Handbook of APPLIED MATHEMATICS

Handbook of APPLIED MATHEMATICS

Selected Results and Methods

Second Edition

edited by
Carl E. Pearson
Professor of Applied Mathematics
University of Washington

VNR VAN NOSTRAND REINHOLD
New York

Library of Congress Catalog Card Number 82-20223
ISBN 0-442-00521-0 (pbk.)

Manufactured in the United States of America

Van Nostrand Reinhold
115 Fifth Avenue
New York, New York 10003

Chapman & Hall
2-6 Boundary Row
London SE1 8HN, England

Thomas Nelson Australia
102 Dodds Street
South Melbourne, Victoria 3205, Australia

Nelson Canada
1120 Birchmount Road
Scarborough, Ontario M1K 5G4, Canada

15 14 13 12 11 10 9 8 7 6 5 4 3 2

Library of Congress Cataloging in Publication Data

Main entry under title:
Handbook of applied mathematics.

Includes bibliographies and index.
1. Mathematics–Handbooks, manuals, etc. I. Pearson, Carl E. II. Title: Applied mathematics.
QA40.H34 1983 510'.2'02 82-20223
ISBN 0-442-23866-5
ISBN 0-442-00521-0 (pbk.)

Preface

Most of the topics in applied mathematics dealt with in this handbook can be grouped rather loosely under the term *analysis*. They involve results and techniques which experience has shown to be of utility in a very broad variety of applications.

Although care has been taken to collect certain basic results in convenient form, it is not the purpose of this handbook to duplicate the excellent collections of tables and formulas available in the *National Bureau of Standards Handbook of Mathematical Functions* (AMS Series 55, U.S. Government Printing Office) and in the references given therein. Rather, the emphasis in the present handbook is on technique, and we are indeed fortunate that a number of eminent applied mathematicians have been willing to share with us their interpretations and experiences.

To avoid the necessity of frequent and disruptive cross-referencing, it is expected that the reader will make full use of the index. Moreover, each chapter has been made as self-sufficient as is feasible. This procedure has resulted in occasional duplication, but as compensation for this the reader may appreciate the availability of different points of view concerning certain topics of current interest.

As editor, I would like to express my appreciation to the contributing authors, to the reviewers, to the editorial staff of the publisher, and to the many secretaries and typists who have worked on the manuscript; without the partnership of all of these people, this handbook would not have been possible.

CARL E. PEARSON

Changes in the Second Edition:

Some material less directly concerned with technique has been omitted or consolidated. Two new chapters, on Integral Equations, and Mathematical Modelling, have been added. Several other chapters have been revised or extended, and known misprints have been corrected.

Contents

Handbook of APPLIED MATHEMATICS

1

Formulas from Algebra, Trigonometry and Analytic Geometry

H. Lennart Pearson*

1.1 THE REAL NUMBER SYSTEM

Readers wishing a logical development of the real number system are directed to the references at the end of the chapter. Here the real numbers are considered to be the set of all terminating and nonterminating decimals with addition, subtraction, multiplication and division (except by zero) defined as usual. Addition and multiplication satisfy

the Commutative Law

$$a + b = b + a \tag{1.1-1}$$

$$ab = ba \tag{1.1-2}$$

the Associative Law

$$a + (b + c) = (a + b) + c \tag{1.1-3}$$

$$a(bc) = (ab)c \tag{1.1-4}$$

the Distributive Law

$$a(b + c) = ab + ac \tag{1.1-5}$$

The real numbers are an ordered set, i.e., given any two real numbers a and b, one

*Prof. H. Lennart Pearson, Dep't. of Mathematics, Illinois Institute of Technology, Chicago, Ill.

of the following must hold:

$$a < b \qquad a = b \qquad a > b \tag{1.1-6}$$

The real numbers fall into two classes, the rational numbers and the irrational numbers. A number is rational if it can be expressed as the quotient of two integers. Division of one integer by another to give a decimal shows that a rational number is either a terminating or repeating decimal. Conversely, any repeating decimal is a rational number, as indicated by the following example:

$$\begin{aligned} &4.328328328 \cdots \\ &= 4 + \frac{328}{10^3} + \frac{328}{10^6} + \frac{328}{10^9} + \cdots \\ &= \frac{4324}{999} \quad \text{by summing the geometric series} \end{aligned}$$

Also note for example $8 = 7.999999 \cdots$.

Non-repeating decimals correspond to irrational numbers.

A set of numbers is said to be bounded above if there exists a number M such that every member of the set is $\leqslant M$. The smallest such M is called the least upper bound of the given set. Similarly a set of numbers is bounded below if there exists a number Q such that every member of the set is $\geqslant Q$, and the greatest lower bound is the largest such Q. Any non-empty set of numbers which is bounded above has a least upper bound, and similarly, if it is bounded below it has a greatest lower bound.

1.2 THE COMPLEX NUMBER SYSTEM

1.2.1 Definition, Real and Imaginary Parts

Two definitions will be given. First, any number of the form $a + ib$ where a and b are any real numbers and $i^2 = -1$ is called a complex number. The number a is called the real part of the complex number and b is called the imaginary part. If $a + ib$ is denoted by the single letter z, then the notation $a = R(z)$, $b = I(z)$ is used.

A second definition, more modern in character, is to define the complex numbers as the set of all ordered pairs (a, b) of real numbers satisfying

$$(a, b) + (c, d) = (a + c, b + d) \tag{1.2-1}$$

$$(a, b)\,(c, d) = (ac - bd, ad + bc) \tag{1.2-2}$$

In particular,

$$(a, 0) + (c, 0) = (a + c, 0)$$

$$(a, 0)\,(c, 0) = (ac, 0)$$

so $(a, 0)$ may be identified with the real number a. By (1.2-2), $(0, 1)(0, 1) = (-1, 0)$ and if i is used for the complex number $(0, 1)$ then

$$a + ib = (a, 0) + (0, 1)(b, 0) = (a, 0) + (0, b) = (a, b)$$

and the two definitions are seen to be equivalent.

1.2.2 Conjugate, Division, Modulus and Argument

The conjugate $\bar{z}$ of the complex number $z = a + ib$ is $\bar{z} = a - ib$.

$$z + \bar{z} = 2R(z) \qquad z - \bar{z} = 2iI(z) \tag{1.2-3}$$

Also,

$$\left.\begin{aligned} \overline{z_1 + z_2} &= \bar{z}_1 + \bar{z}_2 \\ \overline{z_1 z_2} &= (\bar{z}_1)(\bar{z}_2) \\ \overline{\left(\frac{z_1}{z_2}\right)} &= \frac{\bar{z}_1}{\bar{z}_2} \end{aligned}\right\} \tag{1.2-4}$$

Division of one complex number by another is illustrated by example:

$$\frac{3 + 2i}{4 - 3i} = \frac{(3 + 2i)(4 + 3i)}{(4 - 3i)(4 + 3i)} = \frac{6 + 17i}{25} = \frac{6}{25} + i\frac{17}{25}$$

The modulus, or absolute value, of the complex number $z = a + ib$ is $|z| = \sqrt{a^2 + b^2}$ and the argument, or amplitude, of z is arc tan (b/a). Note that

$$z\bar{z} = |z|^2, \quad |z_1 z_2| = |z_1||z_2|, \quad \left|\frac{z_1}{z_2}\right| = \frac{|z_1|}{|z_2|} \tag{1.2-5}$$

1.2.3 The Argand Diagram

A one-to-one correspondence exists between the complex numbers and the points in a plane.

$a + ib \longleftrightarrow$ point in 2-space with coordinates (a, b)

1.2.4 Polar Form, de Moivre's Formula

Introducing polar coordinates in the plane, $x = r \cos\theta$, $y = r \sin\theta$ the complex number $x + iy$ can be written in the polar form:

$$x + iy = r(\cos\theta + i \sin\theta) \tag{1.2-6}$$

where $r = \sqrt{x^2 + y^2}$ is the modulus of z and $\theta = \text{arc tan}\,(y/x)$ is the argument of z.

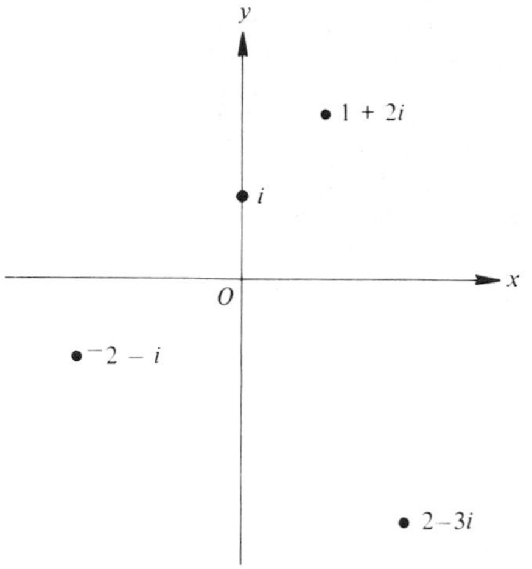

Fig. 1.2-1 Argand diagram: representation of complex numbers in the plane.

Multiplication and division of numbers in polar form yields

$$[r_1(\cos\theta_1 + i\sin\theta_1)]\,[r_2(\cos\theta_2 + i\sin\theta_2)] = r_1r_2[\cos(\theta_1+\theta_2) + i\sin(\theta_1+\theta_2)] \qquad (1.2\text{-}7)$$

$$\frac{r_1(\cos\theta_1 + i\sin\theta_1)}{r_2(\cos\theta_2 + i\sin\theta_2)} = \frac{r_1}{r_2}[\cos(\theta_1-\theta_2) + i\sin(\theta_1-\theta_2)] \qquad (1.2\text{-}8)$$

and extending the multiplication to n equal factors gives de Moivre's Formula:

$$[r(\cos\theta + i\sin\theta)]^n = r^n(\cos n\theta + i\sin n\theta) \qquad (1.2\text{-}9)$$

1.2.5 The nth Roots of a Complex Number

Given $z = r(\cos\theta + i\sin\theta)$,

$$z^{1/n} = r^{1/n}\left[\cos\left(\frac{\theta + 2k\pi}{n}\right) + i\sin\left(\frac{\theta+2k\pi}{n}\right)\right] \qquad k = 0, 1, 2, 3, \ldots, n-1 \quad (1.2\text{-}10)$$

which leads to a more general form of de Moivre's Formula

$$z^{m/n} = r^{m/n}\left[\cos\frac{m}{n}(\theta + 2k\pi) + i\sin\frac{m}{n}(\theta + 2k\pi)\right]$$

$$k = 0, 1, 2, \ldots, n-1; m \text{ and } n \text{ having no factors in common} \quad (1.2\text{-}11)$$

1.3 INEQUALITIES

1.3.1 Rules of Operation, the Triangle Inequality

For a, b and c real numbers,

if $a < b$ and $b < c$ then $a < c$
if $a < b$ then $a + c < b + c$
if $a < b$ then $ac < bc$ if $c > 0$
if $a < b$ then $ac > bc$ if $c < 0$.

Also $|a + b| \leqslant |a| + |b|$ which is the Triangle Inequality, where $|a|$ is a if $a \geqslant 0$ and is $-a$ if $a \leqslant 0$.

1.3.2 The Inequalities of Hölder, Cauchy-Schwarz and Minkowski

Let $a_1, a_2, \ldots, a_n, b_1, b_2, \ldots, b_n$ be any real or complex numbers.

Hölder's Inequality

$$\left|\sum_{k=1}^{n} a_k b_k\right| \leqslant \sum_{k=1}^{n} |a_k b_k| \leqslant \left(\sum_{k=1}^{n} |a_k|^\lambda\right)^{1/\lambda} \left(\sum_{k=1}^{n} |b_k|^\alpha\right)^{1/\alpha}$$

$$\text{where } \lambda > 1 \text{ and } \alpha = \frac{\lambda}{\lambda - 1} \qquad (1.3\text{-}1)$$

Cauchy-Schwarz Inequality

$$\left|\sum_{k=1}^{n} a_k b_k\right|^2 \leqslant \left(\sum_{k=1}^{n} |a_k|^2\right) \left(\sum_{k=1}^{n} |b_k|^2\right) \qquad (1.3\text{-}2)$$

Minkowski Inequality

$$\left(\sum_{k=1}^{n} |a_k + b_k|^\lambda\right)^{1/\lambda} \leqslant \left(\sum_{k=1}^{n} |a_k|^\lambda\right)^{1/\lambda} + \left(\sum_{k=1}^{n} |b_k|^\lambda\right)^{1/\lambda} \quad \text{where } \lambda \geqslant 1 \qquad (1.3\text{-}3)$$

1.4 POWERS AND LOGARITHMS

1.4.1 Rules of Exponents

Let a and b be any positive real numbers and let m and n be positive integers. Then a^n is defined to be the result obtained by multiplying a by itself n times. Then

$$a^n a^m = a^{n+m} \qquad (1.4\text{-}1)$$

$$(a^n)^m = a^{nm} \qquad (1.4\text{-}2)$$

$$(ab)^n = a^n b^n \qquad (1.4\text{-}3)$$

$$\left(\frac{a}{b}\right)^n = \frac{a^n}{b^n} \qquad (1.4\text{-}4)$$

A meaning for a^0 consistent with these rules is obtained by considering $a^n a^0 = a^{n+0} = a^n$, so that $a^0 = 1$; then $a^{-n}a^n = a^{-n+n} = a^0 = 1$ so $a^{-n} = 1/a^n$.

$$\frac{a^n}{a^m} = a^{n-m} \tag{1.4-5}$$

1.4.2 Radicals, Fractional Exponents

Any number a such that $a^n = b$, where n is a positive integer, is called an nth root of b. Any number b has exactly n, n^{th} roots (1.2.5). The principal root of a positive number is defined to be the positive root and the principal n^{th} root (n odd) of a negative number is the negative root. Principal root is not defined when b is negative and n is even. The radical symbol $\sqrt[n]{\;}$ is defined to be the principal n^{th} root, whenever that is defined, and to stand for any one of the n roots if there is no principal root.

Example:

$$\sqrt{4} = 2, \quad \sqrt[5]{-32} = -2, \quad \sqrt{-9} = \pm 3i$$

$a^{1/n}$ is defined to be the same as $\sqrt[n]{a}$

Example:

$$4^{1/2} = 2, \quad (-32)^{1/5} = -2, \quad (-9)^{1/2} = \pm 3i$$

Also

$$a^{m/n} = (\sqrt[n]{a})^m = \sqrt[n]{a^m} \qquad m, n \text{ positive integers} \tag{1.4-6}$$

Finally the definition of a^α, α irrational, will be illustrated by example. Consider 2^π, where $\pi = 3.141592654\ldots$. Then 2^π is defined as the limit of the sequence $2^3, 2^{3.1}, 2^{3.14}, 2^{3.141}, 2^{3.1415}, 2^{3.14159}, 2^{3.141592}, \ldots$, each term of which is defined by the above.

1.4.3 Definitions and Rules of Operation for Logarithms

For any positive number n and any positive number a except 1, there exists a unique real number x such that $n = a^x$. x is called the logarithm of n to the base a. This is written either as above or as $x = \log_a n$. Logarithms have the following properties and rules of operation:

$$\log_a 1 = 0 \qquad \log_a a = 1 \qquad a^{\log_a n} = n \tag{1.4-7}$$

$$\log_a (m \cdot n) = \log_a m + \log_a n \tag{1.4-8}$$

$$\log_a \frac{m}{n} = \log_a m - \log_a n \tag{1.4-9}$$

$$\log_a m^n = n \log_a m \tag{1.4-10}$$

$$\log_a n \cdot \log_b a = \log_b n \tag{1.4-11}$$

1.4.4 Common and Natural Logarithms, the Exponential Function

When the base is 10, the logarithms are called common logarithms, and when the base is $e = 2.718281828459045 \ldots$ the logarithms are called natural logarithms.

$$\begin{aligned} \log_e 10 &= 2.30258509299404568402 \ldots \equiv \ln 10 \\ \log_{10} e &= .43429448190325182765 \ldots \end{aligned} \tag{1.4-12}$$

The inverse function corresponding to the natural logarithm function is called the exponential function, i.e., in symbols e^x.

1.4.5 Log$_e$ z and e^z for z Complex, Principal Value

To generalize these ideas to the complex domain, consider the function $E(z)$ defined by a power series

$$E(z) = 1 + z + \frac{z^2}{2!} + \frac{z^3}{3!} + \cdots \tag{1.4-13}$$

where z is a complex number. This series converges for all z; in particular,

$$E(1) = 1 + 1 + \frac{1}{2!} + \frac{1}{3!} + \cdots = 2.718281828 \ldots = e$$

Direct substitution shows that, if z_1 and z_2 are any complex numbers,

$$E(z_1 + z_2) = E(z_1) \cdot E(z_2) \tag{1.4-14}$$

As a special case of this formula, let x be any real number and n any positive integer.

$$\begin{aligned} E(nx) &= E[(1 + n - 1)x] = E(x)\, E[(n - 1)x] \\ &= E^n(x) \text{ on repeated application of eq. (1.4-14)} \end{aligned}$$

Set $x = 1$

$$E(n) = E^n(1) = e^n$$

For m also a positive integer

$$e^m = E(m) = E\left(n \cdot \frac{m}{n}\right) = \left[E\left(\frac{m}{n}\right)\right]^n$$

so that

$$E\left(\frac{m}{n}\right) = e^{m/n}$$

Again, treating any real x as the limit of a sequence of rational numbers, it follows that $E(x) = e^x$ for any positive real x, and since $E(-x) \cdot E(x) = E(-x + x) = E(0) = 1, E(-x) = 1/E(x) = e^{-x}$. Thus $E(x) = e^x$ for all real x.

In eq. (1.4-13), let $z = iy$, y real, to give $E(iy) = 1 + iy - y^2/2! - i y^3/3! + y^4/4! \cdots = \cos y + i \sin y$ by section 2.2.5.

Then, $E(x + iy) = E(x)\ E(iy) = e^x\ (\cos y + i \sin y)$, from which e^z can be calculated for any complex z. Conversely, given $E(z)$ as any complex number w, not zero, $z = x + iy$ can be found such that $E(z) = w$. This z is called a natural logarithm of w and is written $z = \ln w$. Since $E(z \pm 2n\pi i) = E(z)$, for any integer n, $\ln w$ is a multiple valued function with any two of its values differing by an integral multiple of $2\pi i$.

Examples:

$$\text{Since } E\left(2 + \frac{i\pi}{3}\right) = e^2 \left(\cos \frac{\pi}{3} + i \sin \frac{\pi}{3}\right) = \frac{e^2}{2}(1 + i\sqrt{3}),$$

$$\ln \left[\frac{e^2}{2}(1 + i\sqrt{3})\right] = 2 + i\frac{\pi}{3} \pm 2n\pi i$$

Similarly,

$$\ln 1 = 0 \pm 2n\pi i$$

$$\ln i = \frac{i\pi}{2} \pm 2n\pi i$$

Since $\ln w$ is not single valued, it is convenient to define the principal value of $\ln w$, written $\ln_p w$ as that value for which $-\pi < I(\ln w) \leqslant \pi$.

Examples:

$$\ln_p \frac{e^2}{2}(1 + i\sqrt{3}) = 2 + \frac{i\pi}{3}$$

$$\ln_p i = \frac{i\pi}{2}; \quad \ln_p(-1) = i\pi$$

$$\ln_p(-1-i)(-1-i) = \ln_p(2i) = \ln_p 2 + \frac{i\pi}{2} \neq \ln_p(-1-i) + \ln_p(-1-i)$$

The last example illustrates the ambiguity which arises from $E(\ln w_1 + \ln w_2) = E(\ln w_1) \cdot E(\ln w_2)$ which gives $\ln(w_1 w_2) = \ln w_1 + \ln w_2 \pm 2n\pi i$. Similarly, $\ln(w_1/w_2) = \ln w_1 - \ln w_2 \pm 2n\pi i$.

Next, let w and z be any complex numbers, with $w \neq 0$. Then w^z is defined by

$$w^z = E(z \ln w); \text{ which is also multiple valued.} \tag{1.4-15}$$

Example:

$$i^i = E(i \ln i) = E\left[i\left(\frac{i\pi}{2} \pm 2n\pi i\right)\right] = e^{-\pi/2 \pm 2n\pi}$$

The principal value of w^z, denoted by $(w^z)_p$ is that one value obtained from eq. (1.4-15) by using $\ln_p w$ in place of $\ln w$. For z and w real numbers, using $\ln_p w$ gives results consistent with those of section 1.4.1.

The principal value of $w^{z_1 + z_2}$ is equal to the product of the principal values of w^{z_1} and w^{z_2}. However, $[(w_1 w_2)^z]_p$ is not necessarily equal to $(w_1^z)_p \cdot (w_2^z)_p$, and $(w^{-z})_p$ is not necessarily the same as $1/(w^z)_p$.

Since $\ln 0$ does not exist, 0^z is undefined.

Although eq. (1.4-15) implies that e^z is multiple valued, the convention is to always interpret e^z in the principal value sense. Thus $e^z = E(z)$, always.

1.5 POLYNOMIAL EQUATIONS

1.5.1 Definition, Fundamental Theorem of Algebra

A polynomial of degree n is an expression of the form $a_n x^n + a_{n-1} x^{n-1} + \cdots + a_1 x + a_0$, where $a_n \neq 0, a_{n-1}, \cdots, a_0$ are real or complex numbers and n is a positive integer. When this expression is set equal to 0, the polynomial equation

$$a_n x^n + a_{n-1} x^{n-1} + \cdots + a_1 x + a_0 = 0 \tag{1.5-1}$$

is obtained. For $n = 1, 2, 3, 4, 5$ the equation is called linear, quadratic, cubic, quartic, quintic, respectively.

A polynomial $f(x)$ is divisible by a polynomial $g(x)$ if and only if there exists a unique polynomial $h(x)$ such that $f(x) = g(x) \cdot h(x)$; and $g(x)$, $h(x)$ are called factors of $f(x)$.

A number r such that

$$a_n r^n + a_{n-1} r^{n-1} + a_{n-2} r^{n-2} + \cdots + a_1 r + a_0 = 0 \tag{1.5-2}$$

is called a solution, or root, of the polynomial eq. (1.5-1). Two or more polynomial equations are said to be equivalent if they have the same set of roots. The fundamental theorem of algebra states that every polynomial equation has at least one root, which may be a real or a complex number. Equation (1.5-1) has exactly n roots, not necessarily distinct.

1.5.2 Remainder Theorem, Factor Theorem

The remainder, upon division of the polynomial $f(x) = a_n x^n + a_{n-1} x^{n-1} + \cdots + a_0$ by $ax - b$, is $f(b/a)$.

If r is a root of the equation (1.5-1), then $x - r$ must be a factor of $a_n x^n + a_{n-1} x^{n-1} + \cdots + a_0$. Conversely, if $x - r$ is a factor of the polynomial, then r must be a root of (1.5-1).

1.5.3 Complex Roots, Irrational Roots, Rational Roots

If $a + ib$, $b \neq 0$, is a root of (1.5-1) with real coefficients, then $a - ib$ is also a root; and more generally, if $a + ib$ is a root of multiplicity k, then $a - ib$ must also occur as a root k times.

If $a + b\sqrt{c}$ for a, b, c rational, but $a + b\sqrt{c}$ irrational, is a root of (1.5-1) with rational coefficients, then $a - b\sqrt{c}$ is also a root, and more generally if $a + b\sqrt{c}$ is a root of multiplicity k, then so is $a - b\sqrt{c}$.

If b/a, a and b having no integer factors in common, is a rational root of (1.5-1) with integer coefficients, then a divides a_n and b divides a_0.

1.5.4 Symmetric Functions of the Roots

Division of (1.5-1) by a_n gives an equation of the form

$$x^n + p_{n-1} x^{n-1} + p_{n-2} x^{n-2} + \cdots + p_1 x + p_0 = 0$$

The numbers $p_{n-1}, p_{n-2}, \ldots, p_0$ are given in terms of the roots of (1.5-1) by

$$\left.\begin{aligned} p_{n-1} &= -\text{ (sum of the roots)} \\ p_{n-2} &= +\text{ (sum of the products of the roots two at a time)} \\ p_{n-3} &= -\text{ (sum of the products of the roots three at a time)} \\ &\;\;\vdots \\ p_0 &= (-1)^n \text{ (sum of the products of the roots } n \text{ at a time)} \end{aligned}\right\} \tag{1.5-3}$$

The expressions in the brackets in eqs. (1.5-3) are called the elementary symmetric functions of the roots—symmetric in the sense that an interchange of any two of the roots leaves the expression unchanged.

1.5.5 Descartes' Rule of Signs

In eq. (1.5-1) with real coefficients, the number of positive roots is either the number of variations in sign of the sequence of coefficients, or less than that number by an even integer. The number of negative roots is either the number of variations in sign of the sequence of coefficients found by replacing x by $-x$, or less than that number by an even integer.

1.5.6 Polynomial Equations with Related Roots

To get a polynomial equation each of whose roots is k times the corresponding root of (1.5-1), multiply the coefficients $a_n, a_{n-1}, \ldots, a_0$ by $1, k, k^2, \ldots, k^n$ respectively. In particular, in order to obtain an equation each of whose roots is the negative of the corresponding root of (1.5-1), multiply $a_n, a_{n-1}, \ldots, a_0$ by $1, -1, 1, -1, 1, -1, \ldots, (-1)^n$ respectively.

To obtain a polynomial equation each of whose roots is less by λ than the roots of (1.5-1), divide $a_n x^n + a_{n-1} x^{n-1} + \cdots + a_0$ by $x - \lambda$ giving the quotient $Q_1(x)$ and the remainder R_1; then divide $Q_1(x)$ by $x - \lambda$, giving the quotient $Q_2(x)$ and the remainder R_2; continue this process n times, then the equation $a_n x^n + R_n x^{n-1} + R_{n-1} x^{n-2} + \cdots + R_2 x + R_1 = 0$ is the required equation.

To get a polynomial equation each of whose roots is the reciprocal of the corresponding root of (1.5-1), replace x by $1/x$ giving $a_0 x^n + a_1 x^{n-1} + \cdots + a_{n-1} + a_n = 0$.

To obtain a polynomial equation each of whose roots is the square of the corresponding root of (1.5-1), rewrite (1.5-1) with all the odd powers on one side and all the even powers on the other; square each side, and write $y = x^2$.

1.5.7 Methods of Approximating Roots

All the rational roots of (1.5-1) can be found by applying the last remark of section 1.5.3. Thus, approximation methods are only used in the search for irrational or complex roots.

Newton's Method for approximating a real root is based on the observation that if x_1 is a fairly good approximation to a real root of (1.5-1) then the tangent to the curve $y = a_n x^n + a_{n-1} x^{n-1} + \cdots + a_0$ at $[x_1, y(x_1)]$ will intersect the x axis at a point x_2 which in general will be a better approximation to the root.

The equation of this tangent line is $y - y(x_1) = y'(x_1) \cdot (x - x_1)$. Therefore $x_2 = x_1 - [y(x_1)/y'(x_1)]$, $y'(x_1) \neq 0$. By iteration, successively better approximations $x_3, x_4, x_5 \ldots$ can be obtained. Note that the procedure is not limited to approximating the zeros of polynomial functions.

Horner's Method for approximating an irrational root will be illustrated by ex-

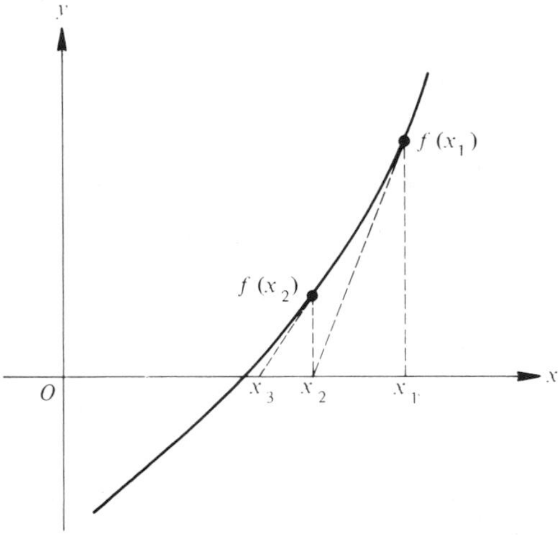

Fig. 1.5-1 Newton's iteration for root-finding, $x_{i+1} = x_i - (f_i/f_i')$.

ample. Find all the roots of the equation: $f(x) = x^4 - x^3 - 2x^2 - 3x - 1 = 0$. By section (1.5.3), ± 1 are the only possible rational roots; substitution shows that neither is a root. Descartes' rule of signs indicates at most one positive root and at most three negative roots. By substituting $x = 0, 1, 2, 3, 4, \ldots$ into $f(x)$, $f(2) < 0$ and $f(3) > 0$, so there is a root r_1 between 2 and 3.

Next, use section 1.5.6 to construct an equation each of whose roots is 2 less than the roots of the given equation. This equation is $x^4 - 3x^3 + 16x^2 + 9x - 7 = 0$. The root of this equation corresponding to r_1 lies between 0 and 1. Now form an equation whose roots are 10 times the roots of the last equation. This equation is $g(x) = x^4 - 30x^3 + 1600x^2 + 9000x - 70{,}000 = 0$ and it must have a root between 0 and 10.

Here $g(4) < 0$, $g(5) > 0$ so there is a root r_2 between 4 and 5. Thus $r_1 = 2 \cdot 4 \ldots$ and the procedure is now iterated, i.e., a new equation is found whose roots are 4 less than the roots of $g(x) = 0$ and again an equation whose roots are 10 times these is constructed. Continuing in this way r_1 is found to be approximately 2.4142 which suggests that perhaps $1 + \sqrt{2}$ is a root. By section 1.5.3, $1 - \sqrt{2}$ would then also be a root and $(x - 1 - \sqrt{2})(x - 1 + \sqrt{2}) = x^2 - 2x - 1$ would be a factor of $f(x)$. Upon division, $x^4 - x^3 - 2x^2 - 3x - 1 = (x^2 - 2x - 1)(x^2 + x + 1)$ so all that remains is to solve the quadratic $x^2 + x + 1 = 0$ whose roots are $(-1 \pm i\sqrt{3})/2$.

If one root is substantially larger in absolute value than any other root, then repeated application of the root-squaring process of Section 1.5.6 further exaggerates the discrepancy, and the multiply squared maximal modulus root will eventually be well approximated by $-p_{n-1}$, because of (1.5-3). This approximation method is associated with the name of Graeffe.

Methods based on function theory are discussed in Chapter 5.

1.5.8 The Linear and Quadratic Equations

The general linear equation is $a_1 x + a_0 = 0$ with solution $x = -(a_0/a_1)$.
The general quadratic equation is $ax^2 + bx + c = 0$ with solutions

$$x = \frac{-b \pm \sqrt{b^2 - 4ac}}{2a}$$

The vanishing of $b^2 - 4ac$, called the discriminant, is a necessary and sufficient condition for equal roots. If a, b, c are all rational numbers, then the roots are real if and only if $b^2 - 4ac > 0$; and if real, rational if and only if the discriminant is the square of an integer. By section (1.5.4), or direct verification, the sum of the roots is $-b/a$ and the product of the roots is c/a.

1.5.9 The Cubic

The roots of a general cubic equation can be obtained by algebraic processes as follows: Division by a_3 gives as the general form of the cubic

$$x^3 + p_2 x^2 + p_1 x + p_0 = 0 \qquad \text{with } p_2 = -\text{ (sum of the roots)} \qquad (1.5\text{-}4)$$

By section 1.5.6 a new equation can be constructed each of whose roots is $-p_2/3$ less than the corresponding root of (1.5-4). This new equation will thus have 0 for the sum of its roots and therefore can be written in the form

$$x^3 + 3q\,x + 2r = 0 \qquad (1.5\text{-}5)$$

Next, let 1, $w = (-1 + i\sqrt{3})/2$, $w^2 = (-1 - i\sqrt{3})/2$ denote the three cube roots of 1, then

$$x^3 + y^3 + z^3 - 3xyz \equiv (x + y + z)(x + wy + w^2 z)(x + w^2 y + wz) \qquad (1.5\text{-}6)$$

is an identity. By (1.5-6), the roots of the equation

$$x^3 - 3yzx + (y^3 + z^3) = 0 \qquad (1.5\text{-}7)$$

are

$$(-y - z), \quad (-wy - w^2 z), \quad (-w^2 y - wz) \qquad (1.5\text{-}8)$$

By setting $-yz = q$, $y^3 + z^3 = 2r$, (1.5-5) and (1.5-7) become the same equation. Substituting $y = -q/z$ into $y^3 + z^3 = 2r$ gives a quadratic equation in z^3 with solutions

$$z^3 = r \pm \sqrt{r^2 + q^3}. \qquad \text{With } z = \sqrt[3]{(r + \sqrt{r^2 + q^3})}$$

$$y = \sqrt[3]{(r - \sqrt{r^2 + q^3})},$$

such that $yz = -q$, the roots of eq. (1.5-5) are then given by eq. (1.5-8).

1.5.10 The Quartic

The general quartic is

$$x^4 + p_3x^3 + p_2x^2 + p_1x + p_0 = 0 \tag{1.5-9}$$

which upon completion of the square for the first two terms can be written

$$\left(x^2 + \frac{p_3}{2}x\right)^2 = \left(\frac{p_3^2}{4} - p_2\right)x^2 - p_1x - p_0$$

Next, add $2(x^2 + (p_3/2)x)y + y^2$ to each side

$$\left(x^2 + \frac{p_3}{2}x + y\right)^2 = \left(\frac{p_3^2}{4} - p_2 + 2y\right)x^2 + (p_3y - p_1)x + y^2 - p_0 \tag{1.5-10}$$

and then choose y so that the right hand side of (1.5-10) is also a perfect square; *i.e.*, set the discriminant equal to 0. This leads to a cubic equation in y, called the resolvent cubic, which can be solved by the method of section 1.5.9. Substituting any root of the cubic into (1.5-10) leads to an equation of the form $(x^2 + (p_3/2)x + y)^2 = (\alpha x + \beta)^2$ which then reduces to solving the two quadratics

$$x^2 + \frac{p_3}{2}x + y = \alpha x + \beta$$

$$x^2 + \frac{p_3}{2}x + y = -(\alpha x + \beta)$$

Abel showed in 1824 that general polynomial equations of degree higher than 4 cannot be solved by purely algebraic methods.

1.5.11 Lagrange Interpolation Formula

The nth degree polynomial which takes on the values $\lambda_1, \lambda_2, \lambda_3, \ldots, \lambda_{n+1}$ at $x_1, x_2, x_3, \ldots, x_{n+1}$ respectively is given by

$$\sum_{i=1}^{n+1} \frac{(x - x_1)\cdots(x - x_{i-1})(x - x_{i+1})\cdots(x - x_{n+1})}{(x_i - x_1)\cdots(x_i - x_{i-1})(x_i - x_{i+1})\cdots(x_i - x_{n+1})} \cdot \lambda_i \tag{1.5-11}$$

This notation means—in the first term of the summation, the factors $(x - x_1)$, $(x_1 - x_1)$ are left out of the numerator and denominator, respectively; in the second term the factors $(x - x_2)$, $(x_2 - x_2)$ are left out, and so on.

1.5.12 Least Squares

The method of least squares consists in finding that polynomial, of prescribed order, which most nearly passes through each of a set of given points (x_i, λ_i), $i = 1, 2, \cdots$

m; the phrase *most nearly* is interpreted in a least squares sense. The case in which the polynomial is to be of first order is most common, and we choose this case as an example.

Let it be required to find coefficients a and b such that the line $y = ax + b$ best fits the given set of points, in the sense that

$$\sum_{i=1}^{m} (ax_i + b - \lambda_i)^2 = \text{minimum}$$

Equating to zero the partial derivatives of the left-hand side with respect to each of a and b, we obtain

$$a \sum x_i^2 + b \sum x_i = \sum \lambda_i x_i$$

$$a \sum x_i + bm = \sum \lambda_i$$

This pair of linear algebraic equations may now be solved for a and b.

1.6 RATIONAL FUNCTIONS AND PARTIAL FRACTIONS

1.6.1 Definition

A rational function is the quotient of two polynomials. If the degree of the numerator is less than degree of the denominator the rational function is said to be proper; otherwise it is improper. If $P_1(x)/P_2(x)$ is improper, by division it can always be written as a polynomial plus a proper fraction.

1.6.2 Partial Fraction Expansions

Given a rational function $P(x)/Q(x)$, writing it as a sum of simpler fractions is called finding the partial fraction expansion, and by section 1.6.1 only proper rational functions need be considered. Let the degree of $Q(x)$ be n, then by section 1.5.2, $Q(x)$ can be factored into n linear factors, real or complex, some of which may be repeated. Then to each nonrepeated factor $x - r_j$ in $Q(x)$ there corresponds a single term $A_j/(x - r_j)$ in the partial fraction expansion, where

$$A_j = \operatorname*{Limit}_{x \to r_j} \frac{(x - r_j)P(x)}{Q(x)}$$

or (1.6-1)

$$A_j = \frac{P(r_j)}{Q'(r_j)}$$

and to a factor $x - r_k$ in $Q(x)$ which is repeated s times, there corresponds s terms

$$\frac{B_s}{(x - r_k)^s} + \frac{B_{s-1}}{(x - r_k)^{s-1}} + \frac{B_{s-2}}{(x - r_k)^{s-2}} + \cdots + \frac{B_2}{(x - r_k)^2} + \frac{B_1}{x - r_k} \qquad (1.6\text{-}2)$$

in the partial fraction expansion, where

$$B_s = \operatorname*{Limit}_{x \to r_k} \frac{(x - r_k)^s P(x)}{Q(x)}$$

and (1.6-3)

$$B_i = \frac{1}{(s - i)!} \operatorname*{Limit}_{x \to r_k} \left\{ \frac{d^{s-i}}{dx^{s-i}} \left[\frac{(x - r_k)^s P(x)}{Q(x)} \right] \right\}$$

$$i = 1, 2, 3, \ldots, s - 1$$

Example:

$$\frac{6x^2 - 5x - 2}{x(x + 1)(x - 2)^2} = \frac{A_1}{x} + \frac{A_2}{x + 1} + \frac{B_2}{(x - 2)^2} + \frac{B_1}{x - 2}$$

with

$$A_1 = \operatorname*{Limit}_{x \to 0} \frac{x(6x^2 - 5x - 2)}{x(x + 1)(x - 2)^2} = \operatorname*{Limit}_{x \to 0} \frac{6x^2 - 5x - 2}{(x + 1)(x - 2)^2} = -\frac{1}{2}$$

similarly, $A_2 = -1$

$$B_2 = \operatorname*{Limit}_{x \to 2} \frac{(x - 2)^2 (6x^2 - 5x - 2)}{x(x + 1)(x - 2)^2} = 2$$

$$B_1 = \frac{1}{1!} \operatorname*{Limit}_{x \to 2} \frac{d}{dx} \left\{ \frac{(6x^2 - 5x - 2)}{x(x + 1)} \right\} = \frac{3}{2}$$

1.7 DETERMINANTS AND SOLUTION OF SYSTEMS OF LINEAR EQUATIONS

1.7.1 Determinants of Orders 2 and 3

The symbol $\begin{vmatrix} a_{11} & a_{12} \\ a_{21} & a_{22} \end{vmatrix}$ is called a determinant of order 2 and stands for the combination $a_{11} a_{22} - a_{21} a_{12}$.

The symbol $\begin{vmatrix} a_{11} & a_{12} & a_{13} \\ a_{21} & a_{22} & a_{23} \\ a_{31} & a_{32} & a_{33} \end{vmatrix}$ is called a determinant of order 3 and stands for the combination $a_{11}a_{22}a_{33} + a_{12}a_{23}a_{31} + a_{13}a_{21}a_{32} - a_{31}a_{22}a_{13} - a_{32}a_{23}a_{11} - a_{33}a_{21}a_{12}$, which can be readily obtained. Write the first two columns of the determinant to its right. Subtract the products obtained by multiplication along the upward arrows from the products obtained by multiplication along the downward arrows.

$$
\begin{vmatrix}
a_{11} & a_{12} & a_{13} \\
a_{21} & a_{22} & a_{23} \\
a_{31} & a_{32} & a_{33}
\end{vmatrix}
\begin{matrix}
a_{11} & a_{12} \\
a_{21} & a_{22} \\
a_{31} & a_{32}
\end{matrix}
$$

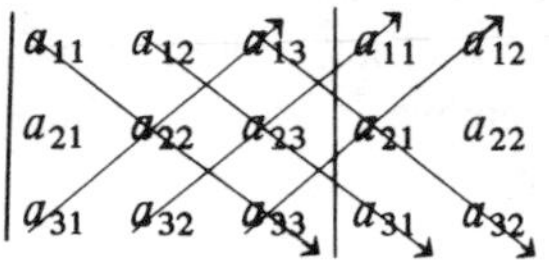

1.7.2 Determinant of Order *n*

The symbol

$$
\begin{vmatrix}
a_{11} & a_{12} & a_{13} & \cdots & a_{1n} \\
a_{21} & a_{22} & a_{23} & \cdots & a_{2n} \\
& & \cdots & & \\
a_{n1} & a_{n2} & a_{n3} & \cdots & a_{nn}
\end{vmatrix}
\tag{1.7-2}
$$

is called a determinant of order *n*, and stands for any one of the combinations

$$
\begin{aligned}
a_{i1}c_{i1} + a_{i2}c_{i2} + \cdots + a_{in}c_{in} && (i = 1, 2, \ldots, \text{or } n) \\
a_{1k}c_{1k} + a_{2k}c_{2k} + \cdots + a_{nk}c_{nk} && (k = 1, 2, \ldots, \text{or } n)
\end{aligned}
\tag{1.7-2}
$$

where c_{ik}, the cofactor of the element a_{ik} is given by

$$
c_{ik} = (-1)^{i+k} m_{ik}
\tag{1.7-3}
$$

where m_{ik}, called the minor of the element a_{ik}, is the determinant of order $n-1$ obtained from the given determinant by removing the row and the column containing a_{ik}. Thus, by iteration of this definition, a determinant of order *n* can be written as a sum of determinants of order 3; the previous rule can then be used.

1.7.3 Properties of Determinants

A determinant is invariant under the interchange of all of its rows for all of its columns, in the same order. (transposition)

If each element in any row (column) is 0, the determinant is 0.

The determinant obtained by multiplying every element of any row (column) by λ, is λ times the original determinant.

If any two rows (columns) of a determinant are interchanged, the new determinant is (-1) times the original.

Two determinants all of whose corresponding rows (columns) but one are identical can be added as illustrated:

$$
\begin{vmatrix}
a_1 & b_1 & c_1 \\
a_2 & b_2 & c_2 \\
a_3 & b_3 & c_3
\end{vmatrix}
+
\begin{vmatrix}
a_1 & d_1 & c_1 \\
a_2 & d_2 & c_2 \\
a_3 & d_3 & c_3
\end{vmatrix}
=
\begin{vmatrix}
a_1 & b_1 + d_1 & c_1 \\
a_2 & b_2 + d_2 & c_2 \\
a_3 & b_3 + d_3 & c_3
\end{vmatrix}
$$

If any row (column) of a determinant is a multiple of any other row (column), the determinant is 0.

The addition to any row (column) of a constant multiple of any other row (column) leaves the determinant unchanged.

The element in the ith row and k^{th} column of the product of the nth order determinant whose general element is a_{ij} with the nth order determinant whose general element is b_{ij} is given by

$$\sum_{j=1}^{n} a_{ij} b_{jk}$$

A determinant each of whose elements is a function of x can be differentiated as follows (prime indicates differentiation):

$$\frac{d}{dx}\begin{vmatrix} a_{11} & a_{12} & \cdots & a_{1n} \\ a_{21} & a_{22} & \cdots & a_{2n} \\ \cdots & \cdots & \cdots & \cdots \\ a_{n1} & a_{n2} & \cdots & a_{nn} \end{vmatrix} = \begin{vmatrix} a'_{11} & a'_{12} & \cdots & a'_{1n} \\ a_{21} & a_{22} & \cdots & a_{2n} \\ \cdots & \cdots & \cdots & \cdots \\ a_{n1} & a_{n2} & \cdots & a_{nn} \end{vmatrix} + \begin{vmatrix} a_{11} & a_{12} & \cdots & a_{1n} \\ a'_{21} & a'_{22} & \cdots & a'_{2n} \\ \cdots & \cdots & \cdots & \cdots \\ a_{n1} & a_{n2} & \cdots & a_{nn} \end{vmatrix} + \cdots + \begin{vmatrix} a_{11} & a_{12} & \cdots & a_{1n} \\ a_{21} & a_{22} & \cdots & a_{2n} \\ \cdots & \cdots & \cdots & \cdots \\ a'_{n1} & a'_{n2} & \cdots & a'_{nn} \end{vmatrix}$$

1.7.4 Cramer's Rule

Consider the system of n linear equations in n unknowns:

$$\begin{aligned} a_{11}x_1 + a_{12}x_2 + \cdots + a_{1n}x_n &= b_1 \\ a_{21}x_1 + a_{22}x_2 + \cdots + a_{2n}x_n &= b_2 \\ \cdots\cdots\cdots\cdots\cdots\cdots\cdots\cdots \\ a_{n1}x_1 + a_{n2}x_2 + \cdots + a_{nn}x_n &= b_n \end{aligned} \tag{1.7-4}$$

If every b_i is equal to 0, (1.7-4) is called a homogeneous system of equations; otherwise it is nonhomogeneous. Let D be the determinant of the coefficients and D_i the determinant obtained from D by replacing the ith column of D with the numbers $b_1, b_2, b_3, \ldots, b_n$. Cramer's Rule states that if $D \neq 0$, then (1.7-4) has a unique solution given by

$$x_i = \frac{D_i}{D} \qquad (i = 1, 2, 3, \ldots, n) \tag{1.7-5}$$

Note that if the system is homogeneous and $D \neq 0$, then the unique solution is the trivial solution $x_i = 0$, $(i = 1, 2, 3, \ldots, n)$. Also note that if $D = 0$, but at least one $D_i \neq 0$, then there can be no solution of the system.

1.7.5 General Systems of Linear Equations

Consider a general system of m equations in n unknowns

$$\begin{array}{l} a_{11}x_1 + a_{12}x_2 + \cdots + a_{1n}x_n = b_1 \\ a_{21}x_1 + a_{22}x_2 + \cdots + a_{2n}x_n = b_2 \\ \cdots\cdots\cdots\cdots\cdots\cdots\cdots\cdots \\ a_{m1}x_1 + a_{m2}x_2 + \cdots + a_{mn}x_n = b_m \end{array} \qquad (1.7\text{-}6)$$

Assuming the equations are so arranged that $a_{11} \neq 0$ the first equation may be used to eliminate x_1 from the remaining $m - 1$ equations. In this new set of $m - 1$ equations either all the coefficients of x_2 are 0 or they are not; if they are not, use one such equation to eliminate x_2 from the remaining $m - 2$ equations; if they are, move on to the first of $x_3, x_4, x_5, \ldots, x_n$ which has a nonzero coefficient in one of the equations. Using this equation to eliminate the unknown from the remaining equations. Continue this process until either no equations remain, or the coefficients of all unknowns in the remaining equations are zero. The resulting system will be of the form

$$\begin{array}{r} a_{11}x_1 + a_{12}x_2 + \cdots + a_{1n}x_n = b_1 \\ c_{22}x_2 + \cdots + c_{2n}x_n = b_{22} \\ \cdots\cdots\cdots\cdots\cdots\cdots \\ \lambda_{rr}x_r + \cdots + \lambda_{rn}x_n = b_{rr} \end{array} \qquad (1.7\text{-}7)$$

with either $r = m$ or $r < m$. In the latter case the rest of the equations will be $0 = b_{r+1,r+1}$, $0 = b_{r+2,r+2}, \ldots, 0 = b_{mm}$, and there will be no solution unless all the b's are indeed 0. If there is a solution, it is obtained by taking $x_{r+1}, x_{r+2}, \ldots, x_n$ to be arbitrary, and solving the last equation for x_r in terms of these arbitrary values. Then each $x_{r-1}, x_{r-2}, \ldots, x_1$ is found by solving each equation of the new system in turn.

Example: Solve the system

$$\begin{aligned} w - 2x + 5y - 3z &= 0 \\ 4w - 8x + 6y - 2z &= 6 \\ -3w + 6x + y + z &= 0 \end{aligned}$$

Eliminating w from the last two equations

$$\begin{aligned} w - 2x + 5y - 3z &= 0 \\ -14y + 10z &= 6 \\ 16y - 8z &= 0 \end{aligned}$$

and eliminating y from the last equation

$$\begin{aligned} w - 2x + 5y - 3z &= 0 \\ -14y + 10z &= 6 \\ 24z &= 48 \end{aligned}$$

Thus $z = 2, y = 1, w = 2x + 1, x$ arbitrary.

The array of coefficients

$$\begin{bmatrix} a_{11} & a_{12} & a_{13} & \cdots a_{1n} \\ a_{21} & a_{22} & a_{23} & \cdots a_{2n} \\ \cdots & \cdots & \cdots & \cdots \\ a_{m1} & a_{m2} & a_{m3} & \cdots a_{mn} \end{bmatrix}$$

taken from eqs. (1.7-6) is an example of an $m \times n$ matrix (see Chapter 16), in this case, called the coefficient matrix. The $m \times (n + 1)$ matrix with the additional column consisting of the numbers $b_1, b_2, \ldots, b_m$ is the augmented matrix corresponding to eqs. (1.7-6). The rank of a matrix is the number of rows (columns) in its largest square submatrix whole determinant is not zero. For example

$$\begin{bmatrix} 1 & 5 & 3 & -15 \\ -1 & 3 & 5 & -9 \\ 1 & 0 & -2 & 0 \end{bmatrix}$$

has rank 2, since each of the four possible determinants of order 3 vanishes, but

$$\begin{vmatrix} 1 & 5 \\ -1 & 3 \end{vmatrix} \neq 0$$

If every b_i in eqs. (1.7-6) is zero, then the equations form a homogeneous system of equations, otherwise it is nonhomogeneous. The existence of solutions can now be characterized in terms of rank as follows:

(1) In a nonhomogeneous system there will be a solution if and only if the rank of the augmented matrix is equal to the rank of the coefficient matrix; if this common rank is equal to the number of unknowns then the solution is unique, if it is less than the number of unknowns then an infinite number of solutions exist.

(2) In a homogeneous system, if the rank of the coefficient matrix is equal to the number of unknowns, then the solution is unique (and trivial) and is $x_i = 0$ for all i. If this rank is less than the number of unknowns, then an infinity of solutions exists.

For example, in the set of equations solved before by elimination, the rank of the augmented matrix equals the rank of the coefficient matrix equals 3, which is less than the number of unknowns; there were indeed an infinite number of solutions.

1.8 PROGRESSIONS

1.8.1 Arithmetic Progressions

An arithmetic progression (A.P.) is a sequence each term of which is obtained from the preceding one by the addition of the same constant. Thus an A.P. has the form

$$a, \quad a+d, \quad a+2d, \quad a+3d \ldots \tag{1.8-1}$$

with the nth term given by

$$a+(n-1)d \tag{1.8-2}$$

and the sum of the first n terms by

$$S_n = \frac{n}{2}\,[2a+(n-1)\,d] \tag{1.8-3}$$

If $t_1, t_2, t_3, \ldots, t_n$ form the first n terms of an A.P., then $t_2, t_3, t_4, \ldots, t_{n-1}$ are called arithmetic means between t_1 and t_n.

A harmonic progression is a sequence whose reciprocals form an A.P.

1.8.2 Geometric Progressions

A geometric progression (G.P.) is a sequence each term of which bears a constant ratio to the preceding term; thus a G.P. has the form

$$a, \quad ar, \quad ar^2, \quad ar^3 \ldots \tag{1.8-4}$$

with the nth term given by

$$ar^{n-1} \tag{1.8-5}$$

and the sum of the first n terms for $r \neq 1$ is

$$S_n = \frac{a(1-r^n)}{1-r} \tag{1.8-6}$$

This last formula can be useful in reverse. For example, in order to divide $1 - x^n$ by $1 - x$, notice that $(1 - x^n)/(1 - x)$ is the sum of a G.P. with first term 1, common ratio x; therefore $(1 - x^n)/(1 - x) = 1 + x + x^2 + x^3 + \cdots + x^{n-1}$. If $T_1, T_2, T_3, \ldots, T_n$ are the first n terms of a G.P., then $T_2, T_3, \ldots, T_{n-1}$ are called geometric means between T_1 and T_n.

1.9 BINOMIAL THEOREM, PERMUTATIONS AND COMBINATIONS

1.9.1 The Binomial Theorem

For any positive integer n,

$$(a + b)^n = a^n + na^{n-1}b + \frac{n(n-1)}{2!}a^{n-2}b^2 + \frac{n(n-1)(n-2)}{3!}a^{n-3}b^3 + \cdots + b^n \quad (1.9\text{-}1)$$

with the general term, the $(r + 1)^{st}$ term $(r = 0, 1, 2, \ldots, n)$, given by

$$\frac{n!}{(n-r)!\,r!}a^{n-r}b^r \quad (1.9\text{-}2)$$

where

$$n! = n \cdot (n-1) \cdot (n-2) \cdot (n-3) \ldots 3 \cdot 2 \cdot 1, \qquad 0! = 1 \quad (1.9\text{-}3)$$

1.9.2 The Binomial Series

The infinite series

$$1 + gx + \frac{g(g-1)}{2!}x^2 + \frac{g(g-1)(g-2)}{3!}x^3 + \frac{g(g-1)(g-2)(g-3)}{4!}x^4 \ldots$$

converges to $(1 + x)^g$ for all x such that $|x| < 1$.

1.9.3 Permutations, Definition and Formulas

The fundamental principle of permutations and combinations is: if one act can be done in n_1 different ways, and upon its completion a second act can be done in n_2

different ways, and upon its completion a third act can be done in n_3 different ways, and so on finitely, then the total number of ways in which all the acts can be done in the order stated is $n_1 \cdot n_2 \cdot n_3 \cdot \ldots$.

Each different arrangement of a certain set of objects selected from a given group of objects is called a permutation of those objects. The total number of different permutations of r objects selected from a set of n objects is

$$P(n, r) = \frac{n!}{(n - r)!} \tag{1.9-4}$$

The total number of different permutations of n objects taken all at a time where $n = n_1 + n_2 + n_3 + \cdots + n_k$, and n_1 are of the same kind, n_2 are identical of a second kind, etc., is $n!/n_1!\,n_2!\ldots n_k!$.

1.9.4 Combinations, Definition and Formulas

Each selection of r objects from a set of n different objects is called a combination of the n objects. The total number of combinations of n objects taken r at a time is given by

$$C(n, r) = \frac{n!}{(n - r)!\,r!} = C(n, n - r) \tag{1.9-5}$$

Notice that the formula for the general term (1.9-2) in the binomial formula can then be written $C(n, r)a^{n-r}b^r$. If a and b are set equal to one in (1.9-1)

$$2^n = 1 + C(n, 1) + C(n, 2) + C(n, 3) + \cdots + C(n, n) \tag{1.9-6}$$

1.10 THE TRIGONOMETRIC FUNCTIONS

1.10.1 Definition

Given any angle θ, $-\infty < \theta < \infty$, place the angle in standard position in a Cartesian coordinate system, *i.e.*, with the vertex of the angle at (0, 0), the initial side on the $+x$-axis and the terminal side wherever it may fall in one of the four quadrants. Positive angles are measured counterclockwise; negative, clockwise.

Choose any point $P(x, y)$ on the terminal side of θ. Let r be the distance of P from 0, *i.e.*, $r = \sqrt{x^2 + y^2}$. Then

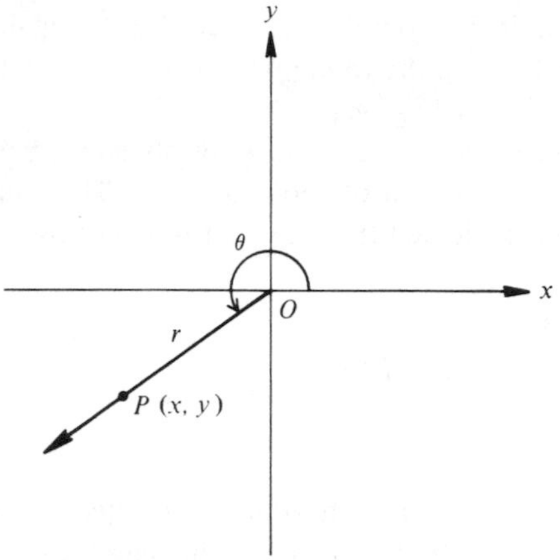

Fig. 1.10-1 Coordinate variables for trigonometric functions.

$$\text{sine } \theta \equiv \sin \theta = \frac{y}{r} \tag{1.10-1}$$

$$\text{cosine } \theta \equiv \cos \theta = \frac{x}{r} \tag{1.10-2}$$

$$\text{tangent } \theta \equiv \tan \theta = \frac{y}{x} \tag{1.10-3}$$

$$\text{cosecant } \theta \equiv \operatorname{cosec} \theta = \frac{r}{y} = \frac{1}{\sin \theta} \tag{1.10-4}$$

$$\text{secant } \theta \equiv \sec \theta = \frac{r}{x} = \frac{1}{\cos \theta} \tag{1.10-5}$$

$$\text{cotangent } \theta \equiv \cot \theta = \frac{x}{y} = \frac{1}{\tan \theta} \tag{1.10-6}$$

In addition to the six trigonometric functions defined above, some classical treatises also use

$$\text{versed sine } \theta \equiv \operatorname{vers} \theta = 1 - \cos \theta \tag{1.10-7}$$

$$\text{coversed sine } \theta \equiv \operatorname{covers} \theta = 1 - \sin \theta \tag{1.10-8}$$

$$\text{haversine } \theta \equiv \operatorname{hav} \theta = \tfrac{1}{2}\,(1 - \cos \theta) \tag{1.10-9}$$

1.10.2 Graphs

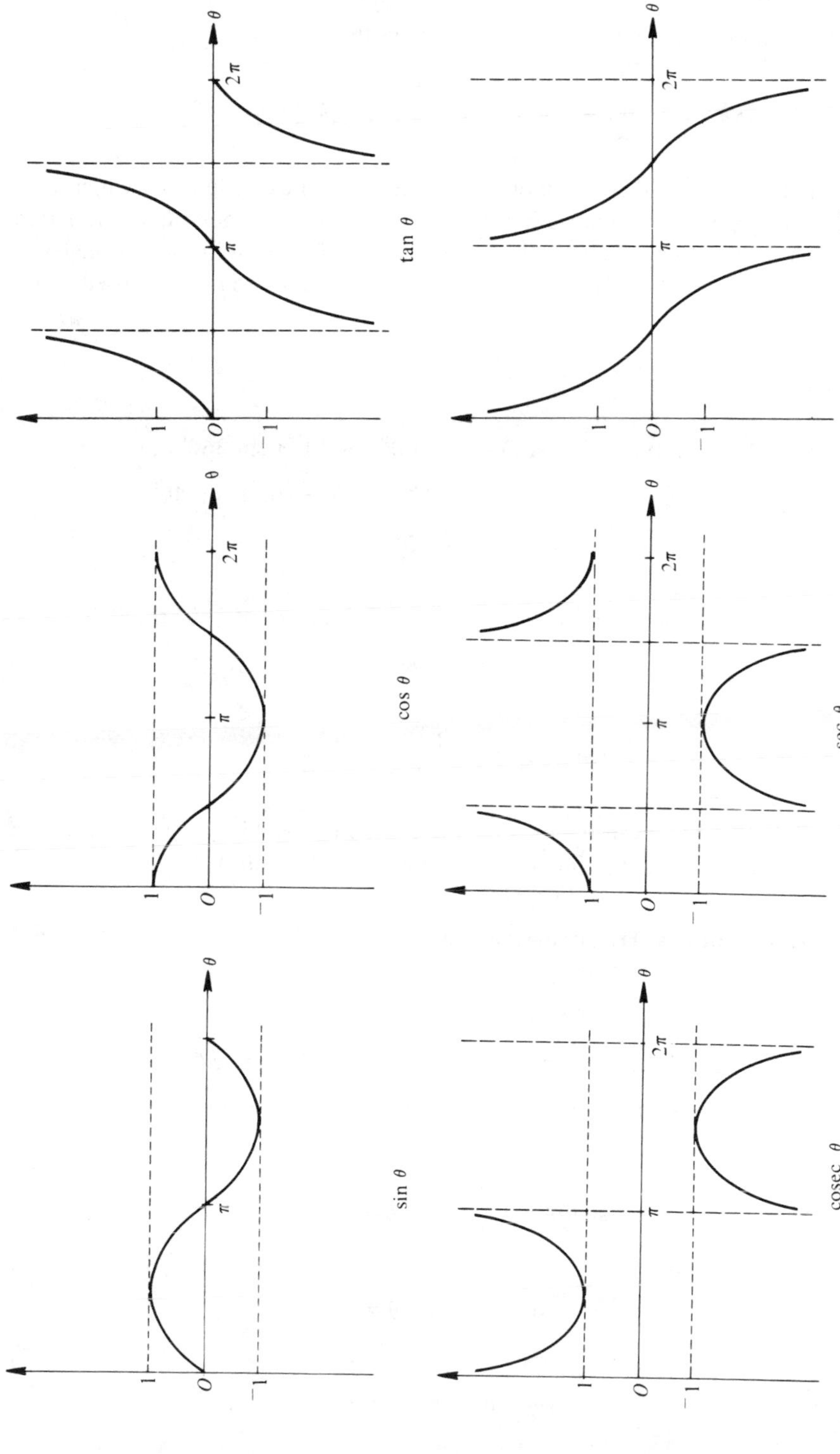

Fig. 1.10-2 Basic trigonometric functions.

1.10.3 Trigonometric Reduction

From the definitions in section (1.10.1) it follows that

$$\sin(2k\pi + \theta) \equiv \sin(k \cdot 360^\circ + \theta) = \sin\theta, \qquad k = 0, \pm1, \pm2, \ldots,$$

and similarly for any other trigonometric function. Thus, reduction, which consists of writing any trigonometric function of any angle as a trigonometric function of an angle in the first quadrant, becomes the question of doing so for angles in quadrants II, III and IV. The identities given in eqs. (1.10-13) can be used to do this.

Examples:

$$\begin{aligned} \cos 330^\circ = \cos(360^\circ - 30^\circ) &= \cos 360^\circ \cos 30^\circ + \sin 360^\circ \sin 30^\circ \\ &= \cos 0^\circ \cos 30^\circ + \sin 0^\circ \sin 30^\circ \\ &= \cos 30^\circ \end{aligned}$$

$$\begin{aligned} \sin 260^\circ = \sin(180^\circ + 80^\circ) &= \sin 180^\circ \cos 80^\circ + \cos 180^\circ \sin 80^\circ \\ &= -\sin 80^\circ \end{aligned}$$

$$\operatorname{cosec} 260^\circ = \frac{1}{\sin 260^\circ} = \frac{-1}{\sin 80^\circ}$$

Further, the identities given in eqs. (1.10-10) make it possible, in the construction of tables of the trigonometric functions, to tabulate only from 0° to 45°.

1.10.4 Basic Identities, Trigonometric Sums

$$\begin{aligned} \sin(90^\circ - \theta) &= \cos\theta & \cos(90^\circ - \theta) &= \sin\theta \\ \operatorname{cosec}(90^\circ - \theta) &= \sec\theta & \sec(90^\circ - \theta) &= \operatorname{cosec}\theta \\ \tan(90^\circ - \theta) &= \cot\theta & \cot(90^\circ - \theta) &= \tan\theta \end{aligned} \tag{1.10-10}$$

$$\tan\theta = \frac{\sin\theta}{\cos\theta} \qquad \cot\theta = \frac{\cos\theta}{\sin\theta} \tag{1.10-11}$$

$$\begin{aligned} \sin^2\theta + \cos^2\theta &= 1 \\ \sec^2\theta - \tan^2\theta &= 1 \\ \operatorname{cosec}^2\theta - \cot^2\theta &= 1 \end{aligned} \tag{1.10-12}$$

$$\begin{aligned}
\sin(A \pm B) &= \sin A \cos B \pm \cos A \sin B \\
\cos(A \pm B) &= \cos A \cos B \mp \sin A \sin B \\
\tan(A \pm B) &= \frac{\tan A \pm \tan B}{1 \mp \tan A \tan B}
\end{aligned} \tag{1.10-13}$$

$$\begin{aligned}
\sin 2A &= 2 \sin A \cos A \\
\cos 2A &= \cos^2 A - \sin^2 A = 2\cos^2 A - 1 = 1 - 2\sin^2 A \\
\tan 2A &= \frac{2 \tan A}{1 - \tan^2 A} \\
\sin 3A &= 3 \sin A - 4 \sin^3 A \\
\cos 3A &= 4 \cos^3 A - 3 \cos A
\end{aligned} \tag{1.10-14}$$

$$\begin{aligned}
\sin \frac{A}{2} &= \pm \sqrt{\frac{1 - \cos A}{2}} \\
\cos \frac{A}{2} &= \pm \sqrt{\frac{1 + \cos A}{2}} \\
\tan \frac{A}{2} &= \frac{1 - \cos A}{\sin A} = \frac{\sin A}{1 + \cos A}
\end{aligned} \tag{1.10-15}$$

$$\begin{aligned}
\sin A + \sin B &= 2 \sin \frac{A+B}{2} \cos \frac{A-B}{2} \\
\sin A - \sin B &= 2 \cos \frac{A+B}{2} \sin \frac{A-B}{2} \\
\cos A + \cos B &= 2 \cos \frac{A+B}{2} \cos \frac{A-B}{2} \\
\cos A - \cos B &= -2 \sin \frac{A+B}{2} \sin \frac{A-B}{2} \\
\tan A \pm \tan B &= \frac{\sin(A \pm B)}{\cos A \cos B}
\end{aligned} \tag{1.10-16}$$

DeMoivre's Formula (1.2-9) and the Binomial Theorem (1.9-1) can be used to write $\sin nA$ and $\cos nA$, for n a positive integer, in terms of powers of $\sin A$ and $\cos A$, or vice versa.

Example:

$$\begin{aligned}
\cos 4A + i \sin 4A &= (\cos A + i \sin A)^4 \\
&= \cos^4 A + 4i \cos^3 A \sin A - 6 \cos^2 A \sin^2 A \\
&\quad - 4i \cos A \sin^3 A + \sin^4 A
\end{aligned}$$

$$\therefore \cos 4A = \cos^4 A - 6\cos^2 A \sin^2 A + \sin^4 A$$

$$\sin 4A = 4 \sin A \cos^3 A - 4 \sin^3 A \cos A$$

and conversely

$$\cos^4 A = \cos 4A + 6\cos^2 A \sin^2 A - \sin^4 A$$

$$= \cos 4A + 6\cos^2 A(1 - \cos^2 A) - (1 - \cos^2 A)^2$$

$$8\cos^4 A = \cos 4A + 8\cos^2 A - 1$$

$$= \cos 4A + 4(\cos 2A + 1) - 1$$

Thus

$$8\cos^4 A = \cos 4A + 4\cos 2A + 3$$

and similarly,

$$8\sin^4 A = \cos 4A - 4\cos 2A + 3$$

$$\sum_{k=1}^{n} C(n,k)\cos kx = 2^n \cos^n \frac{x}{2} \cos \frac{nx}{2} - 1$$

$$\sum_{k=1}^{n} C(n,k)\sin kx = 2^n \cos^n \frac{x}{2} \sin \frac{nx}{2} \qquad (1.10\text{-}17)$$

Also, for any real α and any real $x \neq 2m\pi$ (m an integer)

$$\sum_{k=1}^{n} \cos(\alpha + kx) = \frac{\sin \frac{1}{2} nx}{\sin \frac{1}{2} x} \cos \{\alpha + \tfrac{1}{2}(n+1)x\}$$

$$\sum_{k=1}^{n} \sin(\alpha + kx) = \frac{\sin \frac{1}{2} nx}{\sin \frac{1}{2} x} \sin \{\alpha + \tfrac{1}{2}(n+1)x\} \qquad (1.10\text{-}18)$$

1.10.5 Solution of Triangles, Area, Radii of Inscribed and Circumscribed Circles, Area of a Sector of a Circle

In any triangle let the three angles be A, B, C, and the corresponding sides opposite them be a, b, c. Given certain selections of three parts the other parts can be found. There are four cases to consider.

Case I Given one side and two angles.
The third angle is found from $A + B + C = 180°$ and the other two sides from the Law of Sines

$$\frac{a}{\sin A} = \frac{b}{\sin B} = \frac{c}{\sin C} \qquad (1.10\text{-}19)$$

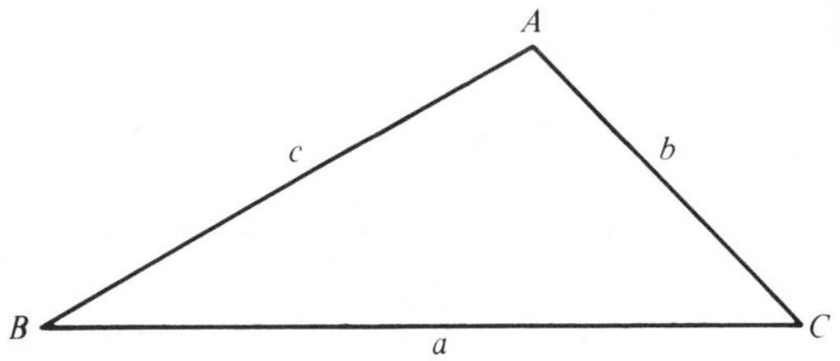

Fig. 1.10-3 Sides and angles of an arbitrary triangle.

Case II Given two sides and the included angle
The third side can be found from the Law of Cosines

$$c^2 = a^2 + b^2 - 2ab \cos C \tag{1.10-20}$$

and the two angles from the Law of Sines, or the Law of Tangents

$$\tan \frac{A - B}{2} = \frac{a - b}{a + b} \cot \frac{C}{2} \tag{1.10-21}$$

together with $(A + B)/2 = 90^\circ - C/2$ can be used to find the missing two angles, and then the Law of Sines to get the third side.

Case III Given two sides and the angle opposite one of them
Assume A, a and $b, a < b$, are given. Then if

(i) $a < b \sin A$, then there is no solution
(ii) $a = b \sin A$, then $B = 90^\circ$ and $C = 90^\circ - A, c^2 = b^2 - a^2$
(iii) $a > b \sin A$, the ambiguous case—two solutions which are found by the Law of Sines and $A + B + C = 180^\circ$

Case IV Given the three sides
The angles may be found by the Law of Cosines, or more conveniently, for logarithmic work, from any one of

$$\sin \frac{A}{2} = \sqrt{\frac{(s - b)(s - c)}{bc}}$$
$$\cos \frac{A}{2} = \sqrt{\frac{s(s - a)}{bc}} \tag{1.10-22}$$

where s is the semiperimeter of the triangle, i.e.,

$$2s = a + b + c \tag{1.10-23}$$

Any one of Mollweide's equations

$$\left.\begin{aligned}\frac{a+b}{c} &= \frac{\cos\frac{1}{2}(A-B)}{\sin\frac{1}{2}C}\\ \frac{a-b}{c} &= \frac{\sin\frac{1}{2}(A-B)}{\cos\frac{1}{2}C}\end{aligned}\right\} \qquad (1.10\text{-}24)$$

may be used to check the results of the above calculations.

The area of any triangle is given by any of the following:

$$\begin{aligned}
A &= \tfrac{1}{2}\,(\text{length of base})\cdot(\text{length of corresponding altitude})\\
A &= \tfrac{1}{2}\,ab\sin C\\
A &= \sqrt{s(s-a)(s-b)(s-c)} \quad \text{Heron's formula}\\
A &= \frac{abc}{4R};\ R \text{ the radius of the circumscribing circle}\\
A &= rs;\ r \text{ the radius of the inscribed circle}\\
A &= 2R^2 \sin A \sin B \sin C\\
A &= r^2 \cot\frac{A}{2}\cot\frac{B}{2}\cot\frac{C}{2}
\end{aligned} \qquad (1.10\text{-}25)$$

R, r are given by

$$R = \frac{a}{2\sin A} \qquad (1.10\text{-}26)$$

$$r = 4R\sin\frac{A}{2}\sin\frac{B}{2}\sin\frac{C}{2} \qquad (1.10\text{-}27)$$

For a sector of a circle, with θ in radians

$$s = r\theta \qquad (1.10\text{-}28)$$

$$\text{Area of sector} = \tfrac{1}{2}\,r^2\theta \qquad (1.10\text{-}29)$$

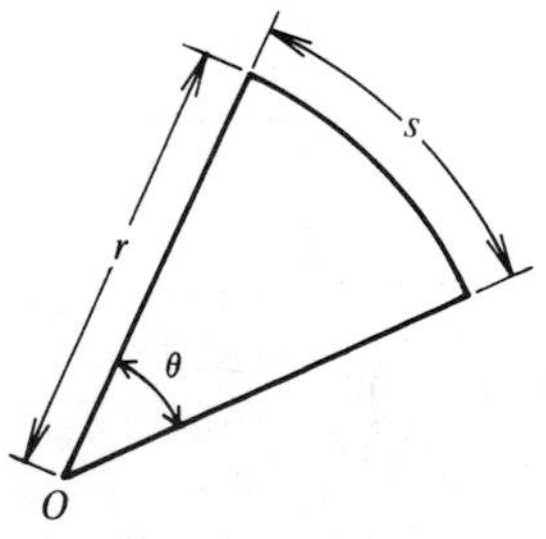

Fig. 1.10-4 Sector of a circle.

1.10.6 Inverse Trigonometric Functions, Graphs, Principal Values

The inverse trigonometric functions are defined by writing $y = \text{arc} \sin x \equiv \sin^{-1} x$ to be the inverse function corresponding to $y = \sin x$ so that

$$\begin{aligned}
&y = \text{arc} \sin x \equiv \sin^{-1} x && \text{implies} \quad x = \sin y \\
&y = \text{arc} \cos x \equiv \cos^{-1} x && \text{implies} \quad x = \cos y \\
&y = \text{arc} \tan x \equiv \tan^{-1} x && \text{implies} \quad x = \tan y \\
&y = \text{arc cosec}\, x \equiv \text{cosec}^{-1} x && \text{implies} \quad x = \text{cosec}\, y \\
&y = \text{arc} \sec x \equiv \sec^{-1} x && \text{implies} \quad x = \sec y \\
&y = \text{arc} \cot x \equiv \cot^{-1} x && \text{implies} \quad x = \cot y
\end{aligned} \tag{1.10-30}$$

Since none of these functions is single valued, it is important in the elementary calculus to restrict the range to the so-called principal values, defined as follows:

$$\begin{aligned}
&y = \text{Arc} \sin x && -\pi/2 \leqslant y \leqslant \pi/2 \\
&y = \text{Arc} \cos x && 0 \leqslant y \leqslant \pi \\
&y = \text{Arc} \tan x && -\pi/2 < y < \pi/2 \\
&y = \text{Arc} \cot x && 0 < y < \pi \\
&y = \text{Arc} \sec x && 0 < y < \pi \\
&y = \text{Arc cosec}\, x && -\pi/2 < y < \pi/2
\end{aligned} \tag{1.10-31}$$

1.10.7 Identities for the Inverse Trigonometric Functions

If no domain is given, then the identity is valid for all x.

$$\begin{aligned}
&\sin(\text{arc} \sin x) = x, \quad \cos(\text{arc} \cos x) = x && |x| \leqslant 1 \\
&\sec(\text{arc} \sec x) = x, \quad \text{cosec}(\text{arc cosec}\, x) = x && |x| \geqslant 1 \\
&\tan(\text{arc} \tan x) = x, \quad \cot(\text{arc} \cot x) = x &&
\end{aligned} \tag{1.10-32}$$

$$\begin{aligned}
&\text{Arc} \sin(\sin x) = x && |x| \leqslant \pi/2 \\
&\text{Arc} \cos(\cos x) = x && 0 \leqslant x \leqslant \pi \\
&\text{Arc} \tan(\tan x) = x && |x| < \pi/2 \\
&\text{Arc cosec}(\text{cosec}\, x) = x && |x| < \pi/2 \\
&\text{Arc} \sec(\sec x) = x && 0 < x < \pi \\
&\text{Arc} \cot(\cot x) = x && 0 < x < \pi
\end{aligned} \tag{1.10-33}$$

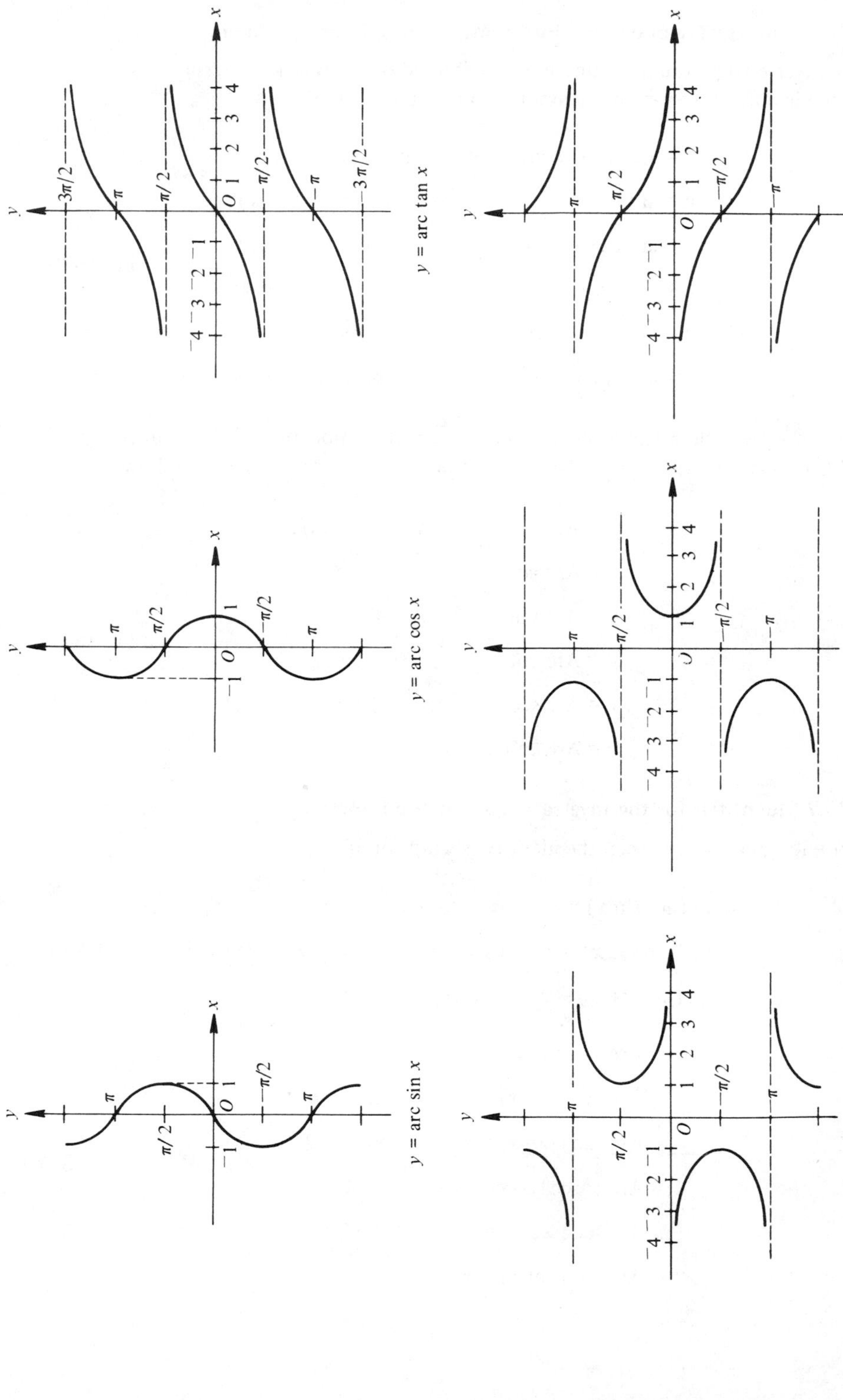

Fig. 1.10-5 Basic inverse trigonometric functions.

$$\begin{aligned}
\text{arc sin}\, x &= (-1)^{\kappa}\, \text{Arc sin}\, x + \kappa\pi \qquad \kappa = 0, \pm 1, \pm 2, \ldots \\
\text{arc cos}\, x &= \pm\, \text{Arc cos}\, x + 2\kappa\pi \\
\text{arc tan}\, x &= \text{Arc tan}\, x + \kappa\pi \\
\text{arc cot}\, x &= \text{Arc cot}\, x + \kappa\pi \\
\text{arc sec}\, x &= \pm\, \text{Arc sec}\, x + 2\kappa\pi \\
\text{arc cosec}\, x &= (-1)^{\kappa}\, \text{Arc sec}\, x + \kappa\pi
\end{aligned} \tag{1.10-34}$$

$$\begin{aligned}
\text{Arc sin}\, x &= \text{Arc cos}\sqrt{1 - x^2} = \text{Arc tan}\, \frac{x}{\sqrt{1 - x^2}} \\
&= \text{Arc cosec}\, \frac{1}{x} = \text{Arc sec}\, \frac{1}{\sqrt{1 - x^2}} = \text{Arc cot}\, \frac{\sqrt{1 - x^2}}{x}
\end{aligned} \tag{1.10-35}$$

$$\begin{aligned}
&\text{Arc sin}\,(-x) = -\,\text{Arc sin}\, x,\ \text{Arc cos}\,(-x) = \pi - \text{Arc cos}\, x \qquad |x| \leqslant 1 \\
&\text{Arc tan}\,(-x) = -\,\text{Arc tan}\, x,\ \text{Arc cot}\,(-x) = \pi - \text{Arc cot}\, x \\
&\text{Arc sec}\,(-x) = \pi - \text{Arc sec}\, x,\ \text{Arc cosec}\,(-x) = -\,\text{Arc cosec}\, x \qquad |x| \geqslant 1
\end{aligned} \tag{1.10-36}$$

$$\begin{aligned}
\text{Arc sin}\, x + \text{Arc sin}\, y &= \text{Arc sin}\,(x\sqrt{1 - y^2} + y\sqrt{1 - x^2}) \\
&\quad \text{if } |\text{Arc sin}\, x + \text{Arc sin}\, y| < \pi/2 \\
&= \pi - \text{Arc sin}\,(x\sqrt{1 - y^2} + y\sqrt{1 - x^2} \\
&\quad \text{if Arc sin}\, x + \text{Arc sin}\, y > \pi/2 \\
&= -\,\pi - \text{Arc sin}\,(x\sqrt{1 - y^2} + y\sqrt{1 - x^2}) \\
&\quad \text{if Arc sin}\, x + \text{Arc sin}\, y < -\,\pi/2
\end{aligned} \tag{1.10-37}$$

$$\text{Arc sin}\, x - \text{Arc sin}\, y = \text{Arc sin}\, x + \text{Arc sin}\,(-y) \tag{1.10-38}$$

$$\begin{aligned}
\text{Arc cos}\, x + \text{Arc cos}\, y &= \text{Arc cos}\,(xy - \sqrt{1 - x^2}\sqrt{1 - y^2}) \\
&\quad \text{if Arc cos}\, x + \text{Arc cos}\, y < \pi \\
&= 2\pi - \text{Arc cos}\,(xy - \sqrt{1 - y^2}) \\
&\quad \text{if Arc cos}\, x + \text{Arc cos}\, y > \pi
\end{aligned} \tag{1.10-39}$$

$$\text{Arc cos}\, x - \text{Arc cos}\, y = \text{Arc cos}\, x + \text{Arc cos}\,(-y) - \pi \tag{1.10-40}$$

$$\begin{aligned}
\text{Arc tan}\, x + \text{Arc tan}\, y &= \text{Arc tan}\, \frac{x + y}{1 - xy} \\
&\quad \text{if } |\text{Arc tan}\, x + \text{Arc tan}\, y| < \pi/2 \\
&= \pi + \text{Arc tan}\, \frac{x + y}{1 - xy} \\
&\quad \text{if Arc tan}\, x + \text{Arc tan}\, y > \pi/2
\end{aligned} \tag{1.10-41}$$

$$= -\pi - \text{Arc tan}\,\frac{x+y}{1-xy}$$
$$\text{if Arc tan}\,x + \text{Arc tan}\,y < -\pi/2$$

$$\text{Arc tan}\,x - \text{Arc tan}\,y = \text{Arc tan}\,x + \text{Arc tan}\,(-y) \qquad (1.10\text{-}42)$$

1.10.8 The Hyperbolic Functions—Graphs, Identities, Inverse Functions

The hyperbolic functions are defined for all real x as:

$$\text{The hyperbolic sine of } x \equiv \sinh x = \frac{e^x - e^{-x}}{2}$$

$$\text{The hyperbolic cosine of } x \equiv \cosh x = \frac{e^x + e^{-x}}{2}$$

$$\text{The hyperbolic tangent of } x \equiv \tanh x = \frac{e^x - e^{-x}}{e^x + e^{-x}}$$

$$\text{The hyperbolic cosecant of } x \equiv \text{cosech}\, x = \frac{1}{\sinh x} \qquad (1.10\text{-}43)$$

$$\text{The hyperbolic secant of } x \equiv \text{sech}\, x = \frac{1}{\cosh x}$$

$$\text{The hyperbolic contangent of } x \equiv \coth x = \frac{1}{\tanh x}.$$

Identities:

$$\sinh x + \cosh x = e^x$$
$$\cosh x - \sinh x = e^{-x} \qquad (1.10\text{-}44)$$

$$\tanh x = \frac{\sinh x}{\cosh x} \qquad \coth x = \frac{\cosh x}{\sinh x} \qquad (1.10\text{-}45)$$

$$\cosh^2 x - \sinh^2 x = 1$$
$$\text{sech}^2 x + \tanh^2 x = 1 \qquad (1.10\text{-}46)$$
$$\text{cosech}^2 x - \coth^2 x = -1$$

$$\sinh (x \pm y) = \sinh x \cosh y \pm \cosh x \sinh y$$
$$\cosh (x \pm y) = \cosh x \cosh y \pm \sinh x \sinh y \qquad (1.10\text{-}47)$$
$$\tanh (x \pm y) = \frac{\tanh x \pm \tanh y}{1 \pm \tanh x \tanh y}$$

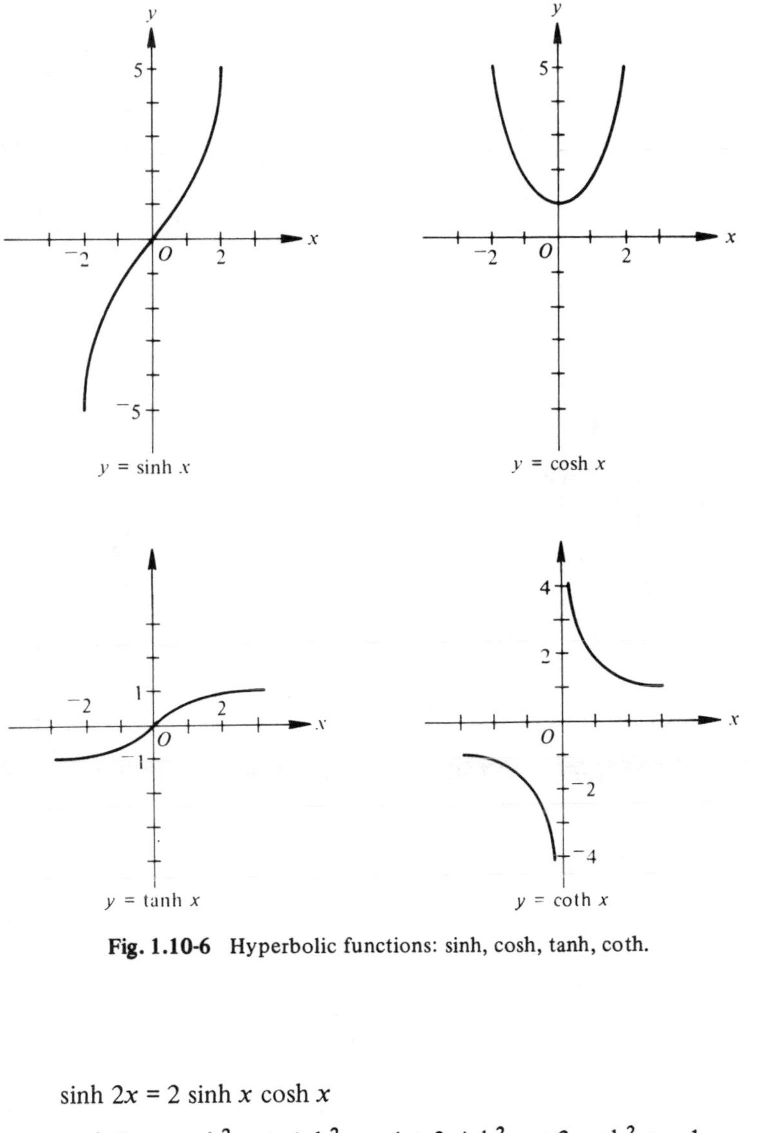

Fig. 1.10-6 Hyperbolic functions: sinh, cosh, tanh, coth.

$$
\begin{aligned}
&\sinh 2x = 2 \sinh x \cosh x \\
&\cosh 2x = \cosh^2 x + \sinh^2 x = 1 + 2 \sinh^2 x = 2 \cosh^2 x - 1 \\
&\tanh 2x = \frac{2 \tanh x}{1 + \tanh^2 x}
\end{aligned}
\tag{1.10-48}
$$

$$
\begin{aligned}
&\sinh \frac{x}{2} = \pm \sqrt{\frac{\cosh x - 1}{2}} \qquad \cosh \frac{x}{2} = \pm \sqrt{\frac{1 + \cosh x}{2}} \\
&\tanh \frac{x}{2} = \frac{\cosh x - 1}{\sinh x} = \frac{\sinh x}{\cosh x + 1}
\end{aligned}
\tag{1.10-49}
$$

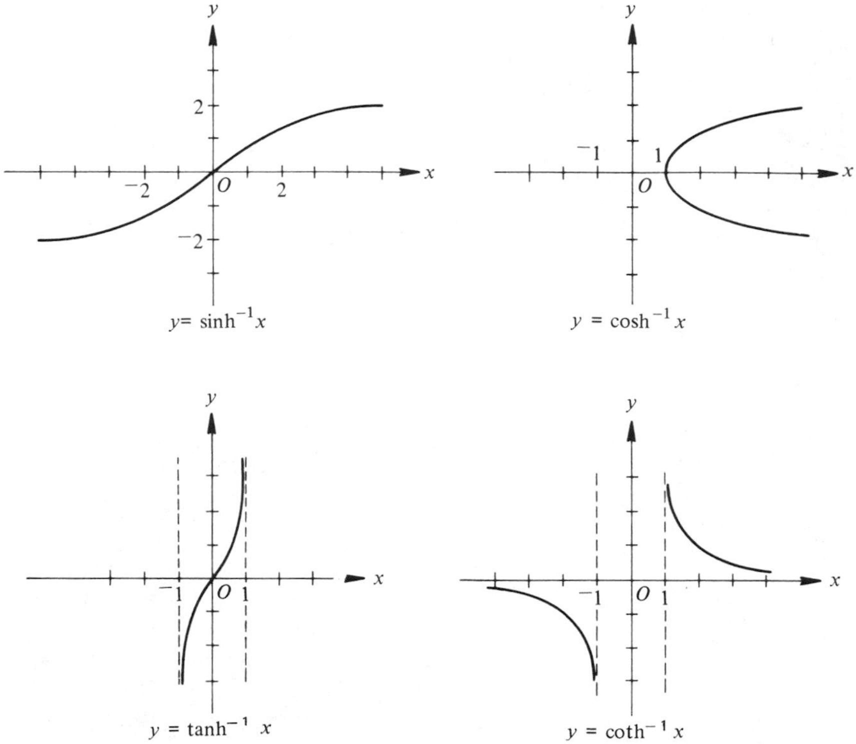

Fig. 1.10-7 Inverse hyperbolic functions.

$$\sinh x + \sinh y = 2 \sinh \frac{x+y}{2} \cosh \frac{x-y}{2}$$

$$\sinh x - \sinh y = 2 \sinh \frac{x-y}{2} \cosh \frac{x+y}{2}$$

$$\cosh x + \cosh y = 2 \cosh \frac{x+y}{2} \cosh \frac{x-y}{2} \tag{1.10-50}$$

$$\cosh x - \cosh y = 2 \sinh \frac{x+y}{2} \sinh \frac{x-y}{2}$$

$$\tanh x \pm \tanh y = \frac{\sinh (x \pm y)}{\cosh x \cosh y}$$

The first two identities give de Moivre's theorem for the hyperbolic functions

$$(\cosh x \pm \sinh x)^n = \cosh nx \pm \sinh nx \tag{1.10-51}$$

and as in section 1.10.4, the Binomial Theorem can be used to write cosh nx and sinh nx for n a positive integer in terms of powers of sinh x and cosh x, and vice versa.

The inverse hyperbolic functions are defined to be the inverse functions corresponding to the hyperbolic functions. Explicitly, $y = \arg\sinh x \equiv \sinh^{-1} x$ implies $x = \sinh y$; and similarly with $y = \arg\cosh x \equiv \cosh^{-1} x$, $y = \arg\tanh x \equiv \tanh^{-1} x$ and $y = \arg\coth x \equiv \coth^{-1} x$. Of these four, only $\cosh^{-1} x$ is multiple valued. The single-valued function Arg cosh x is obtained by restricting its range to $0 \leqslant y < \infty$. These functions are related to certain natural logarithms by

$$\left.\begin{aligned} \arg\sinh x &= \ln\,[x + \sqrt{x^2 + 1}] \\ \operatorname{Arg}\cosh x &= \ln\,[x + \sqrt{x^2 - 1}], \qquad & x \geqslant 1 \\ \arg\tanh x &= \frac{1}{2}\ln\frac{1 + x}{1 - x}, \qquad & |x| < 1 \\ \arg\coth x &= \frac{1}{2}\ln\frac{x + 1}{x - 1}, \qquad & |x| > 1 \end{aligned}\right\} \tag{1.10-51}$$

and satisfy the identities

$$\arg\sinh x \pm \arg\sinh y = \arg\sinh\,[x\sqrt{1 + y^2} \pm y\sqrt{1 + x^2}] \tag{1.10-52}$$

$$|\operatorname{Arg}\cosh x \pm \operatorname{Arg}\cosh y| = \operatorname{Arg}\cosh\,[xy \pm \sqrt{(x^2 - 1)(y^2 - 1)}]$$
$$(x \geqslant 1, y \geqslant 1) \tag{1.10-53}$$

$$\arg\tanh x \pm \arg\tanh y = \arg\tanh\frac{x \pm y}{1 \pm xy}, \qquad |x| < 1, \quad |y| < 1 \tag{1.10-54}$$

1.10.9 Elementary Spherical Trigonometry

Consider a sphere of radius 1. Then a plane intersects the surface of the sphere in a great circle if the plane passes through the center O of the sphere. Let A, B, C be three distinct points on the sphere, not all on the same great circle. The sets of points O, A, B; O, A, C; O, B, C each determine a unique great circle. The spherical triangle formed by taking, in each case, the shorter of the two possible arcs of the great circle is an Euler triangle.

The lengths a, b, c of the arcs $\overparen{BC}$, $\overparen{AC}$, $\overparen{AB}$ are the sides of the spherical triangle and from $s = r\theta$ $(r = 1)$ they are given by angles BOC, AOC, AOB respectively. The angle between two intersecting arcs is taken as the angle between the tangents to the arcs at the point of intersection, and thus the angles α, β, γ of the spherical triangle are determined. For Euler triangles, each side and each angle is less than π.

The sum of the angles is always greater than π and $E = \alpha + \beta + \gamma - \pi$ is called the

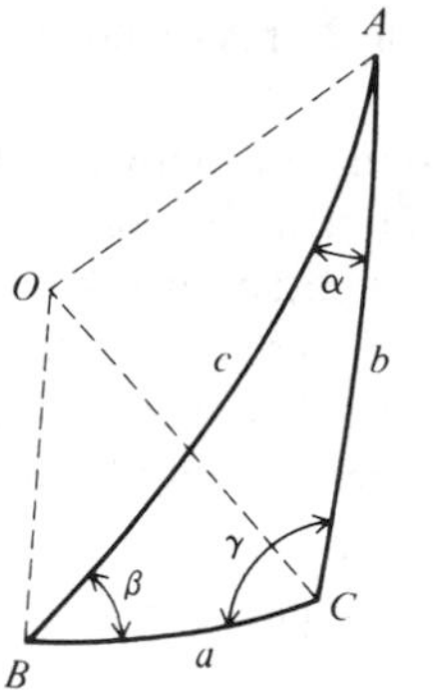

Fig. 1.10-8 Sphere with Euler triangle ABC.

spherical excess. (The area of a spherical triangle on a sphere of radius r is given by Er^2.)

In the solution of spherical triangles, there are two cases.

Case I Right spherical triangles

Let γ be a right angle. Given any other two parts of the triangle, the remaining three can be found by the formulas known as Napier's rule; i.e., consider the scheme of Fig. 1.10-9. Taking any part, the two parts next to it are called adjacent

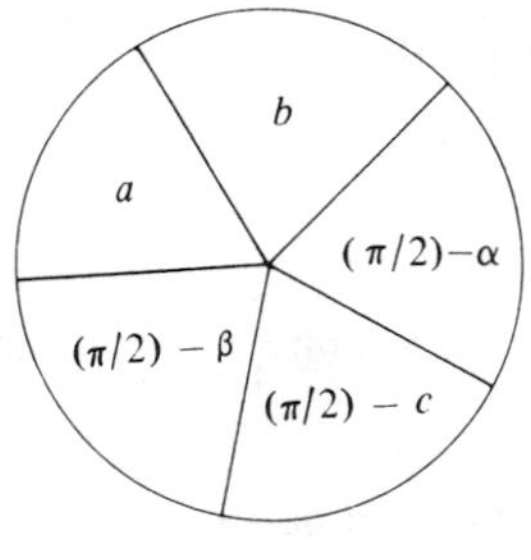

Fig. 1.10-9 Scheme for Napier's rule.

parts, while the other two are called opposite parts. The sine of any part is then equal to either the product of the tangents of the adjacent parts, or the product of the cosines of the opposite parts. Explicitly

$$\begin{aligned} &\sin a = \sin\alpha \sin c && \sin a = \tan b \cot\beta \\ &\sin b = \sin\beta \sin c && \sin b = \tan a \cot\alpha \\ &\cos\alpha = \cos a \sin\beta && \cos\alpha = \tan b \cot c \\ &\cos c = \cos a \cos b && \cos c = \cot\alpha \cot\beta \\ &\cos\beta = \sin\alpha \cos b && \cos\beta = \tan a \cot c \end{aligned} \tag{1.10-55}$$

Case II Oblique spherical triangles
Here there are six possible problems:

(1) Given the three sides
The three angles are found from

$$\tan \frac{\alpha}{2} = \frac{d}{\sin (s - a)} \tag{1.10-56}$$

and corresponding formulas for β, γ; where

$$2s = a + b + c \quad \text{and} \quad d = \sqrt{\frac{\sin (s - a) \sin (s - b) \sin (s - c)}{\sin s}} \tag{1.10-57}$$

or, use

$$\cos a = \cos b \cos c + \sin b \sin c \cos \alpha \qquad \text{(cosine law for sides)} \tag{1.10-58}$$

(2) Given the three angles
The three sides are given by

$$\tan \frac{a}{2} = D \cos (S - \alpha) \tag{1.10-59}$$

and corresponding formulas for b, c; where

$$2S = \alpha + \beta + \gamma \quad \text{and} \quad D = \sqrt{\frac{-\cos S}{\cos (S - \alpha) \cos (S - \beta) \cos (S - \gamma)}} \tag{1.10-60}$$

or, use

$$\cos \alpha = -\cos \beta \cos \gamma + \sin \beta \sin \gamma \cos a \qquad \text{(cosine law for angles)} \tag{1.10-61}$$

(3) Given two sides and the included angle
Use

$$\tan \frac{1}{2} (\alpha - \beta) = \cos \frac{\gamma}{2} \cdot \frac{\sin \frac{1}{2} (a - b)}{\sin \frac{1}{2} (a + b)} \tag{1.10-62}$$

and

$$\tan \frac{1}{2} (\alpha + \beta) = \cot \frac{\gamma}{2} \cdot \frac{\cos \frac{1}{2} (a - b)}{\cos \frac{1}{2} (a + b)} \tag{1.10-63}$$

to get the sum and difference of, and thus the two angles. The third side can be found from either of

$$\tan\frac{c}{2} = \tan\frac{1}{2}(a-b)\cdot\frac{\sin\frac{1}{2}(\alpha+\beta)}{\sin\frac{1}{2}(\alpha-\beta)}$$
$$\tan\frac{c}{2} = \tan\frac{1}{2}(a+b)\cdot\frac{\cos\frac{1}{2}(\alpha+\beta)}{\cos\frac{1}{2}(\alpha-\beta)} \tag{1.10-64}$$

These four formulas are known as Napier's analogies. Alternatively, to get the missing side, use

$$\frac{\sin a}{\sin\alpha} = \frac{\sin b}{\sin\beta} = \frac{\sin c}{\sin\gamma}\ \text{(law of sines)} \tag{1.10-65}$$

(4) Given two angles and the included side
Use Napier's analogies.
(5) Given two sides and an angle opposite one of them
(6) Given two angles and a side opposite one of them
In these two cases, use the Law of Sines and Napier's analogies. (In (5), there may be more than one solution).

Finally, for spherical triangles, the following relations are known as Gauss's equations:

$$\cos\frac{c}{2}\sin\frac{1}{2}(\alpha+\beta) = \cos\frac{\gamma}{2}\cos\frac{1}{2}(a-b)$$
$$\cos\frac{c}{2}\cos\frac{1}{2}(\alpha+\beta) = \sin\frac{\gamma}{2}\cos\frac{1}{2}(a+b)$$
$$\sin\frac{c}{2}\sin\frac{1}{2}(\alpha-\beta) = \cos\frac{\gamma}{2}\sin\frac{1}{2}(a-b) \tag{1.10-66}$$
$$\sin\frac{c}{2}\cos\frac{1}{2}(\alpha-\beta) = \sin\frac{\gamma}{2}\sin\frac{1}{2}(a+b)$$

1.11 ANALYTIC GEOMETRY OF TWO-SPACE

1.11.1 Cartesian, Polar Coordinate Systems

In a plane let $x'ox, y'oy$ be two mutually perpendicular lines intersecting at O, and dividing the plane into four quadrants, as shown in Fig. 1.11-1. To each point in the plane there corresponds an ordered pair of numbers, the x- and y-coordinates of the point, where the x-coordinate (abscissa) is the directed distance of the point from the y-axis and the y-coordinate (ordinate) is the directed distance of the point from the x-axis. The direction of measurement is taken to be positive for distance

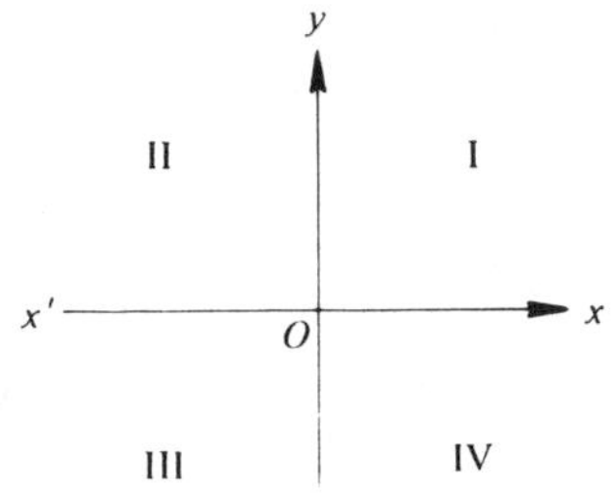

Fig. 1.11-1 Cartesian coordinate system.

measured horizontally to the right or vertically upwards, the opposites are taken to be negative. This one-to-one correspondence between the points of the plane and ordered pairs of real numbers is the rectangular or Cartesian coordinate system in two-space. Other coordinate systems are possible. For example, in polar coordinates each point is specified by the two coordinates (r, θ) where $r = |\overline{OP}|$ and θ is the angle measured positive counterclockwise from the $+x$-axis to the vector **OP**. Note that the polar coordinates of a point are not unique, e.g., $(2, 37°)$ is the same point as $(2, 397°)$, or $(2, -323°)$. The equations connecting polar and rectangular coordinates are:

$$x = r\cos\theta \qquad y = r\sin\theta \tag{1.11-1}$$

$$r = \sqrt{x^2 + y^2} \qquad \theta = \tan^{-1}\frac{y}{x} \tag{1.11-2}$$

1.11.2 Distance Formula, Division of a Line Segment in a Given Ratio

The distance between the two points $P_1(x_1, y_1)$, $P_2(x_2, y_2)$ is

$$d = \sqrt{(x_2 - x_1)^2 + (y_2 - y_1)^2} \tag{1.11-3}$$

That point $P(x, y)$ which divides the segment from $P_1(x_1, y_1)$ to $P_2(x_2, y_2)$ in the ratio r has coordinates

$$x = \frac{x_1 + r x_2}{1 + r} \qquad y = \frac{y_1 + r y_2}{1 + r} \tag{1.11-4}$$

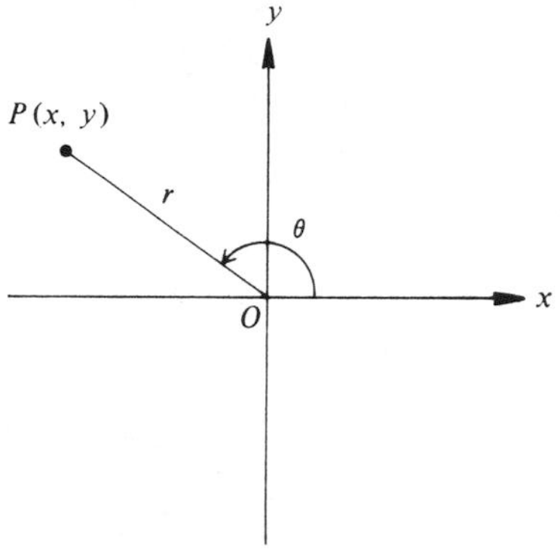

Fig. 1.11-2 Polar coordinates.

and, in particular when $r = 1$, P is the midpoint of $\overline{P_1P_2}$ given by

$$x = \frac{x_1 + x_2}{2}, \qquad y = \frac{y_1 + y_2}{2} \tag{1.11-5}$$

1.11.3 The Straight Line

1.11.3.1 Angle of Inclination, Slope, Intercepts The angle of inclination of a line L is the smallest angle obtained when the $+x$-axis is rotated counterclockwise into coincidence with the line. Thus $0 \leqslant \alpha < 180°$. The slope of L is defined to be $\tan \alpha$ and will be denoted by s. If $P_1(x_1, y_1), P_2(x_2, y_2)$ are two points on L, then

$$s = \frac{y_2 - y_1}{x_2 - x_1} \tag{1.11-6}$$

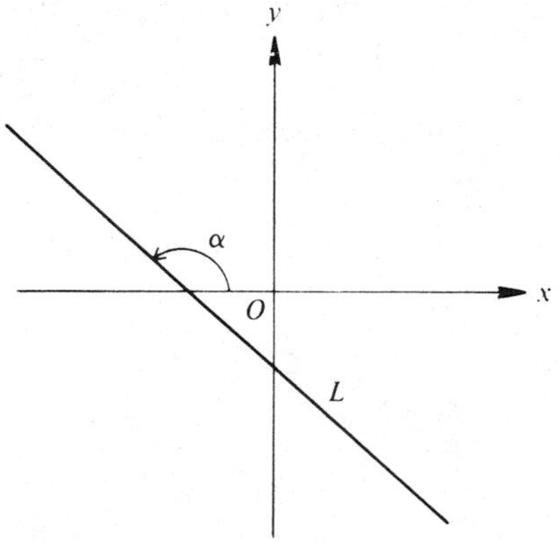

Fig. 1.11-3 Intercept of a line on x- and y-axes.

The x-coordinate of the point where L intersects the x-axis is the x-intercept of L and similarly the y-coordinate of the point where L intersects the y-axis is the y-intercept of L.

1.11.3.2 Standard Forms of the Equation An equation of the form $ax + by + c = 0$; $(a, b) \neq (0, 0)$, always represents a line, and vice versa. Certain standard forms are used, the names implying the information which is given.

(1) The two point form

$$\frac{x_2 - x_1}{x_1 - x} = \frac{y_2 - y_1}{y_1 - y} \tag{1.11-7}$$

with special case

(2) The intercept form

$$\frac{x}{h} + \frac{y}{k} = 1 \tag{1.11-8}$$

where h, k are the x, y intercepts respectively.

(3) The point slope form

$$\frac{y - y_1}{x - x_1} = s \tag{1.11-9}$$

with special case

(4) The slope intercept form

$$y = sx + k \tag{1.11-10}$$

(5) The normal form

$$x \cos \alpha + y \sin \alpha = p \tag{1.11-11}$$

where p is the perpendicular distance from the origin to the line and α is the angle from the $+x$-axis to the line segment along which this perpendicular distance is taken.

(6) The parametric form

$$\begin{aligned} x &= x_1 + t(x_2 - x_1) \\ y &= y_1 + t(y_2 - y_1) \end{aligned} \tag{1.11-12}$$

where $P_1(x_1, y_1), P_2(x_2, y_2)$ are two points on L and $-\infty < t < \infty$.

1.11.3.3 Distance from a Line to a Point The distance between the line $ax + by + c = 0$ and the point $P_1(x_1, y_1)$ is given by

$$\left| \frac{ax_1 + by_1 + c}{\sqrt{a^2 + b^2}} \right| \tag{1.11-13}$$

1.11.3.4 Angle between Two Lines, Parallel and Perpendicular Lines The angle θ between two lines L_1 and L_2 of slopes s_1 and s_2 respectively is given by

$$\tan \theta = \frac{s_2 - s_1}{1 + s_1 s_2} \tag{1.11-14}$$

If L_1 is perpendicular to L_2 then

$$s_1 s_2 = -1 \tag{1.11-15}$$

and if L_1 is parallel to L_2, then $s_1 = s_2$. All lines parallel to or perpendicular to $ax + by + c = 0$ are given by

$$ax + by + q = 0 \tag{1.11-16}$$

$$bx - ay + q = 0 \tag{1.11-17}$$

1.11.3.5 Intersection of Lines, Angle Bisectors The intersection point of the two lines $a_1x + b_1y + c_1 = 0$, $a_2x + b_2y + c_2 = 0$ is found by solving these equations simultaneously. The two lines bisecting the angles of intersection of these two lines have equations

$$\frac{a_1x + b_1y + c_1}{\sqrt{a_1^2 + b_1^2}} = \pm \frac{a_2x + b_2y + c_2}{\sqrt{a_2^2 + b_2^2}} \tag{1.11-18}$$

1.11.3.6 Area of a Triangle, Area of a Polygon The area A of the triangle with vertices $P_1(x_1, y_1)$, $P_2(x_2, y_2)$ and $P_3(x_3, y_3)$ is given by the absolute value of the determinant

$$A = \frac{1}{2} \begin{vmatrix} x_1 & y_1 & 1 \\ x_2 & y_2 & 1 \\ x_3 & y_3 & 1 \end{vmatrix} \tag{1.11-19}$$

The area A of a polygon with vertices $P_1(x_1, y_1)$, $P_2(x_2, y_2)$, . . . , $P_n(x_n, y_n)$ in that order is

$$A = \frac{1}{2}\left\| \begin{vmatrix} x_1 & x_2 \\ y_1 & y_2 \end{vmatrix} + \begin{vmatrix} x_2 & x_3 \\ y_2 & y_3 \end{vmatrix} + \cdots + \begin{vmatrix} x_n & x_1 \\ y_n & y_1 \end{vmatrix} \right\| \qquad (1.11\text{-}20)$$

1.11.4 The Circle

1.11.4.1 General Equation, Parametric Form The equation of the circle with center (h, k) and radius a is

$$(x - h)^2 + (y - k)^2 = a^2 \qquad (1.11\text{-}21)$$

and any equation of the form

$$x^2 + y^2 + Dx + Ey + F = 0 \qquad (1.11\text{-}22)$$

can be written in this form by completion of the square on the terms in x and y. The parametric form of (1.11-21) is

$$\left.\begin{aligned} x &= h + a \cos \theta \\ y &= k + a \sin \theta \end{aligned}\right\} \quad 0 \leqslant \theta < 2\pi \qquad (1.11\text{-}23)$$

The equation of the circle passing through three given, noncollinear points $P_1(x_1, y_1), P_2(x_2, y_2), P_3(x_3, y_3)$ may be written

$$\begin{vmatrix} x^2 + y^2 & x & y & 1 \\ x_1^2 + y_1^2 & x_1 & y_1 & 1 \\ x_2^2 + y_2^2 & x_2 & y_2 & 1 \\ x_3^2 + y_3^2 & x_3 & y_3 & 1 \end{vmatrix} = 0 \qquad (1.11\text{-}24)$$

1.11.4.2 Tangents The equation of the tangent to the circle $(x - h)^2 + (y - k)^2 = a^2$ at the point (x_1, y_1) is

$$(x_1 - h)(x - h) + (y_1 - k)(y - k) = a^2 \qquad (1.11\text{-}25)$$

and the corresponding normal is

$$(y_1 - k)(x - h) - (x_1 - h)(y - k) = 0 \qquad (1.11\text{-}26)$$

The two tangents to $(x - h)^2 + (y - k)^2 = a^2$ with slope s are

$$y - k = s(x - h) \pm a\sqrt{1 + s^2} \qquad (1.11\text{-}27)$$

1.11.5 The Conic Sections

1.11.5.1 General Definition, Test for Type Geometrically, a conic section is a plane section of a right circular cone. If the plane passes through the vertex, then either a point, a straight line, or two straight lines are obtained. If the plane does not pass through the vertex, then if it

(1) is perpendicular to the axis of the cone a circle is obtained;
(2) is parallel to a generator, a parabola is obtained;
(3) intersects only one nappe of the cone and is not perpendicular to the axis, an ellipse is obtained;
(4) intersects both nappes, then a hyperbola is the result; as shown in Fig. 1.11-4.

Equivalently, a conic section is the locus of a point which moves so that the ratio of

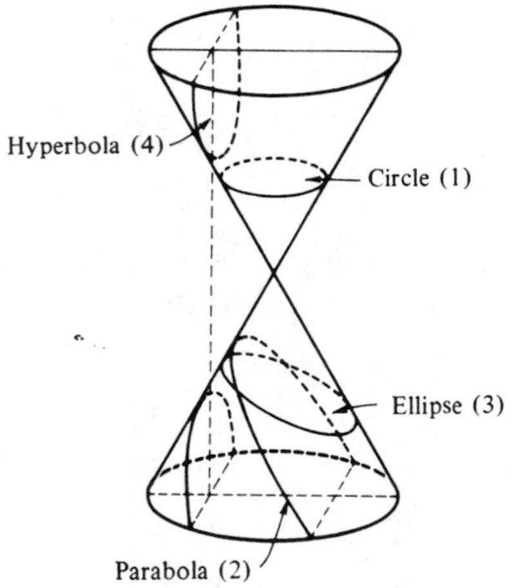

Fig. 1.11-4 Conic sections.

its distance from a fixed point (the focus) to a fixed straight line (the directrix) is a constant (the eccentricity e). If $e = 1$, the conic is a parabola; if $e < 1$, the conic is an ellipse ($e = 0$ corresponds to a circle); if $e > 1$, the conic is a hyperbola. Any equation of the form

$$Ax^2 + Bxy + Cy^2 + Dx + Ey + F = 0 \qquad (1.11\text{-}28)$$

with $(A, B, C) \neq (0, 0, 0)$ corresponds to a conic section and vice versa. The type of conic section is determined by the values of the characteristic $B^2 - 4AC$ and the discriminant

$$\begin{vmatrix} A & \frac{B}{2} & \frac{D}{2} \\ \frac{B}{2} & C & \frac{E}{2} \\ \frac{D}{2} & \frac{E}{2} & F \end{vmatrix}$$

of the eq. (1.11-28) as shown in Table 1.11-1.

Table 1.11-1 Classification of the Conic Sections

Characteristic	Discriminant	Type of Conic
0	$\neq 0$	Non-degenerate parabola
0	0	Degenerate parabola: 2 real or imaginary parallel lines
< 0	$\neq 0$	Non-degenerate ellipse or circle, real or imaginary
< 0	0	Degenerate ellipse: point ellipse or circle
> 0	$\neq 0$	Non-degenerate hyperbola
> 0	0	Degenerate hyperbola: 2 distinct intersecting lines

1.11.5.2 The Parabola Choosing $2a$ to be the distance between the focus and the directrix, the equation of the parabola shown with vertex at (h, k), axis parallel to the x-axis and opening to the right is

$$(y - k)^2 = 4a(x - h) \tag{1.11-29}$$

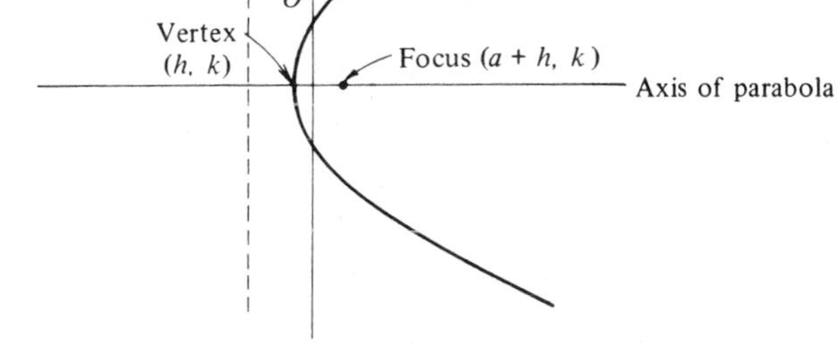

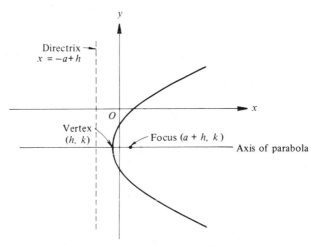

Fig. 1.11-5 The parabola $(y - k)^2 = 4a(x - h)$.

For the parabola opening to the left,

$$(y - k)^2 = -4a(x - h) \tag{1.11-30}$$

and if the axis is parallel to the y-axis, then the corresponding equations are

$$(x - h)^2 = 4a(y - k) \tag{1.11-31}$$

and

$$(x - h)^2 = -4a(y - k) \tag{1.11-32}$$

for opening upwards, and downwards, respectively.

1.11.5.3 The Ellipse Choosing $2a$ to be the length of the major axis, $2b$ to be the length of the minor axis, the equation of the ellipse shown with center at (h, k), major axis parallel to the x-axis is

$$\frac{(x - h)^2}{a^2} + \frac{(y - k)^2}{b^2} = 1 \tag{1.11-33}$$

For the major axis parallel to the y-axis

$$\frac{(y - k)^2}{a^2} + \frac{(x - h)^2}{b^2} = 1 \tag{1.11-34}$$

For any ellipse

$$e = \frac{\sqrt{a^2 - b^2}}{a} \tag{1.11-35}$$

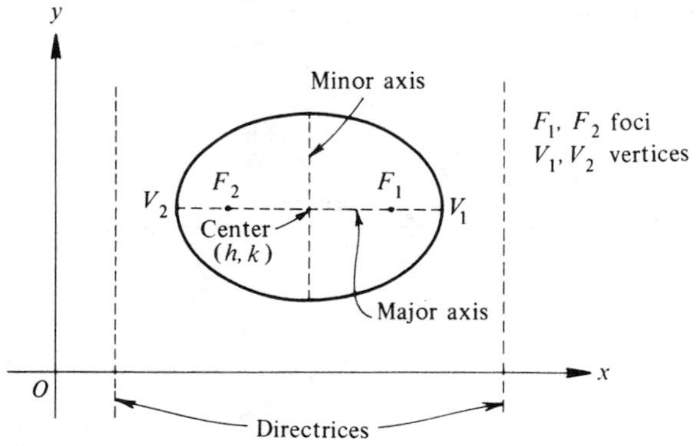

Fig. 1.11-6 The ellipse $(x - h)^2/a^2 + (y - k)^2/b^2 = 1$.

and the distance from the center to either focus is ae, while the distance from the center to either directrix is a/e. Any line segment joining a focus to a point on the ellipse is called a focal radius. An alternative definition comes from the property that for any point on an ellipse, the sum of the two focal radii is a constant, $2a$. The parametric equations of the ellipse of (1.11-33) are

$$\left.\begin{aligned} x &= h + a\cos\theta \\ y &= k + b\sin\theta \end{aligned}\right\} \quad 0 \leqslant \theta < 2\pi \tag{1.11-36}$$

1.11.5.4 The Hyperbola Choosing $2a$ to be the distance between the vertices, *i.e.*, the length of the transverse axis, and

$$b^2 = a^2(e^2 - 1) \tag{1.11-37}$$

the equation of the hyperbola shown with center (h, k), transverse axis parallel to the x-axis is

$$\frac{(x-h)^2}{a^2} - \frac{(y-k)^2}{b^2} = 1 \tag{1.11-38}$$

For the transverse axis parallel to the y axis

$$\frac{(y-k)^2}{a^2} - \frac{(x-h)^2}{b^2} = 1 \tag{1.11-39}$$

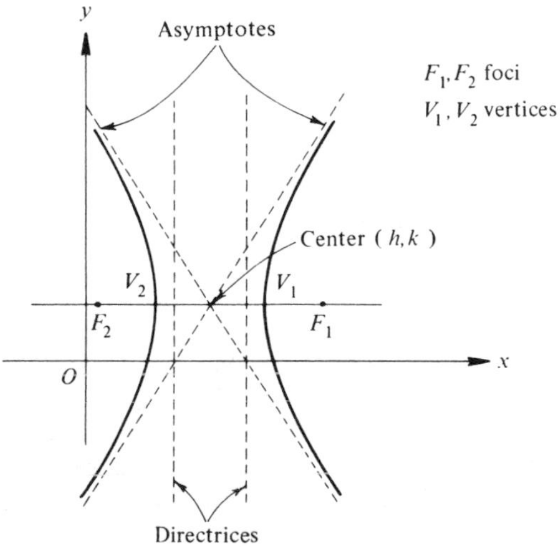

Fig. 1.11-7 The hyperbola $(x-h)^2/a^2 - (y-k)^2/b^2 = 1$.

Even though the hyperbola of (1.11-38) does not intersect the line through the center parallel to the y-axis, $2b$ is called the length of the conjugate axis. The distance from the center to either focus is ae, and the distance from the center to either directrix is a/e. The two asymptotes of the hyperbola (1.11-38) are given by

$$\frac{(x-h)^2}{a^2} - \frac{(y-k)^2}{b^2} = 0 \tag{1.11-40}$$

and similarly for (1.11-39). If $a = b$, the hyperbola is equilateral. Two hyperbolas such that the transverse axis of one is the conjugate axis of the other, and vice versa, are said to be conjugate hyperbolas. For any point on a hyperbola, the difference of the two focal radii is a constant, $\pm 2a$. The parametric equations of the hyperbola of (1.11-38) are

$$\begin{aligned} x &= h + a \sec \theta \\ y &= k + b \tan \theta \end{aligned} \tag{1.11-41}$$

or

$$\begin{aligned} x &= h + a \cosh t \\ y &= k + b \sinh t \end{aligned} \tag{1.11-42}$$

1.11.6 Transformations of Axes

1.11.6.1 Translations Let O' have coordinates (h, k); then the equations relating the coordinates of the same point P under a translation of axes, as shown, are

$$\begin{aligned} x &= x' + h \\ y &= y' + k \end{aligned} \tag{1.11-43}$$

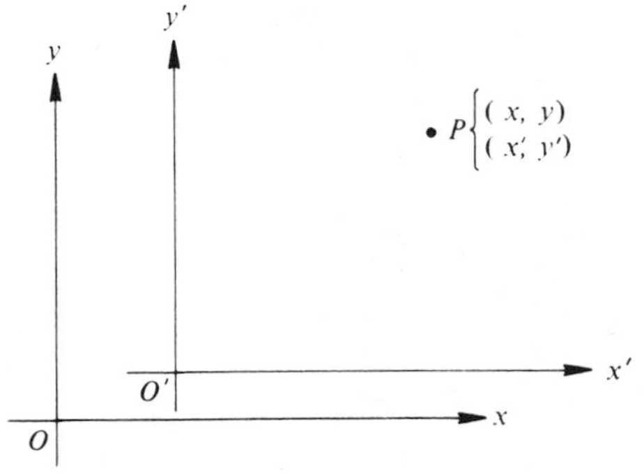

Fig. 1.11-8 Translation of axes $(x, y) \rightarrow (x', y')$.

In the equation of the general conic (1.11-28), if the characteristic $\neq 0$, then substituting (1.11-43) will give equations for h and k which can be solved so that the coefficients of x' and y' will be 0.

1.11.6.2 Direction Angles, Rotation of Axes The angles a vector makes with the $+x$- and $+y$-axes are the direction angles of the vector, and $\cos\alpha$, $\cos\beta$, are called direction cosines for the vector. Consider a rotation of the coordinate system with the origin fixed so that the new axis OX' has a direction angle θ with respect to the OX-axis.

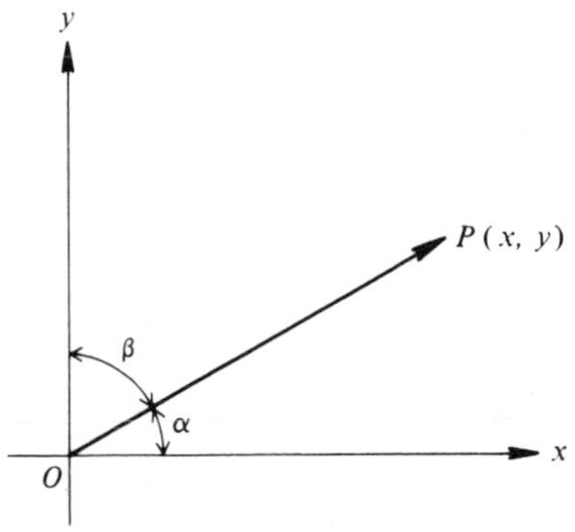

Fig. 1.11-9 Direction angles for vector **OP**.

Then if α_{ij} denotes the cosine of the angle between the x_i and x_j' axes in Fig. 1.11-10, then the equations relating coordinates are

$$\begin{aligned} x_1 &= \alpha_{11}x_1' + \alpha_{12}x_2' \\ x_2 &= \alpha_{21}x_1' + \alpha_{22}x_2' \end{aligned} \qquad (1.11\text{-}44)$$

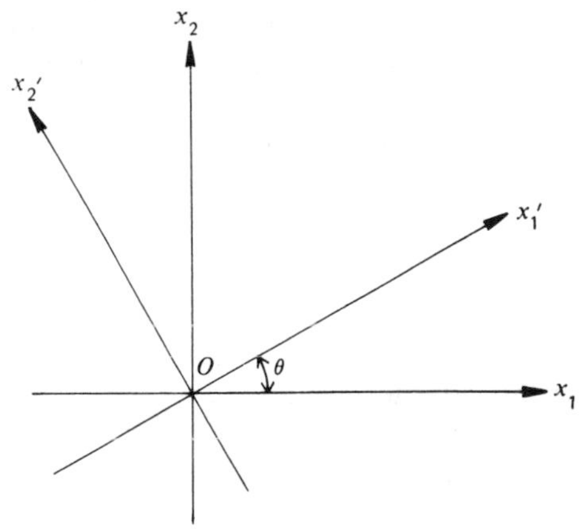

Fig. 1.11-10 Rotation of axes $(x_1, x_2) \to (x_1', x_2')$.

and the inverse transformation is

$$
\begin{aligned}
x'_1 &= \alpha_{11}x_1 + \alpha_{21}x_2 \\
x'_2 &= \alpha_{22}x_1 + \alpha_{22}x_2
\end{aligned}
\qquad (1.11\text{-}45)
$$

both of which sets can be obtained from the multiplication table:

	x'_1	x'_2
x_1	α_{11}	α_{12}
x_2	α_{21}	α_{22}

(1.11-46)

In the particular two-space case under consideration, $\alpha_{11} = \cos\theta$, $\alpha_{12} = \cos(90° + \theta) = -\sin\theta$, $\alpha_{21} = \cos(90° - \theta) = \sin\theta$ and $\alpha_{22} = \cos\theta$, so reverting to $(x_1, x_2) = (x, y)$ the table is

	x'	y'
x	$\cos\theta$	$-\sin\theta$
y	$\sin\theta$	$\cos\theta$

(1.11-47)

In the equation of the general conic (1.11-28), if $A = C$, a rotation through 45° will give an equation in which the coefficient of $x'y'$ is 0, and if $A \neq C$, the relation

$$\tan 2\theta = \frac{B}{A - C} \qquad (1.11\text{-}48)$$

together with the appropriate trigonometric identities will give $\cos\theta$ and $\sin\theta$ such that the coefficient of $x'y'$ is 0. Then completion of squares will lead to the equation of a conic in one of the forms of sections 1.11.5-2 through .4, or to a degenerate case. The discriminant and the characteristic of (1.11-28) are invariant under translation and rotation, as is the quantity $A + C$.

1.11.7 Curves in Polar Coordinates

(1) The line

The relations $r\cos\theta = a$, $r\sin\theta = b$ are the polar forms of the lines $x = a$, $y = b$ respectively and a line through the origin has θ = constant as its polar form. The general line in polar form is obtained from the normal form (1.11-11) by setting $x = r\cos\theta$, $y = r\sin\theta$, *i.e.*,

$$r\cos\theta\cos\alpha + r\sin\theta\sin\alpha = p, \quad \text{or} \quad r\cos(\theta - \alpha) = p \qquad (1.11\text{-}49)$$

(2) The circle

The curve $r = a$ corresponds to the circle center at the origin and radius a, while $r = 2a \cos\theta$, $r = 2a \sin\theta$ are the circles of radius a passing through the origin with centers on the $+x$-, $+y$-axes respectively. The general circle with center in polar coordinates at (r_o, α) and radius a is

$$r^2 + r_o^2 - 2rr_o \cos(\theta - \alpha) = a^2 \tag{1.11-50}$$

(3) The conic

Either of the equations

$$r = \frac{k_1}{k_2 + k_3 \cos\theta}, \qquad r = \frac{k_1}{k_2 + k_3 \sin\theta} \tag{1.11-51}$$

represents a conic with one focus at the origin and a directrix which is either horizontal or vertical; k_1, k_2, k_3 are real numbers and $e = |k_3/k_2|$.

(4) Limaçons

Equations of the form

$$r = a \pm b \cos\theta, \quad r = a \pm b \sin\theta \qquad a \neq b \tag{1.11-52}$$

correspond to limaçons.

(5) Cardioids

This is the special case of (1.11-52) for which $a = b$.

(6) Lemniscates

These are given by either of

$$r^2 = a^2 \cos 2\theta, \quad r^2 = a^2 \sin 2\theta \tag{1.11-53}$$

(7) Rose or petal curves

Equations of the form

$$r = a \sin n\theta, \quad r = a \cos n\theta \tag{1.11-54}$$

where n is an integer correspond to roses–if n is odd the number of petals is n, and if n is even the number of petals is $2n$.

(8) The spiral of Archimedes

This comes from

$$r = a\theta \tag{1.11-55}$$

and is shown in the diagram for $a > 0$. The dotted portion corresponds to θ negative.

(9) The logarithmic spiral
This is given by

$$r = e^{a\theta} \tag{1.11-56}$$

and is shown for $a > 0$. The dotted portion corresponds to $\theta < 0$.

(10) The hyperbolic spiral
The equation for this is

$$r\theta = a \tag{1.11-57}$$

(11) Semicubical parabolas
Semicubical parabolas are represented by equations of the form

$$r = a \tan^2 \theta \sec \theta \tag{1.11-58}$$

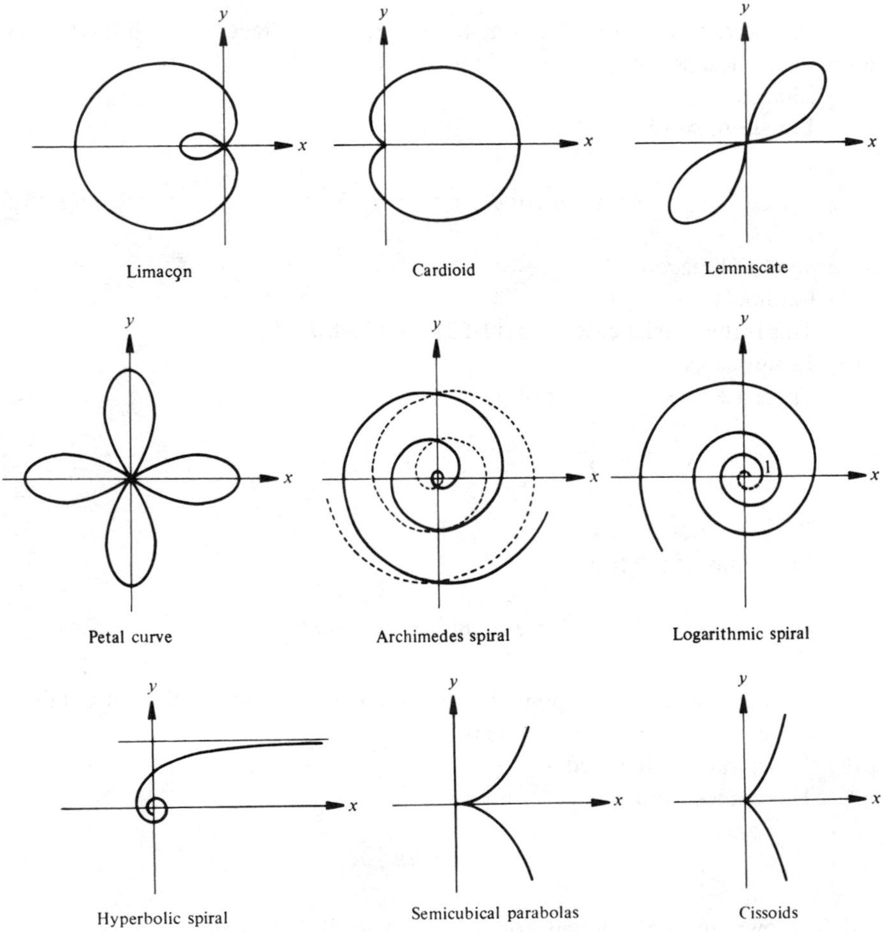

Fig. 1.11-11 Several curves with polar parameters.

(12) Cissoids

This is represented by

$$r = a \sin\theta \tan\theta \qquad (1.11\text{-}59)$$

1.11.8 Curves in Parametric Form

(1) The cycloids

Let a circle of radius a roll along the x-axis without slipping. A fixed point P on the circumference of the circle will describe a path called a cycloid, as shown in Fig. 1.11-12.

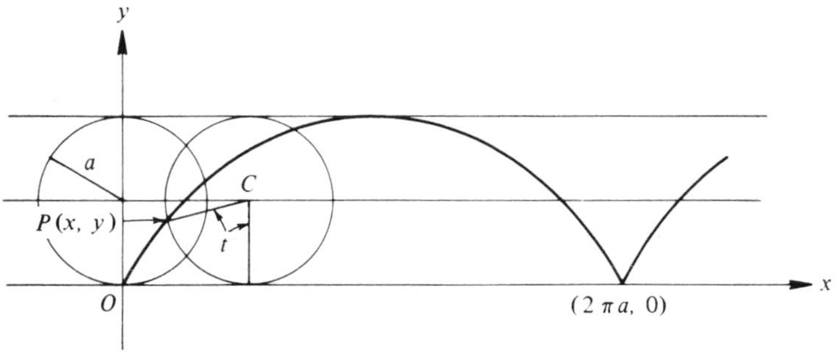

Fig. 1.11-12 The cycloid.

The parametric equations of the cycloid are

$$\begin{aligned} x &= at - a\sin t \\ y &= a - a\cos t \end{aligned} \qquad (1.11\text{-}60)$$

If a point P is, instead, fixed on the radius of the circle (inside the circle with $|\overline{CP}| = q < a$), then the path obtained is a curtate cycloid or trochoid as shown.

The parametric equations of a curtate cycloid are

$$\begin{aligned} x &= at - q\sin t \\ y &= a - q\cos t \end{aligned} \qquad (1.11\text{-}61)$$

If P is fixed on the radius extended (i.e., outside the rolling circle with $|\overline{CP}| = q > a$), then the locus is a prolate cycloid or trochoid as shown.

The parametric equations of a prolate cycloid are (1.11-61).

(2) The epicycloids and the hypocycloids

Fix a point P on the circumference of a circle of radius a and then let this circle roll without slipping on the circumference of a circle of radius A. As the first

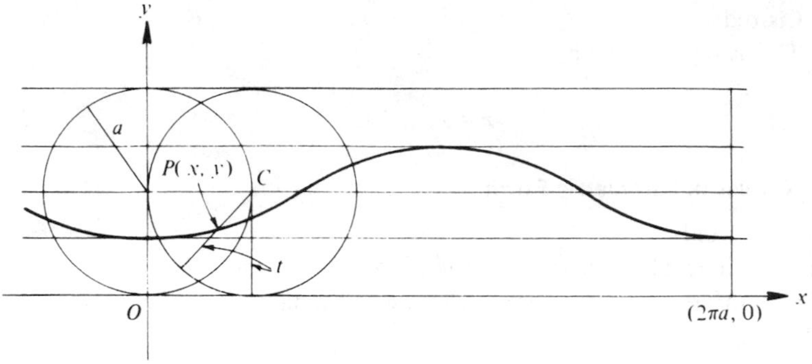

Fig. 1.11-13 Curtate cycloid.

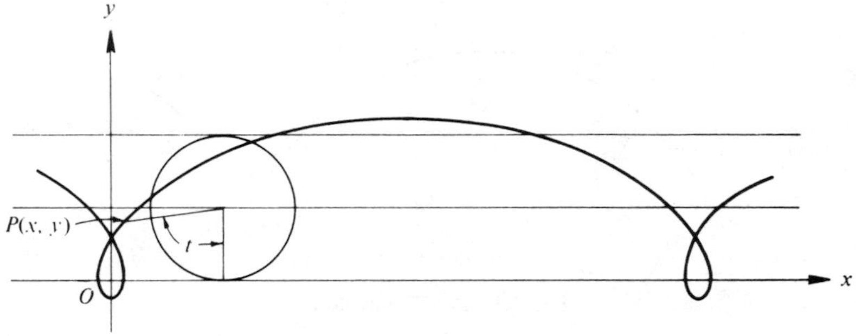

Fig. 1.11-14 Prolate cycloid.

circle rolls on the outside or the inside of the second circle, the corresponding locus is called an epicycloid or a hypocycloid. A typical epicycloid as shown in Fig. 1.11-15.

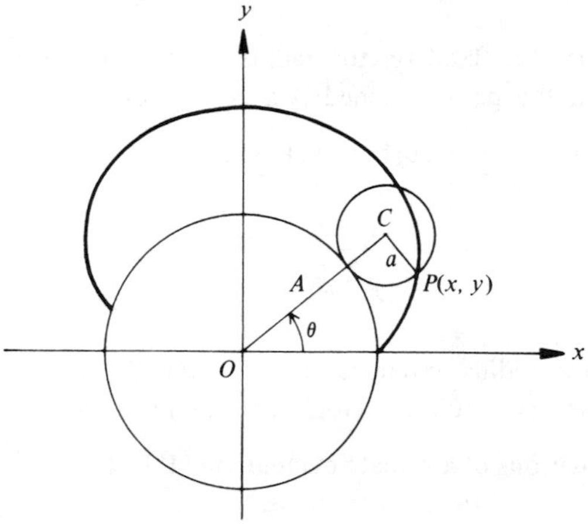

Fig. 1.11-15 Epicycloid.

has parametric equations

$$x = (A + a) \cos \theta - a \cos \left(\frac{A + a}{a}\right) \theta$$
$$y = (A + a) \sin \theta - a \sin \left(\frac{A + a}{a}\right) \theta \qquad (1.11\text{-}62)$$

and a typical hypocycloid has parametric equations

$$x = (A - a) \cos \theta + a \cos \left(\frac{A - a}{a}\right) \theta$$
$$y = (A - a) \sin \theta - a \sin \left(\frac{A - a}{a}\right) \theta \qquad (1.11\text{-}63)$$

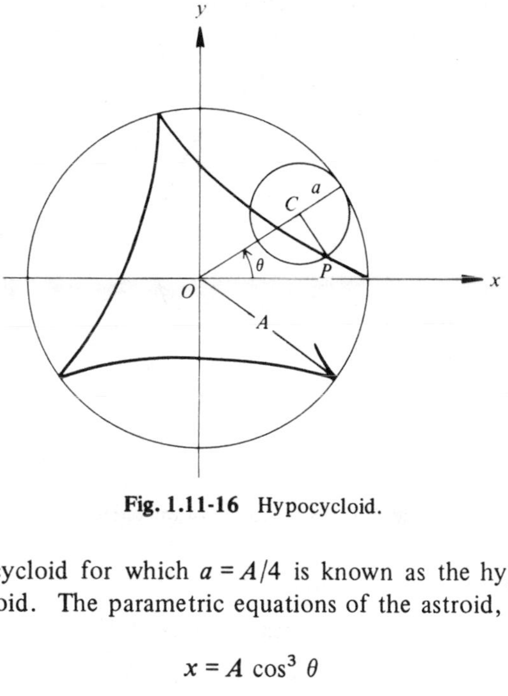

Fig. 1.11-16 Hypocycloid.

The special hypocycloid for which $a = A/4$ is known as the hypocycloid of four cusps, or the astroid. The parametric equations of the astroid, shown below, are

$$x = A \cos^3 \theta$$
$$y = A \sin^3 \theta \qquad (1.11\text{-}64)$$

and the Cartesian coordinate form is

$$x^{2/3} + y^{2/3} = A^{2/3} \qquad (1.11\text{-}65)$$

(3) The involutes of a circle
Wind a string around the circumference of a circle of radius a with one end

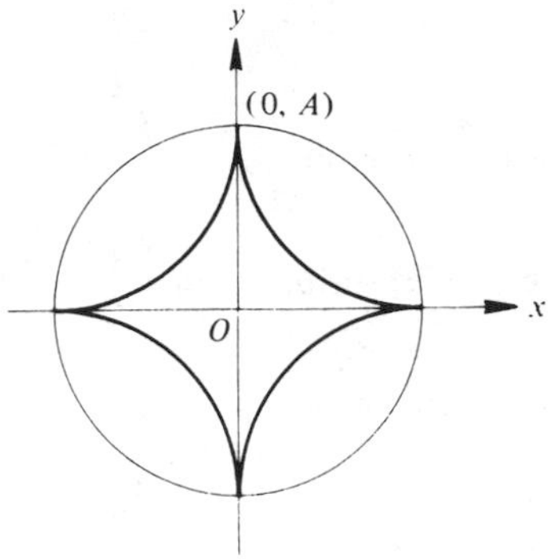

Fig. 1.11-17 Astroid.

attached to the circumference. Keeping the string taut, unwind it; the locus of the free end is called an involute of the circle, as shown in Fig. 1.11-18.

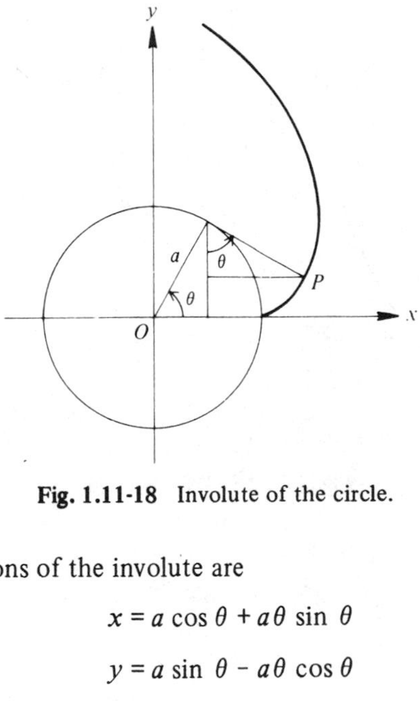

Fig. 1.11-18 Involute of the circle.

The parametric equations of the involute are

$$\begin{aligned} x &= a \cos \theta + a\theta \sin \theta \\ y &= a \sin \theta - a\theta \cos \theta \end{aligned} \tag{1.11-66}$$

1.12 ANALYTIC GEOMETRY OF THREE-SPACE

1.12.1 Coordinate Systems in Three-Space

(1) The Cartesian coordinate system

Let $x'ox$, $y'oy$, $z'oz$ be three mutually perpendicular lines intersecting at the origin O, thus dividing space into eight octants as shown.

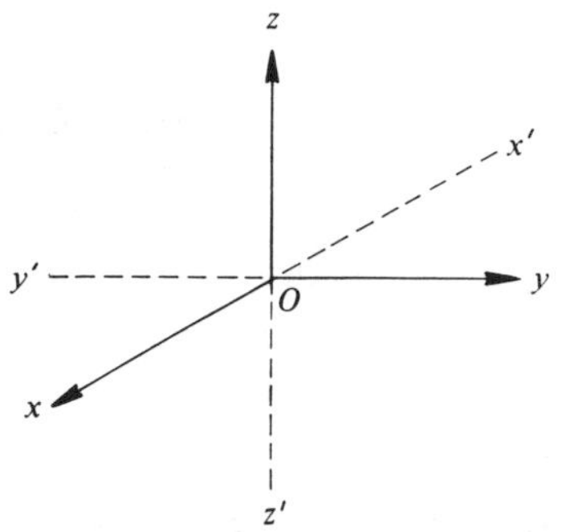

Fig. 1.12-1 Cartesian coordinates in three dimensions.

The reader may visualize the y- and z-axes as being in the plane of the page and the x-axis as pointing towards him. Then to each point there corresponds an ordered triplet (x, y, z) of real numbers; the x-, y-, and z-coordinates of the point. The x-coordinate is the directed distance from the page, taken as positive for points in front of the page; negative otherwise. The y-coordinate is the directed distance from the xz-plane and is taken as positive for points to the right of that plane; negative otherwise. The z-coordinate is the directed distance from the xy-plane and is taken as positive for points above that plane; negative otherwise.

(2) Cylindrical polar coordinates

Here the position of a point is specified by adjoining to the polar coordinates in the plane the z-coordinate of Cartesian coordinates.

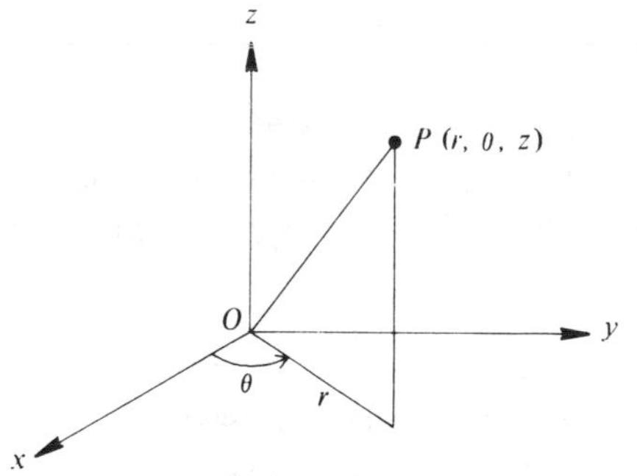

Fig. 1.12-2 Cylindrical polar coordinates (r, θ, z).

The coordinate surfaces are

r = constant—right circular cylinder, axis the z-axis
θ = constant—half plane with one edge the z-axis
z = constant—plane parallel to the xy-plane.

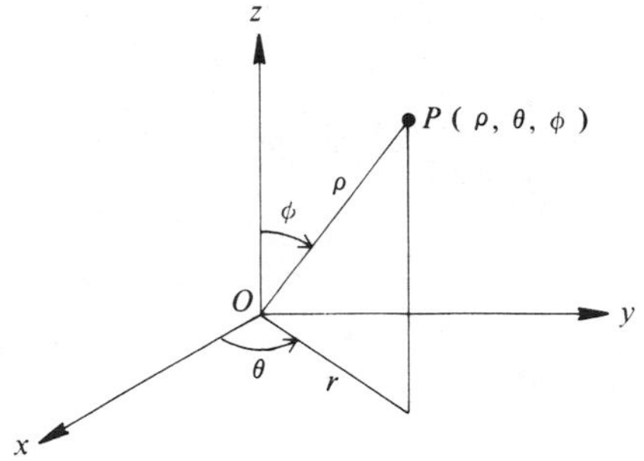

Fig. 1.12-3 Spherical polar coordinates (ρ, θ, ϕ).

The equations connecting cylindrical and rectangular coordinates are

$$\begin{aligned} x &= r \cos \theta \\ y &= r \sin \theta \\ z &= z \end{aligned} \tag{1.12-1}$$

with inverse relations

$$\begin{aligned} r &= \sqrt{x^2 + y^2} \\ \theta &= \tan^{-1} \frac{y}{x} \\ z &= z \end{aligned} \tag{1.12-2}$$

(3) Spherical polar coordinates

These are ρ, the distance of the point from the origin; θ, as in cylindrical polar coordinates; and ϕ, the angle between the $+z$-axis and the vector **OP**.

The coordinate surfaces are

ρ = constant–sphere, center the origin
θ = constant–half plane with one edge the z-axis
ϕ = constant–one nappe of a right circular cone.

The equations connecting spherical polar and Cartesian coordinates are

$$\begin{aligned} x &= \rho \sin \phi \cos \theta \\ y &= \rho \sin \phi \sin \theta \\ z &= \rho \cos \phi \end{aligned} \tag{1.12-3}$$

with inverse relations

$$\rho = \sqrt{x^2 + y^2 + z^2}$$

$$\theta = \tan^{-1} \frac{y}{x} \qquad (1.12\text{-}4)$$

$$\phi = \cos^{-1} \frac{z}{\sqrt{x^2 + y^2 + z^2}}$$

1.12.2 Direction Cosines, Distance Formula

The angles α, β, γ that the vector $\mathbf{P_1P_2}$ makes with the positive x, y, and z directions are direction angles for $\mathbf{P_1P_2}$; and the cosines of these angles are direction cosines for $\mathbf{P_1P_2}$. In terms of the coordinates of P_1 and P_2,

$$\cos\alpha = \frac{x_2 - x_1}{d}, \quad \cos\beta = \frac{y_2 - y_1}{d}, \quad \cos\gamma = \frac{z_2 - z_1}{d} \qquad (1.12\text{-}5)$$

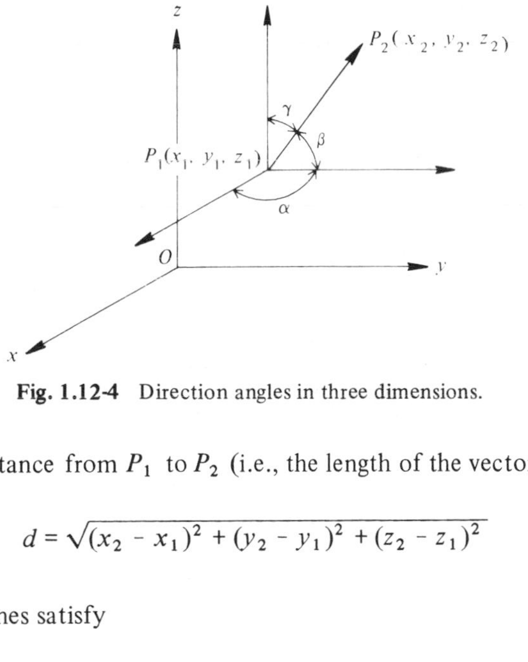

Fig. 1.12-4 Direction angles in three dimensions.

where d is the distance from P_1 to P_2 (i.e., the length of the vector $\mathbf{P_1P_2}$) given by

$$d = \sqrt{(x_2 - x_1)^2 + (y_2 - y_1)^2 + (z_2 - z_1)^2} \qquad (1.12\text{-}6)$$

The direction cosines satisfy

$$\cos^2\alpha + \cos^2\beta + \cos^2\gamma = 1 \qquad (1.12\text{-}7)$$

Direction numbers for a line are the components of any vector on the line, and direction cosines for a line are the components of a unit vector on the line.

1.12.3 Translation, Rotation of Axes

Let the coordinates of O' be (h, k, l); then the relations for a translation of axes are

$$x = x' + h, \quad y = y' + k, \quad z = z' + l \tag{1.12-8}$$

For convenience of notation in studying rotation of axes, replace x, y, z by x_1, x_2, x_3; similarly for x', y', z'. Consider a rotation of axes, with the origin fixed

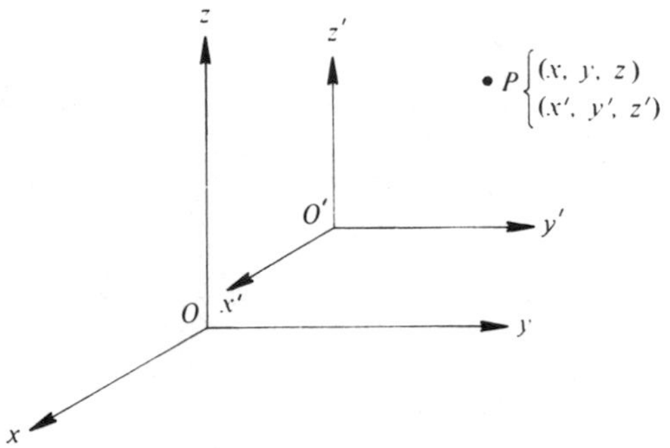

Fig. 1.12-5 Translation of axes in three dimensions.

and the cosine of the angle between the ox_i and ox_j' axes denoted by α_{ij}. Then the equations of transformation of coordinates can be obtained from the multiplication table

	x_1'	x_2'	x_3'
x_1	α_{11}	α_{12}	α_{13}
x_2	α_{21}	α_{22}	α_{23}
x_3	α_{31}	α_{32}	α_{33}

(1.12-9)

Example:

$$x_3 = \alpha_{31}x_1' + \alpha_{32}x_2' + \alpha_{33}x_3'$$

$$x_2' = \alpha_{12}x_1 + \alpha_{22}x_2 + \alpha_{32}x_3$$

etc.

These direction cosines satisfy

$$\sum_{j=1}^{3} \alpha_{ij}\alpha_{kj} = \delta_{ik} \quad \text{and} \quad \sum_{j=1}^{3} \alpha_{ji}\alpha_{jk} = \delta_{ik} \tag{1.12-10}$$

where δ_{ik} is the Kronecker delta, and also

$$\begin{vmatrix} \alpha_{11} & \alpha_{12} & \alpha_{13} \\ \alpha_{21} & \alpha_{22} & \alpha_{23} \\ \alpha_{31} & \alpha_{32} & \alpha_{33} \end{vmatrix} = 1 \tag{1.12-11}$$

1.12.4 The Plane

An equation of the form

$$ax + by + cz + d = 0 \qquad (a, b, c) \neq (0, 0, 0) \tag{1.12-12}$$

always represents a plane, and vice versa. The coefficient vector $[a, b, c]$ is a vector perpendicular to the plane. The equation of the plane through the three points $P_1(x_1, y_1, z_1), P_2(x_2, y_2, z_2), P_3(x_3, y_3, z_3)$ can be written

$$\begin{vmatrix} x - x_1 & y - y_1 & z - z_1 \\ x_2 - x_1 & y_2 - y_1 & z_2 - z_1 \\ x_3 - x_1 & y_3 - y_1 & z_3 - z_1 \end{vmatrix} = 0 \tag{1.12-13}$$

with special case (the intercept form)

$$\frac{x}{h} + \frac{y}{k} + \frac{z}{l} = 1 \tag{1.12-14}$$

The distance between the plane $ax + by + cz + d = 0$ and the point $P(x_1, y_1, z_1)$ is

$$\left| \frac{ax_1 + by_1 + cz_1 + d}{\sqrt{a^2 + b^2 + c^2}} \right| \tag{1.12-15}$$

The cosine of the angle between the two planes $a_1x + b_1y + c_1z + d_1 = 0$ and $a_2x + b_2y + c_2z = d_2 = 0$ is equal to the cosine of the angle between their coefficient vectors and thus can be obtained from the dot product definition.

$$\cos\theta = \frac{a_1a_2 + b_1b_2 + c_1c_2}{\sqrt{a_1^2 + b_1^2 + c_1^2}\sqrt{a_2^2 + b_2^2 + c_2^2}} \tag{1.12-16}$$

1.12.5 The Straight Line

1.12.5.1 Forms of the Equation A straight line in three-space can be considered as the intersection of two nonparallel planes, and hence, the general form of the equation is

$$a_1 x + b_1 y + c_1 z + d_1 = 0$$
$$a_2 x + b_2 y + c_2 z + d_2 = 0 \qquad (1.12\text{-}17)$$

One set of direction numbers for the line defined by eq. (1.12-17) is

$$(b_1 c_2 - b_2 c_1,\ c_1 a_2 - c_2 a_1,\ a_1 b_2 - a_2 b_1) \qquad (1.12\text{-}18)$$

The equation of the line through $P_1(x_1, y_1, z_1)$ and $P_2(x_2, y_2, z_2)$ can be written in

(1) parametric form

$$x = x_1 + t(x_2 - x_1)$$
$$y = y_1 + t(y_2 - y_1) \qquad (1.12\text{-}19)$$
$$z = z_1 + t(z_2 - z_1)$$

where $(x_2 - x_1, y_2 - y_1, z_2 - z_1)$ can be replaced by any set of direction numbers.

(2) symmetric form

If none of $x_2 - x_1, y_2 - y_1, z_2 - z_1$ is 0, then

$$\frac{x - x_1}{x_2 - x_1} = \frac{y - y_1}{y_2 - y_1} = \frac{z - z_1}{z_2 - z_1}$$

and if eg. $y_2 - y_1$ is 0, then the symmetric form would be

$$\frac{x - x_1}{x_2 - x_1} = \frac{z - z_1}{z_2 - z_1}, \quad y = y_1$$

1.12.5.2 Shortest Distance, Angle, Between Two Lines For the lines

$$\frac{x - x_1}{l_1} = \frac{y - y_1}{m_1} = \frac{z - z_1}{n_1}, \quad \frac{x - x_2}{l_1} = \frac{y - y_2}{m_2} = \frac{z - z_2}{n_2}$$

through (x_1, y_1, z_1) and (x_2, y_2, z_2) with direction numbers (l_1, m_1, n_1) and (l_2, m_2, n_2) the shortest distance is the absolute value of

$$\frac{\begin{vmatrix} x_2 - x_1 & y_2 - y_1 & z_2 - z_1 \\ l_1 & m_1 & n_1 \\ l_2 & m_2 & n_2 \end{vmatrix}}{|(l_1, m_1, n_1) \times (l_2, m_2, n_2)|} \qquad (1.12\text{-}21)$$

The angle between two lines is taken as the angle between two vectors, one on each line and is found from the scalar product.

1.12.6 Surfaces and Curves

1.12.6.1 Surfaces of Revolution These are surfaces generated by revolving a plane curve about a fixed line (the axis) in that plane. The equation of such a surface can be found as in this example: let a plane curve, not crossing the y-axis, be

$$f(y, z) = 0, \quad x = 0 \tag{1.12-22}$$

Then the equation of the surface generated by rotation of this arc about the y-axis is

$$f(y, \sqrt{x^2 + z^2}) = 0 \tag{1.12-23}$$

1.12.6.2 Quadric Surfaces These are surfaces whose equations are of the form

$$Ax^2 + By^2 + Cz^2 + 2Dxy + 2Exz + 2Fyz + 2Gx + 2Hy + 2Iz + J = 0 \tag{1.12-24}$$

with $(A, B, C, D, E, F) \neq (0, 0, 0, 0, 0, 0)$.

Special cases:

(1) Cylinders

When one of the variables does not appear in (1.12-24), the surface is a cylinder with generators parallel to the axis of the missing variable. For typical cases, see Fig. 1.12-6.

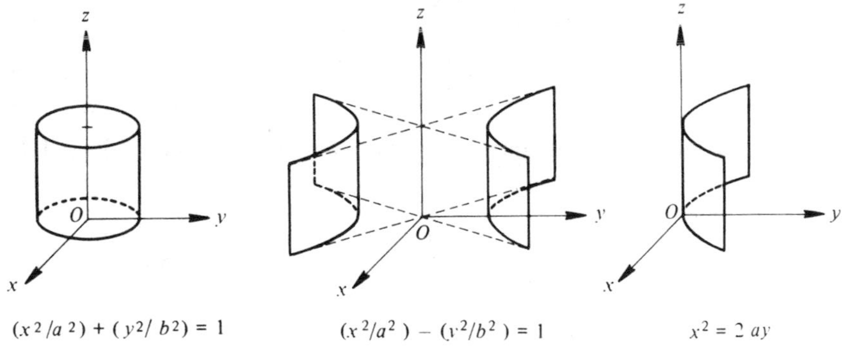

Fig. 1.12-6 Circular, hyperbolic, and parabolic cylinders.

(2) Cones

If in (1.12-24), $G = H = I = J = 0$, the surface is a cone with vertex at the origin. See Fig. 1.12-7.

(3) Spheres

If in eq. (1.12-24), $A = B = C$, $D = E = F = 0$, then the equation is of the form

$$(x^2/a^2) + (y^2/b^2) = (z^2/c^2)$$

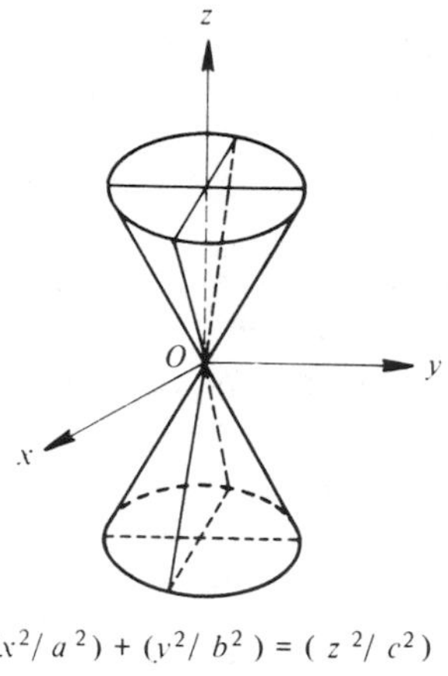

Fig. 1.12-7 Cone.

$$x^2 + y^2 + z^2 + ax + by + cz + d = 0 \qquad (1.12\text{-}25)$$

which by completing the squares, becomes

$$\left(x + \frac{a}{2}\right)^2 + \left(y + \frac{b}{2}\right)^2 + \left(z + \frac{c}{2}\right)^2 = \frac{a^2 + b^2 + c^2 - 4d}{4}$$

This represents a sphere, a single point, or has no real representation, as the right hand side is greater than, equal to, or less than zero.

(4) ellipsoids

Ellipsoids correspond to equations of the form

$$\frac{x^2}{a^2} + \frac{y^2}{b^2} + \frac{z^2}{c^2} = 1 \qquad (1.12\text{-}26)$$

with a, b, c real numbers. This represents surfaces of the form shown in Fig. (1.12-8).

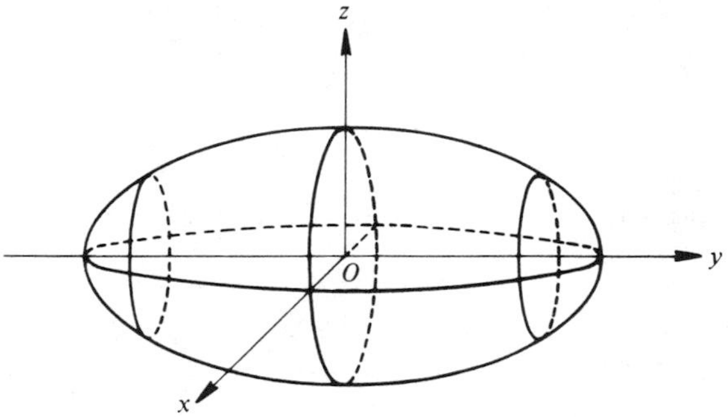

Fig. 1.12-8 Ellipsoid.

In each of the cases $a = b = c$ (sphere), $a = b \neq c$, $b = c \neq a$, $a = c \neq b$, the ellipsoid is a surface of revolution. When such a surface results from a rotation of an ellipse about its major axis, it is called a prolate spheroid; when it results from a rotation about the minor axis, it is called an oblate spheroid.

(5) Elliptic paraboloids
Consider for example

$$\frac{x^2}{a^2} + \frac{y^2}{b^2} = cz \tag{1.12-27}$$

which represents a surface of the form ($c > 0$) shown in Fig. (1.12-9).

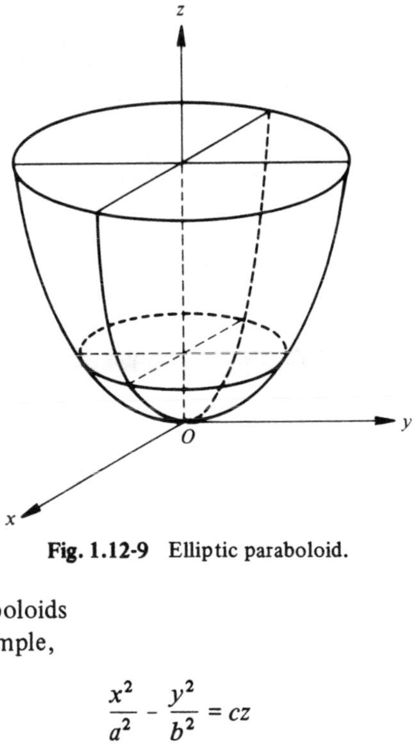

Fig. 1.12-9 Elliptic paraboloid.

(6) Hyperbolic paraboloids
Consider for example,

$$\frac{x^2}{a^2} - \frac{y^2}{b^2} = cz \tag{1.12-28}$$

which represents a surface of the form ($c > 0$) shown in Fig. (1.12-10).

(7) Hyperboloids of one sheet
Consider for example,

$$\frac{x^2}{a^2} + \frac{y^2}{b^2} - \frac{z^2}{c^2} = 1 \tag{1.12-29}$$

representing a surface of the form shown in Fig. (1.12-11).

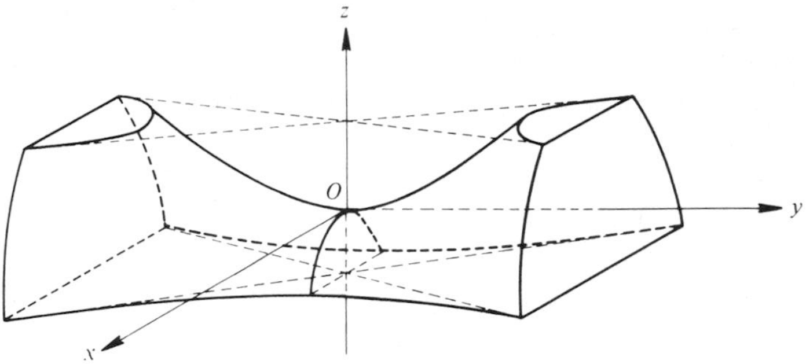

Fig. 1.12-10 Hyperbolic paraboloid.

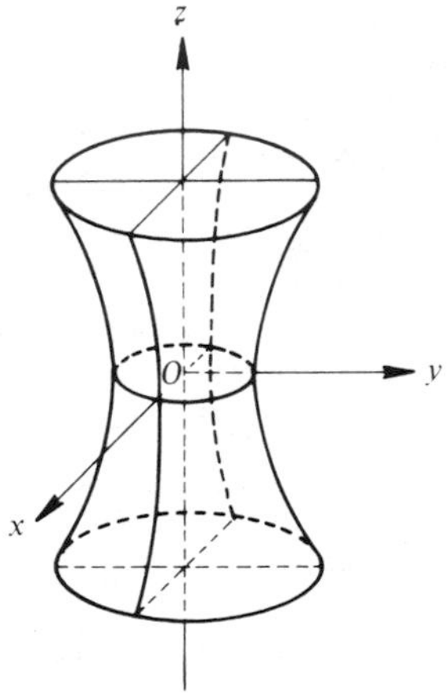

Fig. 1.12-11 Hyperboloid of one sheet.

(8) Hyperboloids of two sheets
Consider for example,

$$\frac{y^2}{b^2} - \frac{x^2}{a^2} - \frac{z^2}{c^2} = 1 \qquad (1.12\text{-}30)$$

representing a surface of the form shown in Fig. (1.12-12).

The general quadric (1.12-24) can be reduced by appropriate coordinate transformations to one of the above eight cases, or to a degenerate case. This is done most conveniently by matrix methods (see Chapter 16), the first step being the elimination of the linear terms (if possible) by a translation of axes. Substitution of (1.12-8) into (1.12-24) gives the following equations for the elimination of the coefficients of x', y', z':

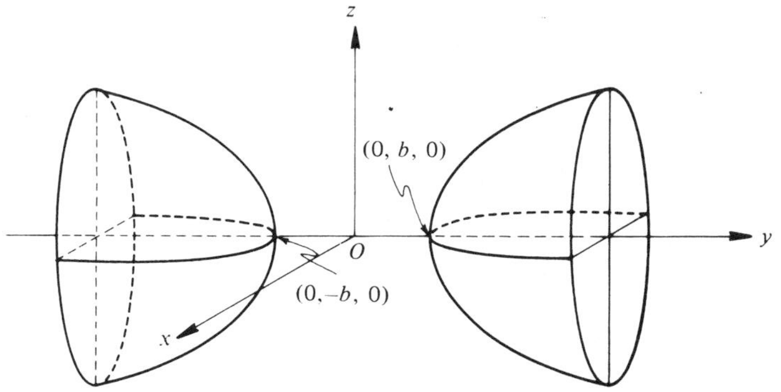

Fig. 1.12-12 Hyperboloid of two sheets.

$$\begin{aligned} Ah + Dk + El + G &= 0 \\ Dh + Bk + Fl + H &= 0 \\ Eh + Fk + Cl + I &= 0 \end{aligned} \qquad (1.12\text{-}31)$$

Considering only the case in which (1.12-31) has a unique solution (for the other cases see references), translation to this point will leave (1.12-24) in the form

$$A(x')^2 + B(y')^2 + C(z')^2 + 2Dx'y' + 2Ex'z' + 2Fy'z' + J' = 0 \qquad (1.12\text{-}32)$$

where J' may or may not be 0. A surface of this form, where a replacement of (x', y', z') by $(-x', -y', -z')$ leaves the equation unchanged, is called a central quadric, and the solution of (1.12-31) gives the center. With the quadratic form

$$A(x')^2 + B(y')^2 + C(z')^2 + 2Dx'y' + 2Ex'z' + 2Fy'z' \qquad (1.12\text{-}33)$$

can be associated the symmetric matrix

$$Q = \begin{pmatrix} A & D & E \\ D & B & F \\ E & F & C \end{pmatrix} \qquad (1.12\text{-}34)$$

such that if X is the column matrix

$$\begin{pmatrix} x' \\ y' \\ z' \end{pmatrix} \qquad (1.12\text{-}35)$$

then the single element of $X^T QX$ = (1.12-33). What is required is a matrix P whose elements satisfy (1.12-10) and (1.12-11) so that the rotation of axes $X = PY$ where Y is the column matrix

$$\begin{pmatrix} x'' \\ y'' \\ z'' \end{pmatrix} \tag{1.12-36}$$

will result in a quadratic form

$$(PY)^T QPY = Y^T P^T QPY \tag{1.12-37}$$

where $P^T QP$ is a diagonal matrix

$$\begin{pmatrix} \lambda_1 & 0 & 0 \\ 0 & \lambda_2 & 0 \\ 0 & 0 & \lambda_3 \end{pmatrix} \tag{1.12-38}$$

with associated quadratic form

$$\lambda_1 (x'')^2 + \lambda_2 (y'')^2 + \lambda_3 (z'')^2 \tag{1.12-39}$$

It turns out that $\lambda_1, \lambda_2, \lambda_3$ are the eigenvalues of the matrix Q and the column vectors of P are corresponding eigenvectors. Depending on the values of $\lambda_1, \lambda_2, \lambda_3$ and J', quadric surfaces as in section 1.12.6.2, or degenerate cases, will be obtained. Consider as an example the quadric

$$7x^2 - 8y^2 - 8z^2 + 8xy - 8xz - 2yz + 9 = 0$$

The eigenvalues obtained from

$$\begin{vmatrix} 7 - \lambda & 4 & -4 \\ 4 & -8 - \lambda & -1 \\ -4 & -1 & -8 - \lambda \end{vmatrix} = 0$$

are -9, -9 and 9. Thus, the quadric after rotation of axes is $-9(x')^2 - 9(y')^2 + 9(z')^2 + 9 = 0$, i.e., a hyperboloid of one sheet. The corresponding eigenvectors satisfying (1.12-10) and (1.12-11) are

$$\begin{bmatrix} 0 \\ \frac{1}{\sqrt{2}} \\ \frac{1}{\sqrt{2}} \end{bmatrix}, \quad \begin{bmatrix} \frac{1}{3} \\ -\frac{2}{3} \\ \frac{2}{3} \end{bmatrix}, \quad \text{and} \quad \begin{bmatrix} \frac{4}{3\sqrt{2}} \\ \frac{1}{3\sqrt{2}} \\ -\frac{1}{3\sqrt{2}} \end{bmatrix}, \quad \ldots$$

which give the rotation matrix.

1.12.6.3 Curves in Three-Space A space curve is most conveniently given in parametric form:

$$x = x(t), \quad y = y(t), \quad z = z(t) \tag{1.12-40}$$

For example, the curve $x = a \cos t, y = a \sin t, z = bt$ is shown in Fig. 1.12-13.

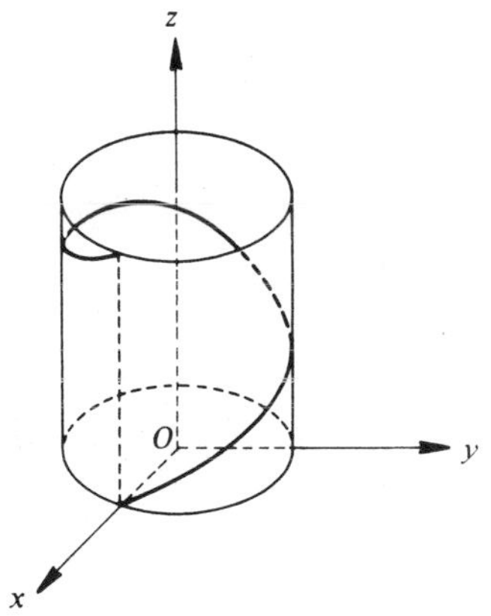

Fig. 1.12-13 The right circular helix.

A curve may also be considered as the intersection of two surfaces in three-space, and in this case the equations of the curve are

$$f(x, y, z) = 0 \qquad g(x, y, z) = 0 \tag{1.12-41}$$

For example, the relations $x^2 + y^2 + z^2 = 4, y + z = 2$ generate the curve shown in Fig. (1.12-14). In a case such as this, where one of the surfaces is a plane, the curve in question will lie in that plane and is called a plane curve.

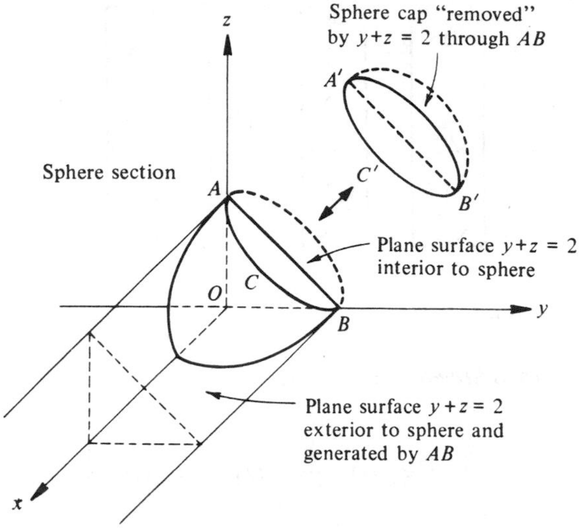

Fig. 1.12-14 Plane curve ACB, intersection of sphere octant with plane surface.

1.12.7 Areas, Perimeters, Volumes, Surface Areas, Centroids, Moments of Inertia, of Standard Shapes

Here A will be used for area, P for perimeter, V for volume, S for surface area, $\bar{x}, \bar{y}, \bar{z}$ for the coordinates of the centroid, I_x, I_y, I_z for the moments of inertia with respect to the X, Y, Z axes respectively, and I_o (a polar moment of inertia) for the moment of inertia of a plane area with respect to an axis through the centroid and perpendicular to the plane of the area. (See sections 2.5.5 and 2.9.2 for definitions.)

1.12.7.1 ***Plane Figures*** The centroid of a figure of area A made up of subregions of areas $A_1, A_2, \ldots, A_n$; where A_i has its centroid at $(\bar{x}_i, \bar{y}_i)$ can be calculated from

$$A\bar{x} = \sum_{i=1}^{n} A_i\bar{x}_i, \quad A\bar{y} = \sum_{i=1}^{n} A_i\bar{y}_i \tag{1.12-42}$$

The moment of inertia of this composite figure with respect to any axis is the sum of the moments of inertia of the separate subareas with respect to the same axis. The transfer formula for moments of inertia states that the moment of inertia with respect to any axis in the plane is equal to the moment of inertia with respect to a parallel centroidal axis plus A multiplied by the square of the distance between the two axes. A corresponding result is valid for polar moments of inertia.

(1) Triangle

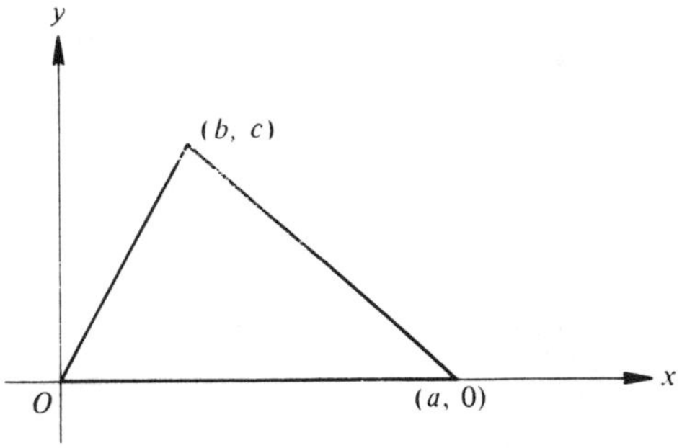

Fig. 1.12-15 Triangle in local coordinate system.

The area and perimeter are given in section 1.10.5. The centroid is the intersection point of the medians, and if the vertices of the triangle are (x_1, y_1), (x_2, y_2), (x_3, y_3), then

$$(\bar{x}, \bar{y}) = \left(\frac{x_1 + x_2 + x_3}{3}, \frac{y_1 + y_2 + y_3}{3}\right) \tag{1.12-43}$$

$$I_x = \frac{ac^3}{12}, \quad I_y = \frac{ac}{12}(a^2 + ab + b^2) \tag{1.12-44}$$

(2) Rectangle ($a = b$ for a square)

$$A = ab, \quad P = 2(a + b) \tag{1.12-45}$$

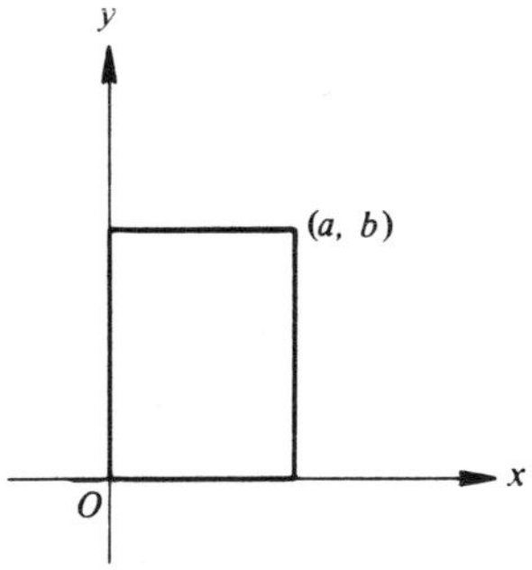

Fig. 1.12-16 Rectangle.

$$(\bar{x}, \bar{y}) = \left(\frac{a}{2}, \frac{b}{2}\right) \tag{1.12-46}$$

$$I_x = \frac{ab^3}{3}, \quad I_y = \frac{a^3 b}{3} \tag{1.12-47}$$

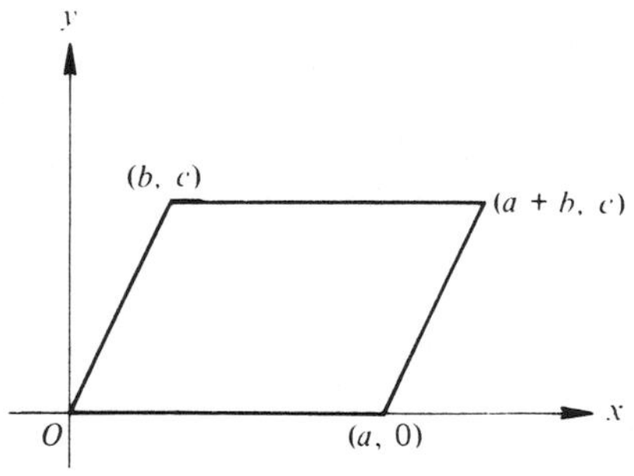

Fig. 1.12-17 Parallelogram.

(3) Parallelogram ($a = b$ for a rhombus)

$$A = ac, \quad P = 2a + 2\sqrt{b^2 + c^2} \tag{1.12-48}$$

The centroid is the intersection point of the diagonals, i.e.,

$$(\bar{x}, \bar{y}) = \left(\frac{a+b}{2}, \frac{c}{2}\right) \tag{1.12-49}$$

$$I_x = \frac{ac^3}{3}, \quad I_y = \frac{ac}{6}(2a^2 + 3ab + 2b^2) \tag{1.12-50}$$

(4) Trapezoid

$$A = \tfrac{1}{2}(a + d - b)c \tag{1.12-51}$$

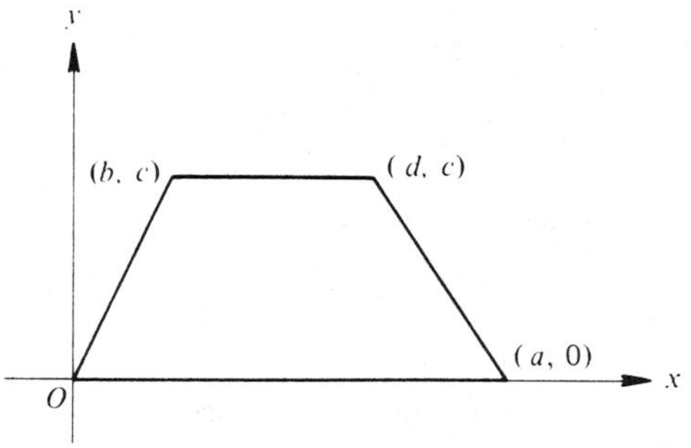

Fig. 1.12-18 Trapezoid.

$$P = a + d - b + \sqrt{b^2 + c^2} + \sqrt{(d - a)^2 + c^2} \tag{1.12-52}$$

and $(\bar{x}, \bar{y})$ by an application of (1.12-42) are

$$(\bar{x}, \bar{y}) = \left(\frac{a^2 + ad + d^2 - b^2}{3(a + d - b)}, \frac{c(a + 2d - 2b)}{3(a + d - b)} \right) \qquad (1.12\text{-}53)$$

$$I_x = \frac{ac^3}{3} - \frac{c^3}{4}(a - d + b) \qquad (1.12\text{-}54)$$

$$I_y = \frac{c}{12}[a(a^2 + ad + d^2) + d^3 - b^3] \qquad (1.12\text{-}55)$$

(5) Polygons in general

The area is found by dividing the polygon into triangles, and the centroid can then be found from (1.12-42). I_x and I_y can be found by integration. In the special case of a regular polygon—all sides and all angles equal—if a is the common length of the n sides and $\alpha = 360/n$ is the central angle, then

$$A = \frac{1}{4} na^2 \cot \frac{\alpha}{2}, \quad P = na \qquad (1.12\text{-}56)$$

The centroid is the center of the polygon.

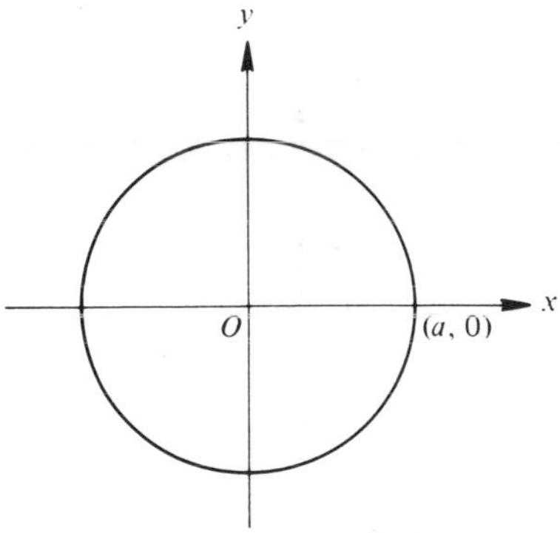

Fig. 1.12-19 Circle.

(6) Circle

$$A = \pi a^2, \quad P = 2\pi a, \quad (\bar{x}, \bar{y}) = (0, 0) \qquad (1.12\text{-}57)$$

$$I_x = I_y = \frac{\pi a^4}{4}, \quad I_o = I_x + I_y = \frac{\pi a^4}{2} \qquad (1.12\text{-}58)$$

(7) Annulus

$$A = \pi(a_1^2 - a_2^2), \quad (\bar{x}, \bar{y}) = (0, 0) \qquad (1.12\text{-}59)$$

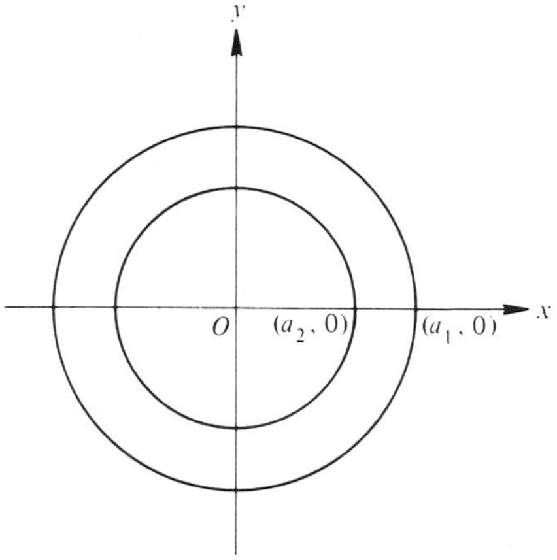

Fig. 1.12-20 Annulus.

$$I_x = I_y = \frac{\pi}{4}(a_1^4 - a_2^4),\quad I_o = \frac{\pi}{2}(a_1^4 - a_2^4) \tag{1.12-60}$$

(8) Sector of a circle (θ in radians)

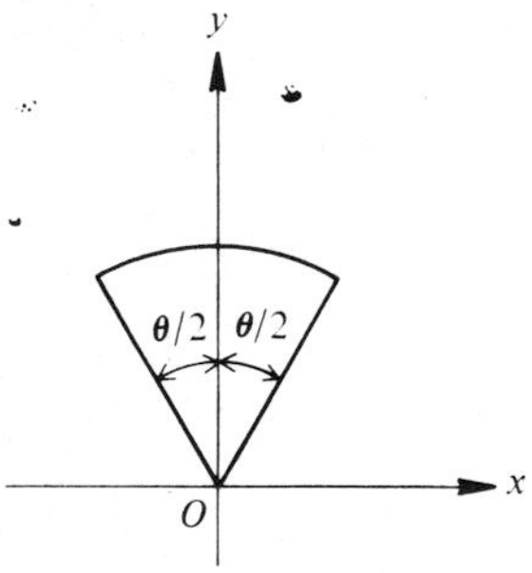

Fig. 1.12-21 Sector of a circle.

$$A = \frac{1}{2}a^2\theta,\quad P = a\theta + 2a \tag{1.12-61}$$

$$(\bar{x}, \bar{y}) = \left(0, \frac{4a}{3\theta}\sin\frac{\theta}{2}\right) \tag{1.12-62}$$

$$I_x = \frac{a^4}{8}(\theta + \sin\theta),\quad I_y = \frac{a^4}{8}(\theta - \sin\theta) \tag{1.12-63}$$

(9) Sector of an annulus

$$A = \frac{1}{2}(a_1^2 - a_2^2)\theta,\quad P = (a_1 + a_2)\theta + 2(a_1 - a_2) \tag{1.12-64}$$

The centroid can be calculated from (1.12-42) and the moments of inertia from composite figures.

(10) Segment of a circle

$$A = \frac{1}{2} a^2(\theta - \sin\theta), \quad P = a\theta + 2a \sin\frac{\theta}{2} \tag{1.12-65}$$

$$(\bar{x}, \bar{y}) = \left(0, \frac{4a \sin^3 \dfrac{\theta}{2}}{3(\theta - \sin\theta)}\right) \tag{1.12-66}$$

$$I_x = \frac{a^4}{16}(2\theta - \sin 2\theta), \quad I_y = \frac{a^4}{48}(6\theta - 8\sin\theta + \sin 2\theta) \tag{1.12-67}$$

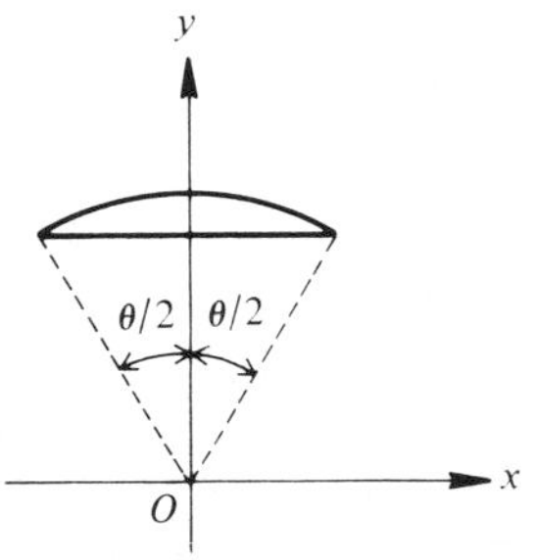

Fig. 1.12-22 Segment of a circle.

(11) Ellipse

$$A = \pi ab, \quad P \text{ is approximately } \pi\left[\frac{3}{2}(a + b) - \sqrt{ab}\right] \tag{1.12-68}$$

$$(\bar{x}, \bar{y}) = (0, 0) \tag{1.12-69}$$

$$I_x = \frac{1}{4}\pi ab^3, \quad I_y = \frac{1}{4}\pi a^3 b, \quad I_o = \frac{1}{4}\pi ab(a^2 + b^2) \tag{1.12-70}$$

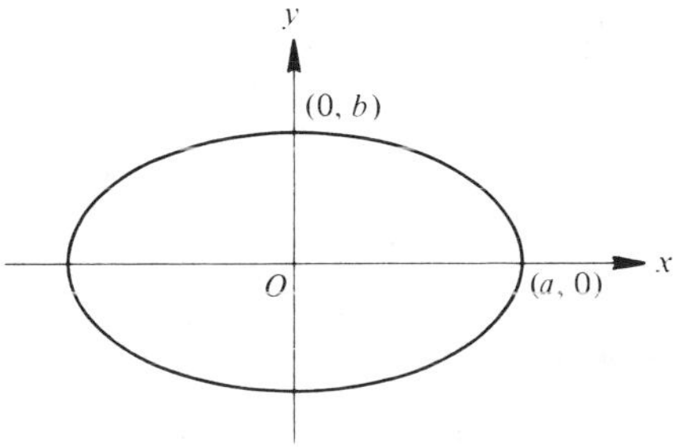

Fig. 1.12-23 Ellipse.

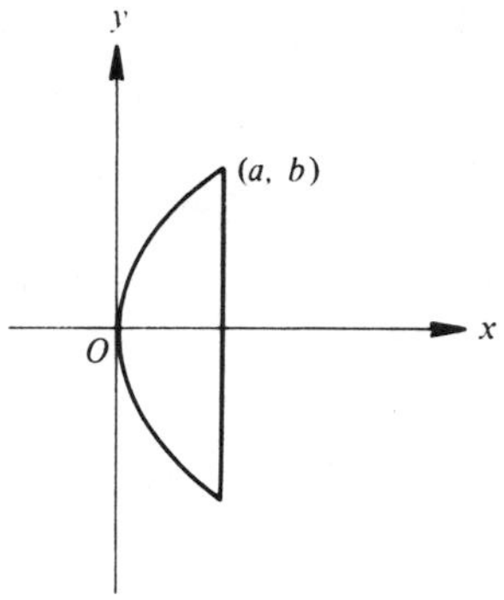

Fig. 1.12-24 Parabolic segment.

(12) Parabolic segment

$$A = \frac{4}{3} ab, \quad (\bar{x}, \bar{y}) = \left(\frac{3}{5} a, 0\right) \tag{1.12-71}$$

$$I_x = \frac{4}{15} ab^3, \quad I_y = \frac{4}{7} a^3 b \tag{1.12-72}$$

1.12.7.2 ***Solids*** The centroid of a solid of volume V made up of subregions of volumes $V_1, V_2, \ldots, V_n$; V_i with centroid at $(\bar{x}_i, \bar{y}_i, \bar{z}_i)$ can be calculated using

$$V\bar{x} = \sum_{i=1}^{n} V_i \bar{x}_i, \quad V\bar{y} = \sum_{i=1}^{n} V_i \bar{y}_i, \quad V\bar{z} = \sum_{i=1}^{n} V_i \bar{z}_i \tag{1.12-73}$$

The statements at the beginning of section (1.12.7.1) for the moments of inertia of composite figures and the transfer formula, hold with 'area' replaced by 'volume'.

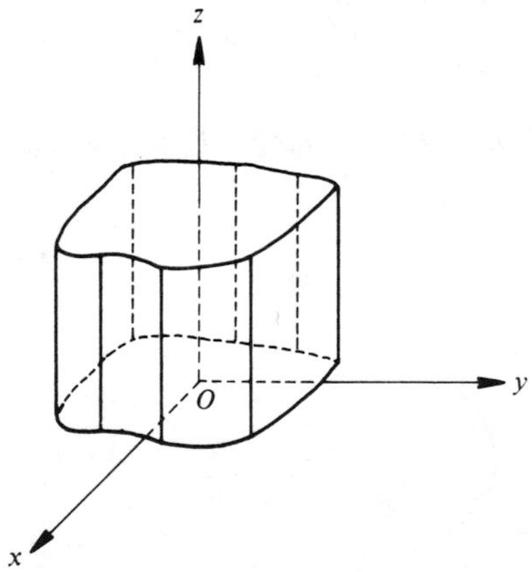

Fig. 1.12-25 Right prism.

(1) Right prism

$$V = (\text{area of base}) \times h \qquad (1.12\text{-}74)$$

$$S = 2(\text{area of base}) + (\text{circumference of base}) \times h \qquad (1.12\text{-}75)$$

The centroid lies on the line perpendicular to the base through its centroid at a height of $h/2$.

$$I_z = (\text{polar moment of base about } z\text{-axis}) \times h \qquad (1.12\text{-}76)$$

Example: Rectangular parallelopiped ($a = b = c$ for a cube)

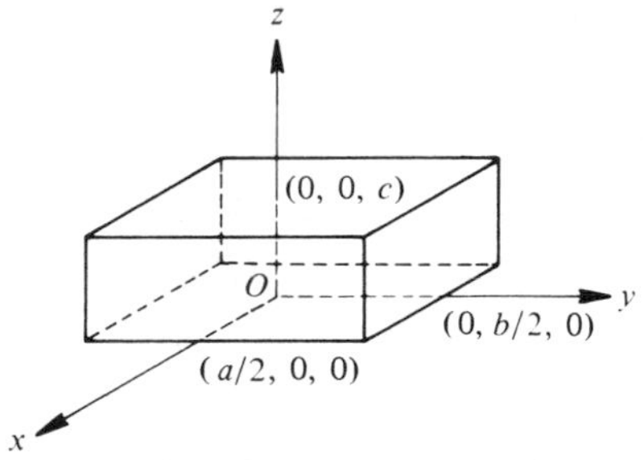

Fig. 1.12-26 Rectangular parallelopiped.

$$V = abc, \quad S = 2(ab + ac + bc) \qquad (1.12\text{-}77)$$

$$(\bar{x}, \bar{y}, \bar{z}) = (0, 0, c/2) \qquad (1.12\text{-}78)$$

$$I_z = \frac{V}{12}(a^2 + b^2) \qquad (1.12\text{-}79)$$

Example: Right circular cylinder of radius a

$$V = \pi a^2 h, \quad S = 2\pi a(a + h) \qquad (1.12\text{-}80)$$

$$(\bar{x}, \bar{y}, \bar{z}) = (0, 0, h/2) \qquad (1.12\text{-}81)$$

$$I_z = \frac{1}{2}\pi a^4 h = \frac{1}{2} V a^2 \qquad (1.12\text{-}82)$$

(2) Frustum of a right circular cone (and the cone itself)

$$V = \frac{1}{3}\pi h(a_1^2 + a_1 a_2 + a_2^2), \quad S = \pi[a_1(l + a_1) + a_2(l + a_2)] \qquad (1.12\text{-}83)$$

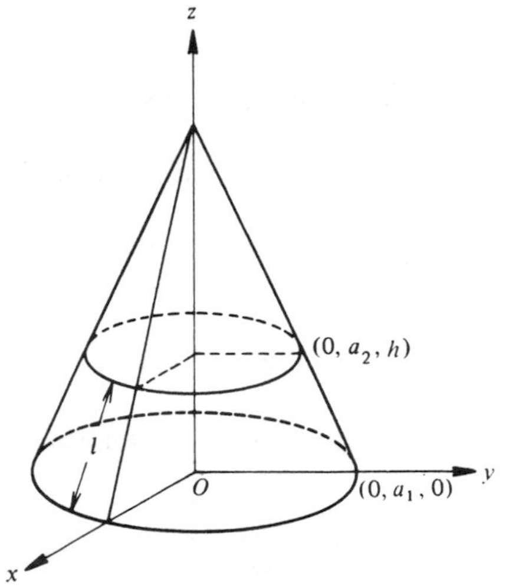

Fig. 1.12-27 Right circular cone.

$$(\bar{x}, \bar{y}, \bar{z}) = \left(0, 0, \frac{h(a_1^2 + 2a_1a_2 + 3a_2^2)}{4(a_1^2 + a_1a_2 + a_2^2)}\right) \tag{1.12-84}$$

$$I_z = \frac{\pi h(a_1^5 - a_2^5)}{10(a_1 - a_2)} \tag{1.12-85}$$

To get the corresponding expressions for a right circular cone set $a_2 = 0$; h is then the height of the cone and l is the slant height. For the right circular cone then, I with respect to an axis through the apex perpendicular to the z-axis is

$$I = \frac{3V}{5}\left(\frac{a_1^2}{4} + h^2\right) \tag{1.12-86}$$

(3) Sphere center at (0, 0, 0); radius a

$$V = \frac{4}{3}\pi a^3, \quad S = 4\pi a^2 \tag{1.12-87}$$

$$(\bar{x}, \bar{y}, \bar{z}) = (0, 0, 0) \tag{1.12-88}$$

$$I_x = I_y = I_z = \frac{8}{15}\pi a^5 \tag{1.12-89}$$

(4) Ellipsoid center (0, 0, 0), semiaxes a, b, c [equation of surface $x^2/a^2 + y^2/b^2 + z^2/c^2 = 1$ see section (1.12.6.2)]

$$V = \frac{4}{3}\pi abc \tag{1.12-90}$$

$$(\bar{x}, \bar{y}, \bar{z}) = (0, 0, 0) \tag{1.12-91}$$

$$I_z = \frac{4}{15}\pi abc(a^2 + b^2) \tag{1.12-92}$$

(5) Ring with circular section (torus)

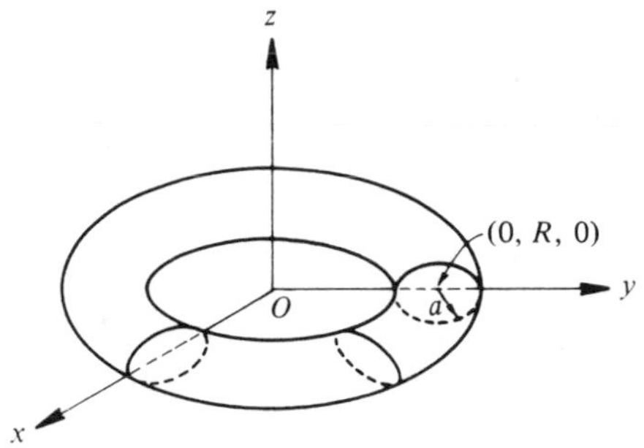

Fig. 1.12-28 Torus with circular section.

$$V = 2\pi^2 Ra^2, \quad S = 4\pi^2 Ra \tag{1.12-93}$$

$$(\bar{x}, \bar{y}, \bar{z}) = (0, 0, 0) \tag{1.12-94}$$

$$I_y = \pi^2 Ra^2 \left(R^2 + \frac{5a^2}{4}\right) \tag{1.12-95}$$

$$I_z = \frac{1}{2}\pi^2 Ra^2(4R^2 + 3a^2) \tag{1.12-96}$$

1.13 REFERENCES AND BIBLIOGRAPHY

1.13.1 References

1-1 Courant, R., and Robbins, H., *What Is Mathematics*?, Oxford University Press, N.Y., 1948. (Sections 1.1, 1.2.)

1-2 Vance, E., *Modern College Algebra*, Addison-Wesley, Reading, Mass., 1967. (Sections 1.3 to 1.9.)

1-3 Churchill, R., *Complex Variables with Applications*, McGraw-Hill, N.Y., 1960. (Sections 1.3 to 1.9.)

1-4 Steinbach, R., and Bellairs, D., *Trigonometry*, Glencoe Press, Beverly Hills, 1971. (Section 1.10.)

1-5 Kells, L.; Kern, W.; and Bland, J., *Plane and Spherical Trigonometry*, McGraw-Hill, N.Y., 1940. (Section 1.10.)
1-6 Morrill, W., *Analytic Geometry*, International, Scranton, Pa., 1964. (Sections 1.11, 1.12.)
1-7 McCrea, W., *Analytic Geometry of Three Dimensions*, Interscience, N.Y., 1948. (Sections 1.11, 1.12.)
1-8 Noble, B., *Applied Linear Algebra*, Prentice-Hall, Englewood Cliffs, N.J., 1969. (Sections 1.11, 1.12.)
1-9 Meriam, J., *Mechanics, Part I*, John Wiley & Sons, N.Y., 1952. (Sections 1.11, 1.12.)

1.13.2 Bibliography

Beaumont, R., Pierce, R., *The Algebraic Foundations of Mathematics*, Addison-Wesley, Reading, Mass., 1963.
Beckenbach, E., and Bellman, R., *Introduction to Inequalities*, Random House, N.Y., 1962.
Brinkman, H., and Klotz, E., *Linear Algebra and Analytic Geometry*, Addison-Wesley, Reading, Mass., 1971.
Chrystal, G., *Textbook of Algebra, vols. I, II*, Chelsea, N.Y., 1952.
Hashisaki, J., and Peterson, J., *Theory of Arithmetic*, John Wiley & Sons, N.Y., 1971.
Hobson, E., *Plane and Advanced Trigonometry*, Dover, N.Y., 1957.
Maxwell, E., *Coordinate Geometry with Vectors and Tensors*, Oxford, N.Y., 1958.
Olmsted, J., *The Real Number System*, Appleton-Century-Crofts, N.Y., 1962.
Salmon, G., *A Treatise on Conic Sections*, Chelsea, N.Y., 1954.
Sigley, D., and Stratton, W., *Solid Geometry*, Dryden, 1956.
Young, J., and Bush, G., *Geometry for Elementary Teachers*, Holden-Day, San Francisco, 1971.

2

Elements of Analysis

H. Lennart Pearson *

This chapter will be limited for the most part to consideration of real functions of a real variable—for the complex variable case, see Chapter 5. The only exceptions will be in certain topics in sequences and series where the extension to complex numbers is immediate, and will be so noted.

2.1 SEQUENCES

2.1.1 Definitions, Convergence

A sequence is an ordered array of infinitely many elements, chosen by a given rule.

Examples:

(1)

$$\frac{1}{2}, -\frac{2}{3}, \frac{3}{4}, -\frac{4}{5}, \ldots, (-1)^{n+1} \frac{n}{n+1}, \ldots$$

(2)

$$x, x^2, x^3, \ldots, x^n, \ldots$$

The sequence of real or complex numbers $\{S_n\}$ *has the limit* L, i.e., it *converges*, if given any $\epsilon > 0$, there exists an $N = N(\epsilon)$ such that $|S_n - L| < \epsilon$ whenever $n > N$. This is written $\lim_{n\to\infty} S_n = L$.

Examples:

(1)

$$\lim_{n\to\infty} n^{1/n} = 1$$

*Prof. H. Lennart Pearson, Dep't. of Mathematics, Illinois Institute of Technology, Chicago, Ill.

(2)

$$\lim_{n \to \infty} \frac{\log n}{n} = 0$$

(3)

$$\lim_{n \to \infty} \left(1 + \frac{1}{n}\right)^n = e = 2.718281828459045 \ldots$$

(4)

$$\lim_{n \to \infty} \left\{ \left(\sum_{k=1}^{n} \frac{1}{k} \right) - \ln n \right\} = \gamma = 0.577215664901532 \ldots \qquad (\gamma \text{ is the Euler constant})$$

If $\{S_n\}$ and $\{T_n\}$ are two sequences of real or complex numbers such that $\lim_{n \to \infty} S_n = L$ and $\lim_{n \to \infty} T_n = M$, then

$$\lim_{n \to \infty} (S_n + T_n) = L + M \tag{2.1-1}$$

$$\lim_{n \to \infty} S_n \cdot T_n = L \cdot M \tag{2.1-2}$$

$$\lim_{n \to \infty} \frac{S_n}{T_n} = \frac{L}{M} \; (M \neq 0) \tag{2.1-3}$$

If in the two real sequences $\{S_n\}$, $\{T_n\}$, $S_n \leqslant S_{n+1}$ and $T_n \geqslant T_{n+1}$, then $\{S_n\}$ is called a *monotonically increasing* sequence, and $\{T_n\}$ is *monotonically decreasing*. If numbers k_1 and k_2 exist such that $S_n < k_1$ and $T_n > k_2$ for all n, then $\{S_n\}$ is said to be *bounded above*, and $\{T_n\}$ is *bounded below*. An application of the Bolzano-Weierstrass Theorem (any infinite bounded set has at least one limit point) establishes convergence for any monotonically increasing sequence which is bounded above, or for any monotonically decreasing sequence which is bounded below. One consequence of the definition of convergence for real or complex sequences is the Cauchy Convergence Principle which states: a necessary and sufficient condition that $\{S_n\}$ be convergent is given any $\epsilon > 0$ there exists an $N = N(\epsilon)$, such that $|S_n - S_m| < \epsilon$, whenever $n, m > N$.

2.1.2 Sequences of Functions; Uniform Convergence

A sequence of real functions $\{f_n(x)\}$ converges on the closed interval $a \leqslant x \leqslant b$, denoted by $[a, b]$, to the limit function $f(x)$, if corresponding to each point x in the interval, given any $\epsilon > 0$, there exists an $N = N(\epsilon, x)$, such that $|f_n(x) - f(x)| < \epsilon$ whenever $n > N$. The corresponding Cauchy Convergence Principle reads: a necessary and sufficient condition that $\{f_n(x)\}$ be convergent on $[a, b]$ is, given any $\epsilon > 0$, there corresponds to each x in the interval an $N = N(\epsilon, x)$, such that $|f_n(x) - f_m(x)| < \epsilon$ whenever $n, m > N$.

In general, as indicated in the above definitions, the N will depend on both ϵ and x. However, if it should happen that given any $\epsilon > 0$ there corresponds an N for $[a, b]$ which is independent of x, then $\{f_n(x)\}$ is said to converge *uniformly* to $f(x)$ on $[a, b]$.

e.g., $\{x^n\}$ converges on $[0, 1]$, and converges uniformly to 0 on any closed subinterval not containing the point 1, but not otherwise.

In this example, each element of the sequence is continuous on $[0, 1]$, but the limit function is not (it is, however, continuous on any subinterval not containing the point 1). If $\{f_n(x)\}$, with each $f_n(x)$ continuous on $[a, b]$, converges uniformly to $f(x)$ on $[a, b]$, then the limit function $f(x)$ will be continuous on $[a, b]$, and

$$\int_a^b f(x)\, dx = \lim_{n \to \infty} \int_a^b f_n(x)\, dx \tag{2.1-4}$$

Further, let $\{f_n(x)\}$ converge to $f(x)$ on $[a, b]$. Then, if $\{f_n'(x)\}$ is a uniformly convergent sequence of continuous functions on $[a, b]$, $f(x)$ has a derivative, given by

$$f'(x) = \lim_{n \to \infty} f_n'(x) \tag{2.1-5}$$

2.2 INFINITE SERIES

2.2.1 Definitions, Convergence

An infinite series is obtained when the terms of a sequence are summed. To every series $\sum_{k=1}^{\infty} a_k$ $\left(\text{henceforth written } \sum a_k\right)$, a_k being real or complex numbers, there corresponds a sequence of partial sums $\{S_n\}$ where $S_n = \sum_{k=1}^{n} a_k$. The series is said to converge to L (the sum of the series) if the corresponding sequence has the limit L; otherwise, the series diverges.

Examples:

(1)

$$\sum ar^{k-1} = \frac{a}{1-r} \qquad \text{for } |r| < 1, \text{ diverges otherwise}$$

(2)

$$\sum \frac{1}{k^p} \qquad \text{converges for } p > 1, \text{ diverges otherwise}$$

By eq. (2.1-1), if $\sum a_k = L$ and $\sum b_k = M$, then $\sum (a_k + b_k) = L + M$. The Cauchy Convergence Principle states: a necessary and sufficient condition for $\sum a_k$ to converge is, given any $\epsilon > 0$, there exists an $N = N(\epsilon)$ such that $|a_{n+1} + a_{n+2} + \cdots + a_m| < \epsilon$, $m > n$, whenever $n > N$. If $\sum a_k$ converges, then $\lim_{k\to\infty} |a_k| = 0$. But $\lim_{k\to\infty} |a_k| = 0$ by itself does not imply the convergence of $\sum a_k$, as the example of the divergent harmonic series $1/k$ shows.

If $\sum |a_k|$ converges, then $\sum a_k$ is called *absolutely convergent*; an absolutely convergent series is convergent. If $\sum a_k$ converges, but $\sum |a_k|$ diverges, then $\sum a_k$ is said to be *conditionally convergent*. Thus, $\sum (-1)^k/k!$ converges absolutely, whereas $\sum (-1)^k/\sqrt{k}$ converges conditionally. The terms of an absolutely convergent series may be rearranged arbitrarily without affecting the sum of the series (provided only that each term appears somewhere in the rearranged series). This statement is not true for conditionally convergent series.

If $\sum a_k$, $\sum b_k$ are convergent series of positive numbers with sums S, T respectively, then the product series

$$a_1 b_1 + (a_2 b_1 + a_1 b_2) + (a_3 b_1 + a_2 b_2 + a_1 b_3) + \cdots$$

converges to $S \cdot T$.

2.2.2 Tests for Convergence

The comparison test:

Let $\sum a_k$, $\sum b_k$ be series of non-negative real numbers with $a_k \leqslant b_k$ for all sufficiently large k; then

(1) if $\sum b_k$ is convergent, so is $\sum a_k$

(2) if $\sum a_k$ is divergent, so is $\sum b_k$

Alternatively, if $\lim_{k\to\infty} a_k/b_k$ exists and is not zero, then the convergence (divergence) of $\sum a_k$ implies the convergence (divergence) of $\sum b_k$.

The integral test:

Let the function $g(x)$ be positive, continuous and decreasing in $N \leqslant x < \infty$, where N is some positive integer. Then $\sum g(k)$ converges (diverges) if the improper integral $\int_N^\infty g(x)\,dx$ converges (diverges).

The alternating series test:

Let $\{a_k\}$, at least for all sufficiently large k, be a decreasing sequence of positive numbers. Then if $\lim_{k\to\infty} a_k = 0$, $\sum (-1)^k a_k$ converges.

The root test:

Let $\sum a_k$ be a series of non-negative real numbers with $\lim_{k\to\infty} (a_k)^{1/k} = p$. Then if $p < 1$, the series is convergent; if $p > 1$, the series diverges; and, if $p = 1$, no conclusion can be drawn from this test.

The ratio test:
For the series of real or complex numbers $\sum a_k$, consider $\lim_{k\to\infty} |a_{k+1}/a_k|$. If this limit is less than one, the series converges absolutely; if the limit is greater than one, or does not exist, the series diverges; and if the limit equals one, the test is inconclusive.

Raabe's test:
Let $\lim_{k\to\infty} k\{1 - |a_{k+1}/a_k|\} = t$ either exist or be $\pm\infty$. Then $\sum a_k$ will be absolutely convergent if $t > 1$, but not if $t < 1$. If $t = 1$, no conclusion can be drawn from this test.

Dirichlet's test:
A series of real numbers of the form $\sum a_k b_k$ converges if
(1) $b_k > 0, b_{k+1} \leqslant b_k, \lim_{k\to\infty} b_k = 0$; and
(2) a number M independent of k exists such that $|a_1 + a_2 + \cdots + a_k| \leqslant M$ for all k.

2.2.3 Series of Functions, Uniform Convergence, Differentiation, Integration

The series of functions $\sum f_k(x)$ converges (uniformly) on the interval $[a, b]$ if the corresponding sequence of partial sums converges (uniformly) on $[a, b]$.

The Weierstrass M-test for uniform convergence:
If to the series $\sum f_k(x)$ defined on an interval there corresponds a convergent series of positive numbers $\sum M_k$ such that $|f_k(x)| < M_k$ for all k and for all x in the interval, then $\sum f_k(x)$ converges uniformly on the interval.

Example: $\sum (\cos nx/n^2)$ converges uniformly in $(-\infty, \infty)$.
If $\sum f_k(x) = S(x)$ converges uniformly on $[a, b]$, and if each $f_k(x)$ is continuous on $[a, b]$, then $S(x)$ is continuous on $[a, b]$, and

$$\int_a^b S(x)\,dx = \sum \int_a^b f_k(x)\,dx \tag{2.2-1}$$

For term by term differentiation, let $\sum f_k(x)$ converge to $S(x)$ on $[a, b]$; then if $\sum f_k'(x)$ is a uniformly convergent series of continuous functions on $[a, b]$,

$$S'(x) = \sum f_k'(x) \tag{2.2-2}$$

2.2.4 Power Series

A series of the form $b_0 + b_1(x - a) + b_2(x - a)^2 + \cdots$ is called a *power series* in $(x - a)$. A translation of axes can be introduced so that the series may be rewritten

$$b_0 + b_1 x + b_2 x^2 + \cdots = \sum_{k=0}^{\infty} b_k x^k \equiv \sum b_k x^k \tag{2.2-3}$$

This series may diverge for all $x \neq 0$, or it may converge absolutely for all x; but more commonly, there will exist a positive number R, the *radius of convergence* such that the series converges absolutely for $|x| < R$, and diverges whenever $|x| > R$. The convergence is uniform on any interval $|x| \leqslant r < R$. (In the case of absolute convergence for all x, the convergence is uniform for any r.)

By section 2.2.3, if $S(x) = \sum b_k x^k$, then

$$\int_a^b S(x)\,dx = \sum b_k \frac{b^{k+1} - a^{k+1}}{k+1}$$

for $-R < a < b < R$.

The series $\sum kb_k x^{k-1}$ obtained by differentiating $\sum b_k x^k$ term by term, has the same radius of convergence as the original series, and hence the same uniform convergence properties. Thus if $S(x) = \sum b_k x^k$, then $S'(x) = \sum kb_k x^{k-1}$, $S''(x) = \sum k(k-1)b_k x^{k-2}$, etc.

Two convergent power series having the same sum for all x in $|x| < r$ must be identical.

2.2.5 Taylor's Formula with Remainder

Let $f(x)$ have continuous derivatives up to the $(n+1)^{\text{st}}$ order in some interval containing the point a. Then

$$f(x) = f(a) + \frac{f'(a)}{1!}(x-a) + \frac{f''(a)}{2!}(x-a)^2 + \cdots + \frac{f^n(a)}{n!}(x-a)^n + R_{n+1}(x) \tag{2.2-4}$$

where

$$R_{n+1}(x) = \frac{f^{(n+1)}(b)}{(n+1)!}(x-a)^{n+1} \qquad a < b < x \tag{2.2-5}$$

or

$$R_{n+1}(x) = \frac{1}{n!}\int_a^x (x-t)^n f^{(n+1)}(t)\,dt \tag{2.2-6}$$

If $\lim\limits_{n\to\infty} R_{n+1} = 0$ for some interval, then from eq. (2.2-4 $f(x)$ has in that interval, the *Taylor series*

$$f(x) = \sum \frac{f^n(a)}{n!}(x-a)^n \tag{2.2-7}$$

That special case for which $a = 0$ is sometimes called the MacLaurin series for $f(x)$.

Example

$$e^x = e^a + e^a(x - a) + \cdots + \frac{e^a(x - a)^n}{n!} + R_{n+1}(x)$$

where

$$R_{n+1}(x) = \frac{e^b}{(n+1)!}(x - a)^{n+1} \qquad a < b < x$$

for all b in (a, x), e^b is bounded, say by M, so

$$|R_{n+1}| < M\frac{|x - a|^{n+1}}{(n+1)!}$$

since $\lim_{n\to\infty} \frac{|x - a|^{n+1}}{(n+1)!} = 0$, $\lim_{n\to\infty} R_{n+1} = 0$, and

$$e^x = e^a + e^a(x - a) + \cdots + \frac{e^a(x - a)^n}{n!} + \cdots \qquad \text{for all } x \qquad (2.2\text{-}8)$$

Other standard Taylor series are

$$\sin x = x - \frac{x^3}{3!} + \frac{x^5}{5!} - \cdots + \frac{(-1)^n x^{2n+1}}{(2n+1)!} + \cdots \qquad \text{for all } x \qquad (2.2\text{-}9)$$

$$\cos x = 1 - \frac{x^2}{2!} + \frac{x^4}{4!} - \cdots + \frac{(-1)^n x^{2n}}{(2n)!} + \cdots \qquad \text{for all } x \qquad (2.2\text{-}10)$$

$$\sinh x = x + \frac{x^3}{3!} + \frac{x^5}{5!} + \cdots + \frac{x^{2n+1}}{(2n+1)!} + \cdots \qquad \text{for all } x \qquad (2.2\text{-}11)$$

$$\cosh x = 1 + \frac{x^2}{2!} + \frac{x^4}{4!} + \cdots + \frac{x^{2n}}{(2n)!} + \cdots \qquad \text{for all } x \qquad (2.2\text{-}12)$$

$$\text{Arc}\sin x = x + \frac{1}{2\cdot 3}x^3 + \cdots + \frac{1\cdot 3\cdot 5\cdots(2n-1)}{2\cdot 4\cdots(2n)(2n+1)}x^{2n+1} + \cdots$$

$$\text{for } -1 \leqslant x \leqslant 1 \qquad (2.2\text{-}13)$$

$$\text{Arc}\tan x = x - \frac{x^3}{3} + \frac{x^5}{5} - \cdots + (-1)^n\frac{x^{2n+1}}{2n+1} + \cdots \qquad \text{for } -1 \leqslant x \leqslant 1 \qquad (2.2\text{-}14)$$

$$\ln(1+x) = x - \frac{x^2}{2} + \frac{x^3}{3} - \cdots + (-1)^n \frac{x^{n+1}}{n+1} + \cdots \qquad \text{for } -1 < x \leqslant 1 \qquad (2.2\text{-}15)$$

$$(1+x)^m = 1 + mx + \frac{m(m-1)}{2!}x^2 + \cdots + \frac{m(m-1)\cdots(m-n+1)}{n!}x^n + \cdots$$

$$\text{for } -1 < x < 1 \qquad (2.2\text{-}16)$$

This last is the Binomial Series, which is also convergent (1) for $x = -1$, if $m > 0$; (2) for $x = 1$, if $m > -1$.

The Bernoulli numbers, B_n, are by definition the coefficients indicated in the MacLaurin series for

$$f(x) = \begin{cases} \dfrac{x}{e^x - 1} & x \neq 0 \\ 1 & x = 0 \end{cases}$$

$$\frac{x}{e^x - 1} = 1 + B_1 x + \frac{B_2}{2!}x^2 + \frac{B_3}{3!}x^3 + \cdots$$

$$\left(B_1 = -\frac{1}{2},\ B_2 = \frac{1}{6},\ B_3 = 0,\ B_4 = -\frac{1}{30},\ B_5 = 0, \cdots\right)$$

The Euler numbers, E_{2n}, are defined by

$$\sec x = E_0 - \frac{E_2}{2!}x^2 + \frac{E_4}{4!}x^4 - \cdots$$

$$(E_0 = 1,\ E_2 = -1,\ E_4 = 5,\ E_6 = -61, \cdots)$$

2.3 FUNCTIONS, LIMITS, CONTINUITY

2.3.1 Definition of Function

Consider a set of ordered pairs of numbers (x, y). If no two different pairs in the set have the same first element, then the dependent variable y is said to be a *single-valued function* of the independent variable x. The set of x values is called the *domain* of the function, the set of corresponding y values is the *range* of the function.

Example:

$$y = f(x) = \sqrt{4 - x^2} \quad \text{with domain } |x| \leqslant 2 \text{ and range } 0 \leqslant y \leqslant 2$$

NOTE: In this example x is not a single-valued function of y. If y is a single-valued function of x and x is also a single-valued function of y, then the function is *one-to-one*.

Functions with a common domain may be added, subtracted, multiplied, and, when the divisor is not zero, divided. Further, if the range of $g(x)$ is contained in the domain of $f(x)$, then $g(x)$ may be substituted into f to obtain the compound or composition function $f(g(x))$.

Example: let $f(x) = 1/x$ and $g(x) = x^3 - 2$
then $f(g(x)) = 1/(x^3 - 2)$ and $g[f(x)] = 1/x^3 - 2$

If f is a one-to-one function, then the inverse function f^{-1} corresponding to f is defined by $f^{-1}[f(x)] = x$.

Example: Arc sin (sin x) = x; $\quad |x| \leqslant \pi/2$

2.3.2 Definition of Limit, Limit Theorems, for Single-Valued Functions

A function f has the limit L at $x = a$ (written $\lim_{x \to a} f(x) = L$) if, given any $\epsilon > 0$, there exists a $\delta > 0$ such that $|f(x) - L| < \epsilon$ whenever $0 < |x - a| < \delta$.

Examples:

(1)

$$f(x) = \begin{cases} 1 & x \neq 2 \\ -1 & x = 2 \end{cases} \qquad \text{then } \lim_{x \to 2} f(x) = 1 \neq f(2)$$

(2)

$$f(x) = \begin{cases} 1 & x \leqslant 2 \\ -1 & x > 2 \end{cases} \qquad \lim_{x \to 2} f(x) \text{ does not exist}$$

If $\lim_{x \to a} f(x) = L$ and $\lim_{x \to a} g(x) = M$, then

$$\lim_{x \to a} [f(x) + g(x)] = L + M \tag{2.3-1}$$

$$\lim_{x \to a} [f(x) \cdot g(x)] = LM \tag{2.3-2}$$

$$\lim_{x \to a} \frac{f(x)}{g(x)} = \frac{L}{M} \qquad (M \neq 0) \tag{2.3-3}$$

If $\lim_{x \to a} g(x) = b$ and $\lim_{x \to b} f(x) = f(b)$, then

$$\lim_{x \to a} f(g(x)) = f(b) \tag{2.3-4}$$

If n is a positive integer, $L > 0$, and $\lim_{x \to a} f(x) = L$, then

$$\lim_{x \to a} \sqrt[n]{f(x)} = \sqrt[n]{L} \tag{2.3-5}$$

A function f has the right hand limit L_1 at $x = a$ (written $\lim_{x \to a+} f(x) = L_1$) if, given any $\epsilon > 0$, there exists a $\delta > 0$ such that $|f(x) - L_1| < \epsilon$ whenever $0 < x - a < \delta$.

A function f has the left hand limit L_2 at $x = a$ (written $\lim_{x \to a-} f(x) = L_2$) if, given any $\epsilon > 0$, there exists a $\delta > 0$ such that $|f(x) - L_2| < \epsilon$ whenever $0 < a - x < \delta$.

A necessary and sufficient condition that $\lim_{x \to a} f(x) = L$ is $\lim_{x \to a+} f(x) = L$ and $\lim_{x \to a-} f(x) = L$.

If there exists a number $A > 0$ such that $|f(x) - L| < \epsilon$ whenever $x > A$, then we write $\lim_{x \to +\infty} f(x) = L$. Similarly, if there exists a number $B < 0$ such that $|f(x) - K| < \epsilon$ whenever $x < B$, then we write $\lim_{x \to -\infty} f(x) = K$.

2.3.3 Some Standard Limits

$$\lim_{x \to +\infty} a^x = +\infty \qquad (a > 1) \tag{2.3-6}$$

$$\lim_{x \to +\infty} a^x = 0 \qquad (0 < a < 1) \tag{2.3-7}$$

$$\lim_{x \to 0} \frac{\sin x}{x} = 1 \qquad (x \text{ in radians}) \tag{2.3-8}$$

$$\lim_{x \to +\infty} \left(1 + \frac{a}{x}\right)^x = e^a \tag{2.3-9}$$

2.3.4 Continuity

A function $f(x)$ is *continuous* at $x = a$ if and only if $\lim_{x \to a} f(x) = f(a)$, and $f(x)$ is continuous in $a < x < b$ if and only if $f(x)$ is continuous at every point in the interval. By section 2.3.2, sums, products, and quotients of continuous functions are continuous; provided that for quotients, the denominator is non-zero in the interval. Further, if $g(x)$ is continuous at a, and $f(x)$ is continuous at $g(a)$, then $f[g(x)]$ is continuous at a.

The functions defined in Chapter 1, the so-called elementary functions of analysis, are continuous within their intervals of definition.

A function f has *left hand continuity* at $x = a$ if $\lim_{x \to a-} f(x) = f(a)$; and f has *right hand continuity* at $x = a$ if $\lim_{x \to a+} f(x) = f(a)$. A necessary and sufficient condition that $f(x)$ be continuous at $x = a$ is $\lim_{x \to a-} f(x) = \lim_{x \to a+} f(x) = f(a)$.

A function f is *sectionally* or *piecewise continuous* on $[a, b]$ if it is continuous at all points of the interval, except for a finite number of points at each of which f has both a finite right hand limit and a finite left hand limit (at the end points only one of these limits will exist).

Let f be continuous on the closed interval $[a, b]$. Then there must exist at least one point x_1, $a \leqslant x_1 \leqslant b$, such that $f(x_1) \geqslant f(x)$ for all x on $[a, b]$; and there must exist at least one point x_2, $a \leqslant x_2 \leqslant b$, such that $f(x_2) \leqslant f(x)$ for all x on $[a, b]$; further if K is any number between the maximum $f(x_1)$ and the minimum $f(x_2)$, then there must be at least one point x_3, $a \leqslant x_3 \leqslant b$, such that $f(x_3) = K$.

2.4 THE DERIVATIVE

2.4.1 Definition and Notation

A single-valued function f is said to be differentiable at $x = a$ if

$$\lim_{h \to 0} \frac{f(a + h) - f(a)}{h} \equiv f'(a) \tag{2.4-1}$$

exists. The number $f'(a)$ is the *derivative* of f at $x = a$. If f has a derivative at each point of $a < x < b$, then f is *differentiable* in (a, b). Among the other notations used for $f'(a)$ are

$$\left.\frac{df}{dx}\right|_{x=a}, \qquad \dot{f}(a), \quad Df(a)$$

If $f'(a)$ does exist, then f must be continuous at $x = a$.

If $\lim\limits_{h \to 0+} [(f(a + h) - f(a))/h]$ exists, then f has a *right hand derivative* at a, denoted by $f'(a+)$; if $\lim\limits_{h \to 0-} [(f(a + h) - f(a))/h]$ exists, then f has a *left hand derivative* at a, denoted by $f'(a-)$.

2.4.2 Differentiation Rules

Let f and g be differentiable in (a, b). Then in (a, b)

$$\frac{d}{dx}[cf(x)] = c\frac{df(x)}{dx} \qquad c \text{ a constant} \tag{2.4-2}$$

$$\frac{d}{dx}[f(x) + g(x)] = \frac{df(x)}{dx} + \frac{dg(x)}{dx} \tag{2.4-3}$$

$$\frac{d}{dx} f(x) \cdot g(x) = f(x) \cdot \frac{dg(x)}{dx} + g(x) \cdot \frac{df(x)}{dx} \tag{2.4-4}$$

$$\frac{(f_1 f_2 f_3 \cdots f_n)'}{f_1 f_2 f_3 \cdots f_n} = \frac{f_1'}{f_1} + \frac{f_2'}{f_2} + \frac{f_3'}{f_3} + \cdots + \frac{f_n'}{f_n} \tag{2.4-5}$$

$$\frac{d}{dx}\frac{f(x)}{g(x)} = \frac{g(x)\dfrac{d f(x)}{dx} - f(x)\dfrac{d g(x)}{dx}}{[g(x)]^2} \tag{2.4-6}$$

$$\frac{d}{dx} f[g(x)] = \frac{d f(g)}{dg} \cdot \frac{d g(x)}{dx} \tag{2.4-7}$$

$$\frac{d^n}{dx^n}[f(x) \cdot g(x)] \equiv \frac{d}{dx}\left(\frac{d}{dx}\left(\frac{d}{dx}\left(\cdots \frac{d}{dx}\left(\frac{d}{dx}(f \cdot g)\right)\right)\right)\right) \quad n \text{ differentiations}$$

$$= f \cdot \frac{d^n g}{dx^n} + C(n, 1)\frac{df}{dx} \cdot \frac{d^{n-1} g}{dx^{n-1}} + C(n, 2)\frac{d^2 f}{dx^2}\frac{d^{n-2} g}{dx^{n-2}} +$$

$$\cdots + C(n, n-1)\frac{d^{n-1} f}{dx^{n-1}}\frac{dg}{dx} + \frac{d^n f}{dx^n} \cdot g(x) \tag{2.4-8}$$

For the definition of $C(n, r)$ see eq. (1.9-5). The previous formula is the *Leibnitz rule* for the nth derivative of a product.

$$\frac{dy}{dx} = \frac{1}{\dfrac{dx}{dy}} \tag{2.4-9}$$

2.4.3 Standard Differentiation Formulas

$$\frac{d}{dx}[u(x)]^n = n u^{n-1}\frac{du}{dx} \qquad (n \text{ any constant})$$

$$\frac{d}{dx}\sin u(x) = \cos u \cdot \frac{du}{dx} \qquad \frac{d}{dx}\cos u(x) = -\sin u \cdot \frac{du}{dx}$$

$$\frac{d}{dx}\tan u(x) = \sec^2 u \cdot \frac{du}{dx} \qquad \frac{d}{dx}\cot u(x) = -\operatorname{cosec}^2 u \cdot \frac{du}{dx}$$

$$\frac{d}{dx}\sec u(x) = \sec u \tan u \frac{du}{dx} \qquad \frac{d}{dx}\operatorname{cosec} u(x) = -\operatorname{cosec} u \cot u \frac{du}{dx}$$

$$\frac{d}{dx}\operatorname{Arc\,sin} u(x) = \frac{1}{\sqrt{1-u^2}}\frac{du}{dx} \qquad \frac{d}{dx}\operatorname{Arc\,cos} u(x) = \frac{-1}{\sqrt{1-u^2}}\frac{du}{dx}$$

$$\frac{d}{dx}\operatorname{Arc\,tan} u(x) = \frac{1}{1+u^2}\frac{du}{dx} \qquad \frac{d}{dx}\operatorname{Arc\,cot} u(x) = \frac{-1}{1+u^2}\frac{du}{dx}$$

$$\frac{d}{dx}\text{Arc sec } u(x) = \frac{1}{u\sqrt{u^2-1}}\frac{du}{dx} \qquad \frac{d}{dx}\text{Arc cosec } u(x) = \frac{-1}{u\sqrt{u^2-1}}\frac{du}{dx}$$

$$\frac{d}{dx}e^{u(x)} = e^u\frac{du}{dx} \qquad \frac{d}{dx}a^{u(x)} \equiv \frac{d}{dx}e^{u\ln a} = a^u \ln a\frac{du}{dx}$$

$$\frac{d}{dx}\ln u(x) = \frac{1}{u}\frac{du}{dx} \qquad \frac{d}{dx}u(x)^{v(x)} \equiv \frac{d}{dx}e^{v\ln u}$$

$$\frac{d}{dx}\sinh u(x) = \cosh u\frac{du}{dx} \qquad \frac{d}{dx}\cosh u(x) = \sinh u\frac{du}{dx}$$

$$\frac{d}{dx}\tanh u(x) = \text{sech}^2 u\frac{du}{dx} \qquad \frac{d}{dx}\coth u(x) = -\text{cosech}^2 u\frac{du}{dx}$$

$$\frac{d}{dx}\text{sech } u(x) = -\text{sech } u \tanh u\frac{du}{dx} \qquad \frac{d}{dx}\text{cosech } u(x) = -\text{cosech } u \coth u\frac{du}{dx}$$

$$\frac{d}{dx}\arg\sinh u(x) = \frac{1}{\sqrt{u^2+1}}\frac{du}{dx} \qquad \frac{d}{dx}\text{Arg}\cosh u(x) = \frac{1}{\sqrt{u^2-1}}\frac{du}{dx} \quad (u > 1)$$

$$\frac{d}{dx}\arg\tanh u(x) = \frac{1}{1-u^2}\frac{du}{dx} \quad (|u| < 1) \qquad \frac{d}{dx}\arg\coth u(x) = \frac{1}{1-u^2}\frac{du}{dx} \quad (|u| > 1)$$

$$\frac{d^n}{dx^n}x^m = \begin{cases} m(m-1)\cdots(m-n+1)x^{m-n} & \text{for all } m \\ n! & \text{for } m = n \end{cases}$$

$$\frac{d^n}{dx^n}\sin x = \sin\left(x + \frac{n\pi}{2}\right) \qquad \frac{d^n}{dx^n}\cos x = \cos\left(x + \frac{n\pi}{2}\right)$$

2.4.4 Geometrical and Physical Applications

Geometrically, $f'(a)$ is the slope of the tangent line to $y = f(x)$ at the point $(a, f(a))$. Physically, $f'(a)$ is the instantaneous rate of change of $f(x)$ with respect to x at $x = a$. Thus, if distance s is expressed as a function of time t by $s = f(t)$, then the velocity at time t is $v(t) = ds/dt$, and the acceleration at time t is $a(t) = dv/dt$.

Let I be the open interval $a < x < b$. If $f'(x) > 0$ for all x in I, then $f(x)$ increases as x increases in I. If $f'(x) < 0$ for all x in I, then $f(x)$ decreases as x increases in I. If $f'(x) = 0$ for all x in I, then $f(x)$ is constant in I. Further, if the differentiable function f has a relative maximum [minimum] at the interior point c, i.e., there exists a $\delta > 0$ such that $f(x) \leqslant f(c)$ [$f(x) \geqslant f(c)$] for all $|x - c| < \delta$, then $f'(c) = 0$. The example $f(x) = x^3$ for which $f'(0) = 0$ but for which $x = 0$ is neither a maximum nor a minimum shows the condition is not sufficient.

First set of sufficient conditions:

1. For c an interior point of (a, b), let f be differentiable in $(c - \delta, c + \delta), \delta > 0$, then if: in $(c - \delta, c), f'(x) > 0; f'(c) = 0$; in $(c, c + \delta), f'(x) < 0$, then f has a relative maximum at c. If the inequalities of the previous sentence are satisfied in reverse, then f has a relative minimum at c.
2. For the end points a and b, let f have a right-hand derivative at a and a left hand derivative at b. Then, if $f'(a+) < 0$, f has a relative maximum at $x = a$; if $f'(b-) > 0$, f has a relative maximum at $x = b$; the inequalities reversed correspond to relative minima.

Second set of sufficient conditions:

Let f have a second derivative in $(c - \delta, c + \delta)$ and let $f'(c) = 0$. Then if

$$f''(c) > 0, \quad f \text{ has a relative minimum at } c$$

$$f''(c) < 0, \quad f \text{ has a relative maximum at } c$$

The condition is not necessary, since $f(x) = x^4$ has a relative (and absolute) minimum at 0, but $f''(0) = 0$.

The second derivative gives further information: if $f''(x) > 0$ in (a, b), then f is concave upwards in the interval, and if $f''(x) < 0$ in (a, b), then f is concave downwards. At a point where f changes concavity (a point of inflection), $f''(x) = 0$.

2.4.5 Mean Value Theorems, Differentials

Let f be differentiable in (a, b) and continuous on $[a, b]$. Then there exists at least one c in (a, b) such that

$$f'(c) = \frac{f(b) - f(a)}{b - a} \tag{2.4-10}$$

If g satisfies the same conditions, and, in addition, $g(a) \neq g(b)$ and $g'(x)$ is never zero, then there exists at least one c in (a, b) such that

$$\frac{f'(c)}{g'(c)} = \frac{f(b) - f(a)}{g(b) - g(a)} \tag{2.4-11}$$

Given $y = f(x)$, let Δx be a change in x and Δy the corresponding change in y. Then the *differentials* dx and dy of x and y are defined by

$$dx = \Delta x; \quad dy = f'(x)\, dx \tag{2.4-12}$$

With this definition, the Leibnitz notation dy/dx for the derivative can be treated as a fraction—the quotient of the two differentials. The differential of y is an approximation to Δy, as shown in Fig. 2.4-1.

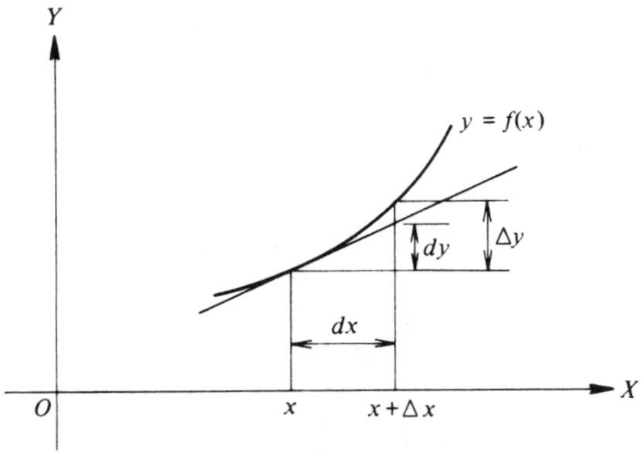

Fig. 2.4-1 The differentials of x and y.

2.4.6 L'Hospital's Rule

If as $x \to a-$, or as $x \to a+$, or as $x \to a$, or as $x \to +\infty$, or as $x \to -\infty$, either

(i) $f(x) \to 0$ and $g(x) \to 0$

(ii) $f(x) \to \infty$ and $g(x) \to \infty$, then

$$\lim \frac{f(x)}{g(x)} = \lim \frac{f'(x)}{g'(x)} \tag{2.4-13}$$

Example:

(1)

$$\lim_{x\to\infty} \frac{e^x}{x^n} = \lim_{x\to\infty} \frac{e^x}{nx^{n-1}} = \lim_{x\to\infty} \frac{e^x}{n(n-1)x^{n-2}}$$

$$= \cdots \lim_{x\to\infty} \frac{e^x}{n(n-1)\cdots(x \text{ to a power} \leqslant 0)}$$

$$= \infty$$

(2)

$$\lim_{x\to\infty} \left(\cos \frac{1}{x}\right)^x$$

here, consider $\lim_{x\to\infty} \ln \left(\cos \frac{1}{x}\right)^x$

$$= \lim_{x \to \infty} x \ln \cos \frac{1}{x} = \lim_{x \to \infty} \frac{\ln \cos \frac{1}{x}}{\frac{1}{x}}$$

$$= \lim_{x \to \infty} - \frac{\sin \frac{1}{x}}{\cos \frac{1}{x}} = 0$$

$$\therefore \lim_{x \to \infty} \left(\cos \frac{1}{x}\right)^x = 1$$

2.4.7 Parametric Equations, Polar Coordinates

Since curves given in parametric or polar form may intersect themselves, such a curve at a particular point may have one tangent, several tangents, or no tangent.

Given $x = x(t), y = y(t), \alpha < t < \beta$; if a tangent exists for $t = t_0$, its slope is

$$\frac{dy}{dx} = \left.\frac{dy/dt}{dx/dt}\right|_{t=t_0} = \frac{y'(t_0)}{x'(t_0)} \tag{2.4-14}$$

and

$$\frac{d^2y}{dx^2} = \frac{\frac{d}{dt}\left(\frac{dy}{dx}\right)}{\frac{dx}{dt}} = \frac{x'(t)\,y''(t) - y'(t)\,x''(t)}{[x'(t)]^3} \tag{2.4-15}$$

For curves given in polar coordinate form $r = f(\theta)$, the coordinate transformation equations $x = r \cos \theta$, $y = r \sin \theta$ give the parameterization $x = f(\theta) \cos \theta, y = f(\theta) \sin \theta$ and the above formulas can now be used.

2.4.8 Curvature

Consider $y = f(x)$ with $f''(x)$ continuous. At any point, let ϕ be the angle between the tangent to the curve and the positive x-axis, and let s be the arc length. Then the *curvature* K is defined by $\mathrm{K} = d\phi/ds$, and is given by

$$\mathrm{K}(x) = \frac{f''(x)}{\{1 + [f'(x)]^2\}^{3/2}} \tag{2.4-16}$$

If the curve is given in parametric form $x = x(t), y = y(t)$, then

$$K(t) = \frac{x'(t)\,y''(t) - x''(t)\,y'(t)}{[(x')^2 + (y')^2]^{3/2}} \qquad (2.4\text{-}17)$$

2.5 THE DEFINITE INTEGRAL

2.5.1 Anti-differentiation

$F(x)$ is an *antiderivative* of $f(x)$ in (a, b) if $F'(x) = f(x)$ in (a, b). We write $\int f(x)\,dx = F(x)$. Thus $\int \cos x\,dx = \sin x$. Note, however, that $\int \cos x\,dx$ also equals $\sin x + \pi^2/2$, which illustrates the following: if F and G are antiderivatives of the same function f on $[a, b]$, then $F(x) = G(x)$ + constant on $[a, b]$. Thus, if F is one antiderivative of f, then the most general antiderivative is given by $\int f(x)\,dx = F(x) + C$.

Antidifferentiation has the properties:

$$\int c\,f(x)\,dx = c\int f(x)\,dx, \qquad c \text{ a constant}$$

$$\int [f(x) + g(x)]\,dx = \int f(x)\,dx + \int g(x)\,dx$$

Notice that each differentiation formula given in section 2.4.3 gives immediately a corresponding antidifferentiation formula.

2.5.2 The Riemann Integral

Let f be a function which is bounded on the interval $[a, b]$. Divide the interval into n subintervals by the points $x_0 = a, x_1, x_2, \ldots, x_{n-1}, x_n = b$ with $x_{i-1} < x_i$, $i = 1, 2, 3, \ldots, n$. Such a subdivision is called a *partition* of $[a, b]$. Let the greatest lower bound and the least upper bound of the values of f on $[x_{i-1}, x_i]$ be denoted by m_i, M_i respectively. Then form the lower sum $s = \sum_{i=1}^{n} m_i\,(x_i - x_{i-1})$ and the upper sum $S = \sum_{i=1}^{n} M_i(x_i - x_{i-1})$. Each partition will in general give different values to each s and S.

Consider all possible partitions and let I be the least upper bound of all lower sums s, and let J be the greatest lower bound of all upper sums S. Then f is *Riemann integrable* on $[a, b]$ if $I = J$; this common value is the Riemann (or definite) integral of f on $[a, b]$, denoted by $\int_a^b f(x)\,dx$.

Any function which never decreases as x increases, or which never increases as x increases, is integrable over any closed interval on which it is defined. Any function continuous on $[a, b]$ is integrable on $[a, b]$. Any function which is bounded and has only a finite number of points of discontinuity on $[a, b]$ is integrable on $[a, b]$.

For generalizations of the Riemann integral, see section 2.5.7—the Riemann-Stieltjes integral, and Chapter 10—the Lebesgue integral.

2.5.3 Properties of the Riemann Integral

1. $\displaystyle\int_a^b [\alpha f(x) + \beta g(x)]\, dx = \alpha \int_a^b f(x)\, dx + \beta \int_a^b g(x)\, dx$ $\quad \alpha, \beta$ any numbers

2. $\displaystyle\int_a^b f(x)\, dx = -\int_b^a f(x)\, dx$

3. if $f < g$ on $[a, b]$, then

$$\int_a^b f(x)\, dx < \int_a^b g(x)\, dx$$

4. $\displaystyle\left|\int_a^b f(x)\, dx\right| \leqslant \int_a^b |f(x|\, dx$

5. if m, M are respectively the greatest lower bound and least upper bound of f on $[a, b]$, then

$$m(b - a) \leqslant \int_a^b f(x)\, dx \leqslant M(b - a)$$

6. if f is continuous on $[a, b]$ and c is any number in $[a, b]$, then

$$\int_a^b f(x)\, dx = \int_a^c f(x)\, dx + \int_c^b f(x)\, dx$$

7. if f is integrable on $[a, b]$, then the function $\displaystyle\int_a^x f(t)\, dt$ is continuous on $[a, b]$, and at any point where f is continuous

$$\frac{d}{dx}\int_a^x f(t)\, dt = f(x)$$

8. let $f(x, y)$ be an integrable function of x for each y, and let $\partial f/\partial y$ be continuous in $a \leqslant x \leqslant b, c \leqslant y \leqslant d$. Then

$$\frac{d}{dy} \int_a^b f(x, y)\, dx = \int_a^b \frac{\partial f}{\partial y}\, dx$$

9. let f be integrable on $[\alpha(x), \beta(x)]$ and let $\alpha(x)$ and $\beta(x)$ be differentiable functions. Then at any point x such that $f[\alpha(x)], f[\beta(x)]$ are continuous

$$\frac{d}{dx} \int_{\alpha(x)}^{\beta(x)} f(t)\, dt = f[\beta(x)] \frac{d\beta}{dx} - f[\alpha(x)] \frac{d\alpha}{dx}$$

2.5.4 Evaluation, Change of Variable, Mean Value Theorem

Let f be integrable on $[a, b]$ and let F be an antiderivative of f for (a, b). Then

$$\int_a^b f(x)\, dx = F(b) - F(a) \equiv F(x)]_a^b \tag{2.5-1}$$

If the substitution $x = g(t)$, where $g(t)$ is differentiable on $[c, d]$, is made in a definite integral, then

$$\int_{g(c)}^{g(d)} f(x)\, dx = \int_c^d f[g(t)]\, g'(t)\, dt \tag{2.5-2}$$

Example:

Find $\displaystyle\int_2^3 \frac{x^3}{(1 + x^2)^3}\, dx$.

Let $x = g(t) = \sqrt{t - 1}$, then $g'(t) = 1/2\sqrt{t - 1}$, and $2 = \sqrt{t - 1}$ gives $t = 5$ and $3 = \sqrt{t - 1}$ gives $t = 10$, so that

$$\int_2^3 \frac{x^3}{(1 + x^2)^3}\, dx = \int_5^{10} \frac{(t - 1)^{3/2}\, dt}{(1 + t - 1)^3\, 2\sqrt{t - 1}}$$

$$= \frac{1}{2} \int_5^{10} \frac{t - 1}{t^3}\, dt = \frac{1}{2} \int_5^{10} (t^{-2} - t^{-3})\, dt = \frac{17}{400}$$

Mean value theorems for integrals

If f is continuous on $[a, b]$, then there exists at least one c in $[a, b]$ such that $\displaystyle\int_a^b f(x)\, dx = f(c) \cdot (b - a)$.

If f and g are continuous on $[a, b]$ and g never changes sign in $[a, b]$, then there exists at least one c in $[a, b]$ such that

$$\int_a^b f(x)\,g(x)\,dx = f(c) \int_a^b g(x)\,dx$$

If g is continuous on $[a, b]$ and if f has a continuous derivative which never changes sign in $[a, b]$, then there exists at least one c in $[a, b]$ such that

$$\int_a^b f(x)\,g(x)\,dx = f(a) \int_a^c g(x)\,dx + f(b) \int_c^b g(x)\,dx$$

2.5.5 Geometrical and Physical Applications

Let f be a non-negative function integrable on $[a, b]$. Then the area A bounded by $x = a, x = b, y = 0$ and $y = f(x)$ is

$$A = \int_a^b f(x)\,dx \tag{2.5-3}$$

In polar coordinates, if a region is bounded by the rays $\theta = \alpha$ and $\theta = \beta$ and by the curve $r = f(\theta)$, then

$$A = \frac{1}{2} \int_\alpha^\beta f^2(\theta)\,d\theta \tag{2.5-4}$$

Let f and g be integrable on $[a, b]$ with $f(x) \geqslant g(x)$. Then the area bounded by $x = a, x = b, y = g(x)$ and $y = f(x)$ is

$$A = \int_a^b [f(x) - g(x)]\,dx \tag{2.5-5}$$

Consider a solid bounded on two sides by the planes $x = a$ and $x = b$. Let the area A of every cross section of the solid perpendicular to the x-axis be known as a function of x. Then the volume of the solid is

$$V = \int_a^b A(x)\,dx \tag{2.5-6}$$

A special case of such a solid is a solid of revolution, generated by rotating the area bounded by $x = a$, $x = b$, $y = 0$ and $y = f(x)$ about the x-axis. In this case,

$A(x) = \pi f^2(x)$, and the volume is

$$V = \pi \int_a^b f^2(x)\,dx \tag{2.5-7}$$

Let C be an arc given by $y = f(x)$, f continuously differentiable in $[\alpha, \beta]$. Then the length L of C is

$$L = \int_\alpha^\beta \sqrt{1 + [f'(x)]^2}\,dx \tag{2.5-8}$$

If C is given in parametric form $x = x(t), y = y(t)$, with $x(t), y(t)$ continuously differentiable in $[\alpha, \beta]$, then

$$L = \int_\alpha^\beta \sqrt{[x'(t)]^2 + [y'(t)]^2}\,dt \tag{2.5-9}$$

If C is given in polar coordinate form $r = f(\theta)$, f continuously differentiable in $[\alpha, \beta]$, then

$$L = \int_\alpha^\beta \sqrt{[f(\theta)]^2 + [f'(\theta)]^2}\,d\theta \tag{2.5-10}$$

Let C be given by any one of the above three descriptions. Then the centroid of C is that point with coordinates $(\bar{x}, \bar{y})$ given by

$$\bar{x} = \frac{\int_\alpha^\beta x\,ds}{\int_\alpha^\beta ds}, \quad \bar{y} = \frac{\int_\alpha^\beta y\,ds}{\int_\alpha^\beta ds} \tag{2.5-11}$$

where ds is chosen as the expression to the right of the integral sign in the expression for arc length which corresponds to the given description of the curve.

With the same choice for ds, if C lies in the upper half plane, then the area S of the surface obtained by rotating C

1. about the x-axis is $S = 2\pi \int_\alpha^\beta y\,ds$

2. about the y-axis is $S = 2\pi \int_{\alpha}^{\beta} x\, ds$.

Consider the plane region bounded by $y = f(x)$ (> 0), $y = 0$, $x = a$ and $x = b$. The coordinates of the centroid of this region are

$$\bar{x} = \frac{\int_a^b x f(x)\, dx}{\int_a^b f(x)\, dx}, \quad \bar{y} = \frac{\frac{1}{2}\int_a^b f^2(x)\, dx}{\int_a^b f(x)\, dx} \tag{2.5-12}$$

Other formulas for $\bar{x}$ and $\bar{y}$ appear in section 2.9.2.

Consider a solid of revolution as described earlier in this section. Its centroid will lie on the x-axis, with

$$\bar{x} = \frac{\int_a^b x[f(x)]^2\, dx}{\int_a^b f^2(x)\, dx} \tag{2.5-13}$$

Theorems of Pappus

1. Let a plane region R lie entirely on one side of a line L in its plane, then the volume of the solid generated by revolving R about L is equal to the product of the area of R and the length of the path described by the centroid of R.
2. If a plane arc C lies wholly on one side of a line L in its plane, then the area of the surface obtained by rotating C about L is equal to the product of the length of C and the length of the path described by the centroid of C.

2.5.6 Approximate Integration

It may happen that eq. (2.5-1) cannot be used to evaluate a definite integral because an antiderivative in closed form does not exist. Various approximation methods exist, two simple ones being:

a. the trapezoidal rule

Let f be continuous in $[a, b]$ and let $x_0 = a, x_1, \ldots, x_n = b$ be a partition into n equal subintervals of common length $(b - a)/n$. Then

$$\int_a^b f(x)\, dx \approx \frac{b-a}{2n}[f(x_0) + 2f(x_1) + 2f(x_2) + \cdots + 2f(x_{n-1}) + f(x_n)] \tag{2.5-14}$$

b. Simpson's rule

Let f be continuous in $[a, b]$ and let $x_0 = a, x_1, x_2, \ldots, x_{2n} = b$ be a partition into $2n$ equal subintervals. Then

$$\int_a^b f(x)\,dx \approx \frac{b-a}{6n}\,[f(x_0) + 4f(x_1) + 2f(x_2) + 4f(x_3) + \cdots$$

$$+ 2f(x_{2n-2}) + 4f(x_{2n-1}) + f(x_{2n})] \quad (2.5\text{-}15)$$

More general rules are considered in Chapter 18.

2.5.7 The Riemann-Stieltjes Integral

Let f and g be functions defined on $[a, b]$ and let $x_0 = a, x_1, x_2, \ldots, x_{n-1}, x_n = b$ be a partition of the interval. Choose any point x_i' such that $x_{i-1} \leqslant x_i' \leqslant x_i$, $i = 1, 2, 3, \ldots, n$. Then

$$\lim_{\substack{n \to \infty \\ \max(x_i - x_{i-1}) \to 0}} \sum_{i=1}^{n} f(x_i')\,[g(x_i) - g(x_{i-1})]$$

if it exists, is the *Riemann-Stieltjes integral*, denoted by $\int_a^b f(x)\,dg(x)$. Notice that for $g(x) = x$, this reduces to the definition of the Riemann integral as given in section 2.5.2; further, if g is continuously differentiable and f is Riemann integrable in $[a, b]$, then

$$\int_a^b f(x)\,dg(x) = \int_a^b f(x)\,g'(x)\,dx$$

As an example, let $f = 1$ on $[a, b]$, then $\int_a^b 1\,dg(x) = g(b) - g(a)$. If $\int_a^b f(x)\,dg(x)$ exists, then so does $\int_a^b g(x)\,df(x)$, and

$$\int_a^b f(x)\,dg(x) = f(b)\,g(b) - f(a)\,g(a) - \int_a^b g(x)\,df(x). \qquad (2.5\text{-}16)$$

2.6 METHODS OF INTEGRATION

Here, only a brief survey of the usual techniques of integration will be given, often by way of an example. Extensive tables of integrals are given in the references at the end of the chapter.

2.6.1 Trigonometric Functions

1. $$\int \sin^5 2x \cos^{1/3} 2x \, dx = \int (\sin^2 2x)^2 \cos^{1/3} 2x \sin 2x \, dx$$

$$= \int (1 - \cos^2 2x)^2 \cos^{1/3} 2x \sin 2x \, dx$$

$$= -\frac{1}{2}\left(\frac{3}{4}\cos^{4/3} 2x - \frac{3}{5}\cos^{10/3} 2x + \frac{3}{16}\cos^{16/3} 2x\right) + C$$

2. $$\int \sin^2 x \cos^4 x dx = \int \left(\frac{1 - \cos 2x}{2}\right)\left(\frac{1 + \cos 2x}{2}\right)^2 dx \qquad \text{by eq. (1.10-15)}$$

$$= \frac{1}{8}\int (1 + \cos 2x - \cos^2 2x - \cos^3 2x) \, dx$$

$$= \frac{1}{8}\left[x + \frac{1}{2}\sin 2x - \int \frac{1 + \cos 4x}{2} dx - \int (1 - \sin^2 2x) \cos 2x \, dx\right]$$

$$= \frac{x}{16} + \frac{1}{48}\sin^3 2x - \frac{1}{64}\sin 4x + C$$

3. $$\int \tan^{-2} x \sec^4 x \, dx = \int \tan^{-2} x (1 + \tan^2 x) \sec^2 x \, dx$$

$$= -\cot x + \tan x + C$$

4. $$\int \sec^{-1/2} x \tan^3 x \, dx = \int (\sec^{-3/2} x \tan^2 x) \sec x \tan x \, dx$$

$$= \int \sec^{-3/2} x (\sec^2 x - 1) \sec x \tan x \, dx$$

$$= \frac{2}{3}\sec^{3/2} x + 2 \sec^{-1/2} x + C$$

5. $$\int \tan^m x\,dx = \int (\tan^{m-2} x)(\sec^2 x - 1)\,dx$$

$$= \frac{\tan^{m-1} x}{m-1} - \int \tan^{m-2} x\,dx$$

which, by iteration, if m is a positive integer, will lead to $\int \tan^0 x\,dx$ if m is even, and for m odd, to $\int \tan x\,dx = \int \frac{\sin x}{\cos x}\,dx = -\ln \cos x + C$.

6. $$\int \frac{\sec x}{2\tan x + \sec x - 1}\,dx = I$$

For a rational function of the trigonometric functions, the substitution $t = \tan(x/2)$ is used. If $t = \tan(x/2)$, then

$$\sin x = \frac{2t}{1+t^2}, \quad \cos x = \frac{1-t^2}{1+t^2} \text{ and } \frac{dx}{dt} = \frac{2}{1+t^2}$$

Then

$$I = \int \frac{dt}{t(t+2)} = \frac{1}{2}\int \left(\frac{1}{t} - \frac{1}{t+2}\right) dt = \frac{1}{2}\ln\left|\frac{t}{t+2}\right| + C = \frac{1}{2}\ln\left|\frac{\tan\frac{x}{2}}{\tan\frac{x}{2}+2}\right| + C$$

2.6.2 Integration by Substitution

If an expression such as $(a^2 - x^2)^{1/2}$ occurs in the integrand, it may be useful to make the substitution $x = a\sin\theta$ (or $x = a\cos\theta$). Similarly when $(a^2 + x^2)^{1/2}$ occurs, try $x = a\tan\theta$, and for $(x^2 - a^2)^{1/2}$, try $x = a\sec\theta$.

Example:

$$\int \frac{1}{(4-x^2)^{3/2}}\,dx; \qquad \text{let } x = 2\sin\theta,\ dx = 2\cos\theta\,d\theta$$

$$= \frac{1}{4}\int \sec^2\theta\,d\theta = \frac{1}{4}\tan\theta + C = \frac{x}{4(4-x^2)^{1/2}} + C$$

In these substitutions, it is sometimes useful to use the results

$$\int \sec x \, dx = \ln | \sec x + \tan x | + C \tag{2.6-1}$$

$$\int \operatorname{cosec} x \, dx = \ln | \operatorname{cosec} x - \cot x | + C \tag{2.6-2}$$

If the integrand contains a single irrational expression $(ax + b)^{p/q}$, the substitution $u^q = ax + b$ may be tried.

Example

$$\int \frac{x - 2}{(3x - 1)^{2/3}} \, dx \qquad \text{let } 3x - 1 = u^3, \quad 3dx = 3u^2 \, du$$

$$= \frac{1}{3} \int (u^3 - 5) \, du = \frac{1}{3} \left(\frac{u^4}{4} - 5u \right) + C$$

$$= \frac{1}{12} (3x - 1)^{4/3} - \frac{5}{3} (3x - 1)^{1/3} + C$$

2.6.3 Integration by Parts

The formula for integration by parts is

$$\int f(x) \frac{dg(x)}{dx} \, dx = f(x) g(x) - \int g(x) \frac{df(x)}{dx} \, dx \tag{2.6-3}$$

An application of the formula replaces one integration problem by another, which may or may not be simpler.

Example:

(1)

$$\int e^{ax} \cos bx \, dx \qquad \text{let } f = e^{ax} \qquad \frac{dg}{dx} = \cos bx$$

$$\frac{df}{dx} = a \, e^{ax} \qquad g = \frac{1}{b} \sin bx$$

$$= \frac{1}{b} e^{ax} \sin bx - \frac{a}{b} \int e^{ax} \sin bx \, dx$$

Here, make a second application of the formula, with

$$f = e^{ax} \qquad dg = \sin bx\, dx$$

$$df = ae^{ax}\, dx \qquad g = -\frac{1}{b}\cos bx$$

$$\int e^{ax} \cos bx\, dx = \frac{1}{b} e^{ax} \sin bx - \frac{a}{b}\left[-\frac{1}{b} e^{ax} \cos bx + \frac{a}{b}\int e^{ax} \cos bx\, dx\right]$$

which can be solved for the required integral, giving

$$\int e^{ax} \cos bx\, dx = \frac{e^{ax}}{a^2 + b^2}\,[b \sin bx + a \cos bx] + C \tag{2.6-4}$$

(2)

$$\int \sin^p x \cos^q x\, dx \qquad q \neq -1,\ \ p + q \neq 0$$

$$\text{let } f = \sin^{p-1} x \qquad dg = \cos^q x \sin x\, dx$$

$$df = (p-1)\sin^{p-2} x \cos x\, dx \qquad g = -\frac{\cos^{q+1} x}{q+1}$$

$$\int \sin^p x \cos^q x\, dx = -\frac{1}{q+1}\sin^{p-1} x \cos^{q+1} x + \frac{p-1}{q+1}\int \sin^{p-2} x \cos^{q+2} x\, dx \tag{2.6-5}$$

$$= -\frac{\sin^{p-1} x \cos^{q+1} x}{p+q} + \frac{p-1}{p+q}\int \sin^{p-2} x \cos^q x\, dx \tag{2.6-6}$$

[by writing $\cos^{q+2} x = \cos^q x\,(1 - \sin^2 x)$]

This last formula, with q set equal to zero, is

$$\int \sin^p x\, dx = -\frac{\sin^{p-1} x \cos x}{p} + \frac{p-1}{p}\int \sin^{p-2} x\, dx$$

a recursion formula for the integration of positive integer powers of sin x.

$$\int_0^{\pi/2} \sin^p x\,dx = -\left.\frac{\sin^{p-1} x \cos x}{p}\right|_0^{\pi/2} + \frac{p-1}{p}\int_0^{\pi/2} \sin^{p-2} x\,dx$$

$$= \frac{p-1}{p}\int_0^{\pi/2} \sin^{p-2} x\,dx \qquad \text{if } p > 1$$

Iteration of this formula for p an integer $\geqslant 2$ gives Wallis' formulas:

$$\int_0^{\pi/2} \sin^{2n} x\,dx = \frac{(2n)!}{2^{2n}(n!)^2}\,\frac{\pi}{2} \tag{2.6-7}$$

$$\int_0^{\pi/2} \sin^{2n+1} x\,dx = \frac{2^{2n}(n!)^2}{(2n)!}\,\frac{1}{2n+1} \tag{2.6-8}$$

2.6.4 Rational Functions

Theoretically, any rational function can be integrated by writing it as a polynomial plus a proper rational function and finding the partial fraction expansion for this latter term by the methods of section 1.6.2.

e.g.,

$$\int \frac{3x^4 + x^3 + 20x^2 + 3x + 31}{(x+1)(x^2+4)^2}\,dx$$

$$= \int \left\{\frac{2}{x+1} + \frac{16-i}{32(x+2i)} + \frac{1}{16(x+2i)^2} + \frac{16+i}{32(x-2i)} + \frac{1}{16(x-2i)^2}\right\} dx$$

$$= \int \left\{\frac{2}{x+1} + \frac{x - \frac{1}{8}}{x^2+4} + \frac{1}{16(x+2i)^2} + \frac{1}{16(x-2i)^2}\right\} dx$$

$$= 2 \ln|x+1| + \frac{1}{2}\ln|x^2+4| - \frac{1}{16}\text{Arc tan}\,\frac{x}{2} - \frac{1}{16(x+2i)} - \frac{1}{16(x-2i)} + C$$

$$= \ln (x+1)^2 (x^2+4)^{1/2} - \frac{1}{16}\text{Arc tan}\,\frac{x}{2} - \frac{x}{8(x^2+4)} + C$$

2.7 IMPROPER INTEGRALS

2.7.1 With Infinite Limits

Let f be integrable on $[a, b]$ for all b. Then by definition

$$\int_a^\infty f(x)\,dx = \lim_{b\to\infty}\int_a^b f(x)\,dx$$

If this limit exists, the improper integral is *convergent*, otherwise, *divergent*. Similarly

$$\int_{-\infty}^b f(x)\,dx = \lim_{a\to-\infty}\int_a^b f(x)\,dx$$

$$\int_{-\infty}^\infty f(x)\,dx = \lim_{a\to-\infty}\int_a^c f(x)\,dx + \lim_{b\to\infty}\int_c^b f(x)\,dx$$

where it is necessary that each improper integral on the right hand side be convergent in order that $\int_{-\infty}^\infty f(x)\,dx$ be convergent.

The *Cauchy Principal Value* of $\int_{-\infty}^\infty f(x)\,dx$ is defined by

$$\text{C.P.V.}\int_{-\infty}^\infty f(x)\,dx = \lim_{\lambda\to\infty}\int_{-\lambda}^{\lambda} f(x)\,dx$$

Examples:

(1)

$$\int_1^\infty \frac{1}{x^p}\,dx = \begin{cases} \dfrac{1}{p-1} & p > 1 \\ \text{diverges for } p \leqslant 1 \end{cases}$$

(2)

$$\int_{-\infty}^\infty x\,dx = \lim_{a\to-\infty}\int_a^1 x\,dx + \lim_{b\to\infty}\int_1^b x\,dx$$

each of which diverges by the first example, but

$$\text{C.P.V.} \int_{-\infty}^{\infty} x\,dx = \lim_{\lambda\to\infty} \int_{-\lambda}^{\lambda} x\,dx = 0$$

Comparison tests for convergence of improper integrals exist analogous to those for infinite series: Let f and g be non-negative with $f \leqslant g$ at least for all $x \geqslant c$. Then if $\int_b^{\infty} g(x)\,dx$ is convergent, so is $\int_a^{\infty} f(x)\,dx$; and if the latter is divergent, so is the former. Also, for f and g non-negative, if $\lim_{x\to\infty} f(x)/g(x)$ exists and is not zero, then either both integrals converge or they both diverge. If the limit should be zero, then if the integral of g converges, so does the integral of f.

Example: $\int_1^{\infty} e^{-x} x^{\alpha-1}\,dx$ converges for all α, by comparison with $\int_1^{\infty} x^{-2}\,dx$

If $\int_a^{\infty} |f(x)|\,dx$ converges, then $\int_a^{\infty} f(x)\,dx$ is called *absolutely convergent*. An absolutely convergent integral is convergent.

Next, consider improper integrals depending on a parameter—such integrals arise, for example, in the Laplace transformation. Let $F(x) = \int_c^{\infty} f(x,y)\,dy$ converge for all x on $[a,b]$; i.e., given any $\epsilon > 0$ there exists a $k_0(\epsilon, x)$ such that

$$\left| F(x) - \int_c^k f(x,y)\,dy \right| < \epsilon \qquad \text{whenever } k \geqslant k_0$$

If k_0 independent of x exists, then the integral converges *uniformly* to F on $[a,b]$.

Let $f(x,y)$ be continuous on $[a,b]$ for $y \geqslant c$. Then, if $F(x) = \int_c^{\infty} f(x,y)\,dy$ converges uniformly on $[a,b]$, $F(x)$ will be continuous on $[a,b]$, and

$$\int_a^b F(x)\,dx = \int_c^{\infty} \int_a^b f(x,y)\,dx\,dy$$

As is the case with infinite series, differentiation requires further restrictions: if $\int_c^{\infty} f(x,y)\,dy$ is convergent to $F(x)$ on $[a,b]$ and $\partial f/\partial x$ is continuous on $[a,b]$ for

$y \geqslant c$, then if $\int_c^\infty \partial f/\partial x \, dy$ is uniformly convergent on $[a, b]$, $dF(x)/dx = \int_c^\infty \partial f/\partial x \, dy$.

2.7.2 Other Improper Integrals

If f is integrable on $[a, c]$, $c < b$ but is not integrable on $[a, b]$, then $\int_a^b f(x)\, dx$ is also said to be an improper integral, and is defined by

$$\int_a^b f(x)\, dx = \lim_{c \to b-} \int_a^c f(x)\, dx$$

Similarly, for f integrable on $[d, b]$ but not on $[a, b]$, $d > a$,

$$\int_a^b f(x)\, dx = \lim_{d \to a+} \int_d^b f(x)\, dx$$

In each case, if the limit exists, the integral is called convergent; otherwise, divergent. For a function f which has an infinite discontinuity at c in the interior of $[a, b]$,

$$\int_a^b f(x)\, dx = \lim_{\lambda \to c-} \int_a^\lambda f(x)\, dx + \lim_{\lambda \to c+} \int_\lambda^b f(x)\, dx$$

where each of the limits on the right must exist.

The Cauchy Principal Value in this last case is defined by

$$\text{C.P.V.} \int_a^b f(x)\, dx = \lim_{\lambda \to 0+} \left\{ \int_a^{c-\lambda} f(x)\, dx + \int_{c+\lambda}^b f(x)\, dx \right\}$$

Example:

$$\int_0^1 \frac{1}{x^p} = \begin{cases} \dfrac{1}{1-p} & p < 1 \\ \text{diverges otherwise} \end{cases}$$

Comparison tests completely analogous to those in the previous section can be stated for these improper integrals.

Example: $\int_0^1 e^{-x} x^{\alpha-1}\, dx$ converges for $\alpha > 0$, since $e^{-x} x^{\alpha-1} < x^{\alpha-1}$ in (0, 1) and $\int_0^1 x^{\alpha-1}\, dx$ converges for $\alpha > 0$.

Improper integrals may be of mixed type, consider for example, the integral defining the *Gamma function* $\Gamma(\alpha)$:

$$\Gamma(\alpha) = \int_0^\infty e^{-x} x^{\alpha-1}\, dx = \int_0^1 e^{-x} x^{\alpha-1}\, dx + \int_1^\infty e^{-x} x^{\alpha-1}\, dx$$

and each of these integrals has been shown to be convergent for $\alpha > 0$ in the examples of section 2.7.

The value of the next improper integral will be used in section 2.9.3.

Consider

$$I = \int_0^\infty e^{-y^2} \cos xy\, dy$$

then

$$\frac{dI}{dx} = -\int_0^\infty e^{-y^2}\, y \sin xy\, dy = \frac{e^{-y^2}}{2} \sin xy \Big|_0^\infty - \frac{x}{2}\int_0^\infty e^{-y^2} \cos xy\, dy$$

so

$$\frac{dI}{dx} = -\frac{x}{2} I \Rightarrow I = A\, e^{-x^2/4}$$

$$I(0) = A = \int_0^\infty e^{-y^2}\, dy = \frac{\sqrt{\pi}}{2} \qquad \text{by eq. (2.9-13)}$$

$$\therefore I = \frac{\sqrt{\pi}}{2} e^{-x^2/4}$$

Let $x = 2b/a, y = at$ $(a > 0)$, then

$$\int_0^\infty e^{-a^2 t^2} \cos 2bt\, dt = \frac{\sqrt{\pi}}{2a} e^{-b^2/a^2} \tag{2.7-1}$$

2.8 PARTIAL DIFFERENTIATION

2.8.1 Functions and Limits for More Than One Variable

Consider a set of ordered $(n + 1)$ - tuplets of real numbers $(x_1, x_2, \ldots, x_n, w)$. If no two different elements of the set have the first n entries identical, then w is a single-valued function of the n independent variables $x_1, x_2, \ldots, x_n$. The set of all $(x_1, x_2, \ldots, x_n)$ values is the domain of the function and the corresponding set of w values is its range.

The definitions of limit and continuity will be given for single-valued functions of two variables; generalizations to three or more variables are immediate.

$\lim_{(x,y)\to(x_0,y_0)} f(x,y) = L$ if given any $\epsilon > 0$, there exists a $\delta > 0$ such that $|f(x,y) - L| < \epsilon$ whenever $(x - x_0)^2 + (y - y_0)^2 < \delta^2$.

The limit theorems given in eqs. (2.3-1), (2.3-2), (2.3-3) extended to functions of several variables, remain valid.

The function $f(x, y)$ is continuous at (x_0, y_0) if and only if $\lim_{(x,y)\to(x_0,y_0)} f(x,y) = f(x_0, y_0)$; $f(x, y)$ is continuous in a region R (a region is an open connected set—see Chapter 5) if it is continuous at each point of R. As in section 2.3.4, sums, products, and quotients of continuous functions are continuous.

2.8.2 Definition and Notations

If w is a single-valued function of $x_1, x_2, \ldots, x_n$; then the derivative obtained by holding $x_1, x_2, \ldots, x_{i-1}, x_{i+1}, \ldots, x_n$ constant and differentiating the resulting function with respect to x_i is called the *partial derivative* of w with respect to x_i, and is denoted by any of $\partial w/\partial x_i$, w_i, $D_i w$. Formally

$$\frac{\partial w}{\partial x_i} = \lim_{h\to 0} \frac{w(x_1, x_2, \ldots, x_{i-1}, x_i + h, x_{i+1}, \ldots, x_n) - w(x_1, x_2, \ldots, x_i, \ldots, x_n)}{h}$$

Example: If $z = e^x \sin y + y \ln x$, then

$$\frac{\partial z}{\partial x} = e^x \sin y + \frac{y}{x} \qquad \frac{\partial z}{\partial y} = e^x \cos y + \ln x$$

Generally, partial differentiation gives functions which can be partially differentiated again, and again, . . . ; yielding partial derivatives of the second, third, . . . orders. For the above example:

$$\frac{\partial}{\partial x}\left(\frac{\partial z}{\partial x}\right) = \frac{\partial^2 z}{\partial x^2} = z_{xx} = D_{xx} z = e^x \sin y - \frac{y}{x^2}$$

$$\frac{\partial}{\partial y}\left(\frac{\partial z}{\partial x}\right) = \frac{\partial^2 z}{\partial y \partial x} = z_{xy} = D_{xy} z = e^x \cos y + \frac{1}{x}$$

$$\frac{\partial}{\partial x}\left(\frac{\partial z}{\partial y}\right) = \frac{\partial^2 z}{\partial x \partial y} = z_{yx} = D_{yx}z = e^x \cos y + \frac{1}{x}$$

$$\frac{\partial}{\partial y}\left(\frac{\partial z}{\partial y}\right) = \frac{\partial^2 z}{\partial y^2} = z_{yy} = D_{yy}z = -\, e^x \sin y$$

Here, the mixed partial derivatives are equal—which is not true in general; however, if the two mixed partials are continuous at the point (x_0, y_0), then they are equal at that point. This result can be generalized to mixed partial derivatives of higher order in any number of variables.

2.8.3 Chain Rule, Exact Differentials

Let $w = f(x_1, x_2, \ldots, x_n)$ and $x_i = x_i(\lambda_1, \lambda_2, \ldots, \lambda_m)$ for $i = 1, 2, 3, \ldots, n$. Then

$$\frac{\partial w}{\partial \lambda_s} = \sum_{j=1}^{n} \frac{\partial f}{\partial x_j}\frac{\partial x_j}{\partial \lambda_s} \qquad (s = 1, 2, \ldots, m) \tag{2.8-1}$$

Example: Let $w = f(r, \theta, z)$ with $r = \sqrt{x^2 + y^2}$, $\theta = \tan^{-1} y/x$, $z = z$. Then

$$\frac{\partial f}{\partial x} = \frac{\partial f}{\partial r}\frac{\partial r}{\partial x} + \frac{\partial f}{\partial \theta}\frac{\partial \theta}{\partial x} + \frac{\partial f}{\partial z}\frac{\partial z}{\partial x}$$

$$= \frac{\partial f}{\partial r}\cos\theta - \frac{\partial f}{\partial \theta}\frac{\sin\theta}{r}$$

$$\frac{\partial f}{\partial y} = \frac{\partial f}{\partial r}\sin\theta + \frac{\partial f}{\partial \theta}\frac{\cos\theta}{r}, \quad \frac{\partial f}{\partial z} = \frac{\partial f}{\partial z}$$

$$\frac{\partial^2 f}{\partial x^2} = \cos\theta\left(\frac{\partial^2 f}{\partial r^2}\frac{\partial r}{\partial x} + \frac{\partial^2 f}{\partial \theta \partial r}\frac{\partial \theta}{\partial x} + \frac{\partial^2 f}{\partial z \partial r}\frac{\partial z}{\partial x}\right) + \frac{\partial f}{\partial r}\frac{\partial}{\partial x}(\cos\theta)$$

$$- \frac{\sin\theta}{r}\left(\frac{\partial^2 f}{\partial r \partial \theta}\frac{\partial r}{\partial x} + \frac{\partial^2 f}{\partial \theta^2}\frac{\partial \theta}{\partial x} + \frac{\partial^2 f}{\partial z \partial \theta}\frac{\partial z}{\partial x}\right) + \frac{\partial f}{\partial \theta}\frac{\partial}{\partial x}\left(-\frac{\sin\theta}{r}\right)$$

$$= \frac{\partial^2 f}{\partial r^2}\cos^2\theta - \frac{\sin\theta\cos\theta}{r}\frac{\partial^2 f}{\partial \theta \partial r} + \frac{\partial f}{\partial r}\frac{\sin^2\theta}{r}$$

$$- \frac{\sin\theta\cos\theta}{r}\frac{\partial^2 f}{\partial r \partial \theta} + \frac{\sin^2\theta}{r^2}\frac{\partial^2 f}{\partial \theta^2} + \frac{\partial f}{\partial \theta}\frac{2}{r^2}\sin\theta\cos\theta$$

Similarly for $\partial^2 f/\partial y^2$, $\partial^2 f/\partial z^2$; giving for the Laplacian of f, i.e., $\nabla^2 f = \partial^2 f/\partial x^2 + \partial^2 f/\partial y^2 + \partial^2 f/\partial z^2$ in cylindrical coordinates:

$$\nabla^2 f = \frac{\partial^2 f}{\partial r^2} + \frac{1}{r}\frac{\partial f}{\partial r} + \frac{1}{r^2}\frac{\partial^2 f}{\partial \theta^2} + \frac{\partial^2 f}{\partial z^2} \tag{2.8-2}$$

For spherical polar coordinates, as defined in section 1.12.1,

$$\nabla^2 f = \frac{\partial^2 f}{\partial \rho^2} + \frac{1}{\rho^2}\frac{\partial^2 f}{\partial \phi^2} + \frac{1}{\rho^2 \sin^2 \phi}\frac{\partial^2 f}{\partial \theta^2} + \frac{2}{\rho}\frac{\partial f}{\partial \rho} + \frac{\cot \phi}{\rho^2}\frac{\partial f}{\partial \phi} \tag{2.8-3}$$

If $w = f(x_1, x_2, \ldots, x_n)$ but each x_i is a function of a single variable, say t, then the chain rule becomes

$$\frac{dw}{dt} = \sum_{j=1}^{n} \frac{\partial f}{\partial x_j}\frac{dx_j}{dt}$$

Recalling the definition of differentials, this gives

$$dw = \sum_{j=1}^{n} \frac{\partial f}{\partial x_j} dx_j \tag{2.8-4}$$

which is the so-called total, or exact, differential of w, and may be used as an approximation to Δw, the actual change in w. More precisely

$$\Delta w = \sum_{j=1}^{n} \frac{\partial f}{\partial x_j} dx_j + \epsilon\left(\sum_{j=1}^{n} |dx_j|\right) \tag{2.8-5}$$

where

$$\lim \epsilon \to 0 \text{ as} \left[\sum_{j=1}^{n} |dx_j|\right]^{1/2} \to 0$$

The expression $P(x, y, z)\, dx + Q(x, y, z)\, dy + R(x, y, z)\, dz$ where the functions and their first partial derivatives are continuous in some region S is an exact differential in S if and only if $\partial P/\partial y = \partial Q/\partial x$, $\partial R/\partial x = \partial P/\partial z$, $\partial Q/\partial z = \partial R/\partial y$. The importance of this last concept is that the line integrals of an exact differential, over two distinct paths joining the same end points are equal, provided that each path can be continuously deformed into the other, without leaving the region.

2.8.4 Taylor's Theorem

Taylor's formula (see section 2.2.5) can be extended to functions of several variables; for example:

$$f(x + h, y + k) = f(x, y) + \frac{df}{1!} + \frac{d^2 f}{2!} + \cdots + \frac{d^n f}{n!} + R_n \tag{2.8-6}$$

where $df, d^2f, \ldots, d^nf$ are the first, second, ..., n^{th} differentials; defined by

$$df = \left(h\frac{\partial}{\partial x} + k\frac{\partial}{\partial y}\right) f$$

$$d^2f = \left(h\frac{\partial}{\partial x} + k\frac{\partial}{\partial y}\right)^{(2)} f = h^2 f_{xx} + 2hk\, f_{xy} + k^2 f_{yy}$$

$$d^3f = \left(h\frac{\partial}{\partial x} + k\frac{\partial}{\partial y}\right)^{(3)} f = h^3 f_{xxx} + 3h^2k\, f_{xxy} + 3hk^2 f_{xyy} + k^3 f_{yyy}$$

$$\vdots$$

$$d^nf = \left(h\frac{\partial}{\partial x} + k\frac{\partial}{\partial y}\right)^{(n)} f$$

and R_n, the remainder, is $d^{n+1}f(x + \theta h, y + \theta k)/(n + 1)!, 0 < \theta < 1$.

2.8.5 Maxima and Minima

A necessary condition that a function of n variables have a relative maximum or minimum at a point is that every first partial derivative of f vanish at that point.

For a function of two variables, if at a point $P_0, f_x = f_y = 0$ and $f_{xx}f_{yy} - f_{xy}^2 > 0$, then f will have a relative maximum at P_0 if $f_{xx}|_{P_0} < 0$, and a relative minimum if $f_{xx}|_{P_0} > 0$. However, if $f_{xx}f_{yy} - f_{xy}^2 < 0$ at P_0, then f has a saddle point at P_0.

Frequently the problem of maximizing (or minimizing) $f(x_1, x_2, \ldots, x_n)$ subject to m side conditions, $(m < n)$

$$g_1(x_1, x_2, \ldots, x_n) = 0, \ldots, g_m(x_1, x_2, \ldots, x_n) = 0 \tag{2.8-7}$$

arises. A useful solution procedure is the method of Lagrange multipliers which is valid if not all of the Jacobians (see section 2.8.6) of the g-functions with respect to m of the variables are zero at the maximum or minimum point in question. Under these conditions, constants $\lambda_1, \ldots, \lambda_m$ will exist such that

$$\frac{\partial f}{\partial x_i} = \lambda_1 \frac{\partial g_1}{\partial x_i} + \lambda_2 \frac{\partial g_2}{\partial x_i} + \cdots + \lambda_m \frac{\partial g_m}{\partial x_i} \qquad i = 1, 2, 3, \ldots, n \tag{2.8-8}$$

The method consists of solving (if possible) the $m + n$ equations in (2.8-7) and (2.8-8) for the unknowns $x_1, \ldots, x_n, \lambda_1, \ldots, \lambda_m$. The points at which extreme values occur will be in this set of solutions.

Example: Show that of all the triangles inscribed in a circle of radius R, the one with the largest perimeter is equilateral.

Let x, y, z be the angles subtended by the three sides at the center of the circle, then the perimeter f is

$$f(x, y, z) = 2R\left(\sin\frac{x}{2} + \sin\frac{y}{2} + \sin\frac{z}{2}\right)$$

The side condition is $g(x, y, z) = x + y + z - 2\pi = 0$. Equations (2.8-8) become

$$R\cos\frac{x}{2} = \lambda,\quad R\cos\frac{y}{2} = \lambda,\quad R\cos\frac{z}{2} = \lambda$$

and the solution of these four equations is $x = y = z = 2\pi/3$, $\lambda = R/2$. Thus, the required triangle is indeed equilateral.

2.8.6 Jacobians, Homogeneous Functions

If the equations $u = f(x, y)$, $v = g(x, y)$ determine x and y as functions of u and v possessing first partial derivatives, then

$$\frac{\partial x}{\partial u} = \frac{g_y}{\begin{vmatrix} f_x & f_y \\ g_x & g_y \end{vmatrix}},\quad \frac{\partial y}{\partial u} = \frac{-g_x}{\begin{vmatrix} f_x & f_y \\ g_x & g_y \end{vmatrix}}$$

with similar formulas for x_v, y_v. The determinant appearing in the denominator is an example of a Jacobian; which is defined in general for n functions of n variables to be

$$\frac{\partial(f_1, f_2, \ldots, f_n)}{\partial(x_1, x_2, \ldots, x_n)} = \begin{vmatrix} \frac{\partial f_1}{\partial x_1} & \frac{\partial f_1}{\partial x_2} & \cdots & \frac{\partial f_1}{\partial x_n} \\ \frac{\partial f_2}{\partial x_1} & \frac{\partial f_2}{\partial x_2} & \cdots & \frac{\partial f_2}{\partial x_n} \\ \cdots & \cdots & \cdots & \cdots \\ \frac{\partial f_n}{\partial x_1} & \frac{\partial f_n}{\partial x_2} & \cdots & \frac{\partial f_n}{\partial x_n} \end{vmatrix} \tag{2.8-9}$$

Examples:

(1) for polar coordinates

$$\frac{\partial(x, y)}{\partial(r, \theta)} = r \tag{2.8-10}$$

(2) for cylindrical polar coordinates

$$\frac{\partial(x, y, z)}{\partial(r, \theta, z)} = r \tag{2.8-11}$$

(3) for spherical polar coordinates

$$\frac{\partial(x, y, z)}{\partial(\rho, \theta, \phi)} = -\rho^2 \sin\phi \qquad (2.8\text{-}12)$$

Note that

$$\frac{\partial(f_1, f_2, \ldots, f_n)}{\partial(x_1, x_2, \ldots, x_n)} \cdot \frac{\partial(x_1, x_2, \ldots, x_n)}{\partial(f_1, f_2, \ldots, f_n)} = 1$$

The function $f(x_1, x_2, \ldots, x_n)$ is *positively homogeneous* of degree k if $f(tx_1, tx_2, \ldots, tx_n) = t^k f(x_1, \ldots, x_n)$ for $t > 0$. For homogeneous functions which are differentiable, Euler's theorem states

$$\sum_{i=1}^{n} x_i \frac{\partial f}{\partial x_i} = kf \qquad (2.8\text{-}13)$$

2.8.7 The Implicit Function Theorem

This theorem states conditions under which the n equations in $n + s$ variables

$$f_i(x_1, \ldots, x_n, y_1, \ldots, y_s) = 0 \qquad i = 1, 2, \ldots, n$$

can be solved for $x_1, \ldots, x_n$ as functions of $y_1, \ldots, y_s$.

Let the functions f_i be defined and have continuous first partial derivatives in an open set S of $n + s$ space. Further, let each $f_i = 0$ at $P_0(x_1^0, \ldots, x_n^0, y_1^0, \ldots, y_s^0)$, and let the Jacobian $\partial(f_1, \ldots, f_n)/\partial(x_1, \ldots, x_n)$ be different from 0 at P_0. Then there will exist a neighborhood Y of $(y_1^0, \ldots, y_s^0)$ and exactly one set of functions $g_1, \ldots, g_n$ defined in Y such that:

(1) $g_i(y_1^0, \ldots, y_s^0) = x_i^0 \qquad i = 1, 2, \ldots, n$
(2) each g_i has continuous first partial derivatives in Y
(3) $f_i(g_1, g_2, \ldots, g_n, y_1, \ldots, y_s) = 0$ for every $(y_1, \ldots, y_s)$ in Y

2.9 MULTIPLE INTEGRALS

2.9.1 Definition and Properties

Let $f(x, y)$ be bounded and continuous in $a < x < b$, $c < y < d$. Then the double integral of f can be defined as

$$\iint\limits_{\substack{a<x<b \\ c<y<d}} f(x, y)\, dA = \int_c^d \left\{ \int_a^b f(x, y)\, dx \right\} dy = \int_a^b \left\{ \int_c^d f(x, y)\, dy \right\} dx$$

where each of the two right hand expressions is an iterated integral, and is usually written without the parentheses. If the domain of f is not a rectangle, but is instead some general bounded set R, then we introduce $f^*(x, y)$ defined by

$$f^*(x,y) = \begin{cases} f(x,y) & (x,y) \text{ in } R \\ 0 & (x,y) \text{ not in } R \end{cases}$$

and let R_1 be any rectangle containing R, then

$$\iint_R f(x,y)\,dA = \iint_{R_1} f^*(x,y)\,dA$$

Example: Find $\iint_R (x^2 + y^3)\,dA$ where R is the region in the first quadrant bounded by $y = x^2$ and $x = y^4$ (see Fig. 2.9-1).

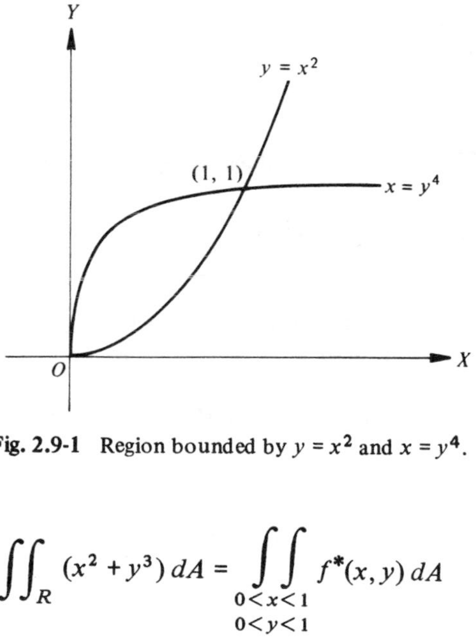

Fig. 2.9-1 Region bounded by $y = x^2$ and $x = y^4$.

$$\iint_R (x^2 + y^3)\,dA = \iint_{\substack{0<x<1 \\ 0<y<1}} f^*(x,y)\,dA$$

where

$$f^*(x,y) = \begin{cases} x^2 + y^3 & (x,y) \text{ in } R \\ 0 & \text{otherwise} \end{cases}$$

$$\iint f^*(x,y)\,dA = \int_0^1 \int_{y^4}^{\sqrt{y}} (x^2 + y^3)\,dx\,dy = \int_0^1 \left[\frac{x^3}{3} + y^3 x\right]_{y^4}^{\sqrt{y}} dy$$

$$= \left[\frac{2}{15} y^{5/2} + \frac{2}{9} y^{9/2} - \frac{y^{13}}{39} - \frac{y^3}{8}\right]_0^1 = \frac{959}{4680}$$

or

$$\iint f^*(x,y)\,dA = \int_0^1 \int_{x^2}^{x^{1/4}} (x^2 + y^3)\,dy\,dx = \frac{959}{4680}$$

The extension of the definition given above for the double integral to triple and higher multiple integrals is immediate.

Multiple integrals satisfy rule of manipulation analogous to those given in section 2.5.3, statements (1), (3), (4), (5), and (6).

2.9.2 Geometrical and Physical Applications

1. Of double integrals

The area A of a plane region R is

$$A = \iint_R 1\,dA \tag{2.9-1}$$

The centroid of this plane area is given by

$$A\bar{x} = \iint_R x\,dA, \quad A\bar{y} = \iint_R y\,dA \tag{2.9-2}$$

The center of gravity of a lamina (a thin sheet of material) of cross section R and density $\rho(x,y)$ is obtained from

$$M\bar{x} = \iint_R x\rho\,dA, \quad M\bar{y} = \iint_R y\rho\,dA \tag{2.9-3}$$

where M, the mass of the lamina, is $\iint_R \rho\,dA$.

The moments of inertia of the above lamina about the x-, y-axes are

$$I_x = \iint_R \rho y^2\,dA, \quad I_y = \iint_R \rho x^2\,dA \tag{2.9-4}$$

and the moment of inertia about an axis through the origin perpendicular to the plane of the lamina is

$$J = \iint_R \rho(x^2 + y^2)\, dA = I_x + I_y \tag{2.9-5}$$

The transfer formula for moments of inertia is stated in section 1.12.7.

Let $z = f(x, y)$ be a non-negative function defined over a finite region R; then the volume of the solid standing under z and above the XY plane is

$$V = \iint_R f(x, y)\, dA \tag{2.9-6}$$

and the surface area of $z = f(x, y)$, (x, y) in R, is

$$S = \iint_R \sqrt{1 + \left(\frac{\partial z}{\partial x}\right)^2 + \left(\frac{\partial z}{\partial y}\right)^2}\, dA \tag{2.9-7}$$

2. Of triple integrals

The volume of a bounded set R is

$$V = \iiint_R 1\, dV \tag{2.9-8}$$

and, if the solid R has density $\rho(x, y, z)$, then the mass of the solid is

$$\iiint_R \rho(x, y, z)\, dV$$

The center of gravity of R is located at $(\bar{x}, \bar{y}, \bar{z})$, where

$$M\bar{x} = \iiint_R x\rho\, dV, \quad M\bar{y} = \iiint_R y\rho\, dV, \quad M\bar{z} = \iiint_R z\rho\, dV \tag{2.9-9}$$

The moment of inertia of R about any axis is

$$I = \iiint_R \rho r^2\, dV \tag{2.9-10}$$

where r is the distance of $P(x, y, z)$ in R from that axis.

2.9.3 Change of Variables or Order of Integration

In $\iint_R f(x,y)\,dA$ let a change of variable be introduced by $x = g(u, v), y = h(u, v)$, where this is a one-to-one mapping of R in the XY-plane into R' of the UV-plane with g, h and their first and second order partial derivatives continuous. Then

$$\iint_R f(x,y)\,dx\,dy = \iint_{R'} f[g(u,v), h(u,v)] \left|\frac{\partial(x,y)}{\partial(u,v)}\right| du\,dv \qquad (2.9\text{-}11)$$

where $\partial(x,y)/\partial(u,v)$ is the Jacobian defined in section 2.8.6, and here may not vanish anywhere in R'. The Jacobian corresponding to a transformation to polar coordinates is r.

If in a triple integral, a change of variables is introduced by $x = g(u, v, w)$, $y = h(u, v, w)$, $z = l(u, v, w)$, then under conditions analogous to those stated above for the double integral

$$\iiint_R f(x,y,z)\,dx\,dy\,dz = \iiint_{R'} f[g,h,l] \left|\frac{\partial(x,y,z)}{\partial(u,v,w)}\right| du\,dv\,dw \qquad (2.9\text{-}12)$$

The Jacobians corresponding to cylindrical and spherical polar coordinates are given by eqs. (2.8-11), (2.8-12).

Examples:

(1) Evaluate $\int_0^\infty e^{-x^2}\,dx = I_1$.

$$I_1^2 = \int_0^\infty e^{-x^2}\,dx \cdot \int_0^\infty e^{-y^2}\,dy = \int_0^\infty \int_0^\infty e^{-(x^2+y^2)}\,dx\,dy$$

$$= \int_0^{\pi/2} \int_0^\infty e^{-r^2}\, r\,dr\,d\theta = -\frac{1}{2}\int_0^{\pi/2} e^{-r^2}\Big|_0^\infty d\theta = \frac{\pi}{4}$$

$$\therefore \int_0^\infty e^{-x^2}\,dx = \frac{\sqrt{\pi}}{2} \qquad (2.9\text{-}13)$$

(2) Evaluate $\int_0^\infty \dfrac{\cos rx}{a^2 + x^2}\,dx = I_2 \qquad (a > 0)$.

Since

$$\int_0^\infty 2z\, e^{-(a^2+x^2)z^2}\, dz = \frac{1}{a^2+x^2}$$

$$I_2 = \int_0^\infty \cos rx \left\{ \int_0^\infty 2z\, e^{-(a^2+x^2)z^2}\, dz \right\} dx$$

$$= \int_0^\infty \int_0^\infty 2z \cos rx\, e^{-(a^2+x^2)z^2}\, dz\, dx$$

$$= \int_0^\infty \int_0^\infty 2z\, e^{-a^2z^2}\, e^{-x^2z^2} \cos rx\, dx\, dz$$

$$= \int_0^\infty 2z\, e^{-a^2z^2} \left(\frac{\sqrt{\pi}}{2z}\, e^{-r^2/4z^2} \right) dz \text{ by eq. (2.7-1)}$$

$$= \sqrt{\pi}\, e^{ar} \int_0^\infty e^{-(az+r/2z)^2}\, dz = \frac{\pi}{2a}\, e^{ar}$$

$$\therefore \int_0^\infty \frac{\cos rx}{a^2+x^2}\, dx = \frac{\pi}{2a}\, e^{ar} \tag{2.9-14}$$

Interchanges in the order of integration for improper double integrals are governed by the theorem:

Let f be positive, and

$$\int_a^c \int_\alpha^\beta f(x,y)\, dy\, dx = \int_\alpha^\beta \int_a^c f(x,y)\, dx\, dy \qquad c < b$$

$$\int_a^b \int_\alpha^\lambda f(x,y)\, dy\, dx = \int_\alpha^\lambda \int_a^b f(x,y)\, dx\, dy \qquad \lambda < \beta,$$

then

$$\int_a^\infty \int_\alpha^\infty f(x,y)\, dy\, dx = \int_\alpha^\infty \int_a^\infty f(x,y)\, dx\, dy$$

if either side is convergent.

2.10 INFINITE PRODUCTS

2.10.1 Definition, Convergence of Infinite Products

If the sequence $P_n = (1 + a_1)(1 + a_2) \cdots (1 + a_n) = \prod_{k=1}^{n} (1 + a_k)$ with the a_k's real or complex numbers and no $a_k = -1$, converges to $P \neq 0$, then the *infinite product* $\prod_{k=1}^{\infty} (1 + a_k)$ is convergent; otherwise divergent.

Example: $\prod_{1}^{\infty} \left(1 + \frac{(-1)^n}{n+1}\right)$ converges, $\prod_{2}^{\infty} \left(1 - \frac{1}{n}\right)$ diverges. If $\prod_{1}^{\infty} (1 + a_k)$ is convergent, then $\lim_{k \to \infty} a_k = 0$.

An infinite product $\prod_{1}^{\infty} (1 + a_k)$ is *absolutely convergent* if $\prod_{1}^{\infty} (1 + |a_k|)$ is convergent; and, as is the case for series, absolute convergence implies convergence.

It is both necessary and sufficient for absolute convergence of $\prod_{1}^{\infty} (1 + a_k)$ that $\sum_{1}^{\infty} a_k$ be absolutely convergent.

An infinite product $\prod_{1}^{\infty} [1 + a_k(x)]$ converges uniformly in an interval if the associated sequence $\{P_n\}$ converges uniformly in the interval, and the product will be uniformly convergent in any interval where $\sum_{1}^{\infty} |a_k(x)|$ is uniformly convergent, see section 2.2.3.

2.11 FOURIER SERIES

Let $f(t)$ be continuous in the interval $-T \leqslant t \leqslant T$, except for a finite number of jump discontinuities. Moreover, let the derivative $f'(t)$ be similarly piecewise continuous. (These conditions may be weakened; see Ref. 2-7.)

Then the Fourier series

$$\frac{a_0}{2} + \sum_{1}^{\infty} \left(a_n \cos \frac{n\pi t}{T} + b_n \sin \frac{n\pi t}{T}\right) \tag{2.11-1}$$

converges to $f(t)$ at each point of continuity, and to $\frac{1}{2}[f(t + 0) + f(t - 0)]$ at each point of discontinuity. (At an end point, the convergence is to $\frac{1}{2}[f(-T + 0) +$

$f(T-0)]$). Here the a_j and b_j coefficients are given by

$$a_n = \frac{1}{T}\int_{-T}^{T} f(t)\cos\frac{n\pi t}{T}\,dt$$
$$b_n = \frac{1}{T}\int_{-T}^{T} f(t)\sin\frac{n\pi t}{T}\,dt \qquad (2.11\text{-}2)$$

For example, the Fourier series for the function $f(t) = |t|$, in the interval $[-1, 1]$, is given by

$$|t| = \frac{1}{2} - \frac{4}{\pi^2}\left[\frac{1}{1^2}\cos\pi t + \frac{1}{3^2}\cos 3\pi t + \frac{1}{5^2}\cos 5\pi t + \cdots\right] \qquad (2.11\text{-}3)$$

Because the trigonometric terms in (2.11-1) are periodic, with period T, the series (2.11-1) also has this period and consequently converges outside $[-T, T]$ to the periodic extension of $f(t)$. Note that the form of (2.11-1) is simplest for the choice $T = \pi$. Note also that a Fourier series for $f(x)$, in the interval $a \leqslant x \leqslant b$, follows from the transformation

$$t = -T + 2T\frac{x-a}{b-a} \qquad (2.11\text{-}4)$$

If $f(t)$ is even in the interval $[-T, T]$, so that $f(-t) = f(t)$, the only the cosine terms in (2.11-1) appear. If $f(t)$ is odd, so that $f(-t) = -f(t)$, then only the sine terms appear. Given $f(t)$ in an interval $[a, b]$, it is often convenient to extend the definition of $f(x)$, in either an odd or an even manner, onto an interval of double length and then obtain a Fourier series expansion over the new interval, involving only sine terms or only cosine terms, respectively.

A Fourier series may also be written in complex form:

$$f(t) = \sum_{-\infty}^{\infty} c_n e^{in\pi t/T} \qquad (2.11\text{-}5)$$

where

$$c_n = \tfrac{1}{2}(a_n - ib_n), \qquad n \geqslant 0$$
$$c_n = \tfrac{1}{2}(a_{-n} + ib_{-n}), \qquad n < 0 \qquad (2.11\text{-}6)$$

Term-by-term integration of a Fourier series in usually permissible; term-by-term differentiation may lead to convergence difficulties. If (2.11-1) is squared and integrated, *Parseval's theorem* results:

$$\frac{1}{T}\int_{-T}^{T}[f(t)]^2\,dt = \frac{a_0^2}{2} + \sum_{1}^{\infty}(a_n^2 + b_n^2) \tag{2.11-7}$$

A related result is that if $f(t)$ is to be approximated by the first N terms of (2.11-1) over the interval $[-T, T]$, in the sense that the integral of the squared discrepancy be least, then the optimal choices for the coefficients a_n and b_n are given by (2.11-2).

A Fourier series expansion for a solution function may be a useful tool in dealing with an ordinary or partial differential equation. Fourier series are also used in the spectral decomposition of time-dependent signals. As a generalization of a Fourier series, functions other than of a trigonometric nature may be used in the expansion; one approach of this kind (Sturm-Liouville theory) is described in Chapter 6.

2.12 REFERENCES AND BIBLIOGRAPHY

2.12.1 References

2-1 Taylor, A. E., and Mann, W. R., *Advanced Calculus*, Xerox, Stamford, 1972. (Sections 2.1, 2.2.)

2-2 Bers, L., *Calculus*, Holt, Rinehart and Winston, New York, 1969. (Sections 2.3 to 2.9.)

2-3 Protter, M. H., and Morrey, C. B., *College Calculus with Analytic Geometry*, Second Edition, Addison-Wesley, Reading, Mass., 1970. (Sections 2.3 to 2.9.)

2-4 Salas, S. L., and Hille, E., *Calculus*, 2nd Ed. Xerox, Stamford, 1974. (Sections 2.3 to 2.9.)

2-5 Gradshteyn, I. S., and Ryzhik, I. M., *Table of Integrals, Sums, Series and Products*, Academic Press, New York, 1966. (Section 2.6.)

2-6 Whittaker, E. T., and Watson, G. N., *Modern Analysis*, Cambridge University Press, New York, 1946. (Section 2.10)

2-7 Jackson, D., *Fourier Series and Orthogonal Polynomials*, Mathematical Association of America, 1943.

2.12.2 Bibliography

Agnew, R. P., *Calculus*, McGraw-Hill, New York, 1962.

Aleksandrov, A. D.; Kolmogorov, A. N.; and Lavrentev, M. A., *Mathematics, Its Content, Methods and Meaning*, MIT Press, Cambridge, Mass., 1969.

Apostol, T. M., *Calculus*, *vol. I, II*, Ginn/Blaisdell, Waltham, 1967.

Apostol, T. M., *Mathematical Analysis*, Second Edition, Addison-Wesley, Reading, Mass., 1974.

Courant, R., and John F., *Introduction to Calculus and Analysis*, Interscience, New York, 1965.

Hardy, G. H., *Pure Mathematics*, Cambridge University Press, New York, 1948.

3

Vector Analysis

Gordon C. Oates*

3.0 INTRODUCTION

In this chapter, elementary concepts of coordinate systems and vector algebra are reviewed. Wherever suitable, the various manipulations are developed without reference to particular coordinate systems.

Development of the appropriate manipulative rules for describing the rate of change of vector quantities leads to what is called vector calculus. Rules for successive operations with vector operators, as well as the development of the several integral theorems involving vector operators, are considered next.

Application of the results of vector calculus to the description of vector fields is then considered, and finally, in the summary, examples of orthogonal curvilinear coordinate systems are given, as are lists of formulas from the vector algebra and vector calculus sections. Detailed examples are included throughout the text to clarify use of the methods.

3.1 COORDINATE SYSTEMS

The concept of a vector as a quantity possessing both magnitude and direction (in contrast to scalar quantities which possess only magnitude) is a familiar one. It will be seen in the following sections that many of the relationships between vector operations can be defined independently of any coordinate system. Often, however, it is desirable to specify a vector in terms of its components, which are simply the projections of the vector on each of the three axes of a reference coordinate system (Fig. 3.1-1). Such axes are formed by extending the unit vector directions of the coordinate system from the reference point. It is most usual in practice to utilize orthogonal coordinates, and for this reason the discussion herein will be restricted to orthogonal coordinate systems. Also, the developments will be restricted

***Prof. Gordon C. Oates, Dep't. of Aeronautics and Astronautics, University of Washington, Seattle, Wash.**

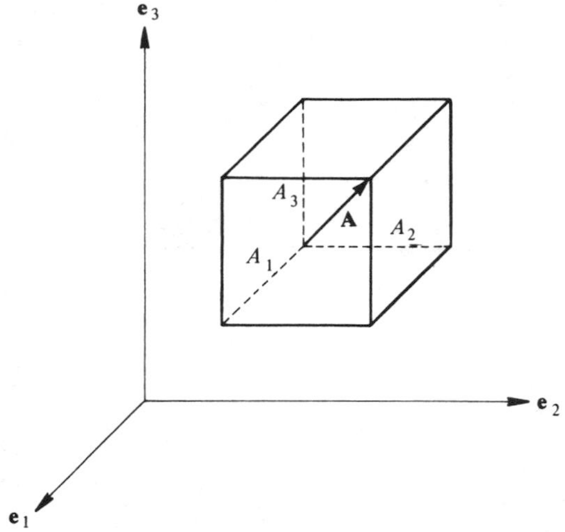

Fig. 3.1-1 Vector and its components–rectangular coordinate system.

to three dimensional vectors. A systematic treatment of more general coordinate systems is found in Chapter 4.

3.1.1 Cartesian Coordinate System

The most familiar coordinate system is the Cartesian coordinate system x_1, x_2, x_3 (Fig. 3.1-2). We consider right-handed systems only; such a system is one in which, if the x_1-axis is rotated into the x_2-axis in the direction shown in the sketch, a right-handed screw rotated in the same direction would tend to advance along the x_3-axis.

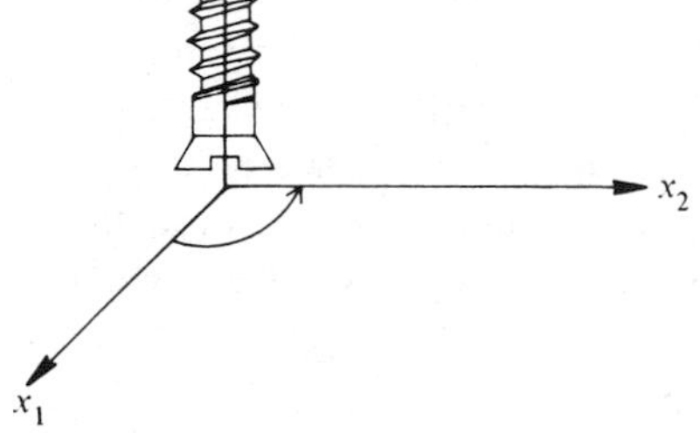

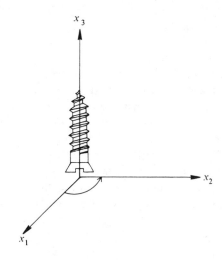

Fig. 3.1-2 Right-handed Cartesian coordinate system.

We also associate with each point a triad of unit vectors **i**, **j** and **k** pointed in the direction of increasing x_1, x_2 and x_3 respectively.

3.1.2 General Orthogonal Curvilinear System

We now consider a generalized coordinate system consisting of three mutually perpendicular surfaces. We define such surfaces by $\xi_i(x_1, x_2, x_3) = \text{constant}$ for $i = 1, 2, 3$. Also, we associate with each point a set of unit vectors $\mathbf{e} \equiv \mathbf{e}_1, \mathbf{e}_2, \mathbf{e}_3$ pointed in the direction of increase of the corresponding coordinate. In general, a differential change in the coordinate ξ_i will not be equal to a differential change in the length element associated with the ξ_i-direction, so the scale factor h_i (a function of position) is introduced, defined such that the differential change in length $d\mathbf{l}$ is given in terms of the differential changes in coordinates and the scale factors by

$$d\mathbf{l} = \sum_{i=1}^{3} h_i \, d\xi_i \, \mathbf{e}_i \tag{3.1-1}$$

Taking the scalar product (see section 3.2.3) of the line element with itself, and utilizing the orthogonality of the system, we see that

$$(dl)^2 = \sum_{i=1}^{3} h_i^2 \, (d\xi_i)^2 \tag{3.1-2}$$

The geometrical relationship between the coordinate surfaces, scale factors, and reference Cartesian coordinate system is shown in Fig. 3.1-3.

The scale factors are easily related to the equations for the coordinate surfaces because of the orthogonality of both the curvilinear and Cartesian systems. The

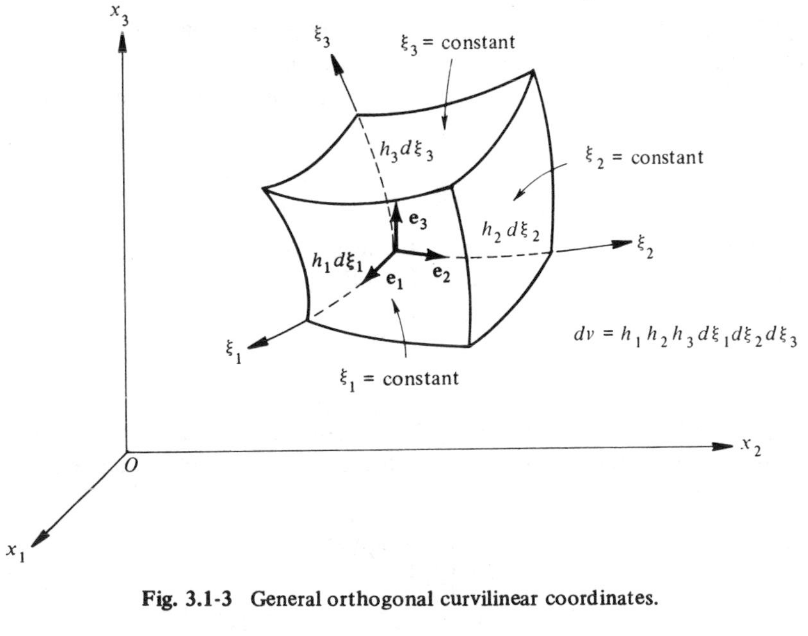

Fig. 3.1-3 General orthogonal curvilinear coordinates.

equations for the coordinate surfaces (in terms of the appropriate constants C_i or K_i) may be written

$$\xi_i(x_1, x_2, x_3) = C_i \tag{3.1-3}$$

or

$$x_i(\xi_1, \xi_2, \xi_3) = K_i \tag{3.1-4}$$

The vector $\mathbf{r}$ with initial point at the origin of the Cartesian coordinate system and terminal point at $P(x, y, z)$ is called the position vector of point P. Thus

$$\mathbf{r} = x_1\mathbf{i} + x_2\mathbf{j} + x_3\mathbf{k} \tag{3.1-5}$$

A change in the position vector due to a change in the coordinate ξ_k has direction $\mathbf{e}_k$ and has magnitude $h_k\,d\xi_k$. Hence

$$\frac{\partial \mathbf{r}}{\partial \xi_k} = h_k \mathbf{e}_k \tag{3.1-6}$$

Taking the scalar product, it follows that

$$h_k^2 = \frac{\partial \mathbf{r}}{\partial \xi_k} \cdot \frac{\partial \mathbf{r}}{\partial \xi_k} = \left\{\frac{\partial x}{\partial \xi_k}\mathbf{i} + \frac{\partial y}{\partial \xi_k}\mathbf{j} + \frac{\partial z}{\partial \xi_k}\mathbf{k}\right\} \cdot \left\{\frac{\partial x}{\partial \xi_k}\mathbf{i} + \frac{\partial y}{\partial \xi_k}\mathbf{j} + \frac{\partial z}{\partial \xi_k}\mathbf{k}\right\}$$

or

$$h_k^2 = \sum_i \left(\frac{\partial x_i}{\partial \xi_k}\right)^2 \tag{3.1-7}$$

A relationship for the scale factors is obtained in terms of the functions ξ_k by considering the gradient (see Section 3.3.7) of the functions. It is shown in Section 3.3.7 that the ith component of the gradient of an arbitrary function F is given by

$$(\nabla F)_i = \frac{1}{h_i}\frac{\partial F}{\partial \xi_i} \qquad (i \text{ not summed})$$

If the function F is now considered to be ξ_k, only the kth component of the gradient will have a nonzero value. It follows that

$$\nabla \xi_k = \frac{1}{h_k}\mathbf{e}_k \tag{3.1-8}$$

Taking the scalar product of this equation with itself, and utilizing the expanded expression for the gradient appropriate for the Cartesian coordinate system, it follows directly that

$$h_k^2 = \left[\sum_i \left(\frac{\partial \xi_k}{\partial x_i}\right)^2\right]^{-1} \tag{3.1-9}$$

3.1.3 Rates of Change of Direction of the Unit Vectors

When it is desired to explore vector calculus, the requirement for describing the rates of change of the unit vectors with position often arises. These rates of change are readily obtained by considering the geometry shown in Figs. 3.1-4 and 3.1-5.

As is indicated in Fig. 3.1-4, the change in the unit vector $\mathbf{e}_1$ due to changing position from ξ_1, ξ_2, ξ_3 to $\xi_1 + d\xi_1$, ξ_2, ξ_3 consists of two parts. The sum of the

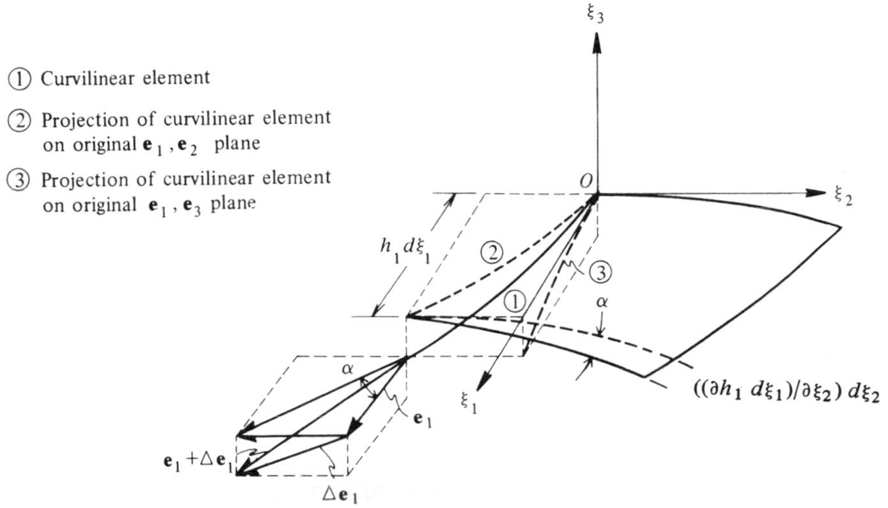

Fig. 3.1-4 Change in unit vector $\mathbf{e}_1$ due to change in ξ_1 to $\xi_1 + d\xi_1$.

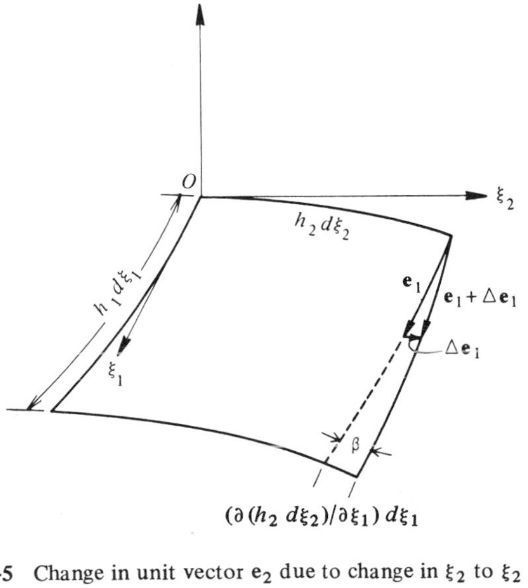

Fig. 3.1-5 Change in unit vector $\mathbf{e}_2$ due to change in ξ_2 to $\xi_2 + d\xi_2$.

changes is denoted $\Delta \mathbf{e}_1$ in the figure. It can be seen that $\Delta \mathbf{e}_1$ consists of two contributions, one in the negative $\mathbf{e}_2$-direction, and one in the negative $\mathbf{e}_3$-direction. The magnitude of the change in the $\mathbf{e}_2$-direction is equal to $\alpha|\mathbf{e}_1| = \alpha$. The angle α is obtained from the geometry as shown in the figure, to give

$$\alpha = \frac{1}{h_2 d\xi_2}\left\{\frac{\partial h_1 d\xi_1}{\partial \xi_2} d\xi_2\right\} = \frac{1}{h_2}\frac{\partial h_1}{\partial \xi_2} d\xi_1$$

A similar change in the $\mathbf{e}_3$-direction gives for the total change in $\mathbf{e}_1$

$$\Delta \mathbf{e}_1 \equiv \frac{\partial \mathbf{e}_1}{\partial \xi_1} d\xi_1$$

$$\Delta \mathbf{e}_1 = \left\{-\frac{\mathbf{e}_2}{h_2}\frac{\partial h_1}{\partial \xi_2} - \frac{\mathbf{e}_3}{h_3}\frac{\partial h_1}{\partial \xi_3}\right\} d\xi_1$$

hence

$$\frac{\partial \mathbf{e}_1}{\partial \xi_1} = -\frac{\mathbf{e}_2}{h_2}\frac{\partial h_1}{\partial \xi_2} - \frac{\mathbf{e}_3}{h_3}\frac{\partial h_1}{\partial \xi_3} \tag{3.1-10}$$

As shown in Fig. 3.1-5, the rate of change of $\mathbf{e}_1$ with ξ_2 involves only one component, given by $\Delta \mathbf{e}_1 = \beta \mathbf{e}_2$, where β is given by $\beta = (d\xi_2/h_1)(\partial h_2/\partial \xi_1)$, so that

$$\frac{\partial \mathbf{e}_1}{\partial \xi_2} = \frac{\mathbf{e}_2}{h_1}\frac{\partial h_2}{\partial \xi_1} \tag{3.1-11}$$

The results of eqs. (3.1-10) and (3.1-11) may obviously be generalized by straightforward permutation of the indices to give the following table:

$$\begin{aligned}
&\frac{\partial \mathbf{e}_1}{\partial \xi_1} = -\frac{\mathbf{e}_2}{h_2}\frac{\partial h_1}{\partial \xi_2} - \frac{\mathbf{e}_3}{h_3}\frac{\partial h_1}{\partial \xi_3}; \quad \frac{\partial \mathbf{e}_1}{\partial \xi_2} = \frac{\mathbf{e}_2}{h_1}\frac{\partial h_2}{\partial \xi_1}; \quad \frac{\partial \mathbf{e}_1}{\partial \xi_3} = \frac{\mathbf{e}_3}{h_1}\frac{\partial h_3}{\partial \xi_1}\\
&\frac{\partial \mathbf{e}_2}{\partial \xi_2} = -\frac{\mathbf{e}_1}{h_1}\frac{\partial h_2}{\partial \xi_1} - \frac{\mathbf{e}_3}{h_3}\frac{\partial h_2}{\partial \xi_3}; \quad \frac{\partial \mathbf{e}_2}{\partial \xi_1} = \frac{\mathbf{e}_1}{h_2}\frac{\partial h_1}{\partial \xi_2}; \quad \frac{\partial \mathbf{e}_2}{\partial \xi_3} = \frac{\mathbf{e}_3}{h_2}\frac{\partial h_3}{\partial \xi_2}\\
&\frac{\partial \mathbf{e}_3}{\partial \xi_3} = -\frac{\mathbf{e}_1}{h_1}\frac{\partial h_3}{\partial \xi_1} - \frac{\mathbf{e}_2}{h_2}\frac{\partial h_3}{\partial \xi_2}; \quad \frac{\partial \mathbf{e}_3}{\partial \xi_1} = \frac{\mathbf{e}_1}{h_3}\frac{\partial h_1}{\partial \xi_3}; \quad \frac{\partial \mathbf{e}_3}{\partial \xi_2} = \frac{\mathbf{e}_2}{h_3}\frac{\partial h_2}{\partial \xi_3}
\end{aligned} \tag{3.1-12}$$

A useful formula that can be verified by direct expansion using eqs. (3.1-12) is:

$$\frac{\partial(h_2 h_3 \mathbf{e}_1)}{\partial \xi_1} + \frac{\partial(h_1 h_3 \mathbf{e}_2)}{\partial \xi_2} + \frac{\partial(h_1 h_2 \mathbf{e}_3)}{\partial \xi_3} = 0 \tag{3.1-13}$$

3.2 VECTOR ALGEBRA

3.2.1 Representation of a Vector by Its Components

As already depicted in Fig. 3.1-1, we will often choose to represent a vector in terms of its components, or projections of the vector on each of the local coordinate directions. Two vectors can be considered equal when each of their respective components are specified to be equal. Thus **A** = **B** if $A_i = B_i$ for $i = 1, 2, 3$. Note that generally when we write $A_i = B_i$, and i is not specified, we mean that the equality holds for all values of i.

Multiplication of a vector **A** by a positive scalar, α, is defined by stating that the resultant vector α**A** is a vector with direction **A** and magnitude α times **A**. Obviously each separate component must also increase in magnitude by the factor α, so if the new vector is denoted **B**, the two expressions $\mathbf{B} = \alpha\mathbf{A}$ and $B_i = \alpha A_i$ are equivalent. If α is negative, the effect of the negative sign is defined to be a reversal of the vector direction.

3.2.2 Vector Addition—The Parallelogram Rule

The sum of vectors is defined according to the parallelogram rule as indicated in Fig. 3.2-1. Thus, the sum of two vectors is the vector obtained by placing the tail of one vector at the tip of the other vector and then connecting the tail of the first

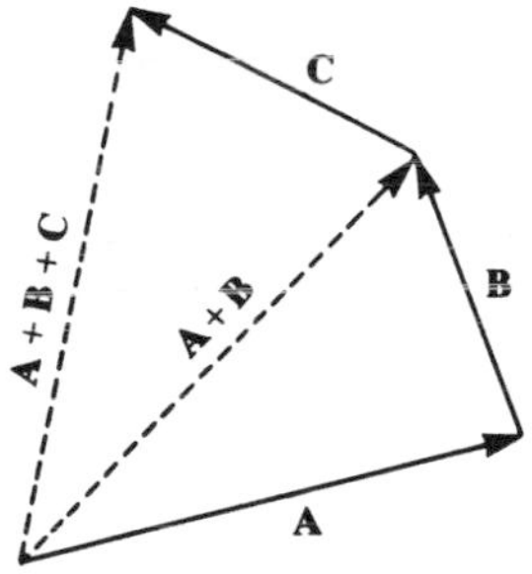

Fig. 3.2-1 Vector addition.

vector to the tip of the second. This process can, of course, be continued to give the sum of several vectors. Clearly, if $\mathbf{D} = \mathbf{A} + \mathbf{B}$, then $D_i = A_i + B_i$.

We may note the results obvious from the geometry of Fig. 3.2-1, that vector addition is both commutative and associative, that is

$$\mathbf{A} + \mathbf{B} = \mathbf{B} + \mathbf{A} \qquad \text{(commutative)} \tag{3.2-1}$$

$$(\mathbf{A} + \mathbf{B}) + \mathbf{C} = \mathbf{A} + (\mathbf{B} + \mathbf{C}) \qquad \text{(associative)} \tag{3.2-2}$$

The extension of these relationships to larger numbers of vectors is obvious. A definition of the magnitude of a vector consistent with the geometrical interpretation is

$$|\mathbf{A}| = \sqrt{A_1^2 + A_2^2 + A_3^2} \tag{3.2-3}$$

We note also that a vector can be written as a vector sum involving its own three components. Thus

$$\mathbf{A} = A_1\mathbf{e}_1 + A_2\mathbf{e}_2 + A_3\mathbf{e}_3 \tag{3.2-4}$$

3.2.3 Scalar Product

Multiplication of a vector by a scalar was defined in the preceding section. We are at liberty to define any number of forms of multiplication of a vector by a vector; two such are particularly useful.

The first form, termed the *scalar product*, results in a scalar. The second form, termed the *vector product*, will be considered in the following section.

The scalar product of two vectors is defined as a number equal to the product of the magnitudes of the two vectors times the cosine of the angle between them. The scalar product is denoted $\mathbf{A} \cdot \mathbf{B}$, so the foregoing definition gives

$$\mathbf{A} \cdot \mathbf{B} = |\mathbf{A}||\mathbf{B}| \cos(A, B) \tag{3.2-5}$$

Geometrically this definition may be interpreted as stating that the scalar product of $\mathbf{A}$ and $\mathbf{B}$ is equal to the magnitude of $\mathbf{A}$ times the projection of $\mathbf{B}$ on $\mathbf{A}$. Such a geometric interpretation makes it clear that $\mathbf{A} \cdot (\mathbf{B} + \mathbf{C}) = \mathbf{A} \cdot \mathbf{B} + \mathbf{A} \cdot \mathbf{C}$, and from eq. (3.2-5) it is obvious that $\mathbf{A} \cdot \mathbf{B} = \mathbf{B} \cdot \mathbf{A}$. Hence, scalar multiplication is both commutative and distributive.

It is worth noting the relationships true for the unit orthogonal vectors, that

$$\mathbf{e}_1 \cdot \mathbf{e}_1 = \mathbf{e}_2 \cdot \mathbf{e}_2 = \mathbf{e}_3 \cdot \mathbf{e}_3 = 1$$

$$\mathbf{e}_1 \cdot \mathbf{e}_2 = \mathbf{e}_2 \cdot \mathbf{e}_1 = \mathbf{e}_2 \cdot \mathbf{e}_3 = \mathbf{e}_3 \cdot \mathbf{e}_2 = \mathbf{e}_1 \cdot \mathbf{e}_3 = \mathbf{e}_3 \cdot \mathbf{e}_1 = 0 \tag{3.2-6}$$

It also follows that, in terms of their components, the scalar product of any two vectors may be written

$$\begin{aligned}\mathbf{A} \cdot \mathbf{B} &= (A_1\mathbf{e}_1 + A_2\mathbf{e}_2 + A_3\mathbf{e}_3) \cdot (B_1\mathbf{e}_1 + B_2\mathbf{e}_2 + B_3\mathbf{e}_3)\\ &= A_1B_1 + A_2B_2 + A_3B_3\\ &= \sum_{i=1}^{3} A_iB_i\end{aligned} \tag{3.2-7}$$

NOTE: ***Summation Convention***—Summations such as that occurring in eq. (3.2-7) occur with such frequency that it is customary to adopt a summation convention in order to simplify the notation in such equations. Thus, whenever a subscript occurs twice in the same term, it will be understood that the subscript is to be summed from 1 to 3. With this convention, eq. (3.2-7) would hence be written.

$$\mathbf{A} \cdot \mathbf{B} = A_i B_i \tag{3.2-8}$$

3.2.4 Vector Product

The "plane" of two vectors is here defined as any plane parallel to that plane containing one of the vectors and parallel to the second vector. With this interpretation, we define the vector product (or cross product) of **A** and **B** as equal to a new vector **C** whose direction is perpendicular to the plane of **A** and **B** and whose magnitude is given by $|\mathbf{C}| = |\mathbf{A}| \cdot |\mathbf{B}| \sin(A, B)$, where (A, B) is taken to be the smaller of the two possible angles. Here the right-hand rule is again invoked to determine the orientation of **C** (Fig. 3.2-2). The vector product is denoted by $\mathbf{C} = \mathbf{A} \times \mathbf{B}$.

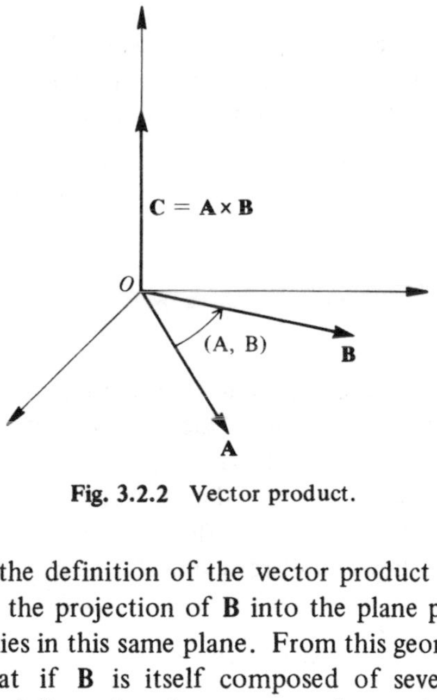

Fig. 3.2.2 Vector product.

It can be seen from the definition of the vector product that $\mathbf{A} \times \mathbf{B}$ has magnitude equal to **A** times the projection of **B** into the plane perpendicular to **A** and that the vector $\mathbf{A} \times \mathbf{B}$ lies in this same plane. From this geometrical interpretation, it may be shown that if **B** is itself composed of several other vectors, say $\mathbf{B} = \mathbf{D} + \mathbf{E} + \cdots$, it then follows that the vector product is distributive, i.e.,

$$\mathbf{A} \times (\mathbf{D} + \mathbf{E} + \cdots) = (\mathbf{A} \times \mathbf{D}) + (\mathbf{A} \times \mathbf{E}) + \cdots \quad \text{(distributive)} \tag{3.2-9}$$

Clearly from the definition of $\mathbf{A} \times \mathbf{B}$ together with Fig. 3.2-2:

$$\mathbf{A} \times \mathbf{B} = -\,\mathbf{B} \times \mathbf{A} \tag{3.2-10}$$

so the vector product is not commutative. The vector product is also not associative, examples being easily constructed to show that $\mathbf{A} \times (\mathbf{B} \times \mathbf{C}) \neq (\mathbf{A} \times \mathbf{B}) \times \mathbf{C}$ in general.

3.2.5 Indicial Notation

It should be clear that vectors and their sums and products have a definite meaning, independent of the particular coordinate system by which they might be described. It is hence possible to prove many of the relationships involving multiple operation with vectors through the use of geometric construction, resorting only to the fundamental definitions given in the preceding sections. While in many cases such proofs are intuitively satisfying, in cases of the more complicated operations the proofs become overly cumbersome, and the intuitive advantage is lost. An alternative approach to the problem of determining the relationships in multiple vector operations is to employ the so-called indicial notation. With the use of this notation, the establishment of otherwise very complex and unwieldy results becomes a matter of routine mechanical algebra.

The symbol, e_{ijk}, is used in our system of indicial notation. This symbol is defined to have the value +1, -1, or 0 depending on the values of the subscripts. More precisely:

(a) e_{ijk} is equal to 0 unless each of the numbers 1, 2 and 3 occurs as a subscript.
(b) e_{ijk} is equal to +1 if the ijk are cyclic; i.e., $e_{123} = e_{231} = e_{312} = +1$.
(c) e_{ijk} is equal to -1 if the ijk are not cyclic; i.e., $e_{132} = e_{321} = e_{213} = -1$.

The utility of this definition is immediately apparent when the formula for the cross product of two vectors is written in indicial notation. Thus, it follows that if $\mathbf{C} = \mathbf{A} \times \mathbf{B}$, the equivalent form is

$$C_i = e_{ijk} A_j B_k \tag{3.2-11}$$

Thus, for example C_1 is obtained by noting that e_{ijk}, in this case, will vanish for all terms except those for which j and k are some combination of 2 and 3. Hence $C_1 = A_2 B_3 - A_3 B_2$. Similarly $C_2 = A_3 B_1 - A_1 B_3$ and $C_3 = A_1 B_2 - A_2 B_1$.

An important further relationship is that existing between the symbol e_{ijk} and the Kronecker delta function δ_{ij}. Recall that δ_{ij} is defined to be zero for $i \neq j$, and to be unity for $i = j$. It follows by direct expansion that

$$e_{ijk} e_{ist} = \delta_{js} \delta_{kt} - \delta_{jt} \delta_{ks} \tag{3.2-12}$$

The very great utility of eqs. (3.2-11) and (3.2-12) will be evident in the next section.

3.2.6 Some Expansion Formulas

It is often useful to manipulate complex groups of vectors involving multiple operations, and the formulas developed in the preceding section lead to rapid determination of the desired relationships. Several examples follow here, and an extensive list is included in section 3.6.2. Note that in order to avoid confusion in handling successive operations, we usually use different subscripts in the indicial group when representing a vector as in eq. (3.2-11).

VECTOR TRIPLE PRODUCT

To Prove:

$$(\mathbf{A} \times \mathbf{B}) \times \mathbf{C} = (\mathbf{A} \cdot \mathbf{C})\, \mathbf{B} - (\mathbf{B} \cdot \mathbf{C})\, \mathbf{A} \qquad (3.2\text{-}13)$$

Proof: Defining $\mathbf{D} = \mathbf{A} \times \mathbf{B}$, we write, using eq. (3.2-11)

$$\begin{aligned} [(\mathbf{A} \times \mathbf{B}) \times \mathbf{C}]_i &= e_{ist} D_s C_t = e_{ist}(e_{sjk} A_j B_k)\, C_t \\ &= e_{sti} e_{sjk} A_j B_k C_t = (\delta_{tj}\delta_{ik} - \delta_{tk}\delta_{ij})\, A_j B_k C_t \\ &= B_i A_j C_j - A_i B_t C_t \equiv [(\mathbf{A} \cdot \mathbf{C})\, \mathbf{B} - (\mathbf{B} \cdot \mathbf{C})\, \mathbf{A}]_i \qquad q.e.d. \end{aligned}$$

NOTE: Similarly, we may show

$$\mathbf{A} \times (\mathbf{B} \times \mathbf{C}) = (\mathbf{A} \cdot \mathbf{C})\, \mathbf{B} - (\mathbf{A} \cdot \mathbf{B})\, \mathbf{C} \qquad (3.2\text{-}14)$$

SCALAR TRIPLE PRODUCT

To Prove:

$$\mathbf{A} \cdot (\mathbf{B} \times \mathbf{C}) = \begin{vmatrix} A_1 & A_2 & A_3 \\ B_1 & B_2 & B_3 \\ C_1 & C_2 & C_3 \end{vmatrix} \qquad (3.2\text{-}15)$$

Proof:

$$\mathbf{A} \cdot (\mathbf{B} \times \mathbf{C}) = A_i(e_{ijk} B_j C_k) = e_{ijk} A_i B_j C_k$$

But $e_{ijk} A_i B_j C_k$ is just the formula for the expansion of the determinant, hence the formula is proved.

NOTE: Also there is the result

$$\mathbf{A} \cdot (\mathbf{B} \times \mathbf{C}) = \mathbf{B} \cdot (\mathbf{C} \times \mathbf{A}) = \mathbf{C} \cdot (\mathbf{A} \times \mathbf{B}) \qquad (3.2\text{-}16)$$

SCALAR PRODUCT OF TWO CROSS PRODUCTS

To Prove:

$$(\mathbf{A} \times \mathbf{B}) \cdot (\mathbf{C} \times \mathbf{D}) = (\mathbf{A} \cdot \mathbf{C})\, (\mathbf{B} \cdot \mathbf{D}) - (\mathbf{A} \cdot \mathbf{D})\, (\mathbf{B} \cdot \mathbf{C})$$

Proof:

$$\begin{aligned}(\mathbf{A}\times\mathbf{B})\cdot(\mathbf{C}\times\mathbf{D}) &= (e_{ijk}A_jB_k)(e_{ist}C_sD_t)\\ &= e_{ijk}e_{ist}A_jB_kC_sD_t\\ &= (\delta_{js}\delta_{kt}-\delta_{jt}\delta_{ks})A_jB_kC_sD_t\\ &= (A_jC_j)(B_kD_k)-(A_jD_j)(B_kC_k)\\ &= (\mathbf{A}\cdot\mathbf{C})(\mathbf{B}\cdot\mathbf{D})-(\mathbf{A}\cdot\mathbf{D})(\mathbf{B}\cdot\mathbf{C}) \qquad \textit{q.e.d.}\end{aligned}$$

Example: As an example of the use of vector algebra techniques consider the problem of finding the length and the location of the end points of the shortest distance joining two lines. Say the lines, PQ and RS are described in terms of the position vectors **P**, **Q**, **R**, and **S** from the origin to the respective points.

Firstly we obtain the equation for a line passing through the ends of a given pair of position vectors. Thus, considering the vector **PQ** joining P to Q, we have $\mathbf{PQ} = \mathbf{Q} - \mathbf{P}$. Similarly the vector joining P to r, r an arbitrary point on the line joining P to Q, is $\mathbf{Pr} = \mathbf{r} - \mathbf{P}$. Now **PQ** and **Pr** are collinear, and hence must be within a scalar multiple of each other. Thus we write $\mathbf{Pr} = t\,\mathbf{PQ}$, or $\mathbf{r} - \mathbf{P} = t(\mathbf{Q} - \mathbf{P})$. The equation for the line joining P to Q may hence be written

$$\mathbf{r}_{PQ} = \mathbf{P} + t(\mathbf{Q} - \mathbf{P})$$

Similarly

$$\mathbf{r}_{RS} = \mathbf{R} + T(\mathbf{S} - \mathbf{R})$$

A general vector, **C**, connecting PQ to RS then has the equation

$$\mathbf{C} = \mathbf{r}_{RS} - \mathbf{r}_{PQ} = \mathbf{R} - \mathbf{P} + T(\mathbf{S} - \mathbf{R}) - t(\mathbf{Q} - \mathbf{P})$$

The magnitude of this vector will be a minimum when **C** is perpendicular to both of the vectors **PQ** and **RS**, giving the two equations for T and t:

$$\mathbf{C}\cdot\mathbf{PQ} = \{\mathbf{R} - \mathbf{P} + T(\mathbf{S} - \mathbf{R}) - t(\mathbf{Q} - \mathbf{P})\}\cdot(\mathbf{Q} - \mathbf{P}) = 0$$

$$\mathbf{C}\cdot\mathbf{RS} = \{\mathbf{R} - \mathbf{P} + T(\mathbf{S} - \mathbf{R}) - t(\mathbf{Q} - \mathbf{P})\}\cdot(\mathbf{S} - \mathbf{R}) = 0$$

It follows immediately that with

$$DEN = \{(\mathbf{S} - \mathbf{R})\cdot(\mathbf{Q} - \mathbf{P})\}^2 - \{(\mathbf{Q} - \mathbf{P})^2(\mathbf{S} - \mathbf{R})^2\}$$

$$T = \frac{1}{DEN}[(\mathbf{Q} - \mathbf{P})^2(\mathbf{S} - \mathbf{R})\cdot(\mathbf{R} - \mathbf{P}) - \{(\mathbf{Q} - \mathbf{P})\cdot(\mathbf{R} - \mathbf{P})\}\{(\mathbf{Q} - \mathbf{P})\cdot(\mathbf{S} - \mathbf{R})\}]$$

$$t = \frac{1}{DEN}[\{(\mathbf{S} - \mathbf{R})\cdot(\mathbf{R} - \mathbf{P})\}\{(\mathbf{S} - \mathbf{R})\cdot(\mathbf{Q} - \mathbf{P})\} - (\mathbf{Q} - \mathbf{P})\cdot(\mathbf{R} - \mathbf{P})(\mathbf{S} - \mathbf{R})^2]$$

Given these values of T and t, the location of the end points follows from the equations for $\mathbf{r}_{PQ}$ and $\mathbf{r}_{RS}$, and the length of the connecting line follows as the magnitude of **C**.

As a numerical example, note that given the points $P(0, 2, 0)$, $Q(2, 0, 0)$, $R(0, 0, 2)$, $S(0, 2, -2)$, the shortest line connecting the lines joining P to Q and R to S is of length 2/3 and has end points located at (4/9, 14/9, 0) and (0, 10/9, -2/9).

3.3 VECTOR CALCULUS

3.3.1 Derivatives of Vectors

In analogy with the definition of the derivative of a scalar, we define the derivative of a vector $d\mathbf{A}/dt$ as the limit of the ratio $\Delta\mathbf{A}/\Delta t$ as $\Delta t \to 0$ (Fig. 3.3-1). The ordinary rules regarding differentiation of products are valid, but obviously care must

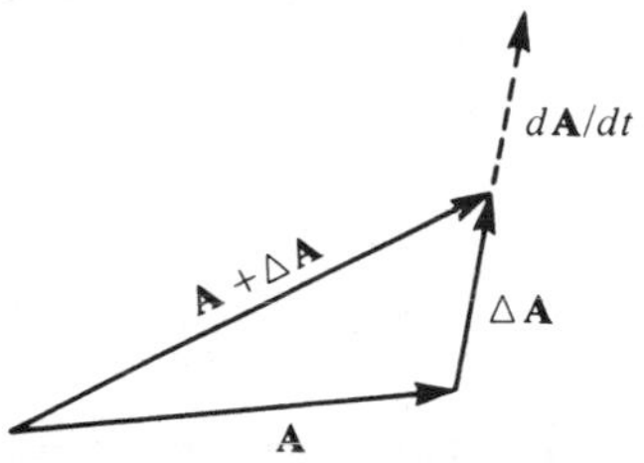

Fig. 3.3-1 Derivative of a vector.

be taken to retain the order of the factors if vector products are involved. Thus, for example

$$\frac{d(\mathbf{A} \times \mathbf{B})}{dt} = \mathbf{A} \times \frac{d\mathbf{B}}{dt} + \frac{d\mathbf{A}}{dt} \times \mathbf{B} = \mathbf{A} \times \frac{d\mathbf{B}}{dt} - \mathbf{B} \times \frac{d\mathbf{A}}{dt} \tag{3.3-1}$$

It should be noted that in a general orthogonal curvilinear system, the unit vectors themselves change direction with position, and hence their rate of change as well as the rate of change of the vector components must be considered.

3.3.2 Divergence

One of the most familiar of vector operators, the divergence is best interpreted by referring directly to the integral definition of the operator. Thus, the divergence of a vector field **F**, denoted $\nabla \cdot \mathbf{F}$ is defined by

$$\nabla \cdot \mathbf{F} = \lim_{\Delta v \to 0} \frac{\iint_s \mathbf{F} \cdot d\mathbf{s}}{\Delta v} \tag{3.3-2}$$

In this expression, the area s is that area enclosing the volume Δv. The vector direction of the elemental vector area $d\mathbf{s}$ is taken to be positive when the element, $d\mathbf{s}$, is directed outward. It is apparent from the definition that the divergence of the vector field is a scalar point function and is a measure of the net outflow of the vector $\mathbf{F}$ at the point where the limiting volume is located.

The expression for the divergence in a variety of coordinate systems is easily obtained directly from the integral definition by considering the volume element Δv to be the elemental volume element of the desired coordinate system. As an example consider the generalized curvilinear coordinate system with volume element as depicted in Fig. 3.3-2.

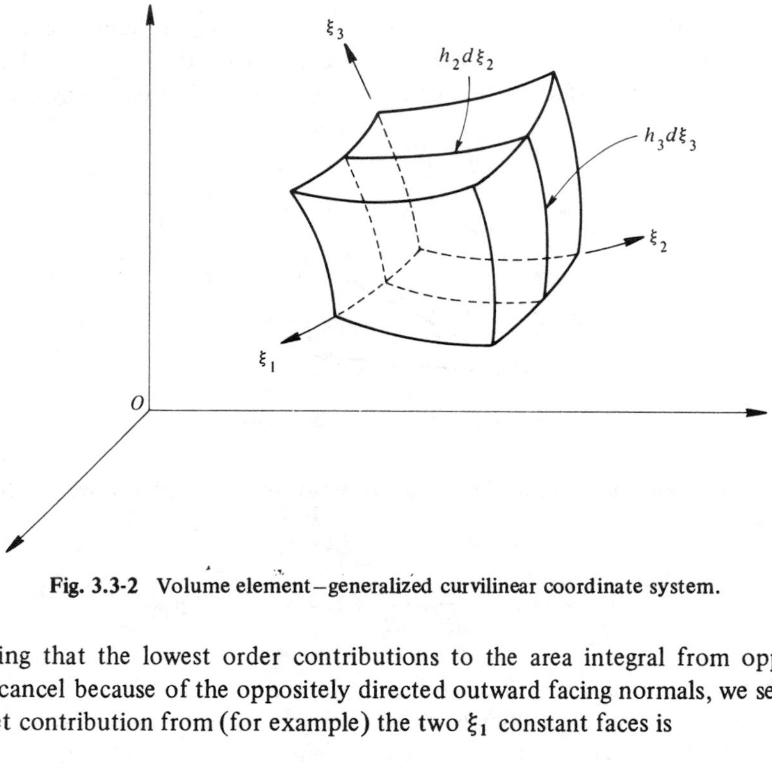

Fig. 3.3-2 Volume element–generalized curvilinear coordinate system.

Noting that the lowest order contributions to the area integral from opposing faces cancel because of the oppositely directed outward facing normals, we see that the net contribution from (for example) the two ξ_1 constant faces is

$$\frac{\partial}{\partial \xi_1}(\mathbf{F}\cdot\mathbf{e}_1\, h_2 h_3\, \Delta\xi_2 \Delta\xi_3)\, \Delta\xi_1 = \frac{\partial(F_1 h_2 h_3)}{\partial \xi_1}\Delta\xi_1 \Delta\xi_2 \Delta\xi_3$$

Adding the contributions of all three pairs of faces and dividing by the infinitesimal volume $\Delta v = h_1 h_2 h_3 \Delta\xi_1 \Delta\xi_2 \Delta\xi_3$ there is obtained

$$\nabla\cdot\mathbf{F} = \frac{1}{h_1 h_2 h_3}\left[\frac{\partial(F_1 h_2 h_3)}{\partial \xi_1} + \frac{\partial(F_2 h_1 h_3)}{\partial \xi_2} + \frac{\partial(F_3 h_1 h_2)}{\partial \xi_3}\right] \tag{3.3-3}$$

3.3.3 Gauss' Divergence Theorem

Consider now the integral over a finite volume, V, of the divergence of the vector $\mathbf{F}$. We may consider the integral as the limit of a summation over small volumes Δv_i, as indicated in Fig. 3.3-3. Thus

$$\iiint_V \nabla \cdot \mathbf{F}\, dv = \lim_{\Delta v_i \to 0} \sum (\nabla \cdot \mathbf{F})_i\, \Delta v_i$$

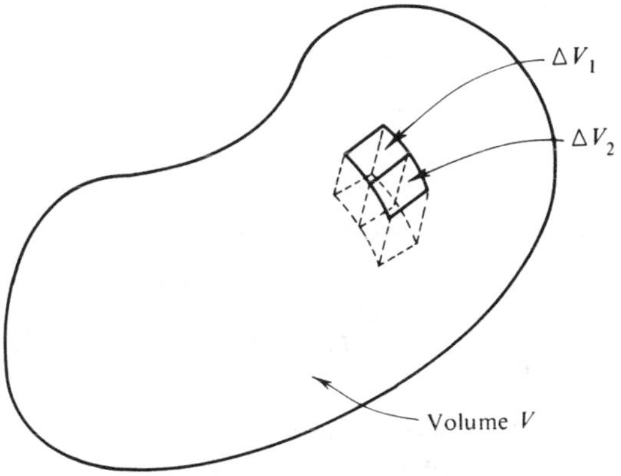

Fig. 3.3-3 Typical volume elements Δv_i comprising V.

From eq. (3.3-2), it follows that the summation over the volumes may be considered the limit of the sum over all the infinitesimal areas of the surface integrals, given by

$$\lim_{\Delta v_i \to 0} \sum (\nabla \cdot \mathbf{F})_i\, \Delta v_i = \lim_{\Delta v_i \to 0} \sum \left(\frac{\iint_{\Delta s} \mathbf{F} \cdot ds}{\Delta v} \right)_i \Delta v_i = \lim_{\Delta s_i \to 0} \sum \iint_{\Delta s_i} \mathbf{F} \cdot ds$$

It is obvious, however, that the contributions from adjacent (internal) surface elements identically cancel, because the value of $\mathbf{F}$ upon such a common surface is the same in each infinitesimal area integral, but the sense of the outward normal is reversed. The result is hence that only the contributions from the exterior surface remain, giving the result that

$$\iiint_V \nabla \cdot \mathbf{F}\, dv = \iint_s \mathbf{F} \cdot ds \tag{3.3-4}$$

This result is known as Gauss' divergence theorem.

The above development is not rigorous, and of course has not indicated any required limitations upon the function **F** or surface s. A rigorous development would show that continuity of **F** and of its divergence throughout V and on the bounding surface is sufficient to ensure validity of the theorem. The surface must be piecewise smooth.

Example: Consider a vector field given by

$$\mathbf{F} = xy^2z\mathbf{i} + y^3z\mathbf{j} + \{(x^2 + y^2) - 2y^2z^2\}\,\mathbf{k}$$

We are to determine the net flux (defined as the value of $\iint F \cdot d\mathbf{s}$ over the given surface) of the field through the surface of the parabaloid $z = 1 - (x^2 + y^2)$, above the x-y plane.

Solution: The area integral over the parabaloid is somewhat awkward to perform, so it is convenient to utilize the divergence theorem to write

$$\iint\limits_{\text{Parabaloid}} \mathbf{F} \cdot d\mathbf{s} = -\iint\limits_{\text{Base}} \mathbf{F} \cdot d\mathbf{s} + \iiint\limits_{v} \nabla \cdot \mathbf{F}\, dv$$

Direct expansion in cartesian coordinates shows that the divergence is zero here, and on the base, the function **F** reduces to $\mathbf{F} = (x^2 + y^2)\,\mathbf{k} = r^2\mathbf{k}$. Then with $d\mathbf{s} = rdrd\theta(-\mathbf{k})$, we obtain

$$\iint\limits_{\text{Parabaloid}} \mathbf{F} \cdot d\mathbf{s} = \int_0^1 \int_0^{2\pi} r^3 drd\theta = \frac{\pi}{2}$$

3.3.4 Curl

The curl measures the tendency of the vector field to "rotate" about a given point. The utility of the operator is apparent in the many examples of vector analysis in which the curl of the vector field emerges as a vital physical property. Perhaps the most familiar example of the use of the curl is in fluid mechanics, where the determination of the curl of the fluid velocity, termed the *vorticity*, is often of prime importance.

The curl of a vector is itself a vector, and its definition is most easily given in terms of the component of the curl in a given direction, say **m** (Fig. 3.3-4). Denoting the curl of **F** as $\nabla \times \mathbf{F}$, we thus define

$$(\nabla \times \mathbf{F})_{\mathbf{m}} = \lim_{\Delta s \to 0} \frac{\int_l \mathbf{F} \cdot d\mathbf{l}}{\Delta s} \tag{3.3-5}$$

In this expression Δs refers to the (sectionally smooth) surface about which the indicated line integral is taken. Δs is in the plane perpendicular to **m**. The direc-

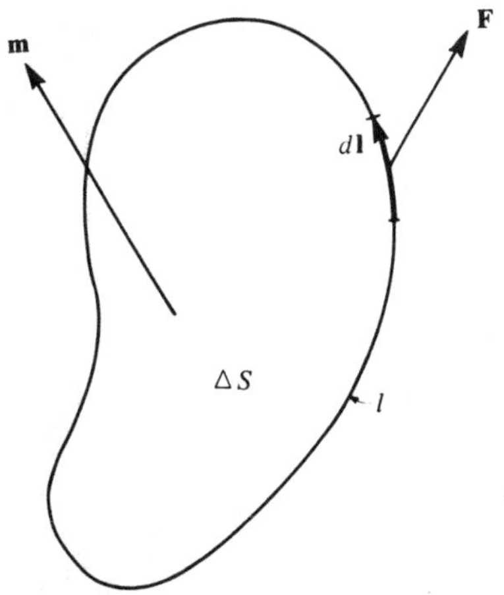

Fig. 3.3-4 Area elements Δs and its normal **m**, with contour l.

tion of the line element is taken to be positive if in traversing the element, a right-handed screw rotated in the same direction would advance in the direction **m.** Alternatively, of course, the positive direction of **m** can be defined in terms of the direction $d\mathbf{l}$.

The expression for the curl in a variety of coordinate systems can be obtained directly from the integral definition by considering the area, Δs, to be the area element of the given coordinate system. Thus, considering the generalized curvilinear coordinate system as an example, the $\mathbf{e}_1$ component of curl follows by considering the contributions to the line integral of each of the four edges shown numbered in Fig. 3.3-5. In this case

$$\int \mathbf{F} \cdot d\mathbf{l} \rightarrow [(\mathbf{F} \cdot \Delta\mathbf{l})_1 + (\mathbf{F} \cdot \Delta\mathbf{l})_3] + [(\mathbf{F} \cdot \Delta\mathbf{l})_2 + (\mathbf{F} \cdot \Delta\mathbf{l})_4]$$

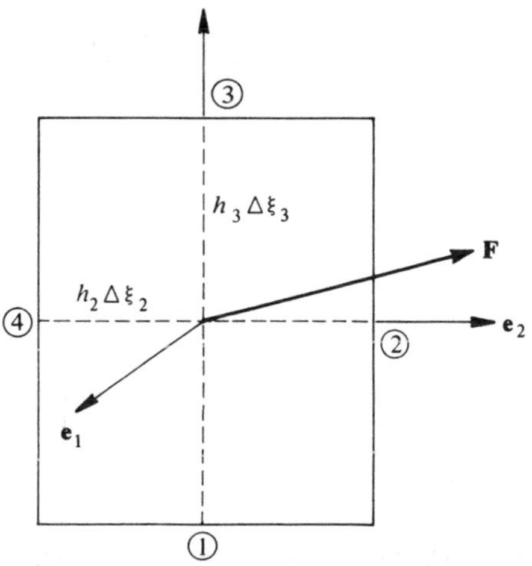

Fig. 3.3-5 Curl component for generalized curvilinear coordinate system.

It can be seen that each of the pairs of terms bracketed contributes because of the change in the value of $\mathbf{F} \cdot \Delta\mathbf{l}$ that exists across the elemental area. In terms of the scale factors and changes in coordinates, the integral becomes

$$\int \mathbf{F} \cdot d\mathbf{l} = -\frac{\partial(F_2 h_2 \Delta\xi_2)}{\partial\xi_3}\Delta\xi_3 + \frac{\partial(F_3 h_3 \Delta\xi_3)}{\partial\xi_2}\Delta\xi_2$$

The $\mathbf{e}_1$ component of the curl is then

$$(\nabla \times \mathbf{F})_{\mathbf{e}_1} = \lim_{\Delta\xi_2\Delta\xi_3 \to 0} \frac{\dfrac{\partial(F_3 h_3)}{\partial\xi_2}\Delta\xi_2\Delta\xi_3 - \dfrac{\partial(F_2 h_2)}{\partial\xi_3}\Delta\xi_2\Delta\xi_3}{h_2 h_3 \Delta\xi_2 \Delta\xi_3}$$

$$= \frac{1}{h_2 h_3}\left(\frac{\partial(F_3 h_3)}{\partial\xi_2} - \frac{\partial(F_2 h_2)}{\partial\xi_3}\right) \qquad (3.3\text{-}6)$$

Similar developments for the other components of the curl lead to (utilizing indicial notation)

$$(\nabla \times \mathbf{F})_i = \frac{e_{ijk}}{h_1 h_2 h_3} h_i \frac{\partial(h_k F_k)}{\partial\xi_j} \qquad (3.3\text{-}7)$$

The expanded form for this expression is included in the summary, section 3.6.3.

3.3.4.1 Volume Integral Form of the Curl It will be of use in the following sections to obtain an expression for the curl in terms of a limiting form of volume integral, because this will lead to easy establishment of the various relationships occurring when operations are performed on sums and products.

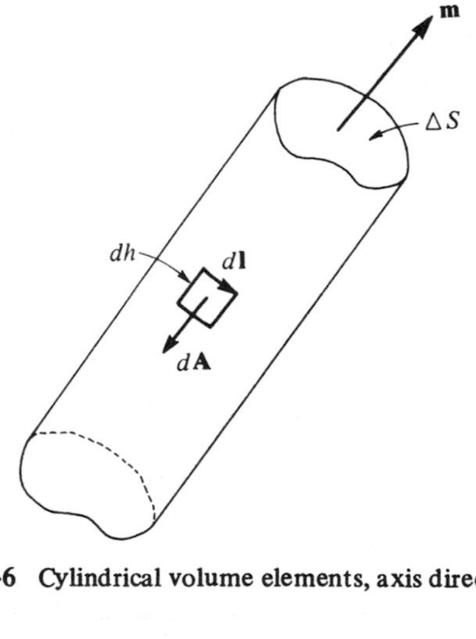

Fig. 3.3-6 Cylindrical volume elements, axis direction **m**.

Consider again the definition of the component of $\nabla \times \mathbf{F}$ in the arbitrary direction $\mathbf{m}$, as given in eq. (3.3-5). We construct a cylinder (which may be of arbitrary cross-sectional shape) with axis in the direction of $\mathbf{m}$, as shown in Fig. 3.3-6. If we consider a small vector area, $d\mathbf{A}$, located on the side of the cylinder we may represent the line element $d\mathbf{l}$ in terms of $\mathbf{m}$ and $d\mathbf{A}$ as

$$d\mathbf{l} = (\mathbf{m} \times d\mathbf{A}) \frac{dl}{dA} = (\mathbf{m} \times d\mathbf{A}) \frac{1}{dh} \tag{3.3-8}$$

Thus, noting that $\mathbf{m} \times \Delta\mathbf{S} = 0$ on the ends of the cylinder we may write

$$(\nabla \times \mathbf{F})_{\mathbf{m}} = \mathbf{m} \cdot (\nabla \times \mathbf{F}) = \lim_{\Delta S \to 0} \frac{\displaystyle\int_l \mathbf{F} \cdot (\mathbf{m} \times d\mathbf{A}) \frac{1}{dh}}{\Delta S}$$

$$= \lim_{\Delta v \to 0} \frac{\displaystyle\iint_A \mathbf{F} \cdot (\mathbf{m} \times d\mathbf{A})}{\Delta v} = \mathbf{m} \cdot \lim_{\Delta v \to 0} \frac{\displaystyle\int_A (d\mathbf{A} \times \mathbf{F})}{\Delta v}$$

The curl, including all components, may hence be written

$$\nabla \times \mathbf{F} = \lim_{\Delta v \to 0} \frac{\displaystyle\iint_A (d\mathbf{A} \times \mathbf{F})}{\Delta v} \tag{3.3-9}$$

3.3.5 Stokes' Theorem

Consider the line integral of the function $\mathbf{F}$ around a finite contour. If a continuous surface is surrounded by this contour, such a surface may be divided up into many small elemental areas. (Figure 3.3-7, note the sense of the positive normal is de-

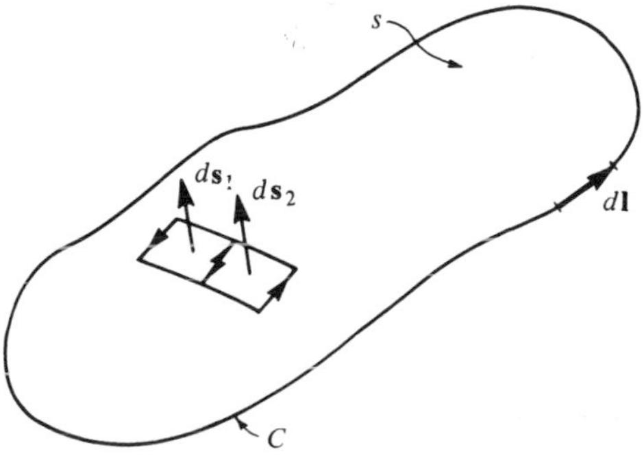

Fig. 3.3-7 Typical area elements $d\mathbf{s}_i$ comprising s.

fined for both the finite and elemental areas as in section 3.3.4). A line integral around each elemental area ds, may be taken, each area contributing $\int_l \mathbf{F} \cdot d\mathbf{l}$ where l is the perimeter of ds. If all such contributions are summed, it is apparent that the contributions of line elements mutual to adjacent areas will cancel, because the sense of $d\mathbf{l}$ is opposite. The only net contribution will arise from the perimeter to the finite area, so that we may write

$$\int_c \mathbf{F} \cdot d\mathbf{l} = \sum_{\text{all areas}} \int_l \mathbf{F} \cdot d\mathbf{l} = \sum \left(\frac{\int_l \mathbf{F} \cdot d\mathbf{l}}{\Delta S} \right) \Delta S$$

But as the size of the area elements is allowed to approach zero the expression in parenthesis becomes the normal component of the curl. Noting also that the summation over the Δs becomes the integral over the area, and that $(\nabla \times \mathbf{F})_n\, ds = (\nabla \times \mathbf{F}) \cdot d\mathbf{s}$, it follows that

$$\iint_s (\nabla \times \mathbf{F}) \cdot d\mathbf{s} = \int_c \mathbf{F} \cdot d\mathbf{l} \tag{3.3-10}$$

This result is known as Stokes' Theorem.

Example: Obtain the surface integral of the curl of the vector field given by $\mathbf{F} = (3x^2 + y)\mathbf{i} + 3xy\mathbf{j} + (xy + z)\mathbf{k}$ over the surface of the parabaloid $z = 1 - (x^2 + y^2)$ that protrudes above the x-y plane.

Solution: The desired result is obtained easily by utilizing Stokes' Theorem and evaluating the contour integral around the circle $r = 1$ on the x-y plane. On this contour $d\mathbf{l} = r d\theta\, \mathbf{e}_\theta$, so noting that $\mathbf{e}_\theta \cdot \mathbf{i} = -\sin\theta$, $\mathbf{e}_\theta \cdot \mathbf{j} = \cos\theta$, we obtain

$$\iint_s (\nabla \times \mathbf{F}) \cdot d\mathbf{s} = \int_c \mathbf{F} \cdot d\mathbf{l} = \int_0^{2\pi} \{(3x^2 + y)(-\sin\theta) + 3xy\cos\theta\}\, d\theta$$

$$= \int_0^{2\pi} \{(3\cos^2\theta + \sin\theta)(-\sin\theta) + 3\cos^2\theta\sin\theta\}\, d\theta$$

$$= -\int_0^{2\pi} \sin^2\theta\, d\theta = -\pi$$

3.3.6 Directional Derivative

It is often of interest to determine the rate of change of a function in a given direction; such a rate of change being termed a directional derivative. In the case of a

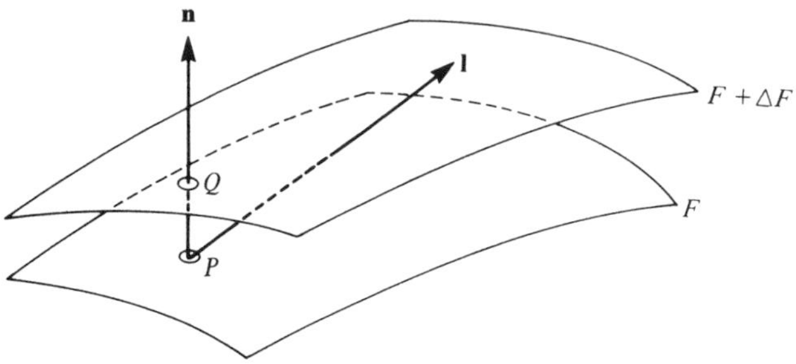

Fig. 3.3-8 Level surfaces and direction of maximum rate of change, **n**.

scalar function, the directional derivative will generally have a maximum in a certain direction, that direction being normal to the surfaces upon which the scalar has a constant value (Fig. 3.3-8). Denoting the unit vector in the direction of the normal to these "level surfaces" by **n**, it is apparent that the directional derivative in an arbitrary direction denoted by the unit vector **l**, is just equal to the scalar product of **n** and **l** times the maximum value of the directional derivative.

The magnitude and direction of this maximum directional derivative is of particular importance, and the directional derivative so defined is termed the gradient of the function.

3.3.7 Gradient

As described in the preceding section, we define the gradient of a scalar function as that vector given in magnitude by the maximum rate of change of the function and in direction by the direction of the maximum rate of change. Referring to Fig. 3.3-8, we see then that the gradient of the scalar function F, denoted ∇F is given by

$$\nabla F = \frac{\partial F}{\partial n}\,\mathbf{n} \tag{3.3-11}$$

The components of the gradient in terms of the generalized coordinate system can be directly obtained from this definition by relating a small change in the normal direction, denoted dn, to a small vector change in an arbitrary direction, $d\mathbf{l}$. With $d\mathbf{l}$ given by eq. (3.1-1), and the relationship $\mathbf{n} \cdot d\mathbf{l} = dn$, it follows that

$$dF = \frac{\partial F}{\partial n}\,dn = \left(\frac{\partial F}{\partial n}\,\mathbf{n}\right) \cdot d\mathbf{l} = (\nabla F) \cdot d\mathbf{l}$$

hence

$$dF = \sum_{i=1}^{3} (\nabla F)_i\, h_i d\xi_i \tag{3.3-12}$$

In these expressions the partial derivative with respect to n indicates the derivative is taken along the direction of **n**, other independent variables (the directions normal to **n**) being held constant. $(\nabla F)_i$ denotes the ith component of ∇F.

From the ordinary rules of calculus

$$dF = \sum_{i=1}^{3} \frac{\partial F}{\partial \xi_i} d\xi_i \tag{3.3-13}$$

so comparison of eqs. (3.3-12) and (3.3-13) shows

$$(\nabla F)_i = \frac{1}{h_i} \frac{\partial F}{\partial \xi_i} \qquad (i \text{ not summed}) \tag{3.3-14}$$

It is appropriate to note here the subscript notation. In this often used notation, a comma is used to indicate differentiation. A benefit of the subscript notation is that it allows indication of the vector character of (for example) the gradient, by using a subscript, just as for other vectors. In this alternative notation, eq. (3.3-14) would be written

$$(\nabla F)_i = \frac{1}{h_i} F_{,i} \qquad (i \text{ not summed}) \tag{3.3-15}$$

3.3.8 The Gradient Theorem

This theorem, which is closely related to Gauss' Divergence Theorem, is easily established directly from eq. 3.3-14. Thus, taking the volume integral of the gradient, we have

$$\iiint_V \nabla F\, dv = \iiint_V (\nabla F)_i \mathbf{e}_i h_1 h_2 h_3\, d\xi_1\, d\xi_2\, d\xi_3$$

$$= \iiint_V \left[\frac{1}{h_1} \frac{\partial F}{\partial \xi_1} \mathbf{e}_1 + \frac{1}{h_2} \frac{\partial F}{\partial \xi_2} \mathbf{e}_2 + \frac{1}{h_3} \frac{\partial F}{\partial \xi_3} \mathbf{e}_3 \right] h_1 h_2 h_3\, d\xi_1\, d\xi_2\, d\xi_3$$

$$= \iint_S F\, [h_2 h_3\, d\xi_2\, d\xi_3 \mathbf{e}_1 + h_1 h_3\, d\xi_1\, d\xi_3 \mathbf{e}_2 + h_1 h_2\, d\xi_1\, d\xi_2 \mathbf{e}_3]$$

$$= \iint_S F\, d\mathbf{s}$$

hence

$$\iiint_V \nabla F\, dv = \iint_S F\, d\mathbf{s} \tag{3.3-16}$$

By taking the limiting form of this expression, the gradient can be expressed as a limiting form of an area integral, thus

$$\nabla F = \lim_{\Delta v \to 0} \frac{\iint_S F \, d\mathbf{s}}{\Delta v} \tag{3.3-17}$$

3.3.9 The Directional Derivative of a Vector—The Convective Operator

As was discussed in section 3.3.6, the directional derivative in the direction **l** of a scalar function F, can be obtained by taking the scalar product of the unit vector **l** with ∇F. If the integral form of ∇F given by eq. (3.3-15) is used, it follows that

$$\mathbf{l} \cdot (\nabla F) = \mathbf{l} \cdot \left\{ \lim_{\Delta v \to 0} \frac{\iint_S F \, d\mathbf{s}}{\Delta v} \right\} = \lim_{\Delta v \to 0} \frac{\iint_S F (\mathbf{l} \cdot d\mathbf{s})}{\Delta v} \tag{3.3-18}$$

Note that in the latter form of the integral, the vector **l** must be considered constant in the limiting process. It is apparent that the relationship of eq. (3.3-18) can be considered an operation on the function F, and in fact such an operation is often denoted $(\mathbf{l} \cdot \nabla) F$. With this interpretation

$$(\mathbf{l} \cdot \nabla) F = \mathbf{l} \cdot (\nabla F) \tag{3.3-19}$$

The operator $(\mathbf{l} \cdot \nabla)$ as defined above, is a scalar operator with the properties of the directional derivative. The operator so defined may also be applied to a vector function, and it can be verified that the operator has the properties of the directional derivative regardless of whether the function being operated upon is a scalar or a vector.

3.3.9.1 The Convective Operator A very slight generalization of the preceding definition is obtained by replacing the unit vector **l** by a more general vector **a**. A development virtually identical to the preceding then indicates that the operator $(\mathbf{a} \cdot \nabla)$, defined by

$$(\mathbf{a} \cdot \nabla) G = \lim_{\Delta v \to 0} \frac{\iint_S G \mathbf{a} \cdot d\mathbf{s}}{\Delta v} \quad \text{where } \mathbf{a} \text{ is considered constant in the limiting process, } G \text{ is any function, scalar or vector} \tag{3.3-20}$$

gives the rate of change of the function G in the direction of **a**, times the magnitude of **a**. This operator is often termed the convective operator because of its frequent use in the equations of fluid mechanics. Thus, for example if the rate of change of some property, G, transported with fluid velocity, $\mathbf{u}$, is desired, it is given by $(\mathbf{u} \cdot \nabla) G$. Note of course that if this were to be the total rate of change of the quantity, the time rate of change (at a point) would have to be zero.

The components of the vector $(\mathbf{a} \cdot \nabla) \mathbf{F}$ in generalized curvilinear components can be obtained fairly readily from the integral definition given by eq. (3.3-20). Thus, consider the infinitesimal volume Δv to be $h_1 h_2 h_3 \Delta\xi_1 \Delta\xi_2 \Delta\xi_3$ indicated in Fig. 3.3-2. In order to obtain the contributions of the faces of the element to the area integral, we consider the net contribution of the pairs of faces formed by constant coordinate values.

Thus, for example, considering the ξ_1 constant faces, we note that at the center of the element

$$\mathbf{a} \cdot d\mathbf{s} = \mathbf{a} \cdot \mathbf{e}_1 \, h_2 h_3 \Delta\xi_2 \Delta\xi_3 \tag{3.3-21}$$

We are interested in the net contribution of the two faces, which is simply

$$\frac{\partial}{\partial \xi_1} (\mathbf{F}\mathbf{a} \cdot d\mathbf{s}) \, \Delta\xi_1 = \frac{\partial \mathbf{F}(\mathbf{a} \cdot \mathbf{e}_1) \, h_2 h_3}{\partial \xi_1} \Delta\xi_1 \Delta\xi_2 \Delta\xi_3 \tag{3.3-22}$$

Similar expressions are obtained for the other pairs of faces to give for the sum of the contributions

$$(\mathbf{a} \cdot \nabla) \mathbf{F} = \lim_{\Delta\xi_1, \Delta\xi_2, \Delta\xi_3, \to 0} \frac{\displaystyle\iint_s \mathbf{F}(\mathbf{a} \cdot d\mathbf{s})}{h_1 h_2 h_3 \Delta\xi_1 \Delta\xi_2 \Delta\xi_3}$$

$$= \frac{1}{h_1 h_2 h_3} \left[\frac{\partial \mathbf{F}(\mathbf{a} \cdot \mathbf{e}_1) \, h_2 h_3}{\partial \xi_1} + \frac{\partial \mathbf{F}(\mathbf{a} \cdot \mathbf{e}_2) \, h_1 h_3}{\partial \xi_2} + \frac{\partial \mathbf{F}(\mathbf{a} \cdot \mathbf{e}_3) \, h_1 h_2}{\partial \xi_3} \right]$$

$$= \frac{1}{h_1 h_2 h_3} \left[\mathbf{F}\mathbf{a} \cdot \left\{ \frac{\partial \mathbf{e}_1 h_2 h_3}{\partial \xi_1} + \frac{\partial \mathbf{e}_2 h_1 h_3}{\partial \xi_2} + \frac{\partial \mathbf{e}_3 h_1 h_2}{\partial \xi_3} \right\} \right.$$

$$\left. + a_1 h_2 h_3 \frac{\partial \mathbf{F}}{\partial \xi_1} + a_2 h_1 h_3 \frac{\partial \mathbf{F}}{\partial \xi_2} + a_3 h_1 h_2 \frac{\partial \mathbf{F}}{\partial \xi_3} \right]$$

In arriving at this form we have utilized the requirement that $\mathbf{a}$ be considered a constant in the limiting process. Noting that the group within brackets is zero, from eq. (3.1-13), we may expand the derivatives of the vector $\mathbf{F}$, utilizing eq. (3.1-12) to give the following expression for the jth component of $(\mathbf{a} \cdot \nabla) \mathbf{F}$:

$$[(\mathbf{a} \cdot \nabla) \mathbf{F}]_j = \frac{a_i}{h_i} \frac{\partial F_j}{\partial \xi_i} + \frac{F_i}{h_i h_j} \left(a_j \frac{\partial h_j}{\partial \xi_i} - a_i \frac{\partial h_i}{\partial \xi_j} \right) \quad (j \text{ not summed}) \tag{3.3-23}$$

The expanded form of this expression is included in the summary, section 3.6.3.

3.3.10 Generalized Form of Gauss' Divergence Theorem

The theorem we seek to prove is

$$\iiint_V [(\mathbf{a}\cdot\nabla)\,\mathbf{F} + \mathbf{F}(\nabla\cdot\mathbf{a})]\,dv = \iint_s \mathbf{F}(\mathbf{a}\cdot d\mathbf{s}) \tag{3.3-24}$$

It may be noted that this theorem reduces to the conventional divergence theorem in the case $\mathbf{F}$ equal a constant.

Proof: We again reference Fig. 3.3-3, and for the moment consider the area integral on the right of eq. (3.3-24) to be taken over the infinitesimal volume element Δv_i with corresponding surface area Δs_i. Clearly, as with the conventional form of the divergence theorem (section 3.3.3) when the contributions of all volume elements are summed, the contributions of adjacent surface elements identically cancel. This follows because the values of both $\mathbf{F}$ and $\mathbf{a}$ are identical at the common interface, but the sense of $d\mathbf{s}$ is reversed. Summing all elements, we then have

$$\lim_{\Delta s_i\to 0}\sum_i \iint_{\Delta s_i} \mathbf{F}(\mathbf{a}\cdot d\mathbf{s}) = \iint_s \mathbf{F}(\mathbf{a}\cdot d\mathbf{s}) \tag{3.3-25}$$

The left side of eq. (3.3-25) is now written in an alternate form. We write the values of $\mathbf{F}$ and $\mathbf{a}$ in each element as $\mathbf{F}_{0i} + \Delta\mathbf{F}_i$ and $\mathbf{a}_{0i} + \Delta\mathbf{a}_i$, where $\mathbf{F}_{0i}$ and $\mathbf{a}_{0i}$ are the values of the functions at the center of element i. Then

$$\sum_i \iint_{\Delta s_i} \mathbf{F}(\mathbf{a}\cdot d\mathbf{s}) = \sum_i \left[\frac{\iint_{\Delta s_i} \mathbf{F}_{0i}(\Delta\mathbf{a}_i\cdot d\mathbf{s})}{\Delta v_i} + \frac{\iint_{\Delta s_i} \Delta\mathbf{F}_i\mathbf{a}_{0i}\cdot d\mathbf{s}}{\Delta v_i}\right]\Delta v_i + \text{H.O.T.}$$

$$= \sum_i \left[\mathbf{F}_{0i}\frac{\iint_{\Delta s_i} \mathbf{a}\cdot d\mathbf{s}}{\Delta v_i} + \frac{\iint_{\Delta s_i} \mathbf{F}\mathbf{a}_{0i}\cdot d\mathbf{s}}{\Delta v_i}\right]\Delta v_i + \text{H.O.T.}$$

In this latter form we have used the result that

$$\iint_{\Delta s_i} \mathbf{a}_{0i}\cdot d\mathbf{s} = \iint_{\Delta s_i} \mathbf{F}_{0i}\mathbf{a}_{0i}\cdot d\mathbf{s} = 0$$

Taking the limit of this expression as $\Delta v_i \to 0$, and utilizing the definitions of eqs. (3.3-2) and (3.3-20) it follows that

$$\lim_{\Delta v_i \to 0} \sum_i \iint_{\Delta s_i} \mathbf{F}(\mathbf{a} \cdot d\mathbf{s}) = \iiint_V [\mathbf{F}(\nabla \cdot \mathbf{a}) + (\mathbf{a} \cdot \nabla)\,\mathbf{F}]\, dv \qquad (3.3\text{-}26)$$

Comparison of eqs. (3.3-25) and (3.2-26) shows that we have proved eq. (3.3-24).

3.3.11 Green's Lemma—The Divergence Theorem in a Plane

Here we consider a two-dimensional vector field that does not vary in a direction perpendicular to the plane containing the two-dimensional vectors. We now consider eq. (3.3-4) applied to the volume element shown in Fig. 3.3-11. There is no

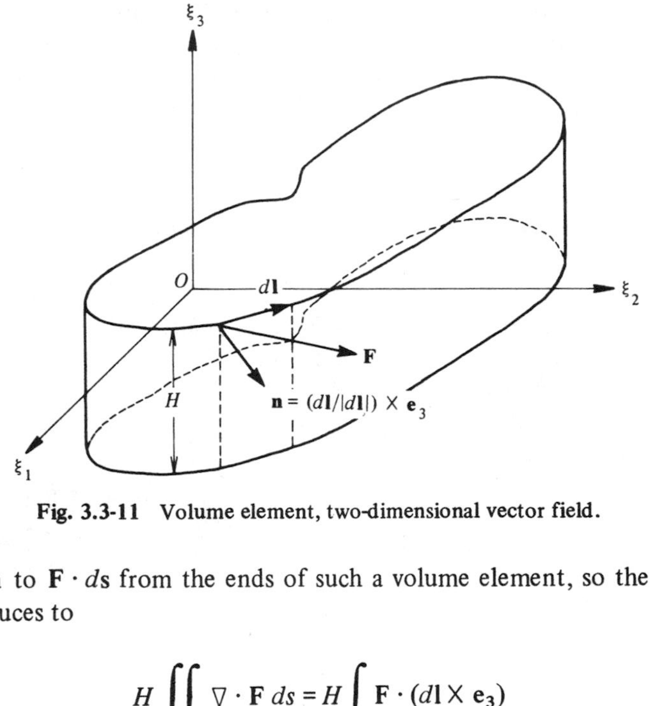

Fig. 3.3-11 Volume element, two-dimensional vector field.

contribution to $\mathbf{F} \cdot d\mathbf{s}$ from the ends of such a volume element, so the divergence theorem reduces to

$$H \iint_s \nabla \cdot \mathbf{F}\, ds = H \int_l \mathbf{F} \cdot (d\mathbf{l} \times \mathbf{e}_3)$$

Now writing $\mathbf{F} = Q\mathbf{e}_1 - P\mathbf{e}_2$, $d\mathbf{l} = h_1 d\xi_1 \mathbf{e}_1 + h_2 d\xi_2 \mathbf{e}_2$ and utilizing eq. (3.3-3) (with $\partial/\partial\xi_3 = 0$) it follows that

$$\int_{\xi_2} \int_{\xi_1} \left[\frac{\partial(h_2 Q)}{\partial \xi_1} - \frac{\partial(h_1 P)}{\partial \xi_2} \right] d\xi_1 d\xi_2 = \oint_l [Q h_2 d\xi_2 + P h_1 d\xi_1] \qquad (3.3\text{-}27)$$

This result is known as Green's Lemma.

Example: As an example of the use of the results developed in this chapter, let us derive the Eulerian forms of the continuity and momentum equations for non-viscous fluid flow. The Eulerian form of the equations describes the flow in terms of the four independent variables, time and the three space coordinates. In deriving the equations we write the expressions for the appropriate quantities considered in terms of a fixed macroscopic volume V, surrounded by surface S.

Solution: Continuity of mass requires that the rate at which mass accumulates in volume V, plus the rate at which mass is convected through the surface S, must be zero. Denoting the mass per volume by ρ, and the fluid velocity by $\mathbf{u}$, we write

$$\frac{\partial}{\partial t}\iiint_V \rho\, dv + \iint_S \rho\mathbf{u}\cdot d\mathbf{s} = 0$$

Noting that $\partial v/\partial t$ is zero for the volume considered, and applying the divergence theorem, eq. (3.3-4) it follows that

$$\iiint_V \frac{\partial \rho}{\partial t}\, dv + \iiint_V \nabla\cdot\rho\mathbf{u}\, dv = 0$$

This result is true for an arbitrary volume V, and hence does not rely on any averaging properties of the integrals. This being the case we may equate the integrands to zero to obtain:

$$\frac{\partial \rho}{\partial t} + \nabla\cdot\rho\mathbf{u} = 0 \qquad \text{continuity equation}$$

The momentum equation may be written in words: The sum of the body forces ($\mathbf{f}$ per unit volume) plus pressure forces equals the rate of accumulation of momentum within the volume plus the rate at which momentum is convected out of the volume.

Noting that $\rho\mathbf{u}$ is the momentum per unit volume, that $\rho\mathbf{u}\cdot d\mathbf{s}$ is the mass flow per second through $d\mathbf{s}$, that $\mathbf{u}$ is the momentum per unit mass, and that the pressure force (inward) on element $d\mathbf{s}$ is $-Pd\mathbf{s}$, the above word equation leads to

$$\iiint_V \mathbf{f}\, dv - \iint_S P d\mathbf{s} = \frac{\partial}{\partial t}\iiint_V \rho\mathbf{u}\, dv + \iint_S \mathbf{u}(\rho\mathbf{u}\cdot d\mathbf{s})$$

Again noting that $\partial v/\partial t$ is zero, and applying eqs. (3.3-4) and (3.3-24) an equation in terms of volume integrals, only, results. Equating integrands, it follows that follows that

$$\mathbf{f} - \nabla P = \frac{\partial \rho \mathbf{u}}{\partial t} + (\rho \mathbf{u} \cdot \nabla)\, \mathbf{u} + \mathbf{u}(\nabla \cdot \rho \mathbf{u})$$

Subtracting the scalar product of **u** and the continuity equation from this gives

$$\mathbf{f} - \nabla P = \rho \left\{ \frac{\partial \mathbf{u}}{\partial t} + (\mathbf{u} \cdot \nabla)\, \mathbf{u} \right\} \qquad \text{momentum equation}$$

3.4 SUCCESSIVE OPERATIONS

It is often required to manipulate expressions involving calculus operations on several variables, or to manipulate expressions involving successive applications of the various operators already considered. The methods for obtaining such relationships will be illustrated by example here. A list of operations on sums and products is given in the summary, section 3.6.6, and a list of successive operations is given in the summary, section 3.6.4.

3.4.1 Operations on Sums and Products

The integral forms for the four operators, divergence, curl, gradient, and convective operator [eqs. (3.3-2), (3.3-9), (3.3-17), (3.3-20)] together with the results of sections 3.2.3 and 3.2.4, show that the four operations are distributive, so that

$$\nabla(\phi + \psi) = \nabla\phi + \nabla\psi \tag{3.4-1}$$

$$(\mathbf{a} \cdot \nabla)(\mathbf{F} + \mathbf{G}) = (\mathbf{a} \cdot \nabla)\, \mathbf{F} + (\mathbf{a} \cdot \nabla)\, \mathbf{G} \tag{3.4-2}$$

$$\nabla \cdot (\mathbf{F} + \mathbf{G}) = \nabla \cdot \mathbf{F} + \nabla \cdot \mathbf{G} \tag{3.4-3}$$

$$\nabla \times (\mathbf{F} + \mathbf{G}) = \nabla \times \mathbf{F} + \nabla \times \mathbf{G} \tag{3.4-4}$$

The same four operators can also operate on appropriate forms of products. The desired expressions can be obtained by reference to the integral definitions of the operators, and by exploiting the technique, already introduced in section 3.3.10, of representing the various functions by their values at the center of the infinitesimal volume plus a small change from the central value existing at the area element of concern (see Fig. 3.4-1). By the use of this artifice the central value, F_0, can be taken outside the area integrals (because it is constant in the limiting process), and in addition, products of the small quantities such as $\Delta F \Delta G$ contribute terms of higher order and may be neglected in the limiting process. Using this technique the required proofs are straightforward. A single example will be given to illustrate the method by considering the curl of a cross product, which is perhaps the most algebraically complex of the proofs.

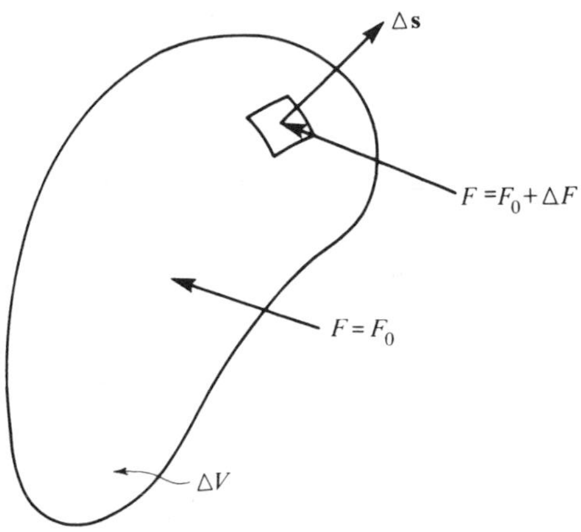

Fig. 3.4-1 Example volume element.

To Prove:

$$\nabla \times (\mathbf{F} \times \mathbf{G}) = (\mathbf{G} \cdot \nabla)\,\mathbf{F} - (\mathbf{F} \cdot \nabla)\,\mathbf{G} + \mathbf{F}(\nabla \cdot \mathbf{G}) - \mathbf{G}(\nabla \cdot \mathbf{F})$$

Proof:

$$\nabla \times (\mathbf{F} \times \mathbf{G}) = \lim_{\Delta v \to 0} \frac{1}{\Delta v} \left[\iint_s d\mathbf{s} \times (\mathbf{F} \times \mathbf{G}) \right]$$

$$= \lim_{\Delta v \to 0} \frac{1}{\Delta v} \left[\iint_s d\mathbf{s} \times (\mathbf{F}_0 \times \mathbf{G}_0) + \iint_s d\mathbf{s} \times (\Delta\mathbf{F} \times \mathbf{G}_0) \right.$$

$$\left. + \iint_s d\mathbf{s} \times (\mathbf{F}_0 \times \Delta\mathbf{G}) + \iint_s d\mathbf{s} \times (\Delta\mathbf{F} \times \Delta\mathbf{G}) \right]$$

In this group, the first term is zero because the value of the area integral (over the closed surface) is zero. The fourth term vanishes in the limiting process because it is of higher order. We then obtain, with the aid of eq. (3.2-14),

$$\nabla \times (\mathbf{F} \times \mathbf{G}) = \lim_{\Delta v \to 0} \frac{1}{\Delta v} \left[\iint_s d\mathbf{s} \times (\Delta\mathbf{F} \times \mathbf{G}_0) + \iint_s d\mathbf{s} \times (\mathbf{F}_0 \times \Delta\mathbf{G}) \right]$$

$$= \lim_{\Delta v \to 0} \frac{1}{\Delta v} \left[\iint_s \Delta\mathbf{F}\, \mathbf{G}_0 \cdot d\mathbf{s} - \iint_s \Delta\mathbf{F} \cdot d\mathbf{s}\, \mathbf{G}_0 \right.$$

$$+ \iint_s \Delta\mathbf{G} \cdot d\mathbf{s}\, \mathbf{F}_0 - \iint_s \Delta\mathbf{G}\, \mathbf{F}_0 \cdot d\mathbf{s}\Bigg]$$

$$= \lim_{\Delta v \to 0} \frac{\iint_s \mathbf{F}\, \mathbf{G}_0 \cdot d\mathbf{s}}{\Delta v} - \mathbf{G} \lim_{\Delta v \to 0} \frac{\iint_s \mathbf{F} \cdot d\mathbf{s}}{\Delta v}$$

$$+ \mathbf{F} \lim_{\Delta v \to 0} \frac{\iint_s \mathbf{G} \cdot d\mathbf{s}}{\Delta v} - \lim_{\Delta v \to 0} \frac{\iint_s \mathbf{G}\, \mathbf{F}_0 \cdot d\mathbf{s}}{\Delta v}$$

$$= (\mathbf{G} \cdot \nabla)\, \mathbf{F} - \mathbf{G}(\nabla \cdot \mathbf{F}) + \mathbf{F}(\nabla \cdot \mathbf{G}) - (\mathbf{F} \cdot \nabla)\, \mathbf{G} \qquad \textit{q.e.d.}$$

It is important to note in this proof, as well as in the proofs leading to the summary of section 3.6.6, that the results are valid in general orthogonal curvilinear coordinates, no restriction to a specific coordinate system being required when the integral definitions of the operators are used.

3.4.2 Successive Application of Vector Operators

There are six ways in which the operations divergence, gradient and curl may be applied twice to a vector or scalar function. In this section we obtain the expansions for these six double operations in terms of the generalized curvilinear coordinates.

DIVERGENCE OF THE GRADIENT OF A SCALAR, THE LAPLACIAN

$$\text{div grad } \phi = \nabla \cdot (\nabla\phi) \equiv \nabla^2 \phi$$

To obtain $\nabla^2\phi$ we successively apply eqs. (3.3-14) and (3.3-3). Hence

$$\nabla^2 \phi = \frac{1}{h_1 h_2 h_3}\left[\frac{\partial}{\partial \xi_1}\left(\frac{h_2 h_3}{h_1}\frac{\partial \phi}{\partial \xi_1}\right) + \frac{\partial}{\partial \xi_2}\left(\frac{h_1 h_3}{h_2}\frac{\partial \phi}{\partial \xi_2}\right) + \frac{\partial}{\partial \xi_3}\left(\frac{h_1 h_2}{h_3}\frac{\partial \phi}{\partial \xi_3}\right)\right] \tag{3.4-5}$$

CURL OF THE GRADIENT OF A SCALAR

$$\text{curl grad } \phi = \nabla \times \nabla\phi$$

Rather than utilizing the expanded form for the gradient, it is more instructive in this case to use the form of eq. (3.3-11). Thus writing $\nabla\phi = \partial\phi/\partial n\, \mathbf{n}$, and utilizing eq. (3.3-5), we obtain for the component of curl grad ϕ in the direction $\mathbf{m}$,

$$(\nabla \times \nabla\phi)_{\mathbf{m}} = \lim_{\Delta s \to 0} \frac{\int_l \frac{\partial \phi}{\partial n}\, \mathbf{n} \cdot d\mathbf{l}}{\Delta s}$$

Now $\mathbf{n} \cdot d\mathbf{l} = dn$, the scalar distance change in the direction $\mathbf{n}$. Thus the line integral around the curve is equal to $\oint \frac{\partial \phi}{\partial n} dn = 0$, and hence

$$\nabla \times \nabla \phi = 0 \tag{3.4-6}$$

GRADIENT OF THE DIVERGENCE OF A VECTOR

$$\text{grad div } \mathbf{F} = \nabla(\nabla \cdot \mathbf{F})$$

Successive application of eqs. (3.3-3) and (3.3-14) yields

$$[\nabla(\nabla \cdot \mathbf{F})]_i = \frac{1}{h_i} \frac{\partial}{\partial \xi_i} \left\{ \frac{1}{h_1 h_2 h_3} \left[\frac{\partial F_1 h_2 h_3}{\partial \xi_1} + \frac{\partial F_2 h_1 h_3}{\partial \xi_2} + \frac{\partial F_3 h_1 h_2}{\partial \xi_3} \right] \right\}$$

$$(i \text{ not summed}) \quad (3.4\text{-}7)$$

DIVERGENCE OF THE CURL OF A VECTOR

$$\text{div curl } \mathbf{F} = \nabla \cdot (\nabla \times \mathbf{F})$$

To evaluate this group we consider the integral form of eq. (3.3-2). Thus

$$\nabla \cdot (\nabla \times \mathbf{F}) = \lim_{\Delta v \to 0} \frac{\iint_s (\nabla \times \mathbf{F}) \cdot d\mathbf{s}}{\Delta v}$$

The area integral is to be taken over a closed area, but for the moment consider the contribution of only a portion of the area, such as that numbered ① in Fig.

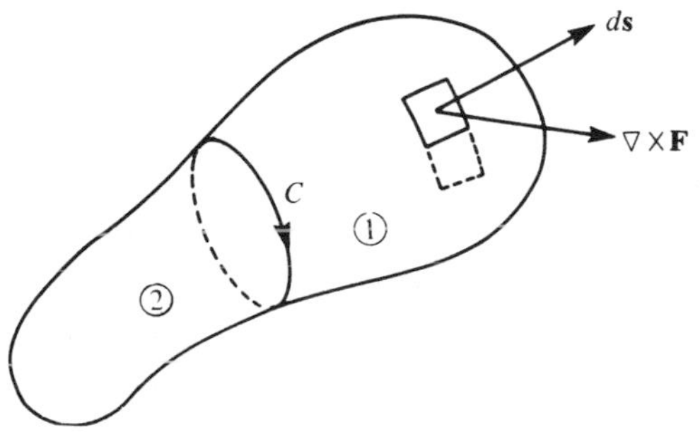

Fig. 3.4-2 Volume element for use in evaluating div curl F.

3.4-2. From Stokes' Theorem, eq. (3.3-10) it follows that

$$\iint_{s_1} (\nabla \times \mathbf{F}) \cdot ds = \int_c \mathbf{F} \cdot d\mathbf{l}$$

Similarly, the integral over area ② gives

$$\iint_{s_2} (\nabla \times \mathbf{F}) \cdot ds = \int_{c'} \mathbf{F} \cdot d\mathbf{l}$$

The curve c' is identical to curve c except that the direction of the line integral is opposite in sense. Thus when the two contributions to the area integral are summed, the result is zero giving

$$\nabla \cdot (\nabla \times \mathbf{F}) = 0 \tag{3.4-8}$$

CURL OF THE CURL OF A VECTOR

$$\text{curl curl } \mathbf{F} = \nabla \times (\nabla \times \mathbf{F})$$

Successive application of eq. (3.3-7) leads to the expression for the ith component of $\nabla \times (\nabla \times \mathbf{F})$

$$[\nabla \times (\nabla \times \mathbf{F})]_i = e_{ijk}\, e_{kst} \frac{h_i}{h_1 h_2 h_3} \frac{\partial}{\partial \xi_j} \left[\frac{h_k^2}{h_1 h_2 h_3} \frac{\partial (h_t F_t)}{\partial \xi_s} \right] \tag{3.4-9}$$

The expanded form of this expression is included in the summary section 3.6.4.

THE LAPLACIAN OF A VECTOR

Strictly speaking, the Laplacian of a vector, i.e.

$$\text{div grad } \mathbf{F} = \nabla \cdot (\nabla \mathbf{F})$$

cannot be defined using vector techniques only, because the gradient of the vector required in the definition is itself, properly a tensor quantity. One approach is to start with cartesian coordinates where the ith component of the Laplacian of a vector is simply the Laplacian of the ith component of the vector. By then considering the operator $\nabla \equiv \mathbf{i}\, \partial/\partial x + \mathbf{j}\, \partial/\partial y + \mathbf{k}\, \partial/\partial z$ to have the properties of a vector, the vector identity is established.

$$\nabla^2 \mathbf{F} = \nabla(\nabla \cdot \mathbf{F}) - \nabla \times (\nabla \times \mathbf{F}) \tag{3.4-10}$$

This may be considered the definition of the Laplacian of a vector.

The expressions for the components of $\nabla^2 \mathbf{F}$ in generalized curvilinear coordinates are very cumbersome and will not be written out here. They may be obtained directly by combining the results of eqs. (3.4-7) and (3.4-9) in eq. (3.4-10).

Examples:

(1) Green's theorem—This theorem results as a special case of Gauss' Divergence Theorem. Thus, consider eq. (3.3-4) when $\mathbf{F}$ is the special vector function given in terms of the two scalar functions ψ and ϕ by $\mathbf{F} = \psi \nabla \phi$. Utilizing the relationship given in the summary, section 3.6.6, for the divergence of a scalar times a vector, it follows that

$$\iiint_v \nabla \cdot (\psi \nabla \phi)\, dv = \iiint_v [\nabla \psi \cdot \nabla \phi + \psi \nabla^2 \phi]\, dv = \iint_s \psi \nabla \phi \cdot ds$$

an alternate form is obtained by taking the difference of the pair of divergences $\nabla \cdot (\psi \nabla \phi) - \nabla \cdot (\phi \nabla \psi)$ to give

$$\iiint_V [\psi \nabla^2 \phi - \phi \nabla^2 \psi]\, dv = \iint_s [\psi \nabla \phi - \phi \nabla \psi] \cdot d\mathbf{s}$$

(2) Steady flow of an incompressible inviscid fluid through a stator row—We consider the effects of the blade forces of the stator row to be represented by a body force $\mathbf{f}$, so utilizing the expanded form of $(\mathbf{u} \cdot \nabla)\, \mathbf{u}$, the momentum equation may be written

$$\rho(\mathbf{u} \cdot \nabla)\, \mathbf{u} = \rho \left\{ \nabla \frac{\mathbf{u}^2}{2} + (\nabla \times \mathbf{u}) \times \mathbf{u} \right\} = \mathbf{f} - \nabla P$$

Taking the scalar product of this equation with $\mathbf{u}$, and noting that $\mathbf{u} \cdot \{(\nabla \times \mathbf{u}) \times \mathbf{u}\} = 0$, it follows that

$$(\mathbf{u} \cdot \nabla) \left(P + \frac{\rho}{2} \mathbf{u}^2 \right) = \mathbf{u} \cdot \mathbf{f}$$

Now in an inviscid fluid, any force resulting from the surface pressures must be perpendicular to the surface, and hence to the relative velocity of the fluid over the surface. In the case of a stator row, as considered here, we hence have $\mathbf{u} \cdot \mathbf{f} = 0$, and consequently $P + \frac{\rho}{2} \mathbf{u}^2$ is conserved along streamlines. With uniform conditions at entry, $P + \frac{\rho}{2} \mathbf{u}^2$ must be constant throughout so that $\nabla \left(P + \frac{\rho}{2} \mathbf{u}^2 \right) = 0$. Subtracting this equation from the first equation of this example, and noting that **f** is zero external to the blade row we find

$$(\nabla \times \mathbf{u}) \times \mathbf{u} = 0$$

Hence when an inviscid fluid with uniform conditions at entry passes through a stator row, if the row imparts vorticity $(\nabla \times \mathbf{u})$, the vorticity must be parallel to the velocity.

3.5 VECTOR FIELDS

3.5.1 Some Preliminary Manipulations

3.5.1.1 Source Points and Field Points We will be concerned with the description of properties of vector fields in terms of certain conditions existing elsewhere in the field. The field point represents the point at which conditions are to be investigated, and the source point represents the point where conditions exist which affect conditions at the field point. For example, an electric current might be flowing at the source point which would affect the magnetic field at the field point. The scalar distance between source and field point is denoted $r(x_i, x_i')$. We will use primed quantities to refer to conditions at the source point, and unprimed quantities to refer to conditions at the field point. The scalar distance $r(x_i, x_i')$ is a symmetric function in terms of x_i and x_i' given by

$$r(x_i, x_i') = \{\mathbf{r} \cdot \mathbf{r}\}^{1/2} = \{(x_i - x_i')\,(x_i - x_i')\}^{1/2}$$

Introducing the symbol ∇' to indicate operation with respect to the source points, and ∇ to indicate operation with respect to the field points it is apparent that when an operation is performed on any function of the scalar distance r, ∇' will be replaced by $-\nabla$. This is because ∇' corresponds to $\partial/\partial x_i'$, and ∇ corresponds to $\partial/\partial x_i$, so that

$$\frac{\partial F(r)}{\partial x_i} = \frac{\partial F}{\partial r}\frac{\partial r}{\partial x_i} = \frac{\partial F}{\partial r}\left(-\frac{\partial r}{\partial x_i'}\right)$$

Hence

$$\nabla' F(r) = -\nabla F(r);\quad (\mathbf{A} \cdot \nabla')F(r) = -(\mathbf{A} \cdot \nabla)F(r);\quad (\nabla')^2 F(r) = \nabla^2 F(r);\quad \text{etc.} \tag{3.5-1}$$

We note also for our use shortly, that if **A** is a constant, we obtain with the results of the summary, section 3.6.6

$$\nabla \times \left(\nabla \times \frac{\mathbf{A}}{r}\right) = \nabla \times \left(\nabla \frac{1}{r} \times \mathbf{A}\right) = (\mathbf{A} \cdot \nabla)\,\nabla \frac{1}{r} - \mathbf{A}\nabla^2 \frac{1}{r} \tag{3.5-2}$$

This latter expression introduces the term $\nabla^2(1/r)$, which we may investigate most easily by introducing spherical coordinates, for which $\xi_1 = r$, $\xi_2 = \theta$, $\xi_3 = \phi$, $h_1 = 1$, $h_2 = r$, $h_3 = r \sin\theta$, $(0 \leqslant r < \infty,\ 0 \leqslant \theta \leqslant \pi,\ 0 \leqslant \phi \leqslant 2\pi)$. In this coordinate system we obtain from eqs. (3.3-14) and (3.4-5)

$$\nabla g(r) = \frac{\partial g}{\partial r}\,\mathbf{e}_r;\quad \text{hence } \nabla\frac{1}{r} = -\frac{1}{r^2}\,\mathbf{e}_r$$

$$\nabla^2 g(r) = \frac{1}{r^2}\frac{\partial}{\partial r}\left(r^2\frac{\partial g}{\partial r}\right);\quad \text{hence } \nabla^2\frac{1}{r} = 0 \text{ except at } r = 0 \tag{3.5-3}$$

The behavior of $\nabla^2(1/r)$ near $r = 0$ can be investigated heuristically by considering the volume integral of $\nabla^2(1/r)$ including the origin. Then, with the divergence theorem:

$$\iiint_V \nabla^2\frac{1}{r}\,dv = \iint_s \nabla\frac{1}{r}\cdot d\mathbf{s} = -\int_0^{2\pi}\int_0^{\pi}\left(\frac{1}{r^2}\right)r^2 \sin\theta\, d\theta\, d\phi$$

$$= -4\pi \tag{3.5-4}$$

It is evident then that $\nabla^2(1/r)$ has the properties of $-4\pi\delta(\mathbf{r})$, where $\delta(\mathbf{r})$ is a three-dimensional version of the Dirac Delta function, defined by the properties

$$\delta(\mathbf{r}) = 0,\ \mathbf{r} \neq 0,\qquad \iiint_V \delta(\mathbf{r})\,dv = 1,\qquad \mathbf{r} = 0 \text{ contained in } V \tag{3.5-5}$$

Note also the corollary

$$\iiint_V F(\mathbf{r})\,\delta(\mathbf{r})\,dv = F(0) \tag{3.5-6}$$

3.5.2 The Scalar and Vector Potentials

It may be shown that a vector field **F** can be described uniquely in terms of its curl and divergence if the source density **C** of its curl, and source density s of its divergence, are zero at infinity. With these restrictions on **C** and s, the two equations

$$\nabla\cdot\mathbf{F} = s \tag{3.5-7}$$

$$\nabla\times\mathbf{F} = \mathbf{C} \tag{3.5-8}$$

lead to a unique solution for **F**.

It is convenient to represent **F** as the sum of two parts, where

$$\mathbf{F} = -\nabla\phi + \nabla\times\mathbf{A} \tag{3.5-9}$$

In this expression ϕ is termed the scalar potential and **A** is termed the vector potential. A convenience of this formulation for **F** lies in the fact that ϕ and **A** may be expressed in terms of the volume integral of s and **C** respectively.

Substitution of eq. (3.5-9) into eqs. (3.5-7) and (3.5-8) gives the following equations for ϕ and **A**.

$$\nabla^2 \phi = -s \tag{3.5-10}$$

$$\nabla \times (\nabla \times \mathbf{A}) = \mathbf{C} \tag{3.5-11}$$

A solution of these equations is given by

$$\phi = \frac{1}{4\pi} \iiint_{V'} \frac{s'}{r} \, dv' \tag{3.5-12}$$

$$\mathbf{A} = \frac{1}{4\pi} \iiint_{V'} \frac{c'}{r} \, dv' \tag{3.5-13}$$

In these expressions, the prime denotes integration with respect to conditions at the source point, and the volume of integration is to include all space. Equation (3.5-12) is easily verified by direct substitution into eq. (3.5-10), with subsequent application of eqs. (3.5-1) and (3.5-5). Thus

$$\nabla^2 \phi = \frac{1}{4\pi} \iiint_{V'} s' \nabla^2 \frac{1}{r} \, dv' = \frac{1}{4\pi} \iiint_{V'} s' \nabla'^2 \frac{1}{r} \, dv'$$

$$= - \iiint_{V'} s' \delta(r) \, dv' = -s$$

Similarly, direct substitution of eq. (3.5-13) into eq. (3.5-11) gives with eq. (3.5-2).

$$\nabla \times (\nabla \times \mathbf{A}) = \frac{1}{4\pi} \iiint_{V'} \nabla \times \left(\nabla \times \frac{C'}{r} \right) dv'$$

$$= \frac{1}{4\pi} \iiint_{V'} \left[(\mathbf{C}' \cdot \nabla) \nabla \frac{1}{r} - \mathbf{C}' \nabla^2 \frac{1}{r} \right] dv'$$

$$= \frac{1}{4\pi} \iiint_{V'} \left[(\mathbf{C}' \cdot \nabla') \nabla' \frac{1}{r} - \mathbf{C}' \nabla'^2 \frac{1}{r} \right] dv'$$

$$= \mathbf{C} + \frac{1}{4\pi} \iiint_{V'} (\mathbf{C}' \cdot \nabla') \nabla' \frac{1}{r} \, dv'$$

The integral in this latter expression may be evaluated through the use of the generalized form of Gauss' Divergence Theorem, eq. (3.3-24) to give

$$\iiint_{V'} (\mathbf{C}' \cdot \nabla')\,\nabla' \frac{1}{r}\,dv' = \iint_{s'} \nabla' \frac{1}{r}\,\mathbf{C}' \cdot ds' - \iiint_{V'} \left(\nabla' \frac{1}{r}\right)(\nabla' \cdot \mathbf{C}')\,dv'$$

Here, the surface integral vanishes because the integral is defined to be taken over all space, and hence the vector source density **C** is zero at the surface. The volume integral also vanishes by eq. (3.4-8). We have thus shown that eq. (3.5-13) satisfies eq. (3.5-11).

3.5.3 The Effect of Finite Boundaries

Equations (3.5-12) and (3.5-13) were shown to be solutions to eqs. (3.5-7) and (3.5-8) when the source densities were prescribed over all space, but it is of interest to investigate the solutions to the equations when finite boundaries are involved. For simplicity we restrict ourselves to the irrotational case, i.e., the case where $\mathbf{A} = 0$.

Green's Theorem (section 3.6.5), with ψ put equal to $1/r$ gives

$$\iiint_{V'} \left[\frac{1}{r}\nabla^2\phi - \phi\nabla'^2 \frac{1}{r}\right] dv' = \iint_{s} \left[\frac{1}{r}\nabla'\phi - \phi\nabla' \frac{1}{r}\right] \cdot ds'$$

With the results of eqs. (3.5-4) to (3.5-6), together with eq. (3.5-10) this may be rearranged to give

$$\phi = \frac{1}{4\pi} \iiint_{V'} \frac{s'}{r}\,dv' + \frac{1}{4\pi} \iint_{s'} \left[\frac{1}{r}\nabla'\phi + \frac{\mathbf{e}_r}{r}\phi\right] \cdot ds \qquad (3.5\text{-}14)$$

The solution for ϕ, when written in this form, may be thought of as the result of contributions from sources within the volume considered, plus the effects of sources external to the boundary. The effects of the sources external to the boundary may be represented in terms of the distribution of the potential and its normal derivative on the bounding surface. Clearly eq. (3.5-12) is just the limiting case of eq. (3.5-14) when the bounding surface is extended to infinity.

We may note here the existence of such techniques for solution as reflection techniques, where the inverse procedure to that implied above is exploited. Thus, for example, the desired boundary conditions at a finite location are generated by placing "apparent" sources external to the boundary.

3.6 SUMMARY

3.6.1 Examples of Orthogonal Curvilinear Coordinate Systems

There are many possible examples of orthogonal curvilinear coordinate systems. The examples here are restricted to those eleven orthogonal coordinate systems that lead to separable forms for the wave equation.

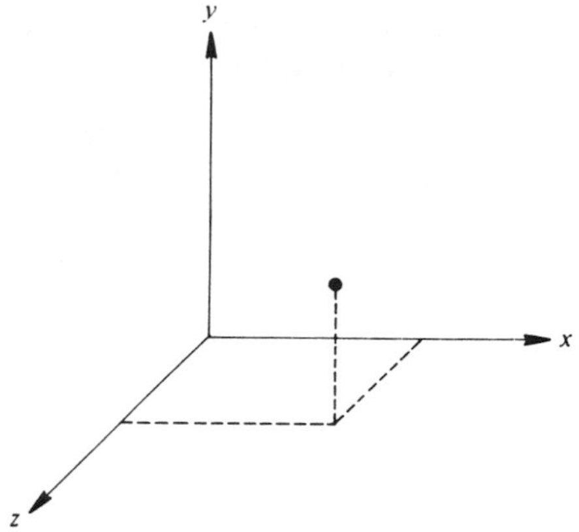

Fig. 3.6-1 Cartesian coordinates.

CARTESIAN COORDINATES

$$\xi_1 = x, \quad \xi_2 = y, \quad \xi_3 = z$$

$$h_1 = h_2 = h_3 = 1$$

Coordinate surfaces are planes. (x = constant, y = constant, z = constant).

CIRCULAR CYLINDER COORDINATES

$$\xi_1 = r, \quad \xi_2 = \theta, \quad \xi_3 = z$$

$$h_1 = 1, \quad h_2 = r, \quad h_3 = 1$$

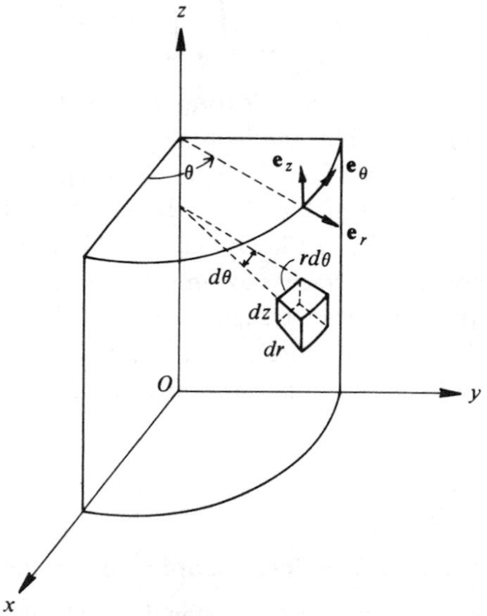

Fig. 3.6-2 Circular cylinder coordinates.

Coordinate surfaces are circular cylinders (r = constant) and planes (z = constant, θ = constant)

SPHERICAL COORDINATES

$$\xi_1 = r, \quad \xi_2 = \theta, \quad \xi_3 = \phi$$

$$h_1 = 1, \quad h_2 = r, \quad h_3 = r \sin\theta$$

Coordinate surfaces are spheres (r = constant), cones (θ = constant) and half planes (ϕ = constant)

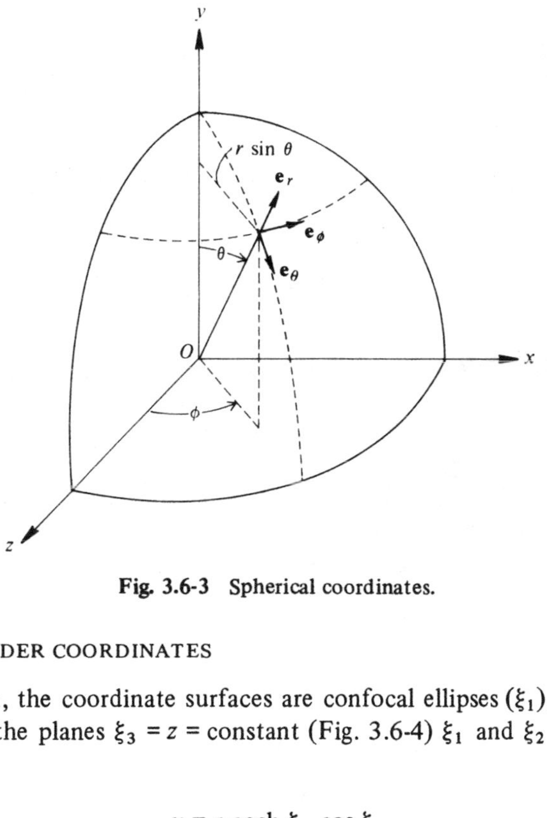

Fig. 3.6-3 Spherical coordinates.

ELLIPTIC CYLINDER COORDINATES

In this system, the coordinate surfaces are confocal ellipses (ξ_1), confocal hyperbolas (ξ_2) and the planes $\xi_3 = z$ = constant (Fig. 3.6-4) ξ_1 and ξ_2 are related to x and y by

$$x = a \cosh \xi_1 \cos \xi_2$$

$$y = a \sinh \xi_1 \sin \xi_2$$

hence

$$\left(\frac{x}{a \cosh \xi_1}\right)^2 + \left(\frac{y}{a \sinh \xi_1}\right)^2 = 1 \qquad \text{(confocal ellipses, } \xi_1 = \text{constant)}$$

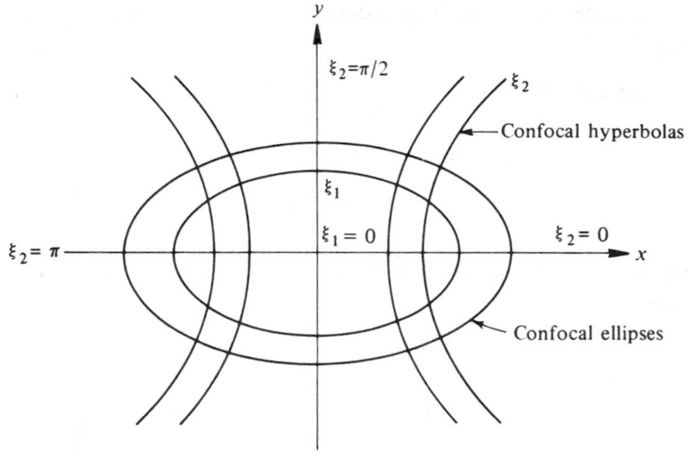

Fig. 3.6-4 Elliptic cylinder coordinates.

$$\left(\frac{x}{a\cos\xi_2}\right)^2 - \left(\frac{y}{a\sin\xi_2}\right)^2 = 1 \qquad \text{(confocal hyperbolas, } \xi_2 = \text{constant)}$$

$$h_1 = h_2 = a\sqrt{\cosh^2\xi_1 - \cos^2\xi_2}, \quad h_3 = 1$$

PROLATE SPHEROIDAL COORDINATES

This system is obtained by rotating the coordinate system of Fig. 3.6-4 about the $\xi_2 = \pi$, $\xi_2 = 0$ axis. The coordinate surfaces are then formed by prolate spheroids (ξ_1 = constant, these are the surfaces formed by rotating the confocal ellipses about the $\xi_2 = \pi$, $\xi_2 = 0$ axis); hyperboloids of two sheets (ξ_2 = constant, these are the surfaces formed by rotating the confocal hyperbolas about the $\xi_2 = \pi$, $\xi_2 = 0$ axis); and planes (ξ_3 = constant). These latter planes are half planes located by their angle from the z-axis. Thus,

$$x = a\cosh\xi_1\cos\xi_2$$

$$y = a\sinh\xi_1\sin\xi_2\sin\xi_3$$

$$z = a\sinh\xi_1\sin\xi_2\cos\xi_3$$

$$\left(\frac{x}{a\cosh\xi_1}\right)^2 + \left(\frac{y}{a\sinh\xi_1}\right)^2 + \left(\frac{z}{a\sinh\xi_1}\right)^2 = 1$$

(Prolate spheroids, ξ_1 = constant)

$$\left(\frac{x}{a\cos\xi_2}\right)^2 + \left(\frac{y}{a\sin\xi_2}\right)^2 - \left(\frac{z}{a\sin\xi_2}\right)^2 = 1$$

(Hyperboloids of two sheets, ξ_2 = constant)

$$\tan \xi_3 = y/z$$

(Half planes, ξ_3 = constant)

$$h_1 = h_2 = a\sqrt{\sinh^2 \xi_1 + \sin^2 \xi_2}, \quad h_3 = a \sinh \xi_1 \sin \xi_2$$

OBLATE SPHEROIDAL COORDINATES

This system is obtained by rotating the coordinate system of Fig. 3.6-4 about the $\xi_2 = \pi/2$ axis. The coordinate surfaces are then formed by oblate spheroids (ξ_1 = constant, these are the surfaces formed by rotating the confocal ellipses about the $\xi_2 = \pi/2$ axis); hyperboloids of one sheet (ξ_2 = constant, these are the surfaces formed by rotating the hyperbolas about the $\xi_2 = \pi/2$ axis); and planes (ξ_3 = constant). These latter planes are half-planes located by their angle from the z-axis. Thus

$$x = a \cosh \xi_1 \sin \xi_2 \sin \xi_3$$

$$y = a \sinh \xi_1 \cos \xi_2$$

$$z = a \cosh \xi_1 \sin \xi_2 \cos \xi_3$$

$$\left(\frac{x}{a \cosh \xi_1}\right)^2 + \left(\frac{y}{a \sinh \xi_1}\right)^2 + \left(\frac{z}{a \cosh \xi_1}\right)^2 = 1$$

(Oblate spheroids, ξ_1 = constant)

$$\left(\frac{x}{a \sin \xi_2}\right)^2 - \left(\frac{y}{a \cos \xi_2}\right)^2 + \left(\frac{z}{a \sin \xi_2}\right)^2 = 1$$

(Hyperboloids of one sheet, ξ_2 = constant)

$$\tan \xi_3 = \frac{x}{z}$$

(Half-planes, ξ_3 = constant)

$$h_1 = h_2 = a\sqrt{\cosh^2 \xi_1 - \sin^2 \xi_2}, \quad h_3 = a \cosh \xi_1 \sin \xi_2$$

PARABOLIC-CYLINDER COORDINATES

In this system the coordinate surfaces are two sets of confocal parabolas, focus on z-axis (ξ_1, ξ_2 = constant) and the planes $\xi_3 = z$ = constant (Fig. 3.6-5). ξ_1 and ξ_2 are related to x and y by

$$x = \tfrac{1}{2}(\xi_2^2 - \xi_1^2)$$

$$y = \xi_1 \xi_2$$

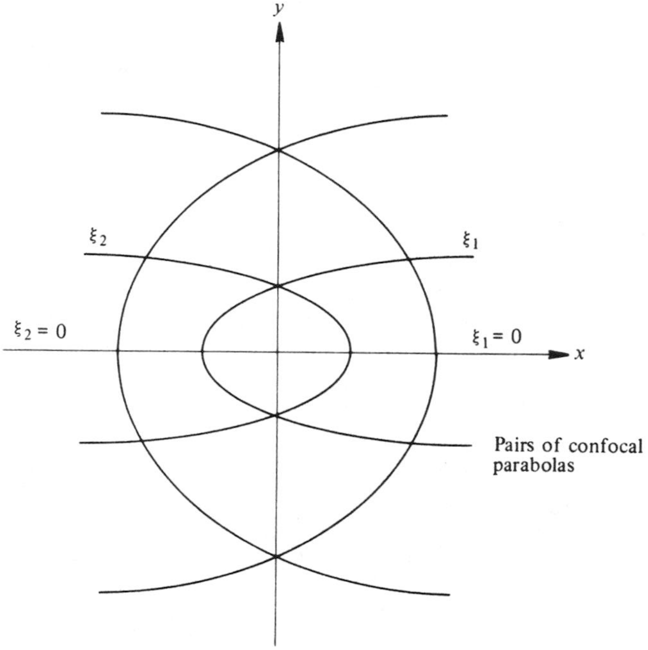

Fig. 3.6-5 Parabolic cylinder coordinates.

hence

$$y^2 = \xi_1^2 (\xi_1^2 + 2x) \qquad \text{(Parabolic cylinders, } \xi_1 = \text{constant)}$$

$$y^2 = \xi_2^2 (\xi_2^2 - 2x) \qquad \text{(Parabolic cylinders, } \xi_2 = \text{constant)}$$

$$h_1 = h_2 = \sqrt{\xi_1^2 + \xi_2^2}, \quad h_3 = 1$$

PARABOLIC COORDINATES

This system is obtained by rotating the coordinate system of Fig. 3.6-5 about the x-axis. (About the $\xi_1 = \xi_2 = 0$ axis.) The coordinate surfaces are then formed by paraboloids of revolution (ξ_1 = constant, ξ_2 = constant) and half-planes, ξ_3 = constant. These latter planes are located by their angle from the z-axis. Thus

$$x = \frac{1}{2} (\xi_2^2 - \xi_1^2)$$

$$y = \xi_1 \xi_2 \sin \xi_3$$

$$z = \xi_1 \xi_2 \cos \xi_3$$

$$y^2 + z^2 = \xi_1^2 (\xi_1^2 + 2x) \qquad \text{(Paraboloids of revolution, } \xi_1 = \text{constant)}$$

$$y^2 + z^2 = \xi_2^2 (\xi_2^2 - 2x) \qquad \text{(Paraboloids of revolution, } \xi_2 = \text{constant)}$$

$$\tan \xi_3 = y/z$$

$$h_1 = h_2 = \sqrt{\xi_1^2 + \xi_2^2}, \quad h_3 = \xi_1 \xi_2$$

CONICAL COORDINATES

The coordinate surfaces for this system consist of spheres ($\xi_1 = r =$ constant), a family of elliptic cones with apex at the origin and axis along the x-axis (ξ_2 = constant) and a family of elliptic cones with apex at the origin and axis along the z-axis (ξ_3 = constant). Clearly the ranges of ξ_2 and ξ_3 must be restricted in order to prevent the cones "doubling" on themselves, or not intersecting the other family at all. The system so defined is related to the Cartesian coordinate system by

$$x^2 = \left(\frac{\xi_1 \xi_2 \xi_3}{bc}\right)^2$$

$$y^2 = \frac{\xi_1^2 (b^2 - \xi_2^2)(\xi_3^2 - b^2)}{b^2 (c^2 - b^2)}$$

$$z^2 = \frac{\xi_1^2 (c^2 - \xi_2^2)(c^2 - \xi_3^2)}{c^2 (c^2 - b^2)}$$

here

$$0 < \xi_2^2 < b^2 < \xi_3^2 < c^2$$

$$x^2 + y^2 + z^2 = r^2 = \xi_1^2 \qquad \text{(Spheres, } \xi_1 = \text{constant)}$$

$$\frac{x^2}{\xi_2^2} - \frac{y^2}{b^2 - \xi_2^2} - \frac{z^2}{c^2 - \xi_2^2} = 0 \qquad \text{(Elliptic cones, } \xi_2 = \text{constant)}$$

$$\frac{x^2}{\xi_3^2} + \frac{y^2}{\xi_3^2 - b^2} - \frac{z^2}{c^2 - \xi_3^2} = 0 \qquad \text{(Elliptic cones, } \xi_3 = \text{constant)}$$

$$h_1 = 1, \quad h_2 = \left[\frac{\xi_1^2 (\xi_3^2 - \xi_2^2)}{(b^2 - \xi_2^2)(c^2 - \xi_2^2)}\right]^{1/2}, \quad h_3 = \left[\frac{\xi_1^2 (\xi_3^2 - \xi_2^2)}{(\xi_3^2 - b^2)(c^2 - \xi_3^2)}\right]^{1/2}$$

ELLIPSOIDAL COORDINATES

The coordinate surfaces in this system consist of ellipsoids centered at the origin (ξ_1 = constant), hyperboloids of one sheet formed by elliptically sweeping the hyperbolas about the z-axis (ξ_2 = constant) and hyperboloids of two sheets, with center at the origin and symmetry about the z-y plane. (ξ_3 = constant). It might perhaps aid the readers geometrical interpretation if the previously described oblate spheroidal coordinates are thought of as a limiting case of ellipsoidal coordinates, wherein rather than sweeping the hyperbolas of one sheet in an ellipse they are swept in a circle. In such a circumstance the hyperbolas of two sheets degenerate to the half planes of the oblate spheroidal coordinate system. Ellipsoidal coordinates are related to the Cartesian system by

$$x^2 = \left(\frac{\xi_1 \xi_2 \xi_3}{ab}\right)^2$$

$$y^2 = \frac{(\xi_1^2 - a^2)(\xi_2^2 - a^2)(a^2 - \xi_3^2)}{a^2(b^2 - a^2)}$$

$$z^2 = \frac{(\xi_1^2 - b^2)(b^2 - \xi_2^2)(b^2 - \xi_3^2)}{b^2(b^2 - a^2)}$$

here

$$0 \leqslant \xi_3^2 < a^2 < \xi_2^2 < b^2 < \xi_1^2 < \infty$$

$$\frac{x^2}{\xi_1^2} + \frac{y^2}{\xi_1^2 - a^2} + \frac{z^2}{\xi_1^2 - b^2} = 1 \qquad \text{(Ellipsoids, } \xi_1 = \text{constant)}$$

$$\frac{x^2}{\xi_2^2} + \frac{y^2}{\xi_2^2 - a^2} - \frac{z^2}{b^2 - \xi_2^2} = 1 \qquad \text{(Hyperboloids of one sheet, } \xi_2 = \text{constant)}$$

$$\frac{x^2}{\xi_3^2} - \frac{y^2}{a^2 - \xi_3^2} - \frac{z^2}{b^2 - \xi_3^2} = 1 \qquad \text{(Hyperboloids of two sheets, } \xi_3 = \text{constant)}$$

$$h_1 = \left[\frac{(\xi_1^2 - \xi_2^2)(\xi_1^2 - \xi_3^2)}{(\xi_1^2 - a^2)(\xi_1^2 - b^2)}\right]^{1/2}, \quad h_2 = \left[\frac{(\xi_2^2 - \xi_3^2)(\xi_1^2 - \xi_2^2)}{(\xi_2^2 - a^2)(b^2 - \xi_2^2)}\right]^{1/2},$$

$$h_3 = \left[\frac{(\xi_1^2 - \xi_3^2)(\xi_2^2 - \xi_3^2)}{(a^2 - \xi_3^2)(b^2 - \xi_3^2)}\right]^{1/2}$$

PARABOLOIDAL COORDINATES

The coordinate surfaces in this system consist of two sets of paraboloids formed by elliptically sweeping the parabolas about the z-axis. Here we denote the set opening in the positive z-direction by ξ_1 = constant, and the set opening in the negative z-direction by ξ_2 = constant. The remaining coordinate surfaces consist of hyperbolic paraboloids (ξ_3 = constant). Note that as in the preceding example, geometrical interpretation can be aided somewhat by considering the previously discussed parabolic coordinates as a limiting case of paraboloidal coordinates, wherein rather than sweeping the parabolas in an ellipse, they are swept in a circle. In such a circumstance the hyperbolic parabolas degenerate to half planes in the parabolic cylinder system. Paraboloidal coordinates are related to Cartesian coordinates by

$$x^2 = 4\frac{(\xi_2 - a)}{(a - b)}(a - \xi_1)(a - \xi_3)$$

$$y^2 = 4\frac{(\xi_2 - b)}{(a - b)}(b - \xi_1)(\xi_3 - b)$$

$$z = \xi_1 + \xi_2 + \xi_3 - a - b$$

here

$$0 < \xi_1 < b < \xi_3 < a < \xi_2$$

$$\frac{x^2}{a - \xi_1} + \frac{y^2}{b - \xi_1} = 4(z - \xi_1) \qquad \text{(Elliptic paraboloid, } \xi_1 = \text{constant)}$$

$$\frac{x^2}{\xi_2 - a} + \frac{y^2}{\xi_2 - b} = -4(z - \xi_2) \qquad \text{(Elliptic paraboloid, } \xi_2 = \text{constant)}$$

$$\frac{x^2}{a - \xi_3} - \frac{y^2}{\xi_3 - b} = 4(z - \xi_3) \qquad \text{(Hyperbolic paraboloid, } \xi_3 = \text{constant)}$$

$$h_1 = \left[\frac{(\xi_2 - \xi_1)(\xi_2 - \xi_3)}{(\xi_2 - a)(\xi_2 - b)}\right]^{1/2}, \quad h_2 = \left[\frac{(\xi_2 - \xi_1)(\xi_3 - \xi_1)}{(a - \xi_1)(b - \xi_1)}\right]^{1/2},$$

$$h_3 = \left[\frac{(\xi_3 - \xi_1)(\xi_2 - \xi_3)}{(a - \xi_3)(\xi_3 - b)}\right]^{1/2}$$

3.6.2 Formulas from Vector Algebra

$$\mathbf{A} \cdot \mathbf{B} = A_i B_i$$

$$\mathbf{A} \times \mathbf{B} = \begin{vmatrix} \mathbf{e}_1 & \mathbf{e}_2 & \mathbf{e}_3 \\ A_1 & A_2 & A_3 \\ B_1 & B_2 & B_3 \end{vmatrix} \equiv e_{ijk} A_j B_k$$

$$\mathbf{A} \cdot (\mathbf{B} \times \mathbf{C}) = \mathbf{B} \cdot (\mathbf{C} \times \mathbf{A}) = \mathbf{C} \cdot (\mathbf{A} \times \mathbf{B}) = \begin{vmatrix} A_1 & A_2 & A_3 \\ B_1 & B_2 & B_3 \\ C_1 & C_2 & C_3 \end{vmatrix} \equiv e_{ijk} A_i B_j C_k$$

$$\mathbf{A} \times (\mathbf{B} \times \mathbf{C}) = (\mathbf{A} \cdot \mathbf{C})\,\mathbf{B} - (\mathbf{A} \cdot \mathbf{B})\,\mathbf{C}$$

$$(\mathbf{A} \times \mathbf{B}) \times \mathbf{C} = (\mathbf{A} \cdot \mathbf{C})\,\mathbf{B} - (\mathbf{B} \cdot \mathbf{C})\,\mathbf{A}$$

$$(\mathbf{A} \times \mathbf{B}) \cdot (\mathbf{C} \times \mathbf{D}) = (\mathbf{A} \cdot \mathbf{C})(\mathbf{B} \cdot \mathbf{D}) - (\mathbf{A} \cdot \mathbf{D})(\mathbf{B} \cdot \mathbf{C})$$

$$(\mathbf{A} \times \mathbf{B}) \times (\mathbf{C} \times \mathbf{D}) = [(\mathbf{A} \times \mathbf{B}) \cdot \mathbf{D}]\,\mathbf{C} - [(\mathbf{A} \times \mathbf{B}) \cdot \mathbf{C}]\,\mathbf{D}$$

3.6.3 Expansion of Operators in General Orthogonal Curvilinear Cooridnates

We use the general orthogonal curvilinear coordinate system (ξ_1, ξ_2, ξ_3). The length element is given in terms of the elemental coordinate changes, unit vectors

$\mathbf{e}_1$, $\mathbf{e}_2$, and $\mathbf{e}_3$, and the scale factors h_1, h_2 and h_3, where

$$d\mathbf{l} = \sum_{i=1}^{3} h_i d\xi_i \mathbf{e}_i$$

$$h_k^2 = \sum_{i=1}^{3} \left(\frac{\partial x_i}{\partial \xi_k}\right)^2 = \left[\sum_{i=1}^{3} \left(\frac{\partial \xi_k}{\partial x_i}\right)^2\right]^{-1}$$

The rates of change of the unit vectors in the coordinate directions are given by

$$\frac{\partial \mathbf{e}_1}{\partial \xi_1} = -\frac{\mathbf{e}_2}{h_2}\frac{\partial h_1}{\partial \xi_2} - \frac{\mathbf{e}_3}{h_3}\frac{\partial h_1}{\partial \xi_3}; \quad \frac{\partial \mathbf{e}_1}{\partial \xi_2} = \frac{\mathbf{e}_2}{h_1}\frac{\partial h_2}{\partial \xi_1}; \quad \frac{\partial \mathbf{e}_1}{\partial \xi_3} = \frac{\mathbf{e}_3}{h_1}\frac{\partial h_3}{\partial \xi_1}$$

$$\frac{\partial \mathbf{e}_2}{\partial \xi_2} = -\frac{\mathbf{e}_1}{h_1}\frac{\partial h_2}{\partial \xi_1} - \frac{\mathbf{e}_3}{h_3}\frac{\partial h_2}{\partial \xi_3}; \quad \frac{\partial \mathbf{e}_2}{\partial \xi_1} = \frac{\mathbf{e}_1}{h_2}\frac{\partial h_1}{\partial \xi_2}; \quad \frac{\partial \mathbf{e}_2}{\partial \xi_3} = \frac{\mathbf{e}_3}{h_2}\frac{\partial h_3}{\partial \xi_2}$$

$$\frac{\partial \mathbf{e}_3}{\partial \xi_3} = -\frac{\mathbf{e}_1}{h_1}\frac{\partial h_3}{\partial \xi_1} - \frac{\mathbf{e}_2}{h_2}\frac{\partial h_3}{\partial \xi_2}; \quad \frac{\partial \mathbf{e}_3}{\partial \xi_1} = \frac{\mathbf{e}_1}{h_3}\frac{\partial h_1}{\partial \xi_3}; \quad \frac{\partial \mathbf{e}_3}{\partial \xi_2} = \frac{\mathbf{e}_2}{h_3}\frac{\partial h_2}{\partial \xi_3}$$

3.6.3.1 Operators in General Orthogonal Curvilinear Coordinates

GRADIENT F:

$$\nabla F = \sum_{i=1}^{3} \frac{1}{h_i}\frac{\partial F}{\partial \xi_i}\mathbf{e}_i$$

DIVERGENCE $\mathbf{F}$:

$$\nabla \cdot \mathbf{F} = \frac{1}{h_1 h_2 h_3}\left[\frac{\partial F_1 h_2 h_3}{\partial \xi_1} + \frac{\partial F_2 h_1 h_3}{\partial \xi_2} + \frac{\partial F_3 h_1 h_2}{\partial \xi_3}\right]$$

CURL $\mathbf{F}$:

$$\nabla \times \mathbf{F} = \frac{1}{h_2 h_3}\left(\frac{\partial F_3 h_3}{\partial \xi_2} - \frac{\partial F_2 h_2}{\partial \xi_3}\right)\mathbf{e}_1 + \frac{1}{h_1 h_3}\left(\frac{\partial F_1 h_1}{\partial \xi_3} - \frac{\partial F_3 h_3}{\partial \xi_1}\right)\mathbf{e}_2 + \frac{1}{h_1 h_2}\left(\frac{\partial F_2 h_2}{\partial \xi_1} - \frac{\partial F_1 h_1}{\partial \xi_2}\right)\mathbf{e}_3$$

alternatively

$$(\nabla \times \mathbf{F})_i = \frac{e_{ijk}}{h_1 h_2 h_3} h_i \frac{\partial (Fh)_k}{\partial \xi_j}$$

CONVECTIVE OPERATOR $(\mathbf{a} \cdot \nabla)\,\mathbf{F}$:

$$(\mathbf{a} \cdot \nabla)\,\mathbf{F} = \mathbf{e}_1 \left[\frac{a_1}{h_1}\frac{\partial F_1}{\partial \xi_1} + \frac{a_2}{h_2}\frac{\partial F_1}{\partial \xi_2} + \frac{a_3}{h_3}\frac{\partial F_1}{\partial \xi_3} + \frac{a_1}{h_1}\left(\frac{F_2}{h_2}\frac{\partial h_1}{\partial \xi_2} + \frac{F_3}{h_3}\frac{\partial h_1}{\partial \xi_3}\right)\right.$$

$$\left. - \frac{a_2}{h_2}\frac{F_2}{h_1}\frac{\partial h_2}{\partial \xi_1} - \frac{a_3}{h_3}\frac{F_3}{h_1}\frac{\partial h_3}{\partial \xi_1}\right]$$

$$+ \mathbf{e}_2 \left[\frac{a_1}{h_1}\frac{\partial F_2}{\partial \xi_1} + \frac{a_2}{h_2}\frac{\partial F_2}{\partial \xi_2} + \frac{a_3}{h_3}\frac{\partial F_2}{\partial \xi_3} + \frac{a_2}{h_2}\left(\frac{F_1}{h_1}\frac{\partial h_2}{\partial \xi_1} + \frac{F_3}{h_3}\frac{\partial h_2}{\partial \xi_3}\right)\right.$$

$$\left. - \frac{a_1}{h_1}\frac{F_1}{h_2}\frac{\partial h_1}{\partial \xi_2} - \frac{a_3}{h_3}\frac{F_3}{h_2}\frac{\partial h_3}{\partial \xi_2}\right]$$

$$+ \mathbf{e}_3 \left[\frac{a_1}{h_1}\frac{\partial F_3}{\partial \xi_3} + \frac{a_2}{h_2}\frac{\partial F_3}{\partial \xi_2} + \frac{a_3}{h_3}\frac{\partial F_3}{\partial \xi_3} + \frac{a_3}{h_3}\left(\frac{F_1}{h_1}\frac{\partial h_3}{\partial \xi_1} + \frac{F_2}{h_2}\frac{\partial h_3}{\partial \xi_2}\right)\right.$$

$$\left. - \frac{a_1}{h_1}\frac{F_1}{h_3}\frac{\partial h_1}{\partial \xi_3} - \frac{a_2}{h_2}\frac{F_2}{h_3}\frac{\partial h_2}{\partial \xi_3}\right]$$

alternatively

$$[(\mathbf{a} \cdot \nabla)\,\mathbf{F}]_j = \frac{a_i}{h_i}\frac{\partial F_j}{\partial \xi_i} + \frac{F_i}{h_i h_j}\left(a_j \frac{\partial h_j}{\partial \xi_i} - a_i \frac{\partial h_i}{\partial \xi_j}\right) \qquad (j \text{ not summed})$$

3.6.4 Successive Operations in General Orthogonal Curvilinear Coordinates

LAPLACIAN OF A SCALAR, DIV GRAD $F \equiv \nabla^2 F$

$$\nabla^2 F = \frac{1}{h_1 h_2 h_3}\left[\frac{\partial}{\partial \xi_1}\left(\frac{h_2 h_3}{h_1}\frac{\partial F}{\partial \xi_1}\right) + \frac{\partial}{\partial \xi_2}\left(\frac{h_1 h_3}{h_2}\frac{\partial F}{\partial \xi_2}\right) + \frac{\partial}{\partial \xi_3}\left(\frac{h_1 h_2}{h_3}\frac{\partial F}{\partial \xi_3}\right)\right]$$

CURL OF THE GRADIENT OF A SCALAR, $\nabla \times \nabla F$

$$\nabla \times \nabla F = 0$$

GRADIENT OF THE DIVERGENCE OF A VECTOR, $\nabla(\nabla \cdot \mathbf{F})$

$$\nabla(\nabla \cdot \mathbf{F}) = \frac{1}{h_i}\frac{\partial}{\partial \xi_i}\left\{\frac{1}{h_1 h_2 h_3}\left[\frac{\partial F_1 h_2 h_3}{\partial \xi_1} + \frac{\partial F_2 h_1 h_3}{\partial \xi_2} + \frac{\partial F_3 h_1 h_2}{\partial \xi_3}\right]\right\}\mathbf{e}_i$$

DIVERGENCE OF THE CURL OF A VECTOR, $\nabla \cdot (\nabla \times \mathbf{F})$

$$\nabla \cdot (\nabla \times \mathbf{F}) = 0$$

CURL OF THE CURL OF A VECTOR, $\nabla \times (\nabla \times \mathbf{F})$

$$\begin{aligned}\nabla \times (\nabla \times \mathbf{F}) = \frac{\mathbf{e}_1}{h_2 h_3}\Bigg\{&\frac{\partial}{\partial \xi_2}\left[\frac{h_3}{h_1 h_2}\left(\frac{\partial F_2 h_2}{\partial \xi_1} - \frac{\partial F_1 h_1}{\partial \xi_2}\right)\right] \\ &- \frac{\partial}{\partial \xi_3}\left[\frac{h_2}{h_1 h_3}\left(\frac{\partial F_1 h_1}{\partial \xi_3} - \frac{\partial F_3 h_3}{\partial \xi_1}\right)\right]\Bigg\} \\ + \frac{\mathbf{e}_2}{h_1 h_3}\Bigg\{&\frac{\partial}{\partial \xi_3}\left[\frac{h_1}{h_2 h_3}\left(\frac{\partial F_3 h_3}{\partial \xi_2} - \frac{\partial F_2 h_2}{\partial \xi_3}\right)\right] \\ &- \frac{\partial}{\partial \xi_1}\left[\frac{h_3}{h_1 h_2}\left(\frac{\partial F_2 h_2}{\partial \xi_1} - \frac{\partial F_1 h_1}{\partial \xi_2}\right)\right]\Bigg\} \\ + \frac{\mathbf{e}_3}{h_1 h_2}\Bigg\{&\frac{\partial}{\partial \xi_1}\left[\frac{h_2}{h_1 h_3}\left(\frac{\partial F_1 h_1}{\partial \xi_3} - \frac{\partial F_3 h_3}{\partial \xi_1}\right)\right] \\ &- \frac{\partial}{\partial \xi_2}\left[\frac{h_1}{h_2 h_3}\left(\frac{\partial F_3 h_3}{\partial \xi_2} - \frac{\partial F_2 h_2}{\partial \xi_3}\right)\right]\Bigg\}\end{aligned}$$

alternatively

$$[\nabla \times (\nabla \times \mathbf{F})]_i = e_{ijk}\, e_{kst} \frac{h_i}{h_1 h_2 h_3} \frac{\partial}{\partial \xi_j}\left[\frac{h_k^2}{h_1 h_2 h_3} \frac{\partial (Fh)_t}{\partial \xi_s}\right]$$

LAPLACIAN OF A VECTOR, $\nabla^2 \mathbf{F}$

This operation is defined by the equation

$$\nabla^2 \mathbf{F} = \nabla(\nabla \cdot \mathbf{F}) - \nabla \times (\nabla \times \mathbf{F})$$

The components of $\nabla^2 \mathbf{F}$ may be obtained by subtracting the components of $\nabla \times (\nabla \times \mathbf{F})$ from $\nabla(\nabla \cdot \mathbf{F})$ as given above.

3.6.5 Integral Theorems

GAUSS' DIVERGENCE THEOREM

$$\iiint_V \nabla \cdot \mathbf{F}\, dv = \iint_s \mathbf{F} \cdot d\mathbf{s}$$

GRADIENT THEOREM

$$\iiint_V \nabla F\, dv = \iint_s F\, d\mathbf{s}$$

STOKES' THEOREM

$$\iint_s (\nabla \times \mathbf{F}) \cdot d\mathbf{s} = \int_c \mathbf{F} \cdot d\mathbf{l}$$

GENERALIZED FORM OF GAUSS' DIVERGENCE THEOREM

$$\iiint_V [(\mathbf{a} \cdot \nabla)\,\mathbf{F} + \mathbf{F}(\nabla \cdot \mathbf{a})]\; dv = \iint_s \mathbf{F}(\mathbf{a} \cdot d\mathbf{s})$$

GREEN'S THEOREMS

$$\iiint_V [\nabla\psi \cdot \nabla\phi + \psi\nabla^2\phi]\; dv = \iint_s \psi\nabla\phi \cdot d\mathbf{s}$$

$$\iiint_V [\psi\nabla^2\phi - \phi\nabla^2\psi]\; dv = \iint_s [\psi\nabla\phi - \phi\nabla\psi] \cdot d\mathbf{s}$$

GREEN'S LEMMA

$$\int_{\xi_2}\int_{\xi_1}\left[\frac{\partial h_2 Q}{\partial \xi_1} - \frac{\partial h_1 P}{\partial \xi_2}\right] d\xi_1 d\xi_2 = \oint_l [Qh_2 d\xi_2 + Ph_1 d\xi_1]$$

3.6.6 Operations on Sums and Products

$$\nabla(\phi + \psi) = \nabla\phi + \nabla\psi$$

$$(\mathbf{a} \cdot \nabla)(\mathbf{F} + \mathbf{G}) = (\mathbf{a} \cdot \nabla)\,\mathbf{F} + (\mathbf{a} \cdot \nabla)\,\mathbf{G}$$

$$\nabla \cdot (\mathbf{F} + \mathbf{G}) = \nabla \cdot \mathbf{F} + \nabla \cdot \mathbf{G}$$

$$\nabla \times (\mathbf{F} + \mathbf{G}) = \nabla \times \mathbf{F} + \nabla \times \mathbf{G}$$

$$\nabla(\phi\psi) = \psi\nabla\phi + \phi\nabla\psi$$

$$\nabla(\mathbf{F} \cdot \mathbf{G}) = (\mathbf{F} \cdot \nabla)\,\mathbf{G} + (\mathbf{G} \cdot \nabla)\,\mathbf{F} + \mathbf{F} \times (\nabla \times \mathbf{G}) + \mathbf{G} \times (\nabla \times \mathbf{F})$$

$$(\mathbf{a} \cdot \nabla)(\phi\mathbf{F}) = \phi[(\mathbf{a} \cdot \nabla)\,\mathbf{F}] + \mathbf{F}[(\mathbf{a} \cdot \nabla)\phi] \equiv \phi[(\mathbf{a} \cdot \nabla)\mathbf{F}] + \mathbf{F}[\mathbf{a} \cdot (\nabla\phi)]$$

$$(\mathbf{a} \cdot \nabla)(\mathbf{F} \times \mathbf{G}) = \mathbf{F} \times [(\mathbf{a} \cdot \nabla)\mathbf{G}] + [(\mathbf{a} \cdot \nabla)\mathbf{F}] \times \mathbf{G}$$

$$\nabla \cdot (\phi\mathbf{F}) = \mathbf{F} \cdot \nabla\phi + \phi\nabla \cdot \mathbf{F}$$

$$\nabla \cdot (\mathbf{F} \times \mathbf{G}) = \mathbf{G} \cdot (\nabla \times \mathbf{F}) - \mathbf{F} \cdot (\nabla \times \mathbf{G})$$

$$\nabla \times (\phi\mathbf{F}) = \nabla\phi \times \mathbf{F} + \phi\nabla \times \mathbf{F}$$

$$\nabla \times (\mathbf{F} \times \mathbf{G}) = (\mathbf{G} \cdot \nabla)\,\mathbf{F} - (\mathbf{F} \cdot \nabla)\,\mathbf{G} + \mathbf{F}\,\nabla \cdot \mathbf{G} - \mathbf{G}\,\nabla \cdot \mathbf{F}$$

3.7 BIBLIOGRAPHY

Batchelor, G. K., *An Introduction to Fluid Dynamics*, Cambridge Univ. Press., New York, 1967.

Moon, P., and Spencer, D. E., *Field Theory Handbook*, Springer-Verlag, Berlin, 1961.

Morse, P. M., and Feshbach, H., *Methods of Theoretical Physics*, McGraw-Hill, New York, 1953.

Panofsky, W. K. H., and Phillips, M., *Classical Electricity and Magnetism*, Addison-Wesley, Reading, Mass., 1956.

Pearson, C. E., *Theoretical Elasticity*, Harvard Univ. Press, Cambridge, Mass., 1959.

Spiegel, M. R., *Vector Analysis*, Schaum's Outline Series, McGraw-Hill, New York, 1959.

Wayland, H., *Differential Equations Applied in Science and Engineering*, Van Nostrand Reinhold, New York, 1958.

4

Tensors

Bernard Budiansky*

4.0 INTRODUCTION

The plan of this chapter is as follows: With reliance on the reader's understanding of ordinary vector algebra and analysis, the vector concept is redeveloped in the spirit of tensor theory for Euclidean three-dimensional space (3-D). This leads to the treatment of higher-order tensors and their behavior in general curvilinear coordinate systems in 3-D. The theory of two-dimensional (2-D) surfaces is discussed next. There follows a section on continuum mechanics, intended to display the power of tensor theory in a particular field of application, and the next section exhibits some interpretations and analogues of tensor results in classical, nontensorial notation. Finally, a very brief excursion into geometry and tensors in n-dimensional hyperspaces concludes the chapter.

The approach, on the whole, is expository, intuitive, and pragmatic, with little attention paid to mathematical pathology. The intent, throughout, is to convey to the reader a sense of how the basic ideas of tensor theory can be developed in a plausible and natural fashion, rather than to compile a comprehensive collection of definitions, postulates and theorems.

Some books on tensors and differential geometry are listed in the bibliography, as is the monograph on mechanics by Sedov, whose polyadic notation for tensors in Euclidean spaces is exploited in the present article.

4.1 VECTORS IN EUCLIDEAN 3-D

4.1.1 Orthonormal Base Vectors

Let $(\mathbf{e}_1, \mathbf{e}_2, \mathbf{e}_3)$ be a right-handed set of three mutually perpendicular vectors of unit magnitude. These *orthonormal* vectors will be used as *base vectors* in the definition of the Cartesian components of arbitrary vectors.

Note that the $\mathbf{e}_i (i = 1, 2, 3)$ may be used to define the Kronecker delta δ_{ij} and the

*Prof. Bernard Budiansky, Div. of Engineering and Applied Physics, Harvard University, Camb., Mass.

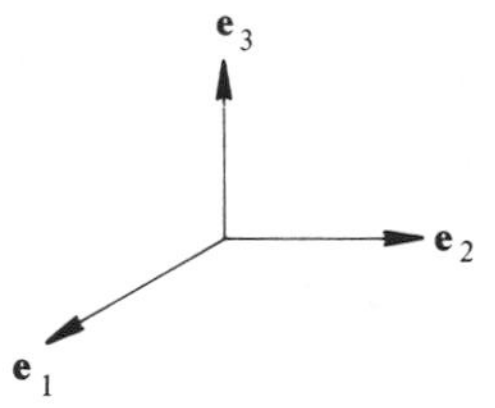

permutation symbol e_{ijk} by means of the equations

$$\mathbf{e}_i \cdot \mathbf{e}_j = \delta_{ij} \tag{4.1-1}$$

and

$$\mathbf{e}_i \times \mathbf{e}_j \cdot \mathbf{e}_k = e_{ijk} \tag{4.1-2}$$

Thus $\delta_{ij} = 1$ for $i = j$; $\delta_{ij} = 0$ for $i \neq j$; $\mathbf{e}_{123} = \mathbf{e}_{312} = \mathbf{e}_{231} = 1$; $\mathbf{e}_{132} = \mathbf{e}_{321} = \mathbf{e}_{213} = -1$; all other $\mathbf{e}_{ijk} = 0$. In turn, a pair of e_{ijk}'s is related to a determinant of δ_{ij}'s by

$$e_{ijk}e_{rst} = \begin{vmatrix} \delta_{ir} & \delta_{is} & \delta_{it} \\ \delta_{jr} & \delta_{js} & \delta_{jt} \\ \delta_{kr} & \delta_{ks} & \delta_{kt} \end{vmatrix} \tag{4.1-3}$$

By setting $r = i$ we recover the $e - \delta$ relation

$$e_{ijk}e_{ist} = \delta_{js}\delta_{kt} - \delta_{jt}\delta_{ks} \tag{4.1-4}$$

4.1.2 Cartesian Components of Vectors; Transformation Rule

The Cartesian components F_i of a vector $\mathbf{F}$, referred to the orthonormal base vectors $\mathbf{e}_i$, are defined, equivalently, by either the projection formula

$$F_i = \mathbf{F} \cdot \mathbf{e}_i \qquad i = 1, 2, 3 \tag{4.1-5}$$

or the composition formula

$$\mathbf{F} = F_i \mathbf{e}_i \tag{4.1-6}$$

[In (4.1-6) the summation convention for repeated indices defined in Chapter 3 is used, as it will be hereafter.]

Suppose a new, rotated set of orthonormal base vectors $\bar{\mathbf{e}}_i (i = 1, 2, 3)$ is introduced. The corresponding new components $\bar{F}_i$ of $\mathbf{F}$ may be computed in terms of the old ones by writing

$$\bar{F}_j = \mathbf{F} \cdot \bar{\mathbf{e}}_j = F_i \mathbf{e}_i \cdot \bar{\mathbf{e}}_j$$

Hence, we have the transformation rule

$$\bar{F}_j = l_{ij} F_i \tag{4.1-7}$$

where

$$l_{ij} \equiv \mathbf{e}_i \cdot \bar{\mathbf{e}}_j \tag{4.1-8}$$

represents the *direction cosine* of the angle between the direction of the old vector $\mathbf{e}_i$ and the new base vector $\bar{\mathbf{e}}_j$. These direction cosines satisfy the useful relations

$$l_{ip} l_{jp} = l_{pi} l_{pj} = \delta_{ij} \tag{4.1-9}$$

4.1.3 General Base Vectors

The idea of defining vector components with respect to a triad of base vectors need not be restricted to the use of orthonormal vectors. Let $\epsilon_1, \epsilon_2, \epsilon_3$ be any three noncoplanar vectors that will play the roles of general base vectors. For consistency and convenience, number these base vectors so that they constitute a right-handed

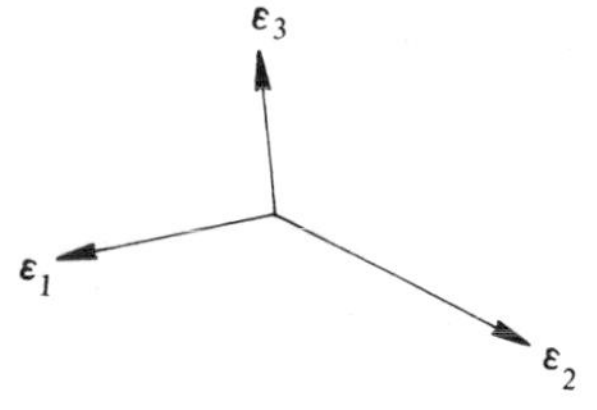

set, meaning that $\epsilon_1 \times \epsilon_2 \cdot \epsilon_3 > 0$. The relations (4.1-1) and (4.1-2) among orthonormal base vectors generalize to the forms

$$\epsilon_i \cdot \epsilon_j = g_{ij} \tag{4.1-10}$$

and

$$\epsilon_i \times \epsilon_j \cdot \epsilon_k = \epsilon_{ijk} \tag{4.1-11}$$

which serve to define the important quantities g_{ij} and ϵ_{ijk}, called the *metric tensor* and the *permutation tensor*, respectively. A relation analogous to (4.1-3) can also be produced. From (4.1-3) the general vector identity

$$(\mathbf{A} \times \mathbf{B} \cdot \mathbf{C})(\mathbf{D} \times \mathbf{E} \cdot \mathbf{F}) = \begin{vmatrix} \mathbf{A}\cdot\mathbf{D} & \mathbf{A}\cdot\mathbf{E} & \mathbf{A}\cdot\mathbf{F} \\ \mathbf{B}\cdot\mathbf{D} & \mathbf{B}\cdot\mathbf{E} & \mathbf{B}\cdot\mathbf{F} \\ \mathbf{C}\cdot\mathbf{D} & \mathbf{C}\cdot\mathbf{E} & \mathbf{C}\cdot\mathbf{F} \end{vmatrix} \tag{4.1-12}$$

can be established; applying this to the base vectors gives

$$\epsilon_{ijk}\epsilon_{rst} = \begin{vmatrix} g_{ir} & g_{is} & g_{it} \\ g_{jr} & g_{js} & g_{jt} \\ g_{kr} & g_{ks} & g_{kt} \end{vmatrix} \tag{4.1-13}$$

Note that the volume v of the parallelopiped having the base vectors as edges equals $\boldsymbol{\epsilon}_1 \times \boldsymbol{\epsilon}_2 \cdot \boldsymbol{\epsilon}_3$; consequently the permutation tensor and the permutation symbol are related by

$$\epsilon_{ijk} = v\, e_{ijk}$$

Denote by g the determinant of the matrix $[g]$ having g_{ij} as its $(i, j)^{\text{th}}$ element; then, by (4.1-13), $(\epsilon_{123})^2 = g = v^2$, so that

$$\epsilon_{ijk} = \sqrt{g}\, e_{ijk} \tag{4.1-14}$$

4.1.4 General Components of Vectors; Transformation Rules

One natural way to define the components of an arbitrary vector $\mathbf{F}$ with respect to the general base vectors $\boldsymbol{\epsilon}_i$ is to follow (4.1-5) and write

$$F_i \equiv \mathbf{F} \cdot \boldsymbol{\epsilon}_i \qquad i = 1, 2, 3 \tag{4.1-15}$$

These components F_i, in context, will always be referred to a specific set of $\boldsymbol{\epsilon}_i$, so that a special notation is usually not needed to distinguish them from the Cartesian components defined by eq. (4.1-5). However, a different kind of component emerges from a generalization of the composition formula (4.1-6); thus we write

$$\mathbf{F} = F^1 \boldsymbol{\epsilon}_1 + F^2 \boldsymbol{\epsilon}_2 + F^3 \boldsymbol{\epsilon}_3 \tag{4.1-16}$$

and this defines components $F^i (i = 1, 2, 3)$ that are not necessarily the same as the F_i. If we retain the summation convention for repeated indices, regardless of whether they are superscripts or subscripts, the formula (4.1-16) is

$$\mathbf{F} = F^i \boldsymbol{\epsilon}_i \tag{4.1-17}$$

For historical reasons (now unimportant) the F_i are called the *covariant* components of $\mathbf{F}$, and the F^i are the *contravariant* components. Clearly, Cartesian components are both covariant and contravariant.

For any given set of general base vectors, the two kinds of components can be related with the help of the metric tensor g_{ij}. Substituting (4.1-17) into (4.1-15) yields

$$F_i = g_{ij} F^j \tag{4.1-18}$$

These relations may be inverted. Use g^{ij} to denote the $(i, j)^{\text{th}}$ element of the inverse of the matrix $[g]$; then

$$g^{ip}g_{pj} = \delta^i_j \tag{4.1-19}$$

where the indices in the Kronecker delta have been placed in superscript and subscript position to conform to their placement on the left-hand side of the equation. Accordingly

$$F^i = g^{ij}F_j \tag{4.1-20}$$

(Note that the existence of the inverse of $[g]$ is guaranteed by the fact that $g = v^2 > 0$, so that $[g]$ is nonsingular.)

It will be seen that, when general vector (and tensor) components enter, the summation convention invariably applies to summation over a repeated index that appears once as a superscript and once as a subscript. For example,

$$\mathbf{F} \cdot \mathbf{F} = (F^i\epsilon_i) \cdot (F^j\epsilon_j) = F^iF^jg_{ij} = F^iF_i$$

The metric tensor g_{ij} can be used to introduce an auxiliary set of base vectors $\epsilon^i (i = 1, 2, 3)$ by means of the definition

$$\epsilon^i = g^{ij}\epsilon_j \tag{4.1-21}$$

It follows that $\epsilon_i \times \epsilon_j = \epsilon_{ijk}\,\epsilon^k$. The ϵ_i and ϵ^i are called, respectively, the covariant and contravariant base vectors, and the following relations are readily established:

$$\epsilon_i = g_{ij}\epsilon^j \tag{4.1-22}$$

$$\epsilon^i \cdot \epsilon_j = \delta^i_j \tag{4.1-23}$$

$$\epsilon^i \cdot \epsilon^j = g^{ij} \tag{4.1-24}$$

$$\mathbf{F} = F_i\epsilon^i \tag{4.1-25}$$

$$F^j = \mathbf{F} \cdot \epsilon^j \tag{4.1-26}$$

Equations (4.1-25) and (4.1-26) show how the use of the contravariant base vectors permits a reversal of the earlier roles of projection and composition in the definition of covariant and contravariant components of a vector.

Consider, finally, the question of calculating the new components of a vector with respect to a new set of base vectors $\bar{\epsilon}_i$; this question generalizes the one answered in section 4.1.2 for rotation of orthonormal base vectors. A direct calculation gives

$$\bar{F}_j = (F^i\epsilon_i) \cdot \bar{\epsilon}_j = F^i(\epsilon_i \cdot \bar{\epsilon}_j) \tag{4.1-27}$$

for the transformed covariant components. We also find easily that

$$\overline{F}_j = F_i(\epsilon^i \cdot \overline{\epsilon}_j) \tag{4.1-28}$$

$$\overline{F}^j = F^i(\epsilon_i \cdot \overline{\epsilon}^j) = F_i(\epsilon^i \cdot \overline{\epsilon}^j) \tag{4.1-29}$$

The relations (4.1-9) generalize to

$$(\epsilon^i \cdot \overline{\epsilon}_p)(\epsilon_j \cdot \overline{\epsilon}^p) = (\epsilon_p \cdot \overline{\epsilon}^i)(\epsilon^p \cdot \overline{\epsilon}_j) = \delta^i_j \tag{4.1-30}$$

4.2 TENSORS IN EUCLIDEAN 3-D

4.2.1 Dyads, Dyadics, and Second-Order Tensors

The mathematical object denoted by **AB**, where **A** and **B** are given vectors, is called a *dyad*. The meaning of **AB** is simply that the operation

$$(\mathbf{AB}) \cdot \mathbf{V}$$

where **V** is any vector, is understood to produce the vector

$$\mathbf{A}(\mathbf{B} \cdot \mathbf{V})$$

A sum of dyads, of the form

$$\mathbf{T} = \mathbf{AB} + \mathbf{CD} + \mathbf{EF} + \cdots$$

is called a *dyadic*, and this just means that

$$\mathbf{T} \cdot \mathbf{V} = \mathbf{A}(\mathbf{B} \cdot \mathbf{V}) + \mathbf{C}(\mathbf{D} \cdot \mathbf{V}) + \mathbf{E}(\mathbf{F} \cdot \mathbf{V}) + \cdots$$

Any dyadic can be expressed in terms of an arbitrary set of general base vectors $\boldsymbol{\epsilon}_i$; since

$$\mathbf{A} = A^i \boldsymbol{\epsilon}_i, \quad \mathbf{B} = B^i \boldsymbol{\epsilon}_i, \quad \mathbf{C} = C^i \boldsymbol{\epsilon}_i, \quad \ldots$$

it follows that

$$\mathbf{T} = A^i B^j \boldsymbol{\epsilon}_i \boldsymbol{\epsilon}_j + C^i D^j \boldsymbol{\epsilon}_i \boldsymbol{\epsilon}_j + E^i F^j \boldsymbol{\epsilon}_i \boldsymbol{\epsilon}_j + \cdots$$

Hence, **T** can always be written in the form

$$\mathbf{T} = T^{ij} \boldsymbol{\epsilon}_i \boldsymbol{\epsilon}_j \tag{4.2-1}$$

in terms of nine numbers T^{ij}.

A dyadic is the same as a *second-order tensor*, and the T^{ij} are called the *contravariant components* of the tensor. These components depend, of course, on the particular choice of base vectors. We re-emphasize the basic meaning of **T** by noting that, for all vectors **V**

$$\mathbf{T} \cdot \mathbf{V} \equiv T^{ij} \epsilon_i (\epsilon_j \cdot \mathbf{V})$$
$$= (T^{ij} V_j) \epsilon_i$$

Thus, if V_j is the jth covariant component of a vector, then $(T^{ij} V_j)$ is the ith contravariant component of another vector. Similarly, we can define the operation

$$\mathbf{V} \cdot \mathbf{T} \equiv T^{ij} (\mathbf{V} \cdot \epsilon_i) \epsilon_j$$
$$= (T^{ij} V_i) \epsilon_j$$

which produces yet another vector having $T^{ij} V_i$ as its contravariant components.

By introducing the contravariant base vectors ϵ^i we can define other kinds of components of the tensor **T**. Thus, substituting

$$\epsilon_i = g_{ip} \epsilon^p, \quad \epsilon_j = g_{jq} \epsilon^q$$

into (4.2-1), we get

$$\mathbf{T} = T_{pq} \epsilon^p \epsilon^q \tag{4.2-2}$$

where the nine quantities

$$T_{pq} \equiv g_{ip} g_{jq} T^{ij} \tag{4.2-3}$$

are called the *covariant* components of the tensor. Similarly, we can define two, generally different, kinds of *mixed* components

$$T^{i}_{\cdot j} \equiv g_{jq} T^{iq} \tag{4.2-4}$$

$$T_{i\cdot}^{\ j} \equiv g_{ip} T^{pj} \tag{4.2-5}$$

that appear in the representation

$$\mathbf{T} = T^{i}_{\cdot j} \epsilon_i \epsilon^j = T_{i\cdot}^{\ j} \epsilon^i \epsilon_j \tag{4.2-6}$$

Dots are often used in the mixed-component forms, as in (4.2-4) and (4.2-5), to remove any typographical ambiguity about which of the two indices, the upper or the lower, comes first.

4.2.2 Transformation Rules

Suppose new base vectors $\bar{\epsilon}_i$ are introduced; what are the new contravariant components $\bar{T}^{ij}$ of $\mathbf{T}$? Substitution of the representations

$$\left.\begin{aligned}\epsilon_i &= (\epsilon_i \cdot \bar{\epsilon}^p)\bar{\epsilon}_p \\ \epsilon_j &= (\epsilon_j \cdot \bar{\epsilon}^q)\bar{\epsilon}_q\end{aligned}\right\} \tag{4.2-7}$$

into (4.2-1) gives

$$\mathbf{T} = T^{ij}(\epsilon_i \cdot \bar{\epsilon}^p)\,(\epsilon_j \cdot \bar{\epsilon}^q)\bar{\epsilon}_p\bar{\epsilon}_q = \bar{T}^{pq}\bar{\epsilon}_p\bar{\epsilon}_q$$

whence

$$\bar{T}^{pq} = T^{ij}(\epsilon_i \cdot \bar{\epsilon}^p)\,(\epsilon_j \cdot \bar{\epsilon}^q) \tag{4.2-8}$$

is the desired transformation rule. Many different, but equivalent, relations are easily derived; for example

$$\bar{T}_{pq} = T^{ij}(\epsilon_i \cdot \bar{\epsilon}_p)\,(\epsilon_j \cdot \bar{\epsilon}_q) \tag{4.2-9}$$

$$\bar{T}^{p}_{\cdot q} = T_{ij}(\epsilon^i \cdot \bar{\epsilon}^p)\,(\epsilon^j \cdot \bar{\epsilon}_q) \tag{4.2-10}$$

and so on.

In many, perhaps most, expositions of tensor theory, the basic definition of a second-order tensor rests on the initial assertion that a transformation law like that given by (4.2-8) must apply to its components. This is, of course, entirely equivalent to the present development, in which the transformation law is derived, and in which the entire framework really relies on an intuitive geometrical perception of the properties of vectors.

4.2.3 Cartesian Components of Second-Order Tensors

In many applications, it suffices to restrict the choice of base vectors to orthonormal ones. As in the case of vectors, the distinction between covariant and contravariant (and mixed) components then disappears, and the tensor components are then usually written with the indices in suffix position. Thus, in terms of the Cartesian base vectors $\mathbf{e}_i$, the tensor $\mathbf{T}$ may be written

$$\mathbf{T} = T_{ij}\mathbf{e}_i\mathbf{e}_j \tag{4.2-11}$$

wherein the T_{ij} are called Cartesian components. The transformation rule for switching to new Cartesian components $\bar{T}_{ij}$ when new Cartesian base vectors $\bar{\mathbf{e}}_i$ are introduced becomes

$$\overline{T}_{ij} = T_{pq} l_{pi} l_{qj} \tag{4.2-12}$$

where the direction cosines l_{ij} are given, as before, by eq. (4.1-8).

4.2.4 Terminology and Grammar

It is customary, and useful, to substitute the phrase "the tensor T^{ij}" for the more precise construction "the contravariant components T^{ij} of the tensor **T**." This is a convenient, harmless shorthand. Similarly, one speaks of "the vector V_i," or "the vector V^j." Perhaps less felicitous is the often-used characterization of T^{ij} as "a contravariant tensor," and of T_{ij} as "a covariant tensor." After all, T^{ij}, T_{ij}, $T_{i\cdot}^{\ j}$ and $T_{\cdot j}^{i}$ are just different kinds of components of the same mathematical object, namely **T**; but here too, the terminology need not be scorned if this is kept in mind. Similarly, if only Cartesian base vectors are involved, it is customary to call T_{ij} a "Cartesian tensor."

This kind of tolerance for sloppy grammar should not, however, be extended to the writing of equations in which modes of representation are mixed. Thus, although we may refer to "the vector V_i" as well as to the same "vector **V**," it would be well beyond the bounds of acceptable mathematical behavior ever to write, for example, $V_i = \mathbf{V}$.

4.2.5 Tensor Operations; Tests for Tensors

We have already noted that if T^{ij} is a tensor and V_i is a vector, then $T^{ij}V_i$ and $T^{ij}V_j$ are vectors. If W_i is another vector, the quantity $T^{ij}V_iW_j$ is a scalar quantity, independent of choice of base vectors. If S^{ij} is another tensor, we also have the following operations:

$$T^{ij}S_{jk} = P^{i}_{\cdot k} \qquad \text{(tensor)}$$

$$T^{ij}S_{ij} \qquad \text{(scalar)}$$

$$g_{ij}T^{ij} \equiv T_i^i \qquad \text{(scalar)}$$

(In the last term the suffix and superscript can be placed one above the other, without ambiguity.)

Some of these relations can serve as diagnostic tools for identifying the tensorial character of two-index quantities. Thus, if it is known that, for all vectors V_i, the operation $T^{ij}V_i$—for *all* base vectors—produces another vector W^j, it must follow that T^{ij} is a tensor. This is proved by showing that T^{ij} transforms like a second-order tensor under changes of base vectors. Since W^j is a vector

$$\begin{aligned}\overline{W}^j &= (\epsilon_q \cdot \bar{\epsilon}^j) T^{pq} V_p \\ &= (\epsilon_q \cdot \bar{\epsilon}^j)(\epsilon_p \cdot \bar{\epsilon}^s) T^{pq} \overline{V}_s\end{aligned}$$

But $\overline{W}^j = \overline{T}^{sj}\overline{V}_s$, and so, since V_s is arbitrary

$$\overline{T}^{sj} = (\epsilon_p \cdot \bar{\epsilon}^s)(\epsilon_q \cdot \epsilon^j)\, T^{pq}$$

which is the correct tensorial transformation rule; hence T^{ij} *is* a tensor.

Other theorems of this type—called *quotient laws*—are easily stated and proved. Thus, T^{ij} is a tensor if:

$T^{ij}X_iY_j$ is a scalar for all vectors X_i, Y_i;
$T^{ij}S_{ij}$ is a scalar for all tensors S_{ij};
$T^{ij}S_{jk}$ is a tensor for all tensors S_{jk};

and so on. On the other hand, if we are given that $T^{ij}X_iX_j$ is a scalar for all vectors X_i, we can conclude that T^{ij} is a tensor *only* if it is known that $T^{ij} = T^{ji}$; tensors of this kind, involving only six independent components, are called *symmetrical* tensors. (Note that symmetry with respect to one set of base vectors implies symmetry for all base vectors.)

4.2.6 The Metric Tensor g_{ij}

Earlier, the quantity $g_{ij} \equiv \epsilon_i \cdot \epsilon_j$ was called the metric *tensor*, and we can now establish its tensor character. One easy way to do this is to note that for all vectors X_i, Y_i, the quantity

$$g_{ij}X^iY^j \equiv X^iY_i = \mathbf{X} \cdot \mathbf{Y}$$

is a scalar—whence the tensor character of g_{ij} follows by the quotient law just discussed. It may be more instructive, however, actually to exhibit, in the form (4.2.2), the tensor having g_{ij} as its covariant components. Consider the tensor

$$\mathbf{g} = \epsilon^i\epsilon_i \tag{4.2-13}$$

to be defined in terms of a particular set of base vectors. Substituting $\epsilon_i = (\epsilon_i \cdot \bar{\epsilon}^p)\bar{\epsilon}_p$ gives

$$\mathbf{g} = \epsilon^i(\epsilon_i \cdot \bar{\epsilon}^p)\bar{\epsilon}_p = \bar{\epsilon}^p\bar{\epsilon}_p$$

So, g has the same form (4.2-13) for *any* choice of base vectors, and rewriting (4.2-13) as

$$\mathbf{g} = g_{ij}\epsilon^i\epsilon^j$$

shows that the $g_{ij} \equiv \epsilon_i \cdot \epsilon_j$ are indeed the covariant components of a tensor. Because its form (4.2-13) is invariant with respect to choice of base vectors, g is called an *isotropic* tensor.

4.2.7 N^{th}-Order Tensors

A third-order tensor, or *triadic*, is the sum of *triads*, as follows:

$$\mathbf{ABC} + \mathbf{DEF} + \mathbf{GHI} + \cdots$$

The meaning of this is that, for any vector **V**, the dot products

$$\mathbf{AB}(\mathbf{C} \cdot \mathbf{V}) + \mathbf{DE}(\mathbf{F} \cdot \mathbf{V}) + \mathbf{GH}(\mathbf{I} \cdot \mathbf{V}) + \cdots$$

or

$$\mathbf{A}(\mathbf{B} \cdot \mathbf{V})\mathbf{C} + \mathbf{D}(\mathbf{E} \cdot \mathbf{V})\mathbf{F} + \mathbf{G}(\mathbf{H} \cdot \mathbf{V})\mathbf{I} + \cdots$$

provide second-order tensors. It is easily established that any third-order tensor can be written

$$\mathbf{T} = T^{ijk}\epsilon_i\epsilon_j\epsilon_k \tag{4.2-14}$$

as well as in the alternative forms

$$T^{ij}_{\cdot\cdot k}\epsilon_i\epsilon_j\epsilon^k$$

$$T^{i}_{\cdot jk}\epsilon_i\epsilon^j\epsilon^k$$

and so on, where indices are lowered via the metric tensor.

The extension to N^{th}-order tensors is now immediate. A general tensor of order N-may be written in the polyadic form

$$\mathbf{T} = \underbrace{T^{ijk}_{\cdot\cdot\cdot lm}\cdots{}^{st}_{\cdot\cdot}}_{N\text{ indices}}\;\underbrace{\epsilon_i\epsilon_j\epsilon_k\epsilon^l\epsilon^m\cdots\epsilon_s\epsilon_t}_{N\text{ base vectors}} \tag{4.2-15}$$

and this means that the dot product of **T** with any vector **V** produces a tensor of order $(N - 1)$. (The dot product with **V** may be with respect to any one of the base vectors; unfortunately, the notations $\mathbf{V} \cdot \mathbf{T}$ and $\mathbf{T} \cdot \mathbf{V}$ are unambiguous only for second-order tensors, and should therefore be avoided for tensors of higher order.) Forms alternative to (4.2-15) are obtained in obvious ways by shifting the up or down locations of repeated indices.

The transformation rule for tensor components is

$$\begin{aligned}\overline{T}^{ijk}_{\cdot\cdot\cdot lm}\cdots{}^{st}_{\cdot\cdot} = T^{abc}_{\cdot\cdot\cdot de}\cdots{}^{fg}_{\cdot\cdot}\,(\epsilon_a \cdot \bar{\epsilon}^i)\,(\epsilon_b \cdot \bar{\epsilon}^j)\\ (\epsilon_c \cdot \bar{\epsilon}^k)\,(\epsilon^d \cdot \bar{\epsilon}_l)\,(\epsilon^e \cdot \bar{\epsilon}_m)\cdots(\epsilon_f \cdot \bar{\epsilon}^s)\,(\epsilon_g \cdot \bar{\epsilon}^t)\end{aligned} \tag{4.2-16}$$

The general form (4.2-15) shows that ordinary vectors are tensors of first order; scalars can be called tensors of zero order. A multitude of operations is possible with tensors of various order; for example

$A^{ijk}B_{kj}$ gives a vector;

$A^{ij}B_{kl\cdot}^{\;\;\;m}$ gives a tensor of fifth order;

$A^{ijk}_{\cdot\cdot\cdot kl}$ gives a tensor of third order;

and so on.

Many fairly obvious theorems of the quotient-law type may be stated and proved for tensors of Nth-order.

4.2.8 The Permutation Tensor ϵ_{ijk}

Choose a particular set of base vectors ϵ_i, and define the third-order tensor

$$\mathbf{E} = (\epsilon_i \times \epsilon_j \cdot \epsilon_k)\epsilon^i\epsilon^j\epsilon^k \tag{4.2-17}$$

Since $\epsilon_i\epsilon^i = \bar{\epsilon}_i\bar{\epsilon}^i$, it follows that $\mathbf{E}$ has the same form as (4.2-17) with respect to *all* sets of base vectors, and so $\epsilon_{ijk} \equiv \epsilon_i \times \epsilon_j \cdot \epsilon_k$ is indeed a tensor. Thus the various indices in the $\epsilon - g$ identities (4.1-12) and (4.1-13) can be shifted to contravariant position. Since the determinant of the g^{ij}'s is $1/g$, it follows that, in contrast to (4.1-14)

$$\epsilon^{ijk} = \frac{1}{\sqrt{g}}\, e^{ijk} \tag{4.2-18}$$

where the permutation symbol e^{ijk} in (4.2-18) has been written with its indices in superscript position, for consistency.

4.2.9 An Example: The Stress Tensor

The concept of *stress* is essentially what led to the invention of tensors (*tenseur* (Fr.) = tensor; *tension* (Fr.) = stress), and the *stress tensor* is the prime example of a

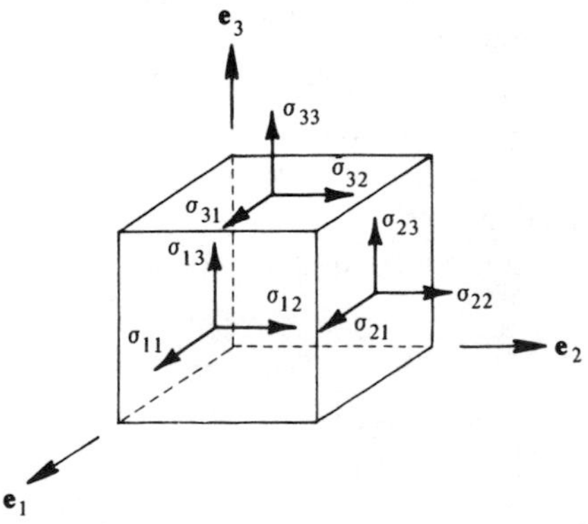

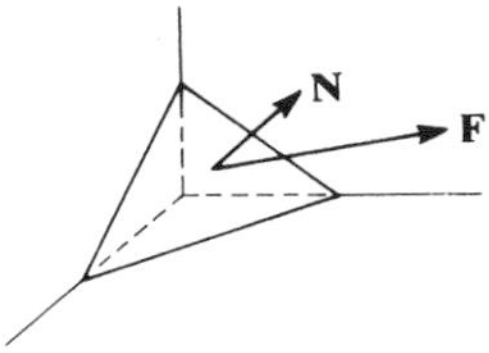

second-order tensor. Introduce Cartesian base vectors $\mathbf{e}_i$, and denote by σ_{ij} (or $\overset{c}{\sigma}_{ij}$) the jth component of the force-per-unit-area acting on the face of the small cube (see sketch) normal to the ith base vector. (The little c emphasizes the fact that the $\overset{c}{\sigma}_{ij}$ are Cartesian components.) Define the tensor

$$\sigma = \overset{c}{\sigma}_{ij}\mathbf{e}_i\mathbf{e}_j \tag{4.2-19}$$

Then, by static equilibrium, it follows from examination of the *Cauchy tetrahedron* (see sketch) that the stress vector **F** (force-per-unit-area) on the area element having normal **N** is just

$$\mathbf{F} = \overset{c}{\sigma}_{ij}(\mathbf{e}_i \cdot \mathbf{N})\mathbf{e}_j \tag{4.2-20}$$

Equation (4.2-20) shows that $\overset{c}{\sigma}_{ij}$ is a Cartesian tensor that transforms, under rotation of axes, according to the rule (4.2-12).

We can now introduce general base vectors ϵ_i, and write σ as

$$\sigma = \sigma^{ij}\epsilon_i\epsilon_j \tag{4.2-21}$$

in terms of the contravariant stress components σ^{ij}. The σ^{ij} lose significance as *physical* components of stress, but they can always be calculated in terms of the Cartesian components $\overset{c}{\sigma}_{ij}$ by means of (4.2.8). The contravariant components F^i of **F** are now given by

$$F^j = \sigma^{ij}N_i \tag{4.2-22}$$

in terms of σ^{ij} and the covariant components $N_i = \mathbf{N} \cdot \epsilon_i$ of the unit normal **N** to the surface area considered.

4.3 GENERAL CURVILINEAR COORDINATES IN EUCLIDEAN 3-D

4.3.1 Coordinate Systems and General Base Vectors

Until now, no hint has been given concerning an appropriate motivation for choosing any particular set of general base vectors. The choice, in applications of tensor analysis, is almost always tied in a special way to the general system of coordinates that is used to locate points in space. (The choice of coordinate system, in turn, is guided by such things as the shape of the region under consideration and the technique of solution to be used for solving boundary-value problems.)

Suppose that the general coordinates are (ξ^1, ξ^2, ξ^3); this means that the position vector $\mathbf{x}$ of a point is a known function of ξ^1, ξ^2, and ξ^3. Then the choice that is usually made for the base vectors is

$$\epsilon^i = \frac{\partial \mathbf{x}}{\partial \xi_i} \tag{4.3-1}$$

For consistency with the right-handedness of the ϵ_i, the coordinates ξ^i must be numbered in such a way that

$$\frac{\partial \mathbf{x}}{\partial \xi^1} \times \frac{\partial \mathbf{x}}{\partial \xi^2} \cdot \frac{\partial \mathbf{x}}{\partial \xi^3} > 0$$

As an example, consider the cylindrical coordinates

$$\begin{cases} \xi^1 = r \\ \xi^2 = \theta \\ \xi^3 = z \end{cases}$$

Then, with $\mathbf{x} = x^i \mathbf{e}_i$ in terms of the Cartesian coordinates x^i and the Cartesian base vectors $\mathbf{e}_i$ we have

$$\begin{cases} x^1 = r \cos \theta \\ x^2 = r \sin \theta \\ x^3 = z \end{cases}$$

and so

$$\begin{cases} \epsilon_1 = (\cos \theta)\mathbf{e}_1 + (\sin \theta)\mathbf{e}_2 \\ \epsilon_2 = (-r \sin \theta)\mathbf{e}_1 + (r \cos \theta)\mathbf{e}_2 \\ \epsilon_3 = \mathbf{e}_3 \end{cases}$$

Note that the magnitude of ϵ_2 depends on r. In general, both the magnitudes and directions of general base vectors now depend on position, and they need not, of course, be orthogonal, as they are for the case of cylindrical coordinates.

4.3.2 Metric Tensor and Jacobian

We have already seen that g_{ij} is a tensor; it will now be shown why it is called the *metric* tensor.

The definition (4.1-10), together with (4.3-1), gives

$$g_{ij} = \frac{\partial \mathbf{x}}{\partial \xi^i} \cdot \frac{\partial \mathbf{x}}{\partial \xi^j} \tag{4.3-2}$$

Now note that

$$d\mathbf{x} = \frac{\partial \mathbf{x}}{\partial \xi^i} d\xi^i \tag{4.3-3}$$

so that an element of arc length ds satisfies

$$(ds)^2 = d\mathbf{x} \cdot d\mathbf{x} = \frac{\partial \mathbf{x}}{\partial \xi^i} \cdot \frac{\partial \mathbf{x}}{\partial \xi^j} d\xi^i d\xi^j = g_{ij} d\xi^i d\xi^j \tag{4.3-4}$$

Thus, with the choice (4.3-1) for the base vectors, g_{ij} provides a set of factors for converting increments in ξ^i to changes in length.

Note that eq. (4.3-3) is the same as

$$d\mathbf{x} = (d\xi^i)\epsilon_i \tag{4.3-5}$$

whence $d\xi^i$ is the general contravariant component of the vector $d\mathbf{x}$. The Jacobian J of the transformation relating Cartesian coordinates and curvilinear coordinates is the determinant of the array $\partial x^i/\partial \xi^j$ $(i, j = 1, 2, 3)$, and the element of volume having the vectors

$$\left(\frac{\partial \mathbf{x}}{\partial \xi^1} d\xi^1\right), \quad \left(\frac{\partial \mathbf{x}}{\partial \xi^2} d\xi^2\right), \quad \left(\frac{\partial \mathbf{x}}{\partial \xi^3} d\xi^3\right)$$

as edges is

$$dV = J\, d\xi^1 d\xi^2 d\xi^3 = (\epsilon_1 \times \epsilon_2 \cdot \epsilon_3)\, d\xi^1 d\xi^2 d\xi^3 = \sqrt{g}\; d\xi^1 d\xi^2 d\xi^3$$

Hence

$$J = \sqrt{g}$$

4.3.3 Transformation Rules for Change of Coordinates

Suppose a new set of general coordinates $\bar{\xi}^i$ is introduced, with the understanding that the relations between ξ^i and $\bar{\xi}^i$ are known, at least in principle. The rule (4.2-16) for changing to new tensor components is (for third order tensors, for example)

$$\bar{T}^{ij}_{\cdot\cdot k} = T^{pq}_{\cdot\cdot r}(\epsilon_p \cdot \bar{\epsilon}^i)(\epsilon_q \cdot \bar{\epsilon}^j)(\epsilon^r \cdot \bar{\epsilon}_k)$$

where $\bar{\epsilon}^i = \bar{g}^{ij}\bar{\epsilon}_j = \bar{g}^{ij}(\partial \mathbf{x}/\partial \bar{\xi}^j)$, $\bar{g}_{ij} = \bar{\epsilon}_i \cdot \bar{\epsilon}_j$, and $\bar{g}^{ij}\bar{g}_{jk} = \delta^i_k$. Then

$$\bar{T}^{ij}_{\cdot\cdot k} = T^{pq}_{\cdot\cdot r}\left(\bar{g}^{is} \frac{\partial \mathbf{x}}{\partial \xi^p} \cdot \frac{\partial \mathbf{x}}{\partial \bar{\xi}^s}\right)\left(\bar{g}^{jt} \frac{\partial \mathbf{x}}{\partial \xi^q} \cdot \frac{\partial \mathbf{x}}{\partial \bar{\xi}^t}\right)\left(g^{ru} \frac{\partial \mathbf{x}}{\partial \xi^u} \cdot \frac{\partial \mathbf{x}}{\partial \bar{\xi}^k}\right)$$

$$= T^{pq}_{\cdot\cdot r}\left(\bar{g}^{is} \frac{\partial \mathbf{x}}{\partial \bar{\xi}^l} \cdot \frac{\partial \mathbf{x}}{\partial \bar{\xi}^s} \frac{\partial \bar{\xi}^l}{\partial \xi^p}\right)\left(\bar{g}^{jt} \frac{\partial \mathbf{x}}{\partial \bar{\xi}^m} \cdot \frac{\partial \mathbf{x}}{\partial \bar{\xi}^t} \frac{\partial \bar{\xi}^m}{\partial \xi^q}\right)\left(g^{ru} \frac{\partial \mathbf{x}}{\partial \xi^u} \cdot \frac{\partial \mathbf{x}}{\partial \xi^n} \frac{\partial \xi^n}{\partial \bar{\xi}^k}\right)$$

and so

$$\bar{T}^{ij}_{\cdot\cdot k} = T^{pq}_{\cdot\cdot r}\left(\frac{\partial\bar{\xi}^i}{\partial\xi^p}\right)\left(\frac{\partial\bar{\xi}^j}{\partial\xi^q}\right)\left(\frac{\partial\xi^r}{\partial\bar{\xi}^k}\right) \tag{4.3-6}$$

Equation (4.3-6) serves as the prototype formula for transformation of any tensor under change of coordinate system.

It may be noted that if at some point in space *all* the components of a tensor vanish with respect to one coordinate system, they vanish at this point for all coordinate systems.

4.4 TENSOR CALCULUS

4.4.1 Gradient of a Scalar

If $f(\xi^1, \xi^2, \xi^3)$ is a scalar function, then

$$\frac{\partial f}{\partial\xi^i} = \frac{\partial f}{\partial x^j}\frac{\partial x^j}{\partial\xi^i} = \left(\frac{\partial f}{\partial x^j}\,\mathbf{e}_j\right)\cdot\frac{\partial\mathbf{x}}{\partial\xi^i}$$

But grad $f \equiv \nabla f = \partial f/\partial x^j\,\mathbf{e}_j$; hence

$$\frac{\partial f}{\partial\xi^i} = (\nabla f)\cdot\epsilon_i \tag{4.4-1}$$

and so, by definition, $\partial f/\partial\xi^i$ is the ith covariant component of ∇f. Hence we can write

$$\nabla f = \frac{\partial f}{\partial\xi^i}\,\epsilon^i$$

An alternative way to conclude that $\partial f/\partial\xi^i$ is a (covariant) vector is to note that

$$df = \frac{\partial f}{\partial\xi^j}\,d\xi^j$$

is a scalar for all $d\xi^j$, recall that $d\xi^j$ is a vector [eq. (4.3-5)], and invoke the appropriate quotient law.

It is notable that the Cartesian components of ∇f, namely $\partial f/\partial x^i$, generalize so neatly to the covariant components $\partial f/\partial\xi^i$ in general coordinates.

4.4-2 Derivative of a Vector; Christoffel Symbols; Covariant Derivative

Consider the partial derivative $\partial\mathbf{F}/\partial\xi^j$ of a vector $\mathbf{F}$. With $\mathbf{F} = F^i\epsilon_i$, we have

$$\frac{\partial\mathbf{F}}{\partial\xi^j} = \frac{\partial F^i}{\partial\xi^j}\,\epsilon_i + F^i\,\frac{\partial\epsilon_i}{\partial\xi^j}$$

Now write

$$\frac{\partial \boldsymbol{\epsilon}_i}{\partial \xi^j} \equiv \Gamma^k_{ij} \boldsymbol{\epsilon}_k \tag{4.4-2}$$

The quantity Γ^k_{ij}—called the *Christoffel symbol of the second kind*—is thus, by definition, the kth contravariant component of the derivative with respect to ξ^j of the ith base vector. Note that

$$\frac{\partial \boldsymbol{\epsilon}_i}{\partial \xi^j} = \frac{\partial^2 \mathbf{x}}{\partial \xi^i \partial \xi^j} = \frac{\partial \boldsymbol{\epsilon}_j}{\partial \xi^i}$$

and therefore

$$\Gamma^k_{ij} = \Gamma^k_{ji}$$

We can now write

$$\frac{\partial \mathbf{F}}{\partial \xi^j} = \left(\frac{\partial F^i}{\partial \xi^j} + F^k \Gamma^i_{kj}\right) \boldsymbol{\epsilon}_i \tag{4.4-3}$$

Introduce the notation

$$\frac{\partial \mathbf{F}}{\partial \xi^j} \equiv F^i_{,j} \boldsymbol{\epsilon}_i \tag{4.4-4}$$

This means that $F^i_{,j}$—called the *covariant derivative* of F^i—is defined as the ith contravariant component of the vector $\partial \mathbf{F}/\partial \xi^j$. Comparing (4.4-4) and (4.4-3) then gives us the formula

$$F^i_{,j} = \frac{\partial F^i}{\partial \xi^j} + F^k \Gamma^i_{kj} \tag{4.4-5}$$

Although $\partial F^i/\partial \xi^j$ is not necessarily a tensor, $F^i_{,j}$ *is* one, for

$$d\mathbf{F} = \frac{\partial \mathbf{F}}{\partial \xi^j} d\xi^j = (F^i_{,j} d\xi^j) \boldsymbol{\epsilon}_i$$

Hence the contravariant components of $d\mathbf{F}$ are $F^i_{,j} d\xi^j$; but since $d\xi^j$ is an arbitrary vector, $F^i_{,j}$ is a (mixed) second-order tensor.

Note that the Christoffel symbols vanish for Cartesian coordinates; hence the $F^i_{,j}$ constitute general tensor components of the Cartesian tensor $\partial \overset{c}{F}_i/\partial x^j$, where $\overset{c}{F}_i$ is the ith Cartesian component of $\mathbf{F}$.

The covariant derivative of F_i, written as $F_{i,j}$, is defined as the ith *covariant* component of $\partial \mathbf{F}/\partial \xi^j$; hence

$$F_{i,j} = g_{ki} F^k_{,j} \tag{4.4-6}$$

A direct calculation of $F_{i,j}$ is more instructive; with $\mathbf{F} = F_i \epsilon^i$, we have

$$\frac{\partial \mathbf{F}}{\partial \xi^j} \equiv F_{i,j} \epsilon^i = \left(\frac{\partial F_i}{\partial \xi^j} \epsilon^i + F_k \frac{\partial \epsilon^k}{\partial \xi^j} \right)$$

Now $\epsilon^k \cdot \epsilon_i = \delta_i^k$, whence

$$\epsilon_i \cdot \frac{\partial \epsilon^k}{\partial \xi^j} = -\epsilon^k \cdot \frac{\partial \epsilon_i}{\partial \xi^j} = -\epsilon^k \cdot (\Gamma^l_{ij} \epsilon_l) = -\Gamma^k_{ij}$$

and therefore

$$\frac{\partial \epsilon^k}{\partial \xi^j} = -\Gamma^k_{ij} \epsilon^i \tag{4.4-7}$$

Consequently

$$F_{i,j} = \frac{\partial F_i}{\partial \xi^j} - F_k \Gamma^k_{ij} \tag{4.4-8}$$

and while this must be the same as (4.4-6) it shows the explicit addition to $\partial F_i/\partial \xi^j$ needed to provide the covariant derivative of F_i.

Other notations are common for covariant derivatives; they are, in approximate order of popularity

$$F_{i|j} \qquad F_{i;j} \qquad D_j F_i \qquad \nabla_j F_i$$

The first of these may actually be more prevalent that $F_{i,j}$ but it is typographically awkward.

Although Γ^k_{ij} is *not* a third-order tensor, the superscript can nevertheless be lowered by means of the operation

$$g_{kp} \Gamma^k_{ij}$$

and the resultant quantity, denoted by $[ij, p]$, is the *Christoffel symbol of the first kind.* The following relations are easily verified:

$$[ij, k] = \epsilon_k \cdot \frac{\partial \epsilon_i}{\partial \xi^j} = \epsilon_k \cdot \frac{\partial \epsilon_j}{\partial \xi^i} = \frac{\partial \mathbf{x}}{\partial \xi^k} \cdot \frac{\partial^2 \mathbf{x}}{\partial \xi^i \partial \xi^j} \tag{4.4-9}$$

$$\frac{\partial g_{ij}}{\partial \xi^k} = [jk, i] + [ik, j] \tag{4.4-10}$$

$$g_{kp}\Gamma^p_{ij} = [ij, k] = \frac{1}{2}\left\{\frac{\partial g_{ik}}{\partial \xi^j} + \frac{\partial g_{jk}}{\partial \xi^i} - \frac{\partial g_{ij}}{\partial \xi^k}\right\} \tag{4.4-11}$$

This last equation shows that the Christoffel symbols really depend only on the metric tensor and its partial derivatives.

4.4.3 Covariant Derivatives of Nth-Order Tensors

Let us work out the formula for the covariant derivatives of $A^{ij}_{\cdot\cdot k}$. Write

$$\mathbf{A} = A^{ij}_{\cdot\cdot k}\,\epsilon_i\epsilon_j\epsilon^k$$

Then, by definition

$$\frac{\partial \mathbf{A}}{\partial \xi^p} \equiv A^{ij}_{\cdot\cdot k,p}\,\epsilon_i\epsilon_j\epsilon^k$$

$$= \frac{\partial}{\partial \xi^p}(A^{ij}_{\cdot\cdot k}\,\epsilon_i\epsilon_j\epsilon^k)$$

and this leads directly to the formula

$$A^{ij}_{\cdot\cdot k,p} = \frac{\partial A^{ij}_{\cdot\cdot k}}{\partial \xi^p} + A^{rj}_{\cdot\cdot k}\Gamma^i_{rp} + A^{ir}_{\cdot\cdot k}\Gamma^j_{rp} - A^{ij}_{\cdot\cdot r}\Gamma^r_{pk} \tag{4.4-12}$$

The general result for N^{th}-order tensors is now obvious, and will not be written out. The tensor character of $A^{ij}_{\cdot\cdot k,p}$ follows from the fact that

$$d\mathbf{A} = (A^{ij}_{\cdot\cdot k,p}\,d\xi^p)\epsilon_i\epsilon_j\epsilon^k$$

Thus $A^{ij}_{\cdot\cdot k,p}d\xi^p$ is a third-order tensor for all $d\xi^p$, whence $A^{ij}_{\cdot\cdot k,p}$ is a fourth-order tensor, by the quotient law. As a matter of convention, the partial derivative $\partial f/\partial \xi^i$ of a scalar is written, without ambiguity, as $f_{,i}$. Similarly, the notation $\mathbf{A}_{,i} \equiv \partial \mathbf{A}/\partial \xi^i$ may be used for tensors $\mathbf{A}$. Also, by convention, all indices following a comma indicate successive covariant differentiation; thus $(A_{ij,p})_{,q} \equiv A_{ij,pq}$.

It is useful to write (4.3-1) as

$$\epsilon_i = \mathbf{x}_{,i}$$

and then

$$\mathbf{x}_{,ij} = \frac{\partial \epsilon_i}{\partial \xi^j} - \epsilon_k\Gamma^k_{ij} = 0 \tag{4.4-13}$$

This result can be reached, alternatively, by reasoning that $\mathbf{x}_{,ij}$ certainly vanishes in a Cartesian coordinate system; hence it must vanish in all coordinate systems. Similar arguments suffice to establish that

$$g_{ij,k} = 0 \tag{4.4-14}$$

and

$$\epsilon_{ijk,l} = 0 \tag{4.4-15}$$

in Euclidean 3-D, because in Cartesian coordinates g_{ij} and ϵ_{ijk} reduce to δ_{ij} and e_{ijk}, partial and covariant differentiation become the same, and $\partial \delta_{ij}/\partial x^k = \partial e_{ijk}/\partial x^l = 0$. It may be noted that (4.4-14) also follows directly from (4.4-10).

The same persistence of equality to zero provides the justification for covariant-differentiation rules of the type

$$(A^{ij}F_k)_{,m} = A^{ij}_{,m}F_k + A^{ij}F_{k,m}$$

This formula holds in Cartesian coordinates; hence it must be right in all coordinate systems, because the vanishing of the difference between the left and right hand side of the equation is invariant with respect to coordinate systems. We can therefore invoke (4.4-14) to justify relations like

$$g_{jk}A^{ij}_{,p} = A^{i}_{\cdot k,p}$$

showing that the operations of multiplying by the metric tensor and performing covariant differentiation are commutative.

4.4.4 Divergence of a Vector

A useful formula for $\nabla \cdot \mathbf{F} \equiv \operatorname{div} \mathbf{F} = F^i_{,i} = (\partial F^i/\partial \xi^i) + F^p \Gamma^i_{ip}$ will be developed for general coordinate systems. We have

$$\begin{aligned}
\Gamma^i_{ip} &= g^{is}\,[ip, s] \\
&= \frac{1}{2} g^{is} \left[\frac{\partial g_{is}}{\partial \xi^p} + \frac{\partial g_{ps}}{\partial \xi^i} - \frac{\partial g_{ip}}{\partial \xi^s}\right] \\
&= \frac{1}{2} g^{is} \frac{\partial g_{is}}{\partial \xi^p}
\end{aligned}$$

But, by determinant theory

$$\frac{\partial g}{\partial \xi^p} = g g^{is} \frac{\partial g_{is}}{\partial \xi^p}$$

Hence

$$\Gamma^i_{ip} = \frac{1}{2g}\frac{\partial g}{\partial \xi^p} = \frac{1}{\sqrt{g}}\frac{\partial(\sqrt{g})}{\partial \xi^p}$$

and therefore

$$\nabla \cdot \mathbf{F} = \frac{1}{\sqrt{g}}\frac{\partial}{\partial \xi^i}(F^i\sqrt{g}) \tag{4.4-16}$$

4.4.5 Riemann-Christoffel Tensor

Since the order of differentiation in repeated partial differentiation of Cartesian tensors is irrelevant (for well-behaved functions), it follows that the indices in repeated covariant differentiation of general tensors in Euclidean 3-D may also be interchanged at will. Thus, identities like

$$\phi_{,ij} = \phi_{,ji} \tag{4.4-17}$$

and

$$f_{k,ij} = f_{k,ji} \tag{4.4-18}$$

must hold. Equation (4.4-17) is easily verified directly, since

$$\phi_{,ij} = \frac{\partial^2 \phi}{\partial \xi^i \partial \xi^j} - \phi_{,p}\Gamma^p_{ij}$$

is obviously symmetrical in i and j. However, the assertion of (4.4-18) in Euclidean 3-D leads to some nontrivial information. By direct calculation it can be shown that

$$f_{k,ij} - f_{k,ji} = R^p_{\cdot kij} f_p \tag{4.4-19}$$

where

$$R^p_{\cdot kij} = \frac{\partial \Gamma^p_{kj}}{\partial \xi^i} - \frac{\partial \Gamma^p_{ki}}{\partial \xi^j} + \Gamma^p_{ri}\Gamma^r_{kj} - \Gamma^p_{rj}\Gamma^r_{ki} \tag{4.4-20}$$

With the help of (4.4-11) it can be shown that R_{pkij}, the *Riemann-Christoffel tensor*, is given by

$$R_{pkij} = \frac{1}{2}\left[\frac{\partial^2 g_{pj}}{\partial \xi^k \partial \xi^i} + \frac{\partial^2 g_{ki}}{\partial \xi^p \partial \xi^j} - \frac{\partial^2 g_{pi}}{\partial \xi^k \partial \xi^j} - \frac{\partial^2 g_{kj}}{\partial \xi^p \partial \xi^i}\right] + g_{rm}[\Gamma^r_{pj}\Gamma^m_{ki} - \Gamma^r_{pi}\Gamma^m_{kj}] \tag{4.4-21}$$

But since the left-hand side of (4.4-19) vanishes for *all* vectors f_k, it follows that

$$R_{pkij} = 0 \tag{4.4-22}$$

in Euclidean 3-D.

Since the components of the Riemann-Christoffel tensor depend only on g_{ij} and its partial derivatives, eqs. (4.4-22) constitute second-order partial differential equations that must be obeyed by the components of the metric tensor. Conversely, it can be shown that if eqs. (4.4-22) are satisfied by functions g_{ij} of the variables (ξ^1, ξ^2, ξ^3) then generalized coordinates ξ^i that correspond to the metric tensor g_{ij} exist.

Although (4.4-22) represents 81 equations, most of them are either identities or redundant, since $R_{pkij} = -R_{pkji} = -R_{kpij} = R_{ijpk}$. Only six distinct nontrivial conditions are specified by (4.4-22), and they may be written as

$$R_{1212} = R_{1223} = R_{1231} = R_{2323} = R_{2331} = R_{3131} = 0 \tag{4.4-23}$$

Since R_{pkij} is antisymmetrical in i and j as well as in p and k, no information is lost if (4.4-22) is multiplied by $\epsilon^{spk}\epsilon^{tij}$. Consequently, a set of six equations equivalent to (4.4-23) is given neatly by

$$S^{st} = 0$$

where S^{st} is the symmetrical, second-order tensor

$$S^{st} \equiv \tfrac{1}{4}\, \epsilon^{spk}\epsilon^{tij} R_{pkij} \tag{4.4-24}$$

The tensor S_{ij} is related simply to the *Ricci tensor* $R_{ij} \equiv R^{p}_{\cdot ijp}$ by

$$R_{ij} = S_{ij} - g_{ij} S^p_p$$

so that (4.4-23) is also equivalent to the assertion $R_{ij} = 0$.

4.4.6 Integral Relations

The familiar divergence theorem (Chapter 3) relating integrals over a volume V and its boundary surface S can obviously be written in tensor notation as

$$\int_V f^i_{,i}\, dV = \int_S f^i N_i\, dS$$

where N_i is the unit outward normal vector to S. Similarly, Stokes' theorem for integrals over a surface S and its boundary line C is just

$$\int_S \epsilon^{ijk} f_{k,j} N_i dS = \oint_C f^k t_k ds$$

where t_k is the unit tangent vector to C, and the usual handedness rules apply for the directions of N_i and t_i.

4.5 THEORY OF SURFACES

4.5.1 Coordinate Systems and Base Vectors

A 2-D surface in Euclidean 3-D is conveniently specified by means of the position vector $\mathbf{x}$ considered as a function of two parameters ξ^1 and ξ^2; to each pair of values of these parameters, the corresponding vector $\mathbf{x}(\xi^1, \xi^2)$ denotes a point on

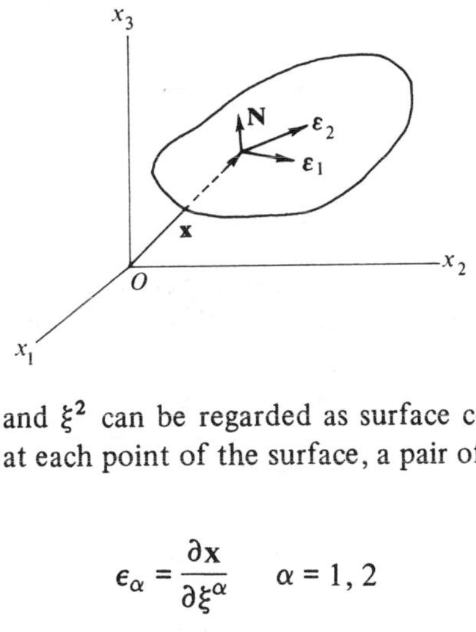

the surface. Thus ξ^1 and ξ^2 can be regarded as surface coordinates. It is convenient, then, to define, at each point of the surface, a pair of base vectors tangent to the surface by

$$\boldsymbol{\epsilon}_\alpha = \frac{\partial \mathbf{x}}{\partial \xi^\alpha} \qquad \alpha = 1, 2 \tag{4.5-1}$$

From now on, Greek indices will be understood to have the range 1, 2.

It will prove convenient to introduce the unit normal vector $\mathbf{N}$ at each point of the surface, so that $\boldsymbol{\epsilon}_1$, $\boldsymbol{\epsilon}_2$, and $\mathbf{N}$ form a right-handed triad of base vectors in 3-D. Further, the coordinate z giving distance from the surface in the direction of $\mathbf{N}$ can also be contemplated.

4.5.2 Surface Vectors and Tensors

With the utilization of the base vectors $\boldsymbol{\epsilon}_\alpha$ and $\mathbf{N}$, an arbitrary vector $\mathbf{F}$ can be written as

$$\mathbf{F} = F^\alpha \boldsymbol{\epsilon}_\alpha + F_N \mathbf{N} \tag{4.5-2}$$

Here, F_N is called the *normal* component of $\mathbf{F}$.

Attention is directed now to a vector field $\mathbf{T}(\xi^1, \xi^2)$, dependent on the surface coordinates, and having zero normal components at (ξ^1, ξ^2); $\mathbf{T}(\xi^1, \xi^2)$ is a field of *surface vectors*, and, at each point of the surface, $\mathbf{T}$ can be written

$$\mathbf{T} = T^\alpha \epsilon_\alpha \tag{4.5-3}$$

The T^α are the surface contravariant components of $\mathbf{T}$. Covariant components T_α are defined by

$$T_\alpha = \mathbf{T} \cdot \epsilon_\alpha \tag{4.5-4}$$

and by defining the surface metric tensor

$$g_{\alpha\beta} \equiv \frac{\partial \mathbf{x}}{\partial \xi^\alpha} \cdot \frac{\partial \mathbf{x}}{\partial \xi^\beta} \equiv \epsilon_\alpha \cdot \epsilon_\beta \tag{4.5-5}$$

and its inverse $g^{\alpha\beta}$, we can derive the relations, analogous to those of 3-D vector analysis

$$T_\alpha = g_{\alpha\beta} T^\beta$$

and

$$T^\alpha = g^{\alpha\beta} T_\beta$$

In fact, most of the theoretical apparatus of vector and tensor algebra developed in sections 4.1, 4.2, and 4.3 carry over to *surface* vectors and tensors defined with respect to the surface base vectors ϵ_α; it is only necessary to suppress the third dimension. Thus, a surface tensor of second order is

$$\mathbf{A} = A^{\alpha\beta} \epsilon_\alpha \epsilon_\beta$$

the transformation rule for change of coordinates in the surface from ξ^α to $\bar{\xi}^\alpha$ is

$$\bar{A}^{\alpha\beta} = A^{\gamma\omega} \frac{\partial \bar{\xi}^\alpha}{\partial \xi^\gamma} \frac{\partial \bar{\xi}^\beta}{\partial \xi^\omega}$$

various quotient laws can be stated and proved; and so on.

However, one thing we can *not* do, in general, is introduce a set of 2-D Cartesian coordinates into an arbitrary surface; this, obviously, is possible only for a *plane*.

4.5.3 2-D Permutation Tensor

In 2-D, the permutation symbol $e_{\alpha\beta}$ (or $e^{\alpha\beta}$) is defined as

$$\begin{aligned} e_{\alpha\beta} &= 1 \quad \text{for } \alpha = 1,\ \beta = 2 \\ &= 0 \quad \text{for } \alpha = \beta \\ &= -1 \quad \text{for } \alpha = 2,\ \beta = 1 \end{aligned}$$

Its values are the same as those of the 3-D permutation symbol $e_{\alpha\beta3}$. If we consider the 3-D coordinate system (ξ^1, ξ^2, z), the appropriate triad of general base vectors becomes $(\boldsymbol{\epsilon}_1, \boldsymbol{\epsilon}_2, \mathbf{N})$, and the corresponding values of the permutation tensor $\epsilon_{\alpha\beta3}$ serve to define the 2-D permutation tensor $\epsilon_{\alpha\beta}$. Thus [see eq. (4.1-11)]

$$\epsilon_{\alpha\beta} \equiv \boldsymbol{\epsilon}_\alpha \times \boldsymbol{\epsilon}_\beta \cdot \mathbf{N} \tag{4.5-6}$$

It follows that

$$\epsilon_{\alpha\beta} = \sqrt{g}\, e_{\alpha\beta}$$

and (4.5-7)

$$\epsilon^{\alpha\beta} = \frac{1}{\sqrt{g}} e^{\alpha\beta}$$

where now $g = g_{11}g_{22} - g_{12}^2$; the $\epsilon - g$ identity (4.1-13) reduces to

$$\epsilon_{\alpha\beta}\epsilon_{\gamma\omega} = g_{\alpha\gamma}g_{\beta\omega} - g_{\alpha\omega}g_{\beta\gamma} \tag{4.5-8}$$

4.5.4 Curvature Tensor

Let P be any one of the single infinity of planes that contain the normal $\mathbf{N}$ at some point p in a surface S (see sketch), and let C be the plane curve of intersection of P

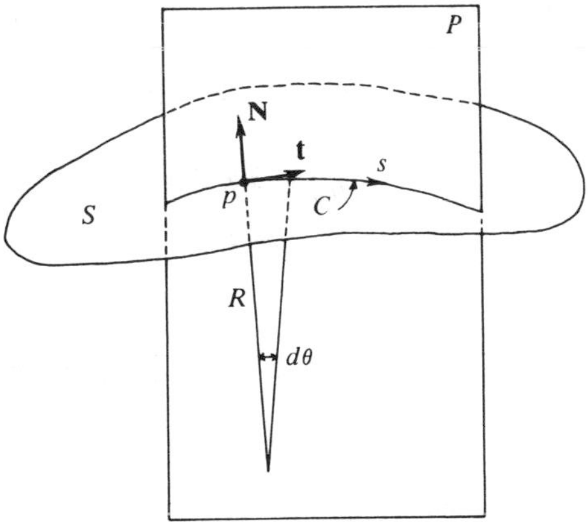

and S. Let s and θ measure arc length and inclination along C, and let $t(s)$ be the unit tangent vector to C, pointing in the direction of increasing s. Then the *curvature* κ of C at p, and its reciprocal R, the *radius of curvature*, are defined by

$$\frac{d\theta}{ds} = \kappa = \frac{1}{R} = -\mathbf{N} \cdot \frac{d\mathbf{t}}{ds} = \mathbf{t} \cdot \frac{d\mathbf{N}}{ds} \tag{4.5-9}$$

Clearly, different values of κ will be associated with different 'normal' sections through S containing $\mathbf{N}$, and the totality of these values would constitute a description of the curvature of the surface itself at p. But, as will now be shown, a much more economical characterization of surface curvatures is possible. Since

$$\mathbf{t} = \frac{d\mathbf{x}}{ds} = \frac{\partial \mathbf{x}}{\partial \xi^\alpha}\frac{d\xi^\alpha}{ds} = t^\alpha \boldsymbol{\epsilon}_\alpha$$

and

$$\frac{d\mathbf{N}}{ds} = \frac{\partial \mathbf{N}}{\partial \xi^\alpha}\frac{d\xi^\alpha}{ds} = \frac{\partial \mathbf{N}}{\partial \xi^\alpha} t^\alpha$$

it follows from (4.5-9) that

$$\kappa = \left(\frac{\partial \mathbf{N}}{\partial \xi^\alpha} \cdot \frac{\partial \mathbf{x}}{\partial \xi^\beta}\right) t^\alpha t^\beta \tag{4.5-10}$$

The quantity

$$b_{\alpha\beta} \equiv \frac{\partial \mathbf{N}}{\partial \xi^\alpha} \cdot \frac{\partial \mathbf{x}}{\partial \xi^\beta} \tag{4.5-11}$$

is a surface tensor, and is called the *curvature tensor.** Thus, a knowledge of the components of $b_{\alpha\beta}$ suffices to determine the curvature of any surface curve produced by a normal section.

Using $t^\alpha = d\xi^\alpha/ds$ and $ds^2 = g_{\gamma\omega} d\xi^\gamma d\xi^\omega$ in (4.5-10) gives

$$\kappa = \frac{b_{\alpha\beta} d\xi^\alpha d\xi^\beta}{g_{\gamma\omega} d\xi^\gamma d\xi^\omega} \tag{4.5-12}$$

This formula displays the ratio of two expressions, quadratic in $d\xi^\alpha$, that are given special names in the literature on differential geometry: the denominator in (4.5-12) is called the *first fundamental form* and the numerator is the *second fundamental form.*

Since $\mathbf{N} \cdot \mathbf{x}_{,\beta} = 0$ it follows that

$$0 = \frac{\partial}{\partial \xi^\alpha}\left(\mathbf{N} \cdot \frac{\partial \mathbf{x}}{\partial \xi^\beta}\right) = \frac{\partial \mathbf{N}}{\partial \xi^\alpha} \cdot \frac{\partial \mathbf{x}}{\partial \xi^\beta} + \mathbf{N} \cdot \frac{\partial^2 \mathbf{x}}{\partial \xi^\alpha \partial \xi^\beta}$$

*This is opposite in sign from the more conventional definition.

Hence, by (4.5-11)

$$b_{\alpha\beta} = -\mathbf{N} \cdot \frac{\partial^2 \mathbf{x}}{\partial \xi^\alpha \partial \xi^\beta} \tag{4.5-13}$$

Consequently, $b_{\alpha\beta} = b_{\beta\alpha}$, and so, in addition to formulas (4.5-11) and (4.5-13), we also have

$$b_{\alpha\beta} = \frac{\partial \mathbf{N}}{\partial \xi^\beta} \cdot \frac{\partial \mathbf{x}}{\partial \xi^\alpha} \tag{4.5-14}$$

4.5.5 Lines-of-Curvature Coordinates; Mean and Gaussian Curvatures

Consider the problem of finding, at any given point of the surface, the largest and smallest curvatures κ given by (4.5-12). Rendering (4.5-12) stationary with respect to $d\xi^\alpha$ leads to

$$(b_{\alpha\beta} - \kappa g_{\alpha\beta})\, d\xi^\beta = 0 \tag{4.5-15}$$

whence, for a nontrivial solution for the extremalizing directions $d\xi^\beta$, the determinantal equation

$$|b_{\alpha\beta} - \kappa g_{\alpha\beta}| = 0 \tag{4.5-16}$$

must be satisfied. The two roots κ_1, κ_2 of (4.5-16) are called the *principal curvatures.* Note that (4.5-16) is equivalent to

$$|b^\alpha_\beta - \kappa \delta^\alpha_\beta| = 0$$

so that κ_1, κ_2 are the eigenvalues of the matrix of numbers b^α_β. (The symmetry of $b_{\alpha\beta}$ makes the ordering of the indices in b^α_β irrelevant, so that the superscript may written directly above the subscript.) For $\kappa = \kappa_1, \kappa_2$, respectively, increments $\overset{(1)}{d}\xi^\beta$ and $\overset{(2)}{d}\xi^\beta$ can be found that satisfy (4.5-15). It is easily verified that for $\kappa_1 \neq \kappa_2$, $g_{\alpha\beta}\overset{(1)}{d}\xi^\alpha \overset{(2)}{d}\xi^\beta = 0$; consequently, the *principal directions* associated with $\overset{(1)}{d}\xi^\alpha$ and $\overset{(2)}{d}\xi^\alpha$ are orthogonal. (The same conclusion can be considered to hold in the case of repeated roots $\kappa_1 = \kappa_2$ of (4.5-16); for then any $d\xi^\alpha$ is in a principal direction, and an orthogonal pair can be chosen arbitrarily.)

Now introduce a new coordinate system $\bar{\xi}^\gamma$ into the surface in such a way that the *coordinate lines* given by $\bar{\xi}^\gamma$ = const. always point in the principal directions. This particular orthogonal coordinate system is called a system of *lines-of-curvature* coordinates. If the curvature tensor $b_{\alpha\beta}$ transforms to $\bar{b}_{\alpha\beta}$ in the lines-of-curvature coordinate, then $\bar{b}_{12} = \bar{b}_{21} = 0$, and

$$\bar{b}_1^1 = \kappa_1 = \frac{1}{R_1}$$
$$\bar{b}_2^2 = \kappa_2 = \frac{1}{R_2} \qquad (4.5\text{-}17)$$

where R_1 and R_2 are the principal radii of curvature.

The *mean curvature* of the surface at any point is defined by the scalar quantity

$$\frac{1}{2} b_\gamma^\gamma = \frac{1}{2}\left(\frac{1}{R_1} + \frac{1}{R_2}\right) \qquad (4.5\text{-}18)$$

Another invariant of the curvature tensor is given by the determinant $|b_\beta^\alpha|$, which by (4.5-16) must equal the product of the principal curvatures. This quantity, called the *Gaussian curvature*, is thus

$$\kappa_G = |b_\beta^\alpha| = b_1^1 b_2^2 - b_2^1 b_1^2 = \frac{1}{R_1 R_2} \qquad (4.5\text{-}19)$$

(Note that the Gaussian curvature really has the dimensions of a curvature squared.)

4.5.6 Gauss and Weingarten Equations

Reverting temporarily to the viewpoint that (ξ^1, ξ^2, z) constitute a 3-D system of coordinates, with base vectors $(\epsilon_1, \epsilon_2, \mathbf{N})$, the following relations hold for derivatives of the three base vectors with respect to ξ^α:

$$\frac{\partial \epsilon_\alpha}{\partial \xi^\beta} = \frac{\partial^2 \mathbf{x}}{\partial \xi^\alpha \partial \xi^\beta} = \frac{\partial \epsilon_\beta}{\partial \xi^\alpha} = \Gamma_{\alpha\beta}^\gamma \frac{\partial \mathbf{x}}{\partial \xi^\gamma} + \Gamma_{\alpha\beta}^3 \mathbf{N} \qquad (4.5\text{-}20)$$

$$\frac{\partial \mathbf{N}}{\partial \xi^\alpha} = \Gamma_{3\alpha}^\omega \frac{\partial \mathbf{x}}{\partial \xi^\omega} + \Gamma_{3\alpha}^3 \mathbf{N} \qquad (4.5\text{-}21)$$

Equation (4.5-20) implies

$$\frac{\partial^2 \mathbf{x}}{\partial \xi^\alpha \partial \xi^\beta} \cdot \mathbf{N} = \Gamma_{\alpha\beta}^3$$

which equals $-b_{\alpha\beta}$ by (4.5-13). Hence, (4.5-20) can be rewritten as

$$\frac{\partial^2 \mathbf{x}}{\partial \xi^\alpha \partial \xi^\beta} = \Gamma_{\alpha\beta}^\gamma \frac{\partial \mathbf{x}}{\partial \xi^\gamma} - b_{\alpha\beta} \mathbf{N} \qquad (4.5\text{-}22)$$

and these are called the *Gauss equations.* Here the surface Christoffel symbols of the second kind $\Gamma^{\gamma}_{\alpha\beta}$ are just

$$\Gamma^{\gamma}_{\alpha\beta} = g^{\gamma\omega}\,[\alpha\beta, \omega]$$

in terms of the surface Christoffel symbols of the first kind $[\alpha\beta, \omega]$, which in turn are defined by

$$[\alpha\beta, \omega] \equiv \frac{\partial^2 \mathbf{x}}{\partial\xi^\alpha \partial\xi^\beta} \cdot \frac{\partial \mathbf{x}}{\partial\xi^\omega} \tag{4.5-23}$$

These relations are just special cases of the basic 3-D results for Christoffel symbols given earlier. Furthermore, the old 3-D relations (4.4-10) and (4.4-11) between the Christoffel symbols and the metric tensor continue to apply if the range of the indices is restricted to 2.

Equation (4.5-21) can be reduced by noting that $\mathbf{N} \cdot \mathbf{N} = 1$ implies $\mathbf{N} \cdot \mathbf{N}_{,\alpha} = 0$, so that $\Gamma^3_{3\alpha} = 0$. Also, from (4.5-21)

$$\frac{\partial \mathbf{N}}{\partial\xi^\alpha} \cdot \frac{\partial \mathbf{x}}{\partial\xi^\beta} = \Gamma^{\omega}_{3\alpha} \frac{\partial \mathbf{x}}{\partial\xi^\omega} \cdot \frac{\partial \mathbf{x}}{\partial\xi^\beta} = g_{\omega\beta}\Gamma^{\omega}_{3\alpha}$$

which, by (4.5-11), equals $b_{\alpha\beta}$. Hence (4.5-21) is just

$$\frac{\partial \mathbf{N}}{\partial\xi^\alpha} = b^\beta_\alpha \frac{\partial \mathbf{x}}{\partial\xi^\beta} \tag{4.5-24}$$

Equations (4.5-24) are the *Weingarten equations.* Note that they imply that in lines-of-curvature coordinates each partial derivative of $\mathbf{N}$ points in the direction of the corresponding surface base vector.

4.5.7 Surface Covariant Differentiation

Consider the partial derivatives of a surface vector $\mathbf{T}(\xi^1, \xi^2)$

$$\frac{\partial \mathbf{T}}{\partial\xi^\alpha} = \frac{\partial}{\partial\xi^\alpha}\left(T^\beta \frac{\partial \mathbf{x}}{\partial\xi^\beta}\right) = \frac{\partial T^\beta}{\partial\xi^\alpha}\frac{\partial \mathbf{x}}{\partial\xi^\beta} + T^\beta \frac{\partial^2 \mathbf{x}}{\partial\xi^\alpha \partial\xi^\beta}$$

Invoking the Gauss equations (4.5-22) gives

$$\frac{\partial \mathbf{T}}{\partial\xi^\alpha} = T^\beta_{,\alpha} \frac{\partial \mathbf{x}}{\partial\xi^\beta} - b_{\alpha\omega}T^\omega \mathbf{N} \tag{4.5-25}$$

where

$$T^\beta_{,\alpha} \equiv \frac{\partial T^\beta}{\partial\xi^\alpha} + \Gamma^\beta_{\alpha\omega}T^\omega \tag{4.5-26}$$

is the surface covariant derivative of T^{β}. It is evident that this covariant derivative has the same form as that in 3-D defined by (4.4-5). The significance of the surface covariant derivative is that it supplies, via (4.5-25), that portion of the partial derivative of **T** that still constitutes a *surface* vector.

With the help of contravariant surface base vectors $\boldsymbol{\epsilon}^{\alpha} = g^{\alpha\beta}\mathbf{x}_{,\beta}$, it is easy to establish, by a calculation similar to that used in 3-D, that

$$\frac{\partial \mathbf{T}}{\partial \xi^{\alpha}} = T_{\beta,\alpha}\boldsymbol{\epsilon}^{\beta} - b_{\alpha}^{\omega} T_{\omega}\mathbf{N} \tag{4.5-27}$$

where

$$T_{\beta,\alpha} \equiv \frac{\partial T_{\beta}}{\partial \xi^{\alpha}} - \Gamma_{\alpha\beta}^{\omega} T_{\omega} \tag{4.5-28}$$

A symbol other than a comma might be used to denote surface (as opposed to 3-D) covariant differentiation if there is any danger of confusion; however, the consistent use of Greek indices for surface components makes separate notation unnecessary. It is also convenient, as in 3-D, to use the comma for a single partial differentiation of scalars, and of unindexed objects like **x** and **N**. Thus, the Weingarten equations can be written

$$\mathbf{N}_{,\alpha} = b_{\alpha}^{\beta}\mathbf{x}_{,\beta} \tag{4.5-29}$$

and the Gauss equations (4.5-22) can be rewritten concisely as

$$\mathbf{x}_{,\alpha\beta} = -b_{\alpha\beta}\mathbf{N} \tag{4.5-30}$$

which follows simply from the replacement of T_{β} in (4.5-28) by $\mathbf{x}_{,\beta}$.

Surface covariant differentiation of higher order surface tensors is defined by formulas entirely analogous to eq. (4.4-12), the only change being that Latin indices with range 3 are replaced by Greek indices with range 2. The geometrical significance of, say, the covariant derivative

$$A^{\alpha\beta}_{\cdot\cdot\gamma,\omega} \equiv \frac{\partial A^{\alpha\beta}_{\cdot\cdot\gamma}}{\partial \xi^{\omega}} + \Gamma^{\alpha}_{\rho\omega}A^{\rho\beta}_{\cdot\cdot\gamma} + \Gamma^{\beta}_{\rho\omega}A^{\alpha\rho}_{\cdot\cdot\gamma} - \Gamma^{\rho}_{\gamma\omega}A^{\alpha\beta}_{\cdot\cdot\rho}$$

is evident from the relation

$$\frac{\partial \mathbf{A}}{\partial \xi^{\omega}} \equiv \frac{\partial}{\partial \xi^{\omega}}\left[A^{\alpha\beta}_{\cdot\cdot\gamma}\boldsymbol{\epsilon}_{\alpha}\boldsymbol{\epsilon}_{\beta}\boldsymbol{\epsilon}^{\gamma}\right] = A^{\alpha\beta}_{\cdot\cdot\gamma,\omega}\boldsymbol{\epsilon}_{\alpha}\boldsymbol{\epsilon}_{\beta}\boldsymbol{\epsilon}^{\gamma} - b_{\omega\alpha}A^{\alpha\beta}_{\cdot\cdot\gamma}\mathbf{N}\boldsymbol{\epsilon}_{\beta}\boldsymbol{\epsilon}^{\gamma}$$
$$- b_{\omega\beta}A^{\alpha\beta}_{\cdot\cdot\gamma}\boldsymbol{\epsilon}_{\alpha}\mathbf{N}\boldsymbol{\epsilon}^{\gamma} - b^{\gamma}_{\omega}A^{\alpha\beta}_{\cdot\cdot\gamma}\boldsymbol{\epsilon}_{\alpha}\boldsymbol{\epsilon}_{\beta}\mathbf{N}$$

which shows how $A^{\alpha\beta}_{\cdot\cdot\gamma,\omega}$ is related to that portion of $\partial\mathbf{A}/\partial\xi^{\omega}$ that is expressible in terms of the surface base vectors $\boldsymbol{\epsilon}_{\alpha} = \partial\mathbf{x}/\partial\xi^{\alpha}$.

The important equation

$$g_{\alpha\beta,\gamma} = 0 \tag{4.5-31}$$

can be verified directly from the definition of covariant differentiation. It is important to observe that it is not possible to invoke the reasoning that was used in Euclidean 3-D to establish that $g_{ij,k} = 0$ vanishes in all coordinate systems simply because it vanishes in Cartesian coordinates; a Cartesian system need not exist in our curved surface. Nevertheless, a variant of this simple argument is accessible, and it depends on the use of the special surface coordinates to be discussed next.

4.5.8 Geodesic Coordinates

Although a Cartesian coordinate system can not be embedded in an arbitrary surface, it is possible to select a surface coordinate system that, in a well-defined sense, is *almost* Cartesian at a given point in the surface. Such a system of *geodesic coordinates*, like Cartesian coordinates, is orthogonal and produces vanishing Christoffel symbols at the point under consideration.

The demonstration of the existence of geodesic coordinates at a point Q on the surface S requires that there be a unique tangent plane P to S through Q. In P, choose 2-D Cartesian coordinates x^{α}, with origin at Q, and then write the position vector to points on S near Q as

$$\mathbf{x} = x^{\alpha}\mathbf{e}_{\alpha} + h(x^1, x^2)\mathbf{N}_0$$

where $\mathbf{N}_0$ is the unit normal to P, and the $\mathbf{e}_{\alpha}$ are orthogonal unit vectors in P. The function h must have the form

$$h(x^1, x^2) = \tfrac{1}{2}\, a_{\alpha\beta}x^{\alpha}x^{\beta} + (\text{terms of order } x^3)$$

where, without loss of generality, $a_{\alpha\beta} = a_{\beta\alpha}$. Thus, near Q, the x^{α} may play the role of surface coordinates ξ^{α}, and so, at Q

$$\frac{\partial^2\mathbf{x}}{\partial\xi^{\alpha}\partial\xi^{\beta}} = \frac{\partial^2\mathbf{x}}{\partial x^{\alpha}\partial x^{\beta}} = a_{\alpha\beta}\mathbf{N}_0$$

whence, by (4.5-23)

$$[\alpha\beta, \omega] = \Gamma^{\omega}_{\alpha\beta} = 0$$

at Q.

It is now possible and easy to prove various rules of covariant differentiation by exploiting the fact that in geodesic coordinates covariant differentiation reduces to partial differentiation. For example, since $g_{\alpha\beta,\gamma}$ vanishes in geodesic coordinates, it must vanish in all surface coordinate systems, by virtue of its tensor structure. Similarly, rules like

$$(f_\alpha d_{\beta\gamma})_{,\omega} = f_{\alpha,\omega} d_{\beta\gamma} + f_\alpha d_{\beta\gamma,\omega}$$

must hold because they are right for partial differentiation, hence they hold in geodesic coordinates, hence in all coordinate systems.

The relation

$$\epsilon_{\alpha\beta,\gamma} = 0 \tag{4.5-32}$$

for the permutation tensor now follows from the ϵ-g identity (4.5-8).

4.5.9 Riemann-Christoffel Tensor

The identity

$$\phi_{,\alpha\beta} = \phi_{,\beta\alpha}$$

holds for all scalars ϕ, as can be verified by a straightforward calculation. But it is *not* possible to assert that for surface vectors f_α

$$f_{\alpha,\beta\gamma} - f_{\alpha,\gamma\beta} = R^{\omega}_{\cdot\alpha\beta\gamma} f_\omega \tag{4.5-33}$$

vanishes, as was the case for (4.4-19) in Euclidean 3-D, wherein Cartesian coordinates could be embedded. Nor can geodesic coordinates be invoked, for

$$R^{\omega}_{\cdot\alpha\beta\gamma} = \frac{\partial \Gamma^{\omega}_{\alpha\gamma}}{\partial \xi^{\beta}} - \frac{\partial \Gamma^{\omega}_{\alpha\beta}}{\partial \xi^{\gamma}} + \Gamma^{\omega}_{\rho\beta}\Gamma^{\rho}_{\alpha\gamma} - \Gamma^{\omega}_{\rho\gamma}\Gamma^{\rho}_{\alpha\beta} \tag{4.5-34}$$

[see (4.4-20)] will vanish only if the derivatives as well as the values of the Christoffel symbols vanish.

Whereas there are six essentially different components of the Riemann-Christoffel tensor in 3-D [see (4.4-23)], there is just one independent component, R_{1212}, in a surface. While R_{1212} must vanish in planes, into which Cartesian coordinates can be embedded, it also vanishes for *developable* surfaces, which are defined as surfaces that can be deformed into planes without stretching. This follows from the observation that R_{1212} depends, via the Christoffel symbols, only on the metric tensor, $g_{\alpha\beta}$; since deformation without extension, by definition, must leave all line elements $ds^2 = g_{\alpha\beta} d\xi^\alpha d\xi^\beta$ unchanged, it follows that $g_{\alpha\beta}$ is unchanged by such a deformation, and that the vanishing of R_{1212} must persist for all surfaces into which planes can be deformed inextensibly.

4.5.10 Codazzi Equations; Gauss' Equation

It is not generally true that six arbitrary functions $g_{\alpha\beta}(\xi^1, \xi^2)$ and $b_{\alpha\beta}(\xi^1, \xi^2)$, represent the metric and curvature tensors of a surface with respect to particular coordinates ξ^α in the surface; there are three independent constraining equations that must be satisfied, and these will now be derived.

From (4.5-29) and (4.5-30)

$$\mathbf{N}_{,\alpha\gamma} = b^{\beta}_{\alpha,\gamma}\mathbf{x}_{,\beta} - b^{\beta}_{\alpha}b_{\beta\gamma}\mathbf{N}$$

But $\mathbf{N}_{,\alpha\gamma} = \mathbf{N}_{,\gamma\alpha}$, whence $b^{\beta}_{\alpha,\gamma} = b^{\beta}_{\gamma,\alpha}$, or

$$b_{\beta\alpha,\gamma} = b_{\beta\gamma,\alpha} \tag{4.5-35}$$

These are the *Codazzi equations*, in which there are just two independent nontrivial relations, namely

$$b_{12,1} = b_{11,2}$$

and

$$b_{21,2} = b_{22,1}$$

We can also calculate that

$$\mathbf{x}_{,\alpha\beta\gamma} = -b_{\alpha\beta,\gamma}\mathbf{N} - b_{\alpha\beta}b^{\omega}_{\gamma}\,\mathbf{x}_{,\omega}$$

and therefore, with the use of the Codazzi equations, it follows that

$$\mathbf{x}_{,\alpha\beta\gamma} - \mathbf{x}_{,\alpha\gamma\beta} = (b^{\omega}_{\gamma}\,b_{\alpha\beta} - b^{\omega}_{\beta}\,b_{\alpha\gamma})\,\mathbf{x}_{,\omega} = R^{\omega}_{\cdot\alpha\beta\gamma}\mathbf{x}_{,\omega}$$

Consequently, the Riemann-Christoffel tensor is related to the curvature tensor by

$$R_{\omega\alpha\beta\gamma} = b_{\omega\beta}\,b_{\alpha\gamma} - b_{\omega\gamma}b_{\alpha\beta} \tag{4.5-36}$$

With the help of (4.5-7), (4.5-8), and (4.5-19) this can be reduced to

$$R_{\omega\alpha\beta\gamma} = \epsilon_{\omega\alpha}\,\epsilon_{\beta\gamma}\kappa_G \tag{4.5-37}$$

in terms of the Gaussian curvature. This clearly displays the single nontrivial component of the Riemann-Christoffel tensor, and we have

$$R_{1212} = g\kappa_G \tag{4.5-38}$$

Equation (4.5-37) [or (4.5-38)] is *Gauss' equation*. It is remarkable in that it shows that the Gaussian curvature of a surface depends *only* on the metric tensor $g_{\alpha\beta}$ and its derivatives.

The two Codazzi equations and Gauss' equation are the conditions constraining the metric and curvature tensors of any surface. Conversely, it can be shown that a real surface having curvature $b_{\alpha\beta}$ and metric $g_{\alpha\beta}$ exists if these three constraining equations are satisfied.

4.5.11 Integral Relations

Let S be a surface bounded by the curve C. The outward *unit surface normal* **n** is perpendicular to C, tangent to S, and points *away* from S. Denote distance along C by s, with the direction of increasing s chosen so that the unit tangent vector

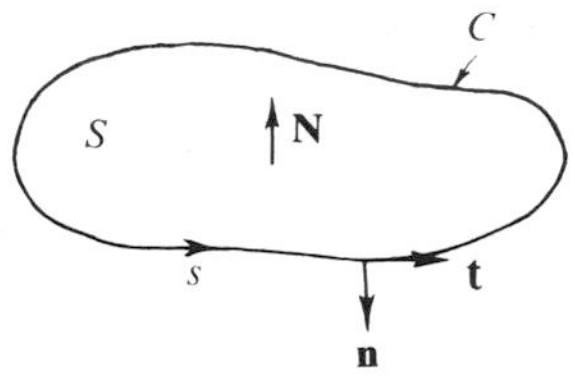

$\mathbf{t} = d\mathbf{x}/ds$ satisfies $\mathbf{t} \times \mathbf{N} = \mathbf{n}$ on C. Note, then, that $n_\alpha = \epsilon_{\alpha\beta} t^\beta$. It will be shown that the *surface divergence theorem*

$$\int_S F^\alpha_{,\alpha} \, dS = \oint_C F^\alpha n_\alpha \, ds \tag{4.5-39}$$

holds.

The absence of a Cartesian coordinate system invalidates the simple proof possible in the analogous Euclidean 3-D case. We can instead use the facts that [see eq. (4.4-16)]

$$F^\alpha_{,\alpha} = \frac{1}{\sqrt{g}} \frac{\partial}{\partial \xi^\alpha} (\sqrt{g} F^\alpha)$$

and

$$dS = \sqrt{g} \, d\xi^1 \, d\xi^2$$

to write

$$\int_S F^\alpha_{,\alpha} \, dS = \int_S \frac{\partial}{\partial \xi^\alpha} (\sqrt{g} F^\alpha) \, d\xi^1 \, d\xi^2$$

which, by Green's theorem, equals

$$\oint_C e_{\alpha\beta}\sqrt{g}\,F^\alpha\, d\xi^\beta$$

But $\sqrt{g}\, e_{\alpha\beta} d\xi^\beta = \epsilon_{\alpha\beta} d\xi^\beta = \epsilon_{\alpha\beta} t^\beta ds = n_\alpha ds$, which establishes the theorem.

An alternative derivation starts from the assertion of Stokes' theorem for surface vectors, which gives directly

$$\int_S \epsilon^{\beta\alpha} G_{\alpha,\beta} dS = \oint_C G_\alpha t^\alpha ds \tag{4.5-40}$$

Then (4.5-39) follows from (4.5-40) if the substitution $G_\alpha = \epsilon_{\rho\alpha} F^\rho$ is made.

4.6 CLASSICAL INTERLUDE

4.6.1 Orthogonal Curvilinear Coordinates (3-D)

It may be useful and instructive to list a few points of contact between tensor theory and the classical analytical apparatus based on orthogonal curvilinear coordinates. If we let (ξ, η, ζ) correspond to (ξ^1, ξ^2, ξ^3) the orthogonality of the coordinate system implies that the arc length ds is governed by

$$ds^2 = \alpha^2 d\xi^2 + \beta^2 d\eta^2 + \gamma^2 d\zeta^2 \tag{4.6-1}$$

in terms of 'scale factors' α, β, and γ, which, in general, depend on (ξ^1, ξ^2, ξ^3). Hence

$$\begin{aligned} g_{11} &= \alpha^2 \\ g_{22} &= \beta^2 \\ g_{33} &= \gamma^2 \end{aligned} \tag{4.6-2a}$$

and $g_{12} = g_{23} = g_{31} = 0$. It follows the g^{ij} also form a diagonal matrix, with

$$\begin{aligned} g^{11} &= \frac{1}{\alpha^2} \\ g^{22} &= \frac{1}{\beta^2} \\ g^{33} &= \frac{1}{\gamma^2} \end{aligned} \tag{4.6-2b}$$

The *unit* vectors $(\mathbf{e}_\xi, \mathbf{e}_\eta, \mathbf{e}_\zeta)$ in the coordinate directions are usually chosen as the base vectors in the classical approach. We have

$$
\begin{aligned}
\mathbf{e}_\xi &= \frac{1}{\alpha}\frac{\partial \mathbf{x}}{\partial \xi} = \frac{\boldsymbol{\epsilon}_1}{\alpha} \\
\mathbf{e}_\eta &= \frac{1}{\beta}\frac{\partial \mathbf{x}}{\partial \eta} = \frac{\boldsymbol{\epsilon}_2}{\beta} \\
\mathbf{e}_\zeta &= \frac{1}{\gamma}\frac{\partial \mathbf{x}}{\partial \zeta} = \frac{\boldsymbol{\epsilon}_3}{\gamma}
\end{aligned}
\tag{4.6-3}
$$

With the use of eq. (4.4-11) the Christoffel symbols in orthogonal coordinates may be identified as

$$
\begin{aligned}
[11,1] &= \alpha\frac{\partial \alpha}{\partial \xi} & \Gamma^2_{11} &= \frac{1}{\alpha}\frac{\partial \alpha}{\partial \xi} \\
[11,2] &= -\alpha\frac{\partial \alpha}{\partial \eta} & \Gamma^2_{11} &= -\frac{\alpha}{\beta^2}\frac{\partial \alpha}{\partial \eta} \\
[12,1] &= \alpha\frac{\partial \alpha}{\partial \eta} & \Gamma^1_{12} &= \frac{1}{\alpha}\frac{\partial \alpha}{\partial \eta} \\
[12,3] &= 0 & \Gamma^3_{12} &= 0
\end{aligned}
\tag{4.6-4}
$$

The rest of the Christoffel symbols can be written down on the basis of these typical results.

The analogues of eq. (4.4-2) for the derivatives of base vectors then become

$$
\begin{aligned}
\frac{\partial \mathbf{e}_\xi}{\partial \xi} &= -\left(\frac{1}{\beta}\frac{\partial \alpha}{\partial \eta}\right)\mathbf{e}_\eta - \left(\frac{1}{\gamma}\frac{\partial \alpha}{\partial \zeta}\right)\mathbf{e}_\zeta \\
\frac{\partial \mathbf{e}_\xi}{\partial \eta} &= \left(\frac{1}{\alpha}\frac{\partial \beta}{\partial \xi}\right)\mathbf{e}_\eta \\
\frac{\partial \mathbf{e}_\xi}{\partial \zeta} &= \left(\frac{1}{\alpha}\frac{\partial \gamma}{\partial \xi}\right)\mathbf{e}_\zeta
\end{aligned}
\tag{4.6-5}
$$

with corresponding results for $\mathbf{e}_\eta$ and $\mathbf{e}_\zeta$.

4.6.2 Physical Components of Vectors and Tensors

In applications the meanings of various components of tensors are best understood if they are evaluated with respect to a set of orthonormal base vectors. When orthonormal curvilinear coordinates are used, tensor components relative to the base vectors $\mathbf{e}_\xi, \mathbf{e}_\eta, \mathbf{e}_\zeta$ are therefore called the *physical* components, and they may be computed easily from a knowledge of the general tensor components. Thus, for example, if we designate the three physical components of a vector **F** by F_ξ, F_η, F_ζ, we have

$$\mathbf{F} = F^i \epsilon_i = F_\xi \mathbf{e}_\xi + F_\eta \mathbf{e}_\eta + F_\zeta \mathbf{e}_\zeta \qquad \text{(No sum on } \xi, \eta, \zeta)$$

Consequently

$$\begin{aligned} F_\xi &= \alpha F^1 = \frac{1}{\alpha} F_1 \\ F_\eta &= \beta F^2 = \frac{1}{\beta} F_2 \\ F_\zeta &= \gamma F^3 = \frac{1}{\gamma} F_3 \end{aligned} \tag{4.6-6}$$

Similarly, the physical components $A_{\xi\zeta}, A_{\xi\eta}$, etc. of a second-order tensor **A** satisfy

$$\mathbf{A} = A^{ij} \epsilon_i \epsilon_j = A_{\xi\zeta} \mathbf{e}_\xi \mathbf{e}_\zeta + A_{\xi\eta} \mathbf{e}_\xi \mathbf{e}_\eta + \cdots \qquad \text{(No sums over } \xi, \eta)$$

Hence

$$\begin{aligned} A_{\xi\xi} &= \alpha^2 A^{11} = \frac{1}{\alpha^2} A_{11} = A_1^1 \\ A_{\xi\eta} &= \alpha\beta A^{12} = \left(\frac{\alpha}{\beta}\right) A^1_{\cdot 2} = \left(\frac{\beta}{\alpha}\right) A_1{}^2_{\cdot} = \left(\frac{1}{\alpha\beta}\right) A_{12} \end{aligned} \tag{4.6-7}$$

and so on.

It is also possible to define physical components with respect to nonorthogonal curvilinear coordinates, but this is less usual, and a distinction must then be made between covariant and contravariant physical components. The unit covariant base vectors with respect to which the physical components are defined are still given by eq. (4.6-3) but are no longer orthogonal. The contravariant physical components F^ξ, F^η, F^ζ of **F** are then defined by the equation

$$\mathbf{F} = F^i \epsilon_i = \frac{1}{\alpha} (F^\xi \epsilon_1) + \frac{1}{\beta} (F^\eta \epsilon_2) + \frac{1}{\gamma} (F^\zeta \epsilon_3)$$

and so

$$\begin{aligned} F^\xi &= \alpha F^1 \\ F^\eta &= \beta F^2 \\ F^\zeta &= \gamma F^3 \end{aligned} \tag{4.6-8}$$

On the other hand, the covariant physical component F_ξ is, by definition

$$F_\xi \equiv \mathbf{F} \cdot \left(\frac{1}{\alpha} \boldsymbol{\epsilon}_1\right)$$

and so

$$\begin{aligned} F_\xi &= \frac{1}{\alpha} F_1 \\ F_\eta &= \frac{1}{\beta} F_2 \\ F_\zeta &= \frac{1}{\gamma} F_3 \end{aligned} \tag{4.6-9}$$

Similarly, we find

$$\begin{aligned} A^{\xi\xi} &= \alpha^2 A^{11} \qquad \text{(No sum over } \xi) \\ A_{\xi\xi} &= \frac{1}{\alpha^2} A_{11} \qquad \text{(No sum over } \xi) \\ A^{\xi\eta} &= \alpha\beta A^{12} \\ A_{\xi\eta} &= \frac{1}{\alpha\beta} A_{12} \end{aligned} \tag{4.6-10}$$

and so on.

4.6.3 Surface Theory in Lines-of-Curvature Coordinates

In applications of surface theory, lines-of-curvature coordinates are usually particularly convenient. If we designate the lines-of-curvature in the surface by (ξ, η), we can invoke eq. (4.6-1) to write

$$ds^2 = \alpha^2 d\xi^2 + \beta^2 d\eta^2 \tag{4.6-11}$$

since the lines-of-curvature are orthogonal. (As in 3-D coordinates, $g_{11} = 1/g^{11} = \alpha^2$, $g_{22} = 1/g^{22} = \beta^2$, and $g_{12} = g^{12} = 0$.)

The orthogonal unit base vectors $\mathbf{e}_\xi$ and $\mathbf{e}_\eta$ in the directions of the coordinate lines are given by the first two of eqs. (4.6-3), and, in addition

$$\mathbf{N} = \mathbf{e}_\xi \times \mathbf{e}_\eta \tag{4.6-12}$$

The radius of curvature R_ξ associated with a normal section in the ξ-direction satisfies

$$\frac{1}{R_\xi} = -\mathbf{N} \cdot \frac{\partial \mathbf{e}_\xi}{\alpha \partial \xi} = \mathbf{e}_\xi \cdot \frac{\partial \mathbf{N}}{\alpha \partial \xi} = \frac{1}{\alpha^2}\left(\frac{\partial \mathbf{x}}{\partial \xi} \cdot \frac{\partial \mathbf{N}}{\partial \xi}\right) = \frac{b_{11}}{\alpha^2} = b_1^1 \tag{4.6-13}$$

and analogous equations hold for the other principal curvature $1/R_\eta$.

Since $b_{12} = 0$, the Weingarten equations (4.5-24) reduce to

$$\begin{aligned} \frac{\partial \mathbf{N}}{\partial \xi} &= \left(\frac{\alpha}{R_\xi}\right) \mathbf{e}_\xi \\ \frac{\partial \mathbf{N}}{\partial \eta} &= \left(\frac{\beta}{R_\eta}\right) \mathbf{e}_\eta \end{aligned} \tag{4.6-14}$$

(In this form, Weingarten's equations are sometimes called the *Rodrigues* formulas.) The 2-D Christoffel symbols are given by eq. (4.6-4), and then, after a little manipulation, Gauss' equations (4.5-22) become

$$\begin{aligned} \frac{\partial \mathbf{e}_\xi}{\partial \xi} &= -\left(\frac{1}{\beta}\frac{\partial \alpha}{\partial \eta}\right) \mathbf{e}_\eta - \left(\frac{\alpha}{R_\xi}\right) \mathbf{N} \\ \frac{\partial \mathbf{e}_\xi}{\partial \eta} &= \frac{1}{\alpha}\frac{\partial \beta}{\partial \xi} \mathbf{e}_\eta \\ \frac{\partial \mathbf{e}_\eta}{\partial \xi} &= \frac{1}{\beta}\frac{\partial \alpha}{\partial \eta} \mathbf{e}_\xi \\ \frac{\partial \mathbf{e}_\eta}{\partial \eta} &= -\left(\frac{1}{\alpha}\frac{\partial \beta}{\partial \xi}\right) \mathbf{e}_\xi - \left(\frac{\beta}{R_\eta}\right) \mathbf{N} \end{aligned} \tag{4.6-15}$$

The two independent Codazzi equations given by (4.5-35) are most easily found in lines-of-curvature classical notation by noting that eqs. (4.6-14) imply

$$\frac{\partial^2 \mathbf{N}}{\partial \xi \partial \eta} = \frac{\partial}{\partial \eta}\left[\frac{\alpha}{R_\xi} \mathbf{e}_\xi\right] = \frac{\partial}{\partial \xi}\left[\frac{\beta}{R_\eta} \mathbf{e}_\eta\right]$$

Then the two scalar components of this result, with the use of eqs. (4.6-5), reduce to

$$\begin{aligned} \frac{\partial}{\partial \eta}\left(\frac{\alpha}{R_\xi}\right) &= \frac{1}{R_\eta}\frac{\partial \alpha}{\partial \eta} \\ \frac{\partial}{\partial \xi}\left(\frac{\beta}{R_\eta}\right) &= \frac{1}{R_\xi}\frac{\partial \beta}{\partial \xi} \end{aligned} \tag{4.6-16}$$

Finally, the Gauss equation (4.5-38) relating the Gaussian curvature to the metric tensor (via R_{1212}) is most efficiently rederived by substituting the first two equations of (4.6-15) into the left and right hand sides, respectively, of the identity

$$\frac{\partial}{\partial \eta}\left(\frac{\partial \mathbf{e}_\xi}{\partial \xi}\right) = \frac{\partial}{\partial \xi}\left(\frac{\partial \mathbf{e}_\xi}{\partial \eta}\right)$$

This leads to the neat form

$$\frac{\partial}{\partial \xi}\left(\frac{1}{\alpha}\frac{\partial \beta}{\partial \xi}\right) + \frac{\partial}{\partial \eta}\left(\frac{1}{\beta}\frac{\partial \alpha}{\partial \eta}\right) + \frac{\alpha\beta}{R_\xi R_\eta} = 0 \tag{4.6-17}$$

for Gauss' equation, which shows explicitly how $\kappa_G = 1/R_\xi R_\eta$ depends only on the metric coefficients α and β.

4.7 AN APPLICATION: CONTINUUM MECHANICS

4.7.1 Equations of Equilibrium

A glimpse will be given of the power of tensor analysis in one area of application. We have already encountered the stress tensor $\boldsymbol{\sigma}$ in section 4.2.9, and now we will derive equations of equilibrium governing its components.

Suppose a body force vector $\mathbf{f}$ per unit volume acts on a body; $\mathbf{f}$ may include inertial forces. Then, on any volume V of the body, bounded by a surface S, equilibrium requires that

$$\int_S \mathbf{F}\, dS + \int_V \mathbf{f}\, dV = 0 \tag{4.7-1}$$

where $\mathbf{F}$ is the vector force per unit area acting on S. For an arbitrary system of curvilinear coordinates, we have,

$$\mathbf{f} = f^j \boldsymbol{\epsilon}_j$$

and eq. (4.2-22) gives

$$\mathbf{F} = F^j \boldsymbol{\epsilon}_j = \sigma^{ij} n_i \boldsymbol{\epsilon}_j$$

where $\mathbf{N} = N^i \boldsymbol{\epsilon}_i$ is the unit outward normal to S. Consequently

$$\int_S \sigma^{ij} \boldsymbol{\epsilon}_j N_i\, dS + \int_V f^j \boldsymbol{\epsilon}_j\, dV = 0$$

Applying the divergence theorem (section 4.4.6) (using, for reasons that will later be apparent, a semicolon rather than a comma to denote covariant differentiation) gives

$$\int_V [f^j \epsilon_j + (\sigma^{ij} \epsilon_j)_{;i}] \, dV = 0$$

But since this holds for all volumes V, and since $\epsilon_{j;i} = \mathbf{x}_{;ji} = 0$, as in eq. (4.4-13), it follows that

$$\sigma^{ij}_{;i} + f^j = 0 \tag{4.7-2}$$

These are the differential equations of force equilibrium for an arbitrary coordinate system.

Similarly, by asserting that the net moment on the body

$$\int_S \mathbf{x} \times \mathbf{F} \, dS + \int_V \mathbf{x} \times \mathbf{f} \, dV$$

must vanish, it can be deduced that $\sigma^{ij} = \sigma^{ji}$, i.e., the stress tensor is symmetrical.

4.7.2 Displacement and Strain

In the study of *deformable* bodies, the concepts of *displacement* and *strain* play important roles. Imagine that the coordinates (ξ^1, ξ^2, ξ^3) identify a *material point* in the body, and that the vector $\mathbf{x}(\xi^1, \xi^2, \xi^3)$ used in the above analysis actually denotes the position of the material point after it has been displaced from its original location $\mathring{\mathbf{x}}(\xi^1, \xi^2, \xi^3)$. (When the coordinates are imagined to be engraved on moving material points of the body, they are called *convected* coordinates.) The displacement vector $\mathbf{U}$ is defined as

$$\mathbf{U} \equiv \mathbf{x}(\xi^1, \xi^2, \xi^3) - \mathring{\mathbf{x}}(\xi^1, \xi^2, \xi^3) \tag{4.7-3}$$

In the study of stress, the base vectors $\epsilon_i \equiv \partial \mathbf{x}/\partial \xi^i$ were chosen on the basis of the geometry of the deformed body in which the stress under consideration existed. It is more natural to define displacement components with respect to base vectors associated with the *undeformed* body; thus

$$\mathbf{U} = u^i \mathring{\epsilon}_i \tag{4.7-4}$$

where

$$\mathring{\epsilon}_i \equiv \frac{\partial \mathring{\mathbf{x}}}{\partial \xi^i} \tag{4.7-5}$$

Different metric tensors are, of course, associated with the two geometries under consideration. Thus, we will write

$$\overset{\circ}{g}_{ij} \equiv \overset{\circ}{\epsilon}_i \cdot \overset{\circ}{\epsilon}_j \tag{4.7-6}$$

and

$$g_{ij} \equiv \epsilon_i \cdot \epsilon_j \tag{4.7-7}$$

Finally, it should be noted that covariant differentiation in the two geometries is different too, and we will use a comma for the undeformed body, keeping the semicolon for the final, deformed state. The Christoffel symbols in the original and final states may be denoted by $\overset{\circ}{\Gamma}{}^i_{jk}$ and Γ^i_{jk}, respectively.

This sets the stage for defining the *Lagrangian strain tensor* η_{ij}, a measure of deformation, by

$$\eta_{ij} \equiv \frac{1}{2}\,(g_{ij} - \overset{\circ}{g}_{ij}) \tag{4.7-8}$$

Note that

$$\begin{aligned} g_{ij} &= \frac{\partial}{\partial \xi^i}\,(\overset{\circ}{\mathbf{x}} + \mathbf{U}) \cdot \frac{\partial}{\partial \xi^j}\,(\overset{\circ}{\mathbf{x}} + \mathbf{U}) \\ &= \overset{\circ}{g}_{ij} + (u^k \overset{\circ}{\epsilon}_k)_{,i} \cdot \overset{\circ}{\epsilon}_j + (u^k \overset{\circ}{\epsilon}_k)_{,j} \cdot \overset{\circ}{\epsilon}_i + (u^k \overset{\circ}{\epsilon}_k)_{,i} \cdot (u^l \overset{\circ}{\epsilon}_l)_{,j} \\ &= \overset{\circ}{g}_{ij} + u^k_{,i}\overset{\circ}{g}_{jk} + u^k_{,j}\overset{\circ}{g}_{ki} + u^k_{,i}u^l_{,j}\overset{\circ}{g}_{kl} \end{aligned}$$

Hence

$$\eta_{ij} = \frac{1}{2}\,[u_{i,j} + u_{j,i} + u_{k,i}u^k_{,j}] \tag{4.7-9}$$

4.7.3 Transformed Equations of Equilibrium

To complete the set of field equations of the mechanics of solid bodies, it is necessary to adjoin to (4.7-2) and (4.7-9) *constitutive* relations among the stresses σ^{ij} and the strains η_{ij} (including, possibly, their rates of change and histories). This will not be pursued further here, but it may be perceived that the use of eq. (4.7-2) could present difficulties associated with the fact that the covariant differentiation therein presupposes a knowledge of the geometry of the deformed body. An alternative set of equilibrium equations can be found that retains the given definition of σ^{ij} with respect to the *deformed* base vectors ϵ_i, but involves covariant differentiation in the *undeformed* body.

First introduce the body force $\mathbf{p}$ per unit *original* volume and then rewrite eq. (4.7-1) as

$$\int_{S_0} \mathbf{F}\left(\frac{dS}{dS_0}\right) dS_0 + \int_{V_0} \mathbf{p}\, dV_0 = 0 \tag{4.7-10}$$

in terms of integrals over the undeformed body. Let dS be the area of the parallelogram bounded by increments $\overset{(1)}{d\mathbf{x}}$ and $\overset{(2)}{d\mathbf{x}}$ tangent to S; then, since

$$\mathbf{N}\, dS = \overset{(1)}{d\mathbf{x}} \times \overset{(2)}{d\mathbf{x}} = \mathbf{x}_{,i} \times \mathbf{x}_{,j} \overset{(1)}{d\xi^i} \overset{(2)}{d\xi^j}$$

it follows that

$$\mathbf{N} \cdot \mathbf{x}_{,k}\, dS \equiv N_k\, dS = \epsilon_{ijk} \overset{(1)}{d\xi^i} \overset{(2)}{d\xi^j}$$

Similarly

$$\overset{\circ}{\mathbf{N}} \cdot \overset{\circ}{\mathbf{x}}_{,k}\, dS_0 = \overset{\circ}{N}_k\, dS_0 = \overset{\circ}{\epsilon}_{ijk} \overset{(1)}{d\xi^i} \overset{(2)}{d\xi^j}$$

in the undeformed body, whence, by eq. (4.1-14)

$$N_k dS = \sqrt{\frac{g}{\overset{\circ}{g}}}\, \overset{\circ}{N}_k\, dS_0 \tag{4.7-11}$$

Hence, substituting $\mathbf{F} = \sigma^{ij} N_i \boldsymbol{\epsilon}_j$ and $\mathbf{p} = p^j \overset{\circ}{\boldsymbol{\epsilon}}_j$ in eq. (4.7-10) gives

$$\int_{S_0} \sqrt{\frac{g}{\overset{\circ}{g}}}\, \sigma^{ij} \boldsymbol{\epsilon}_j \overset{\circ}{N}_i\, dS_0 + \int_{V_0} p^k \overset{\circ}{\boldsymbol{\epsilon}}_k\, dV_0 = 0$$

Introduce the *Kirchhoff stress tensor*

$$\tau^{ij} \equiv \sqrt{\frac{g}{\overset{\circ}{g}}}\, \sigma^{ij}$$

and apply the divergence theorem, this time in the undeformed body. Then

$$\int_{V_0} [(\tau^{ij} \boldsymbol{\epsilon}_j)_{,i} + p^k \overset{\circ}{\boldsymbol{\epsilon}}_k]\, dV_0 = 0$$

and, since $\boldsymbol{\epsilon}_j = [\delta_j^k + u^k_{,j}]\, \overset{\circ}{\boldsymbol{\epsilon}}_k$, we get

$$[\tau^{ij}(\delta_j^k + u^k_{,j}]_{,i} + p^k = 0 \tag{4.7-12}$$

as the desired equilibrium equation involving covariant differentiation in the undeformed body.

In *linearized* theories of continuum mechanics, no distinctions are made between $\boldsymbol{\epsilon}_i$ and $\overset{\circ}{\boldsymbol{\epsilon}}_i$, between $\mathbf{f}$ and $\mathbf{p}$, and between τ^{ij} and σ^{ij}; the nonlinear terms in η_{ij} are dropped, and the linearization of eq. (4.7-12) reduces it to eq. (4.7-2), with the semicolon replaced by a comma.

4.7.4 Equations of Compatibility

Since the six components of the symmetrical strain tensor η_{ij} depend on just three components of displacement, it is evident that the strain components are mathematically constrained in some way. Indeed, there are six so-called *equations of compatibility* governing the η_{ij} that are a direct consequence of the assertion that the Riemann-Christoffel tensor of the *deformed* body must vanish. From eqs. (4.4-21) and (4.4-24), this condition is equivalent to the six equations

$$\epsilon^{spk}\epsilon^{tij}\left[\frac{1}{2}\frac{\partial^2 g_{pi}}{\partial\xi^k\partial\xi^j}+\frac{1}{2}g_{rm}\Gamma^r_{pi}\Gamma^m_{kj}\right]=0 \tag{4.7-13}$$

But, from eq. (4.4-11)

$$g_{rm}\Gamma^r_{pi}\Gamma^m_{kj}=g^{rm}[pi,r][kj,m]$$

where

$$[pi,r]=\frac{1}{2}\left[\frac{\partial g_{pr}}{\partial\xi^i}+\frac{\partial g_{ir}}{\partial\xi^p}-\frac{\partial g_{pi}}{\partial\xi^r}\right]$$

Now let the ξ^i temporarily represent *Cartesian* coordinates; then $\partial\overset{\circ}{g}_{ij}/\partial\xi^p=0$, partial and covariant derivatives are identical, and eq. (4.7-13) can be written

$$\epsilon^{spk}\epsilon^{tij}\left[\eta_{pi,kj}+\tfrac{1}{2}g^{rm}a_{pir}a_{kjm}\right]=0 \tag{4.7-14}$$

where

$$a_{pir}=\eta_{pr,i}+\eta_{ir,p}-\eta_{pi,r}$$

But since this equation holds for Cartesian coordinates, it must, as a bona fide tensor equation, hold for arbitrary coordinates. Therefore, eq. (4.7-14) are the equations of compatibility in arbitrary curvilinear coordinates. Note that g^{rm} is the inverse of the metric tensor g_{ij} in the *deformed* body, and is therefore defined by the condition

$$g^{rm}(\overset{\circ}{g}_{sm}+2\eta_{sm})=\delta^r_s \tag{4.7-15}$$

The permutation tensors, however, may obviously be replaced by the ordinary permutation symbols.

In the linearized theory the quadratic terms in (4.7-14) are dropped, and the equations then apply to the linearized strain tensor.

4.8 TENSORS IN *n*-SPACE

4.8.1 Euclidean *n*-Space

The theoretical framework presented for the study of tensors in Euclidean 3-D generalizes in a straightforward fashion for the treatment of tensors in Euclidean n-space. It is only necessary to think of the position vector $\mathbf{x}$ as having n Cartesian components, and letting the range of all indices be n rather than 3. Then the dot product of two Cartesian vectors would be given by the sum of the n products of their respective Cartesian components, there would be n orthonormal base vectors $\mathbf{e}_i$ and n general base vectors $\boldsymbol{\epsilon}_i = \partial \mathbf{x}/\partial \xi^i$, the second-order metric tensor g_{ij} would still be given by $\boldsymbol{\epsilon}_i \cdot \boldsymbol{\epsilon}_j$—in short, essentially all of the relations in sections 4.1–4.4 continue to make obvious mathematical sense.* In particular, eq. (4.3-6) giving the typical formula for transforming the components of a tensor under a change of variables continue to hold in Euclidean n-space, and its derivation is unaffected by the dimensions of the space.

4.8.2 Riemannian *n*-Space

A Riemannian space of n dimensions is the generalization of a 2-D curved surface, within which it is *not* generally possible to describe points by means of a Cartesian position vector having n components. Recall that the 2-D surface was described by a Cartesian vector $\mathbf{x}(\xi^1, \xi^2)$ having *three* components; in other words, the 2-D surface was *embedded* in a 3-D Euclidean space. But the possibility of this kind of embedding of a Riemannian space in a Euclidean space of more dimensions is not assumed in the usual general treatment of Riemannian geometry and tensors therein.† A brief sketch of this general theory, without much elaboration, will be given.

The starting point is the assumption that points in the space are specified by coordinates $\xi^i (i = 1, 2, \ldots n)$, and that arc length ds is given by the form $ds^2 = g_{ij} d\xi^i d\xi^j$, where now the g_{ij}'s are some functions of the coordinates. The characterization of the geometry of the space resides wholly in the form of $g_{ij}(\xi^1, \xi^2, \ldots \xi^n)$. If $\bar{\xi}^i (i = 1, 2, \ldots n)$ represents a new set of coordinates, then $d\bar{\xi}^i = (\partial \bar{\xi}^i / \partial \xi^j) d\xi^j$, and it is seen that $d\xi^i$ obeys the rules derived in Euclidean spaces for the transformation of the contravariant components of a vector. Furthermore, since ds^2 must remain invariant under change of coordinates, it follows that $\bar{g}_{ij} = g_{kl} \dfrac{\partial \xi^k}{\partial \bar{\xi}^i} \dfrac{\partial \xi^l}{\partial \bar{\xi}^j}$, and so g_{ij} obeys the rules for the transformation of the covariant components of a second-order tensor. The possibility thus presents itself of *defining* vectors and tensors in an n-dimensional Riemannian space simply on the basis of

*Note, however, that the form of the permutation tensor is tied to the dimensionality of the space, since it is an nth-order tensor in n-space. The appropriate definition for the nth-order permutation tensor $\epsilon_{ijk..pq}$, in generalization of (4.1-11), is the value of the n-by-n determinant containing the Cartesian components of ϵ_i in its first row, those of ϵ_j in its second row, and so on. The formula (4.1-13) for the product of two permutation tensors then generalizes to an n-by-n determinant of g_{ij}'s.

†Actually, it can be shown that a Riemannian space of n dimensions can be embedded in a Euclidean space of $\frac{1}{2} n(n+1)$ dimensions.

laws postulated for the transformation of their components—the same laws that were *deduced* for Euclidean spaces. The typical law is that given by eq. (4.3-6). The inverse g^{ij} of the metric tensor is still defined by $g^{ij}g_{jk} = \delta^i_k$, indices on tensors are raised or lowered by means of the metric tensor, and the quotient laws continue to hold.*

Covariant differentiation of tensors is introduced from a new point of view. Equation (4.4-12) showing the general rule for covariant derivatives is assumed as a *definition*, with the Christoffel symbols, in turn, *defined* by eq. (4.4-11) in terms of the metric tensor. In other words, the relations that were *deduced* from definitions involving base vectors in Euclidean space are postulated in Riemannian spaces. The tensor character of covariant derivatives so defined can then be established by verifying that they satisfy the tensor transformation law given by eq. (4.3-6).

It was shown earlier that, through each point on a 2-D surface, a coordinate system exists for which the Christoffel symbols vanish. An analogous theorem for the existence of such geodesic coordinates holds in Riemannian n-space. Accordingly, just as in surface theory, differentiation facts like $g_{ij,k} = 0$, $\epsilon_{ijk..pq,l} = 0$, and $(A_iB_j)_{,k} = A_{i,k}B_j + B_{j,k}A_i$ are readily established by appeal to their validity in geodesic coordinates and the persistence of tensor equations in all coordinate systems.

Finally, we come to the really essential distinction between Euclidean and non-Euclidean spaces. Whereas the interchange of the order of covariant differentiation is valid in Euclidean n-D, the same can not be asserted in a general Riemannian space. (Recall the similar situation in surfaces.) It remains true that $f_{k,ij} - f_{k,ji} = R^p_{\cdot kij} f_p$, with the Riemann-Christoffel tensor R_{pkij} given by eq. (4.4-21); but now it is not necessarily true that R_{pkij} vanishes. If the Riemann-Christoffel tensor *does* vanish, the space under consideration is said to be 'flat'. This means that there is a one-to-one correspondence between points in the space and points in a Euclidean space of the same dimensions and having the identical metric tensor.

In addition to the symmetries

$$R_{pkij} = -R_{pkji} = -R_{kpij} = R_{ijpk}$$

it also follows from eq. (4.4-21) that

$$R_{pkij} + R_{pijk} + R_{pjki} = 0$$

A careful count reveals, then, that in n-space, there are $\frac{1}{12} n^2 (n^2 - 1)$ essentially

*Once again, the permutation tensor demands special consideration, but is easily defined, in generalization of eq. (4.1-14), via the nth-order permutation symbol $e_{ijk..pq}$ by

$$\epsilon_{ijk..pq} = \sqrt{g}\, e_{ijk..pq}$$

The numerical values of the permutation symbol (either +1, 0, or −1) are those taken on by the permutation tensor, in n-D Euclidean space, with respect to orthonormal base vectors. The tensor character of $\epsilon_{ijk..pq}$ in a Riemannian space follows from its satisfaction of the appropriate transformation law.

different, linearly independent, components of R_{pkij}. (Recall that only R_{1212} survived as an independent component, for $n = 2$, in surface theory.)

The Ricci tensor in n-D is still defined by $R_{ij} = R^{p}_{\cdot ijp}$, but it is no longer true, as it was for $n = 3$, that the vanishing of the Ricci tensor is a sufficient condition for flatness.

4.9 BIBLIOGRAPHY

Brillouin, L., *Les Tenseurs*, Dover Publications, New York, 1946.

Bickley, W. G., and Gibson, R. E., *Via Vector to Tensor*, John Wiley & Sons, New York, 1962.

McConnell, A. J., *Applications of the Absolute Differential Calculus*, Blackie and Son, Glasgow, 1931.

Sedov, L. I., *Introduction to the Mechanics of a Continuous Medium*, Addison-Wesley, Reading, Mass., 1965.

Spain, B., *Tensor Calculus*, Oliver and Boyd, Edinburgh, 1953.

Stoker, J. J., *Differential Geometry*, Wiley-Interscience, New York, 1969.

Struik, D. J., *Differential Geometry*, Addison-Wesley, Reading, Mass., 1961.

Synge, J. L., and Schild, A., *Tensor Calculus*, University of Toronto Press, Toronto, 1949.

5

Functions of a Complex Variable

A. Richard Seebass *

5.0 INTRODUCTION

The answers we seek in subjecting physical models to mathematical analysis are most frequently real, but to arrive at these answers we often invoke the powerful theory of analytic functions. This theory often provides major results with little calculation, and the frequency with which we resort to it is a testimony to its utility. In this chapter we summarize the rudiments of complex analysis. Many techniques that rely on complex analysis, e.g., transforms of various types, are treated in subsequent chapters, and consequently, are not mentioned or mentioned only briefly here. Although it is arranged somewhat differently, nearly all of the material presented here will be found in substantially greater detail in Ref. (5-1). A succinct treatment of the basic results of the theory, with some regard for applications, may be found in Ref. (5-2). A less discursive summary of much of this material is contained in Chapter 7 of Ref. (5-3).

5.1 PRELIMINARIES

A function f of a complex variable $z = x + iy$ is an ordered pair of real functions of the real variables x, y. The natural geometric interpretation of such functions and variables motivates the extension of the results of real analysis to complex functions of a complex variable.

5.1.1 The Complex Plane, Neighborhoods and Limit Points

The rules governing complex numbers $x + iy$ are given in sections 1.2 and 1.3. Sequences, series and continuity of functions of z are treated in sections 2.1 to 2.3. Elementary functions of z are discussed in Chapter 1. Each z is uniquely associated

*Prof. A. Richard Seebass, Aerospace and Mechanical Engineering Dep't., University of Arizona, Tucson, Ariz. 85721.

with the ordered pair of real numbers (x, y) through the Argand diagram of section 1.2.3. To this diagram we add the point $z = 1/w$ with $w \to 0$, called the point at infinity. Augmented in this way, the Argand diagram becomes the complex z-plane; this plane can be thought of as the stereographic projection of a unit sphere onto a plane with one pole corresponding to the origin and the other to the point at infinity.

The open region $|z - z_0| < \delta$ is called the *δ-neighborhood* of z_0, unless z_0 is the point at infinity, which has the δ-neighborhood $|1/z| < \delta$. If every neighborhood of z_0 contains a member of a set of points S other than z_0, then z_0 is called a *limit (or accumulation) point* of the set S. A consequence of this definition is that every neighborhood of a limit point of a set contains an infinite number of members of S. The limit point may or may not be a member of S. If every limit point of a set belongs to the set, the set is closed; if some neighborhood of z_0 consists entirely of points of S, z_0 is called an interior point of S; and if all points of S are interior points, S is said to be an open set. A limit point which is not an interior point is a boundary point of S, and the union of all the boundary points is called the boundary of S, while the union of S and its boundary is a closed set called the closure of S.

5.1.2 Domains, Connectedness, and Curves

Open sets for which it is impossible to find two disjoint open sets such that each one contains points of the set and such that all points of the set are in the two open sets are said to be *connected*; any pair of points in a connected set can be joined by a polygonal line consisting only of points of the set. Connected open sets are called *domains*. We are usually concerned with domains of the complex plane. The union of a domain with its boundary points, i.e., its closure, is called a closed domain. A domain is said to be n-connected if its boundary consists of n connected subsets. In particular, if the boundary is a single set we say the domain is simply connected.

A *curve* in the z-plane is defined by $z = x(s) + iy(s)$ where x and y are continuous functions of the real parameter s defined on the closed interval $[a, b]$. If the point z_0 corresponds to more than one s in $[a, b]$ then z_0 is a multiple point. A curve with no multiple points does not intersect itself and is called a simple curve; a curve with a single multiple point corresponding to $s = a$ and $s = b$ is a simple closed curve. Jordan's theorem proves that such a curve divides the plane into two open domains with the curve as their common boundary, a result that is difficult to prove but intuitively correct.

5.1.3 Functions and Line Integrals

We say that $w = u + iv$ is a function of z if to every value of z in a domain D there corresponds one value of w. The continuity of such functions is defined in section 2.3.4, and we note that w is a continuous function of z if and only if $u(x, y)$ and $v(x, y)$ are continuous functions of x and y. The prescription $w = f(z)$ may be thought of as the mapping of a domain of the complex plane into another domain.

Such a mapping is nondegenerate at a point z_0 if the Jacobian of the transformation (see section 5.5) is not zero there. It is one-to-one if distinct points of D map into distinct points of its image. In this case the connectedness of the domain is preserved.

An expression of the form $w = \log z$ does not define a function (in the strict sense), since for $z = re^{i\theta}$ all values of

$$\log r + i(\theta + 2n\pi)$$

with integer n are possible values of the logarithm. In such cases the older literature refers to "multi-valued functions." To define the logarithm as a function it is necessary to specify the choice of n, that is, the branch of $\log z$ must be specified to define the function $w = \log z$. This procedure is discussed more fully in sections 5.2.4 and 5.2.5.

In extending the ideas of calculus to complex functions consider first the integral of a function of a complex variable on a curve $z(s) = x(s) + iy(s)$, $a \leqslant s \leqslant b$, the curve must be rectifiable, that is, of finite length; this is assured if and only if $x(s)$ and $y(s)$ are of bounded variation in $[a, b]$. If $x'(s)$ and $y'(s)$ are continuous then the length L of the curve is given by the real integral

$$L = \int_a^b \{[x'(s)]^2 + [y'(s)]^2\}^{1/2}\, ds = \int_a^b \left|\frac{dz}{ds}\right| ds$$

and the curve is called a smooth curve. In nearly all applied problems we deal with piecewise smooth curves, that is with curves that are smooth at all but a finite number of points. Hereafter we assume that all our curves are piecewise smooth. We assign a direction to such a curve with the positive direction being from a to b, and call the curve the *path* C from a to b. Let $f(z)$ be a function of z defined at all points of C. Consider the sum

$$\sum_{i=1}^{n} f(\zeta_i)(z_i - z_{i-1})$$

where $\{z_i\}$ is an ordered sequence of points on C with $z_0 = a$ and $z_n = b$, and where ζ_i is any point on C between z_i and z_{i-1}. If this sum approaches a limit, independent of the choice of z_i and ζ_i, as $n \to \infty$ and $\max |z_i - z_{i-1}| \to 0$, then this limit is called the *integral of* $f(z)$ *along* C:

$$\int_C f(z)\, dz = \lim_{n\to\infty,\ \max|z_i - z_{i-1}|\to 0} \sum_{i=1}^{n} f(\zeta_i)(z_i - z_{i-1}) \tag{5.1-1}$$

If $f(z)$ is continuous, then with C rectifiable a limit is always assured and the integral defined by (5.1-1) exists. It follows from this definition that the usual rules for definite integrals apply and that we may resort to real variable theory to determine the general conditions under which the integral exists (sections 2.5 and 2.6). For example, if C contains the point at infinity, then the integral must be treated as an improper one (section 2.7.1). Also, it is immediately obvious from (5.1-1) that the integral in the reverse direction is the negative of (5.1-1) and, if $|f(z)|$ is bounded on C by M, then from the triangle inequality (section 1.3)

$$\left|\int_C f(z)\,dz\right| \leqslant \int_C |f(z)|\,|dz| \leqslant ML \tag{5.1-2}$$

Finally, if we have a series defined on C

$$f(z) = \sum_{i=1}^{\infty} f_i(z) \tag{5.1-3}$$

and uniformly convergent there, then, by means of (5.1-2) and the definition of uniform convergence (section 2.2.5), we can conclude that the integral of the sum (5.1-3) is the sum of the integrals of the individual terms.

5.2 ANALYTIC FUNCTIONS

Most functions of a complex variable that arise in analysis are of a special character. They can be represented by a power series about some point and are infinitely differentiable in a neighborhood of that point. Functions that have one of these properties have the other. Indeed, if the first derivative exists in the neighborhood of a point, then all the higher-order derivatives exist there as well. Of the totality of functions of a complex variable this restricted class, the class of *analytic functions*, is of singular utility. This section defines and examines the basic properties of analytic functions.

5.2.1 Differentiability, Cauchy-Riemann Equations, Harmonic Functions

A function of a complex variable $f(z)$ defined in some neighborhood of a point z is said to be *differentiable* at z if the limit as $\zeta - z \to 0$ of

$$\frac{f(\zeta) - f(z)}{\zeta - z}$$

exists and is unique, i.e., if it is independent of how ζ approaches z. This limit, called the derivative of $f(z)$, is denoted by $f'(z)$:

$$f'(z) \equiv \frac{df}{dz} = \lim_{\zeta \to z} \frac{f(\zeta) - f(z)}{\zeta - z} \tag{5.2-1}$$

If a function has a derivative throughout some neighborhood of the point z we say that the function is *analytic* (*regular*, *holomorphic*) there. The function $f(z)$ is analytic 'at infinity' if $f(1/z)$ is analytic at the origin. For multivalued functions we do not insist on a unique value of the derivative at z for all values that the function assumes at z, nor do we insist on a different value of the derivative for differing values of the function. However, once the branch of the function is selected, i.e., once the value at the point is specified, and hence the values in the neighborhood of that point determined, the value of the limit (5.2-1) must be unique. With this definition all the usual rules of differentiation apply (section 2.4.2); in particular, if $f(z)$ is differentiable at z and $\psi(\zeta)$ differentiable at $\zeta = f(z)$, then $\psi[f(z)]$ is also differentiable at z with

$$\frac{d}{dz}\{\psi[f(z)]\} = \psi'[f(z)]\, f'(z)$$

An important consequence of the definition (5.2-1) follows from the requirement that the limit be independent of the path of ζ as $\zeta \to z$. If we let $\zeta - z = \Delta x + i\Delta y$ and consider the two limits, $\Delta x = 0$, $\Delta y \to 0$ and $\Delta y = 0$, $\Delta x \to 0$, then, with $f(z) = u(x, y) + iv(x, y)$, we conclude that

$$u_x = v_y \quad \text{and} \quad u_y = -v_x \tag{5.2-2}$$

and that u and v should be continuous in x and y. Thus for $f(z)$ to be differentiable at z it is necessary that u and v satisfy the *Cauchy-Riemann equations* (5.2-2). It is sufficient that u and v have continuous partial derivatives in a neighborhood of z, and that their derivatives satisfy (5.2-2) there. Analogously, with $\zeta - z = \Delta r + ir\Delta\theta$, we conclude that in polar coordinates (r, θ)

$$u_r = \frac{1}{r} v_\theta \quad \text{and} \quad \frac{1}{r} u_\theta = -v_r$$

As mentioned earlier, the existence of a first derivative of $f(z)$ will be shown to imply the existence of derivatives of all orders. Thus, in turn, we can conclude that if u and v satisfy the Cauchy-Riemann equations then u and v satisfy Laplace's equation

$$u_{xx} + u_{yy} = v_{xx} + v_{yy} = 0$$

Real functions that satisfy Laplace's equation are said to be *harmonic*. Two harmonic functions u and v that satisfy (5.2-2) are called conjugate harmonic functions (see section 5.5).

5.2.2 Cauchy's and Morera's Theorems, Cauchy's Integral and Plemelj Formulas, Maximum Modulus and Liouville's Theorems, Riemann-Hilbert Problem

Much of the utility of complex analysis results from consequences of *Cauchy's theorem*, as modified by Goursat, which states: *If $f(z)$ is analytic in some simply* *connected domain D, then the integral of f(z) along any closed rectifiable path C_o in D is zero:*

$$\int_{C_o} f(z)\,dz = 0 \tag{5.2-3}$$

Hereafter we will use the notation C_o to indicate a *closed path.*

There are many ways to prove this result; the most basic of these is to show that the existence of a derivative insures that the path integral around some arbitrarily small but finite region in D is zero. The path C_o is then constructed from the sum of the paths around the elementary regions that comprise its interior. An obvious consequence of (5.2-3) is that if $f(z)$ is analytic in D then

$$F(z) = \int_{z_0}^{z} f(\zeta)\,d\zeta$$

is independent of the path in D. From the definition (5.2-1) it follows that

$$F'(z) = f(z)$$

It also follows that the indefinite integral

$$\int f(z)\,dz = F(z) + \text{constant}$$

has the same interpretation as it would for real z. *Morera's theorem* supplies a converse to Cauchy's theorem: *If f(z) is continuous in a simply connected domain D and if*

$$\int_{C_o} f(z)\,dz = 0$$

for any closed path in D, then f(z) is analytic in D.

Cauchy and Morera's theorems apply to simply connected domains; they also apply to reducible paths in multiply connected domains. Two paths are *homotopic* if we can continuously deform one into the other without passing outside the domain. A path that can be shrunk to zero without passing outside the domain is homotopic to zero (reducible). Clearly the proof of Cauchy's theorem holds in multiply connected domains for paths that are homotopic to zero. An integral defined on homotopic paths in a multiply connected domain has the same value for each path. A multiply connected domain can be made simply connected by interconnecting the interior boundary paths with each other and with the exterior boundary by

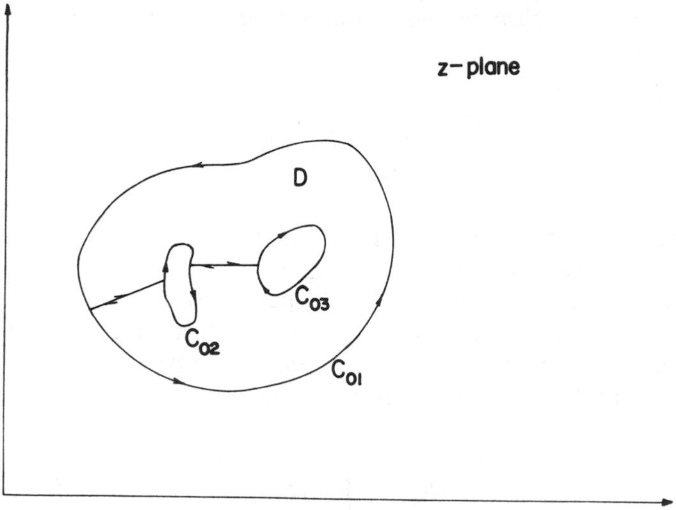

Fig. 5.2-1 Multiply connected domain for (5.2-4).

means of $n - 1$ *cuts* that cannot be crossed by paths of integration. If we integrate along the full boundary of the cut domain, each cut will be traversed twice, but in opposite directions (Fig. 5.2-1); the net result is that the sum of the integrals about the exterior and interior boundaries must be zero

$$\sum_{j=1}^{n} \int_{C_{oj}} f(z)\, dz = 0 \tag{5.2-4}$$

where the C_{oj} form the boundary of D. According to the convention of section 5.1.3, each boundary must be traversed in such a way that D lies on the left, i.e., exterior boundaries are traversed counter-clockwise and interior boundaries clockwise.

A direct consequence of Cauchy's theorem is *Cauchy's integral formula: If $f(z)$ is analytic in a domain D containing a simple closed path C_o and ζ is a point inside C_o, then the value of f at $z = \zeta$ is*

$$f(\zeta) = \frac{1}{2\pi i} \int_{C_o} \frac{f(z)}{z - \zeta}\, dz \tag{5.2-5}$$

The function $f(z)/(z - \zeta)$ is analytic in the doubly connected domain constructed from D by deleting any δ-neighborhood of ζ. Because $f(z)$ is analytic in this neighborhood, the clockwise integral around its boundary is independent of δ and easily deduced to be $-2\pi i f(\zeta)$. Cauchy's integral formula (5.2-5) then follows directly from (5.2-4). For a point in D but outside C_o

$$\frac{1}{2\pi i}\int_{C_o}\frac{f(z)}{z-\zeta}\,dz = 0 \qquad (5.2\text{-}6)$$

To complete the results we need to define the value of the integral for ζ on C_o. Analytic continuation (section 5.2.4) provides the motivation for the proper definition, which is that we interpret the integral as a principal value integral (section 2.7). A simple deformation of the path around ζ then shows that if ζ is on C_o and if C_o is smooth there, then

$$\frac{1}{2\pi i}\int_{C_o}\frac{f(z)}{z-\zeta}\,dz = \frac{1}{2}f(\zeta) \qquad \text{for } \zeta \text{ on } C_o \qquad (5.2\text{-}7)$$

If C has a corner at ζ with interior angle α, then the factor 1/2 on the right-hand side is replaced by $\alpha/2\pi$. It is not necessary that $f(z)$ be analytic at ζ for (5.2-7) to hold, but simply that it satisfy a Lipschitz condition with positive order there.[1] Indeed, *if C is a curve in the complex plane, which need not be closed, and $f(\zeta)$ a function defined on C satisfying a Lipschitz condition of positive order at z, then the function*

$$F(z) = \frac{1}{2\pi i}\int_{C}\frac{f(\zeta)}{\zeta - z}\,d\zeta \qquad (5.2\text{-}8)$$

is analytic everywhere except on C, changes its value discontinuously by $-f(\zeta)$ as z moves across C, from the left to the right, and the principal value of the integral (5.2-8) *with z on C is the mean of the values of* (5.2-8) *with z on each side of C.* The equations which symbolize these results are called the *Plemelj formulas*; they are discussed more fully in Ref. (5-4), pp. 42–55. If C is not closed then at the end points of the curve we need only require that $f(\zeta)$ be integrable.

Cauchy's integral formula (5.2-5) gives an alternate expression for the analytic function $f(z)$. The existence of the derivative of $f(z)$ is assured. Applying the definition (5.2-1) to the integral representation, we find

$$f'(z) = \frac{1}{2\pi i}\int_{C_o}\frac{f(\zeta)}{(\zeta - z)^2}\,d\zeta$$

This procedure can be repeated indefinitely; as a consequence we see that *if $f(z)$ is*

[1] A function $f(t)$ satisfies a *Lipschitz* (*Hölder*) *condition* of order $p\,(>0)$ at t_o if $|f(t) - f(t_o)| < k\,|t - t_o|^p$, where k is any constant, for all t in some neighborhood of t_o.

analytic in D then it has derivatives of all orders in D given by

$$f^n(z) = \frac{n!}{2\pi i} \int_{C_o} \frac{f(\zeta)}{(z - \zeta)^{n+1}} \, d\zeta \tag{5.2-9}$$

and since derivatives of all orders exist, they must all be continuous in D.

Cauchy's integral formula (5.2-5) provides the complex variable analogue of the *mean value theorem* when C_o is specialized to the circular path $|z - \zeta| = R$

$$f(z) = \frac{1}{2\pi} \int_0^{2\pi} f(z + Re^{i\theta}) \, d\theta = \frac{1}{\pi R^2} \int_A f(z + Re^{i\theta}) \, dA$$

Here dA is an element of area of the circle with radius R about z. From this result it follows that

$$|f(z)| \leqslant \frac{1}{\pi R^2} \int_A |f(\zeta)| \, dA$$

and from this inequality we may deduce the *maximum modulus theorem: If $f(z)$ is analytic in a domain D and its boundary C_o, then $|f(z)|$ either attains its maximum value only on C_o or $f(z)$ is constant throughout D.*

Specializing (5.2-9) to a circular path as well, we see that if $M(r)$ is the maximum value of $|f(\zeta)|$ on $|\zeta - f| = r$, then

$$|f^n(z)| \leqslant \frac{n!}{r^n} M(r) \tag{5.2-10}$$

which is known as *Cauchy's inequality.* If $f(z)$ is analytic for all finite values of z, and if $|f(z)| < M$ for all z, then it follows from (5.2-10) that $f'(z)$ must be zero and that $f(z)$ must be a constant. A function that is analytic for all finite values of z is said to be an *entire* or *integral* function. *A bounded entire function is necessarily a constant (Liouville's theorem).*

An important application of the Plemelj formulas is the *Riemann-Hilbert problem*: Determine a function $\mathcal{F}(z)$ that is analytic for all finite z not on a curve C, such that the values of the function as z approaches the curve from the left or right satisfy

$$\mathcal{F}_L(\zeta) = g(\zeta) \, \mathcal{F}_R(\zeta) + h(\zeta) \tag{5.2-11}$$

where $g(\zeta)$ and $h(\zeta)$ are specified functions of ζ defined on C. The function is of course not unique unless we specify its behavior at infinity, and, as we shall, at the end points of C. We have already seen that (5.2-8) provides the solution to the simpler problem

$$F_L(\zeta) - F_R(\zeta) = f(\zeta)$$

Consequently, the function $G(\zeta) = \exp F(\zeta)$, which will also be analytic for all finite ζ if F is, satisfies $G_L(\zeta) = g(\zeta)\, G_R(\zeta)$ where $g(\zeta) = \exp f(\zeta)$. Solving this relation for $g(\zeta)$ and substituting into (5.2-11) we obtain another relation solvable by (5.2-8)

$$\frac{\mathfrak{F}_L(\zeta)}{G_L(\zeta)} - \frac{\mathfrak{F}_R(\zeta)}{G_R(\zeta)} = \frac{h(\zeta)}{G_L(\zeta)}$$

provided $h(\zeta)/G_L(\zeta)$ satisfies the conditions for its application. There is a certain amount of flexibility in the behavior of $\mathfrak{F}(z)$ at infinity and at the end points of C. We could equally well replace $G(z)$ by

$$G(z) = (z - \zeta_1)^{-m}(z - \zeta_2)^{-n} G(z)$$

where m and n are positive integers and ζ_1 and ζ_2 are the end points of C. The choice of m and n determines the behavior (usually singular) of $\mathfrak{F}(z)$ as $z \to \zeta_1, \zeta_2$, and after an entire function has been added to $\mathfrak{F}(z)/G(z)$ its behavior as $z \to \infty$. The arbitrariness in the mathematical solution to (5.2-11) does not arise in applied problems where physical conditions restrict the singular behavior of the solution at the end points of C, as well as the behavior at infinity.

5.2.3 Taylor Series, Radius of Convergence, Schwarz's Lemma

A series of the form

$$\sum_{n=0}^{\infty} a_n(z - z_0)^n \tag{5.2-12}$$

is called a *power series*. It is not hard to prove that if the series (5.2-12) converges for $z = z_i$, then it converges for any z such that $|z - z_0| < |z_i - z_0|$. The least upper bound R of the set $\{|z_i - z_0|\}$ for which (5.2-12) converges is called the radius of convergence of the series. The series (5.2-12) converges uniformly and absolutely for $|z - z_0| < R$ and diverges for $|z - z_0| > R$; the radius of convergence is given by the Cauchy-Hadamard formula

$$R = \liminf_{n\to\infty} \left\{\frac{1}{\sqrt[n]{|a_n|}}\right\} \tag{5.2-13}$$

If the power series (5.2-12) is differentiated term-by-term we obtain a new power series with the same radius of convergence; the derivative of the new series will also have the same radius of convergence, and so forth. Thus the power series (5.2-12) is an analytic function of z for $|z - z_0| < R$.

Consider any closed path C_o about z_0 lying within the circle $|z - z_0| = R$. The function $f(z)$ is analytic in and on C_o and Cauchy's integral formula gives an alternate representation of $f(z)$ for any point z inside C_o,

$$f(z) = \frac{1}{2\pi i} \int_{C_o} \frac{f(\zeta)}{\zeta - z} \, d\zeta = \frac{1}{2\pi i} \int_{C_o} \frac{f(\zeta)}{\zeta - z_0} \left(1 - \frac{z - z_0}{\zeta - z_0}\right)^{-1} d\zeta$$

Since $|(z - z_0)/(\zeta - z_0)|^{-1} < 1$ the series

$$\left(1 - \frac{z - z_0}{\zeta - z_0}\right)^{-1} = 1 + \frac{z - z_0}{\zeta - z_0} + \left(\frac{z - z_0}{\zeta - z_0}\right)^2 + \cdots$$

is uniformly convergent for all points ζ on C_o; integrating term-by-term, we find

$$f(z) = \sum_{n=0}^{\infty} a_n (z - z_0)^n = \sum_{n=0}^{\infty} \frac{f^n(z_0)}{n!} (z - z_0)^n \tag{5.2-14}$$

Both series converge for any z such that $|z - z_0| < R$. The series (5.2-14) is the *Taylor series* for $f(z)$ about $z = z_0$ (section 2.2.6). The remainder after n terms can easily be calculated to be

$$\sum_{i=n}^{\infty} a_i (z - z_0)^i = \frac{(z - z_0)^n}{2\pi i} \int_{C_o} \frac{f(\zeta)}{(\zeta - z_0)^n \, (\zeta - z)} \, d\zeta$$

In many applications it is necessary to know the Taylor series development of an analytic function about some point z_0. The prescription (5.2-14) is seldom the easiest route to the series unless $f(z)$ is easily differentiated repetitively, e.g., e^z. When it is not, the function can often be decomposed into products and sums of functions whose series are known or that can be easily differentiated indefinitely.

There are several immediate consequences of the existence of a Taylor series for analytic functions. If a function is analytic at z_0 and has a zero of order (or multiplicity) p there as well, then

$$\frac{f(z)}{(z - z_0)^p}$$

is analytic there; if $f(z)$ is analytic in a domain D, and if $f(z)$ is not constant inside some simple closed curve lying in D, then the roots of $f(z) = w_0$ are isolated; i.e., each z_0 such that $f(z_0) = w_0$ has a δ-neighborhood of z_0 in which $f(z) \neq w_0$ for $z \neq z_0$; if $f(z)$ is an entire function and if its absolute value increases more slowly than the m-power of $|z|$ as $z \to \infty$, where m is an integer, then $f(z)$ is a polynomial of degree less than m.

Another simple consequence of the existence of a Taylor series is *Schwarz's lemma: If a function $f(z)$ is analytic inside the unit circle $|z| < 1$ and satisfies*

$$f(0) = 0 \quad \textit{and} \quad |f(z)| \leqslant 1 \quad \textit{for} \quad |z| \leqslant 1$$

then $|f(z)| \leqslant |z|$ for any z with $|z| < 1$ and if equality holds at any interior point then it holds everywhere and $f(z) = e^{i\alpha} z$ where α is a real constant. With $f(z) = a_1 z + a_2 z^2 + \cdots$ we see that the function $f(z)/z$ is analytic inside the unit circle. Application of the maximum modulus theorem (section 5.2.2) to this function then gives the result.

5.2.4 Uniqueness Theorem, Analytic Continuation, Monogenic Analytic Functions, Schwarz's Reflection Principle

Perhaps the most important consequence of the existence of the Taylor series representation of an analytic function is the *uniqueness theorem* which is the basis of the powerful principle of *analytic continuation.* The theorem states: *if $f(z)$ is analytic in a domain D and if it vanishes at an infinite sequence of points in D, with a limit point z_L in D, then $f(z)$ vanishes on the entire domain.*

If a function $g(z)$ is defined on a set of points S in a domain D and with a limit point in D, and if $f(z)$ is an analytic function of z in D with $f(z) = g(z)$ on the set S, then $f(z)$ is called the *direct analytic continuation* of $g(z)$. According to the uniqueness theorem if the set S has a limit point then $f(z)$, the analytic continuation of $g(z)$, is unique.

In many applications it is necessary to extend the definition of a function by analytic continuation. This is, in fact, the procedure by which we give meaning to the complex analogues of real functions. Consider, for example, the real function

$$\log x = \int_1^x \frac{dt}{t} \tag{5.2-15}$$

which is defined for all finite real positive values, and the function

$$\log z = \int_{1\,C}^{z} \frac{d\zeta}{\zeta}; \qquad z \neq 0, \infty \tag{5.2-16}$$

where C is any path from 1 to z not passing through $z = 0$ or $z = \infty$. For example, we could proceed along the real axis from 1 to $|z|$ and from $|z|$ along the circular arc $\zeta = |z|\,e^{i\theta}$ to z, where θ varies from 0 to arg (z), i.e., to θ_0 where $z = |z|\,e^{i\theta_0}$. There are two circular arcs from $|z|$ to z; if we choose the arc that does not cross the negative real axis, $\overline{C}$, then

$$\int_{1_{\overline{C}}}^{z} \frac{d\zeta}{\zeta} = \log_p z = \log |z| + i \arg (z), \quad |\arg (z)| < \pi \tag{5.2-17}$$

where we recognize that $\log_p z$ is the principal value of the function $\log z$ (section 1.4.5). The function (5.2-17) is not analytic at the singular points $z = 0$ or ∞; and because its value changes discontinuously on some ray, taken here to be the negative real axis, it is not analytic there either. Yet the function (5.2-15) can be analytically continued by (5.2-16) to any point in the z-plane, with the exception of $z = 0$ and ∞. Indeed, for any two paths not passing through 0 or ∞ we have

$$\int_{1_{C_1}}^{z} \frac{d\zeta}{\zeta} - \int_{1_{C_2}}^{z} \frac{d\zeta}{\zeta} = 2\pi i n$$

where n is the number of times that C_1 encircles the origin counter-clockwise minus the number of times C_2 does likewise; both integrals provide an analytic continuation of the function $\log x$ to complex values. Each integral of the form

$$f(z, D_i) = \int_{1_{C_i}}^{z} \frac{d\zeta}{\zeta}$$

which defines $f(z)$ on a domain D_i, is termed a *function-element* $f(z, D_i)$ of the *monogenic* (or *complete*) *analytic function* defined by the totality of its function elements. Any two function elements of the complete (and multivalued) analytic function $\log z$ defined by (5.2-16) satisfy

$$\log_1 z - \log_2 z = 2\pi i n$$

We observe that two branches of the function $\log z$ differ in value only if their paths for the analytic continuation of (5.2-15) encircle the singular point $z = 0$ to which the function cannot be continued in different directions or a different number of times, i.e., their combined path constitutes a circuit about the origin. It is, in fact, true in general that if there are no singular points between the paths of integration then the result of analytic continuation is the same for each path: *If a function element can be continued analytically along every path in a simply connected domain, then the resulting function is single-valued* (*monodromy theorem*). For doubly connected domains, two analytic continuations of the same function-element coincide if their paths encircle the excluded interior region the same number of times in the same direction. For n-connected domains ($n > 2$) the analytic continuation depends both on the number of times the excluded regions are encircled and on the order in which these encirclements occur. A boundary beyond which no analytic continuation is possible is called a *natural boundary*. Thus the unit circle is the natural boundary of the function

$$f(z) = \sum_{n=0}^{\infty} z^{2^n}$$

which satisfies $f(z) = z + f(z^2)$; since $z = 1$ is a singular point of the series $[f(x) \to \infty$ as $x \to 1$ through real $x < 1]$, it follows from the functional relationship that all points $z = e^{i\theta}$ with $z^{2^p} = 1 (p = 1, 2, \ldots)$ are singular points. These are points of the form $z = e^{i\theta}$ with $\theta = 2\pi q/2^p$ where $(q, p = 1, 2 \ldots)$. Thus every point of the unit circle is the limit point of singular points. But a regular point would have a δ-neighborhood free from singular points; consequently every point is a singular point.

As mentioned above, it is often useful, if not necessary, in practical applications to continue analytically some function under investigation. The straightforward procedure is to construct the Taylor series of the function near its "artificial" boundary; this procedure can, in principle, be continued through a chain of circular domains to the function's natural boundaries. In practice this is often difficult. In many cases we can short-cut this procedure. For example, *if $f(z)$ is defined on a domain D that has a portion of the real axis as one boundary, and if $f(z)$ is continuous up to the boundary and assumes real values for real z then the function*

$$f(z^*) = f^*(z)$$

represents the analytic continuation of $f(z)$ to the values z^* (Schwarz's reflection principle). With certain restrictions this principle can be extended to any curve (see e.g., Ref. 5-5 pp. 221, 222) through conformal mapping (section 5.5). Thus, the analytic continuation of $f(z)$ defined in and real on the unit circle to the domain exterior to $|z| = 1$ is given by $f(1/z^*) = f^*(z)$.

As an example of the process of analytic continuation consider the Gamma function

$$\Gamma(x) = \int_0^{\infty} t^{x-1} e^{-t} \, dt \tag{5.2-18}$$

The integral converges for (real) $x > 0$. (This function was introduced by Euler to interpolate the factorial function $n!$.) Integration by parts shows that $x\Gamma(x) = \Gamma(x + 1)$, and since $\Gamma(1) = 1$, $n! = \Gamma(n + 1)$ for the positive integers. If we define

$$t^{z-1} e^{-t} = e^{-t+(z-1) \log t}$$

with $\log t$ real, then we can replace x in (5.2-18) by the complex variable $z = x + iy$.

The integral

$$\Gamma(z) = \int_0^\infty t^{z-1} e^{-t}\, dt \tag{5.2-19}$$

is absolutely convergent for $x > 0$ and uniformly convergent in every half-plane $x \geqslant \delta > 0$. Now the integral

$$\int_{C_o} \Gamma(z)\, dz = \int_{C_o} dz \int_0^\infty e^{-t} t^{z-1}\, dt = 0$$

about any closed path C_o in $x \geqslant \delta > 0$ (change the order of integration). Thus by Morera's theorem $\Gamma(z)$ given by (5.2-19) represents the analytic continuation of the Gamma function to $Re(z) > 0$. To extend the continuation to the full plane write (5.2-19) as

$$\Gamma(z) = \int_0^1 t^{z-1} \sum_{n=0}^{\infty} (-1)^n \frac{t^n}{n!}\, dt + \int_1^\infty t^{z-1} e^{-t}\, dt; \qquad \mathrm{Re}(z) > 0$$

The series is uniformly convergent on the interval [0, 1] and can be integrated term by term to give *Prym's decomposition*

$$\Gamma(z) = \sum_{n=0}^{\infty} \frac{(-1)^n}{n!(n+z)} + \int_1^\infty t^{z-1} e^{-t}\, dt \tag{5.2-20}$$

The sum converges everywhere but at the points $z = 0, -1, -2, \ldots$. By the uniqueness theorem, (5.2-20) is the analytic continuation of the Gamma function to the complex plane.

5.2.5 Riemann Surfaces and Branch Points

The analytic continuation of a given function along different paths leading to the same point may yield different results. The notion of a *Riemann surface* is useful in depicting the situations that can arise. The Riemann surface of a function $w = f(z)$ is the surface that represents all the values of w for any point in the z-plane; it is sometimes visualized by depicting its multiple sheets one above the other, as in Figs. 5.2.2 and 5.2.3. To each possible result of the continuation of a function along differing paths ending at the same point z_0 we associate one point of the surface. If p_0 is a point corresponding to the function element $f(z) = \sum a_n (z - z_0)^n$, then we define as a δ-neighborhood of p_0 on the Riemann surface all points corresponding to function elements $\sum b_n (z - z_1)^n$ such that $|z_1 - z_0| < \delta$ and such that $\sum b_n(z - z_1)^n$ is a direct analytic continuation of $\sum a_n(z - z_0)^n$. For the purpose of visualizing the surface better, we introduce the notion of a sheet. Suppose that for each z in some region D of the z-plane we can single out a point p of the Riemann surface which lies above z in such a way that the point p' determined in the same manner for z' in D lies in a δ-neighborhood of p provided that $|z' - z|$ is sufficiently small. We say that the collection of points singled out in this way form a *sheet* of the Riemann surface above D.

As an example, consider the expression $w = \log z$, and take D to be the z-plane with the negative real axis removed. A sheet S_n of the Riemann surface of log z above D is defined by taking all the points of the surface corresponding to values of w satisfying

$$(2n - 1)\pi < \text{Im}(w) < (2n + 1)\pi$$

where n is an integer. We can think of the sheets S_n as a pile of copies of the z-plane, cut along the negative real axis, lying above the z-plane. The *edges* of the sheet S_n correspond to function-elements of log z which assign the values log $|x| \pm i\pi n$ to negative real x. Clearly, we should identify one edge of S_n with one edge of S_{n-1} and the other edge of S_n with one edge of S_{n+1}. These sheets make up the spiral ramp-like Riemann surface for the function log z that is pictured in Fig. 5.2.2.

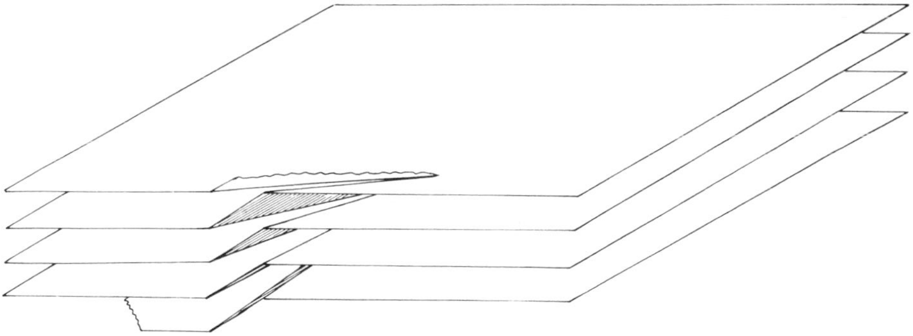

Fig. 5.2-2 Sketch of a portion of the Riemann surface corresponding to log z.

For any monogenic analytic function there is a one-to-one correspondence between values of the function and points of its *Riemann surface*. This surface is made up of at most a denumerably infinite set of sheets on which individual *branches* of the function are defined. These planes are joined along *branch cuts* that connect *branch points*, where the monogenic function fails to be analytic, with each other or with the point at infinity. If $n + 1$ circuits of a branch point carry every branch into itself, and if $n + 1$ is the smallest such integer, we say that the point is a *branch point of order n*. If this never occurs we say the point is a *logarithmic branch point*.

The construction of the Riemann surface of a monogenic analytic function often requires considerable ingenuity. Consider the simple example $w = z^{1/2}$. For each nonzero z there are two possible values, one being the negative of the other. With $z = re^{i\theta}$ and $-\pi < \theta < \pi$, then the values $w = r^{1/2}e^{(1/2)\theta}$ provide a sheet of the Riemann surface with a branch cut along the negative real axis. At the edge $\theta = -\pi$ the value $-i\sqrt{r}$ is to be taken, while along $\theta = \pi$, $+i\sqrt{r}$ is the appropriate value. A second sheet is given by $w = -r^{1/2}e^{(1/2)\theta}$. On the edge above $\theta = -\pi$, w takes the value $i\sqrt{r}$ and the edge of this sheet is to be identified with the edge of the first sheet above $\theta = \pi$. Similarly, the remaining two edges are to be identified with each other. This results in a two-sheeted surface which, while it cannot be constructed from physical sheets of paper without allowing intersections, is sketched in Fig. 5.2-3. The two sheets are only joined in the sense indicated by the arrows. The

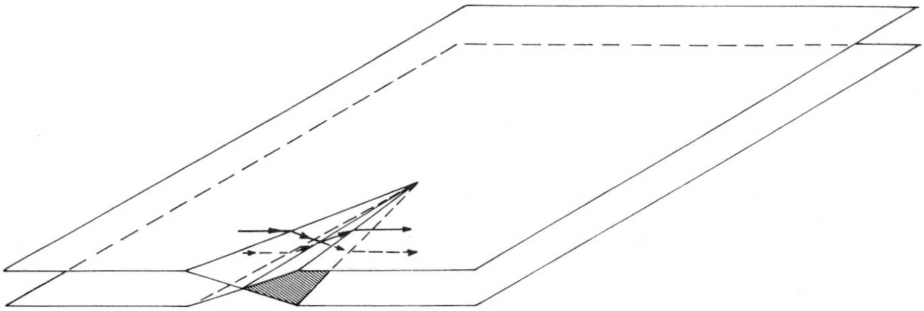

Fig. 5.2-3 Sketch of the Riemann surface corresponding to $z^{1/2}$.

function $z^{1/n}$ has the n-sheeted generalization of this two-sheeted Riemann surface and represents the monogenic analytic continuation of the real function $x^{1/n}, x > 0$.

5.3 SINGULARITIES AND EXPANSIONS

It is the singularities of analytic functions that make them interesting and useful. The term 'singularity' has occurred earlier in contexts where it was clear what the term meant; henceforth we shall be more specific and call any point where a function-element of a monogenic analytic function $f(z)$ ceases or fails to be analytic a *singularity* of $f(z)$. Thus, if $f(z)$ is analytic in a domain D, a singular point is a boundary point z_b of D such that there is no direct analytic continuation of $f(z)$ from D to a domain containing z_b. The Taylor series representation of $f(z)$ about a point z_0 in D has a radius of convergence equal to the distance from z_0 to the nearest singular point, i.e., there is at least one singular point on the circle $|z - z_0| = R$, where R is the radius of convergence of the Taylor series.

A singularity is called an *isolated singularity* of $f(z)$ if $f(z)$ is analytic at every point but the singular point in some δ-neighborhood of the singularity. Note that this means $f(z)$ is single-valued in the annulus about the singular point z_0.

5.3.1 Laurent Series, Poles, Essential Singularities, Singularities of Elementary Functions

We begin our study of singularities with the principal result needed to understand isolated singularities, the Laurent series. If $f(z)$ is analytic and single-valued (e.g., a branch of a multi-valued function) in a domain D with the exception of an isolated singular point at z_0, then the function can be evaluated by (5.2-5) for two circular paths in D, $|z - z_0| = \delta, r$ with $0 < \delta < r$ (see Fig. 5.3-1).

$$f(z) = \frac{1}{2\pi i}\left\{\int_{|z-z_0|=r} \frac{f(\zeta)}{\zeta - z} d\zeta - \int_{|z-z_0|=\delta} \frac{f(\zeta)}{\zeta - z} d\zeta\right\}$$

Following the same procedure we used to arrive at the coefficients for a Taylor series we find that $f(z)$ is given by the *Laurent series*

$$f(z) = \sum_{n=-\infty}^{\infty} a_n (z - z_0)^n \tag{5.3-1}$$

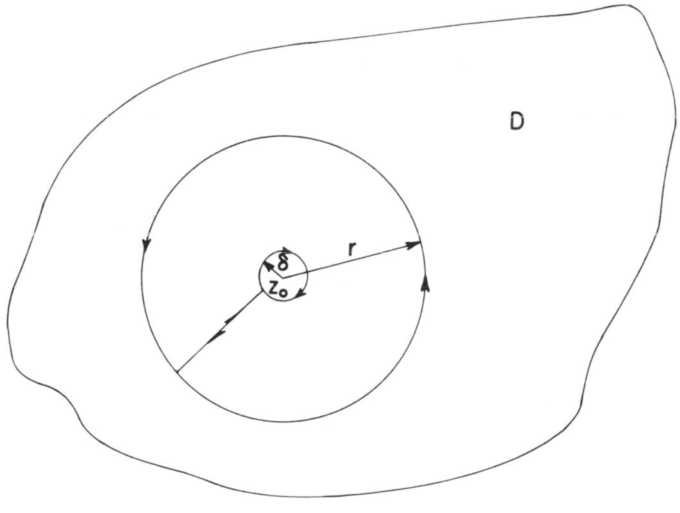

Fig. 5.3-1 Path of integration for the development of the Laurent series.

where

$$a_n = \frac{1}{2\pi i}\int_{C_0} \frac{f(\zeta)}{(\zeta - z_0)^{n+1}} d\zeta \tag{5.3-2}$$

and C_o is any closed curve in D about z_0. The portion of the series consisting of negative powers of $z - z_0$ is called the *principal part* of the series. The coefficient

a_{-1} is called the *residue* at z_0 for $z_0 \neq \infty$; the residue is related to the integral of the values of $f(z)$ about C_o by (5.3-2):

$$\int_{C_o} f(\zeta)\, d\zeta = 2\pi i a_{-1} \tag{5.3-3}$$

If $f(z)$ is analytic or has an isolated singularity at infinity, then $f(z)$ has the Laurent series

$$f(z) = \sum_{n=-\infty}^{\infty} a_n z^n$$

in a neighborhood of the point at infinity; its residue is defined as $-a_{-1}$ and (5.3-3) is satisfied with C_o traversed clockwise. Thus, for example, z^{-1} is analytic at infinity but has the residue -1 there.

If the principal part has a finite number of terms m, then we say that $f(z)$ has a *pole of order m* at z_0; clearly this means that $(z - z_0)^m f(z)$ is analytic at z_0. If the principal part has an infinite number of terms then we say that z_0 is an *isolated essential singularity* of $f(z)$. If we suppose that $|f(z) - c| > \delta$, where c is some complex number, in a neighborhood of an isolated essential singularity z_0, then $[f(z) - c]^{-1}$ has at worst a removable singularity at z_0 and can be expanded in a Taylor series about z_0

$$[f(z) - c]^{-1} = a_0 + a_1(z - z_0) + \cdots$$

Either $a_0 \neq 0$ and $f(z)$ is bounded as $z \to z_0$, which is a contradiction, or $a_0 = a_1 = \cdots = a_m = 0$ and $a_{m+1} \neq 0$, in which case $f(z)$ has a pole of order m, another contradiction. Thus for any c, $f(z) - c$ must be arbitrarily small in some neighborhood of z_0: In every neighborhood of an isolated essential singularity $f(z)$ comes arbitrarily close to every complex number (*Weierstrass' theorem*) and, indeed, one can show that in any neighborhood of z_0 it attains every possible complex value with at most one exception (*Picard's theorem*). *Thus if $f(z)$ has an isolated essential singularity then it has no limit, as $z \to z_0$.*

The nature of an isolated singularity is reflected in the behavior of the function near the singularity. A point z_0 is a removable singularity if and only if $f(z)$ is bounded in $0 < z - z_0 < \delta$, it is a pole of order m if and only if $(z - z_0)^m f(z) \to$ constant as $z \to z_0$; and it is an *essential singularity* if and only if $f(z)$ does not tend to a limit, finite or infinite, as z tends to z_0.

Zeros and (by definition) poles of analytic functions occur at isolated points, i.e., there exists some δ-neighborhood of a zero (pole) that contains no other zero (pole). *If z_L is a limit point of a sequence of zeros of $f(z)$, then $f(z) \equiv 0$ or z_L must be an isolated singularity without a defined limit, i.e., an essential singularity.* An example is $\sin(1/z)$ with zeros $z = \pm 1/n\pi$; $z = 0$ is an isolated essential singularity. In the same way we see that a limit point of poles cannot be a pole and, because no limit is definable, *a limit point of poles must be a nonisolated essential singularity*. An example is $\tan(1/z)$ which has the poles $z = \pm 1/n\pi$.

It must be borne in mind that there are many singularities which are not isolated singularities, such as algebraic and logarithmic branch points (section 5.2.5), points on natural boundaries (section 5.2.5), continuous distributions of poles, etc.

It is, as was mentioned, the singularities of analytic functions that make them interesting. All the elementary functions have singularities of various types. The functions e^z, $\sin z$ and $\cos z$ are entire, i.e., analytic for all finite z, and have an essential singularity at infinity. The functions $\tan z$, $\cot z$, $\sec z$, and $\csc z$ have only poles in the finite plane; the point at infinity is a limit point of these poles and hence an essential singularity. The function $\sin^{-1} z (= i \log(\sqrt{1 - z^2} - iz))$ has branch points of second order at ± 1 and a logarithmic branch point at infinity while $\tan^{-1} z$ has logarithmic branch points at $\pm i$. The function $\log z$ has logarithmic branch points at zero and infinity; $z^{m/n}$ has branch points of finite order at 0 and infinity for integer values of m and n, and z^c has logarithmic branch points there for $c \neq m/n$.

5.3.2 Entire and Meromorphic Functions, Weierstrass' Factorization, Mittag-Leffler Theorem, Phragmén-Lindelöf Theorem

The simplest of the analytic functions are *entire* functions, i.e., functions that are analytic everywhere in the finite plane. The only entire function analytic at infinity is a constant (section 5.2.2). Polynomials are simple examples of entire functions. If an entire function $f(z)$ has no zeros then

$$g(z) = \frac{f'(z)}{f(z)} = \frac{d \log f(z)}{dz}$$

is also entire; consequently $f(z)$ must be of the form

$$f(z) = \exp G(z)$$

where $G(z)$ is an entire function. We can construct entire functions with zeros of given order at given points of the z-plane. Consider an increasing sequence of complex numbers

$$0, z_1, z_2, \ldots, z_n, \ldots,$$

where $|z_n| < |z_{n+1}|$. Then we can construct an entire function with zeros of order m_n at z_n. Obviously the sequence can only have a limit point at $z = \infty$ for the limit point of zeros is an essential singularity. A function with prescribed zeros can be represented by an infinite product of the form

$$z^{m_o} \prod_{n=1}^{\infty} \left(1 - \frac{z}{z_n}\right)^{m_n}$$

if it converges. If it doesn't, convergence can be obtained by multiplying each term by exp $\{P_n(z)\}$ where $P_n(z)$ is a polynomial. In fact, without further information regarding the z_n's the choice

$$P_n(z) = \frac{z}{z_n} + \frac{1}{2}\left(\frac{z}{z_n}\right)^2 + \cdots + \frac{1}{k_n}\left(\frac{z}{z_n}\right)^{k_n}$$

will suffice to guarantee uniform convergence for some choice of integer k_n (see section 2.10.2). Consequently, *an entire function with its only zeros occurring at the points* $0, z_1, \ldots, z_n, \ldots$, *and having the orders* m_n, *is given by the Weierstrass' Factor theorem*

$$f(z) = z^{m_o} \exp g(z) \prod_{n=1}^{\infty} \left\{\left(1 - \frac{z}{z_n}\right) \exp\left[\frac{z}{z_n} + \frac{1}{2}\left(\frac{z}{z_n}\right)^2 + \cdots + \frac{1}{k_n}\left(\frac{z}{z_n}\right)^{k_n}\right]\right\}^{m_n} \tag{5.3-4}$$

where $g(z)$ is an entire function, for a suitable choice of integers k_n; $k_n = n$ or equal to the greatest integer not exceeding $\log n$ will do. If we have further information regarding the z_n's, then we can say more about $P(z)$. For example, if ν is the largest positive integer for which

$$\sum_{n=1}^{\infty} |z_n|^{-\nu} \tag{5.3-5}$$

diverges then we may take

$$P_n = \frac{z}{z_n} + \cdots + \frac{1}{\nu}\left(\frac{z}{z_n}\right)^{\nu}$$

If the series (5.3-5) converges for $\nu = 1$ we can take $P_n = 1$. Thus, for example, the function

$$\prod_{n=1}^{\infty} \left(1 - \frac{z}{n}\right) e^{z/n}$$

is an entire function with zeros at $z = 1, 2, \ldots$.

The reciprocal of an entire function is an analytic function whose only singularities in the finite plane are poles. Such functions are called *meromorphic functions*. The ratio of two polynomials, i.e., a *rational function*, is a meromorphic function. One can use Liouville's theorem to show that any single-valued function with only poles in the finite plane and at infinity is a rational function. The logarithmic derivative of (5.3-4) provides a function with poles at $z = 0, \ldots, z_n, \ldots$, and residues of m_n. Generalizing to functions with prescribed principal parts we have the *Mittag-Leffler theorem: If $f(z)$ is a meromorphic function having poles with the prescribed principal parts*

$$P_n(z) = \sum_{j=1}^{p_n} a_{-j}^n (z - z_n)^{-j}$$

then it is possible to find polynomials $\tilde{P}_n(z)$ and an entire function $g(z)$ such that the series representation

$$f(z) = \sum_{n=0}^{\infty} [P_n(z) + \tilde{P}_n(z)] + g(z) \tag{5.3-6}$$

is uniformly convergent wherever $f(z)$ is analytic. A classic application of the result (5.3-4) is the extension of the factorial function $n!$ to the gamma function $\Gamma(z)$, where $\Gamma(n+1) = n!$ and $z\Gamma(z) = \Gamma(z+1)$. The solution of this functional relationship is a meromorphic function with simple poles at $z = -n, n = 0, 1, \ldots,$ and residues $(-1)^n/n!$; see (5.2-20). The only function that has only these poles and no zeros must be of the form

$$f(z) = \exp\,[-g(z)]\,\frac{1}{z\prod_{n=1}^{\infty}\left(1 + \frac{z}{n}\right)e^{-z/n}}$$

where $g(z+1) - g(z) = 2k\pi i + \gamma$ and γ is Euler's constant[2]; the simplest choice $g(z) = \gamma z$ gives agreement with the definition of the gamma function (5.2-20). See also section 7.2.4.

It is frequently useful to be able to establish bounds on analytic functions that have specified singularities in given domains. There are a number of theorems that provide such results; we discuss two of these very briefly. By applying the mean value theorem (section 5.2.2) to the function $\log f(z)$ where $f(z)$ is meromorphic in $|z| \leqslant R$ with n zeros of orders m_n at $z_1, z_2, \ldots, z_n$ and k poles with order p_k at $z_{(1)}, z_{(2)}, \ldots, z_{(k)}$, with $z_i, z_{(j)} \neq 0$, one deduces *Jensen's formula*

$$\log|f(0)| + \log\left(R^{n-k}\,\frac{|z_{(1)}|^{p_1}\,|z_{(2)}|^{p_2}\cdots|z_{(k)}|^{p_k}}{|z_1|^{m_1}\,|z_2|^{m_2}\cdots|z_n|^{m_n}}\right) = \frac{1}{2\pi}\int_0^{2\pi}\log|f(Re^{i\theta})|\,d\theta \tag{5.3-7}$$

[2] Euler's constant is defined by

$$\lim_{n\to\infty}\left[\sum_{j=1}^{n}(j)^{-1} - \log n\right]$$

relating $|f(0)|$ to the location of the zeros and poles of $f(z)$ and its boundary values. This result can be generalized by using Poisson's formula (5.5-5) to a form that applies to any z with $|z|<R$. See, e.g., Ref. 5-1, p. 75.

We conclude with a result for infinite domains that is a consequence of the mean value theorem applied to the function $f(z)\exp(-\epsilon z^{\beta})$ where $\epsilon>0$ and $\beta<1$, for $\mathrm{Re}(z)\geqslant 0$. The result of this application can be generalized by conformal mapping and some sophisticated reasoning to the *Phragmén-Lindelöf theorem: Consider a simply connected domain D having the point at infinity and 0 among its boundary points C. Let $s(r)$ denote the arc length of the circle $|z|=r$ determined by its intersection with the boundary C of D. If $f(z)$ is analytic in D (except for the point at infinity), and if $f(z)$ satisfies $|f(z)|\leqslant 1$ for z on C and* $\lim_{r\to\infty}\log M(r)/\rho(r)=0$, *where*

$$\rho(r)=\exp\left\{\pi\int_1^r \frac{dt}{s(t)}\right\}$$

and M is the maximum value of $|f(z)|$ on $z=r$, then $f(z)\leqslant 1$ throughout D. An immediate corollary that can be proved directly by the means sketched above is that if D is the angular sector with central angle π/α with $\rho(r)=r^{\alpha}$, then either $|f(z)|\leqslant 1$ throughout the sector or $\log M(r)>(Kr)^{\alpha}$ for some positive K and r greater than some r_0. Thus, for example, the function $f(z)=\exp(z^{\alpha})$ has $|f(z)|\leqslant 1$ on $|\arg z|=\pi/2\alpha$; the logarithm of its maximum modulus is r^{α}. We see that the result leaves little room for refinement. More general results are given in Ref. 5-6, pp. 250–252.

5.4 RESIDUES AND CONTOUR INTEGRALS

One of the most powerful results of complex variable theory is Cauchy's residue theorem, which states that if $f(z)$ is analytic in a domain D, with the exception of an isolated singularity point z_0 in D, then the integral of $f(z)$ on any closed path about z_0 is given by (5.3-3), i.e., the integral is $2\pi i a_{-1}$, where a_{-1} is the residue of the pole at z_0. This result can obviously be extended to a domain with any (finite) number of isolated singular points, and, by using (5.2-7), to allow simple poles on the path of integration. *If $f(z)$ is analytic and single valued in a domain D and its boundary C_o, except at isolated singularities z_i in D with residues a_{-1_i}, and except at simple poles on C_o with residues a_{-1_j}, then*

$$\int_{C_o} f(\zeta)\,d\zeta = 2\pi i\left[\sum_i a_{-1_i}+\sum_j \frac{\alpha_j}{2\pi}a_{-1_j}\right] \tag{5.4-1}$$

where α_j is the interior angle of the path C_o at ζ_j. Hereafter we shall call such integrals *contour integrals*. If an isolated singularity z_0 is a pole of order m, then the residue there can be determined by differentiating the product of function with

$(z - z_0)^m$ $m - 1$ times and taking the limit $z \to z_0$:

$$a_{-1} = \frac{1}{(m-1)!} \lim_{z \to z_0} \frac{d^{m-1}}{dz^{m-1}} [(z - z_0)^m f(z)]$$

There is no direct means of determining the residue when z_0 is an essential singularity.

The residue theorem (5.4-1) often provides an expedient means of evaluating definite integrals or determining their asymptotic behavior, summing series, determining partial function expansions and representing special functions. For integrals its main utility lies in providing an asymptotic evaluation of integrals that cannot be expressed in terms of known functions. Extensive tables of integrals that can be so expressed (Refs. 5-7 and 5-8) have, to a large extent, supplanted their evaluation by (5.4-1) or other means. Contour integral representations of the special function are discussed in Chapter 7; the use of contour integrals in the asymptotic evaluation of the function they represent is detailed in section 12.3. Consequently, these important subjects are not treated here.

5.4.1 Jordan's Lemma, Definite Integrals

Many of the definite integrals that arise are improper ones. Often part of the path of integration includes the point at infinity and not infrequently the integral only exists in a principal-value sense. When the point at infinity lies on the contour of the complex-variable analogue of a real integral, we must determine the requirement on the function's behavior for the integral to exist when the path is a portion of a circular arc with its radius tending toward infinity. Using Jordan's inequality, $\sin\theta \geqslant 2\theta/\pi$, and $|\exp f(z)| = \exp\{Re f(z)\}$, we can deduce *Jordan's lemma: Consider the arc* C_R: $|z| = R$, $|\arg(z - \theta_o)| < \pi/2\mu$; *if*

$$|f(Re^{i\theta})| \leqslant \epsilon(R) \exp[-R^\mu \cos\mu(\theta - \theta_o)]$$

and if

$$\epsilon(R) R^{1-\mu} \to 0 \quad \text{as} \quad R \to \infty$$

then

$$\lim_{R\to\infty} \int_{C_R} f(\zeta)\, d\zeta = 0$$

For example, if $\mu = 1$, and $f(z) = e^{i\nu z} g(z)$, then

$$|f(Re^{i\theta})| \leqslant |g(Re^{i\theta})| \exp(-\nu R \sin\theta)$$

and

$$\lim_{R\to\infty} \int_{C_R} e^{i\nu z} g(z)\, dz = 0$$

provided $g(z) \to 0$ uniformly as $z \to \infty$. Consequently, if $f(z) = P_n(z)/Q_m(z)$, where $P_n(z)$ and $Q_m(z)$ are polynomials of degrees n and m, then we can evaluate integrals involving rational functions. If $Q_m(z)$ has zeros at z_i, $\mathrm{Im}(z_i) > 0$, and simple zeros at x_j, then applying Jordan's lemma we conclude that

$$\int_{-\infty}^{\infty} \frac{P_n(x)}{Q_m(x)} e^{i\nu x}\, dx = 2\pi i \left[\sum_i a_{-1_i} + \frac{1}{2}\sum_j a_{-1_j}\right] \tag{5.4-2}$$

if $m > n$; here a_{-1_i}, a_{-1_j} are the residues of the integrand at the poles corresponding to z_i and x_j.

When some real integrals are to be evaluated by contour integration the complex-valued analogue of the integrand in (5.4-1) may be multivalued; on other occasions it may be convenient to make it multivalued. Consider the integral

$$\int_0^{\infty} x^{\nu} \frac{P_n(x)}{Q_m(x)}\, dx \tag{5.4-3}$$

where ν is not an integer, and its complex variable analogue

$$\int_C z^{\nu} \frac{P_n(z)}{Q_m(z)}\, dz \tag{5.4-4}$$

One choice for C is shown in Fig. 5.4-1. The integrals (5.4-3) and (5.4-4) are well defined provided that $Q_m(0) \neq 0$ and that $0 < \nu + 1 < m - n$. It follows from (5.4-1) that

$$\int_0^{\infty} x^{\nu} \frac{P_n(x)}{Q_m(x)}\, dx - e^{i(\nu+1)\pi} \int_0^{\infty} x^{\nu} \frac{P_n(-x)}{Q_m(-x)}\, dx = 2\pi i \left[\sum_i a_{-1_i} + \frac{1}{2}\sum_j a_{-1_j}\right] \tag{5.4-5}$$

where the a_{-1_i} are the residues at the poles corresponding to the zeros of $Q_m(z)$ for $\mathrm{Im}(z) > 0$ and the a_{-1_j} the residues corresponding to the simple zeros of $Q_m(z)$ for $\mathrm{Im}(z) = 0$. Unless P and Q contain only even powers of x, there are two results. For example, with $P = 1$, $Q = 1 + x$, we find, with $\nu < 1$,

$$\int_0^{\infty} \frac{x^{\nu}}{1+x}\, dx = \frac{\pi}{\sin(\nu+1)\pi}, \quad \int_0^{\infty} \frac{x^{\nu}}{1-x}\, dx = \pi \cot(\nu+1)\pi$$

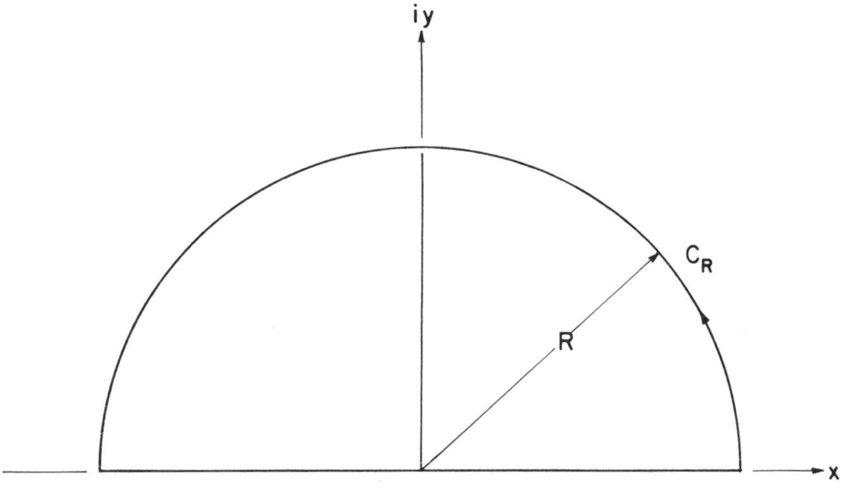

Fig. 5.4-1 Contour for the integral of (5.4-4) is C_R with $R \to \infty$.

As an example where it is convenient to make single-valued integrands multivalued, consider

$$I = \int_0^\infty \frac{x^4}{x^6 - 1}\, dx$$

By using (5.4-5) and taking the limit $\nu \to 0$, we conclude that $I = \pi\sqrt{3}/6$. Often integrands occur that are periodic in θ or $\operatorname{Im}(z)$ and we may take advantage of this periodicity. In other instances we can relate an unknown integral to a known one to arrive at a result. Consider

$$\int_C z^\nu e^{iz}\, dz$$

about the contour shown in Fig. 5.4-2. With $-1 < \nu < 0$, there is no contribution from the circular arcs $|z| = \epsilon, R$ since there are no singularities within the quadrant, and we see that

$$\int_0^\infty x^\nu e^{ix}\, dx = e^{i(\nu+1)\pi/2} \int_0^\infty y^\nu e^{-y}\, dy = e^{i(\nu+1)\pi/2}\, \nu\Gamma(\nu)$$

Finally, we note that integrals of rational functions of $\sin\theta$ and $\cos\theta$ of the form

$$\int_0^{2\pi} F(\sin\theta,\ \cos\theta)\, d\theta$$

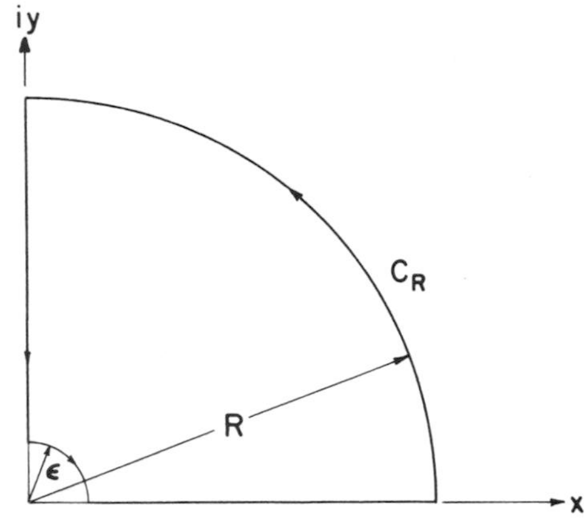

Fig. 5.4-2 Contour for $\int_C z^\nu e^{iz}\,dz$ is C_R with $R \to \infty$.

become integrals of rational functions of z evaluated on the unit circle. (See section 1.10.4.) Thus, e.g., with $\alpha > \beta \geqslant 0$ and real, and $z = e^{i\theta}$

$$\int_0^{2\pi} \frac{d\theta}{\alpha + \beta \cos\theta} = -2i \int_{|z|=1} \frac{dz}{\beta z^2 + 2\alpha z + \beta} = \frac{2\pi}{\sqrt{\alpha^2 - \beta^2}}$$

5.4.2 Partial-Fraction Expansions, Infinite Products, Summation of Series

The Mittag-Leffler theorem of section 5.3.2 gave a method of constructing a partial-fraction expansion of a meromorphic function with prescribed principal parts, but left an entire function undetermined. The residue theorem gives such expansions completely; we illustrate the application here for a meromorphic function with poles of first order. Suppose that $f(z)$ has its (simple) poles at the sequence of points

$$z_1, \ldots, z_n, \ldots$$

where $|z_n| < |z_{n+1}|$, and that the residues there are a_{-1_n}. Consider the contour integral

$$I_n(z) = \frac{1}{2\pi i} \int_{C_{on}} \frac{f(\zeta)}{\zeta(\zeta - z)}\, d\zeta \tag{5.4-6}$$

where C_{on} is a simple closed path enclosing the n poles $z_1, \ldots, z_n$. The integrand of (5.4-6) has simple poles at $z_1, z_2, \ldots, z_n$ and at $\zeta = 0$ and z. Application of the residue theorem yields

$$I_n(z) = \frac{f(z)}{z} - \frac{f(0)}{z} + \sum_{i=1}^{n} \frac{a_{-1_i}}{z_i(z_i - z)}$$

Now if we can choose the contours C_{on} to be equivalent to circles with radii = $O(|z_n|)$, then

$$|I_n| \leqslant \frac{K}{|z_n|} \max |f(\zeta)|_{C_{on}}$$

where K is some constant, and we conclude that if $\max |f(\zeta)|_{C_{on}} = o(|z_n|)$ then $I_n \to 0$ as $n \to \infty$ and

$$f(z) = f(0) + \sum_{i=1}^{\infty} a_{-1_i} \left(\frac{1}{z - z_i} + \frac{1}{z_i} \right) \tag{5.4-7}$$

The function $\sin z$ has simple zeros at $z = \pm n\pi, n = 0, 1, \ldots$ and the function

$$(\sin z)^{-1} - z^{-1}$$

has simple poles with residues $(-1)^n$ at $z = \pm n\pi, n = 1, 2, \ldots$. According to (5.4-7), then

$$\csc z = \frac{1}{z} + \sum_{\substack{n=-\infty \\ n \neq 0}}^{\infty} (-1)^n \left[\frac{1}{z - n\pi} + \frac{1}{n\pi} \right] \tag{5.4-8}$$

We must verify that the successive integrals $I_n \to 0$ as $n \to \infty$. Take C_{on} to be the square contours with corners $(n + \frac{1}{2}) \cdot (\pm 1 \pm i)\pi$; on these contours $(\sin z)^{-1}$ is bounded and hence (5.4-7) holds. Because the series of (5.4-8) with $n \gtrless 0$ converge separately, they may be combined to give

$$\csc z = \frac{1}{z} - 2z \sum_{n=1}^{\infty} \frac{(-1)^n}{n^2 \pi^2 - z^2} \tag{5.4-9}$$

The infinite-product representation of an entire function $f(z)$ can often be deduced from the partial-fraction expansion of its logarithmic derivative; thus, for example, with

$$\frac{d \log (\tan z)}{dz} = 2 \csc (2z)$$

we find that

$$\tan z = z \prod_{n-1}^{\infty} \frac{\left[1 - \left(\frac{z}{n\pi} \right)^2 \right]}{\left[1 - \left(\frac{z}{(n - \frac{1}{2})\pi} \right)^2 \right]}$$

Infinite series of the form $\sum f(n)$, where $f(z)$ is a meromorphic function, can frequently be summed by using the residue theorem with an integrand consisting of the product of $f(z)$ with $\pi \cot \pi z$ or $\pi \csc \pi z$. The latter functions have simple poles at $z = n, n = 0, \pm 1, \ldots$ and residues of $+1$ and $(-1)^n$ there. Consider, for example, a meromorphic function $f(z)$ with simple poles at $z_i (\neq \text{integer})$. Then the integral

$$\int_{C_{okn}} \pi \begin{Bmatrix} \cot \pi z \\ \csc \pi z \end{Bmatrix} f(z)\, dz \tag{5.4-10}$$

about a simple closed contour C_{okn} containing the real integers $k, k+1, \ldots, n$ has the value

$$2\pi i \left[\sum_{j=k}^{n} \begin{Bmatrix} 1 \\ (-1)^j \end{Bmatrix} f(j) + \sum_i a_{-1_i} \pi \begin{Bmatrix} \cot \pi z_i \\ \csc \pi z_i \end{Bmatrix} \right] \tag{5.4-11}$$

where the a_{-1_j} are the residues of the simple poles z_i enclosed by C_{okn}. If the integral (5.4-10) tends toward zero as the sequence of contours C_{okn} has $k \to -\infty$ and $n \to \infty$, then (5.4-11) gives the sum

$$\sum_{n=-\infty}^{\infty} \begin{Bmatrix} 1 \\ (-1)^n \end{Bmatrix} f(n) = - \sum_i a_{-1_i} \pi \begin{Bmatrix} \cot \pi z_i \\ \csc \pi z_i \end{Bmatrix} \tag{5.4-12}$$

Convergence of (5.4-10) is assured, e.g., if $z f(z) \to 0$ as $z \to \infty$. This is the case when C_{okn} is the square with corners $(n + \frac{1}{2}) \cdot (\pm 1 \pm i)$. It follows directly from (5.4-12) that for real α

$$\sum_{n=-\infty}^{\infty} \frac{1}{n^2 + \alpha^2} = \frac{1}{\alpha^2} + 2 \sum_{n=1}^{\infty} \frac{1}{n^2 + \alpha^2} = \frac{\pi}{\alpha} \coth \alpha\pi$$

This procedure applies to functions with higher-order poles as well. Thus the result

$$\sum_{n=-\infty}^{\infty} \frac{1}{(n + \alpha)^2} = \pi^2 \csc^2 \pi\alpha$$

is simply given by the residue of $-\pi \cot \pi z/(z + \alpha)^2$ at the pole $z = -\alpha$ of order two.

5.4.3 Argument Principle, Rouché's Theorem, Local Inverse, Routh-Hurwitz Criterion

In many applied problems, such as the stability of the solutions to a system of linear differential equations, it is necessary to determine the location of the roots of a polynomial. One result that follows directly from (5.4-1) is the argument principle for meromorphic functions. If $f(z)$ is meromorphic in a domain D with boundary C_o and having zeros of order m_i at $z_1, \ldots, z_i, \ldots, z_n$ and poles of order p_j at $z_{(1)}, \ldots, z_{(j)}, \ldots, z_{(m)}$ then function $f'(z)/f(z)$ is analytic in D except at the points corresponding to the zeros and poles of $f(z)$. The residues at the points corresponding to the zeros of $f(z)$ are obviously m_i while those at the points corresponding to the poles are $-p_j$. *Consequently, with $f(z)$ finite and nonzero on C_o*

$$\frac{1}{2\pi i}\int_{C_o}\frac{f'(\zeta)}{f(\zeta)}\,d\zeta = \frac{[\arg f(\zeta)]_{C_o}}{2\pi} = \sum_{i=1}^{n} m_i - \sum_{j=1}^{m} p_j \tag{5.4-13}$$

where $[\ ]_{C_o}$ indicates the change in a complete traversal of C_o. The right-hand side of (5.4-13) is the sum of the number of zeros of $f(z)$ inside C_o, minus the number of poles of $f(z)$ inside C_o, each counted according to their order. By (5.4-13) then, this must be the change in $\arg f(z)/2\pi$ on one complete traversal of C_o; this result is known as the *principle of the argument*.

If we apply (5.4-13) to the sum of two analytic functions $f(z), g(z)$, then

$$[\arg\{f(z)+g(z)\}]_{C_o} = \left[\arg f(z) + \arg\left(1+\frac{g(z)}{f(z)}\right)\right]_{C_o}$$

$$= [\arg f(z)]_{C_o} + \left[\arg\left(1+\frac{g(z)}{f(z)}\right)\right]_{C_o}$$

Now if $|g| < |f|$ on C_o then $|\arg(1+g/f)| < \pi/2$ and a traversal of C_o will return the argument to its original value without an increase or decrease by a factor of 2π. *Thus if $f(z)$ and $g(z)$ are analytic in and on a simple closed curve C_o and if $|g| < |f|$ on C_o, then $f(z)$ and $f(z)+g(z)$ have the same number of zeros, counted according to their order (Rouché's theorem).* As an example consider the two functions αz^n and $\exp z$ where α is real and $\alpha > e$. On $|z| = 1$

$$|\alpha z^n| = \alpha > e \geqslant |\exp z|$$

Thus both αz^n and $\alpha z^n - e^z$ have n zeros in $|z| \leqslant 1$ for $\alpha > e$.

If we write a polynomial of nth degree in the form

$$P_n(z) = a_n z^n\left[1+\frac{a_{n-1}}{a_n z}+\cdots+\frac{a_0}{a_n z^n}\right] = a_n z^n\,[1+f(z)] \tag{5.4-14}$$

With $|z| > R$ then for some R $|f(z)| < 1$, and if $P_n(z)$ has any zeros they will lie inside $|z| < R$. Because $|a_n z^n f(z)| < |a_n z^n|$, $P_n(z)$ and $a_n z^n$ will have the same number of zeros inside $|z| = R$; z^n has one zero of multiplicity n. *Thus any polynomial of degree n will have n zeros, counted according to their order, inside $|z| = R$ for* *some* R. (*Fundamental theorem of algebra.*) Another consequence of Rouché's theorem is *Hurwitz's theorem*: Let $f_n(z)$ be a sequence of functions analytic inside and on C_o that converges uniformly to $f(z) \not\equiv 0$. So long as C_o does not pass through a zero of $f(z)$, then there is an integer N such that every $f_n(z)$ with $n > N$ has the same number of zeros inside C_o as $f(z)$. This follows directly from the uniform convergence of the sequence which implies

$$|f_n(z) - f(z)| < \min |f(z)| \text{ on } C_o \leqslant |f(z)|$$

and Rouché's theorem applied to $f(z)$ and to

$$f(z) + f_n(z) - f(z) = f_n(z)$$

A practical application of Rouché's theorem is the *Routh-Hurwitz criterion* which states: *All the roots z_i of $P_n(z)$ have $Re(z_i) < 0$ if and only if the roots z_k of the $(n - 1)$-degree polynomial*

$$[a_n a_{n-1}^* + a_{n-1} a_n^* - a_n a_n^* z] P_n(z) + a_n^2 z \sum_{i=0}^{n} (-1)^{n-i} a_i^* z^i$$

have $Re(z_k) < 0$ and $Re(a_{n-1}/a_n) > 0$. Successive application of this scheme provides a means of determining whether or not all the roots z_i of $P_n(z)$ have $\text{Re}(z_i) < 0$; this knowledge is usually that sought in stability investigations. Consider, for example, the polynomial $P_3 = z^3 + \alpha z(z + 1) + \beta = 0$, where α and β are real. The above criterion leads to the equations

$$z^3 + \alpha z(z + 1) + \beta = 0,$$

$$\alpha^2 z^2 + (\alpha^2 - \beta) z + \alpha\beta = 0, \quad \text{and}$$

$$(\alpha^2 - \beta) z + \alpha\beta = 0$$

Thus P_3 has three roots with negative real parts if and only if $\alpha, \beta > 0$ and $\beta > \alpha^2$.

Finally, Rouché's theorem supplies a simple proof that an analytic function has a local inverse at points where the derivative is non-zero; this result is needed in the study of conformal mapping (section 5.5). Suppose $w = f(z)$ is analytic at z_0 and $f'(z_0) \neq 0$. The Taylor series development of $f(z)$ about z_0 shows that there is some simple closed contour C_δ about z_0 such that $f(z) = f(z_0)$ only when $z = z_0$ for z in and on C_δ, and with $w_o = f(z_0)$

$$|f(\zeta) - w_o| > \epsilon > 0$$

for ζ on C_δ. Consider those numbers, w_1, such that $|w_1 - w_o| < \epsilon$. Now the function $f(z) - w_o$ has a single zero of first order ($f'(z_0) \neq 0$ in C_δ); by Rouché's theorem so must $f(z) - w_o - w_1 + w_o$. Thus there is only one point, say z_1, where

$f(z_1) = w_1$, and there is a single-valued function $z = g(w)$ with derivative $g'(w_0) = 1/f'(z_0)$. Consequently, if $f'(z_0) \neq 0$, *there is a local single-valued analytic inverse of* $w = f(z)$

$$z = g(w) = \frac{1}{2\pi i} \int_{C_\delta} \frac{\zeta f'(\zeta)}{f(\zeta) - w} \, d\zeta \tag{5.4-15}$$

which can be verified by (5.4-1). Equation (5.4-15) can be used to determine the Taylor series expansion of $g(w)$ about w_o in terms of the derivatives

$$g^{(n)}(w_o) = \left\{ \frac{d^{n-1}}{dz^{n-1}} \left[\frac{z - z_0}{f(z) - w_o} \right]^n \right\}_{z=z_0} \tag{5.4-16}$$

5.5 HARMONIC FUNCTIONS AND CONFORMAL MAPPING

One equation in applied mathematics that is signal, both in its utility and its solvability, is the two-dimensional version of *Laplace's equation*

$$u_{xx} + u_{yy} \equiv \nabla^2 u = 0 \tag{5.5-1}$$

This equation arises in countless applied problems and its close connection with analytic functions renders it uniquely tractable. As noted in section (5.2-1), *if* $u(x, y)$ *is harmonic, i.e., satisfies (5.5-1) in a domain D, then there is a conjugate harmonic function* $v(x, y)$ *such that* $f(z) = u(x, y) + iv(x, y)$ *is an analytic function of z in D.* The converse of this statement is true for simply connected domains. But as the example $u = \log |z|$ shows, the converse is not true for multiply connected domains because v is multivalued for $f(z) = \log z$.

For u to be a single-valued harmonic function in a multiply connected domain D, it is necessary and sufficient for u to be the real part of an analytic function $f(z)$ with

$$\int_{C_o} f'(\zeta) \, d\zeta$$

an imaginary constant or zero for any C_o lying in D.

If we effect a formal change of variables from (x, y) to (z, z^*), then (5.5-1) becomes

$$4u_{zz^*} = 0$$

which we can solve by direct integration. With u and v real functions we have[3]

[3] In the same way it follows that $u = \mathrm{Re}[z^* f(z) + g(z)]$ is the general solution of the biharmonic equation $\nabla^4 u = 0$.

$$u = \frac{f(z) + f^*(z^*)}{2} \quad \text{and} \quad v = \frac{f(z) - f^*(z^*)}{2i} \tag{5.5-2}$$

This heuristic procedure can be justified by a rigorous consideration of the operators $\partial/\partial z$ and $\partial/\partial z^*$. Given a harmonic function, its conjugate and the analytic function they represent are determined uniquely aside from a constant. Thus, given $u(x, y)$, we can use (5.5-2) to write

$$f(z) + f^*(z^*) = 2u\left(\frac{z + z^*}{2}, \frac{z - z^*}{2i}\right) \tag{5.5-3}$$

Direct separation then determines $f(z)$, except for a constant, and v is given by (5.5-2). Consider $u = e^x \cos y$; it follows from (5.5-3) that $f(z) = e^z + c$, c a constant, and from (5.5-2), that $v = e^x \sin y + \text{Im}(c)$.

In studying equations such as (5.5-1), it is sometimes fruitful to introduce new coordinates. If we make a change of coordinates from (x, y) to (ξ, η) then

$$u_{xx} + u_{yy} = \frac{\partial(\xi, \eta)}{\partial(x, y)} [u_{\xi\xi} + u_{\eta\eta}] = 0$$

if and only if $\xi_x = \eta_y$ and $\xi_y = -\eta_x$, where the *Jacobian* of the transformation $\partial(\xi, \eta)/\partial(x, y)$ (see section 2.8.7) is $\xi_x^2 + \xi_y^2$. Because ξ and η satisfy the Cauchy-Riemann conditions, the function $Z = \xi + i\eta$ is an analytic function of z; the Jacobian is positive definite at $z = z_0$ if and only if $f'(z_0) \neq 0$. *Consequently, if u is harmonic in a domain D of the z-plane, then it will be harmonic in the image of D under the mapping $Z = f(z)$ where $f(z)$ is analytic and $f'(z) \neq 0$ in D* (see section 5.1.3).

While Laplace's equation remains unaltered by such a transformation, Poisson's equation

$$\nabla^2 u = p(x, y)$$

and Helmholtz's (reduced wave) equation

$$\nabla^2 u + k(x, y)\, u = 0$$

are altered in detail, but not in form. Some equations can actually be simplified. Thus if

$$\nabla^2 u + \mathbf{q} \cdot \nabla u = 0$$

where $\mathbf{q}$ is a divergence-free vector derivable from a scalar potential, $\mathbf{q} = \nabla\phi$, then the transformation from (x, y) to (ϕ, ψ), where ψ is the conjugate harmonic of ϕ, gives the simpler equation

$$q^2 [\nabla^2 u + u_\phi] = 0$$

in the independent variables ϕ, ψ.

The remainder of this section examines mappings of the form $Z = f(z)$, where f is a meromorphic function, and their applications.

5.5.1 Boundary-Value Problems, Poisson's Formula, Harnack's Inequality

The general boundary-value problem associated with Laplace's equation is to find the function harmonic in a given domain

$$u_{xx} + u_{yy} = 0 \text{ in } D \tag{5.5-4a}$$

assuming prescribed values on its boundary C_o

$$D(\zeta)u + N(\zeta)\frac{\partial u}{\partial n} = B(\zeta) \text{ on } C_o \tag{5.5-4b}$$

where D, N, and B are real functions. When $N \equiv 0$ the boundary-value problem is called a *Dirichlet problem*; when $D \equiv 0$ it is called a *Neumann problem*. In the co-ordinates along and normal to C_o, s and n, the Cauchy-Riemann conditions are $u_s = -v_n$, $u_n = v_s$, where s is positive in the direction of C_o such that D lies to the left and $-n$ is into D; a Dirichlet problem for u or v is equivalent to a Neumann problem for v or u. The uniqueness of solutions to the Dirichlet problem is easily established by using the maximum modulus theorem. If we have a Neumann problem for v, then because

$$\int_{C_o} \frac{\partial u}{\partial s}\, ds = 0, \quad \text{we conclude that} \quad \int_{C_o} \frac{\partial v}{\partial n}\, ds = 0$$

This need not be true for each individual contour in a multiply connected domain, but only for their sum C_o; in such domains u may be multivalued. For an n-connected domain the (multivalued) solution to the Neumann problem is determined to within an arbitrary constant by specifying the values of

$$\int_{C_{o_i}} \frac{\partial v}{\partial n}\, ds$$

on $n - 1$ of the boundaries. The general boundary-value problem (5.5-4a, b) is somewhat more complex (see, e.g., Ref. 5-4, p. 221).

Cauchy's integral formula (5.2-5) supplies the solution to the Dirichlet problem for a circular domain:

$$u + iv = f(z) = \frac{1}{2\pi}\int_0^{2\pi} \frac{f(Re^{i\theta})Re^{i\theta}}{Re^{i\theta} - z}\, d\theta$$

We can combine this expression with the identity

$$\frac{1}{2\pi}\int_0^{2\pi} \frac{f(Re^{i\theta})\,Re^{i\theta}}{Re^{i\theta} - R^2/z^*}\,d\theta = 0$$

to obtain *Poisson's formula*

$$u(r,\theta) = \frac{1}{2\pi}\int_0^{2\pi} u(t)\left[\frac{R^2 - r^2}{R^2 - 2Rr\cos(\theta - t) + r^2}\right] dt \tag{5.5-5}$$

A similar expression obtains for $v(r,\theta)$ and the combined result has the simple form (*Schwarz's formula*)

$$u(r,\theta) + iv(r,\theta) = iv(0) + \frac{1}{2\pi}\int_0^{2\pi} u(t)\left(\frac{Re^{it} + z}{Re^{it} - z}\right) dt$$

If $u > 0$ in $|z| < R$, then from (5.5-5) and the mean-value theorem it follows that $u(r,\theta)$ is restricted by *Harnack's inequality* to the range

$$u(0)\,\frac{R - r}{R + r} \leqslant u(r,\theta) \leqslant u(0)\,\frac{R + r}{R - r}$$

Green's formula, section 3.4.3, relates the integral over a domain of the Laplacian of any two continuous real functions g and h, with continuous second partial derivatives, to their values on the boundary C_o of D

$$\int_{C_o}\left[h\frac{\partial g}{\partial n} - g\frac{\partial h}{\partial n}\right] ds = \int_D [h\nabla^2 g - g\nabla^2 h]\,dx\,dy \tag{5.5-6}$$

This relation can be used to determine a harmonic function $h(x,y)$ that has boundary values (5.5-4b). Now the function $\tilde{g}(z;z_0) = \log|z - z_0|$ is harmonic in the entire finite plane except at the single point $z = z_0$. Consider (5.5-6) for the domain D_o consisting of D with the circle $|z - z_0| = \rho$ removed; then

$$\int_{C_o}\left[h\frac{\partial \tilde{g}}{\partial n} - \tilde{g}\frac{\partial h}{\partial n}\right] ds - \int_0^{2\pi}\left[h\frac{\partial \tilde{g}}{\partial \rho} - \tilde{g}\frac{\partial h}{\partial \rho}\right]\rho\,d\theta = 0$$

Since $\tilde{g}_\rho = \rho^{-1}$ on $|z - z_0| = \rho$, in the limit $\rho \to 0$ we have *the analogue of Cauchy's integral formula for harmonic functions*

$$h(z) = \frac{1}{2\pi} \int_{C_o} \left[h \frac{\partial}{\partial n} \log |\zeta - z| - \log |\zeta - z| \frac{\partial h}{\partial n} \right] ds \tag{5.5-7}$$

If we can find a harmonic function $g(z; z_0)$ that has the local logarithmic behavior of $\tilde{g}(z; z_0)$ and satisfies either $g = 0$ or $g_n = 0$ on C_o, then (5.5-7) will give $h(z)$ in terms of an integral of the product of the boundary values of h_n or h with $g(z; z_0)$ or its normal derivative. Such functions are called *Green's functions* and can frequently be found through conformal mapping. Note that the kernel in the integrand of (5.5-5) is $2\pi g_n(z; z_0)$ for the domain $|z| \leqslant 1$; that is, the Green's function for the Dirichlet problem is

$$g(z; z_0) = \frac{1}{2\pi} \log R \left| \frac{z_0 - z}{R^2 - z_0 z^*} \right| = \frac{1}{4\pi} \log \left[\frac{r^2 + \rho^2 - 2\rho r \cos(\theta - t)}{R^2 + \dfrac{r^2 \rho^2}{R^2} - 2\rho r \cos(\theta - t)} \right] \tag{5.5-8}$$

Applied to the boundary-value problem (5.5-4a, b), eq. (5.5-7) leads to an integral equation for h on C_o; Ref. 5-4 treats such equations.

5.5.2 Conformal Mapping, Riemann's Mapping Theorem, Bilinear Transformations

As we have already noted in the introduction to this section, *mappings of the form* $Z = \xi + i\eta = f(z)$, *where* $f(z)$ *is an analytic function and* $f'(z) \neq 0$, *preserve harmonic functions.* That is, if u is a harmonic function of x and y, it is also a harmonic function of ξ and η. Here we detail some of the general features of such mappings. In section 5.4.3 we showed that these same conditions implied there was a local inverse $z = g(Z)$ in the neighborhood of a point z_0 provided $f'(z_0) \neq 0$; consequently the mapping is locally one-to-one. Consider the image of two continuous curves passing through z_0 under the mapping $Z = f(z)$. *With* $f'(z_0) \neq 0$, *the angle between the two curves is preserved, they are rotated through the angle arg* $f'(z_0)$ *at* $Z = f(z_0)$, *and they are subject to the local linear magnification* $|f'(z_0)|$. These results all follow directly from the existence of a nonzero derivative at z_0. Transformations that preserve the angles between intersecting curves as well as their sense are called *conformal transformations.* (If the angle but not the sense is preserved, the transformation is 'isogonal,' e.g., $f(z^*)$.) Obviously areas are scaled locally by the Jacobian of the map, i.e., by $|f'(z_0)|^2$.

If $Z = f(z)$ and f is an analytic function in D, then D is mapped conformally into its image D' except at the critical points of the transformation where $f'(z) = 0$ or ∞. Suppose that $f'(z)$ has a zero of order n at z_0; then in the neighborhood of z_0 and its image Z_0 we have

$$Z - Z_0 = f(z) - f(z_0) = c(z - z_0)^{n+1} + \cdots$$

where c is some complex constant. In this instance we see that the angle between two curves is multiplied by the factor $(n + 1)$ and the linear magnification is zero.

The mapping is one-to-one onto an $(n+1)$-sheeted Riemann surface for which $Z_0 = f(z_0)$ is a branch point of order n. The mapping is conformal except at z_0. On the other hand, if $f(z)$ is analytic in the domain $0 < \delta \leqslant |z - z_0| < R$, but $f(z)$ has a pole of order n at z_0, then $f'(z_0) = \infty$ there. By examining the mapping $F(z) = 1/f(z)$ near z_0, we can use the above results to determine the local behavior near a pole. In this case $F'(z)$ has a zero of order $n - 1$ and we conclude that $Z = f(z)$ maps a neighborhood of the pole z_0 onto a Riemann surface consisting of n-sheets, with $|Z| > \delta^{-n}$, i.e., with a branch point at $Z = \infty$.

While the conformal mapping of a domain D by $Z = f(z)$ is locally one-to-one, globally it is a one-to-one mapping of D onto the Riemann surface of its inverse function $z = g(Z)$. If this mapping is one-to-one, as it is when the image D' occupies a single sheet of the Riemann surface, then $f(z)$ is said to be *schlicht* (also, simple, univalent, or biuniform). Even the simplest functions have complicated mappings. The inverse function for the mapping

$$Z(z) = \frac{z}{(1 - z)^2}$$

has branch points of order 1 at $Z = -1/4$ and infinity, and a two-sheeted Riemann surface with a branch cut along the negative axis from $-1/4$ to $-\infty$; for the domain $|z| < 1$ the image is the entire cut Z-plane and $Z(z)$ is schlicht in $|z| < 1$.

A central question of conformal mapping theory is what domains can be mapped onto other given domains or, more specifically, what domains can be mapped onto certain standard domains. For simply connected domains the answer is provided by the *Riemann mapping theorem* which states: *Any simply connected domain with at least two boundary points can be mapped conformally and one-to-one onto the disk* $|Z| < 1$; as z tends to points on the boundary of D then the image of z tends toward the boundary $|Z| = 1$. The mapping function $f(z)$ is uniquely determined by setting $f(c) = 0$ and specifying $\arg f'(c)$ where c is an arbitrary point of D. *Any doubly connected domain can be mapped conformally onto the annulus* $1 < |Z| < r_o$; the mapping is unique to within a factor of modulus one and r_o is determined by the mapping. *Any n-connected domain can be mapped onto the Z-plane with n-vertical cuts*; the mapping is uniquely determined by the point z_0 mapped into ∞, the value of

$$\lim_{z \to z_0} \left[f(z) - \frac{\alpha}{z - z_0} \right]$$

where α is a real number, and the value of α (Ref. 5-6).

If we can determine the function $Z = f(z)$ that conformally maps a simply connected domain D of the z-plane into $|Z| < 1$, then we can determine the Green's function for the Dirichlet problem for D from the Green's function for the unit disk (5.5-8).

$$g(z; z_0) = \frac{1}{2\pi} \log \left| \frac{f(z_0) - f(z)}{1 - f(z_0) f^*(z)} \right| \tag{5.5-9}$$

Thus the generalization of Poisson's integral to an arbitrary simply connected domain D with boundary C_o is

$$h(z) = \int_{C_o} h(\zeta) \frac{\partial g(\zeta; z)}{\partial n} \, ds$$

where $g(\zeta; z)$ is given in terms of the mapping of D onto the unit disk by (5.5-9).

While the existence of a conformal mapping is assured, under the conditions discussed above, and while the existence proof is a constructive one, it does not provide a practicable means of determining such mappings. Frequently they can be constructed from a series of elementary transformations; for polygonal boundaries they can be constructed by the method given in the next section. Often they can be found in the tables of conformal transformations given in Refs. 5-9 and 5-10. On occasion, approximate methods must be resorted to; here digital computers prove to be fruitful, with the numerical solution to the numerically mapped problem often having computational advantages over direct numerical solution. In the remainder of this section, and in the following section, we discuss briefly several of the more important transformations. The properties of other transformations, including the elementary functions, are found in Refs. 5-5 and 5-11. Approximate methods of constructing conformal mappings are discussed in Ref. 5-1, pp. 174–180. One of these consists of applying the analogue of (5.5-7) for a boundary point to the harmonic function $h = \text{Im}\,\{\log\,[f(z)/z]\}$, where $f(z)$ maps the interior of the smooth closed curve C_o onto $|Z| < 1$. The result is an integral equation for h that, in general, must be solved numerically.

The most general transformation that maps the entire z-plane onto the entire Z-plane is the *bilinear* (linear fractional, Möbius) *transformation*

$$Z = \frac{az + b}{cz + d} \tag{5.5-10}$$

with $ad - bc \neq 0$ so that (5.5-10) is not constant. The inverse of (5.5-10) is also bilinear, and a bilinear transformation of a bilinear transformation is bilinear (i.e., the bilinear transformations constitute a group). Note that if the points Z_1, Z_2, Z_3, Z_4 correspond to z_1, z_2, z_3, z_4, respectively, then the *cross-ratio equality* holds:

$$\frac{(Z_1 - Z_2)(Z_3 - Z_4)}{(Z_1 - Z_3)(Z_2 - Z_4)} = \frac{(z_1 - z_2)(z_3 - z_4)}{(z_1 - z_3)(z_2 - z_4)}$$

Because $Z'(z) \neq 0$ everywhere, the only critical points of the transformation are $z = \infty$ and $z = -d/c$. With $Z = a/c$ for $z \to \infty$ and $Z \to \infty$ for $z = -d/c$, there is a one-to-one correspondence between the Z-plane and the z-plane, and the mapping is conformal everywhere. The general equation for a circle, $\alpha zz^* + kz + (kz)^* + \gamma = 0$, where α and γ are real and $kk^* > \alpha\gamma$, is invariant in form under (5.5-10), and circles (including the limiting case of a straight line) are mapped onto circles. To determine the map of a given domain D under (5.5-10), it is useful to rewrite the transformation in the form

$$Z - \frac{a}{c} = \frac{bc - ad}{c\zeta}; \qquad \zeta = cz + d$$

We see that the transformation consists of a translation, magnification, and rotation $cz + d$, an inversion, also with rotation and magnification, $(bc - ad)/c\zeta$, and a translation $Z - a/c$. Two special cases of (5.5-10) are

$$Z = \frac{z - a}{z - a^*} e^{i\theta} \quad \text{and} \quad Z = \frac{z - a}{a^* z - 1} e^{i\theta}$$

The first maps $\text{Im}(z) > 0$ onto $|Z| < 1$ if $\text{Im}(a) > 0$; the second maps $|z| < 1$ onto $|Z| < 1$ for $|a| < 1$. Even the maps available from this single transformation are too rich to detail here, and we give only one nontrivial example. For a more complete discussion see, e.g., Refs. 5-2 and 5-12. The region bounded by three circular arcs that are portions of circles that intersect at a single point z_0, as shown in Fig. 5.5-1, can be mapped by

$$Z = \frac{z - z_1}{z - z_0}$$

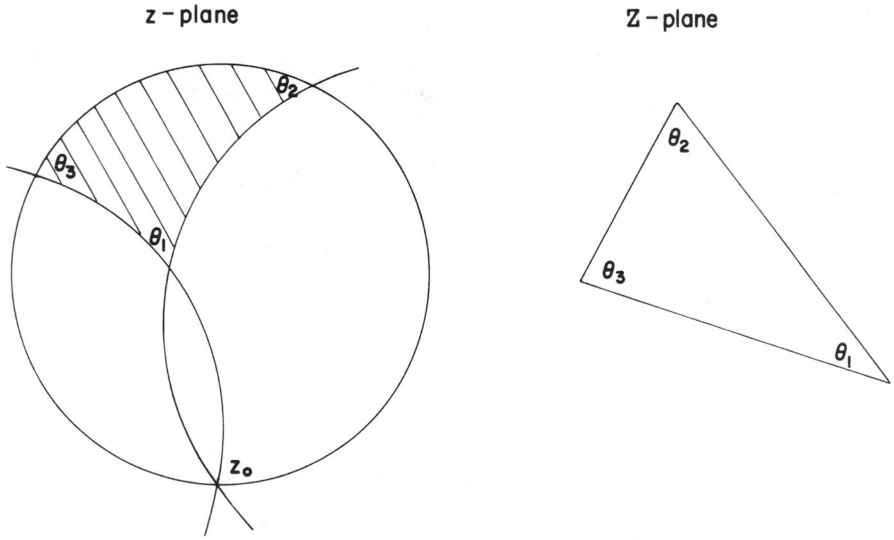

Fig. 5.5-1 Bilinear mapping $Z = (z - z_1)/(z - z_0)$ of region bounded by three circular arcs.

onto a triangular region with interior angles θ_1, θ_2, and θ_3; under this transformation z_0 is mapped to $Z = \infty$ and the circles become straight lines. Since the mapping is conformal $\theta_1 + \theta_2 + \theta_3 = \pi$ for such a configuration. The orientation and size of the resulting triangle can be prescribed by multiplying the transformation by a complex constant. In the next section we examine a class of transformations that map polygons onto the upper-half plane; thus the resulting triangle can be mapped on the upper-half plane, and by another bilinear transformation, mapped onto the unit circle.

5.5.3 Schwarz-Christoffel Transformation, Joukowsky Transformation, Periodic Domains

We have seen that the unit disk can be mapped onto a half plane, and since any simple domain can be mapped onto the unit disk, there is a transformation that will map the interior of a simple closed polygon onto the upper-half plane. Such transformations can be constructed without difficulty; consider a mapping with

$$\frac{dZ}{dz} = K \prod_{i=1}^{N} (z - x_i)^{(\alpha_i/\pi)-1} \tag{5.5-11}$$

where K is a complex constant, the α_i's are real numbers, and the x_i's ordered such that $x_i < x_{i+1}$. Now on the real axis

$$\arg(dZ) - \arg(dx) = \arg(K) + \sum_{i=1}^{n} \left(\frac{\alpha_i}{\pi} - 1\right) \arg(x - x_i)$$

For any x between x_{i-1} and x_i, arg (dZ) is constant and the segment maps int_ _ straight line. Consider now what happens as x traverses an x_i. The argument of dx remains unchanged but arg (dZ) changes by $\pi - \alpha_i$. For $x_i < x < x_{i+1}$, then, arg (dZ) is also constant but it is increased by an amount $\pi - \alpha_i$ over that for $x_{i-1} < x < x_i$. As we traverse the real axis from $-\infty$ to ∞, we traverse a series of straight lines in the Z-plane, changing direction by an amount $\pi - \alpha_i$ each time an x_i is exceeded. These angles correspond to exterior angles of a polygon, as shown in Fig. 5.5-2, and the α_i are its interior angles; note that $\sum_{i=1}^{N} (\pi - \alpha_i) = 2\pi$. To complete the path in the z-plane, let $z = Re^{i\theta}$ and consider $R \to \infty$, with θ ranging from 0 to π. For large R

$$Z \sim Z_0 - \frac{K}{R} e^{-i\theta}$$

and we see that arg $(Z - Z_0)$ decreases from arg $(K) + \pi$ to arg (K) as θ goes from 0 to π, and dZ describes a small semicircular arc of radius $|K|/R$ about Z_0. Thus $z = \infty$ corresponds to a point Z_0 on one side of the polygon (see Fig. 5.5-2). If we choose to put a vertex of the polygon at infinity, i.e., if $x_N = \infty$, then we can redefine K so

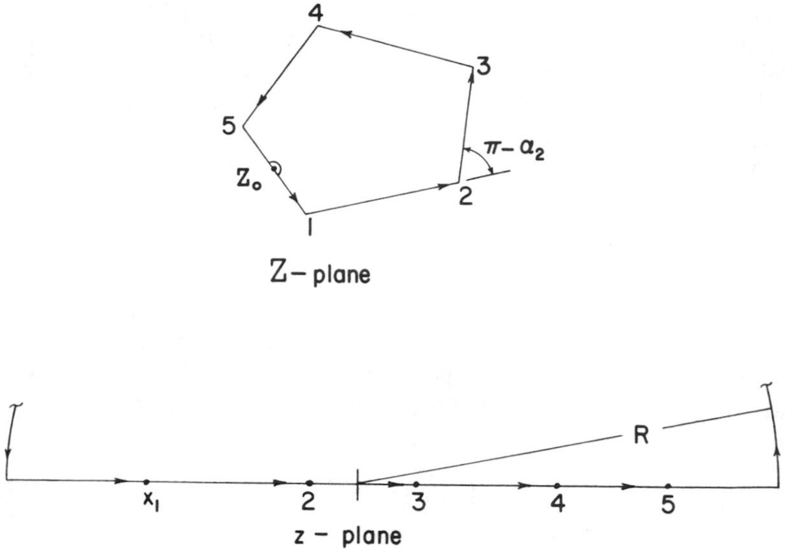

Fig. 5.5-2 The Z- and z-planes for the Schwarz-Christoffel transformation (5.5-12).

that the factor $(z - x_N)$ does not appear in (5.5-11). Quadrature of (5.5-11) gives the *Schwarz-Christoffel formula*

$$Z(z) = K \int_0^z \prod_{i=1}^{N} (\tilde{z} - x_i)^{(\alpha_i/\pi)-1} \, d\tilde{z} + K_1 \tag{5.5-12}$$

The shape of the polygon given by (5.5-12) is determined by the x_i's and α_i's, its scale and orientation by K and its location by K_1. The mapping is conformal except at the critical points x_i. Because $2N$ coordinates are required to specify a polygon with N vertices and there are $2N + 3$ real constants in (5.5-12), comprised of $N\ x_i$'s, $(N - 1)\ \alpha_i$'s $\left[\text{recall } \sum_{i=1}^{N} (\pi - \alpha_i) = 2\pi\right]$ and four constants in K and K_1, any three of the x_i's may be specified arbitrarily.

In general the quadrature (5.5-12) will not be possible in terms of tabulated functions; even the case of a right triangle results in elliptic functions. Still, there are many problems of practical interest that can be solved by utilizing (5.5-12). With Z taken to be the complex potential of hydrodynamics, certain problems involving unknown stream boundaries can be readily solved (Ref. 5-13). Often problems arise in which one or more vertices of the polygon lie at infinity. In such cases the transformation can be constructed by taking the interior angle to be 0 or 2π. For example the degenerate polygon $0 \leqslant \text{Im}(Z) \leqslant Y_0$ can be mapped onto the real

z-axis by

$$Z = i\, Y_0 \log \left(\frac{z - x_2}{x_1 - x_2}\right) \Big/ \log \left(\frac{x_3 - x_2}{x_1 - x_2}\right)$$

where x_1, x_2, and x_3 correspond to the vertices at $0, +\infty + i\, Y_0/2$ and iY_0. Because Y_0 is the only length scale, we may take the x's to be $-1, 0$, and 1 giving the simple transformation

$$Z = \frac{Y_0}{\pi} \log(-z)$$

Another simple transformation with useful properties is the *Joukowsky transformation*, given by

$$Z = \frac{1}{2}\left(z + \frac{1}{z}\right), \qquad z = Z \pm \sqrt{Z^2 - 1} \tag{5.5-13}$$

The ambiguity of the inverse transformation $z(Z)$ is resolved by considering its two-sheeted Riemann surface which is cross-connected along the slit $[-1, 1]$ in the Z-plane. With $z = re^{i\theta}$ and $Z = X + iY$, it follows directly from (5.5-13) that

$$\frac{(2X)^2}{(r + r^{-1})^2} + \frac{(2Y)^2}{(r - r^{-1})} = 1$$

and consequently circles with $|z| = r$ are mapped into ellipses in the Z-plane. The unit circle $r = 1$ becomes the degenerate ellipse corresponding to the cut in the Z-plane; with $r \to 0$ or ∞ the major and minor axes of the ellipse become infinite and we conclude that both the interior and the exterior of $|z| = 1$ map onto the cut Z-plane, i.e., onto the Riemann surface for $z(Z)$. Conversely the inverse transformation $z(Z)$ maps one sheet of the Riemann surface onto the interior of $|z| = 1$ and the other onto its exterior.

Often practical problems arise that are naturally periodic or can be interpreted as problems with periodic boundaries. In such cases mappings of the form Z^n or $Z^{1/n}$ can usually be used to map a periodic domain on the unit circle. Consider for example the function

$$Z = \frac{z}{(1 + z)^2}$$

which is schlict for $|z| < 1$ and maps the unit disk onto the entire Z-plane with a branch cut from $1/4$ to infinity. We can generalize this to

$$Z = \left[\frac{z^n}{(1 + z^n)^2}\right]^{1/n} \tag{5.5-14}$$

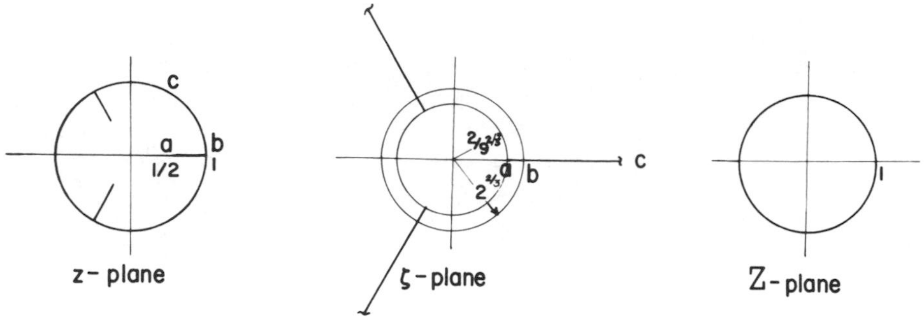

Fig. 5.5-3 Mapping of the unit disk with three cuts onto the unit disk.

Now $\zeta_1 = z^n$ maps the unit disk $|z| < 1$ onto an n-sheeted Riemann surface; $\zeta_2 = \zeta_1/(1 + \zeta_1)^2$ maps this n-sheeted disk onto the n-sheeted plane with n-cuts along the real axis of each sheet from 1/4 to infinity; $Z = \zeta_2^{1/n}$ maps the n-sheeted Riemann surface onto a single sheet with n-cuts running from $|Z| = 1/2^{2n}$ to infinity along $\arg Z = 0, 2\pi/n, 4\pi/n, \ldots$. Thus (5.5-14) maps the Z-plane with n-slits running from $|Z| = 2^{-2n}$ to infinity along the rays $\arg(Z) = 2\pi(k - 1)/n, k = 1, 2, \ldots, n$, onto the disk $|z| < 1$. As an application of this result we may map the domain $|z| < 1$ with slits from $|z| = 1/2$ to $|z| = 1$ on $\arg z = 0, 2\pi/3, 4\pi/3$ onto the unit disk $|Z| < 1$. A simple folding of this domain, shown in Fig. 5.5-3, is accomplished by

$$\zeta = \frac{z}{(1 + z^3)^{2/3}}$$

This gives a ζ-plane with three slits running from $2/9^{2/3}$ to infinity. The previous result can then be used to map the ζ-plane onto the unit disk giving

$$\frac{2Z}{(1 + Z^3)^{2/3}} = \left(\frac{9}{2}\right)^{2/3} \frac{z}{(1 + z^3)^{2/3}}$$

for the complete transformation.

5.6 ACKNOWLEDGMENTS

The author is indebted to Professor C. E. Pearson of the University of Washington and Professors W. H. J. Fuchs and G. S. S. Ludford of Cornell University for their many constructive criticisms of the manuscript and their valuable suggestions for its improvement. He is also grateful to Dr. P. M. Gill of the University of Adelaide for reading the manuscript for accuracy and to Professor P. L. Sachdev of the Indian Institute of Science, Bangalore for reading the manuscript for clarity.

5.7 REFERENCES AND BIBLIOGRAPHY

5.7.1 References

5-1 Carrier, G. F., Krook, M., and Pearson, C. E., *Functions of a Complex Variable*, McGraw-Hill, New York, 1966.

5-2 Phillips, E. G., *Functions of a Complex Variable*, Oliver and Boyd, Edinburgh, 1957.

5-3 Korn, G. A., and Korn, T. M., *Mathematical Handbook for Scientists and Engineers*, Second Edition, McGraw-Hill, New York, 1968.

5-4 Muskhelishvili, N. I., *Singular Integral Equations*, P. Noordhoff, N. V., Groningen, 1953.

5-5 Nevanlinna, R., and Paatero, V., *Introduction to Complex Analysis*, Addison-Wesley, Reading, Mass., 1969.

5-6 Evgrafov, M. A., *Analytic Functions*, W. B. Saunders, Philadelphia, 1966.

5-7 Gradshteyn, I. S., and Ryzhik, I. M., *Table of Series, Products, and Integrals*, Fourth Edition, Academic Press, New York, 1965.

5-8 Erdélyi, A., et. al., *Tables of Integral Transforms* (2 vols.), McGraw-Hill, New York, 1954.

5-9 Kober, H., *Dictionary of Conformal Representations*, Second Edition, Dover, New York, 1957.

5-10 Churchill, R. V., *Complex Variables and Applications*, Second Edition, McGraw-Hill, New York, 1960.

5-11 Sansone, G., and Gerretsen, J., *Lectures on the Theory of Functions of a Complex Variable* (2 vols.), P. Noordhoff, Groningen, 1960.

5-12 Silverman, R. A., *Introductory Complex Analysis*, Prentice-Hall, Englewood Cliffs, 1967.

5-13 Milne-Thomson, L. M., *Theoretical Hydrodynamics*, Fifth Edition, Macmillan, New York, 1968.

5-14 Ahlfors, L. V., *Complex Analysis: An Introduction to the Theory of Analytic Functions of One Complex Variable*, Third Edition, McGraw-Hill, New York, 1978.

5-15 Whittaker, E. T., and Watson, G. N., *A Course of Modern Analysis*, Fourth Edition, Cambridge University Press, Cambridge, 1927.

5-16 Henrici, P., *Applied and Computational Complex Analysis* (2 vols.), John Wiley & Sons, New York, 1974.

5.7.2 Bibliography

Ahlfors, L. V., *Complex Analysis*, Second Edition, McGraw-Hill, New York, 1966.

Carathéodory, C., *Theory of Functions of a Complex Variable* (2 vols.), Chelsea Publishing Co., New York, 1958.

Carrier, G. F., Krook, M., and Pearson, C. E., *Functions of a Complex Variable*, McGraw-Hill, New York, 1966.

Churchill, R. V., *Complex Variables and Applications*, Second Edition, McGraw-Hill, New York, 1960.

Copson, E. T., *An Introduction to the Theory of Functions of a Complex Variable*, Oxford University Press, London, 1935.

Depree, J. D., and Oehring, C. C., *Elements of Complex Analysis*, Addison-Wesley, Reading, 1969.

Evgrafov, M. A., *Analytic Functions*, W. B. Saunders, Philadelphia, 1966.

Fuchs, W. H. J., *Topics in the Theory of Functions of One Complex Variable*, Van Nostrand Reinhold, New York, 1967.

Korn, G. A., and Korn, T. M., *Mathematical Handbook for Scientists and Engineers*, Second Edition, McGraw-Hill, New York, 1968.

Levinson, N., and Redheffer, R. M., *Complex Variables*, Holden-Day, San Francisco, 1970.

Markushevich, A. I., *Theory of Functions of a Complex Variable* (3 vols.), Prentice-Hall, Englewood Cliffs, 1967.

Nevanlinna, R. and Paatero, V., *Introduction to Complex Analysis*, Addison-Wesley, Reading, 1969.

Phillips, E. G., *Functions of a Complex Variable*, Eighth Edition, Oliver & Boyd, Edinburgh, 1957.

Phillips, E. G., *Some Topics in Complex Analysis*, Pergamon Press, Oxford, 1966.

Sansone, G., and Gerretsen, J., *Lectures on the Theory of Functions of a Complex Variable*, 2 vols., P. Noordhoff, Groningen, 1960.

Silverman, R. A., *Introductory Complex Analysis*, Prentice-Hall, Englewood Cliffs, 1967.

Sommer, F., "Functions of Complex Variables," *Mathematics Applied to Physics*, Ed. É. Roubine, Springer-Verlag, New York, 1970.

6

Ordinary Differential and Difference Equations

Edward R. Benton*

6.0 INTRODUCTION

At the heart of many problems in mathematics, physics, and engineering lies the ordinary differential equation or its numerical equivalent, the ordinary finite difference equation. Ordinary differential equations arise not only in countless direct applications, but also occur indirectly, as reductions of partial differential equations (by way of separation of variables or by transform techniques for example; cf. Chaps. 9, 11). Likewise, the probably less familiar difference equations are of inherent interest (in probability, statistics, economics, etc.) but also appear as recurrence relations in connection with differential equations or as numerical approximations to differential equations.

The aim of this chapter is to recount those techniques of some generality and power for solving the three major classical linear problems (the initial-value problem, the boundary-value problem, and the eigenvalue problem). Some attention is also paid to nonlinear equations because of their great interest and relevance.

A first useful step before solving any differential equation problem involves the determination of its classification, because available techniques for solution are catalogued according to a classification scheme.

6.1 BASIC CONCEPTS

6.1.1 Definitions and Classification

A relation of equality involving an unknown function (the *dependent variable*) and one or more of its derivatives is known as a *differential equation*. When the function depends on a single independent variable so that only ordinary derivatives oc-

*Prof. Edward R. Benton, Dep't. of Astro-Geophysics, University of Colorado, Boulder, Col.

cur, then the relation is an *ordinary differential equation* which is written generally as

$$F(x, y, y', y'', \ldots, y^{(n)}) = 0 \tag{6.1-1}$$

Here x is the independent and y the dependent variable. Physically, x normally corresponds to a spatial coordinate or time, but other interpretations are possible. Derivatives of y with respect to x are denoted by primes so that if $y = f(x)$, then $y' = df/dx, y'' = d^2f/dx^2, y^{(n)} = d^n f/dx^n$. Functions of several variables which are determined by equations involving partial derivatives (*partial differential equations*) are treated in Chaps. 8 and 9.

The *order* of a differential equation is that of the highest derivative which occurs; thus eq. (6.1-1) is an nth-order ordinary differential equation. If, as normally happens, integer powers of the unknown function and its derivatives occur, then the power of the highest derivative term is called the *degree* of the differential equation. When the dependence of F on the highest derivative of y is transcendental, rather than rational algebraic, then in effect, all powers occur, and the equation is of infinite degree. Should every nonzero term of the equation involve either the unknown function or its derivatives, the equation is said to be *homogeneous*; otherwise a term dependent on x but independent of $y, y', y'', \ldots, y^{(n)}$ is present and the resulting *inhomogeneous equation* is normally written with the inhomogeneity as the right-hand member (it is referred to as the *forcing function*).

The following examples of ordinary differential equations illustrate this terminology

$$(y''')^2 + (y')^4 + 3y^6 = 0 \tag{6.1-2}$$

$$(1 + x)y' - 2y = -6 \tag{6.1-3}$$

$$y'' + a(x)y' + b(x)y = g(x) \tag{6.1-4}$$

$$(y'')^2 - xy'' + y' = 0 \tag{6.1-5}$$

$$y'' + 2y' - 3y = 0 \tag{6.1-6}$$

$$y'' + \lambda^2 y = 0 \tag{6.1-7}$$

In this set, eq. (6.1-2) is third order, second degree, homogeneous; eq. (6.1-3) is first order, first degree, but inhomogeneous. The other equations are all of second order. In (6.1-4), $g(x)$ is the forcing function. Generally, quantities independent of y which multiply y or its derivatives are the *coefficients* in the equation so eq. (6.1-2) is a *constant coefficient equation* whereas eqs. (6.1-3, 6.1-4) have *variable coefficients.* In eq. (6.1-7) the factor λ^2 is to be regarded as a *parameter*, that is, a quantity independent of x which can take on different (constant) values.

A most important type of classification, especially from the point of view of the solvability of a differential equation, is provided by the criterion of linearity or nonlinearity. If, in eq. (6.1-1), the dependence of F on $y, y', y'', \ldots, y^{(n)}$ is linear then the resulting general n^{th}*-order linear differential equation* is written in the standard form

$$\sum_{j=0}^{n} a_j(x) y^{(j)}(x) = g(x) \tag{6.1-8}$$

where $a_j(x)$ are prescribed variable coefficients, $g(x)$ is the given forcing function, and the zeroth derivative of y, denoted by $y^{(0)}(x)$ is $y(x)$ itself. Comparison of eq. (6.1-8) with eq. (6.1-1) reveals that for linear equations it is necessary that the quantity F in eq. (6.1-1), regarded as a function of the variables, $y, y', y'', \ldots, y^{(n)}$, has a multivariate power series representation containing only the first degree terms in each of these variables. Equations such as eqs. (6.1-2, 6.1-5) that cannot be cast into the form of eq. (6.1-8) are, by definition, *nonlinear*.

Much of the applied mathematics interest in differential equations stems from the challenging problem of constructing *solutions* or *integrals* of some equation valid on an *open* (or *closed*) *interval* $x_1 < x < x_2$ (or $x_1 \leqslant x \leqslant x_2$), by which is meant functions $y = f(x)$ which satisfy the differential equation at every point of the prescribed interval. To do so, y and each of its derivatives which occur in the equation must exist throughout the interval and be such that, upon substitution, the equation is reduced to an identity. In geometry, a functional relationship between a dependent variable y and its independent variable x arises naturally as a means of describing a curve in the x, y-plane. It is therefore convenient to visualize solutions as curves, the phrase *solution curve* or *integral curve* being widely used.

If every term of a homogeneous equation vanishes when $y \equiv 0$, such as for eqs. (6.1-2, 6.1-5, 6.1-6, 6.1-7), then one solution of the equation is the *trivial solution*: $y(x) \equiv 0$. Two solutions which differ by the trivial solution are said to be *identical*. The usual interest is in finding *nontrivial solutions.* A good many techniques have been developed for solving linear differential equations, but most nonlinear equations are either simple transformations of a linear equation, or else they present their own unique difficulties, few generalized methods having been, as yet, found.

Since geometrically, an ordinary differential equation like (6.1-1) is simply a relation between the location y, the slope y', the curvature y'', . . . , of a curve through some point x, there can be many solution curves satisfying a given equation. However, sometimes a single comprehensive formula can be found from which an infinite set of integral curves can be given by assigning different values to a set of constants, whose number is the same as the order of the differential equation. For example, the second order nonlinear differential eq. (6.1-5) admits as a solution any function of the form $y(x) = c_1 x^2 - 4c_1^2 x + c_2$, with c_1 and c_2 arbitrary constants. Notice that y is really a function, not only of x, but also of c_1 and c_2, and the dependence on all three of these quantities is continuous. These solution curves constitute a two-parameter family of parabolas. Such a solution is a *general solution* or *general integral* of the differential equation. In contrast, a *particular solution* or *particular integral* is selected from the general solution by giving a definite set of values to the arbitrary constants (c_1, c_2 in this case). For nonlinear equations, it must not be supposed that the general solution is complete in the sense that it necessarily includes all integrals of the equation. Sometimes a *singular solution* or singular integral exists which cannot be obtained by assigning specific values to the constants in the general solution. For the example of eq. (6.1-5), there is also a singular integral $y = x^3/12$, which is clearly not a parabola. No further attention is given to singular solutions here, but the interested reader can find more material and examples in Refs. 6-16, 6-26.

A problem presented by the general solution is simply that it is too general, usually including multiply infinite numbers of integrals whereas the usual requirement is for a single, or at least finite number of solutions. Since the general differential eq. (6.1-1) can, in principle, be rearranged to yield

$$y^{(n)}(x) = G(x, y, y', y'', \ldots, y^{(n-1)}) \tag{6.1-9}$$

it only specifies how fast the nth derivative of y changes, and even that is known only when $y, y', y'', \ldots, y^{(n-1)}$ are given. Evidently, if a single integral curve is to be determined, these lower order derivatives (or equivalent supplementary information) must be supplied in addition to the differential equation. A straightforward way in which this often occurs is in dynamic situations where x is a time-like variable and the values of $y, y', y'', \ldots, y^{(n-1)}$ are specified at some initial instant $x = x_0$. More generally, n linear equations involving the function and its derivatives evaluated at a single point may be given. Together with these initial conditions the differential equation then constitutes a system termed an *initial-value problem*. Consider, for example, the constant coefficient, second order, linear differential eq. (6.1-6), whose general solution is $y(x) = c_1 e^x + c_2 e^{-3x}$. If we wish to find the integral curve which passes through the origin and has unit slope there, then appropriate initial conditions are $y(0) = 0$, $y'(0) = 1$, and the unique solution is readily shown to be $y(x) = (e^x - e^{-3x})/4$.

In situations where x is a space-like variable extending between two boundaries at $x = x_1$, $x = x_2$, say, then the *boundary-value problem* frequently arises. The extra information needed to associate specific values with the arbitrary constants in the general solution is given as boundary conditions. For example, the integral curve for eq. (6.1-6) which passes through the origin and the point $x = 1, y = 1$ has for boundary conditions $y(0) = 0$, $y(1) = 1$, so that the particular integral curve is described by the solution $y(x) = (e^x - e^{-3x})/(e - e^{-3})$. Just as with differential equations, both initial and boundary conditions are referred to as homogeneous (or inhomogeneous) if the right-hand member is zero (or nonzero).

An extraordinarily important subclass of the boundary-value problem known as the *eigenvalue problem* occurs when a homogeneous equation which includes a parameter is posed along with homogeneous boundary conditions. Suppose, for example, that a solution to eq. (6.1-7) valid in the interval $0 \leqslant x \leqslant 1$ is to be found which vanishes at the end points. The general solution of the second order eq. (6.1-7) is

$$y(x) = c_1 \sin \lambda x + c_2 \cos \lambda x \tag{6.1-10}$$

and in order for $y(0)$ to be 0 clearly c_2 must vanish. Since y is also to be zero at $x = 1$, evidently

$$c_1 \sin \lambda = 0 \tag{6.1-11}$$

One solution of this equation is $c_1 = 0$, but this gives the trivial solution for y. Of far greater significance is the fact that there is a discrete or countably infinite set of numbers λ_n which will reduce eq. (6.1-11) to an identity. These *eigenvalues*, or *characteristic values*, are given, in the present case, by

$$\lambda_n = \pm n\pi, \qquad n = 1, 2, 3, \ldots, \tag{6.1-12}$$

where the subscript has been added to λ to emphasize its dependence on the value of an integer, n. It is concluded that only when the parameter appearing in the original equation takes on very special values is a nontrivial solution possible. The specific solution for any appropriate λ_n is called an *eigenfunction* or *eigensolution*, being given here by

$$y_n(x) = c_n \sin \lambda_n x, \quad \lambda_n = \pi, 2\pi, 3\pi, \ldots \tag{6.1-13}$$

where c_n is an arbitrary constant. This intriguing type of problem is discussed further in section 6.6.

6.1.2 Existence and Uniqueness of Solutions

In most problems which arise as an attempt to model some physical situation, a single solution to a differential equation is required which satisfies given initial or boundary conditions. Consider the typical first order example of eq. (6.1-3) subject to one of the following three initial conditions:

$$(i)\ y(-1) = 0; \quad (ii)\ y(0) = 3; \quad (iii)\ y(-1) = 3 \tag{6.1-14}$$

Since the general solution of eq. (6.1-3) is $y = c(1 + x)^2 + 3$, c = constant, it is immediately evident that problem (*i*) has no solution, problem (*ii*) has just one solution $y(x) = 3$, and problem (*iii*) has the infinite number of solutions given by the general integral. This varied behavior makes it imperative to know when a given initial-value problem has at least one solution (the question of *existence*) and when it has exactly one solution (the question of *uniqueness*). These central issues have occupied mathematicians for many years. Here the important results will simply be quoted, together with references. Although the theorems can readily be generalized for systems of equations, or high order equations, the general flavor is well-illustrated by the simple first order linear initial-value problem

$$y'(x) = g(x, y), \quad y(x_1) = y_1 \tag{6.1-15}$$

The *Existence Theorem* for this case, basically due to Cauchy, can be found proven, for example, in Refs. 6-8, 6-18. It states: *If $g(x, y)$ is continuous at all points of the closed rectangle R: $|x - x_1| \leqslant A$, $|y - y_1| \leqslant B$ and therefore bounded, say $|g(x, y)| \leqslant M$ in R, (A, B, M are positive constants), then the initial-value problem has at least one solution valid in an interval at least as large as $|x - x_1| \leqslant \Gamma$, where the positive constant Γ is the smaller of A and B/M.*

Once existence has been established, the next question is to determine the conditions under which a single solution exists. This is answered by the *Uniqueness Theorem: If $g(x, y)$ and the partial derivative $\partial g/\partial y$ are continuous in the closed rectangle R: $|x - x_1| \leqslant A$, $|y - y_1| \leqslant B$ and therefore bounded, say $|g(x, y)| \leqslant M$, $|\partial g/\partial y| \leqslant N$ in R, (A, B, M, N are positive constants) then the initial-value problem has a unique solution valid in the interval $|x - x_1| \leqslant \Gamma$, where the positive constant Γ is the smaller of A and B/M.*

In essence the theorems require only continuity of g (as a function of x, y) for existence, but uniqueness demands also continuity of $\partial g/\partial y$ (or a Lipschitz condition, defined below). Consider the simple example $y' = (y - a)^{1/2}, y(x_1) = a$. Here $g = (y - a)^{1/2}$ is continuous but $\partial g/\partial y$ fails to exist at $y = a$, so solutions exist but need not be unique. In fact, two solutions which pass through the required point (x_1, a) are $y(x) = a, y(x) = a + \frac{1}{4}(x - x_1)^2$.

One very expedient way to establish both existence and uniqueness is to construct a process which generates the solution. An excellent example of this is the Picard iteration process which consists of evaluating successive approximations to $y(x)$ from the integration formula

$$y_n(x) = y_1 + \int_{x_1}^{x} g(\xi, y_{n-1}(\xi))\, d\xi, \qquad n = 1, 2, 3, \ldots \tag{6.1-16}$$

where $y_0(x) = y_1$. *Picard's theorem* shows that under the slightly different conditions in which $g(x, y)$ is continuous and satisfies a Lipschitz condition (which is weaker than requiring $\partial g/\partial y$ to be continuous)

$$|g(x, y) - g(x, \bar{y})| \leqslant L \cdot (y - \bar{y})$$

with L some positive constant and $\bar{y}$ simply a different value of y, then the Picard process converges in the limit $n \to \infty$ to the unique solution of the initial-value problem. Moreover, the solution function has a continuous first derivative in the interval $|x - x_1| \leqslant \Gamma$, where Γ is the same as before. Of equal importance is the fact that y is also a continuous function of the initial value, y_1.

These and other existence and uniqueness theorems are both proven and illustrated by examples in Chapter 12 of Ref. 6-4.

6.1.3 Systems of Equations

When several functions (the dependent variables) depend on a single independent variable and upon each other, we speak of a *simultaneous system of differential equations*. Most of the foregoing basic concepts and many of the subsequent techniques can, with little extra effort, be extended so as to apply to such systems. This sort of extension is physically relevant to such a scientific problem as, for example, celestial mechanics where the naturally occurring coordinate is time, and where the system consists of several or many bodies subject to the same physical law (Newton's second law) and interacting with each other through their mutual gravitational attractions. An even more elementary, yet still important, example is provided by the equations describing the three-dimensional motion of a single particle of mass m which obeys Newton's second law of motion

$$mx'' = X, \quad my'' = Y, \quad mz'' = Z \tag{6.1-17}$$

where x, y, z, functions of time t, are the Cartesian coordinates of the particle which is acted upon by the force components X, Y, Z which in general are functions of time t, position x, y, z, and velocity x', y', z'. Under the transformation

$$x \to y_1, \quad y \to y_2, \quad z \to y_3, \quad x' \to y_4, \quad y' \to y_5, \quad z' \to y_6 \tag{6.1-18}$$

eq. (6.1-17) reduces to a system of six simultaneous first order, nonlinear ordinary differential equations

$$\begin{aligned} y_1' &= y_4 \\ y_2' &= y_5 \\ y_3' &= y_6 \\ y_4' &= m^{-1} X(t, y_1, y_2, \ldots, y_6) \\ y_5' &= m^{-1} Y(t, y_1, y_2, \ldots, y_6) \\ y_6' &= m^{-1} Z(t, y_1, y_2, \ldots, y_6) \end{aligned} \tag{6.1-19}$$

When, generally, each of the dependent variables $(y_1, y_2, \ldots, y_6)$ appears in each equation, no one equation can be solved independently of the others; then, the system is said to be simultaneous and *coupled*.

Systems of equations also arise in the parametric or implicit description of a curve. Instead of regarding y as a function of the Cartesian coordinate x, it may be advantageous to think of both x and y as functions of another variable z (say, the arc length along the curve). Then the single first order equation $y' = g(x, y)$ is equivalent to the simultaneous system $dx/dz = h(x, y)$, $dy/dz = k(x, y)$ where $k(x, y)/h(x, y) = g(x, y)$.

The foregoing examples are typical of the general system of n first order, ordinary, differential equations

$$
\begin{aligned}
y_1' &= f_1(x, y_1, y_2, \ldots, y_n)\\
y_2' &= f_2(x, y_1, y_2, \ldots, y_n)\\
&\vdots\\
y_n' &= f_n(x, y_1, y_2, \ldots, y_n)
\end{aligned}
\tag{6.1-20}
$$

Notice that the number of equations coincides with the number of dependent variables (if there were greater or lesser numbers of equations than dependent variables, the system would be described as being overspecified or underspecified, respectively). The system eq. (6.1-20) also contains, as a special case, the *uncoupled* or *independent system of equations*; this occurs when each f_j depends only on x and y_j, and is much easier to solve, though less interesting.

The initial-value problem for the system eq. (6.1-20) consists of the system of equations and the initial-conditions

$$y_1(x_1) = a_1, \quad y_2(x_1) = a_2, \ldots, \quad y_n(x_1) = a_n \tag{6.1-21}$$

where $a_1, a_2, \ldots, a_n$ is a set of constants. Thus, each function in the system is given one initial condition.

The existence and uniqueness theorems for the initial-value problem are simple direct extensions of those quoted in section 6.1.2. Continuity of the functions plus satisfaction of a Lipschitz condition guarantees both existence and uniqueness (i.e., there is a single set of functions $y_1, y_2, \ldots, y_n$ satisfying both the system of equations and the initial conditions) in the same sort of interval as before. Also, the Picard process is easily defined and again converges to the unique solution. The details of these theorems can be found in Ref. 6-8.

The initial-value problem for the system of n first order equations in eq. (6.1-20) is also fully equivalent to that for the single n^{th} order eq. (6.1-1) or eq. (6.1-9). To see this, merely introduce the following definitions: $y_1 = y, y_2 = y', \ldots, y_n = y^{(n-1)}$.

Then, the system equivalent to eq. (6.1-9) is

$$
\begin{aligned}
y_1' &= y_2\\
y_2' &= y_3\\
&\vdots\\
y_{n-1}' &= y_n\\
y_n' &= G(x, y_1, y_2, \ldots, y_n)
\end{aligned}
\tag{6.1-22}
$$

and this is clearly of the same type as eq. (6.1-20). Bearing in mind that under appropriate conditions (which are not unduly restrictive) the initial-value problem for the system of n first order equations has a unique solution, the equivalence above shows likewise that the initial-value problem for a single n^{th} order equation admits a unique solution under the same conditions. The differential equation is eq. (6.1-9) and the initial conditions follow from eq. (6.1-21) and the definitions above

$$y(x_1) = a_1, \quad y'(x_1) = a_2, \ldots, y^{(n-1)}(x_1) = a_n \tag{6.1-23}$$

It may be helpful to keep in mind that the initial conditions here specify the function and its first $n - 1$ derivatives at $x = x_1$ while the differential eq. (6.1-9) and its successive derivatives then give, by substitution, the nth and higher order derivatives of y at $x = x_1$. Consequently, under suitable continuity restrictions the function y has a unique Taylor series expansion about $x = x_1$ and therefore exists and is unique.

6.2 FIRST ORDER LINEAR DIFFERENTIAL EQUATIONS

6.2.1 The Principle of Superposition

Several reasons make it sensible to begin a study of differential equations with those which are the simplest of all, the linear equations. Because of this simplicity, highly systematic theory is available for understanding at least the qualitative aspects of any such equation, if indeed, it is not actually possible to construct the general solution. Much of classical analysis can be viewed as an outgrowth of the theory of linear differential equations with variable coefficients. The terminology and basic ideas which enter naturally for linear equations are also carried over into nonlinear analysis, to the extent possible. Moreover, a most useful way to understand a recalcitrant nonlinear problem, is first to solve a linearized version of it. Alternatively, the solution of some nonlinear problems can be made to depend on the solutions of a sequence of linear problems (this is the idea behind certain perturbation methods, see section 6.8.1 for a sketch, and Chapter 14 for detailed exposition).

The fact that the general solution to several important classes of linear differential equations can be written explicitly is largely due to the *principle of superposition*, which states briefly, that solutions of these equations share a group property in that they can be added linearly to one another and the new element remains in the group of solutions. Suppose $y = u(x)$ and $y = v(x)$ are two distinct solutions of the general n^{th} order linear, nonhomogeneous differential equation (6.1-8) so that

$$a_n(x)\, u^{(n)}(x) + a_{n-1}(x)\, u^{(n-1)}(x) + \cdots + a_1(x)\, u'(x) + a_0(x)\, u(x) = g(x)$$

(6.2-1)

$$a_n(x)\, v^{(n)}(x) + a_{n-1}(x)\, v^{(n-1)}(x) + \cdots + a_1(x)\, v'(x) + a_0(x)\, v(x) = g(x)$$

Then it is easily verified by direct substitution that the weighted sum of u and v, say

$$w(x) = (\alpha + \beta)^{-1} [\alpha u(x) + \beta v(x)] \tag{6.2-2}$$

where α, β are any constants, with $\alpha + \beta \neq 0$, satisfies the same equation

$$a_n(x)w^{(n)}(x) + a_{n-1}(x)w^{(n-1)}(x) + \cdots + a_1(x)w'(x) + a_1(x)w(x) = g(x) \tag{6.2-3}$$

Alternatively, if $y = u(x)$, $y = v(x)$ satsify eq. (6.1-8) with $g = g_1$, $g = g_2$, respectively, then $u + v$ satisfies the same equation with $g = g_1 + g_2$. Also, any two solutions of a linear homogeneous equation can be multiplied by arbitrary constants and then added together and the result is still a solution. Thus, elementary solutions can be superimposed to build more complicated ones.

This superposition principle is the single strongest link between all linear differential equations and is a powerful tool of analysis. Such superposition is rarely possible for nonlinear equations.

6.2.2 General Solution of the First Order Linear Differential Equation

The general nonhomogeneous, linear, differential equation of first order is

$$a_1(x)y'(x) + a_0(x)y(x) = g(x) \tag{6.2-4}$$

where a_1, a_0 are the coefficient functions and $g(x)$, the forcing function. Consider first the homogeneous version of this equation, referred to as the *reduced equation*

$$a_1(x)z'(x) + a_0(x)z(x) = 0 \tag{6.2-5}$$

At any point where $z(x) \neq 0$ and $a_1(x) \neq 0$, it is legitimate to write eq. (6.2-5) in the form $z'(x)/z(x) = -a_0(x)/a_1(x)$, and with the proviso that $\int^x I(\xi)\,d\xi$ denotes any function whose derivative is the integrand, $I(x)$, eq. (6.2-5) now yields

$$z(x) = De^{-b(x)} = D \exp\left[-\int^x \frac{a_0(\xi)}{a_1(\xi)} d\xi\right] \tag{6.2-6}$$

where D is an arbitrary constant. As will be emphasized again later, linear equations always admit solutions of characteristically exponential type. Since this solution involves one arbitrary constant, it is the general solution of the reduced equation, customarily being called the *complementary function* or complementary solution of the nonhomogeneous equation. The important point here is that it requires just a single function (the exponential) to describe any solution of the first order linear homogeneous differential equation.

The factor $e^{-b(x)}$ in eq. (6.2-6) has another interpretation which allows it to be used to generate a solution of the nonhomogeneous equation (this feature, where one solution can be used to obtain another, is a recurrent theme for linear equations, and is put to frequent use). In fact, its reciprocal, $e^{b(x)}$, known as an integrating factor, is of just the right structure so that when the homogeneous equation is multiplied by it, an exactly integrable quantity is produced. Since a potential singularity where $a_1(x) = 0$ has already been excluded from consideration, it is convenient to divide eq. (6.2-5) by a_1. Multiplication by the integrating factor then gives

$$e^{b(x)} [z'(x) + b'(x) z(x)] = [e^{b(x)} z(x)]' = 0$$

which is, of course, immediately integrable and again yields eq. (6.2-6). The general solution of the nonhomogeneous equation follows from an application of these same operations to eq. (6.2-4): $[e^{b(x)} y(x)]' = e^{b(x)} g(x)/a_1(x)$. Integration with respect to x, then yields

$$e^{b(x)} y(x) = \int^x e^{b(\xi)} \frac{g(\xi)}{a_1(\xi)} d\xi + C$$

where C is an arbitrary constant. Thus, the general solution of eq. (6.2-4) can be written in terms of integrals (the problem is said to have been *reduced to quadratures*)

$$y(x) = Ce^{-b(x)} + \int^x \frac{g(\xi)}{a_1(\xi)} e^{b(\xi)-b(x)} d\xi \tag{6.2-7}$$

where $b(x) = \int^x [a_0(\xi)/a_1(\xi)]\, d\xi$. This solution is valid on any interval free from zeros of $a_1(x)$. Since any two allowable values of $b(x)$ differ from each other only by a constant ($\ln D$), any of the allowable choices for $b(x)$ in the second term on the right here will do, because D cancels out. In the first term, the addition of $\ln D$ to some allowable value of b merely changes C. Consequently, only one arbitrary constant occurs in eq. (6.2-7) and any solution of eq. (6.2-4) can be found from this formula.

The solution in eq. (6.2-7) is described verbally by saying that the general solution of any first order nonhomogeneous linear differential equation is the sum of the general solution of the reduced equation (the complementary function) and any particular solution of the inhomogeneous equation (this last term in eq. (6.2-7) being referred to as a *particular integral*).

For future reference, it is especially instructive to rewrite the general solution in eq. (6.2-7) in terms of the solution of the reduced equation (6.2-6):

$$y(x) = A\, z(x) + \int^x \frac{z(x)}{z(\xi)} \frac{g(\xi)}{a_1(\xi)} d\xi \tag{6.2-8}$$

where A is an arbitrary constant. This emphasizes again, how one solution, in this case $z(x)$, can be used to build another. Notice especially that a particular integral of the original nonhomogeneous equation, with the non-vanishing factor $a_1(x)$ divided out, can be calculated from a weighted integral of the effective forcing function g/a_1. The weighting function, $z(x)/z(\xi) = e^{b(\xi)-b(x)}$ considered as a function of x, is a solution of the reduced equation. Such a function plays the role of a *Green's function* for this problem, and arises again in a more difficult context in section 6.4.4.

It is a simple matter to utilize eq. (6.2-7) for solving the initial-value problem in which one requires that solution of eq. (6.2-4) which takes on the initial value $y(x_1) = y_1$. The integrals in eq. (6.2-7) are merely converted to appropriate definite integrals and the result is

$$y(x) = y_1 e^{-c(x)} + \int_{x_1}^{x} \frac{g(\xi)}{a_1(\xi)} e^{c(\xi)-c(x)}\, d\xi \tag{6.2-9}$$

where

$$c(\xi) = \int_{x_1}^{\xi} \frac{a_0(\eta)}{a_1(\eta)} d\eta$$

This solution exists and is unique (for a proof see Ref. 6-4) on any interval containing x_1 for which $a_0(x)/a_1(x)$ and $g(x)$ are continuous in the interval. This means that any singularities of the solution function (i.e., locations where it tends to infinity or does not exist) are associated with (and can be found from) singularities of the coefficients and forcing function.

Examples

(a) $$y' + 3y = x^2, \qquad y(0) = 1$$

Here $a_1(x) = 1, a_0(x) = 3, g(x) = x^2, x_1 = 0, y_1 = 1$, so that eq. (6.2-9) yields

$$y(x) = \tfrac{1}{27}\,[25e^{-3x} + 9x^2 - 6x + 2]$$

The complementary solution is Ce^{-3x}; a particular integral is $(9x^2 - 6x + 2)/27$.

(b) $$y' + (\sin x)y = \sin x \cos x, \qquad y(0) = 3$$

Here $a_1(x) = 1$, $a_0(x) = \sin x$, $g(x) = \sin x \cos x$, $x_1 = 0, y_1 = 3$, thus eq. (6.2-9) yields

$$y(x) = e^{-1+\cos x} + 1 + \cos x$$

(c)
$$xy' + y = xe^{x^2}, \quad y(1) = 1$$

Here $a_1(x) = x$, $a_0(x) = 1$, $g(x) = xe^{x^2}$, $x_1 = 1, y_1 = 1$, and eq. (6.2-9) yields

$$y(x) = \frac{1}{x}\left[1 - \frac{1}{2}e + \frac{1}{2}e^{x^2}\right]$$

which is valid for all $x > 0$. The solution becomes singular as $x \to 0$; the fact that $a_1(x)$ vanishes at $x = 0$ means that the basic existence theorem does not hold in any interval including $x = 0$, so the singular behavior as $x \to 0$ is not unexpected.

(d)
$$xy' - 2y = -2, \quad y(0) = 1$$

Here $a_1(x) = 0$ at $x = x_1 = 0$, so that eq. (6.2-9) cannot be applied directly. We can use either the general solution (6.2-7) with an arbitrary constant, or the integrating factor method. We choose the latter, and multiply each term by $1/x^3$ to give

$$\left(\frac{1}{x^2}y\right)' = -\frac{2}{x^3}$$

whence $y = 1 + Cx^2$, where C is a constant. Any choice of C satisfies the initial condition; thus a well-behaved solution exists but is not unique.

Example (c) also illustrates another way in which equations may be solved, namely, by inspection. It is a good general rule to examine each differential equation with an eye towards discovering whether it has any obvious simple properties, before proceeding to the more formal methods of solution. Thus, the trained eye immediately recognizes that this differential equation can be written in the suggestive form $(xy)' = (\frac{1}{2}e^{x^2})'$ from which a single integration gives $y = Ax^{-1} + (2x)^{-1}e^{x^2}$. When A is chosen to satisfy the initial condition, the former result is recovered.

In principle, then, only a single technique is needed to find any solution of the general first order, linear, initial-value problem, and a formal solution, involving nothing more than quadratures, can always be exhibited.

6.3 SECOND-ORDER LINEAR DIFFERENTIAL EQUATIONS WITH CONSTANT COEFFICIENTS

6.3.1 Preliminary Discussion

Second-order differential equations arise very commonly in the physical and social sciences. A general discussion of solution techniques is much more difficult than for the case of a first-order equation; it is therefore useful to begin with a special case which can be solved exactly and which will illustrate some features to be expected in the general case.

We consider a linear, second-order equation with constant coefficients. To provide some background for such a problem, we will adopt the physical context of a forced, damped harmonic oscillator, in either a mechanical or an electric system (however, the solution methods used can be understood without reference to these systems).

6.3.2 Initial-Value Problem for the Damped Harmonic Oscillator

Any physical system which possesses inertia (characterized by mass in a mechanical situation), and which is acted upon by a restoring force (supplied by a spring in the mechanical analogue) when displaced from equilibrium, contains the ingredients necessary to display oscillations. For a linear system, there are two possible natural states; that of rest (corresponding to the trivial solution of an appropriate differential equation) and that of simple harmonic oscillation (a persistent sinusoidal vibration at constant amplitude and frequency). Nature is rarely so primitive, but a surprisingly accurate model for many natural systems is obtained by allowing two more forces to act, one, a damping force proportional to the rate of change of displacement (of frictional origin, as typified by viscous drag) and the other, an external driving force (the forcing function).

Two classically simple physical systems which embrace these fundamental concepts are sketched in Fig. 6.3-1. Here, the initial-value problem is solved for the mass-spring-dashpot system of Fig. 6.3-1a driven by the time-dependent external force $F(x)$, where x is time. The corresponding problem for the *LRC* circuit of Fig. 6.3-1b is solved by the Laplace transform technique in section 6.3.3. Further details on oscillations are given in Chapter 13 as well as in Refs. 6-14, 6-27.

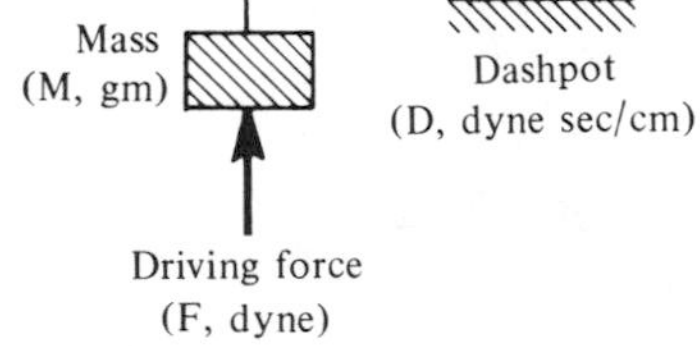

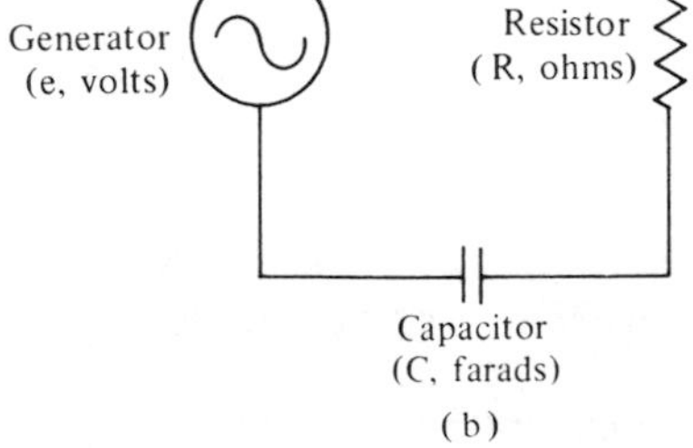

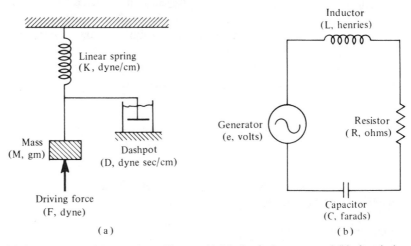

Fig. 6.3.1 The damped harmonic oscillator. (a) Mechanical system; and (b) electrical system.

Consider now the mass in Fig. 6.3-1a where, for the moment, the external force is that due to Newtonian gravitation only. Let $y(x), y'(x)$, and $y''(x)$ denote, respectively, the displacement from equilibrium, the vertical velocity, and the acceleration of the mass at time x. The vertical force balance for the mass (Newton's second law) leads to the constant-coefficient, second order, nonhomogeneous ordinary differential equation $My'' + Dy' + Ky = F(x)$. The force F is external in the sense that it is not specified as a function of the displacement, y, but rather depends only on time.

When possible, it is convenient to rearrange the differential equation so that the coefficient of the highest derivative term is unity. Accordingly, the initial-value problem is usually written in the form

$$y'' + 2\lambda y' + \omega^2 y = f(x); \quad y(0) = a; \quad y'(0) = b \tag{6.3-1}$$

where $2\lambda = D/M$, $\omega^2 = K/M$, $f(x) = M^{-1}F(x)$ and a, b are the initial displacement and velocity of the mass, respectively.

Equation (6.3-1) is linear, inhomogeneous, and has constant coefficients. By virtue of the linearity, the full solution may be again decomposed into the sum of the complementary function, $z(x)$, and a particular integral. To find the former, we consider the reduced equation ($f = 0$), which corresponds physically to the mass being disconnected from any external force, and is therefore referred to as the problem of *free oscillations* (as opposed to the *forced oscillations* which result when $f \neq 0$):

$$z'' + 2\lambda z' + \omega^2 z = 0 \tag{6.3-2}$$

Solutions of purely exponential type (i.e., $z \sim e^{\gamma x}$) are to be expected, because repeated differentiation of such a function, as called for by the differential equation, is tantamount to multiplication by a constant (for example, $d(e^{\gamma x})/dx = \gamma e^{\gamma x}$). Substitution of the trial form $z = Ae^{\gamma x}$ (A = constant) yields $A(\gamma^2 + 2\lambda\gamma + \omega^2)$ $e^{\gamma x} = 0$ whence

$$\gamma^2 + 2\lambda\gamma + \omega^2 = 0 \tag{6.3-3}$$

This polynomial (of the same degree as the differential equation) is called the *characteristic equation* or *auxiliary equation* (this idea recurs again for higher-order systems in section 6.5). Equation (6.3-3) has two roots, γ_1 and γ_2. The general solution of eq. (6.3-2) is now

$$z(x) = A_1 e^{\gamma_1 x} + A_2 e^{\gamma_2 x} \tag{6.3-4}$$

where $\gamma_1 = -\lambda + \eta$, $\gamma_2 = -\lambda - \eta$, $\eta = (\lambda^2 - \omega^2)^{1/2}$ and A_1, A_2 are arbitrary constants. The character of this solution depends upon the relative magnitudes of λ and ω (or, in physical terms, upon the relative strengths of the damping and restoring force). Three distinct possibilities exist:

Case I ($\lambda > \omega$) Then $\eta^2 > 0$, so η is a real quantity, taken as positive without loss of generality. The two roots of eq. (6.3-3), γ_1, γ_2 are then real, distinct, and negative (provided the physically relevant restriction to positive damping, $\lambda > 0$, is made). The solution in eq. (6.3-4) is the sum of two distinct, pure exponential functions, each of which decays to zero as time increases without limit. In this case, $\lambda > \omega$, the motion is said to be *overdamped*, and is not periodic.

Case II ($\lambda = \omega$) When $\eta = 0$, both roots of eq. (6.3-3) coincide, and eq. (6.3-4) no longer gives the general solution of eq. (6.3-2). Use of a limiting process suggests that eq. (6.3-4) should be replaced by

$$z(x) = (A_1 + A_2 x)\, e^{-\lambda x} \tag{6.3-5}$$

and it is readily checked by direct substitution that eq. (6.3-5) does indeed satisfy eq. (6.3-2) when $\gamma_1 = \gamma_2$. With λ positive, the solution again decays to zero as $x \to \infty$, and is non-periodic. Because this case arises when the system is on the verge of oscillatory behavior, the motion is said to be *critically damped.*

Case III ($\lambda < \omega$) When η is purely imaginary, $\eta = i|\eta|$, γ_1, γ_2 form a complex conjugate pair and it is more revealing to write eq. (6.3-4) as

$$z(x) = e^{-\lambda x}\,(B_1 \cos |\eta|\, x + B_2 \sin |\eta|\, x) \tag{6.3-6}$$

where B_1, B_2 are arbitrary real constants. In this case the free oscillation is periodic in time (with period $2\pi/|\eta|$), but its amplitude decays exponentially. The motion is said to be *underdamped.* Note also, the important subcase in which there is no damping, so that $\lambda \equiv 0$. Then the solution becomes

$$z(x) = B_1 \cos \omega x + B_2 \sin \omega x \tag{6.3-7}$$

which describes an undamped, simple harmonic free oscillation at circular frequency ω. For this reason, ω is referred to as the *natural frequency.*

Consider now the problem of finding a particular integral to the nonhomogeneous equation. It is easily verified that the substitution $y(x) = e^{\gamma_1 x} w(x)$ in eq. (6.3-1) yields a linear equation for $w'(x)$, which is solvable by the method of section 6.2.2. We thus obtain

$$y(x) = A_1 e^{\gamma_1 x} + A_2 e^{\gamma_2 x} + \int^x z(x - \xi) f(\xi)\, d\xi \tag{6.3-8}$$

where $z(x) = (\gamma_1 - \gamma_2)^{-1}(e^{\gamma_1 x} - e^{\gamma_2 x})$ when $\gamma_1 \neq \gamma_2$. If $\gamma_1 = \gamma_2$, a slight modification is needed, the appropriate form being

$$y(x) = (A_1 + A_2 x)\, e^{-\lambda x} + \int^x (x - \xi)\, e^{-\lambda(x-\xi)} f(\xi)\, d\xi \tag{6.3-9}$$

It is now a simple matter to evaluate A_1, A_2, for any given initial conditions, the general solution for the initial-value problem of eq. (6.3-1) becoming:

$$y(x) = \frac{1}{\gamma_1 - \gamma_2}\left\{(b - \gamma_2 a)\, e^{\gamma_1 x} - (b - \gamma_1 a)\, e^{\gamma_2 x} + \int_0^x [e^{\gamma_1 (x-\xi)} - e^{\gamma_2 (x-\xi)}]\, f(\xi)\, d\xi\right\} \quad (6.3\text{-}10)$$

where γ_1, γ_2 (presumed distinct) are given by eq. (6.3-4). Should $\gamma_1 = \gamma_2 (= -\lambda = -\omega)$, then the appropriate solution is

$$y(x) = [a + (b + \lambda a)\, x]\, e^{-\lambda x} + \int_0^x (x - \xi)\, e^{-\lambda(x-\xi)} f(\xi)\, d\xi \quad (6.3\text{-}11)$$

An important application is that in which the driving force $F(x)$ is itself oscillatory, at some constant circular frequency Ω, so that

$$f(x) = f_0 \sin(\Omega x + \phi) = \frac{1}{2i} f_0\, [e^{i(\Omega x + \phi)} - e^{-i(\Omega x + \phi)}] \quad (6.3\text{-}12)$$

where $i^2 = -1$, f_0 is the (fixed) amplitude of the external force per unit mass, and ϕ (in radians) is the phase of the driving force. When the free oscillation is underdamped (Case III above, $\lambda < \omega$) the general solution, from eq. (6.3-10), is

$$y(x) = e^{-\lambda x}(A_1 \sin |\eta|\, x + A_2 \cos |\eta|\, x) + f_0 C \sin(\Omega x + \phi - \psi) \quad (6.3\text{-}13)$$

where

$$C = [4\lambda^2 \Omega^2 + (\omega^2 - \Omega^2)^2]^{-1/2},$$

$$A_1 = |\eta|^{-1}\{b + \lambda a - f_0 C[\lambda \sin(\phi - \psi) + \Omega \cos(\phi - \psi)]\}$$

$$A_2 = a - f_0 C \sin(\phi - \psi), \quad \sin \psi = 2\lambda\Omega C, \quad \cos \psi = (\omega^2 - \Omega^2) C$$

As mentioned previously, the solution is the superposition of two parts. The term involving A_1, A_2 constitutes the *transient solution* because, although it helps satisfy the initial conditions of the problem, it ultimately decays to zero (provided the damping coefficient λ, is positive). What remains is called the *steady state solution* (despite its dependence on time); it is a simple harmonic oscillation at the *impressed frequency*, Ω, whose amplitude is that of the driving force (f_0) multiplied by an *amplification factor*, the constant C. The steady state oscillation is shifted in phase by the amount ψ relative to the phase of the driving force.

The conceptually important case of *resonance* occurs when there is no damping

($\lambda = 0$) and the impressed frequency Ω coincides with the natural frequency ω; the amplification factor is then infinite. Physically, the system is being forced to oscillate at precisely the same frequency at which it would oscillate in the absence of forcing. This permits a rapid transfer of energy from the driving force into the system, and leads to an increasing oscillation amplitude. In this singular case, the appropriate initial-value problem is

$$y'' + \omega^2 y = f_0 \sin(\omega x + \phi), \quad y(0) = a, \quad y'(0) = b \tag{6.3-14}$$

and the solution is

$$y(x) = \left(\frac{b}{\omega} + \frac{f_0}{2\omega^2}\cos\phi\right)\sin\omega x + a\cos\omega x - \frac{f_0}{2\omega}x\cos(\omega x + \phi) \tag{6.3-15}$$

Note that the amplitude term $(f_0 x/2\omega)$ grows linearly with time.

A graphical description and further details for this system can be found in Refs. 6-14, 6-27.

6.3.3 Laplace Transform

Electrical circuits can also be used to model linear second-order equations with constant coefficients. We consider such a system, but now for variety we will obtain its solution by a Laplace transform technique (c.f. Chapter 11). In Fig. 6.3-16, let q (in coulombs) be the instantaneous charge on the capacitor plates at any time t, and let $i(t)$ (in amperes) be the electric current in the circuit. Then by Kirchhoff's second law, the sum of the voltage drops around the circuit is zero, and this leads to the first order linear differential equation $L\,di/dt + Ri + q/C = e(t)$, which is again of the form of the second order equation treated in section 6.3.2, because $i = dq/dt$. With the addition of initial conditions, our problem becomes

$$L\frac{di}{dt} + Ri + C^{-1}\int_0^t i(\xi)\,d\xi = e(t), \qquad i(0) = i_0 \tag{6.3-16}$$

where the initial charge on the capacitor, q_0, has been set equal to zero (the generalization to nonzero q_0 merely requires the replacement of e above by $e - q_0/C$).

If the Laplace transform of any function is denoted by a capital letter, e.g., $I(s) = \int_0^\infty e^{-st}\, i(t)\,dt$, then application of the results of Chapter 11 shows that the transformed problem is

$$LsI(s) - Li_0 + RI(s) + \frac{1}{Cs}I(s) = E(s) \tag{6.3-17}$$

where $E(s)$ is the Laplace transform of the applied voltage. The formal solution is

$$I(s) = Li_0 \left(Ls + R + \frac{1}{Cs}\right)^{-1} + \left(Ls + R + \frac{1}{Cs}\right)^{-1} E(s) \qquad (6.3\text{-}18)$$

and the current is the inverse Laplace transform of this function. Because the inversion process is a linear operation, it follows that the current will be a linear superposition of two terms, the first being a free oscillation independent of the forcing and the second a direct result of the forcing. Inversion of this last term, which is a product of two functions whose transforms can be regarded as known, motivates consideration of the *convolution* of two functions, say $g(t), h(t)$, defined by

$$g(t) * h(t) \equiv \int_0^t g(\xi)\, h(t - \xi)\, d\xi = \int_0^t g(t - \xi)\, h(\xi)\, d\xi$$

As noted in Chapter 11, the Laplace transform of the convolution of $g(t)$ and $h(t)$ is simply the product $G(s)\,H(s)$. Consequently, the inverse transform of $I(s)$ is

$$i(t) = Li_0\, k(t) + \int_0^t k(t - \xi)\, e(\xi)\, d\xi \qquad (6.3\text{-}19)$$

where $k(t)$ is the function whose Laplace transform is $(Ls + R + 1/Cs)^{-1}$.

As one simple example of this technique consider the circuit of Fig. 6.3-1b with no charge on the capacitor and no current flowing, initially. The circuit is attached to a constant source of voltage (i.e., a battery). The problem to be solved is $Li' + Ri + C^{-1} \int_0^t i(\xi)\, d\xi = e_0$, $i(0) = i_0 = 0$, and according to eq. (6.3-19) the solution is $i(t) = e_0 \int_0^t k(t - \xi)\, d\xi$. Here $k(t)$ is the inverse Laplace transform of $Cs/(LCs^2 + RCs + 1)$ and is readily calculated (or found in tables) to be $k(t) = [\alpha/L(\alpha - \beta)]\, e^{\alpha t} - [\beta/L(\alpha - \beta)]\, e^{\beta t}$, where α, β are the two roots of $s^2 + R/L\, s + 1/LC = 0$, presumed distinct. Consequently, the solution, found by performing the integration called for above, is $i(t) = [e_0/L(\alpha - \beta)]\, (e^{\alpha t} - e^{\beta t})$.

As another example, where the characteristic equation has a multiple root, and where the forcing is sinusoidal, consider the initial value problem for $y(t)$ given by $y'(t) + 4y(t) + 4 \int_0^t y(\xi)\, d\xi = \sin t$, $y(0) = 1$. The Laplace transform of this equation is $[sY(s) - 1] + 4Y + 4s^{-1}Y = (s^2 + 1)^{-1}$, the solution being

$$Y(s) = \frac{s^2 + 2}{(s^2 + 1)(s + 4 + 4s^{-1})} = \frac{s(s^2 + 2)}{(s^2 + 1)(s + 2)^2}$$

A simple way in which to invert such a rational function of s is to rewrite it in partial fractions

$$Y(s) = \frac{22}{25(s+2)} - \frac{12}{5(s+2)^2} + \frac{3s+4}{25(s^2+1)}$$

The inverse transform is then readily found to be

$$y(t) = \frac{22}{25} e^{-2t} - \frac{12}{5} te^{-2t} + \frac{3}{25}\left(\cos t + \frac{4}{3} \sin t\right)$$

which can be checked by direct substitution.

6.4 SECOND-ORDER LINEAR DIFFERENTIAL EQUATIONS WITH VARIABLE COEFFICIENTS

6.4.1 Standard Forms

The two preceding sections have developed explicit general solutions to the initial-value problems for arbitrary first-order and for constant-coefficient, second-order, linear differential equations. The next stage of complication is to consider second-order, linear equations with variable coefficients. Unfortunately, the ability to display explicit solutions is, in general, lost, but much information is nevertheless available. Equations with polynomial coefficients have been studied with particular diligence, as may be indicated by a partial list of eminent mathematicians who have worked on them: Airy, Bessel, Chebyshev, Gauss, Gegenbauer, Hermite, Laguerre, Legendre, Weber. Some of the special functions which constitute solutions to these equations are treated in Chapter 7. Another important class of second-order equation is that with periodic coefficients, the Mathieu equation being a well-known example (see Chapter 13).

For first-order equations, which require only one extra condition to render the general solution unique, there is clearly no distinction between an initial and a boundary condition. However, such a contrast is notable for second-order equations, where two conditions must be specified. Indeed, the question of existence and uniqueness of solutions requires renewed attention for boundary value problems (a typical theorem is stated below).

Any second-order linear inhomogeneous differential equation can be written in the standard form

$$a_2(x)y''(x) + a_1(x)y'(x) + a_0(x)y(x) = f(x) \tag{6.4-1}$$

In any interval for which $a_2(x) \neq 0$, an often more convenient form is

$$y''(x) + p(x)y'(x) + q(x)y(x) = g(x) \tag{6.4-2}$$

in which $p(x) = a_1(x)/a_2(x), q(x) = a_0(x)/a_2(x), g(x) = f(x)/a_2(x)$. Another equivalent form (valid when $a_2(x) \neq 0$) is found by multiplying eq. (6.4-1) by $r(x)/a_2(x)$ where $r(x) = \exp\left[\int^x a_1(\xi)/a_2(\xi)\,d\xi\right] = \exp\left[\int^x p(\xi)\,d\xi\right]$:

$$[r(x)y'(x)]' + s(x)y(x) = h(x) \tag{6.4-3}$$

in which $s(x) = a_0(x)r(x)/a_2(x), h(x) = r(x)f(x)/a_2(x)$.

Another useful form is obtained when we 'suppress the first derivative.' If $y(x) = w(x)\exp\left[-\frac{1}{2}\int^x p(\xi)\,d\xi\right] = w(x)[r(x)]^{-1/2}$, then eq. (6.4-2) is transformed into

$$w''(x) + t(x)\,w(x) = j(x) \tag{6.4-4}$$

with

$$t(x) = \left[q(x) - \frac{1}{2}p'(x) - \frac{1}{4}p^2(x)\right], \quad j(x) = g(x)\exp\left[\int^x \frac{1}{2}p(\xi)\,d\xi\right]$$

Still another standard form is the reduction of eq. (6.4-1) to a simultaneous system of two first order equations

$$\begin{aligned} a_2(x)u'(x) + a_1(x)u(x) + a_0(x)y(x) &= f(x) \\ y'(x) - u(x) &= 0 \end{aligned} \tag{6.4-5}$$

In what follows, eq. (6.4-2) or eq. (6.4-3) will be used most frequently.

EXISTENCE AND UNIQUENESS THEOREM FOR THE INITIAL-VALUE PROBLEM

In the initial-value problem

$$y''(x) + p(x)y'(x) + q(x)y(x) = g(x), \quad y(x_0) = a, \quad y'(x_0) = b \tag{6.4-6}$$

a, b are any pair of real numbers and $p(x), q(x), g(x)$ are continuous in the interval $x_1 < x < x_2$ which contains x_0. Then there exists a unique continuous solution $y(x)$ valid in $x_1 < x < x_2$. Furthermore, y depends continuously on x_0, a and b. (For a proof, see Ref. 6-4.)

Existence and uniqueness theorems for boundary-value problems are more numerous and involved but far less general (Ref. 6-2). A representative example of some power (which holds also for an important class of nonlinear equation) follows.

EXISTENCE AND UNIQUENESS THEOREM FOR A BOUNDARY-VALUE PROBLEM

In the boundary-value problem

$$y''(x) + p(x)y'(x) + q(x)y(x) = g(x), \quad y(x_1) = a, \quad y(x_2) = b \tag{6.4-7}$$

a, b are any pair of real numbers and $p(x), q(x), g(x)$ are continuous in the interval $x_1 \leqslant x \leqslant x_2$, and therefore bounded, say $|p(x)| \leqslant P$, $|q(x)| \leqslant Q$, $|g(x)| \leqslant G$ for some positive constants P, Q, G. Then if the length of the interval, $x_2 - x_1$, is sufficiently small that $\frac{1}{8}Q(x_2 - x_1)^2 + \frac{1}{2}P(x_2 - x_1) < 1$, a unique solution $y(x)$ to the boundary-value problem exists on the interval $x_1 < x < x_2$.

6.4.2 Linear Independence and the Wronskian

Consider eq. (6.4-2), subject either to initial or boundary conditions. Since the equation is linear, superposition holds, whence the structure of the general solution is anticipated to have the form

$$y(x) = A_1 z_1(x) + A_2 z_2(x) + y_1(x) \tag{6.4-8}$$

where $z_1(x), z_2(x)$ are appropriate nontrivial solutions of the reduced (i.e., homogeneous) equation, and A_1, A_2 are constants to be determined from the initial or boundary conditions; $y_1(x)$, the particular integral, is any solution of the inhomogeneous equation.

As for constant coefficient equations, it is expeditious first to find the complementary solution, i.e., the general solution of

$$z''(x) + p(x)z'(x) + q(x)z(x) = 0 \tag{6.4-9}$$

For eq. (6.4-8) to provide the general solution, $z_1(x)$ and $z_2(x)$ must be essentially different solutions of eq. (6.4-9)—one of them should not simply be a multiple of the other. To illustrate, two solutions of $z'' - 2x^{-1}z' + (4 + 2x^{-2})z = 0$ are $z(x) = x \sin 2x$, $z(x) = x \sin x \cos x$, but by a trigonometric identity, $x(\sin 2x - \frac{1}{2} \sin x \cos x) = 0$ so this particular weighted sum of the two solutions is the trivial solution. They are not sufficiently distinct for both to be useful. This prompts the definition: two functions $z_1(x), z_2(x)$ are said to be *linearly dependent* in an interval $x_1 \leqslant x \leqslant x_2$ if there are two constants, B_1, B_2 not both zero, such that

$$B_1 z_1(x) + B_2 z_2(x) \equiv 0 \quad \text{for} \quad x_1 \leqslant x \leqslant x_2 \tag{6.4-10}$$

Two functions which fail to satisfy this definition are *linearly independent*. Another solution of the sample equation, linearly independent of $x \sin 2x$, is $z(x) = x \cos 2x$.

When the differential equation admits complicated or unfamiliar solutions, it may not be immediately evident whether or not two solutions are linearly dependent or independent. A useful test for this property is as follows. Suppose $z_1(x), z_2(x)$ are two continuous, differentiable solutions of eq. (6.4-9). If they are linearly dependent then two nonzero constants, B_1, B_2 can be found such that eq. (6.4-10) and also its derivative

$$B_1 z_1'(x) + B_2 z_2'(x) = 0 \tag{6.4-11}$$

are both satisfied throughout an interval. With z_1, z_2 given, eqs. (6.4-10), (6.4-11) are two linear, homogeneous algebraic equations for B_1, B_2 and a nontrivial solution for B_1, B_2 exists if and only if the determinant of the coefficient matrix is identically zero throughout the interval. This important determinant is the ***Wronskian determinant*** (or simply the ***Wronskian***, after the Polish mathematician, H. Wronski)

$$W(x) \equiv \begin{vmatrix} z_1(x) & z_2(x) \\ z_1'(x) & z_2'(x) \end{vmatrix} = z_1 z_2' - z_1' z_2 \tag{6.4-12}$$

The consequence is that if $z_1(x), z_2(x)$ are linearly dependent in some interval I, then their Wronskian vanishes identically throughout I. Conversely, if W is zero throughout I, and one of the two functions, say $z_1(x)$, never vanishes therein, then

$$\frac{W(x)}{z_1^2(x)} = \frac{z_1 z_2' - z_1' z_2}{z_1^2} = \left(\frac{z_2}{z_1}\right)' \equiv 0 \quad \text{in } I \tag{6.4-13}$$

But this implies that $z_2 = Cz_1$ with C a constant, which is tantamount to linear dependence.

In addition to its value in deciding whether two solutions are linearly dependent or independent, the Wronskian also enables the construction of a second linearly independent solution when one solution, say $z_1(x)$, is known. If z_1 and z_2 are both to satisfy eq. (6.4-9), then by the equivalence between eq. (6.4-2) and eq. (6.4-3) we have $[r(x)z_1'(x)]' + s(x)z_1(x) = 0$, $[r(x)z_2'(x)]' + s(x)z_2(x) = 0$. If the first (second) of these is multiplied by $z_2(z_1)$ and the resulting equations subtracted (this process being referred to as 'cross multiplication and subtraction', for short), we obtain $[r(z_1 z_2' - z_1' z_2)]' = [rW]' = 0$. Integration of this equation, and combination with eq. (6.4-13) results in Abel's identity

$$W(x) = z_1^2 \left(\frac{z_2}{z_1}\right)' = \frac{C}{r(x)}, \qquad C = \text{constant} \tag{6.4-14}$$

which shows how to evaluate W (apart from the arbitrary multiplicative constant C) directly from the differential equation. The procedure for finding $z_2(x)$, given $z_1(x)$, is first to find $r(x) = \exp\left(\int^x p(\xi)\, d\xi\right)$ from $p(x)$ in eq. (6.4-9) and then evaluate

$$z_2(x) = Cz_1(x) \int^x \frac{d\xi}{z_1^2(\xi) r(\xi)} \tag{6.4-15}$$

Out of the family of solutions described by this important formula, it is convenient to select one, say $\bar{z}_2(x)$, by choosing a definite allowable function $r(x)$, a definite value for C (say $C = 1$) and a particular lower limit for the integration within the interval of interest, say x_0

$$\bar{z}_2(x) = z_1(x) \int_{x_0}^{x} \frac{d\xi}{z_1^2(\xi) r(\xi)} \tag{6.4-16}$$

By direct calculation from Abel's identity, the Wronskian of $z_1(x), \bar{z}_2(x)$ is $1/r(x)$ which has already been presumed not to vanish in the interval. Consequently, it follows that if eq. (6.4-9) has one nontrivial solution, then it has another linearly independent solution. We now prove that the second order eq. (6.4-9) has exactly two linearly independent solutions, so that any solution can be written in the form

$$z(x) = A_1 z_1(x) + A_2 z_2(x) \tag{6.4-17}$$

With $z(x)$ any solution of eq. (6.4-9) and $z_1(x), z_2(x)$ two solutions linearly independent on some interval I, it is always possible to find two non-zero constants A_1, A_2 such that $A_1 z_1(x_0) + A_2 z_2(x_0) = z(x_0)$, $A_1 z_1'(x_0) + A_2 z_2'(x_0) = z'(x_0)$, where z_0 is in I, because the determinant of this linear system is simply $W(z_0)$ which is nonzero. Thus, $z(x)$, and $A_1 z_1(x) + A_2 z_2(x)$ are two solutions of eq. (6.4-9) satisfying the same initial conditions. By the Uniqueness Theorem of section 6.4.1, there is only one such function so the identity of eq. (6.4-17) holds everywhere within I.

Any two linearly independent solutions of the reduced equation form a *solution basis* or *fundamental set* and can be chosen to suit the problem at hand. Notice that two such functions cannot have coincident zeros for at such a point the Wronskian would vanish, which would require C in eq. (6.4-14) to be zero, a result which would violate linear independence. If r does not vanish anywhere in an interval, then it may be shown that the zeros of the functions which form a solution basis separate each other in that interval.

To illustrate the foregoing, consider $xz'' + (2x - 1)z' + (x - 1)z = 0$ which has $z_1 = e^{-x}$ as a solution in $-\infty < x < \infty$. To find a second linearly independent solution, first cast this equation into the form of eq. (6.4-3): $[x^{-1}e^{2x}z'(x)]' + x^{-2}(x - 1)e^{2x}z(x) = 0$ from which it follows, by eq. (6.4-14) that, $W(x) = Cxe^{-2x}$. Equation (6.4-15) then shows that another linearly independent solution is $x^2 e^{-x}$. The general solution is therefore $z(x) = A_1 e^{-x} + A_2 x^2 e^{-x}$.

6.4.3 Variation of Parameters

Once the complementary function has been found, several methods can be utilized to find a particular solution, thereby completing calculation of the general solution. An especially effective technique, due to Lagrange, is the following. With $z_1(x)$, $z_2(x)$ two linearly independent solutions of the reduced equation, look for a solution of the nonhomogeneous eq. (6.4-2) in the form

$$y_1(x) = \alpha(x) z_1(x) + \beta(x) z_2(x) \tag{6.4-18}$$

where the two functions $\alpha(x)$, $\beta(x)$ are to be determined. If we impose

$$\alpha'(x) z_1(x) + \beta'(x) z_2(x) = 0 \tag{6.4-19}$$

then the first derivative of y_1 has exactly the same form as if α and β were constants

$$y_1'(x) = \alpha(x) z_1'(x) + \beta(x) z_2'(x) \tag{6.4-20}$$

Differentiation again and substitution into eq. (6.4-2) leads to $\alpha(z_1'' + pz_1' + qz_1) + \beta(z_2'' + pz_2' + qz_2) + \alpha' z_1' + \beta' z_2' = g$. The two parentheses here vanish, because z_1, z_2 are solutions of the reduced equation, so another equation (in addition to eq. (6.4-19)) which α', β' must satisfy, is

$$\alpha'(x) z_1'(x) + \beta'(x) z_2'(x) = g(x) \tag{6.4-21}$$

Because z_1, z_2 are linearly independent, their Wronskian W does not vanish and hence, eqs. (6.4-19), (6.4-21) admit the solution

$$\alpha(x) = -\int^x \frac{g(\xi)}{W(\xi)} z_2(\xi)\, d\xi, \quad \beta(x) = \int^x \frac{g(\xi)}{W(\xi)} z_1(\xi)\, d\xi$$

The general solution of eq. (6.4-2) is then given by

$$y(x) = A_1 z_1(x) + A_2 z_2(x) - z_1(x) \int^x \frac{g(\xi)}{W(\xi)} z_2(\xi)\, d\xi + z_2(x) \int^x \frac{g(\xi)}{W(\xi)} z_1(\xi)\, d\xi \tag{6.4-22}$$

where $z_1(x), z_2(x)$ are two linearly independent solutions of the reduced eq. (6.4-9) whose Wronskian is $W(x) = z_1(x) z_2'(x) - z_1'(x) z_2(x)$. Notice that conversion to definite integrals, by insertion of specific lower limits of integration, is tantamount to changing the arbitrary constants A_1, A_2. An alternative expression of this solution is

$$y(x) = A_1 z_1(x) + A_2 z_2(x) + \int^x \frac{z_1(\xi) z_2(x) - z_1(x) z_2(\xi)}{z_1(\xi) z_2'(\xi) - z_1'(\xi) z_2(\xi)} g(\xi)\, d\xi \tag{6.4-23}$$

which facilitates a comparison with the corresponding first-order result in eq. (6.2-8).

An illustrative example is the boundary-value problem $(x + 1)y'' + xy' - y = 2(x + 1)^2, y(0) = 0, y(1) = 1$, which, by inspection, admits $z_1 = x$ as one solution of the reduced equation. Application of eq. (6.4-15) leads to another, linearly independent, solution

$$z_2(x) = Cx \int^x (\xi^{-1} + \xi^{-2}) e^{-\xi} \, d\xi = Ce^{-x}$$

With $C = 1$, for convenience, the Wronskian is $W = -(x + 1)e^{-x}$ so the general solution follows from eq. (6.4-22) as $y(x) = A_1 x + A_2 e^{-x} + 2(x^2 - x + 1)$. Satisfaction of boundary conditions, then gives the final solution in the form $y(x) = 2x^2 - (3 - 2e^{-1})x + 2 - 2e^{-x}$.

6.4.4 Green's Function

Two rather different forms in which solutions to differential equations typically can be found are a solution in series (for example, power series, treated in section 6.4.6, or series expansion in eigenfunctions, section 6.6.3) or a solution in integral form. Both have their virtues and disadvantages. Series representations may be relatively easy to construct but frequently converge slowly (or diverge). Integral solutions yield closed forms but are usually difficult to evaluate. Here, we study the integral representation known as the Green's function technique, which systematizes the observation that both first- and second-order linear inhomogeneous equations admit particular solutions in integral form; refer to eqs. (6.2-8), (6.4-23). Our concern is with a second order boundary-value problem of rather general structure

$$[r(x)y'(x)]' + s(x)y(x) = h(x), \quad y(a) = y_a, \quad y(b) = y_b \tag{6.4-24}$$

where $r(x)$ does not vanish in $a \leqslant x \leqslant b$, and y_a, y_b are arbitrary constants.

It is convenient to use superposition in order to replace the given nonhomogeneous boundary conditions by their homogeneous counterparts. To this end let $y(x) = u(x) + v(x)$ where $v(x)$ is any prescribed continuous function whose first and second derivatives exist in $a \leqslant x \leqslant b$ and such that $v(a) = y_a$, $v(b) = y_b$. Then, direct substitution shows that $u(x)$ satisfies the inhomogeneous equation

$$[r(x)u'(x)]' + s(x)u(x) = h(x) - [r(x)v'(x)]' - s(x)v(x) \equiv k(x) \tag{6.4-25}$$

subject to homogeneous boundary conditions $u(a) = u(b) = 0$. This shows that inhomogeneous boundary conditions play the same role as the forcing function, for they can be exchanged for homogeneous conditions provided the forcing function is altered.

The solution of eq. (6.4-25) relies, first, on selection of an appropriate solution basis for the reduced equation. A convenient choice is to make u_1 satisfy the left-hand and u_2 the right-hand homogeneous boundary conditions, and to make each have unit slope at these end points: $u_1(a) = 0, u_1'(a) = 1; u_2(b) = 0, u_2'(b) = 1$. The existence of such a pair of functions is guaranteed by the first theorem of section 6.4.1. For the example of section 6.4.2, this solution basis is

$$u_1(x) = \frac{1}{2a}(x^2 - a^2)e^{-(x-a)}, \quad u_2(x) = \frac{1}{2b}(x^2 - b^2)e^{-(x-b)}$$

which are readily shown to be linearly independent in any interval.

With this choice for u_1, u_2 it follows from eq. (6.4-22) that the complete solution for general $u(x)$ which satisfies the prescribed homogeneous boundary conditions can be written as

$$u(x) = \int_a^x \frac{u_1(\xi)u_2(x)}{W(\xi)r(\xi)} k(\xi)\, d\xi + \int_x^b \frac{u_1(x)u_2(\xi)}{W(\xi)r(\xi)} k(\xi)\, d\xi \tag{6.4-26}$$

This important result is now written in the compact form

$$u(x) = \int_a^b G(x, \xi)k(\xi)\, d\xi \tag{6.4-27}$$

which introduces the *Green's function* as

$$G(x, \xi) = \begin{cases} \dfrac{u_1(x)u_2(\xi)}{r(\xi)[u_1(\xi)u_2'(\xi) - u_1'(\xi)u_2(\xi)]}, & a \leqslant x \leqslant \xi \\[2ex] \dfrac{u_1(\xi)u_2(x)}{r(\xi)[u_1(\xi)u_2'(\xi) - u_1'(\xi)u_2(\xi)]}, & \xi \leqslant x \leqslant b \end{cases} \tag{6.4-28}$$

By Abel's identity, eq. (6.4-14), the common denominator is here a constant. Equation (6.4-27) is the desired integral form; it reveals that the forced solution to a second order linear differential equation subject to homogeneous boundary conditions at $x = a, x = b$, can be found by performing a weighted integration of the forcing function over the interval $a \leqslant x \leqslant b$. The weighting or kernel function here is the Green's function and its form, eq. (6.4-28), shows that it possesses a set of interesting properties (which can in fact be used to define it). Considered as a function of x, G is proportional to one solution, $u_1(x)$, of the reduced equation in $a \leqslant x \leqslant \xi$, and to a different linearly independent solution, $u_2(x)$, in $\xi \leqslant x \leqslant b$. G is continuous at $x = \xi$, but $\partial G/\partial x$ displays a finite discontinuity at $x = \xi$. Indeed, as x approaches ξ from the right (denoted by $x \to \xi+$), $\partial G/\partial x \to u_1(\xi)u_2'(\xi)/W(\xi)r(\xi)$, whereas the approach from the left, $(x \to \xi-)$, gives $\partial G/\partial x \to u_1'(\xi)u_2(\xi)/W(\xi)r(\xi)$; thus there is a discontinuity in slope given by

$$G_x(\xi+, \xi) - G_x(\xi-, \xi) = 1/r(\xi) \tag{6.4-29}$$

where the subscripts on G here denote partial differentiation with respect to x, holding ξ constant. Note further that $G(a, \xi) = G(b, \xi) = 0$, so that as a function of x, G satisfies the same homogeneous boundary conditions as u.

Generally, it should be borne in mind that the Green's function is only determinate when the differential equation, the interval of validity, and suitable homogeneous boundary conditions (not necessarily those illustrated above) are all prescribed.

In deriving Abel's identity, eq. (6.4-14), which plays an important role in both the variation of parameters technique and the Green's function method, we found it useful to work with the fundamental equation in the form of eq. (6.4-3) rather than eq. (6.4-1). The important step required "cross multiplication and subtraction" to produce an integrable quantity, which is impossible for a differential operator of the form

$$\mathcal{L}_1 z = a_2(x) z''(x) + a_1(x) z'(x) + a_0(x) z(x) \tag{6.4-30}$$

unless $a_2' = a_1$ because then $\mathcal{L}_1 z = (a_2 z')' + a_0 z$. However, introduction of the operator $\mathcal{L}_2$ *adjoint* to $\mathcal{L}_1$ by the definition

$$\begin{aligned} \mathcal{L}_2 z &= (a_2 z)'' - (a_1 z)' + a_0 z \\ &= a_2 z'' + (2a_2' - a_1) z' + (a_2'' - a_1' + a_0) z \end{aligned} \tag{6.4-31}$$

renders the following cross-multiplication and subtraction process integrable

$$z_1 \mathcal{L}_1 z_2 - z_2 \mathcal{L}_2 z_1 = [a_2(z_1 z_2' - z_1' z_2) - (a_2' - a_1) z_1 z_2]'$$

Comparison of eqs. (6.4-30), (6.4-31) shows that the two operators $\mathcal{L}_1$, $\mathcal{L}_2$ are identical ($\mathcal{L}_1$ is then said to be *self-adjoint*) if $a_2' = a_1$ and then the cross-multiplication and subtraction process simplifies to $z_1 \mathcal{L}_1 z_2 - z_2 \mathcal{L}_1 z_1 = (a_2 W)'$ where W is the Wronskian of z_1, z_2. An important property of the Green's function for a self-adjoint operator is the reciprocity or symmetry principle which states that $G(x, \xi) = G(\xi, x)$ (for a proof, see Ref. 6-25).

For an instructive example, we construct the Green's function solution for the non-self-adjoint problem $(x+1)y'' + xy' - y = 2(x+1)^2$, $y(0) = y(1) = 0$ valid in the interval $0 \leqslant x \leqslant 1$ (this is the same equation as in section 6.4.3). The more convenient self-adjoint form of this problem, as in eq. (6.4-3), is $[(x+1)^{-1} e^x u']' - (x+1)^{-2} e^x u = 2e^x$, $u(0) = u(1) = 0$ and the solution basis is $u_1(x) = x$, $u_2(x) = \frac{1}{2}(x - e^{1-x})$. The Green's function $G(x, \xi)$ is proportional to $u_1(x)$ in $0 \leqslant x \leqslant \xi$, and to $u_2(x)$ in $\xi \leqslant x \leqslant 1$

$$G(x, \xi) = \begin{cases} A(\xi) x, & 0 \leqslant x \leqslant \xi \\ \frac{1}{2} B(\xi)(x - e^{1-x}), & \xi \leqslant x \leqslant 1 \end{cases}$$

Since G is to be continuous at $x = \xi$, one equation relating A and B is $B(\xi)(\xi - e^{1-\xi}) = 2A(\xi)\xi$. The required discontinuity in the x derivative of G at $x = \xi$ gives another equation as $\frac{1}{2} B(\xi)(1 + e^{1-\xi}) - A(\xi) = (\xi + 1) e^{-\xi}$ from which it follows that

$$G(x, \xi) = \begin{cases} e^{-1}(\xi - e^{1-\xi})x, & 0 \leqslant x \leqslant \xi \\ e^{-1}\xi(x - e^{1-x}), & \xi \leqslant x \leqslant 1 \end{cases}$$

As a result, the solution, from eq. (6.4-26) or (6.4-27) is $u(x) = 2(x - 1)^2 + 2e^{-1}x - 2e^{-x}$, and this is readily verified to satisfy the differential equation and boundary conditions.

An alternative way, both to think about the Green's function as well as to construct it, is provided by saying that $G(x, \xi)$ is that continuous function of x which is the solution of

$$(rG')' + sG = \delta(x - \xi), \quad \text{in} \quad a < x < b \tag{6.4-32}$$

for which $G(a, \xi) = G(b, \xi) = 0$ where the primes here denote partial differentiation with respect to x, holding ξ constant, and $\delta(x - \xi)$ is the Dirac delta function. This latter quantity (also called the unit impulse function) is not a function at all in the conventional sense of that word, because its defining properties are that it be zero at every point but one, the point $x = \xi$, where it is infinite, in just such a way, that its integral is 1. Symbolically,

$$\delta(x - \xi) = 0 \quad \text{for} \quad x \neq \xi, \qquad \int_{-\infty}^{\infty} \delta(x - \xi)\, dx = 1 \tag{6.4-33}$$

where the limits of integration can just as well be $\xi \pm \epsilon$(ϵ any positive number) since δ vanishes outside this region. The Dirac delta function can be usefully thought of as the limit as $\epsilon \to 0$ of the well-defined function $\delta_\epsilon(x - \xi)$ where $\delta_\epsilon(x - \xi) = 0$ for $|x - \xi| > \epsilon$; $\delta_\epsilon(x - \xi) = 1/2\epsilon$, $|x - \xi| \leqslant \epsilon$, for this function is readily seen to satisfy the requirements imposed in eq. (6.4-33).

Because $\delta(x - \xi)$ vanishes nearly everywhere, the solution for G in eq. (6.4-32) will be (as required) a solution of the reduced equation for all x in $a < x < b$ except at $x = \xi$. Furthermore, G will have the correct discontinuity in slope at $x = \xi$, for upon integration with respect to x from $\xi - \epsilon$ to $\xi + \epsilon$, eq. (6.4-32) becomes

$$r(\xi + \epsilon) \left.\frac{\partial G}{\partial x}\right|_{\xi+\epsilon, \xi} - r(\xi - \epsilon) \left.\frac{\partial G}{\partial x}\right|_{\xi-\epsilon, \xi} + \int_{\xi-\epsilon}^{\xi+\epsilon} s(x) G(x, \xi)\, dx = 1$$

and when $\epsilon \to 0$ this again gives eq. (6.4-29) (provided it is recalled that $r(x)$, $s(x)$, and G as a function of x are continuous in $a < x < b$). Consequently, eq. (6.4-32) is equivalent to the former criteria which G must satisfy. A terminology used to describe this behavior is simply that $G(x, \xi)$ represents the response of the system at x to a unit impulse applied at ξ. Equation (6.4-27) then shows how the response to any forcing $k(x)$ can be calculated once the impulse response (i.e., G) is known.

A Green's function technique can be used to solve $\mathcal{L}_1 y = k(x)$, where $\mathcal{L}_1$ is not a self-adjoint operator, without first transforming this operator to self-adjoint form. Consider the former example $\mathcal{L}_1 y = (x+1)y'' + xy' - y = 2(x+1)^2, y(0) = y(1) = 0$, but now define $G(x, \xi)$ by $\mathcal{L}_2 G = \delta(x - \xi)$, $G(0, \xi) = G(1, \xi) = 0$ where $\mathcal{L}_2$, the operator adjoint to $\mathcal{L}_1$, is $\mathcal{L}_2 = (x+1)\, d^2/dx^2 + (2-x)\, d/dx - 2$. Green's function is found to be

$$G(x, \xi) = \begin{cases} e^{-1}(\xi - e^{1-\xi})x(1+x)^{-2} e^x, & 0 \leqslant x \leqslant \xi \\ e^{-1}\xi(1+x)^{-2}(xe^x - e), & \xi \leqslant x \leqslant 1 \end{cases}$$

Cross multiplication and subtraction leads to

$$G\mathcal{L}_1 y - y\mathcal{L}_2 G = \{(x+1)Gy' - y[(x+1)G]' + xyG\}' = G(x, \xi)\, k(x) - y(x)\,\delta(x - \xi)$$

Upon integration with respect to x, from $x = 0$ to $x = 1$, and using the boundary conditions, we find that $y(\xi) = 2\int_0^1 (x+1)^2 G(x, \xi)\, d\xi$. After insertion of Green's function, the former solution is recovered: $y(x) = 2(x-1)^2 + 2e^{-1}x - 2e^{-x}$.

For an example with other homogeneous boundary conditions, we solve $\mathcal{L}y = [(1-x^2)y']' - (1-x^2)^{-1}y = (1+x)^{1/2}$, $y'(0) = y(1) = 0$. The Green's function, which satisfies $\mathcal{L}G = \delta(x - \xi)$, $G_x(0, \xi) = G(1, \xi) = 0$, is found to be

$$G(x, \xi) = \begin{cases} -(1+\xi)^{-1/2}(1-\xi)^{1/2}(1-x^2)^{-1/2}, & 0 \leqslant x \leqslant \xi \\ -(1-\xi^2)^{-1/2}(1+x)^{-1/2}(1-x)^{1/2}, & \xi \leqslant x \leqslant 1 \end{cases}$$

Formation of the quantity $\int_0^1 [y\mathcal{L}G - G\mathcal{L}y]\, dx$ shows that $y(x) = \int_0^1 G(x, \xi)(1+\xi)^{1/2}\, d\xi$ and, with Green's function above, the solution is easily verified to be $y(x) = \frac{4}{3}(1+x)^{-1/2}(1-x) - 2(1+x)^{-1/2}(1-x)^{1/2}$.

6.4.5 Laplace Transform

Although the Laplace transform method is most frequently applied to the solution of linear, initial-value problems for differential equations (of any order) with constant coefficients, it can also be used for such problems where the equation has simple polynomial coefficients. This follows from differentiation of the fundamental transform definition with respect to s. Thus if $Y(s) = \int_0^\infty e^{-sx} y(x)\, dx$, then $Y'(s) = -\int_0^\infty e^{-sx} xy(x)\, dx$, which is the transform of $-xy(x)$. The reason that initial conditions (rather than boundary conditions) must usually be specified, is that the transform of derivatives introduces values of the unknown function and its lower derivatives at the initial point $x = 0$. Nevertheless, it can happen that a boundary-value problem is solvable by this technique.

An illustrative example is the variable coefficient boundary-value problem: $4xy'' + 2y' - (1 + x)y = x^{1/2}$, $y(0) = a$, $y(1) = b$. The Laplace transform of this equation gives $(1 - 4s^2)Y' - (6s + 1)Y = -2a + (\pi^{1/2}/2)s^{-3/2}$ which is a first order differential equation for $Y(s)$, solvable by the technique of section 6.2.2. Notice that the second boundary condition has not yet been needed. The solution of the transformed equation is $Y(s) = 2^{-3/2} C(s - \frac{1}{2})^{-1} (s + \frac{1}{2})^{-1/2} + (a + \pi/2 s^{-1/2})\ (s - \frac{1}{2})^{-1}$ where C is an arbitrary constant. Tables of inverse Laplace transforms readily yield $y(x) = 2^{-3/2}\, Ce^{x/2}\ \mathrm{erf}\,(x^{1/2}) + ae^{x/2} + (\pi/2)^{1/2}\, e^{x/2}\ \mathrm{erf}\,(x/2)^{1/2}$ where the error function (see Chapter 7, section 7.3) is defined by $\mathrm{erf}\,(x) = 2\pi^{-1/2} \displaystyle\int_0^x e^{-\xi^2}\, d\xi$. The boundary condition at $x = 1$ is satisfied by setting $C = 2^{3/2}\ [\mathrm{erf}\,(1)]^{-1} \cdot [be^{-1/2} - a - (\pi/2)^{1/2}\ \mathrm{erf}\,(\frac{1}{2})^{1/2}]$ and this completes the solution.

As a second example, consider $(x^2 - 1)y'' + xy' - 9y = 2 + 7x + 5x^2$, $y(0) = 0$, $y(1) = \alpha$, where α is a constant. The Laplace transform of $y(x)$ obeys the second-order equation $s^2\, Y'' + 3sY' - (8 + s^2)\, Y = -\beta + 2s^{-1} + 7s^{-2} + 10s^{-3}$. Here $\beta = y'(0)$ is an unknown parameter. The form of forcing function suggests seeking a solution in inverse powers of s. A trial solution $Y(s) = As^{-2} + Bs^{-3} + Cs^{-4}$ leads, after Laplace inversion, to $y(x) = \beta x - x^2 - \frac{1}{6}(7 + 8\beta)x^3$. Satisfaction of $y(1) = \alpha$ requires $\beta = -\frac{13}{2} - 3\alpha$ so the desired solution is $y(x) = -(\frac{13}{2} + 3\alpha)x - x^2 + (\frac{15}{2} + 4\alpha)x^3$.

6.4.6 Power Series Solutions

In all the cases treated so far, relatively elementary functions have been sufficient to express the solution in closed form. However, the solutions of many challenging problems are too complicated to be expressible in this way. One alternative is then to delineate the analytic properties and tabulate numerically, classes of special transcendental functions which are defined by those differential equations which occur frequently (refer to Chapter 7). Another methodology, considered now, seeks the solution as an infinite series of elementary functions. Series techniques find very broad application indeed. For the second-order variable coefficient equations of immediate interest, we utilize a power series in the independent variable (i.e., Taylor or MacLaurin series). Subsequently, solution functions will be expanded in a series of eigenfunctions, section 6.6.3, or a series of the powers of a parameter (as in the perturbation expansion, section 6.8.1). Despite its great inherent interest and practical importance, this topic is so large that we cannot do it justice; a cursory treatment is given, but further details can be found in the classic treatises of Forsythe or Ince (Refs. 6-10, 6-16).

Although physical problems are usually formulated in terms of real functions of real variables, it is advantageous at this point to extend the domain of both function and independent variable into the complex plane. This is because the convergence and regularity properties of power series depend on the behavior of the expansion function in the complex, rather than the real plane; for example, the series $1 - x^2 + x^4 - x^6 + \ldots$ fails, for $|x| > 1$, to represent the function $1/(1 + x^2)$, which is well-behaved for all real x, because $1/(1 + x^2)$ has singularities at $x = \pm i$, where $i = \sqrt{-1}$. Consider therefore the second order, variable coefficient, initial-value problem for a function $w = w(z)$

$$w''(z) = p_1(z)\,w'(z) + p_0(z)\,w(z), \quad w(0) = a, \quad w'(0) = b \tag{6.4-34}$$

where the single-valued complex coefficient functions $p_1(z), p_0(z)$ are presumed to have derivatives of all orders (i.e., to be analytic functions of the complex variable z, see Chapter 5) except possibly at certain isolated points $z = z_0, z = z_1, \ldots$ in the complex plane. The restriction to a homogeneous equation here is based on the fact, already demonstrated in sections 6.4.3, 6.4.4 for the real domain, and also true (by the same formal process) for complex functions, that a particular integral of an inhomogeneous equation can be constructed in terms of the linearly independent solutions of the reduced equation. Indeed, all the results on existence, uniqueness, linear independence and the Wronskian in section 6.4.2 carry over immediately to the complex case (for proof, see section 5.2 of Ref. 6-5). If initial conditions are imposed at some point other than the origin, a translation of coordinates brings that point to the origin, so eq. (6.4-34) is perfectly general.

Suppose first, that the origin is an *ordinary point* of the differential equation, meaning that $p_1(z), p_0(z)$ are analytic at $z = 0$. Since both $p_1(z), p_0(z)$ then have convergent MacLaurin expansions about $z = 0$, it is natural to look for the solution function $w(z)$ in the same form

$$w(z) = w(0) + w'(0)z + \frac{1}{2!}\,w''(0)z^2 + \frac{1}{3!}\,w'''(0)z^3 + \cdots \tag{6.4-35}$$

This series is to converge to the solution of eq. (6.4-34) at every point of some neighborhood of $z = 0$ (i.e., for $|z| < R$ where R is some positive number). The first two coefficients are provided by the initial conditions, and higher ones follow from differentiations of eq. (6.4-34), allowable by the prescribed analyticity of p_1, p_0 and the presumed analyticity of w implied by eq. (6.4-35). For example,

$$\begin{aligned} w''(0) &= p_1(0)\,w'(0) + p_0(0)\,w(0) = b p_1(0) + a p_0(0) \\ w'''(0) &= p_1(0)\,w''(0) + p_1'(0)\,w'(0) + p_0(0)\,w'(0) + p_0'(0)\,w(0) \\ &= b\,[p_1^2(0) + p_1'(0) + p_0(0)] + a\,[p_1(0)p_0(0) + p_0'(0)] \end{aligned}$$

Substitution into eq. (6.4-35) shows that $w(z)$ can be decomposed into a superposition of that particular fundamental set of solutions $w_1(z), w_2(z)$ which have unit value and zero slope (w_1) or zero value and unit slope (w_2) at the origin

$$w(z) = a w_1(z) + b w_2(z)$$

where

$$w_1(z) = 1 + \frac{1}{2!} p_0(0) z^2 + \frac{1}{3!} [p_1(0) p_0(0) + p_0'(0)] z^3 + \cdots$$
$$w_2(z) = z + \frac{1}{2!} p_1(0) z^2 + \frac{1}{3!} [p_1^2(0) + p_1'(0) + p_0(0)] z^3 + \cdots \tag{6.4-36}$$

It can be shown that these series are convergent, and it may be verified by direct substitution that $w_1(z), w_2(z)$ are linearly independent solutions of eq. (6.4-34). This proves that $w(z)$ is analytic at an ordinary point of the differential equation, and also provides a means of constructing two linearly independent solutions (and hence the general solution) as power series. Note also the implication that the solution can be nonanalytic only at locations where the coefficients are singular.

In practice, a quicker way to obtain the MacLaurin series representation for $w(z)$ may be to use the *method of undetermined coefficients* in which eq. (6.4-35) is replaced by

$$w(z) = w_0 + w_1 z + w_2 z^2 + w_3 z^3 + \cdots \tag{6.4-37}$$

with similar expressions for $p_1(z), p_0(z)$. Upon substitution into eq. (6.4-34) and regrouping so that all terms of like power in z appear together, there results a power series which equals zero. Such a power series can converge to zero at all points of some neighborhood of the origin, if and only if, each coefficient of a power of z vanishes separately. This provides precisely enough equations to determine, uniquely, the coefficients in eq. (6.4-37).

Consider next the case in which either (or both of) $p_1(z), p_0(z)$ are singular at the origin. In the vicinity of such a *singular point* of the equation, the structure of the solution differs from that just considered. One special case is that for which $p_1(z)$ has a pole while $p_0(z)$ is analytic. As is readily provable (Ref. 6-25), then one solution is still analytic and can be found as before. However, use of the Wronskian technique to find a second linearly independent solution shows that it is not analytic at $z = 0$ but rather has a branch point or pole if $p_1(z)$ has a simple pole at $z = 0$, or has an essential singularity at $z = 0$ should $p_1(z)$ have a double or higher-order pole there.

A situation in which a well-developed theory exists (due mostly to Fuchs and Frobenius) is the case in which the solution function displays, at most, poles or branch points at a finite number of isolated locations. Such behavior is associated with what is termed a *regular singular point* of the equation (as opposed to an *irregular singular point*, for which comparatively little theory exists; but see the cited references). It can be shown that this important category is described by equations of the form

$$w''(z) = r(z) z^{-1} w'(z) + s(z) z^{-2} w(z) \tag{6.4-38}$$

where $r(z)$, $s(z)$ are analytic at the origin. In the notation of eq. (6.4-34), $p_1(z)$ has a simple, and $p_0(z)$ a double pole at $z = 0$. A useful approach is to treat first the easy case where $r(z)$, $s(z)$ are constant, say α, β, respectively. Then eq. (6.4-38) is equivalent to

$$z^2 w''(z) - \alpha z w'(z) - \beta w(z) = 0 \tag{6.4-39}$$

and this Euler equation (treated further in section 6.5.3) admits the general solution

$$w(z) = A_1 z^{\lambda_1} + A_2 z^{\lambda_2} \tag{6.4-40}$$

where, by direct substitution, λ_1, λ_2 are (distinct) roots of the quadratic equation $\lambda(\lambda - 1) - \alpha\lambda - \beta = 0$. Should $\lambda_1 = \lambda_2 = \mu$, say, then the solution is readily verified to be

$$w(z) = z^{\mu} (A_1 + A_2 \ln z) \tag{6.4-41}$$

Thus w is analytic if λ_1, λ_2 are different nonnegative integers, and has poles or branch points otherwise.

This motivates an attempt at a product solution of the problem for nonconstant but analytic $r(z)$, $s(z)$ in the form

$$w(z) = z^{\lambda}(w_0 + w_1 z + w_2 z^2 + \cdots) \tag{6.4-42}$$

where the function in the parenthesis is clearly analytic at $z = 0$; we require $w_0 \neq 0$. This series is now substituted into eq. (6.4-38) and values for $\lambda, w_0, w_1, \ldots$ are sought which make it a solution. (As with any series, its domain of convergence, if any, should subsequently be determined and from a practical viewpoint, the speed of convergence.) When the analytic functions $r(z)$, $s(z)$ are expanded as $r(z) = r_0 + r_1 z + r_2 z^2 + \cdots$; $s(z) = s_0 + s_1 z + s_2 z^2 + \cdots$, and then introduced, along with eq. (6.4-42) into eq. (6.4-38), there results a hierarchy of algebraic recurrence relations (linear difference equations in the terminology of section 6.9) which determine, in principle, the coefficients $w_0, w_1, w_2, \ldots$

$$\begin{aligned}
&w_0 [\lambda(\lambda - 1) - \lambda r_0 - s_0] = 0\\
&w_1 [(\lambda + 1)\lambda - (\lambda + 1) r_0 - s_0] = (\lambda r_1 + s_1) w_0\\
&w_2 [(\lambda + 2)(\lambda + 1) - (\lambda + 2) r_0 - s_0] = (\lambda r_2 + s_2) w_0 + [(\lambda + 1) r_1 + s_1] w_1\\
&\qquad\vdots \qquad\qquad\qquad\qquad \vdots\\
&w_n [(\lambda + n)(\lambda + n - 1) - (\lambda + n) r_0 - s_0] =\\
&\qquad = (\lambda r_n + s_n) w_0 + \cdots + [(\lambda + n - 1) r_1 + s_1] w_{n-1}.
\end{aligned} \tag{6.4-43}$$

Since, by supposition, $w_0 \neq 0$, the first equation here shows that λ must be a root of the *indicial equation*

$$\lambda(\lambda - 1) - \lambda r_0 - s_0 = 0 \tag{6.4-44}$$

If this equation has two distinct roots, not differing by an integer, then by selecting either root and solving eq. (6.4-43) sequentially, we generate two independent solutions to the original equation (6.4-38). Notice, however, that each left hand member of eq. (6.4-43) is obtainable from its immediate forerunner by substituting $\lambda + 1$ for λ. Consequently, if $\lambda_2 = \lambda_1 + n$, where $n > 0$ is an integer, the use of λ_1 in eq. (6.4-43) means that some left hand member of the set of equations will vanish, and the w coefficient in that equation will then be indeterminate. Another special exceptional case where this approach breaks down is when $\lambda_1 = \lambda_2$, so that the indicial equation has coincident roots. Before turning to these awkward situations, we remark that in the case where the indicial roots do not differ by an integer, then the two series solutions obtained from eq. (6.4-42) for $\lambda = \lambda_1, \lambda_2$ are in fact uniformly convergent series within some circular region centered on the origin (for details and proof refer to Ref. 6-5). Furthermore, the two single-valued functions defined by the series in such a region constitute a linearly independent fundamental set of solutions.

In the exceptional circumstance previously excluded, λ_1, say, is greater than λ_2, by a positive integer n (or zero for equal roots). With $\lambda = \lambda_1$, the procedure outlined above still gives one solution, $w_1(z)$ of the form of eq. (6.4-36). The second solution, $w_2(z)$, now follows by application of the Wronskian method, eq. (6.4-15), which in the complex plane becomes

$$w_2(z) = Cw_1(z) \int e^{\int r(z)z^{-1}\,dz} [w_1(z)]^{-2}\,dz \tag{6.4-45}$$

Straightforward substitution and integration shows that the second solution takes on the form

$$\begin{aligned} w_2(z) &= A\,w_1(z)\,[\ln z + h(z)], && \text{when } n = 0 \\ w_2(z) &= A\,[\gamma w_1(z) \ln z + z^{\lambda_2} h(z)], && \text{when } n \neq 0 \end{aligned} \tag{6.4-46}$$

where $h(z)$ is analytic at $z = 0$, A is an arbitrary constant, and γ a fixed constant (which can be zero) whose value is most efficiently found by direct substitution into the differential equation along with $h(z)$ in the form $h_0 + h_1 z + h_2 z^2 + \cdots$, and similarly for $r(z)$, $s(z)$. For $n \neq 0$, the second term on the right is of the same form as when $\lambda_2 - \lambda_1$ is nonintegral. Note further that, with the exception of the case $\gamma = 0$ (described fully in Ref. 6-25), for either $n = 0$ or $n \neq 0$, the logarithmic term (and its attendant branch point at $z = 0$) are present in the general solution (compare with eq. (6.4-41)). Thus, even when both indicial roots are integers, so that the factors z^{λ_1}, z^{λ_2} display poles or are analytic, the general solution still has a branch point.

Except then for the case $\gamma = 0$, it may be concluded that the general solution of a second order equation has a branch point at the location of a regular singularity of the equation. If $p_1(z)$ $[p_0(z)]$ has a pole of order higher than first [second], then one or both of the linearly independent solutions may have a more complicated singularity at this irregular singular point (consult Ref. 6-25 for further details).

So far attention has been focused on a single isolated singular point, but naturally, several such points may occur, and they can be located anywhere in the complex plane, including the *point at infinity*. By definition, the nature of any function $w(z)$ at infinity is taken to be that of $w_1(z_1) = w(1/z_1)$ at $z_1 = 1/z = 0$. Under this transformation, eq. (6.4-34) becomes

$$w_1''(z_1) = -\left[\frac{2}{z_1} + \frac{1}{z_1^2} p_1\left(\frac{1}{z_1}\right)\right] w_1'(z_1) + \frac{1}{z_1^4} p_0\left(\frac{1}{z_1}\right) w_1(z_1)$$

from which it is straightforward to investigate the behavior of the coefficient functions at $z_1 = 0$. For example, $e^{1/z}$ is analytic at infinity, whereas $ze^{-1/z}$ is not; ze^z has an essential singularity, and $(z - a)^3$ a third-order pole with residue $3a^2$, at infinity.

Generally speaking, a transformation of the dependent variable such as

$$w(z) \longrightarrow \frac{(z - z_1)^\sigma (z - z_2)^\eta}{(z - z_3)^{\sigma+\eta}} w(z)$$

will change the indicial roots but not the location of regular singular points (provided z_1, z_2, z_3 are finite; see Ref. 6-5 for details). Another useful construction is a bilinear transformation of the independent variable, $z \to (Az + B)/(Cz + D)$, which can be used to map any three (singular) points into desired standard locations (such as $0, 1, \infty$) without changing the nature or indicial roots at the singular points. By suitable combinations of these transformations the given equation should be manipulated (if possible) into one of the standard forms (given below) in which the singularities have preselected locations and the indicial roots at regular singularities the simplest possible values.

The systematic derivation of standard forms is well-covered in Ref. 6-25, with the following results (where we continue to designate by $w(z)$ the complex function of interest, bearing in mind that it probably represents a transformation of the original problem).

One regular singular point at $z = z_0$

$$w''(z) + 2(z - z_0)^{-1} w'(z) = 0, \quad w(z) = A_1 + A_2(z - z_0)^{-1} \tag{6.4-47}$$

One regular singular point at $z = \infty$

$$w''(z) = 0, \quad w(z) = A_1 + A_2 z \tag{6.4-48}$$

Two regular singular points, at 0 and ∞

$$w''(z) = (\lambda_1 + \lambda_2 - 1)z^{-1}w'(z) - \lambda_1\lambda_2 z^{-2}w(z)$$
$$\begin{aligned} w(z) &= A_1 z^{\lambda_1} + A_2 z^{\lambda_2}, \quad \text{if } \lambda_1 \neq \lambda_2 \\ &= z^{\mu}(A_1 + A_2 \ln z), \quad \text{if } \lambda_1 = \lambda_2 = \mu \end{aligned} \tag{6.4-49}$$

Three regular singular points at 0, 1, ∞ (hypergeometric equation)

$$z(1 - z)w''(z) + [c - (a + b + 1)z]\, w'(z) - abw(z) = 0 \tag{6.4-50}$$

The indicial roots at $z = 0$ are $(0, 1 - c)$, at $z = 1$ are $(0, c - a - b)$, and at $z = \infty$ are (a, b). Solution functions described by this three parameter equation are given further consideration in Chapter 7, section 7.7. A special case, known as the confluent hypergeometric equation, results when the singular points at 1 and ∞ are allowed to merge in a certain way (see Ref. 6-25 for details and solutions)

$$zw''(z) + (c - z)w'(z) - aw(z) = 0 \tag{6.4-51}$$

This equation has a regular singular point at $z = 0$ and an irregular one at $z = \infty$.

One regular singular point $(z = 0)$ and one irregular singular point $(z = \infty)$

$$w''(z) + (1 - \lambda_1 - \lambda_2)z^{-1}w'(z) + [-k^2 + 2\alpha z^{-1} + \lambda_1\lambda_2 z^{-2}]\, w(z) = 0 \tag{6.4-52}$$

The indicial roots at the regular singular point are λ_1, λ_2 and the essential singularity at ∞ is of the form e^{kz} (see Ref. 6-25). The most celebrated example of this problem is the Bessel equation (Chapter 7, sections 7.4, 7.5).

6.5 LINEAR EQUATIONS OF HIGH ORDER AND SYSTEMS OF EQUATIONS

6.5.1 The Nth-Order Linear Equation

For the most part, linear equations of high order are handled by methods that are direct extensions of those already introduced. The simplest high order equation which can easily be solved is the Nth-order linear equation with constant coefficients $a_0, a_1, a_2, \ldots, a_N$ $(a_N \neq 0)$

$$\sum_{n=0}^{N} a_n \frac{d^n y(x)}{dx^n} = f(x) \tag{6.5-1}$$

The general solution contains N arbitrary constants, $C_1, C_2, \ldots, C_N$, that can be chosen to satisfy N initial or N boundary conditions. As for lower order equations, the starting point is consideration of the reduced equation

$$a_N z^{(N)}(x) + a_{N-1} z^{(N-1)}(x) + \cdots + a_2 z''(x) + a_1 z'(x) + a_0 z = 0 \qquad (6.5\text{-}2)$$

or in operator notation

$$\mathcal{L}z = (a_N D^N + a_{N-1} D^{N-1} + \cdots + a_2 D^2 + a_1 D + a_0) z(x) = 0 \qquad (6.5\text{-}3)$$

the symbol D^n signifying (d^n/dx^n). Such equations admit exponential solutions and the trial form $z = e^{\lambda x}$ reduces the differential equation to a problem in algebra, namely that of finding the N roots of the *auxiliary* or *characteristic equation*

$$a_N \lambda^N + a_{N-1} \lambda^{N-1} + \cdots + a_2 \lambda^2 + a_1 \lambda + a_0 = 0 \qquad (6.5\text{-}4)$$

This follows from direct substitution and the observation that $e^{\lambda x}$ never vanishes for finite x. By the fundamental theorem of algebra, eq. (6.5-4) can be rewritten in terms of the roots, λ_i, of the auxiliary equation

$$(\lambda - \lambda_1)(\lambda - \lambda_2)(\lambda - \lambda_3) \cdots (\lambda - \lambda_N) = 0 \qquad (6.5\text{-}5)$$

When the a_i's in eq. (6.5-3) are real, then each λ_i is either real, or some other member of the set is its complex conjugate. If each of the roots is distinct, the general solution of eq. (6.5-2) is

$$z(x) = C_1 e^{\lambda_1 x} + C_2 e^{\lambda_2 x} + \cdots + C_N e^{\lambda_N x} \qquad (6.5\text{-}6)$$

If the two roots λ_1, λ_2, for example, are complex conjugates $(\alpha \pm i\beta)$, it may be convenient to write the associated solutions in the real form $e^{\alpha x}(C_1 \cos \alpha x + C_2 \sin \alpha x)$. If a root, say $\lambda = \mu$, occurs m times, but the other $(N - m)$ roots are simple, then eq. (6.5-5) has the form

$$(\lambda - \lambda_1)(\lambda - \lambda_2) \cdots (\lambda - \lambda_{N-m})(\lambda - \mu)^m = 0 \qquad (6.5\text{-}7)$$

and eq. (6.5-6) is replaced by

$$z(x) = C_1 e^{\lambda_1 x} + C_2 e^{\lambda_2 x} + \cdots + C_{N-m} e^{\lambda_{N-m} x} + (C_{N-m+1} + C_{N-m+2} x + \cdots + C_N x^{m-1}) e^{\mu x} \qquad (6.5\text{-}8)$$

After the general solution of the reduced equation has been found, solution of the complete equation follows by addition of any particular integral. Since any allowable choice of the latter is acceptable, it is convenient to define $y_1(x)$ as that solution of eq. (6.5-1) which satisfies homogeneous initial conditions at $x = x_0$, say

$$y_1(x_0) = y_1'(x_0) = y_1''(x_0) = \cdots = y^{(N-1)}(x_0) = 0 \tag{6.5-9}$$

Direct substitution and careful differentiation then shows that an integral representation of $y_1(x)$ is

$$y_1(x) = a_N^{-1} \int_{x_0}^{x} z(x - \xi) f(\xi)\, d\xi \tag{6.5-10}$$

where $z(x)$ is that solution of the reduced equation which satisfies

$$z(0) = z'(0) = z''(0) = \cdots = z^{(N-2)}(0) = 0, \quad z^{(N-1)}(0) = 1 \tag{6.5-11}$$

To illustrate the foregoing, we solve $y'''' + 5y''' + 15y'' + 5y' - 26y = (157 + 100x)e^{2x}$, $y(0) = 0$, $y'(0) = -2$, $y''(0) = 25$, $y'''(0) = 3$. The reduced equation has roots $1, -2, -2 \pm 3i$ and the complementary function satisfying eq. (6.5-11) is found to be $z(x) = \frac{1}{54}[e^x - 2e^{-2x} + e^{-2x}\cos 3x - e^{-2x}\sin 3x]$. Substitution into eq. (6.5-10) reveals that xe^{2x} is a particular integral so the general solution is $y(x) = xe^{2x} + Ae^x + Be^{-2x} + Ce^{-2x}\cos 3x + De^{-2x}\sin 3x$. When the initial conditions are satisfied, the four constants are found to be $A = 1$, $B = -1$, $C = 0$, $D = -2$ and this completes the solution.

A standard form for the Nth-order variable-coefficient linear equation is

$$b_N(x)y^{(N)}(x) + b_{N-1}(x)y^{(N-1)}(x) + \cdots + b_1(x)y'(x) + b_0(x)y = f(x) \tag{6.5-12}$$

Here the b_j's are continuous, single-valued functions of x in some interval and $b_N(x)$ does not vanish anywhere in the interval. The general theory of such equations can be found in Ref. 6-16.

Linearity assures that the general solution can be expressed as the superposition of the complementary function, $z(x)$, and a particular integral, $y_1(x)$. The complementary function is of the form

$$z(x) = C_1 z_1(x) + C_2 z_2(x) + \cdots + C_N z_N(x) \tag{6.5-13}$$

where $(z_1, z_2, \ldots, z_N)$ is a fundamental set of linearly independent solutions of the reduced equation

$$\mathcal{L}z = 0, \quad \mathcal{L} = b_N(x)\frac{d^N}{dx^N} + b_{N-1}(x)\frac{d^{N-1}}{dx^{N-1}} + \cdots + b_1(x)\frac{d}{dx} + b_0(x) \tag{6.5-14}$$

In general, any N functions $g_1(x), g_2(x), \ldots, g_N(x)$ are *linearly dependent* in an interval I if there are N constants $A_1, A_2, \ldots, A_N$, not all zero, such that $A_1 g_1(x) + A_2 g_2(x) + \cdots + A_N g_N(x) = 0$ in I. Functions not satisfying this requirement are *linearly independent*. If the N functions each possess $N - 1$ continuous derivatives in I, then the *Wronskian determinant* is defined by

$$W(x) = \begin{vmatrix} g_1 & g_2 & g_3 & \cdots g_N \\ g_1' & g_2' & g_3' & \cdots g_N' \\ \cdots & \cdots & \cdots & \cdots \\ g_1^{(N-1)} & g_2^{(N-1)} & g_3^{(N-1)} & \cdots g_N^{(N-1)} \end{vmatrix} \tag{6.5-15}$$

which reduces to the former value, eq. (6.4-12), when $N = 2$. If $g_1, g_2, \ldots, g_N$ are linearly dependent in I then their Wronskian vanishes identically in I. Conversely, if $g_1, g_2, \ldots, g_N$ are linearly independent solutions of the same differential eq. (6.5-14) in I, then their Wronskian does not vanish anywhere in I.

If some, but not all, of the functions $z_1, z_2, \ldots, z_N$, are known, the remaining ones can be found by factoring the differential operator $\mathcal{L}$ into a product of N first order operators and then, in effect, inverting each factor successively. Equivalently, any fundamental set of solutions is expressible in the form $z_1(x) = w_1(x), z_2(x) = w_1(x) \int w_2(x)\, dx, z_3(x) = w_1(x) \int w_2(x) \int w_3(x)\, dx^2, \ldots$ and substitution into the differential equation, results in lower order problems to be solved. We illustrate with a third order example: $(x + 2)z''' + (2x + 3)z'' + xz' - z = 0$. The factored form is $\mathcal{L}_3 \mathcal{L}_2 \mathcal{L}_1 z = 0$ where $\mathcal{L}_1 = (d/dx) + 1$, $\mathcal{L}_2 = (x + 1)(d/dx) - 1$, $\mathcal{L}_3 = (x + 2)(d/dx) + (x + 1)$. $\mathcal{L}_1$ is such that one solution is clearly $z_1(x) = e^{-x}$. With $z_2 = e^{-x} \int w_2\, dx$, we then find that $\mathcal{L}_2 \mathcal{L}_1 z_2 = 0$ provided $(x + 1)w_2' - (x + 2)w_2 = 0$. Thus, $w_2 = (x + 1)e^x$, so a second solution is $z_2 = x$. The expression for z_3 above is found to solve the equation if $(x + 1)w_3' - (x^2 + 2x - 1)w_3 - (x + 1)(x + 3) \int w_3\, dx = 0$. This second order equation for $\int w_3\, dx$ admits the solution $(x + 1)^{-1} e^{-x}$ so a third linearly independent solution is $z_3 = xe^{-x}$.

Once the complementary function has been constructed, a particular integral, $y_1(x)$, is obtainable by the method of *variation of parameters*, which is a direct extension of that given for second order equations in section 6.4.3. Thus, $y_1(x)$ is sought in the form

$$y_1(x) = \alpha_1(x) z_1(x) + \alpha_2(x) z_2(x) + \cdots + \alpha_N(x) z_N(x) \tag{6.5-16}$$

where the N functions $\alpha_i(x)$ are to be determined so that $y_1(x)$ is a solution of eq. (6.5-12). One (nonhomogeneous) relation between these functions and $f(x)$ is provided by substituting eq. (6.5-16) into eq. (6.5-12). The other $N - 1$ consistent relations needed to determine, uniquely, $\alpha_1, \alpha_2, \ldots \alpha_N$, are selected as the homogeneous equations

$$\begin{aligned}
&\alpha_1' z_1 + \alpha_2' z_2 + \cdots + \alpha_N' z_N = 0 \\
&\alpha_1' z_1' + \alpha_2' z_2' + \cdots + \alpha_N' z_N' = 0 \\
&\cdots\cdots\cdots\cdots\cdots\cdots\cdots \\
&\alpha_1' z_1^{(N-2)} + \alpha_2' z_2^{(N-2)} + \cdots + \alpha_N' z_N^{(N-2)} = 0
\end{aligned} \tag{6.5-17}$$

In view of these relations, y_1 in eq. (6.5-16) will satisfy eq. (6.5-12) if

$$\alpha_1' z_1^{(N-1)} + \alpha_2' z_2^{(N-1)} + \cdots + \alpha_N' z_N^{(N-1)} = f(x) \tag{6.5-18}$$

Because $z_1, z_2, \ldots, z_N$ are linearly independent, the N linear equations in eqs. (6.5-17), (6.5-18) determine a unique solution for $\alpha_1', \alpha_2', \ldots, \alpha_N'$ and the required functions then follow upon integration. Further details and examples are given in Ref. 6-16.

6.5.2 Dependence of Solutions on Parameters and Initial Conditions; Stability

A question of considerable theoretical as well as practical importance (for example, in approximation theory) is whether or not the unique solution to an Nth-order initial-value problem is a sensitive function of the initial values. For a large class of differential equations (including N^{th}-order quasi-linear equations wherein the highest derivative term is linear, but all lower order terms can be nonlinear) the answer is that small changes in initial conditions produce correspondingly small changes in the integral curves. The formal result is given in an important theorem (see Refs. 6-8, 6-28, for proof): Suppose $y^{(N)} = F(x, y, y', y'', \ldots, y^{(N-1)})$ is a differential equation satisfied by $y(x)$ where F is a continuous function of its $(N + 1)$ variables in some $(N + 1)$-dimensional space, call it 0. To each point in 0 let there correspond a unique solution function $y = \varphi(x, a, b_0, b_1, b_2, \ldots, b_{N-1})$ satisfying the initial conditions $y(a) = b_0, y'(a) = b_1, y''(a) = b_2, \ldots, y^{(N-1)}(a) = b_{N-1}$. Then φ depends continuously on its $(N + 2)$ variables. Furthermore, if F is not only continuous, but has continuous partial derivatives of order m with respect to its $N + 1$ variables, then φ has continuous partial derivatives of order m (and lower) with respect to its $N + 2$ variables. Finally, if F also depends continuously on some parameters $\lambda_1, \lambda_2, \lambda_3, \ldots, \lambda_n$ then so does φ.

A related question is that of the *stability of a linear differential equation* with constant coefficients. In eq. (6.5-2) let x be time and consider whether or not the solution grows as $x \to \infty$. This equation is said to be *strictly stable* if every member of its fundamental set of solutions decays to zero as $x \to \infty$, and it is *metastable* if some solutions remain bounded but nonzero in this limit. Otherwise, it is *unstable*. The equation is strictly stable if every characteristic root has a negative real part. In this case, $z(x) \to 0$ as $x \to \infty$ and eq. (6.5-10) then shows that y_1 will be bounded as $x \to \infty$ provided $f(x)$ is bounded. Thus, the response of a strictly stable equation to a bounded input $f(x)$ is bounded.

Equation (6.5-2) is metastable when all of the simple roots of its auxiliary equation have nonpositive real parts, at least one such simple root is purely imaginary, and all of the multiple roots have negative real parts. If any root has positive real part, the equation is unstable.

Necessary (but not sufficient; Ref. 6-4) conditions for strict stability (or metastability) are that if eq. (6.5-2) is strictly stable (metastable) then all coefficients of its auxiliary equation must be nonzero and non-negative (non-negative).

Since it is a simple problem, numerically, to compute the roots of an algebraic equation of almost any order, the straightforward way to determine stability is by finding the characteristic roots themselves. However, an alternative procedure which avoids this direct calculation, and is preferable analytically, since it provides both necessary and sufficient conditions for stability, is that associated with the Routh-Hurwitz discriminant. Here one calculates a special sequence of determinants formed from the coefficients in eq. (6.5-2) and, when $a_N > 0$, a necessary and sufficient condition for strict stability is that every member of the sequence be positive. Details are given in Ref. 6-13; see also Chapters 5 and 16.

6.5.3 The Linear Equations of Euler and Laplace

Two high-order, variable-coefficient linear equations of some importance are those associated with the names of Euler and Laplace.

EULER EQUATION

$$b_N(x - x_0)^N y^{(N)}(x) + b_{N-1}(x - x_0)^{N-1} y^{(N-1)}(x) + \cdots + b_1(x - x_0)\, y'(x) + b_0 y(x) = f(x) \quad (6.5\text{-}19)$$

where $b_N, b_{N-1}, \ldots, b_1, b_0, x_0$ are constants. For $N = 2$ this equation was met in section 6.4.6. Each variable coefficient here is a common linear factor raised to the same degree as the order of derivative it multiplies. The substitution $x = x_0 + e^t$ reduces the equation to the constant coefficient form considered in section 6.5.1 for a function $w(t) = y(x_0 + e^t)$

$$a_N w^{(N)}(t) + a_{N-1} w^{(N-1)}(t) + \cdots + a_1 w'(t) + a_0 w(t) = f(x_0 + e^t) \quad (6.5\text{-}20)$$

where the coefficients a_i are related linearly to the coefficients b_j. Since solutions of the reduced equation associated with eq. (6.5-20) are of the form $t^k e^{\lambda t}$ for either real or complex conjugate λ (assuming the b_j's are real) with $k = 0$ for simple roots of the auxiliary equation, the corresponding homogeneous solutions of eq. (6.5-19) are of the form

$$(x - x_0)^\lambda, \quad \text{if } k = 0; \qquad k(x - x_0)^\lambda \ln(x - x_0), \quad \text{if } k \neq 0 \qquad (6.5\text{-}21)$$

LAPLACE EQUATION

Since any variable-coefficient function can be approximated over a sufficiently short interval by a linear function (i.e., a two-term, truncated Taylor expansion), it is useful to consider an equation with linear coefficients. The resulting N^{th}-order homogeneous equation is known as the Laplace equation (e.g., see Ref. 6-17):

$$(a_N + b_N x)\,y^{(N)}(x) + (a_{N-1} + b_{N-1}x)\,y^{(N-1)}(x) + \cdots + (a_1 + b_1 x)\,y'(x) + (a_0 + b_0 x)\,y(x) = 0 \quad (6.5\text{-}22)$$

where the a_j's, b_j's are constants. Naturally, if each $b_j = 0$ then exponential solutions exist. For nonzero b_j, an integral solution can be found in the form

$$y(x) = \int_\alpha^\beta e^{x\xi} f(\xi)\,d\xi \qquad (6.5\text{-}23)$$

where the constants α, β and the function $f(\xi)$ are so chosen as to render the integral a solution. Note the similarity to the integral which defines the Laplace transform. We illustrate by an example: $xy''' + (4 - 3x)\,y'' - 7y' - (2 - 4x)\,y = 0$. Substitution of eq. (6.5-23) into the differential equation leads to

$$\int_\alpha^\beta [G(\xi) + xH(\xi)]\,f(\xi)\,e^{x\xi}\,d\xi = 0$$

where $G(\xi) = 4\xi^2 - 7\xi - 2$, $H(\xi) = \xi^3 - 3\xi^2 + 4$. After integration by parts of the term involving $H(\xi)$, we obtain the alternative form

$$[e^{x\xi}H(\xi)\,f(\xi)]_\alpha^\beta - \int_\alpha^\beta e^{x\xi}\left\{\frac{d}{d\xi}[H(\xi)\,f(\xi)] - G(\xi)\,f(\xi)\right\}d\xi = 0 \quad (6.5\text{-}24)$$

This can be satisfied in two steps by first choosing $f(\xi)$ to nullify, identically, the curly bracket in the integral, and then selecting α, β appropriately to nullify the other term. Thus

$$f(\xi) = [H(\xi)]^{-1} \exp\left[\int \frac{G(\xi)}{H(\xi)}\,d\xi\right] \qquad (6.5\text{-}25)$$

and for this example, $f(\xi) = \xi - 2$, so the first term of eq. (6.5-24) becomes $[(\xi + 1)(\xi - 2)^3\,e^{x\xi}]_\alpha^\beta$. An obvious choice for α, β is $\alpha = -1$, $\beta = 2$, and this gives one solution as

$$y(x) = \int_{-1}^{2} (\xi - 2)\,e^{x\xi}\,d\xi = -x^{-2}e^{2x} + (x^{-2} + 3x^{-1})\,e^{-x}$$

Two more solutions, linearly independent of this one, can, in principle, be found from this one by the technique of section 6.5.1.

6.5.4 Systems of Equations

In many applications, a simultaneous system of coupled equations arises. For linear equations, such a system is equivalent to a single higher order equation, but for nonlinear systems, such a reduction often cannot be made. Furthermore, even for linear equations it is sometimes more convenient to solve a few low order equations rather than a single high order one. Occasionally, a linear system can replace an otherwise nonlinear equation.

This last feature is illustrated by the linear fractional equation

$$\frac{dy}{dx} = \frac{\alpha x + \beta y}{\gamma x + \delta y}, \qquad \alpha\delta - \beta\gamma \neq 0 \tag{6.5-26}$$

Introduction of a new variable, s, upon which both x, y depend, transforms eq. (6.5-26) into a system of constant coefficient linear equations

$$(D - \beta)\, y(s) = \alpha x(s), \quad (D - \gamma)\, x(s) = \delta y(s) \tag{6.5-27}$$

where $D = d/ds$. Operating on the first and second of these with $(D - \gamma)$ and $(D - \beta)$ respectively, we find the equivalent equations

$$\begin{aligned} [(D-\gamma)(D-\beta) - \alpha\delta]\, y &= \left[\frac{d^2y}{ds^2} - (\beta+\gamma)\frac{dy}{ds} - (\alpha\delta - \beta\gamma)\, y\right] = 0 \\ [(D-\beta)(D-\gamma) - \alpha\delta]\, x &= \left[\frac{d^2x}{ds^2} - (\beta+\gamma)\frac{dx}{ds} - (\alpha\delta - \beta\gamma)\, x\right] = 0 \end{aligned} \tag{6.5-28}$$

which shows immediately that the system is second order and that if $\alpha\delta = \beta\gamma$ then dy/dx is a constant, so y is a linear function of x. Either of these equations could be solved directly by the methods of section 6.3, but an instructive alternative is to give the parametric solution of the system eq. (6.5-27). It admits the solution

$$x(s) = A_1 e^{\lambda_1 s} + A_2 e^{\lambda_2 s}, \quad y(s) = \frac{\lambda_1 - \gamma}{\delta} A_1 e^{\lambda_1 s} + \frac{\lambda_2 - \gamma}{\delta} A_2 e^{\lambda_2 s} \tag{6.5-29}$$

if λ_1, λ_2 are distinct roots of the characteristic equation $\lambda^2 - (\beta + \gamma)\lambda - (\alpha\delta - \beta\gamma) = 0$ and A_1, A_2 are the two arbitrary constants appropriate to a second order system. Should $\lambda_1 = \lambda_2 = \mu$ then the solution is

$$x(s) = (A_1 s + A_2)\, e^{\mu s}, \quad y(s) = \delta^{-1}\left[(\mu - \gamma) A_1 s + (\mu - \gamma) A_2 + A_1\right] e^{\mu s} \tag{6.5-30}$$

Many of the ideas developed for low-order equations readily extend to systems of more than two linear, constant coefficient equations where the order of each equation is arbitrary. As an example, consider the following system

$$\begin{aligned} F_1 x(s) + G_1 y(s) + H_1 z(s) &= J_1(s) \\ F_2 x(s) + G_2 y(s) + H_2 z(s) &= J_2(s) \\ F_3 x(s) + G_3 y(s) + H_3 z(s) &= J_3(s) \end{aligned} \tag{6.5-31}$$

where F_j, G_j, H_j are linear differential operators of polynomial form in $D = d/ds$. The order of this system N is that of the highest power of D (which can be considered as an algebraic quantity for this purpose) in the determinant of the coefficients, Δ. Superposition shows that the general solution consists of the complementary function (with N arbitrary constants) plus any particular integral. The characteristic equation, with D treated as algebraic, $\Delta = 0$, has precisely N roots and it is clear that a solution of the reduced system with $J_1 = J_2 = J_3 = 0$ exists in the form $x(s) = A_1 e^{\lambda_1 s} + A_2 e^{\lambda_2 s} + \cdots$ with similar expressions for $y(s)$, $z(s)$. Here λ_1, $\lambda_2, \cdots$ are the *distinct* roots of the characteristic equation and A_1 is a polynomial in s of degree $(m_1 - 1)$ where $m_1 (\geqslant 1)$ is the multiplicity of the root λ_1, and so forth. Substitution of similar expressions into eq. (6.5-31) reduces the problem to the solution of a system of algebraic equations, and then the matrix methods of Chapter 16, section 16.5 can be employed to complete the solution. A fuller discussion together with examples can also be found in Chapter 5 of Ref. 6-4.

6.6 EIGENVALUE PROBLEMS

6.6.1 Preliminary Discussion

A form of ordinary differential equation which arises often in applications is

$$\mathcal{L}\phi(x) = f(x) \tag{6.6-1}$$

where $\mathcal{L}$ is a second order variable coefficient linear differential operator acting on the unknown function $\phi(x)$, and $f(x)$ is the forcing function. One method of solution constructs the integral operator inverse to $\mathcal{L}$ and then operates on eq. (6.6-1) to give the solution in integral form (the Green's function of section 6.4.4 is the kernel of such an operator). An alternative approach, treated now, utilizes the spectral theory of operators to analyze $\mathcal{L}$ itself in terms of its eigenvalues and eigenfunctions; any of a broad class of solution functions are then represented by a series expansion of such eigenfunctions. (Of course, eigenvalue problems are of interest in themselves, also, entirely apart from expansion possibilities.) Since both methods yield solutions to the same problem they are related. For example, the Green's function itself, being a solution of the differential equation, can be expanded in a series of eigenfunctions. This and other connections are treated more fully in Refs. 6-9, 6-11, 6-29.

The selection of an eigenvalue problem for a given operator $\mathcal{L}$ is by no means unique, but it has been found especially useful to expand solutions of differential equations in a series of eigenfunctions of the *Sturm-Liouville system* which consists of the equation

$$[p(x)y'(x)]' + [q(x) + \lambda r(x)]\, y(x) = 0, \qquad \text{in } a \leqslant x \leqslant b \tag{6.6-2}$$

together with homogeneous boundary conditions (of various types) imposed at $x = a$, $x = b$. Note the multiplicity of items required to define a Sturm-Liouville problem: a self-adjoint linear differential operator (the coefficients p, q); a weighting function, $r(x)$; a parameter λ; an interval; and homogeneous boundary conditions.

Sturm-Liouville problems are important, not only because of their contribution to the solution of the boundary value problem for eq. (6.6-1), but also because they arise repeatedly in direct applications, especially one-dimensional vibration problems in continuum or quantum mechanics (the eigenfunctions being the normal modes of oscillation) and as reductions of the important partial differential equations of mathematical physics by separation of variables; then λ is related to the separation constant. See Ref. 6-25 for full coverage of this aspect.

As in section 6.4.1, it may be useful to transform eq. (6.6-2) into other forms, of which the simplest is (Ref. 6-9)

$$\frac{d^2u}{dz^2} + [s(z) + \lambda]\, u(z) = 0, \qquad \text{where } u(z) = (pr)^{1/4} y$$
$$z = \int \left(\frac{r}{p}\right)^{1/2} dx \quad \text{and} \quad s = -(rp)^{-1/4}\,[(rp)^{1/4}]'' + \frac{q}{r} \tag{6.6-3}$$

Another reduction of considerable value is that of Prüfer who introduced an equivalent system of two first-order equations and then solved in polar coordinates, Refs. 6-3, 6-4.

Nontrivial solutions of the homogeneous eq. (6.6-2) which satisfy the boundary conditions are called *eigenfunctions*; we denote them by $y_n(x)$. They are unique only up to a multiplicative constant, C_n, and they exist only when λ takes on special values called *eigenvalues*, λ_n. The complete set of eigenvalues constitutes the *spectrum* of the operator $\mathcal{L}$. When n is an integer (the usual case for finite a, b), the eigenvalues form a denumerable or countably infinite set and the spectrum is said to be *discrete*. If all values of λ in some real interval are eigenvalues (as occurs when $b - a$ is infinite) then the spectrum is *continuous*. In more complicated situations, typified by the Schröedinger equation of quantum mechanics, the spectrum may exhibit both discrete and continuous parts.

The solutions of differential equations that are of general interest are those which are *piecewise continuous*, i.e., those which are continuous in $a \leqslant x \leqslant b$ except for the possibility of a finite number of finite discontinuities within $a < x < b$. Although it is not necessary, the expansion of any such function in a series of eigenfunctions is greatly simplified when the latter form an *orthonormal set.* For this purpose, the inner or *scalar product* of two functions $g(x)$, $h(x)$ relative to the weight function $w(x)$ is denoted and defined by

$$\langle g, h \rangle \equiv \int_a^b w(x) g(x) h(x)\, dx \tag{6.6-4}$$

The scalar product of a function $g(x)$ with itself is called the *norm of the function*, say $N(g)$, and when $N(g) = 1$ then g is a *normalized* function, relative to the weight function w. Any real function $g(x)$ whose norm is bounded is said to be *square integrable* relative to $w(x)$. Two functions whose scalar product vanishes are *orthogonal* on the interval. With this terminology, an *orthonormal* (i.e., both orthogonal and normalized) set of eigenfunctions $y_n(x)$ satisfies the *orthogonality relations*

$$\int_a^b w(x)\, y_m(x)\, y_n(x)\, dx = \delta_{mn} \tag{6.6-5}$$

where the Kronecker delta, δ_{mn}, is 1 when $m = n$ and zero otherwise.

If $\phi(x)$ is any piecewise continuous function on $a \leqslant x \leqslant b$ and $y_n(x)$ denotes an orthonormal set of eigenfunctions with discrete spectrum, then the eigenfunction expansion of $\phi(x)$, if it exists, takes the form

$$\phi(x) = \sum_{n=1}^{\infty} c_n y_n(x) \tag{6.6-6}$$

and the expansion coefficients c_n can be found directly from the scalar product of $\phi(x)$, $y_m(x)$, utilizing eq. (6.6-5). Provided the series above converges uniformly so that the order of integration and summation can be interchanged,

$$c_m = \langle \phi, y_m \rangle = \int_a^b w(x)\, \phi(x)\, y_m(x)\, dx \tag{6.6-7}$$

Alternatively, if c_m is defined by eq. (6.6-7) then the equal sign in eq. (6.6-6) must be interpreted as convergence in the mean square sense, this approximation being referred to again in section 6.8.2. When every piecewise continuous function can be expressed as in eq. (6.6-6), then the basis of eigenfunctions is said to be *complete.* Should the spectrum be continuous, then the series in eq. (6.6-6) is replaced by an integral.

6.6.2 Sturm-Liouville Theory

Consider now *regular* Sturm-Liouville systems in which p, p', q, r are real continuous functions and $p(x), r(x) > 0$ in the closed finite interval $a \leqslant x \leqslant b$. Theorems, proven, for example, in Ref. 6-3, guarantee the existence of a twice continuously differentiable linearly independent solution basis for such a problem.

We first ask what boundary conditions render the eigenfunctions orthogonal. Suppose λ, μ are two distinct eigenvalues of eq. (6.6-2) associated with eigenfunctions $u(x)$, $v(x)$, respectively. Then $(pu')' + (q + \lambda r)\, u = 0$, $(pv')' + (q + \mu r)\, v = 0$. Cross multiplication and subtraction, followed by integration shows that

$$(\lambda - \mu) \int_a^b r(x)\, u(x)\, v(x)\, dx = p(a)\, W(a) - p(b)\, W(b) \tag{6.6-8}$$

where $W(x) = uv' - u'v$ is the Wronskian of u, v. Thus, eigenfunctions belonging to distinct eigenvalues are orthogonal whenever the boundary conditions are such as to nullify the right-hand member. This can happen in several ways. In regular systems, where $p(a) \neq 0$, $p(b) \neq 0$, a common occurrence is for $W(a) = W(b) = 0$ as is implied by boundary conditions such as

$$\alpha y(a) + \beta y'(a) = 0, \quad \gamma y(b) + \delta y'(b) = 0 \tag{6.6-9}$$

where at least one of (α, β) and one of (λ, δ) are nonzero. In a regular *periodic* Sturm-Liouville system $p(a) = p(b)$ and the boundary conditions are

$$y(a) = y(b), \quad y'(a) = y'(b) \tag{6.6-10}$$

which ensures $W(a) = W(b)$ and thence orthogonality. *Singular* Sturm-Liouville problems admit orthogonal eigenfunctions under conditions such that

$$\begin{aligned} &\text{(i)} \;\; W(a) = 0 \text{ and } p(b) = 0;\\ &\text{(ii)} \;\; W(b) = 0 \text{ and } p(a) = 0;\\ &\text{(iii)} \;\; p(a) = p(b) = 0 \end{aligned} \tag{6.6-11}$$

where in (i) and (ii) it is appropriate to require the first or second relation of eq. (6.6-9), respectively, whereas orthogonality is independent of boundary conditions if both endpoints are singularities (iii). Special attention is paid to singular problems in Ref. 6-29.

It is shown in Ref. 6-29 that the real, self-adjoint Sturm-Liouville equation with one-signed weight function $r(x)$ admits only real eigenvalues under any of the boundary conditions above. A useful expression for the eigenvalue associated with the eigenfunction $u(x)$ can be derived by forming the scalar product of u and $\mathcal{L}u$ relative to the weight function $r(x)$

$$\langle u, \mathcal{L}u \rangle = \int_a^b r(x)\, u(x)\, \mathcal{L}u(x)\, dx = \lambda \int_a^b r^2(x)\, u^2(x)\, dx = \lambda \langle u, ru \rangle \quad (6.6\text{-}12)$$

where $\mathcal{L}u = \lambda ru$ since u is an eigenfunction. This shows that

$$\lambda = \langle u, \mathcal{L}u \rangle / \langle u, ru \rangle \quad (6.6\text{-}13)$$

Any operator for which $\langle u, \mathcal{L}u \rangle$ is positive when $r(x)$ is positive is said to be *positive definite.* Equation (6.6-13) implies (since both numerator and denominator are then positive) that positive definite self-adjoint operators have positive eigenvalues. This formula also motivates an approximate method for finding eigenvalues (section 6.8.2).

The transformed Sturm-Liouville eq. (6.6-3) is convenient for investigating the qualitative behavior of eigenfunctions. If, for example, both λ and $s(z)$ are positive, then evidently the curvature of $u(z)$ is opposite in sign to that of $u(z)$, thereby suggesting oscillatory behavior. Indeed, the case of constant s gives sinusoidal oscillations at frequencies which increase as λ increases, the zeros or nodal points moving continuously towards smaller x so that more zeros crowd into a fixed interval as λ increases. The precise results rest on the *Oscillation Theorem* (Refs. 6-3, 6-4, 6-16), which states that the regular Sturm-Liouville system with boundary conditions eq. (6.6-9) possesses an infinite number of eigenvalues which can be ordered (starting with the smallest) as $\lambda_1 < \lambda_2 < \lambda_3 \cdots (\lim_{n \to \infty} \lambda_n = \infty)$ and the eigenfunction $y_n(x; \lambda_n)$ has precisely n simple zeros in the interval $a \leqslant x \leqslant b$. Also, the zeros of real linearly independent eigenfunctions separate each other, so that if y_m has consecutive zeros at x_1, x_2, then y_n vanishes at least once in the open interval $x_1 < x < x_2$ (this *separation theorem* is proved in Ref. 6-16). Certain useful asymptotic properties of the eigenvalues and eigenfunctions, valid for large λ or z, can be proven. For example, if $s(z)$ in eq. (6.6-3) is continuous, and the eigenfunctions are normalized in the interval $0 \leqslant z \leqslant 1$ so that $\int_0^1 u^2(z)\, dz = 1$, then $|u_n(z; \lambda)|$ is bounded in the interval $0 \leqslant z \leqslant 1$, and the bound is independent of both z and λ. Furthermore, an asymptotic representation of the nth eigenfunction $u_n(z)$ in eq. (6.6-3), subject to $u_n(0) = u_n(l) = 0$, is $(2/l)^{1/2} \sin \lambda_n^{1/2} z + \lambda_n^{-1/2}\, 0(1)$, where the asymptotic estimate (as n becomes large) for the nth eigenvalue, presumed positive, is $\lambda_n = n^2 \pi^2 l^{-2} + 0(1)$. The symbol $0(1)$ here denotes a number which remains bounded as $\lambda_n \to \infty$. Further details are given in Ref. 6-9.

6.6.3 Expansions in Eigenfunctions

As is proven in Ref. 6-9, every continuous function $f(x)$ which has piecewise continuous first and second derivatives in $a \leqslant x \leqslant b$ and satisfies the boundary conditions of the Sturm-Liouville eigenvalue system at $x = a, x = b$ can be expanded in an absolutely and uniformly convergent series of eigenfunctions

$$f(x) = \sum_{n=1}^{\infty} c_n y_n(x), \qquad \text{where } c_n = \int_a^b r(x) f(x) y_n(x)\, dx \tag{6.6-14}$$

Furthermore, if $f(x)$ is not continuous but instead has a finite discontinuity at some point interior to the interval, then the series converges to the mean of the left- and right-hand limits at the discontinuity.

A most celebrated eigenfunction expansion is the Fourier series which arises from the simple Sturm-Liouville equation

$$y'' + \lambda y = 0 \tag{6.6-15}$$

which is eq. (6.6-2) with $p(x) = r(x) = 1$, $q(x) = 0$ or eq. (6.6-3) with $s(x) = 0$. If the interval is $-\pi \leqslant x \leqslant \pi$, then boundary conditions $y(-\pi) = 0$, $y(\pi) = 0$ generate the orthonormal set of eigenfunctions $\pi^{-1/2} \sin \lambda^{1/2} x$ with the real, positive, discrete eigenvalue spectrum $\lambda_n = n^2, n = 1, 2, 3, \cdots$. Note that $\lambda = 0$ is not an eigenvalue because the only solution which then satisfies the boundary conditions is the trivial one. Had the boundary conditions been $y'(-\pi) = 0, y'(\pi) = 0$, then an orthonormal set of eigenfunctions is $\pi^{-1/2} \cos \lambda^{1/2} x$ with the same spectrum $\lambda_n = 1, 4, 9, \cdots$. However, $\lambda = 0$ is now also an eigenvalue, because any nonzero constant is a nontrivial solution of $y'' = 0$ which satisfies the boundary conditions. The normalized eigenfunction associated with $\lambda = 0$ is $(2\pi)^{-1/2}$.

Sometimes the boundary conditions are not sufficiently selective to prevent what is called *degeneracy* wherein more than one eigenfunction is associated with an eigenvalue. Thus, eq. (6.6-15) subject only to periodic conditions $y(-\pi) = y(\pi)$, $y'(-\pi) = y'(\pi)$ admits the eigenvalues $\lambda_n = n^2$, $n = 0, 1, 2, 3, \cdots$ but for each λ_n, $\sin \lambda_n^{1/2} x$, $\cos \lambda_n^{1/2} x$ (or any linear combination) are acceptable eigenfunctions. The eigenfunction expansion in this case (discussed thoroughly in Ref. 6-7) is the Fourier series

$$f(x) = \frac{1}{2} a_0 + \sum_{n=1}^{\infty} (a_n \cos nx + b_n \sin nx)$$

where

$$a_n = \frac{1}{\pi} \int_{-\pi}^{\pi} f(x) \cos nx\, dx, \qquad n = 0, 1, 2, \cdots$$

$$b_n = \frac{1}{\pi} \int_{-\pi}^{\pi} f(x) \sin nx\, dx, \qquad n = 1, 2, 3, \cdots \tag{6.6-16}$$

Such a series can represent either a function defined only in the interval $-\pi \leqslant x \leqslant \pi$ for all values of x in that interval, or it can represent a function, periodic with period 2π, for all values of x. It cannot represent a nonperiodic function for all x. A simple example of a nonperiodic finite interval Fourier series is given by eq. (6.6-15) subject to $y(0) = 0$, $y'(\pi) = y(\pi)$. The differential equation and first boundary condition are satisfied by $\sin \lambda^{1/2} x$. The second boundary condition is obeyed if λ is a root of the transcendental equation $\tan \pi\lambda^{1/2} = \lambda^{1/2}$. A result of the preceding section guarantees that the eigenvalues are real (and the eigenfunctions are orthogonal), but for these boundary conditions positive definiteness is not easily determined because

$$\langle y, \mathcal{L}y \rangle = -\int_0^\pi yy''\, dx = -[yy']_0^\pi + \int_0^\pi y'^2(x)\, dx$$

$$= -[y(\pi)]^2 + \int_0^\pi y'^2(x)\, dx$$

For real, positive eigenvalues set $\lambda_n = \pi^{-2}k^2$, (k real), where k is a root of $\tan k = \pi^{-1}k$ (by graphical consideration of the intersections of the curve $\tan k$ with the line $\pi^{-1}k$, this equation is seen to admit an infinite number of roots, k_n, which can be chosen positive for present purposes). Negative eigenvalues would correspond to $\lambda_n = -\pi^{-2}l^2$ where $\tanh l = \pi^{-1}l$ and the only real root of this equation is $l = 0$, but $\lambda = 0$ is not an eigenvalue since the corresponding solution of eq. (6.6-15) cannot satisfy the boundary conditions. Thus, the eigenvalues are positive and the eigenfunction expansion is the nonperiodic Fourier series

$$f(x) = \sum_{n=1}^{\infty} c_n \sin \frac{k_n x}{\pi}, \qquad k_n = \pi \tan k_n$$

$$c_n = 2\,[\pi - \cos^2 k_n]^{-1} \int_0^\pi f(x) \sin \frac{k_n x}{\pi}\, dx \tag{6.6-17}$$

To illustrate how sensitive these problems are to the boundary conditions, consider eq. (6.6-15) subject to $y(0) = 0$, $y'(0) = y(1)$, which falls outside of the general cases treated previously, see eqs. (6.6-9), (6.6-10). The eigenfunctions must again be of the form $\sin \lambda^{1/2} x$ but satisfaction of the second boundary condition requires that $\lambda^{1/2} = \sin \lambda^{1/2}$ and this equation has only one real root $\lambda = 0$ (for which the eigenfunction is $y = x$) and an infinite number of complex roots. Consequently, there are complex eigenvalues. Moreover, these boundary conditions do not nullify the right hand member of eq. (6.6-8) so the eigenfunctions are not orthogonal. Coefficients of the eigenfunction expansion can be found only by considering the adjoint problem (see Ref. 6-11 for details).

An instructive example of a singular Sturm-Liouville system is provided by the equation $[(1 - x^2)y']' + \lambda y = 0$ in $-1 \leqslant x \leqslant +1$, with y and y' bounded at the end-points. In the notation of eq. (6.6-2), $p(x) = (1 - x^2)$, which vanishes at $x = \pm 1$ (hence the singular behavior), $q(x) = 0$, and $r(x) = 1$. Since the operator is real, self-adjoint and r is one signed, the eigenvalues are real. Furthermore, $\mathcal{L}$ is positive definite because

$$\langle u, \mathcal{L}u \rangle = -\int_{-1}^{+1} u(x)\,[(1 - x^2)u'(x)]'\,dx$$

$$= -\,[(1 - x^2)uu']_{-1}^{+1} + \int_{-1}^{+1} (1 - x^2)u'^2(x)\,dx$$

which is greater than zero (provided $u' \not\equiv 0$) since the square bracket vanishes. The eigenfunction $u = \text{constant} \neq 0$ corresponds to the eigenvalue $\lambda = 0$. Apart from this exception, the eigenvalues are positive. One method of solving this problem is by power series expansions about the regular singular points $x = \pm 1$, as in section 6.4.6. A different technique rests on the observation that frequently solutions to variable coefficient differential equations are closely related to the form of the coefficients themselves (polynomial coefficients generate power series solutions, for example). Hence, introduce $u(x) = (x^2 - 1)^n$ and by differentiation obtain $u'(x) = 2nx(x^2 - 1)^{n-1}$ so that a differential equation obeyed by $u(x)$ is $(1 - x^2)u' + 2nxu = 0$. Another differentiation yields $(1 - x^2)u'' + 2(n - 1)xu' + 2nu = 0$. The kth derivative of this last equation gives

$$(1 - x^2)u^{(k+2)} + 2(n - k - 1)xu^{(k+1)} + (2n - k)(k + 1)u^{(k)} = 0$$

This is the desired equation if $k = n$ provided $\lambda = n(n + 1)$ and $y = Cu^{(n)} = C[(x^2 - 1)^n]^{(n)}$, where C is a constant, which can be chosen to normalize these eigenfunctions. The conventional terminology for this problem is that the *Legendre equation*

$$(1 - x^2)\frac{d^2y}{dx^2} - 2x\frac{dy}{dx} + n(n + 1)y = 0 \quad \text{in } -1 \leqslant x \leqslant 1 \tag{6.6-18}$$

admits orthogonal polynomial solutions, $P_n(x)$, as eigenfunctions which are found by evaluating *Rodrigues' formula*

$$P_n(x) = \frac{1}{2^n(n!)}\frac{d^n}{dx^n}[(x^2 - 1)^n] \tag{6.6-19}$$

The constant here makes $P_n(1) = 1$. The first few polynomials are $P_0 = 1, P_1 = x, P_2 = \frac{1}{2}(3x^2 - 1)$ and the explicit formula for the nth degree polynomial is

$$P_n(x) = \sum_{r=0}^{N} (-1)^r \frac{(2n-2r)!}{2^n(r!)\,[(n-r)!]\,[(n-2r)!]} x^{n-2r} \tag{6.6-20}$$

where $N = n/2$ or $N = (n-1)/2$ according as n is even or odd, respectively.

Since each integral power of x, say x^n, can be written as a linear superposition of Legendre polynomials of degree $n, n-1, \ldots, 0$, any function $f(x)$ expandable in MacLaurin series can be expanded in a series of Legendre polynomials

$$f(x) = \sum_{n=0}^{\infty} c_n P_n(x), \qquad c_n = \frac{2n+1}{2} \int_{-1}^{+1} f(x)\, P_n(x)\, dx \tag{6.6-21}$$

the factor $(2n+1)/2$ being required for normalization. Further properties and applications of these functions are developed in the cited references.

Another important expansion is in the eigenfunctions of the Bessel differential equation. The associated Fourier-Bessel series is displayed in Chapter 7, section 7.4.8.

6.7 NONLINEAR ORDINARY DIFFERENTIAL EQUATIONS

6.7.1 Some Solvable First-Order Equations

The techniques developed for solving linear differential equations, seldom carry over unaltered to nonlinear equations. Moreover, wide classes of nonlinear equation are simply unsolvable in exact form. Nontheless, many ingenious approximation techniques (some of which are treated in section 6.8) have been invented to cope with otherwise recalcitrant equations of interest.

As a first step we treat some exactly solvable first-order equations, of the form $y'(x) = f(x,y)$, or

$$P(x,y)\, dx + Q(x,y)\, dy = 0 \tag{6.7-1}$$

When P, Q are not functions of y and x, respectively, then the variables are separable so that integration leads to a general solution in the form

$$\int P(x)\, dx + \int Q(y)\, dy = C \tag{6.7-2}$$

where C is an arbitrary constant of integration. More generally, if $P(x,y) = P_1(x)\, P_2(y)$, $Q(x,y) = Q_1(x)\, Q_2(y)$ then division by $Q_1(x)\, P_2(y)$ (presumed nonzero) results in *separation of variables*

$$\frac{P_1(x)}{Q_1(x)}\, dx + \frac{P_2(y)}{Q_2(y)}\, dy = 0 \tag{6.7-3}$$

The solution is again given by eq. (6.7-2) with $P = P_1/Q_1$, $Q = P_2/Q_2$. Sometimes a change in variables reduces an equation to a form in which the variables separate, an important case being for a *homogeneous equation* which is eq. (6.7-1) with P, Q homogeneous functions of x, y of the same degree. (Roughly speaking, this means that P, Q are such that P/Q is a function of y/x only.) A natural substitution is then $y = vx$ so $dy = x\,dv + v\,dx$. P, Q are, by definition, homogeneous of degree m provided $P(x, vx) = x^m M(v)$, $Q(x, vx) = x^m N(v)$ with M, N independent of x. Under this circumstance the variables separate and the general integral is

$$\int \frac{N(v)\,dv}{M(v) + vN(v)} + \ln x = C \tag{6.7-4}$$

One example of such a homogeneous equation is eq. (6.7-1) with $P(x, y) = ax + by$, $Q(x, y) = cx + ey$ (a, b, c, e constants). These forms can be regarded either as special exact examples, or more generally, as the first two terms of bivariate MacLaurin expansions (therefore valid approximations for arbitrary analytic P, Q). The specific equation is

$$(ax + by)\,dx + (cx + ey)\,dy = 0 \tag{6.7-5}$$

and following substitution of $y = vx$, the general solution is found to be

$$\ln Cx + \int \frac{(c + ev)\,dv}{a + (b + c)\,v + ev^2} = 0 \tag{6.7-6}$$

The general solution of any first order equation is a functional relation between x and y including one arbitrary constant, conveniently written as

$$u(x, y) = C \tag{6.7-7}$$

The differential of this equation, $du = \partial u/\partial x\,dx + \partial u/\partial y\,dy$, is of the form of eq. (6.7-1) if and only if

$$P(x, y) = \frac{\partial u}{\partial x}, \quad Q(x, y) = \frac{\partial u}{\partial y} \tag{6.7-8}$$

Consequently, eq. (6.7-1) can be integrated directly (without any prior manipulation) only if P, Q satisfy the *condition of integrability*

$$\frac{\partial P}{\partial y} = \frac{\partial Q}{\partial x} \tag{6.7-9}$$

each of these partial derivatives being of necessity, by eq. (6.7-8), equal to $\partial^2 u/\partial x \partial y$. If P, Q do not satisfy eq. (6.7-9), the original equation is still reducible to an integrable form but this step requires multiplication by an integrating factor, say $w(x, y)$, so that the equation for determining w is a partial differential equation

$$\frac{\partial}{\partial y}(wP) = \frac{\partial}{\partial x}(wQ) \tag{6.7-10}$$

and this equation is sometimes solvable analytically. An example is $P = y + 2x^2y^4$, $Q = x + 3x^3y^3$. Here, eq. (6.7-10) becomes $y(1 + 2x^2y^3)\, \partial w/\partial y - x(1 + 3x^2y^3)\, \partial w/\partial x - x^2y^3w = 0$, which, by inspection, admits a solution of the form $w(x, y) = w(x^2y^3)$. An integrating factor is then found to be $e^{x^2y^3}$, so, in this case, eq. (6.7-1) has an integral $xye^{x^2y^3} = \text{constant}$. For further properties of integrating factors, see Refs. (6-16, 6-17).

Our final category of first-order nonlinear differential equation exactly solvable is that for which a change in variables reduces the equation to linear form. Two well-known examples are the equations of Bernoulli and Riccati, which are, respectively:

$$y'(x) + p(x)y(x) + q(x)y^n(x) = 0 \tag{6.7-11}$$

$$y'(x) + p(x)y(x) + q(x)y^2(x) = r(x) \tag{6.7-12}$$

Division of the first by y^n suggests the change of variable $u = y^{1-n}$ and the equation for u is then linear

$$u'(x) + (1 - n)p(x)u(x) = -(1 - n)q(x) \tag{6.7-13}$$

The Riccati eq. (6.7-12) is reduced to a second-order linear differential equation by the substitution $y = u'/qu$ which gives

$$u''(x) + [p(x) - q'(x)/q(x)]\, u'(x) - r(x)q(x)u(x) = 0 \tag{6.7-14}$$

and this is solvable by the methods of section 6.4.

As a general rule, an unfamiliar first-order nonlinear differential equation should first be examined from the point of view of the standard types reviewed here. If it is not of the desired form, possibly a transformation of variables will reduce it appropriately. Failing that, it is wise to consult the extensive list of solved differential equations provided by Kamke Ref. 6-18. As a final alternative, approximations, graphical methods, or a numerical solution may be developed.

6.7.2 Second-Order Nonlinear Differential Equations

Nonlinear differential equations of the second order are significantly more difficult to solve than their first-order counterparts. A classical problem is illustrated by the simple undamped pendulum whose instantaneous angular deviation from the vertical $\theta(t)$ obeys the equation

$$\ddot{\theta}(t) + \omega_0^2 \sin \theta(t) = 0 \tag{6.7-15}$$

where a dot signifies a time derivative, $\omega_0^2 = g/l$, g being the constant gravitational acceleration and l the length of the plumb bob. Since $\sin \theta = \theta - \theta^3/3! + \theta^5/5! - \cdots$ the restoring force is effectively of infinite degree in the dependent variable $\theta(t)$. For small enough deflections of the pendulum, a linear approximation suffices ($\sin \theta \doteq \theta$) and it then follows that ω_0 is the angular frequency (in radians/second) of such infinitesimal oscillations. At the next stage of approximation, one takes $\sin \theta \doteq \theta - \theta^3/3!$ and with this sort of a cubic nonlinearity the resulting *Duffing equation* is solvable in implicit form in terms of the incomplete elliptic integral of the first kind (see Ref. 6-20 for details). The exact equation can be handled in much the same way. If $\Omega(t) \equiv \dot{\theta}(t)$ is the pendulum angular velocity, then $\ddot{\theta} = d\Omega/dt = \Omega d\Omega/d\theta$ so eq. (6.7-15) reduces to a first order equation with variables separated: $\Omega d\Omega + \omega_0^2 \sin \theta d\theta = 0$. An integration shows that Ω has the form $\Omega = \pm 2\omega_0(k^2 - \sin^2 \theta/2)^{1/2}$ where $k = \sin \theta_m/2$ and $\theta_m > 0$ is the maximum value of θ achieved in the motion (at which point $\Omega = 0$, on physical grounds). If our interest is in times immediately after that for which $\theta = \theta_m$, Ω is negative so we have to integrate the first order equation $d\theta/dt = -2\omega_0(k^2 - \sin^2 \theta/2)^{1/2}$ and this has variables which separate. It is customary to introduce φ such that $\sin \varphi = k^{-1} \sin \theta/2$ and then the final solution takes on the form

$$\omega_0(t - t_0) = -\int_0^{\varphi} [1 - k^2 \sin^2 \psi]^{-1/2}\, d\psi \tag{6.7-16}$$

where t_0 is the time at which θ (and hence φ) is zero. When $\theta = -\theta_m$, we must switch over to the positive branch of the square root in eq. (6.7-16). The integral which occurs here is the incomplete elliptic integral of the first kind and can be found tabulated in Ref. 6-1 for various values of k, φ. Although the motion is still periodic, as for the linear case, the period τ is now a function of the amplitude, being given by

$$\omega_0 \tau = 4\int_0^{\pi/2} [1 - k^2 \sin^2 \psi]^{-1/2}\, d\psi \tag{6.7-17}$$

If, for example, the pendulum swings to $\pm 60°$, the period is increased by about 7% over the small amplitude period.

It is sometimes advantageous to utilize graphical methods for solution of nonlinear differential equations. Thus, a relatively simple way to solve any second-order equation in which the independent variable is absent is by means of the *phase plane* which is simply a plot of dy/dx versus y. The general equation in this case is

$$f(y, y', y'') = 0 \tag{6.7-18}$$

but in terms of the phase plane variables $y, z \equiv y'$, it becomes $f(y, z, z\, dz/dy) = 0$. Consequently, for given values of y, z (i.e., a given point in the phase plane) the equation defines a slope dz/dy of a solution curve through that point. By taking a small step in phase space in the direction indicated by dz/dy and then finding a new value of the slope, it is possible to sketch out a *trajectory* in the phase plane, this being a curve followed by the system point, in the direction of increasing time. Although the trajectories do not give y as a function of x directly, it is straightforward to determine the value of x associated with a selected point y, z by integrating $dx = dy/z$ along the trajectory. The phase plane trajectories for the pendulum problem can be found plotted in many places, e.g., Ref. 6-27.

6.8 APPROXIMATE METHODS

6.8.1 Perturbation Expansions

Because the mathematical problems presented by modern science cannot usually be solved exactly in closed form utilizing elementary or even transcendental functions, much of applied mathematics is devoted to approximation theory. Moreover, even if an exact solution is available, it is often more expedient to extract the desired information from an approximate representation. One of the most widely used of all such approximate techniques is the perturbation expansion wherein a function dependent on a parameter (usually small, in some sense), as well as an independent variable is expanded in powers of the parameter. Fuller coverage of this powerful method is given in chapter 14, but the simplest ideas are introduced here.

Consider the small vibrations of an undamped harmonic oscillator, such as the pendulum of section 6.7.2, where the length of the pendulum is not constant, but rather oscillates slightly in time at the same frequency as the pendulum would oscillate if the length were fixed. A typical problem for the system is given by

$$y''(x) + (1 + \epsilon \sin x)\, y(x) = 0, \quad y(0) = 1, \quad y'(0) = 0 \tag{6.8-1}$$

where the positive constant ϵ is much smaller than one. Clearly, this problem satisfies the conditions of the theorem quoted in section 6.5.2; furthermore, all derivatives of $(1 + \epsilon \sin x)y$, as a function of ϵ, x, y, exist. Therefore, the dependence of y on ϵ is analytic, being representable by a power series expansion in ϵ (with x-dependent coefficients) which is convergent for all finite ϵ

$$y(x; \epsilon) = y_0(x) + \epsilon y_1(x) + \epsilon^2 y_2(x) + \cdots \tag{6.8-2}$$

This expression is generally referred to as a *perturbation expansion* and it is specifically a *regular* one because of the absence of singularities. Substitution into eq. (6.8-1) and regrouping of terms with like power of ϵ leads to

$$(y_0'' + y_0) + \epsilon(y_1'' + y_1 + y_0 \sin x) + \epsilon^2(y_2'' + y_2 + y_1 \sin x) + \cdots = 0$$

This equation, which cannot be satisfied for all finite ϵ unless each parenthesis vanishes identically, generates an infinite sequential hierarchy of linear, inhomogeneous differential equations for $y_0, y_1, y_2, \ldots$. Evaluation of eq. (6.8-2) and its derivative at $x = 0$, shows that y_0 must satisfy the same initial conditions as y and that $y_1, y_2, \ldots$ obey homogeneous initial conditions. The first three problems are: (i) $y_0'' + y_0 = 0, y_0(0) = 1, y_0'(0) = 0$; (ii) $y_1'' + y_1 = -y_0 \sin x, y_1(0) = y_1'(0) = 0$; (iii) $y_2'' + y_2 = -y_1 \sin x$, $y_2(0) = y_2'(0) = 0$. These problems are solvable by the methods previously described. Correct through order ϵ^2, the solution is

$$y(x; \epsilon) = \cos x - \frac{\epsilon}{3} \sin x\,(1 - \cos x)$$

$$+ \frac{\epsilon^2}{72}(3 \sin^2 x \cos x - 8 \sin^2 x - 16 \cos x - 3x \sin x + 16) + 0(\epsilon^3)$$

where $0(\epsilon^3)$ denotes terms whose ultimate approach to zero is proportional to ϵ^3 as $\epsilon \to 0$. For sufficiently small ϵ, this is probably an adequate description for most purposes, but higher order terms can, if desired, be calculated as follows. The function $y_n(x)$ in eq. (6.8-2) satisfies $y_n'' + y_n = -y_{n-1} \sin x$, $y_n(0) = y_n'(0) = 0$ for $n = 1, 2, 3, \ldots$. The variation of parameters technique for example, section 6.4.3, eq. (6.4-23), then shows that the solution can be found from

$$y_n(x) = \int_0^x y_{n-1}(\xi)\,(\cos x \sin^2 \xi - \sin x \cos \xi \sin \xi)\, d\xi$$

and the integral here can be handled numerically, if necessary.

Suppose next that the small parameter occurs, not in the undifferentiated term, but rather multiplies the most highly differentiated term. Such a situation occurs in a variety of physical fields (such as elasticity, celestial mechanics, plasma physics, fluid dynamics etc.). For a simple example, consider a damped harmonic oscillator whose mass is small in some sense:

$$\epsilon y'' + ay' + by = 0, \quad y(0) = 0, \quad y'(0) = 1$$

where a, b are arbitrary positive constants, and $\epsilon \ll 1$. Substitution of the perturbation expansion, eq. (6.8-2), again leads to a hierarchy but the first problem is now $ay_0' + by_0 = 0, y_0(0) = 0, y_0'(0) = 1$, and this first order equation has no solution which satisfies the two imposed initial conditions. The regular perturbation expansion fails to be useful. Chapter 14 takes up the way to extricate oneself from such a *singular perturbation problem*, so called because $y(x; \epsilon)$ is not well-behaved as $\epsilon \to 0$ (this is consistent with the fact that the theorem of section 6.5.2 is violated in that limit when ϵ multiplies the most highly differentiated term of the differential equation).

As an example of a regular perturbation expansion for an eigenvalue problem (see Refs. 6-6, 6-9, 6-11, for general theory and further details) consider

$$y'' + \lambda y = \epsilon x^2 y \quad \text{in } 0 \leqslant x \leqslant \pi, \qquad y(0) = y(\pi) = 0 \tag{6.8-3}$$

where $0 \leqslant \epsilon \ll 1$. The eigenfunctions and eigenvalues of eq. (6.8-3) both depend on ϵ so we introduce a joint perturbation expansion consisting of eq. (6.8-2) and

$$\lambda = \lambda_0 + \epsilon\lambda_1 + \epsilon^2\lambda_2 + \cdots \tag{6.8-4}$$

where the subscript here does not refer (as it did formerly) to the index of the exact eigenvalue in a discrete spectrum, but rather to the degree of approximation in the perturbation expansion. Substitution into eq. (6.8-3) leads to the first two problems in the form $y_0'' + \lambda_0 y_0 = 0, y_0(0) = y_0(\pi) = 0; y_1'' + \lambda_0 y_1 = -\lambda_1 y_0 + x^2 y_0$, $y_1(0) = y_1(\pi) = 0$. The unperturbed problem (that for y_0) has the solution $y_0(x) = \sin \lambda_0^{1/2} x$, $\lambda_0 = n^2$, $n = 1, 2, 3, \ldots$, so the equation for y_1 is $y_1'' + n^2 y_1 = (x^2 - \lambda_1) \sin nx$ and the general solution is

$$y_1(x) = \left(A + \frac{x^2}{4n^2}\right) \sin nx + \left(B + \frac{2n^2\lambda_1 + 1}{4n^3} x - \frac{x^3}{6n}\right) \cos nx$$

where A, B are arbitrary constants. The boundary conditions are satisfied if $B = 0$ and $\lambda_1 = (\pi^2/3) - (1/2n^2)$. A is left unspecified, but since eigenfunctions are only unique up to multiplicative constants, the term $\epsilon A \sin nx$ in y resulting from $A \sin nx$ in y_1 is not distinct from the term $\sin nx$ in y_0. Thus, without loss of generality we set $A = 0$ and obtain

$$y(x; \lambda, \epsilon) = \left(1 + \frac{\epsilon x^2}{4n^2}\right) \sin nx + \epsilon \frac{\pi^2 x - x^3}{6n} \cos nx + 0(\epsilon^2)$$

$$\lambda = n^2 + \epsilon \left(\frac{\pi^2}{3} - \frac{1}{2n^2}\right) + 0(\epsilon^2)$$

Higher order corrections to the eigenfunctions and eigenvalues follow in similar fashion, but, of course, the calculation rapidly becomes burdensome.

Another situation in which regular perturbation expansions are often successful is for weakly nonlinear equations, a typical example being the following initial-value problem for an otherwise linear oscillator subject to small nonlinear damping

$$y'' + y = -\epsilon y'^2, \quad y(0) = 1, \quad y'(0) = 0 \tag{6.8-5}$$

where $0 < \epsilon \ll 1$. The same form of perturbation expansion as before, eq. (6.8-2), leads to a hierarchy of linear (and therefore, solvable) initial-value problems $y_0'' + y_0 = 0$, $y_0(0) = 1$, $y_0'(0) = 0$; $y_1'' + y_1 = -y_0'^2$, $y_1(0) = y_1'(0) = 0$; $y_2'' + y_2 = -2y_0'y_1'$,

$y_2(0) = y_2'(0) = 0; \ldots$ The total solution is

$$y(x;\epsilon) = \cos x + \frac{\epsilon}{3}(\sin^2 x + 2\cos x - 2)$$

$$+ \frac{\epsilon^2}{18}(3\cos^3 x + 13\cos x + 3x\sin x + 8\sin^2 x - 16) \quad (6.8\text{-}6)$$

An interesting feature of this solution is the appearance of a *secular term* $(\epsilon^2/6)x \sin x$ which grows without bound as $x \to \infty$ and describes a nonperiodic but oscillatory motion. Its origin is in the forcing function for y_2, which, because of the form of y_0 and y_1, contains a term $4/3 \cos x$. This produces resonance, as in section 6.3.2, eqs. (6.3-14), (6.3-15). In some problems where this behavior occurs, such as that for an undamped oscillator with weakly nonlinear spring force, the correct motion is known to be periodic. A special type of expansion, known as an asymptotic expansion, needs then to be introduced and the terms exhibiting unacceptable secular growth are suppressed by a technique usually associated with the names Poincaré, Lighthill, and Kuo (Chapters 12, 13, 14).

Yet another sort of perturbation method relies on an asymptotic expansion from the start to generate approximate solutions to equations of the form $y'' + \epsilon^{-2} f(x) y = 0$. The fuller treatment deserved by this WKB method is given in Chapter 12.

6.8.2 Variational Methods

Many of the basic problems in mathematical physics arise directly from or can be posed in terms of the *calculus of variations*, whose central task is to find that function or functions which renders some *functional* stationary. The fundamental concepts are covered in Chapter 19 and, for example, in Chapter 4 of Ref. 6-9. A functional is a quantity dependent, not on a continuous variable (as is a function), but rather on a function (or functions). An elementary example is provided by the arc length between the origin and some point x_1, y_1 of the x, y-plane measured along a prescribed curve whose equation is $y = y(x)$, the functional being the integral $I[y(x)] = \int_0^x \{1 + [y'(x)]^2\}^{1/2}\, dx$. Admissible functions $y(x)$ are those which pass through the end points $(0, 0)$, (x_1, y_1). Euclidean geometry shows that there is one special function, namely that describing a straight line between the endpoints, for which the integral takes on a minimum value. Generally, functions $y(x)$ which result in the functional becoming a minimum, a maximum, or having an inflection point are said to render the functional stationarv.

The connection between the variational calculus and ordinary differential equations is revealed by studying the functional

$$I(y(x)) = \int_{x_0}^{x_1} [p(x)y'^2(x) - q(x)y^2(x) + 2r(x)y(x)]\, dx \qquad (6.8\text{-}7)$$

It is profitable to ask whether or not we can select, among all functions $y(x)$ such that $y''(x)$ exists and is continuous everywhere in $x_0 \leqslant x \leqslant x_1$, and which satisfy $y(x_0) = a, y(x_1) = b$, that one function which makes stationary the functional, compared to neighboring admissable functions (i.e., those which satisfy the boundary conditions and differ from the solution function by a sufficiently small amount everywhere in the interval of interest). Suppose now that $z(x)$ is the desired function and $w(x)$ is any continuous differentiable function satisfying $w(x_0) = w(x_1) = 0$. If $y(x) = z(x) + \epsilon w(x)$ where ϵ is a constant, then the value of I depends continuously on ϵ and since $z(x)$ makes the integral stationary, we require $[\partial I(\epsilon)/\partial \epsilon]_{\epsilon=0} = 0$. Direct calculation yields

$$\left[\frac{\partial}{\partial \epsilon} I(\epsilon)\right]_{\epsilon=0} = \left\{\frac{\partial}{\partial \epsilon} \int_{x_0}^{x_1} [p(z' + \epsilon w')^2 - q(z + \epsilon w)^2 + 2r(z + \epsilon w)]\, dx\right\}_{\epsilon=0}$$

$$= \left\{\int_{x_0}^{x_1} [2pw'(z' + \epsilon w') - 2qw(z + \epsilon w) + 2rw]\, dx\right\}_{\epsilon=0}$$

$$= 2 \int_{x_0}^{x_1} (pw'z' - qwz + rw)\, dx$$

$$= 2\,(pwz')\Big|_{x_0}^{x_1} - 2 \int_{x_0}^{x_1} [(pz')' + qz - r]\, w\, dx$$

using integration by parts. The integrated term here vanishes because of homogeneous boundary conditions on w, so evidently the last integral must also vanish and this requires

$$(pz')' + qz = r \tag{6.8-8}$$

because otherwise there must (by continuity) exist some sufficiently small subinterval in which $(pz')' + qz - r$ is one-signed. Selection of any continuous w which is the same sign in that subinterval and vanishes elsewhere, provides a contradiction. Recalling that $z(x_0) = a$, $z(x_1) = b$, we then see that the required function that makes the integral stationary is, necessarily, a solution of the self-adjoint boundary value problem consisting of eq. (6.8-8) together with inhomogeneous boundary conditions. It does not follow that the unique solution of the boundary value problem (if it exists) is necessarily that precise function which minimizes the integral. It may for example maximize it, or give it inflection point behavior. Distinguishing between these possibilities can be difficult, but in many applications this degree of sophistication is unnecessary.

This equivalence between solving boundary (or initial) value problems for ordinary differential equations, and searching for the function which renders a functional stationary, provides a basis for a variety of effective approximate techniques which are associated with the *direct method of the variational calculus*. Here one seeks an approximation to some desired function $y(x)$ in the form of a linear superposition of suitably pre-selected trial or comparison functions $\phi_i(x)$ where the ϕ_i's are such that the boundary conditions are satisfied exactly; the approximate method attempts to make the differential equation be obeyed (in contrast to other methods in which the equation is obeyed exactly, but boundary conditions approximately). The starting point is

$$y(x) \approx \sum_{i=1}^{N} c_i \phi_i(x) = c_1 \phi_1(x) + c_2 \phi_2(x) + \cdots + c_N \phi_N(x) \tag{6.8-9}$$

where the manner of determination of the N coefficients c_i is what distinguishes the various methods. If, for example, the ϕ_i's are the first N eigenfunctions of a complete orthonormal set, then the choice $c_i = \langle \phi_i, y \rangle$ is best in a mean square sense, that is, this choice minimizes the mean square error between $y(x)$ and the right hand member of eq. (6.8-9) over the interval of interest (for proof and further details, see Ref. 6-9). Generally, once the ϕ_i's have been selected (the adequacy of the overall approximation depends strongly on a wise choice) the problem is ultimately reduced to one of algebra. Whereas any orthonormal and complete set of eigenfunctions will do for large enough N, it is far more convenient to utilize low degree polynomials, or sometimes Fourier modes. By starting with $N = 1$ and then increasing it, one can also generate a sequence of approximations, which may in fact converge to the exact solution, but proof of convergence is subtle (see, for example, Ref. 6-19).

In the method of Rayleigh and Ritz, one introduces eq. (6.8-9) into the equivalent variational problem and then adjusts the constants $c_1, c_2, \ldots, c_N$ so as to obtain a stationary value for the functional. The corresponding value of the right-hand member of eq. (6.8-9) is then the approximate solution. A simple example is provided by $(x^2 y')' + x^2 y = 0$, in $0 \leqslant x \leqslant \pi$, $y(0) = 1$, $y(\pi) = 0$. Instead of solving directly, we look for that function $y(x)$ which satisfies $y(0) = 1$, $y(\pi) = 0$ and makes stationary

$$I(y(x)) = \int_0^{\pi} x^2 [y'^2(x) - y^2(x)] \, dx \tag{6.8-10}$$

A suitable version of eq. (6.8-9) is $z(x) = (1 - x/\pi)(1 + ax + bx^2 + \cdots)$ which satisfies the boundary conditions for any values of $a, b, \ldots$. We content ourselves with the simplest nontrivial approximation by retaining the term through a only. Then substitution of z for y in eq. (6.8-10) leads to

$$I(a) = \int_0^{\pi} x^2 \left[\left(a - \frac{1}{\pi} - \frac{2ax}{\pi} \right)^2 - \left(1 - \frac{x}{\pi} \right)^2 (1 + ax)^2 \right] dx$$

where the dependence of I on the unknown parameter a is shown. $I(a)$ is stationary when a is such that $\partial I(a)/\partial a = 0$, which yields $a = -7(10 - \pi^2)/4\pi(14 - \pi^2)$. A quadratic approximation to $y(x)$ is then $z(x) \doteq 1 - 0.333x + 0.00473x^2$. In view of the simplicity of the trial function, this compares reasonably well (typical errors are of order 30% or less) with the exact solution in terms of spherical Bessel functions, which is $y(x) = j_0(x) = x^{-1} \sin x$. A better approximation is obtainable by taking $z(x) = [1 - (x/\pi)] \cdot (1 + ax + bx^2)$ and then finding a, b such that $\partial I(a, b)/\partial a = 0$, $\partial I(a, b)/b = 0$. A different application of the Rayleigh-Ritz procedure enables one to find an approximation to the smallest eigenvalue when the eigenfunctions are unknown and unduly difficult to find. The result of eq. (6.6-13) motivates the definition of the *Rayleigh quotient* as the functional

$$R(z(x)) = \frac{\langle z, \mathcal{L}z\rangle}{\langle z, rz\rangle} = \frac{\displaystyle\int_a^b rz\mathcal{L}z\, dx}{\displaystyle\int_a^b r^2 z^2\, dx} \tag{6.8-11}$$

where $\mathcal{L}u = \lambda ru$ is the eigenvalue equation and we suppose that the eigenvalues are non-negative. If η is the smallest eigenvalue and $w(x)$ its associated eigenfunction then R takes on this smallest value when $z(x) = w(x)$ (for proof see Ref. 6-11). When $w(x)$ is unknown, we obtain an approximation to η (and sometimes larger eigenvalues as well) by setting, as in eq. (6.8-9), $z(x) = \phi_1(x) + c_2\phi_2(x) + \cdots + c_N\phi_N(x)$ and then extremalizing R as a function of $c_2, \ldots, c_N$. An estimate of the error in such a scheme is readily obtainable when $\mathcal{L}$ is self-adjoint. Set $z(x) = w(x) + \epsilon\phi(x)$ so that ϵ measures the order of magnitude of the error in the eigenfunction. Direct calculation (using the fact that w is an eigenfunction, $\mathcal{L}w = \lambda rw$, and that $\mathcal{L}$ is self-adjoint) shows that

$$\begin{aligned} R(z) &= \frac{\langle w + \epsilon\phi, \lambda rw + \epsilon\mathcal{L}\phi\rangle}{\langle w + \epsilon\phi, r(w + \epsilon\phi)\rangle} \\ &= \frac{\lambda\langle w, rw\rangle + 2\epsilon\lambda\langle w, r\phi\rangle + \epsilon^2\langle\phi, \mathcal{L}\phi\rangle}{\langle w, rw\rangle + 2\epsilon\langle w, r\phi\rangle + \epsilon^2\langle\phi, \phi\rangle} \\ &= \lambda + \epsilon^2 \frac{\langle\phi, \mathcal{L}\phi\rangle - \lambda\langle\phi, \phi\rangle}{\langle w, rw\rangle + 2\epsilon\langle w, r\phi\rangle + \epsilon^2\langle\phi, \phi\rangle} \end{aligned} \tag{6.8-12}$$

Consequently, the error in estimating the eigenvalue is smaller (of order ϵ^2) than that in approximating the eigenfunction. This means that relatively crude trial functions can give accurate estimates of the eigenvalue, but, conversely, one cannot, with confidence, use the best trial function as a good approximation to the eigenfunction (for another method which circumvents this difficulty, see Ref. 6-6, p. 75). An illustrative example is provided by $y'' + \lambda y = 0, y'(0) = y'(1) = 0$ which admits the exact eigenfunctions $y = \cos \lambda^{1/2}x$, where $\lambda = n^2\pi^2$ and $n = 0, 1, 2, \ldots$. The Rayleigh quotient is

$$R = -\frac{\int_0^1 z z'' \, dx}{\int_0^1 z^2 \, dx} = \frac{\int_0^1 z'^2 \, dx}{\int_0^1 z^2 \, dx}$$

using integration by parts. The simplest trial function z obeying $z'(0) = z'(1) = 0$ is, of course, z = constant and this gives, without error, the eigenvalue 0. To approximate to $\lambda_1 (= \pi^2)$, we introduce $z = a + \frac{1}{2}x^2 - \frac{1}{3}x^3$, ($a$ = constant) so that $z' = x(1 - x)$. The corresponding value of $R(a)$ is found to be $(30a^2 + 5a + \frac{13}{42})^{-1}$ and the value of a which minimizes this is $a = -\frac{1}{12}$. Although $z = -\frac{1}{12} + \frac{1}{2}x^2 - \frac{1}{3}x^3$ is not a very good approximation to $\cos \pi x$, $R(-\frac{1}{12}) \doteq 9.88$ is an excellent approximation to the eigenvalue $\pi^2 \doteq 9.87$.

Another very widely employed method, associated with the name of Galerkin, is not restricted to variational problems but when applied to them yields results very similar to those obtained by the Rayleigh-Ritz technique (c.f. Refs. 6-19, 6-21). Galerkin's method exploits the idea of orthogonality to determine the coefficients in eq. (6.8-9). Suppose, for example, that the functions $\phi_1, \phi_2, \ldots, \phi_N$, were selected as the first N functions of a complete orthonormal set for expressing the solution of $\mathcal{L}y = 0$ subject to homogeneous boundary conditions. The exact solution, achieved as $N \to \infty$, is then such that $\mathcal{L}y$ is orthogonal to each function $\phi_1, \phi_2, \phi_3, \ldots$ over the interval in question. For an approximation based on finite N, the N constants c_i are chosen in such a way that $z = \sum_{i=1}^{N} c_i\phi_i$ is orthogonal to $\phi_1, \phi_2, \ldots, \phi_N$, i.e., there are N orthogonality conditions to be satisfied

$$\int_a^b \phi_i(x) \mathcal{L}z(x) \, dx = 0 \quad \text{for } i = 1, 2, 3, \ldots, N \tag{6.8-13}$$

The situation is clarified by an example: $y'' + y - x = 0$ in $0 \leqslant x \leqslant 1$, $y(0) = y'(1) = 0$. A suitable quadratic polynomial approximation to y, of the form of eq. (6.8-9), is obtained by setting $z'(x) = (1 - x)(a + bx)$. This gives

$$z(x) = a\phi_1(x) + b\phi_2(x) = a\left(x - \frac{x^2}{2}\right) + b\left(\frac{x^2}{2} - \frac{x^3}{3}\right)$$

(Here ϕ_1, ϕ_2 are linearly independent but not orthogonal.) The two orthogonality conditions take the form

$$\int_0^1 \left[-a\left(1 - x + \frac{x^2}{2}\right) + b\left(1 - 2x + \frac{x^2}{2} - \frac{x^3}{3}\right) - x\right]\left(x - \frac{x^2}{2}\right) dx = 0$$

$$\int_0^1 \left[-a\left(1 - x + \frac{x^2}{2}\right) + b\left(1 - 2x + \frac{x^2}{2} - \frac{x^3}{3}\right) - x\right]\left(\frac{x^2}{2} - \frac{x^3}{3}\right) dx = 0$$

which gives $72a + 17b = -75$ and $119a + 58b = -147$. Solving for a, b yields $a \doteq -0.8597$, $b \doteq -0.7706$ so the approximate solution is $z(x) = -0.8597\,x + 0.0446\,x^2 + 0.2569\,x^3$. This compares very well with the exact solution, which is $y(x) = x - \sin x/\cos 1$. For example,

at $x = 0.25$, $z \doteq -0.2081$ whereas $y \doteq -0.2079$

at $x = 0.50$, $z \doteq -0.3866$ whereas $y \doteq -0.3873$

at $x = 0.75$, $z \doteq -0.5115$ whereas $y \doteq -0.5115$

at $x = 1$, $z \doteq -0.5582$ whereas $y \doteq -0.5574$

As a generalization of the Galerkin procedure, the functions to which $\mathcal{L}z$ is made orthogonal need not be the ϕ_i-functions themselves; they can be any set of suitable functions. Popular choices are functions each of which is nonzero only on a subinterval, or delta functions (in this last case we have a collocation procedure; see for example Ref. 6-22).

Variational methods receive further consideration in Chapters 17-20.

6.8.3 Numerical Methods

Problems which are not readily amenable to approximate solution by either perturbation or variational techniques may well yield to a numerical solution. The construction of ingenious numerical methods has proceeded very far and detailed consideration is given in Chapters 18, 19 (see section 18.6 especially). Here we mention only the rather general method of successive approximations due to Picard. It is useful for the general n^{th}-order quasi-linear differential equation or system of n first-order equations. Since it is straightforward to generalize to higher order, we restrict attention to the first order problem

$$y'(x) = f(x, y), \quad y(x_0) = a \tag{6.8-14}$$

The exact problem can be expressed as an equivalent integral equation

$$y(x) = a + \int_{x_0}^{x} f[\xi, y(\xi)]\, d\xi \tag{6.8-15}$$

and Picard's iterative method is based on evaluating the nth approximation to $y(x)$, say $y_n(x)$, by inserting $y_{n-1}(x)$ for y in the integral

$$y_n(x) = a + \int_{x_0}^{x} f(\xi, y_{n-1}(\xi))\, d\xi \tag{6.8-16}$$

In some situations, the integration can be carried out explicitly and then an analytic approximation ensues. However, more generally, the integration can be handled numerically by a wide variety of techniques. The conditions under which this process converges to the exact solution were stated in section 6.1.2.

6.9 ORDINARY DIFFERENCE EQUATIONS

6.9.1 Preliminary Discussion

Our concern up to this point has been with functions y defined on a continuum in x, usually an interval $a \leqslant x \leqslant b$; however, functions arise which are only defined, or only of interest, when x takes on a discrete set of values. Thus, whereas differential equations can be regarded as relating values of the function at points separated by an infinitesimal amount (the differential), *difference equations* are those involving y at values of the argument $x_1, x_2, x_3, \ldots$ where the difference between two successive points is some finite number Δx. Without essential loss of generality, it suffices to consider $\Delta x = 1$. Suppose then that the real number y_n is the value of $y(x)$ when the argument is $x + n$, where x is a fixed number, and $n = 0, 1, 2, 3, \ldots$. The quantity

$$\Delta y(x) \equiv y(x+1) - y(x) \tag{6.9-1}$$

is called the *first forward difference* of $y(x)$ where the adjective 'first' is used because $\Delta y(x)$ requires a knowledge of y at x and only one other point; this is a 'forward' difference because the other point corresponds to a larger value of x. As will be seen, $\Delta y(x)$ is an approximation to the first derivative of y with respect to x. Higher order differences are defined in an obvious way: $\Delta^{k+1}y(x) = \Delta[\Delta^k y(x)]$ so that, for example, $\Delta^2 y(x) = \Delta[\Delta y(x)] = \Delta y(x+1) - \Delta y(x) = y(x+2) - 2y(x+1) + y(x)$. An ordinary difference equation is simply a relation of equality involving the differences of some function of a single discrete variable. Such equations arise in a number of contexts, a few examples being as follows.

In finding the Laplace transform of t^n where n is a positive integer, one is led, after a simple substitution to consider a function $\Gamma(n)$ defined by an integral of the form

$$\Gamma(n) = \int_0^\infty \xi^{n-1} e^{-\xi}\, d\xi, \qquad n = 1, 2, 3, \ldots \tag{6.9-2}$$

which is Euler's gamma function (it can be defined for all positive real n and non-integral negative n but our immediate interest is in just the positive integers). An integration by parts shows that $\Gamma(n)$ satisfies the difference equation $\Gamma(n+1) = n\Gamma(n)$ or in difference notation: $\Delta\Gamma(n) = (n-1)\Gamma(n)$. This is a first order variable coefficient linear homogeneous difference equation. By direct calculation from eq. (6.9-2), $\Gamma(1) = 1$ so by substitution, $\Gamma(2) = \Gamma(1) = 1$, $\Gamma(3) = 2\Gamma(2) = 2$, $\Gamma(4) = 3\Gamma(3) = 3 \cdot 2$ and generally, $\Gamma(n+1) = n!$ for $n = 1, 2, 3, \ldots$.

Another example is the famous problem of the 'gamblers ruin' (which is equivalent to a certain random walk in one dimension). A gambler, initially with n dollars ($n = 0, 1, 2, \ldots$), makes a series of one dollar bets with his opponent, who starts with $N - n$ dollars, the probability of the first gambler's winning each bet being p, a constant (the probability of losing, being $1 - p$). For different n, what is the probability, say $P(n)$ or for notational simplicity, P_n, that the gambler will bankrupt his opponent? At a moment when the gambler has n dollars, so that he eventually bankrupts the opponent with probability P_n, he can either win the next bet, with probability p, thereby increasing his purse to $n + 1$ dollars, or else lose the bet, with probability $1 - p$, thereby decreasing his purse to $n - 1$ dollars. Consequently, the difference equation satisfied by P_n is $P_n = pP_{n+1} + (1 - p)P_{n-1}$, or in difference notation, $p\Delta^2 P_n + (2p - 1)\Delta P_n = 0$, which is a second order, constant coefficient linear, homogeneous difference equation. Two conditions which P_n clearly satisfies are $P_0 = 0$ and $P_N = 1$ and this suffices to determine a unique solution by the methods developed below.

Still another way in which difference equations arise is as recurrence relations in the series solution of ordinary differential equations. A general example was seen in eq. (6.4-43). More specific examples of recurrence relations can be found in Chapter 7.

Finally, an extremely important motivation for studying difference equations is provided by their natural occurrence when derivatives are approximated by finite differences in the numerical solution of differential equations. Indeed derivatives can be written as the limits of differences as follows. The difference notation is first broadened to allow definition on a discrete set of values of x which differ from each other by some arbitrary (not necessarily integer) amount h: $\Delta_h f(x) \equiv f(x + h) - f(x)$. Then clearly, $f'(x) = \lim_{h \to 0} h^{-1} \Delta_h f(x)$. Consequently, an approximation to the first-order differential equation $f'(x) = g(x)$, which is presumably accurate for small enough h, is the difference equation $\Delta_h f(x) = hg(x)$. This intimate relationship between derivatives and differences results in a strong parallelism of methodology for solving these two classes of equation. Much of the preceding terminology carries over directly to the solution of difference equations, so we confine ourselves to a treatment of only a few topics.

6.9.2 First-Order Linear Difference Equations

As for the corresponding differential equation, first-order linear difference equations are readily solved without difficulty. The general problem consists in finding that set of numbers $f(n)$, or f_n for short, such that

$$f_{n+1} - a_n f_n = b_n, \qquad n = \alpha, \alpha + 1, \alpha + 2, \ldots \tag{6.9-3}$$

where α is a constant and a_n, b_n are given sequences of numbers. Clearly, f_n is not yet fully determinate because the difference equation only relates the difference in the function to its value; analogously to an initial condition, we must also know the starting value of f at some point, which, for simplicity, will be taken as $n = \alpha$: suppose $f_\alpha = c$. Because the right hand member of eq. (6.9-3) is independent of f_n this difference equation is *inhomogeneous*. The solution is developed in compact form by introducing the product symbol Π and defining

$$\sigma_{n+1} = \prod_{i=\alpha}^{n} a_i = a_\alpha \cdot a_{\alpha+1} \cdot a_{\alpha+2} \cdots a_n, \qquad \sigma_\alpha = 1 \tag{6.9-4}$$

Then, so long as $\sigma_{n+1} \neq 0$, eq. (6.9-3) is the same as the constant coefficient difference equation

$$g_{n+1} - g_n = d_n, n = \alpha, \alpha + 1, \alpha + 2, \ldots$$

where

$$g_n = f_n/\sigma_n, \qquad d_n = b_n/\sigma_{n+1} \tag{6.9-5}$$

The initial condition $f_\alpha = c$ becomes $g_\alpha = c/\sigma_\alpha = c$ and the difference equation then shows that $g_{\alpha+1} = g_\alpha + d_\alpha = c + d_\alpha, g_{\alpha+2} = g_{\alpha+1} + d_{\alpha+1} = c + d_\alpha + d_{\alpha+1}$ and generally, $g_n = c + \sum_{i=\alpha}^{n-1} d_i$. Reverting to the original notation, we have

$$f_n = c\sigma_n + \sum_{i=\alpha}^{n-1} \sigma_n b_i/\sigma_{i+1}, \quad n = \alpha, \alpha + 1, \ldots \tag{6.9-6}$$

and this is the required solution. As an illustration, consider $f_{n+1} = nf_n + \beta^n, f_1 = 1$, where β is a constant. From eq. (6.9-4), $\sigma_{n+1} = n!$, so the solution is

$$f_n = (n - 1)! + (n - 1)! \sum_{i=1}^{n-1} \beta^i/i!$$

6.9.3 Second-Order Linear Difference Equations

Just as with differential equations, the step from first- to second-order difference equations is a major one. The constant coefficient case is however still completely tractable, and an appropriate starting point is the homogeneous equation

$$f_{n+1} + af_n + bf_{n-1} = 0, \qquad n = 1, 2, 3, \ldots \tag{6.9-7}$$

If f_0, f_1 are known then this equation determines (uniquely) $f_2, f_3, \ldots$ recursively. Any solution dependent on the two parameters f_0, f_1 must then be the most general solution. Since specification of f_0, f_1 is tantamount to specifying f_n and its first difference for $n = 0$, this problem is analogous to the initial-value problem in differential equations. The general solution is readily confirmed to be

$$f_n = Ac_1^n + Bc_2^n \tag{6.9-8}$$

where c_1, c_2 are the two distinct roots of $c^2 + ac + b = 0$ and A, B are arbitrary constants. For definiteness let $c_1 = \frac{1}{2}[-a + (a^2 - 4b)^{1/2}]$, $c_2 = \frac{1}{2}[-a - (a^2 - 4b)^{1/2}]$. Then if $c_1 \neq c_2$, the solution of the initial-value problem is

$$f_n = (c_1 - c_2)^{-1} \; [(f_1 - f_0 c_2)\, c_1^n - (f_1 - f_0 c_1)\, c_2^n] \tag{6.9-9}$$

If instead of initial values, boundary values are given, say f_0 and f_N, then the solution is

$$f_n = (c_1^N - c_2^N)^{-1} \; [(f_N - f_0 c_2^N)\, c_1^n - (f_N - f_0 c_1^N)\, c_2^n] \tag{6.9-10}$$

which reduces to eq. (6.9-9) when $N = 1$. When $c_1 = c_2 = c$, then eq. (6.9-10) must be altered, the correct form (from L'Hospital's rule) being

$$f_n = \frac{n}{N} f_N c^{n-N} - \frac{n - N}{N} f_0 c^n \tag{6.9-11}$$

The appropriate limit for eq. (6.9-9) when $c_1 = c_2 = c$ is obtained from this result by substituting $N = 1$.

The problems of the gamblers ruin is a simple example of a boundary value problem for eq. (6.9-7) and from eq. (6.9-10) the solution is $P_n = [1 - p^{-n} \cdot (1 - p)^n]/[1 - p^{-N}(1 - p)^N]$. When the odds are even, $p = \frac{1}{2}$, so $c_1 = c_2 = 1$; then eq. (6.9-11) shows that $P_n = n/N$.

No general method exists for writing down an explicit solution to the initial-value problem for an arbitrary second-order linear difference equation, but, as for differential equations, the situation is readily resolved if one solution to the homogeneous problem is known. Suppose then that

$$f_{n+1} + a_n f_n + b_n f_{n-1} = c_n, \qquad n = 1, 2, 3, \ldots \tag{6.9-12}$$

with f_0, f_1 given initial values. Let g_n be any known solution of the reduced equation

$$g_{n+1} + a_n g_n + b_n g_{n-1} = 0 \tag{6.9-13}$$

Cross multiplication followed by subtraction of these two equations (so as to eliminate the term in a_n) leads to $f_{n+1}g_n - f_n g_{n+1} + b_n(f_{n-1}g_n - f_n g_{n-1}) = c_n g_n$ and this is a first-order difference equation for the *Casorati determinant* (or Casoratian), K_n, of the two functions f_n, g_n which is defined by

$$K_n \equiv f_n g_{n+1} - f_{n+1} g_n \tag{6.9-14}$$

With this notation, K_n satisfies $K_n - b_n K_{n-1} = c_n g_n$ and with b_n, c_n, and g_n known the solution is obtainable by applying the result of section 6.9.2. With K_n determined, eq. (6.9-14) is then solved as a first-order linear difference equation for f_n, the solution being

$$f_n = g_n \left[\frac{f_0}{g_0} - \frac{K_0}{g_1 g_0} - \frac{K_1}{g_2 g_1} - \cdots - \frac{K_{n-1}}{g_n g_{n-1}} \right] \tag{6.9-15}$$

since eq. (6.9-14) shows that $K_0 = g_1 f_0 - g_0 f_1$, it is clear that f_n depends upon the initial conditions f_0, f_1, and when they are specified, the solution is unique. For the homogeneous case where $c_n \equiv 0$, we obtain $K_n = b_n K_{n-1}$ and eq. (6.9-15) shows that the general solution of the reduced problem is expressible as

$$f_n = g_n \left\{ \frac{f_0}{g_0} + (g_0 f_1 - g_1 f_0) \left[\frac{1}{g_1 g_0} + \frac{b_1}{g_2 g_1} + \frac{b_2 b_1}{g_3 g_2} + \cdots + \frac{b_{n-1} b_{n-2} \cdots b_1}{g_n g_{n-1}} \right] \right\} \tag{6.9-16}$$

To illustrate, the equation $f_{n+1} - (n+2) f_n + n f_{n-1} = 0$ admits $n!$ as a solution and eq. (6.9-16) shows that the general solution is $f_n = f_0(n!) + (f_1 - f_0)(n!) \cdot \sum_{i=1}^{n} (1/i!)$.

The Casoratian introduced above plays the same role for difference equations as the Wronskian does for differential equations. For example, a necessary condition that two functions defined on a discrete set of points be linearly dependent is that their Casoratian vanish identically, and a sufficient condition that they be linearly independent is that their Casoratian does not vanish identically on the set. (See Ref. 6-4 for further properties.) Operational methods, variation of parameters, Green's functions, and eigenvalue problems all have their finite difference analogues (see Refs. 6-6, 6-24 for details). A different connection between difference and differential equations is brought out in the method of *generating functions*, which introduces a function, say $y(z; x)$, so constructed that

$$y(z; x) = \sum_{n=0}^{\infty} f_n(x) z^n \tag{6.9-17}$$

where the f_n coefficients are construed as the solution of an associated finite difference equation or alternatively, $f_n(x)$ is the solution of an ordinary differential equation in the variable x; c.f., Chapter 7 for some examples. Application to the gamblers ruin is given in Ref. 6-6; here we consider a well-known three-term recurrence relation for Legendre polynomials (whose defining differential equation was introduced in section 6.6-3).

$$(n+1)P_{n+1}(x) - (2n+1)xP_n(x) + nP_{n-1}(x) = 0 \tag{6.9-18}$$

where $P_0(x) = 1$, $P_1(x) = x$. Instead of solving directly for P_n by the preceding method, we look for a generating function $y(z;x)$ whose coefficients in a MacLaurin expansion in powers of z are proportional to $P_n(x)$, i.e.,

$$y(z;x) = \sum_{n=0}^{\infty} P_n(x)\, z^n \tag{6.9-19}$$

An effort is now made to find a simple differential equation for $y(z;x)$. Multiplication of eq. (6.9-18) by z^n and summation over n leads to three terms, the first of which can then be manipulated as follows (provided differentiation of the series is allowable):

$$\sum_{n=0}^{\infty} (n+1)P_{n+1}(x)\, z^n = \sum_{n=0}^{\infty} \frac{d}{dz}\left[P_{n+1}(x)\, z^{n+1}\right]$$

$$= \frac{d}{dz}\left\{\left[\sum_{n=0}^{\infty} P_n(x)\, z^n\right] - P_0(x)\right\} = \frac{dy}{dz}$$

upon substitution of eq. (6.9-19). Similar operations applied to the other two terms enable derivation of an equation for y: $(1 - 2xz + z^2)\, dy/dz + y(z - x) = 0$, whose solution is the generating function $y(z;x) = (1 - 2xz + z^2)^{-1/2}$. This proves that the solution of the difference equation (6.9-18) can be written as

$$P_n(x) = \frac{1}{n!}\left\{\frac{d^n}{dz^n}\left[(1 - 2xz + z^2)^{-1/2}\right]\right\}_{z=0} \tag{6.9-20}$$

The results derivable from this expression coincide with those from Rodriques' formula, eq. (6.6-19).

Just as with differential equations, some classes of nonlinear difference equation are reducible by transformation to linear form, and are therefore readily solvable. An important example is the equation of Riccati type

$$f_{n+1}f_n + af_{n+1} + bf_n + c = 0 \tag{6.9-21}$$

The translation $f_n = g_n + A$ renders the problem homogeneous

$$g_{n+1} g_n + (a + A) g_{n+1} + (b + A) g_n = 0 \tag{6.9-22}$$

provided A is a root of $A^2 + (a + b) A + c = 0$. Then if $h_n = 1/g_n$ the problem for h_n is linear

$$(b + A) h_{n+1} + (a + A) h_n + 1 = 0 \tag{6.9-23}$$

Applications of this material as well as further development can be found in the cited references (expecially Refs. 6-4, 6-6, 6-12, 6-23, 6-24).

6.10 REFERENCES

6-1 Abramowitz, M., and Stegun, I., (editors), *Handbook of Mathematical Functions*, National Bureau of Standards, Applied Mathematics Series **55**, Wash., D.C., 1964.

6-2 Bailey, P. B., Shampine, L. F., and Waltman, P. E., *Nonlinear Two Point Boundary Value Problems*, Academic Press, New York, 1968.

6-3 Birkhoff, G., and Rota, G.-C., *Ordinary Differential Equations*, Ginn, Boston, 1962.

6-4 Brand, L., *Differential and Difference Equations*, Wiley, N.Y., 1966.

6-5 Carrier, G. F., Krook, M., and Pearson, C. E., *Functions of a Complex Variable-Theory and Technique*, McGraw-Hill, New York, 1966.

6-6 Carrier, G. F., and Pearson, C. E., *Ordinary Differential Equations*, Blaisdell, Waltham, Massachusetts, 1968.

6-7 Churchill, R. V., *Fourier Series and Boundary Value Problems*, McGraw-Hill, New York, 1941.

6-8 Coddington, E. A., and Levinson, N., *Theory of Ordinary Differential Equations*, McGraw-Hill, New York, 1955.

6-9 Courant, R., and Hilbert, D., *Methods of Mathematical Physics, vol. 1*, Interscience, New York, 1953.

6-10 Forsyth, A. R., *Theory of Differential Equations*, Cambridge Univ. Press, New York, 1906, Dover Publications (Reprint), New York, 1959.

6-11 Friedman, B., *Principles and Techniques of Applied Mathematics*, Wiley, New York, 1956.

6-12 Friedman, B., *Lectures on Applications-Oriented Mathematics*, Holden-Day Inc., California, 1969.

6-13 Gantmacher, F. R., *The Theory of Matrices*, 2 vols., Chelsea, New York, 1959.

6-14 den Hartog, J. P., *Mechanical Vibrations*, McGraw-Hill, New York, 1940.

6-15 Hildebrand, F. B., *Methods of Applied Mathematics*, Prentice-Hall, New York, 1965.

6-16 Ince, E. L., *Ordinary Differential Equations*, Dover Publications, 1944.

6-17 Ince, E. L., *Integration of Ordinary Differential Equations*, Oliver and Boyd, Edinburgh, 1956.

6-18 Kamke, E., *Differentialgleichungen Reeler Funktionen*, Chelsea, London, 1947.

6-19 Kantorovich, L. V., and Krylov, V. I., *Approximate Methods of Higher Analysis* (C. D. Benster, trans.), Interscience, New York, 1958.

6-20 McLachlan, N. W., *Ordinary Non-Linear Differential Equations in Engineering and Physical Sciences*, Oxford, Second Edition, 1958.

6-21 Mikhlin, S. G., *Variational Methods in Mathematical Physics*, Macmillan, New York, 1964.

6-22 Mikhlin, S. G., and Smolitskiy, K. L., *Approximate Methods for Solution of Differential and Integral Equations*, American Elsevier, New York, 1967.

6-23 Miller, K. S., *An Introduction to the Calculus of Finite Differences and Difference Equations*, Henry Holt & Co., New York, 1960.

6-24 Miller, K. S., *Linear Difference Equations*, W. A. Benjamin, New York, 1968.

6-25 Morse, P. M., and Feshbach, H., *Methods of Theoretical Physics*, Parts I, II, McGraw-Hill, New York, 1953.

6-26 Murphy, G. M., *Ordinary Differential Equations and Their Solutions*, D. Van Nostrand, Princeton, New Jersey, 1960.

6-27 Stoker, J. J., *Nonlinear Vibrations in Mechanical and Electrical Systems*, Interscience, New York, 1950.

6-28 Struble, R. A., *Nonlinear Differential Equations*, McGraw-Hill, New York, 1962.

6-29 Titchmarsh, E. C., *Eigenfunction Expansions*, Parts I, 1962, II, 1958, Oxford.

7

Special Functions

Victor Barcilon*

7.0 INTRODUCTION

This chapter on special functions is not meant as a substitute for the handbooks devoted to this subject. Rather, it constitutes an attempt to collect some of the most frequently used formulas involving special functions and to provide a thread for the reader interested in unifying what may seem to be disconnected results.

The desire to present special functions from a unified viewpoint is quite understandable and by no means new. Truesdell's difference-differential equation in Ref. 7-10 and Miller's use of Lie groups are two such recent attempts. In the present chapter, the theory of entire functions is used as the underlying framework to tie apparently unrelated formulas. This theory provides some general techniques for obtaining series, infinite product and integral representations, which can in turn be used to deduce the differential equations or the recurrence and functional relations satisfied by the special function under consideration, as well as its asymptotic expansion. However, it must be said that the useful special functions are too varied and numerous to fit into the straight-jacket of any one unifying framework, and the one presented here is not exempt from drawbacks.

7.0.1 Entire Functions

A function $f(z)$ of the complex variable z is an *entire function* if it is analytic in the finite z-plane.

The entire function $f(z)$ is of *order* ρ if

$$\rho = \overline{\lim_{r \to \infty}} \frac{\log \log M(r)}{\log r} \tag{7.0-1}$$

where $M(r) = \max_{|z|=r} |f(z)|$. The entire function $f(z)$ of positive order ρ is of

*Prof. Victor Barcilon, Dep't. of the Geophysical Sciences, University of Chicago, Chicago, Ill.

type τ if

$$\tau = \overline{\lim_{r \to \infty}} \frac{\log M(r)}{r^\rho} \tag{7.0-2}$$

If $\tau = \infty$, the function is of maximum type. Entire functions of order 1, type τ are referred to as *exponential functions* of type τ. The order and type characterize the growth of the function and appear explicitly in asymptotic expansions. However, it is possible to compute ρ and τ from the series representation. Indeed, if the entire function $f(z) = \sum_{n=0}^{\infty} a_n z^n$ is of finite order ρ, then

$$\rho = \overline{\lim_{n \to \infty}} \frac{n \log n}{\log (1/|a_n|)} \tag{7.0-3}$$

furthermore

$$\tau = \frac{1}{\rho} \overline{\lim_{n \to \infty}} |n!\, a_n|^{\rho/n} \left(\frac{n}{e}\right)^{1-\rho} \tag{7.0-4}$$

If $f(z)$ is an entire function, $f(z)$ and $f'(z)$ are of the same order and type.

If $f(z)$ is an entire function of order ρ, then

$$f(z) = z^m e^{P(z)} \prod_n E\left(\frac{z}{z_n}, k\right) \tag{7.0-5}$$

where $P(z)$ is a polynomial of degree $p \leqslant \rho$, z_n are the zeros and $E(z/z_n, k)$ are the Weierstrass primary factors (Hadamard factorization theorem). This representation is useful for the derivation of results involving zeros of entire functions.

If $f(z) = \sum_0^{\infty} a_n z^n$ is an exponential function of type τ, then the Borel transform $F(z) = \sum_0^{\infty} \frac{n!\, a_n}{z^{n+1}}$ is convergent for $|z| > \tau$. Furthermore, $f(z)$ can be represented thus

$$f(z) = \frac{1}{2\pi i} \int_{|z| = \tau + \epsilon} e^{zu} F(u)\, du \tag{7.0-6}$$

(Pólya representation.)

If $f(z)$ is an exponential function of type τ which is $L_2(-\infty, \infty)$ then

$$f(z) = \int_{-\tau}^{\tau} e^{izt} \hat{f}(t)\, dt \tag{7.0-7}$$

where the Fourier transform $\hat{f} \in L^2(-\tau, \tau)$ and vice-versa: this is the Paley-Wiener theorem. We shall often see that the Paley-Wiener integral representations deduced for $\hat{f} \in L^2(-\tau, \tau)$ are valid even if $\hat{f} \notin L^2(-\tau, \tau)$.

7.0.2 Notations

All the notations and definitions of the special functions used in this chapter are identical to those of Abramowitz and Stegun. The logarithmic function is defined thus

$$\ln z = \ln |z| + i \arg z \qquad |\arg z| < \pi \tag{7.0-8}$$

The Euler constant γ is defined by any one of the four equivalent formulas

$$\begin{aligned} \gamma &= -\int_0^\infty e^{-t} \ln t \, dt \\ &= \int_0^1 \frac{1 - e^{-t} - e^{-1/t}}{t} \, dt \\ &= -\lim_{z \to 0} \left[\int_z^\infty \frac{e^{-t}}{t} \, dt + \ln z \right] \\ &= \lim_{m \to \infty} \left[\sum_{k=1}^{m} \frac{1}{k} - \ln m \right] \end{aligned} \tag{7.0-9}$$

The numerical value of γ is 0.57721. . . .

7.1 EXPONENTIAL INTEGRAL AND RELATED FUNCTIONS

7.1.1 Definition

$$E_1(z) = \int_z^\infty \frac{e^{-t}}{t} \, dt \qquad (|\arg z| < \pi) \tag{7.1-1}$$

The path of integration excludes the origin and does not cross the negative real axis. $E_1(z)$ has a logarithmic branch point at $z = 0, \infty$. The z-plane is cut along the negative real axis.

$$E_1(z) = \int_1^\infty \frac{e^{-zt}}{t} \, dt \qquad (Re\, z > 0) \tag{7.1-2}$$

7.1.2 Series Expansion

Differentiating (7.1-1), expanding in power series, and then integrating term by term, we get (cf. def. of γ)

$$E_1(z) = -\gamma - \ln z - \sum_{n=1}^{\infty} \frac{(-1)^n z^n}{nn!} \qquad (|\arg z| < \pi) \tag{7.1-3}$$

Note that

$$E_1(z) + \gamma + \ln z = \int_0^z \frac{1 - e^{-t}}{t}\, dt \tag{7.1-4}$$

is an exponential function of type 1.

7.1.3 Asymptotic Expansion

Integrating (7.1-1) by parts we deduce that

$$E_1(z) \sim \frac{e^{-z}}{z} \left\{1 - \frac{1}{z} + \frac{2!}{z^2} - \frac{3!}{z^3} + \cdots\right\} \qquad (|\arg z| < \pi) \tag{7.1-5}$$

7.1.4 Sine and Cosine Integrals

$$\mathrm{Si}(z) = \int_0^z \frac{\sin t}{t}\, dt \tag{7.1-6}$$

$$\mathrm{Ci}(z) = \gamma + \ln z + \int_0^z \frac{\cos t - 1}{t}\, dt \tag{7.1-7}$$

Note that $\mathrm{Si}(z)$ and $\mathrm{Ci}(z) - \gamma - \log z$ are entire functions of exponential type 1.

$$\lim_{x \to \infty} \mathrm{Si}(x) = \frac{\pi}{2} \tag{7.1-8}$$

The series expansions are

$$\mathrm{Si}(z) = \sum_{n=0}^{\infty} \frac{(-1)^n z^{2n+1}}{(2n+1)(2n+1)!} \tag{7.1-9}$$

$$\mathrm{Ci}(z) = \gamma + \ln z + \sum_{n=0}^{\infty} \frac{(-1)^n z^{2n}}{2n(2n)!} \tag{7.1-10}$$

In order to obtain the asymptotic expansions of Si(z) and Ci(z), we rewrite (7.1-6) and (7.1-7) as follows:

$$\mathrm{Si}(z) = \frac{\pi}{2} - \int_z^\infty \frac{\sin t}{t}\,dt \qquad (|\arg z| < \pi) \tag{7.1-11}$$

$$\mathrm{Ci}(z) = -\int_z^\infty \frac{\cos t}{t}\,dt \tag{7.1-12}$$

Integrating by parts we get

$$\mathrm{Si}(z) = \frac{\pi}{2} - \frac{\cos z}{z} \sum_{n=0}^{\infty} \frac{(-1)^n\,(2n)!}{z^{2n}} - \frac{\sin z}{z^2} \sum_{n=0}^{\infty} \frac{(-1)^n\,(2n+1)!}{z^{2n}} \tag{7.1-13}$$

$$\mathrm{Ci}(z) = \frac{\sin z}{z} \sum_{n=0}^{\infty} \frac{(-1)^n\,(2n)!}{z^{2n}} - \frac{\cos z}{z^2} \sum_{n=0}^{\infty} \frac{(-1)^n\,(2n+1)!}{z^{2n}} \tag{7.1-14}$$

7.1.5 Paley-Wiener Representation

On account of (7.1-7) and (7.1-9), Si(x)/x is an exponential function of type 1 which is $L^2(-\infty, \infty)$. Therefore it has a Paley-Wiener representation, viz.

$$\frac{\mathrm{Si}(z)}{z} = -\int_0^1 \ln t \cos(zt)\,dt \tag{7.1-15}$$

7.2 GAMMA FUNCTION AND RELATED FUNCTIONS

7.2.1 Definition (Hankel)

$$\Gamma(z) = -\frac{1}{2i \sin \pi z} \int_C (-t)^{z-1} e^{-t}\,dt \tag{7.2-1}$$

where C is shown in Fig. 7.2-1 $\Gamma(z)$ is a meromorphic function with simple poles at $z = -n$ ($n = 0, 1, \ldots$) with residues $(-1)^n/n!$

$$\Gamma(z) = \int_0^\infty e^{-t}\,t^{z-1}\,dt \qquad (\mathrm{Re}\, z > 0) \tag{7.2-2}$$

$$\Gamma(n+1) = n! \qquad (n = 0, 1, 2, \ldots) \tag{7.2-3}$$

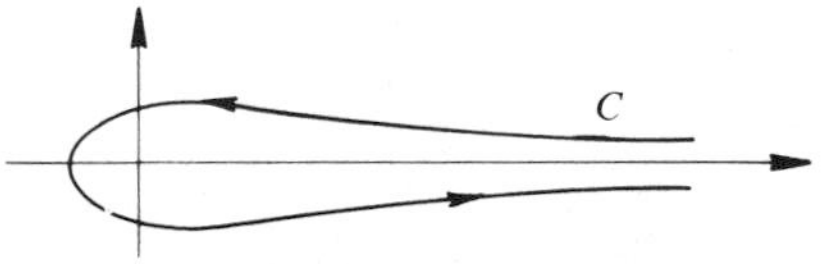

Fig. 7.2-1 Path of integration for integral representation of Γ-function.

7.2.2 Recurrence Relation

Integrating (7.2-1) by parts we get

$$\Gamma(z+1) = z\Gamma(z) \tag{7.2-4}$$

7.2.3 Reflection Formula

$$\Gamma(1-z)\,\Gamma(z) = \frac{\pi}{\sin \pi z} \tag{7.2-5}$$

This formula shows the relationship between the Γ-function and the trigonometric function. The simplest proof consists in proving (7.2-5) for $0 < Re\ z < 1$ and extending the result by analytic continuation.

7.2.4 Multiplication Formulas

$$\Gamma(2z) = (2\pi)^{-(1/2)}\, 2^{2z-(1/2)}\, \Gamma(z)\,\Gamma(z+\tfrac{1}{2}) \tag{7.2-6}$$

$$\Gamma(3z) = (2\pi)^{-1}\, 3^{3z-(1/2)}\, \Gamma(z)\,\Gamma(z+\tfrac{1}{3})\,\Gamma(z+\tfrac{2}{3}) \tag{7.2-7}$$

$$\Gamma(nz) = (2\pi)^{1/2(1-n)}\, n^{nz-(1/2)} \prod_{k=0}^{n-1} \Gamma\left(z+\frac{k}{n}\right) \tag{7.2-8}$$

7.2.5 Power Series

The reader is referred to Ref. 7-1. Series expansion are of limited use because of the poles.

7.2.6 The Function $1/\Gamma(z)$

From (7.2-5) it is clear that $1/\Gamma(z)$ is an entire function. In fact from (7.2-1) and (7.2-5)

$$\frac{1}{\Gamma(z)} = \frac{i}{2\pi}\int_C (-t)^{-z}\, e^{-t}\, dt \tag{7.2-9}$$

The above formula is very useful for deriving integral representations of several special functions.

$$\frac{1}{\Gamma(z)} = z e^{\gamma z} \prod_{n=1}^{\infty}\left(1+\frac{z}{n}\right) e^{-(z/n)} \tag{7.2-10}$$

$1/\Gamma(z)$ is of maximum exponential type.

7.2.7 Asymptotic Expansion

Using the method of steepest descent, (7.2-1) yields

$$\Gamma(z) \sim (2\pi)^{1/2} e^{-z} z^{z-(1/2)} \left[1 + \frac{1}{12z} + \frac{1}{288z^2} + \cdots\right] \qquad (z \to \infty \quad \text{in } |\arg z| < \pi) \tag{7.2-11}$$

$$\Gamma(x+1) = (2\pi)^{1/2} x^{x+(1/2)} \exp\left(-x + \frac{\theta}{12x}\right); \qquad (x > 0, \quad 0 < \theta < 1) \tag{7.2-11$'$}$$

7.2.8 Beta Function

$$B(z, w) = \frac{\Gamma(z)\,\Gamma(w)}{\Gamma(z+w)} \tag{7.2-12}$$

$$\begin{aligned} B(z, w) &= \int_0^1 t^{z-1}(1-t)^{w-1}\, dt; \qquad (Re\, z > 0, \quad Re\, w > 0) \\ &= 2\int_0^{\pi/2} (\sin t)^{2z-1} (\cos t)^{2w-1}\, dt \end{aligned} \tag{7.2-13}$$

7.2.9 Psi (Digamma) Function

$$\psi(z) = \frac{d}{dz} \ln \Gamma(z) = \frac{\Gamma'(z)}{\Gamma(z)} \tag{7.2-14}$$

$\psi(z)$ is a meromorphic function with residue (-1) at $0, -1, -2, \ldots$

$$\begin{aligned} \psi(z) &= -\gamma - \frac{1}{z} + \sum_1^\infty \frac{z}{n(z+n)}; \qquad (z \neq 0, -1, -2, \ldots) \\ &= \lim_{n \to \infty} \left[\ln n - \frac{1}{z} - \cdots - \frac{1}{z+n}\right] \end{aligned} \tag{7.2-15}$$

Special values of psi-function

$$\begin{aligned} \psi(1) &= -\gamma \\ \psi(n) &= -\gamma + \sum_1^{n-1} k^{-1} \qquad n \geq 2 \end{aligned} \tag{7.2-16}$$

Recurrence formula

$$\psi(z+1) = \psi(z) + 1/z \tag{7.2-17}$$

Reflection formula

$$\psi(1-z) = \psi(z) + \pi \cot \pi z \tag{7.2-18}$$

Duplication formula

$$\psi(2z) = \tfrac{1}{2}\psi(z) + \tfrac{1}{2}\psi(z+\tfrac{1}{2}) + \ln 2 \tag{7.2-19}$$

Asymptotic expansion

$$\psi(z) \sim \ln z - \frac{1}{2z} - \frac{1}{12z^2} + \frac{1}{120z^4} + \cdots \quad (z \to \infty \quad \text{in } |\arg z| < \pi) \tag{7.2-20}$$

7.3 ERROR FUNCTION AND RELATED FUNCTIONS

7.3.1 Definition

$$\operatorname{erf}(z) = \frac{2}{\sqrt{\pi}} \int_0^z e^{-t^2}\, dt \tag{7.3.1}$$

$$\operatorname{erfc}(z) = 1 - \operatorname{erf}(z)$$

The function $\operatorname{erf}(z)$ is an entire function of order 2 and type 1.

$$\operatorname{erf}(z) \to 1 \quad \text{as} \quad z \to \infty \quad \text{in } |\arg z| < \frac{\pi}{4} \tag{7.3-2}$$

7.3.2 Series Expansion

Differentiating (7.3-1), expanding in power series and integrating term by term, we get

$$\operatorname{erf}(z) = \frac{2}{\sqrt{\pi}} \sum_{n=0}^{\infty} \frac{(-1)^n z^{2n+1}}{n!(2n+1)} \tag{7.3-3}$$

7.3.3 Asymptotic Expansion

Since

$$\operatorname{erf}(z) = 1 - \frac{2}{\sqrt{\pi}} \int_z^{\infty} e^{-t^2}\, dt \tag{7.3-4}$$

by repeated integration by parts, we get

$$\operatorname{erfc}(z) = 1 - \operatorname{erf}(z) = \frac{e^{-z^2}}{\sqrt{\pi}\, z} \left\{1 - \frac{1}{2z^2} + \cdots\right\} \quad \left(|\arg z| < \frac{3\pi}{4}\right) \tag{7.3-5}$$

7.3.4 Integral Representation

$$\frac{\operatorname{erf}(\sqrt{x})}{\sqrt{x}}$$

which is an exponential function of type 1, admits the following Paley-Wiener like representation:

$$\frac{\operatorname{erf}(\sqrt{x})}{\sqrt{x}} = \frac{1}{\sqrt{\pi}} \int_0^1 \frac{e^{-xt}}{\sqrt{t}}\, dt \tag{7.3-6}$$

7.3.5 Fresnel Integrals—Definition

$$C(z) = \int_0^z \cos \frac{\pi t^2}{2}\, dt \tag{7.3-7}$$

$$S(z) = \int_0^z \sin \frac{\pi t^2}{2}\, dt \tag{7.3-8}$$

Note that

$$C(x) \to \tfrac{1}{2}, \quad S(x) \to \tfrac{1}{2} \qquad (x \to \infty)$$

$C(z)$ and $S(z)$ are entire functions of order 2 type $\frac{\pi}{2}$. Note that

$$C(z) + iS(z) = \frac{1+i}{2} \operatorname{erf}\left[\frac{\sqrt{\pi}}{2}(1-i)\, z\right].$$

7.3.6 Series Expansions

Integrating term by term the series expansion of the integrands we get

$$C(z) = \sum_{k=0}^{\infty} \frac{(-1)^k (\pi/2)^{2k} z^{4k+1}}{(4k+1)(2k)!} \tag{7.3-9}$$

$$S(z) = \sum_{k=0}^{\infty} \frac{(-1)^k (\pi/2)^{2k+1} z^{4k+3}}{(4k+3)(2k+1)!} \tag{7.3-10}$$

7.3.7 Asymptotic Expansions

$$C(z) = \frac{1}{2} - \int_z^{\infty} \cos \frac{\pi t^2}{2}\, dt \tag{7.3-11}$$

$$S(z) = \frac{1}{2} - \int_z^{\infty} \sin \frac{\pi t^2}{2}\, dt \tag{7.3-12}$$

After repeated integrations by parts, we get

$$C(z) \sim \frac{1}{2} + \sin \frac{\pi z^2}{2} \cdot \frac{1}{\pi z}\left[1 + \sum_{m=1}^{\infty} (-1)^m \frac{1 \cdot 3 \cdots (4m-1)}{(\pi z^2)^{2m}}\right] - \cos \frac{\pi z^2}{2}$$

$$\cdot \frac{1}{\pi z}\left[\sum_{m=0}^{\infty} (-1)^m \frac{1 \cdot 3 \cdots (4m+1)}{(\pi z^2)^{2m+1}}\right] \tag{7.3-13}$$

$$S(z) \sim \frac{1}{2} - \cos \frac{\pi z^2}{2} \cdot \frac{1}{\pi z}\left[1 + \sum_{m=1}^{\infty} (-1)^m \frac{1 \cdot 3 \cdots (4m-1)}{(\pi z^2)^{2m}}\right] - \sin \frac{\pi z^2}{2}$$

$$\cdot \frac{1}{\pi z}\left[\sum_{m=0}^{\infty} (-1)^m \frac{1 \cdot 3 \cdots (4m+1)}{(\pi z^2)^{2m+1}}\right] \quad \left(z \to \infty \text{ in } |\arg z| < \frac{\pi}{2}\right) \tag{7.3-14}$$

7.3.8 Integral Representation

$C(\sqrt{z})/\sqrt{z}$ and $S(\sqrt{z})/\sqrt{z}$ which are exponential functions of type $\pi/2$, but not $L^2(-\infty, \infty)$, admit the following Paley-Wiener like representation:

$$\frac{C(\sqrt{z})}{\sqrt{z}} = \frac{1}{\sqrt{2\pi}} \int_0^{\pi/2} \frac{\cos(zt)}{\sqrt{t}}\, dt$$

$$\frac{S(\sqrt{z})}{\sqrt{z}} = \frac{1}{\sqrt{2\pi}} \int_0^{\pi/2} \frac{\sin(zt)}{\sqrt{t}}\, dt \tag{7.3-15}$$

7.4 BESSEL FUNCTIONS

7.4.1 Definition

The Bessel functions $J_\nu(z)$, $Y_\nu(z)$, $H_\nu^{(1)}(z)$ and $H_\nu^{(2)}(z)$ are solutions of

$$z^2 \frac{d^2 w}{dz^2} + z \frac{dw}{dz} + (z^2 - \nu^2)\, w = 0 \tag{7.4-1}$$

7.4.2 Series Representations

$$J_\nu(z) = \left(\frac{z}{2}\right)^\nu \sum_{k=0}^{\infty} \frac{(-1)^k (z/2)^{2k}}{k!\,\Gamma(k+\nu+1)} \tag{7.4-2}$$

$$Y_\nu(z) = \frac{J_\nu(z)\cos\nu\pi - J_{-\nu}(z)}{\sin\nu\pi}, \qquad \nu \neq 0, \pm 1, \ldots \tag{7.4-3}$$

$$Y_n(z) = \lim_{\nu \to n} Y_\nu(z)$$

$$H_\nu^{(1)}(z) = J_\nu(z) + iY_\nu(z), \quad H_\nu^{(2)}(z) = J_\nu(z) - iY_\nu(z) \tag{7.4-4}$$

$J_\nu(z)$ is the Bessel function of the first kind of order ν; $Y_\nu(z)$ is the Bessel function of the second kind of order ν; $H_\nu^{(1)}$, $H_\nu^{(2)}$ are the Hankel functions, or Bessel functions of the third kind. The z-plane is cut along the negative real axis. As a function of z, $z^{-\nu} J_\nu(z)$ is an (even) exponential function of type 1. Therefore

$$J_\nu(z) = \frac{(z/2)^\nu}{\Gamma(\nu+1)} \prod_{k=1}^{\infty} \left(1 - \frac{z^2}{j_{\nu,k}^2}\right) \tag{7.4-5}$$

where $j_{\nu,k}$ are the zeros of $J_\nu(z)$. As a function of ν, $J_\nu(z)$, $Y_\nu(z)$ and $H_\nu(z)$ are exponential functions of maximum type.

$$Y_0(z) = \frac{2}{\pi}\left\{\ln\frac{z}{2} + \gamma\right\} J_0(z) - \frac{2}{\pi}\sum_{k=0}^{\infty} \frac{\{\psi(k+1)+\gamma\}\left(-\dfrac{z^2}{4}\right)^k}{(k!)^2} \tag{7.4-6}$$

$$Y_n(z) = -\frac{(z/2)^{-n}}{\pi}\sum_{k=0}^{n-1} \frac{(n-k-1)!}{k!}\left(\frac{z^2}{4}\right)^k + \frac{2}{\pi}\ln\left(\frac{z}{2}\right) J_n(z)$$

$$-\frac{(z/2)^n}{\pi}\sum_{k=0}^{\infty}\{\psi(k+1)+\psi(n+k+1)\}\frac{\left(-\dfrac{z^2}{4}\right)^k}{k!(n+k)!} \tag{7.4-7}$$

where $\psi(m+1) = -\gamma + 1 + \cdots + (1/m)$ is the psi-function.

7.4.3 Differential Equations—Wronskians

$J_\nu(z)$ and $Y_\nu(z)$ are linearly independent solutions of (7.4-1). The Wronskian

$$W\{J_\nu(z), Y_\nu(z)\} = \frac{2}{\pi z} \tag{7.4-8}$$

$H_\nu^{(1)}(z)$ and $H_\nu^{(2)}(z)$ are also linearly independent solutions of (7.4-1). For $\nu \neq$ integer, $J_\nu(z)$ and $J_{-\nu}(z)$ are also linearly independent solutions and

$$W\{J_\nu(z), J_{-\nu}(z)\} = -\frac{2\sin\nu\pi}{\nu\pi} \tag{7.4-9}$$

$$J_{-n}(z) = (-1)^n J_n(z)$$
$$Y_{-n}(z) = (-1)^n Y_n(z) \tag{7.4-10}$$

If $\mathcal{C}_\nu(z)$ denotes a linear combination of $J_\nu(z)$, $Y_\nu(z)$, $H_\nu^{(1)}(z)$ and $H_\nu^{(2)}(z)$, then the solution of

$$z^2 \frac{d^2 w}{dz^2} + (1 - 2p) z \frac{dw}{dz} + (\lambda^2 q^2 z^{2q} + p^2 - \nu^2 q^2) w = 0 \tag{7.4-11}$$

is

$$w = z^p \mathcal{C}_\nu(\lambda z^q) \tag{7.4-12}$$

7.4.4 Integral Representations

From the definition of the beta function $B(k + \frac{1}{2}, \nu + \frac{1}{2})$ and (7.4-2), we deduce the Poisson representation

$$J_\nu(z) = \frac{(z/2)^\nu}{\sqrt{\pi}\,\Gamma(\nu + \frac{1}{2})} \int_{-1}^{+1} (1 - t^2)^{\nu-(1/2)} \cos zt\, dt \qquad \operatorname{Re} \nu > -\frac{1}{2}, \quad |\arg z| < \pi \tag{7.4-13}$$

$$= \frac{(z/2)^\nu}{\sqrt{\pi}\,\Gamma(\nu + \frac{1}{2})} \int_0^\pi \cos(z \cos\theta) \sin^{2\nu}\theta\, d\theta \tag{7.4-14}$$

For $\operatorname{Re} \nu > 0$, the above formula is a consequence of the Paley-Wiener theorem for the exponential function $z^{-\nu} J_\nu(z)$. (7.4-13) is a special case of the Pólya integral representation in terms of the Borel transform.

Since $J_n(z)$ is also of exponential type and $L_2(-\infty, \infty)$ on the real axis [cf. 7.0.7], we can get another integral representation [Bessel Representation]

$$J_n(z) = \frac{1}{\pi} \int_0^\pi \cos(z \sin\theta - n\theta)\, d\theta \tag{7.4-15}$$

$$= \frac{i^{-n}}{\pi} \int_0^\pi e^{iz\cos\theta} \cos(n\theta)\, d\theta \tag{7.4-16}$$

The above formulas can simply be deduced from the Laurent expansion of the generating function (7.2-24).

Using the Hankel integral representation (7.2-10) of $1/\Gamma(k + \nu + 1)$, we deduce the Sonine-Schläfli representation

$$J_\nu(z) = \frac{1}{\pi}\int_0^\pi \cos(z\sin\theta - \nu\theta)\,d\theta - \frac{\sin(\nu\pi)}{\pi}\int_0^\infty e^{-z\sinh t - \nu t}\,dt \qquad \left(|\arg z| < \frac{\pi}{2}\right) \tag{7.4-17}$$

which generalizes (7.4-15), (7.4-16). As a result of (7.4-17)

$$Y_\nu(z) = \frac{1}{\pi}\int_0^\pi \sin(z\sin\theta - \nu\theta)\,d\theta$$

$$- \frac{1}{\pi}\int_0^\infty \{e^{\nu t} + e^{-\nu t}\cos(\nu\pi)\}\,e^{-z\sinh t}\,dt \qquad \left(|\arg z| < \frac{\pi}{2}\right) \tag{7.4-18}$$

Finally, the following Sommerfeld representations are useful because they are valid for all z. If $\theta = u + iv$ and the contours C_2 and C_3 originate at $u_0 - i\infty$, where $Re\,(ze^{iu_0}) > 0$, and terminate at $-u_0 - \pi + i\infty$ and $\pi - u_0 + i\infty$ then

$$\begin{aligned} J_\nu(z) &= +\frac{1}{2\pi}\int_{C_1} e^{-iz\sin\theta + i\nu\theta}\,d\theta \\ H_\nu^{(1)}(z) &= +\frac{1}{\pi}\int_{C_2} e^{-iz\sin\theta + i\nu\theta}\,d\theta \\ H_\nu^{(2)}(z) &= +\frac{1}{\pi}\int_{C_3} e^{-iz\sin\theta + i\nu\theta}\,d\theta \end{aligned} \tag{7.4-19}$$

C_1 originates at $-u_0 - \pi + i\infty$ and terminates at $\pi - u_0 + i\infty$. These contours are shown in Fig. 7.4-5 for $|\arg z| < \pi$.

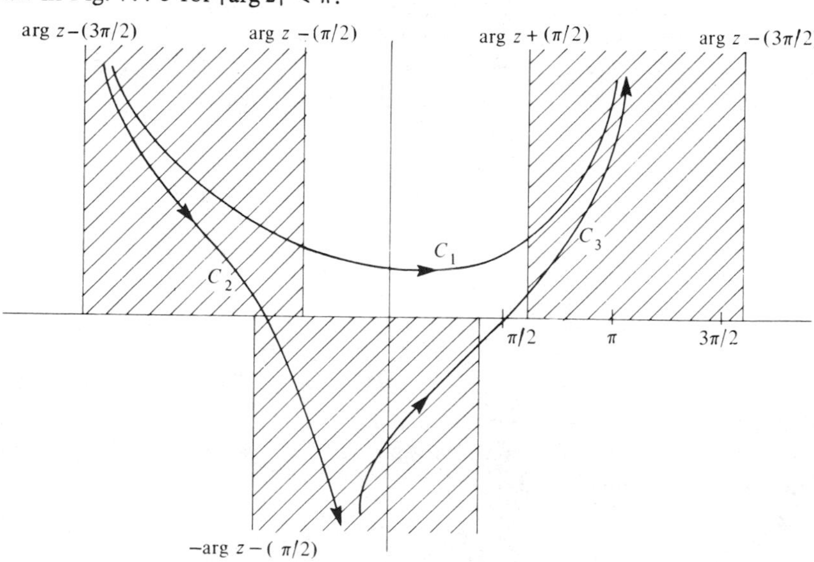

Fig. 7.4-5 Path of integrations for Sommerfeld representations of Bessel functions.

7.4.5 Recurrence Relations

$\mathcal{C}$ denotes $J, Y, H^{(1)}, H^{(2)}$ or any linear combination of these functions

$$\mathcal{C}_{\nu-1}(z) + \mathcal{C}_{\nu+1}(z) = \frac{2\nu}{z} \mathcal{C}_\nu(z) \tag{7.4-20}$$

$$\mathcal{C}_{\nu-1}(z) - \mathcal{C}_{\nu+1}(z) = 2\mathcal{C}'_\nu(z) \tag{7.4-21}$$

In particular

$$J'_0(z) = -J_1(z)$$

$$Y'_0(z) = -Y_1(z)$$

$$\left(\frac{1}{z}\frac{d}{dz}\right)^k z^\nu \mathcal{C}_\nu(z) = z^{\nu-k}\mathcal{C}_{\nu-k}(z), \quad k = 0, 1, \ldots \tag{7.4-22}$$

$$\left(\frac{1}{z}\frac{d}{dz}\right)^k z^{-\nu} \mathcal{C}_\nu(z) = (-1)^k z^{-\nu-k} \mathcal{C}_{\nu+k}(z) \tag{7.4-23}$$

7.4.6 Generating Function

$$\exp\left\{\frac{z}{2}\left(t - \frac{1}{t}\right)\right\} = \sum_{k=-\infty}^{\infty} t^k J_k(z), \qquad t \neq 0 \tag{7.4-24}$$

7.4.7 Asymptotic Approximations

Using Sommerfeld's integral representation and the method of steepest descent, we deduce that for ν fixed and $|z| \to \infty$

$$J_\nu(z) = \sqrt{\frac{2}{\pi z}} \left\{\cos\left(z - \frac{\nu\pi}{2} - \frac{\pi}{4}\right) + e^{|\mathrm{Im}\, z|}\, O\left(\frac{1}{|z|}\right)\right\}, \qquad |\arg z| < \pi \tag{7.4-25}$$

$$Y_\nu(z) = \sqrt{\frac{2}{\pi z}} \left\{\sin\left(z - \frac{\nu\pi}{2} - \frac{\pi}{4}\right) + e^{|\mathrm{Im}\, z|}\, O\left(\frac{1}{|z|}\right)\right\}, \qquad |\arg z| < \pi \tag{7.4-26}$$

$$H_\nu^{(1)}(z) = \sqrt{\frac{2}{\pi z}} \exp\left\{i\left(z - \frac{\nu\pi}{2} - \frac{\pi}{4}\right)\right\} \left\{1 + O\left(\frac{1}{|z|}\right)\right\}, \qquad -\pi < \arg z < 2\pi \tag{7.4-27}$$

$$H_\nu^{(2)}(z) = \sqrt{\frac{2}{\pi z}} \exp\left\{-i\left(z - \frac{\nu\pi}{2} - \frac{\pi}{4}\right)\right\} \left\{1 + O\left(\frac{1}{|z|}\right)\right\}, \qquad -2\pi < \arg z < \pi \tag{7.4-28}$$

Uniform asymptotic approximations for large x and ν are (i) for $x > \nu$

$$\begin{Bmatrix} J_\nu(x) \\ Y_\nu(x) \end{Bmatrix} = \sqrt{\frac{w - \tan^{-1} w}{w}} \left[J_{1/3}(\zeta) \cos\frac{\pi}{6} \mp Y_{1/3}(\zeta) \sin\frac{\pi}{6}\right] + O\left(\frac{1}{\nu^{4/3}}\right) \tag{7.4-29}$$

$$w = \sqrt{\frac{x^2}{\nu^2} - 1}; \qquad \zeta = \nu[w - \tan^{-1} w]$$

(ii) for $x < \nu, 0 < x < \infty$

$$J_\nu(x) = \frac{1}{\pi}\sqrt{\frac{\tanh^{-1} w - w}{w}}\; K_{1/3}(\zeta) + O(\nu^{-4/3})$$

$$Y_\nu(x) = -\sqrt{\frac{\tanh^{-1} w - w}{w}}\; \{I_{1/3}(\zeta) + I_{-1/3}(\zeta)\} + O(\nu^{-4/3}) \tag{7.4-30}$$

$$w = \sqrt{1 - \frac{x^2}{\nu^2}}; \quad \zeta = \nu\{\tanh^{-1} w - w\}$$

7.4.8 Fourier-Bessel and Dini Series

Let $f(x)$ be (i) a real piecewise continuous function; (ii) of bounded variation in every subinterval $[r_1, r_2]$, $0 < r_1 < r_2 < 1$ in $[0, 1]$; (iii) such that $x^{1/2} f(x)$ is $L_1(0, 1)$ then if $\alpha_{\nu k}$ $(k = 1, 2, \ldots)$ are the nonnegative roots of

$$\frac{AJ_\nu(x) + BrJ'_\nu(x)}{x^\nu} = 0; \qquad \nu \geqslant 0; \quad A \cdot B \geqslant 0 \tag{7.4-31}$$

the series

$$\sum_{k=1}^{\infty} C_k J_\nu(\alpha_{\nu k} x) \tag{7.4-32}$$

where

$$C_k = \frac{2\displaystyle\int_0^1 xf(x) J_\nu(\alpha_{\nu k} x)\, dx}{J_\nu'^2(\alpha_{\nu k}) + \{1 - (\nu^2/\alpha_{\nu k}^2)\} J_\nu^2(\alpha_{\nu k})} \tag{7.4-33}$$

converges to $\frac{1}{2}[f(x + 0) + f(x - 0)]$.

7.4.9 Special Cases

For $B = 0$

$$C_k = \frac{2}{J_{\nu+1}^2(\alpha_{\nu k})}\int_0^1 xf(x)\, J_\nu(\alpha_{\nu k} x)\, dx \tag{7.4-34}$$

Furthermore the series converges to 0 at $x = 1$.

For $A = -\nu B$, $\alpha_{\nu 1} = 0$ is a root, since (7.4-25) reduces to $J_{\nu+1}(x)/x^\nu = 0$, and the first term in the series is

$$2(\nu + 1)\,x^{\nu} \int_0^1 x^{\nu+1} f(x)\,dx \qquad (7.4\text{-}35)$$

The other terms correspond to the positive roots of $J_{\nu+1}(x) = 0$. This is the case when $\nu = 0$ and $A = 0$, for which

$$f(x) = C_0 + \sum_{k=1}^{\infty} C_k J_0(j_{1k}x) \qquad (7.4\text{-}36)$$

where j_{1k} are the positive roots of $J_0'(x) \equiv -J_1(x) = 0$, and

$$C_0 = 2\int_0^1 xf(x)\,dx$$

$$C_k = \frac{2\displaystyle\int_0^1 xf(x)\,J_0(j_{1k}x)\,dx}{J_0^2(j_{1k})} \qquad (7.4\text{-}37)$$

7.5 MODIFIED BESSEL FUNCTIONS

7.5.1 Definition

The modified Bessel functions $I_\nu(z)$ and $K_\nu(z)$ are solutions of

$$z^2 \frac{d^2w}{dz^2} + z\frac{dw}{dz} - (z^2 + \nu^2)\,w = 0 \qquad (7.5\text{-}1)$$

7.5.2 Series Representations

$$I_\nu(z) = \left(\frac{z}{2}\right)^{\nu} \sum_{k=0}^{\infty} \frac{(z/2)^{2k}}{k!\,\Gamma(k+\nu+1)} \qquad (7.5\text{-}2)$$

$$K_\nu(z) = \frac{\pi}{2}\,\frac{I_{-\nu}(z) - I_\nu(z)}{\sin \nu\pi}, \qquad \nu \neq 0, \pm 1, \ldots \qquad (7.5\text{-}3)$$

For integral values of ν

$$K_n(z) = \lim_{\nu \to n} K_\nu(z) \qquad (7.5\text{-}4)$$

$$K_0(z) = -\left\{\ln\left(\frac{z}{2}\right) + \gamma\right\} I_0(z) + \sum_{k=0}^{\infty} \{\psi(k+1) + \gamma\}\,\frac{(z/2)^{2k}}{(k!)^2} \qquad (7.5\text{-}5)$$

$$K_n(z) = \frac{1}{2}\left(\frac{z}{2}\right)^{-n} \sum_{k=0}^{n-1} \frac{(n-k-1)!}{k!} (-1)^k \left(\frac{z}{2}\right)^{2k} + (-1)^{n+1} \ln\left(\frac{z}{2}\right) I_n(z)$$

$$+ (-1)^n \frac{1}{2}\left(\frac{z}{2}\right)^n \sum_{k=0}^{\infty} \{\psi(k+1) + \psi(k+n+1)\} \frac{(z/2)^{2k}}{k!\,(n+k)!} \qquad (7.5\text{-}6)$$

$I_\nu(z)$ and $K_\nu(z)$ are analytic in the z-plane cut along the negative real axis; $z^{-\nu}I_\nu(z)$ and $I_n(z)$ are exponential functions of type 1.

As functions of ν, $I_\nu(z)$ and $K_\nu(z)$ are exponential functions of maximum type. Since

$$\left.\begin{aligned} I_\nu(z) &= e^{-\nu(\pi i/2)}\,(J_\nu\, ze^{(i\pi/2)}) \\ K_\nu(z) &= \frac{\pi i}{2}\, e^{+\nu(\pi i/2)}\, H_\nu^{(1)}\,(ze^{(i\pi/2)}) \end{aligned}\right\} \left(-\pi < \arg z \leqslant \frac{\pi}{2}\right) \qquad (7.5\text{-}7)$$

and

$$\left.\begin{aligned} I_\nu(z) &= e^{\nu(3/2)\pi i}\, J_\nu(ze^{-(3/2)\pi i}) \\ K_\nu(z) &= -\frac{\pi i}{2}\, e^{-\nu(\pi i/2)}\, H_\nu^{(2)}(ze^{-(i\pi/2)}) \end{aligned}\right\} \left(-\frac{\pi}{2} < \arg z \leqslant \pi\right) \qquad (7.5\text{-}8)$$

many of the properties of the modified Bessel functions can be deduced from those of the ordinary Bessel functions.

7.5.3 Differential Equations—Wronskians

$I_\nu(z)$ and $K_\nu(z)$ are linearly independent solutions of (7.5-1). The Wronskian

$$W\{K_\nu(z), I_\nu(z)\} = \frac{1}{z} \qquad (7.5\text{-}9)$$

For $\nu \neq$ integer, $I_\nu(z)$ and $I_{-\nu}(z)$ are also linearly independent solutions and

$$W\{I_\nu(z), I_{-\nu}(z)\} = -2\,\frac{\sin(\nu\pi)}{\pi z} \qquad (7.5\text{-}10)$$

If Z denotes a linear combination of I and K, then the solution of

$$z^2 \frac{d^2 w}{dz^2} + (1-2p)\, z\, \frac{dw}{dz} - (\lambda^2 q^2 z^2 - p^2 + \nu^2 q^2)\, w = 0$$

is

$$w = z^p Z_\nu(\lambda z^q) \qquad (7.5\text{-}11)$$

7.5.4 Integral Representations

(See ordinary Bessel functions for pertinent remarks.)

$$I_\nu(z) = \frac{(z/2)^\nu}{\sqrt{\pi}\,\Gamma(\nu + \frac{1}{2})} \int_{-1}^{+1} (1 - t^2)^{\nu-(1/2)} \cosh zt\, dt \qquad |\arg z| < \pi, \quad Re\, \nu > -\frac{1}{2}$$

(7.5-12)

$$K_\nu(z) = \int_0^\infty e^{-z \cosh t} \cosh(\nu t)\, dt \qquad |\arg z| < \frac{\pi}{2} \tag{7.5-13}$$

7.5.5 Recurrence Relations

$$\left.\begin{aligned} I_{\nu-1}(z) - I_{\nu+1}(z) &= \frac{2\nu}{z} I_\nu(z) \\ I_{\nu-1}(z) + I_{\nu+1}(z) &= 2I'_\nu(z) \end{aligned}\right\} \left\{\begin{aligned} K_{\nu-1}(z) - K_{\nu+1}(z) &= -\frac{2\nu}{z} K_\nu & (7.5\text{-}14) \\ K_{\nu-1}(z) + K_{\nu+1}(z) &= -2K'_\nu & (7.5\text{-}15) \end{aligned}\right.$$

$$I'_0(z) = I_1(z), \quad K'_0(z) = -K_1(z)$$

$$\left(\frac{1}{z}\frac{d}{dz}\right)^k z^\nu I_\nu(z) = z^{\nu-k} I_{\nu-k}, \qquad \left(\frac{1}{z}\frac{d}{dz}\right)^k z^\nu K_\nu = (-1)^k z^{\nu-k} K_{\nu-k} \tag{7.5-16}$$

$$\left(\frac{1}{z}\frac{d}{dz}\right)^k z^{-\nu} I_\nu(z) = z^{-\nu-k} I_{\nu+k}, \qquad \left(\frac{1}{z}\frac{d}{dz}\right)^k z^{-\nu} K_\nu = (-1)^k z^{-\nu-k} K_{\nu+k} \tag{7.5-17}$$

7.5.6 Generating Function

$$\exp\left\{\frac{z}{2}\left(t + \frac{1}{t}\right)\right\} = \sum_{k=-\infty}^{\infty} t^k I_k(z) \qquad (t \neq 0) \tag{7.5-18}$$

7.5.7 Asymptotic Expansions for Large Arguments

$$I_\nu(z) \sim \frac{e^z}{\sqrt{2\pi z}} \left\{1 - \frac{4\nu^2 - 1}{8z} + \cdots\right\} \qquad \left(|\arg z| < \frac{\pi}{2}\right) \tag{7.5-19}$$

$$K_\nu(z) \sim \sqrt{\frac{\pi}{2z}}\, e^{-z} \left\{1 + \frac{4\nu^2 - 1}{8z} + \cdots\right\} \qquad \left(|\arg z| < \frac{3\pi}{2}\right) \tag{7.5-20}$$

7.6 ORTHOGONAL POLYNOMIALS

The variable x is assumed real.

7.6.1 Legendre Polynomials

Definition–differential equation

The nth Legendre polynomial $P_n(x)$ is a solution of

$$\frac{d}{dx}\left\{(1-x^2)\frac{dP_n}{dx}\right\}+n(n+1)P_n(x)=0 \tag{7.6-1}$$

If $[\nu]$ denotes the largest integer $\leqslant \nu$ then

$$P_n(x)=\sum_{k=0}^{[n/2]}\frac{(-1)^k(2n-2k)!}{2^n k!(n-k)!(n-2k)!}x^{n-2k} \tag{7.6-2}$$

in particular

$$\begin{aligned}
P_0(x)&=1\\
P_1(x)&=x\\
P_2(x)&=\tfrac{1}{2}[3x^2-1]\\
P_3(x)&=\tfrac{1}{2}[5x^3-3x]\\
P_4(x)&=\tfrac{1}{8}[35x^4-30x^2+3]\\
P_5(x)&=\tfrac{1}{8}[63x^5-70x^3+15x]\\
P_6(x)&=\tfrac{1}{16}[231x^6-315x^4+105x^2-5]
\end{aligned}$$

Rodrigues' formula–recurrence relations

$$P_n(x)=\frac{(-1)^n}{2^n n!}\frac{d^n}{dx^n}[(1-x^2)^n] \tag{7.6-3}$$

$$\begin{aligned}
(n+1)P_{n+1}(x)&=(2n+1)xP_n(x)-nP_{n-1}(x)\\
(1-x^2)\frac{dP_n(x)}{dx}&=-nxP_n(x)+nP_{n-1}(x)
\end{aligned} \tag{7.6-4}$$

Generating function

$$(1-2tx+t^2)^{-1/2}=\sum_{n=0}^{\infty}t^nP_n(x)\qquad |t|<1 \tag{7.6-5}$$

Orthogonality–completeness

$$\int_{-1}^{+1}P_m(x)P_n(x)\,dx=\frac{2}{2n+1}\delta_{mn} \tag{7.6-6}$$

If $f(x) \in L_2(-1, 1)$

$$\sum_{n=0}^{\infty} \left[\frac{2n+1}{2} \int_{-1}^{+1} P_n(x') f(x')\, dx'\right] P_n(x) \tag{7.6-7}$$

converges to $f(x)$ in $L_2(-1, 1)$. In particular, if $f(x)$ is piecewise smooth, then the series is pointwise convergent to $\frac{1}{2}[f(x+0) + f(x-0)]$.

Zeros

The kth zero of $P_n(x)$ is

$$x_k^{(n)} = 1 - \frac{j_{0,k}^2}{2n^2}\left[1 - \frac{1}{n} + O\left(\frac{1}{n^2}\right)\right] \tag{7.6-8}$$

where $j_{0,k}$ is the kth positive root of $J_0(x)$.

7.6.2 Chebyshev Polynomials

Definition–differential equation

The nth Chebyshev polynomial $T_n(x)$ is a solution of

$$(1 - x^2)\frac{d^2 T_n}{dx^2} - x\frac{dT_n}{dx} + n^2 T_n = 0 \tag{7.6-9}$$

If $[\nu]$ denotes the largest integer $\leqslant \nu$, then

$$T_n(x) = \frac{n}{2}\sum_{k=0}^{[n/2]} \frac{(-1)^k (n-k-1)!}{k!(n-2k)!}(2x)^{n-2k} \tag{7.6-10}$$

$$T_n(\cos\theta) = \cos n\theta \tag{7.6-11}$$

in particular

$$T_0(x) = 1$$

$$T_1(x) = x$$

$$T_2(x) = 2x^2 - 1$$

$$T_3(x) = 4x^3 - 3x$$

$$T_4(x) = 8x^4 - 8x^2 + 1$$

$$T_5(x) = 16x^5 - 20x^3 + 5x$$

$$T_6(x) = 32x^6 - 48x^4 + 18x^2 - 1$$

Rodrigues' formula–recurrence relation

$$T_n(x) = \frac{(-1)^n \sqrt{\pi}}{2^{n+1}\Gamma(n+\frac{1}{2})} \cdot \sqrt{1-x^2}\,\frac{d^n}{dx^n}\left[(1-x^2)^{n-(1/2)}\right] \tag{7.6-12}$$

$$\begin{gathered} T_{n+1}(x) = 2xT_n(x) - T_{n-1}(x) \\ (1-x^2)\cdot\frac{dT_n(x)}{dx} = -nxT_n(x) + nT_{n-1}(x) \end{gathered} \tag{7.6-13}$$

Generating function

$$\frac{1-xt}{1-2xt+t^2} = \sum_{n=0}^{\infty} t^n T_n(x) \qquad (|t|<1) \tag{7.6-14}$$

Orthogonality–completeness

$$\int_{-1}^{+1} \frac{T_m(x)\,T_n(x)}{\sqrt{1-x^2}}\,dx = \frac{\pi}{2}\,\delta_{mn} \qquad (n \neq 0) \tag{7.6-15}$$

and

$$\int_{-1}^{+1} \frac{T_m(x)\,T_0(x)}{\sqrt{1-x^2}}\,dx = \pi\delta_{m0} \tag{7.6-16}$$

If $f(x)$ is such that $\int_{-1}^{+1} \frac{|f(x)|^2}{\sqrt{1-x^2}}\,dx$ exists, then the series

$$\frac{1}{\pi}\int_{-1}^{+1} \frac{f(x')\,dx'}{\sqrt{1-x'^2}} + \frac{2}{\pi}\sum_{n=0}^{\infty}\left[\int_{-1}^{+1} \frac{f(x')\,T_n(x')\,dx'}{\sqrt{1-x'^2}}\right] T_n(x) \tag{7.6-17}$$

converges in the mean with weight $(1-x^2)^{-1/2}$ to $f(x)$. In particular, if $f(x)$ is piecewise smooth, then the series converges pointwise to $\frac{1}{2}[f(x+0)+f(x-0)]$.

Zeros

$$x_k^{(n)} = \cos\frac{2k-1}{2n}\,\pi \tag{7.6-18}$$

7.6.3 Jacobi Polynomials

Definition–differential equation

The nth Jacobi polynomial $P_n^{(\alpha,\beta)}(x)$ is a solution of

$$(1-x^2)\frac{d^2P_n^{(\alpha,\beta)}}{dx^2} + [\beta-\alpha-(\alpha+\beta+2)x]\frac{dP_n^{(\alpha,\beta)}}{dx} + n(n+\alpha+\beta+1)P_n^{(\alpha,\beta)} = 0 \quad (7.6\text{-}19)$$

where $\alpha > -1$ and $\beta > -1$.

$$P_n^{(\alpha,\beta)}(x) = \frac{1}{2^n}\sum_{k=0}^{n}\frac{(n+\alpha)!(n+\beta)!}{(n+\alpha-k)!k!(k+\beta)!(n-k)!}\cdot(x-1)^{n-k}(x+1)^k \quad (7.6\text{-}20)$$

Rodrigues' formula–recurrence relation

$$P_n^{(\alpha,\beta)}(x) = \frac{(-1)^n}{2^n n!}\frac{1}{(1-x)^\alpha(1+x)^\beta}\frac{d^n}{dx^n}[(1-x)^{\alpha+n}(1+x)^{\beta+n}] \quad (7.6\text{-}21)$$

$$\begin{aligned} &2(n+1)(n+\alpha+\beta+1)(2n+\alpha+\beta)P_{n+1}^{(\alpha,\beta)}(x) \\ &= \left[(2n+\alpha+\beta+1)(\alpha^2-\beta^2) + \frac{\Gamma(2n+\alpha+\beta+3)}{\Gamma(2n+\alpha+\beta)}x\right]P_n^{(\alpha,\beta)}(x) \\ &\quad - 2(n+\alpha)(n+\beta)(2n+\alpha+\beta+2)P_{n-1}^{(\alpha,\beta)}(x) \end{aligned} \quad (7.6\text{-}22)$$

$$\begin{aligned} &(2n+\alpha+\beta)(1-x^2)\frac{dP_n^{(\alpha,\beta)}}{dx} \\ &= n[\alpha-\beta-(2n+\alpha+\beta)x]P_n^{(\alpha,\beta)}(x) + 2(n+\alpha)(n+\beta)P_{n-1}^{(\alpha,\beta)}(x) \end{aligned} \quad (7.6\text{-}23)$$

Generating function

$$\begin{aligned} &(1-2xt+t^2)^{-1/2}(1-t+\sqrt{1-2xt+t^2})^{-\alpha}(1+t+\sqrt{1-2xt+t^2})^{-\beta} \\ &= \frac{1}{2^{\alpha+\beta}}\sum_{n=0}^{\infty} t^n P_n^{(\alpha,\beta)}(x) \qquad |t| < 1 \end{aligned} \quad (7.6\text{-}24)$$

Orthogonality–completeness

$$\begin{aligned} &\int_{-1}^{+1}(1-x)^\alpha(1+x)^\beta P_m^{(\alpha,\beta)}(x)P_n^{(\alpha,\beta)}(x)\,dx \\ &= \frac{2^{\alpha+\beta+1}}{2n+\alpha+\beta+1}\frac{\Gamma(n+\alpha+1)\Gamma(n+\beta+1)}{n!\Gamma(n+\alpha+\beta+1)}\delta_{mn} \end{aligned} \quad (7.6\text{-}25)$$

If $f(x)$ is such that $\int_{-1}^{1} (1-x)^\alpha (1+x)^\beta |f(x)|^2\,dx$ exists, then the series

$$\sum_{n=0}^{\infty} \left[\int_{-1}^{+1} \left(\frac{2^{\alpha+\beta+1}}{2n+\alpha+\beta+1} \frac{\Gamma(n+\alpha+1)\,\Gamma(n+\beta+1)}{n!\,\Gamma(n+\alpha+\beta+1)} \right)^{-1} \cdot P_n^{(\alpha,\beta)}(x') f(x')\,dx' \right] P_n^{(\alpha,\beta)}(x) \qquad (7.6\text{-}26)$$

converges in the mean with weight $(1-x)^\alpha(1+x)^\beta$ to $f(x)$. In particular, if f is piecewise smooth, then the series converges pointwise to $\frac{1}{2}[f(x+0)+f(x-0)]$.

Zeros

$$\cos^{-1} x_k^{(n)} = \frac{j_{\alpha,k}}{n}[1+o(n)] \qquad (7.6\text{-}27)$$

where $j_{\alpha,k}$ is the kth positive zero of $J_\alpha(x)$.

7.6.4 Laguerre Polynomials

Definition–differential equation

The nth Laguerre polynomial $L_n^{(\alpha)}(x)$ is a solution of

$$x\frac{d^2L_n^{(\alpha)}}{dx^2} + (\alpha+1-x)\frac{dL_n^{(\alpha)}}{dx} + nL_n^{(\alpha)} = 0 \qquad (7.6\text{-}28)$$

where $\alpha > -1$

$$L_n^{(\alpha)}(x) = \sum_{k=0}^{n} \frac{\Gamma(n+\alpha+1)}{\Gamma(k+\alpha+1)} \frac{(-x)^k}{k!(n-k)!} \qquad (7.6\text{-}29)$$

in particular

$$L_0(x) = 1$$

$$L_1(x) = -x + 1$$

$$L_2(x) = \tfrac{1}{2}[x^2 - 4x + 2]$$

$$L_3(x) = \tfrac{1}{6}[-x^3 + 9x^2 - 18x + 6]$$

$$L_4(x) = \tfrac{1}{24}[x^4 - 16x^3 + 72x^2 - 96x + 24]$$

Rodrigues' formula–recurrence relations

$$L_n^{(\alpha)}(x) = \frac{e^x x^{-\alpha}}{n!} \frac{d^n}{dx^n} (x^{n+\alpha} e^{-x}) \tag{7.6-30}$$

$$(n+1) L_{n+1}^{(\alpha)}(x) = (2n + \alpha + 1 - x) L_n^{(\alpha)}(x) - (n + \alpha) L_{n-1}^{(\alpha)}(x) \tag{7.6-31}$$

$$x \frac{dL_n^{(\alpha)}(x)}{dx} = n L_n^{(\alpha)}(x) - (n + \alpha) L_{n-1}^{(\alpha)}(x) \tag{7.6-32}$$

Generating function

$$(1 - t)^{-\alpha-1} \exp\left(\frac{xt}{t-1}\right) = \sum_{n=0}^{\infty} t^n L_n^{(\alpha)}(x) \qquad |t| < 1 \tag{7.6-33}$$

Orthogonality–completeness

$$\int_0^\infty x^\alpha e^{-x} L_m^{(\alpha)}(x) L_n^{(\alpha)}(x)\, dx = \frac{\Gamma(n + \alpha + 1)}{n!} \delta_{mn} \tag{7.6-34}$$

If $f(x)$ is such that $\int_0^\infty x^\alpha e^{-x} |f(x)|^2\, dx$ exists, then the series

$$\sum_{n=0}^{\infty} \frac{n!}{\Gamma(n + \alpha + 1)} \int_0^\infty x'^\alpha e^{-x'} f(x') L_n^{(\alpha)}(x')\, dx' L_n^{(\alpha)}(x) \tag{7.6-35}$$

converges in the mean with weight $x^\alpha e^{-x}$ to $f(x)$. In particular if $f(x)$ is piecewise smooth in every finite subinterval, then the series converges pointwise to $\frac{1}{2}[f(x + 0) + f(x - 0)]$.

Zeros

$$x_k^{(n)} = \frac{j_{\alpha,k}^2}{4n + 2(\alpha + 1)} [1 + O(n^{-2})] \tag{7.6-36}$$

where $j_{\alpha,k}$ is the kth positive zero of $J_\alpha(x)$.

7.6.5 Hermite Polynomials

Definition–differential equation

The nth Hermite polynomial $H_n(x)$ is a solution of

$$\frac{d^2 H_n}{dx^2} - 2x \frac{dH_n}{dx} + 2nH_n = 0 \tag{7.6-37}$$

If $u_n = \exp(-x^2/2) H_n(x)$, then

$$\frac{d^2 u_n}{dx^2} + (2n + 1 - x^2) u_n = 0 \tag{7.6-38}$$

If $[\nu]$ denotes the largest integer $\leqslant \nu$, then

$$H_n(x) = \sum_{k=0}^{[n/2]} \frac{(-1)^k n!}{2^k k!(n-2k)!} (x)^{n-2k} \tag{7.6-39}$$

in particular

$$H_0(x) = 1$$
$$H_1(x) = 2x$$
$$H_2(x) = 4x^2 - 2$$
$$H_3(x) = 8x^3 - 12x$$
$$H_4(x) = 16x^4 - 48x^2 + 12$$
$$H_5(x) = 32x^5 - 160x^3 + 120x$$

Rodrigues' formula–recurrence relations

$$H_n(x) = (-1)^n e^{x^2} \frac{d^n}{dx^n} (e^{-x^2}) \tag{7.6-40}$$

$$H_{n+1}(x) = 2xH_n(x) - 2nH_{n-1}(x) \tag{7.6-41}$$

$$\frac{dH_n(x)}{dx} = 2nH_{n-1}(x) \tag{7.6-42}$$

Generating function

$$e^{2xt-t^2} = \sum_{n=0}^{\infty} \frac{t^n}{n!} H_n(x) \tag{7.6-43}$$

Orthogonality–completeness

$$\int_{-\infty}^{\infty} e^{-x^2} H_m(x) H_n(x)\, dx = 2^n n! \sqrt{\pi}\, \delta_{mn} \tag{7.6-44}$$

If $f(x)$ is such that $\int_{-\infty}^{\infty} e^{-x^2} |f(x)|^2 \, dx$ exists, then the series

$$\sum_{n=0}^{\infty} \frac{1}{2^n n! \sqrt{\pi}} \int_{-\infty}^{\infty} e^{-x'^2} H_n(x') \, dx' \; H_n(x) \tag{7.6-45}$$

converges in the mean with weight e^{-x^2} to $f(x)$. In particular, if $f(x)$ is piecewise smooth in every finite interval, then the series converges pointwise to $\frac{1}{2}[f(x+0) + f(x-0)]$.

7.6.6 Interrelations

$$P_n(x) = P_n^{(0,0)}(x) \tag{7.6-46}$$

$$T_n(x) = \frac{n! \sqrt{\pi}}{\Gamma(n + \frac{1}{2})} P_n^{(-1/2,-1/2)}(x) \tag{7.6-47}$$

$$L_n^{(-1/2)}(x) = \frac{(-1)^n}{2^{2n} n!} H_{2n}(\sqrt{x}) \tag{7.6-48}$$

$$L_n^{(1/2)}(x) = \frac{(-1)^n}{2^{2n+1} n! \sqrt{x}} H_{2n+1}(\sqrt{x}) \tag{7.6-49}$$

7.7 HYPERGEOMETRIC FUNCTIONS AND LEGENDRE FUNCTIONS

7.7.1 Definition

$$F(a, b; c; z) = \frac{\Gamma(c)}{\Gamma(a)\Gamma(b)} \sum_{n=0}^{\infty} \frac{\Gamma(a+n)\Gamma(b+n)}{\Gamma(c+n)} \frac{z^n}{n!} \quad \text{for } |z| < 1 \tag{7.7-1}$$

The function is defined in the remainder of the z-plane by analytic continuation.
Note that $F(a, b; c; z)$ reduces to a polynomial of degree n in z when either a or b equals $-n$. F is an analytic function in the z-plane cut along $(1, \infty)$.

7.7.2 Differential Equation

$F(a, b; c; z)$ is a solution of

$$z(1-z)\frac{d^2 w}{dz^2} + [c - (a+b+1)z]\frac{dw}{dz} - abw = 0 \tag{7.7-2}$$

When all the numbers $c, c-a-b, a-b$ differ from integers, the second solution of

(7.7-2) is

$$\begin{aligned} w_2 &= z^{1-c} F(a - c + 1, b - c + 1; 2 - c; z) \\ &= z^{1-c}(1 - z)^{c-a-b} F(1 - a, 1 - b; 2 - c; z) \end{aligned} \tag{7.7-3}$$

7.7.3 Integral Representation

$$F(a, b; c; z) = \frac{\Gamma(c)}{\Gamma(a)\Gamma(b)} \int_0^1 t^{b-1}(1 - t)^{c-b-1}(1 - tz)^{-a}\, dt$$

$$(Re\, c > Re\, b > 0) \quad (7.7\text{-}4)$$

7.7.4 Gauss' Relations and Linear Transformations

$F(a, b; c; z)$ is related to the six contiguous functions $F(a \pm 1; b; c; z), F(a, b \pm 1; c; z)$ and $F(a, b; c \pm 1; z)$. These are Gauss' relations.

$F(a, b; c; z)$ can also be expressed in terms of hypergeometric functions of the variables $z/(z - 1)$, $1/(z - 1)$, $1/z$, $1 - z$.

See Abramowitz and Stegun or Erdélyi et al.

7.7.5 Relation to Orthogonal Polynomials

$$P_n^{(\alpha,\beta)}(x) = \frac{\Gamma(n + \alpha)}{\Gamma(\alpha)\Gamma(n)} F\left(-n, n + \alpha + \beta + 1; \alpha + 1; \frac{1 - x}{2}\right) \tag{7.7-5}$$

$$T_n(x) = F\left(-n, n; \frac{1}{2}; \frac{1 - x}{2}\right) \tag{7.7-6}$$

$$P_n(x) = F\left(-n, n + 1; 1; \frac{1 - x}{2}\right) \tag{7.7-7}$$

7.7.6 Elementary Functions

$$z^{-1} \ln (1 - z) = F(1, 1; 2; z) \tag{7.7-8}$$

$$\frac{1}{2z} \ln \frac{1 + z}{1 - z} = F\left(\frac{1}{2}, 1; \frac{3}{2}; z^2\right) \tag{7.7-9}$$

$$(1 - z)^{-\alpha} = F(\alpha, \beta; \beta; z) \tag{7.7-10}$$

7.7.7 Legendre Functions

Definition–differential equations

The Legendre functions $P_\nu^\mu(z)$ and $Q_\nu^\mu(z)$ are the solutions of

$$(1 - z^2)\frac{d^2w}{dz^2} - 2z\frac{dw}{dz} + \left[\nu(\nu+1) - \frac{\mu^2}{1-z^2}\right] w = 0 \qquad (7.7\text{-}11)$$

The Legendre functions are analytic in the z-plane cut along $(1, -\infty)$; $|\arg(z \pm 1)| < \pi$, $|\arg z| < \pi$.

7.7.8 Legendre Functions as Hypergeometric Functions

$$P_\nu^\mu(z) = \frac{1}{\Gamma(1-\mu)}\left(\frac{z+1}{z-1}\right)^{\mu/2} F\left(-\nu, \nu+1; 1-\mu; \frac{1-z}{2}\right) \qquad (|1-z| < 2) \qquad (7.7\text{-}12)$$

$$Q_\nu^\mu(z) = e^{i\pi\mu}\frac{\sqrt{\pi}}{2^{\nu+1}}\frac{\Gamma(\nu+\mu+1)}{\Gamma(\nu+\frac{3}{2})}\frac{(z^2-1)^{\mu/2}}{z^{\nu+\mu+1}} F\left(1+\frac{\nu}{2}+\frac{\mu}{2}, \frac{1+\nu+\mu}{2}; \nu+\frac{3}{2}; \frac{1}{z^2}\right) \qquad (|z| > 1) \qquad (7.7\text{-}13)$$

7.7.9 Real Argument $-1 < x < 1$

For real x, the Legendre functions are defined as follows:

$$P_\nu^\mu(x) \equiv \tfrac{1}{2}\left[e^{i\mu(\pi/2)}P_\nu^\mu(x+i0) + e^{-i\mu(\pi/2)}P_\nu^\mu(x-i0)\right] \qquad (7.7\text{-}14)$$

$$Q_\nu^\mu(x) \equiv \frac{e^{-i\mu\pi}}{2}\left[e^{-i\mu(\pi/2)}Q_\nu^\mu(x+i0) + e^{i\mu(\pi/2)}Q_\nu^\mu(x-i0)\right] \qquad (7.7\text{-}15)$$

7.7.10 Special Values

For $\mu = 1, 2, \ldots$

$$P_\nu^m(z) = (z^2-1)^{m/2}\frac{d^m P_\nu(z)}{dz^m}$$
$$Q_\nu^m(z) = (z^2-1)^{m/2}\frac{d^m P_\nu(z)}{dz^m} \qquad (7.7\text{-}16)$$

where $P_\nu(z)$ and $Q_\nu(z)$ stand for $P_\nu^0(z)$ and $Q_\nu^0(z)$. $P_\nu(z)$ satisfies the recurrence relations (7.6-3) and (7.6-4). For $-1 < x < 1$

$$P_\nu^m(x) = (-1)^m(1-x^2)^{m/2}\frac{d^m P_\nu(x)}{dx^m}$$
$$Q_\nu^m(x) = (-1)^m(1-x^2)^{m/2}\frac{d^m Q_\nu(x)}{dx^m} \qquad (7.7\text{-}17)$$

$$P_\nu^{-m}(x) = (-1)^m \frac{\Gamma(\nu - m + 1)}{\Gamma(\nu + m + 1)} P_\nu^m(x) \tag{7.7-18}$$

When ν is equal to an integer n, the $P_n^m(x)$ are referred to as the associated Legendre polynomials. In particular

$$P_1^1(\cos\theta) = \sin\theta$$

$$P_2^1(\cos\theta) = 3\cos\theta\sin\theta$$

$$P_2^2(\cos\theta) = 3\sin^2\theta$$

$$P_3^1(\cos\theta) = \tfrac{3}{2}\sin\theta(5\cos^2\theta - 1)$$

$$P_3^2(\cos\theta) = 15\cos\theta\sin^2\theta$$

$$P_3^3(\cos\theta) = 15\sin^3\theta$$

7.7.11 Spherical Harmonics

$$Y_n^m(\theta,\varphi) = e^{im\varphi} P_n^m(\cos\theta) \qquad (-n \leqslant m \leqslant n) \tag{7.7-19}$$

Orthogonality condition

$$\int_0^{2\pi} d\varphi \int_0^{\pi} d\theta \sin\theta\, \overline{Y_{n_1}^{m_1}(\theta,\varphi)}\, Y_{n_2}^{m_2}(\theta,\varphi) = \frac{4\pi}{2n_1 + 1} \cdot \frac{(n_1 + m_1)!}{(n_1 - m_1)!} \delta_{n_1 n_2} \delta_{m_1 m_2} \tag{7.7-20}$$

Basic equation

$$\left[-\frac{1}{\sin\theta}\frac{\partial}{\partial\theta}\sin\theta\frac{\partial}{\partial\theta} - \frac{1}{\sin^2\theta}\frac{\partial^2}{\partial\varphi^2}\right] Y_n^m(\theta,\varphi) = n(n+1) Y_n^m(\theta,\varphi) \tag{7.7-21}$$

Solutions of Laplace equation $\nabla^2 u = 0$ which are bounded for $\theta = 0, \pi$ can be written as

$$u = \sum_{n=0}^{\infty} \sum_{m=-n}^{n} (A_{mn} r^n + B_{mn} r^{-n-1}) Y_n^m(\theta,\varphi) \tag{7.7-22}$$

7.8 ELLIPTIC INTEGRALS AND FUNCTIONS

7.8.1 Elliptic Integrals of the First and Second Kind

$$F(\phi \backslash \alpha) = \int_0^{\phi} (1 - \sin^2\alpha \sin^2\theta)^{-1/2}\, d\theta \tag{7.8-1}$$

$$E(\phi\backslash\alpha) = \int_0^{\phi} (1 - \sin^2 \alpha \sin^2 \theta)^{1/2}\, d\theta \qquad (7.8\text{-}2)$$

Setting $k^2 = \sin^2 \alpha$ and $w = \sin \phi$ we can also write

$$F(w, k) = \int_0^{w} \frac{dt}{\sqrt{(1 - t^2)(1 - k^2 t^2)}} \qquad (7.8\text{-}3)$$

$$E(w, k) = \int_0^{w} \sqrt{\frac{1 - k^2 t^2}{1 - t^2}}\, dt \qquad (7.8\text{-}4)$$

7.8.2 Complete Elliptic Integrals

$$K(k) = F(1, k) = \int_0^{1} \frac{dt}{\sqrt{(1 - t^2)(1 - k^2 t^2)}} \qquad (7.8\text{-}5)$$

or in terms of the hypergeometric function

$$K(k) = \frac{\pi}{2} F\left(\frac{1}{2}, \frac{1}{2}; 1; k^2\right) \qquad (7.8\text{-}6)$$

Similarly

$$E(k) = E(1, k) = \int_0^{1} \sqrt{\frac{1 - k^2 t^2}{1 - t^2}}\, dt$$

$$= \frac{\pi}{2} F\left(-\frac{1}{2}, \frac{1}{2}; 1; k^2\right) \qquad (7.8\text{-}7)$$

7.8.3 Jacobian Elliptic Function

The Jacobian elliptic function

$$w = \mathrm{sn}\,(z, k) \qquad (7.8\text{-}9)$$

is defined as the inverse of

$$z = \int_0^{w} \frac{dt}{\sqrt{(1 - t^2)(1 - k^2 t^2)}} \qquad (7.8\text{-}10)$$

The other Jacobian elliptic functions can be defined in terms of $\mathrm{sn}\,(z, k)$

$$\mathrm{cn}\,(z, k) = \sqrt{1 - \mathrm{sn}^2\,(z, k)} \qquad (7.8\text{-}11)$$

$$\mathrm{dn}\,(z, k) = \sqrt{1 - k^2\,\mathrm{sn}^2\,(z, k)} \tag{7.8-12}$$

The functions $\mathrm{sn}(z)$, $\mathrm{cn}(z)$, and $\mathrm{dn}(z)$ are doubly periodic meromorphic functions of z. If

$$K' = K(1 - k^2) \tag{7.8-13}$$

then $\mathrm{sn}(z, k)$ has primitive periods $4K$ and $2iK'$. Similarly, $\mathrm{cn}(z, k)$ and $\mathrm{dn}(z, k)$ have respectively primitive periods $(4K, 2K + 2iK')$ and $(2K, 4iK')$. All three functions have poles at

$$z_{mn} = 2mK + (2n + 1)\,iK' \tag{7.8-14}$$

7.8.4 Approximations in Terms of Circular and Hyperbolic Functions

$$\begin{aligned} \mathrm{sn}(z, k) &= \sin z - \frac{k^2}{4}\,(z - \sin z \cos z)\sin z + O(k^4) \\ \mathrm{sn}(z, k) &= \tanh z + \frac{1 - k^2}{4}\,(\sinh z \cosh z - z)\,\mathrm{sech}^2\, z + O\,[(1 - k^2)^2] \end{aligned} \tag{7.8-15}$$

$$\begin{aligned} \mathrm{cn}(z, k) &= \cos z + \frac{k^2}{4}\,(z - \sin z \cos z)\sin z + O(k^4) \\ \mathrm{cn}(z, k) &= \mathrm{sech}\, z - \frac{1 - k^2}{4}\,(\sinh z \cosh z - z)\tanh z\,\mathrm{sech}\, z + O[(1 - k^2)^2] \end{aligned} \tag{7.8-16}$$

$$\begin{aligned} \mathrm{dn}(z, k) &= 1 - \frac{k^2}{2}\sin^2 z + O(k^4) \\ \mathrm{dn}(z, k) &= \mathrm{sech}\, z + \frac{1 - k^2}{4}\,(\sinh z \cosh z + z)\tanh z\,\mathrm{sech}\, z + O[(1 - k^2)^2] \end{aligned} \tag{7.8-17}$$

7.8.5 Weierstrass Elliptic $\wp$-Function

$$\wp(z) = \frac{1}{z^2} + \sum_{m,n=-\infty}^{\infty}{}' \; [(z - z_{mn})^{-2} - z_{mn}^{-2}] \tag{7.8-18}$$

where

$$z_{mn} = 2m\omega_1 + 2n\omega_2 \tag{7.8-19}$$

and a prime on summation sign indicates that the term $m = n = 0$ is omitted.

The function $\wp(z)$ is doubly periodic meromorphic with fundamental periods $(2\omega_1, 2\omega_2)$ and double poles at z_{mn}.

The invariants g_2 and g_3 are

$$g_2 = 60 \sum_{m,n=-\infty}^{\infty}{}' z_{mn}^{-4} \tag{7.8-20}$$

$$g_3 = 140 \sum_{m,n=-\infty}^{\infty}{}' z_{mn}^{-6} \tag{7.8-21}$$

A solution of the following differential equation is $\wp(z; g_2, g_3)$:

$$\wp'^2(z) = 4\wp^3(z) - g_2\wp(z) - g_3 \tag{7.8.22}$$

If $w = \wp(z; g_2, g_3)$ then

$$z = \int_w^\infty (4t^3 - g_2 t - g_3)^{-1/2}\, dt \tag{7.8.23}$$

7.9 OTHER SPECIAL FUNCTIONS

7.9.1 Confluent Hypergeometric Function

$$M(a, b, z) = \frac{\Gamma(b)}{\Gamma(a)} \sum_{n=0}^{\infty} \frac{\Gamma(a+n)}{\Gamma(b+n)} \cdot \frac{z^n}{n!} \tag{7.9-1}$$

If a and b are not negative integers, $M(a, b, z)$ is an entire function of exponential type 1; $M(a, b, z)$ is a solution of

$$z\frac{d^2w}{dz^2} + (b - z)\frac{dw}{dz} - aw = 0 \tag{7.9-2}$$

The second solution is

$$U(a, b, z) = \frac{\pi}{\sin \pi b}\, \frac{M(a, b, z)}{\Gamma(1 + a - b)\Gamma(b)} - z^{1-b}\frac{M(1 + a - b, 2 - b, z)}{\Gamma(a)\Gamma(2 - b)} \tag{7.9-3}$$

As $|z| \to \infty$

$$M(a, b, z) = \frac{\Gamma(b)}{\Gamma(a)} e^z z^{a-b}[1 + O(|z|^{-1})] \qquad (Re\, z > 0) \tag{7.9-4}$$

$$M(a, b, z) = \frac{\Gamma(b)}{\Gamma(b - a)} (-z)^{-a}[1 + O(|z|^{-1})] \qquad (Re\, z < 0) \tag{7.9-5}$$

It is therefore clear that for some domain in the parameter space $(a, b), M(a, b, iy) \in L^2(-\infty, \infty)$ and hence admits a Paley-Wiener integral representation. As a matter of fact, from the integral representation (7.2-13) of the beta function we can deduce from (7.9-1) that

$$M(a, b, z) = \frac{\Gamma(b)}{\Gamma(a)\,\Gamma(b-a)} \int_0^1 e^{zt}\, t^{a-1} (1-t)^{b-a-1}\, dt \tag{7.9-6}$$

From (7.9-6) we can deduce Kummer's functional relation

$$M(a, b, z) = e^z M(b-a, b, -z) \tag{7.9-7}$$

The following special functions can be written in terms of confluent hypergeometric functions:

$$e^z = M(a, a, z) \tag{7.9-8}$$

$$\operatorname{erf} z = \frac{2z}{\sqrt{\pi}} M\left(\frac{1}{2}, \frac{3}{2}, -z^2\right) \tag{7.9-9}$$

$$J_\nu(z) = \frac{e^{iz}\left(\frac{z}{2}\right)^\nu}{\Gamma(\nu+1)} M\left(\nu + \frac{1}{2}, 2\nu + 1, 2iz\right) \tag{7.9-10}$$

$$L_n^{(\alpha)}(x) = \frac{\Gamma(\alpha+n+1)}{\Gamma(\alpha+1)} \frac{1}{n!} M(-n, \alpha+1, x) \tag{7.9-11}$$

7.9.2 Zeroth-Order Angular Prolate Spheroidal Wave Functions

We shall denote by $\Psi_n(c, z)$ a function proportional to the zeroth order angular prolate spheroidal wave function $S_{0n}(c, z)$. Then $\Psi_n(c, z)$ is a solution of

$$(1-z^2)\frac{d^2w}{dz^2} - 2z\frac{dw}{dz} + [\lambda_{0n} - c^2z^2]\, w = 0 \tag{7.9-12}$$

where $\lambda_{0n}(c)$ is determined so that $\Psi_n(c, z)$ is finite at $z = \pm 1$.

The function $\Psi_n(c, z)$ is also a solution of the integral equation

$$\mu_n(c)\Psi_n(c, z) = \int_{-1}^{+1} \frac{\sin c(z-t)}{\pi(z-t)} \Psi_n(c, t)\, dt \tag{7.9-13}$$

The function $\Psi_n(c, z)$ is an exponential function of type c which is $L^2(-\infty, \infty)$. Hence, it admits a Paley-Wiener representation

$$\Psi_n(c, z) = (-i)^n \left[\frac{c}{2\pi\mu_n}\right]^{1/2} \int_{-1}^{+1} e^{ictz}\, \Psi_n(c, t)\, dt \tag{7.9-14}$$

We notice that $\{\Psi_n(c, x); n = 0, 1, 2, \ldots\}$ is complete in $L_2(-1, 1)$ and

$$\int_{-1}^{+1} \Psi_n(c, x)\, \Psi_m(c, x)\, dx = \mu_n(c)\, \delta_{nm} \tag{7.9-15}$$

It is also complete in $L^2(-\infty, \infty)$ for exponential functions which are $L^2(-\infty, \infty)$ and

$$\int_{-\infty}^{\infty} \Psi_n(c, x)\, \Psi_m(c, x)\, dx = \delta_{nm} \tag{7.9-16}$$

7.10 REFERENCES AND BIBLIOGRAPHY

7.10.1 References

7-1 Abramowitz, M., and Stegun, I. A., (editors), *Handbook of Mathematical Functions*, National Bureau of Standards, Wash., D.C., 1964.

7-2 Erdélyi, A., et al., *Higher Transcendental Functions*, 3 vols., McGraw-Hill, New York, 1953.

7-3 Whittaker, E. T., and Watson, G. N., *A Course of Modern Analysis*, Fourth Edition, Cambridge University Press, Cambridge, 1952.

7-4 Boas, R. P., Jr., *Entire Functions*, Academic Press, New York, 1954.

7-5 Muller, C., *Spherical Harmonics*, Springer-Verlag, New York, 1966.

7-6 Buchholz, H., *The Confluent Hypergeometric Function*, Springer-Verlag, New York, 1969.

7-7 Slepian, D., and Pollak, H. O., "Prolate Spheroidal Wave Functions. Fourier Analysis and Uncertainty–I," *Bell System Technical Journal*, **40**, 43–64, 1962.

7.10.2 Bibliography

McBride, E., *Obtaining Generating Functions*, Springer-Verlag, New York, 1971.

Miller, W., *Lie Theory and Special Functions*, Academic Press, New York, 1968.

Truesdell, C., "An Essay Toward a Unified Theory of Special Functions," *Annals of Mathematical Studies*, Princeton University Press, Princeton, 1948.

8

First-Order Partial Differential Equations

Jirair Kevorkian*

8.0 INTRODUCTION

In this chapter we study the solution of first-order partial differential equations of the form

$$F(x_1, x_2, \ldots, x_n, u, u_{x_1}, u_{x_2}, \ldots, u_{x_n}) = 0 \tag{8.0-1}$$

Equation (8.0-1), with appropriate initial conditions, governs the solution of the dependent variable u as a function of the n independent variables x_i. It is a first-order equation, since only the first partial derivatives of u with respect to x_i occur. In general, F will be a nonlinear function of the u_{x_i}. If, however, F is linear in the u_{x_i}, eq. (8.1-1) is said to be quasilinear, and has the most general form

$$\sum_{i=1}^{n} a_i(x_1, x_2, \ldots, x_n, u)\, u_{x_i} = a(x_1, \ldots, x_n, u) \tag{8.0-2}$$

where the coefficients a_i and a do not depend on the u_{x_i} but may involve u. In the special case where these coefficients do not involve u, eq. (8.0-2) is said to be linear and, finally, the simplest case results if the coefficients are constants.

First-order partial differential equations arise immediately in the transformation theory of second-order equations into canonical form, and their solution in this context is quite relevant especially for the case of hyperbolic equations (see Chapter 9 for further discussion). However, more significant and direction applications of first-order partial differential equations can be found in a wide variety of fields. For example, it is a remarkable fact that the Eikonal equation (derived in the next section, and discussed in detail in section 8.4), a nonlinear, first-order, partial differen-

***Prof. Jirair Kevorkian, Dep't. of Aeronautics and Astronautics, and Program in Applied Mathematics, University of Washington, Seattle, Wash.**

tial equation, represents the unifying mathematical bond between such diverse topics as geometrical optics, dynamics, and variational calculus.

To a great extent, the material presented in this chapter is based on the comprehensive treatment which is given in Ref. 8-1.

8.1 EXAMPLES OF FIRST-ORDER PARTIAL DIFFERENTIAL EQUATIONS

8.1.1 Traffic Flow

Let us begin our collection of examples by considering a quasilinear equation describing an all too familiar problem in modern everyday life: the flow of traffic. This is an interesting special illustration of *kinematic waves* discussed in Ref. 8-2. Consider a long road along which distance is measured by x, and let q denote the number of vehicles per unit time. Let k be the concentration; i.e., the number of vehicles per unit distance, and assume there exists a relation of the form

$$q = Q(k, x) \tag{8.1-1}$$

dictated by the human response of the drivers to vehicle concentration (depicted through the dependence of Q upon k) and road condition (x-dependence). For a given road, one can easily measure q as a function of k at any x in the following manner. We draw two lines across the road a distance Δx apart and centered at x. Let T be a moderate time interval long enough to allow many vehicles to pass, and chosen with t in its middle. Then, clearly if one counts the number, n, of vehicles crossing x during the period T we have

$$q = n/T \tag{8.1-2}$$

Moreover, if we denote by dt_i the time each of the n vehicles took to cross the slice of width Δx, then

$$k = \left(\sum_{i=1}^{n} dt_i\right) \Big/ T\Delta x \tag{8.1-3}$$

A series of such measurements covering a wide range of concentrations and locations along the road gives a relation of the form eq. (8.1-1).

Now we apply the law of 'conservation of vehicles' which states that at time $t + \Delta t$ the number of vehicles in a length of road Δx wide centered at x is equal to the number of vehicles which existed in the same slice at time t plus the net inflow during the interval Δt, i.e.

$$k(x, t + \Delta t)\,\Delta x = k(x, t)\,\Delta x + \left[q\left(x - \frac{\Delta x}{2}, t\right)\Delta t - q\left(x + \frac{\Delta x}{2}, t\right)\Delta t\right] \tag{8.1-4}$$

where the bracketed terms in eq. (8.1-4) represent the net inflow.

In the limit as Δx and Δt become vanishingly small, eq. (8.1-4) reduces to

$$\frac{\partial k}{\partial t} + \frac{\partial q}{\partial x} = 0 \tag{8.1-5}$$

Let us introduce the notation

$$c = \frac{\partial Q}{\partial k} \tag{8.1-6}$$

where eq. (8.1-6) is calculated from eq. (8.1-1) with x fixed. If we now multiply eq. (8.1-5) by c we obtain the quasilinear equation

$$\frac{\partial q}{\partial t} + \tilde{c}(q, x)\frac{\partial q}{\partial x} = 0 \tag{8.1-7}$$

where, by combining eq. (8.1-6) with eq. (8.1-1) we regard $\tilde{c}$ as a function of q and x.

The solution of eq. (8.1-7) will be considered in some detail in our discussion of shocks in sections 8.3.2 and 8.3.3.

8.1.2 The Eikonal Equation in Geometrical Optics

We now derive a nonlinear equation which is fundamental in *geometrical optics* and dynamics. For simplicity we first consider the interpretation of this equation in geometrical optics.

In geometrical optics one only considers the propagation boundary of an optical signal in a medium, without regard to the detailed structure of the light wave, i.e., amplitude, phase, and frequency. Thus, given an initial signal, at a later time we distinguish between regions in the medium which have been disturbed, and undisturbed regions which the signal has not yet reached.

The boundary between the disturbed and undisturbed regions can be calculated using a construction due to Huygens. For simplicity, let us consider a two-dimensional disturbance in an isotropic medium; i.e., at each point $P = (x, y)$ the speed of light, c, depends only on x and y and is independent of the direction of propagation. Assume that the initial disturbance is a point source of light at $P_0 = (x_0, y_0)$ and time $t = u_0$ as shown in Fig. 8.1-1.

Light rays emanate radially from P_0 and initially travel with speed $c_0 = c(x_0, y_0)$ so that at time $t = u_0 + \Delta u$ the disturbed region is the interior of the circle $\mathcal{C}_0$ centered at P_0 and having a radius $c_0 \Delta u$. If Δu is sufficiently small, it is consistent to use the value of c at P_0 in this construction. Now, in order to determine the boundary of the disturbed region at a later time, say $t = u_0 + 2\Delta u$ we repeat the previous process regarding all the points on $\mathcal{C}_0$ as signal emitters. For example, consider the two neighboring points P_1 and P_2 on $\mathcal{C}_0$. In general, $c_1 = c(P_1)$ and

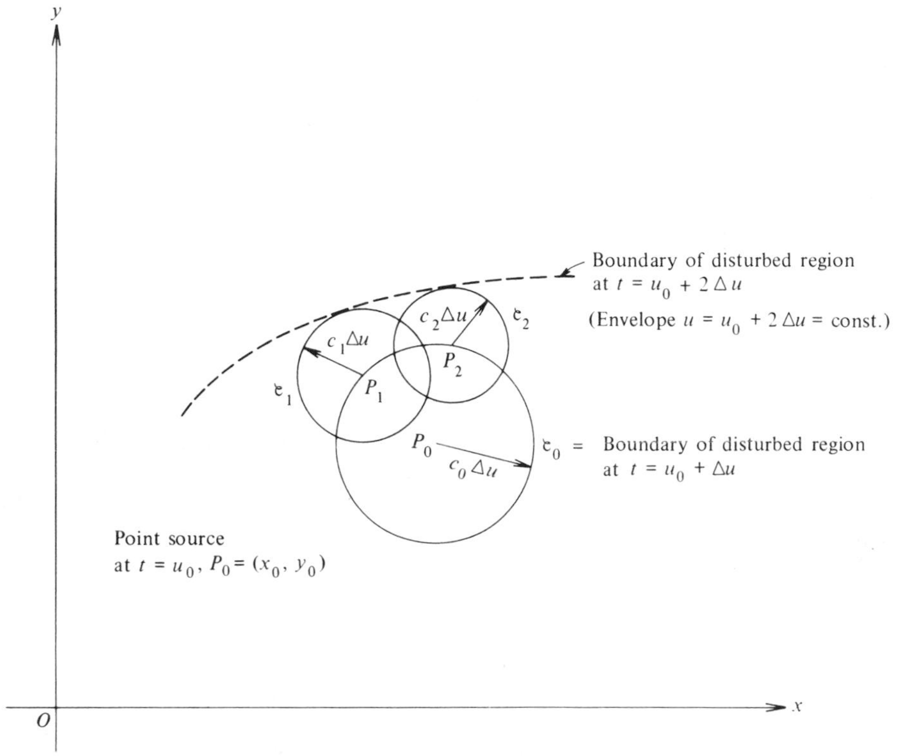

Fig. 8.1-1 Motion of disturbed region.

$c_2 = c(P_2)$ are different so that at time $t = u_0 + 2\Delta u$ the disturbances emitted from P_1 and P_2 have reached the circumferences of the two circles $\mathcal{C}_1$ and $\mathcal{C}_2$. Thus, the boundary of the disturbed region at time $u_0 + 2\Delta u$ is the envelope of all the circles associated with the points on $\mathcal{C}_0$. It is now clear how this construction can be continued indefinitely once we know what the local values of the speed of light, c, are. We also note that the above construction can be carried out starting from an arbitrary initial disturbed region not necessarily a point.

The important geometrical feature of Huygen's construction which will be the key for deriving its analytical description is the fact that the wave fronts are orthogonal to the light rays. Thus, if we denote the infinitesimal distance element along a light ray by ds, we have

$$du = |\text{grad } u| \, ds \tag{8.1-8}$$

in any isotropic medium.

Since du is the time elapsed for light to travel the distance ds, we also have

$$ds = c\,du \tag{8.1-9}$$

Combining eq. (8.1-8) and eq. (8.1-9) gives the *Eikonal Equation*

$$|\text{grad } u|^2 = \frac{1}{c^2} \tag{8.1-10a}$$

which, in a three-dimensional medium, has the form

$$u_x^2 + u_y^2 + u_z^2 = \frac{1}{c^2(x, y, z)} \tag{8.1-10b}$$

We have derived the Eikonal equation using very elementary physical arguments regarding the propagation of an optical signal. These arguments would also apply to any other signal propagation phenomenon. For example, if the acoustic velocity in air is a given function of x, y, z, then eq. (8.1-10) can be interpreted as the equation governing the propagation of acoustic disturbances.*

In Chapter 9, it is shown that the Eikonal equation defines the characteristic surfaces of the wave equation, and the foregoing derivation provides a very useful intuitive description of these surfaces.

In the next section we derive the Eikonal equation using the variational principle of Fermat which states that light rays travel along paths of minimum time. Equation (8.1-10) is also conventionally derived from the wave equation or Maxwell's equations for the case of almost constant speed of light (e.g., see Ref. 8-3 for a comprehensive treatment of the subject). In fact, if one assumes that the speed of light (or index of refraction) in a given medium does not change appreciably over a distance of the order of the wave length of some prescribed initial wave, then the equation governing surfaces of constant optical length obey the Eikonal equation. These surfaces defined by $u(x, y, z)$ = constant are called geometrical wave surfaces or geometrical wave fronts, and the function u is often called the Eikonal.

8.1.3 Hamilton-Jacobi Theory and Variational Calculus

Consider the functional J defined by

$$J = \int_{s_I}^{s_F} L(q_1, \ldots, q_n, \dot{q}_1, \ldots, \dot{q}_n, s)\, ds \tag{8.1-11}$$

where L is a given function of the $2n + 1$ variables q_i, $\dot{q}_i$ and s ($\dot{q}_i$ denotes dq_i/ds). Let the q_i be prescribed functions of s, and let the initial and final values s_I and s_F be fixed. Then J is a number which depends on the particular choice of s_I, s_F and the functions q_i. Let us now vary all the prescribed quantities by arbitrary infini-

*The reader is cautioned that we have adopted a very restricted physical picture in deriving the Eikonal equation. In general, wave propagation phenomena require a more accurate model in which one cannot treat the time as a dependent variable as in eq. (8.1-10).

tesimal amounts and denote the new values by an asterisk, i.e.,

$$J^* = \int_{s_I^*}^{s_F^*} L(q_1^*, \ldots, q_n^*, \dot{q}_1^*, \ldots, \dot{q}_n^*, s)\, ds \tag{8.1-12}$$

If we denote the small difference between a varied quantity and its original value by δ, i.e., $\delta J = J^* - J$, $\delta q = q^* - q$, etc., one can show that

$$\delta J = \left(L - \sum_{i=1}^{n} \frac{\partial L}{\partial \dot{q}_i} \dot{q}_i\right) \delta s_F \Bigg|_{s=s_F} - \left(L - \sum_{i=1}^{n} \frac{\partial L}{\partial \dot{q}_i} \dot{q}_i\right) \delta s_I \Bigg|_{s=s_I} + \left(\sum_{i=1}^{n} \frac{\partial L}{\partial \dot{q}_i} \delta q_i\right)_{s=s_F}$$

$$- \left(\sum_{i=1}^{n} \frac{\partial L}{\partial \dot{q}_i} \delta q_i\right)_{s=s_I} + \int_{s_I}^{s_F} \sum_{i=1}^{n} \left(\frac{\partial L}{\partial q_i} - \frac{d}{ds} \frac{\partial L}{\partial \dot{q}_i}\right) \delta q_i\, ds \tag{8.1-13}$$

The reader is referred to Chapter 20 for more details.

Now, let us consider the following variational principle. We seek functions $q_i(s)$ passing through the fixed end points $s_I, q_i(s_I)$ and $s_F, q_i(s_F)$ such that the value of J is stationary, i.e.,

$$\delta J = 0 \tag{8.1-14a}$$

with

$$\delta s_I = \delta s_F = 0 \tag{8.1-14b}$$

and

$$\delta q_i(s_I) = \delta q_i(s_F) = 0 \tag{8.1-14c}$$

For this variational principle eq. (8.1-15) yields the Euler equations

$$\frac{\partial L}{\partial q_i} - \frac{d}{ds}\left(\frac{\partial L}{\partial \dot{q}_i}\right) = 0 \qquad i = 1, \ldots, n \tag{8.1-15}$$

The solution defined by these equations, together with the prescribed $2n$ conditions on the $q_i(s_I)$ and $q_i(s_F)$, is called an *extremal* and satisfies the variational principle eqs. (8.1-14).

If we interpret L as the *Lagrangian* of a system of particles with s equal to the time and the q_i as the generalized coordinates for the system, then eqs. (8.1-14) is *Hamilton's Principle*, and eq. (8.1-15) are the familiar Lagrange equations in dynamics. See, for example, Ref. 8-4 for more details.

The main goal of this section is to establish the relationship of eq. (8.1-15 [or equivalently the variational principle eqs. (8.1-14)] with a partial differential equation of first order.

To this effect we consider a *Legendre transformation* whereby we replace the Lagrangian L by another function H called a *Hamiltonian* which depends on the q_i and a new set of variables p_i replacing the $\dot{q}_i$ according to

$$p_i = \frac{\partial L}{\partial \dot{q}_i} \qquad i = 1, \ldots, n \tag{8.1-16a}$$

$$H = \sum_{i=1}^{n} p_i \dot{q}_i - L \tag{8.1-16b}$$

The interpretation of the Legendre transformation eqs. (8.1-16) is as follows: We use eq. (8.1-16a) to solve for the $\dot{q}_i$ in terms of the q_i, p_i and s, and this result is then used in eq. (8.1-16b) to define H as a function of the q_i, p_i and s. Clearly, if the determinant of the matrix $\left\{\frac{\partial^2 L}{\partial \dot{q}_i \, \partial \dot{q}_j}\right\}$ is not equal to zero, one can solve for the q_i from (8.1-16a). Note that for the important class of problems to be considered later on, where L is a homogeneous function of order one in the $\dot{q}_i$, the functional determinant of $\left\{\frac{\partial^2 L}{\partial \dot{q}_i \, \partial \dot{q}_j}\right\}$ vanishes and a Legendre transformation does not exist.

Let us now consider only extremals connecting the fixed points $q_i(s_I)$, s_I and $q_i(s_F)$, s_F, i.e., solutions of eq. (8.1-15) subject to the boundary conditions

$$q_i(s_I) = \kappa_i \tag{8.1-17a}$$

$$q_i(s_F) = Q_i \tag{8.1-17b}$$

where the κ_i and Q_i are the prescribed initial and final values.

For the moment, let us fix s_I and the κ_i and consider a succession of values of s_F and corresponding Q_i. This construction will clearly generate a *field of extremals* emanating from the fixed point s_I and κ_i. If we denote

$$P_i = \left(\frac{\partial L}{\partial \dot{q}_i}\right)_{s=s_F} \tag{8.1-18}$$

eq. (8.1-13) then shows that

$$\frac{\partial J}{\partial s_F} = -(H)_{s=s_F} \tag{8.1-19a}$$

$$\frac{\partial J}{\partial Q_i} = P_i \qquad i = 1, \ldots, n \tag{8.1-19b}$$

Equations (8.1-19) hold whether or not a Legendre transformation exists. In the latter case eq. (8.1-19a) becomes equal to zero as will be shown separately later on.

If a Legendre transformation exists eqs. (8.1-19) combine to give

$$\frac{\partial J}{\partial s} + H\left(q_1, \ldots, q_n, \frac{\partial J}{\partial q_1}, \ldots, \frac{\partial J}{\partial q_n}, s\right) = 0 \tag{8.1-20}$$

which is called the *Hamilton-Jacobi equation.* In writing (8.1-20) we have recognized the fact that the end point s_F, Q_i is considered to vary over the allowable field of extremals and hence we have replaced s_F by s and Q_i by q_i.

A result identical to (8.1-20) holds if one varies the initial point holding the end point fixed. Thus, the field of extremals emanating from a given point obeys the Hamilton-Jacobi equation for J regarded as a function of its end point.

In section 8.4 we shall see the role played by the Euler equations in the solution of eq. (8.1-20). See also Chapter 20 for a discussion of the equivalence of Euler's equations with the Hamilton-Jacobi equation in defining the solution of the variational problem eqs. (8.1-14).

For certain problems in geometry as well as in geometrical optics L has the form

$$L = \left(\sum_{i,j=1}^{n} a_{ij}\, \dot{q}_i\, \dot{q}_j\right)^{1/2} \tag{8.1-21}$$

where the matrix a_{ij} is symmetric and its elements depend on the q_i. For example, let $n = 3$ and let the q_i denote Cartesian coordinates in an isotropic medium where the speed of light c is a given function of the q_i. Take L in the form

$$L = \frac{1}{c}\left[\sum_{i=1}^{3}\left(\frac{dq_i}{ds}\right)^2\right]^{1/2} \tag{8.1-22}$$

Thus

$$J = \int_{s_I}^{s_F}\left[\sum_{i=1}^{3}\left(\frac{dq_i}{ds}\right)^2\right]^{1/2}\frac{ds}{c} \tag{8.1-23}$$

is the time elapsed for light to travel from the point s_I to the point s_F along the curve $q_i(s)$, and the variational principle eqs. (8.1-14) is equivalent to Fermat's principle: Between any two points on a light ray traveling through a given medium, the path is one which minimizes the elapsed time.

Note that eq. (8.1-21) is homogeneous of order one and therefore according to *Euler's theorem for homogeneous functions* (see Chapter 2)

$$L = \sum_{i=1}^{n} \dot{q}_i \frac{\partial L}{\partial \dot{q}_i} = \sum_{i=1}^{n} p_i \dot{q}_i \tag{8.1-24}$$

and hence eq. (8.1-16b) implies that $H = 0$.

It is easy to show that if the b_{ij} are the elements of the inverse of the a_{ij} matrix, then

$$\sum_{i,j=1}^{n} b_{ij} p_i p_j = 1 \tag{8.1-25}$$

Therefore, eqs. (8.1-19b) and (8.1-25) combine to give

$$\sum_{i,j=1}^{n} b_{ij} \frac{\partial J}{\partial q_i} \frac{\partial J}{\partial q_j} = 1 \tag{8.1-26}$$

which is a *generalized Eikonal equation.* In particular, for the Lagrangian of eq. (8.1-22) where $b_{ij} = \delta_{ij} c^2$ (δ_{ij} is the Kroneker delta, equal to unity if $i = j$ and equal to zero otherwise) eq. (8.1-26) reduces to

$$\sum_{i=1}^{3} \left(\frac{\partial J}{\partial q_i} \right)^2 = \frac{1}{c^2} \tag{8.1-27}$$

which is precisely the Eikonal eq. (8.1-10) where J denotes u.

8.2 GEOMETRICAL CONCEPTS, QUALITATIVE RESULTS

In preparation for discussing the solution of first order partial differential equations, we review briefly in this section some of the ideas underlying the analytical description of various geometrical structures. We then consider and contrast the geometrical content of first order ordinary and partial differential equations.

8.2.1 Curves, Surfaces, and Manifolds

To begin with, consider a two dimensional Euclidean space in which a curve is to be described. One could introduce Cartesian coordinates, x and y, and then describe the curve by using a function, say $y = f(x)$. This form of description is awkward, if the curve in question requires that f be many-valued (as, for example, for the case of the unit circle $y = \pm\sqrt{1 - x^2}$). To avoid this difficulty, one might use the *implicit representation* $F(x, y) = 0$, which for the case of the unit circle is simply $x^2 + y^2 - 1 = 0$.

A much more convenient representation can be obtained by defining the curve *parametrically.** Thus, if s is a parameter which increases monotonically along the curve (as for instance, if s is the arc length along the curve) then any continuous curve can be defined parametrically by specifying the two functions $x = x(s)$ and $y = y(s)$. For the unit circle, $x(s) = \cos s$ and $y(s) = \sin s$, if we let s be the arc length measured along the circle in the counterclockwise sense from the point $x = 1$, $y = 0$. Note that $x(s) = \cos 2s$, $y(s) = \sin 2s$ where s is not equal to the arc length also defines the *same* circle; in fact, there is no unique parametrization for a given curve.

We also see that solving for s in terms of x from the equation $x = x(s)$ and substituting the result into $y = y(s)$ leads to the alternate description $y = f(x)$. In this case; if s is not a one-valued function of x then, in general, neither is f. Equivalent statements hold if the roles of x and y are interchanged.

The concept of parametric representation of a curve in the plane immediately generalizes to an n-dimensional Euclidean space, where we regard the set of n functions $x_i = x_i(s)$, $i = 1, \ldots, n$ as the Cartesian coordinates of an n-dimensional curve. Note that the implicit form which was useful in a two dimensional space becomes unwieldly in a space of higher dimension, as one would need $n - 1$ functions of the n coordinates to define a curve in an n-dimensional space.

Consider now a surface in a three-dimensional space. With Cartesian components x, y, and z one could, possibly at the price of having to deal with a many-valued function, use one of the coordinates expressed as a function of the other two for describing the surface. For example, the unit sphere $z = \pm\sqrt{x^2 + y^2 - 1}$ requires a two-valued function no matter which two coordinates we choose as independent variables. Or, one could use an implicit definition of a surface in the form $F(x, y, z) = 0$. Again, the parametric representation $x = x(s_1, s_2)$, $y = y(s_1, s_2)$, $z = z(s_1, s_2)$ is more natural. If for the case of the sphere, we let s_1 be the colatitude measured from the z-axis, and s_2 be the longitude measured from the x-axis, the appropriate parametric representation becomes $x = \sin s_1 \cos s_2$, $y = \sin s_1 \sin s_2$, $z = \cos s_1$. We obtain the above most directly by setting the radius equal to unity in the transformation formulas for spherical polar coordinates.

In general, in an n-dimensional space, a set of functions of the form $x_i = x_i(s_1, s_2, \ldots, s_m)$, $i = 1, \ldots, n$ with $m < n$ defines a geometrical structure of dimension m (lower than n); this is called an m-dimensional *manifold*. Thus, a curve is the degenerate case of a one-dimensional 'unifold.'

8.2.2 Families of Manifolds; Envelope Formation

In our study of the structure of the solution of first-order partial differential equations we shall have occasion to work with families of manifolds and associated singular manifolds. These concepts are reviewed next.

Consider the one-parameter *family of curves* in the plane defined in the form

$$x = x(s, \tau) \tag{8.2-1a}$$

*It is pointed out that when calculating line integrals, surface integrals, etc., the use of the parametric representation for curves, surfaces, etc., considerably simplifies the calculation.

$$y = y(x, \tau) \tag{8.2-1b}$$

or implicitly in the form

$$F(x, y, \tau) = 0 \tag{8.2-2}$$

This constitutes a family in the following sense. For any given fixed value of τ, the functions in eq. (8.2-1) provide a parametric representation of a curve in the plane. With this curve one associates a certain fixed value of τ. Now as τ takes on different values, we obtain different curves, and the totality of these curves constitutes a one-parameter (τ) family.

In a similar manner, we define a k-parameter family curves in an n-dimensional space by

$$x_i = x_i(s, \tau_1, \tau_2, \ldots, \tau_k); \qquad i = 1, \ldots, n \tag{8.2-3}$$

In this context it is important to distinguish between the variable s which varies along a curve, and the parameters τ_r which are fixed in order to define a member of the family.

More generally, we can define a k-parameter family of m-dimensional manifolds in an n-dimensional space by

$$x_i = x_i(s_1, \ldots, s_m, \tau_1, \ldots, \tau_k); \qquad i = 1, \ldots, n \tag{8.2-4}$$

Often for a given family of manifolds one can associate a singular manifold. To fix ideas, consider first a one parameter family of curves in the plane defined implicitly in the form eq. (8.2-2). Assume that this family has an envelope defined by the curve $f(x, y) = 0$ and having the qualitative behavior sketched in Fig. (8.2-1).

It is easy to see from the figure that at each of the points e_1, e_2, etc., *both* $F(x, y, \tau)$ and $(\partial F/\partial \tau)(x, y, \tau)$ vanish. Thus, if an envelope exists, it is obtained by eliminating τ from the two equations

$$F(x, y, \tau) = 0 \tag{8.2-5a}$$

$$\frac{\partial F}{\partial \tau}(x, y, \tau) = 0 \tag{8.2-5b}$$

If the family $F = 0$ is given parametrically in the form eq. (8.2-1) then eq. (8.2-5a) implies that as a function of s and τ the function $\varphi(s, \tau) = F[x(s, \tau), y(s, \tau), \tau]$ is identically equal to zero.

This, in turn implies that $\partial\varphi/\partial s = \partial\varphi/\partial\tau = 0$. Thus, at any point through which passes a member of the family $F = 0$, eq. (8.2-5a) implies that

$$\frac{\partial \varphi}{\partial s} = \frac{\partial F}{\partial x}\frac{\partial x}{\partial s} + \frac{\partial F}{\partial y}\frac{\partial y}{\partial s} = 0 \tag{8.2-6a}$$

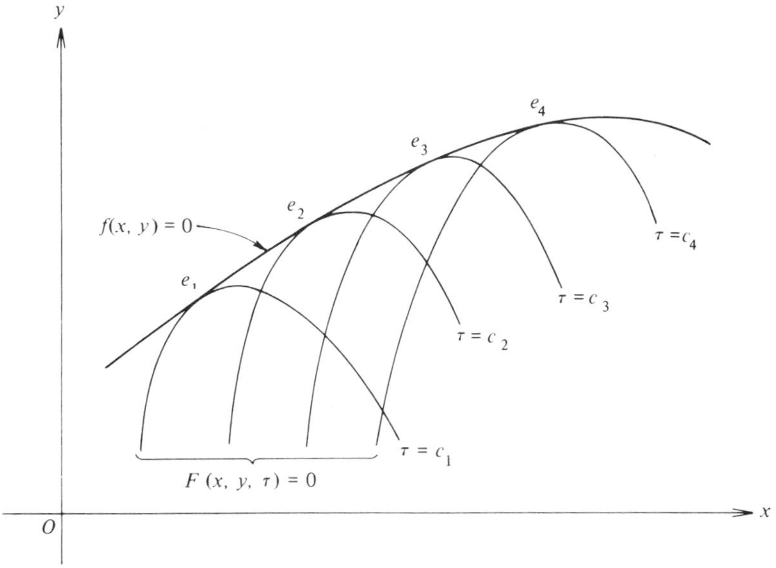

Fig. 8.2-1 Envelope formation.

$$\frac{\partial \varphi}{\partial \tau} = \frac{\partial F}{\partial x}\frac{\partial x}{\partial \tau} + \frac{\partial F}{\partial y}\frac{\partial y}{\partial \tau} + \frac{\partial F}{\partial \tau} = 0 \tag{8.2-6b}$$

Since along the envelope $\partial F/\partial \tau$ is identically zero, eq. (8.2-6) reduces to

$$\Delta(s, \tau) = \frac{\partial x}{\partial s}\frac{\partial y}{\partial \tau} - \frac{\partial y}{\partial s}\frac{\partial x}{\partial \tau} = 0 \tag{8.2-7}$$

which states that the Jacobian of the transformation eq. (8.2-1) vanishes at all points along the envelope. Solving for $\tau(s)$ from eq. (8.2-7) and substituting the result in eq. (8.2-1) gives the parametric form of the envelope.

The reader can also verify that the foregoing formulas may also define loci of *cusps* or *nodes* in a given family. Loci of turning points, cusps or nodes will generally be denoted as *singular.* Finally, it is emphasized that only if a given family has a singular curve can one simultaneously satisfy eqs. (8.2-5a) and (8.2-5b)—or, equivalently, eqs. (8.2-1) and (8.2-7).

As an example, consider the family defined by

$$F(x, y, \tau) = \tau^2 y^2 - 4\tau x + 4 = 0 \tag{8.2-8}$$

This family of parabolas together with its envelopes is shown in Fig. 8.2-2.

Calculating $\partial F/\partial \tau = 0$ gives $\tau = 2x/y^2$ along the envelope, and when this result is substituted into (8.2-8) to eliminate τ we find $y = \pm x$. The same result could have

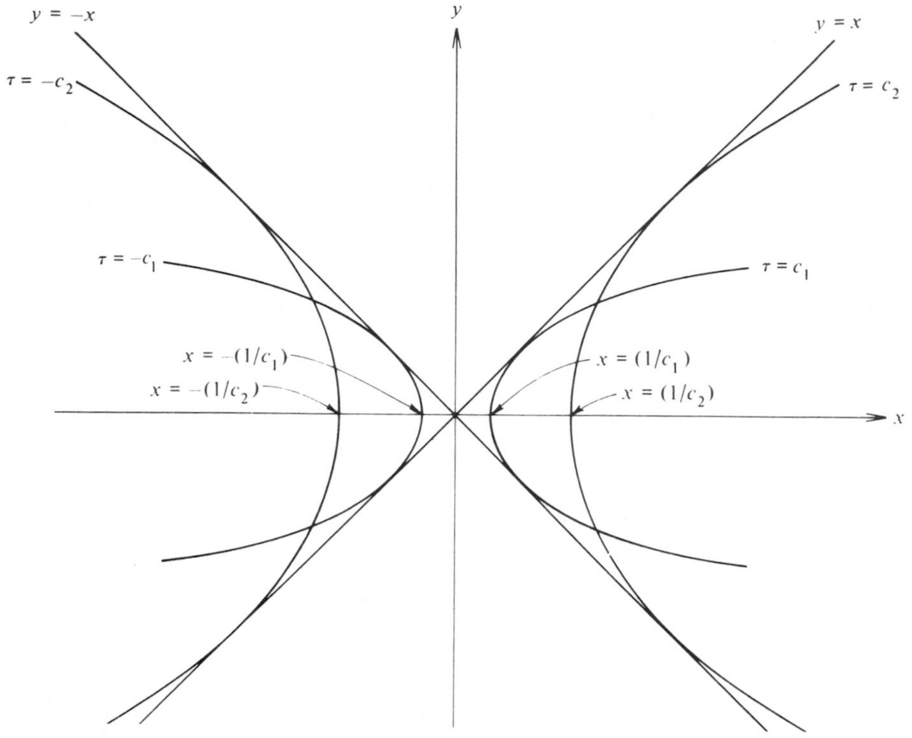

Fig. 8.2-2 Family of parabolas.
$F(x, y, \tau) = \tau^2 y^2 - 4\,\tau x + 4 = 0$

been obtained starting from any one of the possible parametric representations of the family eq. (8.2-8). Consider, for example, the representation

$$x = \frac{1}{\tau} + \frac{\tau s^2}{4} \tag{8.2-9a}$$

$$y = s \tag{8.2-9b}$$

The vanishing of the Jacobian implies that

$$\Delta(s, \tau) = -\left(-\frac{1}{\tau^2} + \frac{s^2}{4}\right) = 0$$

i.e.,

$$\tau = \pm\, 2/s \tag{8.2-10}$$

which when substituted in eqs. (8.2-9) gives again $y = \pm\, x$.

The above ideas generalize to families of manifolds, and particular applications will be given in later sections.

8.2.3 Differential Equations Associated with Families of Functions

Consider the one parameter family of curves in the x, u plane defined by

$$\Phi(x, u, \tau) = 0 \qquad (8.2\text{-}11a)$$

where τ is the parameter. Let us now regard u as a function of x along a given member of the family and calculate the total derivative of Φ with respect to x

$$\frac{d\Phi}{dx} = \frac{\partial \Phi}{\partial x} + \frac{\partial \Phi}{\partial u}\frac{du}{dx} = 0 \qquad (8.2\text{-}11b)$$

Eliminating τ from the pair eqs. (8.2-11) results in a first order (in general, nonlinear), ordinary differential equation of the form

$$F(x, u, u') = 0 \qquad (8.2\text{-}12)$$

where the prime denotes differentiation with respect to x.

For example, if in eq. (8.2-8) we replace y by u and consider it as a prototype for (8.2-11a), we easily calculate for eq. (8.2-11b) the following result

$$-4\tau + 2uu'\,\tau^2 = 0 \qquad (8.2\text{-}13)$$

Discarding the solution $\tau = 0$ which cannot satisfy eq. (8.2-11a) for finite values of x and u, we use the root $\tau = 2/uu'$ in eq. (8.2-8) to calculate

$$v'^2 - 4xv' + 4v = 0 \qquad (8.2\text{-}14)$$

where $v = u^2$. It is easy to verify that the two solutions of eq. (8.2-14) corresponding to the two roots of the quadratic expression for v' give the parabolas in the positive or negative half-planes.

The calculation leading to eq. (8.2-12) can be easily generalized to the case of an nth order ordinary differential equation, by considering an n-parameter family of curves of the form $\Phi(x, u, \tau_1, \ldots, \tau_n) = 0$ and eliminating the τ_i from the above by using the first n derivatives of $\Phi = 0$.

Another interesting result for ordinary differential equations of the form eq. (8.2-12) is the fact that if this equation has a singular solution in the sense defined in section 8.2.2 then this solution can be derived directly from the differential equation without having to calculate the general solution. Thus, if a singular solution exists, it is obtained by eliminating u' from eq. (8.2-12) and

$$\frac{\partial F}{\partial u'}(x, u, u') = 0 \qquad (8.2\text{-}15)$$

It is not surprising that a process of envelope formation for the family eq. (8.2-12), where u' is regarded as a parameter, leads to the singular solution, because the singular solution is also the envelope of the one-parameter family eq. (8.2-11a).

The foregoing ideas for ordinary differential equations set the stage for the necessary generalization to the case of first-order partial differential equations. It would be natural to expect that in order to derive a first-order partial differential equation associated with some given solution, this latter must contain an arbitrary *function* and must represent a family of manifolds rather than curves. As discussed in Ref. 8-1, it turns out that to each n-parameter family of manifolds of dimension n one can associate a first-order (in general nonlinear) partial differential equation in n-independent variables. To illustrate the procedure consider the two-dimensional case and let

$$\Phi(x, y, u, a, b) = 0 \tag{8.2-16}$$

be the two-parameter family of surfaces in x, y, u-space.

By eliminating a and b from the two equations (this can be done if $\Phi_{xa}\Phi_{yb} - \Phi_{ya}\Phi_{xb} \neq 0$)

$$\frac{\partial \Phi}{\partial u}\frac{\partial u}{\partial x} + \frac{\partial \Phi}{\partial x} = 0 \tag{8.2-17a}$$

$$\frac{\partial \Phi}{\partial u}\frac{\partial u}{\partial y} + \frac{\partial \Phi}{\partial y} = 0 \tag{8.2-17b}$$

and substituting the result into eq. (8.2-16) one derives a relation of the form

$$F\left(x, y, u, \frac{\partial u}{\partial x}, \frac{\partial u}{\partial y}\right) = 0 \tag{8.2-18}$$

The generalization of this process to the n-dimensional case is obvious and will not be discussed.

8.2.4 The Complete Integral, Singular Integral, Envelope Formation, Separation of Variables

A solution of the form (8.2-16) involving two *arbitrary** parameters a, b is called a complete integral of eq. (8.2-18). Its role is analogous to that of a general solution for the case of an ordinary differential equation. In fact, we shall show presently that one can use a process of envelope formation to derive a solution of eq. (8.2-18) involving an arbitrary function once the complete integral eq. (8.2-16) is known.

Thus, finding the complete integral is in most cases equivalent to the solution of a

*Note that when (8.2-18) is independent of u the second arbitrary constant in the complete integral is additive, i.e., one can write the complete integral in the form $u = \Phi(x, y, a) + b$.

given initial value problem. As indicated later on, one can sometimes calculate the complete integral very easily, and this fact has far-reaching implications in the Hamilton-Jacobi theory for dynamics (cf. section 8.4.3).

A word of caution is appropriate regarding the generality of a complete integral. As pointed out in Ref. 8-1, not all solutions of a given first-order partial differential equation can be obtained from a complete integral, and in this sense the latter is not as general as the 'general solution' of an ordinary differential equation.

To construct a solution involving an arbitrary function let $b = w(a)$ where w is arbitrary. Now eq. (8.2-16) becomes a one parameter family of surfaces

$$\Phi\,[x, y, u, a, w(a)] = 0 \tag{8.2-19}$$

If these surfaces have an envelope, this latter is obtained by eliminating a from eq. (8.2-19) and

$$\frac{\partial\Phi}{\partial a} + w'(a)\,\frac{\partial\Phi}{\partial b} = 0 \tag{8.2-20}$$

To see the details of this process, and to verify the assertion that this envelope is indeed a solution of eq. (8.2-18) we consider the example discussed in Ref. 8-1.

Let Φ denote the two-parameter family of unit spheres with centers at $x = a$, $y = b$, $u = 0$ in x, y, u-space. Thus

$$\Phi = (x - a)^2 + (y - b)^2 + u^2 - 1 = 0 \tag{8.2-21}$$

In this case eq. (8.2-17) becomes

$$uu_x + (x - a) = 0; \qquad \text{i.e.,}\quad a = x + uu_x \tag{8.2-22a}$$

$$uu_y + (y - b) = 0; \qquad \text{i.e.,}\quad b = y + uu_y \tag{8.2-22b}$$

Substituting the resulting values of a and b into eq. (8.2-21) yields

$$u^2(u_x^2 + u_y^2 + 1) - 1 = 0 \tag{8.2-23}$$

The geometrical interpretation of the envelope formed by the family of unit spheres with $b = w(a)$ is quite straightforward. Clearly, when these unit spheres are centered along the curve $y = w(x)$ their envelope will be a tabular surface of unit radius and axis along the curve $y = w(x)$. Evaluating eq. (8.2-20) for our case gives

$$(x - a) + [y - w(a)]\,w'(a) = 0 \tag{8.2-24}$$

If we let $w = x$, the above yields $a = (x + y)/2$ which when substituted into eq.

(8.2-21) gives the equation for the circular cylinder

$$(x - y)^2 + 2(u^2 - 1) = 0 \tag{8.2-25}$$

with axis along $y = x$.

The reader can show that letting $w(a) = \pm\sqrt{R^2 - a^2}$ will lead to the equation for a torus of large radius R and small radius unity.

The relation of solutions obtained by envelope formation from a complete integral with a given initial value problem will be discussed in section 8.4.1. In fact, as can be seen from the foregoing example, the arbitrary function w is not directly related to a given initial curve. In contrast, for the case of an ordinary differential equation of the type eq. (8.2-12) once an initial point is given, i.e., $u = u_0$ at $x = x_0$, the constant τ can immediately be calculated from the general solution eq. (8.2-11a).

The *singular integral* is the direct analogue of the singular solution of an ordinary differential equation, and is obtained by double envelope formation with respect to both parameters a and b. Thus, we calculate a singular integral by eliminating a and b from eq. (8.2-16) and the solution of

$$\Phi_a = \Phi_b = 0 \tag{8.2-26}$$

In the case of the family of spheres this results in the two parallel planes $u = \pm 1$ as is geometrically obvious. Again, the reader can easily extend all the above ideas to the case of n-dimensions.

8.2.5 Separation of Variables

Let us now turn to a very useful approach for calculating the complete integral, viz., separation of variables. If the assumption that the complete integral can be expressed as

$$u = X(x, a, b) + Y(y, a, b) \tag{8.2-27}$$

allows for the separation of the partial differential equation (8.2-19) in the form

$$F_1\left(X, \frac{\partial X}{\partial x}\right) = F_2\left(Y, \frac{\partial Y}{\partial y}\right) = \text{const.} \tag{8.2-28}$$

then the latter can in principle be solved by quadratures.

As is evident, separability depends on the choice of coordinates. Once such a choice has been made it is a straightforward matter to verify whether or not the equation is separable. Unfortunately, there is no known criterion for finding a coordinate system in terms of which a given partial differential equation is separable. In fact, generally, one cannot even ascertain whether such a coordinate system exists.

Note that the above type of separation of variables is considerably more far-reaching than that encountered for boundary-value problems for higher order equations. In the latter case separation of variables is usually performed in product form, but more importantly its success depends not only on the coordinate system but also on the prescribed boundary conditions. Here, on the other hand, the separability of a given partial differential equation, if feasible, depends only on the choice of coordinates. In section 8.4 we will see that separability of the Hamilton-Jacobi equation is equivalent to the solvability of an associated system of ordinary differential equations, i.e., the existence of a sufficient number of integrals.

As an illustrative example we consider the Eikonal eq. (8.2-10a) in cylindrical polar coordinates with c depending only on the radial distance. Using the expression for the gradient in polar coordinates (see Chapter 3) leads to

$$\left(\frac{\partial u}{\partial r}\right)^2 + \frac{1}{r^2}\left(\frac{\partial u}{\partial \theta}\right)^2 = \frac{1}{c^2(r)} \equiv n^2(r) \tag{8.2-29}$$

We assume that u separates in the form

$$u = R(r, a, b) + \Theta(\theta, a, b) \tag{8.2-30}$$

and substitute it in eq. (8.2-29) to find

$$\left(\frac{\partial R}{\partial r}\right)^2 + \frac{1}{r^2}\left(\frac{\partial \Theta}{\partial \theta}\right)^2 = n^2(r) \tag{8.2-31}$$

The above implies that

$$r^2\left(\frac{\partial R}{\partial r}\right)^2 - r^2 n^2(r) = \left(\frac{\partial \Theta}{\partial \theta}\right)^2 = a^2 = \text{constant} \tag{8.2-32}$$

Therefore, the Eikonal equation with cylindrical symmetry is separable in polar coordinates.

Equation (8.2-32) can immediately be integrated to give

$$\Theta = a\theta \tag{8.2-33a}$$

$$R = \int^r [(a/\xi)^2 + n^2(\xi)]^{1/2}\, d\xi + b \tag{8.2-33b}$$

which determines the complete integral.

The reader may verify that had one started with the equation in Cartesian coordinates with $c = c(\sqrt{x^2 + y^2})$ it would have been impossible to calculate the complete integral by separation of variables.

8.2.6 Geometric Interpretation of a First-Order Equation

Crucial to the discussion of solution methods is an understanding of the geometrical content of a first-order partial differential equation.

Here again, let us review the analogous arguments appropriate for an ordinary differential equation. Consider eq. (8.2-12) in the x, u-plane, and let x_0, u_0 be a point through which passes a solution curve. Evaluating eq. (8.2-12) at x_0, u_0 gives a unique value for $u'(x_0, u_0)$ except if x_0, u_0 is a singular point, in which case u' is undefined. As shown in Chapter 6, one can linearize the problem near a singular point and obtain the solution near that point easily. Everywhere else, eq. (8.2-12) defines a field of tangents in the plane, and a solution curve is one which is everywhere tangent to this field. Incidentally, one could use this property to graphically construct an approximation for the solution (*method of isoclines*).

Understandably, the situation is more complicated for a partial differential equation. The simplest nontrivial case to visualize is the quasilinear equation in two independent variables.

$$a(x,y,u)\,u_x + b(x,y,u)\,u_y = c(x,y,u) \tag{8.2-34}$$

Let us examine the implications of eq. (8.2-34) regarding a solution surface in x, y, u-space. Since possible solution surfaces are functions of the form

$$\Phi = \varphi(x,y) - u = 0 \tag{8.2-35}$$

we wish to see what restrictions are placed upon eq. (8.2-35) by eq. (8.2-34).

Consider the normal to the surface $\Phi = 0$. This is simply the vector **N** with components $\Phi_x = u_x$, $\Phi_y = u_y$, and -1 along the x-, y-, and u-directions respectively. (The sense of **N**, i.e., outward or inward and its magnitude are immaterial for this discussion, since we could have defined the solution surface $\Phi = 0$ with an arbitrary multiplicative constant.)

At any point $P_0 = (x_0, y_0, u_0)$ on a possible solution surface, the three *numbers* $a_0 = a(x_0, y_0, u_0)$, $b_0 = b(x_0, y_0, u_0)$ and $c_0 = c(x_0, y_0, u_0)$ can be calculated for the particular eq. (8.2-34) and this partial differential equation evaluated at P_0 reduces to an *algebraic* equation (in this case also a linear one) between u_x and u_y. Interpreting this relation at P_0 in terms of $\mathbf{N}_0 = \mathbf{N}(x_0, y_0, u_0)$, we see that eq. (8.2-34) defines a one-parameter family of normals emanating from P_0. Moreover, because eq. (8.2-34) is linear the end points of the possible $\mathbf{N}_0$ lie in a *plane* as shown isometrically in Fig. 8.2-3. If we denote the components of these normals by N_1, N_2, N_3, eq. (8.2-34) states

$$a_0 \frac{N_1}{N_3} + b_0 \frac{N_2}{N_3} = c_0 \tag{8.2-36}$$

Now we construct planes perpendicular to this family of normals. These are therefore planes tangent to all the possible solution surfaces passing through P_0.

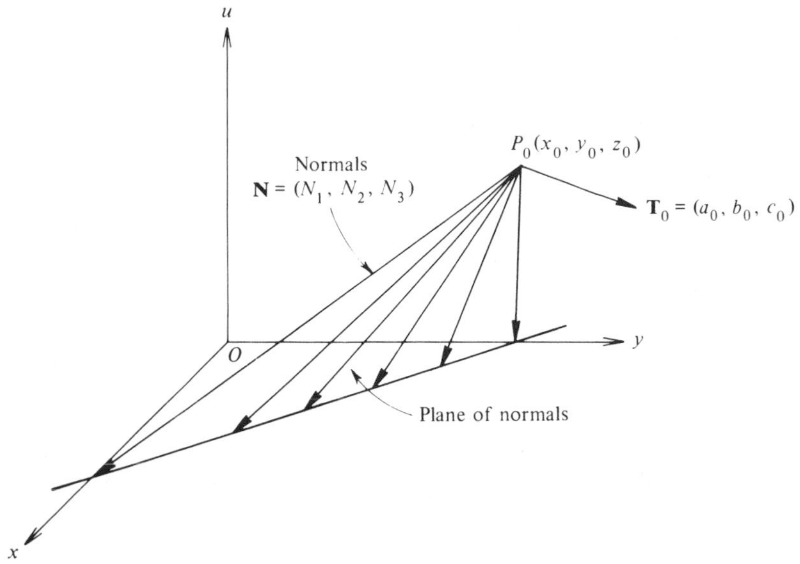

Fig. 8.2-3 Plane of normals to possible solution surfaces for a quasilinear equation.

$$a(x, y, u)\, u_x + b(x, y, u)\, u_y = c(x, y, u)$$

Again, since the equation is linear and hence the normals lie in a plane, it follows that the tangent planes all intersect along a line through the point P_0. Moreover, the coordinate components along this line are proportional to (a_0, b_0, c_0).

To derive this fact analytically, we write eq. (8.2-34) at P_0 in the form

$$a_0 u_x(x_0, y_0, u_0) + b_0 u_y(x_0, y_0, u_0) - c_0 = 0 \tag{8.2-37}$$

which states that the vector $\mathbf{T}_0$ given by

$$\mathbf{T}_0 = (a_0, b_0, c_0) \tag{8.2-38}$$

is perpendicular to all the normals at P_0 and is therefore common to all the planes tangent to the possible solution surfaces at P_0.

The direction $\mathbf{T}_0$, which as we shall see later is of fundamental importance in constructing a solution, is called a *characteristic direction* or *Monge axis*. A curve in x, y, u-space which is everywhere tangent to the characteristic direction is called a *characteristic curve.* The projection of this curve onto the x, y-plane is called a *characteristic ground curve* and is defined by $dx/ds = a, dy/ds = b$. Thus the quasilinear equation, eq. (8.2-34), implies that on a characteristic ground curve $du/ds = c$.

To summarize, we have shown that through any point P_0 there is a one-parameter family of possible planes tangent to a solution and that these planes have the characteristic direction $\mathbf{T}_0$ in common.

With the above geometrical structure one can visualize qualitatively the means for constructing a solution passing through a given curve. Although a graphical approximate construction might be impractical, the steps one would have to follow shed considerable light on the analytical nature of the solution to be discussed in the next section.

Consider a given curve in x, y, u-space denoted by $\mathcal{C}_0$ and given in parametric form with τ as the parameter along $\mathcal{C}_0$. To construct a solution passing through $\mathcal{C}_0$ we calculate the characteristic directions all along $\mathcal{C}_0$, say at the points $\tau = 0$, $\Delta\tau$, $2\Delta\tau$, etc., and if the characteristic directions are not everywhere tangent to $\mathcal{C}_0$ we can proceed along these directions a short distance Δs and obtain a neighboring curve $\mathcal{C}_1$. Continuing this process of "patching" appropriate strips we obtain a solution surface containing $\mathcal{C}_0$ as shown in Fig. 8.2-4.

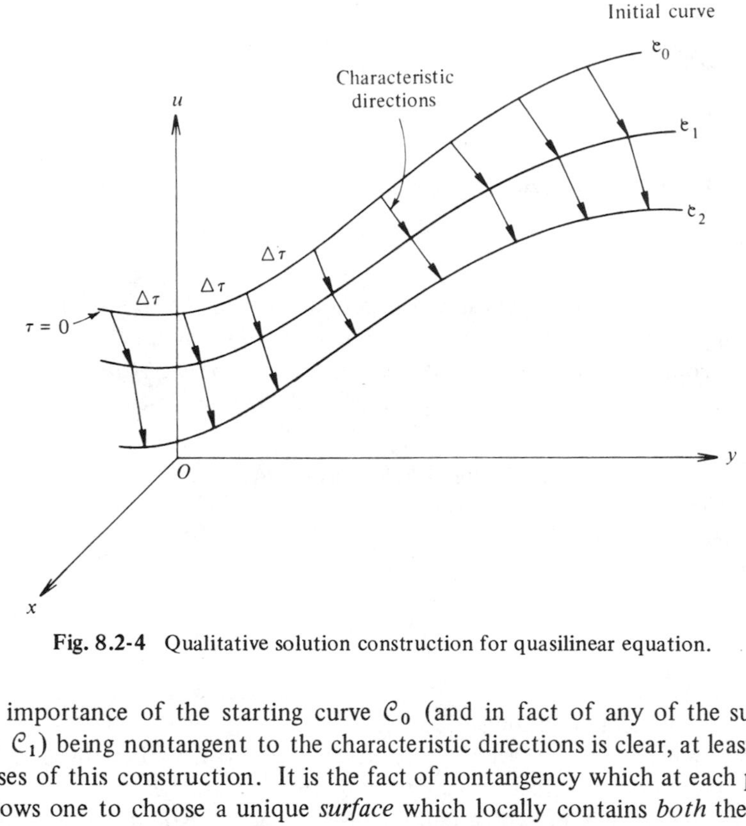

Fig. 8.2-4 Qualitative solution construction for quasilinear equation.

The importance of the starting curve $\mathcal{C}_0$ (and in fact of any of the successive curves $\mathcal{C}_1$) being nontangent to the characteristic directions is clear, at least for the purposes of this construction. It is the fact of nontangency which at each point on $\mathcal{C}_0$ allows one to choose a unique *surface* which locally contains *both* the starting curve and the characteristic. Had the two directions coincided, there would have been available a one-parameter family of possible solution surfaces with no criterion specifying a unique choice.

The geometry is more complicated for the case of a nonlinear equation, primarily because the normals to possible solution surfaces through a point no longer lie in a plane. In fact, at P_0 the one-parameter family of possible normals with components

$(u_x, u_y, -1)$ obeys the *nonlinear* algebraic equation

$$F(x_0, y_0, u_0, u_x, u_y) = 0 \tag{8.2-39}$$

wherein u_y may be a multi-valued function of u_x. Thus, the possible tangent planes no longer intersect along one line but instead they envelope along a curved surface called the *Monge cone.* For purposes of visualization let us reconsider the Eikonal equation in two dimensions with a variable speed of light c

$$u_x^2 + u_y^2 = n^2(x, y) = \frac{1}{c^2} \tag{8.2-40}$$

For this case, the one parameter family of normals describe a circular cone with apex at P_0 and base radius $n_0 = n(x_0, y_0)$. The envelope of planes tangent to possible solutions is also a cone with the same geometry. In this example, which historically motivated the name, the Monge cone is indeed a cone. The reader is cautioned that when (8.2-39) defines some other curved surface, the term *cone* is used only in the local sense. That is, one could find a cone (not necessarily circular) which, over a small neighborhood of interest on a particular solution surface, coincides with the actual Monge surface. Some authors use the term *Monge conoid* to denote this Monge surface.

Let us now study qualitatively the procedure for constructing a solution surface. Starting from a given curve $\mathcal{C}_0$, we calculate the Monge cones associated with each point on the curve. This one-parameter family of cones will, in general, have more than one envelope. For example, for eq. (8.2-40) one has two envelope surfaces each of which can be used to extend the solution from $\mathcal{C}_0$. Thus, we see that for a nonlinear equation one must specify an *initial strip* rather than just an initial curve in order to be able to continue the solution. By a strip *we mean a curve attached to an infinitesimal tangent surface.* In fact, this strip cannot be prescribed arbitrarily but must be the envelope of Monge cones emanating from the points on $\mathcal{C}_0$.

Given this strip we extend it to a neighboring strip by enveloping the Monge cones at the border of the initial strip and requiring the two strips to be joined smoothly. This smoothness requirement precludes the possibility of switching from one brach of the envelope to another since then the derivatives across the juncture of two branches would be discontinuous.

This construction is sketched in Fig. 8.2-5 for an Eikonal equation and for clarity only two points are used along $\mathcal{C}_0$. The cones are sketched using solid lines, and their lines of tangency with the integral surface are indicated by dotted lines.

The strip between $\mathcal{C}_0$ and $\mathcal{C}_1$ is specified by the initial data. In going from $\mathcal{C}_1$ to $\mathcal{C}_2$ we envelope the Monge cones for the points on $\mathcal{C}_1$ and from the two available choices we pick the one which joins the two strips smoothly. Since, $n \neq 0$ the cone angle is finite and this choice is unambiguous.

For a quasilinear equation the Monge cone collapses into a Monge axis, and we see why one does not need to keep track of the appropriate strip.

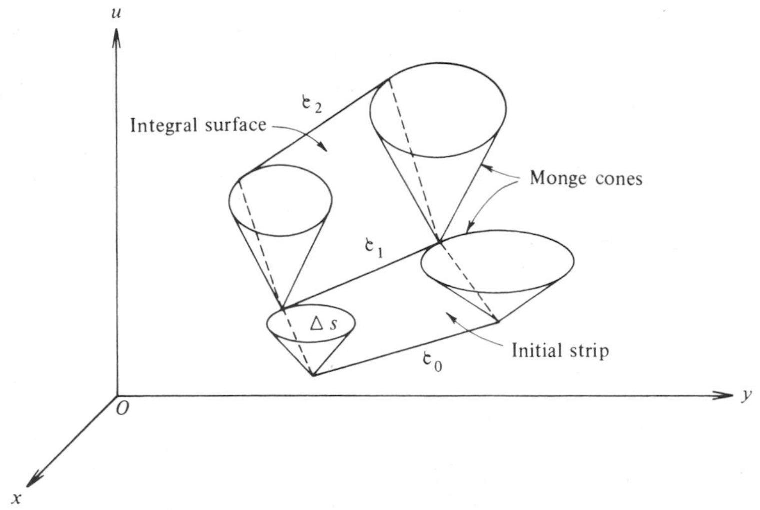

Fig. 8.2-5 Qualitative solution construction for nonlinear equation.

The geometrical interpretations given above generalize to higher dimensions but one losses the possibility of visualization available in three space dimensions.

8.3 QUASILINEAR EQUATIONS

In this section we list without proof the principal results concerning quasilinear equations and discuss the appropriate solution techniques. The reader is referred to Ref. 8-1 for proofs and further discussion.

8.3.1 Two Independent Variables

The geometrical aspects of the solution of eq. (8.2-34) was discussed qualitatively in section 8.2.6. In this subsection we consider the analytical results needed to solve problems and we discuss some illustrative examples in detail.

Throughout this section it will be assumed that the coefficients of eq. (8.2-34) are continuously differentiable functions, and that $a^2 + b^2 \neq 0$ anywhere. Furthermore, we use the term *solution* to denote a one-valued function of x and y which is continuously differentiable and satisfies eq. (8.2-34) everywhere. This precludes the case of discontinuous initial data, as well as solutions with discontinuities, which are termed 'weak' and discussed in sections 8.3.2 and 8.3.3.

The characteristic directions which are formed by the vector field $\mathbf{T} = (a, b, c)$ in x, y, u-space can be connected with a family of curves called *characteristic curves*. These curves are defined by the solution of the following set of first order ordinary differential equations.

$$\frac{dx}{ds} = a(x, y, u) \tag{8.3-1a}$$

$$\frac{dy}{ds} = b(x, y, u) \tag{8.3-1b}$$

$$\frac{du}{ds} = c(x, y, u) \tag{8.3-1c}$$

Note that since the system eq. (8.3-1) is autonomous the solution depends on $s - s_0$ and two other constants. Thus, the constant s_0 may be set equal to zero, and the solution defines a two-parameter family of curves.

An *integral surface* or *solution surface* consists of a surface $u(x, y)$ which at each point has a tangent along the characteristic direction. Thus, it is formed by smoothly joining appropriate characteristic curves. In fact, one can show that a necessary and sufficient condition for a surface to be an integral surface is that it be generated by a one-parameter family of characteristic curves.

An *initial-value problem* consists of finding an integral surface which passes through a given curve $\mathcal{C}_0$ called an initial curve and defined by

$$x = x_0(\tau) \tag{8.3-2a}$$

$$y = y_0(\tau) \tag{8.3-2b}$$

$$u = u_0(\tau) \tag{8.3-2c}$$

This means that out of the two-parameter family of curves obtained by solving eq. (8.3-1) we isolate a one-parameter family

$$x = x(s, \tau) \tag{8.3-3a}$$

$$y = y(s, \tau) \tag{8.3-3b}$$

$$u = u(s, \tau) \tag{8.3-3c}$$

which satisfies the conditions

$$x(0, \tau) = x_0(\tau) \tag{8.3-4a}$$

$$y(0, \tau) = y_0(\tau) \tag{8.3-4b}$$

$$u(0, \tau) = u_0(\tau) \tag{8.3-4c}$$

Thus, we have set $s = 0$ on the initial curve.

If the Jacobian of the transformation eq. (8.3-3a) and eq. (8.3-3b) is not equal to zero everywhere along $\mathcal{C}_0$, there is a small neighborhood of $\mathcal{C}_0$ in which one can solve for s, and τ in terms of x, y and derive a solution surface by substituting this result into eq. (8.3-3c). This result can be generalized as follows.

Equation (8.3-3c) defines a unique solution surface in parametric form in some neighborhood of the given initial curve as long as $\Delta(0, \tau) \neq 0$, where

$$\Delta(s, \tau) = x_s y_\tau - y_s x_\tau \tag{8.3-5}$$

Moreover, this solution can be extended uniquely beyond the initial curve as long as one does not reach a singular curve in the x, y-plane on which $\Delta(s, \tau) = 0$ everywhere. Such a curve is defined implicitly in terms of s, and τ by the relation $\Delta(s, \tau) = 0$.

The solution surface so generated may be visualized geometrically by considering it as an infinitely dense collection of characteristic curves one layer thick. On each of these characteristic curves s varies in both directions to plus and minus infinity from the value zero on $\mathcal{C}_0$. On the other hand, τ remains constant on each characteristic curve.

We now turn to the case where Δ is identically equal to zero on $\mathcal{C}_0$, and this case admits two possibilities: 1) If $\Delta(0, \tau) = 0$ for all τ, and $\mathcal{C}_0$ is characteristic, then one can pass an infinite number of solution surfaces through $\mathcal{C}_0$; and 2) If, however, $\Delta(0, \tau) = 0$, and $\mathcal{C}_0$ is not characteristic, there is no continuously differentiable solution of the initial value problem near $\mathcal{C}_0$. The above statements also hold when $\Delta(s, \tau)$ becomes equal to zero on some other curve away from the initial curve.

If Δ vanishes along a curve $\mathcal{C}$ which is not characteristic then one can show that the projection of $\mathcal{C}$ on the $u = 0$ plane is an envelope of the projection on the same plane of characteristic curves.

For the case of a linear differential equation the vanishing of Δ along a noncharacteristic curve means that the system eq. (8.3-3) describes a cylinder perpendicular to the x, y-plane.

To illustrate the foregoing ideas, consider the following example taken from Ref. 8-1.

$$uu_x + u_y = 1 \tag{8.3-6}$$

The differential equations governing the characteristic curves are

$$\frac{dx}{ds} = u \tag{8.3-7a}$$

$$\frac{dy}{ds} = 1 \tag{8.3-7b}$$

$$\frac{du}{ds} = 1 \tag{8.3-7c}$$

The following solution depending on three constants, x_0, y_0, u_0 is easily obtained

$$x = \frac{s^2}{2} + su_0 + x_0 \tag{8.3-8a}$$

$$y = s + y_0 \tag{8.3-8b}$$

$$u = s + u_0 \tag{8.3-8c}$$

If we regard x_0, y_0, and u_0 as arbitrary functions of a parameter τ, we see that eq. (8.3-8) is in the form eq. (8.3-3) where $x_0(\tau)$, $y_0(\tau)$, and $u_0(\tau)$ represent the initial curve parametrically. We calculate

$$\Delta(s, \tau) = [s + u_0(\tau)]\, y_0'(\tau) - [su_0'(\tau) + x_0'(\tau)] \tag{8.3-9}$$

and study various particular choices of initial curves.

1. Using $x_0 = \tau$, $y_0 = 2\tau$, $u_0 = \tau$ for $\mathcal{C}_0$ gives the following one-parameter family of characteristic curves

$$x = \frac{s^2}{2} + s\tau + \tau \tag{8.3-10a}$$

$$y = s + 2\tau \tag{8.3-10b}$$

$$u = s + \tau \tag{8.3-10c}$$

We calculate the following expression for $\Delta(s, \tau)$

$$\Delta(s, \tau) = s + 2\tau - 1 \tag{8.3-11}$$

and conclude that since $\Delta(0, \tau) = 2\tau - 1$ only vanishes at one point ($\tau = \frac{1}{2}$) on $\mathcal{C}_0$, there exists some neighborhood of $\mathcal{C}_0$ excluding the curve $s + 2\tau - 1 = 0$ for which an integral surface exists. We calculate this surface by solving for s and τ in terms of x and y from eqs. (8.3-10a) and (8.3-10b) and substituting the result in eq. (8.3-10c). This gives

$$s = \frac{(y - 2x)}{(1 - y)} \tag{8.3-12a}$$

$$\tau = \frac{(2x - y^2)}{2(1 - y)} \tag{8.3-12b}$$

$$u = \frac{(2y - y^2 - 2x)}{2(1 - y)} \tag{8.3-12c}$$

It is clear from the above that the line $y = 1$ is a singularity in the transformation $s, \tau \leftrightarrow x, y$ as well as in the solution. Indeed, if we examine the relation eq. (8.3-11) using eqs. (8.3-12a) and (8.3-12b) we find that $\Delta(s, \tau) = 0$ actually defines the line $y = 1$. The exceptional nature of this line is depicted by Fig. 8.3-1 in which we sketch the family of parabolas eq. (8.3-12b) representing the projections of the characteristic curves on the x, y-plane.

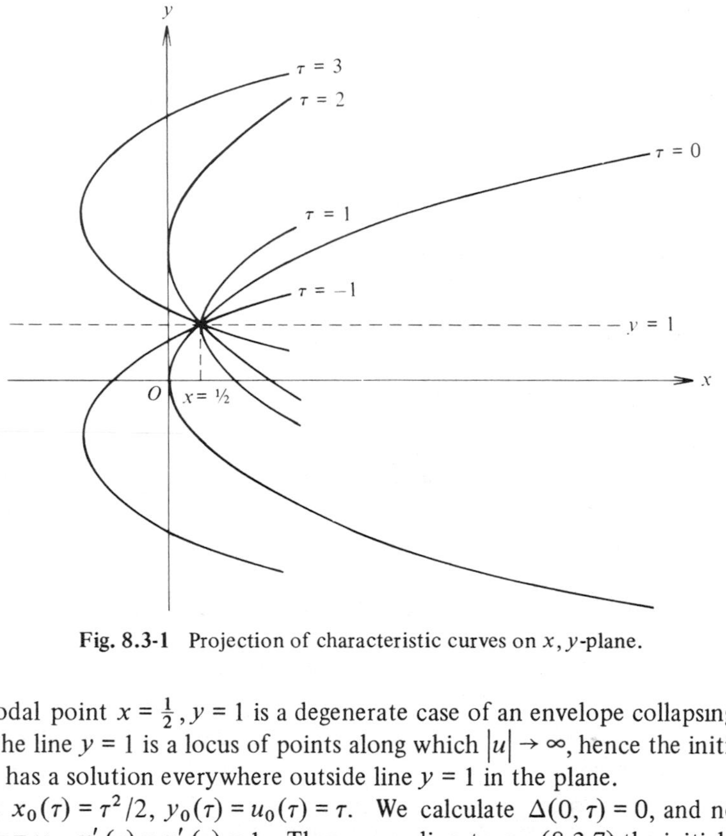

Fig. 8.3-1 Projection of characteristic curves on x, y-plane.

The nodal point $x = \frac{1}{2}, y = 1$ is a degenerate case of an envelope collapsing onto a point. The line $y = 1$ is a locus of points along which $|u| \to \infty$, hence the initial value problem has a solution everywhere outside line $y = 1$ in the plane.

2. Let $x_0(\tau) = \tau^2/2$, $y_0(\tau) = u_0(\tau) = \tau$. We calculate $\Delta(0, \tau) = 0$, and note that $x_0'(\tau) = \tau = u_0$, $y_0'(\tau) = u_0'(\tau) = 1$. Thus, according to eq. (8.3-7) the initial curve is characteristic. In order to demonstrate the nonuniqueness of the solution, we use the above initial values in eq. (8.3-8) to calculate the following expression for the characteristic curves

$$x = \frac{(s + \tau)^2}{2} \tag{8.3-13a}$$

$$y = (s + \tau) \tag{8.3-13b}$$

$$u = (s + \tau) \tag{8.3-13c}$$

Since in eq. (8.3-13) s and τ only occur in the combination $(s + \tau)$ it is clear u is not uniquely defined. In fact, for any function $w(z)$ with $w(0) = 0$, the following defines a solution surface u satisfying eq. (8.3-13):

$$w(u - y) + \frac{u^2}{2} - x = 0 \tag{8.3-14}$$

3. Finally, we consider the initial curve $x_0 = \tau^2$, $y_0 = 2\tau$, $u_0 = \tau$. It is easy to verify that $\Delta(0, \tau) = 0$ but $\mathcal{C}_0$ is not characteristic.

To illustrate the nature of the initial curve we evaluate eq. (8.3-8) for this case

$$x = \frac{s^2}{2} + s\tau + \tau^2 \tag{8.3-15a}$$

$$y = s + 2\tau \tag{8.3-15b}$$

$$u = s + \tau \tag{8.3-15c}$$

Eliminating s from eqs. (8.3-15a) and (8.3-15b) gives the following implicit form for the projection of the one-parameter family of characteristic curves

$$F(x, y, \tau) = \frac{(y - 2\tau)^2}{2} + \tau(y - 2\tau) + \tau^2 - x = 0 \tag{8.3-16}$$

Equation (8.3-16) defines a family of parabolas with envelope, (obtained by eliminating τ from (8.3-16) and the expression $F_\tau = 0$)

$$y^2 = 4x \tag{8.3-17}$$

Thus, the projection of the initial curve on the $u = 0$ plane is the envelope of the corresponding projection of characteristic curves.

Let us now investigate the behavior of the surface u defined by eq. (8.3-15). It is a straightforward calculation to show that eq. (8.3-15) gives

$$u = \frac{y}{2} \pm \left(x - \frac{y^2}{4}\right)^{1/2} \tag{8.3-18}$$

Thus, for values of x and y where u is defined, i.e., $x \geqslant y^2/4$, u is two-valued. More importantly, however we note that u_x and u_y are infinite on the initial curve.

Another interesting example consists of the homogeneous part of eq. (8.3-6) with y interpreted as the time t, and an initial curve corresponding to a wave. More precisely, we study the initial value problem.

$$u_t + uu_x = 0 \tag{8.3-19a}$$

subject to the initial condition

$$u(x, 0) = \sin x \tag{8.3-19b}$$

Integrating eq. (8.3-1) with $a = u$, $b = 1$, $y = t$, $c = 0$ we find the following one-parameter family of characteristic curves satisfying the initial condition eq. (8.3-19b) (which can also be stated in the form $x_0 = \tau$, $t_0 = 0$, $u_0 = \sin \tau$):

$$x = s \sin \tau + \tau \tag{8.3-20a}$$

$$t = s \tag{8.3-20b}$$

$$u = \sin \tau \tag{8.3-20c}$$

The Jacobian of the transformation eqs. (8.3-20a) and (8.3-20b) is

$$\Delta(s, \tau) = x_s t_\tau - t_s x_\tau = -1 - s \cos \tau \tag{8.3-21}$$

Thus, a unique solution exists near the initial curve and is given implicitly by

$$u - \sin (x - ut) = 0 \tag{8.3-22}$$

The transcendental function $u(x, t)$ defined by eq. (8.3-22) cannot be expressed by a finite number of terms.

The curve $\Delta(s, \tau) = 0$ is easily calculated using eqs. (8.3-20a) and (8.3-20b) in the form

$$x = \cos^{-1}\left(\frac{-1}{t}\right) \pm (t^2 - 1)^{1/2} \tag{8.3-23}$$

and is sketched in Fig. 8.3-2 for $0 \leqslant x \leqslant 2\pi$ and $t \geqslant 0$.

Since the solution is a periodic function of x, the behavior in the remainder of the x, t-plane is easily determined.

The branch between A and B on the envelope corresponds to the upper sign in eq. (8.3-23), and the branch AC corresponds to the lower sign.

The point A is a cusp with slope $(dx/dt) = 0$, and is therefore tangent to the characteristic emanating from $x = \pi$, at $t = 0$. As $t \to \infty$, eq. (8.3-23) has the following asymptotic behavior:

$$x = \frac{\pi}{2} + t \text{ on } AB \tag{8.3-24a}$$

$$x = \frac{3\pi}{2} - t \text{ on } AC \tag{8.3-24b}$$

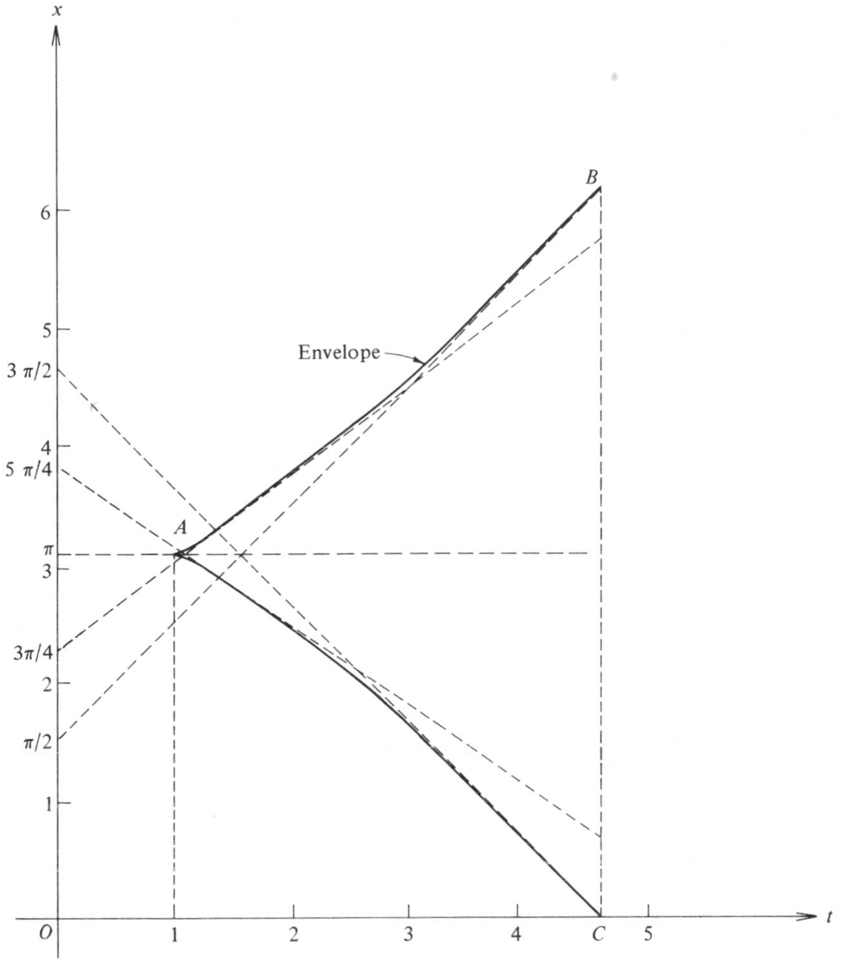

Fig. 8.3-2 Projection of characteristics and envelope for:

$$u_t + uu_x = 0; \quad u(x, 0) = \sin x$$

Thus, the characteristic emanating from $\pi/2$ and $3\pi/2$ at time $t = 0$ become tangent to AB and AC respectively as $t \to \infty$. However, in view of the periodicity of the problem, the points B and C are determined by the intersection with the other envelopes and one calculates the t-coordinate of B and C from $\cos^{-1}(-1/t) = (t^2 - 1)^{1/2}$, i.e., $t \doteq 4.6$. It is clear from this construction that to the right of the curves BAC the solution is no longer one-valued since projections of characteristics cross.

A more vivid qualitative description of the solution can be derived by examining u at $t = 0$, $t = \frac{1}{2}$, $t = 1$, and $t = 2$. This can be done very easily in graphical form by remembering that u remains constant along a characteristic and hence maintains its initial value. Thus, as t increases the portion of the initial wave between $x = \pi/2$ and $x = 3\pi/2$ is squeezed into a progressively narrower region until $t = 1$. This steepens the wave until at $t = 1$ its midpoint becomes vertical. Beyond the envelope BAC eq. (8.3-23) defines a many-valued function.

Figure 8.3-3 gives the wave form in sequence for $t = 0, \frac{1}{2}, 1$, and 2, in which we

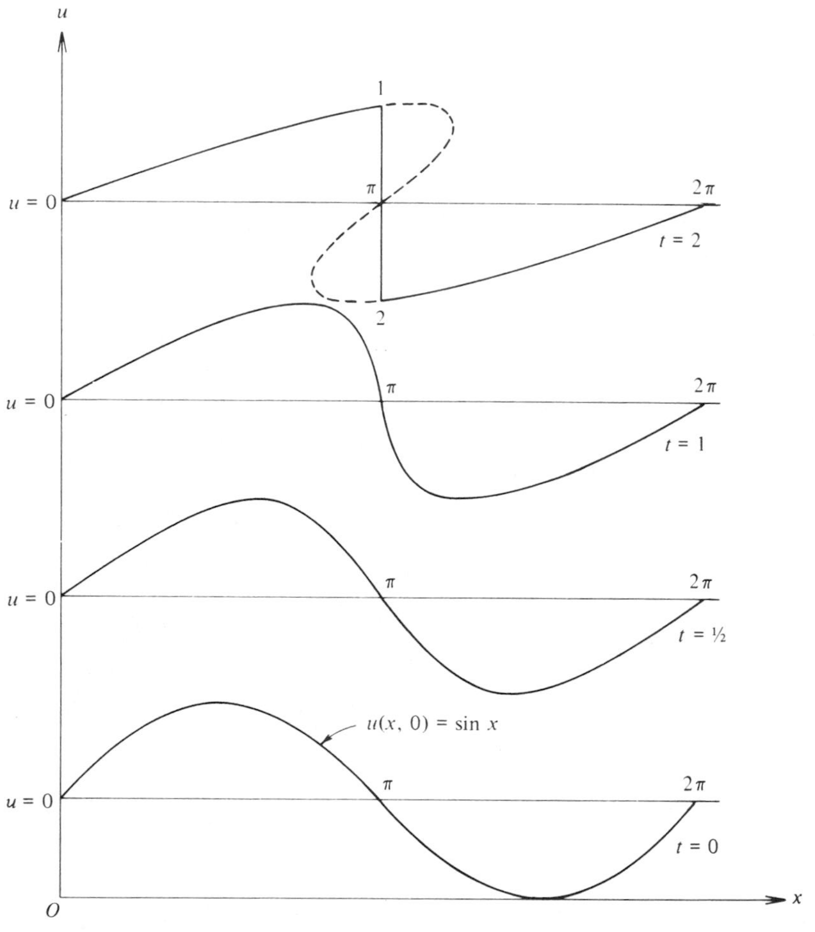

Fig. 8.3-3 Successive solutions of $u_t + uu_x = 0$, $u(x, 0) = \sin x$ at times $t = 0, \frac{1}{2}, 1, 2$ showing many-valuedness when $t > 1$.

use a dotted curve between points 1 and 2 to indicate that the result is not a solution. Following the discussion in sections 8.3.2 and 8.3.3, we shall see that the vertical solid line drawn between points 1 and 2 in this sketch is an appropriate discontinuity.

8.3.2 Weak Solutions, Conservation Laws, Shocks

Many of the results discussed in this section apply equally well to systems of partial differential equations of first order and will be developed in this generality to avoid repetition in Chapter 9.

Consider first a *linear* system of equations in the form

$$L(u) = Au_x + Bu_t + Cu = 0 \tag{8.3-25}$$

in some domain G of the x, t-plane. Here u is the vector $(u_1, u_2, \ldots, u_k)$ and u_x denotes the vector $(u_{1_x}, u_{2_x}, \ldots, u_{k_x})$ etc. The quantities A, B, C are $k \times k$ matrices with components depending only on x and t.

We define the adjoint of the operator L by L^* in the form

$$L^*(u) = -(Au)_x - (Bu)_t + Cu \tag{8.3-26}$$

For any subdomain R of G, let ζ be an *arbitrary, smooth test function which vanishes outside* R. One can then show that for any subdomain R of the solution domain G for u, the following divergence relation or 'conservation law' holds identically (whether or not $L(u) = 0$ as long as u is continuously differentiable)

$$\int_R\int [\zeta L(u) - uL^*(\zeta)]\, dx\, dt = 0 \tag{8.3-27}$$

If $L(u) = 0$, then eq. (8.3-27) reduces to

$$\int_R\int uL^*(\zeta)\, dx\, dt = 0 \tag{8.3-28}$$

It is the validity of the converse of the foregoing statement which is of importance here. One can show that if (8.3-28) is true for a function u which is *continuously differentiable* and for all test functions ζ in all subdomains R of G then $L(u) = 0$ everywhere in G.

The above result allows us to introduce a weaker criterion for a solution by relaxing the requirement that u be continuously differentiable. More precisely, we call u a *weak solution* of $L(u) = 0$ in G if u, together with its first derivatives, is piecewise continuous and eq. (8.3-28) holds for all test functions which vanish outside all subdomains R of G.

Thus, in the subdomains R of G considered there may be discontinuities of u. Let C be the locus in R of such a discontinuity and assume that u is continuously differentiable everywhere else in R. In the case where C divides R into two regions R_1 and R_2 we can easily show that the line integral

$$\int_C \zeta[u]\,(A\varphi_x + B\varphi_t)\,ds = 0 \tag{8.3-29a}$$

where $[u]$ denotes the difference between u evaluated on either side of C, and $\varphi(x, t) = 0$ is the implicit representation of the curve C.

Equation (8.3-29a) being true for all functions ζ implies that

$$[u]\,(A\varphi_x + B\varphi_t) = 0 \tag{8.3-29b}$$

If the vector $[u] \neq 0$, i.e. for the nontrivial case of discontinuous solutions, the matrix $\varphi_x A + \varphi_t B$ has zero determinant. As shown in Chapter 9, this condition determines a characteristic curve of the system eq. (8.3-25). For the case of a single equation, the condition $\varphi_x A + \varphi_t B = 0$ defines the projection of a characteristic curve on the x, t-plane. This is seen from the fact that along any curve $\varphi(x, t) = 0$ we can introduce the parameter s and calculate

$$d\varphi = 0 = \varphi_x \frac{dx}{ds} + \varphi_t \frac{dt}{ds} \tag{8.3-30}$$

Now if $\varphi(x, t) = 0$ is the projection of a characteristic we have, according to eqs. (8.3-1a) and (8.3-1b) that $dx/ds = A$ and $(dt/ds) = B$ and our result follows.

Thus, the important conclusion of the foregoing discussion is the fact that for weak solutions of linear systems, *discontinuities propagate along characteristics*. For a linear system one can a priori calculate the projections of characteristics from eqs. (8.3-1a) and (8.3-1b) without knowledge of u or an initial curve. Moreover, barring singular points of eqs. (8.3-1a) and (8.3-1b) at which $a = b = 0$, the characteristic projections do not cross and therefore discontinuities only arise if the initial data is discontinuous. To summarize our findings for the linear case we see that discontinuities in u arise only if the initial data is discontinuous; any such discontinuity propagates along the particular characteristic $\tau = \tau_0 =$ constant passing through the initial location of the discontinuity and is governed by

$$\frac{d[u]}{ds} = c\{x(s, \tau_0), y(s, \tau_0), u(s, \tau_0)\} \tag{8.3-31}$$

The analogous theory for quasilinear systems is based on the crucial requirement that *the system be a divergence relation or satisfy a conservation law* of the form

$$L(u) = p_t(x, t, u) + q_x(x, t, u) + n(x, t, u) = 0 \tag{8.3-32}$$

It is emphasized that the form (8.3-32) is *necessary* for the ensuing development. Although this requirement is usually met in all problems of physical interest which are naturally governed by conservation laws it is very important to note that *a given set of partial differential equations may have more than one distinct divergence form.* Since, our results on discontinuous solutions will depend on the divergence form with which we start, it follows that no unique jump relations can be derived just from a given system of first order partial differential equations. In general, one has to appeal to some physical argument outside the set (8.3-32) to establish the appropriate set of divergence relations. This is contrasted with the linear case where a unique discontinuity curve and jump condition could be calculated from the system of equations.

Let us consider the very apt example of one-dimensional compressible flow of a perfect gas. Having introduced an equation of state so that the problem involves only the three variables of velocity, density and pressure, the equations of conservation of mass, momentum, and energy have the same continuously differentiable solution as the system in which the equation for conservation of energy is replaced by the equation of conservation of entropy. However, the discontinuous solution of the former set is, in general, not the same as the discontinuous solution of the latter. In this example, when discontinuities are present, we appeal to thermodynamics and note that entropy is actually not conserved across a shock while energy is. Therefore, the appropriate conservation law is apparent. The details of this example are given in section 8.3.3, and for further discussion of the fluid mechanical aspects of hyperbolic systems with discontinuities, Ref. 8-2 is recommended. A general treatment of weak solutions is given in Ref. 8-5.

Actually, in the foregoing as well most physical problems governed by hyperbolic systems of partial differential equations certain terms representing dissipative phenomena have been neglected. These terms, although very small throughout most of the domain of interest, become crucial in regions with large variations of the dependent variables. Thus, a discontinuity in a hyperbolic system is generally an approximation to the physically more realistic occurrence of a small region of rapid variation in the dependent variables. Use of the more accurate equations including the dissipative terms leads to smooth solutions with no ambiguity regarding the appropriate jump phenomena which emerge as regions of rapid variation. Unfortunately, in most problems of interest such an analysis is quite difficult to perform exactly. Approximate techniques have been developed for the study of such 'singular perturbation problems' and Chapter 14 is devoted to this and related topics.

We return now to the system (8.3-32) and again introduce arbitrary smooth test functions ζ which vanish outside subdomains R of G.

The analogue of eq. (8.3.27) is here immediately derived by starting with

$$\int_R\int \zeta L(u)\,dx\,dt = 0 \tag{8.3-33}$$

and applying Gauss' theorem for smooth u to obtain

$$\int_R\int (p\zeta_t + q\zeta_x - n\zeta)\,dx\,dt = 0 \tag{8.3-34}$$

One can show that if u has continuous first derivatives and eq. (8.3-34) holds for all test functions then $L(u) = 0$. Again we use eq. (8.3-34) to define a weak solution in the following way.

A function $u(x, t)$ is a weak solution of eq. (8.3-32) in G if it is piecewise continuous and has piecewise continuous first derivatives, and satisfies eq. (8.3-34) for all test functions ζ in all subdomains R of G.

As before, we can use eq. (8.3-34) and apply Gauss' theorem in the two regions separated by a discontinuity C to obtain

$$\varphi_t [p] + \varphi_x [q] = 0 \tag{8.3-35a}$$

An alternate form for the *jump condition* eq. (8.3-35a) is obtained by calculating the slope $U = dx/dt$ of the discontinuity in the form

$$U[p] = [q] \tag{8.3-35b}$$

It is important to note that the condition governing the *discontinuity C is now coupled with the solution for u* on either side of C, hence in deriving a weak solution both must be considered simultaneously;

8.3.3 Examples of Weak Solutions

Consider first the example of one-dimensional unsteady flow of a perfect compressible gas. As shown, for example, in Ref. 8-2 the equations of conservation of mass, and momentum are respectively (for smooth solutions)

$$\rho_t + (\rho u)_x = 0 \tag{8.3-36a}$$

$$\rho u_t + \rho u u_x + p_x = 0 \tag{8.3-36b}$$

where ρ is the density, u the velocity, and p the pressure.

To complete the description of the above three flow variables one needs another relation. If the flow is adiabatic, the entropy given by p/ρ^γ remains constant along particle paths. Here the constant γ is the ratio of the specific heat at constant pressure divided by the specific heat at constant density. The conservation of entropy along a particle path is given by

$$\left(\frac{p}{\rho^\gamma}\right)_t + u\left(\frac{p}{\rho^\gamma}\right)_x = 0 \tag{8.3-36c}$$

As pointed out earlier, the above set is equivalent to one in which eq. (8.3-36c) is replaced by an energy conservation relation if the variables u, p, ρ are continuously differentiable. The energy conservation relation is usually written in the form

$$\left(\frac{u^2}{2} + \frac{\gamma}{\gamma - 1}\frac{p}{\rho}\right)_t + u\left(\frac{u^2}{2} + \frac{\gamma}{\gamma - 1}\frac{p}{\rho}\right)_x - \frac{p_t}{\rho} = 0 \tag{8.3-37a}$$

and leads to the divergence relation

$$\left(\frac{\rho u^2}{2} + \frac{p}{\gamma - 1}\right)_t + \left[\rho u\left(\frac{u^2}{2} + \frac{\gamma}{\gamma - 1}\frac{p}{\rho}\right)\right]_x = 0 \tag{8.3-37b}$$

if one uses eq. (8.3-36a).

It is a straightforward but tedious calculation to show that if one is allowed to differentiate p, ρ, and u, then eq. (8.3-37) follows from the three eqs. (8.3-36). Thus, for smooth solutions it is immaterial whether one uses eq. (8.3-36c) or eq. (8.3-37) as the third conservation relation.

On the other hand, for solutions with discontinuities, eq. (8.3-36c) is thermodynamically incorrect because the entropy of a gas particle increases as it goes through a shock.

In preparation for calculating the jump conditions, we cast eq. (8.3-36b) into divergence form. This is easily accomplished by adding eq. (8.3-36a) multiplied by u to eq. (8.3-36b) and noting that the result becomes $(\rho u)_t + (\rho u^2 + p)_x = 0$.

To summarize our results, the divergence relations appropriate for calculating the jump conditions for one-dimensional, compressible, unsteady flow are

$$\rho_t + (\rho u)_x = 0 \tag{8.3-38a}$$

$$(\rho u)_t + (\rho u^2 + p)_x = 0 \tag{8.3-38b}$$

$$\left(\frac{\rho u^2}{2} + \frac{p}{\gamma - 1}\right)_t + \left[\rho u\left(\frac{u^2}{2} + \frac{\gamma}{\gamma - 1}\frac{p}{\rho}\right)\right]_x = 0 \tag{8.3-38c}$$

Applying the jump condition in the form eq. (8.3-35) gives the well known *Rankine-Hugoniot* relations

$$\rho_1(u_1 - U) = \rho_2(u_2 - U) \tag{8.3-39a}$$

$$\rho_1(u_1 - U)^2 + p_1 = \rho_2(u_2 - U)^2 + p_2 \tag{8.3-39b}$$

$$\frac{1}{2}(u_1 - U)^2 + \frac{\gamma}{\gamma - 1}\frac{p_1}{\rho_1} = \frac{1}{2}(u_2 - U)^2 + \frac{\gamma}{\gamma - 1}\frac{p_2}{\rho_2} \tag{8.3-39c}$$

Next, we return to the example of traffic flow discussed in section 8.3.2 and study some general results before considering solutions of illustrative initial value problems.

Equation (8.3-19a) is a limiting form of Burger's equation

$$u_t + uu_x = \epsilon u_{xx} \tag{8.3-40}$$

for the case of $\epsilon \to 0$. The right hand side of eq. (8.3-40) represents a dissipative term which for $\epsilon << 1$ is small everywhere unless u_{xx} is large.

As pointed out in Ref. 8-8, one can solve eq. (8.3-40) exactly by noticing that the transformation

$$u = \varphi_x \tag{8.3-41a}$$

$$\varphi = -2\epsilon \log \theta(x, t) \tag{8.3-41b}$$

reduces eq. (8.3-40) to the diffusion equation

$$\theta_t - \epsilon\theta_{xx} = 0 \tag{8.3-42}$$

which is discussed in Chapter 9. Further discussion of Burger's equation appears in Chapter 15.

One can use the exact solution or the approximate solution obtained by singular perturbation methods to show that as $\epsilon \to 0$ discontinuous solutions of eq. (8.3-19a) arise either if the initial data are discontinuous, or if the initial data are smooth but $\Delta(s, \tau)$ becomes zero in the domain of the solution.

Using the above limiting solution as $\epsilon \to 0$ one can show that the following condition is always satisfied across a discontinuity.

$$u_L \leqslant U \leqslant u_R \tag{8.3-43a}$$

Here u_L and u_R denote the values of u to the left and right respectively of the discontinuity curve which propagates with speed U. (It is understood that left and right are with respect to the direction of increasing time in the x, t-plane). Note that the condition $u_L > u_R$ never arises except on an initial curve, because this condition implies that the characteristics on either side of the discontinuity are pointing away from the discontinuity. Clearly, this implies that at an earlier time these characteristics had crossed, and it is at this earlier time that a discontinuity occurred. For more details see the examples discussed later on in this subsection.

For the more general equation of traffic flow where the nonlinear term uu_x is replaced by $a(u)\,u_x$ the condition (8.3-43) becomes

$$u_L \leqslant a(u) \leqslant u_R \tag{8.3-43b}$$

A discontinuous solution which satisfies (8.3-43) is called a *permissible weak solution*, and we shall show how use of such a condition allows one to calculate a unique result. The proof that (8.3-43) determines a unique weak solution is given in Ref. 8-7.

Conditions (8.3-43) are the special cases of the thermodynamic principle that entropy increases across a shock.

To illustrate vividly the fact that the same equation can be put in different divergence forms and hence can lead to different jump relations, we let $v = \log u$ and transform eq. (8.3-19a), i.e.

$$u_t + \left(\frac{u^2}{2}\right)_x = 0 \qquad (8.3\text{-}44a)$$

into

$$v_t + (e^v)_x = 0 \qquad (8.3\text{-}44b)$$

According to eq. (8.3-35), the jump condition for eq. (8.3-44a) is

$$U = \frac{u_R + u_L}{2} \qquad (8.3\text{-}45a)$$

while eq. (8.3-44b) leads to

$$U = (u_R - u_L)/(\log u_R - \log u_L) \qquad (8.3\text{-}45b)$$

Note that eq. (8.3-45a) can always be satisfied whenever $u_L \leqslant u_R$ while eq. (8.3-45b) can never hold true. In fact, it is clear from eq. (8.3-45b) that $U \leqslant u_L$, which is in contradiction with eq. (8.3-43a).

The generation of a shock by appropriate initial conditions can be simply illustrated for eq. (8.3-19a) with the following data at $t = 0$.

$$u(x, 0) = \begin{cases} 0 & \text{if } x > 1 \\ 1 & \text{if } x < 1 \end{cases} \qquad (8.3\text{-}46)$$

Thus, the solution immediately breaks down for $t > x$ and $t > 0$. In this region of the x, t-plane the characteristics emanating from $x < 0$ at $t = 0$ cross with those emanating from $x > 0$. Using the jump condition in the form (8.3-45a) and the fact that $u_L = 0$ and $u_R = 1$ we find that the shock curve starts at $x = 0, t = 0$ and obeys $U = dx/dt = \frac{1}{2}$. Thus, the permissible weak solution of eq. (8.3-19a) subject to (8.3-46) is

$$u = \begin{cases} 0, & x > t/2, \ t \geqslant 0 \\ 1, & x < t/2, \ t \geqslant 0 \end{cases} \qquad (8.3\text{-}47)$$

The above example is the analogue of a uniform shock wave in one-dimensional compressible flow.

The solution eq. (8.3-47) satisfies condition (8.3-43a) since $u_L = 0$, $U = \frac{1}{2}$, and $u_R = 1$. One can also prove that eq. (8.3-47) is the only weak solution satisfying the permissibility condition (8.3-43a).

To illustrate the contradiction one would obtain by trying to derive another weak solution, let us assume that two shocks emanate from $x = t = 0$.

Let U_1 and U_2 denote the two shock speeds with $U_1 > U_2$. Let the solution between the two shocks be given by u_2. The jump conditions become $U_1 = u_2/2$ and $U_2 = (u_2 + 1)/2$ which imply that $U_2 > U_1$ in contradiction with the original hypothesis.

In the previous example the characteristics crossed due to a given initial discontinuity. Let us now take the converse case for which the characteristics diverge from a given initial discontinuity. For example, let

$$u(x, 0) = \begin{cases} 1, & \text{if } x > 0 \\ 0, & \text{if } x < 0 \end{cases} \tag{8.3-48}$$

The following solution immediately results from the characteristic equations.

$$u(x, t) = \begin{cases} 1, & \text{if } 0 \leqslant t < \infty, \ x > t \\ 0, & \text{if } 0 \leqslant t < \infty, \ x < 0 \end{cases} \tag{8.3-48a}$$

However, no information can be derived from the above for the domain $0 \leqslant x \leqslant t$. It is clear that if we introduce any finite number of discontinuities of given speed, condition (8.3-43a) will be violated. For example, if we take the shock $x = t/2$ to separate the region where $u = 1$ from the region when $u = 0$ we have a contradiction because $u_L = 1$, $U = \frac{1}{2}$, and $u_R = 0$. If we introduce two shocks of arbitrary speeds, (8.3-43a) will be violated but less strongly across both shocks.

We thus observe by experimentation that we need an infinite number of discontinuities of infinitesimal strength, which in the limit lead to a solution of the form

$$u = x/t \quad \text{if } 0 \leqslant t \leqslant x \tag{8.3-48b}$$

The above is the analogue of a *simple wave* in compressible flow. Note that the solution defined by eqs. (8.3-48a) and (8.3-48b) is continuous everywhere except at $x = t = 0$, and continuously differentiable everywhere except on the lines $x = t$ and $x = 0$.

So far we have only considered uniform initial data with discontinuities. It is easy to generalize the above results to cases where the initial data is not uniform. Suppose we have an initial condition of the form

$$u(x, 0) = \begin{cases} u_1(x) & \text{if } x > x_0 \\ u_2(x) & \text{if } x < x_0 \end{cases} \tag{8.3-49}$$

1. If $u_1(x_0^+) < u_2(x_0^-)$, a single shock emanates from $x = x_0$, $t = 0$ with initial slope $[u_1(x_0^+) + u_2(x_0^-)]/2$. Barring the occurrence of other discontinuities the shock divides the x, t-plane in two regions in which $u_1(x, t)$ and $u_2(x, t)$ are the continuously differentiable solutions of (8.3-19a) satisfying $u_1(x, 0) = u_1(x)$ and $u_2(x, 0) = u_2(x)$.

The shock speed is obtained by solving the differential equation

$$\frac{dx}{dt} = \frac{u_1(x, t) + u_2(x, t)}{2} \tag{8.3-50}$$

2. If $u_1(x_0^+) > u_2(x_0^-)$ the solution in the triangular region $x_0 + u_1(x_0^+)\, t \leqslant x \leqslant x_0 + u_2(x_0^-)\, t$ is given by

$$u(x, t) = (x - x_0)/t \tag{8.3-51}$$

We now illustrate how solutions of types (1) and (2) interact by listing a series of examples solved without detailed discussion.

If we choose

$$u(x, 0) = \begin{cases} 1 & \text{if } |x| > 1 \\ 0 & \text{if } |x| < 1 \end{cases} \tag{8.3-52}$$

we calculate the solution sketched in Fig. 8.3-4.

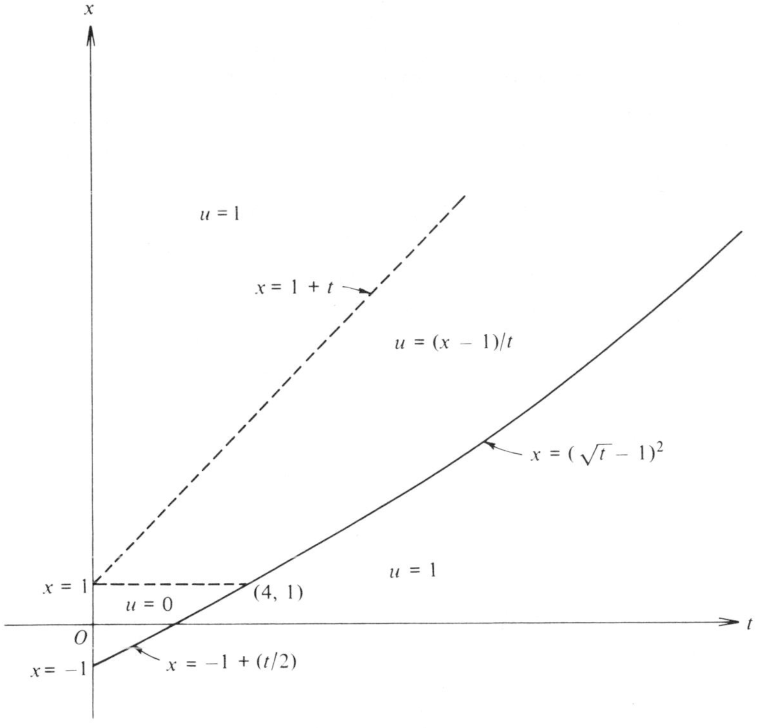

Fig. 8.3-4 Solution of the initial-value problem (8.3-52).

If we choose

$$u(x,0)=\begin{cases}0, & |x|>1\\ 1-x, & 0<x<1\\ -1-x, & -1<x<0\end{cases} \tag{8.3-53}$$

we obtain the solution in Fig. 8.3-5. where, because of the antisymmetry of the initial data with respect to the $x = 0$ axis $u(x, t) = -u(-x, t)$.

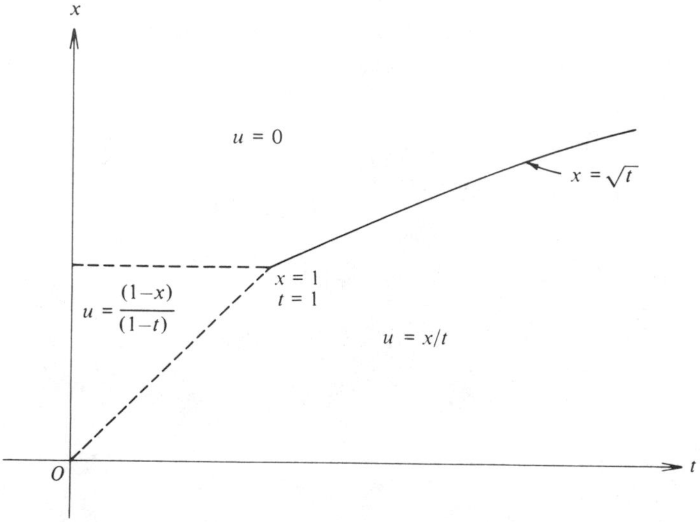

Fig. 8.3-5 Solution of the initial-value problem (8.3-53).

We note that a shock curve starts at the point $x = t = 1$ where all the characteristics emanating from the interval $|x| < 1$ cross.

Finally, the example $u(x, 0) = \sin x$, discussed in section 8.3.1 has the weak solution in which a stationary shock starts at $t = 1$, $x = \pi$ and extends horizontally to $t = \infty$. The effect of this shock is to introduce a jump $u(\pi^+, t) - u(\pi^-, t)$ in the wave at $x = \pi$ (see Fig. 8.3-2 where this is sketched for $t = 2$) and to eliminate the portion of the solution which is many-valued (shown by a dotted curve in Fig. 8.3-3 for $t = 2$).

8.3.4 More than Two Independent Variables

In this section, as in 8.3.1, we list without proof the principal results for equations of the form (8.0-2) (repeated below)

$$\sum_{i=1}^{n} a_i(x_1, \ldots, x_n, u)\, u_{x_i} = a(x_1, \ldots, x_n, u) \tag{8.3-54}$$

The $(n + 1)$ ordinary differential equations

$$\frac{dx_i}{ds} = a_i(x_1, \ldots, x_n, u); \quad i = 1, \ldots, n \tag{8.3-55a}$$

$$\frac{du}{ds} = a(x_1, \ldots, x_n, u) \tag{8.3-55b}$$

define an n-parameter family of characteristic curves which fill the $x_1, \ldots, x_n, u$-space. The above is an n-parameter family because eq. (8.3-55) is autonomous.

We can prove that every integral manifold $u = u(x_1, \ldots, x_n)$ is generated by an $(n - 1)$-parameter family of characteristic curves. Conversely, every manifold generated by an $(n - 1)$-parameter family of characteristic curves is an integral manifold.

An initial value problem consists of passing an integral manifold through a given $(n - 1)$-dimensional initial manifold $\mathcal{C}_0$ defined below

$$x_i = x_i^{(0)}(\tau_1, \ldots, \tau_{n-1}); \quad i = 1, \ldots, n \tag{8.3-56a}$$

$$u = u^{(0)}(\tau_1, \ldots, \tau_{n-1}) \tag{8.3-56b}$$

Thus, an integral manifold is generated from the $(n - 1)$-parameter family of curves

$$x_i = x_i(s, \tau_1, \ldots, \tau_{n-1}); \quad i = 1, \ldots, n \tag{8.3-57a}$$

$$u = u(s, \tau_1, \ldots, \tau_{n-1}) \tag{8.3-57b}$$

which are the solutions of eqs. (8.3-55) and satisfy the initial conditions

$$x_i(0, \tau_1, \ldots, \tau_{n-1}) = x_i^{(0)}(\tau_1, \ldots, \tau_{n-1}); \quad i = 1, \ldots, n \tag{8.3-58a}$$

$$u(0, \tau_1, \ldots, \tau_{n-1}) = u^{(0)}(\tau_1, \ldots, \tau_{n-1}) \tag{8.3-58b}$$

If Δ, defined by

$$\Delta(s, \tau_1, \ldots, \tau_{n-1}) = \frac{\partial(x_1, \ldots, x_n)}{\partial(s, \tau_1, \ldots, \tau_{n-1})} \tag{8.3-59}$$

is not equal to zero initially one can solve the initial value problem uniquely in some neighborhood of the initial manifold in the following manner. Equations (8.3-57a) are solved for the n-quantities $s, \tau_1, \ldots, \tau_{n-1}$ as functions of the $x_1, \ldots, x_n$ and the result is then substituted into (8.3-57b) to obtain the integral manifold in the form $u = u(x_1, \ldots, x_n)$.

This solution can be extended from the neighborhood of the initial manifold to larger values of s uniquely unless a manifold is reached on which Δ becomes equal to zero.

In preparation for discussing the case $\Delta = 0$ we define a *characteristic manifold C*. This is an $(n - 1)$-dimensional manifold in the $(n + 1)$-dimensional space with the property that at each point $(x_1, \ldots, x_n, u)$ on C the characteristic vector

$$T = (a_1, a_2, \ldots, a_n, a) \tag{8.3-60}$$

is tangent to C. This means that T is linearly dependent on the $(n - 1)$-linearly independent tangent vectors T_ν to C. Consequently, there exist $(n - 1)$ constants λ_ν not all equal to zero such that

$$T = \sum_{\nu=1}^{n-1} \lambda_\nu T_\nu \tag{8.3-61}$$

It then follows that the $(n - 1)$-dimensional manifold

$$x_i = x_i(\tau_1, \ldots, \tau_{n-1}); \qquad i = 1, \ldots, n \tag{8.3-62a}$$

$$u = u(\tau_1, \ldots, \tau_{n-1}) \tag{8.3-62b}$$

is characteristic if the following conditions hold for some $(n - 1)$ constants λ_ν not all equal to zero.

$$a_i = \sum_{\nu=1}^{n-1} \lambda_\nu \frac{\partial x_i}{\partial \tau_\nu}; \qquad i = 1, \ldots, n \tag{8.3-63a}$$

$$a = \sum_{\nu=1}^{n-1} \lambda_\nu \frac{\partial u}{\partial \tau_\nu} \tag{8.3-63b}$$

If $\Delta = 0$ everywhere on the initial manifold $\mathcal{C}_0$, then a necessary and sufficient condition for the solvability of eq. (8.3-54) is that $\mathcal{C}_0$ be a characteristic manifold. In this case there exist an infinity of solutions and $\mathcal{C}_0$ is a branch manifold across which any two solutions may be joined smoothly.

The foregoing ideas are illustrated by the example

$$u_x + u_y + u_z = 0 \tag{8.3-64}$$

which has the two-parameter family of characteristic curves

$$x = s + x_0(\tau_1, \tau_2) \tag{8.3-65a}$$

$$y = s + y_0(\tau_1, \tau_2) \tag{8.3-65b}$$

$$z = s + z_0(\tau_1, \tau_2) \tag{8.3-65c}$$

$$u = u_0(\tau_1, \tau_2) \tag{8.3-65d}$$

passing through the initial manifold $x = x_0, y = y_0, z = z_0$, and $u = u_0$.

1. Consider first the initial value problem where $u = xz$ on $y = 0$, i.e., $x_0 = \tau_1$, $y_0 = 0$, $z_0 = \tau_2$, $u_0 = \tau_1 \tau_2$. We calculate $\Delta(s, \tau_1, \tau_2) = -1$, hence a unique solution exists for all x, y, and z. The characteristic curves are

$$x = s + \tau_1 \tag{8.3-66a}$$

$$y = s \tag{8.3-66b}$$

$$z = s + \tau_2 \tag{8.3-66c}$$

$$u = \tau_1 \tau_2 \tag{8.3-66d}$$

Solving the first three of eqs. (8.3-66) gives $s = y$, $\tau_1 = x - y$, $\tau_2 = z - y$, which when substituted into (8.3-66d) provides the solution $u = (x - y)(z - y)$.

2. As an example of a characteristic initial manifold let $x_0 = 2\tau_1, y_0 = \tau_1 + \tau_2$, $z_0 = 2\tau_2$, $u_0 = \tau_1 - \tau_2$. It follows easily from the definitions of x_0, y_0, and z_0 that $\Delta(0, \tau_1, \tau_2) = 0$. To ascertain that $\mathcal{C}_0$ is a characteristic manifold we must find a λ_1 and λ_2 not both equal to zero such that according to eq. (8.3-63)

$$1 = 2\lambda_1 \tag{8.3-67a}$$

$$1 = \lambda_1 + \lambda_2 \tag{8.3-67b}$$

$$1 = 2\lambda_2 \tag{8.3-67c}$$

and

$$0 = \lambda_1 - \lambda_2 \tag{8.3-68}$$

Actually, (8.3-67) is always true if $\Delta = 0$, and the crucial test is eq. (8.3-68). We see that indeed all four conditions are met with $\lambda_1 = \lambda_2 = \frac{1}{2}$.

The characteristic equations for this case are

$$x = s + 2\tau_1 \tag{8.3-69a}$$

$$y = s + \tau_1 + \tau_2 \tag{8.3-69b}$$

$$z = s + 2\tau_2 \tag{8.3-69c}$$

$$u = \tau_1 - \tau_2 \tag{8.3-69d}$$

and can be used to exhibit the nonuniqueness of the solution in the form

$$u = (x - z)/2 + f[y - (x + z)/2] \tag{8.3-70}$$

for any function $f(\xi)$ subject only to $f(0) = 0$.

8.4 NONLINEAR EQUATIONS

As in section 8.3 we begin our discussion by listing without proof the results needed for solving initial-value problems for nonlinear equations with two independent variables and extend these immediately to the case of n independent variables. We then reconsider the concept of a complete integral and study some illustrative examples. The latter part of this section covers the case of more than two independent variables, and the application of these results to dynamics.

8.4.1 Solution of Initial-Value Problem and Relation to Complete Integral

We build on the geometrical description provided in 8.2.6, and list the analytic formulas needed for solving a given initial value problem.

We are concerned here with the nonlinear equation in two variables

$$F(x, y, u, p, q) = 0 \tag{8.4-1}$$

where $p = u_x$ and $q = u_y$. A solution of (8.4-1) is a function $u(x, y)$ which satisfies eq. (8.4-1) everywhere. Moreover, since p and q must be continuously differentiable, the resulting u is a twice continuously differentiable function of x and y.

At each point in x, y, u-space we have a *one-parameter family of characteristic directions* which generate a Monge cone. Curves which at each point are tangent to a characteristic direction are called *focal curves* or Monge curves, and are defined by the three first order equations

$$\frac{dx}{ds} = F_p \tag{8.4-2a}$$

$$\frac{dy}{ds} = F_q \tag{8.4-2b}$$

$$\frac{du}{ds} = pF_p + qF_q \tag{8.4-2c}$$

As pointed out in our geometrical discussion of section 8.2.6, one needs to specify a curve x, y, u (i.e., a focal curve) as well as the associated strip (p and q) in order to isolate the appropriate generator out of the one-parameter family defining the Monge cone.

One can show that in order to have the focal curves embedded in an integral surface (i.e., that in some neighborhood of the projection of the focal curve on the x, y-plane, u is a one-valued, twice continuously differentiable function of x and y), p and q must obey the following equations along the focal curve:

$$\frac{dp}{ds} = -(pF_u + F_x) \tag{8.4-2d}$$

$$\frac{dq}{ds} = -(qF_u + F_y) \tag{8.4-2e}$$

The system of five eqs. (8.4-2) is called the *characteristic system of differential equations.*

Since s is absent from the right-hand sides of eqs. (8.4-2), the solution of this system provides a four-parameter family of functions of s for x, y, u, p, q. It is easy to show that F is constant along any solution of the system (8.4-2); i.e., F is an integral of the characteristic system of differential equations. Thus, a solution of eqs. (8.4-2) which also satisfies the condition $F = 0$ will be a three-parameter family called a *characteristic strip.* Such a strip consists of a *characteristic curve* x, y, u plus an associated *infinitesimal tangent plane* defined by p and q. One can then show that every integral surface is generated by a one-parameter family of characteristic strips.

To see how for a given initial value problem one can isolate a one-parameter family of characteristic strips from the three-parameter family available, consider a given initial strip $\mathcal{C}_0$ defined by

$$x = x_0(\tau) \tag{8.4-3a}$$

$$y = y_0(\tau) \tag{8.4-3b}$$

$$u = u_0(\tau) \tag{8.4-3c}$$

$$p = p_0(\tau) \tag{8.4-3d}$$

$$q = q_0(\tau) \tag{8.4-3e}$$

It is understood that the projection of the curve $x_0(\tau), y_0(\tau), u_0(\tau)$ has no double points. Moreover, we require that

$$F(x_0, y_0, u_0, p_0, q_0) = 0 \tag{8.4-4a}$$

and

$$\frac{du_0}{d\tau} = p_0 \frac{dx_0}{d\tau} + q_0 \frac{dy_0}{d\tau} \tag{8.4-4b}$$

for consistency.

Thus, eqs. (8.4-3) subject to the two conditions eqs. (8.4-4) reduce to three conditions in terms of a parameter τ, and the resulting one-parameter family of characteristic strips can be written in the form

$$x = x(s, \tau) \tag{8.4-5a}$$

$$y = y(s, \tau) \tag{8.4-5b}$$

$$u = u(s, \tau) \tag{8.4-5c}$$

$$p = p(s, \tau) \tag{8.4-5d}$$

$$q = q(s, \tau) \tag{8.4-5e}$$

where $x(0, \tau) = x_0(\tau), y(0, \tau) = y_0(\tau)$, etc.

As in the quasilinear case, we can show that if Δ defined by

$$\Delta(s, \tau) = F_p y_\tau - F_q x_\tau = x_s y_\tau - y_s x_\tau \tag{8.4-6}$$

is not identically zero along the initial strip, there exists some neighborhood of the strip wherein the initial value problem has a unique solution. Moreover, this solution can be extended to larger values of s unless a curve $\Delta(s, \tau) = 0$ is reached.

This solution is obtained by solving eqs. (8.4-5a) and (8.4-5b) for s and τ in terms of x and y and substituting the result into the remainder of the equations to calculate

$$u = u(x, y) \tag{8.4-7a}$$

$$p = p(x, y) \tag{8.4-7b}$$

$$q = q(x, y) \tag{8.4-7c}$$

One can, of course, prove that the resulting $u(x, y), p(x, y)$,and $q(x, y)$ do indeed satisfy the fundamental requirement $p = u_x, q = u_y$ as well as $F = 0$.

If $\Delta = 0$ initially, a solution of eq. (8.4-1) exists if and only if the initial strip is characteristic. In this case there are an infinite number of solutions any two of which may be joined with continuous second derivatives through the initial strip.

If $\Delta = 0$ and the initial strip is not characteristic, then it is a *focal strip*, i.e., a focal curve with an associated infinitesimal surface which is not tangent to an integral surface. In this case, the projection of the focal curve on the x, y-plane is the singular curve of a family of characteristic curves projected on the same plane. The interpretation of this case in geometrical optics is of particular interest, because the light rays are characteristics, and the focal curves associated with these characteristics are the *caustic curves* of the light rays. For further discussion on the physical aspects of caustics the reader is referred to Ref. 8-3.

We next indicate how a complete integral of eq. (8.4-1) defines by a process of envelope formation a three-parameter family of characteristic curves which are equivalent to the solution in complete generality of the system (8.4-2) subject to $F = 0$.

Let $u = \varphi(x, y, a, b)$ be a complete integral of eq. (8.4-1), i.e., a solution of eq. (8.4-1) involving two arbitrary constants a and b such that the determinant $\varphi_{ax} \cdot \varphi_{by} - \varphi_{bx}\varphi_{ay} \neq 0$. This condition insures that a and b are independent, and that the solution does not depend on some function $\gamma(a, b)$, say. Consider the set of five equations

$$u = \varphi(x, y, a, b) \tag{8.4-8a}$$

$$\varphi_a(x, y, a, b) = \lambda s \tag{8.4-8b}$$

$$\varphi_b(x, y, a, b) = \mu s \tag{8.4-8c}$$

$$p = \varphi_x(x, y, a, b) \tag{8.4-8d}$$

$$q = \varphi_y(x, y, a, b) \tag{8.4-8e}$$

where φ is a complete integral. Now eqs. (8.4-8a), (8.4-8d) and (8.4-8e) are self explanatory. Equations (8.4-8b) and (8.4-8c) define a variable s and two parameters λ and μ. If we regard $b = w(a)$, then the latter two equations combine to give the condition for an envelope discussed in eq. (8.2-20) if we interpret

$$w'(a) = -\lambda/\mu \tag{8.4-9}$$

Thus eqs. (8.4-8) together with the condition (8.4-9) define a three-parameter $(a, b, \lambda/\mu)$ family of curves and an associated infinitesimal tangent plane. To see how these curves may be constructed explicitly we envisage the following procedure. One solves for x and y as functions of s and the three-parameters from eqs. (8.4-8b) and (8.4-8c). This is possible since, by hypothesis for a complete integral, $\varphi_{ax}\varphi_{by} - \varphi_{bx}\varphi_{ay} \neq 0$. When the result is substituted into eqs. (8.4-8a), (8.4-8d), and (8.4-8e) one has a three-parameter family of curves

$$x = x(s, a, b, \lambda/\mu) \tag{8.4-10a}$$

$$y = y(s, a, b, \lambda/\mu) \tag{8.4-10b}$$

$$u = u(s, a, b, \lambda/\mu) \tag{8.4-10c}$$

and the associated definition of an infinitesimal tangent plane given by

$$p = p(s, a, b, \lambda/\mu) \tag{8.4-10d}$$

$$q = q(s, a, b. \lambda/\mu) \tag{8.4-10e}$$

The proof that (8.4-10) is in fact a characteristic strip is given in Ref. 8-6. We demonstrate this fact for the example of eq. (8.2-23) later on, and prove it for the Hamilton-Jacobi equation in section 8.4.3. In effect, this construction allows one to calculate the characteristic curves (and the infinitesimal tangent planes attached to them) as the intersection of the complete integral with its arbitrary envelopes.

Thus, the availability of a complete integral completely bypasses the need of solving the characteristic equations (except if the latter have nonunique solutions) subject to a given initial strip. This remarkable result motivates the desirability of calculating the complete integral directly from eq. (8.4-1) by separation of variables as this is generally a much easier task. Of course, once the characteristic equations have been solved, a complete integral is easily obtained as a trivial consequence.

The discussion for the case of n-independent variables does not involve any essential new ideas, so we merely list the characteristic equations here without further comment for the equation

$$F(x_1, x_2, \ldots, x_n, u, p_1, \ldots, p_n) = 0 \tag{8.4-11}$$

$$\frac{dx_i}{ds} = F_{p_i}; \qquad i = 1, 2, \ldots n \tag{8.4-12a}$$

$$\frac{du}{ds} = \sum_{i=1}^{n} p_i F_{p_i} \tag{8.4-12b}$$

$$\frac{dp_i}{ds} = -(F_u p_i + F_{x_i}); \qquad i = 1, 2, \ldots n \tag{8.4-12c}$$

where $p_i = u_{x_i}$.

8.4.2 Examples

(a) Consider the following initial-value problem:

$$F = p^2 + q^2 - \frac{1}{c^2} = 0; \qquad c = \text{const.} \tag{8.4-13a}$$

$$x_0 = y_0 = \tau; \qquad u_0 = \alpha\tau \tag{8.4-13b}$$

To complete the statement of the problem we must also specify p_0 and q_0, and we do so by satisfying eqs. (8.4-4). Thus, for some constant θ we let $p_0 = (\cos\theta)/c$ and $q_0 = (\sin\theta)/c$, it is easy to see that eq. (8.4-4a) is then automatically satisfied. Next we require that θ be the solution of $c\alpha = \sin\theta + \cos\theta$ which specifies possible values of $\theta(\alpha)$ consistent with eq. (8.4-4b).

For this example $\Delta = 2(\cos\theta - \sin\theta)/c$ and it vanishes only if $\theta = \pi/4$ or $5\pi/4$ (i.e., $\alpha = \pm\sqrt{2}/c$). If we exclude these two cases, a unique solution can be calculated as follows.

We integrate eqs. (8.4-2) subject to the prescribed initial conditions and obtain the one-parameter family of characteristic strips.

$$x = \frac{2s}{c}\cos\theta + \tau \tag{8.4-14a}$$

$$y = \frac{2s}{c}\sin\theta + \tau \tag{8.4-14b}$$

$$u = \frac{2s}{c^2} + \frac{(\sin\theta + \cos\theta)}{c}\,\tau \tag{8.4-14c}$$

$$p = p_0 = \frac{\cos\theta}{c} \tag{8.4-14d}$$

$$q = q_0 = \frac{\sin\theta}{c} \tag{8.4-14e}$$

Solving eqs. (8.4-14a) and (8.4-14b) for s and τ in terms of x and y gives

$$s = \frac{c(y - x)}{2(\sin\theta - \cos\theta)} \tag{8.4-15a}$$

$$\tau = \frac{(x\sin\theta - y\cos\theta)}{(\sin\theta - \cos\theta)} \tag{8.4-15b}$$

Substituting the above into (8.4-14c) shows that

$$u = \frac{x\cos\theta + y\sin\theta}{c} \tag{8.4-15c}$$

(b) Another interesting initial-value problem for the Eikonal equation is the calculation of the *integral conoid* which is the wave front generated by a point source of light. For the simple case where c is constant, the location of the source is immaterial so we choose it at the origin. Thus, $x_0 = y_0 = u_0 = 0$. This in turn implies that condition eq. (8.4-4b) is identically satisfied, and we set $p_0 = \sin\tau/c$, $q_0 = \cos\tau/c$ to satisfy eq. (8.4-4a). Here, the initial strip *is the Monge cone at the origin*, and an easy calculation shows that the wave front is given by a continuation of the Monge cone, i.e.

$$u = \frac{(x^2 + y^2)^{1/2}}{c} \tag{8.4-16}$$

Let us now derive the equation governing light rays for the general two dimensional case where $c = c(x, y)$. The characteristic equations are

$$\frac{dx}{ds} = 2p \tag{8.4-17a}$$

$$\frac{dy}{ds} = 2q \tag{8.4-17b}$$

$$\frac{du}{ds} = \frac{2}{c^2} \tag{8.4-17c}$$

$$\frac{dp}{ds} = -\frac{2}{c^3} c_x \tag{8.4-17d}$$

$$\frac{dq}{ds} = -\frac{2}{c^3} c_y \tag{8.4-17e}$$

The differential equation for light rays is obtained by eliminating p and q from the four equations above for x, y, p, and q. The result, which is straightforward to derive, can be expressed with either x or y as the independent variable. Choosing the former representation we find

$$c(x, y) \frac{d^2y}{dx^2} + \left(c_y - c_x \frac{dy}{dx}\right)\left[1 + \left(\frac{dy}{dx}\right)^2\right] = 0 \tag{8.4-18}$$

For the case c = constant, eq. (8.4-18) is trivially solved and gives a two-parameter family of straight lines. If c is not constant, the above is a nonlinear, second order, ordinary differential equation which can be solved in principle (or numerically) once c is specified.

A typical problem would consist of calculating the light rays in a given medium, once an initial wave front were specified in the form $u(x, y) = 0$. In this case, at each point x, y of the curve $u = 0$ one calculates the slope of the light ray as the normal to the wave front.

Thus, the light ray emanating from the point x_0, y_0 with $u(x_0, y_0) = 0$, is the solution of eq. (8.4-18) subject to the initial conditions $y(x_0) = y_0$, $dy(x_0)/dx = u_y(x_0, y_0)/u_x(x_0, y_0)$.

It is interesting to also derive eq. (8.4-18) from Fermat's principle. If we specialize eq. (8.1-23) to the two-dimensional case and use x as our variable of integration by noting that $ds^2 = dx^2 + dy^2$, the Lagrangian becomes

$$L = \frac{\left[1 + \left(\frac{dy}{dx}\right)^2\right]^{1/2}}{c(x, y)} \tag{8.4-19}$$

and Euler's eq. (8.1-15) reduces to eq. (8.4-18).

Many important optical phenomena are governed by equations (8.4-17) or (8.4-18) and can be studied by examination of their solutions. It is beyond the scope of this chapter to consider these topics, and the reader is referred to standard texts on optics, e.g., Ref. 8-3, for further discussion.

(c) Next, we illustrate the relation between the complete integral, and a solution of the characteristic equations by reconsidering the example of eq. (8.2-23).

The characteristic equations are

$$\frac{dx}{ds} = 2u^2 p \tag{8.4-20a}$$

$$\frac{dy}{ds} = 2u^2 q \tag{8.4-20b}$$

$$\frac{du}{ds} = 2u^2(p^2 + q^2) = 2(1 - u^2) \tag{8.4-20c}$$

$$\frac{dp}{ds} = -2pu(1 + p^2 + q^2) = -\frac{2p}{u} \tag{8.4-20d}$$

$$\frac{dq}{ds} = -2qu(1 + p^2 + q^2) = -\frac{2q}{u} \tag{8.4-20e}$$

One can solve for x, y, u, p, q in parametric form in terms of the variable s and three constants by performing the indicated quadratures starting with eq. (8.4-20c) then proceeding to eqs. (8.4-20d) and (8.4-20e), and finally to eqs. (8.4-20a) and (8.4-20b).

A more direct calculation gives the characteristic curves implicitly as the intersections of planes

$$\gamma(x - a) = y - b \tag{8.4-21a}$$

with spheres

$$(x - a)^2 + (y - b)^2 + u^2 = 1 \tag{8.4-21b}$$

where γ, a, b are three arbitrary constants.

We immediately recognize (8.4-21b) as the complete integral (8.2-21). Thus, the characteristic curves are the three-parameter family (a, b, γ) of unit circles centered at $x = a, y = b, u = 0$ in planes normal to the $u = 0$-plane and subtending an angle γ to the x, u-plane.

The geometrical interpretation of the construction discussed following eqs. (8.4-8a) is easily visualized in this case. Although the complete integral only involves *two parameters* a and b, the coordinates of the center of the sphere, we actually are able to generate a *three-parameter* family of curves by envelope formation for the following reason. By specifying the function $b = w(a)$ in our envelope construction, we not only specify b *but also* $w'(a)$, and γ is precisely $\pi/2 - \tan^{-1} w'(a)$, as can be seen from Fig. 8.4-1. Thus, a characteristic curve is the

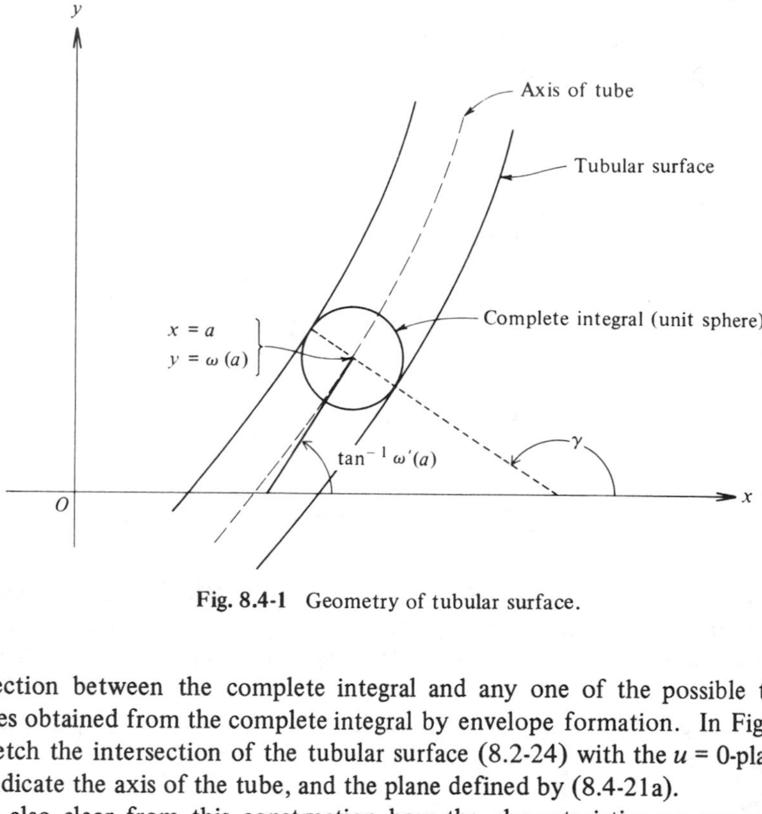

Fig. 8.4-1 Geometry of tubular surface.

intersection between the complete integral and any one of the possible tubular surfaces obtained from the complete integral by envelope formation. In Fig. 8.4-1 we sketch the intersection of the tubular surface (8.2-24) with the $u = 0$-plane; we also indicate the axis of the tube, and the plane defined by (8.4-21a).

It is also clear from this construction how the characteristics are curves across which two solutions may be joined smoothly. For, by merely requiring the two generators $w_1(x)$ and $w_2(x)$ to have the same value and derivative at $x = a$, we can derive two tubular surfaces which branch smoothly out of the circle at $x = a$, $y = w_1(a) = w_2(a)$. One could, for instance, construct a third tubular surface which is generated by w_1 over part of its extent and by w_2 over the remainder.

8.4.3 Hamilton-Jacobi Theory in Dynamics

In section 8.1 we derived the Hamilton-Jacobi equation as the governing equation for the field of extremals from a given point. In this subsection we investigate the properties of eq. (8.1-20) within the context of dynamics. Other applications of our results, for example to variational calculus, are not excluded (cf. Chapter 20) although we choose dynamics as a framework of discussion.

As shown, e.g., in Ref. 8-4, Lagrange's eqs. (8.1-15) which describe a dynamical system with n degrees of freedom are equivalent to Hamilton's equations with the Hamiltonian $H = H(q_1, \ldots, q_n, p_1, \ldots, p_n, t)$ [cf. (8.1-16)]

$$\frac{dq_i}{dt} = H_{p_i}; \qquad i = 1, \ldots, n \tag{8.4-22a}$$

$$\frac{dp_i}{dt} = -H_{q_i}; \qquad i = 1, \ldots, n \tag{8.4-22b}$$

(In dynamical systems described by a Lagrangian, the latter is in general a quadratic form, and therefore a Legendre transformation exists.) One can also derive Hamilton's equations directly from the modified Hamilton principle which is identical to Hamilton's principle (cf. section 8.1.3) when a Legendre transformation exists. In fact, Hamilton's modified principle states that in a dynamical system, motion evolves along a path of coordinates q_i and momenta p_i such that between any two fixed points in space and time the variation of the functional

$$I = \int_{t_I}^{t_F} \left(\sum_{i=1}^{n} p_i \dot{q}_i - H \right) dt \tag{8.4-23}$$

is zero. It is understood that in the integrand of (8.4-23) $\dot{q}_i$ is replaced by the function of q_i and p_i resulting from the Legendre transformation (8.1-16a), and the reader recognizes this integrand as the Lagrangian when (8.1-16b) is used.

In the notation of variational calculus [cf. (8.1-14)] Hamilton's modified principle becomes

$$\delta I = 0 \tag{8.4-24a}$$

with

$$\delta t_I = \delta t_F = 0 \tag{8.4-24b}$$

$$\delta \kappa_i = \delta Q_i = 0 \tag{8.4-24c}$$

Thus, the ordinary differential eqs. (8.4-22) define extremals which, if taken to emanate from a fixed point, describe a field governed by the Hamilton-Jacobi eq. (8.1-20).

$$\frac{\partial J}{\partial t} + H\left(q_1, \ldots, q_n, t, \frac{\partial J}{\partial q_1}, \ldots, \frac{\partial J}{\partial q_n}\right) = 0 \qquad (8.4\text{-}25a)$$

Note that since a Legendre transformation exists $I = J$.

Equation (8.4-25a) is a nonlinear, first-order, partial differential equation for the $n + 1$ independent variables $q_1, \ldots, q_n, t$ and the dependent variable J. If we use our conventional notation $\partial J/\partial q_i = p_i$, $p = \partial J/\partial t$, $J = u$, $x_i = q_i$, eq. (8.4-25a) can be written in the form discussed in section 8.4-1.

$$F = p + H(x_1, \ldots, x_n, t, p_1, \ldots, p_n) = 0 \qquad (8.4\text{-}25b)$$

Two points need be mentioned regarding eq. (8.4-25b). First, we note that $u = J$ does not occur in $F = 0$. More importantly, at this point we view with great caution the fact that we have used the notation p_i for two seemingly different quantities. In eq. (8.4-22b) p_i denotes $\partial L/\partial \dot{q}_i$, and in eq. (8.4-25b) $p_i = \partial J/\partial q_i$. Actually, there is no contradiction as we see by examining the characteristic equations of eq. (8.4-25b). According to eq. (8.4-12) these are

$$\frac{dx_i}{ds} = F_{p_i} = H_{p_i} \qquad (8.4\text{-}26a)$$

$$\frac{dt}{ds} = F_p = 1 \qquad (8.4\text{-}26b)$$

$$\frac{du}{ds} = \sum_{i=1}^{n} p_i F_{p_i} + p \qquad (8.4\text{-}26c)$$

$$\frac{dp_i}{ds} = -(F_{x_i} + p_i F_u) = -H_{x_i} \qquad (8.4\text{-}26d)$$

$$\frac{dp}{ds} = -(F_t + p_i F_u) = -H_t \qquad (8.4\text{-}26e)$$

First we note that eq. (8.4-26b) gives $t = s +$ constant. Therefore, eqs. (8.4-26a) and (8.4-26d) are precisely Hamilton's eqs. (8.4-22) and our notation for p_i is consistent. Equations (8.4-26c) and (8.4-26e) are uncoupled from the remainder of the equations since u does not occur anywhere in F, and p only occurs linearly. In fact, eq. (8.4-26c) gives the variation of u with time and eq. (8.4-26e) gives the result, implicit in eq. (8.4-22) that

$$\frac{dH}{dt} = -\frac{\partial H}{\partial t} \qquad (8.4\text{-}27)$$

and both u, and H may be calculated by quadrature from eqs. (8.4-26c) and (8.4-27) once the solution of Hamilton's equations have been obtained.

We summarize our findings by the statement, *The characteristic equations of the Hamilton-Jacobi partial differential equation contain Hamilton's differential equations.*

We now prove a very important theorem relating the solution of Hamilton's equations and the complete integral of eq. (8.4-25b):

Given a complete integral of eq. (8.4-25b) in the form

$$u = \varphi(x_1, \dots, x_n, t, a_1, \dots, a_n) + a \tag{8.4-28}$$

then the solution of Hamilton's differential equations is contained implicitly in

$$\varphi_{a_i} = b_i = \text{const.}; \qquad i = 1, \dots, n \tag{8.4-29a}$$

$$\varphi_{x_i} = p_i; \qquad i = 1, \dots, n \tag{8.4-29b}$$

This is a special case of the theorem stated without proof in section 8.4.1, and its proof is worth providing. First, however, let us consider the mechanics of obtaining functions $q_i(t)$ and $p_i(t)$ from eq. (8.4-29).

We start by solving eq. (8.4-29a) for the x_i in terms of t and the $2n$ constants a_i, b_i. It is possible to do this because φ is a complete integral; therefore, the determinant of the matrix $\{\varphi_{a_i x_j}\}$ is nonvanishing by hypothesis. We then substitute the expressions just calculated for the x_i into eq. (8.4-29b) and obtain the p_i in terms of t and the $2n$ constants. We also note that this calculation does not involve any quadratures; we merely perform some algebraic operations to obtain $x_i(t, a_1, \dots, a_n, b_1, \dots, b_n)$ and $p_i(t, a_1, \dots, a_n, b_1, \dots, b_n)$ once eq. (8.4-29) are calculated from the complete integral by simple differentiation.

We now prove that this result is a solution of Hamilton's equations involving $2n$ arbitrary constants, hence a general solution of the dynamical problem.

We take the total derivative of eq. (8.4-29a) with respect to t to obtain

$$\varphi_{a_i t} + \sum_{k=1}^{n} \varphi_{a_i x_k} \dot{x}_k = 0; \qquad i = 1, \dots, n \tag{8.4-30}$$

where a dot denotes d/dt.

Next we take the partial derivative with respect to a_i of $\varphi_t + H(x_1, \dots, x_n, t, \varphi_{x_1}, \dots, \varphi_{x_n}) = 0$ to find

$$\varphi_{t a_i} + \sum_{k=1}^{n} H_{p_k} \varphi_{x_k a_i} = 0 \tag{8.4-31}$$

Subtracting eq. (8.4-31) from eq. (8.4-30) and noting that the determinant of $\{\varphi_{x_k a_i}\}$ does not vanish gives eq. (8.4-26a)

To verify eq. (8.4-26d) we calculate the total derivative of eq. (8.4-29b) with respect to t, i.e.

$$\dot{p}_i = \varphi_{x_i t} + \sum_{k=1}^{n} \varphi_{x_i x_k} \dot{x}_k; \quad i = 1, \ldots, n \tag{8.4-32}$$

Now, we differentiate $F = 0$ with respect to x_i to find

$$0 = \varphi_{t x_i} + \sum_{k=1}^{n} H_{p_k} \varphi_{x_k x_i} + H_{x_i}; \quad 1, \ldots, n \tag{8.4-33}$$

Subtracting eq. (8.4-33) from eq. (8.4-32) gives the required result. ■

One of the advantages of having a complete integral is the fact that one can easily solve eq. (8.4-29a) for the constants a_i in terms of the q_i, p_i thus finding explicit formulas for the *n integrals of motion required for the solution of the dynamical system.* The availability of such integrals even in approximate form is of paramount importance for a qualitative understanding of the given motion. Next, we illustrate these ideas by an example for which one can calculate the complete integral exactly. In section 8.4.5 we return to the problem and show how, under certain conditions, one can calculate approximate results in asymptotic form.

Consider the motion of a particle around a spherically symmetric force field. As shown, e.g., in Ref. 8-4, the motion remains in a plane because the angular momentum is conserved. If we then choose polar coordinates in the plane of the motion, the Hamiltonian (which in this example equals the constant energy) becomes

$$H = \frac{1}{2m}\left(p_r^2 + \frac{1}{r^2} p_\theta^2\right) + V(r) \tag{8.4-34}$$

In the above, r and θ are polar coordinates, $V(r)$ is the potential energy and p_r and p_θ are the momenta conjugate to r and θ, i.e., [from eq. (8.1-16a) and the fact that $L = (m/2)(\dot{r}^2 + r^2\dot{\theta}^2) - V$]

$$p_r = m\dot{r} \tag{8.4-35a}$$

$$p_\theta = mr^2\dot{\theta} \tag{8.4-35b}$$

Hamilton's differential equations for this example are

$$\dot{r} = \frac{p_r}{m} \tag{8.4-36a}$$

$$\dot{\theta} = \frac{p_\theta}{mr^2} \tag{8.4-36b}$$

$$\dot{p}_r = \frac{p_\theta^2}{mr^3} - V'(r) \tag{8.4-36c}$$

$$\dot{p}_\theta = 0 \tag{8.4-36d}$$

We see that eqs. (8.3-36a) and (8.4-36b) merely repeat the definitions of p_r and p_θ while eqs. (8.4-36c) reduces to Newton's equation for the radial direction once we replace $\dot{p}_r$ by $m\ddot{r}$. Equation (8.4-36d) which is the first integral of Newton's equation for the circumferential variable states that the magnitude of the angular momentum is conserved.

Although this problem can be solved conventionally from above, we wish to examine the approach using the Hamilton-Jacobi equation which is

$$u_t + \frac{1}{2m}\left(u_r^2 + \frac{1}{r^2}u_\theta^2\right) + V(r) = 0 \tag{8.4-37}$$

The procedure for calculating the complete integral by separation of variables for essentially the same equation was discussed in section 8.2.25. Without repeating the derivation, we find here

$$u = -at + b\theta \pm \int_{r_0}^{r} \{2m[a - V'(\xi)] - b^2/\xi^2\}^{1/2}\, d\xi + c \tag{8.4-38}$$

where r_0 is the value of r for which the quantity in the square root vanishes, and since (8.4-37) is independent of u the third constant c is additive.

Equations (8.4-29a) become, with $a_1 = a, a_2 = b$

$$-t \pm \int_{r_0}^{r} m\{2m[a - V'(\xi)] - b^2/\xi^2\}^{-(1/2)}\, d\xi = -t_0 = \text{constant} \tag{8.4-39a}$$

$$\theta \mp \int_{r_0}^{r} b\xi^{-2}\{2m[a - V'(\xi)] - b^2/\xi^2]^{-(1/2)}\, d\xi = \theta_0 = \text{constant} \tag{8.4-39b}$$

Solving eq. (8.4-39a) gives $r(t)$ and substituting the result in eq. (8.4-39b) gives $\theta(t)$. The reader is referred to texts on dynamics, e.g., Ref. 8-4, for discussions on the physical aspects of this solution, in particular for the case $V = -(k^2/r)$ which corresponds to Newtonian gravitation.

Next we examine the results of eqs. (8.4-29b). Differentiating eq. (8.4-38) with respect to r and setting the result equal to p_r gives

$$\{2m[a - V'(r)] - b^2/r^2\}^{1/2} = p_r \tag{8.4-40a}$$

and the corresponding equation for p_θ is

$$b = p_\theta \tag{8.4-40b}$$

Equation (8.4-40b) directly gives the conservation of angular momentum, and an easy manipulation on (8.4-40a) shows that $a = H =$ const. which is just the conservation of energy.

8.4.4 Canonical Transformations

In preparation for the discussion of a technique of historical interest in dynamics we set down the essential ideas concerning *canonical transformations*, and in so doing show yet another interpretation of the Hamilton-Jacobi equation.

Consider a dynamical system governed by a Hamiltonian, i.e., a set of $2n$ ordinary differential equations derived from a Hamiltonian function

$$H = H(q_1, \ldots, q_n, p_1, \ldots, p_n, t) \tag{8.4-41}$$

such that

$$\frac{dq_i}{dt} = H_{p_i}; \qquad i = 1, \ldots, n \tag{8.4-42a}$$

$$\frac{dp_i}{dt} = -H_{q_i}; \qquad i = 1, \ldots, n \tag{8.4-42b}$$

The reader is reminded that this is a severe restriction on the generality of the dynamical (or other application) system, and excludes, for example, systems involving frictional forces.

We seek a transformation from the variables q_i, p_i and t to a new set q_i^*, p_i^*, t with the property that the Hamiltonian form of eqs. (8.4-42) is preserved. More precisdely, if the transformation is denoted as follows

$$q_i = q_i(q_1^*, \ldots, q_n^*, p_1^*, \ldots, p_n^*, t) \tag{8.4-43a}$$

$$p_i = p_i(q_1^*, \ldots, q_n^*, p_1^*, \ldots, p_n^*, t) \tag{8.4-43b}$$

We calculate a new Hamiltonian by substituting eqs. (8.4-43) for the q_i and p_i throughout the given function H of eq. (8.4-41). The resulting Hamiltonian which depends on the starred variables is denoted by

$$H^* = H^*(q_1^*, \ldots, q_n^*, p_1^*, \ldots, p_n^*, t) \tag{8.4-44}$$

and is used to generate a new set of Hamilton differential equations.

$$\frac{dq_i^*}{dt} = H^*_{p_i^*}; \qquad i = 1, \ldots, n \tag{8.4-45a}$$

$$\frac{dp_i^*}{dt} = -H^*_{q_i^*}; \qquad i = 1, \ldots, n \tag{8.4-45b}$$

Now, a transformation eq. (8.4-43) is said to be canonical *if and only if the solution of eqs. (8.4-45) describes the same motion as the solutions of eqs. (8.4-42)*, i.e., if we use the solution obtained from eqs. (8.4-42) and eqs. (8.4-45), eqs. (8.4-43) are identically satisfied.

The modified variational principle of Hamilton [cf. (8.4-24)] provides us with a very concise representation of a canonical transformation. We use the fact that $\delta I = 0$ in both the q_i, p_i, t and q_i^*, p_i^*, t variables to deduce that for a canonical transformation

$$\sum_{i=1}^{n} p_i \dot{q}_i - H = \sum_{i=1}^{n} p_i^* \dot{q}_i^* - H^* + \frac{dS}{dt} \tag{8.4-46}$$

where S is any one of the four arbitrary functions indicated below obtained by considering all possible combinations of old and new sets of q_i and p_i variables.

$$S = S_1(q_1, \ldots, q_n, q_1^*, \ldots, q_n^*, t) \tag{8.4-47a}$$

$$S = S_2(q_1, \ldots, q_n, p_1^*, \ldots, p_n^*, t) \tag{8.4-47b}$$

$$S = S_3(p_1, \ldots, p_n, q_1^*, \ldots, q_n^*, t) \tag{8.4-47c}$$

$$S = S_4(p_1, \ldots, p_n, p_1^*, \ldots, p_n^*, t) \tag{8.4-47d}$$

Any arbitrary function of the type indicated above generates a canonical transformation when used in eq. (8.4-46), and is called a *generating function*. We show this now for S_1.

Since

$$\frac{dS_1}{dt} = \sum_{i=1}^{n} (S_{1q_i}\dot{q}_i + S_{1p_i}\dot{p}_i) + S_{1t} \tag{8.4-48}$$

eq. (8.4-46) becomes

$$\sum_{i=1}^{n} p_i \dot{q}_i - H = \sum_{i=1}^{n} p_i^* \dot{q}_i^* - H^* + \sum_{i=1}^{n} (S_{1q_i}\dot{q}_i + S_{1p_i}\dot{p}_i) + S_{1t} \tag{8.4-49}$$

which implies that

$$p_i = \frac{\partial S_1}{\partial q_i}; \qquad i = 1, \ldots, n \tag{8.4-50a}$$

$$p_i^* = -\frac{\partial S_1^*}{\partial q_i^*}; \qquad i = 1, \ldots, n \tag{8.4-50b}$$

$$H^* = H + \frac{\partial S_1}{\partial t} \tag{8.4-50c}$$

Equations (8.4-50) define a transformation for each arbitrary S_1 in the sense discussed next. That this transformation is canonical is assured by the fact that eq. (8.4-46) is identically satisfied.

We solve eq. (8.4-50b) for the q_i in terms of the q_i^* and p_i^* to obtain eq. (8.4-34a) then substitute the result into eq. (8.4-50a) to obtain eq. (8.4-43b). Finally, replacing the q_i and p_i in H by the expressions just calculated and also replacing the q_i in $\partial S_1/\partial t$ gives the transformed Hamiltonian H^* in eq. (8.4-50c) in terms of the q_i^*, p_i^* and t.

The reader is cautioned that eqs. (8.4-50) provide only an *implicit* definition of a canonical transformation. The procedure outlined above for deriving an *explicit* transformation of the type eq. (8.4-43) quite often completely obscures the relationship between S_1 and the properties of the transformation, and is, in addition, generally very tedious.

We list below the formulae one derives by the above arguments for the generating functions of form S_2 which are important for the applications discussed in section 8.4.5.

$$p_i = \frac{\partial S_2}{\partial q_i}; \qquad i = 1, \ldots, n \tag{8.4-51a}$$

$$q_i^* = \frac{\partial S_2}{\partial p_i^*}; \qquad i = 1, \ldots, n \tag{8.4-51b}$$

$$H^* = H + \frac{\partial S_2}{\partial t} \tag{8.4-51c}$$

The corresponding formulae for S_3 and S_4 are given in Ref. 8-4 and will not be repeated here.

The reader can verify that the generating function for the *identity transformation* is

$$S_2(q_1, \ldots, q_n, p_1^*, \ldots, p_n^*, t) = \sum_{i=1}^{n} q_i p_i^* \tag{8.4-52}$$

and that all *point transformations* (i.e., transformations $q_i^* = f_i(q_1, \ldots, q_n, t)$ which do not involve the p_i) are canonical and generated by

$$S_2(q_1, \ldots, q_n, p_1^*, \ldots, p_n^*, t) = \sum_{i=1}^{n} f_i(q_1, \ldots, q_n)\, p_i^* \tag{8.4-53}$$

For the reader who has begun to wonder what the foregoing has to do with first-order partial differential equations, we pose the following question: What is the generating function of a canonical transformation for which the transformed Hamiltonian vanishes, hence the Hamilton-Jacobi equation becomes trivial to solve?

We choose to answer the question using a generating function S_2, and we see that according to (8.4-51c) we must solve the equation

$$0 = H + \frac{\partial S_2}{\partial t} \tag{8.4-54}$$

Now, $H = H(q_1, \ldots, q_n, p_1, \ldots, p_n, t)$ and according to eq. (8.4-51b) $p_i = \partial S_2/\partial q_i$. Therefore, the generating function we seek is the solution of the Hamilton-Jacobi equation

$$H^* = \frac{\partial S_2}{\partial t} + H\left(q_1, \ldots, q_n, t, \frac{\partial S_2}{\partial q_1}, \ldots, \frac{\partial S_2}{\partial q_n}\right) = 0 \tag{8.4-55}$$

The motivation for seeking such a transformation S_2 is clear if we examine Hamilton's equations in the transformed variables. Since $H^* = 0$, the p_i^* and q_i^* are $2n$ constants. Therefore, the *transformation relations eq. (8.4-43) are in fact the solution of the dynamical problem!*

This fact puts the procedure for solving a dynamical system on an uncompromisingly consistent basis. For, *one either has to solve Hamilton's differential equations, or the Hamilton-Jacobi equation* no matter what point of view one adopts. This fact had already emerged from our study of the relationship of the characteristic curves (solutions of Hamilton's differential equations) with the complete integral (solution of the Hamilton-Jacobi equation).

Texts on dynamics usually emphasize the present interpretation of the Hamilton-Jacobi equation, whereas texts in mathematics tend to give more importance to the point of view put forward in section 8.4.3. Actually, both points of view are equivalent. In fact, we have seen the fundamental role of the Hamilton-Jacobi equation in dynamics, geometrical optics, and variational calculus. In this chapter we have primarily emphasized the applications in dynamics, with some lesser emphasis on geometrical optics, and even less discussion on variational calculus.

This choice of emphasis was taken because the same theoretical background will be applied to problems in variational calculus and optimization in general in Chapter 20. Thus, to obtain an overall view, the reader is urged to compare and contrast our results and terminology with corresponding ones in other fields.

8.4.5 Canonical Perturbation Theory (von Zeipel's Method)

Until the advent of space flight in 1957 the literature on celestial mechanics had been almost entirely devoted to solution techniques using the Hamiltonian formalism. One might say that the golden age of celestial mechanics ended with Poincaré, who in his celebrated works (Ref. 8-7), discussed with great clarity the principal methods of this field up to his time.

The idea of the method we described here was originally discussed by Poincaré, in volume II of Ref. 8-7. However, since von Zeipel appears to be the first worker to have applied it in Ref. 8-8, the method bears his name in the literature.

Celestial mechanicians continued to use this method (or slight variations thereof) until the recent interest in celestial mechanics by scientists from other disciplines. Consequently, modern and considerably more efficient methods were developed or adapted from other fields by these workers. Nevertheless, a small segment of astronomers still clings to the traditional approach.

These relatively new and more general methods are the techniques based on *multiple scales* or *averaging* described in Chapter 14, and in Ref. 8-9. In a sense, this subsection belongs in Chapter 14. However, since we discuss a technique which has its basis the approximate solution of the Hamilton-Jacobi equation, it is perhaps appropriate to cover it in this closing section of Chapter 8.

To motivate the technique we use the simple example discussed in Ref. 8-9, where the method is compared with that of multiple scales.

Consider the nonlinear oscillator described by

$$\frac{d^2y}{dt^2} + y + \epsilon f(y, t; \epsilon) = 0 \tag{8.4-56}$$

where ϵ is a small parameter and f is periodic in t with period 2π.

We wish to solve (8.4-56) *asymptotically* in the limit $\epsilon \to 0$ and in a form which is *uniformly valid* for large times (cf. Chapter 14 for the idea of a uniformly valid asymptotic expansion).

The idea of the technique is to find a canonical transformation for the Hamiltonian associated with eq. (8.4-56) such that to any desired order in ϵ, the transformed Hamiltonian is *cyclic in all the coordinates*, i.e., the coordinates are absent, hence the momenta are constants. This is a less restrictive requirement than used in section 8.4.4 in re-deriving the Hamilton-Jacobi equation, but leads nevertheless to a solvable problem as we presently show.

The Hamiltonian associated with eq. (8.4-56) is easily found to be

$$H_1(q, p, t; \epsilon) = \frac{p^2 + q^2}{2} + \epsilon g(q, t; \epsilon) \tag{8.4-57a}$$

where we have introduced the notation

$$q = y \tag{8.4-57b}$$

$$p = \dot{y} \tag{8.4-57c}$$

$$g = \int_0^q f(y, t; \epsilon)\, dy \tag{8.4-57d}$$

It is shown in Ref. 8-9 that, if we treat the time as a second coordinate and introduce the canonical transformations

$$w_1 = \frac{[\tan^{-1}(y/\dot{y}) - t]}{2\pi} \tag{8.4-58a}$$

$$w_2 = \frac{t}{2\pi} \tag{8.4-58b}$$

$$J_1 = \pi(y^2 + \dot{y}^2) \tag{8.4-58c}$$

$$J_2 = \pi(y^2 + \dot{y}^2) - 2\pi H_1 \tag{8.4-58d}$$

then the Hamiltonian for the transformed problem becomes

$$H(w_1, w_2, J_1, J_2; \epsilon) = \frac{J_2}{2\pi} + \epsilon g\left[\left(\frac{J_1}{\pi}\right)^{1/2} \sin 2\pi(w_1 + w_2), 2\pi w_2; \epsilon\right] \tag{8.4-59}$$

In the above the J_i are *action variables* and the w_i *are angle variables* (cf. Ref. 8-4 for a discussion of action and angle variables).

Moreover, since g is periodic in t with period 2π, the expression for g in eq. (8.4-59) can be decomposed in the form

$$g = \sigma(w_1, w_2, J_1; \epsilon) + \lambda(w_1, J_1; \epsilon) \tag{8.4-60a}$$

where σ is a periodic function of unit period in both w_1 and w_2 with the property that

$$\langle \sigma \rangle \equiv \int_0^1 \sigma(w_1, w_2, J_1; \epsilon)\, dw_2 = 0 \tag{8.4-60b}$$

and λ and is a periodic function of w_1 with unit period.

The notation $\langle\ \rangle$ of eq. (8.4-60b) will henceforth be adopted to denote the average of a quantity over one period of w_2.

Having seen a concrete example leading to a special Hamiltonian form, we consider the more general problem of Hamiltonians of the form

$$H(w_1, w_2, J_1, J_2; \epsilon) = \frac{J_2}{2\pi} + \epsilon g \tag{8.4-61}$$

where g has the properties of eq. (8.4-60a) but is otherwise arbitrary.

The significance of the decomposition eq. (8.4-60a) which is given explicitly in Ref. 8-9 for eq. (8.4-56) can be seen from examination of Hamilton's differential equations for w_1 and w_2.

$$\frac{dw_1}{dt} = H_{J_1} = \epsilon(\sigma_{J_1} + \lambda_{J_1}) \tag{8.4-62a}$$

$$\frac{dw_2}{dt} = H_{J_2} = \frac{1}{2\pi} \tag{8.4-62b}$$

Thus, w_1 is a quantity of order ϵ, while $w_2 = t/2\pi$ + constant. We see that w_2 is essentially a normalized time variable, in terms of which the period of oscillations of the $\epsilon = 0$ problem are unity. On the other hand w_2 is of order ϵ which implies that periodic functions depending only on w_2 have a period of order $1/\epsilon \gg 1$.

The decomposition of eq. (8.4-60) isolated the *short-periodic terms* σ from the *long-periodic terms* λ. The form of H given in eq. (8.4-59) is analogous to the result one calculates for satellite motions using the *Delauney variables* (cf. Ref. 8-10) in the following sense: The Hamiltonian is proportional to one of the momenta (in this case J_2) if $\epsilon = 0$. When $\epsilon \neq 0$ but small, one can develop H in a power series in ϵ of the form

$$H = H_0(J_2) + \sum_{i=1}^{\infty} H_i(w_1, \ldots, w_n, J_1, \ldots, J_n)\, \epsilon^i \tag{8.4-63}$$

and the H_n are periodic functions of the w_i. In our case $n = 2$, and J_2 is absent from the H_i. For a concrete example of the above, and an application of the von Zeipel method the reader is referred to Ref. 8-11 where the problem of motion of an artificial earth satellite without drag is discussed.

We now return to the problem of eq. (8.4-59) where g satisfies the requirements of (8.4-60) and (8.4-61). The von Zeipel procedure consists of the transformation of H to any given order in ϵ to a form free of all of the angle variables. In our case, we need only remove the w_2 variable to reduce the solution to quadrature. In fact, we require the transformed Hamiltonian to be independent of the transformed w_2-variable. In general, one can in principle, eliminate all the angle variables simultaneously by an appropriate canonical transformation. As a result, the transformed action variables are constants to the order of accuracy retained.

We denote the new variables by an asterisk and introduce a generating function $S(w_1, w_2, J_1^*, J_2^*; \epsilon)$ [cf. eqs. (8.4-51)] which depends on the original angle variables w_1, w_2 and new action variables J_1^*, J_2^* in order to transform H to a form free of w_2^*. In view of the fact that the unperturbed Hamiltonian is already in the proper form, S must reduce to the generating function for the identity transformation when $\epsilon \to 0$, and we assume that it may be developed asymptotically in powers of ϵ as follows:

$$S = w_1 J_1^* + w_2 J_2^* + \epsilon S_1 + \epsilon^2 S_2 + O(\epsilon^3) \tag{8.4-64}$$

where the S_i do not depend on ϵ.

The new variables are related to the original according to eqs. (8.4-51)

$$J_i = S_{w_i} = J_i^* + \epsilon S_{1w_i} + \epsilon^2 S_{2w_i} + O(\epsilon^3) \tag{8.4-65a}$$

$$w_i^* = S_{J_i^*} = w_i + \epsilon S_{1J_i^*} + \epsilon^2 S_{2J_i^*} + O(\epsilon^3) \tag{8.4-65b}$$

and since S is formally independent of t, the new Hamiltonian H^* is obtained from the original Hamiltonian by substituting for the original variables in H the transformed values involving starred quantities [cf. eq. (8.4-51c)]. Before considering the conditions which define the S_i, it is useful to develop the detailed explicit relationship between the original and new variables correct to $O(\epsilon^2)$, and subject to the transformation formulas in (8.4-65).

When the following expansions:

$$J_i = J_i^* + \epsilon F_i^{(1)} + \epsilon^2 F_i^{(2)} + O(\epsilon^3) \tag{8.4-66a}$$

$$w_i = w_i^* + \epsilon G_i^{(1)} + \epsilon^2 G_i^{(2)} + O(\epsilon^3) \tag{8.4-66b}$$

with the $F_i^{(j)}$ and $G_i^{(j)}$ depending upon w_1^*, w_2^*, J_1^*, and J_2^*, are substituted into eqs. (8.4-65) and terms of equal orders are identified, one finds

$$F_i^{(1)} = \frac{\partial S_1}{\partial w_i} \tag{8.4-67a}$$

$$G_i^{(1)} = -\frac{\partial S_1}{\partial J_i^*} \tag{8.4-67b}$$

$$F_i^{(2)} = -\frac{\partial^2 S_1}{\partial w_i \partial w_1}\frac{\partial S_1}{\partial J_1^*} - \frac{\partial^2 S_1}{\partial w_i \partial w_2}\frac{\partial S_1}{\partial J_2^*} + \frac{\partial S_2}{\partial w_i} \tag{8.4-67c}$$

$$G_i^{(2)} = \frac{\partial^2 S_1}{\partial J_i^* \partial w_1}\frac{\partial S_1}{\partial J_1^*} + \frac{\partial^2 S_1}{\partial J_i^* \partial w_2}\frac{\partial S_1}{\partial J_2^*} - \frac{\partial S_2}{\partial J_i^*} \tag{8.4-67d}$$

where all the partial derivatives appearing on the right-hand sides are evaluated at $w_i = w_i^*$.

Thus, eqs. (8.4-66) with the results of (8.4-67) provide the first three terms of the asymptotic expansion of *an explicit canonical transformation* with properties we have not yet specified.

Use of eqs. (8.4-67) in (8.4-61) and the ordering of the results according to powers of ϵ yields the following expression for the new Hamiltonian H^*.

$$H^* = H_0^* + \epsilon H_1^* + \epsilon^2 H_2^* + O(\epsilon^3) \tag{8.4-68a}$$

$$H_0^* = \frac{J_2^*}{2\pi} \tag{8.4-68b}$$

$$H_1^* = \left(\frac{1}{2\pi}\right)\left(\frac{\partial S_1}{\partial w_2}\right) + \sigma + \lambda \tag{8.4-68c}$$

$$\begin{aligned} H_2^* = {} & \frac{1}{2\pi}\left(\frac{\partial S_2}{\partial w_2} - \frac{\partial^2 S_1}{\partial w_1 \partial w_2}\frac{\partial S_1}{\partial J_1^*} - \frac{\partial^2 S_1}{\partial w_2^2}\frac{\partial S_1}{\partial J_2^*}\right) \\ & - \frac{\partial \sigma}{\partial w_1}\frac{\partial S_1}{\partial J_1^*} - \frac{\partial \sigma}{\partial w_2}\frac{\partial S_1}{\partial J_2^*} + \frac{\partial \sigma}{\partial J_1}\frac{\partial S_1}{\partial w_1} \\ & - \frac{\partial \lambda}{\partial w_1}\frac{\partial S_1}{\partial J_1^*} + \frac{\partial \lambda}{\partial J_1}\frac{\partial S_1}{\partial w_1} + \frac{\partial \sigma}{\partial \epsilon} + \frac{\partial \lambda}{\partial \epsilon} \end{aligned} \tag{8.4-68d}$$

where all the terms on the right-hand sides of (8.4-68) are evaluated at $w_i = w_i^*$, $J_i = J_i^*$, and $\epsilon = 0$.

Clearly, the requirement that H_1^* be independent of w_2^* is accomplished by setting

$$\left(\frac{1}{2\pi}\right)\left[\frac{\partial S_1(w_1, w_2, J_1^*, J_2^*)}{\partial w_2}\right] + \sigma(w_1, w_2, J_1^*; 0) = 0 \tag{8.4-69a}$$

in which case H_1^* is simply

$$H_1^* = \lambda(w_1^*, J_1^*; 0) \tag{8.4-69b}$$

Equation (8.4-69a) defines S_1 to within an arbitrary long-periodic term in the form

$$S_1 = \hat{S}_1(w_1, w_2, J_1^*) + S_1^*(w_1, J_1^*, J_2^*) \tag{8.4-70a}$$

where

$$\hat{S}_1 = -2\pi \int_a^{w_2} \sigma(w_1, \eta, J_1^*; 0)\, d\eta \tag{8.4-70b}$$

The lower limit a is an arbitrary function of w_1 and J_1^*, and S_1^* is arbitrary. We note that since $\langle\sigma\rangle = 0$, it is possible to choose the lower limit a in (8.4-70b) such that $\langle\hat{S}_1\rangle = 0$ also.

Use of eqs. (8.4-69) in eq. (8.4-68d) simplifies the expression for H_2^* to

$$H_2^* = \frac{1}{2\pi}\frac{\partial S_2}{\partial w_2} + \frac{\partial\sigma}{\partial J_1}\frac{\partial S_1}{\partial w_1} - \frac{\partial\lambda}{\partial w_1}\frac{\partial S_1}{\partial J_1^*} + \frac{\partial\lambda}{\partial J_1}\frac{\partial S_1}{\partial w_1} + \frac{\partial\sigma}{\partial\epsilon} + \frac{\partial\lambda}{\partial\epsilon} \tag{8.4-71}$$

Removal of the short-periodic terms from H_2^* requires that

$$\frac{1}{2\pi}\frac{\partial S_2}{\partial w_2} + \left(\frac{\partial\sigma}{\partial J_1}\frac{\partial\hat{S}_1}{\partial w_1} - \left\langle\frac{\partial\sigma}{\partial J_1}\frac{\partial\hat{S}_1}{\partial w_1}\right\rangle\right) + \frac{\partial\sigma}{\partial J_1}\frac{\partial S_1^*}{\partial w_1} - \frac{\partial\lambda}{\partial w_1}\frac{\partial\hat{S}_1}{\partial J_1^*} + \frac{\partial\lambda}{\partial J_1}\frac{\partial\hat{S}_1}{\partial w_1} + \frac{\partial\sigma}{\partial\epsilon} = 0 \tag{8.4-72}$$

which in turn defines S_2 by quadrature to within an arbitrary long-periodic function S_2^*. For the purpose of the present comparison, the choice of S_1^*, S_2^*, etc. is immaterial. However, we only consider the case $\partial S_i^*/\partial J_2^* = 0$, which implies that $w_2 = w_2^*$. We note in passing that the judicious choice of S_1^* could simplify the calculation of S_2, etc.

The second-order term in the transformed Hamiltonian now becomes

$$H_2^* = \left\langle\frac{\partial\sigma}{\partial J_1}\frac{\partial S_1}{\partial w_1}\right\rangle - \frac{\partial\lambda}{\partial w_1}\frac{\partial S_1^*}{\partial J_1^*} + \frac{\partial\lambda}{\partial J_1}\frac{\partial S_1^*}{\partial w_1} + \frac{\partial\lambda}{\partial\epsilon} \tag{8.4-73}$$

and this is as far as the von Zeipel procedure is carried out here. With H^* now completely defined to $O(\epsilon^2)$ we find from Hamilton's differential equations for J_1^* and w_1^* that these quantities only depend on $\tilde{t} = \epsilon t$ to $O(\epsilon)$ since

$$\frac{dw_1^*}{d\tilde{t}} = \frac{\partial H_1^*}{\partial J_1^*} + \epsilon\frac{\partial H_2^*}{\partial J_1^*} + O(\epsilon^2) \tag{8.4-74a}$$

$$\frac{dJ_1^*}{d\tilde{t}} = -\frac{\partial H_1^*}{\partial w_1^*} - \epsilon\frac{\partial H_2^*}{\partial w_1^*} + O(\epsilon^2) \tag{8.4-74b}$$

Furthermore, since $H_1^* + \epsilon H_2^*$ is also independent of J_2^* the dependence on w_2^* and J_2^* is completely isolated. The Hamiltonian $H_1^* + \epsilon H_2^*$ is an integral to $O(\epsilon)$ of the system (8.4-74), thus reducing the solution for w_1^* and J_1^* in terms of $\tilde{t}$ from (8.4-74) to quadrature.

The solutions for w_2^* and J_2^* are simply

$$w_2^* = \frac{t}{2\pi} + \text{constant} + O(\epsilon^3) \tag{8.4-75a}$$

$$J_2^* = \text{constant} + O(\epsilon^3) \tag{8.4-75b}$$

and the above, together with the solutions one could obtain for w_1^* and J_1^* in terms of t, represent a complete solution to the problem when used in (8.4-66).

With this example, the reader can readily generalize the procedure to systems of n degrees of freedom. One eliminates the angle variables to any desired order by a transformation as described above. The final result is that the transformed Hamiltonian becomes free of all angle variables; therefore the transformed momenta are integrals of motion.

More precisely, if the final transformed Hamiltonian H^* is of the form

$$H^* = H^*(J_1^*, J_2^*, \ldots, J_n^*) \tag{8.4-76}$$

Hamilton's equations give

$$\frac{dw_i^*}{dt} = H_{J_i^*}^* = \gamma_i^* = \text{constant}, \qquad i = 1, \ldots, n \tag{8.4-77a}$$

$$\frac{dJ_i^*}{dt} = 0, \qquad i = 1, \ldots, n \tag{8.4-77b}$$

Thus, the solution can be expressed in the form

$$J_i^* = J_i^*(0) = \text{constant}, \qquad i = 1, \ldots, n \tag{8.4-78a}$$

$$w_i^* = \gamma_i^* t + w_i^*(0), \qquad i = 1, \ldots, n \tag{8.4-78b}$$

which involves the $2n$ arbitrary constants $J_i^*(0)$ and $w_i^*(0)$.

Of course, all such statements are valid in an asymptotic sense to some given order in ϵ, and in general it is difficult to prove the convergence of the procedure to the limit of an infinite number of terms.

The elegance of the von Zeipel method and its principal advantage hinge on the fact that one reduces the given dynamical system to a 'solved' one by deriving n integrals, which are the transformed momenta of eq. (8.4-78a). Our example eq. (8.4-75b), shows that J_2^* is a constant to $O(\epsilon^2)$. By inverting eq. (8.4-66a) we may express J_2^* in terms of w_1, w_2, J_1, and J_2 in the form

$$J_2^* = J_2 - \epsilon \frac{\partial S_1}{\partial w_2} + \epsilon^2 \left[\frac{\partial^2 S_1}{\partial w_1 \partial J_1^*} \frac{\partial S_1}{\partial w_1} + \frac{\partial^2 S_1}{\partial w_2 \partial J_2^*} \frac{\partial S_1}{\partial w_2} - \frac{\partial S_2}{\partial w_2} \right] = \text{constant} \tag{8.4-79}$$

and this defines *a formal integral* to $O(\epsilon^2)$, because all the terms on the right-hand side are known functions of w_1, w_2, J_1, J_2.

Unfortunately, much of the results in the literature based on the von Zeipel method stop at the stage of calculating a transformed Hamiltonian free of angle variables. Although this means that the solution is available in principle, a quantitative solution is far from being at hand. Much more work needs to be done in order to express the coordinate and velocity components as functions of time and the initial conditions.

Recent modifications to the von Zeipel technique use Lie series to calculate an explicit canonical transformation directly. Thus, although one might be able to bypass part of the tedious calculations mentioned above, one still cannot derive the solution in terms of useful variables.

It is for this reason that methods based on the Hamiltonian formalism, although elegant in concept, are usually impractical for application.

In contrast, multiple-scale, or averaging techniques provide efficiently and with a minimum of obscurity, results in the form of the coordinates as functions of time and arbitrary initial conditions. A case in point is the fact that one can solve eq. (8.4-56) directly to obtain an expansion for $y(t, \epsilon)$ without any transformations. Moreover, these techniques apply to non-Hamiltonian systems arising, for example, when frictional forces are present. For the above reasons the more modern techniques are preferable.

8.5 REFERENCES

8-1 Courant, R., and Hilbert, D., *Methods of Mathematical Physics*, vol. II, Wiley-Interscience, New York, 1962.

8-2 Whitman, G. B., *Linear and Nonlinear Waves*, Wiley-Interscience, New York, 1974.

8-3 Born, M., and Wolf, E., *Principles of Optics*, Fourth Edition, Pergamon Press, Elmsford, New York, 1970.

8-4 Goldstein, H., *Classical Mechanics*, 2nd ed., Addison-Wesley, Reading, Mass., 1980.

8-5 Lax, P. D., "Hyperbolic Systems of Conservative Laws," *Communications on Pure and Applied Mathematics*, **10**, 537, 1957.

8-6 Garabedian, P., *Partial Differential Equations*, John Wiley & Sons, Inc., New York, 1964.

8-7 Poincaré, H., *Les Méthodes Nouvelles de la Mécanique Céleste*, three vols., 1892, 1893, 1899, republished by Dover Publications, New York, 1957.

8-8 von Zeipel, H., "Recherche sur le Mouvement des Petites Planetes," *Ark. Astron. Nat. Fys.*, 11–13, 1916.

8-9 Kevorkian, J., and Cole, J. D., *Perturbation Methods in Applied Mathematics*, Springer-Verlag, New York, 1981.

8-10 Brouwer, D., and Clemence, G. M., *Methods of Celestial Mechanics*, Academic Press, New York, 1961.

8-11 Brouwer, D., "Solution of the Problem of Artificial Satellite Theory Without Drag," *The Astronomical Journal*, **64**, 378, 1959.

9

Partial Differential Equations of Second and Higher Order

Carl E. Pearson*

9.0 SURVEY OF CONTENTS

We will use the abbreviation PDE for partial differential equation and ODE for ordinary differential equation.

Chapter 8 deals with PDE's of first order; here we consider those of higher order. The two chapters are largely independent of one another, and it is not necessary to read Chapter 8 in order to use most of Chapter 9.

The first section of this chapter describes, with examples, methods which may be used for the careful derivation of PDE's governing a particular kind of problem. The other sections are concerned with a discussion of general properties of PDE's and with practical solution techniques. Most of the techniques are illustrated by means of simple examples; emphasis is on basic ideas, rather than possible pathologies. Some topics–perturbations, wave equations, variational methods–are touched on only briefly in this chapter, because of their coverage in other chapters.

Throughout this chapter, we take it for granted that the various functions occurring are continuously differentiable to as high an order as is appropriate for the particular context.

9.1 DERIVATION EXAMPLES

In this section, we give two examples of the derivation of the PDE's governing the behavior of a system. The first of these deals with a simple diffusion problem, and concentrates attention on alternative methods of careful derivation; the second con-

*Prof. Carl E. Pearson, Dep'ts. of Aeronautics and Astronautics, and of Applied Mathematics, University of Washington, Seattle, Wash.

siders the application of the conservation integral method to a nonlinear membrane equilibrium problem.

9.1.1 A Diffusion Problem

Consider the problem of heat conduction along a thin bar of material, of circular cross-sectional area A (cm^2), so oriented that its central axis coincides with the x-axis. The sides of the bar are thermally insulated, and the initial and boundary conditions are to be such that the temperature φ depends only on position x (cm) and time t (secs)–i.e., the temperature is uniform over any cross section. We write $\varphi = \varphi(x, t)$, and measure it (°C) relative to any chosen reference level (e.g., room temperature). Denote the bar density by ρ (gm/cm^3), the specific heat by c (cals/gm-deg), and the thermal conductivity by k (cals/sec-cm-deg). We permit ρ, c, and k to be functions of x.

Our first method of derivation is in terms of a conservation-law integral. Let $x = a$ and $x = b$, with $b > a$, denote two fixed positions along the bar. Then, by the meaning of an integral, the rate of change of thermal energy contained between a and b is

$$\frac{\partial}{\partial t}\int_a^b c\rho A\varphi\, dx = \int_a^b c\rho A\varphi_t\, dx \tag{9.1-1}$$

where (here, and in future) we use a subscript to denote partial differentiation. We use the principle of conservation of energy to equate this expression to the net rate at which heat energy is entering that part of the bar between a and b

$$A\,[k(b)\,\varphi_x(b,t) - k(a)\,\varphi_x(a,t)] = A\int_a^b (k\,\varphi_x)_x\, dx \tag{9.1-2}$$

Thus, cancelling the constant A

$$\int_a^b [c\rho\varphi_t - (k\varphi_x)_x]\, dx = 0 \tag{9.1-3}$$

Within the constraints of the model used, this equation is exact. We appeal now to the fact that if $f(x)$ is a continuous function defined over a certain interval, and if $\int_a^b f(x)\, dx = 0$ for all choices of (a, b) within that interval, then* $f(x) \equiv 0$, to de-

*The proof of this statement is almost obvious. If there is some point x_0 at which $f(x) \neq 0$, then by continuity there is some interval surrounding x_0 within which $f(x)$ has the same sign as $f(x_0)$; a choice of a and b within this interval would then contradict the hypothesis.

duce that

$$\varphi_t = \frac{1}{c\rho}(k\,\varphi_x)_x \tag{9.1-4}$$

The essential steps leading to eq. (9.1-4) were: 1) a formulation of the physical model; 2) an application of an exact conservation law in integral form; and 3) a mathematical deduction from that conservation law. Notice that it was also assumed, implicitly, that all functions are sufficiently nonpathological that the various integrals exist; that the orders of partial differentiation and integration may be interchanged [as in eq. (9.1-1)] ; and that the integrand of eq. (9.1-3) is continuous.

Whenever a PDE has been derived on the basis of some model, a number of checks should be made. If physical quantities are involved, the dimensions of all terms must be consistent; in eq. (9.1-4), for example, we verify that each side has the dimensions of (°C/sec). Secondly, the equation should 'look reasonable'. Here, for example, we can examine the special case in which k is constant; the sign of φ_t is then the same as the sign of φ_{xx}, and this is in accordance with physical expectation. Thirdly, there are often special cases in which the solution is known at once; thus in eq. (9.1-4), if we consider the case in which $\varphi(x, 0)$ = constant, we find $\varphi_t = 0$ as we should.

A second method of exact derivation again makes use of the conservation law, but avoids the integral form. We use the mean value theorem of the integral calculus to replace the expression (9.1-1) by

$$\int_a^b c(x)\,\rho(x)\cdot A\cdot\varphi_t(x,t)\,dx = A\cdot c(\xi)\,\rho(\xi)\,\varphi_t(\xi,t)\cdot(b-a)$$

where all quantities are evaluated at some point ξ satisfying $a \leqslant \xi \leqslant b$. The basic equality is then written in the form (again after a cancellation of A)

$$c(\xi)\,\rho(\xi)\,\varphi_t(\xi,t)\,(b-a) = k(b)\,\varphi_x(b,t) - k(a)\,\varphi_x(a,t) \tag{9.1-5}$$

Again, eq. (9.1-5) is exact, although we do not of course know the value of ξ. We now divide both sides by $(b-a)$, and permit $b \to a$ (so that $\xi \to a$ also). On the assumption that the right-hand side limit does exist, we obtain

$$c(a)\,\rho(a)\,\varphi_t(a,t) = (k\,\varphi_x)_x\big|_{x=a}$$

and, replacing a by x, this equation coincides with eq. (9.1-4).

Frequently, a more casual form of derivation is used. We consider two neighboring points along the bar, say at x and $x + \Delta x$, and write

$$c\rho A\,\Delta x\,\varphi_t = (kA\,\varphi_x)\big|_{x+\Delta x} - (kA\,\varphi_x)\big|_x$$

where it is assumed that Δx is sufficiently small that it is not necessary to specify the exact value of x at which φ_t is evaluated. Dividing by Δx, and proceeding to the limit, $\Delta x \to 0$, again yields eq. (9.1-4).

In using the "casual method" for more complicated problems, involving perhaps noncartesian coordinates, it may be difficult to be sure that all relevant terms are correctly included. It has been the author's experience that the most reliable and effective method is the integral conservation method. A useful tool in this method, which provides the mechanism for altering boundary terms to interior terms [as in eq. (9.1-2)] is Gauss' divergence theorem of Chapter 3, which reads

$$\int_S B_i n_i \, dS = \int_V B_{i,i} \, dV \tag{9.1-6}$$

where B_i is the ith component of a vector field defined over some region V, with surface S; the unit normal vector at each surface element of S is (n_i), and the summation convention of Chapters 3 and 4 (from 1 to 2 in two dimensions, and from 1 to 3 in three dimensions) is used.

9.1.2 A Membrane Problem

A portion of a thin membrane (of negligible thickness) is sketched in Fig. (9.1-1); denote its equation by $z = z(x, y)$. We postulate that the membrane material (e.g., a soap film) is such that it cannot sustain shear. Consequently, if we draw some line element dl in the surface of the membrane, the internal membrane force acting across dl must be in the plane of the membrane and perpendicular to dl. We denote

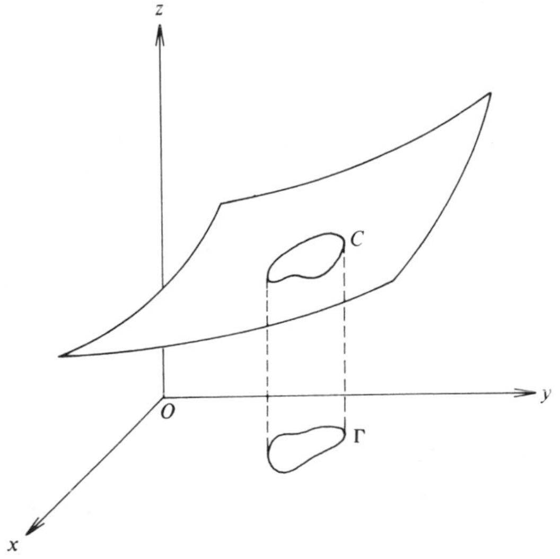

Fig. 9.1-1 Membrane surface.

this vector force, per unit length of dl, by $\mathbf{T}$ (dynes/cm); we can expect it to be a function of x and y. We draw, in imagination, some closed curve C in the surface of the membrane, and denote by Γ the orthogonal projection of C onto the x, y-plane; we now apply the laws of mechanics to that part of the membrane contained within C.

It is first necessary to compute the vector direction of $\mathbf{T}$ at any point on C. Let the line element dl of C have projection components dx and dy, parallel to the x- and y-axes respectively; then $dz = z_x dx + z_y dy$, so that the three vector components of $\mathbf{dl}$ are $(dx, dy, z_x\, dx + z_y\, dy)$. Next, the vector with components $(1, 0, z_x)$ is easily seen to be tangential to the surface at the point (x, y); so is the vector with components $(0, 1, z_y)$; the cross product of these two vectors has components $(-z_x, -z_y, 1)$, and is necessarily perpendicular to the surface. Finally, the cross product of this normal vector with the vector $\mathbf{dl}$ yields a vector in the direction of $\mathbf{T}$; dividing by the magnitude of this cross product vector so as to obtain a unit vector, we can therefore write the three components of $\mathbf{T}$ as

$$\mathbf{T} = \frac{T}{dl \cdot \sqrt{1 + z_x^2 + z_y^2}} \{z_x z_y\, dx + (1 + z_y^2)\, dy, \quad -(1 + z_x^2)\, dx - z_x z_y\, dy, \quad -z_y\, dx + z_x\, dy\}$$

where dl is the length of $\mathbf{dl}$, and where T is the magnitude of $\mathbf{T}$.

The mass-acceleration laws of mechanics then yield

$$\int_C \mathbf{T}\, dl = 0$$

i.e.

$$\int_\Gamma \frac{T}{\sqrt{1 + z_x^2 + z_y^2}} (z_x z_y\, dx + [1 + z_y^2]\, dy) = 0 \tag{9.1-7}$$

and similarly for the other two components. Use of Stokes' theorem [which coincides with the divergence theorem of eq. (9.1-6) for the two-dimensional case], and of the fact that eq. (9.1-7) must hold for any curve Γ, gives

$$[\tau(1 + z_y^2)]_x = (\tau z_x z_y)_y \tag{9.1-8}$$

where $\tau = T/\sqrt{1 + z_x^2 + z_y^2}$. Similarly

$$\begin{aligned} (\tau\, z_x z_y)_x &= [\tau(1 + z_x^2)]_y \\ (\tau\, z_x)_x &= -(\tau\, z_y)_y \end{aligned} \tag{9.1-9}$$

Equations (9.1-8) and (9.1-9) provide a set of three coupled equations for τ and z. They are easily manipulated to show that

$$\ln \tau + \tfrac{1}{2} \ln (1 + z_x^2 + z_y^2) = \text{constant}$$

whence T = constant. Using this result, the second of eqs. (9.1-9) then leads to

$$z_{xx} + z_{yy} + z_x^2 z_{yy} + z_y^2 z_{xx} - 2\, z_x z_y z_{xy} = 0 \qquad (9.1\text{-}10)$$

as the nonlinear equation governing the shape of the membrane surface. Incidentally, the result T = constant is in accordance with physical expectation for the surface tension in a soap film. Moreover, since a soap film is known to attain a surface shape of minimal area, we would expect that the problem of determining that surface shape, which for a given boundary space curve has minimal surface area, to lead to the same PDE. In fact, to illustrate the variational principle technique for constructing a PDE, we now formulate this problem.

Let C_0 be some given closed space curve, whose projection on the x, y-plane is the closed curve C_1. It is required to find that surface, of boundary C_0, which has least surface area. Since an area element dA_0 of the surface will have a projected area of $dA_1 = dA_0/\sqrt{1 + z_x^2 + z_y^2}$, our problem is to find $z(x, y)$ so as to minimize

$$I = \int_{A_1} (dA_1)\sqrt{1 + z_x^2 + z_y^2}$$

where the integral is over the area A_1 enclosed by C_1, in the x, y-plane. This is a conventional calculus of variations problem, and the procedure $\delta I = 0$, subject to $\delta z = 0$ on C_1, leads again to eq. (9.1-10).

9.2 THE SECOND-ORDER LINEAR EQUATION IN TWO INDEPENDENT VARIABLES

9.2.1 Classification

Let $A, B, \ldots, F$, be given functions of (x, y), and consider the linear PD operator

$$L = A \frac{\partial^2}{\partial x^2} + 2B \frac{\partial^2}{\partial x \partial y} + C \frac{\partial^2}{\partial y^2} + D \frac{\partial}{\partial x} + E \frac{\partial}{\partial y} + F \qquad (9.2\text{-}1)$$

A problem which arises frequently in the sciences, and which has been much studied, is to find a function $\varphi(x, y)$ such that

$$L\varphi = G \qquad (9.2\text{-}2)$$

in a region R of the x, y-plane, where $G(x, y)$ is specified, and such that certain conditions are satisfied on a part of the boundary of R. The basic properties of the operator eq. (9.2-1) are largely determined by the relative magnitudes of the coefficients A, $2B$, C, of the second-order terms; depending on these relative magnitudes, a coordinate transformation can be effected which will reduce the second-order portion of L to one of three standard forms. The criterion which determines the appropriate standard form is the sign of the *discriminant* $B^2 - AC$. We assume that this discriminant has the same sign (or vanishes) throughout R (if this is not the case, we divide R into subregions in each of which this is the case, and then designate any one of these subregions by R). The three cases are:

Case I $B^2 - AC > 0$ in R

Equation (9.2-2) is then said to be *hyperbolic*. There exist two families of curves, called *characteristics*; we term generic members of these two families C_+- and C_--curves. A C_+ curve is one for which*

$$\frac{dy}{dx} = \frac{B + \sqrt{B^2 - AC}}{A} \tag{9.2-3}$$

while a C_--curve is one for which

$$\frac{dy}{dx} = \frac{B - \sqrt{B^2 - AC}}{A} \tag{9.2-4}$$

Since A, B, C are given functions of (x, y), each of eqs. (9.2-3) and (9.2-4) is an ordinary differential equation, which is in general solvable (at least numerically). Thus we can find that curve in the x, y-plane, for which eq. (9.2-3) is satisfied, and which passes through some chosen initial point (x_0, y_0). By choosing different initial points, we obtain the other members of the C_+-family. Similarly, the members of the C_--family may be obtained.

Next, let $\xi(x, y)$ be some function whose contour curves (defined by ξ = constant) coincide with the curves of the family C_+, and let $\eta(x, y)$ be a function whose contour curves coincide with the curves of the family C_-. Then (ξ, η) may be chosen as new independent variables,** and the function $\varphi = \varphi(\xi, \eta)$ will satisfy

$$\varphi_{\xi\eta} + D_1\varphi_\xi + E_1\varphi_\eta + F_1\varphi = G_1 \tag{9.2-5}$$

where $D_1, \ldots, G_1$ are certain functions of (ξ, η). Equation (9.2-5) is one of the two standard, or *canonical,* forms for the hyperbolic case. The second canonical

*If $A = 0$, we use $dx/dy = (B \pm \sqrt{B^2 - AC})/C$. If both A and C vanish, eq. (9.2-2) is already in canonical form.

**The Jacobian $\xi_x\eta_y - \xi_y\eta_x$ cannot vanish, for if it did the contour curves of ξ and η would be tangent, and the condition $B^2 - AC > 0$ prevents this—as follows from eqs. (9.2-3) and (9.2-4).

form is obtained by the transformation $\xi = \alpha + \beta$, $\eta = \alpha - \beta$, which leads to

$$\varphi_{\alpha\alpha} - \varphi_{\beta\beta} + D_2\varphi_\alpha + E_2\varphi_\beta + F_2\varphi = G_2 \tag{9.2-6}$$

As an example, the equation

$$(1 + x^2)\,\varphi_{xx} + 2\varphi_{xy} + 3\varphi = 0 \tag{9.2-7}$$

is hyperbolic everywhere. From eqs. (9.2-3) and (9.2-4), members of the C_+ and C_- families have slopes $2/(1 + x^2)$ and 0 respectively, and suitable ξ and η functions are

$$\xi = y - 2\tan^{-1} x, \quad \eta = y$$

In terms of these (ξ, η) variables, eq. (9.2-7) becomes

$$\varphi_{\xi\eta} - \left[\tan\left(\frac{\eta - \xi}{2}\right)\right]\varphi_\xi - \frac{3}{4}\left[\sec^2\left(\frac{\eta - \xi}{2}\right)\right]\varphi = 0 \tag{9.2-8}$$

Case II $B^2 - AC = 0$ in R

Equation (9.2-2) is said to be *parabolic*. There is now only one family of real characteristics; each member of this family has slope

$$\frac{dy}{dx} = \frac{B}{A}$$

(if $A = 0$, so is B from $B^2 - AC = 0$, and the equation is already in canonical form). If $\xi(x, y)$ is a function whose contour curves coincide with these characteristics, and if $\eta(x, y)$ is any other function such that $\eta_x \xi_y - \eta_y \xi_x$ is nonzero in R, then in terms of ξ- and η-coordinates eq. (9.2-2) takes the *canonical form*

$$\varphi_{\eta\eta} + D_2\varphi_\xi + E_2\varphi_\eta + F_2\varphi = G_2 \tag{9.2-9}$$

where D_2, E_2, F_2, G_2 are certain functions of (ξ, η).

For example, the equation

$$x^2\varphi_{xx} + 2x\varphi_{xy} + \varphi_{yy} + y\varphi_x - \varphi = 0 \tag{9.2-10}$$

is parabolic in the region R defined by $x > 0$. The characteristics are given by $y = \ln x + \text{const.}$, and we can take $\xi = y - \ln x$, $\eta = y$, to obtain

$$\varphi_{\eta\eta} + \varphi_\xi\,[1 - \eta \exp(\xi - \eta)] - \varphi = 0 \tag{9.2-11}$$

Case III $B^2 - AC < 0$ in R

Equation (9.2-2) is said to be *elliptic*. There are no real characteristics. It is, however, possible to find a new ξ, η-coordinate system such that eq. (9.2-2) takes the canonical form

$$\varphi_{\xi\xi} + \varphi_{\eta\eta} + D_3\varphi_\xi + E_3\varphi_\eta + F_3\varphi = G_3 \tag{9.2-12}$$

where D_3, E_3, F_3, G_3 are certain functions of (ξ, η). The conditions which must be satisfied by the functions $\xi(x,y)$ and $\eta(x,y)$ are

$$\begin{aligned} A\,\xi_x^2 + 2B\,\xi_x\xi_y + C\,\xi_y^2 &= A\,\eta_x^2 + 2B\,\eta_x\eta_y + C\,\eta_y^2 \\ A\,\xi_x\eta_x + B(\xi_x\eta_y + \xi_y\eta_x) + C\,\xi_y\eta_y &= 0 \end{aligned} \tag{9.2-13}$$

or, with $i = \sqrt{-1}$

$$A(\xi_x + i\,\eta_x)^2 + 2B(\xi_x + i\,\eta_x)(\xi_y + i\,\eta_y) + C(\xi_y + i\,\eta_y)^2 = 0$$

One solution is

$$\frac{\xi_x + i\,\eta_x}{\xi_y + i\,\eta_y} = -\frac{B + i\sqrt{AC - B^2}}{A} \tag{9.2-14}$$

(note that A cannot vanish). In turn, eq. (9.2-14) requires that the *Beltrami equations* be satisfied.

$$\xi_x = \frac{B\,\eta_x + C\,\eta_y}{\sqrt{AC - B^2}} \qquad \xi_y = -\frac{A\,\eta_x + B\,\eta_y}{\sqrt{AC - B^2}} \tag{9.2-15}$$

The existence of a solution is proved in Ref. 9-1. If A, B, C are analytic, or if they may be sufficiently closely approximated by analytic functions, then a process of analytic continuation may be used. We illustrate by means of an example.

Consider

$$x^2\,\varphi_{xx} + 2\,\varphi_{xy} + \varphi_{yy} = 0 \tag{9.2-16}$$

which is elliptic in the region R defined by $x > 1$. The equation of a complex characteristic satisfies

$$\frac{dy}{dx} = \frac{1 + i\sqrt{x^2 - 1}}{x^2}$$

and the solution curves are

$$\psi(x,y) = y + \frac{1}{x} + i\left[\frac{\sqrt{x^2 - 1}}{x} - \ln\left(x + \sqrt{x^2 - 1}\right)\right] = \text{constant}$$

If we choose ξ and η to be the real and imaginary parts, respectively, of ψ, then along these curves we have

$$\frac{dy}{dx} = -\frac{\psi_x}{\psi_y} = -\frac{\xi_x + i\,\eta_x}{\xi_y + i\,\eta_y}$$

so that ξ and η satisfy eq. (9.2-14). In terms of ξ and η, eq. (9.2-16) becomes

$$\varphi_{\xi\xi} + \varphi_{\eta\eta} + \frac{2x}{x^2 - 1}\varphi_\xi - \frac{x(x^2 - 2)}{(x^2 - 1)^{3/2}}\varphi_\eta = 0 \qquad (9.2\text{-}17)$$

where x is that function of η defined by

$$\eta = \frac{\sqrt{x^2 - 1}}{x} - \ln\left(x + \sqrt{x^2 - 1}\right)$$

In some cases, inspection is a useful alternative. For example, in the *Tricomi equation*

$$u_{xx} + x\,u_{yy} = 0$$

the substitution $\eta = \eta(x)$, with an appropriate choice for the function $\eta(x)$, is effective for the case $x > 0$.

9.2.2 Boundary Conditions

Any completely-posed problem which leads to a PDE will include, as part of the problem statement, one or more conditions which must be satisfied on a portion of a boundary curve. For prototype equations of the kinds considered in section (9.2.1), physically appropriate boundary or initial conditions are as follows (with y replaced by time, t, in case I and II):

Case I Wave equation (hyperbolic case)

$$\varphi_{tt} - \varphi_{xx} = 0 \qquad (9.2\text{-}18)$$

A typical "initial value problem" is one in which eq. (9.2-18) is to hold for $t > 0$, $0 < x < L$, with $\varphi(x, 0)$ and $\varphi_t(x, 0)$ specified for $0 < x < L$, and with $\varphi(0, t)$, $\varphi(L, t)$ specified for $t > 0$. (Alternatively, the boundary data at $x = 0$, $x = L$ could involve information concerning φ_x).

Case II Diffusion equation (parabolic case)

$$\varphi_t = \varphi_{xx} \qquad (9.2\text{-}19)$$

A typical problem would require eq. (9.2-19) to be valid on $0 < x < L, t > 0$, with $\varphi(x, 0)$ specified for $0 < x < L$, and with $\varphi(0, t)$ and $\varphi(L, t)$ specified for $t > 0$. (Alternatively, the data at $x = 0, x = L$ could again involve φ_x).

Case III Potential Equation (elliptic case)

$$\varphi_{xx} + \varphi_{yy} = 0 \tag{9.2-20}$$

Typically, eq. (9.2-20) is to hold in a region R; on the boundary Γ of R, the value of φ is specified. Alternatively, values of the normal derivative $\partial\varphi/\partial n$, or of some combination of φ and $\partial\varphi/\partial n$ could be specified on Γ.

In general, these are the kinds of boundary conditions encountered in practice for more general equations of each kind. They have the property that a problem involving such boundary conditions, specifically appropriate to the kind of PDE, is *properly posed*, in the sense that (a) a solution exists, (b) it is unique, and (c) the solution depends continuously on the boundary data.

It is usually the case that a problem involving a PDE of one kind, and boundary conditions appropriate to a PDE of another kind, is not properly posed. For example, consider $u_{xx} + u_{yy} = 0$ in the region $y > 0$, with $u(x, 0) = 0$ and $u_y(x, 0) = f(x)$ where $f(x)$ is specified. The reflection principle of potential theory (section 9.9) permits us to extend the domain of harmonicity to the lower half plane, including the x-axis itself; however, a harmonic function is necessarily analytic in (x, y), so that only in the very special case in which $f(x)$ is analytic can a solution be found. Moreover, even if $f(x)$ is analytic, the problem is not properly posed, as the Hadamard example $f(x) = (\sin Nx)/N^2$, N integral, shows—the solution $u = (\sinh Ny \sin Nx)/N^3$ does not $\to 0$ as $N \to \infty$, even though $f(x) \to 0$.

There is nevertheless a "standard" boundary-value problem in terms of which eq. (9.2-2) is often discussed, irrespective of type. It deals with analytic boundary data, and concerns itself only with the existence of a solution, not with whether or not the problem is properly posed. Consider eq. (9.2-2), and let each of the coefficients $A, B, \ldots, G$ be analytic functions of (x, y) in R, with A nonzero. Let (x_0, y_0) be a point in R, and let $\varphi(x_0, y) = f(y)$ and $\varphi_x(x_0, y) = g(y)$ be prescribed analytic functions of y for y in some neighborhood of y_0. Then the *Cauchy-Kowalewski theorem* states that there exists a unique solution of eq. (9.2-2), satisfying these initial conditions, and analytic in some neighborhood of (x_0, y_0). By a suitable transformation of variables, this result may be framed in terms of prescribed data on an analytic curve in the (x, y) plane; a similar result also holds for higher-order and possibly nonlinear systems (see Ref. 9-2, p. 39; Ref. 9-3, p. 6; or Ref. 9-4, p. 14, for a more general formulation and for a proof).

As an illustration of a situation to which the Cauchy-Kowalewski theorem applies, consider again the problem $u_{xx} + u_{yy} = 0$, $u(x, 0) = 0$, $u_y(x, 0) = f(x)$, where $f(x)$ is analytic. Since a harmonic function is the real part of an analytic function $F(z)$ of a complex variable $z = x + iy$ (cf. section 9.9), we can write $F = u + iv$, with

$F'(z) = u_x + iv_x = u_x - iu_y = -if(x)$ for $y = 0$; thus $F'(z) = -if(z)$, $F(z) = -i\int_0^z f(z)\,dz$, and $u = \text{Re}\{F(z)\}$ then provides the unique analytic solution.

In general, a situation in which φ and its normal derivative $\partial\varphi/\partial n$ are prescribed along some curve in the x, y-plane is termed a *Cauchy data* problem. We have remarked that, even though solutions to special problems of this kind can sometimes be found for other kinds of equations, it is only in the hyperbolic case that such problems are well-posed. Even in the hyperbolic case, an exception must be made for the case in which the initial data curve is a characteristic; in this case, the PDE reduces to a consistency condition for the Cauchy data along the curve, and if this condition is satisfied there will be an infinity of solutions to the initial-value problem. Curves which are not characteristic are sometimes said to be *free* (section 9.11).

If the region of interest is infinite, then some condition at infinity must be imposed; this condition is frequently that the solution vanish, or in any event be bounded, at infinity. If there are discontinuities in the boundary data, the solution to even a properly-posed problem may not be unique without some further constraint (arising, for example, from a physical requirement that the energy be finite) on the order of singularity permissible in the neighborhood of the discontinuity. For example, let φ be harmonic in the region $x > 0$, $y > 0$, with $\varphi(x, 0) = 0$ for $x > 0$, $\varphi(0, y) = 1$ for $y > 0$. A bounded solution is $\varphi = (2/\pi)\tan^{-1}(y/x)$; any harmonic function of the form $r^{-2n}\sin 2n\,\theta$, where r and θ are polar coordinates and n is a positive integer, could be added to this solution without affecting the boundary conditions or the behavior at infinity.

9.3 MORE GENERAL EQUATIONS

Equation (9.2-2) may be generalized in several ways—there can be more independent variables, the order may be higher, the equation may be nonlinear, and there may be a set of coupled equations rather than a single equation. One example would be that of three-dimensional diffusion, which might lead to the equation

$$\varphi_t = (k\,\varphi_x)_x + (k\,\varphi_y)_y + (k\,\varphi_z)_z + g \tag{9.3-1}$$

where $\varphi = \varphi(x, y, z, t)$ is a function of the three space variables (x, y, z) and of time t, where k is a given function of (x, y, z, φ), and where g is a function of (x, y, z, t). If k does not depend on φ, then the equation is linear, and superposition methods (series, transforms, Green's functions) are often applicable. If k does depend on φ, the equation is nonlinear, and recourse to approximate methods (perturbations, variations, numerical methods) is often necessary. As a second example, the much-studied *Navier-Stokes equations* for an ideal gas read

$$\rho\left(\frac{\partial u_i}{\partial t} + u_{i,j}\,u_j\right) = -p_{,i} + \mu\,u_{i,jj} + \frac{1}{3}\,\mu\,u_{j,ji}$$

$$\frac{\partial \rho}{\partial t} + (\rho\,u_j)_{,j} = 0$$

$$p = \rho\,R\,T \tag{9.3-2}$$

$$\rho\,c_v\left(\frac{\partial T}{\partial t} + T_{,j}\,u_j\right) = (k\,T_{,i})_{,i} - p\,u_{i,i} - \frac{2}{3}\,\mu(u_{k,k})^2 + \mu(u_{i,j} + u_{j,i})u_{i,j}$$

where $u_i (i = 1, 2, 3)$ are the three velocity components, ρ the density, T the temperature, and p the pressure; we use the index notation of Chapter 3 (with coordinates x_1, x_2, x_3, and time t), so that a repeated index indicates summation from 1 to 3. The first of these equations is to hold for each choice of i. The quantities R, c_v, μ, k, are the gas constant, the specific heat at constant volume, the viscosity, and the conductivity, respectively. There are six equations in the six unknowns $u_1, u_2, u_3, \rho, p, T$. Clearly the system is nonlinear. To make any progress on a system as complicated as this, it is necessary to examine special cases—e.g., one- or two-dimensional flow, incompressible flow, or nonviscous non-heat-conducting flow.

Another source of complexity arises in the boundary conditions. Not only may the boundary in a multidimensional problem be geometrically complicated, but the specified boundary conditions may be nonlinear. For example, a heat radiation condition may read $\partial\varphi/\partial n = A(\varphi^4 - \varphi_0^4)$ where A, φ_0 are constants. Again, approximation methods are probably necessary.

In general, any PDE system may be rewritten as a coupled set of first-order equations. The idea is simple, and may be illustrated for the general second-order equation

$$F(x, y, \varphi, \varphi_x, \varphi_y, \varphi_{xx}, \varphi_{xy}, \varphi_{yy}) = 0 \tag{9.3-3}$$

where F is some function of the eight arguments listed. We define $p = \varphi_x$, $q = \varphi_y$, $r = \varphi_{xx}$, $s = \varphi_{xy}$, and obtain

$$F(x, y, \varphi, p, q, r, s, q_y) = 0$$

$$\varphi_x = p$$

$$\varphi_y = q$$

$$p_x = r$$

$$q_x = s$$

as a set of five equations for the five unknowns φ, p, q, r, s. A different approach to this kind of substitution is described in Ref. 9-3, p. 8.

The reverse process—that of replacing a set of coupled first-order equations by a single equivalent high-order equation—is not generally feasible. However, there are cases in which this kind of substitution is both feasible and useful. The most common case is probably that of 'irrotationality'; if three functions $A_1(x_1, x_2, x_3)$, $A_2(x_1, x_2, x_3)$ and $A_3(x_1, x_2, x_3)$ satisfy the conditions $A_{i,j} = A_{j,i}$ for all choices of i, j, then these conditions may be replaced by the statement that there exists some function $\psi(x_1, x_2, x_3)$ such that $A_i = \psi_{,i}$, for all i. The function ψ is termed a 'potential'. A set of PDE's involving the A_i may now be rewritten in terms of ψ. For example, if $A_{i,j} = A_{j,i}$ for all i and j, and if $A_{i,i} = 0$, then this set of four equations reduces to the single equation $\psi_{,ii} = 0$. Generalizations of this idea have been much studied in elasticity; see Refs. 9-5, 9-6.

As remarked in section 9.2.2, a generalized version of the Cauchy-Kowalewski theorem for the case of analytic Cauchy data holds for sets of equations (of analytic structure) of the kinds considered here; see the references listed there. Also, as in section 9.2.2, a given problem is not necessarily well-posed unless the kind of boundary data given is appropriate to the scientific context in which the PDE system arises.

When we depart from the equations considered in section 9.2, it is difficult to provide a formal classification scheme, and only rarely feasible to transform the equation into some canonical form. For certain cases, there is however some generally-accepted nomenclature. Thus it is natural to term the equations

$$\varphi_{xx} + \varphi_{yy} + \varphi_{zz} = 0$$

$$\varphi_{tt} - \varphi_{xx} - \varphi_{yy} = 0$$

$$\varphi_t - \varphi_{xx} - \varphi_{yy} = 0$$

elliptic, hyperbolic, and parabolic, respectively. More generally, consider an equation for $\varphi(x_1, x_2, \ldots, x_n)$ involving the second-order terms

$$\sum_{i,j=1}^{n} a_{ij}\varphi_{ij} + \cdots = 0 \qquad (9.3\text{-}4)$$

where φ_{ij} means $\partial^2\varphi/\partial x_i \partial x_j$, and where the a_{ij} are functions of $(x_1, x_2, \ldots, x_n)$. At a point P, let the a_{ij} attain values a_{ij}^0. By means of a suitable local affine transformation from the x_i-variables to a set of y_i-variables (i.e., each y_i is a linear function of the x_i), the second-order terms as evaluated at P become

$$\sum_{i=1}^{n} \lambda_i \frac{\partial^2 \varphi}{\partial y_i^2}$$

where $\lambda_i = +1, -1$, or 0. If all $\lambda_i = +1$ (or all $\lambda_i = -1$), eq. (9.3-4) is termed *elliptic*. If all λ_i except exactly one have the same sign, eq. (9.3-4) is termed *hyperbolic*.

If several λ_i are equal to + 1, and several are equal to − 1, we say the equation is *ultrahyperbolic*. If one or more of the λ_i vanish, we say eq. (9.3-4) is *parabolic*. All these designations refer to the character of the PDE at point P. Of course, if the a_{ij} are constants, then the canonical transformation appropriate for P is applicable at all points.

On occasion, a function satisfying a high-order PDE may be expressed in terms of simpler functions. For example, let $\varphi(x, y, z)$ satisfy the biharmonic equation

$$\Delta\Delta\varphi = \left(\frac{\partial^2}{\partial x^2} + \frac{\partial^2}{\partial y^2} + \frac{\partial^2}{\partial z^2}\right)^2 \varphi = 0$$

in a suitable region. Then φ must have the form $\varphi = \alpha + x\beta$, where α and β are harmonic (Ref. 9-6, p. 120).

9.4 SERIES SOLUTIONS

A series solution is often feasible for a linear PDE where the boundary conditions are prescribed on coordinate surfaces. Although the series may be tedious to calculate if many terms are required for accuracy (a computer may be used of course), it does have the advantage of exactness. A disadvantage is that if a function φ is obtained in such a series form, and if φ_x or φ_{xx} is desired, then term-by-term differentiation will generally result in slower convergence and, perhaps, in no convergence at all. (In such cases, it may be possible to explicitly sum the more troublesome component of the series.)

The basic idea is an outgrowth of Fourier series expansion techniques. Instead of using sines and cosines, however, we use functions particularly adapted to the PDE and to the boundary geometry. To find such functions, we use the separation of variables idea in the PDE operator to produce ODE's and adjoin homogeneous boundary conditions so as to obtain Sturm-Liouville problems (cf. Chapter 6). The advantage of this procedure is that the eigenfunctions of such problems are complete, so that expansions of other functions in terms of them are known to be feasible; moreover, the fact that we start with the PDE operator means that the resulting eigenfunctions are naturally associated with the problem in which they are eventually to be used. (We remark that the PDE problem itself need not have homogeneous boundary conditions.)

To illustrate, let $\varphi(x, t)$ satisfy the equation

$$\varphi_t = (x^2\varphi_x)_x + g(x, t) \tag{9.4-1}$$

in $1 < x < 2$, $t > 0$, and let $\varphi(x, 0) = f(x)$, $\varphi(1, t) = 0$, $\varphi(2, t) = t$. Here $f(x)$ and $g(x, t)$ are prescribed functions of (x, t). Although we might notice that the transformation $x = e^y$ would simplify the PDE, it would lead to a non-self-adjoint situation which is sometimes inconvenient; we therefore proceed directly. We want to write

$$\varphi = \sum \alpha_n(t)\, \beta_n(x) \tag{9.4-2}$$

and to find suitable $\beta_n(x)$-functions we consider $\psi = \alpha(t)\, \beta(x)$ in an equation involving the PDE operator alone, i.e., in

$$\psi_t = (x^2 \psi_x)_x$$

This yields

$$\frac{\alpha'}{\alpha} = \frac{(x^2 \beta')'}{\beta}$$

which is possible only if each side is a constant, say $-\lambda$. Thus we are led to consider

$$(x^2 \beta')' + \lambda\beta = 0 \tag{9.4-3}$$

and, since we observe that boundary conditions involve a prescription of φ at $x = 1$ and $x = 2$, we obtain a Sturm-Liouville problem by adjoining to the (self-adjoint) ODE (9.4-3) the homogeneous boundary conditions $\beta(1) = \beta(2) = 0$.* The solutions of this Sturm-Liouville problem are

$$\begin{aligned} \beta_n(x) &= \frac{1}{\sqrt{x}} \sin\left[\frac{\pi n \ln x}{\ln 2}\right] \\ \lambda_n &= \frac{1}{4} + \left(\frac{\pi n}{\ln 2}\right)^2 \\ n &= 1, 2, \ldots \end{aligned} \tag{9.4-4}$$

Since this set of $\beta_n(x)$-functions is complete, the expansion (9.4-2) is certainly possible, in the usual mean-square sense

$$\lim_{N\to\infty} \int_1^2 \left[\varphi - \sum_{n=1}^{N} \alpha_n(t)\, \beta_n(x)\right]^2 dx = 0 \tag{9.4-5}$$

for any fixed choice of t. The series need not, of course, converge at each point x in the interval $[1, 2]$; moreover, term-by-term differentiation need not be permissible. For that reason, we do not substitute the series (9.4-2) directly into the PDE.

We observe instead that, by virtue of the usual eigenfunction orthogonality rela-

*Had the boundary conditions involved a prescription of $\partial\phi/\partial x$, the natural homogeneous conditions to choose for the Sturm-Liouville problem would have been $\partial\varphi/\partial x = 0$ at the ends.

tion (Chapter 6)

$$\alpha_n(t) = \frac{1}{\ln\sqrt{2}} \int_1^2 \varphi(x, t)\, \beta_n(x)\, dx \tag{9.4-6}$$

(the weight function being unity in this case). If eq. (9.4-1) is multiplied by $\beta_n(x)$ and integrated with respect to x from $x = 1$ to $x = 2$ we therefore obtain—after two integrations by parts, and use of eq. (9.4-3)

$$\int_1^2 \beta_n(x)\, \varphi_t(x, t)\, dx = -\,[x^2 \beta_n' \varphi]_1^2 - \lambda_n \int_1^2 \varphi(x, t)\, \beta_n(x)\, dx + \int_1^2 g(x)\, \beta_n(x)\, dx$$

or

$$\alpha_n'(t) = \left[\frac{2\sqrt{2}\,(-1)^{n+1}\,\pi n}{(\ln 2)^2}\right] t - \lambda_n \alpha_n(t) + g_n(t) \tag{9.4-7}$$

where $g_n(t)$ is the nth coefficient in the expansion of $g(x, t)$ in terms of the $\beta_n(x)$. Thus

$$\alpha_n = \int_0^t e^{-\lambda_n (t-\tau)} \left[\frac{2\sqrt{2}\,(-1)^{n+1}\,\pi n \tau}{(\ln 2)^2} + g_n(\tau)\right] d\tau + f_n e^{-\lambda_n t} \tag{9.4-8}$$

where $f(x) = \sum f_n \beta_n(x)$.

The final expansion (9.4-2) obtained by this method is valid in the sense of eq. (9.4-5) despite the fact that it does not converge at $x = 2$ (since each $\beta_n(2) = 0$, whereas $\varphi(2, t) = t$). The fact that the series (9.4-2) converges only in the mean-square sense means that the series (9.4-2) is almost certainly not term-by-term differentiable, so that a substitution of eq. (9.4-2) directly into eq. (9.4-1) would have led to disaster. The present method, sometimes called a *finite transform method*, avoids this difficulty.

An alternative technique would be to "strengthen" the convergence of eq. (9.4-2) by expanding, not $\varphi(x, t)$, but rather an associated function $\Omega(x, t)$ defined by

$$\Omega(x, t) = \varphi(x, t) - (x - 1)\, t$$

Thus $\Omega(1, t) = \Omega(2, t) = 0$, and Ω satisfies the PDE

$$\begin{aligned} \Omega_t &= (x^2 \Omega_x)_x + g_1(x, t) \\ g_1(x, t) &= g(x, t) + 2xt + 1 - x \end{aligned} \tag{9.4-9}$$

with $\Omega(x, 0) = f(x)$ as before. If we now write

$$\Omega = \sum \alpha_n^{(1)}(t)\, \beta_n(x) \tag{9.4-10}$$

and proceed as before, the fact that Ω and each $\beta_n(x)$ satisfies the same homogeneous conditions at $x = 1$, $x = 2$ means that term-by-term differentiation is (at least in this case) permissible, and a substitution of eq. (9.4-10) into eq. (9.4-9) will give a result essentially equivalent to that of eq. (9.4-2) as previously obtained. In fact, the results may be made identical by explicitly summing that part of eq. (9.4-2) which involves the end condition at $x = 2$, i.e., that part involving the first portion of each integrand in eq. (9.4-8). (If one had a solution in the form of eq. (9.4-2), with α_n given by eq. (9.4-8), and wanted to differentiate term-by-term, this same summation procedure could be used as a preliminary step).

If more variables, say $x_1, x_2, \ldots, x_n$, are involved, the situation is essentially similar; one expands any unknown function (or functions) φ in the form $\varphi = \left[\sum\sum\cdots\sum \alpha_1(x_1)\, \alpha_2(x_2) \ldots \alpha_n(x_n)\right]$, where suitable expansion functions $\alpha_i(x_i)$ are determined from associated Sturm-Liouville problems. For example, let $\varphi(x, r, \theta, t)$ satisfy the wave equation

$$\varphi_{tt} = \varphi_{xx} + \varphi_{rr} + \frac{1}{r}\varphi_r + \frac{1}{r^2}\varphi_{\theta\theta} + f(x, r, \theta, t)$$

where f is a given function, and let certain values for φ be prescribed on the boundary of the cylindrical region $0 < x < l$, $0 < r < a$, $0 < \theta < 2\pi$, $0 < t$, with φ and φ_t also prescribed throughout the region at $t = 0$. Then suitable expansion functions are given by

$$\varphi = \sum_{n=1}^{\infty}\sum_{m=0}^{\infty}\sum_{j=1}^{\infty} [a_{nmj}(t)\cdot \cos m\theta + b_{nmj}(t)\cdot \sin m\theta]\cdot \sin\frac{n\pi x}{l} J_m\left(\zeta_{mj}\frac{r}{a}\right)$$

where ζ_{mj} is the jth zero of the Bessel function J_m. In this problem, one of the Sturm-Liouville problems has periodic end conditions, and another has a finiteness condition at one end of the interval. As before, it might be convenient, although it is not essential, to carry out a preliminary change of dependent variable which will make the boundary conditions homogeneous.

A large collection of examples of series expansions will be found in Refs. 9-7, 9-8, 9-9.

9.5 TRANSFORM METHODS

The use of transform methods for PDE's is discussed in detail in Chapter 11. We consider here only one simple example, in order to compare the results so obtained with those obtained by series expansions. We conclude with some general remarks on the applicability of transform methods.

Consider again eq. (9.4-1), with the boundary conditions there given. Denote the Laplace transform, with respect to time, of any function $u(x, t)$ by the corresponding uppercase letter $U(x, s)$, so that

$$U(x, s) = \int_0^\infty e^{-st}\, u(x, t)\, dt$$

Multiply each side of eq. (9.4-1) by e^{-st} and integrate with respect to t from 0 to ∞ (this process is often termed 'taking the transform of the equation') to obtain

$$-f(x) + s\Phi(x, s) = (x^2 \Phi_x)_x + G(x, s) \tag{9.5-1}$$

(We have assumed that $\varphi(x, t)$ is sufficiently well-behaved that the processes of integration with respect to t and differentiation with respect to x can be interchanged). At this point, we can write down the formal solution of eq. (9.5-1), subject to the conditions $\Phi(1, s) = 0$, $\Phi(2, s) = 1/s^2$; this transform expression can then be inverted to obtain again the solution of section 9.4. However, the transform form often permits a useful approximation to be made, and to illustrate this fact we consider the special case $f = g = 0$, so that eq. (9.5-1) leads to

$$\Phi(x, s) = \frac{\sqrt{2}\, \sinh\left(\frac{1}{2}\sqrt{1 + 4s}\, \ln x\right)}{s^2\sqrt{x}\, \sinh\left(\frac{1}{2}\sqrt{1 + 4s}\, \ln 2\right)} \tag{9.5-2}$$

For sufficiently large values of s, this expression may be expanded as

$$\Phi = \frac{\sqrt{2}}{s^2\sqrt{x}}\, e^{-(1/2)\sqrt{1+4s}\,\ln(2/x)} \{1 - e^{-\sqrt{1+4s}\,\ln x} + e^{-\sqrt{1+4s}\,\ln 2} - e^{-\sqrt{1+4s}\,(\ln x + \ln 2)} + \cdots\} \tag{9.5-3}$$

and this series may now be inverted term by term; the result is a series of error-function-type expressions. However, the point of interest is that the inverted series converges very rapidly for small values of t [since a property of Laplace transforms is that the small-t behavior of $\varphi(x, t)$ is linked to the large-s behavior of $\Phi(x, s)$]. In contrast, the corresponding series of section 9.4 converges very slowly for small values of t; thus the solution techniques of this section and of section 9.4 are complementary to each other. In the present case, inversion of eq. (9.5-3) yields

$$\varphi(x, t) \sim \sqrt{\frac{2}{x}}\left[\left(t + \frac{1}{2}\ln^2\frac{2}{x}\right)\operatorname{erfc}\left(\frac{1}{2\sqrt{t}}\ln\frac{2}{x}\right) - \sqrt{\frac{t}{\pi}}\left(\ln\frac{2}{x}\right)\exp\left(\frac{-\ln^2\frac{2}{x}}{4t}\right)\right] \tag{9.5-4}$$

valid for t small.

We now list some general remarks concerning transform methods:

1. The use of a Laplace, Fourier, Mellin, or other transform represents an extension of the idea of series expansions of section 9.4 to the situation in which an interval is infinite.

2. As with the "finite transforms" of section 9.4, a main motivation is that a PDE can be turned into an ODE, or at least into a PDE with a smaller number of independent variables.

3. The method is generally restricted to linear equations, where the coefficients do not depend on the variable with respect to which the transform is being taken [except perhaps as a low-order polynomial, in which case a differential equation in the transform variable occurs–thus if $F(s)$ is the transform of $f(t)$, then the transform of $tf(t)$ is $-d/ds\, F(s)$].

4. In inverting a transform expression, powerful approximate or asymptotic methods are sometimes available. An example was given above; a more complete discussion will be found in Chapters 11 and 12, and in Ref. 9-10.

5. In taking the transform of an equation, it is implicitly assumed that transforms of the dependent functions do exist. This means in turn that any singularity at infinity or near a boundary discontinuity is assumed not to be too strong–thus the transform method acts as a filter, in that it rules out solutions which are too badly behaved.

6. In certain problems, the inversion formula for a transform may involve a complex-variable path passing through singularities of the integrand. A far-field radiation condition may provide guidance; alternatively, the introduction of artificial dissipation may be useful. See, for example, Ref. 9-10, p. 337.

9.6 THE PERTURBATION IDEA

If a PDE problem is close, in some sense, to another problem whose solution is known, then it may be possible to reformulate the given problem so as to take advantage of this fact. For example, we can easily solve (by series, say) the equation

$$\varphi_{xx} + \varphi_{yy} = g(x, y) \tag{9.6-1}$$

in the region $0 < x < 1$, $0 < y < 1$ (where g is a given function), subject to $\varphi = f$ on the boundary, where f is a given function of boundary position. If we instead encounter the problem

$$\varphi_{xx} + (1 + \tfrac{1}{4} x^5)\, \varphi_{yy} = g \tag{9.6-2}$$

with the same boundary conditions, then we could use the solution $\varphi^{(0)}$ of eq. (9.6-1) as an approximation for the term $\frac{1}{4} x^5\ \varphi_{yy}$, and rewrite eq. (9.6-2) as

$$\varphi_{xx} + \varphi_{yy} \cong [g - \tfrac{1}{4} x^5\ \varphi_{yy}^{(0)}] \tag{9.6-3}$$

This equation has the same form as eq. (9.6-1), and since the right-hand side is a known function of (x, y), we can solve it by the same means (still subject to the condition $\varphi = f$ on the boundary) to find a function $\varphi^{(1)}$. We could now set up an iterative procedure, and find $\varphi^{(2)}$ as the solution of the problem

$$\varphi_{xx} + \varphi_{yy} = [g - \tfrac{1}{4} x^5 \varphi_{yy}^{(1)}]$$

$\varphi = f$ on boundary, and so on. It is clear that nonlinear situations—for example

$$\varphi_{xx} + (1 + \tfrac{1}{10} \varphi)\, \varphi_{yy} = g$$

could be handled similarly; the basic requirement is that the "perturbing" terms $\frac{1}{4}x^5$ or $\frac{1}{10}\varphi$ be sufficiently small that the iterative process converge rapidly. A formal alternative to the iterative process is to replace eq. (9.6-2) with the equation

$$\varphi_{xx} + (1 + \epsilon x^5)\, \varphi_{yy} = g \tag{9.6-4}$$

and to assume φ to be analytic in the small parameter ϵ, so that

$$\varphi(x, y, \epsilon) = \psi^{(0)}(x, y) + \epsilon \psi^{(1)}(x, y) + \epsilon^2 \psi^{(2)}(x, y) + \cdots \tag{9.6-5}$$

to substitute this expression into eq. (9.6-4), and to collect powers of ϵ so as to obtain

$$\begin{aligned} \psi_{xx}^{(0)} + \psi_{yy}^{(0)} &= g \\ \psi_{xx}^{(1)} + \psi_{yy}^{(1)} &= -x^5 \psi_{yy}^{(0)} \\ \psi_{yy}^{(2)} + \psi_{yy}^{(2)} &= -x^5 \psi_{yy}^{(1)} \\ \cdots\cdots&\cdots\cdots \end{aligned}$$

These equations may be solved in sequence, subject to the boundary conditions $\psi^{(0)} = f$, $\psi^{(j)} = 0$ for $j > 0$, on the boundary. Of course, we eventually set $\epsilon = \frac{1}{4}$ in the series (9.6-5).

It is worth pointing out that the "perturbation" may occur in a boundary condition, as well as in the PDE itself. A simple example occurs in the problem of finding $\varphi(r, \theta)$, harmonic in the region $0 \leqslant r < 1 + \epsilon f(\theta)$, $0 \leqslant \theta \leqslant 2\pi$ with $\varphi(1 + \epsilon f(\theta), \theta) = g(\theta)$. Here f and g are given functions of θ, and $|\epsilon| \ll 1$. We write

$$\varphi(r, \theta, \epsilon) = \varphi^{(0)}(r, \theta) + \epsilon \varphi^{(1)}(r, \theta) + \epsilon^2 \varphi^{(2)}(r, \theta) + \cdots \tag{9.6-6}$$

and find at once that each $\varphi^{(j)}$-function must be harmonic. The crucial point here is in the boundary conditions for the $\varphi^{(j)}$. The condition $\varphi(1 + \epsilon f, \theta) = g$ can be rewritten

$$g = \varphi(1, \theta) + (\epsilon f)\, \varphi_r(1, \theta) + \tfrac{1}{2} (\epsilon f)^2\, \varphi_{rr}(1, \theta) + \cdots \tag{9.6-7}$$

and if the series (9.6-6) is substituted into eq. (9.6-7) we obtain the set of boundary conditions

$$\varphi^{(0)}(1,\theta) = g$$
$$\varphi^{(1)}(1,\theta) = -f(\theta)\cdot\varphi_r^{(0)}(1,\theta)$$
$$\varphi^{(2)}(1,\theta) = -f(\theta)\cdot\varphi_r^{(1)}(1,\theta) - \tfrac{1}{2}f^2(\theta)\cdot\varphi_{rr}^{(0)}(1,\theta)$$
$$\cdots\cdots\cdots\cdots$$

Thus each $\varphi^{(j)}$-function is determined by a calculation involving the unit circle only, and the desired solution (9.6-6) may be computed.

Although the perturbation idea is simple, and exceedingly useful (even in cases in which ϵ is surprisingly large), there are many subtleties inherent in the perturbation method, and we now refer the reader to Chapter 14.

9.7 CHANGE OF VARIABLE

It is hardly necessary to point out that a given PDE may sometimes by simplified by an adroit change in either the dependent or independent variable. For example, the constant a^2 in the equation $\varphi_t = a^2\varphi_{xx}$ can be suppressed (except perhaps in the boundary conditions) by the replacement $x = a\xi$, where (ξ, y) are the new independent variables. In the equation $\varphi_t = \varphi_{xx} + A\varphi_x + B\varphi$ (where A, B, are constants), the substitution $\varphi = e^{\alpha x+\beta t}\psi$ leads to an equation for ψ in which the terms in ψ_x and ψ will not appear, if the constants α and β are chosen appropriately.

A more subtle situation arises in connection with the nonlinear *Burger's equation*

$$u_t + uu_x = \epsilon u_{xx} \tag{9.7-1}$$

which to some extent models the one-dimensional flow of a viscous fluid. Here ϵ is a positive constant. A transformation of variables which reduces this equation to a linear diffusion equation was found by Hopf. We rewrite eq. (9.7-1) as

$$u_t = (\epsilon u_x - \tfrac{1}{2}u^2)_x \tag{9.7-2}$$

This equation states that the vector field whose components are u and $(\epsilon u_x - \frac{1}{2}u^2)$ in the x, t-plane is irrotational, so that a potential function $Q(x, t)$ must exist such that

$$u = Q_x; \qquad \epsilon u_x - \tfrac{1}{2}u^2 = Q_t$$

Replacing u by Q_x in the second of these equations, we obtain

$$Q_t = \epsilon Q_{xx} - \tfrac{1}{2}Q_x^2 \tag{9.7-3}$$

Although this equation is still nonlinear, the nonlinearity is quadratic, and an exponential transformation is sometimes effective in such cases. Here in fact we define

$$\psi(x, t) = \exp\left\{-\frac{1}{2\epsilon} Q(x, t)\right\} \tag{9.7-4}$$

to obtain the familiar diffusion equation

$$\psi_t = \epsilon\psi_{xx} \tag{9.7-5}$$

Observe that, from eq. (9.7-4), ψ is always nonzero. To obtain u itself, from any solution ψ of eq. (9.7-5), we use $u = -2\epsilon\psi_x/\psi$.

We consider now two often-used kinds of coordinate transformations; a third kind, connected with characteristics, is discussed in section 9.11.

9.7.1 Conformal Mapping

The Laplacian operator $\Delta = \partial^2/\partial x^2 + \partial^2/\partial y^2$ is of frequent occurrence; problems involving it can sometimes be simplified by use of the close connection between this operator and complex variable methods.

The simplest case is that in which it is required to find a function $\varphi(x, y)$, harmonic in a region R of the x, y-plane (i.e., $\Delta\varphi = 0$ in R), with φ prescribed (say $\varphi = g$) on the boundary Γ of R. Define $z = x + iy$ where $i = \sqrt{-1}$, and let $w = f(z)$ be a complex-valued function of z, analytic in R, with $w = u(x, y) + iv(x, y)$. We can think of the functional relationship $w = f(z)$ as a mapping from the x, y-plane into the u, v-plane; let the region R be mapped into a region R'. A basic property of this mapping is that it is *conformal*, except at critical points; this means that, except where $f'(z) = 0$, the angle of intersection of any pair of arcs at a point z_0 in R is the same (in both magnitude and direction) as the angle of intersection of the image arcs at $w_0 = f(z_0)$ in R'. Moreover, the Jacobian $|\partial(u, v)/\partial(x, y)|$ is nonzero if $f' \neq 0$, so that the transformation is locally one-to-one. However, the property of immediate interest is that, if $\Omega(u, v)$ denotes the value of $\varphi(x, y)$ as evaluated at that point (x, y) which under the mapping corresponds to (u, v), then Ω is harmonic in R'. Since the boundary values $\varphi = g$ may also be thought of as being carried along by the transformation, the result is to transform the problem $\Delta\varphi = 0$ in R, $\varphi = g$ on Γ, into the problem $\Delta\Omega = 0$ in R', $\Omega = g$ on Γ'. If R' is a particularly simple region, then this kind of transformation is clearly worthwhile.

An immediate question is whether or not a transformation into a 'particularly simple' region R' exists, and if so, how hard it is to find. The first part of this question is answered by the *Riemann mapping theorem*, which states that if R is simply-connected and has more than one boundary point, then there exists a transformation mapping R conformally onto the standard region R' defined by $|w| < 1$, with any of the following additional conditions satisfied (Ref. 9-14):

a. a given point z_0 in R maps into the point $w_0 = 0$ in R', and a chosen direction through z_0 maps into the direction of the positive w-axis at $w_0 = 0$;

b. a given point z_0 in R maps into $w_0 = 0$, and a given point z_1 on Γ maps into a chosen point w_1 on the boundary $|w| = 1$ of R';

c. three arbitrary points z_1, z_2, z_3 on Γ map into three given points w_1, w_2, w_3 (but chosen in the same sense) on the boundary $|w| = 1$ of R'.

For multiply-connected regions, the connectivity must generally be maintained in the mapping; it may not be possible to arbitrarily specify the ratios of lengths characteristic of the corresponding standard domain.

Some mappings have been extensively studied. Among these are the *bilinear transformations* $[w = (az + b)/(cz + d)]$ which map circles or straight lines into circles or striaght lines; *Joukowsky transformations* $(w = z + 1/z)$; and *Schwarz-Christoffel transformations*, which map the exterior or interior of any polygon into a circle. A 'dictionary' of many standard transformations of this kind is given in Ref. 9-11.

It frequently happens that a transformation is available which "almost" maps a given region R into the unit circle. In such cases, it is possible to contrive a transformation, determinable at least approximately, which maps the "almost circle" into a circle, so that the composition of the two transformations produces the desired result. See Ref. 9-10, p. 174. We refer also to Ref. 9-10, Chapter 4, for a number of instructive examples of conformal mappings, including the practically important case of degenerate polygons, in which "slits" of various kinds occur.

Problems involving Poisson's equation $[\Delta\varphi = h(x, y)]$, more complicated boundary conditions $(\varphi + \alpha\partial\varphi/\partial n = \beta)$, and higher-order equations (e.g., $\Delta\Delta\varphi = 0$) can also be usefully treated by means of conformal mapping. In some cases, physical objects—e.g., sources or dipoles—retain their physical interpretation under the mapping; in others, certain contour integrals (e.g., circulation) play an important physical role. A general discussion will be found in Ref. 9-10, Chapter 4; for applications to fluid mechanics, see Ref. 9-12, and for applications to elasticity see Ref. 9-13.

9.7.2 Legendre Transformation

The most common changes of variable are probably those in which the independent variables are replaced by a new set of independent variables, or in which the dependent variables are replaced by a new set of dependent variables. It may however be useful to set aside this restriction. A simple example is given by $u_x + uu_y = 0$; think of this equation as defining $x(u, y)$ rather than $u(x, y)$; then $x(u, y)$ satisfies the linear equation $ux_y = 1$.

A useful tool in these kinds of manipulations is afforded by the *Legendre transformation*. Let u be a function of x and y, and define $p = u_x, q = u_y$. Then

$$du = p\,dx + q\,dy \tag{9.7-6}$$

Define now

$$w = px + qy - u \tag{9.7-7}$$

so that

$$dw = x\,dp + y\,dq \tag{9.7-8}$$

and it follows that, if we think of w as being a function of p and q (which in a sense involve u)

$$\left.\frac{\partial w}{\partial p}\right|_q = x; \quad \left.\frac{\partial w}{\partial q}\right|_p = y \tag{9.7-9}$$

Of course, p and q will not be suitable independent variables unless the Jacobian determinant $|\partial(p,q)/\partial(x,y)|$ is nonzero, i.e., unless $u_{xx}u_{yy} - u_{xy}^2 \neq 0$ (i.e., the function $u = u(x,y)$ must not represent a *developable surface*).

A first example of the use of this Legendre transformation is in the problem

$$y + u_x^2 + u_y^2 x = 0 \tag{9.7-10}$$

We observe that x and y occur linearly here, but that p and q do not. This suggests that we make p and q the new independent variables, and define $w(p,q)$ as in eq. (9.7-7). Then eq. (9.7-10) becomes

$$w_q + p^2 + q^2 w_p = 0$$

which is indeed a linear equation.

Instead of changing to variables (p,q), it might be useful to try (y,p) as new independent variables. A suitable Legendre transformation would be given by

$$v = px - u$$

for then

$$dv = x\,dp - q\,dy$$

so that $v_p = x$, $v_y = -q$. We could similarly define $r = qy - u$, if we were aiming at new variables (x,q).

Turning to second-order situations, consider again eq. (9.7-7), for $w(p,q)$. Since $x = w_p$, we have (by taking $\partial/\partial x$ or $\partial/\partial y$ of both sides)

$$\begin{aligned} 1 &= w_{pp}p_x + w_{pq}q_x \\ &= w_{pp}u_{xx} + w_{pq}u_{xy} \end{aligned}$$

and

$$0 = w_{pp}u_{xy} + w_{pq}u_{yy}$$

Similarly, starting with $y = w_q$, we obtain

$$\begin{aligned} 0 &= w_{pq}u_{xx} + w_{qq}u_{xy} \\ 1 &= w_{pq}u_{xy} + w_{qq}u_{yy} \end{aligned}$$

Thus, with $D = u_{xx}u_{yy} - u_{xy}^2 \neq 0$, we have $w_{pp} = u_{yy}/D$, $w_{pq} = -u_{xy}/D$, and $w_{qq} = u_{xx}/D$.

As an example of a second-order equation, consider the equation of *minimal surface area*

$$(1 + u_y^2)\, u_{xx} + (1 + u_x^2)\, u_{yy} - 2u_{xy}u_xu_y = 0 \tag{9.7-11}$$

which the above transformation changes into the linear equation

$$(1 + q^2)\, w_{qq} + (1 + p^2)\, w_{pp} + 2\, pq\, w_{pq} = 0 \tag{9.7-12}$$

In problems of fluid mechanics, a change of this kind, in which velocities (components of the potential gradient) become new independent coordinates is termed a *hodograph transformation*. Of course, the two velocity components need not be used directly; the velocity magnitude and flow direction angle are equivalent variables.

We remark finally that, from an equation like (9.7-8), one can deduce useful auxiliary relationships. Thus with x and y considered to be functions of p and q, eq. (9.7-8) implies at once that $x_q = y_p$. In more extensive calculations of this kind, a convenient auxiliary tool is the identity that results from a functional relationship of the form $f(x, y, z) = 0$; the identity reads

$$\left.\frac{\partial x}{\partial y}\right|_z \left.\frac{\partial y}{\partial z}\right|_x \left.\frac{\partial z}{\partial x}\right|_y = -1$$

9.8 GREEN'S FUNCTIONS

By the term *Green's function* we generally mean the solution of a linear PDE whose nonhomogeneous term has been replaced by a product of delta functions, so as to represent the effect of a concentrated forcing function. A superposition technique may then enable one to write down, at least formally, the solution of more general problems. We begin by reviewing delta functions and Green's identities.

9.8.1 Delta Function

Consider the sequence of functions $\delta(x, \epsilon)$ defined, for various values of the parameter $\epsilon > 0$, by

$$\delta(x, \epsilon) = \begin{cases} 0, & \text{for } |x| > \epsilon \\ \dfrac{1}{2\epsilon}, & \text{for } |x| < \epsilon \end{cases} \tag{9.8-1}$$

Let $f(x)$ be any continuous function in the interval $(-1, 1)$. Define

$$I = \int_{-1}^{1} f(x)\,\delta(x, \epsilon)\,dx = \int_{-\epsilon}^{\epsilon} f(x) \cdot \frac{1}{2\epsilon}\,dx \tag{9.8-2}$$

so that, if M and m denote the maximum and minimum values of $f(x)$ in $[-\epsilon, \epsilon]$, it is clear that

$$m \leqslant I \leqslant M$$

and that, as $\epsilon \to 0$

$$I \to f(0)$$

It is often convenient to use a symbolic shorthand for this kind of limiting process, and this is done in terms of the *Dirac delta function* $\delta(x)$. We interpret the equation

$$\int_{-1}^{1} f(x)\,\delta(x)\,dx = 0 \tag{9.8-3}$$

to mean that $\delta(x)$ is replaced by $\delta(x, \epsilon)$ as defined by eq. (9.8-1), that the integration is performed, and that we then permit $\epsilon \to 0$ in the result of the integration. Although we can think of $\delta(x)$ in a picturesque manner as being a function which vanishes everywhere except at the origin, where its value is infinite in such a way that "area" under the curve is unity (so that in effect we get a unit impulse at the origin), the above interpretation–in which we first integrate and then let $\epsilon \to 0$– avoids mathematical awkwardness. (We remark that any meaningful use of the delta function will eventually involve an integration.) Of course, the limits of integration in eq. (9.8-2) need not be $(-1, 1)$; it is necessary only that they include the origin.

It is convenient to use a function somewhat better behaved then that defined by eq. (9.8-1). One choice is

$$\delta(x, \epsilon) = \frac{1}{2\sqrt{\pi\epsilon}} \exp\left(-\frac{x^2}{4\epsilon}\right) \tag{9.8-4}$$

which again leads to eq. (9.8-2), but which has the advantage of differentiability everywhere. We could in fact now consider

$$J = \int_{-1}^{1} f(x)\,\delta'(x, \epsilon)\,dx$$

$$= [f(x) \cdot \delta(x, \epsilon)]_{-1}^{1} - \int_{-1}^{1} f'(x)\,\delta(x, \epsilon)\,dx$$

$$\to -f'(0) \quad \text{as } \epsilon \to 0$$

More generally, if ξ is some fixed value of x, we can write—subject to the above interpretation—

$$\int_\alpha^\beta f(x)\,\delta(x-\xi)\,dx = \begin{cases} 0, & \xi \text{ not in } [\alpha, \beta] \\ f(\xi), & \xi \text{ in } \quad (\alpha, \beta) \end{cases}$$

$$\int_\alpha^\beta f(x)\,\delta'(x-\xi)\,dx = \begin{cases} 0, & \xi \text{ not in } [\alpha, \beta] \\ -f'(\xi), & \xi \text{ in } \quad (\alpha, \beta) \end{cases}$$

Similarly

$$\int_\alpha^\beta \int_\gamma^\tau f(x,y)\,\delta(x-\xi)\delta(y-\eta)\,dx\,dy = f(\xi, \eta)$$

if the point (ξ, η) is inside the region of integration, and zero otherwise.

An example from ODE's may be worthwhile. Let $g(x, \xi)$ be the solution of

$$\begin{aligned} g_{xx} - g &= \delta(x-\xi) \\ g(0, \xi) &= g(1, \xi) = 0 \end{aligned} \tag{9.8-5}$$

where the parameter ξ satisfies the condition $0 < \xi < 1$. Except at $x = \xi$, g is to be continuously differentiable; we require also $g(\xi-, \xi) = g(\xi+, \xi)$, so that g is to be continuous in x across $x = \xi$. An integration between $(\xi - 0)$ and $(\xi + 0)$ of eq. (9.8-5) shows that the increase in slope, g_x, across the point $x = \xi$ is given by

$$g_x(\xi+, \xi) - g_x(\xi-, \xi) = 1$$

These conditions define g uniquely as

$$g = \frac{-1}{\sinh 1} \begin{cases} \sinh x \sinh(1-\xi), & 0 < x < \xi \\ \sinh(1-x)\sinh \xi, & \xi < x < 1 \end{cases}$$

If instead of eq. (9.8-5), we encounter

$$\begin{aligned} y'' - y &= f(x) \\ y(0) = A; &\quad y(1) = B \end{aligned} \tag{9.8-6}$$

then we can multiply eq. (9.8-5) by y, eq. (9.8-6) by g, subtract, and integrate from 0 to 1 to obtain

$$\int_0^1 (y g_{xx} - g y_{xx})\,dx = \int_0^1 y(x) \cdot \delta(x-\xi)\,dx - \int_0^1 g(x, \xi) f(x)\,dx \tag{9.8-7}$$

i.e.

$$y(\xi) = Bg_x(1, \xi) - Ag_x(0, \xi) + \int_0^1 g(x, \xi) f(x)\, dx \tag{9.8-8}$$

Thus if $g(x, \xi)$ is known, the solution of eq. (9.8-6) for any $f(x)$ can be obtained by quadratures. This idea carries over to PDE's, as we shall see.

A more formal and systematic treatment of the delta function will be found in Ref. 9-15.

9.8.2 Green's Identities

Let A_i be the ith component of a vector field defined over a region V with boundary S. Using $A_{i,j}$ to denote $\partial A_i/\partial x_j$, where x_i is the ith Cartesian coordinate, and a repeated index to denote summation (cf., Chapter 3), Gauss' divergence theorem reads

$$\int_V A_{i,i}\, dV = \int_S A_i n_i\, dS \tag{9.8-9}$$

in either two or three dimensions. Here n_i is the ith component of the unit outward normal vector. We assume, of course, that S possesses a tangent plane at each point.

Next, let φ be any scalar, and define $A_i = \varphi_{,i}$. Then eq. (9.8-9) gives

$$\int_V \varphi_{,ii}\, dV = \int_S \frac{\partial \varphi}{\partial n}\, dS \tag{9.8-10}$$

As a generalization, let φ and ψ be two scalars, and define $A_i = \varphi\psi_{,i}$ to obtain

$$\int_V \varphi\psi_{,ii}\, dV + \int_V \varphi_{,i}\psi_{,i}\, dV = \int_S \varphi \frac{\partial \psi}{\partial n}\, dS \tag{9.8-11}$$

an equation which is termed *Green's first identity*. Interchanging φ and ψ, and subtracting the result so obtained from eq. (9.8-11) gives *Green's second identity*

$$\int_V (\varphi\psi_{,ii} - \psi\varphi_{,ii})\, dV = \int_S \left(\varphi \frac{\partial \psi}{\partial n} - \psi \frac{\partial \varphi}{\partial n}\right) dS \tag{9.8-12}$$

The above results are valid in either two or three dimensions; to proceed further, we must specialize. Consider the three-dimensional case, and let (ξ_i) be some fixed point inside V. Define $r > 0$ by

$$r^2 = (x_i - \xi_i)(x_i - \xi_i) \tag{9.8-13}$$

(still using the summation convention), and in eq. (9.8-12) let $\psi = 1/r$. Then

$$\psi_{,ii} = -4\pi\delta(x_1 - \xi_1) \cdot \delta(x_2 - \xi_2) \cdot \delta(x_3 - \xi_3) \tag{9.8-14}$$

so that eq. (9.8-12) leads to *Green's third identity*

$$4\pi\varphi(\xi_1, \xi_2, \xi_3) = \int_S \left[\frac{1}{r}\frac{\partial\varphi}{\partial n} - \varphi\frac{\partial}{\partial n}\left(\frac{1}{r}\right)\right] dS - \int_V \frac{1}{r}\varphi_{,ii}\, dV \tag{9.8-15}$$

(The volume integral converges, since near the point (ξ_i) the element of volume is proportional to r^2.)

In two dimensions, we use $\ln r$ rather than $1/r$ to obtain similarly

$$2\pi\varphi(\xi_1, \xi_2) = \int_S \left[\left(\ln\frac{1}{r}\right)\frac{\partial\varphi}{\partial n} - \varphi\frac{\partial}{\partial n}\ln\frac{1}{r}\right] dS - \int_V \left(\ln\frac{1}{r}\right)\varphi_{,ii}\, dV \tag{9.8-16}$$

Each of eqs. (9.8-15) and (9.8-16) relate the value of any function φ at an interior point to values of φ and $\partial\varphi/\partial n$ on the boundary, as well as to an integral over the interior of its Laplacian.

9.8.3 Green's Functions for the Laplacian Operator

As in section 9.8.2, let (ξ_i) be a point within the three-dimensional region V. Let $g(x_i, \xi_i)$ denote that function satisfying

$$\begin{gathered} \Delta g = g_{,ii} = 4\pi\delta(x_1 - \xi_1)\delta(x_2 - \xi_2)\delta(x_3 - \xi_3) \\ g = 0 \quad \text{for } (x_i) \text{ on } S \end{gathered} \tag{9.8-17}$$

where the Laplacian operator Δ means $(\partial^2/\partial x_1^2 + \partial^2/\partial x_2^2 + \partial^2/\partial x_3^2)$. In view of eq. (9.8-14), $g = -(1/r) +$ a harmonic function. Replacing ψ in eq. (9.8-12) by g, we obtain

$$4\pi\varphi(\xi_1, \xi_2, \xi_3) = \int_V g\Delta\varphi\, dV + \int_S \varphi\frac{\partial g}{\partial n}\, dS \tag{9.8-18}$$

Thus, once the *Green's function* g has been found for our region, any problem of the form

$$\begin{aligned} \Delta\varphi &= \text{prescribed in } V \\ \varphi &= \text{prescribed on } S \end{aligned} \tag{9.8-19}$$

can be solved by quadratures via eq. (9.8-18).

Since the specifying conditions (9.8-17) are simpler than those of (9.8-19), it may be easier to find g (say, by series expansions) than to find φ directly. In any event, once g has been determined, eq. (9.8-18) enables us to solve eq. (9.8-19) for a sequence of functions φ. In some cases, g is particularly easy to find by use of the symmetry of the problem. For example, consider a sphere of radius R centered at the origin of a spherical coordinate system with coordinates (r, θ, γ) (i.e., $x_1 = r \sin\theta \cos\gamma$, etc.). Denote the coordinates of the fixed point by (ρ, α, β). The 'image point' corresponding to this fixed point has coordinates $(R^2/\rho, \alpha, \beta)$, and the reader may verify that

$$g(r, \theta, \gamma; \rho, \alpha, \beta) = \frac{R}{\rho r_2} - \frac{1}{r_1} \tag{9.8-20}$$

where r_1 and r_2 denote the distances from (r, θ, γ) to (ρ, α, β) and $(R^2/\rho, \alpha, \beta)$ respectively. Thus, by use of eq. (9.8-18), the problem (9.8-19) is solved for the sphere.

It may be shown that $g(x_i; \xi_i)$ is symmetric in its two sets of variables–i.e., $g(x_1, x_2, x_3; \xi_1, \xi_2, \xi_3) = g(\xi_1, \xi_2, \xi_3; x_1, x_2, x_3)$.

If the problem (9.8-19) is replaced by the problem

$$\begin{aligned} \Delta\varphi &= \text{prescribed in } V \\ \frac{\partial\varphi}{\partial n} &= \text{prescribed on } S \end{aligned} \tag{9.8-21}$$

then, on the assumption that the consistency condition (9.8-10) is satisfied, we define a Green's function $h(x_i, \xi_i)$ by

$$\begin{aligned} \Delta h &= 4\pi\delta(x_1 - \xi_1)\delta(x_2 - \xi_2)\delta(x_3 - \xi_3) \\ \frac{\partial h}{\partial n} &= K \quad \text{on } S \end{aligned} \tag{9.8-22}$$

where the constant K is obtained by setting $\varphi = h$ in eq. (9.8-10). The function h is only defined within an arbitrary function of (ξ_1, ξ_2, ξ_3), and this arbitrary function may be chosen so that h is symmetric in its two sets of arguments. Equation (9.8-12), with ψ replaced by h, yields

$$4\pi\varphi(\xi_1, \xi_2, \xi_3) = K \int_S \varphi\, dS - \int_S h \frac{\partial \varphi}{\partial n} dS + \int_V h \Delta\varphi\, dV \tag{9.8-23}$$

A topic closely related to the subject of Green's functions for Laplacian operators is that of surface and volume distributions of charges and dipoles; see, for example, section 9.9 and Refs. 9-3, 9-27.

9.8.4 Other Green's Function Examples

Let $u(x_1, x_2)$ denote the displacement of an elastically-restrained membrane, so that

$$\Delta u - \alpha^2 u = f \tag{9.8-24}$$

where α is a real constant, and where $f(x_1, x_2)$ is a prescribed loading function. We define a Green's function $g(x_1, x_2; \xi_1, \xi_2)$, where (ξ_1, ξ_2) is a point inside the region R of interest, by

$$\begin{gathered} \Delta g - \alpha^2 g = \delta(x_1 - \xi_1)\,\delta(x_2 - \xi_2) \\ g = 0 \text{ for } (x_1, x_2) \text{ on } C \end{gathered} \tag{9.8-25}$$

where C is the boundary of R. The usual "multiply and subtract technique" gives

$$\int_R [g\Delta u - u\Delta g]\, dA = \int_R fg\, dA - u(\xi_1, \xi_2)$$

or

$$u(\xi_1, \xi_2) = \int_R fg\, dA + \int_C u \frac{\partial g}{\partial n} ds \tag{9.8-26}$$

The singular part of g will here have the form $-(1/2\pi)K_0(\alpha r)$ where r is the distance between (ξ_i) and (x_i), and where K_0 is the modified Bessel function (which has a logarithmic singularity at $r = 0$). If α were purely imaginary, as is the case for the reduced wave equation, then a similar technique would apply. More precisely, set $\varphi = \text{Re}\{\psi e^{i\omega t}\}$ in the wave equation $\varphi_{tt} - c^2\Delta\varphi = m$, where $m = \text{Re}\{-fe^{i\omega t}\}$, to obtain

$$\Delta\psi + \frac{\omega^2}{c^2} = \frac{1}{c^2} f \tag{9.8-27}$$

Taking into account the condition that any wave at infinity must be outgoing, an appropriate Green's function would be

$$g = \frac{i}{4} H_0^{(2)}\left(r \frac{\omega}{c}\right) + p \tag{9.8-28}$$

where p is a well-behaved function satisfying

$$\Delta p + \frac{\omega^2}{c^2} p = 0$$

The singular parts of the three-dimensional equivalents of these two Green's functions would involve $e^{-\alpha r}/r$ and $e^{-i\omega r/c}/r$, respectively.

We remark in passing that when a region extending to infinity is of interest for the three-dimensional version of eq. (9.8-27), and if no energy is incident from infinity, then ψ must satisfy the *Sommerfeld radiation condition*

$$R\left(\frac{\partial \psi}{\partial R} + i \frac{\omega}{c} \psi\right) \longrightarrow 0 \quad \text{as} \quad R \longrightarrow \infty \tag{9.8-29}$$

where R is the radius of a large sphere. We remark also, as a general matter of technique, that if the multiply and subtract technique is to be used in conjunction with the divergence theorem for a non-self-adjoint equation, then the Green's function for the adjoint equation should be used. For the operator

$$L\varphi = A\varphi_{xx} + 2B\varphi_{xy} + C\varphi_{yy} + D\varphi_x + E\varphi_y + F\varphi \tag{9.8-31}$$

the adjoint operator is defined by

$$L^*\varphi = (A\varphi)_{xx} + (2B\varphi)_{xy} + (C\varphi)_{yy} - (D\varphi)_x - (E\varphi)_y + F\varphi \tag{9.8-31}$$

Some other special Green's functions are of interest; we give here the final results of their use, and refer to Refs. 9-2, 9-7, and 9-16 for details and for further examples:

a. $\varphi_t = a^2 \varphi_{xx}$ in $-\infty < x < \infty$, $0 < t$, with $\varphi(x, 0) = f(x)$. Then

$$\varphi(x, t) = \frac{1}{2a\sqrt{\pi t}} \int_{-\infty}^{\infty} f(\xi) \exp\left\{-\frac{(x-\xi)^2}{4a^2 t}\right\} d\xi \tag{9.8-32}$$

b. $\varphi_{tt} - c^2 \Delta\varphi = m$. Here c is constant, m is a given function of position and time. Let $[\varphi]_R$ denote the value of φ at a time earlier than t by an amount required for signal propagation (at velocity c) over the distance r between the observation point and the element of integration. Then *Kirchhoff's formula* reads

$$\varphi(\xi_1, \xi_2, \xi_3, t) = \frac{1}{4\pi c^2} \int_V \frac{[m]_R}{r} \, dV$$

$$+ \frac{1}{4\pi} \int_S \left\{ \frac{1}{r} \left[\frac{\partial \varphi}{\partial n} \right]_R - [\varphi]_R \frac{\partial}{\partial n} \left(\frac{1}{r} \right) + \frac{1}{rc} \left[\frac{\partial \varphi}{\partial t} \right]_R \frac{\partial r}{\partial n} \right\} dS \quad (9.8\text{-}33)$$

c. $\varphi_{tt} - c^2 \Delta\varphi = m$, as in (b). Then

$$\varphi(x_1, x_2, x_3, t) = \int_0^{t+} d\tau \left[\int_{V_0} m(\xi, \tau) g(x, t; \xi, \tau) \, dV_0 \right.$$

$$\left. - c^2 \int_{S_0} \left\{ \varphi \frac{\partial g}{\partial n_0} - g \frac{\partial \varphi}{\partial n_0} \right\} dS_0 \right] - \int_{V_0} \left[\varphi \frac{\partial g}{\partial \tau} - g \frac{\partial \varphi}{\partial \tau} \right]_{\tau=0} dV_0 \quad (9.8\text{-}34)$$

where the subscript '0' indicates the ξ-variable region. Here

$$g = \frac{1}{4\pi c^2 r} \delta \left[t - \tau - \frac{r}{c} \right] \quad (9.8\text{-}35)$$

where, as before, r is the distance between (ξ_i) and (x_i). In one space-dimension and time, an alternative formulation of the wave-equation result is in terms of a *Riemann function*; see Ref. 9-2, p. 449.

The method of integral transforms (Chapter 11) is often very efficient in the problem of determining at least the singular part of a Green's function.

9.8.5 Integral Equations

Green's function methods may be used to transform a given PDE problem into the form of an integral equation. We give here only one simple illustrative example, and refer to Refs. 9-3, 9-7, and 9-17 for further details and more examples.

Let S be a closed surface enclosing a volume V. Let $\Delta\varphi = 0$ in V, with $\varphi = f$ (a prescribed function) on S. Then φ may be written as

$$\varphi = \int_S k \frac{\partial}{\partial n} \left(\frac{1}{r} \right) dS$$

where the function k satisfies the integral equation

$$\varphi(P) + 2\pi k(P) = \int_S k \frac{\partial}{\partial n} \left(\frac{1}{r} \right) dS \quad (9.8\text{-}36)$$

P is any point on S (we assume each point of S possesses a tangent plane), and r is the distance from dS to P. Equation (9.8-36) involves only S, and so is of lower dimensionality than the original problem; on the other hand, there may be accuracy problems in any numerical procedure as applied to eq. (9.8-36), because of the nature of the integrand.

9.9 POTENTIAL THEORY (See also Section 9.8)

We deal here with Laplace's equation $\Delta\varphi = 0$, in either two or three dimensions; properties in a higher number of dimensions are similar. The operator Δ means $(\partial^2/\partial x^2 + \partial^2/\partial y^2)$ or $(\partial^2/\partial x^2 + \partial^2/\partial y^2 + \partial^2/\partial z^2)$. A function φ satisfying $\Delta\varphi = 0$ is said to be *harmonic*. A problem in which $\Delta\varphi = 0$ inside a region V, with φ specified on the boundary S of V, is termed a *Dirichlet problem*. If $\partial\varphi/\partial n$ is specified on S, the problem is said to be a *Neumann problem*, and if a linear combination of φ and $\partial\varphi/\partial n$ is specified on S, it is called a *mixed problem*. A Neumann problem for a harmonic function is consistent, and has a solution, only if $\int_S (\partial\varphi/\partial n)\, dS = 0$.

In two dimensions, the value of a harmonic function at a point P is equal to the average of its values on either (a) the circumference of any circle centered on P, or (b) the area of any disk centered on P, provided of course that the circle or disk lies entirely within the region of harmonicity. This *mean value theorem* holds also in three dimensions, with an expression in terms of spheres rather than circles.

If φ is harmonic in a region, then the maximum value of φ, and also the minimum value of φ, occur on the boundary of the region. A consequence of this maximum property is that the Dirichlet problem is unique, and that the Neumann problem is unique within an additive constant.

A *fundamental solution* of $\Delta\varphi = 0$ is a solution which depends only on the distance r from some chosen source or origin point. In two dimensions, a fundamental solution is $\ln r$; in three dimensions, one is $(1/r)$.

The solution of the Dirichlet problem for a circle of radius R is given by *Poisson's integral*

$$\varphi(r,\theta) = \frac{1}{2\pi}\int_0^{2\pi} \frac{\varphi(R,\alpha)(R^2 - r^2)\, d\alpha}{R^2 + r^2 - 2Rr\cos(\alpha - \theta)} \tag{9.9-1}$$

where (r, θ) are polar coordinates. In three dimensions, the equivalent formula (using spherical coordinates r, θ, γ) is

$$\varphi(\rho,\alpha,\beta) = \frac{1}{4\pi}\int_0^{2\pi} d\gamma \int_0^{\pi} d\theta \cdot \frac{R(R^2 - \rho^2)\varphi(R,\theta,\gamma)\sin\theta}{[R^2 + \rho^2 - 2R\rho\{\sin\theta\sin\alpha\cos(\gamma - \beta) + \cos\theta\cos\alpha\}]^{3/2}} \tag{9.9-2}$$

Similar formulas for the Neumann problem, and also for the region outside a circle or sphere, are available (Ref. 9-18).

From eq. (9.9-1) or (9.9-2) we obtain *Harnack's inequality*, valid within any circle (sphere) within which φ is harmonic

$$\frac{R-r}{R+r}\varphi(0,\theta) \leqslant \varphi(r,\theta) \leqslant \frac{R+r}{R-r}\varphi(0,\theta) \tag{9.9-3}$$

$$\frac{R(R-\rho)}{(R+\rho)^2}\varphi(0,\alpha,\beta) \leqslant \varphi(\rho,\alpha,\beta) \leqslant \frac{R(R+\rho)}{(R-\rho)^2}\varphi(0,\alpha,\beta) \tag{9.9-4}$$

A consequence of these results is that a function which is harmonic and bounded in all space is necessarily a constant.

Another consequence of eq. (9.9-1) or (9.9-2) is that a harmonic function is analytic–i.e., expressible as a power series expansion with respect to any interior point in the region; the series converges in some neighborhood of that point.

If a function is continuous in a region, and satisfies the mean value theorem in that region, then the function is harmonic. (It is in fact enough if, for each point P in V, there is *some* sphere centered on P for which the mean value theorem is satisfied.)

If $\varphi(x, y)$ is harmonic in a portion of the upper half plane adjacent to a portion of the x-axis, and if $\varphi = 0$ on that x-axis portion, then the *Schwarz reflection principle* permits us to extend φ harmonically into the lower half plane, by defining $\varphi(x, -y) = -\varphi(x, y)$. A similar result holds for a three-dimensional region bounded by a portion of a plane, and also for regions bounded by arcs of circles or portions of spheres.

If a sequence of harmonic functions converges uniformly in a region, the limit function is harmonic.

In two dimensions, there is a close relationship between functions of a complex variable $z = x + iy$ and harmonic functions. The real or imaginary parts of any analytic $f(z)$ are harmonic, and, conversely, given any harmonic function $u(x, y)$, an analytic function $f(z)$ exists with the property that $u = \text{Re}\{f(z)\}$. If $f = u + iv$, then the harmonic function v is said to be conjugate to u. A Dirichlet problem for u becomes a Neumann problem for v. A harmonic function remains harmonic under conformal mapping; cf. section 9.7.1.

In three dimensions, the potential of a source of intensity α is defined to be α/r, where r is the distance from an observation point to that source. The potential of a density distribution of intensity $\rho(x_1, x_2, x_3)$ is given by

$$\varphi(x_1, x_2, x_3) = \int_V \frac{\rho(\xi_1, \xi_2, \xi_3)}{r}\, dV \tag{9.9-5}$$

where the integration is over ξ_i-space, and where $r^2 = (x_i - \xi_i)(x_i - \xi_i)$. The potential of a surface distribution of intensity σ is similarly given by

$$\varphi(x_1, x_2, x_3) = \int_S \frac{\sigma}{r} \, dS \tag{9.9-6}$$

By use of the results of section 9.8, it follows from eq. (9.9-5) that

$$\Delta\varphi = -4\pi\rho \tag{9.9-7}$$

A *dipole* is defined to be the result of permitting two sources, of opposite sign, to approach one another while their intensity grows. A dipole of strength α, oriented along a unit vector (n_i), has the potential $(\alpha \cos \omega)/r^2$ where r is the magnitude of the radius vector from the dipole to the observation point, and ω is the angle between this radius vector and the vector (n_i). If a portion of a surface carries a dipole distribution of (position-dependent) intensity k, normal to the surface, then the discontinuity in potential across this surface is $4\pi k$. Source and dipole distribution provide convenient concepts in the formulation of potential problems in terms of integral equations; see section 9.8.5 and Ref. 9-27.

A survey of information pertinent to potential theory will be found in Refs. 9-2, 9-3, and, more specifically in 9-18. We remark that the maximum properties of harmonic functions are special cases of more general properties of this kind applicable to general elliptic operators, and sometimes to other kinds of PDE's, see Ref. 9-19.

9.10 EIGENVALUE PROBLEMS

A typical eigenvalue problem is to find a nontrivial function $u(x, y)$ satisfying

$$u_{xx} + u_{yy} + \lambda p u = 0 \tag{9.10-1}$$

in a region R, subject to $u = 0$ on the boundary Γ of R. Here p is a given function, and λ is to be some constant. By the phrase nontrivial we mean that u is not to be identically zero in R. In this case, there are only certain values of the parameter λ for which the problem has a solution; such values of λ are termed *eigenvalues*, and the corresponding solutions u are termed *eigenfunctions*.

Let u_m correspond to λ_m, and u_n to λ_n. An often-occurring property is that of orthogonality. Here we write

$$\Delta u_m + \lambda_m p u_m = 0$$

$$\Delta u_n + \lambda_n p u_n = 0$$

multiply the first of these equations by u_n, the second by u_m, and integrate over R, using the divergence theorem to obtain

$$(\lambda_m - \lambda_n) \int_R p u_m u_n \, dA = 0 \tag{9.10-2}$$

Thus if $\lambda_m \neq \lambda_n$, we deduce that u_m and u_n are *orthogonal with respect to the weight function* $p(x, y)$. Frequently, p will be one-signed throughout R; we now restrict our attention to this case, and for definiteness take $p > 0$ in R.

If λ_m and u_m are complex-valued, then we can set $\lambda_n = \lambda_m^*$ and $u_n = u_m^*$ to deduce that $\int_R p u_m u_m^* \, dA = 0$, which is a contradiction; thus all eigenvalues of the problem (9.10-1), with $p > 0$, are real, and all eigenfunctions (which involve an arbitrary multiplicative constant anyway) can be taken to be real. Moreover, by appropriate choice of these arbitrary constants, the u_n can be *normalized* in the sense that

$$\int p u_n^2 \, dA = 1 \tag{9.10-3}$$

If we multiply the equation

$$\Delta u_n + \lambda_n p u_n = 0$$

through by u_n, and integrate over R, we are led to

$$\lambda_n = \frac{\int_R [(\partial u_n/\partial x)^2 + (\partial u_n/\partial y)^2] \, dA}{\int_R p u_n^2 \, dA} \tag{9.10-4}$$

which shows that $\lambda_n > 0$, all n. The form of eq. (9.10-4) suggests a variational procedure; it can be shown that the eigenfunctions are those functions satisfying $u = 0$ on Γ and which make the *Rayleigh quotient* $\left[\int_R (u_x^2 + u_y^2) \, dA\right] \Big/ \left[\int_R p u^2 \, dA\right]$ stationary; cf. section 9.12.

An eigenvalue problem can usually be rephrased as an integral equation problem. In the present case, let g satisfy $\Delta g = \delta(x - \xi)\delta(y - \eta)$ (cf. section 9.8), with $g = 0$ on Γ; a multiply and subtract process leads to

$$u(\xi, \eta) + \lambda \int_A p(x, y) g(x, y; \xi, \eta) u(x, y) \, dx \, dy = 0$$

which is an integral equation eigenvalue problem. An extensive collection of results concerning integral equation eigenvalue problems is available; see Ref. 9-2, vol. 1, and 9-17.

Perturbation methods are often applied to eigenvalue problems. Thus if $\varphi = 0$ on Γ, with

$$\Delta\varphi + [\lambda - \epsilon g(x, y)]\, p\varphi = 0 \tag{9.10-5}$$

in R, where $p(x, y)$ and $g(x, y)$ are given functions, and where $|\epsilon| \ll 1$, then it is natural to expand about the solution to the problem (9.10-1). We write for the nth eigenfunction and eigenvalue, say

$$\begin{aligned} \lambda &= \lambda_0 + \epsilon\lambda_1 + \epsilon^2\lambda_2 + \cdots \\ \varphi &= \varphi_0 + \epsilon\varphi_1 + \epsilon^2\varphi_2 + \cdots \end{aligned} \tag{9.10-6}$$

where, for $\epsilon = 0$, λ_0 corresponds to our previous λ_n and φ_0 to our previous u_n. Substitution into eq. (9.10-5) leads to a set of equations, which may be solved in the usual way. Frequently, undetermined constants will occur in this perturbation process, as a result of the fact that φ is arbitrary within a multiplicative constant. Scaling of φ will remove this indeterminacy.

It is also possible to perturb the region, in a way similar to that of section 9.6. Again expansions of the form (9.10-6) are used, and all quantities are evaluated on the undisturbed boundary. An interesting example, involving numerical calculation considerations, is given in Ref. 9-20.

Although this simple example has illustrated many of the common properties and ideas in connection with eigenvalue problems arising in PDE's, there are many cases of practical interest in which the results are not so straightforward. For example, the Schrödinger equation

$$\Delta u + \frac{k}{r}u + \lambda u = 0$$

(with $k > 0$ a constant, and u finite at $r = 0$ and at ∞, and where Δ is the spherical coordinate Laplacian operator) has a rather different eigenvalue structure, in that every positive value of λ is an eigenvalue. The negative eigenvalues, however, form a discrete set.

For some problems, general theorems relating to upper or lower bounds on eigenvalues are available. A particularly useful general technique is related to Courant's maximum-minimum principle; see Ref. 9-2, vol. 1. For other methods leading to bounds, see Refs. 9-19, 9-21, and 9-27.

9.11 CHARACTERISTICS

9.11.1 Definition

There are various ways in which a *characteristic curve* or *surface* may be defined, all essentially equivalent. We will illustrate the situation for the second-order equation already considered in section 9.2.1.

$$A\varphi_{xx} + 2B\varphi_{xy} + C\varphi_{yy} + D\varphi_x + E\varphi_y + F\varphi = G \tag{9.11-1}$$

where $A, B, \ldots, G$ are functions of (x, y) only. We noted in section 9.2.1 that, in the hyperbolic case, there were two families of curves, $\xi(x, y) = \text{constant}$ and $\eta(x, y) = \text{constant}$, such that eq. (9.11-1) took a particularly simple form when rewritten in terms of (ξ, η) as independent variables. This is one way of introducing characteristics–as contour curves of a coordinate system in which the PDE attains a canonical form.

Alternatively, we could ask for that curve Γ defined by $\xi(x, y) = 0$ along which the prescription of Cauchy data (i.e., φ and its normal derivative $\partial\varphi/\partial n$) is inadequate, in the sense that a Taylor series cannot be constructed so as to determine φ–at least for the analytic data case–in a neighborhood of the curve. If we embed this special curve in a family of curves $\xi(x, y) = \text{constant}$, and construct a nontangential second family of curves $\eta(x, y) = \text{constant}$, so as to have available a ξ, η-coordinate system, then a prescription of Cauchy data along Γ is equivalent to a prescriptionof φ_ξ and φ_η on Γ, i.e., as functions of η. Differentiation yields $\varphi_{\eta\eta}$, $\varphi_{\eta\eta\eta}, \ldots, \varphi_{\xi\eta}, \varphi_{\xi\eta\eta}, \ldots$ along the curve, but to obtain $\varphi_{\xi\xi}$ we have to use the PDE itself, which reads

$$\varphi_{\xi\xi}(A\xi_x^2 + 2B\xi_x\xi_y + C\xi_y^2) + \text{other terms} = 0 \tag{9.11-2}$$

where all "other terms" are known on Γ. Only if the coefficient of $\varphi_{\xi\xi}$ is nonzero is it possible to solve for $\varphi_{\xi\xi}$, and if this is indeed the case, all higher derivatives of φ on the curve Γ can be obtained–e.g., by differentiating eq. (9.11-2) with respect to ξ to obtain $\varphi_{\xi\xi\xi}$ as a function of η, and so on. But this process is not feasible if the coefficient vanishes. Thus we see that if we define a characteristic curve to be one along which Cauchy data is inadequate, we are again led to the families of curves encountered in section 9.2-1.

Another definition is very closely related to the preceding one. If the curve Γ defined by $\xi(x, y)$ is such that the coefficient of $\varphi_{\xi\xi}$ in eq. (9.11-2) vanishes, then the remaining terms imply a constraint on the given Cauchy data. Thus we could define a characteristic to be a curve along which Cauchy data may not be arbitrarily prescribed, or, equivalently, a curve along which the given PDE reduces to a statement involving rates of change along that curve only.

Perhaps the most fundamental definition of a characteristics, however, is in terms of curves across which a solution may possess a derivative discontinuity. Suppose that we had asked for a curve Γ, defined by $\xi(x, y) = 0$, such that on each side of Γ the function φ and its derivatives are continuous, but such that across Γ there is a discontinuity in the second transverse derivative. Specifically, if we denote the function φ on two sides of Γ by φ^+ and φ^-, and if again we embed the curve Γ in a ξ, η-coordinate system, we require

$$\varphi^+ = \varphi^- \text{ across } \Gamma$$

$$(\text{and so } \varphi_\eta^+ = \varphi_\eta^-, \quad \varphi_{\eta\eta}^+ = \varphi_{\eta\eta}^-, \ldots, \text{ on } \Gamma)$$

$$\varphi_\xi^+ = \varphi_\xi^- \text{ across } \Gamma$$

$$\varphi_{\xi\xi}^+ \neq \varphi_{\xi\xi}^- \text{ across } \Gamma$$

Writing eq. (9.11-2) for each of φ^+ and φ^-, as evaluated at two neighboring points on opposite sides of Γ, subtracting the two equations, and permitting the points to approach one another, we obtain

$$(\varphi_{\xi\xi}^+ - \varphi_{\xi\xi}^-)(A\xi_x^2 + 2B\xi_x\xi_y + C\xi_y^2) = 0 \tag{9.11-3}$$

since all other terms are continuous by hypothesis. Thus if $\varphi_{\xi\xi}^+ \neq \varphi_{\xi\xi}^-$, we are again led to the same condition on Γ as before. The same result would have been obtained had we asked for that curve Γ across which $\varphi, \varphi_\xi, \varphi_{\xi\xi}$ were continuous, but $\varphi_{\xi\xi\xi}$ was discontinuous–for merely differentiate eq. (9.11-2) with respect to ξ, and use the same method. In general, then, if *any* transverse derivative of order higher than the first is to be discontinuous across Γ, Γ must be a characteristic.

Physically, this last definition is related to the idea of signal propagation. Let one of the x, y-variables be space-like, and the other time-like. Then if a function has a different functional form on two sides of a curve Γ, we can think of Γ as being a trajectory in space-time which separates these two different forms for φ–i.e., a path traced out by a signal front. For the present example, the existence of a characteristic is thus equivalent to the requirement that signal propagation be possible. (We exclude however shock waves, in which φ or φ_ξ could be discontinuous–see Chapter 8.)

These various definitions carry over to more complicated situations. For example, we could let the coefficients $A, B, \ldots,$ of eq. (9.11-1) depend on φ; the characteristic condition would then involve the given data for φ, and so would pertain to a curve in x, y, φ-space rather than to a curve in x, y-space. In such a case, one sometimes distinguishes in nomenclature between the characteristic space curve and its projection on the x, y-plane, the latter being termed the *characteristic trace*. When more dependent or independent variables occur, there may be more than two families of characteristics, and the characteristics will be surfaces (manifolds) rather than curves.

As a simple example, consider the wave equation in two space dimensions (x, y), and time t

$$\varphi_{tt} - c^2(\varphi_{xx} + \varphi_{yy}) = 0 \tag{9.11-4}$$

where we take $c = c(x, y)$. We ask for a surface M $\xi(x, y, t) = 0$, such that in terms of an embedding (ξ, η, τ) coordinate system, $\varphi_{\xi\xi}$ may be discontinuous across M. We find easily that the condition defining M is

$$\xi_t^2 - c^2(\xi_x^2 + \xi_y^2) = 0 \tag{9.11-5}$$

which is the eikonal equation of Chapter 8. Solutions of eq. (9.11-5) may be constructed with the help of the characteristic curves for that equation, or its rays, as discussed in Chapter 8; the characteristics of eq. (9.11-5) are termed the *bicharacteristics* of eq. (9.11-4).

A further discussion of characteristics, together with a number of examples, will be found in Refs. 9-2, 9-3, 9-22, and 9-27. We turn now to an illustrative example involving a coupled pair of nonlinear equations.

9.11.2 Method of Characteristics

A pair of equations arising in shallow water theory (Ref. 9-23) is as follows:

$$\begin{aligned} u_t + uu_x + 2cc_x &= 0 \\ 2c_t + 2uc_x + cu_x &= 0 \end{aligned} \tag{9.11-6}$$

where $u(x, t)$ is the fluid velocity, and $c(x, t)$ is a function related to the wave height.

We try to find a curve Γ, defined by $\xi(x, t) = 0$, in the x, t-plane such that any of the previous characteristic conditions is fulfilled. We choose in particular the discontinuity condition of section 9.11.1, and, in terms of an embedding ξ, η-coordinate system as described in that section, we obtain

$$\begin{aligned} u_\xi(\xi_t + u\xi_x) + c_\xi(2c\xi_x) + \text{other terms} &= 0 \\ u_\xi(c\xi_x) + 2c_\xi(\xi_t + u\xi_x) + \text{other terms} &= 0 \end{aligned} \tag{9.11-7}$$

where the "other terms" involve u_η or ξ_η. If either of u_ξ or c_ξ can be discontinuous across Γ, with everything else continuous, then u_ξ and c_ξ cannot be uniquely determined by eq. (9.11-7), so that the characteristic condition becomes that the coefficient determinant must vanish

$$\begin{vmatrix} \xi_t + u\xi_x & 2c\xi_x \\ c\xi_x & 2(\xi_t + u\xi_x) \end{vmatrix} = 0$$

or that

$$\xi_t + (u \pm c)\,\xi_x = 0 \tag{9.11-8}$$

Thus there are two families of characteristic curves

$$\begin{aligned} C_+ \text{ family:} \quad \frac{dx}{dt} &= u + c \text{ for } \Gamma \\ C_- \text{ family:} \quad \frac{dx}{dt} &= u - c \text{ for } \Gamma \end{aligned} \tag{9.11-9}$$

Since the fluid velocity is u, we observe that c represents the velocity of signal propagation relative to the fluid.

Along a C_+ curve, we have

$$\frac{du}{dt} = u_t + (u + c)u_x$$

$$\frac{dc}{dt} = c_t + (u + c)\, c_x$$

so that, by use of eq. (9.11-6)

$$u + 2c = \text{constant along a } C_+ \text{ curve} \tag{9.11-10}$$

Similarly

$$u - 2c = \text{constant along a } C_- \text{ curve} \tag{9.11-11}$$

These equations permit a convenient numerical solution scheme to be implemented. Suppose that u and c are prescribed at $t = 0$—i.e., along a portion of the x-axis in the x, t-plane as shown in Fig. (9.11-1). In particular, suppose we know u

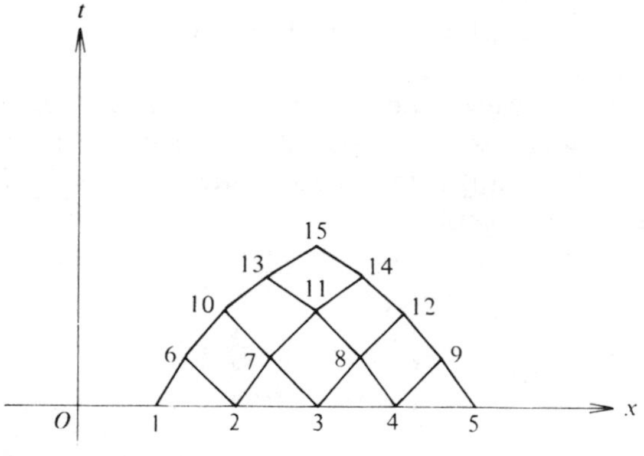

Fig. 9.11-1 Method of characteristics.

and c at the points marked 1, 2, 3, 4, 5 in the figure. Since the slope $u \pm c$ of a C_+ or C_- characteristic is thereby known at each of these points, we can draw short straight-line sections through these points in the characteristic directions, so as to intersect at points 6, 7, 8, 9 as shown. Along the line 6-1, $u + 2c$ = constant, so that $u + 2c$ is known at point 6. Similarly, using the constancy of $u - 2c$ along line 6-2, $u - 2c$ is known at point 6. Thus each of u and c can be calculated at point 6, and similarly for points 7, 8, 9. Knowing u and c at these points, we can draw new local characteristic lines, intersecting at points 10, 11, 12, and proceed in the same way up to point 15.

Refining this process by means of a closer spacing of mesh points, it follows that if u and c are prescribed along the portion 1-5 of the x-axis, we will be able to determine u and c throughout the curvilinear triangle formed by this portion of the x-axis and the two characteristics through the ends 1 and 5. It is clear that the information prescribed at any one point on the x-axis, say the point 2, can have an effect on the solution only in that region of the x, t-plane lying between the two characteristics emanating from that point. This region would be called the domain of influence of point 2. Conversely, the solution at any point in the x, t-plane, say the point 15, will depend only on the initial data as prescribed along that part of the x-axis lying between the intersection points with the two characteristics through the point; this part of the x-axis would be called the *domain of dependence* of the point 15.

A minor difficulty will occur if we try to extend the solution region so as to include a physical boundary. For example, in Fig. (9.11-2), let a boundary start at

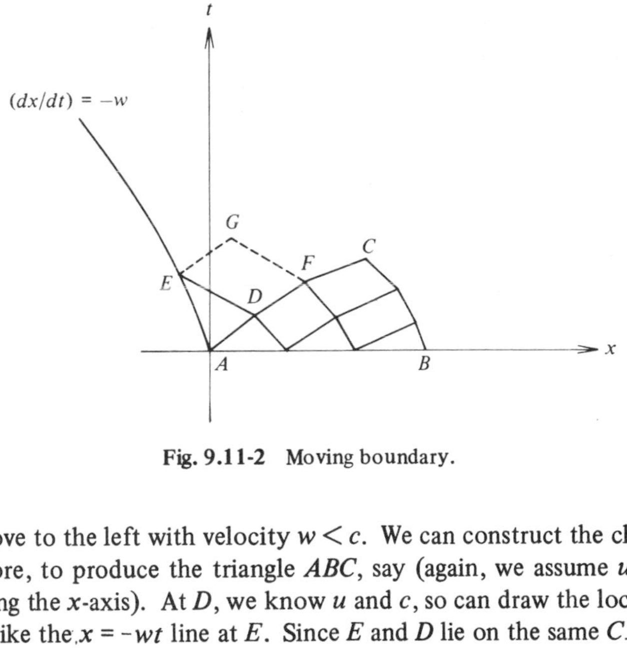

Fig. 9.11-2 Moving boundary.

$x = 0$ and move to the left with velocity $w < c$. We can construct the characteristics mesh as before, to produce the triangle ABC, say (again, we assume u and c to be specified along the x-axis). At D, we know u and c, so can draw the local C_- characteristic to strike the $x = -wt$ line at E. Since E and D lie on the same C_- characteristic, $u_E - 2c_E$ is known. But u_E is known at E, since the fluid in contact with the wall must move with the velocity $-w$. Thus $u_E = -w$, so that our knowledge of $u_E - 2c_E$ permits us to determine c_E also. We can then draw the C_+ characteristic from E (shown dotted), to intersect the C_- characteristic from F at G; it is now clear how we proceed to fill up the rest of the x, t-plane (to the right of the boundary curve) with mesh points.

This kind of graphical construction constitutes the *method of characteristics*. Of course, it may not be possible to obtain the conditions holding along the characteristic in integrated form, as in eqs. (9.11-10) and (9.11-11), but these equations represent only a convenience and not a necessity. The crucial items are the step-by-step procedure, and the consideration of the regions of influence and dependence. The method of characteristics is rarely used for practical solution purposes if there are more than two independent variables, see, however, Ref. 9-2, vol. 2, Chapter 6. We will now pursue the present example a little further.

9.11.3 Simple Wave

In some problems of physical interest, straight line characteristics play an important role. This is in fact the case for the present problem. Let one member, say Γ, of a C_+-family be straight; then the slope $u + c$ is constant along Γ. But $u + 2c$ is also constant along Γ, so that each of u and c must be individually constant. If we now consider an adjoining C_+ characteristic, say Γ', along which $u + 2c$ is constant, and realize that $u - 2c$ on this characteristic must be the same as on Γ since a point on Γ' may be linked to a point on Γ via a C_- characteristic, it follows that each of u and c is again constant along Γ', so that Γ' is straight. Thus a *straight-line characteristic is necessarily embedded in a family of straight-line characteristics*. (Of course, this theorem holds for the present problem, but not necessarily for other problems.) Such a situation is termed a *simple wave*.

An example in which a simple wave occurs is that in which we consider the region $x > 0$, with $u(x, 0) = 0$, $c(x, 0) = c_0$ (constant) as initial condition. Let $u(0, t)$ be prescribed, say $u(0, t) = f(t)$. The characteristics emanating from points on the x-axis are straight lines, with slopes $\pm c_0$, as shown in Fig. 9.11-3. The boundary characteristic to this region is Γ_0; since it is straight, adjoining characteristics ($\Gamma_1, \Gamma_2, \ldots$) must also be straight, although we do not as yet know their

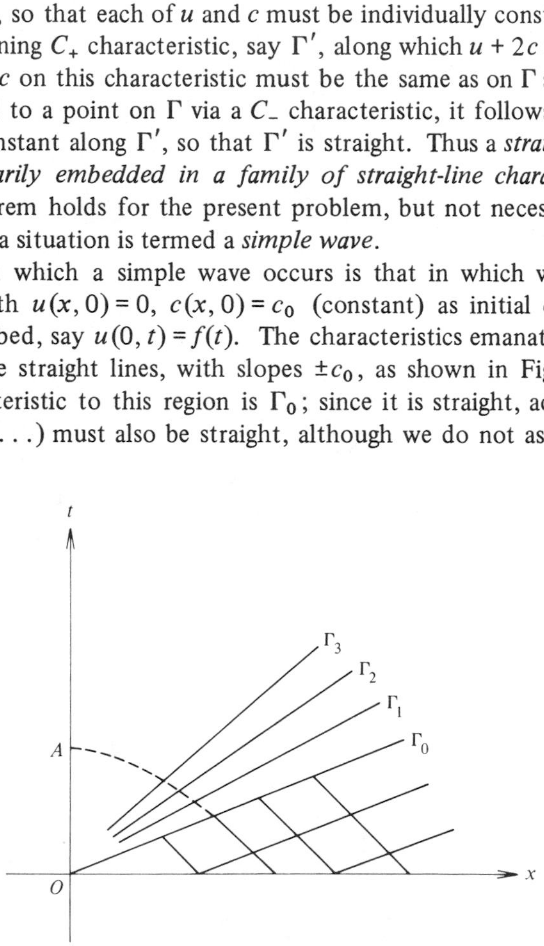

Fig. 9.11-3 Simple wave.

directions. A typical C_- characteristic will continue beyond the constant-state triangle as some sort of curved line; one such is shown dotted in the figure. At the point A, u is known via $u_A = f(t_A)$; moreover, $u_A - 2c_A$ must equal $0 - 2c_0$ because of the linkage to the constant-state region via the dotted curve. Thus $c_A = c_0 + \frac{1}{2} f(t_A)$ is also determinable. At this point the C_+ characteristic emanating from A has slope $u_A + c_A = c_0 + \frac{3}{2} f(t_A)$ and so can be drawn in its entirety. We thus can construct the simple wave filling up the rest of the region in the x, t-plane, and thereby avoid the mesh approach of section 9.11.2.

A similar situation holds if the boundary at $x = 0$ is made to move, as in Fig. 9.11-2; if the initial state is one of constancy, then a simple wave is appropriate. We remark however that the solution technique may lead to situations in which characteristics of the same family intersect one another; since the information at such an intersection point would be self-contradictory, the solution can be expected to break down near such a point, perhaps via the formation of a shock as described in Chapter 8. See also Refs. 9-22, 9-23, and 9-27.

9.12 VARIATIONAL METHODS

Any PDE problem can be restated as a variational problem. One method for doing this is as follows. Let a function satisfy a PDE (linear or nonlinear) of the form $L(\varphi) = 0$, in a region $\boldsymbol{R}$, and let the boundary or initial conditions be specified for a portion Γ of the boundary of R in the form $M(\varphi) = 0$. Then the given problem is clearly equivalent to that of minimizing the expression

$$\int_R [L(\varphi)]^2 \, dR + \int_\Gamma [M(\varphi)]^2 \, dS \tag{9.12-1}$$

More generally, we could weight this combination of integrals differently, we could use absolute values rather than squares in the integrands, and we could raise these absolute values to any power. Any system of PDE's could be treated similarly.

For many of the equations of classical interest, other variational principles are available; frequently these principles involve lower order derivatives than those occurring in (9.12-1), and in turn this results in simpler Euler equations.

A classical example is that of the Dirichlet integral, which for the two-dimensional case is

$$I = \int_A (\varphi_x^2 + \varphi_y^2) \, dA \tag{9.12-2}$$

The problem of finding that function φ, satisfying a given boundary condition $\varphi = f$ on the boundary of A, which minimizes I, is equivalent to the problem of solving $\Delta\varphi = 0$ in A, $\varphi = f$ on Γ.

Many problems may be formulated in variational terms by means of basic principles of physics—e.g., Hamilton's principle of Chapter 8 and 20. In fact, it may be that this kind of approach is particularly useful in deriving the governing PDE and boundary conditions—as in shell theory in elasticity.

However, our interest here is in using the variational approach as the vehicle for an approximation method. A well-known idea is that of the *Rayleigh-Ritz* method, which may be exemplified for eq. (9.12-2). We use our physical intuition or experience to approximate the solution φ of the problem $\delta I = 0$ by

$$\psi = \psi_0 + c_1 \psi_1 + c_2 \psi_2 \tag{9.12-3}$$

where ψ_0, ψ_1, ψ_2 are chosen functions satisfying $\psi_0 = f$ on Γ, $\psi_1 = \psi_2 = 0$ on Γ, and where c_1, c_2 are as-yet-undetermined constants. For any choice of c_1, c_2, ψ as defined by eq. (9.12-3) satisfies the boundary condition on φ; the purpose of inserting c_1 and c_2 into the expression for ψ is to provide for some flexibility in our approximation. We now replace φ in eq. (9.12-2) by ψ, and adjust c_1, c_2 so as to make I stationary. In this way, we make the approximate solution share a property that the exact solution φ possesses, within the constraints imposed by the form (9.12-3). An advantage of the use of the variational approach via eq. (9.12-2) is that only first derivatives of the ψ-functions enter into the calculation; it is not necessary to worry too much about the character of the second derivatives of ψ. On the other hand, even though we may be able to approximate φ rather well by ψ, the derivatives of φ may not be well approximated by those of ψ.

More generally, of course, we can adjoin more functions $\psi_3, \psi_4, \ldots, \psi_n$ to eq. (9.12-3), each multiplied by its appropriate constant c_i. The problem of making $I(\psi)$ stationary leads then to a set of n linear equations for the c_i. We could also incorporate the c_i in a nonlinear manner into the function ψ, in which case the stationary condition would lead to a set of nonlinear algebraic equations for the c_i.

A modification of this technique would be to replace the variational principle by a new one in which the boundary condition is obtained as a consequence of the principle—i.e., as a natural boundary condition (Chapter 20). For example, let the above boundary condition $\varphi = f$ on Γ be replaced by the condition $\partial\varphi/\partial n + \varphi = g$ on Γ, where g is prescribed. Then an appropriate reformulation of the variational principle is to make stationary the combination

$$J = \int_A \frac{1}{2} (\varphi_x^2 + \varphi_y^2)\, dA + \int_\Gamma \frac{1}{2} (\varphi - g)^2\, ds \tag{9.12-4}$$

where no restrictions are imposed on φ. In using the Rayleigh-Ritz method, with an approximating function

$$\psi = c_0 \psi_0 + c_1 \psi_1 + \cdots + c_n \psi_n$$

the chosen ψ_j-functions need now not satisfy any assigned boundary conditions. Another way of handling prescribed auxiliary (e.g., boundary) conditions is by means of Lagrange multipliers; see Chapter 20.

Very similar methods may be used for eigenvalue problems. A useful tool is the Rayleigh quotient idea, which may be exemplified for the problem

$$\begin{gathered}\Delta\varphi + \lambda\varphi = 0 \text{ in } R \\ \varphi = 0 \text{ on } \Gamma, \text{ the boundary of } R\end{gathered} \tag{9.12-5}$$

As in section 9.10, we find that each eigenvalue λ, and its corresponding eigenfunction φ, satisfy the equation

$$\lambda = \frac{\int_R (\varphi_x^2 + \varphi_y^2)\, dA}{\int_R \varphi^2\, dA} \tag{9.12-6}$$

If we now ask for that function φ which makes the "Rayleigh quotient" occurring on the right-hand side of eq. (9.12-6) stationary, subject to $\varphi = 0$ on Γ, we find that the solutions of this problem are the eigenfunctions of problem (9.12-5), and that the corresponding values of the quotient are the eigenvalues.

A method which is closely related to the variational idea is that of *Galerkin*. To illustrate it, we consider again the problem $\Delta\varphi = 0$ in R, $\varphi = f$ on Γ, where f is some function prescribed on the boundary Γ of the two-dimensional region R. We choose approximation functions ψ_0, ψ_1, ψ_2, in the form of eq. (9.12-3), satisfying the boundary conditions there specified. We now form the expression $\Delta\psi$, and adjust c_1 and c_2 so as to make $\Delta\psi$ orthogonal to each of two chosen functions, Ω, and Ω_2, in the sense

$$\int (\Delta\psi)\Omega_j dA = 0$$

for $j = 1, 2$. If the functions Ω_1 and Ω_2 are chosen to be the same as ψ_1 and ψ_2, we recover the Rayleigh-Ritz method (in this example); if Ω_1 and Ω_2 are each products of delta functions (section 9.8.1), the method is termed the *collocation method*. Another common idea is to make each Ω_j zero, except within a certain subregion of R, where it equals unity.

Many variants of these ideas are possible. For example, the 'constants' c_i could be made functions of one or more of the variables, in which case the variational or Galerkin idea would lead to ODE's or PDE's in the c_j's. At this point, however, we refer to Chapter 20 and to Refs. 9-24, 9-25, and 9-26, where many examples are displayed.

9.13 NUMERICAL TECHNIQUES

Several of the approximation techniques discussed in sections 9.4 (series solutions), 9.5 (transforms), 9.6 (perturbations), 9.8.5 (integral equations), 9.11 (characteristics), and 9.12 (variational methods) are amenable to numerical implementation on a computer–in whole or in part. In this section, however, we will emphasize those methods which reply on the replacement of a PDE problem by one involving finite differences. One of the simplest situations is that in which "time-stepping" is used to solve a diffusion problem in one space dimension; in section 9.13.1, this problem is chosen as a vehicle for the exploration of accuracy and stability. Section 9.13.2 considers extensions to a higher number of dimensions, as well as the effects of a more complicated form of equation. Boundary value problems are discussed in section 9.13.3 (relaxations) and in section 9.13.4 (finite elements). Wave-type problems, and hyperbolic systems in general, are considered in section 9.13.5.

9.13.1 Diffusion Equation

Suppose that we seek a function $\phi(x, t)$, depending on distance x and time t, such that

$$\phi_t = a^2 \phi_{xx} \tag{9.13-1}$$

for x lying within a space interval $(0, L)$ and for all values of $t > 0$; the diffusivity a^2 is taken as constant. As is appropriate for this kind of problem, we adjoin the initial condition $\phi(x, 0) = f(x)$ and certain end conditions which, for simplicity, we choose as $\phi(0, t) = 0$, $\phi(L, t) = 0$. The function $f(x)$ is prescribed for x in $(0, L)$; to avoid a minor end point difficulty, we will here require $f(0) = f(L) = 0$.

Instead of trying to determine ϕ for each possible x-value and for each possible t-value, we will content ourselves with a knowledge of an approximation to ϕ at certain discrete x-values and discrete t-values. Specifically, divide the interval $(0, L)$ into m equal subintervals, and define

$$x_j = jh, \quad j = 0, 1, \ldots m \tag{9.13-2}$$

with $h = L/m$. Also, define

$$t_n = nk, \quad n = 0, 1, 2, \ldots \tag{9.13-3}$$

where k is some convenient time interval. Finally, denote by $\phi_j^{(n)}$ the approximate (as determined numerically) value of ϕ at x_j and at t_n, so that (hopefully)

$$\phi_j^{(n)} \simeq \phi(x_j, t_n) \tag{9.13-4}$$

where ϕ as before denotes the exact solution of eq. (9.13-1). We now recall that

$$\phi_t(x, t) = \lim_{\delta t \to 0} \frac{\phi(x, t + \delta t) - \phi(x, t)}{\delta t} \tag{9.13-5}$$

and that

$$\begin{aligned}\phi_{xx}(x,t) &= \lim_{\delta x\to 0} \frac{\phi_x(x+\frac{1}{2}\delta x, t) - \phi_x(x-\frac{1}{2}\delta x, t)}{\delta x} \\ &= \lim_{\delta x\to 0} \frac{1}{\delta x}\left\{\frac{\phi(x+\delta x,t)-\phi(x,t)}{\delta x} - \frac{\phi(x,t)-\phi(x-\delta x,t)}{\delta x}\right\} \\ &= \lim_{\delta x\to 0} \frac{\phi(x+\delta x,t) - 2\phi(x,t) + \phi(x-\delta x,t)}{(\delta x)^2}\end{aligned} \tag{9.13-6}$$

These equations suggest that, if h and k are adequately small

$$\phi_t(x_j, t_n) \cong \frac{\phi_j^{(n+1)} - \phi_j^{(n)}}{k} \tag{9.13-7}$$

and

$$\phi_{xx}(x_j, t_n) \cong \frac{\phi_{j+1}^{(n)} - 2\phi_j^{(n)} + \phi_{j-1}^{(n)}}{h^2} \tag{9.13-8}$$

We are thus led to the replacement of eq. (9.13-1) by

$$\phi_j^{(n+1)} - \phi_j^{(n)} = \left(a^2 \frac{k}{h^2}\right)(\phi_{j+1}^{(n)} - 2\phi_j^{(n)} + \phi_{j-1}^{(n)}) \tag{9.13-9}$$

To use eq. (9.13-9) in practice, we would use the given initial condition to require $\phi_j^{(0)} = f(x_j)$ for all j-values, and the end conditions in the form $\phi_0^{(n)} = 0$, $\phi_m^{(n)} = 0$. Equation (9.13-9), with $n = 0$, permits us to determine $\phi_j^{(1)}$ for all mesh points j satisfying $1 \leqslant j \leqslant m - 1$; the choices $n = 1, 2, \ldots$ in sequence then yield $\phi_j^{(2)}$, $\phi_j^{(3)}, \ldots$. This repetitive process is easily programmed for a computer.

In investigating the error associated with this kind of *finite-difference* process, it is useful to examine a problem for which the exact solution is available. Let $f(x) = \sin(N\pi x/L)$, where N is some chosen integer, so that

$$\phi(x,t) = \exp\left(-\frac{N^2\pi^2a^2}{L^2}t\right)\sin\frac{N\pi x}{L} \tag{9.13-10}$$

The corresponding solution of eq. (9.13-9) is

$$\phi_j^{(n)} = \left\{1 - 4\frac{a^2k}{h^2}\sin^2\frac{N\pi h}{2L}\right\}^n \sin\frac{N\pi hj}{L} \tag{9.13-11}$$

These two solutions are easily verified by direct substitution. We now observe that if $N\pi h/L$ is small, the first factor in eq. (9.13-11) can be approximated by

$$\left\{1 - 4\frac{a^2 k}{h^2}\left(\frac{N\pi h}{2L}\right)^2\right\}^n$$

and, for small k, the first portion of the binomial expansion of this result is indeed a good approximation to the first factor of eq. (9.13-10). (The second factors of eqs. (9.13-10) and (9.13-11) agree exactly at the mesh point values $x = x_j$). Thus for appropriate choices of h and k, the approximate solution (9.13-11) can be very satisfactory.

However, suppose that the choices for h and k are such that $a^2 k/h^2 > \frac{1}{2}$. Then it may be possible to find a value for N such that

$$1 - 4\frac{a^2 k}{h^2} \sin\left(\frac{N\pi h}{2L}\right) < -1 \tag{9.13-12}$$

so that the first factor in eq. (9.13-11) would grow in magnitude, in an unbounded manner, as n increases. This behavior is in remarkable contrast to that of the first factor in eq. (9.13-10), which decays in time; the numerical solution would here be entirely unacceptable. Moreover, any prescription of $f(x)$ may contain Fourier components of the form sin $(N\pi x/L)$ for a variety of values of N (and such Fourier components can be inserted by inadvertent roundoff in the computer, in any event), so that if the *stability condition*

$$\frac{a^2 k}{h^2} < \frac{1}{2} \tag{9.13-13}$$

is violated, we can expect to encounter Fourier components for which eq. (9.13-12) is satisfied and whose amplitude in consequence will increase indefinitely. This stability analysis method, in which a harmonic decomposition of an error is considered, is termed the *von Neumann method.* An alternative technique examines the nature of the matrix operator which, when applied to the mesh point values at time level n, yields those at level $n + 1$; see Refs. 9-28 and 9-38.

This example—which in effect deals with a particular Fourier component of $f(x)$—makes it plausible that if eq. (9.13-13) is satisfied, then as $h \to 0$, the approximate solution $\phi_j^{(n)}$ will approach the exact solution $\phi(x_j, t_n)$. For a formal statement and proof of the relevant theorem, see Ref. 9-28. We note, however, that eq. (9.13-13) is very restrictive, in that a halving of the mesh spacing h, undertaken to improve accuracy, would require a quadrupling of the number of time steps required to cover a given time interval.

If next the diffusivity a^2 is a function of x, rather than a constant, we would simply use $a^2(x_j)$ instead of a^2 in eqs. (9.13-9) and (9.13-13); this latter equation would be required to hold for all j. Also, if the end condition at $x = 0$ required $\phi(0, t) = p(t)$, with p a prescribed function of t, we would simply use $\phi_0^{(n)} = p(t_n)$

at each step. If more generally $A\phi(0, t) + B\phi_x(0, t) = p(t)$, where A and B are constants, then the end condition

$$A\phi_0^{(n)} + B\frac{\phi_1^{(n)} - \phi_0^{(n)}}{h} = p(t_n)$$

could be adjoined to eq. (9.13-9). A similar extension can be made at $x = L$.

The use of Taylor's formula shows (Chapter 18 or Ref. 9-28) that eq. (9.13-7) is accurate to the first order, and eq. (9.13-8) to the second order, in powers of k and h, respectively. Since the right-hand side of eq. (9.13-7) would give a second-order-accurate expression for $\phi_t(x_j, \frac{1}{2}[t_n + t_{n+1}])$, a more accurate version of (9.13-9) would be obtained by averaging values of the right-hand side, evaluated at time t_n and t_{n+1}

$$\phi_j^{(n+1)} - \phi_j^{(n)} = \left(a^2 \frac{k}{h^2}\right)\left\{\frac{\phi_{j+1}^{(n)} - 2\phi_j^{(n)} + \phi_{j-1}^{(n)}}{2} + \frac{\phi_{j+1}^{(n+1)} - 2\phi_j^{(n+1)} + \phi_{j-1}^{(n+1)}}{2}\right\} \tag{9.13-14}$$

This *Crank-Nicholson method* requires that a set of algebraic equations be solved at each time step (note, however, that the coefficient matrix is tridiagonal, so that Gaussian elimination is effective); thus the method is *implicit* rather than *explicit*, as in eq. (9.1-9). An analysis similar to that which led to eq. (9.13-13) shows that eq. (9.13-14) is stable for all choices of k. Although the stability condition (9.13-13) can now be discarded, it is, of course, still necessary to choose k small relative to the time scale over which the boundary conditions are changing.

Many other finite difference approximations to eq. (9.13-1) are available. Two others are:

a) centered time difference equation

$$\frac{\phi_j^{(n+1)} - \phi_j^{(n-1)}}{2k} = a^2 \frac{\phi_{j+1}^{(n)} - 2\phi_j^{(n)} + \phi_{j-1}^{(n)}}{h^2}$$

This equation is *consistent* (as our previous approximations are) in the sense that as h and $k \to 0$, eq. (9.13-14) $\to$ eq. (9.13-1). It is applicable only for $n = 2, 3, \ldots$, so that some other method would be necessary in order to find $\phi_j^{(1)}$. Despite its second-order accuracy, eq. (9.13-14) is, unfortunately, unstable for all choices of h and k.

b) DuFort-Frankel equation

$$\frac{\phi_j^{(n+1)} - \phi_j^{(n-1)}}{2k} = a^2 \frac{\phi_{j+1}^{(n)} - \phi_j^{(n+1)} - \phi_j^{(n-1)} + \phi_{j-1}^{(n)}}{h^2} \tag{9.13-15}$$

This equation is always stable, but it is consistent with eq. (9.13-1) only if the ratio $(k/h) \to 0$ as h and $k \to 0$.

9.13.2 Extensions

For the equivalent of eq. (9.13-1) in two space dimensions x and y

$$\phi_t = a^2(\phi_{xx} + \phi_{yy}) \tag{9.13-16}$$

it is appropriate to replace the one-dimensional finite difference Laplacian

$$(\phi_{j+1}^{(n)} - 2\phi_j^{(n)} + \phi_{j-1}^{(n)})/h^2$$

that occurs in eqs. (9.13-9) and (9.13-14) by its two-dimensional equivalent

$$(\phi_{i,j+1}^{(n)} - 2\phi_{i,j}^{(n)} + \phi_{i,j-1}^{(n)})/h^2 + (\phi_{i+1,j}^{(n)} - 2\phi_{i,j}^{(n)} + \phi_{i-1,j}^{(n)})/h^2 \tag{9.13-17}$$

Here the (x, y) region of interest is covered by a square grid, and $\phi_{i,j}^{(n)}$ denotes the approximation to $\phi(x, y, t)$ at time nk and at the mesh point denoted by (i, j) (see Fig. 9.13-1). The stability criterion for the two-dimensional form of eq. (9.13-9) is that $(a^2 k/h^2) < \frac{1}{4}$; again, this is undesirably restrictive.

The two-dimensional form of eq. (9.13-14) uses the values of the Laplacian of ϕ at times (n) and $(n + 1)$, and so is again implicit. The coefficient matrix is no longer tridiagonal, so that the calculation of the $\phi^{(n+1)}$ values is now more troublesome. An alternative method, associated with the names of Peaceman, Rachford, and Douglas, is to use two different formulas for alternate time steps. In the first step, the first term of (9.13-17) is computed at time $(n + 1)$ and the second at time n; for the next time step, the roles are reversed. The resulting *alternating-difference-implicit method* (*ADI method*) is unconditionally stable, and although still implicit

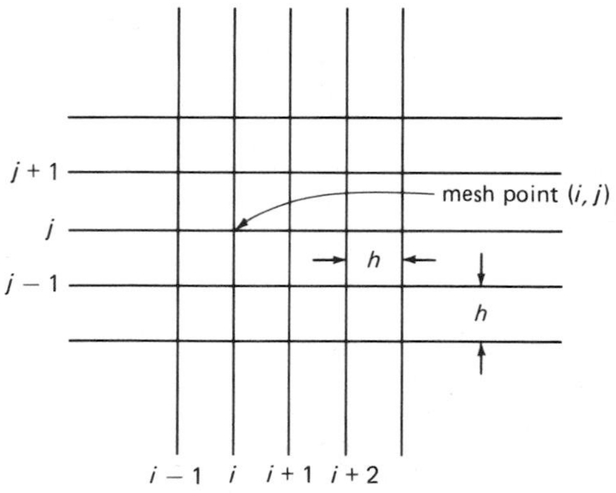

Fig. 9.13-1 Mesh geometry.

it has the computational advantage that only tridiagonal matrices appear. Extensions to three dimensions are possible (Ref. 9-27). A similar technique, due to Bagrinovskii and Godunov, consists in replacing $\phi_{xx} + \phi_{yy}$ by $2\phi_{xx}$ in the first step and by $2\phi_{yy}$ in the second step, using an implicit formulation in each step; the method is unconditionally stable, involves only tridiagonal matrices, and is directly extensible to the three-dimensional case. Other "splittings" are described in Ref. 9-38, Chapter 2.

A further complication arises if additional terms are present in eq. (9.13-1) or (9.13-16). Consider for example

$$\phi_t = a\phi_{xx} + b\phi_{yy} + c\phi_{xy} + d\phi_x + e\phi_y + f\phi + g$$

where $a, b, \ldots, g$ are functions of x, y, and t. For the case $a > 0, b > 0, 4ab > c^2$, stability for the explicit case (analogue of eq. (9.13-9)) usually requires at least that $(a + b)k/h^2 < \frac{1}{2}$; the equivalent of the implicit eq. (9.13-14) is generally unconditionally stable. Some care may, however, be necessary to ensure accuracy in regions in which the coefficients are altering rapidly; a locally dense mesh may be necessary, together with some interpolation procedure connecting this fine mesh to the mesh used elsewhere.

So far, we have not discussed the matter of boundary values for higher-dimensional problems. There is an obvious difficulty in that a rectangular mesh will rarely fit exactly the time boundary of the region, as is indicated for example in Fig. 9.13-2. One approach is to replace the true boundary by a boundary consisting of mesh sides (as is shown dotted); the value of ϕ at any boundary mesh point must be obtained by interpolation from its local values on the true boundary. Again, a locally dense mesh could be useful near boundary segments where ϕ (or the boundary shape) is changing rapidly. It may, however, be more efficient to use a finite element approach in such problems; see Section 9.13.4.

A topic of considerable interest is that of convergence—as h and $k \to 0$, does the

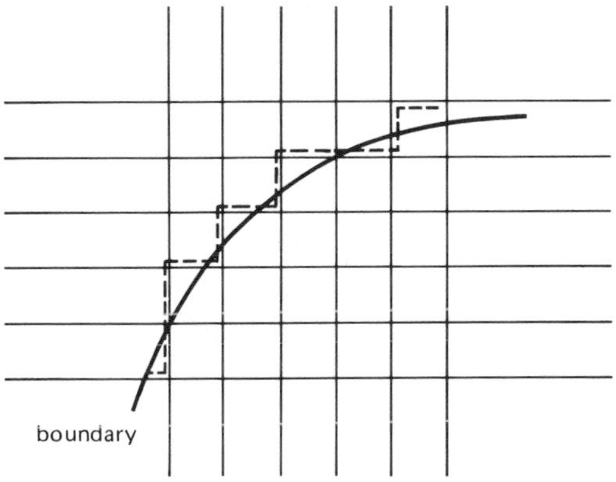

Fig. 9-13-2 Mesh points near boundary.

solution of the finite difference approximation approach the exact solution of the original PDE? For nonlinear equations, little theoretical information is available, and recourse to experimentation with different choices of h and k is usually necessary. For linear equations, a theorem of Lax guarantees convergence for stable algorithms which are consistent with the PDE; roundoff error may, however, still be troublesome (see Refs. 9-28 and 9-35).

9.13.3 Boundary Value Problems (Elliptic PDE)

A typical problem would be to find $\phi(x, y)$ so that

$$\phi_{xx} + \phi_{yy} = f(x, y) \tag{9.13-18}$$

in some region R of the (x, y) plane, where values of ϕ are given for all boundary points of R. Again, we cover R with a square grid (as in Figs. 9.13-1 and 9.13-2), and we replace the exact boundary of R by an approximate boundary fitted to the mesh; values of ϕ at boundary points of the mesh are obtained by interpolation from boundary data. It is now necessary to determine ϕ at interior points so as to satisfy the finite difference equivalent of eq. (9.13-18). Proceeding as in the preceding sections, we obtain

$$\phi_{i,j+1} + \phi_{i,j-1} + \phi_{i+1,j} + \phi_{i-1,j} - 4\phi_{i,j} = h^2 f_{i,j} \tag{9.13-19}$$

where $\phi_{i,j}$ denotes the approximation to ϕ at the (i, j) mesh point and where h is the mesh spacing. One equation of the form (9.13-19) is obtained for each choice of (i, j) corresponding to an interior mesh point; the totality of these equations represents a set of linear algebraic equations for the unknown $\phi_{i,j}$ values.

Direct solution of this equation set is feasible if the total number of equations is less than some hundreds; beyond this number, *relaxation* is more efficient. The basic idea behind the relaxation process is that of iterative correction—solution values are first guessed and then repeatedly corrected, point by point, in some systematic manner. A primitive relaxation scheme, called *Jacobi relaxation*, consists in the computation of a new $\phi_{i,j}$ value at each point (i, j), such that (if the $\phi_{i,j}$ values at surrounding points were unaltered), eq. (9.13-19) would be satisfied exactly at that point. Such a calculation is made for each point; upon completion (and only then), the old values of the $\phi_{i,j}$ at all points are replaced by the new values. It is generally more efficient, however, to use *Gauss-Seidel relaxation*, in which old values are replaced by new values as soon as they are calculated.

It is still more efficient to use a *successive overrelaxation* (*SOR*) *method*, in which $\phi_{i,j}$ values are overcorrected. Thus if we denote by $\phi_{i,j}^{(\text{old})}$ and by $\phi_{i,j}^{(\text{new})}$ the old and new values of $\phi_{i,j}$, the formula used to compute $\phi_{i,j}^{(\text{new})}$ is

$$\phi_{i,j}^{(\text{new})} = \phi_{i,j}^{(\text{old})} + \frac{\alpha}{4}\left[\phi_{i-1,j}^{(\text{new})} + \phi_{i,j+1}^{(\text{new})} + \phi_{i+1,j}^{(\text{old})} + \phi_{i,j-1}^{(\text{old})} - 4\phi_{i,j}^{(\text{old})} - h^2 f_{i,j}\right] \tag{9.13-20}$$

It is assumed here that the relaxation sweep is from left to right and from the top down; thus ϕ values corresponding to points to the left of or above (i, j) will already have been given these "new" values. The quantity α is a relaxation parameter. For a region which is approximately square, with N mesh points along a side, the optimal value of α is given by

$$\alpha = 2 - 2 \sin (\pi/N) \tag{9.13-21}$$

With this choice for α, about $3N/4$ relaxation sweeps are required to reduce an initial error in the guessed values of ϕ by a factor of 100; in contrast, the Jacobi method would require about N^2 sweeps. To estimate α, eq. (9.13-21) can be used, with N chosen in accordance with (for example) the total area of R. Alternatively, and particularly if R has a complicated shape, the convergence rates corresponding to various values of α may be determined experimentally for a problem of known exact solution (e.g., choose some ϕ, determine f so as to satisfy eq. (9.13-18), and record the values of ϕ on the boundary of R).

It is clear that a number of refinements can be made. For example, a coarse mesh could be used in an initial relaxation process, in order to obtain improved starting values for the fine mesh iteration. Also, boundary refinements can be increasingly incorporated as the relaxation proceeds. Different values of the relaxation parameter can be used for different portions of R, and this parameter can in fact be altered during the course of the relaxation process. Extensions to higher dimensions or to situations in which some combination of ϕ and its normal derivative are specified on the boundary are straightforward in principle but often tedious in practice.

An extensive literature on relaxation processes exists; Refs. 9-29 and 9-30 are particularly informative. We note, incidentally, that the finite difference equation corresponding to a relaxation process can be interpreted as time-stepping for a related diffusion equation; see, for example, Ref. 9-39, p. 33.

9.13.4 Finite Elements

We begin with a simple example illustrating the finite element method, and then describe some extensions.

Consider a region R (Fig. 9.13-3) of the (x, y) plane, and let boundary values of a function $\phi(x, y)$ be prescribed. We want to determine ϕ in R such that $\phi_{xx} + \phi_{yy} = 0$ in R, and such that ϕ takes on the prescribed boundary values. Rather than use a rectangular mesh as before, we divide R into a number of triangular subregions as shown and number the nodes in some manner. Our purpose is to try to determine, at least approximately, the values of ϕ at all interior nodes, subject, of course, to the condition that values of ϕ at boundary nodes (e.g., at nodes, 6, 7, and 8 in the figure) correspond to the prescribed boundary data. In this example, we will approximate ϕ by a linear function in x and y over each triangle, where the coefficients of this function are determined by the nodal values for that triangle. For example, inside the triangle defined by node numbers 1, 2, and 3, we use the approximation

$$\phi \cong \bar{\phi} = ax + by + c \tag{9.13-22}$$

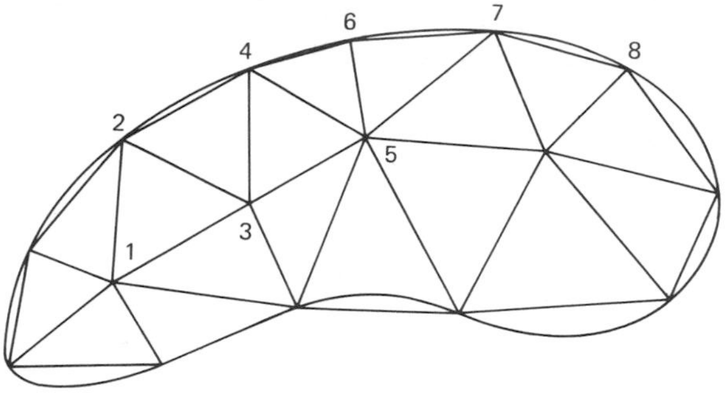

Fig. 9.13-3 Finite elements.

where

$$a = \frac{1}{2A} [\bar{\phi}_1 (y_2 - y_3) + \bar{\phi}_2 (y_3 - y_1) + \bar{\phi}_3 (y_1 - y_2)]$$

$$b = \frac{1}{2A} [\bar{\phi}_1 (x_3 - x_2) + \bar{\phi}_2 (x_1 - x_3) + \bar{\phi}_3 (x_2 - x_1)]$$

$$c = \frac{1}{2A} [\bar{\phi}_1 (x_2 y_3 - x_3 y_2) + \bar{\phi}_2 (x_3 y_1 - x_1 y_3) + \bar{\phi}_3 (x_1 y_2 - x_2 y_1)]$$

Here A is the area of the triangle, given by

$$2A = (x_2 - x_1)(y_3 - y_1) - (x_3 - x_1)(y_2 - y_1)$$

and (x_i, y_i) are the coordinates of node i (at which node $\bar{\phi}$ has the value $\bar{\phi}_i$). Observe that the coefficients a, b, and c are linear functions of the nodal values $\bar{\phi}_1$, $\bar{\phi}_2$, and $\bar{\phi}_3$.

Next, we make use of the Rayleigh-Ritz method of Section 9.12, in which the condition that $\phi_{xx} + \phi_{yy} = 0$ in R is replaced by the equivalent variational condition that $\delta I = 0$ where

$$I = \int_R (\phi_x^2 + \phi_y^2)\, dA$$

To evaluate I, we replace the integration over R by a sum of integrals over the triangular subdomains and evaluate each such integral by use of the local approximation (9.13-22). For the triangle with nodes (1, 2, 3), since $\bar{\phi}_x = a$ and $\bar{\phi}_x = b$, we obtain $(a^2 + b^2)A$ as the contribution of this subregion to I, where a and b are given in terms of $\bar{\phi}_1$, $\bar{\phi}_2$, and $\bar{\phi}_3$ as above. Adding together all such triangular-region contributions, we obtain an approximation to I which is quadratic in the $\bar{\phi}_i$ values.

We then determine the interior nodal $\overline{\phi}_i$ values such that this quadratic form is stationary (here, a minimum); i.e., we set $\partial I/\partial\overline{\phi}_i = 0$ for each such $\overline{\phi}_i$. This process results in a set of linear algebraic equations which can be solved to yield the $\overline{\phi}_i$ values. It is clear, incidentally, that in computational practice the contributions to the coefficients of the resulting linear equation set arising from each triangle can be generated very systematically.

Note that the composite function approximating ϕ consists of "elements" which are different linear functions over each triangle, but which are continuous across triangle boundaries. Note also that although second derivatives arise in the original PDE problem, only first derivatives of the approximating function were encountered in the variational principle. It is clear that generalizations are possible—more general shapes than triangles (e.g., rectangles or trapezoids) could have been used, and more complicated functional elements (defined by additional parameters) could have been devised, provided that continuity across boundaries could have been maintained. The advantages of the finite element method are that even complicated regions R can usually be fitted well by triangles or other shapes, that the computational process is straightforward, that lower derivatives than those occurring in the original PDE are used, and that the coefficient matrix of the final algebraic equation set is sparse and (usually) well-conditioned. The main disadvantage is that it may eventually be necessary to solve a very large system of algebraic equations (numbered perhaps in the thousands), and this can be time-consuming. Sometimes it is useful to divide the original problem into parts, which are solved individually and then connected; also, iterative methods can be applied to the equation set.

The variational formulation for the example problem considered here was well known. In other cases, as in nonlinear problems, it may not be easy to derive an equivalent variational principle; a Galerkin approach may then be feasible. Suppose that the PDE reads $M(\phi) = 0$, where M is some PDE operator, and where ϕ, as before, is the function sought. Then we can construct an approximate solution, say $\overline{\phi}$, as before, and determine any coefficients (e.g., nodal values of $\overline{\phi}$) such that

$$\int_R M(\overline{\phi})\,\psi_j\,dA = 0 \tag{9.13-23}$$

for $j = 1, 2, \ldots$, where the ψ_j are some set of suitable functions, equal in number to the total number of coefficients in $\overline{\phi}$. (In a sense, eq. (9.13-23) makes $M(\overline{\phi})$ "orthogonal" to each of the ψ_j; the exact solution ϕ makes $M(\phi) = 0$, and 0 is orthogonal to all functions, not just the ψ_j.) As a special case, the ψ_j could be chosen as those functions used in the finite element depiction of ϕ. It is important to note that if second derivatives occur in $M(\overline{\phi})$, the piecewise representation of $\overline{\phi}$ would have to be continuously differentiable across boundaries. However, it may be possible to use the divergence theorem in eq. (9.13-23) so as to reduce the orders of the derivatives occurring in $M(\overline{\phi})$, at the cost of introducing derivatives of the ψ_j.

At this point, it seems clear that the finite element method can be applied to equations of evolution also. Consider for example, the diffusion problem of eq. (9.13-16). Again, the region R can be subdivided into triangles, with ϕ approximated by a piecewise linear function $\bar{\phi}$ defined by the nodal values $\bar{\phi}_j$, each of which is now considered to be a function of time. To obtain a Galerkin formulation, we write

$$\int_R [\bar{\phi}_t - a^2(\bar{\phi}_{xx} + \bar{\phi}_{yy})]\, \psi_j\, dA = 0$$

which is equivalent to

$$\int_R [\bar{\phi}_t \psi_j + a^2\{\bar{\phi}_x (\psi_j)_x + \bar{\phi}_y (\psi_j)_y\}]\, dA - a^2 \int_C \psi_j \frac{\partial \bar{\phi}}{\partial n}\, dl = 0$$

where C is the boundary of R, and $\partial/\partial n$ denotes a normal derivative. In this form, only first derivatives occur, so that a piecewise linear representation for $\bar{\phi}$ is adequate.

In recent years, the finite element method has attracted much theoretical attention; an excellent survey will be found in Ref. 9-31. For discussions of applications, with examples, see Refs. 9-32 and 9-33; sample computer programs will be found in Ref. 9-34.

9.13.5 Wave Problems

For the prototype wave problem

$$\phi_{tt} = c^2 \phi_{xx} \tag{9.13-24}$$

satisfied by a function $\phi(x, t)$ in, say, $0 < x < L$, $t > 0$, with $c > 0$ constant, the method of Section 9.13.1 leads to the finite difference equivalent (with similar notation)

$$\frac{\phi_j^{(n+1)} - 2\phi_j^{(n)} + \phi_j^{(n-1)}}{k^2} = c^2 \frac{\phi_{j+1}^{(n)} - 2\phi_j^{(n)} + \phi_{j-1}^{(n)}}{h^2} \tag{9.13-25}$$

The customary initial data on ϕ and ϕ_t would determine $\phi_j^{(0)}$ and $\phi_j^{(1)}$ values (these latter from $\phi_j^{(1)} - \phi_j^{(0)} = k\phi_t(jh, 0)$, for example). Equation (9.13-25) would then be applied for $n = 2, 3, \ldots$, in sequence, for all j-values other than those corresponding to $x = 0$ and $x = L$; at these two points, prescribed boundary data would be used. This numerical process is stable if the condition $(ck/h) < 1$ is satisfied, and this *Courant-Friedrichs-Lewy* (*CFL*) *condition* may be interpreted as requiring the domain of dependence (Section 9.11.2) of the finite-difference equation to contain that of the PDE.

To avoid this stability constraint, which limits the permissible size of time step, the right-hand side of eq. (9.13-25) may be replaced by the average of two similar terms, one computed for time level $(n+1)$ and the other for time level $(n-1)$. Extensions to more space dimensions and to situations in which first-order derivatives occur may be made in much the same way as for the diffusion equation of Sections 9.9.4 and 9.9.5. Also, it may be convenient to rewrite eq. (9.13-24) (or its generalization) as a coupled pair of first-order equations and to use a Lax-Wendroff method of the kind to be discussed below. These various topics are dealt with in Refs. 9-28 and 9-35.

For a nonlinear wave equation, or system of equations, it may be necessary to deal with a shock wave (Sections 8.3.2 and 8.3.3), where the solution can become discontinuous. In such cases, it is usually desirable to write the governing equations in conservation form (Section 8.3.2), the intuitive reasoning being that a conservation law continues to hold across a shock, whereas the PDE does not. As an example, the first-order PDE

$$\phi_t + \phi\phi_x = 0 \tag{9.13-26}$$

could be written in the conservation form

$$\phi_t + \tfrac{1}{2}(\phi^2)_x = 0 \tag{9.13-27}$$

As a second example, the divergence forms for the equations governing the one-dimensional flow of a compressible gas have been given in eqs. (8.3-38). The Lax-Wendroff approach to the solution of eq. (9.13-27) achieves accuracy by use of a Taylor expansion of the form

$$\phi_j^{(n+1)} = \phi_j^{(n)} + k\left(\frac{\partial \phi}{\partial t}\right)_j^{(n)} + \frac{1}{2}k^2\left(\frac{\partial^2 \phi}{\partial t^2}\right)_j^{(n)}$$

Here $\phi_t = -\frac{1}{2}(\phi^2)_x$, and

$$\begin{aligned}\phi_{tt} &= -(\tfrac{1}{2}\phi^2)_{xt} = -(\phi\phi_t)_x \\ &= \tfrac{1}{2}[\phi(\phi^2)_x]_x\end{aligned} \tag{9.13-28}$$

A suitable finite-difference equivalent of the right-hand side of eq. (9.13-28) would be given by

$$\frac{1}{h}\left\{\phi_{j+1/2}^{(n)}\left[\frac{(\phi_{j+1}^{(n)})^2 - (\phi_j^{(n)})^2}{h}\right] - \phi_{j-1/2}^{(n)}\left[\frac{(\phi_j^{(n)})^2 - (\phi_{j-1}^{(n)})^2}{h}\right]\right\} \tag{9.13-29}$$

It is customary here to replace the term $\phi_{j+1/2}^{(n)}$ by $\frac{1}{2}(\phi_{j+1}^{(n)} + \phi_j^{(n)})$ in order to avoid midmesh point values.

The Lax-Wendroff approach to the more general problem (again in conservation form)

$$\phi_t + [f(\phi)]_x = 0 \tag{9.13-30}$$

where now ϕ is a column vector of dependent variables, is similar. Equation (9.13-29) will involve the Jacobian matrix of f with respect to the elements of f (for details, see Refs. 9-28 and 9-35); it may be preferable to use Richtmeyer's two-step version of the Lax-Wendroff method, in which Jacobian evaluations are avoided.

For the example problem of eq. (9.13-27), Richtmeyer's approach is to first compute quantities $\phi_{j+1/2}^{(n+1/2)}$ given by

$$\phi_{j+1/2}^{(n+1/2)} = \frac{1}{2}(\phi_{j+1}^{(n)} + \phi_j^{(n)}) - \frac{k}{4h}[(\phi_{j+1}^{(n)})^2 - (\phi_j^{(n)})^2]$$

and then obtain $\phi_j^{(n+1)}$ by

$$\phi_j^{(n+1)} = \phi_j^{(n)} - \frac{k}{2h}[(\phi_{j+1/2}^{(n+1/2)})^2 - (\phi_{j-1/2}^{(n+1/2)})^2]$$

The accuracy of these formulations is of second order, and the stability constraints are similar to the CFL condition. The generalization of the two-step method to systems of equations, of the form of eq. (9.13-30), is immediate. It should also be remarked that different two-step methods are available; one could for example use a left-hand space difference in the first time step and a right-hand difference in the second. See Ref. 9-28, Chapter 12 and Ref. 9-38, Chapter 4.

An alternative approach to nonlinear hyperbolic systems is to deliberately incorporate an artificial viscosity term into the PDE. For example, eq. (9.13-26) could be replaced by

$$\phi_t + \phi\phi_x = \epsilon\phi_{xx} \tag{9.13-31}$$

where the small positive constant ϵ plays the role of a viscosity coefficient. The effect of the added term $\epsilon\phi_{xx}$ is to smooth out abrupt transitions in the solution function, so that a shock wave transition now occupies several adjoining mesh points. (Even in the Lax-Wendroff-type approach described above, it may be useful to introduce an artificial viscosity term, to moderate oscillations near a shock front.) It may also be useful to increase the density of the mesh spacing near a shock, the dense region being made to move along with the shock. Some sample calculations involving artificial viscosity are given in Ref. 9-28.

9.13.6 Concluding Remarks

PDE problems met in practice are frequently nonlinear and often involve sets of coupled equations. They may exhibit a combination of the features of diffusion, potential, and wave-type problems, and the general character may change from region to region. Little general information is available, but a good deal of experience has accumulated in special areas of practical interest—as indicated, for example, in Refs. 9-28, 9-24, 9-36, and 9-34. Reference 9-39 gives both background considerations and computer programs for a plasma problem of considerable practical interest.

Some general suggestions may be made. First of all, the mesh spacing should be reasonably small compared with the distance scale over which dependent variables change significantly. Second, if the problem is a complicated one, it may be very useful to first obtain computational experience with a simpler problem of the same general nature and, in fact, preferably one for which the exact solution is known. Stability, accuracy, and convergence can then be investigated reliably and economically before returning to the original problem. It may also be useful to remove singularities, perhaps by local analytical subtraction. Finally, it may be worthwhile to consider unconventional methods. For example, one could replace the spatial Laplacian operator in a diffusion problem by a finite difference operator, but leave the time derivative alone; the result of this hybrid approach is to obtain a set of coupled ordinary differential equations for the mesh point values, solvable, for example, by a Runge-Kutta algorithm. As a second example, a spectral decomposition technique can be used even in nonlinear problems, provided one uses the fast Fourier transform as an economical device to carry out the transitions between time and frequency domains required to evaluate time-dependent nonlinearities. Spectral methods have been successfully applied to problems of both potential and time evolution character; Refs. 9-37 and 9-39 are particularly informative. Also, of course, computational methods based on characteristics, integral equations, or numerical perturbations—all representing direct numerical implementations of topics treated earlier in this and the preceding chapter—are often effective. As a final remark, it is hardly necessary to remind the reader that if a sequence of numerical studies, for various geometrics or parameter values, is required, a prior nondimensionalization of the problem may result in computer economy.

9.14 REFERENCES

9-1 Ahlfors, L. V., "Conformality with Respect to Riemannian Metrics," *Ann. Acad. Sci. Fenn.*, Series A1, No. 206, 1955.

9-2 Courant, R., and Hilbert, D., *Methods of Mathematical Physics*, vols. 1–2, Wiley-Interscience, New York, 1962.

9-3 Garabedian, P. R., *Partial Differential Equations*, Wiley, New York, 1964.

9-4 Petrovsky, I. G., *Lectures on Partial Differential Equations*, Wiley-Interscience, New York, 1954.

9-5 Marguerre, K., "Ansätze zur Lösung der Grundgleichungen der Elastizitätstheorie," *Z.A.M.M.*, 35, 242, 1955.

9-6 Pearson, C., *Theoretical Elasticity*, Harvard University Press, Cambridge, Mass., 1959.

9-7 Morse, P., and Feshbach, H., *Methods of Theoretical Physics*, vols. 1–2, McGraw-Hill, New York, 1953.

9-8 Carslaw, H. J., and Jaeger, J. C., *Conduction of Heat in Solids*, Oxford University Press, New York, 1959.

9-10 Carrier, G., Krook, M., and Pearson, C., *Functions of a Complex Variable*, McGraw-Hill, New York, 1966.

9-11 Kober, H., *Dictionary of Conformal Representations*, Dover, New York, 1952.

9-12 Milne-Thomson, L. M., *Theoretical Hydrodynamics*, Macmillan, New York, 1960.

9-13 Muskhelishvili, N. I., *Some Basic Problems of the Mathematical Theory of Elasticity*, Noordhoff, Groningen, Netherlands, 1953.

9-14 Carathéodory, C., *Conformal Representation*, Cambridge University Press, Cambridge, 1932.

9-15 Schwarz, L., *Théorie des Distributions*, vols. 1–2, Hermann et Cie, Paris, 1950.

9-16 Sommerfield, A., *Partial Differential Equations in Physics*, Academic Press, New York, 1949.

9-17 Tricomi, F. G., *Integral Equations*, Wiley-Interscience, New York, 1957.

9-18 Kellogg, O. D., *Foundations of Potential Theory*, Dover, New York, 1953.

9-19 Protter, M. H., and Weinberger, H. F., *Maximum Principles in Differential Equations*, Prentice-Hall, New Jersey, 1967.

9-20 Hutchinson, J. W., and Niordson, F. I., *Designing Vibrating Membranes*, Report No. 12, 1971, DCAMM, Technical University of Denmark.

9-21 Polya, G., and Szegö, G., "Isoperimetric Inequalities in Mathematical Physics," *Ann. of Math. Studies*, No. 27, Princeton University Press, Princeton, 1951.

9-22 Courant, R., and Friedrichs, K. O., *Supersonic Flow and Shock Waves*, Wiley-Interscience, New York, 1948.

9-23 Stoker, J. J., *Water Waves*, Wiley-Interscience, New York, 1957.

9-24 Kantorovich, L., and Krylov, V., *Approximate Methods in Higher Analysis*, Wiley-Interscience, New York, 1958.

9-25 Mikhlin, S., *Variational Methods in Mathematical Physics*, Pergamon, New York, 1964.

9-26 Strang, G., and Fix, G., *An Analysis of the Finite Element Method*, Prentice-Hall, New Jersey, 1973.

9-27 Carrier, G. F., and Pearson, C. E., *Partial Differential Equations*, Academic Press, New York, 1976.

9-28 Richtmeyer, R. D., and Morton, K. W., *Difference Methods for Initial Value Problems*, 2 ed., Wiley, New York, 1967.

9-29 Varga, R. S., *Matrix Iterative Analysis*, Prentice-Hall, New Jersey, 1962.

9-30 Brikhoff, G., *The Numerical Solution of Elliptic Equations*, SIAM Regional Conference Series No. 1, 1971.

9-31 Strang, G., and Fix, G., *An Analysis of the Finite Element Method*, Prentice-Hall, New Jersey, 1973.

9-32 Mitchell, A. R., and Wait, R., *The Finite Element Method in Partial Differential Equations*, Wiley, New York, 1977.

9-33 Zienkiewicz, O. C., *The Finite Element Method in Engineering Science*, McGraw-Hill, London, 1971.

9-34 Huebner, K. H., *The Finite Element Method for Engineers*, Wiley, New York, 1975.

9-35 Smith, G. D., *Numerical Solution of Partial Differential Equations: Finite Difference Methods*, 2 ed., Oxford University Press, Oxford, 1978.

9-36 Wirz, H. J., and Smolderen, J. J. (eds.), *Numerical Methods in Fluid Dynamics*, McGraw-Hill, New York, 1978.

9-37 Orszag, S., *Numerical Simulation of Incompressible Flows*, Studies in Applied Math, 50, 1971, p. 293.

9-38 Mitchell, A. R., and Griffiths, D. F., *The Finite Difference Method in Partial Differential Equations*, Wiley, New York, 1980.

9-39 Bauer, F., Betancourt, O., and Garabedian, P., *A. Computational Method in Plasma Physics*, Springer-Verlag, New York, 1978.

9-40 Gottlieb, D., and Orszag, S. A., *Numerical Analysis of Spectral Methods*, SIAM Regional Conference No. 26, 1977.

10

Integral Equations

Donald F. Winter*

10.1 INTRODUCTION

An integral equation is a functional equation in which the unknown variable ϕ appears under an integral sign. In the case of a single independent variable x, a form commonly encountered can be written as

$$g(x)\,\phi(x) = f(x) + \lambda \int_a^b F[x, \xi; \phi(\xi)]\, d\xi \qquad (a \leqslant x \leqslant b) \qquad (10.1\text{-}1)$$

In this expression, $f(x)$ and $g(x)$ are known functions, and $\phi(x)$ is to be determined. The form of the functional F is known, λ is a parameter, and the range of integration $[a, b]$ is specified.

Simple examples of readily solvable integral equations arise from the standard integral transforms. For example, the Fourier and Mellin transforms described in Chapter 11 qualify as integral equations, if the functions in the integrands are considered to be unknown. In each case, the solution is given by the inversion formula. To the extent that integral transform pairs represent special cases of integral equations and corresponding solutions, it may be said that the research of Laplace in the latter part of the 18th century was probably the earliest in the field. However, the theory of integral equations received its real impetus from the work of Niels Henrik Abel (1802–1829). Abel proposed and solved (by two different methods) the following problem: imagine a bead sliding down a frictionless wire under the influence of gravity. Suppose the wire follows a plane (x, y) curve of the form $x = \phi(y)$, with $\phi(0) = 0$. The bead begins its descent from the point (x, y) where $y > 0$ and arrives at the origin at a time t, which is a prescribed function of y: $t = f(y)$, say. The problem is to determine the form of the wire corresponding to a given $f(y) = t$. If g denotes acceleration due to gravity, elementary considerations lead to the expression

*Prof. Donald F. Winter, Dep'ts. of Oceanography and Applied Mathematics, University of Washington, Seattle, Wash. 98195.

$$f(y) = \frac{1}{\sqrt{2g}} \int_0^y \left\{ \frac{[\phi'(y_0)]^2 + 1}{y - y_0} \right\}^{1/2} dy_0$$

An equivalent form

$$f(y) = \frac{1}{\pi} \int_0^y \frac{\psi(y_0)}{\sqrt{y - y_0}} \, dy_0$$

is known as *Abel's integral equation,* and for $f(0) = 0$, its solution is given by

$$\psi(y) = \int_0^y \frac{f'(y_0)}{\sqrt{y - y_0}} \, dy_0$$

It can be said of most boundary value problems involving either ordinary or partial differential equations that exact solutions cannot be found. The same is true of problem formulations leading to integral equations, and as a consequence recourse to approximate or numerical solutions is often necessary. This chapter presents a brief survey of exact and approximate solution techniques applicable to various types of integral equations that arise in practice.

This discussion is selective in character and is intended to provide an overview of the subject. The reader who requires elaboration of specific topics is referred to the several comprehensive treatises on integral equations that have appeared in recent years. With few exceptions, theorems are stated herein without proof, and solution procedures are sketched with a minimum of derivational detail, because several presentations of theoretical material are readily available. For example, discussions of linear integral equations accessible to nonspecialists are to be found in Refs. 10-1 to 10-5. Ref. 10-6 is a comprehensive treatise on numerical methods for solving integral equations of several types. This last work also contains an extensive bibliography, citing many journal articles and reports treating various aspects of the subject.

10.2 DEFINITIONS AND CLASSIFICATIONS

When the integrand in eq. (10.1-1) can be written in the form

$$F[x, \xi; \phi(\xi)] = k(x, \xi)\, \phi(\xi)$$

the integral equation is said to be *linear*. The factor $k(x, \xi)$ is called the kernel of the equation and is a function of x and ξ throughout the rectangle R defined by $a \leqslant x \leqslant b$, $a \leqslant \xi \leqslant b$. If the derivative of the dependent variable appears in an integral equation, the expression is called an *integrodifferential equation.* In many problems of practical importance, the unknown function depends on two or more independent variables, and, in such cases, multidimensional integral equations can arise. If more than a single dependent variable is to be determined, the solution of a set of coupled integral equations will be required (see Refs. 10-7 and 10-8).

A one-dimensional linear integral equation can be written in the general form

$$g(x)\,\phi(x) = f(x) + \lambda \int_a^b k(x, \xi)\,\phi(\xi)\,d\xi \tag{10.2-1}$$

The functions $f(x)$ and $g(x)$ are known, λ is a parameter, a and b are specified integration limits, and the kernel $k(x, \xi)$ is defined over the region R.

If $g(x) \equiv 0$ in eq. (10.2-1) and if a and b are constants, the equation is called a *Fredholm integral equation of the first kind*

$$f(x) + \lambda \int_a^b k(x, \xi)\,\phi(\xi)\,d\xi = 0 \tag{10.2-2}$$

Here the function $\phi(\xi)$ is to be determined. If $f(x) \equiv 0$ and $g(x) = 1$ in eq. (10.2-1), the equation is said to be a *homogeneous Fredholm integral equation of the second kind*

$$\phi(x) = \lambda \int_a^b k(x, \xi)\,\phi(\xi)\,d\xi \tag{10.2-3}$$

If $f(x) \neq 0$ and $g(x) = 1$, the equation is an *inhomogeneous Fredholm equation of the second kind*

$$\phi(x) = f(x) + \lambda \int_a^b k(x, \xi)\,\phi(\xi)\,d\xi \tag{10.2-4}$$

If the upper limit of the range of integration in eq. (10.2-2) is replaced by x, the equation is termed a *Volterra integral equation of the first kind*

$$f(x) + \lambda \int_a^x k(x, \xi)\,\phi(\xi)\,d\xi = 0 \tag{10.2-5}$$

Similarly, if $b = x$ in eqs. (10.2-3) and (10.2-4), the equations are called *Volterra integral equations of the second kind*. Corresponding to eq. (10.2-4), we have

$$\phi(x) = f(x) + \lambda \int_a^x k(x, \xi)\,\phi(\xi)\,d\xi \tag{10.2-6}$$

and this Volterra equation is said to be *homogeneous* if $f(x) \equiv 0$. The Volterra integral equations may be considered special cases of Fredholm equations with $k(x, \xi) = 0$ for $\xi > x$. Linear Volterra and Fredholm integral equations have been the subject of intensive study in the past. It turns out that the parameter λ plays an important role in the underlying theory, and for this reason it cannot simply be absorbed into k. The role of λ is exemplified by the observation that when the kernel is a bounded integrable function over R and the range of integration is finite, the solution of the Volterra equation is an entire function of λ, whereas the solution of the corresponding Fredholm equation is a meromorphic function of λ.

Nonlinear integral equations assume such a variety of forms that there is currently no conventional system of classification. An expression of the general type (10.1-1), with the range [0, 1], is often referred to in the literature as an *Urysohn equation*. One particular Urysohn equation, which has received a fair amount of attention, is an obvious extension of the linear Fredholm equation

$$\phi(x) = f(x) + \lambda \int_0^1 k(x, \xi)\, [\phi(\xi)]^n \, d\xi \tag{10.2-7}$$

where n is an integer greater than 1. Other special cases of the Urysohn equation arise in the theory of radiative transfer

$$\phi(x) = f(x) + \lambda \int_0^1 k(x, \xi)\, F[\xi, \phi(\xi)] \, d\xi$$

known as the *Hammerstein equation*, and the so-called *H-equation*

$$\phi(x) = 1 + \lambda x \phi(x) \int_0^1 \frac{g(\xi)}{x + \xi} \phi(\xi) \, d\xi$$

where $g(\xi)$ is a given (usually simple) function of ξ. A rather different type is exemplified by the *equation of Bratu*

$$\phi(x) = \lambda \int_0^b g(x, \xi)\, e^{\phi(\xi)} \, d\xi \tag{10.2-8}$$

where the kernel is a Green's function of the form

$$g(\xi, x) = \frac{1}{b} \begin{cases} \xi(b - x), & 0 < \xi < x \\ x(b - \xi), & x < \xi < b \end{cases} \tag{10.2-9}$$

We shall consider both eqs. (10.2-7) and (10.2-8) in section 10.7, where techniques for solving nonlinear integral equations are discussed.

A nonlinear integrodifferential equation of the form

$$\frac{1}{\phi}\frac{d\phi}{dt} = \left(1 - \frac{\phi}{\phi_m}\right) + \lambda \int_a^t k(t - t_0)\,\phi(t_0)\,dt_0 \tag{10.2-10}$$

where ϕ_m is constant, is referred to as the nonlinear *integrodifferential equation of Volterra*, who first studied equations of this sort in connection with the growth of populations influenced both by intraspecific competition and hereditary factors. Although the parameter λ is again displayed explicitly in each of the examples above, the role it plays in the solutions of nonlinear integral equations may be entirely different from the linear case.

The ease (or difficulty) of solving a particular integral equation depends to a large extent on the nature of the kernel and the inhomogeneous term. To simplify the discussion, throughout most of this chapter, we shall consider $\phi(x)$ to be a real function of a real variable x; exceptions to this rule will be obvious from the context. The ensuing discussion of linear integral equations is facilitated by introducing some definitions at this point. We shall say that a given function $f(x)$ is $L_p\,[a, b]$ in the interval $[a, b]$ if it is measurable (i.e., Lebesque integrable) in $[a, b]$ and satisfies the inequality

$$\int_a^b |f(x)|^p\,dx < \infty$$

Suppose the sign of $f(x)$ is such that $\int_a^b f(x)\,dx > 0$. A kernel $k(x, \xi)$ is said to be *definite* if the constant c defined by

$$\int_a^b dx \int_a^b dt\; k(x, t) f(x) f(t) = c$$

is either always positive or always negative when f is $L_2\,[a, b]$. In the former case, the kernel is said to be *positive-definite*; in the latter, the kernel is *negative-definite*. If $c \geqslant 0$ or $c \leqslant 0$, the kernel is *semidefinite* (some writers use the term *indefinite*).

As already mentioned, if $f(x)$ is identically zero in (10.2-4) or (10.2-6), the integral equation is said to be homogeneous. Obviously, every homogeneous equation has the trivial solution $\phi(x) = 0$. However, suppose that a homogeneous linear equation has at least one nonzero solution $\phi_1(x)$. Then if c_1 is any scalar, $c_1\phi_1(x)$ is also a solution, and thus there is no loss in generality if a suitable constant has been absorbed into ϕ such that ϕ is *normalized* in the sense

$$\int_a^b [\phi(x)]^2\,dx = 1$$

In the sequel we will generally make this assumption. Moreover, if $\phi_n(x)$ ($n = 1, 2, \ldots, m$) are solutions of a homogeneous linear integral equation, then all linear combinations of $c_n \phi_n(x)$ are also solutions. Therefore, by application of the Gram-Schmidt procedure described in Chapter 17, if several linearly independent solutions $\phi_n(x)$ are known, these may be considered orthogonal and normalized over the interval $[a, b]$ in the sense

$$\int_a^b \phi_m(x)\,\phi_n(x)\,dx = \delta_{mn} \tag{10.2-11}$$

where δ_{mn} is the usual Kronecker delta.

A value of the parameter λ_1 for which a homogeneous linear integral equation possesses a nontrivial $L_2[a, b]$ solution $\phi_1(x)$ is called an *eigenvalue* of the kernel $k(x, \xi)$, and the function $\phi_1(x)$ is an *eigenfunction* of the kernel associated with the eigenvalue λ. More generally, there may correspond to each eigenvalue of the kernel a number of linearly independent (orthonormal) eigenfunctions $\phi(x)$.

Suppose that $f(x)$ and $k(x, \xi)$ are both nonzero L_2 functions, and that each is continuous and bounded in the appropriate domain. In subsequent discussions we shall have occasion to refer to *iterated kernels*, which are denoted by $k_n(x, \xi)$, and to a *resolvent kernel* $\Gamma(x, \xi; \lambda)$. The iterated kernels associated with the kernel $k(x, \xi)$ are defined by the sequence of functions

$$k_1(x, \xi) = k(x, \xi)$$

$$k_2(x, \xi) = \int_a^b k(x, u_1)\,k(u_1, \xi)\,du_1$$

$$\ldots$$

$$k_n(x, \xi) = \int_a^b \underset{(n-1)}{\cdots} \int_a^b k(x, u_1)\,k(u_1, u_2) \ldots k(u_{n-1}, \xi)\,du_1\,du_2 \ldots du_{n-1} \tag{10.2-12}$$

Under the stated conditions, orders of integration can be interchanged, giving the recursion relationship

$$k_n(x, \xi) = \int_a^b k(x, u)\,k_{n-1}(u, \xi)\,du \tag{10.2-13}$$

and (more generally)

$$k_n(x, \xi) = \int_a^b k_m(x, u)\,k_{n-m}(u, \xi)\,du$$

for the nth iterated kernel associated with $k(x, \xi)$. The resolvent kernel $\Gamma(x, \xi; \lambda)$ is defined by the series expression

$$\Gamma(x, \xi; \lambda) = \sum_{n=1}^{\infty} \lambda^n k_n(x, \xi) \tag{10.2-14}$$

over its range of convergence. A kernel is *symmetric* if it satisfies the relation $k(x, \xi) = k(\xi, x)$. An *antisymmetric* kernel has the form $k(x, \xi) = -k(\xi, x)$, if k is real.

Further nomenclature is frequently encountered in the literature. In particular, if $k(x, \xi)$ is a complex $L_2[a, b]$ kernel, then $k^*(x, \xi) = \overline{k(\xi, x)}$ is also an $L_2[a, b]$ kernel and is sometimes referred to as the *Hermitian adjoint* of $k(x, \xi)$—the overbar denotes the complex conjugate. If $k(x, \xi) = k^*(x, \xi)$, then $k(x, \xi)$ is called a *Hermitian kernel*. Evidently, if $k(x, \xi)$ is a real-valued function, the kernel is Hermitian if and only if it is symmetric. A *polar kernel* can be written as

$$k(x, \xi) = r(\xi)\, g(x, \xi) \tag{10.2-15}$$

where $g(x, \xi)$ is symmetric; this form frequently arises in practice.

A kernel that can be expressed as a terminating sum of products of functions $\alpha_i(x)$ and $\beta_i(\xi)$ is called a *degenerate kernel*

$$k(x, \xi) = \sum_{i=1}^{N} \alpha_i(x)\, \beta_i(\xi) = \sum_{i=1}^{M} \sum_{j=1}^{M} c_{ij}\, \psi_i(x)\, \psi_j(\xi) \tag{10.2-16}$$

The second form of eq. (10.2-16) is obtained by expressing each of the α_i and β_i in terms of a linear combination of a set of M functions ψ_i, which are orthonormal over the region R. It may be noted that a symmetric kernel $k(x, \xi)$ can be approximated uniformly by symmetric degenerate kernels; thus, if $\tilde{k}(x, \xi)$ approximates the symmetric kernel $k(x, \xi)$, then the kernel

$$\tfrac{1}{2}\,[\tilde{k}(x, \xi) + \tilde{k}(\xi, x)]$$

also approximates $k(x, \xi)$.

An integral equation is said to be *singular* if it is characterized by one or both of the following properties

1. the kernel k has one or more singularities in the region R
2. either a or b (or both) is infinite.

The kernels of integral equations derived by the Green's function procedure or by transform methods are often singular. When a singular kernel is of the form

$$k(x, \xi) = \frac{g(x, \xi)}{|x - \xi|^{\beta}} \qquad (|g(x, \xi)| \leqslant M) \tag{10.2-17}$$

where $g(x, \xi)$ is a (bounded) smooth function in R, and β is a real positive constant <1, it can be shown that the nth iterated kernel $k_n(x, \xi)$ satisfies the inequality

$$k_n(x, \xi) \leqslant \frac{cM^n}{|x - \xi|^{n\beta - n + 1}}$$

where c is a finite constant. Hence, $k_n(x, \xi)$ will be bounded if $\beta < 1 - 1/n$, and it follows that, for any finite real constant $\beta < 1$, there exists a value of n such that the iterated kernel $k_n(x, \xi)$ is nonsingular. A kernel of the form (10.2-17) is said to be *weakly singular*. The examples considered in the next section will illustrate several of the definitions and concepts given above.

10.3 ORIGIN OF INTEGRAL EQUATIONS

Integral equations are encountered in a wide variety of scientific fields: statistics, acoustics, electromagnetic theory, fluid mechanics, radiology, cardiology, psychology, and population dynamics, to cite just a few examples. Quantitative considerations of certain types of problems lead most naturally to a conventional boundary value problem in the form of an ordinary differential equation or a partial differential equation, together with appropriate boundary conditions. This is to be expected when the process by which a variable property ϕ changes as a function of x (say) depends only on the behavior of ϕ at points in the immediate neighborhood of x or at isolated points elsewhere. However, if the character of ϕ at x depends on the behavior of ϕ over a range of x, then the quantification of the process will produce an integral equation or an integrodifferential equation. One example of the latter type of problem is the nonlinear integrodifferential equation of Volterra. If, in eq. (10.2-10), ϕ is a measure of the number of individuals growing in a bounded community, then the first term on the right-hand side represents the growth rate dependence on the instantaneous population size. The integral term describes hereditary factors and represents a weighted effect of the population size over some time interval in the past. Other examples are presented below.

Example 1: Consider the spreading of a thin oil slick over a porous plane medium. Suppose that the slick covers an area $A(t)$ at time t, where $t = 0$ corresponds to the beginning of the spill. Assume the thickness h of the oil slick to be constant (observed to be approximately the case for crude oil at low temperature). Denote by $Q(t)$ the volume rate of spill entering the center of the slick. Suppose the substrate material is such that the rate at which oil seeps into the medium at any point is a function of the time elapsed since the slick first overran the point. Conservation arguments lead immediately to an integral equation for dA/dt of the form

$$h\frac{dA}{dt} = Q(t) - \gamma \int_0^t k(t - t_0)\frac{dA}{dt_0}\, dt_0 \tag{10.3-1}$$

where γ is a constant. In this case, $A(t)$ can be determined directly by application of the Laplace transform method, once $k(t - t_0)$ is specified.

Example 2: Problems involving radiative energy exchange often lead to integral equations. The *Milne problem* of radiative transfer in a plane-stratified semi-infinite medium is a frequently discussed example (see, e.g., Ref. 10.8). A somewhat different example is provided by a consideration of radiation in a semi-infinite cylindrical hole of unit diameter drilled in a medium of constant temperature T. Let the center of the hole coincide with the x-axis in a cylindrical coordinate system (r, θ, x), with θ denoting the polar angle. The surface of the hole is defined by $r = \frac{1}{2}$. Suppose that the material of the hole is a gray body of emissivity ϵ, and let $B(x)$ denote the total flux emitted from an element dS_1 of the surface. When radiative equilibrium prevails, the total emitted flux is equal to $\epsilon\sigma T^4$ plus reflected radiation received from the rest of the surface. If $I(x)$ denotes the total radiation incident per unit area at the location of dS_1, the equilibrium condition can be expressed as

$$B(x) = \epsilon\sigma T^4 + (1 - \epsilon) I(x) \tag{10.3-2}$$

The incident radiative flux dI at dS_1 contributed by an area element dS_2 at $(\frac{1}{2}, \theta, \xi)$ is given by

$$dI = B(\xi) \frac{a^2}{\pi R^4} dS_2$$

where a, R, and dS_2 are shown in Fig. 10.3-1. Making reference to the illustration, one can derive the following relationships from elementary geometric considerations:

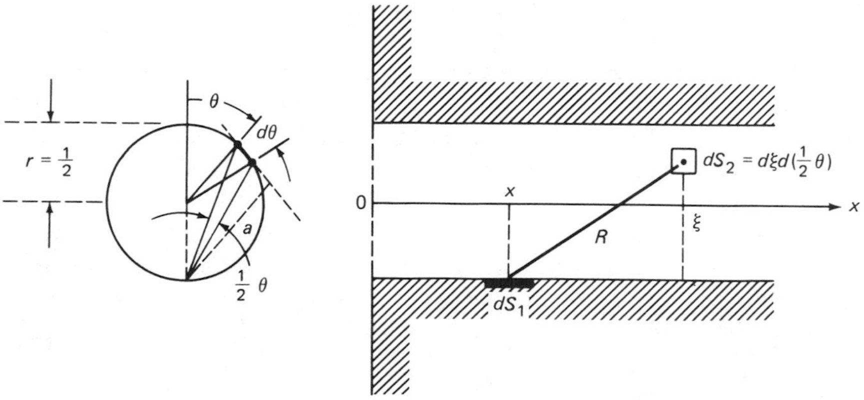

Fig. 10.3-1 Radiation in a cylindrical hole.

$$a = \cos^2\left(\tfrac{1}{2}\theta\right)$$
$$R^2 = (x - \xi)^2 + \cos^2\left(\tfrac{1}{2}\theta\right)$$

and

$$dS_2 = \tfrac{1}{2}\, d\xi\, d\theta$$

Here, use has been made of the fact that the surface of the hole is $r = \frac{1}{2}$. Now, dI can be expressed as

$$dI = \frac{1}{\pi} B(\xi) \cos^4\left(\frac{1}{2}\theta\right) \frac{d(\frac{1}{2}\theta)\, d\xi}{R^4}$$

The total incident flux at dS_1 is the integral of dI over the entire surface of the cylindrical hole

$$I(x) = \frac{1}{\pi} \int_0^\infty B(\xi) \left\{ \int_0^\pi \frac{\cos^4\left(\frac{1}{2}\theta\right)}{\left[(x - \xi)^2 + \cos^2\left(\frac{1}{2}\theta\right)\right]^2}\, d\left(\frac{1}{2}\theta\right) \right\} d\xi.$$

The integration over θ can be carried out, giving the result

$$I(x) = \int_0^\infty k(x - \xi)\, B(\xi)\, d\xi \qquad (10.3\text{-}3)$$

where the kernel is given by the expression

$$k(x - \xi) = 1 - |x - \xi| \frac{(x - \xi)^2 + \frac{3}{2}}{\left[(x - \xi)^2 + 1\right]^{3/2}}$$

Combining eqs. (10.3-2) and (10.3-3), we obtain an integral equation for the emitted radiative flux as a function of distance x from the entrance to the hole

$$B(x) = \epsilon \sigma T^4 + (1 - \epsilon) \int_0^\infty k(x - \xi)\, B(\xi)\, d\xi$$

Example 3: A boundary value problem involving two or three space dimensions can sometimes be reduced to a one- or two-dimensional problem by means of an integral equation formulation. For example, consider the propagation of small-amplitude, shallow water waves in an infinite sea of uniform depth d. For an incompressible fluid, the motion is governed by equations conserving mass and horizontal momentum. Consider waves generated by a point-source disturbance at (x_s, y_s), and suppose that the waves are monochromatic of angular frequency ω. Assuming time dependence of the form $e^{-i\omega t}$, we write the periodic velocity components, u and v, and wave height η as complex functions of x and y. The real part of the product of each with $e^{-i\omega t}$ represents the required real variable, i.e.

$$u = Re\{u^*(x, y)\, e^{-i\omega t}\}, \quad v = Re\{v^*(x, y)\, e^{-i\omega t}\}, \quad \eta = Re\{h^*(x, y)\, e^{-i\omega t}\}$$

The linearized equations governing the motion can be written as

$$-i\omega u^* - fv^* = -gh_x^*$$

$$-i\omega v^* + fu^* = -gh_y^*$$

$$u_x^* + v_y^* = \frac{i\omega}{d} h^*$$

where f is the Coriolis parameter and the subscripts indicate partial derivatives. We set $h^* = h_i^* + \phi(x, y)$, where h_i^* represents the incident wave which originates from a point-source at (x_s, y_s). It follows that the variable ϕ satisfies the inhomogeneous Helmholtz equation (with a point-source term represented by delta functions)

$$\nabla^2\phi + k^2\phi = A_i \delta(x - x_s)\, \delta(y - y_s)$$

where $k^2 = (\omega^2 - f^2)/gd > 0$, and of the two linearly independent solutions, we choose the one that satisfies the radiation condition at infinity. Now consider the scattering of waves by a cylindrical island of radius a, centered at the origin. If polar coordinates are used, the wave height at any point is determined by the equation

$$\nabla^2\phi + k^2\phi = A_i \frac{1}{r} \delta(r - r_s)\, \delta(\theta - \theta_s)$$

which is to be solved in the region $r \geqslant a$, subject to the radiation condition and the requirement of zero normal flow at $r = a$

$$-i\omega\phi_r + \frac{f}{a}\phi_\theta = 0 \qquad (r = a)$$

At this stage of the proceedings, an expression for ϕ could be found by separating variables. The final result is a rather cumbersome expression involving Bessel and Hankel functions.

Alternatively, consider the free-space Green's function that satisfies the equation

$$\nabla^2 g + k^2 g = \frac{1}{r}\delta(r - r_0)\, \delta(\theta - \theta_0)$$

and the radiation condition. The solution is well known (Ref. 10-9)

$$g(r, \theta; r_0, \theta_0) = -\frac{i}{4} H_0^{(1)}(kR)$$

where

$$R = [r^2 + r_0^2 - 2rr_0 \cos(\theta - \theta_0)]^{1/2}$$

The standard Green's function procedure now leads to the result

$$\phi(r_0, \theta) = -A_i \frac{i}{4} H_0^{(1)}(kR_s) - \int_0^{2\pi} \left[\frac{if}{\omega a} g\phi_\theta + \phi g_r\right] a\, d\theta \qquad (10.3\text{-}4)$$

where $R_s = R(r = r_s)$, and the boundary condition has been used to eliminate the normal derivative of ϕ from the integrand. Note that the inhomogeneous term represents the incident wave field at the point of observation (r_0, θ_0), while the integral represents the wave scattered by the island. Obviously, if ϕ were known on the boundary, the scattered wave field could be calculated. The variable $\phi(a, \theta)$ is, in fact, the solution of the integral equation obtained from eq. (10.3-4) by allowing the observation point to approach a point on the boundary. This integral equation involves only the independent variable θ, and the dimension of the problem is thereby reduced.

Example 4: There is a close relationship between ordinary differential and integral equations. We begin with a simple example where the solution is known, in order to illustrate some of these relationships in a transparent way.

Let $\phi(x)$ satisfy

$$\phi_{xx} - \lambda\phi = r(x) \qquad (10.3\text{-}5)$$

for x in $[a, b]$, with $\phi(a) = A$ and $\phi(b) = B$, where A, B, and $b > a$ are given real numbers. As before, λ is a parameter, and we take $r(x)$ to be some prescribed function of x. Define a Green's function $g(x; \xi)$ by

$$g_{xx} = \delta(x - \xi) \qquad (10.3\text{-}6)$$

for $a < \xi < b$, where $g(a; \xi) = g(b; \xi) = 0$. Here $\delta(x - \xi)$ is the usual delta function. In this simple problem, g may be found explicitly; it is

$$g(x, \xi) = -\frac{1}{b - a}\begin{cases}(x - a)(b - \xi), & a < x < \xi \\ (\xi - a)(b - x), & \xi < x < b\end{cases}$$

Next, multiply each side of eq. (10.3-5) by g, each side of eq. (10.3-6) by ϕ, subtract, and integrate with respect to x from a to b to obtain

$$\int_a^b (\phi_{xx}g - \phi g_{xx})\,dx = \lambda \int_a^b g\phi\,dx + \int_a^b rg\,dx - \phi(\xi)$$

or, since

$$\phi_{xx}g - \phi g_{xx} = (\phi_x g - \phi g_x)_x$$

$$[\phi_x g - \phi g_x]_a^b = \lambda \int_a^b g\phi\,dx + \int_a^b rg\,dx - \phi(\xi)$$

Rearranging, and inserting boundary values, we obtain the equation

$$\phi(\xi) = f(\xi) + \lambda \int_a^b g(x, \xi)\,\phi(x)\,dx$$

where

$$f(\xi) = Bg_x(b, \xi) - Ag_x(a, \xi) + \int_a^b r(x)\,g(x, \xi)\,dx$$

This is a Fredholm integral equation of the second kind. We note, of course, that $g_x(b; \xi)$ and $g_x(a; \xi)$ are obtainable from the explicit expression for g. Using the symmetry of g (for this problem), and interchanging the coordinates x and ξ, we find that the integral equation can be written in more conventional form as

$$\phi(x) = f(x) + \lambda \int_a^b g(x, \xi)\,\phi(\xi)\,d\xi \tag{10.3-7}$$

Corresponding to a given $r(x)$ (which determines the form of $f(x)$), eq. (10.3-7) will usually have a unique solution of a form to be discussed in the sequel, and this will coincide with the solution of eq. (10.3-5). We observe incidentally that the solution process for eq. (10.3-7) uses only that equation; it is not necessary to adjoin any boundary conditions satisfied by $\phi(x)$ (of course, boundary conditions for $\phi(x)$ were incorporated into the derivation of eq. (10.3-7)). A second observation is that integral equation theory singles out certain values of the parameter λ for which there may not be a solution. In this example, such a result corresponds to a well-known result in differntial equation theory. To illustrate, denote the eigenvalues and the corresponding eigenfunctions of the problem

$$\phi_{xx} - \lambda\phi = 0; \qquad \phi(a) = \phi(b) = 0$$

by λ_n and ϕ_n. Now suppose that in eq. (10.3-5) the parameter λ is equal to one of these, say $\lambda = \lambda_N$. Moreover, suppose that $A = B = 0$. Then differential equation theory states that eq. (10.3-5) will have a solution if and only if the prescribed function $r(x)$ is orthogonal to the corresponding eigenfunction $\phi_N(x)$, i.e., if $\int_a^b r\phi_N\,dx = 0$. For this case, eq. (10.3-7) reads

$$\phi(x) = f(x) + \lambda_N \int_a^b g(x, \xi)\,\phi(\xi)\,d\xi$$

where

$$f(x) = \int_a^b g(x, \xi)\,r(\xi)\,d\xi$$

Moreover, since $\phi_N(x)$ is a solution of the problem for the case $r(x) = 0$ (which implies $f(x) = 0$), we see that $\phi_N(x)$ must satisfy the homogeneous problem

$$\phi_N(x) = \lambda_N \int_a^b g(x, \xi)\,\phi_N(\xi)\,d\xi$$

Thus the exceptional value of λ for the differential equation problem is also an exceptional value for the integral equation problem, in the sense that the homogeneous equation possesses a nontrival solution. This example makes plausible the assertion that an integral equation such as (10.3-7) need not have a solution for all values of λ, and this we will later see to be the case.

Example 5: It seems clear that the use of a Green's function, and the "multiply and subtract" technique of Example (4), may be extended to more complicated situations. One example is afforded by the case of harmonic shallow water motion in a channel of constant depth but linearly varying width, which leads to an equation governing the velocity amplitude $\phi(x)$ of the form

$$L(\phi) \equiv (x\phi_x)_x - \frac{1}{x}\phi = -\lambda x\phi$$

for $0 < x < 1$, with $\phi(0) = \phi(1) = 0$. Here, λ is a parameter related to the wavelength.

Define $g(x, \xi)$ by

$$L(g) = \delta(x - \xi)$$

with $g = 0$ at $x = 0$, 1. Proceeding as in Example (4), we are led to the integral equation

$$\phi(\xi) = -\lambda \int_0^1 xg(x, \xi)\, \phi(x)\, dx \tag{10.3-8}$$

This homogeneous integral equation can be expected to have nontrivial solutions only for particular values of λ (specifically, the requirement is that $J_1(\sqrt{\lambda}) = 0$).

Although g is symmetric, the kernel xg in this equation is not. Since integral equations with symmetric kernels usually have simpler properties, it is useful to note that the device of multiplication by $\sqrt{\xi}$ will replace this last integral equation by one with a symmetric kernel, in terms of a new unknown function $\sqrt{x}\, \phi(x)$

$$\sqrt{\xi}\, \phi(\xi) = \lambda \int_0^1 \sqrt{x\xi}\, g(x, \xi)\, [\sqrt{x}\, \phi(x)]\, dx$$

Example 6: Differential equations of initial value type can frequently be recast as Volterra integral equations. To illustrate, a certain model of phytoplankton growth leads to the equation

$$\phi_{xx} + (\lambda^2 e^{-x} - \tfrac{1}{4})\, \phi = 0$$

to be solved subject to the initial conditions $\phi(0) = 1$, $\phi'(0) = 0$. One integration from 0 to x gives

$$\phi_x + \int_0^x (\lambda^2 e^{-\xi} - \tfrac{1}{4})\, \phi(\xi)\, d\xi = 0$$

(where $\phi'(0) = 0$ has been used to set the constant of integration equal to 0), and a second integration gives

$$\phi(x) + \int_0^x \left[\int_0^\eta (\lambda^2 e^{-\xi} - \tfrac{1}{4})\, \phi(\xi)\, d\xi \right] d\eta = 1$$

Integrating by parts, we obtain the Volterra equation

$$\phi(x) + \int_0^x (x - \xi)(\lambda^2 e^{-\xi} - \tfrac{1}{4})\, \phi(\xi)\, d\xi = 1 \tag{10.3-9}$$

Here, the parameter λ plays a somewhat different role than in the case of a Fredholm equation. It turns out that if ϕ is to be bounded, then λ must be a root of the equation

$$2\lambda J_0(2\lambda) - J_1(2\lambda) = 0$$

10.4 NONSINGULAR LINEAR INTEGRAL EQUATIONS

Solutions of nonsingular linear integral equations of the Fredholm and Volterra type can be expressed in various forms. In any given problem, the form of the solution is determined by the ease with which the expression can be established. Inhomogeneous linear equations of the second kind are solved analytically by three different standard procedures: (1) the method of Fredholm, which gives $\phi(x)$ in terms of an integral of the ratio of two power series in λ; (2) successive substitution; and (3) the Hilbert-Schmidt method in which $\phi(x)$ is expressed as a series involving the eigenfunctions $\phi_n(x)$ of the corresponding homogeneous equation. We begin with Fredholm's method for solving eq. (10.2-4).

10.4.1 Fredholm's Determinant, First Minor, the Fundamental Fredholm Relations

Consider the possibility of finding an approximate solution of the inhomogeneous Fredholm equation of the second kind, eq. (10.2-4), by replacing the integral with a finite sum. Subdivide the range $[a, b]$ into $n - 1$ equal segments, each of length δ; denote the jth mesh point by ξ_j, with $j = 1, 2, \ldots, n$. A natural approximation of eq. (10.2-4) is the form

$$\phi_i - \lambda\delta \sum_{j=1}^{n} k_{ij}\phi_j = f(x_i) \qquad (i = 1, 2, \ldots, n) \tag{10.4-1}$$

where ϕ_i is an approximation to $\phi(x_i)$, and k_{ij} is the value of the kernel at the point (x_i, ξ_j) in the rectangle R. The system (10.4-1) will have a unique solution for ϕ_i if the determinant of coefficients $D_n(\lambda)$ for ϕ_i is nonzero. The element in the ith row and jth column of that determinant is $\delta_{ij} - \lambda\delta k_{ij}$, where δ_{ij} is the Kronecker delta. In the arguments that follow, the kernel $k(x, \xi)$ is assumed to be a real function of x and ξ, and it is either continuous throughout the rectangle $R\ (a \leqslant x, \xi \leqslant b)$ or it is continuous in the subdomain $a \leqslant \xi \leqslant x \leqslant b$ and equal to zero when $\xi > x$. (It is possible to extend most of the developments in this section to complex functions of real variables and to relax the requirements on continuity to some degree.) The determinant of coefficients $D_n(\lambda)$ can be expanded in powers of λ

$$D_n(\lambda) = \begin{vmatrix} 1 - \lambda\delta k_{11} & -\lambda\delta k_{12} \cdots & -\lambda\delta k_{1n} \\ -\lambda\delta k_{21} & 1 - \lambda\delta k_{22} \cdots & -\lambda\delta k_{2n} \\ & \cdots & \\ -\lambda\delta k_{n1} & \cdots & 1 - \delta\lambda k_{nn} \end{vmatrix}$$

$$= 1 - \lambda \sum_{i=1}^{n} \delta k_{ii} + \frac{\lambda^2}{2!} \sum_{i=1}^{n} \sum_{j=1}^{n} \delta^2 \begin{vmatrix} k_{ii} & k_{ij} \\ k_{ji} & k_{jj} \end{vmatrix} - \cdots$$

Returning to the linear system (10.4-1) with the determinant of coefficients $D_n(\lambda)$ given above, we denote by $D_n(x_i, x_j; \lambda)$ the cofactor in $D_n(\lambda)$ of the term involving k_{ij}. Again, expanding in powers of λ, we have

$$D_n(x_i; x_j; \lambda) = \lambda\delta k_{ij} - \lambda^2 \sum_{p=1}^{n} \delta^2 \begin{bmatrix} k_{ij} & k_{ip} \\ k_{pj} & k_{pp} \end{bmatrix} + \cdots$$

Moreover, if $D_n(\lambda) \neq 0$, the linear system for ϕ_i has the solution

$$\phi_i = \frac{1}{D_n(\lambda)} \sum_{j=1}^{n} D_n(x_i, x_j; \lambda) f(x_j)$$

This expression strongly suggests that if we let $n \to \infty$ and $\delta \to 0$, the limiting form may provide a solution of the inhomogeneous equation. That this is indeed the case was first shown by Fredholm.

The limiting forms of $D_n(\lambda)$ and $D_n(x, \xi; \lambda)$ are readily obtained from the definitions and are expressible as power series in λ. Thus, we have for $D(\lambda)$ the expansion

$$D(\lambda) = 1 + \sum_{j=1}^{\infty} a_j \lambda^j \tag{10.4-2}$$

where

$$a_j = \frac{(-1)^j}{j!} \int_a^b \underset{(j)}{\cdots} \int_a^b \begin{bmatrix} k(\xi_1, \xi_1) \cdots k(\xi_1, \xi_j) \\ \cdots \\ k(\xi_j, \xi_1) \cdots k(\xi_j, \xi_j) \end{bmatrix} d\xi_1\, d\xi_2 \cdots d\xi_j$$

Denote the determinant in the integrand by Δ_j. By assumption, the kernel $k(x, \xi)$ is continuous and therefore bounded in R by a constant M (say). Since the kernels are real, so are the determinants Δ_j, and it can be inferred from Hadamard's theorem that $|\Delta_j| \leqslant M^j j^{j/2}$. An upper bound can now be established for a_j; specifically

$$|a_j| \leqslant \frac{M^j}{j!} j^{j/2} (b-a)^j$$

and therefore eq. (10.4-2) is an absolute and uniformly convergent series in integer powers of λ. $D(\lambda)$ is called *Fredholm's determinant*, or simply the determinant of $k(x, \xi)$.

In like manner, the limiting form of $\delta^{-1} D_n(x_i, \xi_j; \lambda)$ is expressible as a series in powers of λ; for all x_i and ξ_j in R, we obtain

$$D(x, \xi; \lambda) = \lambda k(x, \xi) + \sum_{j=1}^{\infty} b_j(x, \xi) \lambda^{j+1} \tag{10.4-3}$$

where

$$b_j(x,\xi) = \frac{(-1)^j}{j!} \int_a^b \underset{(j)}{\cdots} \int_a^b \begin{bmatrix} k(x,\xi) & k(x,\xi_1) & \cdots k(x,\xi_j) \\ k(\xi_1,\xi) & k(\xi_1,\xi_1) & \cdots k(\xi_1,\xi_j) \\ & \cdots & \\ k(\xi_j,\xi) & & \cdots k(\xi_j,\xi_j) \end{bmatrix} d\xi_1\, d\xi_2 \cdots d\xi_j$$

For real nonsingular kernels, the series is absolutely and uniformly convergent in integer powers of λ, for all (x, ξ) in R. The quantity $D(x, \xi; \lambda)$ is called *Fredholm's first minor*.

Referring to the determinant in the integrand of the expression for $b_j(x, \xi)$, we note that the minor of $k(x, \xi)$ is the determinant in the integrand of the expression for a_j (eq. (10.4-2)). After some determinant manipulations, which follow naturally from that observation (see, for example, Ref. 10-5), one arrives at an expression of the form

$$D(x,\xi;\lambda) = \lambda D(\lambda)\, k(x,\xi) + \sum_{j=1}^{\infty} g_j(x,\xi)\lambda^{j+1} \tag{10.4-4}$$

where

$$g_j(x,\xi) = \frac{(-1)^{j+1}}{(j-1)!} \int_a^b \underset{(j)}{\cdots} \int_a^b k(\xi_0,\xi)$$

$$\begin{bmatrix} k(x,\xi_0) & k(x,\xi_1) \cdots k(x,\xi_{j-1}) \\ k(\xi_1,\xi_0) & k(\xi_1,\xi_{j-1}) \\ & \cdots \\ k(\xi_{j-1},\xi_0) & k(\xi_{j-1},\xi_{j-1}) \end{bmatrix} d\xi_0\, d\xi_1 \cdots d\xi_{j-1}$$

A comparison of eqs. (10.4-3) and (10.4-4) leads immediately to the important result

I. $$D(x,\xi;\lambda) = \lambda D(\lambda)\, k(x,\xi) + \lambda \int_a^b k(\xi_1,\xi)\, D(x,\xi_1;\lambda)\, d\xi_1 \tag{10.4-5}$$

which is *Fredholm's first fundamental relation.* A related result is obtained by a different expansion of the integrand in eq. (10.4-3)

II. $$D(x,\xi;\lambda) = \lambda D(\lambda)\, k(x,\xi) + \lambda \int_a^b k(x,\xi_1)\, D(\xi_1,\xi;\lambda)\, d\xi_1 \tag{10.4-6}$$

This is *Fredholm's second fundamental relation*; the two relations form the theoretical basis of Fredholm's solution of the inhomogeneous integral equation that bears his name. A nontrivial consequence of the fundamental relations is the expression

$$\int_a^b D(x, x; \lambda)\, dx = -\lambda \frac{dD(\lambda)}{d\lambda} \tag{10.4-7}$$

The Fredholm relations suggest a particular form for the solution of the inhomogeneous Fredholm equation. Thus, suppose the coordinates x and ξ in eq. (10.2-4) are replaced by ξ and ξ_1, respectively, so that, with a minor rearrangement of terms, the equation reads

$$f(\xi) = \phi(\xi) - \lambda \int_a^b k(\xi, \xi_1)\, \phi(\xi_1)\, d\xi_1 \tag{10.4-8}$$

Now if the parameter λ is such that $D(\lambda) \neq 0$, we can multiply eq. (10.4-8) by $D(x, \xi; \lambda)/D(\lambda)$ and integrate from a to b. Using the first of Fredholm's relations, we are led to the result

$$\int_a^b f(\xi) \frac{D(x, \xi; \lambda)}{D(\lambda)}\, d\xi = \lambda \int_a^b k(x, \xi)\, \phi(\xi)\, d\xi$$

$$= \phi(x) - f(x)$$

Hence, if Fredholm's equation has a solution, it can be expressed in the form

$$\phi(x) = f(x) + \int_a^b \frac{D(x, \xi; \lambda)}{D(\lambda)} f(\xi)\, d\xi \tag{10.4-9}$$

It may be verified by substitution of eq. (10.4-9) into eq. (10.2-4) that this last form is indeed a solution of the inhomogeneous Fredholm equation of the second kind.

10.4.2 The Homogeneous Fredholm Equation

An important corollary of the previous considerations is that the homogeneous equation (obtained by setting $f(x) \equiv 0$) has no continuous solution except $\phi(x) = 0$, unless the parameter λ is such that the Fredholm determinant is zero. (When the parameter λ is not a root of $D(\lambda) = 0$, it is said to be a *regular value*.) However, if λ_1 is a simple root of $D(\lambda) = 0$, and $D(x, \xi_0; \lambda_1) \neq 0$ for some value of ξ_0 in $[a, b]$, then by comparison with the second of the Fredholm relations, eq. (10.4-6), the function

$$\phi(x) = D(x, \xi_0; \lambda_1)$$

is a continuous nonzero solution of the homogeneous equation

$$\phi(x) = \lambda_1 \int_a^k k(x, \xi)\, \phi(\xi)\, d\xi \tag{10.4-10}$$

More generally, suppose that λ_1 is a root of $D(\lambda) = 0$ of multiplicity r and that $D(x, \xi; \lambda_1) \neq 0$ for all (x, ξ) in R. Since $D(x, \xi; \lambda)$ is an integral function of λ, it can be expanded in a Taylor series about λ_1

$$D(x, \xi; \lambda) = \sum_{n=m}^{r} c_n(x, \xi)\, (\lambda - \lambda_1)^n \qquad (m \geqslant 1) \tag{10.4-11}$$

If we consider $D(x, \xi; \lambda)$ to be a function of the complex variable λ, analytic in the neighborhood of λ_1, then by Cauchy's integral formula, the coefficients of the Taylor series expansion are given by

$$c_n(x, \xi) = \frac{1}{2\pi i} \int_\gamma \frac{D(x, \xi; \lambda)}{(\lambda - \lambda_1)^{n+1}}\, d\lambda$$

where the contour γ is a simple closed curve enclosing the pole at λ_1. Suppose that λ is taken to be a circle of radius ρ centered at λ_1 and that M is an upper bound of $|D(x, \xi; \lambda)|$ for λ on γ. It follows that

$$|c_n(x, \xi)| \leqslant M\rho^{-n}$$

and hence the series (10.4-11) is absolutely and uniformly convergent for $|\lambda - \lambda_1| < \rho$. Substitution of the series (10.4-11) into eq. (10.4-7) for $D(\lambda)$ demonstrates that $D'(\lambda) = 0$ has a root λ_1 of order $r \geqslant m$. Consequently, $D(\lambda)$ has a zero at λ_1 at least of order $m + 1$, and $D(\lambda)$ has therefore an expansion of the form

$$D(\lambda) = a'_{m+1}(\lambda - \lambda_1)^{m+1} + a'_{m+2}(\lambda - \lambda_1)^{m+2} + \cdots$$

If this last expression is substituted into eq. (10.4-7), together with the expansion for $D(x, \xi; \lambda)$, and the coefficients of like powers of $\lambda - \lambda_1$ are equated, one obtains

$$c_n(x, \xi) = \lambda_1 \int_a^b k(x, \xi_1)\, c_n(\xi_1, \xi)\, d\xi_1$$

Now, if there exist values of $\xi = \xi_0$ in $[a, b]$ such that $c_n(x, \xi_0) \neq 0$, it follows that

$$\phi_n(x) = c_n(x, \xi_0)$$

is an eigenfunction associated with the eigenvalue λ_1.

By way of summary to this point, if λ_1 is a simple root of $D(\lambda) = 0$, then λ_1 is an eigenvalue that has associated with it a single eigenfunction $\phi_1(x)$. If λ_1 is a root of $D(\lambda) = 0$ of multiplicity, r, then there exist $q\,(\leqslant r)$ linearly independent eigenfunctions $\phi_{11}(x), \phi_{21}(x), \ldots, \phi_{q1}(x)$ associated with λ_1. In other words, there corresponds to each eigenvalue a finite number of linear independent eigenfunctions. Each eigenvalue of the homogeneous equation corresponds to a pole of the ratio $D(x, \xi; \lambda)/D(\lambda)$. Moreover, this ratio is a meromorphic function of λ in the complex λ-plane, because only a finite number of poles lie within any circle of finite radius ρ. It can also be proved that if the Fredholm determinant has an infinite number of roots, then the modulus of successive values of λ_n approach infinity.

When the kernel is complex, these same remarks hold for the *transposed* homogeneous equation defined by

$$\psi(x) = \bar{\lambda} \int_a^b \overline{k(\xi, x)}\, \psi(\xi)\, d\xi$$

Moreover, if λ_1 is an eigenvalue of $k(x, \xi)$, where k is complex, then $\overline{\lambda_1}$ is an eigenvalue of the associated kernel $\overline{k(\xi, x)}$. In fact, the homogeneous equation [eq. (10.4-10)] and the transposed equation have the same number of eigenvalues.

Kernels for which eigenvalues and eigenfunctions have been determined are limited in number (the same can be said for differential equations as well). In many instances, however, the Fredholm determinant is a terminating polynomial whose roots are easily established. Table I displays selected elementary kernels, associated intervals $[a, b]$, and corresponding Fredholm determinants from which eigenvalues and eigenfunctions are readily calculated. Although kernels of the simple types listed in the table are not often encountered in practice, they can be useful as approximating kernels.

More frequently, $D(\lambda)$ is expressible as an infinite series in powers of λ, and, in such cases, the number of eigenvalues is usually infinite. (The Heaviside step function $U(x - \xi)$ in the range $[0, 1]$ is an example of a kernel with no real eigenvalues, although $D(\lambda) = e^{-\lambda}$ is expressible as a nonterminating series of integral powers of λ.)

10.4.3 The Inhomogeneous Fredholm Equation of the Second Kind

Three different standard procedures have been developed to solve linear inhomogeneous nonsingular integral equations of the form of eq. (10.2-4)

$$\phi(x) = f(x) + \lambda \int_a^b k(x, \xi)\, \phi(\xi)\, d\xi \tag{10.2-4}$$

where $f(x) \neq 0$. The first method, due to Fredholm, has already been previewed in Section 10.4.1; it gives the solution in terms of a ratio of two power series in λ,

Table I Representative Fredholm Determinants of Simple Form.

$k(x, \xi)$	Interval	$D(\lambda)$
$k(x)$ or $k(\xi)$	$[a, b]$	$1 - \lambda \int_a^b k(\xi)\, d\xi$
$x + \xi$	$[0, b]$	$1 - \lambda b^2 - \frac{1}{12}\lambda^2 b^4$
$x^2 + \xi^2$	$[0, b]$	$1 - \lambda\left(\frac{2b^3}{3}\right) - \lambda^2\left(\frac{4b^6}{45}\right)$
$x\xi + \xi^2$	$[0, b]$	$1 - \lambda\left(\frac{2b^3}{3}\right) - \lambda^2\left(\frac{b^6}{72}\right)$
$x\xi$	$[0, b]$	$1 - \lambda\left(\frac{b^3}{3}\right)$
$\sin x \sin \xi$	$[0, b]$	$1 - \lambda\left(\frac{b}{2} - \frac{\sin 2b}{4}\right)$
$e^{x+\xi}$	$[0, b]$	$1 - \lambda\left(\frac{e^{2b}}{2} - \frac{1}{2}\right)$
$k_1(x)\, k_2(\xi)$	$[a, b]$	$1 - \lambda \int_a^b k_1(\xi)\, k_2(\xi)\, d\xi$

each having an infinite radius of convergence under rather general conditions. The series in the numerator involves x, but the denominator is independent of x. The second method, originally due to Liouville (1837) and later elaborated by Neumann (1870), involves successive substitutions and leads to a form that is related to and generalized by the theory of Fredholm. A third method due to Hilbert and Schmidt gives $\phi(x)$ as a series involving the eigenfunctions $\phi_n(x)$ of the corresponding homogeneous equation; that is, a solution is developed in the form

$$\phi(x) = \sum_{n=1}^{\infty} c_n \phi_n(x) \tag{10.4-12}$$

The method of successive substitutions is simple in concept, yielding a series in integral powers of λ, but it often involves laborious integrations. The Fredholm and Hilbert-Schmidt methods are somewhat more elaborate, but the forms of the solutions are at the same time more generally valid. In each case, the presentation below will be brief, for the procedures draw largely on the results developed in the previous sections.

10.4.3.1 The Fredholm Method Let $f(x)$ be a given nonzero continuous function in $[a, b]$. If $k(x, \xi)$ is continuous in R, and if the parameter λ is not a zero of the Fredholm determinant $D(\lambda)$, then eq. (10.2-4) has a unique continuous solution given by

$$\phi(x) = f(x) + \lambda \int_a^b \frac{D(x, \xi; \lambda)}{D(\lambda)} f(\xi)\, d\xi \tag{10.4-13}$$

As shown in section 10.4.1, this result is readily obtained by multiplying the original equation by $D(x, \xi; \lambda)\, d\xi$, integrating from a to b, and using the first of Fredholm's fundamental relations. Here, $D(\lambda)$ and $D(x, \xi; \lambda)$ are the series given by eqs. (10.4-2) and (10.4-3), respectively. Both series are absolutely convergent for all λ and $D(x, \xi; \lambda)$ converges uniformly in R.

If λ_1 is a root of order $r \geqslant 1$ of $D(\lambda) = 0$, then the inhomogeneous equation has a continuous solution if and only if $f(x)$ is orthogonal to all the eigenfunctions of the transposed homogeneous equation. Thus, if $D(\lambda_1) = 0$, and if the kernel is (generally) complex, the transposed homogeneous equation

$$\psi(x) = \overline{\lambda_1} \int_a^b \overline{k(\xi, x)}\, \psi(\xi)\, d\xi \tag{10.4-14}$$

will be satisfied by a complete set of r eigenfunctions $\psi_j(x), j = 1, 2, \ldots, r$. If the function $f(x)$ is orthogonal to all the $\psi_j(x)$, then the inhomogeneous equation has solutions of the form

$$\phi(x) = f(x) + \lambda \int_a^b \frac{D^*(x, \xi, x_1, \xi_1, \ldots, x_r, \xi_r; \lambda_1)}{D^*(x_1, \xi_1, \ldots, x_r, \xi_r; \lambda_1)} f(\xi)\, d\xi + \sum_{k=1}^{r} c_j \phi_j(x)$$

where c_j are arbitrary constants, $(x_1, \xi_1, \ldots, x_r, \xi_r)$ are constants in $[a, b]$, and where the numerator and denominator of the ratio in the integrand are generalizations of the Fredholm first minor and determinant, respectively. As a practical matter, however, the Fredholm form in this case is too complicated to be of much use. The reader may consult Ref. 10-10 for further discussion.

Example: Consider the equation

$$\phi(x) = \tfrac{1}{2} + \sin(2x) + \pi^{-1} \int_0^\pi \cos^2(x - \xi)\, \phi(\xi)\, d\xi$$

We first replace π^{-1} by λ and calculate the Fredholm determinant: $D(\lambda) = (1 - \frac{1}{2}\pi\lambda)(1 - \frac{1}{4}\pi\lambda)^2$. Because π^{-1} is not a zero of $D(\lambda)$, the solution is given by eq.

(10.4-13). In this case, the first minor can be manipulated into the form

$$D(x, \xi; \pi^{-1}) = D(\pi^{-1})\,[1 + \cos 2(x - \xi)]$$

where $D(\pi^{-1}) = \frac{9}{32}$; the solution of the integral equation is

$$\phi(x) = 1 + \tfrac{4}{3} \sin 2x$$

Example: In the case of a degenerate kernel, the Fredholm equation can be written in the form

$$\phi(x) = f(x) + \lambda \sum_{i=1}^{N} \alpha_i(x)\, c_i$$

where

$$c_i = \int_a^b \beta_i(\xi)\, \phi(\xi)\, d\xi$$

Define the auxiliary quantities

$$k_{ij} = \int_a^b \alpha_i(x)\, \beta_j(x)\, dx \qquad \text{and} \qquad g_j = \int_a^b \beta_j f(x)\, dx$$

Multiplying the original integral equation by $\beta_j(x)\,dx$ and integrating, we obtain a system of linear equations for c_j

$$c_i - \lambda \sum_{i=1}^{N} k_{ij} c_i = g_j \qquad j = 0, 1, \ldots, N$$

The determinant of coefficients is the Fredholm determinant

$$D(\lambda) = \begin{bmatrix} 1 - \lambda k_{11} & -\lambda k_{12} \cdots & -\lambda k_{1N} \\ -\lambda k_{21} & 1 - \lambda k_{22} & \\ & \cdots\cdots\cdots & \\ -\lambda k_{N1} & \cdots & 1 - \lambda k_{NN} \end{bmatrix}$$

and when $D(\lambda) \neq 0$, the system can be solved for all c_j in a straightforward way, because the values of g_j are known. In the first example considered above, the kernel can be expressed as a three-term degenerate kernel by using the appropriate trigonometric identities.

10.4.3.2 Method of Successive Substitution Again, suppose the kernel $k(x, \xi)$ is real and continuous in R, with absolute upper bound M. Let $f(x)$ be a nonzero continuous function in $[a, b]$, and let λ be a regular value. Successive substitution into eq. (10.2-4) of the expression for ϕ itself leads to a series of the form

$$\phi(x) \doteq f(x) + \lambda \int_a^b k(x, \xi) f(\xi)\, d\xi + \lambda^2 \int_a^b \int_a^b k(x, \xi)\, k(\xi, \xi_1) f(\xi_1)\, d\xi_1\, d\xi$$

$$+ \cdots + \lambda^N \int_a^b \underset{(N)}{\cdots} \int_a^b k(x, \xi)\, k(\xi, \xi_1) \ldots k(\xi_{N-2}, \xi_{N-1}) f(\xi_{N-1})\, d\xi_{N-1} \ldots d\xi$$

(10.4-15)

Under the stated conditions, the solution is continuous and unique, and the series converges absolutely and uniformly in R for values of λ such that

$$|\lambda| < M^{-1}(b - a)^{-1}$$

As the kernel is required to be continuous, orders of integration can be interchanged in each term of the series (10.4-15), thereby allowing the introduction of the iterated kernels $k_n(x, \xi)$, to give the more compact result

$$\phi(x) \doteq f(x) + \sum_{n=1}^{N} \lambda^n \int_a^b k_n(x, \xi) f(\xi)\, d\xi + R_N(x)$$

As N is increased without limit, $R_N(x)$ approaches zero, and the solution can be expressed in terms of the resolvent kernel $\Gamma(x, \xi; \lambda)$

$$\phi(x) = f(x) + \lambda \int_a^b \Gamma(x, \xi; \lambda) f(\xi)\, d\xi \qquad (10.4\text{-}16)$$

where

$$\Gamma(x, \xi; \lambda) = \sum_{n=1}^{\infty} \lambda^{n-1} k_n(x, \xi)$$

Expansions of the form (10.4-16) generally have radii of convergence greater than $M^{-1}(b - a)^{-1}$, as can be demonstrated both by simple examples and by analytic continuation. A comparison of eqs. (10.4-13) and (10.4-16) suggests the identification

$$\Gamma(x, \xi; \lambda) = \frac{D(x, \xi; \lambda)}{D(\lambda)}$$

In fact, the ratio $D(x, \xi; \lambda)/D(\lambda)$ represents an analytic continuation of the resolvent kernel $\Gamma(x, \xi; \lambda)$ and is meromorphic throughout the complex λ-plane.

An inhomogeneous Volterra equation can also be solved by successive substitution. When $k(x, \xi)$ and $f(x)$ satisfy the same conditions as stated above, the method of successive substitution give the series

$$\phi(x) \doteq f(x) + \lambda \int_a^x k(x, \xi) f(\xi)\, d\xi + \lambda^2 \int_a^x k(x, \xi) \int_a^\xi k(\xi, \xi_1)\, d\xi_1\, d\xi$$

$$\cdots + \lambda^N \int_a^x k(x, \xi) \int_a^\xi k(\xi, \xi_1) \ldots \int_a^{\xi_{N-1}} k(\xi_{N-2}, \xi_{N-1})\, d\xi_{N-1} \ldots d\xi_1\, d\xi$$

Here, the series is absolutely and uniformly convergent and represents a unique solution.

10.4.3.3 The Hilbert-Schmidt Method for Symmetric Kernels When the kernel is symmetric ($k(x, \xi) = \overline{k(\xi, x)}$ if it is complex), the corresponding eigenvalues and eigenfunctions are characterized by many useful properties, and the Fredholm equations have solution of elegant form. For the most part, this discussion will assume the kernel is real.

Before stating the theorem that is central to the Hilbert-Schmidt method, we record a few definitions and lemmas that provide essential background to the theorem. Consider a system of functions $\phi_n(x)$, orthonormal over the range $[a, b]$ and a function $f(x)$ that is $L_2\,[a, b]$. If, for any positive number ϵ, there exists a corresponding number M, such that

$$P_N \equiv \int_a^b \left[f(x) - \sum_{n=1}^{N} a_n \phi_n(x) \right]^2 dx < \epsilon \qquad \text{for } N > M$$

we say that the series in the integrand *converges in the mean* to $f(x)$ and write

$$f(x) = \underset{N\to\infty}{\text{l.i.m.}} \sum_{n=1}^{N} a_n \phi_n(x) \tag{10.4-17}$$

The quantity P_N is a minimum if and only if the coefficients a_n are the *Fourier coefficients* of $f(x)$ defined by

$$a_n \equiv \int_a^b f(x)\, \phi_n(x)\, dx$$

A function $f(x)$ that is $L_2\,[a, b]$ can be represented in the form

$$f(x) = \sum_{n=1}^{\infty} a_n \phi_n(x)$$

where $\{\phi_n(x)\}$ is a given orthonormal set on $[a, b]$, if and only if Parseval's equation holds

$$\int_a^b f^2(x)\,dx = \sum_{n=1}^{\infty} a_n^2$$

Moreover, if Parseval's equation holds for any function $f(x)$ in L_2 (or in a subset of L_2), then the orthonormal system S of functions $\phi_n(x)$ is said to be *complete* with respect to $f(x)$ in L_2 (or in the corresponding subset of L_2). If Parseval's equation holds for two different functions $f_1(x)$ and $f_2(x)$, then it is satisfied by an linear combination of $f_1(x)$ and $f_2(x)$. A system of $L_2[a, b]$ functions is said to be *closed* in $L_2[a, b]$ if there is no other function that is orthogonal to all the members of the system S.

From the definition of the iterated kernels and the recursion relations, eqs. (10.2-12) and (10.2-13), it is easily seen that when the kernels are symmetric, the following relation holds

$$k_{m+n}(x, x) = \int_a^b k_m(x, \xi)\,k_n(x, \xi)\,d\xi$$

The *trace* A_n of the symmetric kernel $k(x, \xi)$ is defined by the expression

$$A_n \equiv \int_a^b k_n(x, x)\,dx \qquad (n = 1, 2, \ldots) \tag{10.4-18}$$

where $k_1(x, \xi) \equiv k(x, \xi)$. From the definition and the recursion relation above, it can be shown that

$$A_{2n}^2 \leqslant A_{2n-2}A_{2n+2} \qquad (n = 2, 3, \ldots)$$

and hence the traces of even index are all positive. Moreover, division of each side of this last inequality by $A_{2n}A_{2n+2}$ gives

$$\frac{A_{2n}}{A_{2n+2}} \leqslant \frac{A_{2n-2}}{A_{2n}}$$

which must approach a finite limit as n increases without bound. Clearly, the limit value of the sequence of ratios is the radius of convergence of the series

$$\sum_{n=1}^{\infty} A_{2n}\lambda^{2n}$$

and furthermore the radius of convergence is equal to the square of the modulus of the lowest eigenvalue of the kernel whose traces of even index are A_{2n}; that modulus is

$$|\lambda_1| \leqslant \sqrt{\frac{A_{2n}}{A_{2n+2}}} \qquad \text{for } n = 1, 2, \ldots \tag{10.4-19}$$

Hence, we have the important result that if the L_2 symmetric kernel under consideration is nonzero, then there exists at least one eigenvalue λ_1 whose modulus satisfies the inequality above (see Ref. 10-2 for a more rigorous proof of this conclusion). The ambiguity in the sign of λ_1 is removed if the kernel is positive-definite, because, in such case, the eigenvalues are all positive. Moreover, if the eigenvalues are well separated at the lower end of the spectrum, then eq. (10.4-19) can provide an estimate of $|\lambda_1|$ that is sufficiently accurate for numerical iteration.

We proceed now to the construction of a sequence of *shortened kernels*, which process ultimately leads to an expansion theorem for symmetric $k(x, \xi)$ and thence to the Hilbert-Schmidt theorem. Since every symmetric L_2 kernel has at least one eigenvalue λ_1 with a corresponding normalized eigenfunction ϕ_1, we may consider a second shortened symmetric kernel

$$k^{(2)}(x, \xi) = k(x, \xi) - \frac{\phi_1(x)\,\phi_1(\xi)}{\lambda_1}$$

If this kernel is nonzero, it will have at least one eigenvalue λ_2 and a corresponding normalized eigenfunction ϕ_2. Although λ_2 may be equal to λ_1, the eigenfunctions will not be the same. This process may be repeated, generating thereby a sequence of shortened kernels, the $(N + 1)$th member of the sequence being given by

$$k^{(N+1)}(x, \xi) = k(x, \xi) - \sum_{n=1}^{N} \frac{\phi_n(x)\,\phi_n(\xi)}{\lambda_n}$$

If there exists a value of N for which $k^{(N+1)}(x, \xi) = 0$, the process is terminated, and the symmetric kernel $k(x, \xi)$ is of degenerate form and has only a finite number of eigenvalues

$$k(x, \xi) = \sum_{n=1}^{N} \frac{\phi_n(x)\,\phi_n(\xi)}{\lambda_n} \tag{10.4-20}$$

Equation (10.4-20) is called the *bilinear formula* of the symmetric kernel $k(x, \xi)$. If the construction can be continued indefinitely, the bilinear formula implies the existence of an infinite number of eigenvalues. In either case, the eigenvalues may be considered ordered in absolute value, so that $|\lambda_1| \leqslant |\lambda_2| \leqslant \cdots$.

The bilinear formula cannot immediately be extended to an infinite number of terms since the sum may not converge. However, if the series

$$\sum_{n=1}^{\infty} \frac{\phi_n(x)\,\phi_n(\xi)}{\lambda_n}$$

converges uniformly, or if to any positive number ϵ there corresponds a positive integer M such that

$$\int_a^b \int_a^b \left[\sum_{n=N+1}^{\infty} \frac{\phi_n(x)\,\phi_n(\xi)}{\lambda_n}\right]^2 dx\, d\xi < \epsilon \qquad \text{for } N > M$$

then the symmetric kernel with eigenvalues λ_n and orthonormal eigenfunctions $\phi_n(x)$ has the expansion

$$k(x, \xi) = \sum_{n=1}^{\infty} \frac{\phi_n(x)\,\phi_n(\xi)}{\lambda_n} \tag{10.4-21}$$

If $k(x, \xi)$ is a definite symmetric kernel, continuous over the rectangle R or if the kernel has a finite number of eigenvalues, all of the same sign, we are assured that the expansion (10.4-21) is absolutely and uniformly convergent in R. Moreover, the Riesz-Fisher theorem and Bessel's inequality can be used to show that the bilinear formula converges in the mean to the L_2 symmetric kernel $k(x, \xi)$ in the sense of eq. (10.4-17) even if the system $\{\phi_n\}$ is incomplete.

Another consequence of the foregoing developments is relevant to the Hilbert-Schmidt theorem: in particular, any function $f(x)$ that is $L_2\,[a, b]$ is orthogonal to a definite symmetric kernel if and only if

$$\int_a^b k(x, \xi) f(\xi)\, d\xi = 0 \tag{10.4-22}$$

For if $\phi_n(x)$ is any orthogonal eigenfunction of the definite symmetric kernel $k(x, \xi)$, then we can write

$$\int_a^b \phi_n(x)\left[\int_a^b k(x, \xi) f(\xi)\, d\xi\right] dx = \int_a^b f(\xi)\left[\int_a^b k(x, \xi)\,\phi_n(x)\, dx\right] d\xi$$

$$= \lambda_n^{-1} \int_a^b f(\xi)\,\phi_n(\xi)\, d\xi = 0$$

provided eq. (10.4-22) holds. Conversely, if $f(\xi)$ is orthogonal to $\phi_n(x)$ for $n = 1, 2, \ldots,$ then for any integer N

$$\int_a^b k(x, \xi) f(\xi)\, d\xi = \int_a^b \left[k(x, \xi) - \sum_{n=1}^{N} \frac{\phi_n(x)\,\phi_n(\xi)}{\lambda_n}\right] f(\xi)\, d\xi$$

Hence, for N sufficiently large, we have

$$\left[\int_a^b k(x, \xi) f(\xi)\, d\xi\right]^2 \leqslant \left\{\int_a^b \left[k(x, \xi) - \sum_{n=1}^{N} \frac{\phi_n(x)\,\phi_n(\xi)}{\lambda_n}\right]^2 d\xi\right\}\left\{\int_a^b f^2(\xi)\, d\xi\right\}$$

Since the series in the integrand of the first factor on the right converges in the mean under the stated conditions, the right-hand side of this last inequality can be made arbitrarily small by choosing N sufficiently large. This last result is significant, for it states that the orthogonality of an L_2 function to the eigenfunctions of a definite symmetric kernel can be demonstrated without actually knowing the explicit form of the $\phi_n(x)$.

We are now in a position to state the Hilbert-Schmidt theorem: Let $g(\xi)$ be a function that is $L_2\,[a, b]$ and $k(x, \xi)$ by a symmetric L_2 kernel with corresponding orthonormal eigenfunctions $\phi_n(x)$. Then if $f(x)$ can be represented as the transform of $g(x)$,

$$f(x) = \int_a^b k(x, \xi)\, g(\xi)\, d\xi$$

then $f(x)$ can also be represented by the Fourier expansion

$$f(x) = \sum_{n=1}^{\infty} a_n \phi_n(x)$$

where the Fourier coefficients are given by

$$a_n = \int_a^b f(x)\, \phi_n(x)\, dx$$

for $n = 1, 2, \ldots$. Furthermore, if $k(x, \xi)$ is square-integrable, then the expansion is absolutely and uniformly convergent. Proofs of the Hilbert-Schmidt theorem are to be found in a number of texts (see Refs. 10-1, 10-3, and 10-4).

An important corollary of the Hilbert-Schmidt theorem is that if the symmetric kernel $k(x, \xi)$ is $L_2\,[a, b]$, then all the associated iterated kernels $k_m(x, \xi)$ have the absolutely convergent expansions

$$k_m(x, \xi) = \sum_{n=1}^{\infty} \frac{\phi_n(x)\, \phi_n(\xi)}{\lambda_n^m} \qquad (m \geqslant 2) \tag{10.4-23}$$

Moreover, if $k(x, \xi)$ is square-integrable, then each of the series is uniformly convergent.

An expression for the L_2 solution of the Fredholm integral equation of the second kind with a symmetric L_2 kernel can be obtained as a direct consequence of the Hilbert-Schmidt theorem when λ is a regular value. Thus, if $f(x)$ is a function $L_2\,[a, b]$, then $g(x) = \phi(x) - f(x)$ is an $L_2\,[a, b]$ function, and, since $\phi - f$ appears in the integral equation as the transform of an L_2 function, it has the Fourier series representation

$$\phi(x) - f(x) = \sum_{n=1}^{\infty} c_n \phi_n(x)$$

and

$$c_n = \int_a^b [\phi(x) - f(x)]\, \phi_n(x)\, dx = d_n - a_n$$

Here, d_n is the nth Fourier coefficient of ϕ, and a_n is the nth Fourier coefficient of f. Moreover, the L_2 solution ϕ also satisfies the relations

$$d_n = \int_a^b \phi_n(x)\, \phi(x)\, dx = \int_a^b \phi_n(x) \left[f(x) + \lambda \int_a^b k(x, \xi)\, \phi(\xi)\, d\xi \right] dx$$

$$= a_n + \frac{\lambda}{\lambda_n} \int_a^b \phi_n(\xi)\, \phi(\xi)\, d\xi$$

so that with eq. (10.4-21)

$$d_n = a_n + \frac{\lambda}{\lambda_n} d_n$$

It follows that if $\lambda \neq \lambda_n$, the solution of the inhomogeneous Fredholm equation of the second kind with a symmetric L_2 kernel can be represented in the form

$$\phi(x) = f(x) + \lambda \sum_{n=1}^{\infty} \frac{a_n}{\lambda_n - \lambda} \phi_n(x) \tag{10.4-24}$$

Moreover, if $k(x, \xi)$ is definite and continuous over R, the series in eq. (10.4-24) is absolutely and uniformly convergent in $[a, b]$.

If λ_1 is an eigenvalue of multiplicity r, the solutions exist only if

$$\int_a^b f(x)\, \phi_n(x)\, dx = 0$$

for $n = 1, 2, \ldots r$. In such case, the solution can be written

$$\phi(x) = f(x) + \lambda_1 \sum_{n=r+1}^{\infty} \frac{a_n}{\lambda_n - \lambda_1} \phi_n(x) + \sum_{n=1}^{r} c_n \phi_n(x)$$

where the c_n are arbitrary constants.

Successful implementation of the Hilbert-Schmidt method requires the determination of the eigenvalues and eigenfunctions of the kernel (or the transposed kernel if $k(x, \xi)$ is complex). In general, such determinations must be carried out approximately or numerically. We have already recorded an upper bound of the modulus of the lowest eigenvalue, developed during the course of the discussion leading to the bilinear formula (see eq. (10.4-19). A lower bound of $|\lambda_1|$ can be established by setting $\xi = x$ in eq. (10.4-23) and integrating from a to b; since the $\phi_n(x)$ are orthonormal

$$A_m = \int_a^b k_m(x, x)\, dx = \sum_{n=1}^{\infty} \frac{1}{\lambda_n^m} \qquad \text{for } m \geqslant 2 \tag{10.4-25}$$

Hence, the traces with even indices can be used to generate lower bounds of the modulus of the lowest eigenvalue $|\lambda_1|$

$$A_{2m} = \sum_{n=1}^{\infty} \frac{1}{\lambda_n^{2m}} \geqslant \frac{1}{\lambda_1^{2m}}$$

If to each eigenvalue there corresponds a single eigenfunction, then for sufficiently large m, we have approximately

$$|\lambda_1| \geqslant (A_{2m})^{-1/2m} \tag{10.4-26}$$

A lower bound of the modulus of the second lowest eigenvalue can be estimated from

$$|\lambda_2| \geqslant \frac{1}{|\lambda_1|}\left(\frac{1}{2} B_{2m}\right)^{-1/2m} \tag{10.4-27}$$

where $B_{2m} = A_{2m}^2 - A_{4m}$ (see Ref. 10-2). In each case, the formula provides not only a lower bound of the eigenvalue, but also an estimate that is often accurate within a few percent for integers $m = 1$ and $m = 2$. In any event, we have the result

$$(A_{2m})^{-1/2m} \leqslant |\lambda_1| \leqslant \sqrt{\frac{A_{2m}}{A_{2m+2}}} \qquad \text{for } m = 1, 2, \ldots \tag{10.4-28}$$

When the eigenvalues are well separated at the lower end of the spectrum, the approximations above usually provide satisfactory initial estimates for numerical iteration.

If $k(x, \xi)$ is a polar kernel, there may be a decided advantage to converting to a symmetric form. In many cases, the traces with even indices may be calculated from

$$A_{2m} = 2\int_a^b \int_a^x k_m^2(x, \xi)\, d\xi\, dx$$

where $k_n(x, \xi)$ is the expression appropriate for $\xi < x$. In some instances, however, the polar form is $L_2[a, b]$, whereas the symmetric form is not, in which case this labor-saving device will not work.

Example: The problem of small-amplitude, periodic shallow water waves in a channel of uniform depth and triangular plan view was expressed in the form of an integral equation [eq. (10.3-8)]; setting $g(x, \xi) = -k(x, \xi)$, we have

$$\phi(\xi) = \lambda \int_0^1 k(x, \xi)\, \phi(x)\, dx$$

where

$$k(x, \xi) = \frac{1}{2} x \begin{cases} x\left(\dfrac{1}{\xi} - \xi\right), & 0 < x < \xi \\ \xi\left(\dfrac{1}{x} - x\right), & \xi < x < 1 \end{cases}$$

The eigenfunctions of the kernel are

$$\phi_k(x) = J_1(\lambda_k^{1/2} x)$$

and the eigenvalues λ_k are the roots of the equation

$$J_1(\lambda_k^{1/2}) = 0$$

The lowest eigenvalue (obtained from a table) is $\lambda_1 = 14.682 \ldots$. An estimate of the value of λ_1 can be obtained by calculating the trace A_2 corresponding to the kernel $k(x, \xi)$. We first compute $k_2(x, \xi)$ and set $\xi = x$ to obtain

$$k_2(x, \xi) = \tfrac{1}{8} x^5 - \tfrac{1}{8} x^3 - \tfrac{1}{4} x^3 \ln x$$

The second trace of $k(x, \xi)$ is then calculated from

$$A_2 = \int_0^1 k_2(x, x)\, dx = \tfrac{1}{192}$$

Hence, a first estimate (and lower bound) of the lowest eigenvalue is $\lambda_1 = 13.856 \ldots$.

Example: The Green's function satisfying the differential equation

$$g_{xx}(x, \xi) = \delta(x - \xi)$$

in the range [0, 1], subject to the boundary conditions $g(0, \xi) = g(1, \xi) = 0$, is of the form

$$g(x, \xi) = -\begin{cases} x(1 - \xi), & 0 < x < \xi \\ \xi(1 - x), & \xi < x < 1 \end{cases} \tag{10.4-29}$$

The corresponding bilinear formula is

$$g(x, \xi) = -\sum_{n=1}^{\infty} \frac{(\sqrt{2} \sin n\pi x)(\sqrt{2} \sin n\pi\xi)}{(n\pi)^2}$$

and the eigenvalues are $\lambda_n = (n\pi)^2$, $n = 1, 2, \ldots$. For the purpose of illustration, we construct estimates of the two lowest eigenvalues of the kernel $k(x, \xi) = -g(x, \xi)$. First, we compute the traces $A_2 = 1.1111 \times 10^{-2}$, $A_4 = 1.0582 \times 10^{-4}$, $A_6 = 1.0822 \times 10^{-6}$, $A_8 = 1.1107 \times 10^{-8}$, and the quantities $B_2 = 35.4965$ and $B_4 = 385.463$. For $m = 3$ in eqs. (10.4-26) and (10.4-28), and $m = 2$ in eq. (10.4-27), we have

$$9.8692 < \lambda_1 < 9.8708 \quad \text{and} \quad \lambda_2 \geqslant 39.0954,$$

which may be compared with the exact results $\lambda_1 = 9.86960\ldots$ and $\lambda_2 = 39.4784\ldots$.

There is an alternative approach to the generation of eigenvalues and eigenfunctions of a symmetric kernel that suggests a different procedure for approximating the eigenvalues of a kernel (see, e.g., Ref. 10-1). Consider the quadratic integral form

$$J(\phi, \phi) = \int_a^b \int_a^b k(x, \xi)\, \phi(x)\, \phi(\xi)\, dx\, d\xi$$

where ϕ is any function that is at least piecewise continuous in $[a, b]$, subject to the normalization condition

$$\int_a^b \phi(x)\, \phi(x)\, dx = 1$$

The lowest positive eigenvalue λ_1 of the symmetric kernel $k(x, \xi)$ is the reciprocal of the maximum value attained by the quadratic integral under the normalization condition. Moreover, $J(\phi, \phi)$ is maximized when $\phi(x)$ is the first eigenfunction $\phi_1(x)$ of $k(x, \xi)$ (for a proof of this assertion see Ref. 10-1). The second lowest eigenvalue is the reciprocal of the maximum attained by $J(\phi, \phi)$ subject to the normalization condition and to the orthogonality constraint

$$\int_a^b \phi(x)\,\phi(x)\,dx = 1 \qquad \text{and} \qquad \int_a^b \phi(x)\,\phi_1(x)\,dx = 0$$

In general, the positive eigenvalues λ_n ($n = 2, 3, \ldots$), ordered as an increasing sequence, are defined as the reciprocals of the maximum values attained by the quadratic forms subject to the conditions

$$\int_a^b \phi(x)\,\phi(x)\,dx = 1 \qquad \text{and} \qquad \int_a^b \phi(x)\,\phi_j(x)\,dx = 0 \qquad \text{for } j = 1, 2, \ldots, n-1$$

If $J(\phi, \phi)$ can assume negative values, the ordered set of negative eigenvalues can be generated by an analogous process involving minimization of the quadratic integral forms. The final result of both processes taken together is a complete sequence of eigenvalues and corresponding eigenfunctions of the kernel that are orthogonal on the interval $[a, b]$. If the kernel has only a finite number of eigenvalues, the $k(x, \xi)$ is degenerate, and the procedure leads to the bilinear formula.

The foregoing procedure suggests an estimation method based on successive approximations. Thus, let $\phi^{(n-1)}(x)$ denote the $(n-1)$th estimate of the eigenfunction corresponding to the lowest eigenvalue λ_1. If $\lambda_1^{(n-1)}$ is the $(n-1)$th estimate of λ_1, then the nth estimate of the eigenfunction is obtained from

$$\phi_1^{(n)}(x) = \lambda_1^{(n-1)} \int_a^b k(x, \xi)\,\phi^{(n-1)}(\xi)\,d\xi \tag{10.4-30}$$

subject to the normalization condition

$$\int_a^b [\phi_1^{(n)}]^2\,dx = 1 \tag{10.4-31}$$

The hope is, of course, that as n increases without limit $\phi_1^{(n)}$ and $\lambda_1^{(n)}$ will approach the true eigenfunction and eigenvalue. Repeated application of eq. (10.4-30), beginning with a first estimate of the lowest eigenfunction $\phi_1^{(1)}(x)$, together with the orthogonality condition, ultimately leads to the expression

$$1 = [\lambda_1^{(n+1)}]^2 [\lambda_1^{(n)}]^2 \ldots [\lambda_1^{(1)}]^2 \int_a^b \int_a^b k_{2n+2}(u, v)\,\phi_1^{(1)}(u)\,\phi_1^{(1)}(v)\,du\,dv$$

from which it follows that

$$[\lambda_1^{(n+1)}]^2 = \frac{\displaystyle\int_a^b \int_a^b k_{2n}(u, v)\,\phi_1^{(1)}(u)\,\phi_1^{(1)}(v)\,du\,dv}{\displaystyle\int_a^b \int_a^b k_{2n+2}(u, v)\,\phi_1^{(1)}(u)\,\phi_1^{(1)}(v)\,du\,dv}$$

where $\phi_1^{(1)}(x)$ is a first estimate of the lowest eigenfunction. In particular, if $\phi_1^{(1)}(x)$ is taken to be $k(x, x_0)$, then

$$\lambda_1^{(n)} \doteq \frac{k_{2n}(x_0, x_0)}{k_{2n+2}(x_0, x_0)}$$

10.4.4 Fredholm Equations of the First Kind

Fredholm integral equations of the first kind have the general form

$$f(x) = \lambda \int_a^b k(x, \xi)\, \phi(\xi)\, d\xi \tag{10.4-32}$$

where $f(x)$ is a given arbitrary function of x, $k(x, \xi)$ is the kernel, and $\phi(x)$ is to be determined. We note in passing that Fredholm equations of the second kind can be considered equations of the first kind with the singular kernel $\lambda^{-1}\delta(\xi - x) - k(x, \xi)$. However, there is little advantage in this interpretation because most Fredholm equations of the first kind have no solution. Moreover, even if a solution can be found, it may not be unique.

To illustrate the difficulties associated with Fredholm equations of the first kind, consider the case in which $k(x, \xi)$ is a nonsingular degenerate kernel

$$k(x, \xi) = \sum_{i=1}^{N} \alpha_i(x)\, \beta_i(\xi)$$

Substitution into eq. (10.4-32) shows that $f(x)$ must be expressible in the form

$$f(x) = \sum_{i=1}^{N} c_i \alpha_i(x)$$

in the interval $[a, b]$, where the Fourier coefficients are

$$c_i = \int_a^b \beta_i(\xi)\, \phi(\xi)\, d\xi$$

Obviously, if $f(x)$ has no such expansion, then the integral equation has no solution.

In general, in order for eq. (10.4-32) to have a solution, $f(x)$ and $k(x, \xi)$ must satisfy very restrictive conditions. In fact, if we are to find Fredholm equations of the first kind that have solutions under less stringent conditions, we must look to those that are singular. We turn our attention to that class of integral equations in the next section.

10.5 SINGULAR LINEAR INTEGRAL EQUATIONS

An integral equation is said to be singular either if the kernel is singular within the range of integration or if one or both limits are infinite. No general theory is available for singular integral equations, but procedures have been developed for dealing with several special cases, some of which are discussed elsewhere in this handbook. Five solution methods are dealt with in this section: (1) the generating function method for Fredholm equations of the first kind; (2) transform techniques; (3) complex variables methods; (4) the iterated kernel method for weakly singular kernels; and (5) the treatment of logarithmic kernels.

10.5.1 The Generating Function Method

A special case of importance for Fredholm equations of the first kind arises when the kernel can be written in the form

$$k(x, \xi) = \sum_{n=0}^{\infty} w(\xi)\, \psi_n(\xi)\, x^n \tag{10.5-1}$$

where $\{\psi_n(\xi)\}$ is a complete orthogonal set on $[a, b]$ with respect to $w(\xi)$

$$\int_a^b w(\xi)\, \psi_m(\xi)\, \psi_n(\xi)\, d\xi = h_n \delta_{mn}$$

The kernel $k(x, \xi)$ is then the generating function of the set $\{\psi_n(\xi)\}$ (some illustrations are given in the table below). In such cases, the solution of the integral equa-

Table II Singular Kernels and Generating Functions.

Kernel	Series	Range	Polynomial
$\dfrac{J_\alpha(2\sqrt{x\xi})}{(x\xi)^{\alpha/2}}$	$= \sum_{n=0}^{\infty} \dfrac{1}{\Gamma(\alpha+n+1)} L_n^{(\alpha)}(\xi)\, e^{-x} x^n$	$0 < x, \xi < 0$ $\alpha > -1$	Generalized Laguerre
$\dfrac{1}{\sqrt{1 - 2x\xi + x^2}}$	$= \sum_{n=0}^{\infty} P_n(\xi)\, x^n$	$-1 < x, \xi < +1$	Legendre
$\dfrac{1 - x\xi}{1 - x\xi + x^2}$	$= \sum_{n=0}^{\infty} T_n(\xi)\, x^n$	$-1 < x, \xi < +1$	Chebyshev, first kind
$1 - \frac{1}{2} \ln \lvert 1 - 2x\xi + \xi^2 \rvert$	$= \sum_{n=1}^{\infty} \dfrac{1}{n} T_n(\xi)\, x^n$	$-1 < x, \xi < +1$	Chebyshev, first kind
$\dfrac{1}{1 - x\xi + \xi^2}$	$= \sum_{n=0}^{\infty} U_n(\xi)\, x^n$	$-1 < x, \xi < +1$	Chebyshev, second kind
$e^{-(x-\xi)^2}$	$= \sum_{n=0}^{\infty} \dfrac{e^{-\xi^2}}{n!} H_n(\xi)\, x^n$	$-\infty < x, \xi < \infty$	Hermite

tion can be expressed as

$$\phi(x) = \frac{1}{\lambda} \sum_{n=0}^{\infty} \frac{f^{(n)}(0)}{h_n n!} \psi_n(x) \tag{10.5-2}$$

Example: The singular integral equation

$$f(x) = \int_0^{\infty} J_0(2\sqrt{x\xi})\, \phi(\xi)\, d\xi \tag{10.5-3}$$

can be solved by observing that the kernel is expressible in terms of the following generating function relation

$$J_0(2\sqrt{x\xi}) = \sum_{n=0}^{\infty} \frac{1}{n!} L_n(\xi)\, e^{-x} x^n$$

where $L_n(\xi)$ is the Laguerre polynomial of order n. The orthogonality relation is

$$\int_0^{\infty} e^{-\xi} L_m(\xi)\, L_n(\xi)\, d\xi = \delta_{mn} \tag{10.5-4}$$

Assume the solution has an expansion in terms of appropriately weighted Laguerre polynomials

$$\phi(x) = e^{-x} \sum_{m=0}^{\infty} c_m L_m(x) \tag{10.5-5}$$

Substituting eq. (10.5-5) into eq. (10.5-3) and using the orthogonality relation, we obtain

$$e^x f(x) = \sum_{n=0}^{\infty} c_n \frac{1}{n!} x^n$$

and therefore

$$c_n = \frac{d^n}{dx^n} [e^x f(x)] \Big|_{x=0}$$

This last result, together with eq. (10.5-5), gives the solution.

10.5.2 Method of Integral Transforms

Several different types of singular integral equations are amenable to solution by transform techniques. Chapter 11 describes transform methods in general terms, and some selected types of singular integral equations are presented in that discussion. The transform pairs used below are as defined in Chapter 11. In most cases, the transform technique gives the solution of the original equation in terms of a quadrature in the complex plane. In each of the inversion expressions given below, it is essential that the several transformed functions appearing in the integrand have a common region of analyticity. We first discuss singular equations with an infinite range of integration.

Several types of inhomogeneous equations of the second kind can be solved by application of the Mellin transform. For instance, the solution of the integral equation

$$\phi(x) = f(x) + \lambda \int_0^\infty k\left(\frac{x}{\xi}\right) \phi(\xi) \frac{d\xi}{\xi} \qquad (x > 0)$$

can be expressed in terms of the inverse Mellin transform

$$\phi(x) = \frac{1}{2\pi i} U(x) \int_{\sigma - i\infty}^{\sigma + i\infty} x^{-s} \frac{F(s)}{1 - \lambda K(s)} \, ds$$

In this equation and those that follow, $s = \sigma + i\tau$ lies entirely within the strip of analyticity. The Mellin transform can also be used to solve a similar integral equation of the second kind with a kernel of the form $k(x\xi)$

$$\phi(x) = f(x) + \lambda \int_0^\infty k(x\xi) \phi(\xi) \, d\xi \qquad (x > 0)$$

The solution is given as an integral along a line in the complex s-plane that lies within the strip of common analyticity of the several functions in the integrand

$$\phi(x) = \frac{1}{2\pi i} U(x) \int_{\sigma - i\infty}^{\sigma + i\infty} x^{-s} \frac{F(s) + \lambda K(s) F(1 - s)}{1 - \lambda^2 K(s) K(1 - s)} \, ds$$

The related integral equation of the first kind

$$f(x) = \lambda \int_0^\infty k(x\xi) \phi(\xi) \, d\xi \qquad (x > 0)$$

is also solvable in terms of the inverse Mellin transform

$$\phi(x) = \frac{1}{2\pi i} U(x) \int_{\sigma - i\infty}^{\sigma + i\infty} x^{-s} \frac{F(1-s)}{\lambda K(1-s)} \, ds$$

The Laplace transform pair is a special case of this last example, with $\sigma = \frac{1}{2}$ and $K(s) = \Gamma(s)$. The Laplace transform can also be used to express the solution of the equation

$$\phi(x) = f(x) + \lambda \int_{x}^{\infty} k(\xi - x)\, \phi(\xi)\, d\xi$$

as an integral in the complex plane

$$\phi(x) = \frac{1}{2\pi i} U(x) \int_{\sigma_0 - i\infty}^{\sigma_0 + i\infty} e^{px} \frac{F(p)}{1 - \lambda K(-p)} \, dp$$

Certain integral equations with displacement kernels $k(x - \xi)$ can often be treated by Fourier transform techniques. For example, consider the homogeneous equation

$$\phi(x) = \lambda \int_{-\infty}^{\infty} k(x - \xi)\, \phi(\xi)\, d\xi$$

Because the range of integration is infinite, we expect the eigenvalue spectrum to be continuous. Application of the Fourier transform operator to this equation yields

$$[1 - \lambda\sqrt{2\pi}\, K(w)]\; \Phi(w) = 0$$

where the convolution theorem has been used. Clearly, $\Phi(w)$ can differ from zero only at the roots w_j of

$$1 - \lambda\sqrt{2\pi}\, K(w) = 0$$

Suppose that w_j is of multiplicity r and that there are N such zeros. Let $\exp(c_1 |x|)\, k(x)$ be $L_1[-\infty, \infty]$ and $\exp(-c_2 |x|) f(x)$ be $L_2[-\infty, \infty]$, where $c_1 > c_2 > 0$. The solution of the homogeneous equation can then be expressed formally as

$$\phi(x) = \sum_{j=1}^{N} \sum_{q=1}^{r} c_{jq} x^{q-1} e^{-iw_j x}$$

where the c_{jq} are constants.

Similarly, the Fourier transform can be used to solve an integral equation of the first kind with an infinite range of integration and a displacement kernel $k(x - \xi)$

$$f(x) = \lambda \int_{-\infty}^{\infty} k(x - \xi)\, \phi(\xi)\, d\xi$$

where $f(x)$ is $L_2\,[-\infty, \infty]$ and $k(x)$ is $L_1\,[-\infty, \infty]$. There exists a solution $L_2\,[-\infty, \infty]$ if and only if $F(w)/K(w)$ is $L_2\,[-\infty, \infty]$, and, under the stated conditions, the solution is

$$\phi(x) = \frac{1}{2\pi} \int_{-\infty}^{\infty} e^{-iwx} \frac{F(w)}{\lambda K(w)}\, dw$$

When the range of integration is $[0, \infty]$ and the kernel is of the displacement type, the generalized Fourier transform method is the most effective. Since this is one of the special cases most frequently discussed, we will only outline the procedure here. Consider the homogeneous equation

$$\phi(x) = \lambda \int_{-\infty}^{\infty} k(x - \xi)\, \phi(\xi)\, d\xi \tag{10.5-6}$$

where $k(x)$ is such that $K(w)$ is analytic in the strip $\tau_0 < Im(w) < \tau_1$. Further, suppose $\Phi_+(w)$ is analytic for $Im(w) > \tau_0'$ and $\Phi_-(w)$ is analytic for $Im(w) < \tau_1'$. When $\tau_0' < \tau_1$ and $\tau_1' > \tau_0$, the regions of analyticity of Φ_+ and K overlap, as do the analytic domains of Φ_- and K. Let the real constants a and b satisfy the inequalities $\tau_0' < a < \tau_1$ and $\tau_0 < b < \tau_1'$. After some algebraic rearrangement, eq. (10.5-6) can be expressed in terms of the appropriate generalized Fourier inversions

$$\int_{-\infty + ia}^{\infty + ia} dw\, e^{-iwx}\, [1 - \lambda\sqrt{2\pi}\, K(w)]\, \Phi_+(w) + \int_{-\infty + ib}^{\infty + ib} dw\, e^{-iwx}\, \Phi_-(w) = 0$$

It can be inferred that the integrands must each be analytic in the strip $b \leqslant Im(w) < a$, and therefore

$$[1 - \lambda\sqrt{2\pi}\, K(w)]\, \Phi_+(w) = -\Phi_-(w) \tag{10.5-7}$$

Suppose the factor Φ_+ has n simple zeros w_j in the strip and that it is possible to find a factorization of the form

$$1 - \lambda\sqrt{2\pi}\, K(w) = N_+(w)\, N_-(w) \tag{10.5-8}$$

where $N_+(w)$ is analytic for $Im(w) \geqslant \tau_0'' > b$, and $N_-(w)$ is analytic for $Im(w) \leqslant \tau_1'' < a$. We can then write

$$N_+(w)\,\Phi_+(w) = -\frac{\Phi_-(w)}{N_-(w)} = P(w)$$

where $P(w)$ is analytic throughout the strip of common analyticity. It follows that

$$\phi_+(x) = \frac{1}{\sqrt{2\pi}} \int_{-\infty+ia}^{\infty+ia} e^{-iwx} \frac{P(w)}{N_+(w)}\, dw \tag{10.5-9}$$

Now, $P(w)$ can be at most a polynomial in w of finite degree. Specifically, $P(w)$ may be determined by the analyticity condition together with an additional requirement; typically that $\Phi_+(w) \to 0$ as $|w| \to \infty$. Then, by closing the contour in the appropriate part of the complex plane, the integral may be evaluated by the Cauchy residue theorem. The factorization procedure is called the *Wiener-Hopf technique*. For further elaboration and examples, the reader may consult Chapter 11.

Fourier transforms can also be used to treat Volterra equations. For example, the inhomogeneous Volterra equation with a difference kernel

$$\phi(x) = f(x) + \lambda \int_0^x k(x - \xi)\, \phi(\xi)\, d\xi \qquad (x > 0) \tag{10.5-10}$$

has an $L_2\,[0, \infty]$ solution under certain circumstances. Specifically, suppose that $\exp(-c_1 x) f(x)$ is $L_2\,[0, \infty]$ and that $\exp(-c_1 x)\, k(x)$ is $L_1\,[0, \infty]$ for $c_1 > 0$. Then there exists a unique solution $\phi(x)$ such that $\exp(-c_2 x)\, \phi(x)$ is $L_2\,[0, \infty]$, where $c_2 > 0$. Application of a Fourier transform gives $\phi(x)$ as

$$\phi(x) = \lim_{\sigma \to \infty} \frac{1}{\sqrt{2\pi}} \int_{-\sigma+i\tau_0}^{\sigma+i\tau_0} e^{-iwx} \frac{F(w)}{1 - \lambda\sqrt{2\pi}\, K(w)}\, dw$$

for τ_0 sufficiently large, the Laplace transform gives equivalent form

$$\phi(x) = \frac{1}{2\pi i} U(x) \int_{\sigma_0-i\infty}^{\sigma_0+i\infty} e^{px} \frac{F(p)}{1 - \lambda K(p)}\, dp$$

for sufficiently large σ_0.

Example: Consider eq. (10.3-1), the inhomogeneous Volterra equation for the time rate of change of the area of an oil slick spreading over a permeable medium; the equation is of the form (10.5-10). From physical considerations, a plausible form for $k(t)$ is $t^{-1/2}$. Assume that the volume rate of spill is constant and equal to Q_0. If eq. (10.3-1) is then divided by h, Q_0/h set equal to q_0, and γ/h denoted by λ, the equation reads

$$\frac{dA}{dt} = q_0 - \lambda \int_0^t (t - t_0)^{-1/2} \frac{dA}{dt_0} \, dt_0$$

Application of the Laplace transform gives

$$\frac{dA}{dt} = \frac{q_0}{2\pi i} U(t) \int_{\sigma_0 - i\infty}^{\sigma_0 + i\infty} e^{pt} \frac{1}{p + \lambda\sqrt{\pi p}} \, dp$$
$$= q_0 U(t) \exp(\tau^2) \operatorname{erfc}(\tau)$$

where $\tau = \lambda\sqrt{\pi t}$. This last expression can be integrated to give an expression for the area at time t

$$A(t) = \frac{q_0}{\pi\lambda^2} U(t) \left[\exp(\tau^2) \operatorname{erfc}(\tau) + \frac{2}{\sqrt{\pi}} \tau - 1\right]$$

The related singular Volterra equation of the first kind can also be solved by the use of a Fourier or Laplace transform. Let $\exp(-cx) f(x)$ be $L_2[0, \infty]$ and $\exp(-cx) k(x)$ be $L_1[0, \infty]$, with $c > 0$. Then the equation

$$f(x) = \lambda \int_0^x k(x - \xi) \phi(\xi) \, d\xi \qquad (x > 0) \tag{10.5-11}$$

has a solution such that $\exp(-cx) \phi(x)$ is $L_2[0, \infty]$ if and only if

$$\int_{-\infty}^{\infty} \left| \frac{F(w)}{K(w)} \right|^2 dw < M$$

where $w = \sigma + i\tau_0$, for τ_0 sufficiently large. When these conditions are satisfied

$$\phi(x) = \frac{1}{2\pi} \int_{-\infty + i\tau_0}^{\infty + i\tau_0} e^{-iwx} \frac{F(w)}{\lambda K(w)} \, dw$$

Example: Abel's integral equation is of the form of eq. (10.5-11), with $k(x - \xi) = (x - \xi)^{-1/2}$ and $\lambda = 1$. In this case

$$K(w) = \frac{1}{\sqrt{2}} \frac{1}{(-iw)^{1/2}}$$

and we choose the branch such that $(-iw)^{-1/2}$ is real for $\tau > 0$ ($w = \sigma + i\tau$). We can then write formally

$$\phi(x) = \frac{1}{\sqrt{2}\,\pi} \int_{-\infty + i\tau_0}^{+\infty + i\tau_0} e^{-iwx} (-iw)^{1/2} F(w)\, dw$$

If $\phi(x)$ is assumed to be $L_2\,[0, \infty]$, then the integral of $\phi(x)$ can be manipulated into the form

$$I(x) = \frac{1}{\pi^{3/2}} \int_0^{\infty} f(\xi)\, d\xi \int_{-\infty + i\tau_0}^{\infty + i\tau_0} \frac{e^{-iw(\xi - x)} - e^{iw\xi}}{(-iw)^{1/2}}\, dw$$

The inner integral vanishes if $\xi > x$. If $0 < \xi < x$, the contribution from $e^{iw\xi}$ is zero, and appropriate deformation of the contour gives

$$I(x) = \frac{1}{\pi} \int_0^x \frac{f(\xi)}{\sqrt{x - \xi}}\, d\xi$$

and the problem is reduced to a differentiation and a quadrature.

10.5.3 Complex Variables Methods

An integral equation of the general form

$$g(x)\,\phi(x) = f(x) + \lambda \,⨍_a^b \frac{\phi(\xi)}{x - \xi}\, d\xi \qquad a < x < b \tag{10.5-12}$$

is said to have a "Cauchy kernel." Here, the integral is to be interpreted as the principal value. If $g(x) \equiv 0$, the problem can be expressed as a quadrature by the Riemann-Hilbert method. The Riemann-Hilbert problem and integral equations of the type (10.5-12) over the range [0, 1] are described in some detail in Chapter 11. We supplement that discussion with a few examples with the range [-1, +1]. Thus, the solution of the integral equation of the first kind

$$f(x) = ⨍_{-1}^{+1} \frac{\phi(\xi)}{x - \xi}\, d\xi \tag{10.5-13}$$

can be expressed in the form

$$\phi(x) = -\frac{1}{\pi^2} \sqrt{\frac{1 - x}{1 + x}} \int_{-1}^{+1} \frac{f(\xi)\sqrt{1 + \xi}}{\sqrt{1 - \xi}\,(\xi - x)}\, d\xi + \frac{c_0}{\pi\sqrt{1 - x^2}}$$

Under certain conditions, a more explicit form can be obtained. If, after setting $x = -\cos\theta$, the inhomogeneous term can be shown to have a uniformly converging Fourier series expansion of the form

$$f(-\cos\theta) = -\frac{1}{2}\sum_{n=1}^{\infty} a_n \frac{\sin(n\theta)}{\sin\theta}$$

then the series

$$\phi(x) = \frac{1}{2\pi\sqrt{1-x^2}}\left[\frac{1}{2}a_0 + \sum_{n=1}^{\infty} a_n \cos(nx)\right]$$

converges uniformly and represents a solution of eq. (10.5-13). In the special case $f(x) = 1$, the solution is

$$\phi(x) = \frac{x^2 - 1 + c_0}{\pi\sqrt{1-x^2}}$$

An equation of similar form is

$$f(x) = \int_0^1 \frac{\phi(\xi)}{|x-\xi|^{\alpha}}\,d\xi \qquad (0 < x < 1, 0 < \alpha < 1)$$

Here, the kernel has an algebraic singularity and the solution of the equation can be given in terms of quadratures

$$\phi(x) = \frac{\sin(\alpha\pi)}{2\pi}\frac{d}{dx}\int_0^x \frac{f(\xi)}{(x-\xi)^{1-\alpha}}\,d\xi - \frac{\cos^2\left(\frac{\alpha\pi}{2}\right)}{\pi^2}\frac{d}{dx}\int_0^x \frac{[\xi(1-\xi)]^{1/2(1-\alpha)}}{(x-\xi)^{1-\alpha}}\,d\xi$$

$$\cdot\int_0^1 \frac{f(\xi')\,[\xi'(1-\xi')]^{1/2(\alpha-1)}}{\xi'-\xi}\,d\xi'$$

In the special case $f(x) = 1$, the solution takes the simple form

$$\phi(x) = \frac{\cos\left(\frac{\alpha\pi}{2}\right)}{\pi}(x - x^2)^{1/2(\alpha-1)}$$

Recall that in defining a kernel with a weak singularity of the form $|x - \xi|^{-\beta}$, it was observed that the nth iterated kernel $k_n(x, \xi)$ would be nonsingular if $\beta \leqslant 1 - 1/n$. By way of illustration, consider the weakly singular equation

$$\phi(x) = f(x) + \lambda \int_a^b \frac{g(x, \xi)}{|x - \xi|^{1/2}} \phi(\xi)\, d\xi$$

where $g(x, \xi)$ is analytic in R. With $\beta = \frac{1}{2}$, the kernel $k_2(x, \xi)$ will be nonsingular. We proceed as follows: change the variable x to u, multiply by $k(x, u)\, du$, and integrate from a to b. If we define the function $f_2(x)$ by

$$f_2(x) = f(x) + \lambda \int_a^b \frac{g(x, u)}{|x - u|^{1/2}} f(u)\, du$$

then the iterated equation takes the form

$$\phi(x) = f_2(x) + \lambda^2 \int_a^b k_2(x, \xi)\, \phi(\xi)\, d\xi$$

where $k_2(x, \xi)$ is nonsingular. Note that Abel's equation has a weakly singular kernel and therefore can be solved by iteration (more easily than by the Fourier transform method in fact).

If, in eq. (10.5-12), the Cauchy kernel is replaced by a logarithmic displacement kernel, i.e., if

$$f(x) = \int_{-1}^{+1} \ln |x - \xi|\, \phi(\xi)\, d\xi \qquad (-1 < x < 1) \tag{10.5-14}$$

the function $\phi(x)$ can be found by "Carleman's method," a technique that is analogous to the kernel factorization of the Wiener-Hopf procedure. The solution can be written as

$$\phi(x) = \frac{1}{\pi^2 \sqrt{1 - x^2}} \left[\int_{-1}^{+1} \frac{\sqrt{1 - \xi^2}\, f'(\xi)}{\xi - x}\, d\xi - \frac{1}{\ln 2} \int_{-1}^{+1} \frac{f(\xi)}{\sqrt{1 - \xi^2}}\, d\xi \right]$$

this last being referred to as "Carleman's formula." The following generalization of eq. (10.5-14)

$$f(x) = \int_{-1}^{+1} [u(x - \xi) \ln |x - \xi| + v(x - \xi)]\, \phi(\xi)\, d\xi$$

where $u(x)$ and $v(x)$ are polynomials, is discussed in Ref. 10-7.

10.6 APPROXIMATE SOLUTION OF INTEGRAL EQUATIONS

Most approximation procedures fall into one or more of the following three categories: (1) kernel replacement, (2) numerical quadrature, and (3) expansion methods. The literature in this aspect of the subject is quite extensive. For those who are interested primarily in "numbers," the development of approximate solutions with the aid of high-speed computers represents an attractive alternative to the Fredholm method and the Hilbert-Schmidt method. If an integral equation does not yield to conventional techniques, Ref. 10-6 is a good resource; that work is devoted entirely to numerical solution of integral equations and contains an extensive bibliography on the subject. If the integral equation is singular or nonlinear, use of an approximate or numerical procedure may be essential, although methods are not well developed for such problems.

10.6.1 Kernel Replacement

After formulating a problem as an integral equation, it may turn out that the kernel is rather complicated and the prospects for solving the integral equation exactly are poor. In such event, consideration may be given to replacing the exact kernel with an approximate one that will yield (at the least) some useful information regarding the asymptotic behavior of $\phi(x)$.

This approach may be especially helpful in problems of the Wiener-Hopf variety, where it is required to perform the factorization $1 - \lambda\sqrt{2\pi}K(w) = N_+(w)N_-(w)$. It may happen that a factored form is difficult to interpret, if indeed it can be found at all. We consider an example from Ref. 10-7.

$$1 = \int_0^\infty k(x - \xi)\,\phi(\xi)\,d\xi \qquad (x > 0) \tag{10.6-1}$$

where the kernel is given by

$$k(x) = \frac{1}{\pi}\int_0^\infty \frac{ue^{-u}}{x^2 + u^2}\,du$$

The form of the equation suggests the Wiener-Hopf approach, and, although the kernel is a somewhat complicated function (it can be expressed in terms of $Ei(ix)$), its Fourier transform has the simple form $(1 + |w|)^{-1}$. Following the standard Wiener-Hopf method, we define

$$\begin{aligned} \phi_+(x) &= \phi(x) \qquad (x > 0) \\ &= 0 \qquad\quad\;\; (x < 0) \end{aligned}$$

and set

$$f_-(x) + U(x) = \int_{-\infty}^{\infty} k(x - \xi)\, \phi_+(\xi)\, d\xi$$

where

$$\begin{aligned} f_-(x) &= 0 \qquad (x > 0) \\ &= f(x) \qquad (x < 0) \end{aligned}$$

and $U(x)$ is the Heaviside step function. Taking the Fourier transform of eq. (10.6-1), we write

$$F_-(w) - \frac{1}{\sqrt{2\pi}} \frac{1}{iw} = \frac{1}{\sqrt{2\pi}} \frac{1}{1 + |w|} \Phi_+(w)$$

where $F_-(w)$ and $\Phi_+(w)$ are both unknown. As is often the case in the Wiener-Hopf method, the factorization of $(1 + |w|)^{-1}$ is not straightforward. However, if $1 + |w|$ is expressed as $1 + \sqrt{w^2 + \epsilon^2}$, the required factorization can be carried out, and, in the limit as ϵ is allowed to approach zero, we have

$$\frac{1}{1 + |w|} = N_+(w) N_-(w) = \left[\frac{1}{\sqrt{1 + w}} \exp\left(-\frac{i}{\pi} Q(w)\right)\right] \left[\frac{1}{\sqrt{1 + w}} \exp\left(\frac{i}{\pi} Q(w)\right)\right]$$

where $Q(w) = \int_0^w [\ln z/(1 - z^2)]\, dz$. The first factor is analytic in the upper half plane, $-\frac{1}{2}\pi < \arg w < \frac{3}{2}\pi$, and the second factor is analytic in the lower half plane, $-\frac{3}{2}\pi < \arg w < \frac{1}{2}\pi$. The expression for $\phi_+(x)$ can then be written as the inversion

$$\phi_+(x) = \frac{1}{2\pi} \int_{-\infty}^{\infty} e^{-iwx} \frac{1}{iN_+(w) N_-(0)}\, dw$$

where the path of integration passes above the singularity at the origin. In this case, it does not appear feasible to try to carry out contour integration to obtain the exact solution. However, the asymptotic behavior for large and small $x\,(x > 0)$ can be inferred. Thus, for large positive x, $\phi_+(x) \sim 1 + 0(x^{-1})$, while for small positive values of x, the behavior is $\phi_+(x) \sim 1/\sqrt{\pi x} + 0(x^{3/2} \ln x)$.

Since this seems to be the best we can do without a good deal more labor, we consider replacing the given kernel with an approximate kernel $\tilde{k}(x - \xi)$, chosen to give the same asymptotic behavior. Consider the choice

$$\tilde{k}(x) = \frac{1}{\pi} K_0(x)$$

where $K_0(x)$ is the modified Bessel function of order zero. The Fourier transform of $\tilde{k}(x)$ is

$$\tilde{K}(w) = \frac{1}{\sqrt{2\pi}} \frac{1}{\sqrt{1 + w^2}}$$

Since $\lim_{|w| \to \infty} K(w) = \lim_{|w| \to \infty} \tilde{K}(w)$, we anticipate adequate representation of the true behavior of $\phi_+(x)$ for small x. Also, $K(0) = \tilde{K}(0)$, and $K'(0) = \tilde{K}'(0)$, so that the form of the solution should have the correct behavior for large x. The corresponding Wiener-Hopf problem is now easily solved; the final result is

$$\tilde{\phi}(x) = \operatorname{erf} x^{1/2} + \frac{1}{\sqrt{\pi x}} e^{-x}$$

To compare with the asymptotic behavior deduced above, we observe that $\tilde{\phi}_+(x) \sim 1 + O(1/\sqrt{x})$ and $\tilde{\phi}_+(x) \sim 1/\sqrt{\pi x} + O(x^{1/2})$, for large and small x, respectively, demonstrating that our expectations were justified for the asymptotic forms. For additional discussion and examples, see Ref. 10-11.

Kernel replacement for a finite range of integration may also be productive. By way of an example, we consider the Fredholm equation

$$\phi(x) = x + \tfrac{1}{2} \int_{-1}^{+1} (1 - |x - \xi|)\, \phi(\xi)\, d\xi \qquad (-1 < x < 1)$$

whose exact solution can easily be found

$$\phi(x) = \sec(1) \sin x$$

Since the range is $[-1, +1]$, we can generate an approximate solution by expanding kernel in a terminating series involving Legendre polynomials. Taking two terms only, we write

$$\begin{aligned} k(x, \xi) = (1 - |x - \xi|) &\doteq a_0(x) P_0(\xi) + a_1(x) P_1(\xi) \\ &= a_0(x) + a_1(x)\, \xi \end{aligned}$$

and determine $a_0(x)$ and $a_1(x)$ by preserving the area and first moment of the kernel

$$\int_{-1}^{+1} k(x, \xi)\, d\xi = 2a_0(x) = 1 - x^2$$

and

$$\int_{-1}^{+1} k(x, \xi)\, \xi\, d\xi = \tfrac{2}{3}\, a_1(x) = x - \tfrac{1}{3} x^3$$

Straightforward calculations lead ultimately to the result

$$\phi(x) \doteq \tfrac{11}{6} x - \tfrac{5}{18} x^3$$

which is a fair approximation, despite the crude representation of the kernel.

10.6.2 Numerical Quadrature

If the inhomogeneous term and the kernel of a nonsingular linear integral equation are well behaved functions, and if λ is a regular value, then the use of a quadrature rule, together with an interpolation formula, can be expected to give good results. Simpson's rule or an n-point Gaussian rule are recommended, although the trapezoidal rule may be adequate for sufficiently smooth functions.

We begin with a basic quadrature rule, which can be written as

$$\int_{\xi_i}^{\xi_{i+1}} g(\xi)\, d\xi \doteq \sum_{l=0}^{m} w_l g(z_l) \tag{10.6-1}$$

where w_l is a weight factor characteristic of the rule. A quadrature rule has precision r if it is exact for a polynomial of degree r.

We next consider a subdivision of the range of integration

$$a \leqslant \xi_0 < \xi_1 < \cdots < \xi_j < \cdots < \xi_n \leqslant b$$

A repeated rule can be constructed by successive application of the basic rule. For simplicity, suppose the interval $[a, b]$ is divided into n subintervals, each of length $\delta = \xi_{j+1} - \xi_j$. Then a repeated rule over $[a, b]$ allows an approximate representation of the integral of $g(\xi)$

$$\int_a^b g(\xi)\,d\xi \doteq \sum_{j=0}^{n} w_j \frac{1}{q} \sum_{p=1}^{q} g\left[\xi_p + \frac{1}{n}(\xi_j - a)\right]$$
$$\equiv \sum_{j=0}^{n} w_j g(z_j) \tag{10.6-2}$$

These ideas can be applied to nonsingular linear integral equations in the following way. Suppose the integral term of the inhomogeneous Fredholm equation can be approximated by a quadrature rule

$$\phi(x) - \lambda \sum_{j=0}^{n} w_j k(x, \xi_j)\, \phi(\xi_j) - f(x) = r(x) \tag{10.6-3}$$

where $r(x)$ denotes the residual. If x is discretized in the same way as ξ (that is, $x_i = \xi_i$ for $i = 0, 1, \ldots, n$) and if we require the equation to be represented exactly at each discrete value of x, the result is a set of $n + 1$ linear equations for $\tilde{\phi}_j, j = 0, 1, \ldots, n$, where the tilde has been added to emphasize that the value of $\phi(x_j)$ is only approximate. This procedure is referred to as the *Nyström method.*

Introducing the matrices $[k] = [k(x_i, \xi_j]$, $[D] = \text{diag}(w_0, w_1, \ldots, w_n)$, $[\tilde{\phi}] = [\tilde{\phi}_0, \tilde{\phi}_1, \ldots, \tilde{\phi}_n]^T$, and $[f] = [f_0, f_1, \ldots, f_n]^T$, we can write the equation set compactly as

$$[\tilde{\phi}] - \lambda[k]\;[D]\;[\tilde{\phi}] = [f]$$

This system may be processed by a linear equation solver to give a set of values $\tilde{\phi}_j \doteq \tilde{\phi}(x_j)$. An approximate representation of $\phi(x)$ can then be constructed using an appropriate interpolant. When the trapezoidal rule or Simpson's rule is used, eq. (10.6-3) is the natural interpolating formula. If the Nyström method seems appropriate for the problem at hand, reference may be made to Ref. 10-11, which contains FORTRAN programs for solving nonsingular linear nonhomogeneous Fredholm equations of the second kind by Simpson's rule and by the n-point Gauss-Legendre rule. In the latter case, n is successively set equal to integer powers of 2 up to $n = 256$, beyond which number a repeated rule is used. Both programs are well documented and contain error analyses.

Since the nonsingular homogeneous equation has solutions only for discrete λ_i, it will be necessary to estimate the eigenvalues for kernels of general form. If $f(x) = 0$, the Nyström procedure leads to the linear system

$$\{[I] - \tilde{\lambda}[k]\;[D]\}\;[\tilde{\phi}] = [0]$$

which is a matrix eigenvalue problem whose solution gives $n + 1$ approximate eigenvalues $\tilde{\lambda}_0, \tilde{\lambda}_1, \ldots, \tilde{\lambda}_n$ corresponding to $n + 1$ approximate eigenvectors $\tilde{\phi}_0, \tilde{\phi}_1, \ldots, \tilde{\phi}_n$. The choice of method for generating the eigenfunctions and eigenvectors will depend on the nature of the kernel; Ref. 10-6 may be consulted for details concerning the choice of procedure.

If an estimate of the lowest eigenvalue can be obtained, that estimate can be improved by iterative methods, and subsequent eigenvalues determined by standard techniques. Estimate of the lowest eigenvalue can be calculated by various methods, two of which have been mentioned in connection with the Hilbert-Schmidt method in Section 10.4.3. Reference 10-2 may be consulted for further approximate techniques.

Linear equations of the second kind, which are singular because the range of integration is $[0, \infty]$ or $[-\infty, +\infty]$, may be solved numerically by Gauss-Laguerre or Gauss-Hermite quadrature, provided the kernel and the inhomogeneous term are relatively well-behaved functions.

If the range $[a, b]$ is finite and the kernel is logarithmic or weakly singular, then a product integration technique will usually give more accurate results than the standard Nyström method. In this procedure, the assumption is made that $\phi(x)$ is well behaved and that the kernel can be split into two factors $g(x, \xi)$ and $h(x, \xi)$, with the singular behavior incorporated entirely in the factor $h(x, \xi)$. For simplicity, subdivide the range $[a, b]$ into n segments, each of equal length $\delta = (b - a)/n$, and define a linear interpolation of $g(x, \xi)\,\phi(\xi)$ over the interval $\xi_{j-1} \leqslant \xi \leqslant \xi_j$

$$[g(x, \xi)\,\phi(\xi)] \doteq \frac{1}{\delta}\,[(\xi_j - \xi)\,g(x, \xi_{j-1})\,\phi(\xi_{j-1}) + (\xi - \xi_{j-1})\,g(x, \xi_j)\,\phi(\xi_j)]$$

for $j = 1, 2, \ldots, n$ and $a \leqslant x \leqslant b$. The integral term in the Fredholm equation can then be approximated by

$$\int_a^b k(x, \xi)\,\phi(\xi)\,d\xi \doteq \sum_{j=0}^{n} v_j(x)\,g(x, \xi_j)\,\phi(\xi_j)$$

where the weights $v_j (j = 1, 2, \ldots, n - 1)$ are given by the integrals

$$v_j = \frac{1}{\delta}\left[\int_{\xi_{j-1}}^{\xi_j} h(x, \xi)\,(\xi - \xi_j)\,d\xi + \int_{\xi_j}^{\xi_{j+1}} h(x, \xi)\,(\xi_{j+1} - \xi)\,d\xi\right]$$

and at the end points

$$v_0 = \frac{1}{\delta}\int_{\xi_0}^{\xi_1} h(x, \xi)\,(\xi_1 - \xi)\,d\xi, \quad v_n = \frac{1}{\delta}\int_{\xi_{n-1}}^{\xi_n} h(x, \xi)\,(\xi - \xi_{n-1})\,d\xi$$

The weight factors corresponding to the logarithmic displacement kernel can be computed analytically (see Ref. 10-11). For weakly singular algebraic kernels, the weights must be evaluated numerically, but the calculations are straightforward. As in the nonsingular case, the integral equation can be represented approximately by a linear system of the form

$$\tilde{\phi}(x_i) - \lambda \sum_{j=0}^{n} v_{ji} g(x_i, \xi_j) \tilde{\phi}(\xi_j) = f(x_i) \qquad (10.6\text{-}4)$$

whose solution, together with the weights v_{ji} and an interpolation formula, gives the numerical solution of the Fredholm equation. Thus, the foregoing procedure extends the Nyström method to weakly singular and logarithmic displacement kernels.

10.6.3 Expansion Methods

In the expansion method, $\phi(x)$ is represented approximately as a weighted sum of $n + 1$ linearly independent functions, $u_i(x)$, continuous on the interval $[a, b]$. In the case of the inhomogeneous Fredholm equation, we write

$$\sum_{j=0}^{n} a_j u_j(x) - \lambda \sum_{j=0}^{n} a_j \int_a^b k(x, \xi)\, u_j(\xi)\, d\xi - f(x) = r(x) \qquad (10.6\text{-}5)$$

where the unknown variable $\phi(x)$ is expressed as the series

$$\tilde{\phi}(x) = \sum_{j=0}^{n} a_j u_j(x)$$

The approximate functional form can be constructed once the coefficients are determined. The values of a_j are determined by multiplying eq. (10.6-5) successively by functions $v_i(x), i = 0, 1, \ldots, n$, and then requiring the inner product of $r(x)$ and $v_i(x)$ to be zero for each value of i. The choice $v_i(x) = \delta(x - x_i)$ is called the *collocation method.* If the v_i are set equal to x^i for $i = 0, 1, \ldots, n$, the procedure is referred to as the *method of moments*. The choice $w_i u_i(x)$, where the $u_i(x)$ are $L_2[a, b]$ and $w(x)$ is an appropriate weight factor, is called the *Galerkin method.* In this latter case, there is some advantage in selecting $\{u_i(x)\}$ to be a set orthogonal on $[a, b]$ with respect to $w(x)$. In all cases except for collocation there are quadratures to be performed, which can be done numerically if necessary. In any event, experience suggests that expansion methods are most likely to succeed when the functions $u_i(x)$ can be chosen so as to represent the important features of the solution (if indeed they are known). As a general rule, there is little advantage to be gained by choosing an expansion procedure over Nyström's method.

10.7 NONLINEAR INTEGRAL EQUATIONS

Procedures for solving nonlinear integral equations are extensions of techniques developed for linear equations: successive approximations, expansion methods, and numerical quadrature combined with an algorithm for solving nonlinear algebraic equations. We shall touch only briefly on some representative methods, using the Urysohn equation as an example.

Consider the nonlinear equation

$$\phi(x) = f(x) + \lambda \int_0^1 F[x, \xi, \phi(\xi)]\, d\xi \tag{10.7-1}$$

where $f(x)$ is bounded by $|f|$ in $[a, b]$, and $F(x, \xi, \phi)$ is integrable and bounded in absolute value by K in R: $0 < x, \xi < 1, |\phi| < b$. Furthermore, suppose that F satisfies the Lipschitz condition $|F(x, \xi, \phi) - F(x, \xi, \phi_0)| < L|\phi - \phi_0|$ in R. Then the method of successive approximations provides a means for generating an approximate solution of eq. (10.7-1)

$$\phi^{(0)}(x) = f(x) - f(0)$$

$$\phi^{(1)}(x) = f(x) + \lambda \int_0^1 F[x, \xi, \phi^{(0)}(\xi)]\, d\xi \tag{10.7-2}$$

$$\cdots$$

$$\phi^{(m)}(x) = f(x) + \lambda \int_0^1 F[x, \xi, \phi^{(m-1)}(\xi)]\, d\xi$$

for $\lambda < \max(L, M)$, where $M = K + |f(0)|\,|\lambda|^{-1}$.

A somewhat similar iterative procedure, based on Newton's method, begins with a more arbitrary first estimate of the solution, $\phi^{(1)}(x)$. If $\phi^{(1)}$ is substituted into eq. (10.7-1), there remains a residual $r_1(x)$

$$\phi^{(1)}(x) - f(x) - \lambda \int_0^1 F[x, \xi, \phi^{(1)}(\xi)]\, d\xi = r_1(x) \tag{10.7-3}$$

Using Newton's method, we can obtain a second estimate $\phi^{(2)}(x)$ from the expression

$$\phi^{(2)}(x) = \phi^{(1)}(x) - y_1(x)$$

where $y_1(x)$ is the solution of the linear integral equation

$$y_1(x) = r_1(x) + \lambda \int_0^1 F_\phi[x, \xi, \phi^{(1)}(\xi)]\, y_1(\xi)\, d\xi$$

where the subscript ϕ denotes differentiation with respect to ϕ. The process is repeated by replacing $\phi^{(1)}$ in eq. (10.7-3) with $\phi^{(2)}$ and calculating a second residual $r_2(x)$. At the mth stage of the iteration, we obtain

$$\phi^{(m+1)}(x) = \phi^{(m)}(x) - y_m(x) \tag{10.7-4}$$

where y_m is the solution of

$$y_m(x) = r_m(x) + \lambda \int_0^1 F_\phi[x, \xi, \phi^{(m)}(\xi)]\, y_m(\xi)\, d\xi \tag{10.7-5}$$

with the inhomogeneous term given by

$$r_m(x) = \phi^{(m)}(x) - f(x) - \lambda \int_0^1 F[x, \xi, \phi^{(m)}(\xi)]\, d\xi \tag{10.7-6}$$

The foregoing iterative technique can be combined with numerical quadrature to avoid cumbersome integrations. Thus, let the range of integration be subdivided in a manner appropriate to a chosen quadrature formula: $0 \leqslant \xi_1 < \xi_2 < \cdots < \xi_j < \cdots < \xi_n \leqslant 1$, with weight factors w_j. Suppose the range of x is subdivided in the same way, and let $\tilde{\phi}_i^{(m+1)} = \tilde{\phi}^{(m+1)}(x_i)$ denote $(m+1)$th approximation of $\phi(x_i)$. We can then employ the following iterative procedure, for $i = 0, 1, 2, \ldots, n$

$$\tilde{\phi}_i^{(m+1)} = \tilde{\phi}_i^{(m)} - \tilde{y}_i^{(m)} \tag{10.7-7}$$

where

$$\tilde{y}_i^{(m)} = \tilde{r}_i^{(m)} + \lambda \sum_{j=0}^{n} w_j F_\phi[x_i, \xi_j, \tilde{\phi}_j^{(m)}]\, \tilde{y}_j^{(m)} \tag{10.7-8}$$

with the values $\tilde{r}_i^{(m)}$ calculated from

$$\tilde{r}_i^{(m)} = \tilde{\phi}_i^{(m)} - f_i - \lambda \sum_{j=0}^{n} w_j F[x_i, \xi_j, \tilde{\phi}_j^{(m)}] \tag{10.7-9}$$

To illustrate the techniques described above, we shall consider two of the nonlinear integral equations mentioned in section 2. First we examine the homogeneous Urysohn equation

$$\psi(x) = \lambda \int_0^1 k(x, \xi)\, [\psi(\xi)]^n\, d\xi$$

Recall that if $n = 1$, the equation has only the trivial solution $\phi(x) = 0$, unless the parameter λ is an eigenvalue of the kernel. The situation is quite different in the nonlinear case. When n is a positive integer equal to or greater than two, we may set $\psi(x) = \lambda^m \phi(x)$, where $m = -(n - 1)^{-1}$, to obtain

$$\phi(x) = \int_0^1 k(x, \xi)\, [\phi(\xi)]^n \, d\xi \qquad (10.7\text{-}10)$$

which demonstrates that λ can be set equal to unity without loss of generality.

Suppose that $k(x, \xi)$ is a degenerate kernel of the form

$$k(x, \xi) = \sum_{i=1}^{N} \alpha_i(x)\, \beta_i(\xi) \qquad (10.7\text{-}11)$$

where $\alpha_i(x)$ is a set of functions linearly independent on $[0, 1]$. Equation (10.7-10) then takes the form

$$\phi(x) = \sum_{i=1}^{N} \alpha_i(x)\, c_i \qquad (10.7\text{-}12)$$

where

$$c_i = \int_0^1 \beta_i(\xi)\, [\phi(\xi)]^n \, d\xi$$

Substitution of eq. (10.7-12) into eq. (10.7-10) gives

$$\sum_{i=1}^{N} \alpha_i(x)\, c_i - \sum_{i=1}^{N} \alpha_i(x) \int_0^1 \beta_i(\xi) \left[\sum_{j=1}^{N} \alpha_j(\xi)\, c_j\right]^n d\xi = 0$$

Since the functions $\alpha_i(x)$ are linearly independent by assumption, we can equate the coefficients of $\alpha_i(x)$ to obtain a set of N algebraic equations of degree n for the constants c_i, $i = 1, 2, \ldots, N$. In general, there will exist $N^n - 1$ solutions in addition to the trivial one $\phi(x) = 0$. Moreover, it can be shown that if N is even, at least one of the solutions will be real. The solution of the nonlinear algebraic equations may be obtained by an iterative technique, such as Newton's method.

Example: The extension to the inhomogeneous problem is straightforward when $f(x)$ has an expansion in terms of $\alpha_i(x)$. By way of example, consider the nonlinear Urysohn equation

$$\phi(x) = \tfrac{3}{4}x + \int_0^1 x\xi\,[\phi(\xi)]^2\,d\xi \qquad (10.7\text{-}13)$$

whose solution is $\phi(x) = x$. This problem is one of three examined by R. H. Moore to illustrate the use of Newton's method in solving equations of the form $P(\phi) = 0$, where P is a nonlinear operator on a Banach space (as, for example, a nonlinear integral operator) [Ref. 10-12]. Moore showed that with two-point Gaussian quadrature, Newton's method converged to 10-digit accuracy in five iterations, beginning with the initial estimate $\phi^{(0)}(x) = \frac{1}{2}$. Of course, the exact solution is very easy to obtain in this case because the degenerate kernel is of the simplest form and the inhomogeneous term is proportional to $\alpha_1(x) = x$. This same problem may be solved (for the purpose of illustration) by the iterative method described above. Beginning with the estimate $\phi^{(0)} = \frac{1}{2}$, we calculate the residual $r_1(x) = \frac{1}{2} - 7x/8$ from eq. (10.7-6), and $y_1(x)$ from eq. (10.7-5), to obtain $\phi^{(1)}(x) = 13x/16$ from eq. (10.7-4). A second iteration gives $\phi^{(2)}(x) = 0.9852x$, which is close to the exact solution.

Example: The equation of Bratu is a homogeneous nonlinear integral equation of the form

$$\phi(x) = \lambda \int_0^b k(x, \xi)\, e^{\phi(\xi)}\, d\xi \qquad (10.7\text{-}14)$$

where $k(x, \xi)$ is the Green's function

$$k(x, \xi) = \frac{1}{b}\begin{cases} \xi(b - x), & 0 < \xi < x \\ x(b - \xi), & x < \xi < b \end{cases}$$

This integral equation is the equivalent of the nonlinear boundary value problem

$$\phi_{xx} + \lambda e^{\phi} = 0 \qquad 0 < x < b$$

subject to $\phi(0) = \phi(b) = 0$. A first integral of the equation can be written

$$[\phi_x(x)]^2 = 2\lambda(\alpha - e^{\phi})$$

$$\alpha = 1 + \frac{1}{2\lambda}[\phi_x(0)]^2$$

Integrating once again, we obtain the implicit result

$$x = \frac{1}{\sqrt{2\lambda}} \int_0^{\phi} \frac{1}{\sqrt{\alpha - \exp(\phi)}}\, d\phi \tag{10.7-15}$$

It can be shown that the equation of Bratu has two real and distinct solutions for $0 < \lambda < \lambda_c$, where

$$\lambda_c = \frac{(1.8745\ldots)^2}{b^2}$$

and that the graphs of these solutions are positive, concave downward, passing through $\phi(0) = \phi(b) = 0$. The curves are symmetric about $x = x_1$, where

$$x_1 = \sqrt{\frac{2}{\lambda\alpha}}\ \ln\left[\sqrt{\alpha} + \sqrt{\alpha - 1}\right] \tag{10.7-16}$$

There are two values of $\phi_x(0)$ corresponding to a given value of b, where

$$b < b_{\max} = \frac{1.8745\ldots}{\sqrt{\lambda_c}}$$

and therefore two values of x_1 from eq. (10.7-16). As $\lambda \to \lambda_c$, the two solutions merge to a single unique limit. If λ exceeds λ_c, Bratu's equation has no solution.

The iterative numerical quadrature procedure given by eqs. (10.7-7) through (10.7-9) is appropriate for finding approximate solutions of eq. (10.7-4). If $\lambda < \lambda_c$ and $b < b_{\max}$, a five-point Gaussian quadrature rule is adequate, with initial estimates of the form $\phi^{(1)}(x) = cx(b - x)$. By way of illustration, if $\lambda = 1$ and $b = \frac{5}{3}$, the iterative numerical method gives the two distinct solutions to 10-digit accuracy with $c = \frac{3}{5}$ and $c = \frac{8}{3}$ in four iterations. The same procedure applied to the Urysohn equation (10.7-13) also converges with 10-digit accuracy in four iterations beginning with $\phi^{(1)} = 1$. Additional examples are to be found in Ref. 10-12.

10.8 REFERENCES

10-1 Courant, R., and Hilbert, D., *Methods of Mathematical Physics*, vol. 1, Wiley-Interscience, New York, 1962.

10-2 Mikhlin, S. G., *Integral Equations*, 2nd ed., Pergamon, New York, 1964.

10-3 Smithies, F., *Integral Equations*, 2nd ed., Cambridge University Press, Cambridge, 1962.

10-4 Tricomi, F. G., *Integral Equations*, Wiley-Interscience, New York, 1957.

10-5 Whittaker, E. T., and Watson, G. N., *A Course of Modern Analysis*, Cambridge University Press, Cambridge, 1978.
10-6 Baker, C. T. H., *The Numerical Treatment of Integral Equations*, Clarendon Press, Oxford, 1971.
10-7 Carrier, G. F., Krook, M., and Pearson, C. E., *Functions of a Complex Variable*, McGraw-Hill, New York, 1966.
10-8 Titchmarsh, E. C., *Introduction to the Theory of Fourier Integrals*, Oxford University Press, Oxford, 1937.
10-9 Morse, P., and Feshbach, H., *Methods of Theoretical Physics*, vol. 1., McGraw-Hill, New York, 1953.
10-10 Lovitt, W. V., *Linear Integral Equations*, Dover, New York, 1950.
10-11 Atkinson, K. E., *A Survey of Numerical Methods for the Solution of Fredholm Integral Equations of the Second Kind*, Society for Industrial and Applied Mathematics, Philadelphia, 1976.
10-12 Moore, R. H., *Newton's Method and Variations*, in *Nonlinear Integral Equations*, P. M. Anselone, ed., University of Wisconsin Press, Madison, 1964.

11

Transform Methods

Gordon E. Latta*

11.0 INTRODUCTION

The object of this chapter is to introduce the integral transforms of most common occurrence in applied mathematics, and to illustrate typical applications by means of examples. The treatment is not meant to be exhaustive, nor is it particularly well-motivated, although we frequently use a specific boundary-value problem, or integral equation, to introduce a particular transform or technique.

Proofs are either omitted entirely, or the key steps are indicated. Reference material for this chapter may be found in Chapters 5, 7, and in Refs. 11-1 and 11-2.

11.1 FOURIER'S INTEGRAL FORMULA

Fourier deduced the following formula by considering Fourier series on the interval $(-L, L)$ and formally letting $L \to \infty$.

$$f(x) = \frac{1}{\pi} \int_0^{\infty} d\lambda \int_{-\infty}^{\infty} f(t) \cos \lambda (x - t)\, dt \qquad (11.1\text{-}1)$$

Since this formula is the starting point for all transform methods, it is important to know sufficient conditions for its validity.

(a) Dirichlet conditions: on each finite interval, $f(x)$ is piecewise differentiable (having a finite number of finite discontinuities), and $\int_{-\infty}^{\infty} |f|\, dx < \infty$. In this case, the right side of (11.1-1) converges to $\frac{1}{2}\,[f(x) + 0) + f(x - 0)]$, and so, to $f(x)$ at each point of continuity.

*Prof. Gordon E. Latta, Dep't. of Mathematics, Naval Postgraduate School, Monterey, Calif. 93940.

(b) If $f(x)$ belongs to the class $L_2(-\infty, \infty)$ of Lebesgue integrable functions, then then (11.1-1) holds a.e. (almost everywhere), in the sense of mean convergence.

(c) Fourier's formula is valid for other classes of functions as well, but it is not easy to characterize these classes. Thus it is clear that for

$$f(x) = \begin{cases} \dfrac{1}{\sqrt{x}}, & x > 0 \\ 0, & x < 0 \end{cases}$$

(11.1-1) is valid, although this $f(x)$ does not fall into any of the above categories.

In the remainder of this chapter, we concern ourselves with variants of (11.1-1), as well as with other formulas, and will tacitly assume the validity of the formulas for the $f(x)$ given.

11.1.1 Fourier Transforms

On replacing $\cos \lambda(x - t)$ by its exponential equivalent, we may rewrite (11.1-1) as

$$f(x) = \frac{1}{2\pi} \int_{-\infty}^{\infty} e^{-i\lambda x}\, d\lambda \int_{-\infty}^{\infty} e^{i\lambda t} f(t)\, dt \tag{11.1-2}$$

Following Cauchy, we define the Fourier transform of $f(x)$ by

$$F(\lambda) = \frac{1}{\sqrt{2\pi}} \int_{-\infty}^{\infty} e^{i\lambda t} f(t)\, dt \tag{11.1-3}$$

and then (11.1-2) reads

$$f(x) = \frac{1}{\sqrt{2\pi}} \int_{-\infty}^{\infty} e^{-i\lambda x} F(\lambda)\, d\lambda \tag{11.1-4}$$

Modern terminology has become fairly well standardized in this regard, although a number of references still use different normalizations for the Fourier transform, such as $y(\lambda) = \int_{-\infty}^{\infty} e^{2\pi i\lambda t} f(t)\, dt$. We shall use (11.1-3) to denote the Fourier transform of $f(t)$ with (11.1-4) being the inversion formula.

Applying (11.1-2) to even functions, $f(-x) = f(x)$, (11.1-3) becomes

$$F(\lambda) = \sqrt{\frac{2}{\pi}} \int_{0}^{\infty} \cos \lambda(t) f(t)\, dt$$

while (11.1-4) reads

$$f(x) = \sqrt{\frac{2}{\pi}} \int_0^\infty \cos \lambda x \, F(\lambda) \, d\lambda$$

Instead of regarding these formulas as particular cases of Fourier transforms, it is convenient to take $f(x)$ arbitrary on $(0, \infty)$, and define

$$F_c(\lambda) = \sqrt{\frac{2}{\pi}} \int_0^\infty \cos \lambda t \, f(t) \, dt \tag{11.1-5}$$

$$f(x) = \sqrt{\frac{2}{\pi}} \int_0^\infty \cos \lambda x \, F_c(\lambda) \, d\lambda \tag{11.1-6}$$

These are called *(Fourier) cosine transforms.*

Similarly, for odd functions, $f(-x) = -f(x)$, we obtain the Fourier sine transform pair

$$F_s(\lambda) = \sqrt{\frac{2}{\pi}} \int_0^\infty \sin \lambda t \, f(t) \, dt \tag{11.1-7}$$

$$f(x) = \sqrt{\frac{2}{\pi}} \int_0^\infty \sin \lambda x \, F_s(\lambda) \, d\lambda \tag{11.1-8}$$

Examples:

(1)

$$f(x) = \begin{cases} e^{-x}, & 0 < x < \infty \\ 0, & -\infty < x < 0 \end{cases}$$

$$F(\lambda) = \frac{1}{\sqrt{2\pi}} \int_0^\infty e^{i\lambda x - x} \, dx = \frac{1}{\sqrt{2\pi}\,(1 - i\lambda)}$$

$$F_c(\lambda) = \sqrt{\frac{2}{\pi}} \left(\frac{1}{1 + \lambda^2} \right) = F(\lambda) + F(-\lambda)$$

$$F_s(\lambda) = \sqrt{\frac{2}{\pi}} \left(\frac{\lambda}{1 + \lambda^2} \right) = \frac{F(\lambda) - F(-\lambda)}{i}$$

(2) From (7.4.4)

$$J_0(x) = \frac{2}{\pi} \int_0^1 \frac{\cos xt \, dt}{\sqrt{1 - t^2}}$$

Using (11.1-5), (11.1-6), we have

$$\sqrt{\frac{\pi}{2}} J_0(x) = \sqrt{\frac{2}{\pi}} \int_0^1 \cos \lambda x \frac{d\lambda}{\sqrt{1 - \lambda^2}}$$

$$\int_0^\infty J_0(x) \cos \lambda x \, dx = \begin{cases} \dfrac{1}{\sqrt{1 - \lambda^2}}, & \lambda < 1 \\ 0, & \lambda > 1 \end{cases}$$

(3) Tables of transforms list numerous identities connecting the transforms of $f(x)$, $f'(x)$, $f(kx)$, and so on. We take such results to be self-evident and refer to them only as the need arises. Thus

$$F(\lambda) = \frac{1}{\sqrt{2\pi}} \int_{-\infty}^\infty e^{i\lambda t} f(t) \, dt$$

implies

$$\frac{1}{\sqrt{2\pi}} \int_{-\infty}^\infty e^{i\lambda t} f'(t) \, dt = -i\lambda F(\lambda)$$

for smooth enough $f(x)$. On the other hand, it is precisely these elementary results which lead to powerful methods for solving certain boundary-value problems. Consider the problem for $\varphi(x,y)$ in $-\infty < x < \infty$, $y > 0$, $\Delta\varphi \equiv \varphi_{xx} + \varphi_{yy} = 0$, $\varphi(x, 0) = f(x)$, $\varphi(x,y) \to 0$ as $|x|$, $y \to \infty$. Let $\Phi(\lambda, y)$ be the Fourier transform of $\varphi(x,y)$:

$$\Phi(\lambda, y) = \frac{1}{\sqrt{2\pi}} \int_{-\infty}^\infty e^{i\lambda x} \varphi(x,y) \, dx$$

$$\Phi_{yy} = \frac{1}{\sqrt{2\pi}} \int_{-\infty}^\infty e^{i\lambda x} (-\varphi_{xx}) \, dx = \lambda^2 \Phi(\lambda, y)$$

Thus, the transform $\Phi(\lambda, y)$ satisfies a differential equation in one fewer independent variable; we have 'transformed out' one variable. We have

$$\Phi(\lambda, y) = A(\lambda)e^{|\lambda|y} + B(\lambda)e^{-|\lambda|y}$$

From $\varphi(x,y) \to 0$ as $|x|, y \to \infty$, we infer that

$$\Phi(\lambda, y) \to 0 \text{ as } y \to \infty, \quad \text{so that } A(\lambda) \equiv 0$$

and

$$\Phi(\lambda, 0) = \frac{1}{\sqrt{2\pi}} \int_{-\infty}^{\infty} e^{i\lambda x} \varphi(x, 0)\, dx = F(\lambda), \quad \text{so that } B(\lambda) = F(\lambda)$$

Thus

$$\Phi(\lambda, y) = F(\lambda)e^{-|\lambda|y} \tag{11.1-9}$$

and we have a formula for the solution of our boundary-value problem

$$\varphi(x,y) = \frac{1}{\sqrt{2\pi}} \int_{-\infty}^{\infty} e^{-i\lambda x} F(\lambda)e^{-|\lambda|y}\, d\lambda \tag{11.1-10}$$

It is customary to express the solution of the problem in terms of the given boundary data $f(x)$, instead of in terms of its transform. We have

$$\begin{aligned}
\varphi(x,y) &= \frac{1}{\sqrt{2\pi}} \int_{-\infty}^{\infty} e^{-i\lambda x} e^{-|\lambda|y}\, d\lambda \, \frac{1}{\sqrt{2\pi}} \int_{-\infty}^{\infty} f(t)e^{i\lambda t}\, dt \\
&= \frac{1}{2\pi} \int_{-\infty}^{\infty} f(t)\, dt \int_{-\infty}^{\infty} e^{-|\lambda|y} e^{-i\lambda(x-t)}\, d\lambda \\
&= \frac{y}{\pi} \int_{-\infty}^{\infty} \frac{f(t)\, dt}{(x-t)^2 + y^2} \quad \text{(Poisson's formula)}
\end{aligned}$$

(4) Let us use the sine transform on the example (3). We call

$$\Phi(x, \lambda) = \sqrt{\frac{2}{\pi}} \int_{0}^{\infty} \sin \lambda y \, \varphi(x,y)\, dy$$

then

$$\Phi_{xx} = \sqrt{\frac{2}{\pi}} \int_0^\infty \sin \lambda y \, (-\varphi_{yy}) \, dy$$

$$\Phi_{xx} = \lambda^2 \Phi - \sqrt{\frac{2}{\pi}} \lambda f(x)$$

We now need a formula for the solution to this ordinary (constant coefficient) differential equation. In fact, we could use Fourier transforms on this problem, or use the known fundamental solution for the problem (see section 11.6.1).

$$\Phi(x, \lambda) = \frac{1}{2} \sqrt{\frac{2}{\pi}} \int_{-\infty}^{\infty} f(t) e^{-\lambda |x-t|} \, dt$$

Using the inversion formula for the sine transform, we again recover Poisson's formula.

(5) While we are at it, consider applying the cosine transform to this same example. Let

$$\psi(x, \lambda) = \sqrt{\frac{2}{\pi}} \int_0^\infty \cos \lambda y \, \varphi(x, y) \, dy$$

then

$$\psi_{xx} - \lambda^2 \psi = \sqrt{\frac{2}{\pi}} \, \varphi_y(x, 0)$$

In this case, the problem for the transform contains an unknown boundary term. We could obtain a representation for $\psi(x, \lambda)$, invert, and obtain an integral equation for the unknown $\varphi_y(x, 0)$. However, these formulations are in general more difficult to solve than the original problem. Thus some transforms are better suited to a given problem than others, and a certain amount of experience is useful in suggesting the more likely one(s) to try.

(6) As a rule of thumb, when the cosine and sine transforms are applicable to a given differential equation, we use the sine transform when the boundary data involve only even-order derivatives, and the cosine transform for odd-order derivatives. The most common cases include quarter-plane (space) problems.

11.1.2 Convolution Integrals, Parseval's Equation

As in ex. (3) of section 11.1.1, we frequently encounter a product of Fourier transforms to invert. Let $F(\lambda)$, $G(\lambda)$ be the Fourier transforms of $f(x), g(x)$; then

$$\frac{1}{\sqrt{2\pi}}\int_{-\infty}^{\infty} e^{-i\lambda x} F(\lambda)\, G(\lambda)\, d\lambda = \frac{1}{\sqrt{2\pi}}\int_{-\infty}^{\infty} e^{-i\lambda x} G(\lambda)\, d\lambda \,\frac{1}{\sqrt{2\pi}}\int_{-\infty}^{\infty} e^{-i\lambda t} f(t)\, dt$$

$$= \frac{1}{\sqrt{2\pi}}\int_{-\infty}^{\infty} f(t)\, dt \,\frac{1}{\sqrt{2\pi}}\int_{-\infty}^{\infty} e^{-i\lambda(x-t)}\, G(\lambda)\, d\lambda$$

$$= \frac{1}{\sqrt{2\pi}}\int_{-\infty}^{\infty} f(t)\, g(x-t)\, dt \qquad (11.1\text{-}11)$$

This integral is called a *convolution integral*, and we obtain the transform pair

$$F(\lambda)\, G(\lambda) \quad \text{and} \quad \frac{1}{\sqrt{2\pi}}\int_{-\infty}^{\infty} f(t)\, g(x-t)\, dt$$

Conversely, integrals of the convolution type have particularly simple Fourier transforms. Thus certain integral equations (mostly arising from boundary-value problems for constant coefficient partial differential equations) can be solved by transform methods (see 11.4).

With $G(\lambda) = \overline{F(\lambda)}$, the same procedure used in deriving (11.1-11) yields

$$\frac{1}{\sqrt{2\pi}}\int_{-\infty}^{\infty} |F(\lambda)|^2\, d\lambda = \frac{1}{\sqrt{2\pi}}\int_{-\infty}^{\infty} |f(t)|^2\, dt \qquad (11.1\text{-}12)$$

upon setting $x = 0$. This is called *Parseval's equation*, and is entirely analogous to the Fourier series equation of the same name.

Examples:

(1)

$$f(x) = \begin{cases} e^{-x}, x > 0 & F(\lambda) = \dfrac{1}{\sqrt{2\pi}}\dfrac{1}{1-i\lambda} \\ 0, \quad x < 0 & \end{cases}$$

$$\int_{-\infty}^{\infty} |F(\lambda)|^2\, d\lambda = \frac{1}{2\pi}\int_{-\infty}^{\infty} \frac{d\lambda}{1+\lambda^2} = \frac{1}{2} = \int_0^{\infty} e^{-2x}\, dx = \int_{-\infty}^{\infty} |f|^2\, dx$$

(2)

$$y''(x) - k^2 y(x) = f(x); \quad -\infty < x < \infty$$

$$y \to 0 \text{ as } |x| \to \infty$$

Taking Fourier transforms

$$-(\lambda^2 + k^2)\, Y(\lambda) = F(\lambda)$$

$$Y(\lambda) = -\frac{F(\lambda)}{\lambda^2 + k^2}$$

$$y(x) = -\frac{1}{2k}\int_{-\infty}^{\infty} e^{-k|x-t|} f(t)\, dt$$

The integrand $G(x, t) = -(1/2k)e^{-k|x-t|}$ is called the *Green's function* for the problem (see 11.6-1).

11.2 LAPLACE TRANSFORMS

In (11.1-3), we restrict $f(x)$ to be zero for $x < 0$. (Note that if $f(0) \neq 0$, there is automatically a discontinuity at $x = 0$). Then let $\lambda = is$, and we are led to the results

$$F(s) = \int_0^{\infty} e^{-sx} f(x)\, dx \tag{11.2-1}$$

$$f(x) = \frac{1}{2\pi i}\int_{-i\infty}^{i\infty} e^{sx} F(s)\, ds, \quad x > 0 \tag{11.2-2}$$

Although it is not often stressed, (11.1-4) also yields

$$\frac{1}{2\pi i}\int_{-i\infty}^{i\infty} e^{xs} F(s)\, ds = \begin{cases} \frac{1}{2} f(0), & x = 0 \\ 0, & x < 0 \end{cases} \tag{11.2-3}$$

The integral in (11.2-1) is convergent (by assumption) for s purely imaginary. If we now regard $s = \sigma + i\tau$ as a complex variable, then

$$F(\sigma + i\tau) = \int_0^{\infty} e^{-\sigma x} e^{-ix\tau} f(x)\, dx$$

converges for $\sigma = 0$, and hence for all $\sigma \geqslant 0$, and tends to zero as $\sigma \to \infty$. Further, the formal derivative

$$F'(\sigma + i\tau) = -\int_0^{\infty} x f(x)\, e^{-ix\tau - \sigma x}\, dx$$

converges (absolutely) for $\sigma > 0$. In other words, $F(s)$ is an analytic function of s for $R(s) > 0$; in the right-half plane; $R(s)$ = real part of s.

We now combine the strictly real variable formulas (11.2-1), (11.2-2) with Cauchy's theorem and the theory of analytic functions. Thus, the path of integration [the imaginary axis for (11.2-2)] can be shifted to the right to *any* line L: $R(s) = \sigma_0$. Hence

$$F(s) = \int_0^\infty e^{-sx} f(x)\, dx, \quad R(s) > 0$$

$$\frac{1}{2\pi i} \int_L e^{xs} F(s)\, ds = \begin{cases} f(x), & x > 0 \\ 0, & x < 0 \end{cases} \tag{11.2-4}$$

where L is any line $R(s) = \sigma_0 > 0$.

A second extension now follows immediately. If $f(x)$ is locally integrable, $e^{-kx}|f(x)|$ integrable on $(0, \infty)$ for some $k > 0$, then

$$\int_0^\infty e^{-sx} f(x)\, dx = F(s), \quad R(s) > k \tag{11.2-5}$$

$$\frac{1}{2\pi i} \int_L e^{xs} F(s)\, ds = f(x), \quad x > 0 \tag{11.2-6}$$

with L any path (line) in $R(s) > k$.

Examples:

(1)

$$f(x) = \sin x$$

$$F(s) = \frac{1}{1 + s^2}, \quad R(s) > 0$$

(11.2-6) becomes

$$\frac{1}{2\pi i} \int_L e^{xs} \frac{ds}{s^2 + 1}$$

which can readily be evaluated by the residue theorem, "closing the contour to the left," yielding

$$\text{residue}\,\Big|_{s=i} + \text{residue}\,\Big|_{s=-i} = \frac{e^{ix}}{2i} - \frac{e^{-ix}}{2i} = \sin x$$

(2) For suitable smooth $f(x)$, with Laplace transform $F(s)$, $F(s) = L\{f(x)\}$, we have

$$L\{f'(x)\} = sF(s) - f(0)$$

$$L\{f''(x)\} = s^2F(s) - sf(0) - f'(0), \text{ and so on}$$

Accordingly, the Laplace transform is particularly well suited to treating initial-value problems. Consider the constant coefficient problem

$$\begin{gathered} y'' + ay' + by = f(x) \\ y(0) = y_0, \quad y'(0) = y_0' \end{gathered} \tag{11.2-7}$$

With

$$Y(s) = L\{y(x)\}$$

we obtain

$$(s^2 + as + b)Y(s) = F(s) + y_0 s + y_0' + ay_0 \tag{11.2-8}$$

a purely algebraic problem, with $Y(s)$ analytic for $R(s)$ greater than that of the roots of $s^2 + as + b = 0$, and for $F(s)$ analytic.

(3) Transforms *act like filters* (to borrow a phrase from Ref. 11-3): when applied to a differential equation, they can "filter out" whole classes of undesired solutions. Consider the Laplace transform of the Bessel function $y = J_0(x)$, satisfying

$$\begin{gathered} xy'' + y' + xy = 0 \\ y(0) = 1, \quad y'(0) \text{ finite} \end{gathered}$$

We find

$$(s^2 + 1)\,Y'(s) + sY(s) = 0 \tag{11.2-9}$$

using

$$L\{xy\} = -\frac{d}{ds}Y(s)$$

and so

$$Y(s) = \frac{c}{\sqrt{1 + s^2}} \tag{11.2-10}$$

Note that the specific initial value $y(0) = 1$ cancels out in (11.2-9), and some other device is needed to evaluate c in (11.2-10). (See 11.9).

(4) The convolution theorem (1.1-11) for Laplace transforms becomes

$$\frac{1}{2\pi i}\int_L e^{xs}\,F(s)\,G(s)\,ds = \int_0^x f(t)\,g(x-t)\,dt$$

$$= \int_0^x g(t)\,f(x-t)\,dt \qquad (11.2\text{-}11)$$

Thus, for example $1/(1+s^2)^2$ is the transform of $\int_0^x \sin t \sin (x-t)\,dt = (\sin x - x\cos x)/2$. Again, from $\int_0^\infty x^{a-1}\,e^{-sx}\,dx = \Gamma(a)/s^a$, we have $1/s^{a+b}$ is the Laplace transform of

$$\frac{1}{\Gamma(a)\,\Gamma(b)}\int_0^x t^{a-1}\,(x-t)^{b-1}\,dt = \frac{x^{a+b-1}}{\Gamma(a+b)}$$

from the beta-function integral.

11.2.1 Generalized Fourier Transforms

The analytic function ideas of (11.2) are readily applied to the Fourier transform (occasionally referred to as the two-sided Laplace transform). We define

$$F_+(w) = \frac{1}{\sqrt{2\pi}}\int_0^\infty e^{ixw}\,f(x)\,dx \qquad (11.2\text{-}12)$$

$$F_-(w) = \frac{1}{\sqrt{2\pi}}\int_{-\infty}^0 e^{ixw}\,f(x)\,dx \qquad (11.2\text{-}13)$$

valid for $f(x)$ of exponential type, i.e., for $e^{-k|x|}\,|f(x)|$ integrable on $(-\infty, \infty)$ for some $k > 0$. With $w = u + iv$, then (11.2-12) is analytic for $I(w) > k$, (11.2-13) is analytic for $I(w) < -k$. If $k = 0$, the ordinary Fourier transform is revealed as the sum of the boundary values of two analytic functions, each analytic (and tending to zero as $w \to \infty$) in a half-plane. If $k < 0$, these half planes overlap, and we speak of a common 'strip' of analyticity for $F_+(w)$ and $F_-(w)$.

Example:

(1)

$$f(x) = e^{|x|}$$

$$F_+(w) = -\frac{1}{\sqrt{2\pi}\,(1 + iw)}$$

$$F_-(w) = \frac{1}{\sqrt{2\pi}\,(iw - 1)}$$

Then

$$f(x) = \frac{1}{\sqrt{2\pi}} \int_{ai-\infty}^{ai+\infty} e^{-ixw}\, F_+(w)\, dw + \frac{1}{\sqrt{2\pi}} \int_{bi-\infty}^{bi+\infty} e^{-ixw}\, F_-(w)\, dw \tag{11.2-14}$$

where the lines $v = a$, $v = b$ lie in the respective half-planes of analyticity; here

$$a > 1, \quad b < -1$$

Thus

$$\frac{1}{\sqrt{2\pi}} \int_{ai-\infty}^{ai+\infty} e^{-ixw}\, F_+(w)\, dw = -\frac{1}{2\pi i} \int_{ai-\infty}^{ai+\infty} e^{-ixw} \frac{dw}{w - i} = \begin{cases} 0, & x < 0 \\ e^x, & x > 0 \end{cases}$$

An important theorem in applying these generalized transforms is the following result: If $F_+(w)$, $v \geqslant a$ and $F_-(w)$, $v \leqslant -b$ are generalized Fourier transforms, and if

$$\frac{1}{\sqrt{2\pi}} \int_{ai-\infty}^{ai+\infty} e^{-ixw}\, F_+(w)\, dw + \frac{1}{\sqrt{2\pi}} \int_{bi-\infty}^{bi+\infty} e^{-ixw}\, F_-(w)\, dw \equiv 0$$

then $F_+(w)$ is actually analytic in the larger domain $v \geqslant -b$, $F_-(w)$ for $v \leqslant a$, and in this strip (it is assumed $a, b > 0$)

$$F_+(w) + F_-(w) \equiv 0$$

11.2.2 Mellin Transforms

Combining the ideas of (11.1) and (11.2.1), we define, using (11.1-3)

$$F(s) = \int_0^\infty x^{s-1} f(x)\, dx \tag{11.2-15}$$

where we have replaced e^t by x, $i\lambda$ by s, and $(1/\sqrt{2\pi}) f(t)$ by $f(x)$. This $F(s)$ is called the *Mellin transform* of $f(x)$: $F(s) = M\{f(x)\}$. In (11.2-15), $\int_0^1 x^{s-1} f(x)\, dx$ corresponds to F_+ of the generalized Fourier transform, and is analytic in a right half-plane $R(s) > \sigma_0$. Correspondingly $\int_1^\infty x^{s-1} f(x)\, dx$ is analytic in a left half-plane $R(s) < \sigma_1$. For most applications of the Mellin transform it is assumed that these half-planes overlap, and we write

$$F(s) = \int_0^\infty x^{s-1} f(x)\, dx, \quad \sigma_0 < R(s) < \sigma_1 \tag{11.2-15}$$

$$f(x) = \frac{1}{2\pi i} \int_L x^{-s} F(s)\, ds \tag{11.2-16}$$

the inversion formula following from (11.1-4), where L is any line $R(s) = \sigma = \text{const}$ in the strip of analyticity $\sigma_0 < \sigma < \sigma_1$.

Examples:

(1)

$$f(x) = e^{-x}$$

$$F(s) = \Gamma(s), \quad \sigma > 0$$

$$\frac{1}{2\pi i} \int_L x^{-s}\, \Gamma(s)\, ds = \sum_0^\infty \frac{(-1)^n x^n}{n!} = e^{-x}$$

This last integral is evaluated by residues, shifting L to the left, the process valid for $|x| < 1$. An appeal to analytic continuation is needed to verify the result for $x > 1$.

(2) Consider $F(s) = \Gamma(s)$, $-1 < \sigma < 0$. As above, $f(x) = e^{-x} - 1$. This example shows that the strip of analyticity is an integral part of the definition of $F(s)$.

(3) The convolution formula for Mellin transforms can take several useful forms.

$$\frac{1}{2\pi i}\int_L F(s)\,G(s)x^{-s}\,ds = \int_0^\infty g(t)\,f\left(\frac{x}{t}\right)\frac{dt}{t}$$

$$\frac{1}{2\pi i}\int_L F(s)\,G(1-s)x^{-s}\,ds = \int_0^\infty f(xt)\,g(t)\,dt \tag{11.2-17}$$

Precisely as the Fourier transform is useful in simplifying convolution integrals, the Mellin transform is useful in treating integrals with 'product' kernels. For example (and ignoring the circular reasoning) consider (11.1-7) as an integral equation.

$$\sqrt{\frac{2}{\pi}}\int_0^\infty f(x)\sin\lambda x\,dx = g(\lambda)$$

We regard $g(\lambda)$ as given, and wish to solve for the unknown $f(x)$. Formally applying (11.2-17) and using

$$M\{\sin x\} = \Gamma(s)\sin\frac{\pi s}{2}$$

$$\int_0^\infty \lambda^{s-1}\,g(\lambda)\,d\lambda = G(s) = \sqrt{\frac{2}{\pi}}\int_0^\infty x^{-s}f(x)\,dx\;\Gamma(s)\sin\frac{\pi s}{2}$$

$$= \frac{2}{\pi}\Gamma(s)\sin\frac{\pi s}{2}\,F(1-s)$$

Hence $F(s) = \sqrt{2/\pi}\;\Gamma(s)\sin(\pi s/2)\,G(1-s)$, and so

$$f(x) = \sqrt{\frac{2}{\pi}}\int_0^\infty \sin\lambda x\,g(\lambda)\,d\lambda, \quad \text{which is (11.1-8)}$$

(4) From

$$\frac{1}{2\pi i}\int_L \frac{\Gamma(s)\,\Gamma(a-s)\,\Gamma(b-s)}{\Gamma(c-s)}x^{-s}\,ds$$

$$= \sum_0^\infty \frac{(-1)^n}{n!}\frac{\Gamma(a+n)\,\Gamma(b+n)}{\Gamma(c+n)}x^n, \quad |x| < 1$$

valid for $a, b > 0$ and $0 < R(s) < a, b$ for L, we see that

$$M\{{}_2F_1(a,b;c;-x)\} = \frac{\Gamma(s)\,\Gamma(a-s)\,\Gamma(b-s)}{\Gamma(c-s)}\,\frac{\Gamma(c)}{\Gamma(a)\,\Gamma(b)}$$

$$0 < R(s) < a, b \quad (11.2\text{-}18)$$

In fact, (11.2-18) is valid much more generally: in performing the inversion as indicated above, we merely deform the path L so as to keep the poles of $\Gamma(a - s)$, $\Gamma(b - s)$ to the right of the path, while the poles of $\Gamma(s)$ are on the left. The constants a, b can be complex.

11.3 LINEARITY, SUPERPOSITION, REPRESENTATION FORMULAS

The superposition principle for solutions of linear differential equations is of the utmost importance. If L is a linear differential operator, and $\varphi_1, \varphi_2, \ldots$ are solutions of $L\varphi = 0$, then so is $C_1\varphi_1 + C_2\varphi_2 + \ldots$ a solution for arbitrary constants $C_1, C_2 \ldots$. Passing to the limit as n, the number of terms, tends to infinity, we obtain an infinite series or definite integral representations for particular solutions (hopefully of the particular solution we seek).

Examples:

(1)

$$\Delta\phi = \phi_{xx} + \phi_{yy} = 0 \quad \text{in the upper half-plane } \phi(x, 0) = f(x)$$

We can readily find elementary solutions which contain a free parameter, which we can use for superposition.

$$\varphi = e^{\alpha x + \beta y}, \quad \text{where } \alpha^2 + \beta^2 = 0$$

or

$$\varphi = e^{i\lambda x}\, e^{-\lambda y}$$

or

$$\varphi = e^{-\lambda y}\, e^{i\lambda(x - x')} f(x')$$

or

$$\varphi = e^{-\lambda y} \cos \lambda(x - x') f(x')$$

Superposing on x', we may obtain the integral

$$\int_{-\infty}^{\infty} e^{-\lambda y} \cos \lambda(x - x') f(x')\, dx'$$

and further superposing on λ

$$\varphi = \int_0^\infty e^{-\lambda y}\, d\lambda \int_{-\infty}^\infty \cos \lambda\,(x - x')f(x')\, ds'$$

$$= y \int_{-\infty}^\infty \frac{f(x')\, dx'}{(x - x')^2 + y^2}$$

If we now let $y \downarrow 0$, the right side becomes $\pi f(x)$, so that

$$\phi(x,y) = \frac{y}{\pi} \int_{-\infty}^\infty \frac{f(x')\, dx'}{(x - x')^2 + y^2}$$

Poisson's formula is suggested as a representation of the solution of the given boundary-value problem. Even more illuminating is the suggested result

$$\pi f(x) = \lim_{y \to 0} \int_0^\infty e^{-\lambda y}\, d\lambda \int_{-\infty}^\infty \cos \lambda (x - x') f(x')\, dx'$$

which is, of course, Fourier's integral formula.

For the same boundary-value problem, we could equally well obtain elementary solutions in polar coordinates.

$$\Delta\phi = \phi_{rr} + \frac{1}{r}\phi_r + \frac{1}{r^2}\phi_{\theta\theta} = 0; \quad r > 0, 0 < \theta < \pi$$

We have $r^{-\lambda} \sin \lambda\theta, r^{-\lambda} \cos \lambda\theta$ as solutions for all λ. Guided by (11.2-16), we try a superposition of the form

$$\varphi = \frac{1}{2\pi i} \int_L r^{-\lambda}\, [A(\lambda) \cos \lambda\theta + B(\lambda) \sin \lambda\theta]\, d\lambda \tag{11.3-1}$$

Now, for $\theta = 0, \varphi(r, 0) = f(r)$, we have

$$f(r) = \frac{1}{2\pi i} \int_L r^{-\lambda} A(\lambda)\, d\lambda \tag{11.3-2}$$

so that $A(\lambda)$ is the Mellin transform of $f(r)$. Likewise, on $\theta = \pi, \varphi = f(-r)$

$$f(-r) = \frac{1}{2\pi i} \int_L r^{-\lambda} \, [A(\lambda) \cos \pi\lambda + B(\lambda) \sin \pi\lambda] \, d\lambda$$

and (11.2-15) solves for $B(\lambda)$.

(2) Consider the axially-symmetric potential problem $\Delta\phi = \phi_{xx} + \phi_{yy} + \phi_{zz} = 0$ in the upper half-space, $\phi \to 0$ as $R = \sqrt{x^2 + y^2 + z^2} \to \infty$, and $\phi(x, y, 0) = f(r)$, $r^2 = x^2 + y^2$. From symmetry considerations, a solution $\phi(r, z)$ exists

$$\phi_{rr} + \frac{1}{r} \phi_r + \phi_{zz} = 0$$

Elementary solutions (product solutions) are

$$\phi = e^{-\lambda z} J_0(\lambda r)$$

and superposition leads to

$$\phi(\lambda, z) = \int_0^\infty a(\lambda) J_0(\lambda r) e^{-\lambda z} \, d\lambda \qquad (11.3\text{-}3)$$

[Note that in forming the superposition, we keep in mind the conditions at infinity, and any boundedness conditions in the finite space. Thus we use only $\lambda \geqslant 0$ here, since $e^{-\lambda z}$ is unbounded for $\lambda < 0$; similarly we ignore $N_0(\lambda r)$, since it is unbounded at $r = 0$].

Equation (11.3-3) will formally solve the boundary-value problem provided

$$\phi(r, 0) = f(r) = \int_0^\infty a(\lambda) J_0(\lambda r) \, d\lambda \qquad (11.3\text{-}4)$$

Applying the Mellin transform to (11.3-4), and using

$$\int_0^\infty x^{s-1} J_0(x) \, dx = \frac{2^{s-1} \Gamma\left(\frac{s}{2}\right)}{\Gamma\left(1 - \frac{s}{2}\right)}; \quad 0 < R(s) < \frac{3}{2}$$

we obtain

$$F(s) = \frac{2^{s-1} \Gamma\left(\frac{s}{2}\right)}{\Gamma\left(1 - \frac{s}{2}\right)} \int_0^\infty \lambda^{-s} a(\lambda) \, d\lambda$$

$$= \frac{2^{s-1} \Gamma\left(\frac{s}{2}\right)}{\Gamma\left(1 - \frac{s}{2}\right)} A(1 - s)$$

Thus

$$A(s) = F(1 - s)\,\frac{\Gamma\left(\frac{1}{2} + \frac{s}{2}\right) 2^{+s}}{\Gamma\left(\frac{1}{2} - \frac{s}{2}\right)}$$

and so

$$a(\lambda) = \int_0^\infty \lambda r\, J_0(\lambda r)\, f(r)\, dr \tag{11.3-5}$$

The conditions for validity of these manipulations may be obtained for individual cases; see also section 11.3.1.

11.3.1 Hankel Transforms

As suggested by ex. (2) of section 11.3, we have the self-reciprocal formulas

$$F(\lambda) = \int_0^\infty \sqrt{\lambda x}\, J_0(\lambda x)\, f(x)\, dx \tag{11.3-6}$$

$$f(x) = \int_0^\infty \sqrt{\lambda x}\, J_0(\lambda x)\, F(\lambda)\, d\lambda \tag{11.3-7}$$

Sufficient conditions for their validity are essentially the same as for the Fourier transform. Thus if $\int_0^\infty |f|\, dx < \infty$ and $f(x)$ is piecewise smooth (Dirichlet conditions), then (11.3-6), (11.3-7) hold at each point of continuity.

If we rewrite the sine and cosine transforms, using $J_{1/2}(x) = \sqrt{2/\pi x}\,\sin x$, and $J_{1/2}(x) = \sqrt{2/\pi x}\,\cos x$, we find

$$F(\lambda) = \int_0^\infty \sqrt{\lambda x}\, J_\nu(\lambda x)\, f(x)\, dx$$

$$f(x) = \int_0^\infty \sqrt{\lambda x}\, J_\nu(\lambda x)\, F(\lambda)\, d\lambda \tag{11.3-8}$$

are reciprocal (transform) formulas, for $\nu = 0, \pm\frac{1}{2}$. These are three examples of a more general class of formulas. E. C. Titchmarsh calls $k(x)$ a generalized Fourier kernel if

$$g(\lambda) = \int_0^\infty k(\lambda x) f(x)\, dx$$
$$f(x) = \int_0^\infty k(\lambda x) g(\lambda)\, d\lambda \tag{11.3-9}$$

imply each other. In terms of their Mellin transforms, (11.3-9) becomes

$$F(s) = K(s)\, G(1 - s), \quad G(s) = K(s)\, F(1 - s)$$

and so

$$K(s) K(1 - s) \equiv 1 \tag{11.3-10}$$

must hold for a generalized Fourier kernel.

Many particular cases are known, although the most useful ones stem from

$$k(x) = \sqrt{x}\, J_\nu(x); \quad K(s) = \frac{2^{s-(1/2)} \Gamma\left(\frac{s}{2} + \frac{\nu}{2} + \frac{1}{4}\right)}{\Gamma\left(\frac{\nu}{2} - \frac{s}{2} + \frac{3}{4}\right)}$$

which give the formulas (11.3-8) for arbitrary ν (although some restriction, such as $R(\nu) > -3/2$, is necessary to guarantee convergence of the integrals). These formulas are called the *Hankel transform*. Such kernels arise in axially-symmetric potential problems of higher dimension.

COMMENT: An ever-increasing number of special transforms appear in the literature based on a nonsymmetrical variant of these generalized Fourier kernels. We see that

$$g(\lambda) = \int_0^\infty k(\lambda x) f(x)\, dx$$
$$f(x) = \int_0^\infty h(\lambda x) g(\lambda)\, d\lambda \tag{11.3-11}$$

provide transform and inversion formula provided that the Mellin transforms of the kernels satisfy

$$H(s)\,K(1-s) \equiv 1 \tag{11.3-12}$$

One merely finds a special case, and then studies conditions for its validity.

11.3.2 Kontorovich-Lebedev Transforms

In ex. (1) of section 11.3, we used polar coordinates and the Mellin formulas to solve Laplace's equation in the upper half-plane. (Fourier transforming on x was even simpler, although, as seen in 11.4, there are some problems where the Mellin formulas yield a tremendously simpler solution. With an eye to the future, we seek similar simplification for other boundary value problems).

Let us try the same technique for the upper half-plane problem

$$\Delta\phi - \phi = 0, \qquad (\Delta\phi = \phi_{xx} + \phi_{yy})$$

$$\phi(x, 0) = f(x), \qquad \phi \to 0 \text{ as } r = \sqrt{x^2 + y^2} \to \infty$$

Product solutions of the differential equation are

$$\phi = R(r)\,e^{i\lambda\theta}; \quad r^2R'' + rR' - (\lambda^2 + r^2)\,R = 0$$

The function $R = K_\lambda(r)$, the modified Bessel function with imaginary argument, behaves like $r^{-\lambda}$ for small r, and suggests a generalization of the Mellin formula

$$\phi(r, \theta) = \frac{1}{2\pi i}\int_L K_\lambda(r)\,[A(\lambda)\cos\lambda\theta + B(\lambda)\sin\lambda\theta]\,d\lambda \tag{11.3-13}$$

(of 11.3-1), where we anticipate that L will be a path (line) of integration parallel to the imaginary axis.

To satisfy the boundary conditions, we require

$$\phi(r, 0) = f(r) = \frac{1}{2\pi i}\int_L K_\lambda(r)\,A(\lambda)\,d\lambda \tag{11.3-14}$$

with a similar result for $\theta = \pi$. Hence our problem reduces to the inversion of (11.3-14). By means of the formulas

$$\begin{aligned} K_{ix}(r) &= \frac{1}{\operatorname{ch}\dfrac{\pi x}{2}}\int_0^\infty \cos(x\theta)\cos(r\operatorname{sh}\theta)\,d\theta \\ &= \frac{1}{\operatorname{sh}\dfrac{\pi x}{2}}\int_0^\infty \sin(x\theta)\sin(r\operatorname{sh}\theta)\,d\theta \end{aligned} \tag{11.3-15}$$

and taking the path of integration in (11.3-14) to be the imaginary axis, we have

$$f(r) = \frac{1}{\pi} \int_0^\infty K_{ix}(r)\, g(x)\, dx \tag{11.3-16}$$

and

$$g(x) = \frac{2x}{\pi} \operatorname{sh}(\pi x) \int_0^\infty K_{ix}(r)\, f(r)\, \frac{dr}{r} \tag{11.3-17}$$

In order to obtain this last integral, it has been necessary to use a $(C, 1)$ summability, that is (11.3-17) is to be interpreted as a $(C, 1)$ integral.

The pair (11.3-16), (11.3-17) is called the *Kontorovich-Lebedev transform.*

Example: Rewriting (11.3-13), we have

$$\phi(r, \theta) = \frac{1}{\pi} \int_0^\infty K_{ix}(r)\, [a(x) \operatorname{ch}(x\theta) + b(x) \operatorname{sh}(x\theta)]\, dx \tag{11.3-18}$$

Let us take $\phi \equiv 1$ on the boundary. (We note that $\phi(x, y) \equiv \phi(y)$ now, $\phi_{yy} - \phi = 0$, and so $\phi = e^{-y}$.) From (11.3-18), $\theta = 0$, we have

$$1 = \frac{1}{\pi} \int_0^\infty K_{ix}(r)\, a(x)\, dx$$

so that (11.3-17) gives

$$a(x) = \frac{2x}{\pi} \operatorname{sh}(\pi x) \int_0^\infty K_{ix}(r) \frac{dr}{r}$$

$$= \frac{2x}{\pi} \operatorname{sh}(\pi x) \int_0^\infty \frac{dr}{r} \frac{1}{\operatorname{sh}\left(\frac{\pi x}{2}\right)} \int_0^\infty \sin(x\theta) \sin(r \operatorname{sh}\theta)\, d\theta, \text{ from (11.3-15)}$$

$$= 2x \operatorname{ch}\left(\frac{\pi x}{2}\right) \int_0^\infty \sin(x\theta)\, d\theta$$

$$= 2 \operatorname{ch}\left(\frac{\pi x}{2}\right), \quad \text{evaluated } (C, 1)$$

Similarly, $b(x) = -2\,\text{sh}\,(\pi x/2)$, and

$$\phi(r,\theta) = \frac{2}{\pi}\int_0^\infty K_{ix}(r)\,\text{ch}\left(\frac{\pi}{2} - \theta\right) x\,dx$$

$$= e^{-r\cos(\pi/2-\theta)} = e^{-y}$$

11.4 THE WIENER-HOPF TECHNIQUE

So far, as we have introduced various transforms, we have illustrated their use by applying them to boundary-value problems in which a routine application of the transform yielded an immediate solution of the problem. The domains have been infinite, or semi-infinite, and the boundary conditions have been of the first kind (function prescribed) or of the second kind (normal derivative prescribed). More challenging, and so more interesting, are the problems involving boundary conditions of the third kind, in which the function is prescribed on part of the boundary, and a derivative on the remainder of the boundary.

These problems can usually be recast as an integral equation (with a difference kernel or a product kernel in simpler cases), and have been solved only in special cases. As an illustration, we consider

$$\Delta\phi - \phi = 0; \quad \Delta\phi = \phi_{xx} + \phi_{yy} \text{ in the upper half plane}$$

$$\begin{aligned} \phi(x,0) &= e^{-x}; \quad x > 0 \\ \phi_y(x,0) &= 0; \quad\;\; x < 0 \end{aligned} \tag{11.4-1}$$

with $\phi \to 0$ as $x, y \to \infty$.

Applying the Fourier transform

$$\Phi(w,y) = \frac{1}{\sqrt{2\pi}}\int_{-\infty}^\infty e^{ixw}\phi(x,y)\,dx$$

we find

$$\Phi_{yy} - (1 + w^2)\,\Phi = 0$$

$$\Phi(w,y) = A(w)\,e^{-y\sqrt{1+w^2}}, \quad \text{using the condition at } \infty$$

$$\Phi_y(w,y) = -\sqrt{1+w^2}\,A(w)\,e^{-y\sqrt{1+w^2}} \tag{11.4-2}$$

$$\Phi(w,0) = \frac{-\Phi_y(w,0)}{\sqrt{1+w^2}}$$

We note here, using the convolution theorem for Fourier transforms

$$\phi(x, 0) = -\frac{1}{\pi}\int_{-\infty}^{\infty} K_0(|x - t|)\, \phi_y(t, 0)\, dt \tag{11.4-3}$$

as an identity connecting the boundary values of $\phi(x, 0)$, $\phi_y(x, 0)$.

For our boundary-value problem, we have $\phi_y(x, 0) = 0$, $x < 0$; define $\phi_y(x, 0) = -\pi f(x), x > 0$. Then

$$\int_0^{\infty} K_0(|x - t|)\, f(t)\, dt = e^{-x}; \quad x > 0 \tag{11.4-4}$$

This is an integral equation (singular) of the first kind for the unknown $f(x)$; its solution clearly solves the full problem.

A closer look at (11.4-4) shows that there are two unknown functions involved; $f(x)$ for $x > 0$, and $g(x) = \int_0^{\infty} K_0(|x - t|)\, f(t)\, dt$ for $x < 0$. In terms of these unknowns, (11.4-2) becomes

$$G_-(w) + \frac{1}{\sqrt{2\pi}\,(1 - iw)} = \frac{\pi F_+(w)}{\sqrt{1 + w^2}} \tag{11.4-5}$$

where G_-, F_+ are the generalized Fourier transforms. Thus, there are likewise two unknown functions in the transform plane, and a single equation to determine them. The procedure for solving this equation, using analytic continuation techniques, has become known as the Wiener-Hopf method. Actually, Wiener and Hopf gave a prescription for solving the singular Fredholm equation of the second kind

$$f(x) = \int_0^{\infty} k(x - t)\, f(t)\, dt, \qquad x > 0 \tag{11.4-6}$$

which transforms into

$$F_+(w) + F_-(w) = K(w)\sqrt{2\pi}\, F_+(w) \tag{11.4-7}$$

again a single equation for two unknown functions.

We sketch in three illustrative cases. First of all, from (11.4-4) we assume $f(x) = O(e^{kx})$, $k < 1$, as $x \to \infty$, in order to guarantee the existence of the integral. Then $F_+(w)$ is analytic for $v > k$, and $\to 0$ as $w \to \infty$ there. Further the definition of $g(x)$ shows that $g(x) = O(e^x)$ as $x \to -\infty$, so that $G_-(w)$ is analytic for $v < 1$.

Accordingly, all the functions occurring in (11.4-5) are single-valued and analytic in the strip $k < v < 1$. We take the branch cuts for $\sqrt{1+w^2}$ from i to ∞ and $-i$ to $-\infty$ along the imaginary axis. Following the line suggested by E. T. Copson, we characterize $F_+(w)$ by analytic continuation, as follows.

(a) Equation (11.4-5) immediately continues $F_+(w)$ to be single-valued, analytic for $v > -1$.

(b) Continuing $F_+(w)$ around $w = -i$, as (11.4-5) shows $F_+(w)$ defined in the full plane, we find $F_+^*(w)$, the continuation, satisfies $F_+^*(w) = -F_+(w)$. Hence

$$F_+(w) = \frac{\phi(w)}{\sqrt{w+i}} \tag{11.4-8}$$

and $\phi(w)$ is single-valued in the full plane.

(c)
$$\pi\phi(w) = \sqrt{w-i}\left\{(w+i)\,G_-(w) + \frac{i}{\sqrt{2\pi}}\right\} \tag{11.4-9}$$

From (11.4-9) we observe that $\phi(w)$ is analytic at $w = -i$, and so is entire.

(d) From (11.4-8), $|\phi(w)| = o(|w|^{1/2})$ as $w \to \infty$ in the upper half-plane, while (11.4-9) shows $|\phi| = o(|w|^{3/2})$ at ∞ in the lower half-plane.

Liouville's theorem now yields $\phi(w)$ is a polynomial of degree $\leqslant (\frac{3}{2})$, and so at most of degree 1. Since $|\phi| = o(|w|^{1/2})$ in the upper half-plane, $\phi(w) \equiv$ constant. Finally (11.4-9) yields

$$\begin{aligned} \pi\phi(-1) &= \sqrt{2i}\,\frac{i}{\sqrt{2\pi}} \\ \phi(w) &= \sqrt{-2i}\,\frac{i}{\pi\sqrt{2\pi}} \end{aligned} \tag{11.4-10}$$

This approach is elegant, although we can see formidable difficulties in performing the analytic continuation in more complicated cases.

The second example outlines the Wiener-Hopf lemma. The basic weapon is the application of Cauchy's integral formula to a function analytic in a strip $a < v < b$, vanishing appropriately at ∞. If $H(w)$ is such a function, then $H(w) = H_+(w) + H_-(w)$ uniquely, with H_+ analytic and $o(1)$ in $v > a$, H_- analytic and $o(1)$ in $v < b$. We take the limiting case of a rectangle lying in $a < v < b$ as the right and left ends approach ∞. Hence

$$H_+(w) = \frac{1}{2\pi i}\int_{\alpha i-\infty}^{\alpha i+\infty} \frac{H(t)\,dt}{t-w}, \quad a < \alpha, \quad I(w) > \alpha$$

Next, given $L(w)$, $H(w) = \log L(w)$; if $H(w)$ satisfies the above requirements, then

$$L(w) = e^{H_+(w)} \cdot e^{H_-(w)} \tag{11.4-11}$$

and $L(w)$ is 'factored' into a product (quotient) of bounded zero-free functions, each well-behaved in a full half-plane.

These ideas are now applied to (11.4-7)

$$F_+(w) \{1 - \sqrt{2\pi}\, K(w)\} = -F_-(w) \tag{11.4-12}$$

where we assume $K(w)$ analytic in a strip $a < v < b$. The key step is to factor $1 - \sqrt{2\pi}\, K(w) = N_+(w)\, N_-(w)$, obtaining

$$F_+(w)\, N_+(w) = \frac{-F_-(w)}{N_-(w)} \tag{11.4-13}$$

Each side of (11.4-12) then represents the same entire function, which we can write down provided it is bounded by a power of w as $w \to \infty$. (By Liouville's theorem it is a polynomial.)

The following lemma describes in detail how to factor $1 - \sqrt{2\pi}\, K(w)$.

Lemma (*Wiener-Hopf*): Let $\phi(w)$ be analytic in the strip $-1 < v < 1$ and let $\int_{\infty}^{\infty} |\phi(u + iv)|^2 \, du < M = M(\alpha)$ in any interior strip $-1 < -\alpha \leqslant v \leqslant \alpha < 1$. In any interior strip, $-1 < -\beta \leqslant v \leqslant \beta < 1$, $1 - \phi(w)$ has only a finite number of zeros, $w_1, \cdots, w_n$. Then $1 - \phi(w) = (w - w_1) \cdots (w - w_n)\, \phi_+(w)/\phi_-(w)$, where ϕ_+ is analytic and zero-free in $v \leqslant \beta$, ϕ_- is analytic and zero-free in $v \geqslant -\beta$; and in their respective half-planes of analyticity

$$|\phi_+| > K|w|^{-(n/2)-k}, \quad |\phi_-| > K|w|^{(n/2)-k}$$

where k is a positive integer depending only on $\phi(w)$.

COMMENT: To factor $1 - \phi(w)$, we want $\log\,[1 - \phi(w)]$ to be single-valued, analytic, and $\to 0$ as $w \to \infty$ in the strip. Clearly, we must first remove the zeros $(w - w_1) \cdots (w - w_n)$ of $1 - \phi(w)$. We have

$$1 - \phi(w) = (w - w_1) \cdots (w - w_n) \frac{\psi(w)}{(1 + w^2)^{n/2}} \tag{11.4-14}$$

so that $\psi(w) \to 1$ as $w \to \infty$. This still does not guarantee that $\log \psi(w) \to 0$ as $w \to \infty$ in the strip. To achieve this result, we set

$$\psi(w) = \left(\frac{w - i}{w + i}\right)^k \chi(w) \tag{11.4-15}$$

choosing k so that the variation in $\log \chi(w)$ is zero along the whole strip. Now, $\chi(w)$ factors as in (11.4-11).

In (11.4-12) we now have

$$F_+(w)\,(w - w_1)\cdots(w - w_n)\,\frac{1}{(1 + w^2)^{n/2}}\left(\frac{w - i}{w + i}\right)^k \frac{\Omega_+(w)}{\Omega_-(w)} = -F_-$$

or

$$F_+(w)\,\frac{(w - w_1)\cdots(w - w_n)}{(w + i)^{n/2 + k}}\,\Omega_+ = -F_-\,\Omega_-(w - i)^{n/2 - k}$$

Hence, each side is entire, a polynomial of degree $\leqslant n/2 - k$, the constants of which can be evaluated by substitution into the given integral equation.

Example:

$$k(x) = \lambda\, e^{-|x|}, \qquad \lambda < \tfrac{1}{2}$$

$$1 - \sqrt{2\pi}\, K(w) = \frac{w^2 + 1 - 2\lambda}{w^2 + 1}$$

$$F_+\,\frac{(w^2 + 1 - 2\lambda)}{w + i} = -F_-(w - i)$$

Each side is constant, so that

$$F_+(w) = \frac{c(w + i)}{w^2 + 1 - 2\lambda}$$

As a final example, from Oseen flow past a semi-infinite flat plate, we obtain a relation in the transform plane valid for real w only; there is no strip of analyticity. The boundary-value problem for the velocity vector (u, v) and the pressure, p, is

$$\left.\begin{aligned} \nu(u_{xx} + v_{yy}) &= U u_x + p_x \\ \nu(u_{xx} + v_{yy}) &= U v_x + p_y \\ u_x + v_y &= 0 \end{aligned}\right\} -\infty < x < \infty, \quad y > 0$$

with

$$v = 0 \text{ on } y = 0$$

$$u = u_0 \text{ on } y = 0, \quad x > 0 \quad \text{(the plate)}$$

$$u, v, p \to 0 \text{ as } x, y \to \infty$$

Denoting $u(x, 0) = g(x)$ for $x < 0$, and $u_y(x, 0) = f(x)$ for $x > 0$, we obtain

$$G_-(w) + \frac{iu_0}{\sqrt{2\pi}\, w} = \frac{iv}{u}\left(1 - \sqrt{1 - \frac{iu}{vw}}\right) \operatorname{sgn}(w)\, F_+(w) \tag{11.4-16}$$

where

$$\operatorname{sgn}(w) = \begin{cases} 1, & w > 0, \\ -1, & w < 0, \end{cases}$$

and the quantity $\sqrt{1 - iu/vw} \to 1$ as $w \to \pm\infty$.

NOTE: One way to create a strip of analyticity is to imbed this problem in a one-parameter family of problems, replacing $|w|$ by $\sqrt{w^2 + \epsilon^2}$, $\operatorname{sgn} w = |w|/w \to \sqrt{w^2 + \epsilon^2}/w$ and use the strip $-\epsilon < v < \epsilon$. Alternately, we may proceed directly.

Equation (11.4-16), while not possessing a strip in which every term is analytic nevertheless is a relation between the boundary values of analytic functions (F_+ in the upper, and G_- in the lower). There is another difficulty: $\sqrt{1 - iu/vw}$ is to tend to one as $w \to \pm\infty$ along the real axis. On the other hand, the analytic function $\sqrt{1 - iu/vw}$, defined to tend to 1 as $w \to +\infty$, is now a specific branch, and we have no control over its nature as $w \to -\infty$. We can choose the branch cuts, however, so that this specific branch does have the desired behavior as $w \to -\infty$. To this end, we cut, parallel to the imaginary axis, from $w = 0$ to $w = -i\infty$, and $w = iu/v$ to $i\infty$.

In the cut plane, on the real axis, (11.4-16) reads

$$\begin{aligned} G_-(w) + \frac{iu_0}{\sqrt{2\pi}} &= -\frac{iv}{u}\left(1 - \sqrt{1 - \frac{iu}{vw}}\right) F_+(w) \text{ on the negative axis} \\ G_-(w) + \frac{iu_0}{\sqrt{2\pi}} &= \frac{iv}{u}\left(1 - \sqrt{1 - \frac{iu}{vw}}\right) F_+(w) \text{ on the positive axis} \end{aligned} \tag{11.4-17}$$

Starting from the requirement that $F_+(w)$ is single-valued, analytic in $v > 0$, we see that (11.4-17) permits the continuation of $F_+(w)$ to the full plane, and further, that the continued value around $w = 0$ is $-F_+(w)$. Hence

$$F_+(w) = \frac{\phi(w)}{\sqrt{w}} \tag{11.4-18}$$

and $\phi(w)$ is single-valued everywhere. We obtain from (11.4-17) a single relation valid on the full real axis

$$G_-(w) + \frac{iu_0}{\sqrt{2\pi}} = -\frac{iv}{u}\left(\frac{1}{\sqrt{w}} - \frac{1}{w}\sqrt{w - \frac{iu}{v}}\right)\phi(w) \tag{11.4-19}$$

Finally, assuming the local integrability of $f(x)$, $g(x)$, and their vanishing at ∞, we have $wG_-(w)$, $wF_+(w) \to 0$ as $w \to 0$.

Thus, $\phi(w)$ is single-valued everywhere and can be singular only at $w = 0$, where it can have, at worst, a pole; since $wF_+(w) \to 0$, $\phi(w)$ must be analytic at $w = 0$, and $\phi(w) \equiv$ const. from Liouville's theorem.

11.4.1 Dual Integral Equations

Consider the mixed boundary-value problem for the three-dimensional, axially-symmetric potential $\phi(r, z)$:

$$\phi_{rr} + \frac{1}{r}\phi_r + \phi_{zz} = 0 \quad \text{outside } z = 0, r \leqslant 1$$

with

$$\begin{aligned} &\phi(r, 0) = 1, \quad r < 1 \\ &\phi \to 0 \text{ as } r, z \to \infty \end{aligned} \tag{11.4-20}$$

(This is the conductor potential of a circular disc.)

Invoking symmetry, uniqueness, $\phi(r, z)$ is even in z for > 1. Hence, equivalent boundary conditions are

$$\begin{aligned} &\phi(r, 0) = 1, \quad r < 1 \\ &\phi_z(r, 0) = 0, \quad r > 1 \end{aligned} \tag{11.4-21}$$

As in (2), section 11.3, we have

$$\phi(r, z) = \int_0^\infty f(\lambda)\, J_0(\lambda r)\, e^{-\lambda z}\, d\lambda \tag{11.4-22}$$

with

$$\begin{aligned} \phi(r, 0) &= \int_0^\infty f(\lambda)\, J_0(\lambda r)\, d\lambda = 1, \quad r < 1 \\ -\phi_z(r, 0) &= \int_0^\infty \lambda f(\lambda)\, J_0(\lambda r)\, d\lambda = 0, \quad r > 1 \end{aligned} \tag{11.4-23}$$

The pair of equations (11.4-23) are called dual integral equations; their simultaneous solution is required.

The following solution, given by Titchmarsh[11-4], involves analytic continuation, and several of the ideas used in the Wiener-Hopf method. From the Hankel formulas (11.3-8), and denoting $\phi_z(r, 0) = -h(r)$, $r < 1$, the second integral in (11.4-23) yields

$$f(\lambda) = \int_0^1 r\,h(r)\,J_0(\lambda r)\,dr$$

Hence we may assume $f(\lambda)$ has a Mellin transform, analytic in a strip $0 < R(s) < 3/2$. Applying convolution formula for the Mellin transform to (11.4-23), we obtain

$$\begin{aligned} &\frac{1}{2\pi i}\int_L F(s)\,\frac{2^{-s}\Gamma\left(\frac{1}{2}-\frac{s}{2}\right)}{\Gamma\left(\frac{1}{2}-\frac{s}{2}\right)}\,r^{s-1}\,ds = 1, \qquad r < 1 \\ &\frac{1}{2\pi i}\int_L F(s)\,\frac{2^{1-s}\Gamma\left(1-\frac{s}{2}\right)}{\Gamma\left(\frac{1}{2}\,s\right)}\,r^{s-2}\,ds = 0, \qquad r > 1 \end{aligned} \tag{11.4-24}$$

where L is a line in $0 < R(s) < 3/2$.

We next 'factor' the integrands of (11.4-24) defining

$$F(s) = \frac{2^s\Gamma\left(\frac{s}{2}\right)}{\Gamma\left(\frac{1}{2}-\frac{s}{2}\right)}\,G(s) \tag{11.4-25}$$

and obtaining

$$\begin{aligned} &\frac{1}{2\pi i}\int_L \frac{\Gamma\left(\frac{s}{2}\right)}{\Gamma\left(\frac{1}{2}+\frac{s}{2}\right)}\,G(s)\,r^{s-1}\,ds = 1, \qquad r < 1 \\ &\frac{1}{2\pi i}\int_L \frac{\Gamma\left(1-\frac{s}{2}\right)}{\Gamma\left(\frac{1}{2}-\frac{s}{2}\right)}\,G(s)\,r^{s-1}\,ds = 0, \qquad r > 1 \end{aligned} \tag{11.4-26}$$

The object of the factoring is to achieve $\Gamma(s/2)/\Gamma\left(\frac{1}{2}+\frac{s}{2}\right)$ in the first integral, analytic for $R(s) > 0$, and having no zeros there. Similarly in the second integral, $\Gamma\left(1 - \frac{s}{2}\right) \Big/ \Gamma\left(\frac{1}{2} - \frac{s}{2}\right)$ is analytic, zero-free for $R(s) < 1$.

In the next steps, the path of integration will be shifted between two lines L_1, L_2 in the strip $0 < R(s) < 1$. Specifically, let z be any point in the strip, and take L_1 to lie to the left of z, L_2 to the right of z.

Take $L = L_2$ for the first of the integrals (11.4-26), multiply by r^{-z} and integrate $0 \leqslant r \leqslant 1$.

$$\frac{1}{2\pi i} \int_{L_2} \frac{\Gamma\left(\frac{s}{2}\right)}{\Gamma\left(\frac{1}{2} + \frac{s}{2}\right)} G(s) \frac{ds}{s - z} = \frac{1}{1 - z} \tag{11.4-25}$$

Now shift to L_1, using Cauchy's theorem.

$$\frac{1}{2\pi i} \int_{L_1} \frac{\Gamma\left(\frac{s}{2}\right)}{\Gamma\left(\frac{1}{2} + \frac{s}{2}\right)} G(s) \frac{ds}{s - z} = \frac{1}{1 - z} - \frac{\Gamma\left(\frac{z}{2}\right)}{\Gamma\left(\frac{1}{2} + \frac{z}{2}\right)} G(z) \tag{11.4-26}$$

This equation effects the analytic continuation of $G(z)$ to the right half-plane, since the integral on the left is analytic for z to the right of L_1, and so for $R(z) > 0$. Instead of the analytic characterization of $G(z)$, Titchmarsh obtains the solution by formula, with minimal extra assumptions. From (11.4-26), we have $G(z) - \Gamma\left(\frac{1}{2} + \frac{z}{2}\right) \Big/ \Gamma\left(\frac{z}{2}\right)(1 - z)$ is analytic for $R(z) > 0$, since $\Gamma\left(\frac{z}{2}\right) \Big/ \Gamma\left(\frac{1}{2} + \frac{z}{2}\right)$ is zero-free there. Accordingly

$$\frac{1}{2\pi i} \int_{L_2} \left\{ G(s) - \frac{\Gamma\left(\frac{s}{2} + \frac{1}{2}\right)}{\Gamma\left(\frac{s}{2}\right)(1 - s)} \right\} \frac{ds}{s - z} = 0 \tag{11.4-27}$$

provided $G(s)$ is suitably behaved at ∞.

In a similar way, the second integral of (11.4-26) is multiplied by r^{-z}, integrated 1 to ∞, and L is taken to be L_1.

$$\frac{1}{2\pi i} \int_{L_1} \frac{\Gamma\left(1 - \frac{s}{2}\right)}{\Gamma\left(\frac{1}{2} - \frac{s}{2}\right)} \frac{G(s)\, ds}{s - z} = 0$$

Shifting to L_2, we find that $\Gamma\left(1-\frac{s}{2}\right)\Gamma\left(\frac{1}{2}-\frac{s}{2}\right)G(s)$, and hence $G(s)$ are analytic for $R(s) < 1$. Thus

$$\begin{aligned}\frac{1}{2\pi i}\int_{L_1}\frac{G(s)\,ds}{s-z} &= 0\\ \frac{1}{2\pi i}\int_{L_2}\frac{G(s)\,ds}{s-z} &= G(z)\end{aligned} \qquad (11.4\text{-}28)$$

The solution of the problem is now contained in (11.4-27) and (11.4-28).

$$G(z)=\frac{1}{2\pi i}\int_{L_2}\frac{\Gamma\left(\frac{1}{2}+\frac{s}{2}\right)}{\Gamma\left(\frac{s}{2}\right)(1-s)}\,\frac{1}{s-z}\,ds=\frac{1}{\sqrt{\pi}\,(1-z)}$$

$$F(s)=\frac{2^{s-1}\Gamma\left(\frac{s}{2}\right)}{\sqrt{\pi}\,\Gamma\left(\frac{3}{2}-\frac{s}{2}\right)}$$

$$f(\lambda)=\frac{2}{\pi}\frac{\sin\lambda}{\lambda}$$

Titchmarsh further discusses the dual pair

$$\int_0^\infty y^\alpha f(y)\,J_\nu(xy)\,dy=g(x), \qquad 0<x<1$$

$$\int_0^\infty f(y)\,J_\nu(xy)\,dy=0, \qquad x>1$$

by the same methods.

11.5 ABEL'S INTEGRAL EQUATION, FRACTIONAL INTEGRALS, WEYL TRANSFORMS

The n-fold iterated integral

$$\begin{aligned}I^n f(x) &= \int_a^x\int_a^{x_1}\cdots\int_a^{x_{n-1}} f(x_n)\,dx_n\,dx_{n-1}\cdots dx_1\\ &= \frac{1}{(n-1)!}\int_a^x (x-t)^{n-1} f(t)\,dt\end{aligned}$$

is readily generalized to

$$I^\alpha f(x) = \frac{1}{\Gamma(\alpha)} \int_a^x (x - t)^{\alpha - 1} f(t)\, dt \tag{11.5-1}$$

In this form $I^\alpha f(x)$ exists for integrable $f(x)$, $R(\alpha) > 0$ and is called a *fractional integral*, or the Riemann-Liouville integral.

We have $I^\alpha [I^\beta f(x)] = I^{\alpha+\beta} f(x)$ for $R(\alpha) > 0, R(\beta) > 0$, (see ex. (4); section 11.2), and $I^0 f(x) = f(x)$ for $f(x)$ continuous.

As a function of the complex number α, $I^\alpha f(x)$ is analytic for $R(\alpha) > 0$ and can be continued to be left half-plane for smooth enough $f(x)$ by means of

$$I^\alpha f(x) = \frac{(x - a)^\alpha}{\Gamma(\alpha + 1)} f(a) + I^{\alpha+1} f'(x) \tag{11.5-2}$$

on integration by parts. We have

$$I^{(-n)} f(x) = f^{(n)}(x) = \frac{d^n}{dx^n} f(x) \tag{11.5-3}$$

Since $I^{1/2}(I^{1/2} f(x)) = If(x) = \int_a^x f(t)\, dt$, it is clear that $I^{1/2} f(x)$ is a fractional integral. (M. Riesz generalized these integrals to a multidimensional form and used the analytic continuation properties to solve the wave equation in an arbitrary number of dimensions. The analogues of (11.5-3) exist in odd dimensions and lead to Huygen's principle.)

Analogous results are suggested by iterating integrals $\int_x^\infty f(t)\, dt$. The n-fold iterated integral

$$W^n f(x) = \frac{1}{(n - 1)!} \int_x^\infty (t - x)^{n-1} f(t)\, dt$$

and

$$W^\alpha f(x) = \frac{1}{\Gamma(\alpha)} \int_x^\infty (t - x)^{\alpha - 1} f(t)\, dt \tag{11.5-4}$$

is the so-called *Weyl transform*. In general, $W^\alpha f(x)$ is analytic only in a strip $0 < R(\alpha) < k$. Within this strip, $W^\beta [W^\alpha f] = W^{\alpha+\beta} f$ for $R(\alpha) > 0$, $R(\beta) > 0$, and $W^0 f(x) = f(x)$ for $f(x)$ continuous.

For differentiable $f(x)$, suitably behaved at ∞, $W^{\alpha}f(x) = -W^{\alpha+1}f'(x)$ extends $W^{\alpha}f$ into the left half-plane.

Examples:

(1) *Abel's integral equation* is

$$g(x) = \int_0^x (x - t)^{-\alpha} f(t)\,dt, \quad 0 < \alpha < 1$$

From the foregoing section

$$g(x) = \Gamma(1 - \alpha)\, I^{1-\alpha} f(x)$$

$$I^{\alpha} g(x) = \Gamma(1 - \alpha)\, I f(x)$$

$$f(x) = \frac{1}{\Gamma(1 - \alpha)} \frac{d}{dx} I^{\alpha} g(x)$$

$$f(x) = \frac{\sin \pi\alpha}{\pi} \frac{d}{dx} \int_0^x (x - t)^{\alpha - 1} g(t)\,dt$$

the solution also obtained by Abel.

(2) Since the lower limit for the Riemann-Liouville integral can be any fixed number, the limits of integration are frequently indicated as $\overset{x}{\underset{a}{I}}{}^{\alpha} f(x) = 1/\Gamma(\alpha) \times \int_a^x (x - t)^{\alpha-1} f(t)\,dt$.

Consider

$$\overset{x}{\underset{0}{I}}{}^{\alpha} J_0(k\sqrt{x}) = \sum_0^{\infty} \frac{(-1)^n \left(\frac{k}{2}\right)^{2n}}{(n!)^2} I^{\alpha}(x^n)$$

$$= \sum_0^{\infty} \frac{(-1)^n \left(\frac{k}{2}\right)^{2n}}{\Gamma(n + \alpha + 1) \cdot n!} x^{n+\alpha} \tag{11.5-5}$$

$$= \left(\frac{2}{k}\right)^{\alpha} x^{\alpha/2} J_{\alpha}(k\sqrt{x})$$

(3) The Laplace transform is ideally suited to evaluate these fractional integrals. Consider

$$L\{\overset{x}{\underset{0}{I}}{}^{\alpha} J_0(k\sqrt{x})\} = \frac{1}{s^{\alpha}} L\{J_0(k\sqrt{x})\} = \frac{1}{s^{\alpha}} \frac{e^{-k^2/4s}}{s}$$

$$\mathop{I^{\alpha}}_{0}^{x} J_0(k\sqrt{x}) = \left(\frac{2}{k}\right)^{\alpha} x^{\alpha/2} J_{\alpha}(k\sqrt{x})$$

(4)

$$W^{\alpha}(J_0(k\sqrt{x})) = \left(\frac{2}{k}\right)^{\alpha} x^{\alpha/2} J_{-\alpha}(k\sqrt{x}) \tag{11.5-6}$$

One way to evaluate Weyl transforms is to use the convolution theorem for the Fourier transform. Thus, the Fourier transform of

$$W^{\alpha}f(x) \text{ is } \frac{F(w)}{(e^{\pi i/2}w)^{\alpha}}$$

where $F(w)$ is the Fourier transform of $f(x)$, and

$$\frac{1}{(e^{\pi i/2}w)^{\alpha}} = \begin{cases} \dfrac{e^{-(\pi i/2)\alpha}}{w^{\alpha}}, & w > 0 \\[2ex] \dfrac{e^{(\pi i/2)\alpha}}{|w|^{\alpha}}, & w < 0 \end{cases}$$

11.5.1 Dual Integral Equations Revisited

We consider again the dual integral equation (11.4-23)

$$\int_0^{\infty} f(\lambda) J_0(\lambda r)\, d\lambda = 1, \qquad r < 1 \tag{11.5-7}$$

$$\int_0^{\infty} \lambda f(\lambda) J_0(\lambda r)\, d\lambda = 0, \qquad r > 1 \tag{11.5-8}$$

Replacing r by $\sqrt{r}$ in (11.5-7), and applying $\mathop{I}_{0}^{r}{}^{1/2}$, using (11.5-5), we obtain

$$\int_0^{\infty} f(\lambda) \sqrt{\frac{2}{\lambda}}\, r^{1/4} J_{1/2}(\lambda\sqrt{r})\, d\lambda = 2\sqrt{\frac{r}{\pi}}, \quad \sqrt{r} < 1$$

or

$$\int_0^{\infty} f(\lambda) \frac{\sin \lambda r}{\lambda} d\lambda = r, \qquad r < 1$$

or

$$\int_0^\infty f(\lambda)\cos\lambda r\,d\lambda = 1, \qquad r<1 \tag{11.5-9}$$

Similarly, in (11.5-8), with r replaced by $\sqrt{r}$, and applying $\overset{\infty}{\underset{r}{W}}{}^{1/2}$, we obtain, with (11.5-6)

$$\int_0^\infty f(\lambda)\cos\lambda r\,d\lambda = 0, \qquad r>1 \tag{11.5-10}$$

Now from (11.5-9), (11.5-10) and the cosine transform

$$f(\lambda) = \frac{2}{\pi}\int_0^1 \cos\lambda r\,dr = \frac{2}{\pi}\frac{\sin\lambda}{\lambda}$$

as previously obtained.

This result is a special case of a more general situation. The function $f(\lambda)$ appearing in (11.5-7) and (11.5-8) determines a three-dimensional, axially-symmetric potential, having mixed boundary data. The same $f(\lambda)$ also determines a two-dimensional potential with unmixed boundary data, namely

$$\phi(x,y) = \int_0^\infty f(\lambda)\cos\lambda x\, e^{-\lambda y}\,d\lambda$$

$$\phi(x,0) = \begin{cases} 1, & 0<x<1 \\ 0, & x>1, \quad \text{and } \phi(-x,y)=\phi(x,y). \end{cases}$$

We have thus a method of 'descent,' in which the higher dimensional mixed problem is trivially solved (by formula) from the related lower dimensional problem.

More generally, the dual equations arising from the mixed axially-symmetric potential in $\beta + 3$ dimensions,

$$\phi(r,z) = \int_0^\infty f(\lambda)\,(\lambda r)^{-\beta/2} J_{\beta/2}(\lambda r)\, e^{-\lambda z}\,d\lambda$$

with

$$\begin{aligned} &\int_0^\infty f(\lambda)\,(\lambda r)^{-\beta/2} J_{\beta/2}(\lambda r)\,d\lambda = g(r), \qquad r<1 \\ &\int_0^\infty \lambda f(\lambda)\,(\lambda r)^{-\beta/2} J_{\beta/2}(\lambda r)\,d\lambda = h(r), \qquad r>1 \end{aligned} \tag{11.5-11}$$

can be solved immediately from the Hankel formula after applying $I^{1/2}$, $W^{1/2}$ to (11.5-11) with $r \to \sqrt{r}$. We obtain

$$\int_0^\infty f(\lambda)\,(\lambda r)^{-\beta/2-1/2}\,J_{\beta/2-1/2}(\lambda r)\,d\lambda \quad \text{known for } 0 < r < \infty$$

See (11.3-8).

11.6 POISSON'S FORMULA, SUMMATION OF SERIES

If $f(x)$ is of bounded variation on $(0, \infty)$, continuous, and tends to zero as $x \to \infty$, let $F_c(\lambda)$ be its cosine transform; then

$$\sqrt{\beta}\left\{\frac{1}{2}F_c(0) + \sum_1^\infty F_c(n\beta)\right\} = \sqrt{\alpha}\left\{\frac{1}{2}f(0) + \sum_1^\infty f(n\alpha)\right\} \tag{11.6-1}$$

where $\alpha\beta = 2\pi$, $\alpha > 0$.

Examples:

(1) The function $f(x) = e^{-x}$ has $F_c(\lambda) = \sqrt{2/\pi}\; 1/1 + \lambda^2$ and yields

$$\sqrt{\alpha}\left(\frac{1}{2} + \sum_1^\infty e^{-n\alpha}\right) = \frac{\sqrt{\alpha}}{2}\coth\frac{\alpha}{2}$$

$$= \frac{2}{\sqrt{\alpha}}\left(\frac{1}{2} + \sum_1^\infty \frac{1}{1 + \dfrac{4\pi^2}{\alpha^2}n^2}\right)$$

or equivalently

$$\sum_{-\infty}^\infty \frac{1}{n^2 + a^2} = \frac{\pi}{a}\coth \pi a$$

(2) The function $f(x) = e^{-1/2x^2}$ has $F_c(\lambda) = e^{-1/2\lambda^2}$ so that

$$\sqrt{\alpha}\left(\frac{1}{2} + \sum_1^\infty e^{-1/2\alpha^2 n^2}\right) = \sqrt{\beta}\left(\frac{1}{2} + \sum_1^\infty e^{-1/2\beta^2 n^2}\right)$$

a result useful in the asymptotic evaluation of theta functions.

Comparable formulas hold in terms of the sine transform, and the Mellin transform. Among other things, they offer one method of transforming infinite series into other forms.

11.6.1 Fundamental Solutions, Green's Functions

The transforms we have studied so far all involve infinite or semi-infinite intervals. Accordingly, in using transforms to simplify boundary-value problems, we can treat only special domains, infinite in extent.

There is, however, one important ingredient used in the solution of general boundary value problems which we can obtain by transform methods, namely, the fundamental solution. This solution of the differential equation(s) has the so-called characteristic singularity for the problem. The *Green's function* is then the fundamental solution plus a regular (nonsingular) solution.

The results which we discuss here are best suited to constant coefficient problems, but do apply to some variable coefficient cases.

Consider a scalar partial differential equation

$$L\varphi(x) = f(x) \text{ in } R_n, \quad x = x_1, \ldots, x_n \tag{11.6-2}$$

Here, L is a polynomial in $p_i = \partial/\partial x_i$. We may write $L(p_i)\,\varphi(x) = f(x)$. The fundamental solution, $\gamma(x)$, is defined by the requirement that

$$\varphi(x) = \int_{R_n} \gamma(x - t) f(t)\, dt \tag{11.6-3}$$

solves (11.6-2) for all $f(x)$ of compact support, and subject to appropriate behavior at ∞ (such as vanishing, radiation condition, etc). We regard $\gamma(x)$ as a (Schwarzian) distribution, and in the following, $\delta = \delta(x)$ is the delta-function. Equation (11.6-3) is the convolution product

$$\varphi = \gamma \otimes f \tag{11.6-4}$$

Distributionally, $L\varphi = \delta \otimes f$, and $f = \delta \otimes f$, so

$$\begin{aligned} \delta \otimes f = L\varphi &= L\delta \otimes \varphi \\ &= L\delta \otimes \gamma \otimes f \\ &= (L\delta \otimes \gamma) \otimes f \end{aligned}$$

for all distributions f, and thus

$$L\delta \otimes \gamma = \delta, \quad \text{or } L\gamma = \delta \tag{11.6-5}$$

The fundamental solution, $\gamma(x)$, is thus the distribution (of order 0) satisfying

$$L\gamma(x) = 0, \quad x \neq 0$$

$$L\gamma \otimes f = f \text{ for all } f \text{ of compact support}$$

Applying the n-fold Fourier transform, we replace $\partial/\partial x_i$ by $-i\lambda_i$ in polynomial L, obtaining

$$L(-i\lambda_i, \ldots, -i\lambda_n)\,\bar{\gamma}(\lambda) = 1 \tag{11.6-6}$$

where $\bar{\gamma}(\lambda)$ is the n-fold Fourier transform of $\gamma(x)$.

We must emphasize that this is a purely formal result, as multidimensional distributions are not necessarily the cross product of n one-dimensional distributions, among other things. However, in most cases of interest, the distributions $\gamma(x)$ are actually smooth functions for large x, and the relation (11.6-6) remains valid.

For example, take $L\varphi = \varphi_{xx} + \varphi_{yy} - \varphi$.

$$L(p) = p_1^2 + p_2^2 - 1$$

$$L(-i\lambda) = -\lambda_1^2 - \lambda_2^2 - 1$$

$$\bar{\gamma} = -\frac{1}{\lambda_1^2 + \lambda_2^2 - 1}$$

$$\gamma = -\frac{1}{2\pi} K_0(\sqrt{x^2 + y^2})$$

These results find frequent application in obtaining representations (via integral equations) for the solution of boundary-value problems. To solve $L\varphi = 0$ in a domain D, φ prescribed on ∂D, the boundary of D, we consider $L\varphi = f$ in R_n where f is a distribution having support on ∂D. Then $\varphi = \gamma \otimes f$ is a representation, with f to be determined from the boundary data. For example, for $\varphi_{xx} + \varphi_{yy} - \varphi = 0$ in the upper half-plane, $\varphi(x, 0) = g(x)$. We may try $L\varphi = f(x)\,\delta(y)$ or $L\varphi = h(x)\,\delta'(y)$, yielding

$$\varphi = \gamma \otimes f = -\frac{1}{2\pi}\int_{-\infty}^{\infty} K_0(\sqrt{(x-t)^2 + y^2})\, f(t)\, dt$$

or

$$= \frac{1}{2\pi}\int_{-\infty}^{\infty} \frac{\partial}{\partial y} K_0(\sqrt{(x-t)^2 + y^2})\, h(t)\, dt$$

For systems of partial differential equations

$$A \cdot J = F$$

where A is a square matrix of differential operators, $m \times m$, with J, F being m vectors, the fundamental solution matrix can be found as follows: We have $A = A(p)$, a polynomial matrix in $p_i = \partial/\partial x_i$; adj A is the cofactor matrix and

$$A \cdot \text{adj}\, A = \det A \cdot I \tag{11.6-7}$$

where det A is a scalar differential operator. Divide each individual column of adj A by any common (polynomial) factor, obtaining B

$$A \cdot B = D$$

where D is a diagonal matrix of scalar operators

$$D = (\theta_{ii})$$

Let $\theta_{ii}\, \gamma_i = \delta$, that is, γ_i are the m fundamental solutions of the scalar operators θ_{ii}. Then

$$C = \begin{pmatrix} \gamma_1 & & & 0 \\ & \gamma_2 & & \\ & & \ddots & \\ 0 & & & \gamma_m \end{pmatrix}, \qquad DC = \begin{pmatrix} \delta \\ \vdots \\ \delta \end{pmatrix}$$

and $B \cdot C$ is the fundamental solution matrix for A.

$$\left[A \int_{R_n} B \cdot C \cdot F\, dt = AB \int_{R_n} C \cdot F\, dt \right.$$
$$= D \int_{R_n} C \cdot F\, dt$$
$$= D\delta \otimes C \otimes F$$
$$= \delta \otimes F$$
$$\left. = F \right]$$

As previously, we may use Fourier transforms to obtain the γ_i.

NOTE: We obtain a decomposition of the solution vector J into (at most) m portions, one corresponding to each scalar operator θ_{ii} (or to each γ_i). A well-known instance is longitudinal and transverse waves in linearized hydrodynamical flows.

11.7 HILBERT TRANSFORMS, RIEMANN-HILBERT PROBLEM

As has been already observed, the transforms we have been using are either analytic functions, or are the boundary values of analytic functions. When we recall that if an analytic function on an arc or merely at a sequence of points converging to a point of analyticity is known, then we know the analytic function everywhere, we see that there must be other techniques, similar to the Wiener-Hopf, for using transforms to solve boundary-value problems.

In this section, we consider boundary values, on all or a portion of the real axis, of analytic functions. (Similar results will pertain for curved arcs.) For the functions involved, $f(z)$, or $f(x)$, defined on $(-\infty, \infty)$ or merely on $(0, 1)$, we assume that $\int_{-\infty}^{\infty} |f(x+iy)|^p \, dx$, and $\int_0^1 |f(x)|^p \, dx$ are finite, and $p > 1$; that is, these are L_p functions.

As a starting point, for $f(z)$ in $L_p(-\infty, \infty)$ uniformly in $y > 0$, Cauchy's formula yields

$$f(z) = \frac{1}{2\pi i} \int_{-\infty}^{\infty} \frac{f(t)\,dt}{t-z}, \qquad y > 0 \tag{11.7-1}$$

With $f(x + i0) = u(x) + iv(x)$, and using the Cauchy singular integral, we have

$$u(x) + iv(x) = \frac{1}{2\pi i} \fint_{-\infty}^{\infty} \frac{u(t) + iv(t)}{t - x} dt + \frac{1}{2} \{u(x) + iv(x)\}$$

The symbol $\fint_{-\infty}^{\infty} \varphi(t)\,dt/(t-x)$ is the Cauchy principal value integral.

Thus

$$u(x) = \frac{1}{\pi} \fint_{-\infty}^{\infty} \frac{v(t)\,dt}{t-x}$$
$$-v(x) = \frac{1}{\pi} \fint_{-\infty}^{\infty} \frac{u(t)\,dt}{t-x} \tag{11.7-2}$$

The equations (11.7-2) define the *Hilbert transform* and its inverse.

Conversely, if $v(x)$ is $L_p(-\infty, \infty)$, and $u(x)$ is defined by the first of equations (11.7-2), then $u(x)$ is also L_p, and the second equation in (11.7-2) follows. Furthermore, the combination $u + iv$ is the boundary value of an analytic function $f(z)$, analytic and L_p in the upper half-plane.

Denoting

$$H\{v(x)\} = \frac{1}{\pi} \fint_{-\infty}^{\infty} v(t)\,dt/(t-x),$$

we have

$$u(x) = H\{v(x)\}, \quad v(x) = -H\{u(x)\}$$

and

$$f(z) = \frac{1}{2\pi i}\int_{-\infty}^{\infty} \frac{v(t)\,dt}{t - z} \tag{11.7-3}$$

yields

$$f(x + i0) = \frac{1}{2i} H\{v(x)\} + \frac{1}{2} v(x)$$

$$= \frac{1}{2i}\{u(x) + iv(x)\}$$

A completely equivalent restatement of the above results is the following: necessary and sufficient conditions that $f(x)$ represent the boundary values (on the real axis) of a function analytic in the upper half-plane is that

$$H\{f(x)\} = if(x) \tag{11.7-4}$$

By the same token, $H\{f\} = -if$ characterizes the L_p boundary values of lower half-plane analytic functions.

Corollary: If $f(x)$ is an arbitrary L_p-function, then

$$f(x) = \tfrac{1}{2}\{f(x) + iH\{f(x)\}\} + \tfrac{1}{2}\{f(x) - iH\{f(x)\}\}$$

gives a unique decomposition of $f(x)$ into the sum of boundary values of two analytic functions, each analytic and L_p in a half-plane.

It now follows that every L_p-function $F(w)$ is a Fourier transform, and that, for example

$$\frac{1}{2}[F(w) - iH\{F(w)\}] = \frac{1}{\sqrt{2\pi}}\int_0^{\infty} e^{ixw} f(x)\,dx = F_+(w)$$

Example:

(1)

$$f(x) = \begin{cases} 0, & |x| > 1 \\ \operatorname{sgn} x, & |x| < 1 \end{cases}$$

$$f_+(x) = \frac{1}{2}[f(x) - iH\{f(x)\}] = \frac{-i}{2\pi}\log\left(\frac{x^2 - 1}{x^2}\right)$$

On semi-infinite intervals, and finite intervals we have the following self-explanatory result. Given $h(x)$ in L_p,

$$F(z) = \frac{1}{2\pi i} \int_0^1 \frac{h(t)\,dt}{t - z} \tag{11.7-5}$$

defines an analytic function in the full plane, cut from $z = 0$ to $z = 1$. The boundary values from above and below are

$$F_+(x) = F(x + i0) = \frac{1}{2\pi i} \fint_0^1 \frac{h(t)\,dt}{t - x} + \frac{1}{2} h(x)$$

$$F_-(x) = F(x - i0) = \frac{1}{2\pi i} \fint_0^1 \frac{h(t)\,dt}{t - x} - \frac{1}{2} h(x)$$

$$F_+(x) + F_-(x) = \frac{1}{\pi i} \fint_0^1 \frac{F_+(t) - F_-(t)}{t - x}\,dt \tag{11.7-6}$$

We note further that $F(z) = O(1/z)$ as $z \to \infty$. Conversely, if $F(z)$ is analytic in the cut plane, $O(1/z)$ at infinity, then (11.7-6) holds.

(2) Many simple examples can be constructed from functions analytic in the plane cut from 0 to ∞, such as $\sqrt{z}$, $\log z, \ldots$ by combining them with the mapping $\zeta = z/1 - z$. Consider $\sqrt{z/z - 1}$. It is analytic in the cut plane, but not $O(1/z)$ at ∞. However $\sqrt{(z/z - 1)} - 1$ is. With $F(z) = \sqrt{z/z - 1} - 1$

$$F_+(x) = -i\sqrt{\frac{x}{1 - x}} - 1, \quad F_-(x) = i\sqrt{\frac{x}{1 - x}} - 1$$

Hence (11.7-6) yields

$$-2 = -\frac{2i}{\pi i} \fint_0^1 \sqrt{\frac{t}{1 - t}} \frac{dt}{t - x}, \quad \text{or} \quad \frac{1}{\pi} \fint_0^1 \sqrt{\frac{t}{1 - t}} \frac{dt}{t - x} = 1$$

Similarly, $(z/z - 1)^\alpha - 1$ gives

$$\frac{1}{\pi} \fint_0^1 \left(\frac{t}{1 - t}\right)^\alpha \frac{dt}{t - x} = \csc \pi\alpha - \cot \pi\alpha \left(\frac{x}{1 - x}\right)^\alpha$$

(3) Suppose we wish to evaluate a specific singular integral, for instance

$$\frac{1}{\pi}\int_0^1 \frac{\sqrt{t(1-t)}}{(t-x)(1-\sigma t)}\,dt, \quad 0<\sigma<1$$

From (11.7-6), we want $F(z)$ so that

$$F_+(x) - F_-(x) = \frac{\sqrt{x(1-x)}}{1-\sigma x} \qquad (11.7\text{-}7)$$

and are led to the simplest example of a more general problem: to construct $F(z)$ given a relation between its boundary values. These are called *Hilbert problems* or *Riemann-Hilbert problems.*

In simple cases, such as here, we can deduce $F(z)$ by analytic continuation, or perhaps even guess in an educated way.

$$F_1(z) = \frac{\sqrt{z(z-1)}}{1-\sigma z}$$

satisfies

$$F_{1_+}(x) - F_{1_-}(x) = \frac{2i\sqrt{x(1-x)}}{1-\sigma x}$$

but is not analytic at $z = 1/\sigma$, nor $O(1/z)$ at ∞.
However

$$F(z) = -\frac{i}{2}\frac{\sqrt{z(z-1)}}{1-\sigma z} + \frac{i}{2}\frac{\sqrt{\frac{1}{\sigma}\left(\frac{1}{\sigma}-1\right)}}{1-\sigma z} - \frac{i}{2\sigma}$$

clearly does the job. We have

$$\frac{1}{\pi}\int_0^1 \frac{\sqrt{t(1-t)}}{(t-x)(1-\sigma t)}\,dt = \frac{1}{\sigma}\left\{1 - \frac{\sqrt{1-\sigma}}{1-\sigma x}\right\} \qquad (11.7\text{-}8)$$

(4) Consider (11.7-6) being given $F_+(x) + F_-(x)$. For example

$$\frac{1}{\pi}\int_0^1 \frac{h(t)\,dt}{t-x} = 1.$$

Now $F_+(x) + F_-(x) = 1$, and we have a singular integral equation to solve (another Hilbert problem). Actually, we only require $h(x)$ which is $i[F_+(x) - F_-(x)]$.

General formulas for solving such problems are given in (5) and (6). Here we use analytic continuation. We have

$$F_+(x) + F_-(x) = 1$$

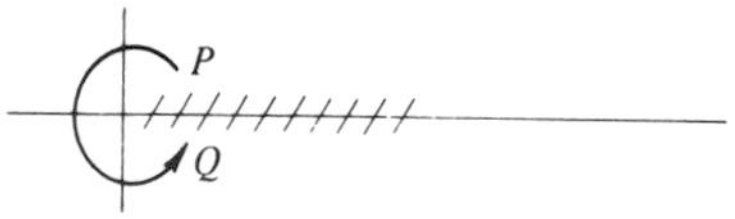

We use this equation to deduce the multivalued nature of $F(z)$ at $z = 0, z = 1$. With z at P, on the top of the cut, $F(z) = F_+$, while at Q, we have $F^*(z) = F_-$, the continuation around $z = 0$.

$$F^*(z) = 1 - F(z), \quad F^{**}(z) = 1 - F^*(z) = F(z)$$

Hence $F(z)$ is a single-valued function of $\sqrt{z}$. Similar results pertain at $z = 1$. Now set $F(z) = A + \sqrt{z}\,B; A, B$ single-valued at $z = 0$.

$$F^* = A - \sqrt{z}\,B = 1 - (A + \sqrt{z}\,B)$$

$$F(z) = \tfrac{1}{2} + \sqrt{z}\,B(z)$$

$$= \tfrac{1}{2} + \sqrt{z(z-1)}\,C(z)$$

Then $C_+ - C_- = 0$, $C(z)$ is single-valued in the full cut plane (including the cut).

Now

$$F(z) = \frac{1}{2\pi i}\int_0^1 \frac{h(t)\,dt}{t - z}$$

cannot have a pole at $z = 0, z = 1$

$$F(z) = \tfrac{1}{2} + \sqrt{z(z-1)}\,C(z)$$

shows that $C(z)$ can be singular only at $z = 0, z = 1$, having at worst a first-order pole there, while $F(z) = O(1/z)$ at ∞ gives

$$C(z) = -\frac{1}{2z} + O\left(\frac{1}{z^2}\right) \quad \text{as} \quad z \to \infty$$

Hence

$$C(z) = \frac{a}{z} + \frac{b}{z-1}, \quad a + b = -\frac{1}{2}$$

$$F(z) = \frac{1}{2} + \sqrt{z(z-1)}\left\{\frac{a}{z} - \frac{(\frac{1}{2}+a)}{z-1}\right\}$$

$$F_+ - F_- = 2i\sqrt{x(1-x)}\left(\frac{a}{x} - \frac{\frac{1}{2}+a}{x-1}\right) = \frac{2i}{\sqrt{x(1-x)}}\left(\frac{1}{2}x + a\right)$$

$$h(x) = \frac{1}{\sqrt{x(1-x)}}(x + 2a)$$

We note the presence of an eigensolution, $1/\sqrt{x(1-x)}$. See also (2).

(5) We consider here the general solution of the so-called airfoil equation

$$\frac{1}{\pi} \int_0^1 \frac{u(t)\,dt}{t-x} = g(x) \tag{11.7-9}$$

of which (4) is a special case.

By taking real and imaginary parts of a product $f(z) \cdot g(z)$ of two functions analytic in $I(z) > 0$, and using (11.7-2)

$$H\{u\,Hv + v\,Hu\} = Hu\,Hv - uv \tag{11.7-10}$$

Using

$$v(x) = \begin{cases} \sqrt{x(1-x)}, & 0 < x < 1 \\ 0, & \text{elsewhere} \end{cases}$$

$$\frac{1}{\pi} \int_0^1 \frac{u(t)\,dt}{t-x} = g(x) = Hu, \quad u \equiv 0 \text{ off } (0, 1)$$

we have $Hv = \frac{1}{2} - x$ from (11.7-8), and then (11.7-10) gives

$$\frac{1}{\pi} \int_0^1 \left(\frac{1}{2} - t\right) \frac{u(t)\,dt}{t-x} + \frac{1}{\pi} \int_0^1 \frac{\sqrt{t(1-t)}\,g(t)\,dt}{t-x} = \left(\frac{1}{2} - x\right) g(x) - \sqrt{x(1-x)}\,u(x)$$

or

$$u(x) = \frac{\frac{1}{\pi}\int_0^1 u(t)\,dt}{\sqrt{x(1-x)}} - \frac{1}{\pi\sqrt{x(1-x)}} \int_0^1 \frac{\sqrt{t(1-t)}\,g(t)\,dt}{t-x} \qquad (11.7\text{-}11)$$

Thus there is an eigensolution; if for example, $g(x)$ is smooth, and $u(x)$ is to be finite at $x = 0$ (or at $x = 1$) we can determine the solution uniquely.

(6) The general Hilbert problem we are concerned with is

$$a(x)\,F_+(x) + b(x)\,F_-(x) = r(x) \qquad (11.7\text{-}12)$$

for $F(z)$, single-valued and analytic in the cut plane, cut from $z = 0$ to $z = 1$.

As with all linear problems, the general solution of (11.7-12) will consist of a particular solution plus the general solution of the homogeneous ($r \equiv 0$) case. A particular solution can be found by the following method, due to Carleman.

Let

$$F(z) = e^{T(z)}\,U(z) \qquad (11.7\text{-}13)$$

$$a(x)\,e^{T_+}\,U_+ + b(x)\,e^{T_-}\,U_- = r(x)$$

Determine $T(z)$:

$$a(x)\,e^{T_+} = -b(x)\,e^{T_-}$$

$$T_+ - T_- = \log\left(-\frac{b}{a}\right) \qquad (11.7\text{-}14)$$

We have from (11.7-5), (11.7-6)

$$T(z) = \frac{1}{2\pi i}\int_0^1 \log\left(-\frac{b}{a}\right)\frac{dt}{t-z} \qquad (11.7\text{-}15)$$

(Of course, this solution is not unique; we may add an arbitrary single-valued function.) Next

$$U_+ - U_- = \frac{r(x)}{ae^{T_+}} = \frac{r(x)}{-be^{T_-}} = \frac{r(x)}{\sqrt{-abe^{T_+ + T_-}}}$$

In the last form, $T_+ + T_-$ is readily obtained from (11.7-15), and then

$$U(z) = \frac{1}{2\pi i}\int_0^1 \frac{r(t)\, e^{-1/2(T_+ + T_-)}}{\sqrt{-ab}} \frac{dt}{t - z} \tag{11.7-16}$$

Completely analogous results hold for the semi-infinite interval.

(7) The Hilbert matrix $(b_{mn}) = (1/m + n + \frac{1}{2})$ occurs in several applied areas. If $\{a_n\}, -\infty < n < \infty$ is L_2; i.e., $\Sigma\, |a_n|^2 < \infty$, then

$$b_m = \frac{1}{\pi}\sum_{-\infty}^{\infty} \frac{a_n}{n + m + \frac{1}{2}} \tag{11.7-17}$$

$\{b_m\}$ is L_2, and further

$$\frac{1}{\pi}\sum_{-\infty}^{\infty} \frac{b_m}{m + n + \frac{1}{2}} = a_n \tag{11.7-18}$$

Equations (11.7-18), (11.7-17) may be considered as a pair of discrete transforms, acting on sequences. They are closely related to finite Hilbert transforms, which may be seen as follows:

Let

$$\phi(x) = \sum_0^\infty a_n x^n, \quad \psi(x) = \sum_0^\infty a_{-(n+1)}\, x^n$$

$$\xi(x) = \sum_0^\infty b_m x^m, \quad \eta(x) = \sum_0^\infty b_{-(m+1)}\, x^m$$

Then (11.7-17) may be rewritten

$$\xi(x) = \frac{1}{\pi}\int_0^1 \frac{\phi(t)\, dt}{\sqrt{t}\,(1 - xt)} - \frac{1}{\pi}\int_0^1 \frac{\sqrt{t}\,\psi(t)\, dt}{t - x} \tag{11.7-19}$$

$$\eta(x) = \frac{1}{\pi}\int_0^1 \frac{\phi(t)\, dt}{\sqrt{t}\,(t - x)} - \frac{1}{\pi}\int_0^1 \frac{\sqrt{t}\,\psi(t)\, dt}{1 - xt} \tag{11.7-20}$$

From (11.7-18) we get a similar pair, with ϕ, ψ interchanged with ξ, η. We have, in effect, a pair of singular integral equations together with their inversion formulas.

(8) Dual Series—This example is not intended to indicate the right way to solve the boundary-value problem, but to indicate a procedure which we might have to face occasionally.

Consider

$$\Delta \phi = \phi_{rr} + \frac{1}{r}\phi_r + \frac{1}{r^2}\phi_{\theta\theta} = 0, \qquad r < 1$$

$$\phi(1, \theta) = f(\theta), \qquad 0 < \theta < \pi$$

$$\phi_r(1, \theta) = 0, \qquad -\pi < \theta < 0$$

Starting from the eigenfunction representation

$$\phi(r, \theta) = \frac{a_0}{2} + \sum_1^\infty r^n (a_n \cos n\theta + b_n \sin n\theta)$$

we have

$$\frac{a_0}{2} + \sum_1^\infty a_n \cos n\theta + b_n \sin n\theta = f(\theta), \qquad 0 < \theta < \pi,$$

$$\sum_1^\infty n\, a_n \cos n\theta + n\, b_n \sin n\theta = 0, \qquad -\pi < \theta < 0 \tag{11.7-21}$$

Equivalent dual series are

$$\sum_1^\infty (a_n \cos n\theta + b_n \sin n\theta) = f(\theta) - \frac{a_0}{2}, \qquad 0 < \theta < \pi$$

$$\sum_1^\infty (a_n \sin n\theta + b_n \cos n\theta) = \lambda, \qquad 0 < \theta < \pi \tag{11.7-22}$$

where λ is a constant of integration, to be determined. To illustrate, we take $f(\theta) = \cos\theta$. Multiplying equations (11.7-22) by $\sin m\theta$, $\cos m\theta$, $m \geqslant 1$, and integrating, we obtain

$$\sum_1^\infty a_{2n}\left(\frac{1}{n - m - \frac{1}{2}}\right) = \frac{a_0}{2m + 1}$$

$$\sum_0^\infty a_{2n+1}\left(\frac{1}{n + \frac{1}{2} - m}\right) = -\frac{1}{2m + 1} - \frac{1}{2m - 1} \tag{11.7-23}$$

With $\sum_1^\infty a_{2n} x^n = x\phi(x)$, the first becomes

$$\int_0^1 \frac{\sqrt{t}\,\phi(t)\,dt}{t-x} = \frac{a_0}{2}\int_0^1 \frac{dt}{\sqrt{t}\,(1-xt)} = -\frac{a_0}{2}\int_0^1 \frac{dt}{\sqrt{t}\,(t-x)}$$

Hence

$$\sqrt{x}\,\phi(x) = -\frac{a_0}{2\sqrt{x}} + \frac{c}{\sqrt{x(1-x)}} \tag{11.7-24}$$

and since $\phi(x)$ is analytic at $x = 0$,

$$\phi(x) = -\frac{a_0}{2x}\left(1 - \frac{1}{\sqrt{1-x}}\right) \tag{11.7-25}$$

On the other hand, since $\sum_1^\infty a_n^2 < \infty$ for Fourier series, $\phi(x)$ must be $L_2(0, 1)$. Hence $a_0 = 0, a_{2n} \equiv 0, n = 1, 2, \ldots.$

Similar arguments yield

$$\psi(x) = \sum_0^\infty a_{2n+1}\, x^n = \frac{1 + x - \sqrt{1-x}}{2x}$$

$$a_1 = \tfrac{3}{4}, \quad a_{2n+1} = \tfrac{1}{2}(-1)^n \binom{\frac{1}{2}}{n+1}, \qquad n = 1, 2, \ldots$$

Finally

$$b_{2m+1} \equiv 0, \quad b_{2m} = (-1)^{m+1}\binom{\frac{1}{2}}{m}, \qquad m = 1, 2, \ldots$$

11.7.1 Avoiding the Wiener-Hopf

We have already seen examples of using transform methods on boundary value problems for which one approach required the use of Wiener-Hopf methods, analytic continuation, in order to obtain the solution, while a different approach led to a solution by formula. We were then able to solve more complicated problems for the second approach by now using the Wiener-Hopf approach.

For example, $\Delta\,\phi = \phi_{xx} + \phi_{yy} = 0$ in the upper half-plane, with mixed data

$$\phi(x, 0) = f(x), \qquad x > 0$$
$$\phi_y(x, 0) = g(x), \qquad x < 0$$

together with Fourier transforms, leads to the Wiener-Hopf case. On the other hand, using the Mellin formulas, the solution can be written down immediately, by formula.

The doubly mixed problem, $\Delta\phi = 0, y > 0$

$$\phi(x, 0) = f(x), \qquad 0 < x < 1,$$
$$\phi_y(x, 0) = h(x), \qquad x < 0, \ \ x > 1$$

leads to dual integral equations which can be solved by the methods of (11.4-1), while from the Fourier transform approach, we obtain the equivalent of

$$\int_0^1 k(x - t)\, u(t)\, dt = r(x)$$

an equation which has not been solved except in one or two very special cases.

We might try the same reasoning to $\Delta\phi - \phi = 0, y > 0$. The mixed problem

$$\phi(x, 0) = f(x), \qquad x > 0$$
$$\phi_y(x, 0) = g(x), \qquad x < 0$$

does not require the Weiner-Hopf if we use the Kontorovich-Lebedev transform. Perhaps we can invoke analytic continuation to solve the doubly mixed problem by using this transform.

Unfortunately, the integral appearing in the Kontorovich-Lebedev transform

$$f(r) = \frac{1}{2\pi i} \int_L K_\lambda(r) A(\lambda)\, d\lambda$$

yields information only about the even part of $A(it)$ when λ is restricted to the imaginary axis, and analytic continuation does not appear available.

Another way in which we can avoid the Wiener-Hopf is illustrated below.

$$\Delta\varphi - \varphi = 0, \qquad -\infty < x < \infty, \ \ y > 0$$

$$\left.\begin{aligned} \varphi(x, 0) &= e^{-x}, \qquad x > 0 \\ \varphi_y(x, 0) &= 0, \qquad x < 0 \end{aligned}\right. \tag{11.7-24}$$

As previously obtained, this problem can be recast as the integral equation

$$\int_0^\infty K_0(|x - t|)\, f(t)\, dt = e^{-x}, \qquad x > 0$$

or

$$\frac{1}{2}\int_{-\infty}^{\infty} e^{-ixw}\, F_+(w)\, \frac{dw}{\sqrt{1+w^2}} = e^{-x}, \qquad x>0$$

with

$$F_+(w) = \int_0^{\infty} f(t)\, e^{ixt}\, dt \tag{11.7-25}$$

Now multiply by e^{ixz} and integrate from 0 to ∞, valid for $I(z)>0$.

$$\int_{-\infty}^{\infty} \frac{F_+(w)\, dw}{\sqrt{1+w^2}\,(w-z)} = -\frac{2}{z+i} \tag{11.7-26}$$

We may now close the contour in the upper half-plane, and let $z \to iy+0, y>1$

$$2 \fint_1^{\infty} \frac{F_+(it)\, dt}{\sqrt{t^2-1}\,(t-y)} = -\frac{2}{y+1} \tag{11.7-27}$$

This equation is a simple Hilbert problem, solvable by formula from (11.7-11). Of course the special form of the transform of $K_0(|x|)$ namely $\sqrt{\pi/2}\,(1/\sqrt{1+w^2}$, makes this approach work. Perhaps the same method might lead to a solution of a doubly mixed case.

Consider

$$\int_{-1}^{1} K_0(|x-t|)\, f(t)\, dt = 1, \qquad |x|<1$$

$$\frac{1}{2}\int_{-\infty}^{\infty} \frac{e^{-ixw}}{\sqrt{1+w^2}} F(w) \left\{ \frac{e^{i(z-w)} - e^{-i(z-w)}}{z-w} \right\} dw = \frac{e^{iz} - e^{-iz}}{z}$$

The function $F(w) = \int_{-1}^{1} e^{ixw} f(x)\, dx$ is entire, even in w. Taking $I(z)>0$, and closing the contours of the integrals in (11.7-28) in the lower and upper half-planes, respectively, we obtain, for $\phi(x) = e^{-x} F(ix)/\sqrt{x^2-1}$

$$e^{x} \fint_1^{\infty} \frac{\phi(t)\, dt}{t-x} + e^{-x} \int_1^{\infty} \frac{\phi(t)\, dt}{t+x} = -\frac{2 \sinh x}{x} \tag{11.7-29}$$

or

$$\cosh \sqrt{x} \, \fint_1^\infty \frac{\phi(\sqrt{t})\, dt}{t - x} + \sqrt{x} \sinh \sqrt{x} \, \fint_1^\infty \frac{\phi(\sqrt{t})\, dt}{(t - x)\sqrt{t}} = -\frac{2 \sinh \sqrt{x}}{\sqrt{x}} \tag{11.7-30}$$

in terms of

$$F(z) = \frac{1}{2\pi i} \int_1^\infty \frac{\phi(\sqrt{t})\, dt}{t - z}, \quad G(z) = \frac{1}{2\pi i} \int_1^\infty \frac{\phi(t)\, dt}{\sqrt{t}\,(t - z)}$$

we have

$$F_+ - F_- = \sqrt{x}\,(G_+ - G_-)$$

$$e^{\sqrt{x}}\,(F_+ + F_- + \sqrt{x}\,(G_+ + G_-)) + e^{-\sqrt{x}}\,(F_+ + F_- - \sqrt{x}\,G_+ - \sqrt{x}\,G_-) = \frac{2i}{\pi\sqrt{x}}\left(e^{\sqrt{x}} - e^{-\sqrt{x}}\right) \tag{11.7-31}$$

a particular Hilbert-Riemann problem. This problem has not yet been solved, although it appears feasible.

11.8 FINITE TRANSFORMS

Although the end result of this section is essentially the same as Fourier series, eigenfunction expansions, there is some point in regarding these objects as transforms in their own right, and since the range of integration is finite, they are termed finite transforms. For example

$$F(k) = \int_0^\pi f(x) \sin kx\, dx \tag{11.8-1}$$

is the finite Fourier sine transform of $f(x)$. Regarding k as a positive integer variable, and using Fourier sine series, we can invert (11.8-1) to yield

$$f(x) = \frac{2}{\pi} \sum_1^\infty F(k) \sin kx \tag{11.8-2}$$

valid for $f(x) \in L_2(0, \pi)$ for the Lebesgue integral, or for the Dirichlet conditions on $f(x)$, in which case (11.8-2) converges to $\frac{1}{2}\,[f^*(x + 0) + f^*(x - 0)]$ at each x, where $f^*(x)$ is the odd periodic extension of $f(x)$.

Clearly, for each eigenfunction expansion of the type arising from separation of variables for a boundary value problem, we can construct a similar finite transform and its inversion formula. These transforms find their most immediate application in the solution of non homogeneous problems. For example, consider

$$u_t = u_{xx}, \quad 0 < x < L$$
$$u(0,t) = e^{-t}, \quad u(x,0) = 1, \quad u(L,t) = 0. \tag{11.8-3}$$

If the left-hand boundary condition had been $u(0,t) = 0$, then separation of variables and superposition would have given

$$u(x,t) = \sum_1^\infty a_n \sin \frac{n\pi x}{L} e^{-(n^2\pi^2/L^2)t}$$

To deal directly with the problem (11.8-3), we use the finite sine transform of $u(x,t)$ over $0 < x < L$, obtaining

$$U(k,t) = \int_0^L u(x,t) \sin \frac{k\pi x}{L} dx \tag{11.8-4}$$

and by requirement, $u(x,t)$ is piecewise smooth.

We find

$$\begin{aligned}\frac{\partial U(k,t)}{\partial t} &= \int_0^L u_{xx} \sin \frac{k\pi x}{L} dx \\ &= \left[u_x \sin \frac{k\pi x}{L} - \frac{uk\pi}{L} \cos \frac{k\pi x}{L}\right]_0^L - \frac{k^2\pi^2}{L^2} U(k,t) \\ &= \frac{k\pi}{L} e^{-t} - \frac{k^2\pi^2}{L^2} U(k,t)\end{aligned} \tag{11.8-5}$$

$$U(k,0) = \int_0^L \sin \frac{k\pi x}{L} dx = \frac{L}{k\pi}\{1 - (-1)^k\} \tag{11.8-6}$$

Hence

$$U(k,t) = \frac{\frac{k\pi}{L}}{\left(\frac{k^2\pi^2}{L^2} - 1\right)} e^{-t} - \frac{\frac{k\pi}{L}}{\left(\frac{k^2\pi^2}{L^2} - 1\right)} e^{-(k^2\pi^2/L^2)t} + \frac{L}{k\pi}\{1 - (-1)^k\} e^{-(k^2\pi^2/L^2)t} \tag{11.8-7}$$

(The special case $k\pi/L = 1$ for some k is handled separately.) Finally

$$u(x, t) = \frac{2}{L} \sum_{1}^{\infty} U(k, t) \sin \frac{k\pi x}{L} \tag{11.8-8}$$

One useful aspect of these ideas is that in simple cases, we can transform the original problem into vastly easier ones. Thus, here, we see from (11.8-8) and (11.8-7), that

$$u(x, t) = e^{-t}\, h(x) + v(x, t)$$

where

$$v_t = v_{xx}, \quad v = 0 \text{ at } x = 0, x = L$$

and so we have

$$v_t = v_{xx}, \quad v = 0 \text{ at } x = 0, x = L$$

$$h''(x) + h(x) = 0, \quad h(0) = 1, h(L) = 0$$

$$v(x, 0) = 1 - h(x) = 1 - \frac{\sin (L - x)}{\sin L}$$

Of particular importance is that the best suited finite transform for a given problem is the one which arises naturally in the corresponding homogeneous problem (provided there is one at all). For further remarks on finite transforms, see Chapter 9.

11.9 ASYMPTOTIC RESULTS

The results suggested by the title are of great importance in several applications; they are likewise fairly complicated. We content ourselves with a few examples, and refer to Chapter 12 for further treatment.

For the Laplace transform, there is a rule of thumb which shows that for $F(s) = \int_0^\infty e^{-st} f(t)\, dt$, the values of $F(s)$ as $s \to \infty$ are determined by those of $f(x)$ for $x \to 0$, while $F(s)$ for $s \to 0$ is determined by $f(x)$ as $x \to \infty$.

These ideas are covered by Abelian and Tauberian asymptotics respectively. Thus, if

$$f(t) \sim \sum_{0}^{\infty} a_n t^{\lambda_n}, \qquad -1 < \lambda_0 < \lambda_1 < \cdots \tag{11.9-1}$$

as $t \to 0$, then

$$F(s) \sim \sum_0^\infty a_n \frac{\Gamma(1+\lambda_n)}{s^{1+\lambda_n}} \quad \text{as} \quad s \to \infty \tag{11.9-2}$$

The full asymptotic series need not exist; if only a finite number of terms of (11.9-1) hold, the corresponding finite number of terms occurs in (11.9-2). Thus, the analytic case (or merely the continuous case) is contained in these formulas. Most frequently a single term is all that is used (the so-called dominant term). For example, from (11.2-9),

$$Y(s) = \int_0^\infty e^{-sx} J_0(x)\,dx = c/\sqrt{1+s^2}\,.$$

Since $J_0(x) \sim 1$ as $x \to 0$, $Y(s) \sim 1/s$ as $s \to \infty$, so that $c = 1$.
Again

$$\int_0^\infty \frac{e^{-sx}}{1+x}\,dx \sim \sum_0^\infty \frac{(-1)^n\, n!}{x^{n+1}}$$

even though the series clearly diverges for each x.

The results for 'small s' and 'infinite t' are not nearly so immediate. Thus if $F(s)$ is analytic for $R(s) > \sigma$, $F(s)$ has only poles in the finite plane, occurring at $\{s_n\}$, $s_0 > s_1 > \cdots$, $f(t)$ real, residues $c_0, c_1, \ldots$ real, then $f(t) \sim \sum_0^\infty c_n e^{s_n t}$ as $t \to \infty$.

Another fragmentary result is if

$$F(s) = 1/s\,[a_0 + b_0\sqrt{s} + a_1 + b_1 s^{3/2} + \cdots], \quad |s| < 1$$

then

$$f(t) \sim a_0 + \sum_n b_n \frac{t^{-[n+(1/2)]}}{\Gamma(\frac{1}{2} - n)} \quad \text{as} \quad t \to \infty.$$

11.10 OPERATIONAL FORMULAS

All the foregoing results, as well as most of the rest of this book, are solidly based on firm analytical procedures and can be used with confidence when the appropriate conditions are satisfied. Of course, we usually ignore the tedious justifications step by step, and rather 'analyze': We assume adequate behavior for our integrals and analysis in order to obtain the answer. It is then a simpler matter to check that our analysis is indeed valid for this answer.

One further step involves all the above except that, for some reason or other, it is impractical to verify our assumptions (the answer may involve an unknown function, non linear equation, etc.) and we adopt the purely formal procedure. This aspect of analysis is frequently encountered in perturbation problems (whose stability is unknown), but for which a good formal procedure yields satisfactory looking answers. If no other procedure is available, we are then stuck with the pure formalism.

Another result of a similar nature is the use of operational formulas or methods. These frequently enable us to drastically short cut the labor involved in a problem (very worthwhile), and are usually based on correct analytical procedures, although not necessarily obviously so. Thus, Heaviside operational calculus involving the Laplace transform is a very fast and powerful method for obtaining solutions to certain initial-value problems. Similar operational formulas are available for a variety of inversion formulas to various transforms.

A lesser known, but extremely useful operational formula (used by Ramanujan, Carathéodory) appears to have originated with Lucas in the following manner:

$$\frac{x}{e^x - 1} = \sum_0^\infty \frac{(-1)^n B_n x^n}{n!} \tag{11.10-1}$$

defines the generating function for the Bernoulli numbers. These coefficients occur in a number of important applications (the Euler-Maclaurin summation formula, power series for $\tan x, \ldots$). There is no simple formula (computationally) for B_n itself. Lucas wrote B_n as B^n (superscript, not power) $x/e^x - 1 = \sum_0^\infty (-1)^n B^n x^n / n!$, clearly indistinguishable from e^{-Bx}, provided we interpret the result only in terms of the power series.

We write $x/e^x - 1 \stackrel{\circ}{=} e^{-Bx}$, operational equality

$$-\frac{x}{e^{-x} - 1} = \frac{xe^x}{e^x - 1} \stackrel{\circ}{=} e^{(1-B)x} \stackrel{\circ}{=} e^{Bx}$$

Hence

$$B^n \stackrel{\circ}{=} B_n \stackrel{\circ}{=} (1 - B)^n \stackrel{\circ}{=} \sum_{r=0}^{n} \binom{n}{r} (-1)^r B_r \tag{11.10-2}$$

a correct result, and useful computationally.

To illustrate some useful applications, let us denote by M the class of functions $f(x)$, analytic in a neighborhood of the real positive axis, having Mellin transforms valid in $\sigma_0 < R(s) < \sigma_1$. Thus $\sin x, e^{-x}, 1/1 + x, J_0(x), \ldots$ belong to M. It is this class of functions which arises in applications most often. For $f(x)$ in M,

$$f(x) = \sum_0^\infty \frac{(-1)^n a_n x^n}{n!} \stackrel{\circ}{=} e^{-ax}$$

although the series may have a finite radius of convergence. We define

$$\phi(s) = \frac{1}{\Gamma(s)} \int_0^\infty x^{s-1} f(x)\, dx \stackrel{\circ}{=} a^{-s} \tag{11.10-3}$$

and think of $\phi(s)$ as the 'coefficient function.' We have $\phi(-n) = a^n \stackrel{\circ}{=} a_n$ and $\phi(s)$ is analytic in a full left hand plane, $O(e^{(\pi/2)\sigma})$ there. This result readily follows from the following analysis:

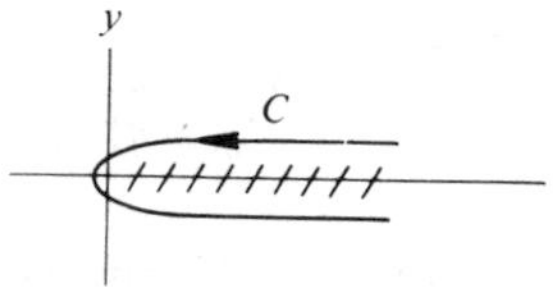

The function $\frac{1}{2\pi i} \int_c z^{s-1} f(z)\, dz$ is analytic in s in a full left half-plane.

$$\frac{1}{2\pi i} \int_c z^{s-1} f(z)\, dz = \frac{e^{2\pi is} - 1}{2\pi i} \int_0^\infty x^{s-1} f(x)\, dx$$

$$= \frac{e^{i\pi s}}{\Gamma(1-s)\,\Gamma(s)} \frac{1}{} \int_0^\infty x^{s-1} f(x)\, dx$$

Thus

$$\phi(s) = e^{-i\pi s}\, \Gamma(1-s) \frac{1}{2\pi i} \int_c z^{s-1} f(z)\, dz \tag{11.10-4}$$

Further define $||a||$ as the radius of convergence of a power series $\sum_0^\infty a_n t^n$. Consider now $\int_0^\infty \cos \lambda x\, f(x)\, dx$ for $f(x) \in M$. We have

$$\int_0^\infty \cos \lambda x\, f(x)\, dx \stackrel{\circ}{=} \int_0^\infty e^{-ax} \cos \lambda x\, dx = \frac{a}{a^2 + \lambda^2}$$

$$\frac{a^2}{a^2+\lambda^2}=\frac{1}{\lambda^2}\sum_0^\infty \frac{(-1)^n a^{2n+1}}{\lambda^{2n}} \overset{\circ}{=} \sum_0^\infty \frac{(-1)^n a_{2n+1}}{\lambda^{2n+2}} \overset{\circ}{=} \sum_0^\infty \frac{(-1)^n \phi(-2n-1)}{\lambda^{2n+2}} \tag{11.10-5}$$

It is not hard to replace the operational steps by Mellin formulas and the convolution formulas. The result is surprising, but very useful. The integral is an analytic function of $\{a_n\}$; the analytic function can be found by using the special $f(x)$ of M, viz. e^{-ax}. Thus, we have obtained the cosine transform of *all of M*.

Of course, $\sum_0^\infty (-1)^n a_{2n+1}/\lambda^{2n+2}$ converges only for $|\lambda| > \|a\|$. What happens if this quantity is infinite! We could try

$$\frac{a}{a^2+\lambda^2}=\sum_0^\infty \frac{(-1)^n \lambda^{2n}}{a^{2n+1}} \overset{\circ}{=} (?) \sum_0^\infty \frac{(-1)^n \lambda^{2n}}{a^{2n+1}} \tag{11.10-6}$$

Unfortunately, $\phi(s)$ need not be analytic for all s. If it is, then (11.10-6) is valid.

Another application—consider $f(x) = J_0(x)$.

$$J_0(x)=\sum_0^\infty \frac{(-1)^n \left(\frac{x}{2}\right)^{2n}}{(n!)^2} \overset{\circ}{=} e^{-ax}$$

$$a_{2n+1}=0,\quad a_{2n}=\frac{(-1)^n \Gamma(2n+1)}{2^{2n}\{\Gamma(n+1)\}^2},\quad n=0,1,2,\ldots$$

We then know $\phi(-n)$ for all n. Can we obtain an analytic formula for $\phi(-n)$? The problem is that $(-1)^n = \cos n\pi = e^{in\pi} = e^{-in\pi} = \cdots$ each coming from a different analytic function. The problem is solved by means of Carlson's theorem, guaranteeing unique interpolation for $\phi(n)$ to $\phi(s)$ provided $\phi(s) = o(e^{\pi\sigma/2})$ and $f(x)$ is in M. There is now no choice. We must use $(-1)^n = \cos n\pi$, $\phi(-2n) = \sqrt{\pi}/\Gamma(1+n)\,\Gamma(\frac{1}{2}-n)$

$$\phi(-n)=\frac{\sqrt{\pi}}{\Gamma\left(1+\frac{n}{2}\right)\Gamma\left(\frac{1}{2}-\frac{n}{2}\right)} \tag{11.10-7}$$

and so

$$\phi(s)=\frac{\sqrt{\pi}}{\Gamma\left(1-\frac{s}{2}\right)\Gamma\left(\frac{1}{2}+\frac{s}{2}\right)} \tag{11.10-8}$$

Thus we can construct the Mellin transform for each $f(x)$ in M knowing its power series expansion. Again

$$f(x) = \frac{1}{1+x} = \sum_0^\infty \frac{(-1)^n a_n x^n}{n!} = \sum_0^\infty (-1)^n x^n$$

$$a_n = n!, \quad \phi(-n) = \Gamma(1+n), \quad \phi(s) = \Gamma(1-s)$$

One more formula is worth noting. If the operational answer to a problem is $g(a)$, how do we express it if a power series is not clearly indicated?

If

$$g(a) = \sum_0^\infty \frac{(-1)^n a_n a^n}{n!}$$

$$\psi(s) = \frac{1}{\Gamma(s)} \int_0^\infty x^{s-1} g(x)\, dx \qquad \text{(The coefficient function for } g(x))$$

Then

$$g(a) = \frac{1}{2\pi i} \int_L \Gamma(s)\, \phi(s)\, \psi(s)\, ds \tag{11.10-9}$$

closed to the left.

This same formula holds more generally. An extension to $g(x)$ not in M is

$$\psi_\delta(s) = \frac{1}{\Gamma(s)} \int_0^\infty e^{-\delta x} x^{s-1} g(x)\, dx$$

and then

$$g(a) = \lim_{\delta \to 0} \frac{1}{2\pi i} \int_L \Gamma(s)\, \phi(s)\, \psi_\delta(s)\, ds \tag{11.10-10}$$

For example

$$g(a) = a^n, \quad \psi_\delta(s) = \frac{\Gamma(s+n)}{\Gamma(s)\, \delta^{s+n}}$$

$$g(a) = \lim_{\delta \to 0} \frac{1}{2\pi i} \int_L \Gamma(s+n)\, \phi(s)\, \delta^{-s-n}\, ds$$

$$= \lim_{\delta \to 0} \left(\sum_0^\infty \left\{ \frac{(-1)^k}{k!} \phi(-n-k)\, \delta^k \right\} \right)$$

$$= \lim_{\delta \to 0} \left(\phi(-n) - \delta\, \phi(-n-1) + \tfrac{1}{2}\, \phi(-n-2)\, \delta^2 \pm \cdots \right.$$

$$= \phi(-n)$$

as previously.

11.11 REFERENCES

11-1 Erdelyi, A. (ed.), *Higher Transcendental Functions*, vols. 1-3, McGraw-Hill, New York, 1953.

11-2 Erdelyi, A. (ed.), *Tables of Integral Transforms*, vols. 1 and 2, McGraw-Hill, New York, 1954.

11-3 Carrier, G.; Krook, M.; and Pearson, C. E., *Functions of a Complex Variable*, McGraw-Hill, New York, 1966.

11-4 Titchmarsh, E. C., *Theory of Fourier Integrals*, 2nd ed., Oxford University Press, New York, 1948.

12

Asymptotic Methods

Frank W. J. Olver*

12.1 DEFINITIONS

12.1.1 Asymptotic and Order Symbols

Let $f(x)$ and $\phi(x)$ be functions defined on a point set $\mathbf{X}$ and c be a limit point of $\mathbf{X}$. If $f(x)/\phi(x) \to 1$ as $x \to c$, then we say that $f(x)$ is *asymptotic* to $\phi(x)$and write

$$f(x) \sim \phi(x) \qquad (x \to c \text{ in } \mathbf{X})$$

The point c is called the *distinguished point* of this *asymptotic relation*; c need not belong to $\mathbf{X}$. The set $\mathbf{X}$ may be real or complex; in the latter event it is required that $f(x)/\phi(x)$ approaches its limit uniformly with respect to $\arg x$.

In a similar way, if $f(x)/\phi(x) \to 0$ as $x \to c$ then we write

$$f(x) = o\{\phi(x)\} \qquad (x \to c \text{ in } \mathbf{X})$$

And if $|f(x)/\phi(x)|$ is bounded as $x \to c$, then

$$f(x) = O\{\phi(x)\} \qquad (x \to c \text{ in } \mathbf{X})$$

Lastly, if $|f(x)/\phi(x)|$ is bounded in the *whole* of $\mathbf{X}$, then we write

$$f(x) = O\{\phi(x)\} \qquad (x \in \mathbf{X})$$

Examples:

$$\sinh x \sim x \qquad (x \to 0 \text{ in any point set})$$

$$\sin\left(n\pi + \frac{1}{n}\right) = O\left(\frac{1}{n}\right) \qquad (n \to \infty \text{ through integer values})$$

*Prof. Frank W. J. Olver, Institute for Fluid Dynamics and Applied Mathematics, University of Maryland, College Park, Md. 20742.

$$e^{ix}/(1+x) = O(x^{-1}) \qquad (x \in \text{real line})$$

$$e^{-x} = o(1) \qquad (x \to \infty \text{ in the sector } |\arg x| \leqslant \tfrac{1}{2}\pi - \delta, \text{ where } \delta > 0)$$

The last relation is invalid in the open sector $|\arg x| < \frac{1}{2}\pi$ owing to lack of uniformity with respect to $\arg x$.

The symbol $o\{\phi(x)\}$ or, more briefly, $o(\phi)$ may be used to denote an *unspecified* function with the property stated in the second paragraph. This use is generic, that is, $o(\phi)$ need not denote the same function f at each occurrence. The distinguished point is understood to be the same, however. Similarly for $O(\phi)$. Thus, for example

$$o(\phi) + o(\phi) = o(\phi); \quad o(\phi) = O(\phi); \quad O(\phi)O(\psi) = O(\phi\psi)$$

Relations of this kind are not necessarily reversible. For example, $O(\phi) = o(\phi)$ is false. Another instructive example is supplied by

$$\{1 + o(1)\} \cosh x - \{1 + o(1)\} \sinh x \qquad (x \to +\infty)$$

This expression is $o(e^x)$, not $\{1 + o(1)\}e^{-x}$ because the $o(1)$ terms may represent different functions.

12.1.2 Integration and Differentiation of Asymptotic and Order Relations

Asymptotic and order relations may be integrated subject to obvious restrictions on the convergence of the integrals involved. Suppose, for example, that in the interval (a, ∞) the function $f(x)$ is continuous and $f(x) \sim x^\nu$ as $x \to \infty$, where ν is a real or complex constant. Then as $x \to \infty$*

$$\int_x^\infty f(t)\,dt \sim -\frac{x^{\nu+1}}{\nu+1} \qquad (\operatorname{Re}\nu < -1)$$

and

$$\int_a^x f(t)\,dt \sim \begin{cases} \text{a constant} & (\operatorname{Re}\nu < -1) \\ \ln x & (\nu = -1) \\ x^{\nu+1}/(\nu+1) & (\operatorname{Re}\nu > -1) \end{cases}$$

These results are extendible to complex integrals in a straightforward manner.

Differentiation is permissible only with extra conditions. For example, let $f(z)$ be holomorphic for all sufficiently large $|z|$ in a given sector **S**, and

$$f(z) = O(z^\nu) \qquad (z \to \infty \text{ in } \mathbf{S})$$

*Except where otherwise stated, proofs of all results quoted in the present chapter will be found in Olver 1974a.

where ν is a fixed real number. Then

$$f^{(m)}(z) = O(z^{\nu - m}) \qquad (z \to \infty \text{ in } \mathbf{S}')$$

where $\mathbf{S}'$ is any sector properly interior to $\mathbf{S}$ and having the same vertex. This result also holds if the symbol O is replaced in both places by o.

12.1.3 Asymptotic Solution of Transcendental Equations

Let $f(\xi)$ be a strictly increasing function of the real variable ξ in an interval (a, ∞), and

$$f(\xi) \sim \xi \qquad (\xi \to \infty)$$

Then for $u > f(a)$ the equation $f(\xi) = u$ has a unique root $\xi(u)$ in (a, ∞), and

$$\xi(u) \sim u \qquad (u \to \infty)$$

As an example, consider the equation

$$x^2 - \ln x = u$$

In the notation just given we may take $\xi = x^2$, $f(\xi) = \xi - \frac{1}{2}\ln \xi$, and $a = \frac{1}{2}$. Then $\xi(u) \sim u$ as $u \to \infty$, implying that

$$x = u^{1/2}\{1 + o(1)\} \qquad (u \to \infty)$$

Higher approximations can be found by successive resubstitutions. Thus

$$x^2 = u + \ln x = u + \ln\,[u^{1/2}\{1 + o(1)\}] = u + \tfrac{1}{2}\ln u + o(1)$$

and thence

$$x = u^{1/2}\left\{1 + \frac{\ln u}{4u} + o\left(\frac{1}{u}\right)\right\}$$

and so on.

The same procedure can be used for complex variables, provided that the function $f(\xi)$ is analytic and the parameter u restricted to a ray or sector properly interior to the sector of validity of the relation $f(\xi) \sim \xi$.

12.1.4 Asymptotic Expansions

Let $f(x)$ be a function defined on a real or complex unbounded point set $\mathbf{X}$, and $\sum a_s x^{-s}$ a formal power series (convergent or divergent). If for each positive integer n

$$f(x) = a_0 + \frac{a_1}{x} + \frac{a_2}{x^2} + \cdots + \frac{a_{n-1}}{x^{n-1}} + O\left(\frac{1}{x^n}\right) \qquad (x \to \infty \text{ in } \mathbf{X})$$

then $\sum a_s x^{-s}$ is said to be the *asymptotic expansion* of $f(x)$ as $x \to \infty$ in $\mathbf{X}$, and we write

$$f(x) \sim a_0 + \frac{a_1}{x} + \frac{a_2}{x^2} + \cdots \qquad (x \to \infty \text{ in } \mathbf{X})$$

It should be noticed that the symbol $\sim$ is being used in a different (but not inconsistent) sense from that of 12.1.1.

A necessary and sufficient condition that $f(x)$ possesses an asymptotic expansion of the given form is that

$$x^n\left\{f(x) - \sum_{s=0}^{n-1} \frac{a_s}{x^s}\right\} \to a_n$$

as $x \to \infty$ in $\mathbf{X}$, for each $n = 0, 1, 2, \ldots$. In the special case when $\sum a_s x^{-s}$ converges for all sufficiently large $|x|$, the series is automatically the asymptotic expansion of its sum in any point set.

In a similar manner, if c is a finite limit point of a set $\mathbf{X}$ then

$$f(x) \sim b_0 + b_1(x - c) + b_2(x - c)^2 + \cdots \qquad (x \to c \text{ in } \mathbf{X})$$

means that the difference between $f(x)$ and the nth partial sum of the right-hand side is $O\{(x - c)^n\}$ as $x \to c$ in $\mathbf{X}$.

Asymptotic expansions having the same distinguished point can be added, subtracted, multiplied, or divided in the same way as convergent power series. They may also be integrated. Thus if $\mathbf{X}$ is the interval $[a, \infty)$ where $a > 0$, and $f(x)$ is continuous with an asymptotic expansion of the above form as $x \to \infty$, then

$$\int_a^x f(t)\,dt \sim A + a_0 x + a_1 \ln x - \frac{a_2}{x} - \frac{a_3}{2x^2} - \frac{a_4}{3x^3} - \cdots \qquad (x \to \infty)$$

where

$$A = \int_a^\infty \left\{f(t) - a_0 - \frac{a_1}{t}\right\} dt - a_0 a - a_1 \ln a$$

The last integral necessarily converges because the integrand is $O(t^{-2})$ as $t \to \infty$.

Differentiation of an asymptotic expansion is legitimate when it is known that the derivative $f'(x)$ is continuous and its asymptotic expansion exists. Differentiation is also legitimate when $f(x)$ is an analytic function of the complex variable x, provided that the result is restricted to a sector properly interior to the sector of validity of the asymptotic expansion of $f(x)$.

If the asymptotic expansion of a given function exists, then it is unique. On the other hand, corresponding to any prescribed sequence of coefficients $a_0, a_1, a_2, \ldots,$ there exists an infinity of analytic functions $f(x)$ such that

$$f(x) \sim \sum_{s=0}^{\infty} \frac{a_s}{x^s} \qquad (x \to \infty \text{ in } \mathbf{X})$$

The point set $\mathbf{X}$ can be, for example, the real axis or any sector of finite angle in the complex plane. Lack of uniqueness is demonstrated by the null expansion

$$e^{-x} \sim 0 + \frac{0}{x} + \frac{0}{x^2} + \cdots \qquad (x \to \infty \text{ in } |\arg x| \leqslant \tfrac{1}{2}\pi - \delta)$$

where δ is a positive constant not exceeding $\frac{1}{2}\pi$.

12.1.5 Generalized Asymptotic Expansions

The definition of an asymptotic expansion can be extended in the following way. Let $\{\phi_s(x)\}$, $s = 0, 1, 2, \ldots,$ be a sequence of functions defined on a point set $\mathbf{X}$ such that for every s

$$\phi_{s+1}(x) = o\{\phi_s(x)\} \qquad (x \to c \text{ in } \mathbf{X})$$

Then $\{\phi_s(x)\}$ is said to be an *asymptotic sequence* or *scale.* Additionally, suppose that $f(x)$ and $f_s(x)$, $s = 0, 1, 2, \ldots,$ are functions such that for each nonnegative integer n

$$f(x) = \sum_{s=0}^{n-1} f_s(x) + O\{\phi_n(x)\} \qquad (x \to c \text{ in } \mathbf{X})$$

Then $\sum f_s(x)$ is said to be a *generalized asymptotic expansion of $f(x)$ with respect to the scale* $\{\phi_s(x)\}$, and we write

$$f(x) \sim \sum_{s=0}^{\infty} f_s(x); \quad \{\phi_s(x)\} \text{ as } x \to c \text{ in } \mathbf{X}$$

Some, but by no means all, properties of ordinary asymptotic expansions carry over to generalized asymptotic expansions.

12.2 INTEGRALS OF A REAL VARIABLE

12.2.1 Integration by Parts

Asymptotic expansions of a definite integral containing a parameter can often be found by repeated integrations by parts. Thus for the Laplace transform

$$Q(x) = \int_0^\infty e^{-xt} q(t)\, dt$$

assume that $q(t)$ is infinitely differentiable in $[0, \infty)$, and for each s

$$q^{(s)}(t) = O(e^{\sigma t}) \qquad (0 \leqslant t < \infty)$$

where σ is an assignable constant. Then for $x > \sigma$

$$Q(x) = \frac{q(0)}{x} + \frac{q'(0)}{x^2} + \cdots + \frac{q^{(n-1)}(0)}{x^n} + \epsilon_n(x)$$

where n is an arbitrary nonnegative integer, and

$$\epsilon_n(x) = \frac{1}{x^n} \int_0^\infty e^{-xt} q^{(n)}(t)\, dt$$

With the assumed conditions

$$|\epsilon_n(x)| \leqslant \frac{A_n}{x^n} \int_0^\infty e^{-xt+\sigma t}\, dt = \frac{A_n}{x^n(x - \sigma)} \qquad (x > \sigma)$$

A_n being assignable. Thus $\epsilon_n(x) = O(x^{-n-1})$, and

$$Q(x) \sim \sum_{s=0}^\infty \frac{q^{(s)}(0)}{x^{s+1}} \qquad (x \to \infty)$$

An example is furnished by the incomplete Gamma function:

$$\Gamma(\alpha, x) = e^{-x} x^\alpha \int_0^\infty e^{-xt}(1 + t)^{\alpha-1}\, dt \sim e^{-x} x^{\alpha-1} \sum_{s=0}^\infty \frac{(\alpha - 1)(\alpha - 2) \cdots (\alpha - s)}{x^s}$$

as $x \to \infty$, α being fixed. The $\sim$ sign is now being used in the sense that

$$\Gamma(\alpha, x)/(e^{-x} x^{\alpha-1})$$

has the sum as its asymptotic expansion, as defined in 12.1.4. In the present case a straightforward extension of the analysis shows that if α is real and $n \geqslant \alpha - 1$, then the nth error term of the asymptotic expansion is bounded in absolute value by the first neglected term and has the same sign.

12.2.2 Watson's Lemma

Let $q(t)$ now denote a real or complex function of the positive real variable t having a finite number of discontinuities and infinities, and the property

$$q(t) \sim \sum_{s=0}^{\infty} a_s t^{(s+\lambda-\mu)/\mu} \qquad (t \to 0+)$$

where λ and μ are constants such that $\operatorname{Re}\lambda > 0$ and $\mu > 0$. Assume also that the Laplace transform of $q(t)$ converges throughout its integration range for all sufficiently large x. Then formal term-by-term integration produces an asymptotic expansion, that is

$$\int_0^{\infty} e^{-xt}\, q(t)\, dt \sim \sum_{s=0}^{\infty} \Gamma\left(\frac{s+\lambda}{\mu}\right) \frac{a_s}{x^{(s+\lambda)/\mu}} \qquad (x \to \infty)$$

This result is known as *Watson's lemma* and is one of the most frequently used methods for deriving asymptotic expansions. It should be noticed that by permitting $q(t)$ to be discontinuous the case of a finite integration range $\int_0^b$ is automatically included.

Example: Consider

$$\int_0^{\infty} e^{-x\cosh\tau}\, d\tau = e^{-x} \int_0^{\infty} \frac{e^{-xt}}{(2t+t^2)^{1/2}}\, dt$$

Since

$$(2t+t^2)^{-1/2} = \sum_{s=0}^{\infty} (-)^s \frac{1\cdot 3\cdot 5\cdots(2s-1)}{s!\, 2^{2s+(1/2)}}\, t^{s-(1/2)} \qquad (0<t<2)$$

the above result is applied with $\lambda = \frac{1}{2}$ and $\mu = 1$, to give

$$\int_0^{\infty} e^{-x\cosh\tau}\, d\tau \sim e^{-x}\sqrt{\frac{\pi}{2x}} \sum_{s=0}^{\infty} (-)^s \frac{1^2\cdot 3^2\cdot 5^2\cdots(2s-1)^2}{s!\,(8x)^s} \qquad (x \to \infty)$$

12.2.3 Riemann-Lebesgue Lemma

Let a be finite or $-\infty$, b be finite or $+\infty$, and $q(t)$ continuous in (a, b) save possibly at a finite number of points. Then

$$\int_a^b e^{ixt}\, q(t)\, dt = o(1) \qquad (x \to \infty)$$

provided that this integral converges uniformly at a, b, and the exceptional points, for all sufficiently large x. This is the *Riemann-Lebesgue lemma.*

It should be noticed that if the given integral converges absolutely throughout its range, then it also converges uniformly, since x is real. On the other hand, it may converge uniformly but not absolutely. For example, if $0 < \delta < 1$ then by integration by parts it can be seen that

$$\int_0^\infty \frac{e^{ixt}}{t^\delta}\, dt$$

converges uniformly at both limits for all sufficiently large x, but converges absolutely only at the lower limit.

12.2.4 Fourier Integrals

Let a and b be finite, and $q(t)$ infinitely differentiable in $[a, b]$. Repeated integrations by parts yield

$$\int_a^b e^{ixt} q(t)\, dt = \sum_{s=0}^{n-1} \left(\frac{i}{x}\right)^{s+1} \{e^{iax} q^{(s)}(a) - e^{ibx} q^{(s)}(b)\} + \epsilon_n(x)$$

where

$$\epsilon_n(x) = \left(\frac{i}{x}\right)^n \int_a^b e^{ixt} q^{(n)}(t)\, dt$$

As $x \to \infty$ we have $\epsilon_n(x) = o(x^{-n})$, by the Riemann-Lebesgue lemma. Hence the expansion just obtained is asymptotic in character.

A similar result applies when $b = \infty$. Provided that each of the integrals

$$\int_a^\infty e^{ixt} q^{(s)}(t)\, dt \qquad (s = 0, 1, \ldots)$$

converges uniformly for all sufficiently large x, we have

$$\int_a^\infty e^{ixt} q(t)\,dt \sim \frac{ie^{iax}}{x} \sum_{s=0}^{\infty} q^{(s)}(a) \left(\frac{i}{x}\right)^s \qquad (x \to \infty)$$

Whether or not b is infinite, an error bound is supplied by

$$|\epsilon_n(x)| \leqslant x^{-n}\, \mathcal{V}_{a,b}\{q^{(n-1)}(t)\}$$

where $\mathcal{V}$ is the *variational operator*, defined by

$$\mathcal{V}_{a,b}\{f(t)\} = \int_a^b |f'(t)\,dt|$$

12.2.5 Laplace's Method

Consider the integral

$$I(x) = \int_a^b e^{-xp(t)} q(t)\,dt$$

in which x is a positive parameter. The peak value of the factor $e^{-xp(t)}$ is located at the minimum t_0, say, of $p(t)$. When x is large this peak is very sharp, and the overwhelming contribution to the integral comes from the neighborhood of t_0. It is therefore reasonable to approximate $p(t)$ and $q(t)$ by the leading terms in their ascending power-series expansions at t_0, and evaluate $I(x)$ by extending the integration limits to $-\infty$ and $+\infty$, if necessary. The result is *Laplace's approximation* to $I(x)$.

For example, if t_0 is a simple minimum of $p(t)$ which is interior to (a, b) and $q(t_0) \neq 0$, then

$$I(x) \doteq \int_a^b e^{-x\{p(t_0)+(1/2)(t-t_0)^2 p''(t_0)\}} q(t_0)\,dt$$

$$\doteq q(t_0) e^{-xp(t_0)} \int_{-\infty}^{\infty} e^{-(1/2)x(t-t_0)^2 p''(t_0)}\,dt = q(t_0) e^{-xp(t_0)} \sqrt{\frac{2\pi}{xp''(t_0)}}$$

In circumstances now to be described, approximations obtained in this way are asymptotic representations of $I(x)$ for large x.

By subdivision of the integration range and change of sign of t (if necessary) it can always be arranged for the minimum of $p(t)$ to be at the left endpoint a. The other endpoint b may be finite or infinite. We further assume that:

a. $p'(t)$ and $q(t)$ are continuous in a neighborhood of a, save possibly at a.
b. As $t \to a$ from the right

$$p(t) - p(a) \sim P(t-a)^{\mu}; \quad q(t) \sim Q(t-a)^{\lambda-1}$$

and the first of these relations is differentiable. Here P, μ, Q, and λ are constants such that $P > 0$, $\mu > 0$, and $\mathrm{Re}\,\lambda > 0$.

c. $\int_a^b e^{-xp(t)} q(t)\,dt$ converges absolutely throughout its range for all sufficiently large x.

With these conditions

$$\int_a^b e^{-xp(t)} q(t)\,dt \sim \frac{Q}{\mu} \Gamma\left(\frac{\lambda}{\mu}\right) \frac{e^{-xp(a)}}{(Px)^{\lambda/\mu}} \qquad (x \to \infty)$$

Example: Consider

$$I(x) = \int_0^\infty e^{x\tau - (\tau-1)\ln\tau}\,d\tau$$

The maximum value of the integrand is located at the root of the equation

$$x - 1 - \ln\tau + (1/\tau) = 0$$

For large x the relevant root is given by

$$\tau \sim e^{x-1} = \zeta$$

say; compare 12.1.3. To apply the Laplace approximation the location of the peak needs to be independent of the parameter x. Therefore we take $t = \tau/\zeta$ as new integration variable, giving

$$I(x) = \zeta^2 \int_0^\infty e^{-\zeta p(t)} q(t)\,dt$$

where

$$p(t) = t(\ln t - 1); \quad q(t) = t$$

The minimum of $p(t)$ is at $t = 1$, and Taylor-series expansions at this point are

$$p(t) = -1 + \tfrac{1}{2}(t-1)^2 - \tfrac{1}{6}(t-1)^3 + \cdots; \quad q(t) = 1 + (t-1)$$

In the notation introduced above we have $p(a) = -1$, $P = \frac{1}{2}$, $\mu = 2$ and $Q = \lambda = 1$. Hence

$$\int_1^\infty e^{-\xi p(t)}\, q(t)\, dt \sim \left(\frac{\pi}{2\xi}\right)^{1/2} e^{\xi}$$

On replacing t by $2 - t$, we find that the same asymptotic approximation holds for the corresponding integral over the range $0 \leqslant t \leqslant 1$. Addition of the two contributions and restoration of the original variable yields the required approximation

$$I(x) \sim (2\pi)^{1/2}\, e^{3(x-1)/2} \exp\left(e^{x-1}\right) \qquad (x \to \infty)$$

12.2.6 Asymptotic Expansions by Laplace's Method

The method of 12.2.5 can be extended to derive asymptotic expansions. Suppose that in addition to the previous conditions

$$p(t) \sim p(a) + \sum_{s=0}^{\infty} p_s (t-a)^{s+\mu}; \quad q(t) \sim \sum_{s=0}^{\infty} q_s (t-a)^{s+\lambda-1}$$

as $t \to a$ from the right, and the first of these expansions is differentiable. Then

$$\int_a^b e^{-xp(t)}\, q(t)\, dt \sim e^{-xp(a)} \sum_{s=0}^{\infty} \Gamma\left(\frac{s+\lambda}{\mu}\right) \frac{a_s}{x^{(s+\lambda)/\mu}} \qquad (x \to \infty)$$

where the coefficients a_s are defined by the expansion

$$\frac{q(t)}{p'(t)} \sim \sum_{s=0}^{\infty} a_s v^{(s+\lambda-\mu)/\mu} \qquad (v \to 0+)$$

in which $v = p(t) - p(a)$. By reversion and substitution the first three coefficients are found to be

$$a_0 = \frac{q_0}{\mu p_0^{\lambda/\mu}}; \quad a_1 = \left\{\frac{q_1}{\mu} - \frac{(\lambda+1)p_1 q_0}{\mu^2 p_0}\right\} \frac{1}{p_0^{(\lambda+1)/\mu}}$$

$$a_2 = \left[\frac{q_2}{\mu} - \frac{(\lambda+2)p_1 q_1}{\mu^2 p_0} + \{(\lambda+\mu+2)p_1^2 - 2\mu p_0 p_2\} \frac{(\lambda+2)q_0}{2\mu^3 p_0^2}\right] \frac{1}{p_0^{(\lambda+2)/\mu}}$$

In essential ways Watson's lemma (12.2.2) is a special case of the present result.

12.2.7 Method of Stationary Phase

This method applies to integrals of the form

$$I(x) = \int_a^b e^{ixp(t)}\, q(t)\, dt$$

and resembles Laplace's method in some respects. For large x the real and imaginary parts of the integrand oscillate rapidly and cancel themselves over most of the range. Cancellation does not occur, however, at (i) the endpoints (when finite) owing to lack of symmetry; and at (ii) the zeros of $p'(t)$, because $p(t)$ changes relatively slowly near these *stationary points.*

Without loss of generality the range of integration can be subdivided in such a way that the stationary point (if any) in each subrange is located at the left endpoint a. Again the other endpoint b may be finite or infinite. Other assumptions are:

a. In (a, b), the functions $p'(t)$ and $q(t)$ are continuous, $p'(t) > 0$, and $p''(t)$ and $q'(t)$ have at most a finite number of discontinuities and infinities.

b. As $t \to a+$

$$p(t) - p(a) \sim P(t - a)^{\mu}; \quad q(t) \sim Q(t - a)^{\lambda - 1}$$

the first of these relations being twice differentiable and the second being differentiable. Here P, μ, and λ are positive constants, and Q is a real or complex constant.

c. $q(t)/p'(t)$ is of bounded variation in the interval (k, b) for each $k \in (a, b)$ if $\lambda < \mu$, or in the interval (a, b) if $\lambda \geqslant \mu$.

d. As $t \to b-$, $q(t)/p'(t)$ tends to a finite limit, and this limit is zero when $p(b) = \infty$.

With these conditions $I(x)$ converges uniformly at each endpoint for all sufficiently large x. Moreover,

$$I(x) \sim e^{\lambda \pi i/(2\mu)} \frac{Q}{\mu} \Gamma\left(\frac{\lambda}{\mu}\right) \frac{e^{ixp(a)}}{(Px)^{\lambda/\mu}} \qquad (x \to \infty)$$

if $\lambda < \mu$, or

$$I(x) = -\lim_{t \to a+} \left\{\frac{q(t)}{p'(t)}\right\} \frac{e^{ixp(a)}}{ix} + \lim_{t \to b-} \left\{\frac{q(t) e^{ixp(t)}}{p'(t)}\right\} \frac{1}{ix} + o\left(\frac{1}{x}\right) \qquad (x \to \infty)$$

if $\lambda \geqslant \mu$.

Example: The Airy integral of negative argument

$$\mathrm{Ai}(-x) = \frac{1}{\pi} \int_0^\infty \cos\left(\frac{1}{3} v^3 - xv\right) dv \qquad (x > 0)$$

The stationary points of the integrand satisfy $v^2 - x = 0$, and the only root in the range of integration is $\sqrt{x}$. To render the location of the stationary point independent of x, we substitute $v = \sqrt{x}\,(1 + t)$, giving

$$\mathrm{Ai}(-x) = \frac{x^{1/2}}{\pi} \int_{-1}^{\infty} \cos\,\{x^{3/2} p(t)\}\, dt$$

where $p(t) = -\frac{2}{3} + t^2 + \frac{1}{3} t^3$. With $q(t) = 1$, it is seen that as $t \to \infty$ the ratio $q(t)/p'(t)$ vanishes and its variation converges. Accordingly, the given conditions are satisfied. For the range $0 \leqslant t < \infty$ we have $p(0) = -\frac{2}{3}$, $\mu = 2$, and $P = Q = \lambda = 1$. The role of x is played here by $x^{3/2}$, and we derive

$$\int_0^{\infty} \exp\,\{ix^{3/2} p(t)\}\, dt \sim \frac{1}{2}\pi^{1/2} e^{\pi i/4} x^{-3/4} \exp\left(-\frac{2}{3} i x^{3/2}\right)$$

The same approximation is found to hold for $\int_{-1}^{0}$, and on taking real parts we arrive at the desired result:

$$\mathrm{Ai}(-x) = \pi^{-1/2} x^{-1/4} \cos\left(\tfrac{2}{3} x^{3/2} - \tfrac{1}{4}\pi\right) + o(x^{-1/4}) \qquad (x \to \infty)$$

As in the case of Laplace's method, the method of stationary phase can be extended to the derivation of asymptotic expansions; see Erdélyi 1956, section 2.9, and Olver 1974b.

12.3 CONTOUR INTEGRALS

12.3.1 Watson's Lemma

The result of 12.2.2 can be extended to complex values of the large parameter. Again, let $q(t)$ be a function of the positive real variable t having a finite number of discontinuities and infinities, and the property

$$q(t) \sim \sum_{s=0}^{\infty} a_s t^{(s+\lambda-\mu)/\mu} \qquad (t \to 0+)$$

with $\mathrm{Re}\,\lambda > 0$ and $\mu > 0$. Assume also that the abscissa of convergence (section 11.2) of the transform

$$Q(z) = \int_0^{\infty} e^{-zt} q(t)\, dt$$

is finite or $-\infty$. Then

$$Q(z) \sim \sum_{s=0}^{\infty} \Gamma\left(\frac{s+\lambda}{\mu}\right) \frac{a_s}{z^{(s+\lambda)/\mu}}$$

as $z \to \infty$ in the sector $|\arg z| \leqslant \frac{1}{2}\pi - \delta \; (< \frac{1}{2}\pi)$, the power $z^{(s+\lambda)/\mu}$ having its principal value.

When $q(t)$ is an analytic function of t, the region of validity of the last expansion can often be increased by rotation of the integration path about the origin. The general result is as follows. In addition to the foregoing conditions, assume that $q(t)$ is holomorphic in the sector $\alpha_1 < \arg t < \alpha_2$, α_1 being negative and α_2 positive. Assume also that for each $\delta \in (0, \frac{1}{2}\alpha_2 - \frac{1}{2}\alpha_1)$ the given asymptotic expansion of $q(t)$ for small $|t|$ applies in the sector $\alpha_1 + \delta \leqslant \arg t \leqslant \alpha_2 - \delta$, and $q(t) = O(e^{\sigma|t|})$ as $t \to \infty$ in the same sector. Here σ is an assignable constant. Then $Q(z)$ can be continued analytically into the sector $-\alpha_2 - \frac{1}{2}\pi < \arg z < -\alpha_1 + \frac{1}{2}\pi$, and the given asymptotic expansion for large z holds when $-\alpha_2 - \frac{1}{2}\pi + \delta \leqslant \arg z \leqslant -\alpha_1 + \frac{1}{2}\pi - \delta$.

Example: Consider

$$Q(z) = \int_0^{\infty} e^{-zt} \ln(1 + \sqrt{t})\, dt$$

The singularities of $q(t) \equiv \ln(1 + \sqrt{t})$ are given by $\sqrt{t} = -1$, hence $q(t)$ is holomorphic in the sector $|\arg t| < 2\pi$. Within the unit circle

$$q(t) = \sum_{s=1}^{\infty} (-)^{s-1} \frac{t^{s/2}}{s}$$

With $\lambda = \mu = 2$ we derive

$$Q(z) \sim \frac{1}{2z} \sum_{s=1}^{\infty} (-)^{s-1} \frac{\Gamma(\frac{1}{2}s)}{z^{s/2}} \qquad \left(z \to \infty, \; |\arg z| \leqslant \tfrac{5}{2}\pi - \delta\right)$$

δ being any positive constant less than $\frac{5}{2}\pi$.

12.3.2 Laplace's Method

Extensions of the results of 12.2.5 and 12.2.6 to complex variables necessitate great care in the choice of branches of the many-valued functions which are used. Let $\mathcal{P}$ denote the path for the contour integral

$$I(z) = \int_a^b e^{-zp(t)} q(t)\, dt$$

and assume that the endpoint a is finite. The other endpoint b may be finite or infinite. Also, let ω denote the angle of slope of $\mathcal{P}$ at a, that is, the limiting value of $\arg(t-a)$ as $t \to a$ along $\mathcal{P}$. The functions $p(t)$ and $q(t)$ are assumed to be holomorphic in an open domain **T** which contains $\mathcal{P}$, with the possible exception of the endpoints a and b.

Further assumptions are:

a. In the neighborhood of a there are convergent series expansions

$$p(t) = p(a) + \sum_{s=0}^{\infty} p_s(t-a)^{s+\mu}; \quad q(t) = \sum_{s=0}^{\infty} q_s(t-a)^{s+\lambda-1}$$

in which $p_0 \neq 0$, $\operatorname{Re}\lambda > 0$, and $\mu > 0$. When μ or λ is nonintegral (and this can only happen when a is a boundary point of **T**) the branches of $(t-a)^\mu$ and $(t-a)^\lambda$ are determined by

$$(t-a)^\mu \sim |t-a|^\mu e^{i\mu\omega}; \quad (t-a)^\lambda \sim |t-a|^\lambda e^{i\lambda\omega}$$

as $t \to a$ along $\mathcal{P}$, and by continuity elsewhere.

b. The parameter z is confined to a sector or single ray, given by $\theta_1 \leqslant \theta \leqslant \theta_2$, where $\theta = \arg z$ and $\theta_2 - \theta_1 < \pi$.

c. $I(z)$ converges at b absolutely and uniformly for all sufficiently large $|z|$.

d. $\operatorname{Re}\{e^{i\theta} p(t) - e^{i\theta} p(a)\}$ attains its minimum on $\mathcal{P}$ at $t = a$ (and nowhere else).

The last condition is crucial; it demands that the endpoint a is the location of the peak value of the integrand when $|z|$ is large.

With the foregoing conditions

$$I(z) \sim e^{-zp(a)} \sum_{s=0}^{\infty} \Gamma\left(\frac{s+\lambda}{\mu}\right) \frac{a_s}{z^{(s+\lambda)/\mu}}$$

as $z \to \infty$ in $\theta_1 \leqslant \arg z \leqslant \theta_2$. In this expansion the branch of $z^{(s+\lambda)/\mu}$ is

$$|z|^{(s+\lambda)/\mu} e^{i\theta(s+\lambda)/\mu}$$

and the coefficients a_s are determined by the method and formulas of 12.2.6, with the proviso that in forming the powers of p_0, the branch of $\arg p_0$ is chosen to satisfy

$$|\arg p_0 + \theta + \mu\omega| \leqslant \tfrac{1}{2}\pi$$

This choice is always possible, and it is unique.

12.3.3 Saddle-Points

Consider now the integral $I(z)$ of 12.3.2 in cases when the minimum value of $\mathrm{Re}\,\{zp(t)\}$ on the path $\mathcal{P}$ occurs not at a but an interior point t_0, say. For simplicity, assume that $\theta(\equiv \arg z)$ is fixed, so that t_0 is independent of z. The path may be subdivided at t_0, giving

$$I(z) = \int_{t_0}^{b} e^{-zp(t)}\, q(t)\, dt - \int_{t_0}^{a} e^{-zp(t)}\, q(t)\, dt$$

For large $|z|$ the asymptotic expansion of each of these integrals can be found by application of the result of 12.3.2, the role of the series in Condition (a) being played by the Taylor-series expansions of $p(t)$ and $q(t)$ at t_0. If $p'(t_0) \neq 0$, then it transpires that the asymptotic expansions of the two integrals are exactly the same, and on subtraction only the error terms are left. On the other hand, if $p'(t_0) = 0$ then the μ of Condition (a) is an integer not less than 2; in consequence, *different* branches of $p_0^{1/\mu}$ are used in constructing the coefficients a_s, and the two asymptotic expansions no longer cancel.

Cases in which $p'(t_0) \neq 0$ can be handled by deformation of $\mathcal{P}$ in such a way that on the new path the minimum of $\mathrm{Re}\,\{zp(t)\}$ occurs either at one of the endpoints or at a zero of $p'(t)$. As indicated in the preceding paragraph, the asymptotic expansion of $I(z)$ may then be found by means of one or two applications of the result of 12.3.2. Thus the zeros of $p'(t)$ are of great importance; they are called *saddle-points*. The name derives from the fact that if the surface $|e^{p(t)}|$ is plotted against the real and imaginary parts of t, then the tangent plane is horizontal at a zero of $p'(t)$, but in consequence of the maximum-modulus theorem this point is neither a maximum nor a minimum of the surface. Deformation of a path in the t-plane to pass through a zero of $p'(t)$ is equivalent to crossing a mountain ridge *via* a pass.

The task of locating saddle-points for a given integral is generally fairly easy, but the construction of a path on which $\mathrm{Re}\,\{zp(t)\}$ attains its minimum at an endpoint or saddle-point may be troublesome. An intelligent guess is sometimes successful, especially when the parameter z is confined to the real axis. Failing this, a partial study of the conformal mapping between the planes of t and $p(t)$ may be needed.

In constructing bounds for the error terms in the asymptotic expansion of $I(z)$ it is advantageous to employ integration paths along which $\mathrm{Im}\,\{zp(t)\}$ is constant. On the surface $|e^{zp(t)}|$ these are the paths of steepest descent from the endpoint or saddle-point. In consequence, the name *method of steepest descents* is often used. For the purpose of deriving asymptotic expansions, however, use of steepest paths is not essential.

Example: Bessel functions of large order—An integral of Schläfli for the Bessel function of the first kind is given by

$$J_\nu(\nu \operatorname{sech} \alpha) = \frac{1}{2\pi i} \int_{\infty - \pi i}^{\infty + \pi i} e^{-\nu p(t)}\, dt$$

where

$$p(t) = t - \operatorname{sech} \alpha \sinh t$$

Let us seek the asymptotic expansion of this integral for fixed positive values of α and large positive values of ν.

The saddle-points are the roots of $\cosh t = \cosh \alpha$ and are therefore given by $t = \pm\alpha, \pm\alpha \pm 2\pi i, \ldots$. The most promising is α, and as a possible path we consider that indicated in Figure 12.3-1. On the vertical segment we have $t = \alpha + i\tau$ where $-\pi \leqslant \tau \leqslant \pi$, and therefore

$$\operatorname{Re}\{p(t)\} = \alpha - \tanh \alpha \cos \tau > \alpha - \tanh \alpha \qquad (\tau \neq 0)$$

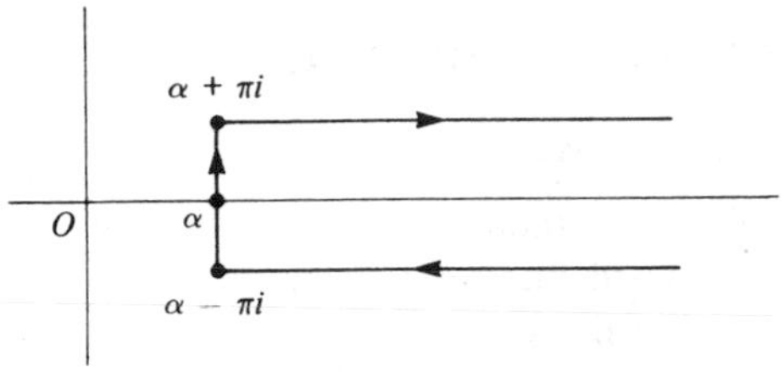

Fig. 12.3-1 *t*-plane.

On the horizontal segments $t = \alpha \pm \pi i + \tau$ where $0 \leqslant \tau < \infty$, and

$$\operatorname{Re}\{p(t)\} = \alpha + \tau + \operatorname{sech} \alpha \sinh(\alpha + \tau) \geqslant \alpha + \tanh \alpha$$

Clearly $\operatorname{Re}\{p(t)\}$ attains its minimum on the path at α, as required.

The Taylor series for $p(t)$ at α is given by

$$p(t) = \alpha - \tanh \alpha - \tfrac{1}{2}(t-\alpha)^2 \tanh \alpha - \tfrac{1}{6}(t-\alpha)^3 - \tfrac{1}{24}(t-\alpha)^4 \tanh \alpha + \cdots$$

In the notation of 12.3.2, we have $\mu = 2, p_0 = -\frac{1}{2}\tanh \alpha, p_1 = -\frac{1}{6}, p_2 = -\frac{1}{24}\tanh \alpha$, and $\lambda = q_0 = 1$. On the upper part of the path $\omega = \frac{1}{2}\pi$, and since $\theta = 0$ the correct choice of branch of $\arg p_0$ is $-\pi$. The formulas of 12.2.6 yield

$$a_0 = (\tfrac{1}{2}\coth \alpha)^{1/2} i; \quad a_1 = \tfrac{1}{3}\coth^2 \alpha; \quad a_2 = (\tfrac{1}{2} - \tfrac{5}{6}\coth^2 \alpha)(\tfrac{1}{2}\coth \alpha)^{3/2} i$$

Hence from 12.3.2

$$\int_{\alpha}^{\infty + \pi i} e^{-\nu p(t)}\, dt \sim e^{-\nu(\alpha - \tanh \alpha)} \sum_{s=0}^{\infty} \Gamma\left(\frac{s+1}{2}\right) \frac{a_s}{\nu^{(s+1)/2}}$$

The corresponding expansion for $\int_\alpha^{\infty - \pi i}$ is obtained by changing the sign of i throughout. Combination of the results yields

$$J_\nu(\nu \operatorname{sech} \alpha) \sim \frac{e^{-\nu(\alpha - \tanh \alpha)}}{(2\pi\nu \tanh \alpha)^{1/2}} \left\{1 + \left(\frac{1}{8} \coth \alpha - \frac{5}{24} \coth^3 \alpha\right) \frac{1}{\nu} + \cdots\right\}$$

This is *Debye's expansion.* No expression for the general term is available; the easiest way of calculating higher terms is *via* differential-equation theory (12.8.2).

Conformal mapping was not required in this example because a suitable path was easily guessed. For the corresponding problem with complex ν and α, however, the mapping is almost unavoidable.

12.4 FURTHER METHODS FOR INTEGRALS

12.4.1 Logarithmic Singularities

Watson's lemma, Laplace's method, and the method of stationary phase can be extended in a straightforward manner to cases in which the integrand has a logarithmic singularity at the saddle-point.

For example, with the conditions of 12.2.2

$$\int_0^\infty e^{-xt} q(t) \ln t \, dt \sim \sum_{s=0}^\infty \Gamma'\left(\frac{s+\lambda}{\mu}\right) \frac{a_s}{x^{(s+\lambda)/\mu}} - \ln x \sum_{s=0}^\infty \Gamma\left(\frac{s+\lambda}{\mu}\right) \frac{a_s}{x^{(s+\lambda)/\mu}}$$

In other words, formal differentiation of the general result of 12.2.2 with respect to the exponent λ (or μ) is legitimate. Such differentiations may be repeated any number of times.

12.4.2 Generalizations of Laplace's Method

The underlying idea of Laplace's method may be applied to integrals in which the parameter x enters in a more general way than in sections 12.2 and 12.3. Consider the integral

$$I(x) = \int_a^b \exp\{-xp(t) + x^\alpha r(t)\} q(t) \, dt$$

in which $p(t)$ and $q(t)$ satisfy the conditions of 12.2.5, $r(t)$ is independent of x, and α is a constant. What kind of behavior can be permitted in $r(t)$ at $t = a$ without changing the result already obtained for the case $r(t) = 0$? A sufficient condition is, in fact

$$r(t) \sim R(t - a)^\nu \qquad (t \to a+)$$

where R and ν are constants such that $R \neq 0, \nu \geqslant 0$, and $\nu > \mu\alpha$. In the case $R > 0$ we must also have $\alpha < 1$.

When $\nu \leqslant \mu\alpha$ the term $x^\alpha r(t)$ may no longer be treated as a negligible perturbation. The case $\nu < \mu\alpha$ can be handled simply by interchanging the roles of $p(t)$ and $r(t)$, and regarding x^α instead of x as the large parameter.

The case $\nu = \mu\alpha$ is more interesting, because $I(x)$ can no longer be approximated satisfactorily in terms of elementary functions. The simplest integral having the same character is *Faxén's integral*

$$\mathrm{Fi}(\xi, \eta; y) = \int_0^\infty \exp(-\tau + y\tau^\xi)\tau^{\eta-1}\, d\tau \qquad (0 \leqslant \mathrm{Re}\, \xi < 1, \mathrm{Re}\, \eta > 0)$$

This is used as approximant in the following general result.

Let

$$I(x) = \int_0^b \exp\{-xp(t) + x^{\nu/\mu} r(t) + s(x, t)\} q(x, t)\, dt$$

in which b is finite, and

a. In the interval $(0, b]$ the functions $p'(t)$ and $r(t)$ are continuous and $p'(t) > 0$.

b. As $t \to 0+$

$$p(t) = p(0) + Pt^\mu + O(t^{\mu_1}); \quad p'(t) = \mu P t^{\mu-1} + O(t^{\mu_1 - 1}); \quad r(t) = Rt^\nu + O(t^{\nu_1})$$

where $P > 0$, $\mu_1 > \mu > \nu \geqslant 0$, and $\nu_1 > \nu$.

c. For all sufficiently large x the functions $s(x, t)$ and $q(x, t)$ are continuous in $0 < t \leqslant b$, and

$$|s(x, t)| \leqslant Sx^\gamma t^\sigma; \quad |q(x, t) - Qt^{\lambda-1}| \leqslant Q_1 x^\beta t^{\lambda_1 - 1}$$

where $S, \gamma, \sigma, Q, \lambda, Q_1, \beta$, and λ_1 are independent of x and t, and*

$$\sigma \geqslant 0; \quad \lambda > 0; \quad \lambda_1 > 0; \quad \gamma < \min(1, \sigma/\mu); \quad \beta < (\lambda_1 - \lambda)/\mu$$

Then

$$I(x) = \frac{Q}{\mu} \mathrm{Fi}\left(\frac{\nu}{\mu}, \frac{\lambda}{\mu}; \frac{R}{P^{\nu/\mu}}\right) \frac{e^{-xp(0)}}{(Px)^{\lambda/\mu}} \left\{1 + O\left(\frac{1}{x^{\tilde{\omega}/\mu}}\right)\right\} \qquad (x \to \infty)$$

where $\tilde{\omega} = \min(\mu_1 - \mu, \sigma - \mu\gamma, \lambda_1 - \lambda - \mu\beta, \nu_1 - \nu)$.

*None of γ, β, or $\lambda_1 - \lambda$ is required to be positive.

12.4.3 Properties of Faxén's Integral

Commonly needed pairs of values of the parameters are $\xi = \eta = \frac{1}{2}$ and $\xi = \eta = \frac{1}{3}$. For these cases

$$\mathrm{Fi}\left(\tfrac{1}{2}, \tfrac{1}{2};\ y\right) = \sqrt{\pi}\, e^{y^2/4} \left\{1 + \mathrm{erf}\left(\tfrac{1}{2} y\right)\right\}$$

$$\mathrm{Fi}\left(\tfrac{1}{2}, \tfrac{1}{2};\ y\right) + \mathrm{Fi}\left(\tfrac{1}{2}, \tfrac{1}{2};\ -y\right) = 2\sqrt{\pi}\, e^{y^2/4}$$

$$\mathrm{Fi}\left(\tfrac{1}{3}, \tfrac{1}{3};\ y\right) = 3^{2/3}\, \pi\, \mathrm{Hi}\left(3^{-1/3} y\right)$$

and

$$e^{-\pi i/6}\, \mathrm{Fi}\left(\tfrac{1}{3}, \tfrac{1}{3};\ y e^{\pi i/3}\right) + e^{\pi i/6}\, \mathrm{Fi}\left(\tfrac{1}{3}, \tfrac{1}{3};\ y e^{-\pi i/3}\right) = 3^{2/3}\, 2\pi\, \mathrm{Ai}\left(-3^{-1/3} y\right)$$

Here $\mathrm{Hi}(x)$ denotes *Scorer's function*, defined by

$$\mathrm{Hi}(x) = \frac{1}{\pi} \int_0^\infty \exp\left(-\tfrac{1}{3} t^3 + xt\right) dt$$

and Ai is the Airy integral.

Example: Parabolic cylinder functions of large order—An integral representation for the parabolic cylinder function is supplied by*

$$U\left(n + \frac{1}{2}, y\right) = \frac{e^{-y^2/4}}{\Gamma(n+1)} \int_0^\infty e^{-yw - (w^2/2)}\, w^n\, dw \qquad (n > -1)$$

We seek an asymptotic approximation for large positive n and fixed y.

The integrand attains its maximum at $w = -\frac{1}{2} y + \sqrt{\frac{1}{4} y^2 + n}$. Since this is asymptotic to $\sqrt{n}$ for large n we make the substitution $w = \sqrt{n}\,(1 + t)$; compare the example at the end of 12.2.5. Accordingly

$$U\left(n + \frac{1}{2}, y\right) = \exp\left(-y\sqrt{n} - \frac{1}{2} n - \frac{1}{4} y^2\right) \frac{n^{(n+1)/2}}{\Gamma(n+1)} \int_{-1}^\infty e^{-np(t) - yt\sqrt{n}}\, dt$$

where

$$p(t) = t + \tfrac{1}{2} t^2 - \ln(1 + t) = t^2 - \tfrac{1}{3} t^3 + \cdots \qquad (t \to 0)$$

*This notation is due to Miller (1955). In the older notation of Whittaker, $U(n + \frac{1}{2}, y)$ is denoted by $D_{-n-1}(y)$.

The general result of 12.4.2 is applied with $x = n$, $r(t) = -yt$, $s(x, t) = 0$, and $q(x, t) = 1$. Thus $\mu = 2, \mu_1 = 3, \nu = 1, \widetilde{\omega} = 1$, and

$$\int_0^b e^{-np(t)-yt\sqrt{n}}\,dt = \frac{\mathrm{Fi}(\frac{1}{2}, \frac{1}{2}; -y)}{2\sqrt{n}}\left\{1 + O\left(\frac{1}{\sqrt{n}}\right)\right\}$$

for any fixed value of the positive number b. Similarly,

$$\int_{-b}^0 e^{-np(t)-yt\sqrt{n}}\,dt = \frac{\mathrm{Fi}(\frac{1}{2}, \frac{1}{2}; y)}{2\sqrt{n}}\left\{1 + O\left(\frac{1}{\sqrt{n}}\right)\right\}$$

provided that $0 < b < 1$. The contributions from the tails $\int_b^\infty$ and $\int_{-1}^{-b}$ are exponentially small when n is large, hence by addition and use of Stirling's approximation (7.2-11) we derive the required result

$$U\left(n + \frac{1}{2}, y\right) = \frac{\exp\,(-y\sqrt{n} + \frac{1}{2}n)}{\sqrt{2}\,n^{(n+1)/2}}\left\{1 + O\left(\frac{1}{\sqrt{n}}\right)\right\} \qquad (n \to \infty)$$

12.4.4 More General Kernels

Watson's lemma (12.2.2 and 12.3.1) may be regarded as an inductive relation between two asymptotic expansions; thus

$$q\,(t) \sim \sum_{s=0}^{\infty} a_s t^{(s+\lambda-\mu)/\mu} \qquad (t \to 0+)$$

implies

$$\int_0^\infty e^{-xt} q(t)\,dt \sim \sum_{s=0}^{\infty} \Gamma\left(\frac{s+\lambda}{\mu}\right)\frac{a_s}{x^{(s+\lambda)/\mu}} \qquad (x \to +\infty)$$

provided that $\mathrm{Re}\,\lambda > 0$, $\mu > 0$, and the integral converges. Similar induction of series occurs for integrals in which the factor e^{-xt} is replaced by a more general kernel $g(xt)$. Thus

$$\int_0^\infty g(xt) q\,(t)\,dt \sim \sum_{s=0}^{\infty} G\left(\frac{s+\lambda}{\mu}\right)\frac{a_s}{x^{(s+\lambda)/\mu}} \qquad (x \to +\infty)$$

in which $G(\alpha)$ denotes the Mellin transform of $g(t)$:

$$G(\alpha) = \int_0^\infty g(\tau)\tau^{\alpha-1}\,d\tau$$

Special cases include

$$\int_0^\infty \mathrm{Ai}(xt)q(t)\,dt \sim \sum_{s=0}^\infty 3^{-(s+\lambda+2\mu)/(3\mu)}\Gamma\left(\frac{s+\lambda}{\mu}\right)\left\{\Gamma\left(\frac{s+\lambda+2\mu}{3\mu}\right)\right\}^{-1}\frac{a_s}{x^{(s+\lambda)/\mu}}$$

and

$$\int_0^\infty K_0(xt)q(t)\,dt \sim \frac{1}{4}\sum_{s=0}^\infty \left\{\Gamma\left(\frac{s+\lambda}{2\mu}\right)\right\}^2 a_s\left(\frac{2}{x}\right)^{(s+\lambda)/\mu}$$

where K_0 is the modified Bessel function. It is assumed that for large t, $q(t)$ is $O(e^{\sigma t^{3/2}})$ for the Ai kernel and $O(e^{\sigma t})$ for the K_0 kernel, σ being an assignable constant.

12.4.5 Bleistein's Method

Let

$$I(\alpha, x) = \int_0^k e^{-xp(\alpha,t)}q(\alpha, t)t^{\lambda-1}\,dt$$

where k and λ are positive constants (k possibly being infinite), α is a variable parameter in the interval $[0, k)$, and x is a large positive parameter. Assume that $\partial^2 p(\alpha, t)/\partial t^2$ and $q(\alpha, t)$ are continuous functions of α and t, and also that for given α the minimum value of $p(\alpha, t)$ in $[0, k)$ is attained at $t = \alpha$, at which point $\partial p(\alpha, t)/\partial t$ vanishes but both $\partial^2 p(\alpha, t)/\partial t^2$ and $q(\alpha, t)$ are nonzero. For large x Laplace's method gives

$$I(\alpha, x) \sim e^{-xp(\alpha,\alpha)}q(\alpha, \alpha)\alpha^{\lambda-1}\left\{\frac{x}{2\pi}\left[\frac{\partial^2 p(\alpha, t)}{\partial t^2}\right]_{t=\alpha}\right\}^{-1/2}$$

if $\alpha \neq 0$, or

$$I(\alpha, x) \sim \frac{1}{2}e^{-xp(0,0)}q(0, 0)\Gamma\left(\frac{1}{2}\lambda\right)\left\{\frac{x}{2}\left[\frac{\partial^2 p(0, t)}{\partial t^2}\right]_{t=0}\right\}^{-\lambda/2}$$

if $\alpha = 0$.

Whether or not $\lambda = 1$, the first of these approximations does not reduce to the second as $\alpha \to 0$. This abrupt change means that the first approximation is nonuniform for arbitrarily small values of α.

To obtain a uniform approximation, we introduce a new integration variable w, given by

$$p(\alpha, t) = \tfrac{1}{2} w^2 - aw + b$$

where a and b are functions of α chosen in such a way that the endpoint $t = 0$ corresponds to $w = 0$, and the stationary point $t = \alpha$ corresponds to the stationary point $w = a$. Thus

$$b = p(\alpha, 0); \quad a = \{2p(\alpha, 0) - 2p(\alpha, \alpha)\}^{1/2}$$

and

$$w = \{2p(\alpha, 0) - 2p(\alpha, \alpha)\}^{1/2} \pm \{2p(\alpha, t) - 2p(\alpha, \alpha)\}^{1/2}$$

the upper or lower sign being taken according as $t >$ or $< \alpha$. The relationship between t and w is one-to-one, and because

$$\frac{dw}{dt} = \pm \frac{1}{\{2p(\alpha, t) - 2p(\alpha, \alpha)\}^{1/2}} \frac{\partial p(\alpha, t)}{\partial t}$$

the relationship is free from singularity at $t = \alpha$.

Transformation to w as variable gives

$$I(\alpha, x) = e^{-xp(\alpha,0)} \int_0^{\kappa} \exp\left\{-x\left(\frac{1}{2} w^2 - aw\right)\right\} f(\alpha, w) w^{\lambda-1}\, dw$$

where

$$f(\alpha, w) = q(\alpha, t)\left(\frac{t}{w}\right)^{\lambda-1} \frac{dt}{dw}$$

and $\kappa = \kappa(\alpha)$ is the value of w at $t = k$. The factor $f(\alpha, w)$ is expanded in a Taylor series centered at the peak value $w = a$ of the exponential factor. This series has the form

$$f(\alpha, w) = \sum_{s=0}^{\infty} \phi_s(\alpha)(w - a)^s$$

in which the coefficients $\phi_s(\alpha)$ are continuous at $\alpha = 0$. The required uniform expansion is then obtained in a similar manner to Laplace's method: κ is replaced by

∞ and the series integrated term by term. Thus with the notation

$$F_s(y) = \int_0^\infty \exp\left(-\frac{1}{2}\tau^2 + y\tau\right)(\tau - y)^s \tau^{\lambda-1}\, d\tau$$

we derive

$$I(\alpha, x) \sim \frac{e^{-xp(\alpha,0)}}{x^{\lambda/2}} \sum_{s=0}^{\infty} \phi_s(\alpha) \frac{F_s(a\sqrt{x})}{x^{s/2}} \qquad (x \to \infty)$$

in the sense that this series is a generalized asymptotic expansion with respect to an appropriate scale (12.1.5).

Example: Let

$$I(\alpha, x) = \int_0^{\pi/2} e^{x(\cos\theta + \theta \sin\alpha)}\, d\theta$$

where $0 \leqslant \alpha < \frac{1}{2}\pi$, and x is a large positive parameter. In the present notation

$$p(\alpha, \theta) = -\cos\theta - \theta\sin\alpha; \quad \partial p(\alpha, \theta)/\partial\theta = \sin\theta - \sin\alpha$$

The minimum of $p(\alpha, \theta)$ in the range of integration is $\theta = \alpha$. Since this approaches an endpoint as $\alpha \to 0$ we have exactly the situation described above.

The appropriate transformation is given by

$$\cos\theta + \theta\sin\alpha = 1 + aw - \tfrac{1}{2}w^2$$

where

$$a = \sqrt{2}\,(\cos\alpha + \alpha\sin\alpha - 1)^{1/2}$$

Thus

$$w = a \pm \sqrt{2}\,\{\cos\alpha + (\alpha - \theta)\sin\alpha - \cos\theta\}^{1/2} \qquad (\theta \gtrless \alpha)$$

The new integral is

$$I(\alpha, x) = e^x \int_0^{\kappa} \exp\left\{-x\left(\frac{1}{2}w^2 - aw\right)\right\} \frac{d\theta}{dw}\, dw$$

where

$$\kappa = a + \sqrt{2}\,\{\cos\alpha + (\alpha - \tfrac{1}{2}\pi)\sin\alpha\}^{1/2}$$

and

$$\frac{d\theta}{dw} = \frac{w - a}{\sin\theta - \sin\alpha} = \sum_{s=0}^{\infty} \phi_s(\alpha)(w - a)^s$$

In the last expansion the first three coefficients are given by

$$\phi_0(\alpha) = \frac{1}{(\cos\alpha)^{1/2}};\quad \phi_1(\alpha) = \frac{\sin\alpha}{3\cos^2\alpha};\quad \phi_2(\alpha) = \frac{5 - 2\cos^2\alpha}{24(\cos\alpha)^{7/2}}$$

Write

$$X_s(\alpha, x) = \Gamma(\tfrac{1}{2}s + \tfrac{1}{2}) + (-)^s\gamma\{\tfrac{1}{2}s + \tfrac{1}{2}, x(\cos\alpha + \alpha\sin\alpha - 1)\} \qquad (s = 0, 1, \ldots)$$

where $\gamma(\alpha, x)$ is the incomplete Gamma function

$$\gamma(\alpha, x) = \int_0^x e^{-t}\,t^{\alpha-1}\,dt \qquad (\mathrm{Re}\,\alpha > 0)$$

Then we may express the required asymptotic expansion in the form

$$I(\alpha, x) = \frac{e^{x(\cos\alpha+\alpha\sin\alpha)}}{(2x)^{1/2}}\left\{\sum_{s=0}^{n-1} \phi_s(\alpha)X_s(\alpha, x)\left(\frac{2}{x}\right)^{s/2} + O\left(\frac{1}{x^{n/2}}\right)\right\} \qquad (x \to \infty)$$

where n is an arbitrary nonnegative integer. The O-term is uniform in any interval $0 \leqslant \alpha \leqslant \alpha_0$ for which α_0 is a constant less than $\pi/2$.

For fixed α and large x the incomplete Gamma function can be approximated in terms of elementary functions; compare 12.2.1. Then the uniform asymptotic expansion reduces to either the first or second Laplace approximation given at the beginning of this subsection, depending whether $\alpha > 0$ or $\alpha = 0$. This is, of course, to be expected, both in the present example and in the general case.

12.4.6 Method of Chester, Friedman, and Ursell

Let

$$I(\alpha, x) = \int_{\mathcal{P}} e^{-xp(\alpha,t)}\,q(\alpha, t)\,dt$$

be a contour integral in which x is a large parameter, and $p(\alpha, t)$ and $q(\alpha, t)$ are analytic functions of the complex variable t and continuous functions of the parameter α. Suppose that $\partial p(\alpha, t)/\partial t$ has two zeros which coincide for a certain value $\hat{\alpha}$, say, of α, and at least one of these zeros is in the range of integration. The problem of obtaining an asymptotic approximation for $I(\alpha, x)$ which is uniformly valid for α in a neighborhood of $\hat{\alpha}$ is similar to the problem treated in 12.4.5. In the present case we employ a cubic transformation of variables, given by

$$p(\alpha, t) = \tfrac{1}{3}w^3 + aw^2 + bw + c$$

The stationary points of the right-hand side are the zeros $w_1(\alpha)$ and $w_2(\alpha)$, say, of the quadratic $w^2 + 2aw + b$. The values of $a = a(\alpha)$ and $b = b(\alpha)$ are chosen in such a way that $w_1(\alpha)$ and $w_2(\alpha)$ correspond to the zeros of $\partial p(\alpha, t)/\partial t$. The other coefficient, c, is prescribed in any convenient manner.

The given integral becomes

$$I(\alpha, x) = e^{-xc} \int_{\mathcal{Q}} \exp\left\{-x\left(\frac{1}{3}w^3 + aw^2 + bw\right)\right\} f(\alpha, w)\, dw$$

where $\mathcal{Q}$ is the w-map of the original path $\mathcal{P}$, and

$$f(\alpha, w) = q(\alpha, t)\frac{dt}{dw} = q(\alpha, t)\frac{w^2 + 2aw + b}{\partial p(\alpha, t)/\partial t}$$

With the prescribed choice of a and b, the function $f(\alpha, w)$ is analytic at $w = w_1(\alpha)$ and $w = w_2(\alpha)$ when $\alpha \neq \hat{\alpha}$, and at the confluence of these points when $\alpha = \hat{\alpha}$. For large x, $I(\alpha, x)$ is approximated by the corresponding integral with $f(\alpha, w)$ replaced by a constant, that is, by an Airy or Scorer function, depending on the path $\mathcal{Q}$.

Example: Let us apply the method just described to the integral

$$A(\alpha, x) = \int_0^\infty e^{-x(\operatorname{sech}\alpha \sinh t - t)}\, dt$$

in which $\alpha \geqslant 0$ and x is large and positive. The integrand has saddle-points at $t = \alpha$ and $-\alpha$. The former is always in the range of integration, and it coincides with the latter when $\alpha = 0$. We seek an asymptotic expansion of $A(\alpha, x)$ which is uniform for arbitrarily small values of α.

By symmetry, the appropriate cubic transformation has the form

$$\operatorname{sech}\alpha \sinh t - t = \tfrac{1}{3}w^3 - \zeta w$$

The stationary points of the right-hand side are $w = \pm\zeta^{1/2}$. Since they are to correspond to $t = \pm\alpha$, the value of the coefficient ζ is determined by

$$\tfrac{2}{3}\zeta^{3/2} = \alpha - \tanh\alpha$$

Then

$$A(\alpha, x) = \int_0^\infty \exp\left\{-x\left(\frac{1}{3}w^3 - \zeta w\right)\right\}\frac{dt}{dw}\,dw$$

The peak value of the exponential factor in the new integrand occurs at $w = \zeta^{1/2}$. We expand t in a Taylor series at this point, in the form

$$t = \alpha + \sum_{s=0}^{\infty}\frac{\phi_s(\alpha)}{s+1}(w - \zeta^{1/2})^{s+1}$$

The coefficients $\phi_s(\alpha)$ can be found, for example, by repeatedly differentiating the equation connecting t and w and then setting $t = \alpha$ and $w = \zeta^{1/2}$. In particular

$$\phi_0(\alpha) = \left(\frac{4\zeta}{\tanh^2\alpha}\right)^{1/4}; \quad \phi_1(\alpha) = \frac{2 - \{\phi_0(\alpha)\}^3}{3\phi_0(\alpha)\tanh\alpha}$$

It is easily verified that each of these expressions tends to a finite limit as $\alpha \to 0$.

The desired asymptotic expansion is now obtained by termwise integration. Thus

$$A(\alpha, x) \sim \pi\sum_{s=0}^{\infty}\phi_s(\alpha)\frac{\mathrm{Qi}_s(x^{2/3}\zeta)}{x^{(s+1)/3}} \qquad (x \to \infty)$$

where

$$\mathrm{Qi}_s(y) = \frac{1}{\pi}\int_0^\infty \exp\left(-\frac{1}{3}t^3 + yt\right)(t - y^{1/2})^s\,dt \qquad (s = 0, 1, \ldots)$$

These integrals are related to Scorer's function (12.4.3) by

$$\mathrm{Qi}_0(y) = \mathrm{Hi}(y); \quad \mathrm{Qi}_1(y) = \mathrm{Hi}'(y) - y^{1/2}\,\mathrm{Hi}(y); \quad \mathrm{Qi}_2(y) = \pi^{-1} - 2y^{1/2}\,\mathrm{Qi}_1(y)$$

and

$$\mathrm{Qi}_s(y) = (-)^s\pi^{-1}y^{(s-2)/2} - 2y^{1/2}\,\mathrm{Qi}_{s-1}(y) + (s-2)\,\mathrm{Qi}_{s-3}(y) \qquad (s \geqslant 3)$$

If α is restricted to a finite interval $[0, \alpha_0]$, then the error on truncating the asymptotic expansion of $A(\alpha, x)$ at its nth term is $O\{x^{-(n+1)/3}\,\mathrm{Qi}_n(x^{2/3}\zeta)\}$ uniformly with respect to α, provided that n is even. For α ranging over the infinite interval $[0, \infty)$ this error is uniformly $O\{(1+\zeta)^{(n+1)/4}x^{-(n+1)/3}\,\mathrm{Qi}_n(x^{2/3}\zeta)\}$, in this case provided that n is even and nonzero.

12.5 SUMS AND SEQUENCES

12.5.1 Bernoulli Polynomials

The polynomials $B_0(x), B_1(x), \ldots$, defined by the generating function

$$\frac{te^{xt}}{e^t - 1} = \sum_{s=0}^{\infty} B_s(x)\frac{t^s}{s!} \qquad (|t| < 2\pi)$$

are called the *Bernoulli polynomials*. Their values at $x = 0$ are the *Bernoulli numbers* $B_s = B_s(0)$.

The first few Bernoulli numbers are given by

$$B_0 = 1 \quad B_1 = -\tfrac{1}{2} \quad B_2 = \tfrac{1}{6} \quad B_3 = 0 \quad B_4 = -\tfrac{1}{30}$$

$$B_5 = 0 \quad B_6 = \tfrac{1}{42} \quad B_7 = 0 \quad B_8 = -\tfrac{1}{30} \quad B_9 = 0$$

the only nonvanishing B_s of odd subscript being, in fact, B_1. Corresponding polynomials are

$$B_0(x) = 1; \qquad B_1(x) = x - \tfrac{1}{2}$$

$$B_2(x) = x^2 - x + \tfrac{1}{6}; \qquad B_3(x) = x(x - \tfrac{1}{2})(x - 1)$$

$$B_4(x) = x^4 - 2x^3 + x^2 - \tfrac{1}{30}; \qquad B_5(x) = x(x - \tfrac{1}{2})(x - 1)(x^2 - x - \tfrac{1}{3})$$

$$B_6(x) = x^6 - 3x^5 + \tfrac{5}{2}x^4 - \tfrac{1}{2}x^2 + \tfrac{1}{42}$$

Important properties include

$$B_s(x) = \sum_{j=0}^{s} \binom{s}{j} B_{s-j}x^j$$

$$B_s'(x) = sB_{s-1}(x); \quad \int_0^1 B_s(x)\,dx = 0 \quad (s \geqslant 1); \quad B_s(1 - x) = (-)^s B_s(x)$$

$$\sum_{j=1}^{n} j^s = \frac{1}{s+1}\{B_{s+1}(n+1) - B_{s+1}\} \qquad (s \geqslant 1)$$

$$\sum_{j=1}^{\infty} \frac{1}{j^{2s}} = (-)^{s-1}\frac{(2\pi)^{2s}B_{2s}}{2(2s)!} \qquad (s \geqslant 1)$$

and

$$\int_0^\infty \frac{x^{2s-1}}{e^{2\pi x} - 1}\, dx = (-)^{s-1} \frac{B_{2s}}{4s} \qquad (s \geqslant 1)$$

When $s \geqslant 1$, the only zeros of $B_{2s+1}(x)$ in the interval $[0, 1]$ are $0, \frac{1}{2}$, and 1, and the only zeros of $B_{2s}(x) - B_{2s}$ in the same interval are 0 and 1. Also,

$$|B_{2s}(x)| \leqslant |B_{2s}| \qquad (0 \leqslant x \leqslant 1)$$

12.5.2 The Euler-Maclaurin Formula

If a, m, and n are integers such that $a < n$ and $m > 0$, and $f^{(2m)}(x)$ is absolutely integrable over the interval (a, n), then

$$\sum_{j=a}^{n} f(j) = \int_a^n f(x)\,dx + \frac{1}{2} f(a) + \frac{1}{2} f(n) + \sum_{s=1}^{m-1} \frac{B_{2s}}{(2s)!} \{f^{(2s-1)}(n) - f^{(2s-1)}(a)\}$$
$$+ \int_a^n \frac{B_{2m} - B_{2m}(x - [x])}{(2m)!} f^{(2m)}(x)\,dx$$

Here $[x]$ denotes the integer in the interval $(x - 1, x]$; in consequence, as a function of x, $B_{2m}(x - [x])$ is periodic and continuous, with period 1.

The formula just given is the *Euler-Maclaurin formula*. Its uses include numerical quadrature, numerical summation of slowly convergent series, and asymptotic approximation of sums of series. Another version of the formula is furnished by

$$\sum_{j=a}^{n} f(j) = \int_a^n f(z)\,dz + \frac{1}{2} f(a) + \frac{1}{2} f(n) - 2 \int_0^\infty \frac{\operatorname{Im}\{f(a + iy)\}}{e^{2\pi y} - 1}\, dy$$
$$+ \sum_{s=1}^{m} \frac{B_{2s}}{(2s)!} f^{(2s-1)}(n) + 2 \frac{(-)^m}{(2m)!} \int_0^\infty \operatorname{Im}\{f^{(2m)}(n + i\vartheta_n y)\} \frac{y^{2m}\,dy}{e^{2\pi y} - 1}$$

ϑ_n being some number in the interval $(0, 1)$. This second form is valid with the conditions:

a. $f(z)$ is continuous throughout the strip $a \leqslant \operatorname{Re} z \leqslant n$, and holomorphic in its interior.

b. $f(z)$ is real on the intersection of the strip with the real axis.

c. $f(z) = o(e^{2\pi|\operatorname{Im} z|})$ as $\operatorname{Im} z \to \pm\infty$ in $a \leqslant \operatorname{Re} z \leqslant n$, uniformly with respect to $\operatorname{Re} z$.

d. $\displaystyle\int_0^\infty \frac{\text{Im}\,\{f(a+iy)\}}{e^{2\pi y}-1}\,dy$ converges

e. $f^{(2m)}(z)$ is continuous on the line Re $z = n$.

Example: Let us seek the asymptotic expansion of the sum

$$S(n) = \sum_{j=1}^{n} j \ln j$$

for large n. Setting $f(x) = x \ln x$, we have

$$\int f(x)\,dx = \frac{1}{2}x^2 \ln x - \frac{1}{4}x^2; \quad f'(x) = \ln x + 1; \quad f^{(s)}(x) = (-)^s \frac{(s-2)!}{x^{s-1}} \quad (s \geqslant 2)$$

The first form of the Euler-Maclaurin formula leads to

$$S(n) = \frac{1}{2}n^2 \ln n - \frac{1}{4}n^2 + \frac{1}{2}n \ln n + \frac{1}{12}\ln n + C$$

$$- \sum_{s=2}^{m-1} \frac{B_{2s}}{2s(2s-1)(2s-2)n^{2s-2}} - R_m(n)$$

where m is an arbitrary integer exceeding unity,

$$C = \frac{1}{4} - \frac{1}{720} - \frac{1}{12}\int_1^\infty \frac{B_4(x-[x])}{x^3}\,dx$$

and

$$R_m(n) = \int_n^\infty \frac{B_{2m} - B_{2m}(x-[x])}{2m(2m-1)x^{2m-1}}\,dx$$

The final result in 12.5.1 shows that $B_{2m} - B_{2m}(x-[x])$ is bounded in absolute value by B_{2m} and has the same sign. Hence

$$|R_m(n)| \leqslant \frac{|B_{2m}|}{2m(2m-1)(2m-2)n^{2m-2}} \quad (m \geqslant 2)$$

Since the last quantity is $O(1/n^{2m-2})$ as $n \to \infty$, the expansion for $S(n)$ is an asymptotic expansion, complete with error bound.

A numerical estimate for the constant C can be found by summing directly the first few terms of the series $\sum j \ln j$, and using the bound for $|R_m(n)|$. For example, if $m = 4$ and $n = 5$, then we have

$$|R_4(5)| \leqslant \frac{(1/30)}{8 \cdot 7 \cdot 6} \frac{1}{5^6} \doteqdot 0.6 \times 10^{-8}$$

Direct summation yields $S(5) = 18.27449823$, and subtracting the values of the known terms in the expansion of $S(5)$, we find that $C = 0.24875449$, correct to eight decimal places.

An analytical expression for C can be derived from the second form of the Euler-Maclaurin formula given above. The result is expressible as

$$C = \frac{\gamma + \ln(2\pi)}{12} - \frac{\zeta'(2)}{2\pi^2}$$

where γ denotes Euler's constant, $\zeta(z)$ is the Riemann Zeta function

$$\zeta(z) = \sum_{s=1}^{\infty} \frac{1}{s^z} \qquad (\mathrm{Re}\, z > 1)$$

and $\zeta'(2)$ the derivative of $\zeta(z)$ at $z = 2$.

12.5.3 Asymptotic Expansions of Entire Functions

The asymptotic behavior, for large $|z|$, of entire functions defined by their Maclaurin series

$$f(z) = \sum_{j=0}^{\infty} a_j z^j$$

can sometimes be found by expressing the sum as a contour integral and applying the methods of sections 12.2 to 12.4.

Example: Consider the function

$$f(\rho, x) = \sum_{j=0}^{\infty} \left(\frac{x^j}{j!}\right)^{\rho}$$

for large positive values of x, where ρ is a constant in the interval $(0, 4]$. From the residue theorem it follows that

$$\sum_{j=0}^{n-1} \left(\frac{x^j}{j!}\right)^{\rho} = \frac{1}{2i} \int_{\mathcal{C}} \left\{\frac{x^t}{\Gamma(t+1)}\right\}^{\rho} \cot(\pi t)\, dt$$

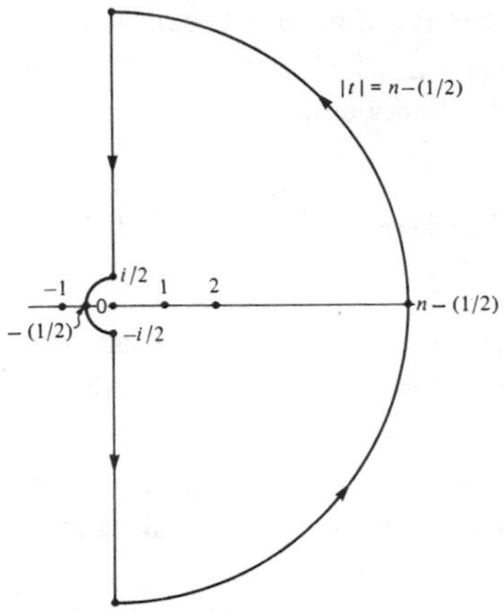

Fig. 12.5-1 *t*-plane: contour $\mathcal{C}$.

where $\mathcal{C}$ is the closed contour depicted in Figure 12.5-1. Now

$$\frac{\cot(\pi t)}{2i} = -\frac{1}{2} - \frac{1}{e^{-2\pi i t} - 1} = \frac{1}{2} + \frac{1}{e^{2\pi i t} - 1}$$

Hence

$$\sum_{j=0}^{n-1} \left(\frac{x^j}{j!}\right)^\rho = \int_{-1/2}^{n-(1/2)} \left\{\frac{x^t}{\Gamma(t+1)}\right\}^\rho dt - \int_{\mathcal{C}_1} \left\{\frac{x^t}{\Gamma(t+1)}\right\}^\rho \frac{dt}{e^{-2\pi i t} - 1}$$
$$+ \int_{\mathcal{C}_2} \left\{\frac{x^t}{\Gamma(t+1)}\right\}^\rho \frac{dt}{e^{2\pi i t} - 1}$$

where $\mathcal{C}_1$ and $\mathcal{C}_2$ are respectively the upper and lower halves of $\mathcal{C}$.

By means of Stirling's approximation (7.2-11) it is verifiable that the integrals around the large circular arcs vanish as $n \to \infty$, provided that $\rho \leqslant 4$ (which we have assumed to be the case). Also, $|x^{t\rho}| \leqslant 1$ when $x \geqslant 1$ and $\operatorname{Re} t \leqslant 0$. Hence

$$f(\rho, x) = \int_0^\infty \left\{\frac{x^t}{\Gamma(t+1)}\right\}^\rho dt + O(1) \quad (x \geqslant 1)$$

The asymptotic behavior of the last integral can be found by use of Stirling's approximation and Laplace's method in the manner of the example treated in 12.2.5. The final result is given by

$$f(\rho, x) \sim \frac{e^{\rho x}}{\rho^{1/2}(2\pi x)^{(\rho-1)/2}} \qquad (x \to \infty)$$

12.5.4 Coefficients in a Maclaurin or Laurent Expansion

Let $f(t)$ be a given analytic function and

$$f(t) = \sum_{n=-\infty}^{\infty} a_n t^n \qquad (0 < |t| < r)$$

a Laurent expansion. What is the asymptotic behavior of a_n as n approaches ∞ or $-\infty$? More specially, what is the asymptotic behavior of the sequence of coefficients in a Maclaurin series?

Problems of this kind can be brought within the scope of sections 12.3 and 12.4 by use of the Cauchy formula

$$a_n = \frac{1}{2\pi i} \int_{\mathcal{C}} \frac{f(t)}{t^{n+1}} \, dt$$

in which $\mathcal{C}$ is a simple closed contour encircling $t = 0$. However, in cases when $f(t)$ has finite singularities other than $t = 0$, the method of the next subsection often yields the required approximation in an easier way.

12.5.5 Method of Darboux

In the complex t-plane let r be the distance of the nearest singularity of $f(t)$ from the origin, and suppose that a 'comparison' function $g(t)$ can be found with the properties:

a. $g(t)$ is holomorphic in $0 < |t| < r$.
b. $f(t) - g(t)$ is continuous in $0 < |t| \leqslant r$.
c. The coefficients in the Laurent expansion

$$g(t) = \sum_{n=-\infty}^{\infty} b_n t^n \qquad (0 < |t| < r)$$

have known asymptotic behavior.

Then by allowing the contour in Cauchy's formula to expand, we deduce that

$$a_n - b_n = \frac{1}{2\pi i} \int_{|t|=r} \frac{f(t) - g(t)}{t^{n+1}} \, dt = \frac{1}{2\pi r^n} \int_0^{2\pi} \{f(re^{i\theta}) - g(re^{i\theta})\} \, e^{-ni\theta} \, d\theta$$

Application of the Riemann-Lebesgue lemma (12.2.3) to the last integral yields

$$a_n = b_n + o(r^{-n}) \qquad (n \to \infty)$$

This is a first approximation to a_n. Often this result is refinable in two ways. First, if $f^{(m)}(t) - g^{(m)}(t)$ is continuous on $|t| = r$ then the integral for $a_n - b_n$ may be integrated m times by parts to yield the stronger result

$$a_n = b_n + o(r^{-n}n^{-m}) \qquad (n \to \infty)$$

Secondly, it is unnecessary for $f(t) - g(t)$—or $f^{(m)}(t) - g^{(m)}(t)$—to be continuous on $|t| = r$; it suffices that the integrals involved converge uniformly with respect to n.

Example: Legendre polynomials of large order—The standard generating function for the Legendre polynomials is given by

$$\frac{1}{(1 - 2t\cos\alpha + t^2)^{1/2}} = \sum_{n=0}^{\infty} P_n(\cos\alpha)t^n \qquad (|t| < 1)$$

Let the left-hand side be denoted by $f(t)$. The only singularities of this function are branch-points at $t = e^{\pm i\alpha}$. To insure that these points do not coincide we restrict $0 < \alpha < \pi$ in what follows.

Let $(e^{i\alpha} - t)^{-1/2}$ be the branch of this square root which is continuous in the t-plane cut along the outward-drawn ray through $t = e^{i\alpha}$ and takes the value $e^{-i\alpha/2}$ at $t = 0$. Similarly, let $(e^{-i\alpha} - t)^{-1/2}$ denote the conjugate function. Then $f(t)$ can be factorized as

$$f(t) = (e^{i\alpha} - t)^{-1/2}(e^{-i\alpha} - t)^{-1/2} \qquad (|t| < 1)$$

If $t \to e^{\mp i\alpha}$ from within the unit circle, then

$$(e^{i\alpha} - t)^{-1/2} \to e^{-\pi i/4}(2\sin\alpha)^{-1/2} \qquad (t \to e^{-i\alpha})$$

$$(e^{-i\alpha} - t)^{-1/2} \to e^{\pi i/4}(2\sin\alpha)^{-1/2} \qquad (t \to e^{i\alpha})$$

Accordingly, in the notation used above we set

$$g(t) = e^{-\pi i/4}(2\sin\alpha)^{-1/2}(e^{-i\alpha} - t)^{-1/2} + e^{\pi i/4}(2\sin\alpha)^{-1/2}(e^{i\alpha} - t)^{-1/2}$$

The coefficient of t^n in the Maclaurin expansion of $g(t)$ is

$$b_n = \left(\frac{2}{\sin\alpha}\right)^{1/2}\binom{-\frac{1}{2}}{n}\cos\left\{\left(n + \frac{1}{2}\right)\alpha + \left(n - \frac{1}{4}\right)\pi\right\}$$

By Darboux's method a first approximation to $P_n(\cos\alpha)$ is given by

$$P_n(\cos\alpha) = b_n + o(1) \qquad (n \to \infty)$$

Since, however, by Stirling's approximation

$$b_n = \{2/(\pi n \sin\alpha)\}^{1/2} \cos(n\alpha + \tfrac{1}{2}\alpha - \tfrac{1}{4}\pi) + O(n^{-3/2})$$

this estimate for $P_n(\cos\alpha)$ reduces effectively to $o(1)$.

An improved result is obtainable by observing that the integral of $f'(t) - g'(t)$ around the unit circle converges uniformly with respect to n. Accordingly, we may integrate once by parts and apply the Riemann-Lebesgue lemma to obtain

$$P_n(\cos\alpha) = b_n + o(n^{-1})$$

By synthesizing a function $g(t)$ which matches the behavior of $f(t)$ *and* its first m derivatives at $t = e^{\mp i\alpha}$, we can extend this result into

$$P_n(\cos\alpha) = \left(\frac{2}{\sin\alpha}\right)^{1/2} \sum_{s=0}^{m-1} \binom{-\frac{1}{2}}{s} \binom{s-\frac{1}{2}}{n} \frac{\cos\alpha_{n,s}}{(2\sin\alpha)^s} + O\left(\frac{1}{n^{m+(1/2)}}\right) \qquad (n \to \infty)$$

where $\alpha_{n,s} = (n - s + \frac{1}{2})\alpha + (n - \frac{1}{2}s - \frac{1}{4})\pi$, and m is an arbitrary positive integer.

12.5.6 Haar's Method

Let $f(t)$ be given by an inverse Laplace transform (section 11.2)

$$f(t) = \frac{1}{2\pi i} \int_{c-i\infty}^{c+i\infty} e^{pt} F(p)\, dp$$

and $g(t)$ a comparison function having a known transform

$$g(t) = \frac{1}{2\pi i} \int_{c-i\infty}^{c+i\infty} e^{pt} G(p)\, dp$$

and known asymptotic behavior for large positive t. By subtraction

$$f(t) - g(t) = \frac{e^{ct}}{2\pi} \int_{-\infty}^{\infty} e^{itv} \{F(c + iv) - G(c + iv)\}\, dv$$

If the last integral converges uniformly at each limit for all sufficiently large t, then the Riemann-Lebesgue lemma (12.2.3) shows that

$$f(t) = g(t) + o(e^{ct}) \qquad (t \to \infty)$$

If, in addition, the corresponding integrals with F and G replaced by their derivatives $F^{(j)}$ and $G^{(j)}$, $j = 1, 2, \ldots, m$, converge uniformly, then by repeated integrations by parts and use again of the Riemann-Lebesgue lemma, we derive

$$f(t) = g(t) + o(t^{-m} e^{ct}) \qquad (t \to \infty)$$

This method for approximating a given function $f(t)$ is due to Haar, and is analogous to Darboux's method for sequences. The best results are obtained by translating the integration contour to the left to make the value of c as small as possible.

Example: Bessel functions of large argument—For $t > 0$ and $\nu > -\frac{1}{2}$, the Bessel function $J_\nu(t)$ is representable by

$$f(t) = \frac{1}{2\pi i}\int_{c-i\infty}^{c+i\infty} \frac{e^{tp}\,dp}{(p^2+1)^{\nu+(1/2)}} \qquad (c > 0)$$

where $f(t) = \pi^{1/2}(\frac{1}{2}t)^\nu J_\nu(t)/\Gamma(\nu+\frac{1}{2})$, and $(p^2+1)^{\nu+(1/2)}$ has its principal value. To approximate $f(t)$ for large t we deform the path into two loop integrals*

$$\int_{-\infty+i}^{(i+)} + \int_{-\infty-i}^{(-i+)}$$

In the first of these the factor $(p^2+1)^{-\nu-(1/2)}$ is replaced by its expansion in ascending powers of $p - i$. Then using Hankel's integral for the reciprocal of the Gamma function (7.2-9), we derive

$$\frac{1}{2\pi i}\int_{-\infty+i}^{(i+)} \frac{e^{tp}\,dp}{(p^2+1)^{\nu+(1/2)}}$$

$$= \frac{1}{2^{\nu+(1/2)}e^{(2\nu+1)\pi i/4}} \sum_{s=0}^{n-1} \binom{-\nu-\frac{1}{2}}{s} \frac{1}{(2i)^s} \frac{e^{it}}{\Gamma(\nu+\frac{1}{2}-s)\,t^{s-\nu+(1/2)}} + \epsilon_n(t)$$

Here n is an arbitrary integer, and

$$\epsilon_n(t) = \frac{1}{2\pi i}\int_{-\infty+i}^{(i+)} e^{tp}\, O\{(p-i)^{n-\nu-(1/2)}\}\,dp$$

the O-term being uniform on the loop path.

*The notation $\int_a^{(b+)}$ means that the integration path begins at a, encircles the singularity at b once in the positive sense, and returns to its starting point without encircling any other singularity of the integrand.

If we restrict $n > \nu - \frac{1}{2}$, then the path in the last integral may be collapsed onto the two sides of the cut through $p = i$ parallel to the negative real axis. Thence it follows that $\epsilon_n(t)$ is $O(1/t^{n-\nu+(1/2)})$ as $t \to \infty$. Similar analysis applies to the other loop integral, and combination of the results gives the required expansion

$$J_\nu(t) = \left(\frac{2}{\pi t}\right)^{1/2} \sum_{s=0}^{n-1} \binom{-\nu - \frac{1}{2}}{s} \frac{\Gamma(\nu + \frac{1}{2})}{\Gamma(\nu + \frac{1}{2} - s)} \frac{\cos\{t - (\frac{1}{2}s + \frac{1}{2}\nu + \frac{1}{4})\pi\}}{(2t)^s} + O\left(\frac{1}{t^{n+(1/2)}}\right)$$

From the standpoint of Haar, the role of $F(p)$ is played here by $(p^2 + 1)^{-\nu-(1/2)}$ and that of $G(p)$ by

$$\sum_{s=0}^{n-1} \frac{1}{2^{\nu+(1/2)}} \binom{-\nu - \frac{1}{2}}{s} \left\{\frac{(p-i)^{s-\nu-(1/2)}}{e^{(2\nu+1)\pi i/4}(2i)^s} + \frac{(p+i)^{s-\nu-(1/2)}}{e^{-(2\nu+1)\pi i/4}(-2i)^s}\right\}$$

12.6 THE LIOUVILLE-GREEN (OR JWKB) APPROXIMATION

12.6.1 The Liouville Transformation

Let

$$\frac{d^2 w}{dx^2} = f(x) w$$

be a given differential equation, and $\xi(x)$ any thrice-differentiable function. On transforming to ξ as independent variable and setting

$$W = \left(\frac{d\xi}{dx}\right)^{1/2} w$$

we find that

$$\frac{d^2 W}{d\xi^2} = \left[\dot{x}^2 f(x) - \frac{1}{2}\{x, \xi\}\right] W$$

Here the dot signifies differentiation with respect to ξ, and $\{x, \xi\}$ is the *Schwarzian derivative*

$$\{x, \xi\} = -2\dot{x}^{1/2} \frac{d^2}{d\xi^2}(\dot{x}^{-1/2}) = \frac{\dddot{x}}{\dot{x}} - \frac{3}{2}\left(\frac{\ddot{x}}{\dot{x}}\right)^2$$

properties of which include

$$\{x, \xi\} = -\left(\frac{dx}{d\xi}\right)^2 \{\xi, x\}; \quad \{x, \zeta\} = \left(\frac{d\xi}{d\zeta}\right)^2 \{x, \xi\} + \{\xi, \zeta\}$$

The foregoing change of variables is called the *Liouville transformation*. If we now prescribe

$$\xi = \int f^{1/2}(x)\, dx$$

then $\dot{x}^2 f(x) = 1$

$$\frac{d^2 W}{d\xi^2} = \left[1 - \frac{1}{2}\{x, \xi\}\right] W$$

and

$$\{x, \xi\} = \frac{5f'^2(x) - 4f(x)f''(x)}{8f^3(x)} = \frac{2}{f^{3/4}} \frac{d^2}{dx^2}\left(\frac{1}{f^{1/4}}\right)$$

Neglect of the Schwarzian enables the equation in W and ξ to be solved exactly, and this leads to the following general solution of the original differential equation:

$$Af^{-1/4}(x) \exp\left\{\int f^{1/2}(x)\, dx\right\} + Bf^{-1/4}(x) \exp\left\{-\int f^{1/2}(x)\, dx\right\}$$

where A and B are arbitrary constants. This is the *Liouville-Green* or *LG approximation*, also known as the *JWKB approximation*. The expressions

$$f^{-1/4} \exp\left(\pm \int f^{1/2}\, dx\right)$$

are called the *LG functions*.

In a wide range of circumstances, described in following subsections, neglect of the Schwarzian is justified and the LG approximation accurate. An important case of failure is immediately noticeable, however. At a zero of $f(x)$ the Schwarzian is infinite, rendering the LG approximation meaningless. Zeros of $f(x)$ are called *transition points* or *turning points* of the differential equation. The reason for the names is that on passing through a zero of $f(x)$ on the real axis, the character of each solution changes from oscillatory to monotonic (or vice-versa). Satisfactory approximations cannot be constructed in terms of elementary functions in the neighborhood of a transition point; see section 12.8 below.

12.6.2 Error Bounds: Real Variables

In stating error bounds for the LG approximation, it is convenient to take the differential equation in the form

$$\frac{d^2 w}{dx^2} = \{f(x) + g(x)\}\, w$$

It is assumed that in a given finite or infinite interval (a_1, a_2), $f(x)$ is a positive, real, twice-continuously differentiable function, and $g(x)$ is a continuous, real or complex function. Then the equation has twice-continuously differentiable solutions

$$w_1(x) = f^{-1/4}(x) \exp\left\{\int f^{1/2}(x)\,dx\right\} \{1 + \epsilon_1(x)\}$$

$$w_2(x) = f^{-1/4}(x) \exp\left\{-\int f^{1/2}(x)\,dx\right\} \{1 + \epsilon_2(x)\}$$

with the error terms bounded by

$$|\epsilon_j(x)|,\ \tfrac{1}{2} f^{-1/2}(x)\,|\epsilon_j'(x)| \leqslant \exp\{\tfrac{1}{2}\,\mathcal{V}_{a_j,x}(F)\} - 1 \qquad (j = 1, 2)$$

Here $\mathcal{V}$ denotes the variational operator defined in 12.2.4, and $F(x)$ is the *error-control function*

$$F(x) = \int \left\{\frac{1}{f^{1/4}} \frac{d^2}{dx^2}\left(\frac{1}{f^{1/4}}\right) - \frac{g}{f^{1/2}}\right\} dx$$

The foregoing result applies whenever the $\mathcal{V}_{a_j,x}(F)$ are finite.

A similar result is available for differential equations with solutions of oscillatory type. With exactly the same conditions, the equation

$$\frac{d^2 w}{dx^2} = \{-f(x) + g(x)\}\, w$$

has twice-continuously differentiable solutions

$$w_1(x) = f^{-1/4}(x) \exp\left\{i \int f^{1/2}(x)\,dx\right\} \{1 + \epsilon_1(x)\}$$

$$w_2(x) = f^{-1/4}(x) \exp\left\{-i\int f^{1/2}(x)dx\right\} \{1 + \epsilon_2(x)\}$$

such that

$$|\epsilon_j(x)|,\ f^{-1/2}(x)\,|\epsilon_j'(x)| \leqslant \exp\{\mathcal{V}_{a,x}(F)\} - 1 \qquad (j = 1, 2)$$

Here a is an arbitrary point in the closure of (a_1, a_2)—possibly at infinity—and the solutions $w_1(x)$ and $w_2(x)$ depend on a. When $g(x)$ is real, $w_1(x)$ and $w_2(x)$ are complex conjugates.

12.6.3 Asymptotic Properties with Respect to the Independent Variable

We return to the equation

$$\frac{d^2w}{dx^2} = \{f(x) + g(x)\}\, w$$

The error bounds of 12.6.2 immediately show that

$$w_1(x) \sim f^{-1/4} \exp\left(\int f^{1/2}\, dx\right) \qquad (x \to a_1+)$$

$$w_2(x) \sim f^{-1/4} \exp\left(-\int f^{1/2}\, dx\right) \qquad (x \to a_2-)$$

These results are valid whether or not a_1 and a_2 are finite, and also whether or not f and $|g|$ are bounded at a_1 and a_2. All that is required is that the error-control function $F(x)$ be of bounded variation in (a_1, a_2).

A somewhat deeper result, not immediately deducible from the results of 12.6.2, is that when $\left|\int f^{1/2}\, dx\right| \to \infty$ as $x \to a_1$ or a_2, there exist solutions $w_3(x)$ and $w_4(x)$ with the complementary properties

$$w_3(x) \sim f^{-1/4} \exp\left(\int f^{1/2}\, dx\right) \qquad (x \to a_2-)$$

$$w_4(x) \sim f^{-1/4} \exp\left(-\int f^{1/2}\, dx\right) \qquad (x \to a_1+)$$

The solutions $w_1(x)$ and $w_2(x)$ are unique, but not $w_3(x)$ and $w_4(x)$. At a_1, $w_1(x)$ is said to be *recessive* (or *subdominant*), whereas $w_4(x)$ is *dominant*. Similarly for $w_2(x)$ and $w_3(x)$ at a_2.

Example: Consider the equation

$$\frac{d^2 w}{dx^2} = (x + \ln x)w$$

for large positive values of x. We cannot take $f = x$ and $g = \ln x$ because $\int gf^{-1/2}\,dx$ would diverge at infinity. Instead, set $f = x + \ln x$ and $g = 0$. Then for large x, $f^{-1/4}(f^{-1/4})''$ is $O(x^{-5/2})$, consequently $\mathcal{V}(F)$ converges. Accordingly, there is a unique solution $w_2(x)$ such that

$$w_2(x) \sim (x + \ln x)^{-1/4} \exp\left\{-\int (x + \ln x)^{1/2}\,dx\right\} \qquad (x \to \infty)$$

and a nonunique solution $w_3(x)$ such that

$$w_3(x) \sim (x + \ln x)^{-1/4} \exp\left\{\int (x + \ln x)^{1/2}\,dx\right\} \qquad (x \to \infty)$$

These asymptotic forms are simplifiable by expansion and integration; thus

$$w_2(x) \sim x^{-(1/4)-\sqrt{x}} \exp(2x^{1/2} - \tfrac{2}{3}x^{3/2}); \quad w_3(x) \sim x^{-(1/4)+\sqrt{x}} \exp(\tfrac{2}{3}x^{3/2} - 2x^{1/2})$$

12.6.4 Convergence of $\mathcal{V}(F)$ at a Singularity

Sufficient conditions for the variation of the error-control function to be bounded at a finite point a_2 are given by

$$f(x) \sim \frac{c}{(a_2 - x)^{2\alpha+2}}; \quad g(x) = O\left\{\frac{1}{(a_2 - x)^{\alpha-\beta+2}}\right\} \qquad (x \to a_2-)$$

provided that c, α, and β are positive constants and the first relation is twice differentiable.

Similarly, when $a_2 = \infty$ sufficient conditions for $\mathcal{V}(F)$ to be bounded are

$$f(x) \sim cx^{2\alpha-2}; \quad g(x) = O(x^{\alpha-\beta-2}) \quad (x \to \infty)$$

again provided that c, α, and β are positive and the first relation is twice differentiable. When $\alpha = \frac{3}{2}$ we interpret the last condition as $f'(x) \to c$ and $f''(x) = O(x^{-1})$; when $\alpha = 1$ we require $f'(x) = O(x^{-1})$ and $f''(x) = O(x^{-2})$.

12.6.5 Asymptotic Properties with Respect to Parameters

Consider the equation

$$\frac{d^2w}{dx^2} = \{u^2 f(x) + g(x)\}\, w$$

in which u is a large positive parameter. If we again suppose that in a given interval (a_1, a_2) the function $f(x)$ is positive and $f''(x)$ and $g(x)$ are continuous, then the result of 12.6.2 may be applied with $u^2 f(x)$ playing the role of the previous $f(x)$. On discarding an irrelevant factor $u^{-1/2}$ it is seen that the new differential equation has solutions

$$w_j(u, x) = f^{-1/4}(x) \exp\left\{(-)^{j-1} u \int f^{1/2}(x)\, dx\right\} \{1 + \epsilon_j(u, x)\} \qquad (j = 1, 2)$$

where

$$|\epsilon_j(u, x)|, \; \frac{1}{2uf^{1/2}(x)} \left|\frac{\partial \epsilon_j(u, x)}{\partial x}\right| \leqslant \exp\left\{\frac{\mathcal{V}_{a_j,x}(F)}{2u}\right\} - 1$$

the function $F(x)$ being defined exactly as before. Since $F(x)$ is independent of u, the error bound is $O(u^{-1})$ for large u and fixed x. Moreover, if $F(x)$ is of bounded variation in (a_1, a_2), then the error bound is $O(u^{-1})$ uniformly with respect to x in (a_1, a_2). The differential equation may have a singularity at either endpoint without invalidating this conclusion as long as $\mathcal{V}(F)$ is bounded at a_1 and a_2.

Thus the LG functions represent asymptotic solutions in the neighborhood of a singularity (as in 12.6.3), and uniform asymptotic solutions for large values of a parameter. This *double* asymptotic property makes the LG approximation a remarkably powerful tool for approximating solutions of linear second-order differential equations.

Example: Parabolic cylinder functions of large order—The parabolic cylinder functions satisfy the equation

$$\frac{d^2w}{dx^2} = \left(\frac{1}{4}x^2 + a\right) w$$

a being a parameter. In the notation of 12.6.2, we take $f(x) = \frac{1}{4}x^2 + a$ and $g(x) = 0$. Referring to 12.6.4, we see that $\mathcal{V}(F)$ is finite at $x = +\infty$. Hence there exist solutions which are asymptotic to $f^{-1/4}e^{\pm\xi}$ for large x, where

$$\xi = \int \left(\frac{1}{4}x^2 + a\right)^{1/2} dx$$

On expansion and integration, we find that

$$\xi = \tfrac{1}{4}x^2 + a \ln x + \text{constant} + O(x^{-2}) \qquad (x \to \infty)$$

Hence the asymptotic forms of the solutions reduce to constant multiples of $x^{a-(1/2)}e^{x^2/4}$ and $x^{-a-(1/2)}e^{-x^2/4}$. The principal solution $U(a, x)$ is specified (uniquely) by the condition

$$U(a, x) \sim x^{-a-(1/2)}e^{-x^2/4} \qquad (x \to \infty)$$

How does $U(a, x)$ behave as $a \to +\infty$? Making the transformations $a = \frac{1}{2}u$ and $x = (2u)^{1/2}t$, we obtain

$$\frac{d^2w}{dt^2} = u^2(t^2 + 1)w$$

A solution of this equation which is recessive at infinity is given by

$$w(u, t) = (t^2 + 1)^{-1/4}e^{-u\hat{\xi}(t)}\{1 + \epsilon(u, t)\}$$

where

$$\hat{\xi}(t) = \int (t^2 + 1)^{1/2}\, dt = \frac{1}{2}t(t^2 + 1)^{1/2} + \frac{1}{2}\ln\,\{t + (t^2 + 1)^{1/2}\}$$

The error term is bounded by

$$|\epsilon(u, t)| \leqslant \exp\,\{\mathcal{V}_{t,\infty}(F)/(2u)\} - 1$$

with

$$F(t) = \int (t^2 + 1)^{-1/4}\,\{(t^2 + 1)^{-1/4}\}''\, dt = -\frac{t^3 + 6t}{12(t^2 + 1)^{3/2}}$$

The solutions $w(u, t)$ and $U(\frac{1}{2}u, \sqrt{2u}\,t)$ must be in constant ratio as t varies, since both are recessive at infinity. The value of the ratio may be found by comparing the asymptotic forms at $t = +\infty$. Thus we arrive at the required approximation:

$$U(\tfrac{1}{2}u, \sqrt{2u}\,t) = 2^{(u-1)/4}e^{u/4}u^{-(u+1)/4}(t^2 + 1)^{-1/4}e^{-u\hat{\xi}(t)}\{1 + \epsilon(u, t)\}$$

This result holds for positive u and all real values of t, or, on returning to the original variables, positive a and all real x. For fixed u (not necessarily large) and large positive t, we have $\epsilon(u, t) = O(t^{-2})$. On the other hand, since $\mathcal{V}_{-\infty,\infty}(F) < \infty$ we have $\epsilon(u, t) = O(u^{-1})$ for large u, uniformly with respect to $t \in (-\infty, \infty)$. These estimates illustrate the doubly asymptotic nature of the LG approximation.

Incidentally, the result of the example in 12.4.3 is obtainable from the present more general result by setting $u = 2n + 1$, $t = y/\sqrt{4n+2}$ and expanding $\hat{\xi}(t)$ for small t.

12.6.6 Error Bounds: Complex Variables

Let $f(z)$ and $g(z)$ be holomorphic in a complex domain $\mathbf{D}$ in which $f(z)$ is non-vanishing. Then the differential equation

$$\frac{d^2w}{dz^2} = \{f(z) + g(z)\}\, w$$

has solutions which are holomorphic in $\mathbf{D}$, depend on arbitrary (possibly infinite) reference points a_1 and a_2, and are given by

$$w_j(z) = f^{-1/4}(z) \exp\{(-)^{j-1}\xi(z)\}\{1 + \epsilon_j(z)\} \qquad (j = 1, 2)$$

where

$$\xi(z) = \int f^{1/2}(z)\, dz$$

and

$$|\epsilon_j(z)|, \quad |f^{-1/2}(z)\epsilon_j'(z)| \leqslant \exp\{\mathcal{V}_{a_j,z}(F)\} - 1$$

Here $F(z)$ is defined as in 12.6.2, with $x = z$.

In contrast to the case of real variables, the present error bounds apply only to subregions $\mathbf{H}_j(a_j)$ of $\mathbf{D}$. These subregions comprise the points z for which there exists a path $\mathcal{P}_j$ in $\mathbf{D}$ linking a_j with z, and along which $\mathrm{Re}\{\xi(z)\}$ is nondecreasing $(j = 1)$ or nonincreasing $(j = 2)$. Such a path is called *ξ-progressive*. In the bound $\exp\{\mathcal{V}_{a_j,z}(F)\} - 1$ the variation of $F(z)$ has to be evaluated along a ξ-progressive path. Parts of $\mathbf{D}$ excluded from $\mathbf{H}_j(a_j)$ are called *shadow zones.* The solutions $w_j(z)$ exist and are holomorphic in the shadow zones, but the error bounds do not apply there.

Asymptotic properties of the approximation with respect to z in the neighborhood of a singularity, or with respect to large values of a real or complex parameter, carry over straightforwardly from the case of real variables.

12.7 DIFFERENTIAL EQUATIONS WITH IRREGULAR SINGULARITIES

12.7.1 Classification of Singularities

Consider the differential equation

$$\frac{d^2 w}{dz^2} + f(z)\frac{dw}{dz} + g(z)\,w = 0$$

in which the functions $f(z)$ and $g(z)$ are holomorphic in a region which includes the punctured disc $0 < |z - z_0| < a$, z_0 and a being given finite numbers.

If both $f(z)$ and $g(z)$ are analytic at z_0, then z_0 is said to be an *ordinary point* of the differential equation. In this event all solutions are holomorphic in the disc $|z - z_0| < a$.

If z_0 is not an ordinary point, but both $(z - z_0)f(z)$ and $(z - z_0)^2 g(z)$ are analytic at z_0, then this point is said to be a *regular singularity* or *singularity of the first kind.* In this case independent solutions can be constructed in series involving fractional powers of $z - z_0$ and also, possibly, $\ln(z - z_0)$. The series converge when $|z - z_0| < a$; compare 6.4.6.

Lastly, if z_0 is neither an ordinary point nor a regular singularity, then it is said to be an *irregular singularity*, or *singularity of the second kind.* If an integer r exists such that $(z - z_0)^{r+1}f(z)$ and $(z - z_0)^{2r+2}g(z)$ are both analytic at z_0, then the least value of r is said to be the *rank* of the singularity. By analogy, a regular singularity is sometimes said to have zero rank.

In the neighborhood of an irregular singularity it is usually impossible to find convergent series expansions for the solutions in terms of elementary functions. Instead, asymptotic expansions are employed. From section 12.6, especially 12.6.4,* it can be seen that the LG functions furnish asymptotic approximations at an irregular singularity. The purpose of the present section is to extend these approximations into asymptotic expansions for singularities of finite rank. We begin with the simplest and commonest case in applications.

12.7.2 Singularities of Unit Rank

Without loss of generality the singularity is assumed to be at infinity: a finite singularity z_0 can always be projected to infinity by taking $(z - z_0)^{-1}$ as new independent variable. Thus we consider the differential equation of 12.7.1 with

$$f(z) = \sum_{s=0}^{\infty} \frac{f_s}{z^s}; \quad g(z) = \sum_{s=0}^{\infty} \frac{g_s}{z^s}$$

these series converging for sufficiently large $|z|$. Not all of the coefficients f_0, g_0, and g_1 vanish, otherwise the singularity would be regular.

*The symbols f and g are now being used differently.

Formal series solutions in descending powers of z can be constructed in the form

$$w = e^{\lambda z} z^{\mu} \sum_{s=0}^{\infty} \frac{a_s}{z^s}$$

Substituting in the differential equation and equating coefficients, we obtain in turn

$$\lambda^2 + f_0 \lambda + g_0 = 0$$

$$(f_0 + 2\lambda)\,\mu = -(f_1 \lambda + g_1)$$

and

$$(f_0 + 2\lambda)\, s a_s = (s - \mu)(s - 1 - \mu)\, a_{s-1} + \{\lambda f_2 + g_2 - (s - 1 - \mu) f_1\} a_{s-1}$$
$$+ \{\lambda f_3 + g_3 - (s - 2 - \mu) f_2\} a_{s-2} + \cdots + \{\lambda f_{s+1} + g_{s+1} + \mu f_s\} a_0$$

The first of these equations yields two possible values

$$\lambda_1, \lambda_2 = -\tfrac{1}{2} f_0 \pm (\tfrac{1}{4} f_0^2 - g_0)^{1/2}$$

for λ, called the *characteristic values.* The next equation determines the corresponding values μ_1 and μ_2, of μ. Then the values of a_0, say $a_{0,1}$ and $a_{0,2}$ in the two cases, may be assigned arbitrarily and higher coefficients $a_{s,1}$ and $a_{s,2}$, $s = 1, 2, \ldots$, determined recursively. The process fails if, and only if, $\lambda_1 = \lambda_2$, that is, $f_0^2 = 4g_0$. This case is treated below.

In general the formal series diverge. Corresponding to each, however, there is a unique solution $w_j(z), j = 1, 2$, of the differential equation with the property

$$w_j(z) \sim e^{\lambda_j z} z^{\mu_j} \sum_{s=0}^{\infty} \frac{a_{s,j}}{z^s}$$

as $z \to \infty$ in the sector

$$|\arg\{(\lambda_2 - \lambda_1) z\}| \leqslant \tfrac{3}{2}\pi - \delta \quad (j = 1); \quad |\arg\{(\lambda_1 - \lambda_2) z\}| \leqslant \tfrac{3}{2}\pi - \delta \quad (j = 2)$$

δ being an arbitrary positive constant.

In the sector $|\arg\{(\lambda_2 - \lambda_1)z\}| < \pi/2$, the solution $w_1(z)$ is recessive, and the two branches of $w_2(z)$ are dominant. These roles are interchanged in $|\arg\{(\lambda_1 - \lambda_2)z\}| < \pi/2$. Although both $w_1(z)$ and $w_2(z)$ can be continued analytically to any range of $\arg z$, the given sectors of validity of the asymptotic expansions are maximal (unless the expansions happen to converge).

The case in which the characteristic values λ_1 and λ_2 are equal can be handled by the preliminary transformation

$$w = e^{-f_0 z/2} W; \quad t = \sqrt{z}$$

In the new equation either the singularity at $t = \infty$ is regular, or it is irregular with unequal characteristic values (and therefore amenable to the foregoing analysis).

Example: Bessel's equation—Bessel functions of real or complex order ν satisfy

$$\frac{d^2 w}{dz^2} + \frac{1}{z}\frac{dw}{dz} + \left(1 - \frac{\nu^2}{z^2}\right) w = 0$$

In the present notation $f_1 = g_0 = 1$, $g_2 = -\nu^2$, and all other coefficients vanish. The equations for λ and μ yield $\lambda_1 = -\lambda_2 = i$, and $\mu_1 = \mu_2 = -\frac{1}{2}$. With $a_{0,1} = a_{0,2} = 1$ it is found that

$$a_{s,1} = i^s A_s(\nu); \quad a_{s,2} = (-i)^s A_s(\nu)$$

where

$$A_s(\nu) = (4\nu^2 - 1^2)(4\nu^2 - 3^2) \cdots \{4\nu^2 - (2s - 1)^2\}/(s!\, 8^s)$$

Renormalizing $w_1(z)$ and $w_2(z)$ by the factors $(2/\pi)^{1/2} \exp\{\mp(\frac{1}{2}\nu + \frac{1}{4})\pi i\}$, we obtain solutions $H_\nu^{(1)}(z)$ and $H_\nu^{(2)}(z)$ with the properties

$$H_\nu^{(1)}(z) \sim \left(\frac{2}{\pi z}\right)^{1/2} e^{i\zeta} \sum_{s=0}^{\infty} i^s \frac{A_s(\nu)}{z^s} \qquad (z \to \infty \text{ in } -\pi + \delta \leqslant \arg z \leqslant 2\pi - \delta)$$

$$H_\nu^{(2)}(z) \sim \left(\frac{2}{\pi z}\right)^{1/2} e^{-i\zeta} \sum_{s=0}^{\infty} (-i)^s \frac{A_s(\nu)}{z^s} \qquad (z \to \infty \text{ in } -2\pi + \delta \leqslant \arg z \leqslant \pi - \delta)$$

where δ is an arbitrary positive constant, and $\zeta = z - \frac{1}{2}\nu\pi - \frac{1}{4}\pi$. These are the Hankel functions of order ν; $H_\nu^{(1)}(z)$ is recessive at infinity in the upper half-plane, or, more precisely when $0 < \arg z < \pi$; $H_\nu^{(2)}(z)$ is recessive when $-\pi < \arg z < 0$.

12.7.3 Stokes' Phenomenon

The functions $H_\nu^{(1)}(z)$ and $H_\nu^{(2)}(z)$ introduced in the last example can be continued analytically to any range of arg z. Appropriate asymptotic expansions can be constructed by means of the continuation formulas

$$H_\nu^{(1)}(ze^{m\pi i}) = -\operatorname{cosec}(\nu\pi)\, [\sin\{(m-1)\nu\pi\}\, H_\nu^{(1)}(z) + e^{-\nu\pi i} \sin(m\nu\pi)\, H_\nu^{(2)}(z)]$$

$$H_\nu^{(2)}(ze^{m\pi i}) = \operatorname{cosec}(\nu\pi)\, [e^{\nu\pi i} \sin(m\nu\pi)\, H_\nu^{(1)}(z) + \sin\{(m+1)\nu\pi\}\, H_\nu^{(2)}(z)]$$

in which m is an integer. For example, if we take $m = 2$ in the first formula, substitute in the right-hand side by means of the expansions of 12.7.2, and then replace z by $ze^{-2\pi i}$, we arrive at

$$H_\nu^{(1)}(z) \sim \left(\frac{2}{\pi z}\right)^{1/2} \left\{e^{i\xi} \sum_{s=0}^{\infty} i^s \frac{A_s(\nu)}{z^s} + (1 + e^{-2\nu\pi i})\, e^{-i\xi} \sum_{s=0}^{\infty} (-i)^s \frac{A_s(\nu)}{z^s}\right\}$$

valid when $z \to \infty$ in $\pi + \delta \leqslant \arg z \leqslant 3\pi - \delta$.

It will be observed that two different asymptotic expansions are available for $H_\nu^{(1)}(z)$ in the sector $\pi + \delta \leqslant \arg z \leqslant 2\pi - \delta$, namely, the expansion just given and the original expansion of 12.7.2. In this sector, however, $e^{-i\xi}$ is exponentially small at infinity compared with $e^{i\xi}$, hence the *whole* contribution of the series it multiplies is absorbable, in Poincaré's sense, in any of the error terms associated with the first series.* Accordingly, there is no inconsistency.

Extensions to other phase ranges may be found in the same manner by taking other values of m. In each case an expansion of the form

$$\left(\frac{2}{\pi z}\right)^{1/2} \left\{\alpha_m(\nu)\, e^{i\xi} \sum_{s=0}^{\infty} i^s \frac{A_s(\nu)}{z^s} + \beta_m(\nu)\, e^{-i\xi} \sum_{s=0}^{\infty} (-i)^s \frac{A_s(\nu)}{z^s}\right\}$$

is obtained, where $(m-1)\pi + \delta \leqslant \arg z \leqslant (m+1)\pi - \delta$, and $\alpha_m(\nu)$ and $\beta_m(\nu)$ are independent of z. The need for *discontinuous* changes in these coefficients as arg z is continuously increased (or decreased) is called the *Stokes phenomenon*. It is not confined to solutions of Bessel's differential equation.

12.7.4** Singularities of Higher Rank

When the differential equation of 12.7.1 has a singularity at infinity of rank $k + 1$, the coefficients $f(z)$ and $g(z)$ can be expanded in series

$$f(z) = z^k \sum_{s=0}^{\infty} \frac{f_s}{z^s}; \quad g(z) = z^{2k} \sum_{s=0}^{\infty} \frac{g_s}{z^s}$$

which converge for large $|z|$, at least one of f_0, g_0, and g_1 being nonzero.

Provided that $f_0^2 \neq 4g_0$, formal series solutions of the form

$$e^{\xi_j(z)} \sum_{s=0}^{\infty} \frac{a_{s,j}}{z^s} \qquad (j = 1, 2)$$

*For numerical purposes, however, the second series should be retained when $\frac{3}{2}\pi \leqslant \arg z \leqslant 2\pi$.

**Proofs of results in this subsection are given in Olver and Stenger 1965.

can be constructed. Here

$$\xi_j(z) = z^{k+1} \sum_{s=0}^{k} \frac{(-)^{j-1}\phi_s - \frac{1}{2} f_s}{(k+1-s)\, z^s} + \{(-)^{j-1}\phi_{k+1} - \tfrac{1}{2} f_{k+1} - \tfrac{1}{2} k\} \ln z$$

the coefficients ϕ_s being defined by the expansion

$$\{\tfrac{1}{4} f^2(z) + \tfrac{1}{2} f'(z) - g(z)\}^{1/2} = z^k \sum_{s=0}^{\infty} \frac{\phi_s}{z^s}$$

The other coefficients $a_{s,j}$ may be calculated recursively by substituting in the differential equation and equating coefficients, the value of $a_{0,j}$ being arbitrary.

Let the z-plane be divided into $2k + 2$ sectors of equal angle, as indicated in Figure 12.7-1 in the case $k = 3$. In any closed sector lying properly within* the union

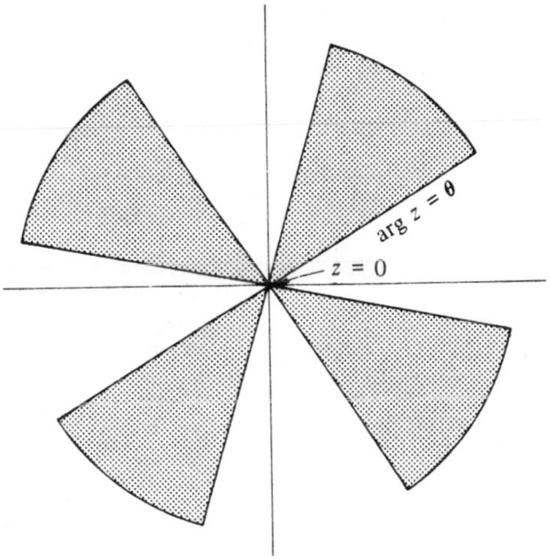

Fig. 12.7-1 z-plane, $k = 3$: $\theta = (\frac{1}{2}\pi - \arg \phi_0)/(k+1)$.

of a shaded sector and the two adjacent unshaded sectors there is a solution of the differential equation having the formal series, with $j = 1$, as its asymptotic expansion. This solution is recessive in the shaded sector and dominant in the abutting sectors. Similarly in any closed sector within the union of an unshaded sector and the two adjacent shaded sectors there is a solution having the formal series, with $j = 2$, as its asymptotic expansion, and this solution is recessive in the unshaded sector and dominant in the abutting sectors.

Again, the exceptional case $f_0^2 = 4g_0$ is amenable to an appropriate preliminary transformation of independent and dependent variables.

*Except for the common vertex.

12.8 DIFFERENTIAL EQUATIONS WITH A PARAMETER

12.8.1 Classification and Preliminary Transformations

In this section we discuss asymptotic solutions of differential equations of the form

$$\frac{d^2w}{dz^2} = \{u^2 f(z) + g(z)\} w$$

in which u is a large real or complex parameter, and z ranges over a real interval or complex domain **D**, say. Equations of this type occur frequently in mathematical physics. The form of the asymptotic solutions depends on the nature of the *transition points* in **D**, that is, points at which $f(z)$ or $g(z)$ is singular, or $f(z)$ vanishes; compare 12.6.1.

In the case in which **D** is free from transition points—which we shall call henceforth *Case I*—it was shown in section 12.6 that the LG functions

$$f^{-1/4}(z) \exp\left\{\pm u \int f^{1/2}(z)\, dz\right\}$$

furnish asymptotic solutions with uniform relative error $O(u^{-1})$ as $|u| \to \infty$. In 12.8.2 we construct asymptotic expansions in descending powers of u, the initial terms of which are the LG approximations.

Later subsections treat cases in which **D** contains a single transition point z_0, say. If, at z_0, $f(z)$ has a pole of order $m \geqslant 2$ and $g(z)$ is either analytic or has a pole of order less than $\frac{1}{2}m + 1$, then the LG approximation or the expansions of 12.8.2 may be used. In *Case II*, treated in 12.8.4, z_0 is a simple zero of $f(z)$ and an analytic point of $g(z)$; in *Case III*, treated in 12.8.6, z_0 is a simple pole of $f(z)$ and $(z - z_0)^2 g(z)$ is analytic there.

Basically the same approach is made in all cases. First, the Liouville transformation (12.6.1) is applied. This introduces new variables W and ξ, related by

$$W = \dot{z}^{-1/2} w$$

the dot denoting differentiation with respect to ξ. Then

$$\frac{d^2W}{d\xi^2} = \{u^2 \dot{z}^2 f(z) + \psi(\xi)\} W$$

where

$$\psi(\xi) = \dot{z}^2 g(z) + \dot{z}^{1/2} \frac{d^2}{d\xi^2}(\dot{z}^{-1/2})$$

The transformation is now prescribed in such a way that: (i) ξ and z are analytic functions of each other at the transition point (if any); (ii) the approximating differential equation obtained by neglect of $\psi(\xi)$, or part of $\psi(\xi)$, has solutions which are functions of a single variable. The actual prescriptions are as follows:

Case I:

$$\dot{z}^2 f(z) = 1, \quad \text{giving} \quad \xi = \int f^{1/2}(z)\, dz$$

Case II:

$$\dot{z}^2 f(z) = \xi, \quad \text{giving} \quad \frac{2}{3}\xi^{3/2} = \int_{z_0}^{z} f^{1/2}(t)\, dt$$

Case III:

$$\dot{z}^2 f(z) = \frac{1}{\xi}, \quad \text{giving} \quad 2\xi^{1/2} = \int_{z_0}^{z} f^{1/2}(t)\, dt$$

The transformed differential equation becomes

$$\frac{d^2 W}{d\xi^2} = \{u^2 \xi^m + \psi(\xi)\} W$$

with $m = 0$ (Case I), $m = 1$ (Case II), or $m = -1$ (Case III).

In Cases I and II approximate solutions of the new equation are obtained by neglecting $\psi(\xi)$. In Case I this is the LG approximating procedure used in section 12.6. In Case II the approximants are Airy functions. In Case III the basic approximating equation is

$$\frac{d^2 W}{d\xi^2} = \left(\frac{u^2}{\xi} + \frac{\rho}{\xi^2}\right) W$$

where ρ is the value of $\xi^2 \psi(\xi)$ at $\xi = 0$. The solutions are expressible in terms of modified Bessel functions of order $\pm\sqrt{1 + 4\rho}$ and argument $2u\sqrt{\xi}$.

12.8.2 Case I: No Transition Points

The standard form of differential equation is given by

$$\frac{d^2 W}{d\xi^2} = \{u^2 + \psi(\xi)\} W$$

The variable ξ ranges over a bounded or unbounded complex domain Δ, being the map of the original z-domain $\mathbf{D}$. The function $\psi(\xi)$ is holomorphic in Δ.

A formal series solution can be constructed in the form

$$W = e^{u\xi} \sum_{s=0}^{\infty} \frac{A_s(\xi)}{u^s}$$

This gives

$$\frac{dW}{d\xi} = ue^{u\xi} \sum_{s=0}^{\infty} \frac{A_s(\xi) + A'_{s-1}(\xi)}{u^s}; \quad \frac{d^2 W}{d\xi^2} = u^2 e^{u\xi} \sum_{s=0}^{\infty} \frac{A_s(\xi) + 2A'_{s-1}(\xi) + A''_{s-2}(\xi)}{u^s}$$

Satisfaction of the given differential equation requires

$$2A'_s(\xi) = -A''_{s-1}(\xi) + \psi(\xi)A_{s-1}(\xi) \qquad (s = 0, 1, \ldots)$$

Thus $A_0(\xi) = \text{constant}$, which we take to be unity without loss of generality, and higher coefficients are found recursively by

$$A_{s+1}(\xi) = -\frac{1}{2}A'_s(\xi) + \frac{1}{2}\int \psi(\xi)A_s(\xi)\,d\xi \qquad (s = 0, 1, \ldots)$$

the constants of integration being arbitrary. Each coefficient $A_s(\xi)$ is holomorphic in Δ.

A second formal solution is obtainable by replacing u by $-u$ throughout. In general both formal series diverge. However, corresponding to any positive integer n there exist solutions $W_{n,j}(u, \xi)$, $j = 1, 2$, which are holomorphic in Δ, depend on arbitrary reference points α_j, and are given by

$$W_{n,1}(u, \xi) = e^{u\xi} \sum_{s=0}^{n-1} \frac{A_s(\xi)}{u^s} + \epsilon_{n,1}(u, \xi)$$

$$W_{n,2}(u, \xi) = e^{-u\xi} \sum_{s=0}^{n-1} (-)^s \frac{A_s(\xi)}{u^s} + \epsilon_{n,2}(u, \xi)$$

where the error terms are bounded by

$$|\epsilon_{n,j}(u, \xi)|, \quad \left|\frac{\partial \epsilon_{n,j}(u, \xi)}{u\,\partial\xi}\right| \leqslant 2\left|e^{(-)^{j-1} u\xi}\right| \exp\left\{\frac{2\mathcal{V}_{\alpha_j,\xi}(A_1)}{|u|}\right\} \frac{\mathcal{V}_{\alpha_j,\xi}(A_n)}{|u|^n}$$

These bounds apply at each point ξ of Δ which can be linked to α_j by a path $\mathcal{Q}_j$ lying in Δ and such that $\operatorname{Re}(uv)$ is nondecreasing ($j = 1$) or nonincreasing ($j = 2$) as v passes along $\mathcal{Q}_j$ from α_j to ξ. The variations in the error bounds must be evaluated along $\mathcal{Q}_j$. The reference points α_j can be at infinity, subject to convergence of the variations.

In the context of the present result, the condition on the path $\mathcal{Q}_j$ is called the *monotonicity condition*, and admissible paths are said to be *($u\xi$)-progressive*; compare 12.6.6. Again, points of Δ excluded by the monotonicity condition are called shadow zones; the zones depend on j, $\arg u$, and the choice of α_j.

Example: Modified Bessel functions of large order–The functions $z^{1/2}I_\nu(\nu z)$ and $z^{1/2}K_\nu(\nu z)$ satisfy

$$\frac{d^2w}{dz^2} = \left\{\nu^2 \frac{1 + z^2}{z^2} - \frac{1}{4z^2}\right\} w$$

The preceding theory will now be applied to derive uniform asymptotic expansions for large positive real values of ν.

The appropriate Liouville transformation is given by

$$\xi = \int \frac{(1 + z^2)^{1/2}}{z} dz; \quad w = \left(\frac{z^2}{1 + z^2}\right)^{1/4} W$$

Then

$$\frac{d^2W}{d\xi^2} = \{\nu^2 + \psi(\xi)\} W$$

where

$$\psi(\xi) = z^2(1 - \tfrac{1}{4} z^2)(1 + z^2)^{-3}$$

Integration yields

$$\xi = (1 + z^2)^{1/2} + \ln z - \ln\{1 + (1 + z^2)^{1/2}\}$$

it being convenient to take the arbitrary constant of integration to be zero. The mapping between the planes of z and ξ is indicated in Figs. 12.8-1 and 12.8-2. We take **D** to be the sector $|\arg z| < \pi/2$. Its map Δ comprises the union of the sector $|\arg \xi| < \pi/2$ and the strip $|\operatorname{Im} \xi| < \pi/2$.

From the above results it follows that the transformed differential equation has solutions

$$W_{n,1}(\nu, \xi) = e^{\nu\xi}\left\{\sum_{s=0}^{n-1} \frac{A_s}{\nu^s} + \eta_{n,1}(\nu, \xi)\right\}$$

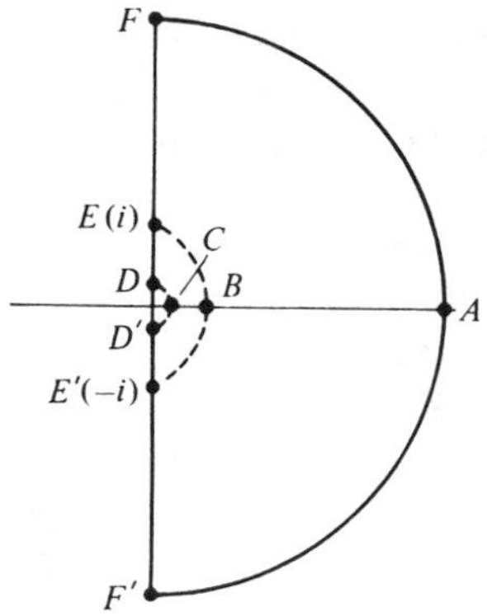

Fig. 12.8-1 z-plane: domain D.

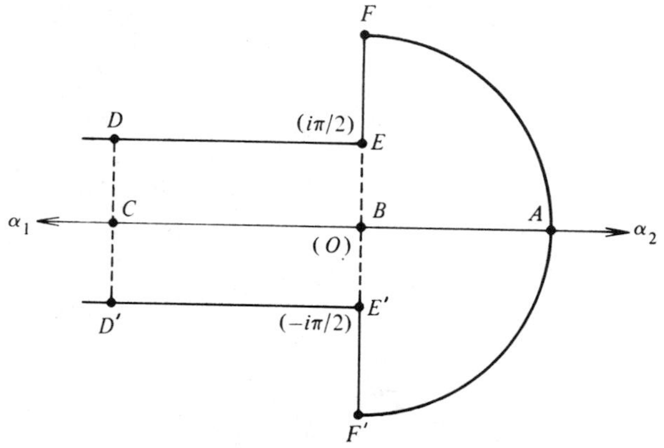

Fig. 12.8-2 ξ-plane: domain Δ.

$$W_{n,2}(\nu,\xi)=e^{-\nu\xi}\left\{\sum_{s=0}^{n-1}(-)^s\frac{A_s}{\nu^s}+\eta_{n,2}(\nu,\xi)\right\}$$

With the reference points α_1 and α_2 taken to be $-\infty$ and $+\infty$ respectively, the present error terms are bounded by

$$|\eta_{n,1}(\nu,\xi)|\leqslant 2\exp\left\{\frac{2\mathcal{V}_{-\infty,\xi}(A_1)}{\nu}\right\}\frac{\mathcal{V}_{-\infty,\xi}(A_n)}{\nu^n}$$

$$|\eta_{n,2}(\nu,\xi)|\leqslant 2\exp\left\{\frac{2\mathcal{V}_{\xi,\infty}(A_1)}{\nu}\right\}\frac{\mathcal{V}_{\xi,\infty}(A_n)}{\nu^n}$$

The condition on the variational paths is that Re ξ is monotonic; from Figs. 12.8-1 and 12.8-2 it is clear that the whole of Δ is admissible both for $j = 1$ and $j = 2$. In other words, shadow zones are absent in the present case.*

On reverting to the original variables we find that the recurrence relation for the coefficients $A_s(\xi)$ becomes

$$A_{s+1} = -\frac{1}{2}\frac{z}{(1+z^2)^{1/2}}\frac{dA_s}{dz} + \frac{1}{8}\int \frac{z(4-z^2)}{(1+z^2)^{5/2}} A_s\, dz$$

From this equation it can be deduced that A_s is a polynomial in $p \equiv (1+z^2)^{-1/2}$ of degree $3s$. If the integration constants are chosen in such a way that A_s vanishes at $z = \infty$ when $s \geqslant 1$, then we obtain

$$A_0 = 1; \quad A_1 = (3p - 5p^3)/24; \quad A_2 = (81p^2 - 462p^4 + 385p^6)/1152$$

To identify $I_\nu(\nu z)$ and $K_\nu(\nu z)$ in terms of the solutions just constructed, we note that as $z \to 0$, that is, $\xi \to -\infty$, both $I_\nu(\nu z)$ and $(1+z^2)^{-1/4} W_{n,1}(\nu, \xi)$ are recessive; hence their ratio is independent of z. Similarly, by letting $z \to +\infty$, that is, $\xi \to +\infty$, we see that the ratio of $K_\nu(\nu z)$ and $(1+z^2)^{-1/4} W_{n,2}(\nu, \xi)$ is independent of z. In both cases the ratio may be found by considering asymptotic forms as $z \to +\infty$, ν being fixed. The final expansions are

$$I_\nu(\nu z) = \frac{1}{1+\eta_{n,1}(\nu,\infty)}\frac{e^{\nu\xi}}{(2\pi\nu)^{1/2}(1+z^2)^{1/4}}\left\{\sum_{s=0}^{n-1}\frac{A_s}{\nu^s} + \eta_{n,1}(\nu,\xi)\right\}$$

$$K_\nu(\nu z) = \left(\frac{\pi}{2\nu}\right)^{1/2}\frac{e^{-\nu\xi}}{(1+z^2)^{1/4}}\left\{\sum_{s=0}^{n-1}(-)^s\frac{A_s}{\nu^s} + \eta_{n,2}(\nu,\xi)\right\}$$

valid when $\nu > 0$, $|\arg z| < \pi/2$, and n is any positive integer.

If z is restricted to the sector $|\arg z| \leqslant (\pi/2) - \delta\ (< \pi/2)$, then the error bounds show that

$$I_\nu(\nu z) \sim \frac{e^{\nu\xi}}{(2\pi\nu)^{1/2}(1+z^2)^{1/4}}\sum_{s=0}^{\infty}\frac{A_s}{\nu^s}$$

$$K_\nu(\nu z) \sim \left(\frac{\pi}{2\nu}\right)^{1/2}\frac{e^{-\nu\xi}}{(1+z^2)^{1/4}}\sum_{s=0}^{\infty}(-)^s\frac{A_s}{\nu^s}$$

as $\nu \to \infty$, uniformly with respect to z.

*This would not be so if ν were complex.

12.8.3 Auxiliary Functions for the Airy Functions

For negative values of x, the Airy functions $\mathrm{Ai}(x)$ and $\mathrm{Bi}(x)$ are of oscillatory character, the period and amplitude of the oscillation diminishing as $x \to -\infty$. On the other hand, for increasing positive x both functions are positive and monotonic, $\mathrm{Ai}(x)$ tending rapidly to zero and $\mathrm{Bi}(x)$ to infinity. To make a combined assessment of the magnitudes of both functions applicable to both negative and positive arguments, we introduce a *weight function* $E(x)$, *modulus function* $M(x)$, and *phase function* $\theta(x)$.

The weight function is defined by

$$E(x) = \sqrt{\mathrm{Bi}(x)/\mathrm{Ai}(x)} \quad (c \leqslant x < \infty); \qquad E(x) = 1 \quad (-\infty < x \leqslant c)$$

Here $c = -0.36605\ldots$ is the negative root of the equation

$$\mathrm{Ai}(x) = \mathrm{Bi}(x)$$

having smallest absolute value. The function $E(x)$ is continuous and nondecreasing.

The modulus and phase functions are defined by

$$E(x)\,\mathrm{Ai}(x) = M(x) \sin\theta(x); \quad E^{-1}(x)\,\mathrm{Bi}(x) = M(x)\cos\theta(x)$$

where $E^{-1}(x) = 1/E(x)$. In consequence

$$M(x) = \sqrt{2\,\mathrm{Ai}(x)\,\mathrm{Bi}(x)}; \quad \theta(x) = \tfrac{1}{4}\pi \quad (x \geqslant c)$$

or

$$M(x) = \sqrt{\mathrm{Ai}^2(x) + \mathrm{Bi}^2(x)}; \quad \theta(x) = \tan^{-1}\{\mathrm{Ai}(x)/\mathrm{Bi}(x)\} \quad (x \leqslant c)$$

the branch of the inverse tangent being $\pi/4$ at $x = c$, and continuous elsewhere. The modulus is a slowly changing function with the property

$$M(x) \sim \pi^{-1/2}|x|^{-1/4} \qquad (x \to \pm\infty)$$

The phase is a nonincreasing function.

The following constant is required in the next subsection:

$$\lambda = \sup_{(-\infty,\infty)} \{\pi|x|^{1/2}M^2(x)\} = 1.04\ldots$$

12.8.4 Case II: Simple Turning Point

The standard form of differential equation for Case II is

$$\frac{d^2W}{d\xi^2} = \{u^2\xi + \psi(\xi)\}\,W$$

For simplicity, it is supposed here that the variables are real, u being positive, and also that $\psi(\xi)$ is infinitely differentiable in a finite or infinite interval (α, β) which contains the turning point $\xi = 0$.

The basic approximating equation is

$$\frac{d^2 W}{d\xi^2} = u^2 \xi W$$

solutions of which are $\mathrm{Ai}(u^{2/3}\xi)$ and $\mathrm{Bi}(u^{2/3}\xi)$. A formal series solution of the given differential equation can be constructed in the form

$$W = \mathrm{Ai}(u^{2/3}\xi) \sum_{s=0}^{\infty} \frac{A_s(\xi)}{u^{2s}} + \frac{\mathrm{Ai}'(u^{2/3}\xi)}{u^{4/3}} \sum_{s=0}^{\infty} \frac{B_s(\xi)}{u^{2s}}$$

with $A_0(\xi) = 1$. Differentiating and equating coefficients, we find that

$$B_s(\xi) = \frac{1}{2\xi^{1/2}} \int_0^{\xi} \{\psi(v)\, A_s(v) - A_s''(v)\} \frac{dv}{v^{1/2}} \qquad (\xi > 0)$$

$$B_s(\xi) = \frac{1}{2(-\xi)^{1/2}} \int_{\xi}^{0} \{\psi(v)\, A_s(v) - A_s''(v)\} \frac{dv}{(-v)^{1/2}} \qquad (\xi < 0)$$

and

$$A_{s+1}(\xi) = -\frac{1}{2} B_s'(\xi) + \frac{1}{2} \int \psi(\xi)\, B_s(\xi)\, d\xi$$

These equations determine recursively sequences of functions which are infinitely differentiable throughout (α, β), *including* $\xi = 0$.

Again, the formal series diverges in general, but corresponding to each nonnegative integer n there exists an infinitely differentiable solution $W_{2n+1,1}(u, \xi)$, such that

$$W_{2n+1,1}(u, \xi) = \mathrm{Ai}(u^{2/3}\xi) \sum_{s=0}^{n} \frac{A_s(\xi)}{u^{2s}} + \frac{\mathrm{Ai}'(u^{2/3}\xi)}{u^{4/3}} \sum_{s=0}^{n-1} \frac{B_s(\xi)}{u^{2s}} + \epsilon_{2n+1,1}(u, \xi)$$

with

$$|\epsilon_{2n+1,1}(u, \xi)| \leq 2 \frac{M(u^{2/3}\xi)}{E(u^{2/3}\xi)} \exp\left\{\frac{2\lambda \mathcal{V}_{\xi,\beta}(|\xi|^{1/2} B_0)}{u}\right\} \frac{\mathcal{V}_{\xi,\beta}(|\xi|^{1/2} B_n)}{u^{2n+1}}$$

E, M, and λ being defined as in 12.8.3. For positive or negative ξ, the bound for $\epsilon_{2n+1,1}(u, \xi)$ is essentially $\mathrm{Ai}(u^{2/3}\xi)O(u^{-2n-1})$, except near the zeros of $\mathrm{Ai}(u^{2/3}\xi)$.

A second solution is furnished by

$$W_{2n+1,2}(u, \xi) = \mathrm{Bi}(u^{2/3}\xi) \sum_{s=0}^{n} \frac{A_s(\xi)}{u^{2s}} + \frac{\mathrm{Bi}'(u^{2/3}\xi)}{u^{4/3}} \sum_{s=0}^{n-1} \frac{B_s(\xi)}{u^{2s}} + \epsilon_{2n+1,2}(u, \xi)$$

with

$$|\epsilon_{2n+1,2}(u, \xi)| \leqslant 2E(u^{2/3}\xi)\, M(u^{2/3}\xi) \exp\left\{\frac{2\lambda \mathcal{V}_{\alpha,\xi}(|\xi|^{1/2}B_0)}{u}\right\} \frac{\mathcal{V}_{\alpha,\xi}(|\xi|^{1/2}B_n)}{u^{2n+1}}$$

Analogous results are also available for complex variables; see Olver 1974a, Chapter 11.

12.8.5 Auxiliary Functions for Bessel Functions

For nonnegative real values of ν and positive values of x, a weight function $E_\nu(x)$ is defined by

$$E_\nu(x) = \{-Y_\nu(x)/J_\nu(x)\}^{1/2} \quad (0 < x \leqslant X_\nu); \quad E_\nu(x) = 1 \quad (X_\nu \leqslant x < \infty)$$

where $x = X_\nu$ is the smallest root of the equation

$$J_\nu(x) + Y_\nu(x) = 0$$

$E_\nu(x)$ is a continuous, positive, nonincreasing function of x.

Corresponding modulus and phase functions are defined by

$$J_\nu(x) = E_\nu^{-1}(x)\, M_\nu(x) \cos\theta_\nu(x); \quad Y_\nu(x) = E_\nu(x)\, M_\nu(x) \sin\theta_\nu(x)$$

thus

$$M_\nu(x) = \{2|Y_\nu(x)|J_\nu(x)\}^{1/2}; \quad \theta_\nu(x) = -\tfrac{1}{4}\pi \quad (0 < x \leqslant X_\nu)$$

$$M_\nu(x) = \{J_\nu^2(x) + Y_\nu^2(x)\}^{1/2}; \quad \theta_\nu(x) = \tan^{-1}\{Y_\nu(x)/J_\nu(x)\} \quad (x \geqslant X_\nu)$$

the branch of the inverse tangent being chosen to make $\theta_\nu(x)$ continuous everywhere.

12.8.6 Case III: Simple Pole

The final form of differential equation to be considered in the present section is

$$\frac{d^2W}{d\xi^2} = \left\{\frac{u^2}{4\xi} + \frac{\nu^2 - 1}{4\xi^2} + \frac{\psi(\xi)}{\xi}\right\} W$$

in which u is again a large positive parameter, ν is a nonnegative real constant, and $\psi(\xi)$ is infinitely differentiable in a finite or infinite interval (α, β) containing $\xi = 0$.

For $\xi \in (0, \beta)$ a formal series solution is supplied by

$$W = \xi^{1/2} I_\nu(u\xi^{1/2}) \sum_{s=0}^{\infty} \frac{A_s(\xi)}{u^{2s}} + \frac{\xi}{u} I_{\nu+1}(u\xi^{1/2}) \sum_{s=0}^{\infty} \frac{B_s(\xi)}{u^{2s}}$$

where I_ν is the modified Bessel function, and the coefficients are determined recursively by $A_0(\xi) = 1$

$$B_s(\xi) = -A_s'(\xi) + \frac{1}{\xi^{1/2}} \int_0^\xi \left\{ \psi(v)\, A_s(v) - \left(\nu + \frac{1}{2}\right) A_s'(v) \right\} \frac{dv}{v^{1/2}}$$

and

$$A_{s+1}(\xi) = \nu B_s(\xi) - \xi B_s'(\xi) + \int \psi(\xi)\, B_s(\xi)\, d\xi$$

Each coefficient tends to a finite limit as $\xi \to 0$.

Corresponding to any nonnegative integer n, there exists a solution $W_{2n+1,1}(u, \xi)$ of the given differential equation of the form

$$W_{2n+1,1}(u, \xi) = \xi^{1/2} I_\nu(u\xi^{1/2}) \sum_{s=0}^{n} \frac{A_s(\xi)}{u^{2s}} + \frac{\xi}{u} I_{\nu+1}(u\xi^{1/2}) \sum_{s=0}^{n-1} \frac{B_s(\xi)}{u^{2s}} + \epsilon_{2n+1,1}(u, \xi)$$

with

$$|\epsilon_{2n+1,1}(u, \xi)| \leqslant \lambda_1(\nu)\, \xi^{1/2} I_\nu(u\xi^{1/2}) \exp\left\{ \frac{\lambda_1(\nu)}{u} \mathcal{V}_{0,\xi}(\xi^{1/2} B_0) \right\} \frac{\mathcal{V}_{0,\xi}(\xi^{1/2} B_n)}{u^{2n+1}}$$

Here $\lambda_1(\nu)$ is the (finite) constant defined by

$$\lambda_1(\nu) = \sup_{x \in (0,\infty)} \{2x I_\nu(x) K_\nu(x)\}$$

For each n an independent second solution is given by

$$W_{2n+1,2}(u, \xi) = \xi^{1/2} K_\nu(u\xi^{1/2}) \sum_{s=0}^{n} \frac{A_s(\xi)}{u^{2s}} - \frac{\xi}{u} K_{\nu+1}(u\xi^{1/2}) \sum_{s=0}^{n-1} \frac{B_s(\xi)}{u^{2s}} + \epsilon_{2n+1,2}(u, \xi)$$

where K_ν is the second modified Bessel function, and

$$|\epsilon_{2n+1,2}(u,\xi)| \leqslant \lambda_1(\nu)\,\xi^{1/2}K_\nu(u\xi^{1/2})\exp\left\{\frac{\lambda_1(\nu)}{u}\,\mathcal{V}_{\xi,\beta}(\xi^{1/2}B_0)\right\}\frac{\mathcal{V}_{\xi,\beta}(\xi^{1/2}B_n)}{u^{2n+1}}$$

For negative ξ the solutions $W_{2n+1,1}(u,\xi)$ and $W_{2n+1,2}(u,\xi)$ are no longer appropriate since they have branch-points at $\xi = 0$ and become complex when continued to negative values. The coefficients $A_s(\xi)$ and $B_s(\xi)$ are free from singularity at $\xi = 0$, however, and may be continued to negative ξ; thus

$$B_s(\xi) = -A_s'(\xi) + \frac{1}{|\xi|^{1/2}}\int_\xi^0 \left\{\psi(v)\,A_s(v) - \left(\nu + \frac{1}{2}\right)A_s'(v)\right\}\frac{dv}{|v|^{1/2}}$$

and $A_{s+1}(\xi)$ is related to $B_s(\xi)$ by the same formula as for positive ξ.

Solutions of the given differential equation for $\xi \in (\alpha, 0)$ are given by

$$W_{2n+1,3}(u,\xi) = |\xi|^{1/2}J_\nu(u|\xi|^{1/2})\sum_{s=0}^{n}\frac{A_s(\xi)}{u^{2s}} - \frac{|\xi|}{u}J_{\nu+1}(u|\xi|^{1/2})\sum_{s=0}^{n-1}\frac{B_s(\xi)}{u^{2s}} + \epsilon_{2n+1,3}(u,\xi)$$

$$W_{2n+1,4}(u,\xi) = |\xi|^{1/2}Y_\nu(u|\xi|^{1/2})\sum_{s=0}^{n}\frac{A_s(\xi)}{u^{2s}} - \frac{|\xi|}{u}Y_{\nu+1}(u|\xi|^{1/2})\sum_{s=0}^{n-1}\frac{B_s(\xi)}{u^{2s}} + \epsilon_{2n+1,4}(u,\xi)$$

where n is again an arbitrary nonnegative integer, and

$$|\epsilon_{2n+1,3}(u,\xi)| \leqslant \lambda_3(\nu)|\xi|^{1/2}\frac{M_\nu(u|\xi|^{1/2})}{E_\nu(u|\xi|^{1/2})}\exp\left\{\frac{\lambda_2(\nu)}{u}\,\mathcal{V}_{\xi,0}(|\xi|^{1/2}B_0)\right\}\frac{\mathcal{V}_{\xi,0}(|\xi|^{1/2}B_n)}{u^{2n+1}}$$

$$|\epsilon_{2n+1,4}(u,\xi)| \leqslant \lambda_4(\nu)|\xi|^{1/2}E_\nu(u|\xi|^{1/2})\,M_\nu(u|\xi|^{1/2}) \times \exp\left\{\frac{\lambda_2(\nu)}{u}\,\mathcal{V}_{\alpha,\xi}(|\xi|^{1/2}B_0)\right\}\frac{\mathcal{V}_{\alpha,\xi}(|\xi|^{1/2}B_n)}{u^{2n+1}}$$

Here E_ν and M_ν are defined as in 12.8.5, and $\lambda_2(\nu)$, $\lambda_3(\nu)$, $\lambda_4(\nu)$ denote the suprema of the functions

$$\pi x M_\nu^2(x), \quad \pi x|J_\nu(x)|E_\nu(x)\,M_\nu(x), \quad \pi x|Y_\nu(x)|M_\nu(x)/E_\nu(x)$$

respectively, for $x \in (0, \infty)$. Each is finite.

Again, analogous results are available for complex u and ξ.

12.9 ESTIMATION OF REMAINDER TERMS

12.9.1 Numerical Use of Asymptotic Approximations

When a realistic analytical bound for the error term in a given expansion is unavailable, it is unsafe to infer the size of the error term simply by inspecting the rate of numerical decrease of the terms in the series. Even in the case of a convergent power-series expansion this cannot be done: the tail has to be majorized analytically —for example, by a geometric progression—before final accuracy can be guaranteed. For a divergent asymptotic expansion the situation is much worse. First, it is impossible to majorize the tail. Secondly, the series represents an infinite class of functions, and the error term depends on which particular member of the class we have in mind.

In cases where the asymptotic variable x, say, is real and positive and the distinguished point is at infinity, the wanted function should be computed by an independent (preferably non-asymptotic) method at the smallest value of x it is intended to apply the asymptotic approximation. If the results are in agreement to S significant figures, then it is likely (but not *certain*) that the approximation will be accurate to at least S significant figures for all greater values of x.

For a complex variable z, both $|z|$ and arg z have to be considered in appraising accuracy. Suppose that an asymptotic approximation is valid as $z \to \infty$ in any closed sector within $\theta_1 < \arg z < \theta_2$, but not within a larger sector. Then the accuracy of the approximation deteriorates severely as the rays $\arg z = \theta_1, \theta_2$ are approached. In consequence, numerical work should be confined to a sector $\theta_1' \leqslant \arg z \leqslant \theta_2'$ lying well within $\theta_1 < \arg z < \theta_2$, and independent evaluations made at $\arg z = \theta_1'$ and θ_2' for the smallest value of $|z|$ intended to be used.

When the regions of validity in the complex plane are not sectors error appraisal is more complicated. Basically, however, the guiding principle is to keep a safe distance from the true boundaries of the region of validity.

12.9.2 Converging Factors

Let $f(z)$ be a function of z having the asymptotic expansion

$$f(z) \sim a_0 + \frac{a_1}{z} + \frac{a_2}{z^2} + \cdots$$

as $z \to \infty$ in a sector **S**: $\theta_1 \leqslant \arg z \leqslant \theta_2$. As a special case **S** can degenerate into the positive or negative real axis. Suppose that successive terms in the series diminish in absolute value until the $(n+1)^{\text{th}}$ term is reached, thereafter they increase. Clearly $n = n(z)$, where $n(z)$ is a discontinuous function of $|z|$ which is independent of arg z. Generally the expansion yields its greatest accuracy when truncated at $n(z)$ terms. We write

$$R_n(z) = f(z) - \sum_{s=0}^{n-1} \frac{a_s}{z^s}$$

and call $R_{n(z)}(z)$ the *optimum remainder term*, whether or not it is actually the least.

Now define

$$C(z) = R_{n(z)}(z)/\{a_{n(z)}z^{-n(z)}\}$$

so that

$$f(z) = \sum_{s=0}^{n(z)-1} \frac{a_s}{z^s} + C(z)\frac{a_{n(z)}}{z^{n(z)}}$$

If we have a way of assessing $C(z)$ when $|z|$ is large, then the magnitude of the optimum remainder term can be estimated. In some cases it is actually possible to construct an asymptotic expansion for $C(z)$ in descending powers of z or $n(z)$. In these fortunate circumstances $C(z)$ can be calculated to several significant figures, considerably increasing the attainable accuracy in the computed value of $f(z)$. For this reason $C(z)$ is called a *converging factor*.

Example: The exponential integral—From (7.1-2) we have

$$E_1(z) = e^{-z}\int_0^\infty \frac{e^{-zt}}{1+t}\,dt \qquad (|\arg z| < \tfrac{1}{2}\pi)$$

Since

$$\frac{1}{1+t} = 1 - t + t^2 - \cdots + (-)^{n-1}t^{n-1} + \frac{(-)^n t^n}{1+t}$$

for any nonnegative integer n, it follows that

$$E_1(z) = e^{-z}\left\{\sum_{s=0}^{n-1} u_s(z) + R_n(z)\right\}$$

where

$$u_s(z) = (-)^s s!\, z^{-s-1}$$

and

$$R_n(z) = (-)^n \int_0^\infty \frac{e^{-zt}t^n}{1+t}\,dt$$

The series $\sum u_s(z)$ is the asymptotic expansion of $e^z E_1(z)$ for large $|z|$, and the term of smallest absolute value is $u_{[|z|]}(z)$, unless $|z|$ is an integer in which event there are two equally small terms $u_{|z|-1}(z)$ and $u_{|z|}(z)$.

We write $\theta = \arg z$ and seek an asymptotic approximation to

$$R_n\{(n+\zeta)e^{i\theta}\} \equiv (-)^n \int_0^\infty \frac{\exp\{-(n+\zeta)e^{i\theta}t\}\, t^n}{1+t}\, dt$$

for large n and fixed values of ζ and θ. The saddle-point of the integrand is given by

$$\frac{d}{dt}\{t \exp(-e^{i\theta}t)\} = 0$$

that is, by $t = e^{-i\theta}$. The path of integration is made to pass through this point by rotation through an angle $-\theta$. Then setting $t = \tau e^{-i\theta}$, we obtain

$$R_n\{(n+\zeta)e^{i\theta}\} = (-)^n e^{-i(n+1)\theta} \int_0^\infty \frac{e^{-(n+\zeta)\tau}\tau^n}{1+\tau e^{-i\theta}}\, d\tau$$

And although this result has been derived on the assumption that $|\theta| < \pi/2$, it is easily extended to $|\theta| < \pi$ by further rotation of the path and appeal to analytic continuation. The desired asymptotic expansion is then found by Laplace's method (12.2.6) to be

$$R_n\{(n+\zeta)e^{i\theta}\} \sim (-)^n(1-\alpha)e^{-n-\zeta-i(n+1)\theta}\left(\frac{2\pi}{n}\right)^{1/2}$$
$$\times \left\{1 + \frac{\zeta^2 - 2\zeta + 2\alpha\zeta + \frac{1}{6} - 2\alpha + 2\alpha^2}{2n} + \cdots\right\}$$

as $n \to \infty$, uniformly with respect to ζ in any compact set and $\theta \in [-\pi+\delta, \pi-\delta]$. Here $\alpha \equiv 1/(1+e^{i\theta})$, and δ is an arbitrary positive constant less than π.

To derive the corresponding expansion for the converging factor

$$C_n(z) \equiv R_n(z)/u_n(z)$$

we make use of Stirling's series (7.2-11) for $n!$. In this way we obtain

$$C_n\{(n+\zeta)e^{i\theta}\} \sim (1-\alpha)\left\{1 + \frac{\alpha(\zeta - 1 + \alpha)}{n} + \cdots\right\}$$

again as $n \to \infty$, uniformly with respect to bounded ζ and $\theta \in [-\pi+\delta, \pi-\delta]$. To apply this result for an assigned value of z, we take $\theta = \arg z$, $n = [|z|]$, and $\zeta = |z| -$

$[|z|]$, so that $(n + \zeta)e^{i\theta} = z$. Truncating the expansion for $C_n\{(n + \zeta)e^{i\theta}\}$ at its first term, for example, we conclude that if the original expansion for $E_1(z)$ is truncated at its $[|z|]^{\text{th}}$ term, then the remainder term is approximately equal to the first neglected term multiplied by $1 - \alpha$, that is, $1/(1 + e^{-i\theta})$. For positive real z this becomes $\frac{1}{2}$.

As a numerical illustration, take $z = 5$. Then

$$u_0(5) = 0.2;\quad u_1(5) = -0.04;\quad u_2(5) = 0.016;\quad u_3(5) = -0.0096;\quad u_4(5) = 0.00768$$

whence

$$u_0(5) + u_1(5) + \cdots + u_4(5) = 0.17408$$

compared with the correct value $e^5E_1(5) = 0.170422176\ldots.$ From the asymptotic expansion for the converging factor, we calculate

$$C_5(5) \sim \tfrac{1}{2}\left(1 - \tfrac{1}{4}\cdot\tfrac{1}{5} + \cdots\right) = 0.475$$

on neglecting terms beyond the second. Since $u_5(5) = -0.00768$, the estimate $C_5(5)\,u_5(5)$ for the remainder term is $-0.00365\ldots.$ Impressively, this equals the discrepancy between the partial sum $u_0(5) + u_1(5) + \cdots + u_4(5)$ and $e^5E_1(5)$, to within one unit of the fifth decimal place.

12.9.3 Euler's Transformation

Another way of increasing the accuracy obtainable from an asymptotic expansion is to transform it into a new series in which the initial terms decrease at a faster rate. Then it is often the case that the optimum remainder term is smaller for the new series than for the original series. It might even happen that the new series converges.

The most frequently used transformation is due to Euler, and is as follows. Let $a_0, a_1, a_2, \ldots$ be a given sequence and $b_0, b_1, b_2, \ldots$ a derived sequence, defined by

$$b_s = k^s\,[\Delta^s(a_jk^{-j})]_{j=0}$$

Here k is an arbitrary number and Δ the forward difference operator: $\Delta v_j = v_{j+1} - v_j$, $\Delta^2 v_j = \Delta v_{j+1} - \Delta v_j$, and so on. Suppose that $f(z)$ is an analytic function of the complex variable z such that

$$f(z) \sim \sum_{s=0}^{\infty} \frac{a_s}{z^{s+1}}$$

as $z \to \infty$ in a given sector **S**. Then

$$f(z) \sim \sum_{s=0}^{\infty} \frac{b_s}{(z - k)^{s+1}} \qquad (z \to \infty \text{ in } \mathbf{S})$$

With $k = 1$ and $z = -1$, the transformation reduces to

$$\sum_{s=0}^{\infty} (-)^s a_s \sim \sum_{s=0}^{\infty} (-)^s \frac{\Delta^s a_0}{2^{s+1}}$$

Example: Let us consider again the asymptotic expansion of the function $e^5 E_1(5)$ of 12.9.2. Application of Euler's transformation is delayed until the smallest term is reached; thus $a_s = (-)^s u_{s+5}(5)$. Relevant forward differences are given in units of the fifth decimal place in the accompanying table.

s	a_s	Δa_s	$\Delta^2 a_s$	$\Delta^3 a_s$	$\Delta^4 a_s$	$\Delta^5 a_s$	$\Delta^6 a_s$
0	-0.00768	- 154	- 214	- 192	- 280	- 434	-804
1	-0.00922	- 368	- 406	- 472	- 714	-1238	
2	-0.01290	- 774	- 878	-1186	-1952		
3	-0.02064	-1652	-2064	-3138			
4	-0.03716	-3716	-5202				
5	-0.07432	-8918					
6	-0.16350						

The first few terms of the transformed series are

$$\sum_{s=0}^{\infty} (-)^s \frac{\Delta^s a_0}{2^{s+1}} = 10^{-5} (-384 + 38 - 27 + 12 - 9 + 7 - 6 + \cdots)$$

Truncation at the sixth term gives -0.00363, then addition to $u_0(5) + u_1(5) + \cdots + u_4(5)$ yields 0.17045, agreeing with the value of $e^5 E_1(5)$ to within 0.00003; compare 12.9.2. Even closer agreement is attainable by working with more terms and more decimal places, and applying a second Euler transformation.

From the analytical standpoint the numerical procedure is equivalent to expanding the remainder term

$$R_5(z) = e^z E_1(z) - \sum_{s=0}^{4} u_s(z)$$

as an asymptotic series in descending powers of $z + 5$, and truncating the new series at the optimum stage.

12.10 REFERENCES AND BIBLIOGRAPHY

12.10.1 References

12-1 Olver, F. W. J., *Asymptotics and Special Functions*, Academic Press, New York, 1974a. (Sections 12.1.2 and 12.8.4.)

12-2 Erdélyi, A., *Asymptotic Expansions*, Dover, New York, 1956. (Section 12.2.7.)

12-3 Olver, F. W. J., "Error Bounds for Stationary Phase Approximations," *SIAM Journal on Math. Anal.*, **5**, 19–29, 1974b. (Section 12.2.7.)

12-4 Miller, J. C. P., *Tables of Weber Parabolic Cylinder Functions*, Her Majesty's Stationery Office, London, 1955. (Section 12.4.3.)

12-5 Olver, F. W. J., and Stenger, F., "Error Bounds for Asymptotic Solutions of Second-Order Differential Equations Having an Irregular Singularity of Arbitrary Rank," *SIAM Journal on Numer. Anal.*, Series B, **2**, 244–249, 1965. (Section 12.7.4.)

12.10.2 Bibliography

Berg, L., *Asymptotische Darstellungen und Entwicklungen*, VEB Deutscher Verlag der Wissenschaften, Berlin, 1968.

Bleistein, N., and Handelsman, R. A., *Asymptotic Expansions of Integrals*, Holt, New York, 1975.

de Bruijn, N. G., *Asymptotic Methods in Analysis*, Wiley-Interscience, New York, second edition, 1961.

Copson, E. T., *Asymptotic Expansions*, Cambridge University Press, Cambridge, 1965.

Erdélyi, A., *Asymptotic Expansions*, Dover, New York, 1956.

Evgrafov, M. A., *Asymptotic Estimates and Entire Functions*, Translated by A. L. Shields, Gordon and Breach, New York, 1961.

Feshchenko, S. F., Shkil', N. I., and Nikolenko, L. D., *Asymptotic Methods in the Theory of Linear Differential Equations*, American Elsevier, New York, 1967.

Jeffreys, H., *Asymptotic Approximations*, Clarendon Press, Oxford, 1962.

Lauwerier, H. A., *Asymptotic Expansions*, Mathematisch Centrum, Amsterdam, 1966.

Olver, F. W. J., *Asymptotics and Special Functions*, Academic Press, New York, 1974.

Sirovich, L., *Techniques of Asymptotic Analysis*, Springer-Verlag, New York, 1971.

Wasow, W., *Asymptotic Expansions for Ordinary Differential Equations*, Wiley-Interscience, New York, 1965.

Wong, R., "Error bounds for asymptotic expansions of integrals," *SIAM Rev.*, **22**, 401, 1980.

13

Oscillations

Richard E. Kronauer*

13.0 INTRODUCTION

The scope of this chapter will be limited to systems of finite dimension; i.e., those which can be described by a limited number of ordinary differential equations of finite order. Any such set of equations can, of course, be brought into a canonical set of first-order equations by augmenting the number of variables, but for most of the analysis here it is more convenient to deal with the equations in the form in which they are generated.

A class of systems can be constructed mathematically, and closely approximated physically, for which the motions continue to oscillate unabated once they are initially excited. Such systems are conservative, and they possess an integral of the motion which can be identified physically as the energy. There are several non-conservative mechanisms which can be added to otherwise conservative systems and which change the energy as the motion proceeds. The most common is simple dissipation, which always removes energy and thereby causes the oscillations to decay. Another is continuing excitation, which may either supply or remove energy depending on how it is applied. A third is parametric variation–i.e., time variation of the coefficients in the differential equations–which may also increase or decrease the energy, often with great effect. The strength of the nonconservative elements can be measured in terms of the fractional change which occurs in the amplitude of an oscillation over one cycle. For weakly nonconservative systems, the analysis can be performed in two stages: a preliminary conservative analysis followed by a nonconservative perturbation. This is the procedure which will be adopted here.

13.1 LAGRANGE EQUATIONS

13.1.1 Nonlinear Systems

We begin by introducing a vector of generalized coordinates, q, of dimension n, which together with its time derivative, $\dot{q}$, defines the state space for the system.

*Prof. Richard E. Kronauer, Div. of Engineering and Applied Physics, Harvard University, Camb., Mass.

Utilizing the method of Lagrange equations first developed in classical mechanics, Ref. 13-1, we assume that a scalar Lagrangian, $L(q, \dot{q}, t)$ is formulated and that the differential equations are given by

$$\frac{d}{dt}\left(\frac{\partial L}{\partial \dot{q}_i}\right) - \frac{\partial L}{\partial q_i} = f_i(t) \qquad i = 1, 2, \ldots, n \tag{13.1-1}$$

Here f is a vector of 'forces' which may include both internal dissipation (not encompassed by the Lagrangian) and applied excitation. The Lagrange equations specifically include parametric variation through the explicit dependence of L on the independent variable, t.

13.1.2 Linear Systems

If we assume that the motion of the system is confined to a local region of state space and redefine the origin of q and $\dot{q}$ so as to lie within the region, all sufficiently regular functions L may be Taylor-expanded

$$L = L_0 + a^T(t)\dot{q} + b^T(t)q + \tfrac{1}{2}\dot{q}^T A'(t)\dot{q} + \tfrac{1}{2}\dot{q}^T B'(t)q + \tfrac{1}{2}q^T C'(t)q + \cdots \tag{13.1-2}$$

where a and b are vectors and A', B', C' are square matrices, all of which may have an arbitrary imposed time dependence. The corresponding Lagrange equations are

$$\dot{a} - b + \tfrac{1}{2}(A' + A'^T)\ddot{q} + \tfrac{1}{2}(\dot{A}' + \dot{A}'^T)\dot{q} + \tfrac{1}{2}(B' - B'^T)\dot{q} \\ + \tfrac{1}{2}\dot{B}'q - \tfrac{1}{2}(C' + C'^T)q + \cdots = f \tag{13.1-3}$$

These may be simplified to

$$\dot{a} - b + A\ddot{q} + \dot{A}\dot{q} + B\dot{q} + \tfrac{1}{2}\dot{B}'q + Cq + \cdots = f \tag{13.1-4}$$

where

$$\begin{aligned} A &= \tfrac{1}{2}(A' + A'^T) \\ B &= \tfrac{1}{2}(B' - B'^T) \\ C &= -\tfrac{1}{2}(C' + C'^T) \end{aligned} \tag{13.1-5}$$

Thus, for systems which are amenable to the Lagrangian formulation the matrices of coefficients for the linear terms in the differential equations have necessary symmetry or skew-symmetry properties, with the sole exception of the term $\frac{1}{2}\dot{B}'q$ for which the matrix may have both skew and symmetric components.

Of special interest are linear systems with constant coefficients. For these the equations become

$$-b + A\ddot{q} + B\dot{q} + Cq = f \tag{13.1-6}$$

Provided the matrix C is not singular, a datum shift in the q coordinates can be used to eliminate b, leaving simply

$$A\ddot{q} + B\dot{q} + Cq = f \tag{13.1-7}$$

In applied mechanics the matrix A represents system inertia, C represents stiffness, and B represents gyroscopic coupling. This latter is particularly interesting since $B\dot{q}$ are velocity-dependent forces which are conservative because of the skewness property of B. In electricity, if the q_j are identified as mesh charges, then A represents inductance, C represents capacitance, and B represents 'gyrators'; the f_j are voltages.

If f is set equal to zero in eq. (13.1-7), i.e., if there are no dissipative or excitative mechanisms, then the system possesses a scalar invariant H, given by

$$H = \tfrac{1}{2}\dot{q}^T A\dot{q} + \tfrac{1}{2} q^T Cq = T + V \tag{13.1-8}$$

which, in mechanics, is the Hamiltonian and has the further interpretation as the total system energy. The first term in H is the kinetic energy, T, while the second is the potential energy, V. The invariance property

$$\dot{H} = 0 \tag{13.1-9}$$

follows from the symmetry properties of A, B, and C which give

$$\begin{aligned} A\ddot{q} + B\dot{q} + Cq &= 0 \\ \ddot{q}^T A - \dot{q}^T B + q^T C &= 0 \end{aligned} \tag{13.1-10}$$

and hence

$$\dot{q}^T A\ddot{q} + \ddot{q}^T A\dot{q} + \dot{q}^T Cq + q^T C\dot{q} = 0 \tag{13.1-11}$$

It is of some interest to note that the quadratic terms in the Lagrangian can be identified with T and V, thus

$$\tfrac{1}{2}\dot{q}^T A'\dot{q} + \tfrac{1}{2}\dot{q}^T B'q + \tfrac{1}{2}q^T C'q = T + \tfrac{1}{2}\dot{q}^T B'q - V \tag{13.1-12}$$

The first two terms are often combined into a single quantity known as the *kinetic coenergy* (Ref. 13-2)

$$T_{CO} = T + \tfrac{1}{2}\dot{q}^T B'q \tag{13.1-13}$$

so that

$$\text{quadratic terms in } L = T_{CO} - V$$

13.2 CONSERVATIVE LINEAR SYSTEMS, DIRECT COUPLED

In this section we restrict study to linear systems with constant coefficients and no gyroscopic coupling

$$A\ddot{q} + Cq = f \tag{13.2-1}$$

The added forces, f, may be of two kinds: 1) linear dissipation in the form $-D\dot{q}$ where D is a symmetric matrix, and 2) applied excitation, $f'(t)$. Dissipative effects will be considered to be small and will be handled by perturbation methods in section 13.2.4. The effect of excitation can be found by integration once the solutions of the homogeneous equations

$$A\ddot{q} + Cq = 0 \tag{13.2-2}$$

are known. This is routine and will be touched upon briefly in section 13.2.2. The properties of the homogeneous solutions are fundamental and will be studied first.

13.2.1 Eigenvalues, Eigenvectors, Orthogonality

The solutions of eq. (13.2-2) are taken in the form

$$q = \lambda e^{i\omega t} \tag{13.2-3}$$

where λ is a complex vector. The differential equations become algebraic,

$$(-\omega^2 A + C)\lambda = 0 \tag{13.2-4}$$

which for nontrivial λ require

$$|\omega^2 A + C| = 0$$

This characteristic equation will have n roots for ω^2 which we denote by ω_k^2 and $2n$ eigenvalues, $\pm\omega_k$. If all roots are distinct, the matrix $(-\omega_k^2 A + C)$ is of rank $n - 1$ for each root and there will be a distinct eigenvector $\lambda_{(k)}$, uniquely determined up to an arbitrary multiple. The case of repeated roots is discussed later.

For any root we may write

$$\lambda_{(k)}^{*T}(-\omega_k^2 A + C)\lambda_{(k)} = 0 \tag{13.2-6}$$

where the superscript (*) denotes complex conjugate. Because of the symmetry of A and C,

$$\lambda_{(k)}^{*T} A \lambda_{(k)} = \text{real}, \quad \lambda_{(k)}^{*T} C\lambda_{(k)} = \text{real} \tag{13.2-7}$$

Therefore each

$$\omega_k^2 = \frac{\lambda_{(k)}^{*T} C \lambda_{(k)}}{\lambda_{(k)}^{*T} A \lambda_{(k)}} \tag{13.2-8}$$

is real. Since the eigenvectors are found from the equations

$$(-\omega_k^2 A + C)\lambda_{(k)} = 0 \tag{13.2-9}$$

for which all the coefficients are real, it follows that the ratios of the elements of each eigenvector must be real. It becomes convenient then to speak of these eigenvectors as real, since if one element is real all elements are real, although a complex factor is in general required on each to match a given set of real initial values of q and $\dot{q}$.

If $\omega_k^2 > 0$, that eigensolution may be spoken of as stable in the sense that for a bounded initial condition the solution remains bounded for all time. The system as a whole may be described as stable if $\omega_k^2 > 0$ for all k. A sufficient condition to ensure this is positive definiteness for both A and C. That such a condition is not necessary can be seen by the elementary counterexample where A and C are each negative definite. In mechanics, the inertia matrix A *is* positive definite, in which case the positive definiteness of C is both necessary and sufficient for stability.

A form of orthogonality between eigenvectors is easily established. Since

$$(-\omega_k^2 A + C)\lambda_{(k)} = 0$$
$$\lambda_{(j)}^T (-\omega_j^2 A + C) = 0$$

premultiplication of the first by $\lambda_{(j)}^T$ and postmultiplication of the second by $\lambda_{(k)}$ leads to

$$(\omega_k^2 - \omega_j^2)\lambda_{(j)}^T A \lambda_{(k)} = 0 \tag{13.2-10}$$

Then for distinct eigenvalues

$$\lambda_{(j)}^T A \lambda_{(k)} = 0 \tag{13.2-11}$$

which is the orthogonality condition. An equivalent, but not independent condition is

$$\lambda_{(j)}^T C \lambda_{(k)} = 0 \tag{13.2-12}$$

The case of repeated roots can be treated quite easily. If two roots ω_j^2 and ω_k^2 are equal, the matrix $(-\omega_j^2 A + C)$ will be of rank $n - 2$, and two linearly independent vectors $\lambda_{(j)}$ and $\lambda'_{(j)}$ can be found. The second of these could be denoted $\lambda_{(k)}$, but then the orthogonality condition would not in general hold between $\lambda_{(j)}$

and $\lambda_{(k)}$. However, if we let

$$\lambda_{(k)} = \lambda'_{(j)} + \alpha\lambda_{(j)} \tag{13.2-13}$$

where α is a scalar, then we may set

$$\lambda^T_{(j)} A\lambda_{(k)} = \alpha\lambda^T_{(j)} A\lambda_{(j)} + \lambda^T_{(j)} A\lambda'_{(j)} = 0 \tag{13.2-14}$$

and choose α so that orthogonality is obtained. The extension to cases of multiplicity higher than two is straightforward.

In all subsequent subsections of section 13.2 we will assume that the roots ω_k^2 have been ordered in ascending value (thus ω_1^2 is the most negative if negative roots exist) and that in the event of repeated roots the vectors have been constructed to be orthogonal.

13.2.2 Normal Coordinates

Consider a time-invariant coordinate transformation

$$q = Gx \tag{13.2-15}$$

where the transformation matrix columns are the eigenvectors

$$G = [[\lambda_{(1)}]\ [\lambda_{(2)}]\ \cdots\ [\lambda_{(n)}]] \tag{13.2-16}$$

Here each of the vectors will be taken to be real with a magnitude yet to be specified. The eqs. (13.2-1) are now

$$AG\ddot{x} + CGx = f(t) \tag{13.2-17}$$

Premultiply this by G^T and a set of uncoupled equations result

$$\lambda^T_{(k)} A\lambda_{(k)}\ddot{x}_k + \lambda^T_{(k)} C\lambda_{(k)} x_k = \phi_k(t) \tag{13.2-18}$$

where

$$\phi_k(t) = \lambda^T_{(k)} f \tag{13.2-19}$$

If we assume that $\lambda^T_{(k)} A\lambda_{(k)} \neq 0$ (a necessary condition for all the roots to be finite) we may choose the size of each $\lambda_{(k)}$ so that

$$\lambda^T_{(k)} A\lambda_{(k)} = \pm 1 \tag{13.2-20}$$

In the event that A is positive definite, all of these will be +1. With the size of

$\lambda_{(k)}$ so chosen, it follows from eq. (13.2-8) that

$$\lambda_{(k)}^T C \lambda_{(k)} = \pm \omega_k^2 \tag{13.2-21}$$

and eqs. (13.2-18) are reduced to canonical form

$$\ddot{x}_k + \omega_k^2 x_k = \pm \phi_k(t) \tag{13.2-22}$$

The x_k are the normal coordinates. They cannot be determined unless the complete set of eigenvectors is known. In multidimensional systems this could require extensive computation, but it is a task which generally need not be done. In subsections to follow we will use simply the knowledge of the existence of a normal coordinate transformation to demonstrate various important system properties.

For any imposed excitation $f(t)$, hence imposed $\phi_k(t)$, the particular solution to eq. (13.2-22) is a convolution integral. If we denote the solution of the kth equation by $z_k(t)$ the particular integral of the q vector is given by

$$\underset{\text{particular}}{q(t)} = Gz = \sum_k \lambda_{(k)} z_k(t) \tag{13.2-23}$$

From eq. (13.2-19) we see that the individual components of ϕ depend on individual eigenvectors so that any component $z_k(t)$ can be determined from the corresponding eigenvector alone. Also, the contribution to q from that z_k is seen from eq. (13.2-23) to require only the single eigenvector. Thus the response of q to excitation can be assembled piecemeal from the knowledge of individual eigensolutions and does not require knowledge of the complete set to proceed.

13.2.3 Constraints and Bounds on Eigenvalues

Following Rayleigh, we introduce the concept of 'reduced' potential and kinetic energies

$$\begin{aligned} \overline{V} &= \tfrac{1}{2} y^T C y \\ \overline{T} &= \tfrac{1}{2} y^T A y \end{aligned} \tag{13.2-24}$$

where the y is a real nondimensional vector. The Rayleigh quotient, $\mathcal{R}$, is defined by

$$\mathcal{R} = \overline{V}/\overline{T} \tag{13.2-25}$$

and the stationary values of $\mathcal{R}$ found from

$$\begin{aligned} \frac{\partial \mathcal{R}}{\partial y_k} &= 0 = \frac{1}{\overline{T}} \left(-\mathcal{R} \frac{\partial \overline{T}}{\partial y_k} + \frac{\partial \overline{V}}{\partial y_k} \right) \\ 0 &= (-\mathcal{R} A + C) y \end{aligned} \tag{13.2-26}$$

are seen to correspond to the roots of the secular equation, ω_k^2. The associated y are, of course, the eigenvectors.

Now we assume that A is a positive definite matrix, as it is in most cases of practical physical interest. Furthermore we will assume that a transformation to normal coordinates has been effected. This means that $\mathcal{R}$ can be written

$$\mathcal{R} = \frac{\sum_{k=1}^{n} \omega_k^2 y_k^2}{\sum_{k=1}^{n} y_k^2} \tag{13.2-27}$$

and in accordance with section 13.2.1 the ω_i^2 are arranged in ascending sequence.

A linear constraint on the system is defined by an n-dimensional constraint vector, h, in the form

$$h^T q = 0 \tag{13.2-28}$$

and if a set of m independent constraints is applied they may be represented by an $m \times n$ constraint matrix, H, where an element H_{ij} is the jth element of the ith constraint, and

$$Hq = 0 \tag{13.2-29}$$

These constraints transform readily into an equivalent set in normal coordinates

$$HGx = 0 = H'x \tag{13.2-30}$$

or, in terms of the variables, y, used to form $\mathcal{R}$

$$H'y = 0 \tag{13.2-31}$$

We will denote the Rayleigh quotient of the system subject to m constraints by ${}^m\mathcal{R}$ and examine first some bounds on its minimum value. By simply partitioning eqs. (13.2-31) into

$$\sum_{j=1}^{j=k} H'_{ij} y_j = - \sum_{j=k+1}^{j=n} H'_{ij} y_j \qquad 1 \leq i \leq m \tag{13.2-32}$$

we observe that, if for any k, the coefficient matrix on the l.h.s. has a rank less than k, at least one nontrivial vector y can be found for which $y_j = 0, j \geq k + 1$. This is always the case for $k \geq m + 1$. If we denote by $(k_{\min} + 1)$ the lowest integer for which the rank of the l.h.s. coefficient matrix is less than k we can construct an ${}^m\mathcal{R} \leqslant \omega^2_{k_{\min}+1}$. Then it follows

$$ {}^m\mathcal{R}_{\min} \leqslant \omega^2_{k_{\min}+1}; \qquad k_{\min} \leqslant m \tag{13.2-33} $$

This implies that no set of m constraints can raise the minimum eigenvalue above the $(m+1)^{\text{st}}$ eigenvalue of the original system. One set of constraints which will produce the maximum effect is simply

$$ \begin{aligned} H'_{ij} &= 1 \text{ for} \quad i=j, \; i,j \leqslant m \\ H'_{ij} &= 0 \quad \text{otherwise} \end{aligned} \tag{13.2-34} $$

since these are constraints which selectively suppress the eigenvalues sequentially from the lowest. Other less restricted forms of constraint can also have maximum effect, but sufficiency conditions for this depend on the entire set of eigenvalues. Two necessary conditions for maximum effect can however be easily stated:

1. The square matrix H'_{ij}, $1 \leqslant i,j \leqslant m$, must be of rank m.
2. The vector $H'_{i,(m+1)}$ must be null.

The first is an immediate consequence of eq. (13.2-33). If the second condition does not hold, an ${}^m\mathcal{R} < \omega^2_{m+1}$ can be constructed by the choice: $y_j = 0$ for $j > m+1$ and $y_{m+1} = 1$ (which leads to nonzero y_j for $j < m+1$). From eqs. (13.2-30) and (13.2-16) we observe that the elements of the vector $H'_{i,(m+1)}$ are the inner products of the original m constraint vectors (for the q coordinates) and the $(m+1)^{\text{st}}$ eigenvector. Thus if we wish m constraints to produce the maximum effect in raising the lowest eigenvalue of a system, each constraint must be orthogonal to the $(m+1)^{\text{st}}$ eigenvector.

The effect of constraints on eigenvalues other than the lowest can be found by a simple extension of the fundamental result

$$ {}^m\mathcal{R}_{\min} \leqslant \omega^2_{m+1} \tag{13.2-35} $$

If we denote the eigenvalues of the system with m constraints as ${}^m\omega_k^2$, ordered in ascending sequence, we can show

$$ \omega_k^2 \leqslant {}^m\omega_k^2 \leqslant \omega^2_{k+m} \tag{13.2-36} $$

The right inequality follows from the idea that if we add $k-1$ additional and correctly chosen constraints to the already m-constrained system we can generate a new system with ${}^m\omega_k^2$ as its lowest eigenvalue. However, the composite set of $m+k-1$ constraints which have been applied to the original system can in no way raise the minimum eigenvalue above ω^2_{k+m}. The left inequality is demonstrated by two successive inequalities. If a set of $k-1$ properly chosen constraints is applied to the original system, the lowest eigenvalue can be raised to ω_k^2. Now the arbitrary m constraints are added as well. Let the resulting minimum eigenvalue be denoted by ω'^2. From the application sequence described it is clear that ω'^2 cannot be below the minimum attained by the $k - 1$ constraints alone

$$\omega_k^2 \leqslant \omega'^2$$

However, the final system cannot depend on the sequence of application and must therefore be the same as that resulting from the application of $k - 1$ constraints to the m-constrained system for which

$$\omega'^2 \leqslant {}^m\omega_k^2$$

which completes the arguments.

There are several useful applications of these results. For example, if two systems are joined together the joining mechanism represents a constraint on the total variable set of the unjoined systems. Therefore the eigenvalues of the joined system are bracketed by the composite eigenvalue spectrum of the component systems. If the component systems have two eigenvalues closely matched, one eigenvalue of the joined system will be very closely constrained. If a double root exists in one component system, this must necessarily be an eigenvalue of the joined system, etc. Another application lies in the use of constraints to raise eigenvalues so as to avoid dangerous resonances. If only m elementary constraints of the form $q_k = 0$ for a few selected k are physically feasible, the foregoing analysis says that they should be placed at coordinates for which the $(m + 1)^{\text{st}}$ eigenvector has very small amplitude.

It is worth noting that all of the properties of constrained systems have been derived independent of the dimension, n, of the coordinate field. They are therefore applicable to continuous conservative systems which have a discrete eigenvalue spectrum. Furthermore, they apply to systems with negative roots ω_k^2 (unstable roots). If there are only a few unstable roots and many stable ones, the application of constraints is a stabilizing influence.

13.2.4 Parameter Perturbations

Parameter perturbations may be classified as ordinary or special depending on whether the eigenvalue changes induced are much smaller than, or comparable to, the original spacing between eigenvalues. Dealing with ordinary perturbations first, we augment the eqs. (13.2-2) by four incremental matrices to give

$$(A + \epsilon\widetilde{A})\ddot{q} + \epsilon(\widetilde{B} + \widetilde{D})\dot{q} + (C + \epsilon\widetilde{C})q = 0 \tag{13.2-37}$$

Here ϵ is a small scalar establishing the strength of the perturbations. $\widetilde{D}$ is a symmetric matrix and represents dissipation while $\widetilde{B}$ is skew-symmetric and represents gyroscopic coupling. If $\widetilde{A}$ and $\widetilde{C}$ represent conservative elements they will be symmetric, although the analysis will encompass skew (nonconservative) elements in these as well. It will be assumed that A, the original system matrix, is positive definite.

A transformation to normal coordinates yields

$$\sum_j \{(I_{lj} + \epsilon \hat{A}_{lj})\ddot{x}_j + \epsilon(\hat{B}_{lj} + \hat{D}_{lj})\dot{x}_j + (\omega_l^2 I_{lj} + \epsilon \hat{C}_{lj})x_j\} = 0 \qquad (13.2\text{-}38)$$

where

$$\hat{A} = G^T \tilde{A} G; \quad \hat{B} = G^T \tilde{B} G; \quad \text{etc.} \qquad (13.2\text{-}39)$$

The matrix of solutions to eqs. (13.2-38) is taken in the form

$$x_{j(k)} = (I_{jk} + \epsilon \Gamma_{j(k)})e^{(i\omega_k + \epsilon s_k)t} \qquad (13.2\text{-}40)$$

with $\Gamma_{k(k)} = 0$ (by scaling). Substitution of these solutions gives, for the terms of order ϵ

$$-\omega_k^2(\hat{A}_{lk} + \Gamma_{l(k)}) + i\omega_k(2s_k I_{lk} + \hat{B}_{lk} + \hat{D}_{lk}) + \omega_l^2 \Gamma_{l(k)} + \hat{C}_{lk} = 0 \quad (13.2\text{-}41)$$

The components of the perturbation eigenvectors, $\Gamma_{l(k)}$, are found from the equations for which $l \neq k$

$$\Gamma_{l(k)} = \frac{1}{\omega_k^2 - \omega_l^2}(-\omega_k^2 \hat{A}_{lk} + i\omega_k(\hat{B}_{lk} + \hat{D}_{lk}) + \hat{C}_{lk}) \qquad (13.2\text{-}42)$$

while the perturbation eigenvalues are found from the equations for which $l = k$:

$$-2s_k = i\omega_k \hat{A}_{kk} + \hat{D}_{kk} + \frac{1}{i\omega_k}\hat{C}_{kk} \qquad (13.2\text{-}43)$$

The interpretation of these results is straightforward. All of the off-diagonal elements of $\hat{A}, \hat{B}, \hat{C}, \hat{D}$ serve to couple the normal coordinates, with $\hat{B}$ and $\hat{D}$ making imaginary contributions to $\Gamma_{l(k)}$. The strength of the coupling between coordinates diminishes as the difference between the original eigenvalues for the coordinates is increased. Only the diagonal elements of $\hat{A}$, $\hat{D}$, and $\hat{C}$ affect the eigenvalues. Because of the symmetry of the transformations, eqs. (13.2-39), these elements are derived only from the symmetrical parts of $\tilde{A}, \tilde{D}$, and $\tilde{C}$. If $\tilde{A}$ is positive semi-definite then all the $\hat{A}_{kk}$ will be either positive or zero. The effect on any eigenvalue for positive ϵ will be to lower its magnitude (or leave it unchanged) regardless of the sign of ω_k^2. With $\hat{C}$ the effect depends on the sign of ω_k^2. The nonconservative $\hat{D}$ makes a real contribution to the time exponent. If $\hat{D}$ is positive definite (and $\epsilon > 0$) the solutions (13.2-40) are seen to decay in time. It is especially interesting to note that while the skew parts of $\tilde{A}$ and $\tilde{C}$ are also nonconservative, in this perturbation analysis they make no contribution to the growth or decay of oscillations.

It is evident from eq. (13.2-42) that if two eigenvalues are close to one another, the perturbation eigenvector will not be small. To handle this special case, provision must be made at the start for the strong interaction. We assume two eigenvalues ω_r^2 and ω_{r+1}^2 to be close, i.e.,

$$\omega_{r+1}^2 = \omega_r^2(1 + \epsilon\gamma) \tag{13.2-44}$$

where γ is a 'detuning' parameter, and take the rth solution in the form

$$x_{j(r)} = (\chi_{j(r)} + \epsilon\Gamma_{j(r)})e^{(i\omega_r + \epsilon s_r)t} \tag{13.2-45}$$

where

$$\chi_{j(r)} \neq 0 \qquad \text{only for } j = r,\ r+1$$

$$\Gamma_{j(r)} = 0 \qquad \text{for } j = r,\ r+1$$

If this is substituted into eq. (13.2-38), the rth equation, to terms of order ϵ, is

$$\chi_{r(r)}[2i\omega_r s_r - \omega_r^2 \hat{A}_{rr} + i\omega_r \hat{D}_{rr} + \hat{C}_{rr}] + \chi_{r+1(r)}[-\omega_r^2 \hat{A}_{r(r+1)} + i\omega_r(\hat{B}_{r(r+1)} + \hat{D}_{r(r+1)}) + \hat{C}_{r(r+1)}] = 0 \tag{13.2-46}$$

The $(r+1)^{\text{st}}$ equation, to similar order, is

$$\chi_{r(r)}[-\omega_r^2 \hat{A}_{(r+1)r} + i\omega_r(\hat{B}_{(r+1)r} + \hat{D}_{(r+1)r}) + \hat{C}_{(r+1)r})] + \chi_{r+1(r)}[2\gamma\omega_r^2 + 2i\omega_r s_r - \omega_r^2 \hat{A}_{(r+1)(r+1)} + i\omega_r \hat{D}_{(r+1)(r+1)} + \hat{C}_{(r+1)(r+1)}] = 0 \tag{13.2-47}$$

Together these two equations determine the perturbation eigenvalue, s_r, and two associated eigenvectors.

The most interesting question in regard to these equations is that of system stability, i.e., sign of $\text{Re}(s_r)$. To obtain the answer we first observe that by considering the symmetrical components of $\hat{A}$ and $\hat{C}$ alone it is possible to redefine the components of χ and revise the value of γ so as to eliminate all of these perturbation components. Without pursuing the algebra here we will assume that this has been done so that $\hat{A}$ and $\hat{C}$ are skew. The only remaining algebraic difficulties arise from $\hat{D}$. With only a slight sacrifice in generality we will assume that the dissipation is diagonal and equal in the two equations, thus separating it from the skew matrix, $\hat{B}$. In summary

$$\hat{A}_{rr} = \hat{A}_{(r+1)(r+1)} = 0 \qquad \hat{A}_{r(r+1)} = -\hat{A}_{(r+1)r}$$
$$\hat{C}_{rr} = \hat{C}_{(r+1)(r+1)} = 0 \qquad \hat{C}_{r(r+1)} = -\hat{C}_{(r+1)r} \tag{13.2-48}$$
$$\hat{D}_{r(r+1)} = \hat{D}_{(r+1)r} = 0 \qquad \hat{D}_{rr} = \hat{D}_{(r+1)(r+1)}$$

The characteristic equation for s_r is then simply

$$\omega_r^2(2s_r + \hat{D}_{rr})(2s_r + \hat{D}_{rr} - 2i\gamma\omega_r) - [-\omega_r^2\hat{A}_{r(r+1)} + i\omega_r\hat{B}_{r(r+1)} + \hat{C}_{r(r+1)}]^2 = 0 \tag{13.2-49}$$

for which the solution is

$$2s_r = -\hat{D}_{rr} + i\gamma\omega_r \pm [-\gamma^2\omega_r^2 + (-\omega_r\hat{A}_{r(r+1)} + i\hat{B}_{r(r+1)} + \hat{C}_{r(r+1)}/\omega_r)^2]^{1/2} \tag{13.2-50}$$

From this we conclude that the simultaneous occurrence of gyroscopic coupling, $\hat{B}$, and nonconservative elements, $\hat{A}$ or $\hat{C}$, will make a positive real contribution to one root $s_r + \hat{D}_{rr}/2$. The system may still not be unstable, however, if the dissipation is sufficiently large. Gyroscopic coupling in the absence of nonconservative $\hat{A}$ or $\hat{C}$ will produce no instability, but acts instead to increase the separation between eigenvalues. Nonconservative $\hat{A}$ or $\hat{C}$ acting by themselves reduce the separation between eigenvalues, and if they are sufficiently strong they will give a positive real contribution to one of them. Then once again the relative size of dissipation controls the question of stability.

In summary it can be seen that the skew elements of $\hat{A}$, $\hat{B}$, and $\hat{C}$ which have no effect on eigenvalues in ordinary perturbations become very important when a system has close eigenvalues to begin with. Perturbation instability requires the presence of nonconservative perturbation elements and is greatly enhanced by gyroscopic coupling.

13.3 SYSTEMS WITH GYROSCOPIC COUPLING

13.3.1 Eigenvalue Properties

The system to be studied is the homogeneous form of eqs. (13.1-7)

$$A\ddot{q} + B\dot{q} + Cq = 0 \tag{13.3-1}$$

We take the solutions to be

$$q = \lambda e^{st} \tag{13.3-2}$$

and obtain

$$(s^2 A + sB + C)\lambda = 0 \tag{13.3-3}$$

with the associated characteristic equation

$$\left|s^2 A + sB + C\right| = 0 \tag{13.3-4}$$

Since the transpose of a determinant has the same value, the symmetry properties of the component matrices leads to

$$0 = \left|s^2 A^T + sB^T + C^T\right| = \left|s^2 A - sB + C\right| \tag{13.3-5}$$

from which we observe that for every root s_j of eq. (13.3-4) there is a corresponding negative root, $-s_j$. Since the matrices A, B, and C are real, the s_j also occur in complex conjugate pairs. Thus, any complex roots actually occur in sets of four-negative and conjugate pairs. For the system to be stable (no root with a positive real part) all roots must therefore be imaginary.

We denote the eigenvector associated with the eigenvalue s_j as $\lambda_{(j)}$ so that

$$(s_j^2 A + s_j B + C)\lambda_{(j)} = 0 \tag{13.3-6}$$

Multiplying this by the conjugate transpose of $\lambda_{(j)}$ gives

$$\lambda_{(j)}^{*T}(s_j^2 A + s_j B + C)\lambda_{(j)} = 0 \tag{13.3-7}$$

Let

$$\begin{aligned} \lambda_{(j)}^{*T} A \lambda_{(j)} &= a \\ \lambda_{(j)}^{*T} B \lambda_{(j)} &= ib \\ \lambda_{(j)}^{*T} C \lambda_{(j)} &= c \end{aligned} \tag{13.3-8}$$

where a, b, and c are all real due to the matrix symmetries. The roots s_j of eq. (13.3-7) satisfy

$$s_j = i\left[\frac{-b \pm \sqrt{b^2 + 4ac}}{2a}\right] \tag{13.3-9}$$

and it can be seen that for s_j to be stable it is necessary for

$$4ac \geqslant -b^2 \tag{13.3-10}$$

Now if A and C are both positive definite matrices, both a and c will be positive for all possible eigenvectors so all eigenvalues will be stable. That is, conditions which are sufficient to insure stability in the absence of gyroscopic coupling, B, are sufficient with an arbitrary B. Furthermore, eq. (13.3-10) shows that b acts to improve stability margins, and can produce stable eigenvalues when the product $ac < 0$. This is 'spin stabilization' of an otherwise unstable system, and will be examined more closely below.

If a symmetric dissipation matrix is added to eqs. (13.3-1) to give

$$A\ddot{q} + (B + D)\dot{q} + Cq = 0 \tag{13.3-11}$$

then the eigenvalues no longer occur in negative pairs—but they do arise in conjugate pairs. If we let

$$\lambda_{(j)}^{*T} D \lambda_{(j)} = d \tag{13.3-12}$$

this will be a positive real quantity for dissipative mechanisms. Then, analogous to eq. (13.3-9), we have

$$s_j = \frac{-(d + ib) \pm \sqrt{(d + ib)^2 - 4ac}}{2a} \tag{13.3-13}$$

It is a simple algebraic exercise to show that, if $d > 0$ and $ac < 0$, the real component of the radical is larger than d. Thus one of the s_j will be unstable. The dissipation destroys the ability of gyroscopic coupling to stabilize.

13.3.2 Spin Stabilization

There is one further aspect of this physically useful phenomenon which is not covered in the general results above. Consider an example of two gyroscopically-coupled equations

$$\begin{aligned} \ddot{q}_1 - q_1 + b\dot{q}_2 &= 0 \\ \ddot{q}_2 + cq_2 - b\dot{q}_1 &= 0 \end{aligned} \tag{13.3-14}$$

where b is the coupling parameter. By itself, the coordinate q_1 is unstable, while q_2 is stable for $c > 0$. The characteristic equation for this set is

$$s^4 - (1 - c - b^2)s^2 - c = 0 \tag{13.3-15}$$

from which

$$2s^2 = (1 - c - b^2) \pm \sqrt{(1 - c - b^2)^2 + 4c} \tag{13.3-16}$$

Now, for $c > 0$ the radical is a positive real and larger than $1 - c - b^2$. Therefore, two of the eigenvalues are real and one of these is unstable. That is, gyroscopic coupling cannot produce system stability by coupling an unstable motion to a stable one. The coupled motions must both be unstable. For $c < 0$, the threshold level for b at which stability is found is

$$b \geqslant 1 + (-c)^{1/2} \tag{13.3-17}$$

13.4 MATHIEU-HILL SYSTEMS

Mathieu-Hill systems (M.-H. systems) are represented by sets of linear differential equations with periodic coefficients. The mathematical complexity possible is very great, so only a limited view can be taken here. The aim is to present a few results with some clarity. The differential equations are taken in the form

$$A(t)\ddot{q} + B(t)\dot{q} + C(t)q = 0 \tag{13.4-1}$$

All of the $n \times n$ coefficient matrices are assumed to have a common period, τ; i.e.

$$A(t + \tau) = A(t) \quad \text{etc.} \tag{13.4-2}$$

and the differential order of the system will be assumed to be the same at all time

$$|A(t)| \neq 0 \tag{13.4-3}$$

13.4.1 Floquet Theory

We suppose that $2n$ independent vector solutions to eqs. (13.4-1) have been generated, which we denote as $\eta_{(j)}(t)$. After one period the differential equations return to their previous condition, so the solutions with shifted argument must also constitute a complete set. Furthermore, we must be able to express them as linear combinations of the original solutions

$$\eta_{(j)}(t + \tau) = \sum_k \beta_{jk}\eta_{(k)}(t) \tag{13.4-4}$$

The square matrix β is a transition matrix. A time-invariant linear transformation of the $\eta_{(j)}$ can be used to diagonalize the transition matrix and yield a set of 'normal' solutions defined by

$$w_{(j)}(t + \tau) = \alpha_j w_{(j)}(t) \tag{13.4-5}$$

The α_j are the eigenvalues of β and are either real or occur in conjugate pairs. The $w_{(j)}$ are eigenvectors in the set of η functions. If we define

$$\alpha_j = e^{s_j \tau} \tag{13.4-6}$$

Then a normal solution can be written

$$w_j = e^{s_j \tau} \phi_{(j)}(t) \tag{13.4-7}$$

where $\phi_{(j)}$ is a vector of periodic functions. It may be observed that systems with constant coefficients, which are a subclass of M.-H. systems, have constants for the functions $\phi_{(j)}$.

Special consideration is required if two or more eigenvalues, α_j, are equal. One possible behavior is that independent vectors $w_{(j)}$ can be found, equal in number to the multiplicity of the eigenvalue. If these do not exist, a new kind of temporal behavior does. This new form can be found by imagining that slightly different values of the coefficients in eqs. (13.4-1) will cause the eigenvalues to be distinct, but close. The new form corresponds to the difference between the two distinct solutions, which can be found by differentiating the normal solution parametrically (in α_j or s_j). Formally

$$\begin{aligned} w'_{(j)} = \frac{\partial w_{(j)}}{\partial s_j} &= t e^{s_j t} \phi_{(j)}(t) + e^{s_j t} \frac{\partial \phi_{(j)}}{\partial s_j} \\ &= e^{s_j t} [t \phi_{(j)}(t) + \phi'_{(j)}(t)] \end{aligned} \tag{13.4-8}$$

If the multiplicity of the eigenvalue is greater than two, further derivatives in s_j may be required, leading to functions like $e^{s_j t} t^2 \phi_{(j)}(t)$, etc.

There is a large class of M.-H. systems arising from the analysis of stability for periodic solutions to nonlinear systems. Returning to the original Lagrangian formulation, eqs. (13.1-1) we define the 'momentum' vector, p, by

$$p_i = \frac{\partial L}{\partial \dot{q}_i} \tag{13.4-9}$$

So that the unforced equations are

$$\dot{p}_i - \frac{\partial L}{\partial q_i} = 0 \tag{13.4-10}$$

We now assume that a periodic solution (period τ) has been found for this system, which we denote by $\widetilde{q}(t)$ and for which the corresponding momentum vector is $\widetilde{p}(t)$. For the stability analysis we let

$$q = \widetilde{q} + u, \quad p = \widetilde{p} + v \tag{13.4-11}$$

With a slight modification this important result can also be demonstrated for systems which arise in other contexts. One such is the class of system

$$A\ddot{u} + B\dot{u} + C(t)u = 0 \tag{13.4-18}$$

where A is a symmetric constant matrix, B is a skew constant matrix and C is a symmetric matrix of period τ. By introducing

$$v \triangleq A\dot{u} + \tfrac{1}{2}Bu \tag{13.4-19}$$

the function (13.4-14) can once again be shown to be invariant and the occurrence of eigenvalues in reciprocal pairs follows as before.

Another result special to the variational system, eqs. (13.4-12), can be obtained by noting that one of its solutions is simply $u = \dot{\tilde{q}}$, $v = \dot{\tilde{p}}$ and therefore is purely periodic (i.e., $\alpha = 1$). This solution corresponds to the difference between two points infinitesimally displaced in time along the original solution. From the reciprocal pair property, another normal solution must have $\alpha = 1$, so this eigenvalue is repeated. The second solution generally has a linear increase with time and corresponds to another periodic solution, adjacent in phase space to the original one but with a slightly different period.

A stable normal solution must have $|\alpha_r| \leqslant 1$, and a stable system requires that all normal solutions be stable. For those systems with reciprocally paired eigenvalues, stability is found only if $|\alpha_r| = 1$ for all solutions. Thus none of the solutions may decay in time, since if a decaying solution exists, a paired growing solution must exist as well. This result is to be expected from the conservative property of such systems.

13.4.2 The Single Mathieu-Hill Equation

If we have a single equation

$$a'(t)\ddot{u} + b'(t)\dot{u} + c'(t)u = 0 \tag{13.4-20}$$

where all coefficients have period τ and $a'(t) \neq 0$, we may first divide by $a'(t)$ and then use the transformation

$$\begin{aligned} u &= q(t)e^{-\int_0^t (b'/2a')\,dt} \\ c &= \frac{c'}{a'} - \left(\frac{b'}{2a'}\right)^2 - \frac{d}{dt}\left(\frac{b'}{2a'}\right) \end{aligned} \tag{13.4-21}$$

to give

$$\ddot{q} + c(t)q = 0 \tag{13.4-22}$$

and the variational equations are

$$v_i = \sum_j \left(\frac{\partial^2 L}{\partial \dot{q}_i \partial q_j} u_j + \frac{\partial^2 L}{\partial \dot{q}_i \partial \dot{q}_j} \dot{u}_j \right)$$
$$\dot{v}_i = \sum_j \left(\frac{\partial^2 L}{\partial q_i \partial q_j} u_j + \frac{\partial^2 L}{\partial q_i \partial \dot{q}_j} \dot{u}_j \right) \tag{13.4-12}$$

where the derivatives of L are evaluated on the periodic solution. These equations may be written

$$v = B(t)u + A(t)\dot{u}$$
$$\dot{v} = C(t)u + B^T(t)\dot{u} \tag{13.4-13}$$

All the matrices have period τ, and A and C are symmetric. Clearly v could be easily eliminated, but we will not do so here.

Let the vectors $u_{(r)}(t)$ and $v_{(r)}(t)$ represent one solution of the variational equations and $u_{(s)}(t)$, $v_{(s)}(t)$ another independent solution. Following Poincaré, Ref. 13-3, we observe that the scalar

$$u_{(r)}^T v_{(s)} - v_{(r)}^T u_{(s)} \tag{13.4-14}$$

is a constant. This can be demonstrated by differentiation and the use of eqs. (13.4-13) with the associated matrix symmetry properties. If these solutions are each normal solutions, then it follows that $u_{(r)}(t + \tau) = \alpha_r u_{(r)}(t)$ and $u_{(s)}(t + \tau) = \alpha_s u_{(s)}(t)$ (similarly for v) so that

$$(\alpha_r \alpha_s - 1)\,(u_{(r)}^T v_{(s)} - v_{(r)}^T u_{(s)}) = 0 \tag{13.4-15}$$

Thus, normal solutions must either be orthogonal in the sense

$$u_{(r)}^T v_{(s)} - v_{(r)}^T u_{(s)} = 0 \tag{13.4-16}$$

or the eigenvalues must be related by

$$\alpha_r \alpha_s = 1 \tag{13.4-17}$$

For a dynamical system of dimension n, there are $2n$ independent normal solutions. If we take the solution $u_{(r)}$ as a reference, there can be no more than $2n - 1$ normal solutions $u_{(s)}$ which satisfy the orthogonality condition, eq. (13.4-16), and one of these is the solution $u_{(r)}$ itself. Thus there can be no more than $2n - 2$ independent normal solutions orthogonal to $u_{(r)}$. There must therefore be at least one normal solution for which $\alpha_s = \alpha_r^{-1}$, for every α_r. That is, the eigenvalues of the variational M.-H. system occur in reciprocal pairs.

which is Hill's equation. If $c(t) = \delta + \epsilon \cos 2t$ so

$$\ddot{q} + (\delta + \epsilon \cos 2t)q = 0 \tag{13.4-23}$$

we have Mathieu's equation (with the conventional choice $\tau = \pi$). This falls in the class of eqs. (13.4-18), and therefore its two eigenvalues are a reciprocal pair. Further, since complex eigenvalues are conjugate, complex eigenvalues are stable. For instability the eigenvalues must be real.

There is an important distinction between the cases of positive and negative real eigenvalues. In the negative case, if the normal solutions are written as in eq. (13.4-7), the exponents s_j each have an imaginary part $-i$. This means that $w_{(j)}$ can be factored into a real exponential function and a periodic function with period 2π (double the period of the coefficients in the differential equation). In the case of positive eigenvalues the $w_{(j)}$ are the product of a real exponential function and a function of period π.

It is customary to display the stability properties of Mathieu's equation in a parametric diagram, Fig. 13.4-1. Here the δ, ϵ pairs corresponding to stable solutions are shown unshaded, while the two kinds of shading represent unstable solutions with either positive or negative real eigenvalues. Limitations of space force us to omit the analytic details by which the boundaries defining the regions are obtained. (See Ref. 13-4.) Only a brief explanation of the results is possible.

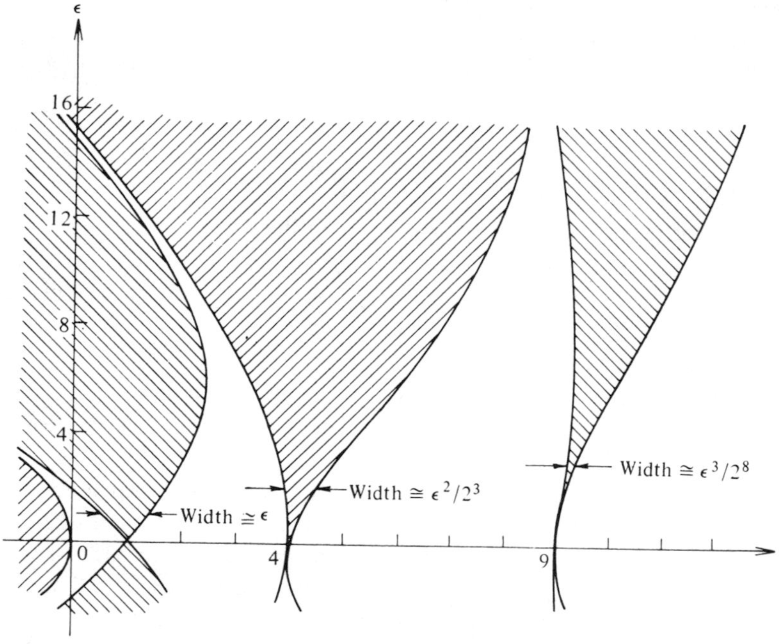

Fig. 13.4-1

When $\epsilon = 0$ and $\delta > 0$ the solutions are known to be sinusoids (stable). The unstable parameter regimes intersect this axis at integer-square values where the frequencies of the solutions are integer multiples of the half-frequency of the equation coefficient, $c(t)$. For small ϵ, the only unstable region for which the 'width' in δ is linear in ϵ is at $\delta = 1$. This represents a resonance condition whose implications will be discussed below. The boundaries separating stable and unstable regimes correspond to repeated eigenvalue pairs of either +1 or −1. With the exception of the points at $\epsilon = 0$ one periodic solution exists on any boundary arc, and these are called Mathieu functions. The second solution on each arc exhibits a linear time growth. Along each arc the number of zeros for the periodic function, lying in the interval $0 \leqslant t < 2\pi$, is fixed. This is because the eigenfunctions are continuous in the parameters, and if zeros were to be gained or lost at some bifurcation value of these parameters, the eigenfunction at that condition would have to exhibit both a zero value and zero derivative at some point in the periodic interval. Such a function would be zero everywhere. The zero count is therefore fixed for each arc and different for each pair of arcs bounding a stable region. These arcs, then, cannot intersect, and although the stable regions become very narrow at large ϵ they do not terminate.

One unusual feature of the stability diagram is the presence of stable regions for $\delta < 0$. The values of ϵ necessary to produce this 'parametric stabilization' are not small, but it is remarkable that it should happen at all. This phenomenon has an important application in the 'strong focusing' of beams of charged particles.

13.4.3 Mathieu-Hill Perturbations

Ordinarily the addition of weak terms of the Mathieu-Hill type to an oscillatory system will produce small effects. The analysis is conventional and will not be repeated here. There are, however, resonance cases where small M.-H. terms produce effects which are cumulatively large. The simplest example is afforded by the Mathieu equation (13.4-23), for $\delta > 0$ and $\epsilon \ll 1$. A straightforward perturbation would take the solution to be

$$q = q_0 + \epsilon q_1 + \cdots \tag{13.4-24}$$

for which

$$q_0 = \lambda \cos(\delta^{1/2} t + \phi) \tag{13.4-25}$$

is found, and the differential equation for q_1 is

$$\begin{aligned} \ddot{q}_1 + \delta q_1 &= -\lambda \cos 2t \cos(\delta^{1/2} t + \phi) \\ &= -\frac{\lambda}{2} [\cos(2t + \delta^{1/2} t + \phi) + \cos(2t - \delta^{1/2} t - \phi) \end{aligned} \tag{13.4-26}$$

Difficulties arise when $2 - \delta^{1/2} \cong \delta^{1/2}$, or more precisely

$$1 - \delta^{1/2} = O(\epsilon) \tag{13.4-27}$$

so that the particular integral of q_1 is $O(\epsilon^{-1})$.

We know from the results of section 13.4.2 that when $\delta^{1/2} - 1$ is small a possible form of the solution is the product of an exponential and a function of period 2π. The form of expansion adopted here can represent this kind of function only for a limited extent in time, and a different approach is required. One method of analysis is to introduce the exponential factor ad hoc into the generating function. A more general method is to recognize that the exponential time dependence is a process which takes place in a different time scale, and introduce an auxiliary independent variable to describe that dependence. This is a 'two-time' analysis and has the capability of dealing with weakly nonlinear oscillations as well as M.-H. perturbations. The details will be developed in sections 13.5 and 13.6.

The need for this more powerful form of analysis arises because the product of the primary system response, q_0, with the periodic coefficient produces a 'combination frequency' component near resonance. For the Mathieu equation $\delta \cong 1$ is the only such resonance case. However, if the equation were of the Hill type where the periodic coefficient contains a full set of Fourier terms, then combination frequency components will be resonant for δ close to any square integer. In multidimensional systems another form of resonance can be found. Suppose that ω_j is the set of system eigenfrequencies and ω_0 is the frequency of the M.-H. perturbations. Then combination frequencies $\omega_j \pm \omega_0$ can arise and these can be resonant with any other eigenfrequency, ω_k. Allowing further for the periodic coefficients to contain harmonics of ω_0 the possible resonance interactions become

$$\omega_j \pm \omega_k = n_0 \tag{13.4-28}$$

where n is any integer. In a system of any complexity these can be legion. Factors such as dissipation and detuning (deviation from precise resonance) act to inhibit the number of resonance interactions actually observed. A quantitative treatment of their effect is given in sections 13.5 and 13.6.

13.5 OSCILLATIONS WITH WEAK NONLINEARITIES

13.5.1 Introduction

A classic example which demonstrates the mathematical problem is the Van der Pol equation

$$\ddot{q} + q + \epsilon(q^2 - 1)\dot{q} = 0 \tag{13.5-1}$$

which will be considered here only for $\epsilon << 1$. As in section 13.4.3, if a straightforward expansion

$$q = q_0 + \epsilon q_1 + \epsilon^2 q_2 + \cdots \tag{13.5-2}$$

is attempted, it yields

$$q_0 = \lambda \cos(t + \phi), \qquad \lambda \neq 0 \tag{13.5-3}$$

and

$$\ddot{q}_1 + q_1 = (-q_0^2 + 1)\dot{q}_0 = \left(\frac{\lambda^3}{4} - \lambda\right) \sin(t + \phi) + \frac{\lambda^3}{4} \sin 3(t + \phi) \tag{13.5-4}$$

There is no bounded solution for q_1 unless λ is chosen to be 2. Even if this is done, a similar problem arises with the next term in the expansion and no bounded solution can be found for q_2 within the confines of this procedure.

The shortcomings of the procedure lie in its inability to encompass two types of actual solution behavior. One is a gradual growth or decay of the solution. (The Van der Pol equation is known to possess a stable periodic solution with an amplitude close to 2, and all nontrivial initial conditions, however large or small, will lead asymptotically to the periodic motion.) The second is a change in the frequency of oscillation. (The angular frequency of the periodic Van der Pol solution differs from unity by an amount proportional to ϵ^2.)

Any suitable analysis must take account of changes in amplitude and phase of oscillations which are slow but persistent. The fact that they are slow is the basis for the expansions. Poincaré, Ref. 13-3, developed expansions suitable for determining periodic solutions. Krylov and Bogoliubov, Ref. 13-5, initiate the 'method of averaging' which permitted treatment of transients as well. Cole and Kevorkian, Ref. 13-6, introduced the method of 'two-timing' which, in the present context, leads to the same results as the method of averaging (see Ref. 13-7 for a comparison) but more easily.

No attempt will be made here to deal with the difficult question of the time-regime of validity of solutions generated by these expansions. The method of averaging appears better suited to attack the question, but the strongest results available (e.g. Ref. 13-8) show regions of validity which are generally much smaller than those for which we expect the solutions to be qualitatively correct.

In this extended sense the analysis by either method must be regarded as heuristic. Further remarks on this subject will be found at the close of section 13.6, and in Chapter 14.

13.5.2 Asymptotic Expansion for the Single Oscillator

In section 13.2.2, x was used to denote the vector of normal coordinates. In the treatment of weak nonlinear effects it will be assumed that a transformation to normal coordinates has been made so that the linearized equations are uncoupled.

For the single oscillator, which is the subject of study throughout section 13.5, it will be understood that x is a scalar.

We take the differential equation in the form

$$\ddot{x} + x + \epsilon\,[f(x, \dot{x}, t) + \nu x] = 0 \tag{13.5-5}$$

which assumes a definition of the independent variable so that the angular frequency is normalized to unity. The function of the term $\epsilon\nu x$ will be described below. Two new independent variables are introduced, which are related to t through

$$\begin{aligned} \tilde{t} &= t(1 + a_2\epsilon^2 + a_3\epsilon^3 + \cdots) \\ \hat{t} &= t(\epsilon + b_2\epsilon^2 + b_3\epsilon^3 + \cdots) \end{aligned} \tag{13.5-6}$$

where $\tilde{t}$ is a 'fast' time and $\hat{t}$ is a 'slow' time. The a_i and b_i are expansion coefficients to be determined. The solution is expanded in the series

$$x(\tilde{t}, \hat{t}) = x_0(\tilde{t}, \hat{t}) + \epsilon x_1(\tilde{t}, \hat{t}) + \epsilon^2 x_2(\tilde{t}, \hat{t}) + \cdots \tag{13.5-7}$$

and its time derivatives are

$$\dot{x} = \frac{dx}{dt} = \frac{\partial x}{\partial \tilde{t}}\frac{d\tilde{t}}{dt} + \frac{\partial x}{\partial \hat{t}}\frac{d\hat{t}}{dt} \tag{13.5-8}$$

$$= (x_{0,\tilde{t}} + \epsilon x_{1,\tilde{t}} + \cdots)(1 + a_2\epsilon^2 + \cdots) + (x_{0,\hat{t}} + \epsilon x_{1,\hat{t}} + \cdots)(\epsilon + b_2\epsilon^2 + \cdots)$$

and

$$\ddot{x} = \frac{d^2x}{dt^2} = \frac{\partial^2 x}{\partial \tilde{t}^2}\left(\frac{d\tilde{t}}{dt}\right)^2 + 2\frac{\partial^2 x}{\partial \tilde{t}\,\partial \hat{t}}\left(\frac{d\tilde{t}}{dt}\cdot\frac{d\hat{t}}{dt}\right) + \frac{\partial^2 x}{\partial \hat{t}^2}\left(\frac{d\hat{t}}{dt}\right)^2$$

$$\begin{aligned} &= (x_{0,\tilde{t}\tilde{t}} + \epsilon x_{1,\tilde{t}\tilde{t}}\cdots)(1 + a_2\epsilon^2\cdots)^2 + (x_{0,\tilde{t}\hat{t}} + \epsilon x_{1,\tilde{t}\hat{t}} + \cdots)(1 + a_2\epsilon^2\cdots) \\ &\quad \cdot(\epsilon + b_2\epsilon^2\cdots) + (x_{0,\hat{t}\hat{t}} + \epsilon x_{1,\hat{t}\hat{t}}\cdots)(\epsilon + b_2\epsilon^2\cdots)^2 \end{aligned} \tag{13.5-9}$$

where the subscripts following the comma denote partial differentiation. The expansion of f is performed about a generating solution based on the leading terms in the x and dx/dt series

$$f(x, \dot{x}, t) = f(x_0, x_{0,\tilde{t}}, \tilde{t}) + f_{,x}\,[x - x_0] + f_{,\dot{x}}\,[\dot{x} - x_{0,\tilde{t}}] \tag{13.5-10}$$

where the derivatives of f are evaluated at $x = x_0$, $\dot{x} = x_{0,\tilde{t}}$, $t = \tilde{t}$. Replacing t by $\tilde{t}$ in these functions means that we are taking f to be specified in a time variable which depends on the solution to be generated. The dependence on t is determined post facto.

Substitution of these expansions into eq. (13.5-5) gives the sequence of partial differential equations

$$x_{0,\tilde{t}\tilde{t}} + x_0 = 0$$

$$x_{1,\tilde{t}\tilde{t}} + x_1 = -f(x_0, x_{0,\tilde{t}}, \tilde{t}) - \nu x_0 - 2x_{0,\tilde{t}\hat{t}}$$

$$x_{2,\tilde{t}\tilde{t}} + x_2 = -f_{,x}x_1 - f_{,\dot{x}}[x_{0,\hat{t}} + x_{1,\tilde{t}}] - 2a_2x_{0,\tilde{t}\tilde{t}} - 2b_2x_{0,\tilde{t}\hat{t}} - x_{0,\hat{t}\hat{t}} - \nu x_1 - 2x_{1,\tilde{t}\hat{t}} \tag{13.5-11}$$

etc.

The first of these integrates to

$$x_0 = \lambda(\hat{t}) \cos [\tilde{t} + \phi(\hat{t})] \tag{13.5-12}$$

which is inserted into the second to yield

$$x_{1,\tilde{t}\tilde{t}} + x_1 = f[\lambda \cos (\tilde{t} + \phi), -\lambda \sin (\tilde{t} + \phi), \tilde{t}] - \nu\lambda \cos (\tilde{t} + \phi) + 2\dot{\lambda} \sin (\tilde{t} + \phi) + 2\lambda\dot{\phi} \cos (t + \phi) \tag{13.5-13}$$

Here the superscript dot on λ and ϕ denotes differentiation in their only time argument, $\hat{t}$. The functions λ and ϕ are undetermined by the differential equation for x_0. Equation (13.5-13) will generate x_1 solutions which grow without limit in $\tilde{t}$ if its r.h.s. contains any resonant excitation, and this would of course violate the premise of the expansion procedure. The last two terms can be used to eliminate resonant excitation from the r.h.s. (suppress secular terms in x_1) by the freedom to choose λ and ϕ. This process yields coupled differential equation for λ and ϕ as we shall see, but first the nature of the explicit time dependence of f must be considered.

If f has no explicit independence on $\tilde{t}$, the original system is autonomous and the function on the r.h.s. of eq. (13.5-13) clearly has a fundamental period in $\tilde{t}$ identical with that of x_0. The interesting cases of explicit time dependence are those which are also periodic and have frequencies equal or close to some multiple of the x_0 frequency. The standard procedure is then to bring the frequency of the imposed (explicit) periodicity into exact agreement with the appropriate harmonic of the oscillator by adjusting the frequency of the oscillation. This is the purpose of the linear term $\epsilon\nu x$ in eq. (13.5-5). It represents the amount by which the (square) of the oscillator frequency must be altered to produce resonance with the imposed periodicity, and will therefore be called the detuning parameter. Thus we will assume that whatever explicit time dependence exists in f is periodic, with a fundamental period of 2π.

The suppression of secular terms in x_1 is effected by setting

$$\dot{\lambda} = \frac{1}{2\pi}\int_0^{2\pi} f[\lambda \cos(\tilde{t}+\phi), -\lambda \sin(\tilde{t}+\phi), \tilde{t}] \sin(\tilde{t}+\phi)\, d\tilde{t} \stackrel{\Delta}{=} g_1(\lambda, \phi)$$

$$(13.5\text{-}14)$$

$$\lambda\dot{\phi} = \frac{1}{2\pi}\int_0^{2\pi} f[\lambda \cos(\tilde{t}+\phi), -\lambda \sin(\tilde{t}+\phi), \tilde{t}] \cos(\tilde{t}+\phi)\, d\tilde{t} \stackrel{\Delta}{=} g_2(\lambda, \phi)$$

Both λ and ϕ are constant during the integration in $\tilde{t}$. After this integration we have a set of autonomous first-order ordinary differential equations. They may be nonlinear and very complicated, but to them we may apply the wide variety of techniques developed for such two dimensional systems (Ref. 13-9). Of special interest are the equilibrium solutions, $\dot{\lambda} = 0 = \dot{\phi}$, which represent synchronization with the imposed periodicity, and limit cycles which represent a steady combination of 'free' and 'forced' oscillations.

Only for very few problems of interest can analytic solutions for λ and ϕ be obtained, and the extension of the expansion to higher order terms cannot generally be carried out. In principle, with λ and ϕ known, the procedure is to generate a solution for x_1 in two parts: a particular integral derived from all the remaining terms on the r.h.s., and a complementary function in the form $[\lambda_1 \cos(\tilde{t}+\phi) + \lambda_2 \sin(\tilde{t}+\phi)]$. The complementary function is then available in the x_2 equation to suppress secular terms, leading to two coupled differential equations in slow time for λ_1 and λ_2. Unlike eqs. (13.5-14), these are not autonomous but contain functions of $\hat{t}$ obtained from λ and ϕ. One case which can be studied in some detail is that where eqs. (13.5-14) have a stable singular point (synchronized solution) and the approach to it is locally exponential in $\hat{t}$. The differential equation set for λ_1 and λ_2 can be shown to possess, asymptotically, the same exponential eigenvalues and are therefore resonant with the $\hat{t}$-functions obtained from λ and ϕ. This can produce secular terms in λ_1 and λ_2 and the adjustable coefficients, a_2 and b_2, are a means of suppressing them. Depending on the example they may not always work, and if they do not the assumed form of expansion is inadequate. What this implies is that the problem at hand contains a third intrinsic time scale which must be accommodated by a third variable, say $\hat{\hat{t}}$, of order $\epsilon^2 t$. Even if the coefficients a_2 and b_2 do work, should the equations for λ and ϕ possess more than one singular point a different set a_2, b_2 will be required for each point.

For the remainder of section 13.5 we will restrict our attention to the first order solution described by eqs. (13.5-14). One common feature of all such solutions is that a reversal of sign of λ is equivalent to an increment in ϕ of π, so that only $\lambda \geqslant 0$ need be considered. In a wide variety of problems, f is a sum of three fundamental kinds of functions:

1. autonomous nonlinearities: $f(x, \dot{x})$;
2. resonant excitation: $F \cos \tilde{t}$; and
3. linear Mathieu: $x \cos 2\tilde{t}$.

We will consider these alone and in combination.

13.5.3 The Autonomous Oscillator-Examples

For an autonomous f, the integrands in eqs. (13.5-14) have the composite variable $(\tilde{t} + \phi)$ as their sole argument so g_1 and g_2 are independent of ϕ. In order for the expansion procedure to be valid the integrals should be of the same magnitude as λ itself. This limits the size of f, but f need not be continuous. For many problems f arises in the form

$$f(x, \dot{x}) = x^m \dot{x}^n \qquad m, n \geqslant 0 \tag{13.5-15}$$

Sometimes this is just a local approximation to a more complicated function. If such a representation is employed and $m + n > 1$, there will be a limit of some $|\lambda|$, depending on ϵ, beyond which the nonlinear contributions to eq. (13.5-5) are no longer weak. It must be understood that eqs. (13.5-14) are not to be used outside of this limit.

With f in the form (13.5-15) and integers for m, n we have

for m even, n odd

$$g_1 = -\lambda^{m+n}\left[\frac{m!}{\left(\frac{m}{2}\right)!}\cdot\frac{(n+1)!}{\left(\frac{n+1}{2}\right)!}\cdot\frac{1}{\left(\frac{m+n+1}{2}\right)!\,2^{m+n+1}}\right]$$

$$g_2 = 0$$

for m odd, n even

$$g_1 = 0$$

$$g_2 = \lambda^{m+n}\left[\frac{(m+1)!}{\left(\frac{m+1}{2}\right)!}\,\frac{n!}{\left(\frac{n}{2}\right)!}\cdot\frac{1}{\left(\frac{m+n+1}{2}\right)!\,2^{m+n+1}}\right]$$

for $m + n$ even

$$g_1 = 0; \quad g_2 = 0$$

Since nonlinearities with m odd and n even make no contribution to $\dot{\lambda}$ they may be classified as conservative, while those with m even and n odd are nonconservative. Those with $m + n$ even are completely ineffective. All these descriptions, however, apply only to the first order approximation.

Since for any autonomous f both g_1 and g_2 are independent of ϕ the eqs. (13.5-14) are integrable in succession. Furthermore, ν may be set equal to zero since there is no externally established period. For the Van der Pol eq. (13.5-1), eqs. (13.5-14) are

$$\dot{\lambda} = -\frac{\lambda^3}{8} + \frac{\lambda}{2}; \quad \dot{\phi} = 0 \tag{13.5-17}$$

The first equation has two singular points at $\lambda = 0$ and $\lambda = 2$. The latter is stable while the origin is unstable. The frequency oscillation for $\lambda = 2$ is the same as for the linearized system. If a conservative term of the 'Duffing' type is added to the Van der Pol equation to give

$$\ddot{x} + x + \epsilon\,[(x^2 - 1)\,\dot{x} + x^3] = 0 \tag{13.5-18}$$

the eqs. (13.5-14) are

$$\dot{\lambda} = -\frac{\lambda^3}{8} + \frac{\lambda}{2}; \quad \dot{\phi} = \frac{3}{8}\lambda^3 \tag{13.5-19}$$

The amplitude of stable oscillation is the same as for Van der Pol, but the frequency of these oscillations is now increased.

13.5.4 Resonant Excitation-Examples

For resonant excitation, $f = -F \cos \tilde{t}$, the contributions to g_1 and g_2 are

$$g_1 = -\frac{F}{2} \sin \phi; \quad g_2 = -\frac{F}{2} \cos \phi \tag{13.5-20}$$

Consider the forced linear oscillator with dissipation and detuning

$$\ddot{x} + x + \epsilon\,[\mu \dot{x} + \nu x - F \cos t] = 0 \tag{13.5-21}$$

Eqs. (13.5-14) are now

$$\begin{aligned} \dot{\lambda} &= -\frac{\mu\lambda}{2} - \frac{F}{2} \sin \phi \\ \lambda\dot{\phi} &= \frac{\lambda\nu}{2} - \frac{F}{2} \cos \phi \end{aligned} \tag{13.5-22}$$

and for $\mu > 0$ these possess the stable equilibrium solution

$$\lambda^2 = \frac{F^2}{\mu^2 + \nu^2}; \quad \tan \phi = -\frac{\mu}{\nu}$$

which is a system response synchronized with the excitation. It agrees with the exact asymptotic solution for the forced linear oscillator to order ϵ.

Suppose now that a general autonomous nonlinearity is added to the excited oscillator to give

$$\ddot{x} + x + \epsilon\,[f(x, \dot{x}) + \mu\dot{x} + \nu x - F\cos t] = 0 \tag{13.5-23}$$

The λ, ϕ equations become

$$\begin{aligned} \dot{\lambda} &= g_1(\lambda) - \frac{\mu\lambda}{2} - \frac{F}{2}\sin\phi \\ \dot{\phi} &= \frac{g_2(\lambda)}{\lambda} + \frac{\nu}{2} - \frac{F}{2\lambda}\cos\phi \end{aligned} \tag{13.5-24}$$

where $g_1(\lambda)$ and $g_2(\lambda)$ are the functions arising from $f(x, \dot{x})$ as in section 13.5.3. If these equations have equilibrium solutions they are described by

$$\begin{aligned} \frac{F}{2}\sin\phi &= g_1 - \frac{\mu\lambda}{2} \\ \frac{F}{2}\cos\phi &= g_2 + \frac{\nu\lambda}{2} \end{aligned} \tag{13.5-25}$$

$$\left(g_1 - \frac{\mu\lambda}{2}\right)^2 + \left(g_2 + \frac{\nu\lambda}{2}\right)^2 - \left(\frac{F}{2}\right)^2 = 0$$

The stability of any equilibrium solution is found by forming the Jacobian matrix, J, of the r.h.s. of eqs. (13.5-24) at the equilibrium point. The necessary and sufficient conditions for stability are

$$\begin{aligned} -Tr(J) &> 0 \\ |J| &> 0 \end{aligned} \tag{13.5-26}$$

For eqs. (13.5-24) these conditions may be shown to be equivalent to

$$\begin{aligned} &\frac{1}{\lambda}\frac{d}{d\lambda}(g_1\lambda) < \mu \\ &\frac{1}{\lambda}\frac{d}{d\lambda}\left[\left(g_1 - \frac{\mu\lambda}{2}\right)^2 + \left(g_2 + \frac{\nu\lambda}{2}\right)^2\right] > 0 \end{aligned} \tag{13.5-27}$$

A specific example which displays a special kind of behavior peculiarly character-

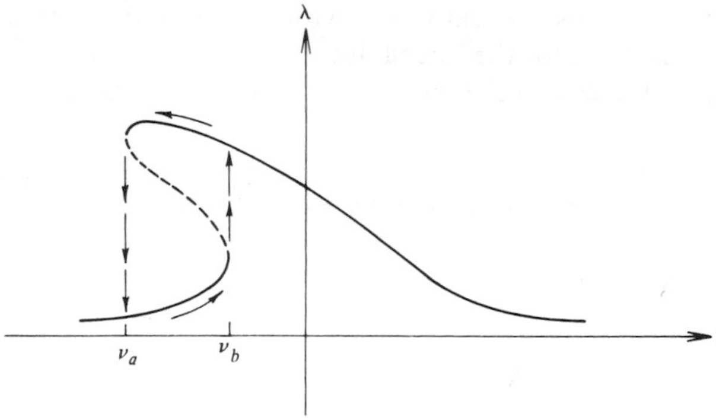

Fig. 13.5-1

istic of nonlinear oscillations is $f(x, \dot{x}) = x^3$. Then

$$g_1 = 0; \quad g_2 = \tfrac{3}{8}\lambda^3 \tag{13.5-28}$$

and the equilibria are given by

$$(\mu\lambda)^2 + (\tfrac{3}{4}\lambda^3 + \nu\lambda)^2 - F^2 = 0 \tag{13.5-29}$$

It is customary to represent these in a 'response diagram' as in Fig. 13.5-1, where F and μ are held fixed and the detuning, ν, is an adjustable parameter. For a range of $\nu < 0$ (and small dissipation) there will be three equilibrium solutions, with the middle amplitude unstable. Note that $\nu < 0$ means the linear oscillator frequency is below the excitation frequency. In this case there are two different asymptotic synchronized states, and the one to which the system goes depends on the initial conditions. A typical phase portrait with λ and ϕ as polar coordinates is sketched in Fig. 13.5-2. The initial conditions in the shaded region lead to the low level state. These are generally low amplitude conditions, but critically paired λ, ϕ can do so for large λ as well.

For two critical values of $\nu < 0$, one of the stable equilibrium points coalesces with the unstable equilibrium, leaving a unique asymptotic state. It is now possible to imagine a very slow time sequence in which the frequency of the excitation begins at a value below the oscillator frequency (so $\nu > 0$) and is raised to be above it. According to Fig. 13.5-1 the synchronized motion will follow the upper arc of the response curve until ν_a is reached, whereupon a transition to the lower arc will take place in the time scale $\hat{t}$. If the excitation frequency is very slowly reduced, a reverse transition occurs at ν_b. It might be noted that if the change of excitation frequency were to take place somewhat faster, in a time scale comparable to $\hat{t}$, a first

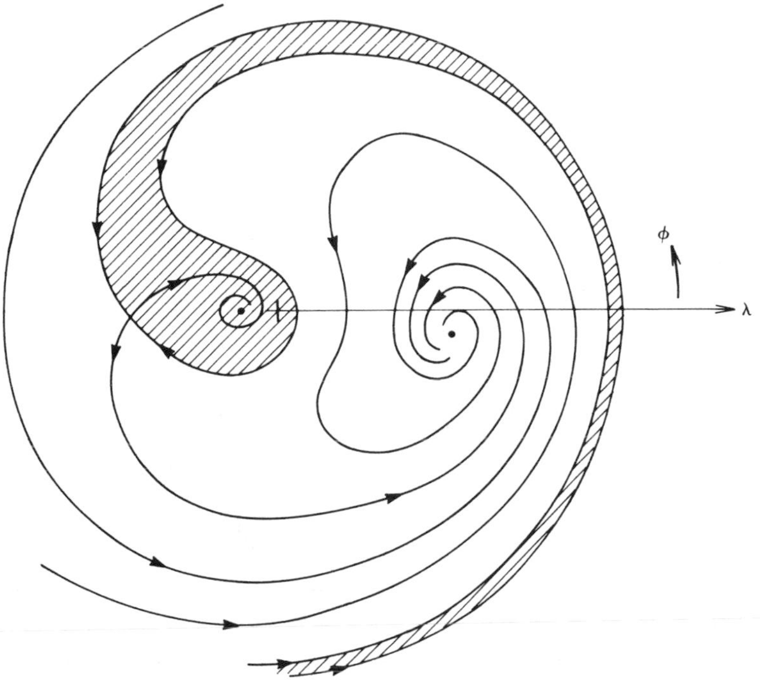

Fig. 13.5-2

approximation to the behavior is obtained by inserting the appropriate $\nu(\hat{t})$ in eqs. (13.5-24).

One other example which has received much attention in the literature is the externally excited Van der Pol oscillator. A wide variety of synchronized and non-synchronized behaviors can be found. For details, see Refs. 13-10 and 13-11.

13.5.5 Parametric Excitation-Examples

There are many types of parametric terms which make resonant contributions, for example $x \cos 2\tilde{t}, x^2 \cos \tilde{t}, x^2 \cos 3\tilde{t}, x^3 \cos 2\tilde{t}$ etc. Limitations of space permit us only to consider the linear, Mathieu, form $x \cos 2\tilde{t}$. This contributes

$$g_1 = \frac{\lambda}{4} \sin 2\phi; \quad g_2 = \frac{\lambda}{4} \cos 2\phi \tag{13.5-30}$$

If we examine first the Mathieu equation with detuning and dissipation,

$$\ddot{x} + x + \epsilon \left[x \cos 2t + \nu x + \mu \dot{x}\right] = 0 \tag{13.5-31}$$

we find

$$\dot{\lambda} = -\frac{\mu\lambda}{2} + \frac{\lambda}{4}\sin 2\phi$$
$$\lambda\dot{\phi} = \frac{\nu\lambda}{2} + \frac{\lambda}{4}\cos 2\phi \tag{13.5-32}$$

which are homogeneous in λ. The second equation may be integrated independently. For $\nu < \frac{1}{2}$ it has two stationary values ($\dot{\phi} = 0$)

$$\cos\phi = -2\nu \tag{13.5-33}$$

for which

$$\frac{\dot{\lambda}}{\lambda} = \frac{1}{2}\left[-\mu \pm \left(\frac{1}{4} - \nu^2\right)^{1/2}\right] \tag{13.5-34}$$

The solutions thus defined have a fixed phase relative to the parametric excitation and an exponential amplitude dependence. For $\mu = 0$ one of these is stable and the other unstable. They are the two normal solutions of Mathieu's equation in the unstable region near $\delta = 1$, Fig. 13.4-1. Returning to the notation of section 13.4.2, to terms of order ϵ, $s_j = \epsilon\dot{\lambda}/\lambda$ and $\delta - 1 = \epsilon\nu$, so for $\mu = 0$ eqs. (13.5-33) and (13.5-34) are

$$\cos 2\phi = 2(1 - \delta)/\epsilon$$
$$s_j = \pm\frac{1}{2}\left[\frac{\epsilon^2}{4} - (\delta - 1)^2\right]^{1/2} \tag{13.5-35}$$

Two important features should be noted. First, at any fixed ϵ (fixed level of parametric excitation) the detuning range for which unstable solutions exist is $\epsilon/2 \leqslant \delta - 1 \leqslant \epsilon/2$ with the maximum instability being at $\delta = 1$. Second, both the growing and decaying solutions have a fixed phase with respect to the excitation. This means that in physical applications where it is desired to use the positive growth effect of the parametric excitation to offset system dissipation, as in eq. (13.5-34), the phase of the excitation relative to the oscillations present in the system must be carefully regulated. It will be seen in section (13.6) that in systems of two or more oscillators, parametric excitation need not be synchronized to be effective.

In an autonomous oscillator, such as that of Van der Pol, the equilibrium amplitude can be seen to be the result of a balance between nonconservative effects as in eq. (13.5-17). For systems with parametric excitation, since the system growth rate depends on detuning, conservative nonlinear elements which alter oscillator frequency can indirectly limit solution growth. The problem may be addressed in general terms. We assume that the system has autonomous nonlinearities as well as dissipation, detuning and the Mathieu term

$$\ddot{x} + x + \epsilon\,[f(x,\dot{x}) + \nu x + \mu\dot{x} + \gamma x \cos 2t] = 0 \tag{13.5-36}$$

The eqs. (13.5-14) are

$$\begin{aligned} \dot{\lambda} &= g_1(\lambda) - \frac{\mu\lambda}{2} + \frac{\gamma\lambda}{4}\sin 2\phi \\ \dot{\phi} &= \frac{g_2(\lambda)}{\lambda} + \frac{\nu}{2} + \frac{\gamma}{4}\cos 2\phi \end{aligned} \tag{13.5-37}$$

where g_1 and g_2 are the contributions of $f(x, \dot{x})$. If equilibrium (synchronized) solutions exist (other than $\lambda = 0$) they are given by

$$\begin{aligned} -\frac{\gamma}{4}\sin 2\phi &= \frac{g_1}{\lambda} - \frac{\mu}{2} \\ -\frac{\gamma}{4}\cos 2\phi &= \frac{g_2}{\lambda} + \frac{\nu}{2} \end{aligned} \tag{13.5-38}$$

or

$$\left(\frac{g_1}{\lambda} - \frac{\mu}{2}\right)^2 + \left(\frac{g_2}{\lambda} + \frac{\nu}{2}\right)^2 - \frac{\gamma^2}{16} = 0 \tag{13.5-39}$$

Just as in section 13.5.4, the stability of an equilibrium point is found from the Jacobian matrix of the r.h.s. of eqs. (13.5-37) and analogous to eqs. (13.5-27) we have

$$\begin{aligned} &\frac{1}{\lambda}\frac{d}{d\lambda}(g_1\lambda) < \mu \\ &\lambda\frac{d}{d\lambda}\left[\left(\frac{g_1}{\lambda} - \frac{\mu}{2}\right)^2 + \left(\frac{g_2}{\lambda} + \frac{\nu}{2}\right)^2\right] > 0 \end{aligned} \tag{13.5-40}$$

Again an interesting example is afforded by $f(x, \dot{x}) = x^3$ for which

$$g_1 = 0; \quad g_2 = \tfrac{3}{8}\lambda^3 \tag{13.5-41}$$

Equilibrium points are given by

$$\frac{3}{4}\lambda^2 = -\nu \pm \left(\frac{\gamma^2}{4} - \mu^2\right)^{1/2} \tag{13.5-42}$$

and for a real solution

$$\gamma > 2\mu; \quad \nu < \left(\frac{\gamma^2}{4} - \mu^2\right)^{1/2} \tag{13.5-43}$$

The first of these means that for synchronized oscillations the Mathieu effect must be at least large enough to offset the linear dissipation. The second implies that there is a limit to the permitted positive detuning—but no limit to negative detuning. This is because the effect of the cubic nonlinearity (with the positive sign given to it) is to raise the oscillator frequency. If

$$\nu < -\left(\frac{\gamma^2}{4} - \mu^2\right)^{1/2} \tag{13.5-44}$$

there are two real solutions. Applying the stability criteria we find that the lower amplitude equilibrium is unstable while the larger amplitude equilibrium is stable. The phase portrait for eqs. (13.5-37) with ν satisfying (13.5-44) is sketched qualitatively in Fig. 13.5-3. It can be seen that all initial conditions of sufficiently small amplitude decay to $\lambda = 0$ (and a small fraction of the large amplitude initial conditions as well). Only if $|\lambda|$ exceeds a critical value will the system approach stable synchronized oscillations. A system with an initial condition threshold such as this is said to be 'hard excited.' The explanation of this behavior is that for small λ the nonlinearity does not raise the oscillator frequency enough to bring it within the range for which the Mathieu term is effective in producing solution growth.

One other related example is parametric synchronization of the Van der Pol oscil-

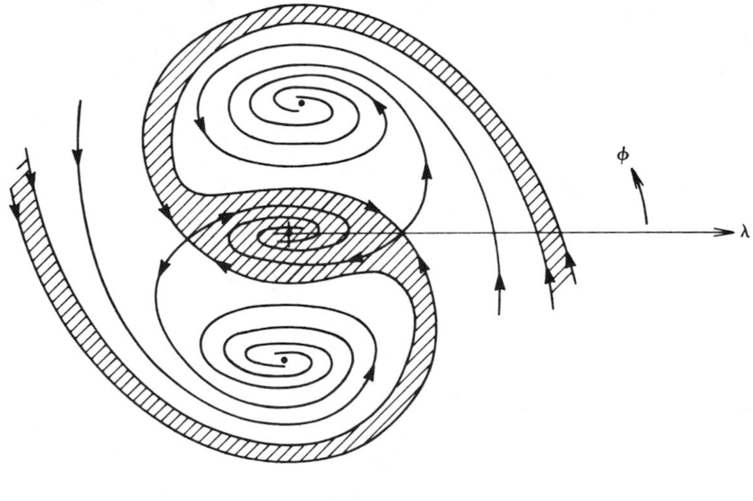

Fig. 13.5-3

lator. In eq. (13.5-36) we set $\mu = 0$ and take

$$f(x, \dot{x}) = (x^2 - 1)\,\dot{x} \tag{13.5-45}$$

so

$$g_1 = \frac{\lambda}{2} - \frac{\lambda^3}{8}; \quad g_2 = 0 \tag{13.5-46}$$

The origin $\lambda = 0$ is always an unstable equilibrium point while the other equilibria are

$$\frac{\lambda^2}{4} = 1 \pm \left(\frac{\gamma^2}{4} - \nu^2\right)^{1/2} \tag{13.5-47}$$

Real solutions exist only for $\gamma > 2\nu$ and they occur in pairs if $\gamma^2 - 4\nu^2 < 4$. In this case the larger λ is stable. If $\gamma^2 - 4\nu^2 > 4$ the single equilibrium is stable. Thus for $\gamma > 2\nu$ all initial conditions lead to synchronized oscillations. For $\gamma < 2\nu$ there is no stable equilibrium and the asymptotic limit set is a limit cycle surrounding the origin in the polar phase plane. This corresponds to a free running oscillator with some modulation due to the parameter excitation.

13.5.6 Explicit Integration in Certain Cases

If the autonomous nonlinearities $f(x, \dot{x})$ are such as to contribute only to $g_2(\lambda)$ [i.e., $g_1(\lambda) = 0$] and dissipation is absent ($\mu = 0$), then a general combination of autonomous nonlinearity, resonant excitation, and linear parametric excitation gives

$$\begin{aligned} \dot{\lambda} &= -\frac{F}{2}\sin\phi + \frac{\gamma\lambda}{4}\sin 2\phi \\ \dot{\phi} &= -\frac{F}{2\lambda}\cos\phi + \frac{\gamma}{4}\cos 2\phi + \frac{g_2(\lambda)}{\lambda} \end{aligned} \tag{13.5-48}$$

These equations possess a first integral

$$\bar{I}(\lambda, \phi) = -\frac{F\lambda}{2}\cos\phi + \frac{\gamma\lambda^2}{8}\cos 2\phi + \int_0^{\lambda} g_2 d\lambda = \text{const.} \tag{13.5-49}$$

whereby the solution of eqs. (13.5-48) may be reduced to quadrature. For each choice of $\bar{I}$ a closed contour is determined in the polar phase diagram and the system is 'conservative' in the λ, ϕ variables. We will postpone further discussion of this subject to section 13.6-1.

13.6 OSCILLATORS COUPLED BY WEAK NONLINEARITY

13.6.1 A Fundamental Invariant

In section 13.5 some indication was given of the wide variety of system behavior to be found in a single oscillator with nonlinearity. Extension of analysis to several oscillators results in overwhelming complexity unless some restrictions are imposed. The most useful single restriction is to make the system of oscillators conservative. A wide variety of new phenomena, peculiar to the multidimensional system can in this way be concisely described, and these provide a starting point for the selective addition of nonconservative elements.

We assume that the linearized system has been cast in normal coordinates, x_j, and that small detuning corrections have been included in each oscillator to make the frequencies of the system 'susceptible to coupling.' By this we mean that the frequencies of groups of oscillators satisfy relations of the form

$$\sum_j \beta_{kj}\omega_j = 0 \tag{13.6-1}$$

where the coefficients β_{kj} are integers. Now, if large integers are used, a collection of given frequencies will have no unique representation such as this. However, it will be shown that if a power series is used to describe the nonlinearity, no term in the series of order p can be at all effective in coupling unless

$$p \geqslant \sum_j |\beta|_{kj} \tag{13.6-2}$$

and even if this is satisifed, large numerical values of p are not very effective. The truly interesting cases are those in which the coefficients are low integers and only a few oscillators take part in any single coupling. In practice such considerations greatly reduce the options, and certain of the oscillators may not participate in any coupling.

The Lagrangian of the system is written

$$L = \tfrac{1}{2}\sum_j m_j \dot{x}_j^2 - \tfrac{1}{2}\sum_j m_j\omega_j^2 x_j^2 + \epsilon\left[l'(x_j, \dot{x}_j) - \tfrac{1}{2}\sum_j m_j\omega_j^2\nu_j x_j^2\right] \tag{13.6-3}$$

where the terms multiplied by the small parameter consist of the nonlinear function of all the x_j and $\dot{x}_j$ and the detunings used to bring the frequencies to the form (13.6-1). For convenience we will combine these into a single function

$$l(x_j, \dot{x}_j) = l'(x_j, \dot{x}_j) - \tfrac{1}{2}\sum_j m_j\omega_j^2\nu_j x_j^2 \tag{13.6-4}$$

which we will refer to as the 'perturbation Lagrangian.' The system differential equations are

$$\ddot{x}_j + \omega_j^2 x_j + \epsilon f_j(x_j, \dot{x}_j) = 0 \qquad (13.6\text{-}5)$$

where

$$f_j = \frac{1}{m_j}\left[\frac{d}{dt}\left(\frac{\partial l}{\partial \dot{x}_j}\right) - \frac{\partial l}{\partial x_j}\right] \qquad (13.6\text{-}6)$$

The individual f_j may clearly be functions of all the coordinates, and these are what couple the system of equations.

The expansion in ϵ proceeds as in section 13.5.2

$$\begin{aligned} x_j(\tilde{t}, \hat{t}) &= x_{j0}(\tilde{t}, \hat{t}) + \epsilon x_{j1}(\tilde{t}, \hat{t}) + \cdots \\ \tilde{t} &= t(1 + a_2\epsilon^2 + \cdots) \\ \hat{t} &= t(\epsilon + b_2\epsilon^2 + \cdots) \end{aligned} \qquad (13.6\text{-}7)$$

which lead to the equations

$$x_{j0,\tilde{t}\tilde{t}} + \omega_j^2 x_{j0} = 0 \qquad (13.6\text{-}8)$$

$$x_{j1,\tilde{t}\tilde{t}} + \omega_j^2 x_{j1} = -f_j(x_{j0}, x_{j0,\tilde{t}}) - 2x_{j0,\tilde{t}\hat{t}} \qquad (13.6\text{-}9)$$

etc.

The solution to eqs. (13.6-8) is taken in the form

$$x_{j0}(\tilde{t}, \hat{t}) = \lambda_j(\hat{t}) \cos[\omega_j\tilde{t} + \phi_j(\hat{t})] \qquad (13.6\text{-}10)$$

for which

$$x_{j0,\tilde{t}} = -\omega_j\lambda_j \sin(\omega_j\tilde{t} + \phi_j) \qquad (13.6\text{-}11)$$

$$x_{j0,\tilde{t}\hat{t}} = -\omega_j\dot{\lambda}_j \sin(\omega_j\tilde{t} + \phi_j) - \omega_j\lambda_j\dot{\phi}_j \cos(\omega_j\tilde{t} + \phi_j) \qquad (13.6\text{-}12)$$

For simplicity we will introduce the notation

$$\begin{aligned} f_{j0} &= f_j(x_{j0}, x_{j0,\tilde{t}}) \\ &= f_j[\lambda_j \cos(\omega_j\tilde{t} + \phi_j), -\lambda_j\omega_j \sin(\omega_j\tilde{t} + \phi_j)] \end{aligned} \qquad (13.6\text{-}13)$$

so that the conditions for suppression of secular terms in (13.6-9) may be written

$$\dot{\lambda}_j = \frac{1}{2\pi}\int_0^{2\pi/\omega_j} f_{j0} \sin(\omega_j \tilde{t} + \phi_j)\, d\tilde{t}$$

$$\lambda_j \dot{\phi}_j = \frac{1}{2\pi}\int_0^{2\pi/\omega_j} f_{j0} \cos(\omega_j \tilde{t} + \phi_j)\, d\tilde{t} \qquad (13.6\text{-}14)$$

Unlike eqs. (13.5-14) which they formally resemble, the integrand functions f_{j0} are not simply periodic functions with period $2\pi/\omega_j$. Here they may have a wide variety of frequencies. The integrals in (13.6-14) will consequently have a set of terms with dependence on fast time $\tilde{t}$ as well as slow time contributions. We do not mean to include these in $\dot{\lambda}$ and $\dot{\phi}$ so we revise the notation to

$$\dot{\lambda}_j = \omega_j^{-1}\, \overline{f_{j0} \sin(\omega_j \tilde{t} + \phi_j)}$$

$$\lambda_j \dot{\phi}_j = \omega_j^{-1}\, \overline{f_{j0} \cos(\omega_j \tilde{t} + \phi_j)} \qquad (13.6\text{-}15)$$

where the superscript bar means identification of that component of the indicated function which is independent of $\tilde{t}$. If it happens that one of the frequency component arising in f_{j0} actually has a very low frequency (compared to ω_j) this is an indication that the original choice of coupling relations (13.6-1) overlooked something significant. A new choice of detuning parameters must be made so that the very low frequency is brought to zero frequency and thereby contributes to $\dot{\lambda}_j$ and $\dot{\phi}_j$.

If any oscillator is completely uncoupled from the others it will be evidenced by $\dot{\lambda}_j = 0 = \dot{\phi}_j$. In some cases this can be seen in advance by the frequency ω_j being absent from any of the coupling relations. It can also occur by the absence of the proper nonlinearity, even though a group of oscillators has a suitable frequency relationship. For any uncoupled oscillator its solution is given independently of the others and will be ignored here. Thus, any oscillator which is retained in the set under study will be assumed to have at least one coupling mechanism.

We now revert to the definition of f_j, eq. (13.6-6). The leading term in an expansion of the r.h.s. can be written

$$f_{j0} = \frac{1}{m_j}\left[\frac{\partial}{\partial \tilde{t}}\left(\left.\frac{\partial l}{\partial \dot{x}_j}\right|_0\right) - \left.\frac{\partial l}{\partial x_j}\right|_0\right] \qquad (13.6\text{-}16)$$

where the notation $|_0$ means that the function in question is evaluated with $x_j = x_{j0}$ and $\dot{x}_j = x_{j0,\tilde{t}}$. This can be used in eq. (13.6-15) along with (13.6-10) and (13.6-11) to give

$$\dot{\lambda}_j = \frac{-1}{m_j \lambda_j \omega_j^2}\, \overline{\left[\frac{\partial}{\partial \tilde{t}}\left(\left.\frac{\partial l}{\partial \dot{x}_j}\right|_0\right) - \left.\frac{\partial l}{\partial x_j}\right|_0\right] x_{j0,\tilde{t}}}$$

$$\lambda_j \dot{\phi}_j = \frac{1}{m_j \lambda_j \omega_j}\, \overline{\left[\frac{\partial}{\partial \tilde{t}}\left(\left.\frac{\partial l}{\partial \dot{x}_j}\right|_0\right) - \left.\frac{\partial l}{\partial x_j}\right|_0\right] x_{j0}} \qquad (13.6\text{-}16)$$

The operator superscript bar is identical to integration, as far as the selected component of f_{j0} is concerned, so a transformation equivalent to a parts integration may be used to give

$$\dot{\lambda}_j = -\frac{1}{m_j \lambda_j \omega_j^2}\left[-\overline{\left.\frac{\partial l}{\partial \dot{x}_j}\right|_0 x_{j0,\tilde{t}\tilde{t}}} - \overline{\left.\frac{\partial l}{\partial x_j}\right|_0 x_{j0,\tilde{t}}}\right]$$
$$\lambda_j \dot{\phi}_j = \frac{1}{m_j \lambda_j \omega_j}\left[-\overline{\left.\frac{\partial l}{\partial \dot{x}_j}\right|_0 x_{j0,\tilde{t}}} - \overline{\left.\frac{\partial l}{\partial x_j}\right|_0 x_{j0}}\right] \tag{13.6-17}$$

We next consider a function

$$\bar{l}(\lambda_j, \phi_j) = \overline{l(x_{j0}, x_{j0,\tilde{t}})} \tag{13.6-18}$$

That is, we go to the original perturbation Lagrangian, substitute the zeroth order solutions for x_j and $\dot{x}_j$, and then identify that component which is independent of $\tilde{t}$. The 'averaged perturbation Lagrangian,' as we will call it, is an explicit function of the λ_j and ϕ_j. We will proceed in a formal way to evaluate its partial derivatives in these arguments

$$\frac{\partial \bar{l}}{\partial \lambda_j} = \overline{\left.\frac{\partial l}{\partial x_j}\right|_0 \frac{\partial x_{j0}}{\partial \lambda_j}} + \overline{\left.\frac{\partial l}{\partial \dot{x}_j}\right|_0 \frac{\partial x_{j0,\tilde{t}}}{\partial \lambda_j}}$$
$$\frac{\partial \bar{l}}{\partial \phi_j} = \overline{\left.\frac{\partial l}{\partial x_j}\right|_0 \frac{\partial x_{j0}}{\partial \phi_j}} + \overline{\left.\frac{\partial l}{\partial \dot{x}_j}\right|_0 \frac{\partial x_{j0,\tilde{t}}}{\partial \phi_j}} \tag{13.6-19}$$

Through the use of eqs. (13.6-10) and (13.6-11) it is easy to relate eqs. (13.6-19) to eq. (13.6-17) thus

$$\frac{\partial \bar{l}}{\partial \lambda_j} = -m_j \omega_j \lambda_j \dot{\phi}_j$$
$$\frac{\partial \bar{l}}{\partial \phi_j} = m_j \omega_j \lambda_j \dot{\lambda}_j \tag{13.6-20}$$

These are powerful results. First we see that $\bar{l}$ is an invariant of the entire system

$$\frac{d\bar{l}}{d\hat{t}} = \sum_j \left(\frac{\partial \bar{l}}{\partial \lambda_j}\dot{\lambda}_j + \frac{\partial \bar{l}}{\partial \phi_j}\dot{\phi}_j\right) = 0 \tag{13.6-21}$$

Further, if we introduce a new set of amplitude variables

$$R_j(\hat{t}) = \frac{m_j \omega_j \lambda_j^2(\hat{t})}{2} \tag{13.6-22}$$

the eqs. (13.6-20) are in canonical Hamiltonian form

$$\begin{aligned} \frac{\partial \bar{l}(R_j, \phi_j)}{\partial R_j} &= -\dot{\phi}_j \\ \frac{\partial \bar{l}(R_j, \phi_j)}{\partial \phi_j} &= \dot{R}_j \end{aligned} \qquad (13.6\text{-}23)$$

To summarize, for the entire coupled system, the averaged perturbation Lagrangian is the Hamiltonian of the slow-time equations for R_j and ϕ_j. It is the only nonlinear function which needs to be evaluated.

There is an important extension of this result which is extremely easy to make. Throughout the previous derivation all of the oscillators which participated in the exchange process were autonomous. However, when the invariance of $\bar{l}$ was demonstrated, eq. (13.6-21), the terms cancelled in pairs

$$\frac{\partial \bar{l}}{\partial \lambda_j} \dot{\lambda}_j + \frac{\partial \bar{l}}{\partial \phi_j} \dot{\phi}_j = 0 \qquad (13.6\text{-}24)$$

for each oscillator. Thus, periodic external excitation can be added to the system by the simple expedient of including fictitious oscillators of appropriate frequency which are coupled to the real oscillators by appropriate terms in the perturbation Lagrangian. We then imagine these fictitious oscillators to have their response amplitudes and phases established by whatever forces may be required. The fictitious oscillators will then not be included in the invariant, but the $\bar{l}$ formed for all of the real oscillators will still be an invariant—and also the Hamiltonian generator of the slow-time equations as in (13.6-23).

Three elementary examples will show the procedure. If the perturbation Lagrangian has added to it a term $x_1 \cos \omega_0 t$, the differential equation for x_1 will have the added term $\cos \omega_0 t$ (simple excitation). If the Lagrangian term is $x_1^2 \cos \omega_0 t$, the added term in the x_1 equation is $2x_1 \cos \omega_0 t$ (Mathieu excitation). Finally a term in the Lagrangian of the form $x_1 x_2 \cos \omega_0 t$ contributes $x_2 \cos \omega_0 t$ to the x_1 equation and $x_1 \cos \omega_0 t$ to the x_2 equation, which gives a parametric coupling between the x_1 and x_2 equations. The first two of these examples are represented in the invariant for the single oscillator displayed in section 13.5.6. The third example will be discussed below.

13.6.2 Further Invariants

Before dealing with generalities, we will examine one of the simplest systems—three oscillators with the frequency relation

$$\omega_1 + \omega_2 - \omega_3 = 0; \quad 0 < \omega_1 < \omega_2 < \omega_3 \qquad (13.6\text{-}25)$$

The perturbation Lagrangian will be taken to be

$$l = x_1 x_2 x_3 + \nu x_1^2 \tag{13.6-26}$$

and for simplicity all of the m_i will be set equal to unity. Only one detuning term is required in l to bring the three oscillators to the precise tuning condition. With the solutions taken in the form (13.6-10), we have

$$\bar{l} = \frac{\lambda_1 \lambda_2 \lambda_3}{4} \cos(\phi_1 + \phi_2 - \phi_3) + \frac{\nu \lambda_1^2}{2} \tag{13.6-27}$$

and if we transform to R-variables as in (13.6-22)

$$\bar{l} = \left[\frac{R_1 R_2 R_3}{2\omega_1 \omega_2 \omega_3}\right]^{1/2} \cos(\phi_1 + \phi_2 - \phi_3) + \frac{\nu R_1}{\omega_1} \tag{13.6-28}$$

We next introduce the definitions.

$$\begin{aligned} g(R_1, R_2, R_3) &\equiv \left[\frac{R_1 R_2 R_3}{2\omega_1 \omega_2 \omega_3}\right]^{1/2} \\ \theta &\equiv \phi_1 + \phi_2 - \phi_3 \end{aligned} \tag{13.6-29}$$

so

$$\bar{l} = g \cos\theta + \frac{\nu R_1}{\omega_1} \tag{13.6-30}$$

According to (13.6-23) we obtain the differential equations

$$\begin{aligned} \dot{R}_1 &= -g \sin\theta \frac{d\theta}{d\phi_1} = -g \sin\theta \\ \dot{R}_2 &= -g \sin\theta \frac{d\theta}{d\phi_2} = -g \sin\theta \\ R_3 &= -g \sin\theta \frac{d\theta}{d\phi_3} = g \sin\theta \\ \dot{\theta} &= \dot{\phi}_1 + \dot{\phi}_2 - \dot{\phi}_3 = \frac{-g}{2}\left[\frac{1}{R_1} + \frac{1}{R_2} - \frac{1}{R_3}\right] \cos\theta - \frac{\nu}{\omega_1} \end{aligned} \tag{13.6-31}$$

From the first three of these we can immediately write down two invariants

$$\begin{aligned} k_1 &= R_1 - R_2 = \text{const.} \\ k_2 &= R_1 + R_3 = \text{const.} \end{aligned} \tag{13.6-32}$$

Combinations of these are also invariant, and if we form

$$\omega_3 k_2 - \omega_2 k_1 = \frac{\omega_1^2 \lambda_1^2}{2} + \frac{\omega_2^2 \lambda_2^2}{2} + \frac{\omega_3^2 \lambda_3^2}{2} = U \tag{13.6-33}$$

we obtain an invariant which can be interpreted as the total energy of the oscillators (to order ϵ).

The two invariants (13.6-32) together with $\overline{l}$ constitute a complete set for eqs. (13.6-31) so their solution is reduced to quadrature. Similar results can be obtained for n oscillators coupled by a single nonlinear mechanism [i.e., with a single relevant frequency relation, eq. (13.6-1)]. For this there will be only one relevant (composite) phase variable, there will be $n + 1$ equations like (13.6-31), and there will be $n - 1$ invariants of the form (13.6-32) which, with $\overline{l}$, constitute a complete set.

Now let us suppose there are two nonlinear mechanisms and two corresponding frequency relations. If the two mechanisms involve independent sets of oscillators, the perturbation Lagrangian can be partitioned and the behavior of each set is reduced to quadrature as before. However, if one or more of the oscillators take part in both mechanisms then all are coupled together. If the total number of oscillators is n, the number of equations like (13.6-31) will be $n + 2$ (since there are two phase variables). The number of constraints like eq. (13.6-32) are now $n - 2$, since the equations for $\dot{R}_i$ involve the two phases. These together with $\overline{l}$ enable the system of differential equations to be reduced to third order. Whether any other invariants can be found is an open question. Extending this to the case of m fully-interacting mechanisms, our present knowledge of invariants permits a reduction of the slow-time differential equations to order $2m - 1$, and no further. (Note that for n oscillators, $m \leqslant n - 1$.)

13.6.3 Synchronous and Asynchronous Nonlinearities

For each frequency relation, eq. (13.6-1), there is a corresponding phase variable

$$\theta_k = \sum_j \beta_{kj} \phi_j \tag{13.6-34}$$

and as we saw by example above, a nonlinear term in l of the form

$$l = \prod_j x_j^{\beta_{kj}} \tag{13.6-35}$$

will produce an $\overline{l}$ which is a function of θ_k. It can be seen from eqs. (13.6-23) that only those parts of l which involve phase contribute to $\dot{R}_j$; that is, effect energy exchange between oscillators. Now if one of the oscillators concerned in a particular

θ_k has its oscillation controlled by imposed forces and all of the other oscillators in that set have a very small amount of damping, the entire set may be brought to a synchronized state. Thus nonlinearities which make contributions to $\bar{l}$ involving θ_k may be described as synchronous. Nonlinearities which make phase-independent contributions are asynchronous. The role of a particular nonlinearity can depend on the context. For example, $x_1^2 x_2^2$ in the Lagrangian will always give an asynchronous contribution to $\bar{l}$ in the form $\lambda_1^2 \lambda_2^2/4$, but if it should happen that $\omega_1 = \omega_2$ there will be a synchronous contribution as well.

In order for any nonlinear term of the form

$$l = \prod_j x_j^{\gamma_j} \tag{13.6-36}$$

To make a synchronous contribution for some θ_k, it is necessary that each component ϕ_j appear to the correct multiple. This implies

$$\gamma_j \geqslant |\beta|_{kj} \tag{13.6-37}$$

If we denote by p the total degree of the term, then

$$p = \sum_j \gamma_j \geqslant \sum_j |\beta|_{kj}$$

which was cited earlier as eq. (13.6-2). It should be noted that the degree of nonlinearity appearing in the differential equations is $p - 1$. (As an example to illustrate that an inequality is correct in eq. (13.6-7), for the case $\omega_1 + \omega_2 - \omega_3 = 0$ the nonlinearity $x_1^3 x_2 x_3$ is synchronous.)

There are other important results which space limitations prevent us from examining in detail. One of these is that the numerical size of the coupling decreases with increasing p. This means that energy exchanges are weaker if many oscillators are involved or if the frequency relations are complicated. For any given θ_k, the most effective nonlinearity will be the minimal form (13.6-35). Another is that the conclusions regarding minimum p for synchronization of oscillations and effectiveness of nonlinearities applies to nonconservative nonlinearities as well. A more thorough discussion in a special application can be found in Ref. 13-12. It must be emphasized that all of these results are confined to an expansion truncated at the first power of ϵ. Expansion to higher order permits many more, but weaker, interactions (Ref. 13-13).

13.6.4 Coupled Oscillators-Nonconservative Examples

Several interesting examples can be found within the framework of the collection of three oscillators introduced in section 13.6.2. Their free (conservative) oscillations were completely specified by the three invariants (13.6-27) and (13.6-32).

We now consider the effect of adding small dissipation to each oscillator and periodic excitation to the one with highest frequency, ω_3. The differential equations are taken as

$$\begin{aligned}
&\ddot{x}_1 + \omega_1^2 x_1 + \epsilon(x_2 x_3 + \mu_1 \dot{x}_1 - 2\nu_1 x_1) = 0 \\
&\ddot{x}_2 + \omega_2^2 x_2 + \epsilon(x_1 x_3 + \mu_2 \dot{x}_2) = 0 \\
&\ddot{x}_3 + \omega_3^2 x_3 + \epsilon(x_1 x_2 + \mu_3 \dot{x}_3 - 2\nu_3 x_3 - (2\omega_3)^{1/2} F \cos \omega_3 t) = 0
\end{aligned} \tag{13.6-38}$$

where the μ_i are the damping coefficients. Two detuning coefficients, ν_1 and ν_3, are now required. The first is used to adjust the sum $\omega_1 + \omega_2$ to be equal to ω_3, while the second is used to account for the natural frequency of the third oscillator being different from ω_3. The full set of slow-time equations is

$$\begin{aligned}
\dot{R}_1 &= -g \sin\theta - \mu_1 R_1 \\
\dot{R}_2 &= -g \sin\theta - \mu_2 R_2 \\
\dot{R}_3 &= g \sin\theta - \mu_3 R_3 - R_3^{1/2} F \sin\phi_3 \\
\dot{\phi}_1 &= \frac{-g\cos\theta}{2R_1} - \frac{\nu_1}{\omega_1} \\
\dot{\phi}_2 &= \frac{-g\cos\theta}{2R_2} \\
\dot{\phi}_3 &= \frac{-g\cos\theta}{2R_3} - \frac{\nu_3}{\omega_3} - \frac{F}{2R_3^{1/2}}\cos\phi_3
\end{aligned} \tag{13.6-39}$$

where g and θ are as defined in eq. (13.6-29). We may reduce this set to five by forming one equation for $\dot{\theta}$ and dropping the two equations for $\dot{\phi}_1$ and $\dot{\phi}_2$.

Of special interest are the equilibrium solutions, defined by $\dot{R}_i = 0 = \dot{\phi}_3 = \dot{\theta}$, which denote synchronized forced oscillations. One such is given by

$$R_1 = 0 = R_2; \quad R_3 = F^2 \Big/ \left(\mu_3^2 + \left(\frac{2\nu_3}{\omega_3}\right)^2\right) \tag{13.6-40}$$

which is simply the isolated response of the excited oscillator. More interesting are the cases where $R_1, R_2 \neq 0$. From the first two eqs. (13.6-39) we have

$$-g \sin\theta = \mu_1 R_1 = \mu_2 R_2 \tag{13.6-41}$$

while from the third and fourth,

$$\dot{\phi}_1 + \dot{\phi}_2 = 0 = -\frac{g\cos\theta}{2}\left(\frac{1}{R_1} + \frac{1}{R_2}\right) - \frac{\nu_1}{\omega_1} \tag{13.6-42}$$

Substitution of (13.6-41) into (13.6-42) yields

$$\tan\theta = \frac{(\mu_1 + \mu_2)}{2}\frac{\omega_1}{\nu_1} \tag{13.6-43}$$

while from (13.6-41) and (13.6-29)

$$\mu_1\mu_2 R_1 R_2 = g^2 \sin^2\theta = \frac{R_1 R_2 R_3}{2\omega_1\omega_2\omega_3}\sin^2\theta \tag{13.6-44}$$

This permits a specification of R_3.

$$R_3 = 2\mu_1\mu_2\omega_1\omega_2\omega_3 \cos^2\theta = 2\mu_1\mu_2\omega_1\omega_2\omega_3\left[1 - \frac{2\nu_1}{\omega_1(\mu_1 + \mu_2)}\right] \tag{13.6-45}$$

Equation (13.6-45) is an especially curious result, because it shows R_3 to have a value specified by the system parameters, independent of the strength of the excitation, F. That is, the magnitude of R_3 is set by the requirements of the two indirectly-excited oscillators [only the first, second, fourth, and fifth equations of (13.6-39) have been used].

From the third and sixth of (13.6-39), along with (13.6-41) and (13.6-42) we find

$$F^2 R_3 = (\mu_2 R_2 + \mu_3 R_3)^2 + \left(-R_2 \frac{\mu_2}{(\mu_1 + \mu_2)}\frac{2\nu_1}{\omega_1} + R_3\frac{2\nu_3}{\omega_3}\right)^2 \tag{13.6-46}$$

which relates F to R_2 [and through eq. (13.6-41) to R_1], since R_3 is already set. Only real and positive R_2 correspond to physical solutions. A critical value of F, F_c, is that for which $R_2 = 0$, given by

$$F_c^2 R_3 = (\mu_3 R_3)^2 + \left(\frac{2\nu_3 R_3}{\omega_3}\right)^2 \tag{13.6-47}$$

which is, of course identical in form to eq. (13.6-40). The distinction is that here R_3 takes on a particular value, from eq. (13.6-45). For all positive μ_2, μ_3 there is no real solution for R_2 when $F = 0$ (other than $R_2 = 0$ cited earlier). If the composite parameter

$$\sigma = \mu_1\mu_3 - \frac{\mu_2}{(\mu_1 + \mu_2)}\frac{4\nu_1\nu_3}{\omega_1\omega_3} \tag{13.6-48}$$

is positive, as F^2 is increased one real positive R_2 is found for $F^2 > F_c^2$ corresponding to stable synchronized oscillations. Thus F_c may be regarded as a threshold level for sustaining coupled oscillations. For $\sigma < 0$ (very lightly damped, or strongly detuned systems) two real positive R_2 are found when

$$F_c^2 > F^2 > 4R_3\left(\frac{\nu_3}{\omega_3} + \frac{\mu_2\mu_3}{(\mu_1+\mu_2)}\frac{\nu_1}{\omega_1}\right)^2\bigg/\left[1 - \left(\frac{2\nu}{(\mu_1+\mu_2)\omega_1}\right)^2\right] \qquad (13.6\text{-}49)$$

Only the smaller of these is stable, and it goes to zero as $F^2 \to F_c^2$. Thus, for $F^2 > F_c^2$ and $\sigma < 0$, no synchronized oscillations will be found.

It can be shown that if the periodic excitation had been applied to one of lower frequency oscillators (close to the appropriate resonance frequency) it is impossible to sustain oscillations in the other two.

A quite different example is obtained when the oscillator with frequency ω_3 is assumed to be forced in such a way that its response is constrained to be

$$x_3 = \lambda_3 \cos \omega_3 t \qquad (13.6\text{-}50)$$

with λ_3 constant. Under these conditions the differential equations (13.6-38) reduce to

$$\begin{aligned} \ddot{x}_1 + \omega_1^2 x_1 + \epsilon(x_2\lambda_3 \cos \omega_3 t + \mu_1\dot{x}_1 - 2\nu_1 x_1) &= 0 \\ \ddot{x}_2 + \omega_2^2 x_2 + \epsilon(x_1\lambda_3 \cos \omega_3 t + \mu_2\dot{x}_2) &= 0 \end{aligned} \qquad (13.6\text{-}51)$$

which are a linear Mathieu-Hill system. The 'slow' time equations are

$$\begin{aligned} \dot{R}_1 &= -g \sin\theta - \mu_1 R_1 \\ \dot{R}_2 &= -g \sin\theta - \mu_2 R_2 \\ \dot{\theta} = \dot{\phi}_1 + \dot{\phi}_2 &= -\frac{g\cos\theta}{2}\left(\frac{1}{R_1} + \frac{1}{R_2}\right) - \frac{\nu_1}{\omega_1} \end{aligned} \qquad (13.6\text{-}52)$$

with g as defined in (13.6-29). These are most concisely interpreted by a change to the variables

$$U = R_1 + R_2; \quad \Psi = \tan^{-1}(R_2/R_1)^{1/2} \qquad (13.6\text{-}53)$$

and a redefinition of parameters to

$$\begin{aligned} &a = (R_3/2\omega_1\omega_2\omega_3)^{1/2} \\ &K = \nu_1/2\omega_1 a; \quad \rho = (\mu_1 - \mu_2)/4a; \quad \sigma = (\mu_1 + \mu_3)/4a \end{aligned} \qquad (13.6\text{-}54)$$

which gives

$$\dot{\Psi} = a(-\sin\theta \cos 2\Psi + \rho \sin 2\Psi)$$
$$\dot{\theta} = -2a(\cos\theta \csc 2\Psi + K) \tag{13.6-55}$$
$$\dot{U} = -2Ua(\sin\theta \sin 2\Psi + \sigma + \rho \cos 2\Psi)$$

In this form the first two equations may be solved separately. Since both R_1 and R_2 are positive, we take $0 \leqslant \Psi \leqslant \pi/2$.

Stationary solutions given by $\dot{\Psi} = 0 = \dot{\theta}$ correspond to 'normal' solutions of the M.-H. system since they represent fixed relative phases and amplitudes between the x_1 and x_2 oscillations. The growth or decay exponent of these oscillations is given by $\dot{U}/2U$. These normal solutions have

$$\sin\theta = \rho \tan 2\Psi; \quad \cos\theta = -K \sin 2\Psi \tag{13.6-56}$$

which lead to

$$\sin^4\theta + \sin^2\theta\,(\rho^2 + K^2 - 1) - \rho^2 = 0 \tag{13.6-57}$$

This equation has one solution $0 \leqslant \sin^2\theta \leqslant 1$ for all $\rho \neq 0$. Of the four possible values of θ, only two satisfy the requirements $0 \leqslant \Psi \leqslant \pi/2$. The growth exponents are

$$\dot{U}/2U = -a\sigma - a\sin\theta \csc 2\Psi \tag{13.6-58}$$

The first term on the right represents decay due to the average dissipation in the two oscillators. The second term takes on opposite sign for the two permitted θ, and is the effect of the parametric excitation contributing to growth in one normal solution and decay in the other.

In this way we have found what appear to be two normal solutions, but which are in fact four (a complete set). The reason is that each solution here has no fixed phase relation to the parametric excitation (only the sum $\theta = \phi_1 + \phi_2$ is specified) and therefore may be arbitrarily shifted in time. In the case of the single Mathieu equation each normal solution is phase-linked to the excitation. It is this phase independence which makes the two-oscillator M.-H. system so useful in physical applications.

The special case $\rho = 0$ which corresponds to matched dissipation in the two oscillators is curious since the first two of eqs. (13.6-55) acquire an invariant

$$c = \tan 2\Psi \cos\theta + K \sec 2\Psi \tag{13.6-59}$$

For $K < 1$ all phase plane trajectories terminate at the singularities, which are nodes. For $K > 1$ the singularities are centers and the trajectories are either closed about them or extend to $\pm\infty$ in θ. Also for $K > 1$ the only solution to eq. (13.6-57) is $\sin^2 \theta = 0$ and, from eq. (13.6-58), we see that all normal solutions decay just as they would in the absence of parametric excitation. Even when the dissipation in the oscillators in slightly mismatched $(0 < \rho << 1)$, it follows from eq. (13.6-57) that $\sin \theta$ will be small for $K > 1$, so $K = 1$ represents a limit to significant effects of the parametric excitation on the growth exponent.

Finally, we remark that if the parametric excitation had been applied through one of the low frequency oscillators (e.g., the motion x_2 is constrained while x_3 is left free), this mechanism can make no net input to the strength of oscillations in the remaining two oscillators. That is, parametric coupling at a frequency which is at or close to the *difference* between the resonances of two oscillators cannot sustain oscillations in them. This analysis is left to the reader.

13.6.5 Validity of Solutions

The basic question in these asymptotic expansions is how well does the truncated solution represent the true solution. If the question is posed quantitatively by forming a function which bounds the difference between the two solutions, the results thus far obtained suggest that the time interval for validity is $O(\epsilon^{-m})$ which the expansion is carried to $O(\epsilon^m)$, Ref. 13-8. This result is very plausible, since both the method of averaging and the multi-time method are designed to accommodate slow but persistent changes in amplitude and phase of oscillations and, in general, refinements to these 'drift rates' will be found at each stage in the expansion.

A different view of validity is concerned with the qualitative behavior of solutions. Such questions as whether oscillations grow or decay, whether energy exchanges between oscillators can occur, or whether synchronization with excitation is possible are all related to the nature of the limit sets of the system. It may then be asked whether the solutions to the truncated expansion represent completely the limit sets of the true solutions. For certain simple cases (e.g., the free-running Van der Pol oscillator) results are well-established, but very little is known in general.

A useful concept in dealing with qualitative behavior is the 'criticality' (inverse to 'structural stability,' Ref. 13-16) of the system. A conservative system is critical in that changes in certain parameters, however small, alter the limit sets of the system. The single conservative oscillator provides a classic example: The addition of Van der Pol terms of arbitrary small size produces a stable limit cycle. Generally the addition of nonconservative elements makes a system less critical, although the last example in section 13.6.4 shows that a conservative subsystem may be imbedded in an otherwise nonconservative system when dissipation coefficients are matched. Many examples in section 13.5 and 13.6 have shown the effect of nonconservative elements in preventing the existence of certain limit behaviors to be enhanced by deviations from precise frequency relations (i.e., detuning).

In order for a new limit behavior to arise at later stages in the expansion process it is necessary that the system be sufficiently close to some critical state. By comparing the strength of the nonlinear mechanism acting to sustain a new limit behavior with the nonconservative and detuning effects which prevent it from arising, it may be possible to eliminate some, and perhaps all, alternatives to the limit sets already described at the existing stage in the expansion.

The study of strongly nonlinear oscillations presents a very difficult challenge. Where weakly nonlinear oscillations have modulations which are psuedo-continuous in the slow-time variable, computer studies (Refs. 13-14 and 13-15) show that strongly nonlinear systems may exhibit great changes from cycle to cycle. Some of these can be explained as subharmonics which are themselves slowly modulated, while others defy simple explanations and resemble random phenomena. This suggests that an entirely different line of mathematical inquiry will be required.

13.7 REFERENCES AND BIBLIOGRAPHY

13.7.1 References

13-1 Goldstein, H., *Classical Mechanics*, Addison-Wesley, Reading, Mass., 1953.

13-2 Crandall, S. H., et al., *Dynamics of Mechanical and Electromechanical Systems*, McGraw-Hill, New York, 1968.

13-3 Poincaré, H., *Les Methodes Nouvelles de la Mécanique Céleste*, 3 vols., Gauthiers-Villars, Paris, 1892–99. Also in translation, *New Methods of Celestial Mechanics*, 3 vols., NASA TTF-450, 451, 452, CFSTI, Springfield, Virginia 22151.

13-4 Whittaker, E. I., and Watson, G. N., *Modern Analysis*, Chapter XIX, Cambridge University Press, Cambridge, 1915.

13-5 Krylov, N. and Bogoliubov, N., "Introduction to Nonlinear Mechanics," Translated from Russian by S. Lefschetz, *Annals of Mathematical Studies*, No. 11, Princeton, 1947.

13-6 Cole, J. D., and Kevorkian, J., "Uniformly Valid Asymptotic Approximation for Certain Non-Linear Differential Equations," *Proceedings of the International Symposium on Non-Linear Differential Equations and Non-Linear Mechanics*, Academic Press, New York, 1963. pp. 113–120.

13-7 Morrison, J. A., "Comparison of the Modified Method of Averaging and the Two-Variable Expansion Procedure," *SIAM Review*, **8** (1), Jan. 1966.

13-8 Mahoney, J. J., "On the Validity of Averaging Methods for Systems with Periodic Solutions," *Report No. 1*, 1970 Fluid Mechanics Research Institute, University of Essex, England.

13-9 Andronov, A. A.; Witt, A. A.; and Khaikin, S. E., *Theory of Oscillators*, (translated from Russian), Pergamon Press, New York, 1966.

13-10 Cartwright, M. L., "Forced Oscillations in Nearly Sinusoidal Systems," *Journal of the Institute of Electrical Engineers*, (London), **96** (3), 88–96, 1948.

13-11 Hayashi, C., *Nonlinear Oscillations in Physical Systems*, Chapters 12 and 13, McGraw-Hill, New York, 1964.

13-12 Kronauer, R. E., and Musa, S. A., "Necessary Conditions for Subharmonic and Superharmonic Synchronization in Weakly Nonlinear Systems," *Quarterly of Applied Mathematics*, **24** (2), July 1966.

13-13 Gambill, R., and Hale, J., "Subharmonic and Ultraharmonic Solutions for Weakly Nonlinear Systems," *Journal of Rational Mechanical Analysis*, **5**, 353–394, 1956.

13-14 Hénon, M., and Heiles, C., "The Applicability of the Third Integral of Motion: Some Numerical Experiments," *Astronomical* Journal, **69** (1), 73–79, 1964.

13-15 Barbanis, B., "On the Isolating Character of the Third Integral in a Resonance Case," *Astronomical Journal*, **71** (6), 415–424, 1966.

13-16 Lefschetz, S., *Differential Equations: Geometric Theory*, Pure and Applied Mathematics, vol. VI, Wiley-Interscience, New York, 1957.

13.7.2 Bibliography

Bogoliubov, N. N., and Mitropolsky, Y. A., *Asymptotic Methods in the Theory of Nonlinear Oscillations*, (translated from Russian), Gordon and Breach, New York, 1961.

Hale, J. K., *Ordinary Differential Equations*, Pure and Applied Mathematics, vol. XXI, Wiley-Interscience, New York, 1969.

Minorsky, N., *Nonlinear Oscillations*, Van Nostrand Reinhold, New York, 1962.

Stoker, J. J., *Nonlinear Vibrations*, Wiley-Interscience, New York, 1950.

14

Perturbation Methods

G. F. Carrier*

14.1 INTRODUCTION

Frequently, one encounters a problem in which the function sought is not an elementary function of its arguments. When those arguments include a small parameter, ϵ, it is sometimes advantageous to seek a representation of the function $f(x_i; \epsilon)$ in the form

$$f(x_i; \epsilon) = \sum_n f_n(x_i)\, \epsilon^n \tag{14.1-1}$$

More generally, of course, when the simpler description given by eq. (14.1-1) fails to be useful, it can be advantageous to write

$$f(x_i; \epsilon) = \sum_n f_n(x_i; \epsilon) g_n(\epsilon) \tag{14.1-2}$$

Series of the form (14.1-2) [which include those having the form of (14.1-1), of course] may be convergent or asymptotic in character but, from the point of view of those who use mathematics to study scientific phenomena or technological questions, the important requirement is that a very few terms of the series provide an approximate description of the function which has all of the accuracy which the investigation requires.

Procedures which provide useful series having the form of eq. (14.1-1) or (14.1-2) frequently are called perturbation methods. Ordinarily, this terminology is applied only to problems in which differentiation and/or integrations with regard to ϵ do not appear in the statement of the problem; in the following sections we will describe several procedures which are useful in such problems and we will introduce and illustrate their use, using appropriate informative examples.

*Prof. George F. Carrier, Div. of Engineering and Applied Physics, Harvard University, Camb., Mass.

14.2 PERTURBATION METHODS FOR ORDINARY DIFFERENTIAL EQUATIONS

Among the ordinary differential equations which arise frequently, many require the use of perturbation methods. Furthermore, in addition to their own importance, the ideas and techniques which are useful in more intricate systems can be displayed profitably using ordinary differential equations as a vehicle.

14.2.1 A Simple Example

Suppose we want to find $u(x, \beta; \epsilon)$ in $0 \leqslant x < \infty$ when it is required that

$$u''(x, \beta; \epsilon) - (1 + \epsilon x e^{-\beta x}) u(x, \beta; \epsilon) = 0 \tag{14.2-1}$$

and that

$$u(0, \beta; \epsilon) = 1, \ \ u'(0, \beta; \epsilon) = -1 \tag{14.2-2}$$

where $\beta > 1$, where $0 < \epsilon << 1$, and where $u' \equiv \partial u/\partial x$. Suppose also that we are primarily interested in the behavior of u as x becomes very large. We know that for each β and each x, u is an entire function of ϵ, Ref. 14-1, and it is natural to attempt to construct a representation of u in the form

$$u = u_0(x, \beta) + \epsilon u_1(x, \beta) + \epsilon^2 u_2(x, \beta) + \cdots \tag{14.2-3}$$

and to substitute this supposed answer into eqs. (14.2-1) and (14.2-2). When this is done, we find

$$(u_0'' - u_0) + \epsilon(u_1'' - u_1 - xe^{-\beta x} u_0) + \epsilon^2(u_2'' - u_2 - xe^{-\beta x} u_1) + \cdots = 0 \tag{14.2-4}$$

$$u_0(0, \beta) + \epsilon u_1(0, \beta) + \cdots = 1 \tag{14.2-5}$$

$$u_0'(0, \beta) + \epsilon u_1'(0, \beta) + \cdots = -1 \tag{14.2-6}$$

Here, and elsewhere, we display all of the arguments of u and of u_i whenever it seems informative to do so and we omit such arguments elsewhere.

If the power series on the left side of eq. (14.2-4) is to converge to zero for all ϵ, the coefficient of each power of ϵ must vanish; furthermore, corresponding statements must be true for eqs. (14.2-5) and (14.2-6). It follows that

$$u_0'' - u_0 = 0 \tag{14.2-7}$$

with

$$u_0(0, \beta) = 1; \ \ u_0'(0, \beta) = -1$$

so that

$$u_0 = e^{-x} \tag{14.2-8}$$

It also follows that

$$u_1'' - u_1 = xe^{-(\beta+1)x} \tag{14.2-9}$$

with

$$u_1(0, \beta) = u_1'(0, \beta) = 0$$

so that, with some slightly cumbersome algebra, we find

$$u_1 = \frac{(\beta^2 + 2\beta)x + 2(\beta + 1)}{(\beta^2 + 2\beta)^2} e^{-(\beta+1)x} - \frac{1}{2\beta^2} e^{-x} + \frac{\beta^2}{2(\beta^2 + 2\beta)^2} e^x \tag{14.2-10}$$

As we noted earlier, one always hopes that a very few terms of an expansion of this sort will be needed to describe the solution of a given problem with all necessary accuracy. Otherwise the series may be rather useless (contemplate, for example, the direct use of $s = \sum_{n=0}^{\infty} (-\epsilon x)^n/n!$ for the calculation of $e^{-\epsilon x}$ when $\epsilon = 10^{-4} << 1$ but $x = 10^7$). In the problem under study here, however, the function u_1 does not directly encourage this hope because the final term of (14.2-10), $[\beta^2/2(\beta^2 + 2\beta)^2]e^x$, becomes extremely large as x gets large—so much so that, however small ϵ may be, there is always some x_0 such that $\epsilon u_1 > u_0$ in $x > x_0$. Nevertheless, one can see from the equation for u_2, i.e.

$$u_2'' - u_2 = xe^{-\beta x} u_1 \tag{14.2-11}$$

that the most rapidly growing term in u_2 also will be a constant multiple of e^x (the non-homogeneous term implies only a term of order $xe^{(1-\beta)x}$), and there is reason to hope that $\epsilon^2 u_2/(u_0 + \epsilon u_1)$ is large compared to ϵ for *no* positive value of x. In fact, using the boundary conditions, $u_2(0, \beta) = u_2'(0, \beta) = 0$ and eq. (14.2-11) we find

$$u_2 = (Nx^2 + Bx + C)e^{-(2\beta+1)x} + (Ax + D)e^{-(\beta+1)x} + (Ex + F)e^{(1-\beta)x} + Le^x + Me^{-x} \tag{14.2-12}$$

where

$$4N = [(\beta^2 + \beta)(\beta^2 + 2\beta)]^{-1}$$

$$B = \frac{1}{2\beta(\beta^2 + 2\beta)^2} + \frac{(2\beta + 1)N}{\beta^2 + \beta}$$

$$C = \frac{(2\beta + 1)B - N}{2(\beta^2 + \beta)}$$

$$A = -[2\beta^3(\beta + 2)]^{-1}$$

$$D = -\frac{(\beta + 1)}{[(\beta^2 + 2\beta)^2\beta^2]}$$

$$E = [2(\beta + 2)^2(\beta^2 - 2\beta)]^{-1}$$

$$F = 2(\beta - 1)(\beta^2 - 2\beta)^{-1}E$$

$$2L = \beta D + 2\beta C - A - B + F(\beta - 2) - E$$

$$2M = A + B + E - (\beta + 2)D - 2(\beta + 1)C - \beta F$$

Specifically, it seems that, for small ϵ, the quantity $u_0 + \epsilon u_1 + \epsilon^2 u_2$ differs very little indeed from $u_0 + \epsilon u_1$.

There are two reasons which underlie the inclusion of the foregoing lengthy and otherwise useless expressions. They provide firm support for the statement that, unless the first few terms of a perturbation series suffice, the series is relatively useless. However, the recipes for E and F also suggest that the coefficients of later terms in the series can be such that the assertion "ϵ multiplies a quantity of order unity" may be incorrect or, at least, delicate to justify. In this particular problem, the difficulty is minor; when $\beta - 2$ is very small, the quantity $(Ex + F)e^{(1-\beta)x} + Me^{-x}$ is uniformly bounded in x and β, even in the limit $\beta = 2$. In later terms of the series, other values of β are superficially anomalous but it is true that $[u - (u_0 + \epsilon u_1)]/u \ll 1$ for all x. Thus, u is superbly approximated for all x by

$$u \simeq u^* \equiv e^{-x} + \epsilon u_1 \tag{14.2-13}$$

whenever $\epsilon \ll 1$.

In conventional terminology, this result can be stated in the form: u is *uniformly* approximated by u^*. Notice that in this problem u is *not* uniformly approximated by u_0.

More commonly, in problems wherein the perturbation series is convergent, and is at all useful, the function sought *is* uniformly approximated by u_0 and the reader may find it instructive to solve eq. (14.2-1) with any initial conditions other than those for which $u(0)/u'(0) = -1$.

14.2.2 A Homogeneous Problem

The foregoing procedure requires some minor modification when one is treating a homogeneous linear problem. Suppose now that

$$u'' + \lambda[1 + \epsilon \sin x]u = 0 \tag{14.2-14}$$

in $0 < x < N\pi$ (N is an even integer and $0 < \epsilon << 1$) and that

$$u(0) = u(N\pi) = 0 \tag{14.2-15}$$

It is well known that nontrivial solutions of such a problem exist only for a certain discrete countable set of values of λ which we denote by $\lambda^{(n)}$, and it is to be expected that the numbers $\lambda^{(n)}$ will differ for different values of ϵ. Thus, we denote them by $\lambda^{(n)}(\epsilon)$ and our objective is to find $\lambda^{(n)}(\epsilon) - \lambda^{(n)}(0)$. It is also clear that $\lambda^{(n)}$ and the nontrivial solution u associated with $\lambda^{(n)}$ [which we denote by $u^{(n)}(x, \epsilon)$] depend on N as well as on ϵ, but we will not insert this dependence into the notation.

Since both $u^{(n)}$ and $\lambda^{(n)}$ depend on ϵ, it is natural to seek a description of each as a power series in ϵ; in doing so, we suppress the superscript n until it is needed again.

Thus, we write

$$u = u_0(x) + \epsilon u_1(x) + \cdots \tag{14.2-16}$$

$$\lambda = \lambda_0 + \epsilon\lambda_1 + \cdots \tag{14.2-17}$$

With the substitution of these series into eqs. (14.2-14) and (14.2-15) we obtain

$$(u_0'' + \lambda_0 u_0) + \epsilon(u_1'' + \lambda_0 u_1 + \lambda_1 u_0 + \lambda_0 u_0 \sin x) + \epsilon^2(u_2'' + \lambda_0 u_2 + \lambda_1 u_1 + \lambda_2 u_0 + \lambda_0 u_1 \sin x + \lambda_1 u_0 \sin x) + \cdots = 0 \tag{14.2-18}$$

and

$$u_j(0) = u_j(N\pi) = 0 \qquad \text{for } 0 \leqslant j \tag{14.2-19}$$

In particular, then

$$u_0'' + \lambda_0 u_0 = 0$$

with

$$u_0(0) = u_0(N\pi) = 0$$

so that the eigenfunctions (with arbitrary but obviously convenient normalization) and eigenvalues associated with this zeroth order problem are

$$u_0^{(n)}(x) = \sin\frac{nx}{N} \tag{14.2-20}$$

and

$$\lambda_0^{(n)} = n^2/N^2 \tag{14.2-21}$$

In first order

$$u_1'' + \lambda_0 u_1 = -\lambda_1 u_0 - \lambda_0 u_0 \sin x \tag{14.2-22}$$

and

$$u_1(0) = u_1(N\pi) = 0 \tag{14.2-23}$$

It follows (after an uncomfortable amount of algebra and trigonometry) that only if $\lambda_1 = 0$ can eq. (14.2-22) have a solution which obeys the boundary conditions of eq. (14.2-23). With $\lambda_1 = 0$, u_1 becomes (unless $n = N/2$)

$$u_1 = -\frac{n^2}{2N(2n+N)} \cos\left[\left(\frac{n}{N}+1\right)x\right] + \frac{-n^2}{2N(2n-N)} \cos\left[\left(\frac{n}{N}-1\right)x\right] + \frac{2n^3}{N(4n^2-N^2)} \cos\left(\frac{n}{N}x\right) + A_1 \sin\frac{nx}{N} \tag{14.2-24}$$

where no constraint on the choice of A_1 has appeared. However, if

$$u = u_0 + \epsilon u_1 + \cdots$$

is a solution, so is the function $u^* = (1 - \epsilon A_1)[u_0 + \epsilon u_1 + \cdots]$ and this function, u^*, can be written in the form

$$u^* = u_0 + \epsilon(u_1 - A_1 u_0) + \cdots$$

and we note that the term in u^* of order ϵ is the same as that in

$$u = u_0 + \epsilon u_1 + \cdots \tag{14.2-25}$$

except that in u^* the coefficient of $\sin nx/N$ is zero. Thus, there is no loss of generality in taking $A_1 = 0$ immediately; it is merely one step in the sequence of decisions which determines the particular choice for the normalization of u.

In the exceptional case noted above i.e., $n = N/2$, the recipe for u_1 becomes

$$u_1 = \frac{1}{16}\left(\cos\frac{x}{2} - \cos\frac{3x}{2}\right) - \frac{1}{8} x \sin\frac{x}{2} \tag{14.2-26}$$

instead of that given by eq. (14.2-24).

The equations which implicitly describe u_2 [as implied by eqs. (14.2-18) and (14.2-19)] are

$$u_2'' + \lambda_0 u_2 = -\lambda_2 u_0 - \lambda_0 u_1 \sin x \tag{14.2-27}$$

and

$$u_2(0) = u_2(N\pi) = 0 \tag{14.2-28}$$

and the right-hand side of eq. (14.2-27) can be written as a linear combination of the functions sin $(A_p x)$ where

$$A_p = \frac{n}{N} + p \qquad \text{for } p = -2, -1, 0, 1, 2$$

No solution of eq. (14.2-27) and (14.2-28) exists unless, in that right-hand side, the coefficient of sin $(A_0 x)$ vanishes; however, that coefficient vanishes only if

$$\lambda_2^{(n)} = \frac{-n^4}{2N^2(4n^2 - N^2)} \tag{14.2-29}$$

or, when $n = N/2$

$$\lambda_2^{(N/2)} = -\frac{5}{128} \tag{14.2-30}$$

Thus, the perturbation procedure has produced a useful estimate of $\lambda(\epsilon) - \lambda(0)$ *but only when* $\epsilon^2 n^4 \ll 2N^2(N^2 - 4n^2)$. For other values of n, the contributions $\epsilon^k \lambda_k^{(n)}$ are not uniformly dominated by $\lambda_0^{(n)}$ nor is the domination of $\epsilon^k u_k$ by u_0 uniform in n. Thus, for such n, the result at best is unreliable; at worst it can be a very bad estimate. We shall look later at methods for dealing with nonuniformities in the validity of such expansions.

14.2.3 A Nonlinear Application

The foregoing use of a regular perturbation technique can be adapted to many nonlinear problems. One of the more interesting of these is associated with the Van der Pol oscillator and displays facets of the perturbation process which did not arise in sections 14.2.1 and 14.2.2.

There are vacuum tube circuits, Ref. 14-2, for which

$$u''(t) - \epsilon[1 - u^2(t)]u'(t) + \omega_0^2 u(t) = F(t) \tag{14.2-31}$$

where u is the (nondimensional) grid voltage of the circuit at time t, ϵ and ω_0 are numbers which depend on the circuit parameters and $F(t)$ is the externally provided input signal. An interesting question asks: When $0 < \epsilon << 1$ and when $F(t) = B' \cos 3\omega t$, are there values of B' and of ω such that $u(t)$ is periodic and has period $2\pi/\omega$? That is, are there values of B' and ω such that the oscillation will synchronize with this particular subharmonic of the input signal?

It is known that, when $\epsilon << 1$ and $F \equiv 0$, a periodic oscillation does occur and that its period is very nearly given by $2\pi/\omega_0$. It is natural, therefore, to anticipate that synchronization is most likely to occur if ω is not too far from ω_0. Accordingly, it is useful to study first the situation in which $|\omega - \omega_0| << \omega$. In fact, it is of crucial importance that one describe ω_0^2 in the form of $\omega^2 - \epsilon c$. With this substitution,* and with the replacement of F by $A \cos(3\omega t + \phi)$ [it is slightly more convenient to adjust the phase of F by the inclusion of the constant ϕ than to provide the corresponding adjustment in the phase of u], eq. (14.2-31) becomes

$$u'' + \omega^2 u = A \cos(3\omega t + \phi) + \epsilon(1 - u^2)u' + \epsilon c u \tag{14.2-32}$$

The reader can verify readily that, with

$$u = u_0 + \epsilon u_1 + \epsilon^2 u_2 + \cdots$$

and with the requirement that each of $u_0, u_1, u_2, \ldots$ have period $2\pi/\omega$

$$u_0 = A_0 \cos(3\omega t + \phi) + \alpha \cos \omega t \tag{14.2-33}$$

where $8\omega^2 A_0 = -A$ and that

$$u_1'' + \omega^2 u_1 = (1 - u_0^2)u_0' + c u_0 \tag{14.2-35}$$

The function u_1 can be periodic only if $(1 - u_0^2)u_0' + cu_0$ is orthogonal to $\sin \omega t$ and to $\cos \omega t$, a requirement which is met only if

$$4c = -A_0 \omega \alpha \sin \phi \tag{14.2-36}$$

and

$$4 - 2A_0^2 - \alpha^2 = A_0 \alpha \cos \phi \tag{14.2-37}$$

Equations (14.2-36) and (14.2-37) imply that u_1 is periodic only if α^2 and A_0^2 lie on an ellipse described by

$$\alpha^4 - \alpha^2(8 - 3A_0^2) + 4\left[4\frac{c^2}{\omega^2} + (2 - A_0^2)^2\right] = 0 \tag{14.2-38}$$

*The reader should note, by trying it, that unless this description of ω_0 is used, it is very difficult to make progress.

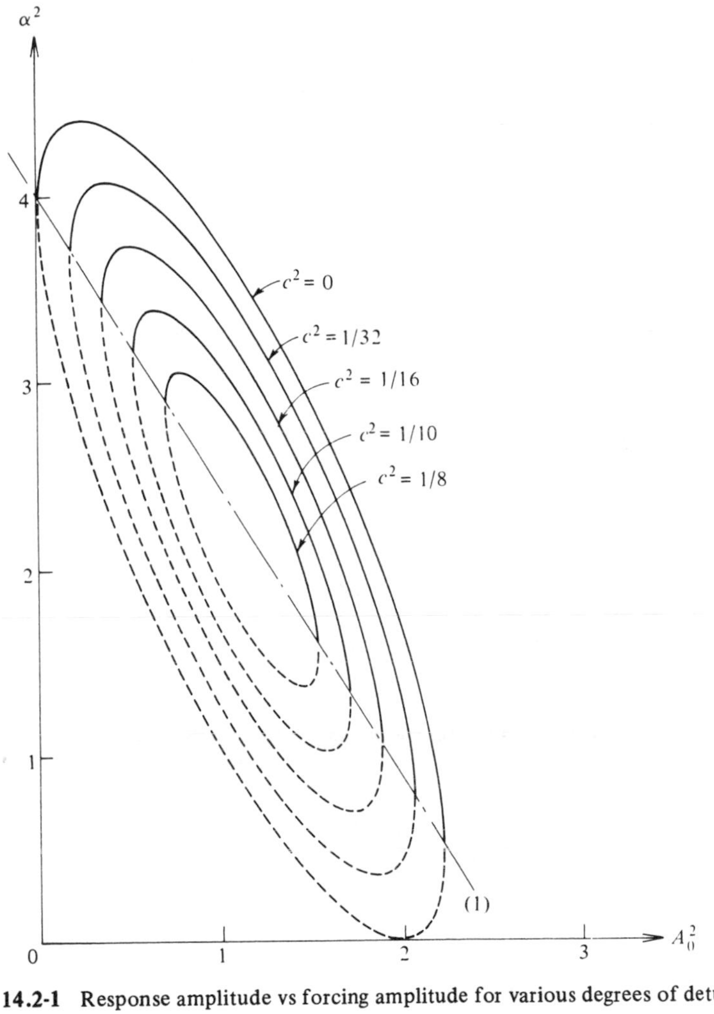

Fig. 14.2-1 Response amplitude vs forcing amplitude for various degrees of detuning.

There is one ellipse for each c in $0 < c < \omega/7$ as indicated in Fig. 14.2-1. Note that synchronization occurs for a given detuning, c, only if A^2 lies in a range of values limited both from above and from below.

One question which remains unanswered at this point asks whether either of the two solutions for each c is stable and would be seen in a real experiment. The analysis of that problem so closely resembles some aspects of the problem of section 14.4.4 that we omit the stability analysis here; it suffices to note that the (solid) upper branch of each curve in Fig. 14.2-1 denotes the stable solution and that a comprehensive analysis of this whole problem can be found in Ref. 14-3.

14.3 PARTIAL DIFFERENTIAL EQUATIONS

The method we have been presenting can be equally useful in the analysis of the solutions of partial differential equations. The procedures are so similar to the foregoing that we will not select examples in which the simplest use of method is fully successful but will choose illustrations which require variations on the method and which lead into alternative procedures. The first of these rapidly reduces to one involving a system of ordinary differential equations.

14.3.1 A Refraction Problem

The refraction of waves by a nonhomogeneous medium is illustrated rather nicely by the problem whose differential equation is

$$\phi_{xx} + \phi_{yy} + k^2 F(x, y, \epsilon)\phi = 0 \qquad \text{in } |x| < \infty, 0 < y < 2\pi \tag{14.3-1}$$

where

$$F = \begin{cases} 1 & \text{in } x < 0 \\ 1 - \epsilon \sin x \cos y & \text{in } 0 < x < L \\ 1 & \text{in } x > L \end{cases} \tag{14.3-2}$$

For simplicity of algebra, we take L to be an integral multiple of 2π. The wave potential is $\phi(x, y, \epsilon)e^{i\omega t}$ and, in $x > L$, it consists* of a given monochromatic wave $\varphi_i = Ie^{ikx}$ along with whatever waves are reflected back towards $x \to \infty$. In $x < 0$ it is a collection of waves travelling in the negative x-direction. More precisely, there must be no rightward propagating waves in $x < 0$ and, in $x > L$, only the incident wave is propagating to the left. It will be easier to give mathematical recipes for these requirements later in the analysis. The boundary conditions along $y = 0$ and $y = 2\pi$ are

$$\phi_y(x, 0, \epsilon) = \phi_y(x, 2\pi, \epsilon) = 0 \tag{14.3-3}$$

The Fourier representation

$$\phi(x, y, \epsilon) = \sum_{n=0}^{\infty} \phi^{(n)}(x, \epsilon) \cos ny$$

is not only mathematically appropriate but it should provide a readily interpretable description. Its use leads directly to

$$\phi_{xx}^{(n)} - n^2\phi^{(n)} + k^2\phi^{(n)} - \left(\frac{\epsilon k^2}{2}\right)(\phi^{(n+1)} + \phi^{(n-1)}) f(x) = 0 \tag{14.3-4}$$

*There is a small arithmetic advantage when one lets the transmitted wave emerge to the left at $x = 0$.

for $n > 1$, but for $n = 0, 1$, we have

$$\phi_{xx}^{(0)} + k^2\phi^{(0)} - \left(\frac{\epsilon k^2}{2}\right)\phi^{(1)} f(x) = 0 \tag{14.3-5}$$

and

$$\phi_{xx}^{(1)} + (k^2 - 1)\phi^{(1)} - (\epsilon k^2)[\phi^{(0)} + \tfrac{1}{2}\,\phi^{(2)}]\, f(x) = 0 \tag{14.3-6}$$

where

$$f(x) = \begin{cases} 0 & x < 0 \\ \sin x & 0 < x < L \\ 0 & x > L \end{cases} \tag{14.3-7}$$

The most interesting value of k is unity. For that value it is clear that, in $x > L$

$$\begin{aligned} \phi^{(0)} &= Ie^{ix} + \mathrm{Re}^{-ix} \\ \phi^{(1)} &= \Phi_1 \ \text{(a const)} \\ \phi^{(2)} &= \Phi_2 e^{-x\sqrt{3}} \\ \phi^{(n)} &= \Phi_n e^{-x\sqrt{n^2-1}} \end{aligned}$$

i.e.

$$\begin{aligned} \phi^{(0)\prime}(L, \epsilon) + i\phi^{(0)}(L, \epsilon) &= 2iIe^{iL} \\ \phi^{(0)\prime}(L, \epsilon) - i\phi^{(0)}(L, \epsilon) &= -2i\mathrm{Re}^{-iL} \end{aligned} \tag{14.3-8}$$

$$\begin{aligned} \phi^{(1)\prime}(L, \epsilon) &= 0 \\ \phi^{(n)\prime}(L, \epsilon) &= -\sqrt{n^2 - 1}\,\phi^{(n)}(L, \epsilon) \end{aligned} \tag{14.3-9}$$

Correspondingly, at $x = 0$

$$\begin{aligned} \phi^{(0)\prime}(0, \epsilon) - i\phi^{(0)}(0, \epsilon) &= 0 \\ \phi^{(0)\prime}(0, \epsilon) + i\phi^{(0)}(0, \epsilon) &= 2iT \\ \phi^{(1)\prime}(0, \epsilon) &= 0 \\ \phi^{(n)\prime}(0, \epsilon) &= \sqrt{n^2 - 1}\,\phi^{(n)}(0, \epsilon) \end{aligned} \tag{14.3-10}$$

Since $\phi^{(n)}$ and $\phi^{(n)\prime}$ must be continuous at $x = 0$ and at $x = L$, eqs. (14.3-8), (14.3-9) and (14.3-10) are correct statements concerning the end point values of the functions $\phi^{(n)}(x, \epsilon)$ in the domain $0 < x < L$.

In these equations, I, R, and T are the intensities of the incident, reflected and transmitted waves, respectively, and R/I, and T/I, are the reflection and transmission coefficients. Any one of I, R, T, may be regarded as given. We use T as the given quantity so that the first two equations of eqs. (14.3-10) are the boundary conditions on $\phi^{(0)}$ for its determination in $0 < x < L$. It follows that eqs. (14.3-8) are not boundary conditions but merely equations to determine I and R when the $\phi^{(n)}$ have been found.

An attempt at the fully conventional procedure now follows but, in an effort to minimize the complications of notation, we replace $\phi^{(0)}, \phi^{(1)}, \phi^{(2)}, \ldots$ by $u, v, w, p, q, r \ldots$ respectively. With this notation, we try

$$u(x, \epsilon) = u_0(x) + \epsilon u_1(x) + \epsilon^2 u_2(x) + \cdots$$
$$v = v_0(x) + \epsilon v_1(x) + \cdots$$

Equations (14.3-4), (14.3-5), etc., imply that

$$u_0'' + u_0 = 0; \quad v_0'' = 0; \quad w_0'' - 3w_0 = 0, \cdots$$

With the boundary conditions (14.3-9) and (14.3-10), these imply

$$u_0 = Te^{ix}; \quad v_0 = V_0 = \text{const}; \quad w_0 = p_0 = \cdots 0$$

The equation for v_1 is

$$v_1'' - u_0 \sin x = 0$$

This equation has no solution for which $v_1'(0) = v_1'(L) = 0$ and our procedure has failed! However, there is one place in the procedure where an arbitrary (and incorrect) choice was made. When we chose to describe u, v, w, as power series in ϵ, we wrote each series in the form

$$\phi^{(n)} = \sum_{m=m_{\min}}^{\infty} \epsilon^m \phi_m^{(n)}$$

where each $m_{\min} = 0$, even though there is no *a priori* assurance that, for example, $\phi^{(1)}$ is bounded as $\epsilon \to 0$. As a matter of fact, the phenomenon is a steady-state one, and $\phi^{(1)}$ (i.e., v) must be large enough so that the energy being fed into v by u (i.e., because of the interaction term, $\frac{1}{2} u \sin x$, in the equation for v) is balanced by the energy being fed back from v into u, w, etc. As it happens (and one can rarely find this out except by trying various alternatives and seeing which one works) this balance is achieved when

$$\phi^{(0)} \equiv u = \sum_{n=0}^{\infty} \epsilon^n u_n(x)$$

$$\phi^{(1)} \equiv v = \sum_{n=-1}^{\infty} \epsilon^n v_n(x) \tag{14.3-11}$$

$$\phi^{(2)} \equiv w = \sum_{n=0}^{\infty} \epsilon^n w_n(x)$$

and, for larger m

$$\phi^{(m)} = \sum_{n=m-2}^{\infty} \epsilon^n \phi_n^{(m)}(x)$$

When *this* set of descriptions is used, the equation for v_{-1} becomes

$$v''_{-1} = 0 \tag{14.3-12}$$

and its solution is

$$v_{-1}(x) = b = \text{const.} \tag{14.3-13}$$

The equations for $\phi_0^{(n)}$ are

$$u''_0 + u_0 - \frac{b}{2}\sin x = 0$$

$$v''_0 = 0 \tag{14.3-14}$$

$$w''_0 - 3w_0 - \frac{b}{2}\sin x = 0$$

and their solutions are

$$u_0 = a_0 \cos x + b_0 \sin x - \tfrac{1}{4} bx \cos x$$

$$v_0 = c_0 x + d_0 \tag{14.3-15}$$

$$w_0 = f_0 e^{-x\sqrt{3}} + g_0 e^{(x-L)\sqrt{3}} - \frac{b}{8}\sin x$$

The boundary conditions imply

$$a_0 = T;\quad b_0 = iT + \frac{b}{4}$$

$$f_0 = -g_0 = \frac{-b}{16\sqrt{3}};\quad c_0 = 0 \tag{14.3-16}$$

but, at this stage of the development, no further information is available to help determine b. Accordingly, we turn to the equations for $\phi_1^{(n)}$, i.e.

$$
\begin{aligned}
u_1'' + u_1 &= \left(\frac{v_0}{2}\right) \sin x \\
v_1'' &= \left(u_0 + \frac{w_0}{2}\right) \sin x \\
w_1'' - 3w_1 &= \left(\frac{v_0}{2}\right) \sin x \\
p_1'' - 8p_1 &= \left(\frac{w_0}{2}\right) \sin x
\end{aligned}
\tag{14.3-17}
$$

The solutions are

$$
\begin{aligned}
u_1 &= a_1 \cos x + b_1 \sin x - \tfrac{1}{4} d_0 x \cos x \\
v_1 &= -\left(\frac{a_0}{8}\right) \sin 2x + c_1 x + d_1 + \frac{b}{32} (x \sin 2x + \cos 2x) \\
&\quad + \frac{f_0}{16} e^{-x\sqrt{3}} (\sin x + \sqrt{3} \cos x) + \frac{g_0}{16} e^{\sqrt{3}(x-L)} (\sin x - \sqrt{3} \cos x) \\
&\quad + \frac{1}{8}\left(b_0 - \frac{b}{16}\right)(2x^2 + \cos 2x) \qquad (14.3\text{-}18) \\
w_1 &= f_1 e^{-x\sqrt{3}} + g_1 e^{(x-L)\sqrt{3}} - \left(\frac{d_0}{8}\right) \sin x \\
p_1 &= \cdots
\end{aligned}
$$

The boundary conditions on v_1 lead to the statements

$$
\begin{aligned}
\frac{-a_0}{4} + c_1 - \left(\frac{f_0}{8}\right)(1 - e^{-L\sqrt{3}}) &= 0 \\
\frac{-a_0}{4} + c_1 + \frac{bL}{16} + \frac{f_0}{8}(1 - e^{-L\sqrt{3}}) + \frac{L}{2}\left(b_0 - \frac{b}{16}\right) &= 0
\end{aligned}
\tag{14.3-19}
$$

and, ultimately, to

$$
\begin{aligned}
b &= \frac{32\sqrt{3}\,LTi}{(1 - e^{-L\sqrt{3}}) - 10\sqrt{3}\,L} \\
I_0 &= T + \frac{iL^2 T}{\frac{5}{2}L - \frac{1}{4\sqrt{3}}(1 - e^{-L\sqrt{3}})}
\end{aligned}
\tag{14.3-20}
$$

and to

$$\frac{R_0}{I_0} = \frac{L^2}{L^2 - \dfrac{5iL}{2} + \dfrac{i}{4\sqrt{3}}(1 - e^{-L\sqrt{3}})} \tag{14.3-21}$$

The procedure can be carried further into the higher orders and refined estimates of R/I can be determined. It is interesting to note, when this is done, that

$$\frac{R}{I} = \frac{R_0}{I_0} + O(\epsilon^2)$$

However, two other observations are more interesting! One notes that the intensity of the $v \cos y$ contribution to ϕ is

$$v \simeq -\frac{16iT}{5\epsilon} \simeq \frac{8I_0}{\epsilon L}$$

when $L >> 1$.

The other notes that u_1 has a term of the form $x \cos x$, u_2 has one of the form $x^2 \cos x$, v_1 has a $c_1 x$, v_2 has an x^2 and so on. Thus, for example, the orders of magnitude of successive terms of u are

$$T, \epsilon LT, \epsilon^2 L^2 T, \ldots$$

Clearly, we again are faced with a situation in which, when $\epsilon L << 1$, the foregoing analysis is adequate but when $\epsilon L = O(1)$, it is not. In section 14.4.3 we will use a continuing analysis of this problem for $\epsilon L = O(1)$ to illustrate one technique which is very useful when non-uniformities of this type arise. Before proceeding to such complications, however, we turn to a different subtlety of this problem.

14.3.2 The Detuned Case

Suppose now that, in the problem of section 14.3.1, we wish to find R/I when $k^2 - 1 = O(1)$. In this situation, it happens that the coupling (i.e., the energy feed from u to v) is weak and the expansion which will succeed is

$$u = \sum_{n=0}^{\infty} \epsilon^n u_n(x)$$

$$v = \sum_{n=1}^{\infty} \epsilon^n v_n(x) \tag{14.3-22}$$

$$w = \sum_{n=2}^{\infty} \epsilon^n w_n(x)$$

The equation of order zero for u is

$$u_0'' + k^2 u_0 = 0 \tag{14.3-23}$$

with

$$u_0'(0) - iku_0(0) = 0$$

and*

$$u_0'(0) + iku_0(0) = 2ikT$$

so that

$$u_0 = T \cos kx + iT \sin kx \tag{14.3-24}$$

The equations of first order in ϵ are

$$\begin{aligned} u_1'' + k^2 u_1 &= 0 \\ v_1'' - (1 - k^2) v_1 &= k^2 u_0 \sin x \end{aligned} \tag{14.3-25}$$

and the boundary conditions are

$$u_1(0) = u_1'(0) = v_1'(0) - \sqrt{1 - k^2}\, v_1(0) = v_1'(L) + \sqrt{1 - k^2}\, v_1(L) = 0$$

where, when $k^2 < 1$, $\sqrt{1 - k^2}$ is positive; when $k^2 > 1$, $\sqrt{1 - k^2}$ is a positive imaginary number.

It follows that

$$u_1 = 0$$

$$v_1 = c_1 e^{\sqrt{1-k^2}x} + d_1 e^{-\sqrt{1-k^2}x} - \frac{Tk^2}{4(1+k)} \sin(1+k)x - \frac{Tk^2}{4(1-k)} \sin(1-k)x$$
$$- \frac{iTk^2}{4(1-k)} \cos(1-k)x + \frac{iTk^2}{4(1+k)} \cos(1+k)x \tag{14.3-26}$$

where

$$c_1 = \frac{Tk^2}{4\sqrt{1-k^2}} \left[1 + \frac{ki}{\sqrt{1-k^2}}\right] e^{ikL - L\sqrt{1-k^2}} \tag{14.3-27}$$

*The k appears here with T because the transmitted wave now has the form Te^{ikx}.

and

$$d_1 = -\frac{Tk^2}{4\sqrt{1-k^2}} + \frac{ik^3T}{4(1-k^2)} \tag{14.3-28}$$

Thus, there is still no reflected wave to order ϵ, and we must continue the analysis and find u_2. The procedure is the routine one and the result is

$$\begin{aligned} u_2 &= a_2 \cos kx + b_2 \sin kx + \frac{k^2 \cos x}{4\sqrt{1-k^2}} [-c_1 e^{\sqrt{1-k^2}x} + d_1 e^{-\sqrt{1-k^2}x}] \\ &- \frac{Tx\, k^3 \sin kx}{16k(1-k^2)} + \frac{iTx\, k^3 \cos kx}{16k(1-k^2)} - \frac{Tk^4}{16(1-k)^2} \cos[(2-k)x] \\ &- \frac{Tk^4}{64(1-k)^2} \cos[(2+k)x] + \frac{iTk^4}{64(1-k)^2} \sin[(2-k)x] \\ &- \frac{iTk^4}{64(1-k)^2} \sin[(2+k)x] \end{aligned} \tag{14.3-29}$$

where

$$a_2 = \frac{k^2(c_1 - d_1)}{4\sqrt{1-k^2}} + \frac{k^4}{64(1+k^2)} T + \frac{k^4}{64(1-k)^2} T \tag{14.3-30}$$

and

$$b_2 = \frac{k(c_1 + d_1)}{4} - \frac{iTk^4}{16(1-k^2)} - \frac{(2-k)iTk^3}{64(1-k)^2} - \frac{(2+k)iTk^4}{64(1+k)^2} \tag{14.3-31}$$

In particular

$$\frac{R}{I} = -e^{2ikL} \frac{\epsilon^2 [u_{2,x}(L) - iku_2(L)]}{2ikTe^{ikL} + \epsilon^2 [u_{2,x}(L) + iku_2(L)]} + O(\epsilon^3) \tag{14.3-32}$$

In the limit $k^2 - 1 \to 0$, this result has no validity; in fact, we can expect eq. (14.3-32) to provide a useful estimate only when $|\epsilon^2 u_2| \ll u_0$. Since

$$u_2(L) \simeq TL^2/16\sqrt{k^2-1}$$

as $k^2 \to 1$, we can expect eq. (14.3-32) to be useful only when

$$\epsilon^2 L^2 \ll 16\sqrt{k^2-1}$$

It would be difficult to infer the reflection coefficient in $0 < |k^2 - 1| \leqslant O(\epsilon^4 L^4)$ from the analysis of sections 14.3.1 and 14.3.2, but one can use the procedure of section 14.2.3 modified in a way which is suggested by eq. (14.3-32) to obtain R/I for that frequency range.

In section 14.2.3, the discrepancy between the frequency and unity was described by writing

$$\omega^2 - 1 = \epsilon\sigma$$

and the perturbation series which emerged gave an interpretable and valid description of the frequency dependence. In this problem, however, if that description were adopted, that is, if we wrote $k^2 - 1 = \epsilon\alpha$, we would find it impossible to construct the series for any value of α except zero. One might anticipate this by drawing the appropriate inference from eq. (14.2-32). In that equation, {recalling that $u_2(L) = O[(k^2 - 1)^{1/2}]$ and $u_2'(L) = O(1)$} it is clear that the u_2 contribution in the denominator and the u_0 contribution in that same denominator become comparable in size only when $k^2 - 1 = O(\epsilon^4)$. Thus, it seems appropriate to skip over any other descriptions of k of the form $k^2 - 1 = \epsilon^n\alpha$ and just use $k^2 - 1 = \alpha\epsilon^4$.

Once this crucial decision has been made, the procedure, in character, is a duplication of that of section 14.3.1. Equations (14.3-15) and (14.3-16) remain unchanged, but the right-hand side of eqs. (14.3-20) become $ib\sqrt{\alpha}$ and $-ib\sqrt{\alpha}$, respectively, instead of zero. The constant b then takes the value

$$b = \frac{32\sqrt{3}\,LT}{-i(1 - e^{-L\sqrt{3}}) + 10\sqrt{3}\,iL - 128\sqrt{3\alpha}}$$

and the reflection coefficient is given by

$$\frac{R}{I} = \frac{iL^2}{iL^2 + \dfrac{5}{2}L - \dfrac{1}{4\sqrt{3}}(1 - e^{L\sqrt{3}}) + 32i\sqrt{\alpha}} + O(\epsilon^2)$$

This result is valid when $|\epsilon^4\alpha| << 1$ whereas eq. (14.3-32) is valid when $\alpha >> L^4$ (i.e., $k^2 - 1 >> \epsilon^4 L^4$). The two recipes are in agreement when $|\epsilon^4\alpha| << 1$ and $\alpha >> L^4$.

A more complete account of this problem can be found in Ref. 14-4.

14.4 MULTISCALING METHODS

We have seen in sections 14.2.2 and 14.3.1 that, when an ordinary perturbation method is attempted in connection with some problems, it fails to be uniformly valid over the full range of an independent variable or a parameter. One technique which frequently is effective in dealing with such a difficulty is called the multiscaling method. Again, it is best described with the help of illustrative examples of which several now follow.

14.4.1 The Oscillator with Cubic Damping

The advantages of the multiscaling technique are rather uniquely highlighted when one studies the problem in which (again, primes denote differentiation with respect to t)

$$u''(t, \epsilon) + \epsilon[u'(t, \epsilon)]^3 + u(t, \epsilon) = 0 \quad \text{in } t > 0 \tag{14.4-1}$$

with

$$u(0, \epsilon) = 1, \quad \text{and} \quad u'(0, \epsilon) = 0$$

When $0 < \epsilon << 1$, it is clear that one can expect an oscillation in which u becomes zero at intervals in t of about 2π, and it is clear that the oscillation will decay slowly. Accordingly, one might imagine that the methods of section 14.2 would apply and that we could write

$$\begin{aligned} u(t, \epsilon) &= u_0(t) + \epsilon u_1(t) + \cdots \\ u(0, \epsilon) &= u_0(0) + \epsilon u_1(0) + \cdots = 1 \\ u'(0, \epsilon) &= u_0'(0) + \epsilon u_1'(0) + \cdots = 0 \end{aligned} \tag{14.4-2}$$

and proceed, successively, to find $u_0, u_1, \ldots$.

When one carries out this process, he finds

$$\begin{aligned} u_0 &= \cos t \\ u_1 &= -\frac{3}{8} t \cos t + \frac{\sin 3t}{32} + \frac{9}{32} \sin t \\ u_2 &= \frac{27}{128} t^2 \cos t + \cdots \end{aligned}$$

If we now collect those contributions from each of $u_0, u_1, u_2, \ldots$ which is largest for large t, we see that there is really a series in powers of ϵt, i.e.

$$u = \cos t - \frac{3}{8} \epsilon t \cos t + \frac{27}{128} (\epsilon t)^2 \cos t + \cdots \tag{14.4-3}$$

From this description of u it is very difficult to decide (a) whether and where the series might converge or (b) at what rate u decays. In fact, for $\epsilon t > 1$, this series description of u is quite useless.

However, this description does remind us, rather forcefully, of the fact that there are really two time scales in the phenomenon. We already know that the oscillation occurs on a time scale of order 2π and we should be confident that the decay occurs at a slower rate and especially, in fact, at a rate which depends upon ϵ. Equation (14.4-3) suggests that the decay will be described by a function of ϵt, and, in the multiscaling process, we take advantage of this suggestion.

In that procedure, we agree to describe u as a function of the *two* variables t and ϵt (we call the latter τ) as well as of ϵ. More explicitly, we write

$$u(t, \epsilon) = w(t, \tau, \epsilon) \tag{14.4-4}$$

With this substitution, eq. (14.4-1) becomes*

$$w_{tt} + 2\epsilon w_{t\tau} + \epsilon^2 w_{\tau\tau} + \epsilon(w_t + \epsilon w_\tau)^3 + w = 0 \tag{14.4-5}$$

and the boundary conditions become

$$w(0, 0, \epsilon) = 1; \quad w_t(0, 0, \epsilon) + \epsilon w_\tau(0, 0, \epsilon) = 0 \tag{14.4-6}$$

We now can write

$$w(t, \tau, \epsilon) = w_0(t, \tau) + \epsilon w_1(t, \tau) + \epsilon^2 w_2(t, \tau) + \cdots \tag{14.4-7}$$

and, using the usual procedure

$$w_{0,tt} + w_0 = 0$$

so that

$$w_0 = A_0(\tau) \cos t + B_0(\tau) \sin t \tag{14.4-8}$$

where

$$A_0(0) = 1 \quad \text{and} \quad B_0(0) = 0$$

Furthermore

$$w_{1,tt} + w_1 = -2w_{0,t\tau} - (w_{0,t})^3 \tag{14.4-9}$$

Since there is no further information available which pertains directly to A_0 and B_0, it now emerges that $A_0(\tau)$ and $B_0(\tau)$ will be determined uniquely only when some arbitrary constraint on w_1 is demanded. The need for this constraint arises because we provided for a lot of flexibility in the choice of a description of u when we allowed ϵ to appear in w_0 (through its role in defining τ) as well as in the coefficient of w_1. We can try to use that flexibility to insist that ϵw_1 be dominated

*A subscript t or τ denotes differentiation with regard to that variable.

uniformly by w_0 and, in fact, as the procedure is continued, that ϵw_n be uniformly dominated by w_{n-1}.

To accomplish that end we must choose A_0 and B_0 so that the right-hand side of eq. (14.4-9) is orthogonal to $\cos t$ and to $\sin t$. That choice is most readily made by writing the nonhomogeneous term in the form

$$-[2B_0' + \tfrac{3}{4} B_0(A_0^2 + B_0^2)] \cos t + [2A_0' + \tfrac{3}{4} A_0(A_0^2 + B_0^2)] \sin t + \tfrac{1}{4} B_0(3A_0^2 - B_0^2) \cos 3t + \tfrac{1}{4} A_0(3B_0^2 - A_0^2) \sin 3t$$

and noting that we must choose A_0 and B_0 according to

$$A_0' + \tfrac{3}{8} A_0(A_0^2 + B_0^2) = 0 \tag{14.4-10}$$

and

$$B_0' + \tfrac{3}{8} B_0(A_0^2 + B_0^2) = 0 \tag{14.4-11}$$

with

$$A_0(0) = 1; \quad B_0(0) = 0 \tag{14.4-12}$$

The solutions of these equations are given by

$$A_0 = \left(1 + \frac{3\tau}{4}\right)^{-1/2}; \quad B_0 = 0 \tag{14.4-13}$$

Furthermore, according to eq. (14.3-6), the solution of eq. (14.4-9) is then given by

$$w_1 = A_1(\tau) \cos t + B_1(\tau) \sin t + \frac{1}{32} A_0^3(\tau) \sin 3t \tag{14.4-14}$$

where A_1 and B_1 obey the initial conditions

$$A_1(0) = 0; \quad B_1(0) = \frac{9}{32} \tag{14.4-15}$$

$A_1(\tau)$ and $B_1(\tau)$ must be so chosen that w_2 is uniformly dominated by w_1. The equation for w_2 is

$$w_2'' + w_2 = -w_{0,\tau\tau} - 2w_{1,t\tau} - 3w_{0,t}^2 (w_{1,t} + w_{0,\tau}) \tag{14.4-16}$$

$$= -A_0'' \cos t - 2 B_1' \cos t + 2 A_1' \sin t - \frac{9}{16} A_0^2 A_0' \cos 3t$$

$$+\frac{3}{4}A_0^2\left[(A_0'+B_1)(\cos 3t-\cos t)-A_1(\sin 3t-3\sin t)\right.$$

$$\left.-\frac{3}{32}A_0^3(2\cos 3t-\cos t-\cos 5t)\right] \qquad (14.4\text{-}17)$$

The choice, $A_1 = 0$, eliminates the sin t contributions from the nonhomogeneous term and B_1 must obey the equation

$$2B_1' + A_0'' + \tfrac{3}{4}A_0^2(A_0' + B_1) - \tfrac{9}{128}A_0^5 = 0 \qquad (14.4\text{-}18)$$

if terms in cos t are to be absent on the right hand side of eq. (14.4-16), i.e.

$$\left(\frac{B_1}{A_0}\right)' = -\frac{45}{216}\left(1+\frac{3}{4}\tau\right)^{-2}$$

or

$$B_1 = \tfrac{3}{64}A_0 + \tfrac{15}{64}A_0^3 \qquad (14.4\text{-}19)$$

In this analysis, we have gone to the trouble of finding B_1 because it might have turned out that B_1/A_0 increased without limit as $t \to \infty$. In many applications the coefficient functions (i.e., the functions of τ) develop in that unfortunate way and further remedies must be adopted. In this problem, however, successive steps of the sort illustrated above lead to a series which, for all t, is dominated by, and well approximated by, the first term

$$w_0 = (1 + \tfrac{3}{4}\epsilon t)^{-1/2}\cos t \qquad (14.4\text{-}20)$$

14.4.2 An Oscillator with Three Time Scales

A closely related problem is one in which there are three time scales, i.e.

$$u'' + \epsilon(u')^3 + c\epsilon^2 u' + u = 0 \qquad (14.4\text{-}21)$$

with

$$u(0,\epsilon) = 1; \quad u'(0,\epsilon) = 0$$

It is easily inferred that, early in the oscillation, the cubic damping term will control the decay rate and the amplitude will again resemble $(1 + \frac{3}{4}\epsilon t)^{-1/2}$. However, once the decay has reduced the amplitude to a small multiple of $\sqrt{\epsilon}$, the linear damping contributions will have become the predominant decay mechanism and the decay will evolve like $e^{-\epsilon^2 ct/2}$.

One might imagine from the foregoing qualitative description of the phenomenon that he should write

$$u(t, \epsilon) = w(t, \tau, \beta, \epsilon)$$

where $\tau = \epsilon t$ and $\beta = \epsilon^2 t$, but this is not only unnecessary, it is also very inefficient. The important thing to remember is that not all multiscale problems require multiscale methods; it is profitable to use such methods only when alternative, intrinsically simpler methods lead to descriptions which are nonuniformly valid.

Accordingly, we proceed optimistically by rewriting eq. (14.4-21) in the form

$$u'' + \epsilon(u')^3 + \epsilon a u' + u = 0 \tag{14.4-22}$$

where $a = \epsilon c$, with

$$u(0, \epsilon) = 1; \quad u'(0, \epsilon) = 0$$

and then define w as in the foregoing section, i.e.

$$u(t, \epsilon) = w(t, \tau, \epsilon) \tag{14.4-23}$$

where $\tau = \epsilon t$.

In doing this, we anticipate that we need the second variable τ for exactly the same reasons that we needed it in section 14.4.1, but that we might be lucky enough so that no nonuniformity arises in connection with the linear damping.

There is one other feature of this problem, however, which we must allow for or repetitions will become necessary. One knows that, when

$$q'' + \epsilon a q' + q = 0$$

the zero spacings of q are not precisely 2π. In fact, of course, q is given by

$$q = A e^{-\epsilon a t/2} \cos mt$$

where

$$m = \left(1 - \frac{\epsilon^2 a^2}{4}\right)^{1/2} \simeq 1 - \frac{\epsilon^2 a^2}{8}$$

and it will be no surprise if this shift in the faster time scale is present in the problem of eq. (14.4-22). Accordingly, we *don't* use eq. (14.4-23) but rather, we use

$$u(t, \epsilon) = w(s, \tau, \epsilon) \tag{14.4-24}$$

where

$$t = s(1 + \epsilon^2 \lambda_2 + \epsilon^4 \lambda_4 + \cdots)$$

The procedure now follows precisely that of section 14.4.1. We have

$$w = w_0(s, \tau) + \epsilon w_1(s, \tau) + \cdots$$

with

$$w_{0,ss} + w_0 = 0$$
$$w_{1,ss} + w_1 = -2 w_{0,s\tau} - [a + w_{0,s}^2] w_{0,s}$$
$$w_0(0, 0) = 1; \quad w_{0,s}(0, 0) = 0$$
$$w_1(0, 0) = 0; \quad w_{1,s}(0, 0) + w_{0,\tau}(0, 0) = 0$$

The zeroth-order solution is

$$w_0 = A_0(\tau) \cos s + B_0(\tau) \sin s$$
$$A_0(0) = 1; \quad B_0(0) = 0$$

and, using these in the equation w_1, we obtain

$$\begin{aligned} w_{1,ss} + w_1 = & -\cos s\{2B_0'(\tau) + B_0(\tau) [a + \tfrac{3}{4} (A_0^2 + B_0^2)]\} \\ & + \sin s\{2A_0'(\tau) + A_0(\tau) [a + \tfrac{3}{4} (A_0^2 + B_0^2)]\} \\ & + \tfrac{1}{4} B_0(3A_0^2 - B_0^2) \cos 3s + \tfrac{1}{4} A_0(3B_0^2 - A_0^2) \sin 3s \quad (14.4\text{-}25) \end{aligned}$$

where

$$w_1(0, 0) = 0; \quad w_{1,s}(0, 0) = -A_0'(0)$$

Thus, w_1 will be as small as w_0, uniformly in t, only if

$$2B_0' = -B_0 [a + \tfrac{3}{4} (A_0^2 + B_0^2)]$$
$$2A_0' = -A_0 [a + \tfrac{3}{4} (A_0^2 + B_0^2)]$$

with

$$A_0(0) = 1; \quad B_0(0) = 0$$

From these, it follows that $B_0 \equiv 0$ and that

$$A_0^2 = e^{-a\tau} \Big/ \left[1 + \frac{3}{4a}(1 - e^{-a\tau})\right] \equiv e^{-\epsilon^2 ct} \Big/ \left[1 + \frac{3}{4\epsilon c}(1 - e^{-\epsilon^2 ct})\right] \quad (14.4\text{-}26)$$

Clearly, this is consistent with the qualitative description given earlier in this section.

It is now readily inferred from eq. (14.4-25) that

$$w_1 = A_1(\tau) \cos s + B_1(\tau) \sin s + \frac{A_0^3}{32} \sin 3s \qquad (14.4\text{-}27)$$

When we put eq. (14.4-27) into the equation for w_2 and demand that its right-hand side contain no terms of the form $M(\tau) \cos s$ or $M(\tau) \sin s$, we obtain

$$2B_1' + 3(a + \tfrac{3}{4}A_0^2)(B_1 + A_0') - 2\lambda_2 A_0 + A_0'' + \tfrac{9}{128}A_0^5 = 0 \qquad (14.4\text{-}28)$$

and

$$2A_1' + (a + \tfrac{3}{4}A_0^2)A_1 = 0$$

with

$$A_1(0) = 0, \quad \text{and} \quad B_1(0) + \tfrac{3}{32} + A_0'(0) = 0$$

Hence $A_1 = 0$ but B_1 must satisfy eq. (14.4-28) in which λ_2 must be so chosen that B_1 is uniformly not larger than A_0. But we see with a little study (see Ref. 14-5 for details), that this uniformity arises only when $\lambda_2 = a^2/8$.

Accordingly, we find in this problem that the two time scales are the decay time associated with the early attenuation and that distortion of unity which describes the 'frequency' of the oscillation. It is not unusual that distortions of t or of ϵt or of both are needed if a two-scale procedure is to be fully valid!

14.4.3 More Refraction

We noted in section 14.3.2 that, when $\epsilon L \geqslant O(1)$, the expansions obtained in that analysis were useless and that a cure for the nonuniformity with regard to L was needed. The presence of terms in $(\epsilon x)^n$, for each index n, suggests again that it may be fruitful to describe u, v, w, etc. as functions of the two 'independent variables', x and $\tau \equiv \epsilon x$. When the problem is so formulated (retaining the use of $u(x, \tau, \epsilon)$, $v(x, \tau, \epsilon)$, . . . where $u(x, \epsilon)$, $v(x, \epsilon)$, . . . , were used before) the problem has the statement

$$u_{xx} + 2\epsilon u_{x\tau} + \epsilon^2 u_{\tau\tau} + k^2 u + \left(\frac{\epsilon k^2}{2}\right) v f(x) = 0$$

$$v_{xx} + 2\epsilon v_{x\tau} + \epsilon^2 v_{\tau\tau} + (k^2 - 1) v + \epsilon k^2 \left[u + \frac{w}{2}\right] f(x) = 0 \qquad (14.4\text{-}29)$$

$$w_{xx} + 2\epsilon w_{x\tau} + \epsilon^2 w_{\tau\tau} + (k^2 - 4) w + \left(\frac{\epsilon k^2}{2}\right) [v + p] f(x) = 0$$

and we will keep things a bit simpler by analyzing only the case $k = 1$.

The boundary conditions (with $k \equiv 1$) are

$$
\begin{gathered}
u_x(0, 0, \epsilon) + \epsilon u_\tau(0, 0, \epsilon) - iu(0, 0, \epsilon) = 0 \\
u_x(0, 0, \epsilon) + \epsilon u_\tau(0, 0, \epsilon) + iu(0, 0, \epsilon) = 2i(T_0 + \epsilon T_1) \\
v_x(0, 0, \epsilon) + \epsilon v_\tau(0, 0, \epsilon) = v_x(L, L^*, \epsilon) + \epsilon v_\tau(L, L^*, \epsilon) = 0 \\
w_x(0, 0, \epsilon) + \epsilon w_\tau(0, 0, \epsilon) + \sqrt{3}\, w(0, 0, \epsilon) = 0 \\
w_x(L, L^*, \epsilon) + \epsilon w_\tau(L, L^*, \epsilon) - \sqrt{3}\, w(L, L^*, \epsilon) = 0
\end{gathered} \tag{14.4-30}
$$

In the foregoing equations L^* denotes ϵL and T is written as $T_0 + \epsilon T_1$ in anticipation of the fact that the intensity of the transmitted wave will be small compared to that of the incident wave whenever L^* is of order unity (i.e., in those cases where the range of τ is of order unity; when $\tau \ll 1$, there is no need for this expansion!) Because of this we will be forced to take $T_0 = 0$, and we might as well provide for that contingency now.

One might anticipate, in view of eq. (14.3-11), that the expansion for v would begin with a term of order ϵ^{-1}. Any attempt to construct the series in that form fails; as we can see when we've gotten the answer, the dependence of v_0 (the first term in the series for v) on L^* is such that $v_0(x, \tau, \epsilon)$ approaches the behavior implied in $v_{-1}(x, \epsilon)$ as L gets small enough.

Thus, we write

$$
\begin{gathered}
u = \sum_{n=0}^{\infty} \epsilon^n u_n(x, \tau) \\
v = \sum_{n=0}^{\infty} \epsilon^n v_n(x, \tau) \\
w = \sum_{n=1}^{\infty} \epsilon^n w_n(x, \tau)
\end{gathered} \tag{14.4-31}
$$

and we substitute these representations into eqs. (14.4-1) and (14.4-2) to obtain boundary value problems for each of the $u_n, v_n, \ldots$.

The zeroth-order problem is

$$
u_{0,xx} + u_0 = 0; \quad v_{0,xx} = 0 \tag{14.4-32}
$$

with

$$
\begin{gathered}
u_{0,x}(0, 0) - iu_0(0, 0) = 0 \\
u_{0,x}(0, 0) + iu_0(0, 0) = 2iT_0 \\
v_{0,x}(0, 0) = 0; \quad v_{0,x}(L, L^*) = 0
\end{gathered} \tag{14.4-33}
$$

These imply that

$$u_0 = b_0(\tau)\cos x + c_0(\tau)\sin x$$
$$v_0 = d_0(\tau)x + e_0(\tau) \qquad (14.4\text{-}34)$$

where

$$c_0(0) - ib_0(0) = 0$$
$$c_0(0) + ib_0(0) = 2iT_0 \qquad (14.4\text{-}35)$$
$$d_0(0) = d_0(L^*) = 0$$

No further constraints are available until the analysis of the contributions of order ϵ so we proceed to that task. We have

$$u_{1,xx} + u_1 = \frac{d_0}{2}x\sin x + \left(\frac{e_0}{2} + 2b_0'\right)\sin x - 2c_0'\cos x \qquad (14.4\text{-}36)$$

$$v_{1,xx} = \frac{b_0}{2}\sin 2x - \frac{c_0}{2}\cos 2x + \frac{c_0}{2} - 2d_0' \qquad (14.4\text{-}37)$$

$$w_{1,xx} - 3w_1 = \frac{d_0}{2}x\sin x + \frac{e_0}{2}\sin x \qquad (14.4\text{-}38)$$

where the prime denotes $d/d\tau$.

Once again, we want to use the flexibility which the extra variable τ provides to permit us to insist that the intensities of u_1, v_1, not be larger with increasing x than those of u_0, v_0. Accordingly, we must suppress all of the terms on the right-hand side of eq. (14.4-36) and the last two terms in eq. (14.4-37); i.e.

$$d_0 = e_0 + 4b_0' = c_0' = c_0 - 4d_0' = 0$$

Thus, eqs. (14.4-34) reduce to

$$u_0 = b_0(\tau)\cos x \qquad (14.4\text{-}39)$$

$$v_0 = e_0(\tau) \qquad (14.4\text{-}40)$$

$$e_0 + 4b_0' = 0 \qquad (14.4\text{-}41)$$

We can now calculate u_1, v_1 and w_1; we obtain

$$u_1 = b_1(\tau)\cos x + c_1(\tau)\sin x \qquad (14.4\text{-}42)$$

$$v_1 = d_1(\tau)x + e_1(\tau) - \frac{b_0}{8} \sin 2x \tag{14.4-43}$$

$$w_1 = f_1(\tau)e^{\sqrt{3}(x-L)} + g_1(\tau)e^{-x\sqrt{3}} - \frac{e_0}{8} \sin x \tag{14.4-44}$$

To find b_1, c_1, d_1, e_1, f_1, and g_1, we must suppress the secular terms in the equations for $u_2, v_2, \ldots$. Without writing out the full equations, we record that this suppression implies

$$d_1 = 0 \tag{14.4-45}$$

$$e_1 + 4b_1' = 0 \tag{14.4-46}$$

$$2c_1' + b_0'' + \frac{b_0}{32} = 0 \tag{14.4-47}$$

$$c_1 - \frac{e_0}{16} - 4d_1' - 2e_0'' = 0 \tag{14.4-48}$$

$$f_1' = g_1' = 0 \tag{14.4-49}$$

Equations (14.4-41), (14.4-47) and (14.4-48) now imply that

$$16c_1'' - c_1 = 0$$

whereupon

$$c_1 = \sigma_1 \cosh \frac{\tau}{4} + \lambda_1 \sinh \frac{\tau}{4}$$

Equations (14.4-47) and (14.4-48) now yield

$$b_0 = \delta_0 \sin \frac{\tau}{\sqrt{32}} + \gamma_0 \cos \frac{\tau}{\sqrt{32}} - \frac{16}{3} \sigma_1 \sinh \frac{\tau}{4} - \frac{16}{3} \lambda_1 \cosh \frac{\tau}{4}$$

$$e_0 = \frac{-\delta_0}{\sqrt{2}} \cos \frac{\tau}{\sqrt{32}} + \frac{\gamma_0}{\sqrt{2}} \sin \frac{\tau}{\sqrt{32}} + \frac{16}{3} \sigma_1 \cosh \frac{\tau}{4} + \frac{16}{3} \lambda_1 \sinh \frac{\tau}{4}$$

The boundary conditions of order unity require that

$$u_{0,\tau}(0,0) + u_{1,x}(0,0) - iu_1(0,0) = 0$$

$$u_{0,\tau}(0,0) + u_{1,x}(0,0) + iu_1(0,0) = 2iT_1$$

$$v_{1,x}(0,0) + v_{0,\tau}(0,0) = 0$$

$$v_{1,x}(L,L^*) + v_{0,\tau}(L,L^*) = 0$$

$$w_{1,x}(0,0) - \sqrt{3}\, w_1(0,0) = 0$$

$$w_{1,x}(L,L^*) + \sqrt{3}\, w_1(L,L^*) = 0$$

and it then follows that

$$\gamma_0 = \lambda_1 = 0 \quad (\text{whence } T_0 = 0)$$

$$\sigma_1 = \frac{3iT_1 \sin(L^*/\sqrt{32})}{8\sqrt{2}\sinh(L^*/4) - \sin(L^*/\sqrt{32})}$$

$$\delta_0 = \frac{64iT_1 \sinh(L^*/4)}{8\sqrt{2}\sinh(L^*/4) - \sin(L^*/\sqrt{32})}$$

$$f_1 = \frac{e_0(L^*)}{16\sqrt{3}}$$

$$g_1 = -\frac{e_0(0)}{16\sqrt{3}}$$

$$b_1(0) = T_1$$

If we call the reflected wave by the name Re^{-ix} and the incident wave Ie^{+ix} then,

$$Ie^{iL} + \mathrm{Re}^{-iL} = u_0(L,L^*) + \epsilon u_1(L,L^*) + \cdots$$

$$i(Ie^{iL} - \mathrm{Re}^{-iL}) = u_{0,x}(L,L^*) + \epsilon[u_{0,\tau}(L,L^*) + u_{1,x}(L,L^*)] + \cdots$$

so that

$$2iI = e^{-iL}\{ib_0(L^*) + \epsilon[b_0'(L^*) + c_1(L^*) + ib_1(L^*)] + O(\epsilon^2)$$

$$2iR = e^{iL}\{ib_0(L^*) - \epsilon[b_0'(L^*) + c_1(L^*) - ib_1(L^*)] + O(\epsilon^2)$$

and the reflection coefficient, R/I, has a magnitude close to unity *except where $b_0(L^*)$ is zero or very small.* When b_0 is zero, i.e., when $L^* = m\pi\sqrt{32}$, R/I becomes

$$R/I \simeq \frac{-e^{2iL}}{(16\sqrt{3}\, i - 1)}$$

Note that this peculiar 'passing band' could not possibly be detected (analytically—not experimentally) using the ordinary perturbation method; it is an artifact of the particular transcendental dependence of R/I on ϵL and the multiscaling method is remarkably useful in detecting such phenomena.

One could push this problem further in many directions and, for different parameter ranges, representations constructed by various ordinary perturbation methods and by multiscaling methods could be blended together to describe the whole range of phenomena. However, the arithmetic, both in the foregoing and in the extensions, is so cumbersome that a thorough digestion of the foregoing is probably more profitable than a long proliferation of the procedures.

14.4.4 A Vibrating String Problem

A mathematical problem which suitably models the behavior of an elastic string, the distance between whose end points varies from the nominal separation, L, by a small multiple of $f(t)$ is

$$u''(t,\epsilon) + \beta\epsilon u'(t,\epsilon) + [1 + \epsilon f(t)]\, u(t,\epsilon) + \epsilon u^3(t,\epsilon) = 0 \qquad (14.4\text{-}50)$$

The function $u(t, \epsilon)\sin(\pi x/L)$ describes the lateral displacement of the string and the cubic contribution represents the consequence of the stretching of the string as it bows. All quantities have been normalized for convenience. The problem is doubly interesting as a vehicle for perturbation methods when both β, which characterizes the damping in the system, and ϵ are small. An interesting initial value problem asks: What is $u(t,\epsilon)$ when $f(t) = \sin 2t$ and when $u(0,\epsilon) = 1$ and $u'(0,\epsilon) = 0$?

Once again we could initiate a conventional perturbation process in ϵ but again we would find a series whose successive terms were of order $(\epsilon t)^n$. Thus, without documenting the details of that effort here, we can expect again that it will be helpful to write

$$u(t,\epsilon) = w(t,\tau,\epsilon) = \sum_{n=0} \epsilon^n w_n(t,\tau) \qquad (14.4\text{-}51)$$

where $\tau = \epsilon t$.

Again, the equation for w_0 is

$$w_{0,tt} + w_0 = 0 \qquad (14.4\text{-}52)$$

and the boundary conditions on w_0, as implied by

$$w(0,0,\epsilon) = 1; \quad w_t(0,0,\epsilon) + \epsilon w_\tau(0,0,\epsilon) = 0 \qquad (14.4\text{-}53)$$

require that

$$w_0 = A_0(\tau)\cos t + B_0(\tau)\sin t \qquad (14.4\text{-}54)$$

with

$$A_0(0) = 1, \quad \text{and } B_0(0) = 0 \qquad (14.4\text{-}55)$$

The equation for w_1 is

$$w_{1,tt} + w_1 = -\beta w_{0,t} - 2w_{0,t\tau} - w_0 \sin 2t - w_0^3 \tag{14.4-56}$$

and the right-hand side can be manipulated into the form

$$\begin{aligned}\text{RHS} = &-\left[2B_0' + \left(\beta + \frac{1}{2}\right)B_0 + \frac{3}{4}A_0\left(A_0^2 + B_0^2\right)\right] \cos t \\ &+\left[2A_0' + \left(\beta - \frac{1}{2}\right)A_0 - \frac{3}{4}B_0\left(A_0^2 + B_0^2\right)\right] \sin t \\ &+\left[\frac{B_0}{2} - \frac{A_0^3}{4} + \frac{3A_0B_0^2}{4}\right]\cos 3t + \left[\frac{B_0^3}{4} - \frac{3A_0^2B_0}{4} - \frac{A_0}{2}\right]\sin 3t\end{aligned} \tag{14.4-57}$$

In order that w_1 not grow faster with time than does w_0, the first two bracketted quantities in eq. (14.4-57) must vanish. The relationship between the quantities A_0 and B_0 which meet this requirement can be studied via their trajectories in the phase plane, and it is particularly simple to infer useful information when $\beta = 0$. Thus, we now insert β into the arguments of A_0, B_0; i.e., they become $A_0(\tau, \beta)$, $B_0(\tau, \beta)$ and we further introduce the terminology, $A_0(\tau, 0) = A(\tau)$ and $B_0(\tau, 0) = B(\tau)$. We then have

$$2B' + \tfrac{1}{2}B + \tfrac{3}{4}A\,(A^2 + B^2) = 0 \tag{14.4-58}$$

$$2A' - \tfrac{1}{2}A - \tfrac{3}{4}B\,(A^2 + B^2) = 0 \tag{14.4-59}$$

These collapse to the form

$$\frac{dB}{dA} = -\frac{B + \frac{3}{2}A\,(A^2 + B^2)}{A + \frac{3}{2}B\,(A^2 + B^2)} \tag{14.4-60}$$

the exact integrals of which are

$$AB + \tfrac{3}{8}(A^2 + B^2)^2 + k = 0 \qquad k \leqslant \tfrac{1}{6} \tag{14.4-61}$$

These curves are sketched (schematically only) in Fig. 14.4-1. It is convenient later to know that the area, A_k, within Γ_k, for $k > 0$, is

$$A_k = \frac{4}{3}\int_{\theta_0}^{\pi/4} \sqrt{\sin^2 2\theta - 6k}\, d\theta \tag{14.4-62}$$

where $\theta_0 = \frac{1}{2}\,\text{arc sin}\,(3k/2)$.

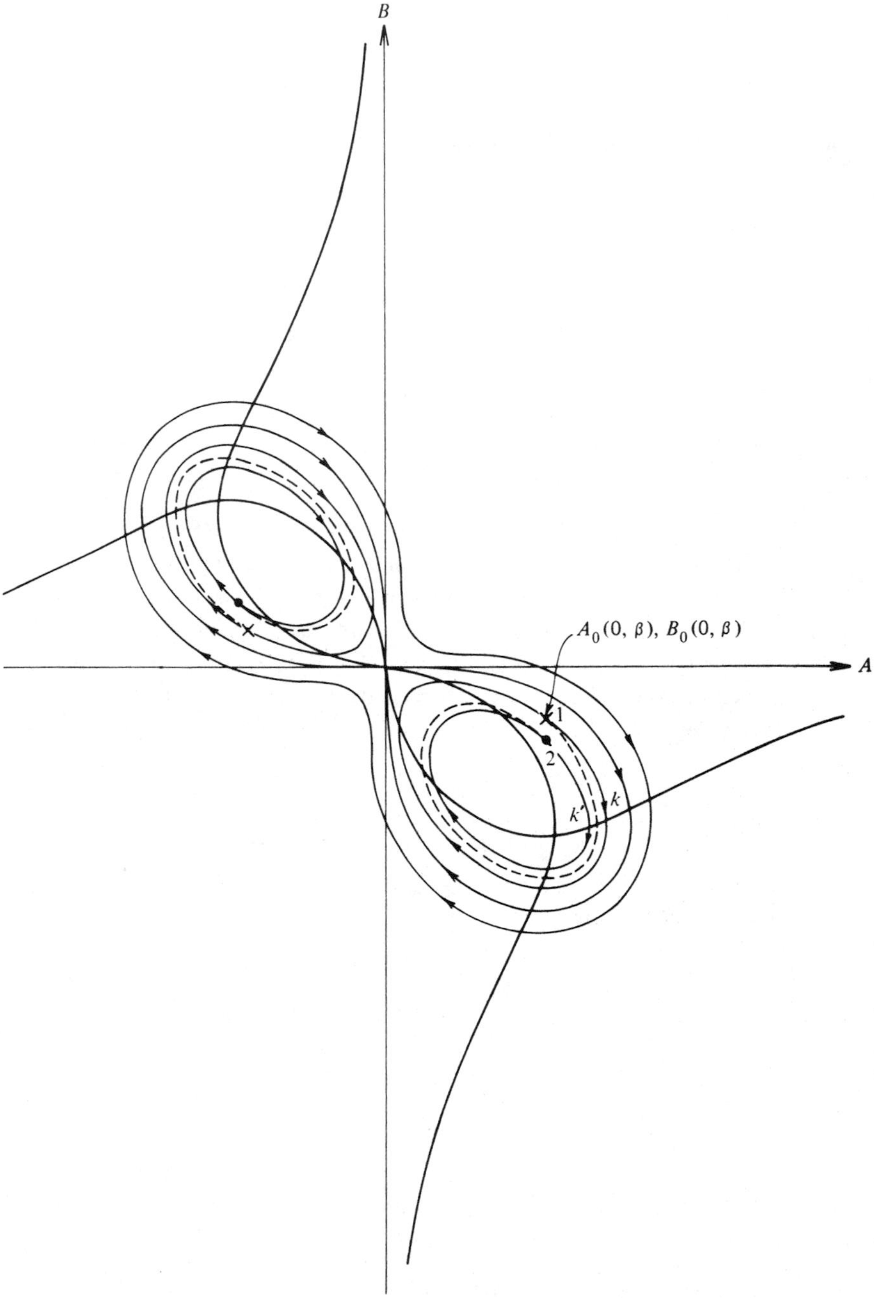

Fig. 14.4-1 Phase plane for string oscillation. Solid curves are for $\beta = 0$; dotted curves are for $\beta > 0$.

Equations (14.4-58) and (14.4-59) with the help of (14.4-61) also can be manipulated to reveal that the time, $T(k)$, needed to transverse the curve Γ_k is given by

$$T(k) = \int_{x_0}^{x_1} \frac{4\,dx}{\sqrt{x^2 - (2k + \frac{3}{4}x^2)^2}} \tag{14.4-63}$$

where $x_0, x_1 = \frac{2}{3}\,[1 \mp \sqrt{1 - 6k}]$.

At this point it is advantageous to introduce another very informal perturbation procedure. It is clear from the physics of the problem, and it is equally clear following a mathematical analysis of the singular points of the equations

$$2B_0' + (\beta + \tfrac{1}{2})B_0 + \tfrac{3}{4}A_0\,(A_0^2 + B_0^2) = 0 \tag{14.4-64}$$

and

$$2A_0' + (\beta - \tfrac{1}{2})A_0 - \tfrac{3}{4}B_0\,(A_0^2 + B_0^2) = 0 \tag{14.4-65}$$

that each of the integral curves of these equations in the A_0, B_0-plane must spiral inward either to the singular point at $A_0 = -B_0 = \sqrt{1/3}$ or that at $A_0 = -B_0 = -\sqrt{1/3}$. The question of interest asks: What are $A_0(\tau, \beta)$ and $B_0(\tau, \beta)$ when $\beta << 1$? To answer this, we do *not* attempt a formal perturbation process with β playing the role of a small parameter. Rather, we note that, with $A_0(0, \beta)$, $B_0(0, \beta)$ as shown in Fig. 14.4-1 (where it is denoted by 1) the first trajectory of $A_0(\tau, \beta)$, $B_0(\tau, \beta)$ will lie extremely close to the integral curve for $\beta = 0$ on which $A_0(0, \beta)$, $B_0(0, \beta)$ lies. However, in time $T(k)$, it will have spiralled in to a position (called 2 in Fig. 14.4-1) on a trajectory $\Gamma_{k'}$ (with $k' > k$) when it has completed that first trajectory.

We denote by $k(0)$, the value of k associated with the curve on which point 1 lies and, by $k(\tau_1)$, the value of k for the curve on which point 2 lies and we note that, from eqs. (14.4-64) and (14.4-65), we have

$$4\beta(A_0 dB_0 - B_0 dA_0) = d\,[A_0 B_0 + \tfrac{3}{8}(A_0^2 + B_0^2)^2] \tag{14.4-66}$$

That is, the change in k which is accomplished in time $T(k)$ is just 4β times the area inside the curve Γ_k. That is

$$k'(\tau) \simeq 4\beta \frac{A_k}{T_k}$$

Thus we can calculate the time history of the entire trajectory by writing

$$\tau = \frac{1}{4\beta} \int_{k_\mu}^{k} \frac{T(k)}{A(k)}\,dk \tag{14.4-67}$$

Note carefully how incredibly more messy it would have been to continue *any* formal perturbation process in the problem, for example, where $\beta = \epsilon^2$.

14.5 BOUNDARY LAYERS

There is another category of problems which incorporate a small parameter; it is characterized by the fact that the most highly differentiated term in the differential equation is multiplied by that parameter. The immediate consequence is that any expansions of the foregoing kinds are immediately doomed to failure because each member of the series is governed by a differential equation of order n accompanied by boundary conditions appropriate to an equation of order greater than n. Thus, a distinctly different procedure is required.

14.5.1 An Ordinary Differential Equation

Let

$$\epsilon u'' - (3 - x^2)u = -x \quad \text{in } 0 < x < 1 \tag{14.5-1}$$

with

$$u(0) = u(1) = 0 \tag{14.5-2}$$

and with

$$0 < \epsilon \ll 1$$

For sufficiently small ϵ it seems plausible, initially, that $\epsilon u''$ would play a role of negligible importance and, that, to a first order of approximation, u could be replaced by u_0 where

$$(3 - x^2)u_0 = x \tag{14.5-3}$$

This, of course, is equivalent to the adoption of an expansion

$$u = u_0 + f_1(\epsilon)u_1 + \cdots \tag{14.5-4}$$

followed by the usual formalities.

The function u_0 so obtained does vanish at $x = 0$, but it fails to do so at $x = 1$. Furthermore, there is no hope that subsequent terms in the series can help! Therefore, we must conclude that $\epsilon u''(x)$ *is* important and that its implications must be displayed through some different technique. But $\epsilon u''$ can be competitive with, say $-x$, in the balance described by eq. (14.5-1) only if u'' is huge. In fact, $\epsilon u''$ can be of order unity near $x = 1$ only if u depends intrinsically on $(x - 1)/\sqrt{\epsilon}$ rather than on x itself. Noting this, it seems plausible that an appropriate description of

u might be given by

$$u = u_0 + w_0 \left[\frac{(x-1)}{\sqrt{\epsilon}}\right] \equiv \frac{x}{3-x^2} + w_0(\eta) \tag{14.5-5}$$

so we substitute this recipe into eq. (14.5-1) and see what emerges. We obtain

$$w_0''(\eta) + \epsilon u_0''(x) - (3-x^2)w_0 = (3-x^2)u_0 - x \tag{14.5-6}$$

where, by the definition of u_0, the right-hand side vanishes and where, as we already know $\epsilon u_0''$ is very small throughout the domain.

Ignoring $\epsilon u_0''$ again, we must seek a solution of

$$w_0''(\eta) - (3-x^2)w_0 = 0 \tag{14.5-7}$$

which, when substituted into eq. (14.5-5) will allow us to obey the boundary conditions. The function w_0 which does so must have the value $-\frac{1}{2}$ at $\eta = 0$ and it will clearly change *very* rapidly with x as x decreases. It follows that we should seek a function such that

$$w_0(0) = -\tfrac{1}{2}; \quad w_0(-1/\sqrt{\epsilon}) << 1$$

or, in a slightly relaxed vein

$$w_0(0) = -\tfrac{1}{2}; \quad w_0(-\infty) = 0$$

However, it is clear that, if w decreases rapidly enough as x decreases from unity, its behavior will be of real interest only when $x \simeq 1$, i.e., when $3 - x^2 \simeq 2$. Accordingly, we replace eq. (14.5-7) by

$$w_0'' - 2w_0 = 0 \tag{14.5-8}$$

and we obtain

$$w = -\tfrac{1}{2}e^{+\sqrt{2}\,\eta} = -\tfrac{1}{2}e^{(x-1)\sqrt{2/\epsilon}} \tag{14.5-9}$$

The result

$$u = \frac{x}{3-x^2} - \frac{1}{2}e^{(x-1)\sqrt{2/\epsilon}} \tag{14.5-10}$$

is an excellent approximation of the true solution. Henceforth, we shall refer to this very informal sort of analysis as the *boundary-layer technique*.

To avoid unwarranted complacency it is necessary, quickly, to look at the problem

$$\epsilon u'' + (3 - x^2)u = x \quad \text{in } 0 < x < 1 \tag{14.5-11}$$

with $u(0) = u(1) = 0$, and to try to treat it in the same way.

Once again, we get

$$u_0 = \frac{x}{3 - x^2}$$

but this time, with $u = u_0 + w(\eta)$, we get

$$0 = w'' + (3 - x^2)w \simeq w'' + 2w \tag{14.5-12}$$

and we find that eq. (14.5-12) has *no* solution which decays rapidly with distance from $\eta = 0$.

The first of our boundary-value problems *is* a boundary-layer problem and the second *is not*! The technique which we are introducing here *is* useful when the 'steep' behavior of the function sought is highly localized; ordinarily it is *not* useful when the steep behavior is not localized!

A correct generalized conclusion which we can draw from these two examples is: **When a boundary-value problem is characterized by the fact that its most highly differentiated term is multiplied by a very small coefficient, it may be of boundary-layer character, and the boundary-layer technique *may* work but there is *no* general, *a priori*, assurance that it necessarily is of that character or that the technique *will* work!**

We now return to the first problem and, using the foregoing as a guide, we construct a perturbation series suitable for this problem.

Since the solution clearly involves two length scales (just as it did in the multiscaling problems), it is clear that the solution should be described as a function of the two appropriate variables, x and η. This time, however, the slow variable is less of a modulation (as it was previously) than a local augmentation so we write

$$\begin{aligned} u &= U(x, \epsilon) + w(\eta, \epsilon) \\ &= u_0(x) + \epsilon^{k_1} u_1(x) + \epsilon^{k_2} u_2(x) + \cdots \\ &\quad + w_0(\eta) + \epsilon^{\nu_1} w_1(\eta) + \epsilon^{\nu_2} w_2(\eta) + \cdots \end{aligned} \tag{14.5-13}$$

where k_1, k_2, ν_1, etc. have been used because, with $\sqrt{\epsilon}$ already appearing in the definition of η, the series may involve nonintegral powers of ϵ. We require of each $w_i(\eta)$ that it tend rapidly (exponentially) to zero as $\eta \to -\infty$ and we substitute eq. (14.5-13) into (14.5-1) as follows:

$$\epsilon[u_0'' + \epsilon^{k_1} u_1'' + \epsilon^{k_2} u_2'' + \cdots] - (3 - x^2)\,[u_0 + \epsilon^{k_1} u_1 + \cdots]$$
$$+ [w_0'' + \epsilon^{\nu_1} w_1'' + \epsilon^{\nu_2} w_2'' + \cdots] - [2 - 2\sqrt{\epsilon}\,\eta - \epsilon\eta^2]\,[w_0 + \epsilon^{\nu_1} w_1 + \cdots] + x = 0 \tag{14.5-14}$$

It seems plausible to choose $k_n = n$ at this point and write

$$u_0 = \frac{x}{(3 - x^2)} \tag{14.5-15}$$

$$u_n = \frac{u_{n-1}''}{(3 - x^2)} \tag{14.5-16}$$

. . . .

The series so obtained for U is asymptotic in character, it vanishes at $x = 0$ and, at $x = 1$, it has the asymptotic description

$$U(1) = \tfrac{1}{2} + \tfrac{5}{4}\,\epsilon + \cdots \tag{14.5-17}$$

The terms in the series for w, if calculated in the same sequential way would not be properly dominated by the early terms and we must do something better. For the present problem it suffices to write $w(\eta) = W(x, \epsilon)$ so that

$$W'' - \frac{1}{\epsilon}(3 - x^2)\,W = 0 \tag{14.5-18}$$

and use the classical substitution

$$W = -U(1)\,e^{(1/\sqrt{\epsilon})\int_1^x V(x', \epsilon)\,dx'}$$

where $V(x, \epsilon) = V_0(x) + \sqrt{\epsilon}\,V_1(x) + \cdots$

We obtain

$$V_0 = \sqrt{3 - x^2} \tag{14.5-19}$$

$$V_1 = -\tfrac{1}{4}\,[\ln(3 - x^2)]' \tag{14.5-20}$$

so that

$$w(\eta, \epsilon) = \frac{U(1)}{(2 - 2\eta\sqrt{\epsilon} - \epsilon\eta^2)^{1/4}} \exp\left\{\frac{1}{\sqrt{\epsilon}} \int_1^{1+\eta\sqrt{\epsilon}} \sqrt{3 - s^2}\,ds + \sqrt{\epsilon}\,(\cdots)\right\} \tag{14.5-21}$$

Ordinarily, in boundary layer problems, the details of the construction cannot be carried out in so naive a manner but also ordinarily, *it is rarely that one needs for any scientific purpose any terms beyond the $u_0 + w_0$ yielded by the crude analysis.* Our purpose here was merely to show that $u_0 + w_0$ was indeed the first term of a perturbation series whose subsequent terms could be constructed systematically. Occasionally, in what follows, we refer to this construction as an 'additive' expansion in contrast to the next procedure which is usually called the method of matched asymptotic expansions.

Once again we study eqs. (14.5-1) and (14.5-2) but this time we construct two series; one (hopefully) will represent $u(x, \epsilon)$ accurately in some thin region (R_1) near $x = 1$ and the other will represent it in an overlapping region, (R_2) extending to $x = 0$.

In R_2 then, we expect a smooth dependence of u on x and we write

$$u = u_0(x) + \epsilon^{k_1} u_1(x) + \cdots \tag{14.5-22}$$

whereas in R_1 we again expect u to be smooth in η so we write

$$u(x, \epsilon) \equiv \psi(\eta, \epsilon) = \psi_0(\eta) + \epsilon^{\nu_1} \psi_1(\eta) + \cdots \tag{14.5-23}$$

Substituting eq. (14.5-23) into eq. (14.5-1) we obtain

$$\sum_{n=0} \epsilon^{\nu_n} \psi_n'' - [2 - 2\sqrt{\epsilon}\,\eta - \epsilon\eta^2] \sum_{n=0} \epsilon^{\nu_n} \psi_n = -1 - \sqrt{\epsilon}\,\eta \tag{14.5-24}$$

and with $\nu_0 = 0$

$$\psi_0'' - 2\psi_0 = -1$$

so that

$$\psi_0 = \tfrac{1}{2} + A_0 e^{-\eta\sqrt{2}} + B_0 e^{\eta\sqrt{2}} \tag{14.5-25}$$

Similarly, with $\nu_1 = \frac{1}{2}$

$$\psi_1'' - 2\psi_1 = -\eta - 2\eta\psi_0$$

and

$$\psi_1 = \eta + \left(\frac{\eta^2}{2\sqrt{2}} + \frac{\eta}{4}\right) A_0 e^{-\eta\sqrt{2}} - \left(\frac{\eta^2}{2\sqrt{2}} - \frac{\eta}{4}\right) B_0 e^{\eta\sqrt{2}} + A_1 e^{-\eta\sqrt{2}} + B_1 e^{\eta\sqrt{2}} \tag{14.5-26}$$

Since $\psi(0, \epsilon) = 0$, $A_0 + B_0 = -\frac{1}{2}$ and $A_1 + B_1 = 0$.

In R_2, the series again starts as it did in eq. (5.1-15) with

$$u_0 = \frac{x}{(3 - x^2)}$$

$$u_1 = \frac{u_0''}{(3 - x^2)}$$

$$\cdots\cdots\cdots\cdots$$

It is to be expected that the region of common validity of the two series will be one in which η is large and negative but $(x - 1)$ is small and negative; with this expectation, we write

$$x = 1 + \eta\sqrt{\epsilon}$$

in the expressions for u_0, u_1, etc. and expand each in a power series in $\sqrt{\epsilon}$. Then we equate the series $\psi_0 + \epsilon\psi_1 + \cdots$ to this expanded form of u to obtain

$$\begin{aligned} &\frac{1}{2} + A_0 e^{-\eta\sqrt{2}} - \left(\frac{1}{2} + A_0\right) e^{\eta\sqrt{2}} \\ &\quad + \sqrt{\epsilon}\left[\eta + \left(\frac{\eta^2}{2\sqrt{2}} + \frac{\eta}{4}\right) A_0 e^{-\eta/\sqrt{2}} + \left(\frac{\eta^2}{2\sqrt{2}} - \frac{\eta}{4}\right)\left(\frac{1}{2} + A_0\right) e^{\eta\sqrt{2}} \right. \\ &\quad \left. + A_1 e^{-\eta\sqrt{2}} - A_1 e^{\eta\sqrt{2}}\right] + \cdots \\ &= u = \frac{1}{2} + \eta\sqrt{\epsilon} + \frac{5\eta^2\epsilon}{4} + \epsilon\left[\frac{5}{4} + \cdots\right] \end{aligned} \qquad (14.5\text{-}27)$$

Clearly, this equality can be true over a negative range of η, say $-\eta = 0(\epsilon^{-1/4})$, only if A_0 and A_1 each vanish. Having chosen all the A's in this way, the reader can verify that

1. The two series are seen to be identical when enough of the . . . are filled in.
2. The function ψ is the sum of the function w introduced in eq. (14.5-13) and the power series expansion of the function U also introduced in that equation.
3. The answer given by eq. (14.5-10) is more readily used than are the equations for ψ_0 and w_0, i.e.

$$\psi_0 = \frac{1}{2}(1 - e^{\eta\sqrt{2}}); \quad u_0 = \frac{x}{3 - x^2} \qquad (14.5\text{-}28)$$

merely because eq. (14.5-10) is a single recipe for the result whereas the user must himself blend the two recipes of eqs. (14.5-28).

14.5.2 General Remarks

Before going on to other illustrations which display other subtleties, the following comments may be helpful:

In general–

1. The expansions one obtains in boundary layer problems are asymptotic in character.
2. It would be very difficult to establish that the series obtained *are* truly asymptotic representations of the function sought.
3. It doesn't matter whether such a demonstration can be made! What matters is that the investigator find either a single truncated series which, with reasonable plausibility, approximates well the function sought, or that he find two (or several) such truncated series which do have common regions of validity and which, compositely, do approximate the function over the domain of interest.
4. It doesn't matter whether one uses an informal blending procedure, such as that used in the foregoing, or a highly formalized procedure, such as that found in Ref. 14-2.

Here, we shall use only informal methods; both because it is the technique preferred by the writer and also because the literature on singular perturbation theory already abounds in formalities.

14.5.3 Problems Which Require Matching

There are many problems in which the foregoing additive process would be clumsy at best. One which is frequently encountered is commonly referred to as the WKB problem.

Let, for example

$$\epsilon u''(x) - Q(x)\, u(x) = 0 \quad \text{in} \quad -\infty < x < \infty \tag{14.5-29}$$

with $u(\infty) = 0$ and $0 < \epsilon \ll 1$.

where

$$Q'(x) > 0 \qquad \text{for all } x$$

$$Q(0) = 0 \qquad \text{and } Q'(0) = 1$$

The function u will clearly exhibit two scales, that on which Q varies and (near any given x not too close to the origin) that given by $\sqrt{\epsilon/Q}$.

The multiscaling process could be invoked to cope with this two-scale character but a simpler initiation of the analysis is equally useful. We merely write

$$u = e^{\int w(x,\epsilon)\, dx}$$

so that

$$\epsilon[w^2 + w_x] = Q \tag{14.5-30}$$

Then, with

$$w \sim \epsilon^{-1/2} \sum_{n=0}^{\infty} \epsilon^{n/2} w_n(x)$$

we obtain

$$w_0 = \pm\sqrt{Q}$$

$$w_1 = \{\ln [Q^{-1/4}]\}'$$

$$\cdots\cdots\cdots\cdots\cdots\cdots$$

and, since there are two linearly independent solutions

$$u = Q^{-1/4}\left[A \exp\left\{\epsilon^{-1/2}\int_{x_0}^{x} \sqrt{Q}\,dx + O(\epsilon^{1/2})\right\} + B \exp\left\{-\epsilon^{-1/2}\int_{x_0}^{x} \sqrt{Q}\,dx + O(\epsilon^{1/2})\right\}\right] \tag{14.5-31}$$

The function so described has meaning in $x > 0$ and, under our boundary conditions (with arbitrary normalization)

$$u \sim u_1 = Q^{-1/4} \exp\left\{-\epsilon^{-1/2}\int_{0}^{x} \sqrt{Q}\,dx\right\} \quad \text{in } x > x_1 \tag{14.5-32}$$

In $x < 0$, u must be given by

$$u \sim u_2 = (-Q)^{-1/4}\,[Ce^{i\epsilon^{-1/2}\int_0^x \sqrt{-Q}\,dx} + De^{-i\epsilon^{-1/2}\int_0^x \sqrt{-Q}\,dx}] \quad \text{in } x < -x_1 \tag{14.5-33}$$

Only when x is so close to zero that $\epsilon(Q')^2$ is comparable in size* to Q^3, do eqs. (14.5-32) and (14.5-33) fail to be good asymptotic approximations. Thus eqs.

*One sees this by calculating u'' using eq. (14.5-32) or (14.5-33) and comparing the largest 'left over' with Qu.

(14.5-32) and (14.5-34) are adequate descriptions when, respectively, $x \gg \epsilon^{1/3}$, $-x \gg \epsilon^{1/3}$.

It might seem, superficially, that one could write, say

$$u \sim u_1 + F\left(\frac{x}{\epsilon^\nu}\right) \tag{14.5-34}$$

where $x > 0$ and proceed as in section (14.5.1) to find F so that (14.5-34) was uniformly valid. For this to be useful, however, F would have to be singular at the origin in just the right way to cancel the singularity of u_1 and this does not sound too promising. Accordingly, we anticipate that for small x the scale will be small and we write

$$u(x, \epsilon) = w(\eta, \epsilon) \quad \text{where } \eta = \frac{x}{\epsilon^\nu} \tag{14.5-35}$$

With this substitution, eq. (14.5-29) becomes

$$\epsilon^{1-2\nu} w'' - (\epsilon^\nu \eta + a_2 \epsilon^{2\nu} \eta^2 + \cdots)\, w = 0 \tag{14.5-36}$$

and the choice $\nu = \frac{1}{3}$ implies

$$w'' - \eta w \pm O(\epsilon^{1/3}) = 0 \tag{14.5-37}$$

The general solution of eq. (14.5-37) is

$$w = E \text{ Ai}(\eta) + F \text{ Bi}(\eta) \tag{14.5-38}$$

where Ai and Bi are the Airy functions.

The asymptotic behavior of Ai (η) for $\eta \gg 1$ is given by

$$\text{Ai} \sim \left(\frac{\eta^{-1/4}}{2\sqrt{\pi}}\right) \exp\left(\frac{-2\eta^{3/2}}{3}\right) = \left(\frac{\epsilon^{1/12} x^{-1/4}}{2\sqrt{\pi}}\right) \exp\left(-\frac{2}{3}\epsilon^{-1/2} x^{3/2}\right) \tag{14.5-39}$$

whereas Bi grows exponentially with $x^{3/2}$ in $\epsilon^{1/3}\, x \gg 1$.

When $\epsilon^{1/3} \ll x \ll 1$, u_1 (and hence u itself) is well approximated by

$$u_1 \sim x^{-1/4} \exp\left(-\tfrac{2}{3}\epsilon^{-1/2} x^{3/2}\right).$$

In the same region, w (and hence u) is well approximated by the same expression provided only that

$$F = 0 \quad \text{and} \quad E = 2\sqrt{\pi}\, \epsilon^{-1/12} \tag{14.5-40}$$

The asymptotic behavior of w in $-\eta \gg 1$ is

$$w \sim 2\,\epsilon^{-1/12}(-\eta)^{-1/4} \sin\left[\frac{2}{3}(-\eta)^{3/2} + \frac{\pi}{4}\right]$$

$$= 2(-x)^{-1/4} \sin\left[\frac{2}{3}\,\epsilon^{-1/2}(-x)^{3/2} + \frac{\pi}{4}\right] \tag{14.5-41}$$

and this blends with u_2 [see eq. (14.5-33)] only if C and D are so chosen that

$$u_2 = 2[-Q]^{-1/4} \sin\left[\epsilon^{-1/2} \int_0^x \sqrt{-Q(x')}\, dx' + \frac{\pi}{4}\right] \tag{14.5-42}$$

Equations (14.5-32), (14.5-38), (14.5-40), and (14.5-42) now provide, compositely, a description of u. The individual descriptions apply, respectively, in

$$x \gg \epsilon^{1/3};\quad |x| \ll 1;\quad -x \gg \epsilon^{1/3}$$

and it is clear that, for $\epsilon \ll 1$, they cover the entire line $-\infty < x < \infty$.

Ordinarily, one gains little by going further than this but it would be too bad to conclude this discussion without noting that the asymptotic behavior (that for large positive ϵx^3 and that for large negative ϵx^3) is such that $\left[\text{with } p(x) = \int_0^x \sqrt{Q(x')}\, dx'\right]$

$$y(x, \epsilon) = [(p(x)]^{1/6} Q(x)^{-1/4}\ \mathrm{Ai}\ \{\epsilon^{-1/3}\, [\tfrac{3}{2}\, p(x)]^{2/3}\} \tag{14.5-43}$$

has the same behavior for small ϵ and for all x as does the composite description given above.

Another elementary problem in which the additive process is not effective and in which matching must be used is

$$\epsilon u_{yy} - (1 - y^2)\, u_x = -y^2 \tag{14.5-44}$$

in $x > 0$ and $-1 < y < 1$, with

$$u(x, \pm 1) = 0 \text{ and } u(0, y) = 0$$

The obvious tentative solution which should be valid away from the boundary layers is

$$u \simeq u_0 = \frac{xy^2}{1 - y^2} \tag{14.5-45}$$

but near $|y| = 1$, it is clear that eq. (14.5-45) is useless.

The fact that u_0 is unbounded as $|y| \to 1$ implies that a solution of the form $u_0(x, y) + w(x, \eta)$ (where, as before, η, is a suitable distorted replacement of y) would be most inconvenient to construct since the singularity of u_0 at $y = 1$ would have to be balanced by an equivalent but oppositely signed singularity of w. Thus it may be better in this problem to use a blending technique wherein we describe u by u_0 for large enough $|y| - 1$ and by other functions, w, for smaller values of $|y| - 1$.

The equation which should suffice for small values of $1 + y$ can be written in terms of w where

$$u(x, y) \equiv \epsilon^{-\gamma} w(x, \eta)$$

and where

$$\eta = \frac{(y + 1)}{\epsilon^{\alpha}}$$

With this substitution, eq. (14.5-44) becomes

$$\epsilon^{1-2\alpha-\gamma} w_{\eta\eta} - 2\eta\epsilon^{\alpha-\gamma} w_x \simeq -1 \tag{14.5-46}$$

Thus, we choose

$$\gamma = \alpha = \tfrac{1}{3}$$

so that w and its derivatives will each be of order unity. With those choices and with the simple form of (14.5-46), we can look for a similarity solution for w, i.e.

$$w = x^{2/3} f\left(\frac{\eta}{x^{1/3}}\right) = x^{2/3} f(\beta)$$

With this definition, eq. (14.5-46) becomes

$$f'' + \tfrac{2}{3}\beta^2 f' - \tfrac{4}{3}\beta f = -1 \tag{14.5-47}$$

It is clear that eq. (14.5-47) has a solution, f_0, which has the asymptotic form

$$f_0 \sim \frac{1}{2\beta} + \frac{1}{4\beta^4} + \cdots$$

and it has one homogeneous solution, f_H, which decays with increasing β like

$$e^{-2\beta^3/9}$$

The function

$$f^* = f_0 + A f_H \qquad (14.5\text{-}48)$$

can be so chosen (i.e., A can be so chosen) that $f^*(0) = 0$ and, with that value of A, eq. (14.5-48) provides the required function. The numerically obtained description, f^*, of the solution of eq. (14.5-47) is given in Fig. 14.5-1.

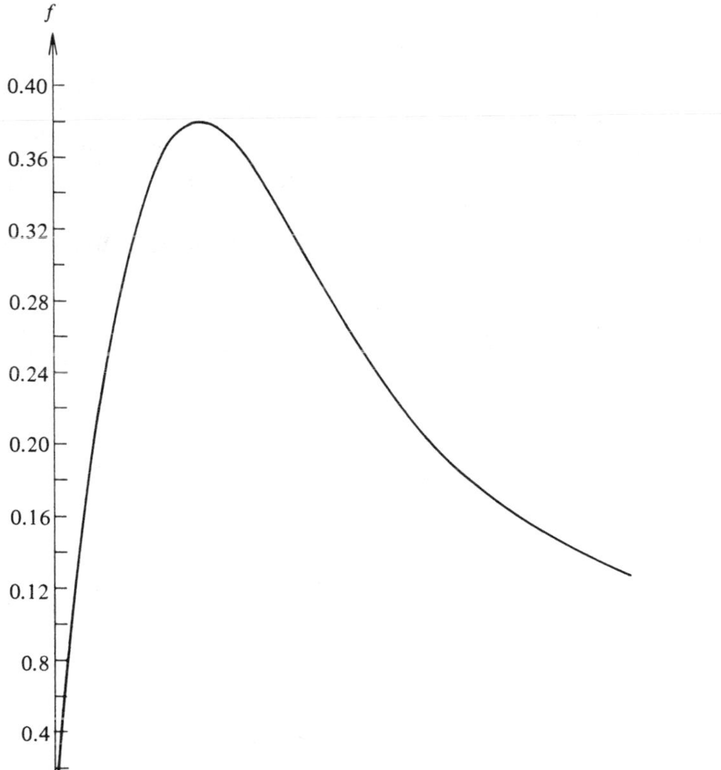

Fig. 14.5-1 f vs β with: $f_{max} = 0.38$ at $\beta = 1.025$, $f'(0) = 0.8$, and $f(\beta) \sim 1/2\beta + 1/4\beta^4 + \cdots$.

Once again, y can be written as

$$y = \eta\epsilon^{1/3} - 1 = \beta(\epsilon x)^{1/3} - 1$$

and the dominant part of u_0 for $\beta \ll (\epsilon x)^{-1/3}$ can be written

$$u_0 \simeq \frac{x^{2/3}}{2\beta\epsilon^{1/3}}$$

which is the same as the asymptotic behavior of $\epsilon^{-1/3} w$ when $\beta \gg 1$.

Thus the region of common validity lies in

$$(\epsilon x)^{-1/3} \gg \frac{y+1}{(\epsilon x)^{1/3}} \gg 1$$

i.e., in $1 \gg (y+1) \gg (\epsilon x)^{1/3}$. Thus, compositely, u_0, and w_1, and another boundary layer identical to w except that it lies along $y = 1$, describe an excellent asymptotic approximation to $u(x, y)$ in $0 < x \ll \epsilon^{-1/3}$.

At this point one should not be trapped into believing the argument which says: "Since, in the foregoing problem, the investigator had to resort to numerical integration anyway, the use of the boundary layer ideas has no advantage over a direct numerical attack on the original problem." The argument, of course, is silly! In a direct numerical attack, one must solve a partial differential equation over a domain containing many mesh points and he must do it again and again for each value of ϵ. Furthermore, the grid must become finer and finer (in some subregion) as the selected value of ϵ becomes smaller. Alternatively, when the boundary layer approach *is* used only one numerical integration is needed; the differential equation to be so solved is ordinary and the delineation of the result applies to all $\epsilon \ll 1$.

14.5.4 A Multilayer Problem

For our look at a multilayer problem, let us suppose that

$$(x^2 + \epsilon w)\, w' + w - (2x^3 + x^2) = 0 \tag{14.5-49}$$

with $w(1) = 1 + e$. An attempt to expand w in the form

$$w \sim \sum_{n=0} \epsilon^n w_n(x)$$

leads immediately to

$$w_0 = e^{1/x} + x^2 \tag{14.5-50}$$

a function which misbehaves rather strongly near the origin.

If, as in section 14.5.3, we define new variables to aid in identifying the contributions from each term of eq. (14.5-49) which are important near $x = 0$, we can write

$$x = \epsilon^{\nu} y$$

and

$$w(x) = \epsilon^{\mu}\phi(y) \tag{14.5-51}$$

We obtain

$$\epsilon^{\nu+\mu}y^2\phi' + \epsilon^{1+2\mu-\nu}\phi\phi' + \epsilon^{\mu}\phi - \epsilon^{2\nu}(y^2 + 2\epsilon^{\nu}y) = 0$$

Small values of x must correspond to values of y of order unity whereas large values of w will correspond to $\phi = O(1)$. Accordingly,

$$\mu < 0 \quad \text{and} \quad \nu > 0$$

It follows that $\nu - \mu = 1$ and that

$$\phi\phi' + \phi \simeq 0 \tag{14.5-52}$$

Thus

$$\phi \simeq a - y$$

i.e.

$$w \sim -\epsilon^{-1}x + a\epsilon^{\mu} \tag{14.5-53}$$

near $x = 0$.

In no way can this function blend with that of eq. (14.5-50) and we must conclude that either (1) the method isn't applicable or (2) there must be an intermediate region in which a transition from the behavior of eq. (14.5-50) to that of eq. (14.5-53) is accomplished. One could expect the need for this transitional behavior via the following argument. The behavior described in eq. (14.5-50) is a result of a balance among w, x^2w', and the nonhomogeneous term. The behavior indicated in (14.5-53) arises from a balance between $\epsilon ww'$ and w. Somewhere in between x^2w' must gradually relinquish its importance and $\epsilon ww'$ must assert its role, and it is likely that this will require an analysis which includes contributions from both x^2w', $\epsilon ww'$ and w.

We denote the 'location' of the transition zone by ϵ^{α} and write

$$x = \epsilon^{\alpha} + \epsilon^{\beta}z$$

$$w(x) = \epsilon^{\gamma} v(z)$$

where we expect $0 < \alpha < \beta$ and $\gamma < 0$.

With these substitutions, eq. (14.5-49) becomes

$$(\epsilon^{2\alpha+\gamma-\beta} + \epsilon^{1+2\gamma-\beta} v)\, v' + \epsilon^{\gamma} v + \cdots = 0$$

The choice $2\alpha = \beta = 1 + \gamma$ is consistent with the foregoing discussion and

$$(1 + v)\, v' + v = 0 \tag{14.5-54}$$

so that

$$\ln v + v = c - z \tag{14.5-55}$$

For large z

$$v \sim e^{c-z} \tag{14.5-56}$$

whereas for small x, eq. (14.5-50) says

$$w = \epsilon^{\gamma} v = e^{1/x} + x^2 \simeq \exp\,[\epsilon^{-\alpha} - z] + O(\epsilon^{2\alpha})$$

So, with $\epsilon^{\gamma} = e^{p}$ (where $p = \epsilon^{-\alpha}$) and with $c = 0$, eqs. (14.5-50) and (14.5-55) are mutually consistent.

Since, $\gamma \ln \epsilon = \epsilon^{-\alpha}$ and $\gamma + 1 = 2\alpha$, so that

$$(1 - 2\alpha) \ln \frac{1}{\epsilon} = \left(\frac{1}{\epsilon}\right)^{\alpha} \tag{14.5-57}$$

we see that, for *very* small ϵ

$$\alpha \simeq \frac{(\ln \ln\,[1/\epsilon])}{\ln\,[1/\epsilon]}$$

For negative and moderately large z, eq. (14.5-55) (with $c = 0$), implies that

$$v \sim -z$$

or

$$w \sim -\epsilon^{\gamma-\beta}(x - \epsilon^{\alpha}) \tag{14.5-58}$$

Hence, with $\beta = 1 + \gamma$ and $a = 1$ and $\mu = 2\gamma + \alpha - 1$, eq. (14.5-53) and (14.5-55) are consistent. Thus there are three descriptions of w which, compositely describe its behavior in $0 < x < 1$.

i.e.

$$w \sim e^{1/x} + x^2 \text{ for large enough } x$$

$$\ln(\epsilon^{-\gamma} w) + (\epsilon^{-\gamma} w) \sim -\epsilon^{-\beta}(x - \epsilon^{\alpha}) \text{ for } x \text{ near } \epsilon^{\alpha} \tag{14.5-59}$$

and

$$w \sim -\epsilon^{-1} x + \epsilon^{\mu} \quad \text{for smaller } x$$

where α is that small positive number for which $(1 - 2\alpha) \ln 1/\epsilon = (1/\epsilon)^{\alpha}$, $\gamma = 2\alpha - 1$, $\beta = 2\alpha$, and $\mu = \alpha - 1$.

The foregoing problem deserves further comment. Not only does it illustrate situations in which the function exhibits more than two scales of behavior* but it also illustrates those problems in which one or more of the scales involve a transcendental dependence on ϵ. It is probably difficult in such cases to continue the construction of the series of which w_0, v, and ϕ are ostensibly the first terms but it is important to note that the analysis still proceeds from the basic ideas of singular perturbation theory and that this kind of truncated analysis can be extremely useful.

14.5.5 A Hyperbolic Problem

A rather peculiar set of results emerges when one studies the problem in which

$$\epsilon(u_{yy} - u_{xx}) + (au_x + u_y) = 0 \quad \text{in } y > 0 \tag{14.5-60}$$

with $u(x, 0) = f(x)$, $u_y(x, 0) = g(x)$ in $0 < x < L$.

The simplest possibility is that there is a thin boundary layer near $y = 0$ and that, in the rest of the triangle $(0, 0; L, 0; L/2, L/2)$, u is well approximated by $w(x, y)$ where

$$aw_x + w_y = 0 \tag{14.5-61}$$

If this expectation were realized w would be given by

$$w = F(x - ay) \tag{14.5-62}$$

and we could seek a description of u in the form

$$u = F(x - ay) + G\left(x, \frac{y}{\epsilon^{\mu}}\right) \tag{14.5-63}$$

*We will see another in section 14.5.6.

It would then follow that (with $\eta = y/\epsilon^{\mu}$)

$$\epsilon^{1-2\mu}G_{\eta\eta} - \epsilon\, G_{xx} + aG_x + \epsilon^{-\mu}G_\eta = \epsilon(1 - a^2)\, F''$$

so that, with $\mu = 1$

$$G_{\eta\eta} + G_\eta = 0 \tag{14.5-64}$$

and

$$G = A(x)\, e^{-\eta} \tag{14.5-65}$$

Thus

$$u = F(x - ay) + A(x)\, e^{-\eta} \tag{14.5-64}$$

The boundary conditions then imply

$$F(x) + A(x) = f(x) \text{ and } -aF'(x) - \epsilon^{-1}A(x) = g(x) \tag{14.5-65}$$

It is quickly seen that we should have written

$$u = F(x - ay) + \epsilon G(x, \eta) \tag{14.5-66}$$

instead of (14.5-63), but this change is accommodated by permitting A to be of order ϵ. In fact, using eqs. (14.5-65), we have

$$F(x) \simeq f(x)$$

and

$$A(x) \simeq -\epsilon g(x) - \epsilon a f'(x)$$

so that

$$u \sim f(x - ay) - \epsilon\,[g(x) + af'(x)]\; e^{-y/\epsilon} \tag{14.5-67}$$

Whenever $|a| < 1$, and $L \ll \epsilon^{-1}$ this result is completely adequate but when, for example, $a > 1$, then part of the region (triangle) in which u should be determined is not 'covered' by eq. (14.5-67) simply because $f(x)$ is not defined in $x < 0$.

We can initiate an improved analysis by inspecting the exact solution of the problem wherein

$$f(x) = \sin kx; \quad g(x) \equiv 0; \quad \text{in } 0 < x < \infty$$

The solution of eq. (14.5-60) with these boundary conditions is

$$u \simeq \operatorname{Im}\left\{\frac{1}{1-A}\left[e^{\epsilon(a^2-1)k^2 y} - Ae^{-y/\epsilon}\right] e^{ik(x-ay)}\right\} \tag{14.5-68}$$

where $A = ika\epsilon + \epsilon^2 k^2$.

Above the boundary layer the solution is almost that which was given by eq. (14.5-67) supplemented by the analytic continuation of eq. (14.5-67) into the region between $y = x$ and $y = x/a$. The 'almost' comes in because the factor $e^{\epsilon k^2(a^2-1)y}$ implies a slow growth with y which the simplest theory didn't detect. Equation (14.5-68) is such, in fact, that, in $y \gg \epsilon$, we might expect u to be better described by

$$u = H[x - ay, \epsilon(a^2 - 1)y] = H(z, s)$$

where the $a^2 - 1$ is inserted for future convenience.

When this is substituted into eq. (14.5-60) we obtain

$$H_{zz} + H_s = 0 \tag{14.5-69}$$

The description of u in the whole triangle then becomes [see eq. (14.5-65)]

$$u = G + H \tag{14.5-70}$$

and the function obtained in this way is in excellent agreement with eq. (14.5-68).

It is now of interest to see what eqs. (14.5-65), (14.5-69), and (14.5-70) imply when

$$f(x) = \frac{1}{\sqrt{x}}; \quad g(x) = 0 \quad \text{in } x > 0$$

The boundary conditions require that

$$H(x, 0) + A(x) = \frac{1}{\sqrt{x}}$$

and

$$\epsilon(a^2 - 1)\, H_s(x, 0) - aH_x(x, 0) - \frac{A(x)}{\epsilon} = 0$$

So, except in $x = O(\epsilon)$, $H \simeq 1/\sqrt{x}$ and we can try

$$H = z^{-1/2}\, P\left(\frac{z}{\sqrt{s}}\right) = z^{-1/2} P(\beta)$$

with $P(\infty) = 1$.

The equation for $P(\beta)$ is

$$P'' - \left(\frac{1}{2}\beta + \frac{1}{\beta}\right)P' + \left(\frac{3}{4\beta^2}\right)P = 0$$

and

$$P = P_0\,\beta e^{\beta^2/8}K_{1/4}\left(\frac{\beta^2}{8}\right);\quad P_0 = \frac{1}{2\sqrt{\pi}}$$

so that

$$H = \frac{1}{2\sqrt{\pi}}\,s^{-1/4}e^{\beta^2/8}\left[\beta^{1/2}K_{1/4}\left(\frac{\beta^2}{8}\right)\right]$$

This result implies that along $x = ay$

$$H = \frac{1}{2\sqrt{\pi}}\,[\epsilon(a^2 - 1)y]^{-1/4}\,\Gamma\left(\frac{1}{4}\right)$$

and along $y = x$

$$H \sim \frac{(\epsilon y)^{-1/4}}{2}\,e^{(a-1)^2 y/4\epsilon}$$

This result is so peculiar that the reader may wish to verify it by extending the boundary conditions to read

$$f(x) = \begin{cases} 0 & x \leqslant 0 \\ x^{-1/2} & x > 0 \end{cases};\quad g(x) = 0;\quad \text{in } -\infty \leqslant x < \infty$$

using the Fourier transform in x to obtain an integral representation of u, and estimating the discrepancy between the result so obtained and H. The results of so doing are gratifying.

14.5.6 A Relaxation Oscillation

When the parameter in the Van der Pol equation, (14.2-31) of section 14.2.3 is large compared to unity rather than small, the phenomena and the mathematics required to describe them are very different from those of that section. In particular, with a simple change of variable, the homogeneous form of that equation can be written

$$\epsilon u''(t) - (1 - u^2)\,u'(t) + u(t) = 0 \qquad (14.5\text{-}71)$$

and it is of interest to find the periodic function which satisfies it. Since the differential equation is of such a form that the solution may exhibit boundary layers, we study first the behavior of u in any region where $\epsilon u''$ is of negligible importance. In such regions $u \simeq w(t)$ where

$$-(1 - w^2)\, w' + w = 0 \tag{14.5-72}$$

and

$$\ln w - \frac{w^2}{2} - t = \text{const} \tag{14.5-73}$$

Without loss of generality one can choose the constant so that, nominally, $w(0) = 1$. That is

$$\ln w - \frac{(w^2 - 1)}{2} = t \tag{14.5-74}$$

This function is graphed in Fig. 14.5-2.

There is, of course, an alternative solution in which $\ln(-w)$, rather than $\ln w$ appears. That solution is relevant in regions where $w < 0$ but the symmetry of the problem is such that $u(t + T) = -u(t)$, where T is the *half-period* of the oscillation and we need not deal explicitly with the alternative solution.

When t is close to zero and w is close to unity, w' is large and w'' is larger. In

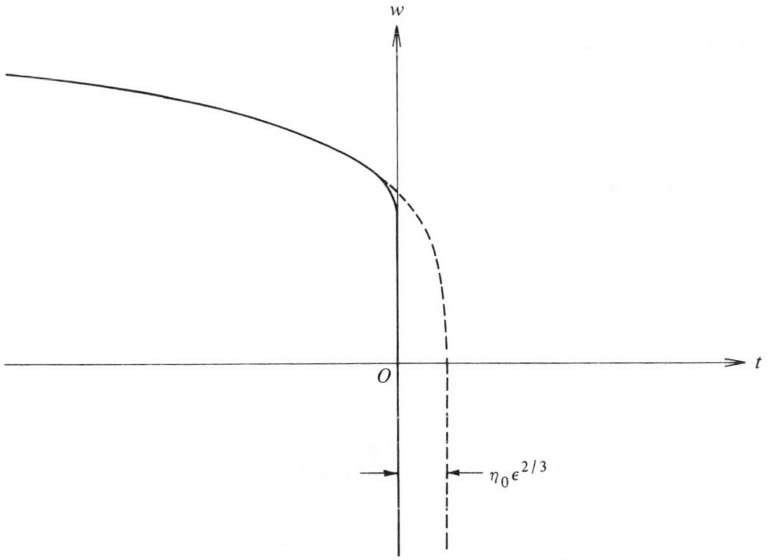

Fig. 14.5-2 Plot of $w(t)$ vs t (solid) and $1 + \epsilon^{1/3} q(\eta)$ vs t (dotted).

fact, as $w \to 1$, $\epsilon w''/w \to \infty$, and it is clear that in eq. (14.5-71) $\epsilon u''$ is important in the neighborhood of $u = 1$. It would be nice if, once again, we could write

$$u \simeq w(t) + f(\tau)$$

where τ denotes some stretched variable [analogous to the $(x - 1)/\sqrt{\epsilon}$ of section 14.5.1] but unfortunately, w is singular at $t = 0$ and u almost surely is not. This implies that f would also have a (cancelling) singularity at $t = 0$ and this makes the additive process cumbersome or even inadequate. Therefore, we seek an approximation to u which is adequate in the region near $t = 0$ and which will blend with w in some region of common validity. As is indicated by the graph of w (Fig. 14.5-2), we can expect u to be of order unity in the region near $t = 0$, and, in view of the steepening of w, we can expect the time scale to be short. Thus, the evident (but inadequate as we shall see) set of variables to try to use is

$$u \simeq v(\xi) \tag{14.5-75}$$

where

$$\xi = t\epsilon^{-k} \tag{14.5-76}$$

and where we expect k to be positive.

When eqs. (14.5-75) and (14.5-76) are substituted into eq. (14.5-71), we obtain

$$\epsilon^{1-2k} v'' - \epsilon^{-k}(1 - v^2)\, v' + v = 0 \tag{14.5-77}$$

and the choice $k = 1$ suggests that a solution of

$$v'' - (1 - v^2)\, v' \simeq 0 \tag{14.5-78}$$

might provide the needed results.

Equation (14.5-78) can be integrated once to give

$$v' - v + \frac{v^3}{3} = \text{const} \tag{14.5-79}$$

which is equivalent to

$$\epsilon w' - w + \frac{w^3}{3} = \text{const} \tag{14.5-80}$$

It is clear that there is nowhere a region in which the w and w' of eq. (14.5-80) bear any resemblence to those of eq. (14.5-72).

The method hasn't failed. The difficulty lies in the fact that, in the region where eq. (14.5-72) is valid the balance implied by eq. (14.5-71) is primarily a balance among the last two terms. Alternatively, when (14.5-78) is valid, the balance lies between the first two terms. The lack of matching occurs because there is a transitional region in which the importance of the third term fades out as the first term assumes more importance.

In order to describe this transition we can anticipate that

$$u = 1 + \epsilon^{\alpha} q(\eta) \tag{14.5-81}$$

where $\eta = \epsilon^{\beta} t$, $\alpha > 0$, and $\beta < 0$. That is, we expect u to stay near unity and the time scale to be shorter than that of $w(t)$.

When eq. (14.5-81) is substituted into eq. (14.5-71), we obtain

$$\epsilon^{1+\alpha+2\beta} q'' + (2\epsilon^{\alpha} q + \epsilon^{2\alpha} q^2)\, \epsilon^{\alpha+\beta} q' + 1 + \epsilon^{\alpha} q = 0 \tag{14.5-82}$$

Since α and β are to be so chosen that q, q', q'', are each of order unity in size, the largest contributions from each of the three original terms are, respectively, $\epsilon^{1+\alpha+2\beta}$, $2\epsilon^{2\alpha+\beta} qq'$, and 1. Thus, if $1 + \alpha + 2\beta = 2\alpha + \beta = 0$ that is, if $\alpha = \frac{1}{3}$ and $\beta = -\frac{2}{3}$, the dominant terms in eq. (14.5-82) form the equation

$$q'' + 2qq' + 1 = 0 \tag{14.5-83}$$

An integral of this equation is

$$q' + q^2 + \eta - C = 0 \tag{14.5-84}$$

When we define $q = \phi'/\phi$, this Ricatti equation becomes

$$\phi'' + (\eta - C)\, \phi = 0 \tag{14.5-85}$$

and its solution is

$$\begin{aligned} \phi &= (\eta - C)^{1/2} \{A J_{1/3} [\tfrac{2}{3} (\eta - C)^{3/2}] + B J_{1/3} [\tfrac{2}{3} (\eta - C)^{3/2}]\} \\ &\equiv (C - \eta)^{1/2} \{A' K_{1/3} [\tfrac{2}{3} (C - \eta)^{3/2}] + B' I_{1/3} [\tfrac{2}{3} (C - \eta)^{3/2}]\} \end{aligned} \tag{14.5-86}$$

We now must find a range of η (and therefore of t) in which eqs. (14.5-74) and (14.5-86) describe the same functional behavior. Those values of η must lie in a region where $\epsilon w'' \ll w$, but $w = 1 + \epsilon^{1/3} q(\eta)$ and $\eta = \epsilon^{-2/3} t$ so eq. (14.5-74) implies that (for $w - 1 \ll 1$),

$$\begin{aligned} q &\simeq (-\eta)^{1/2} \\ q' &\simeq -\tfrac{1}{2} (-\eta)^{-1/2} \\ q'' &\simeq -\tfrac{1}{4} (-\eta)^{-3/2} \end{aligned}$$

and

$$\epsilon w''(t) \simeq q''(\eta) \simeq -\tfrac{1}{4}(-\eta)^{-3/2} \tag{14.5-87}$$

Thus, eq. (14.5-74) can be valid (i.e., can be reasonably accurate) only in

$$-\eta \gg 1 \tag{14.5-88}$$

On the other hand, the asymptotic behavior of the Hankel functions of eq. (14.5-86) is such that only for $B' \equiv 0$ is $\phi'/\phi \sim (C-\eta)^{1/2}$. For all other values of the constant B', $\phi'/\phi \sim -(C-\eta)^{1/2}$. Furthermore, eq. (14.5-86) is valid whenever $\epsilon^{1/3}q \ll 1$, that is, whenever

$$\epsilon^{1/3}(C-\eta)^{1/2} \ll 1 \tag{14.5-89}$$

Therefore, if we choose $B' \equiv 0$ and $C \equiv 0$, eqs. (14.5-88) and (14.5-89) are both true in the region $1 \ll -\eta \ll \epsilon^{-2/3}$ and this defines the region in which the two descriptions, eqs. (14.5-86) and (14.5-74), have a common behavior.

When $B' = 0$, the constants A and B are equal and since only ϕ'/ϕ is of interest, the constant A cancels out so that

$$\begin{aligned}\phi &= (-\eta)^{1/2}K_{1/3}\left[\tfrac{2}{3}(-\eta)^{3/2}\right]\\ &\equiv \eta^{1/2}\left\{J_{1/3}\left[\tfrac{2}{3}(\eta)^{3/2}+J_{-1/3}\left[\tfrac{2}{3}\eta^{3/2}\right]\right\}\right.\end{aligned} \tag{14.5-90}$$

The function $q = \phi'/\phi$ has a singularity at the zeroes of $J_{1/3} + J_{-1/3}$, so the region in which eq. (14.5-90) is relevant to the phenomenon falls short of the point $\eta = \eta_0$, the smallest zero of this Airy function.

Near η_0,

$$q \simeq \frac{1}{\eta - \eta_0} \tag{14.5-91}$$

and we must find still another description of u which is valid near and beyond η_0 and which blends with the behavior of eq. (14.5-91). To do this, we look again at eq. (14.5-79) and compare it to (14.5-84), writing

$$v = 1 + \epsilon^{1/3}q \quad \text{and} \quad \xi = \epsilon^{-1/3}(\eta - \eta_0) \tag{14.5-92}$$

Note that we have redefined ξ [see eq. (14.5-76)] only by a translation of that coordinate. With this substitution, eq. (14.5-79) is

$$\epsilon^{2/3}q' - (1 + \epsilon^{1/3}q) + \left(\frac{1}{3} + \epsilon^{1/3}q + \epsilon^{2/3}q^2 + \frac{\epsilon}{3}q^3\right) = C' \tag{14.5-93}$$

i.e., when $q \simeq (\eta - \eta_0)^{-1}$, eq. (14.5-93) agrees with eq. (14.5-91) to terms of order $\epsilon^{1/3}$ when $C' = -\frac{2}{3}$.

Actually, this problem is delicate enough so that one should, and can, do better than this. The better calculation emerges when we compare eq. (14.5-93) to eq. (14.5-84), recalling that $C = 0$ in the latter. In that comparison, C' must be chosen so that, near $\eta = \eta_0$

$$q' + q^2 + \eta = q' + q^2 - (C' + \tfrac{2}{3})\,\epsilon^{-2/3} + O(\epsilon^{1/3})$$

and this implies that $C' = -\epsilon^{2/3}\eta_0 - \frac{2}{3}$.

Thus, eq. (14.5-93) becomes

$$v' - v + \frac{v^3}{3} = \frac{-2}{3} - \epsilon^{2/3}\eta_0 \qquad (14.5\text{-}94)$$

and

$$\frac{3dv}{(v+a)(v-b)(v-b^*)} = -d\xi \qquad (14.5\text{-}95)$$

where $a = 2 + \epsilon^{2/3}\eta_0/3$ and $b = 1 + \epsilon^{2/3}\eta_0/6 + i\,\epsilon^{1/3}\sqrt{\eta_0}$. Thus

$$\frac{1}{3}\ln(v+a) + \frac{1}{1-v} - \frac{1}{3}\ln(1-v) = -\xi \qquad (14.5\text{-}96)$$

The reader will note that some liberties have been taken in the passage from eq. (14.5-95) to (14.5-96) but, by carrying the more correct arithmetic he can also note that the answer is not affected to the order of interest in that answer. In particular, eq. (14.5-96) is consistent near $\xi = 0$ (which is the same as η near η_0) with eq. (14.5-91). Furthermore, eq. (14.5-96) implies that, as $\xi \to \infty$, $v(\xi) \to -a$. We know, however, that the influence of the third term of eq. (14.5-71) is to initiate a return of u (and hence v) toward positive values. No artifacts of that third term have survived in eq. (14.5-96) but this can be rectified by refining the calculation in the neighborhood of $u = -2$.

Thus, we write

$$u(t) = -a + \epsilon y(\xi) = -2 - \epsilon^{2/3}\frac{\eta_0}{3} + \epsilon y(\xi)$$

we have already defined ξ.

The equation for y becomes

$$y'' + (a^2 - 1)\,y' - a = 0$$

so that

$$y = \frac{a\xi}{a^2 - 1} + B\, e^{-(a^2-1)\,\xi} + N \tag{14.5-97}$$

where B and N must be of order unity. The blending of eq. (14.5-97) with (14.5-96) requires that $B = 3/e$ and any N of order unity changes the time, t, at which eq. (14.5-97) exhibits the same behavior as $-w(t)$, by an amount of order ϵ.

Thus, with errors of order ϵ from the source (and an error of order $\epsilon \ln \epsilon$ from an earlier approximation), we combine the foregoing (with $N = 0$) to obtain for the half-period

$$T = \tfrac{3}{2} - \ln 2 + \tfrac{3}{2}\, \epsilon^{2/3}\, \eta_0 \tag{14.5-98}$$

In substance, of course, this is the time required for w to proceed from the value a to the point near unity, where the description $u = 1 + \epsilon^{1/3} q$ becomes appropriate, plus the time required for q to 'reach' its asymptote at $t = \epsilon^{2/3}\eta_0$. The time required to go from there to $-a$ is of order $|\epsilon \ln \epsilon|$ and is too small to be estimated at all in the foregoing analysis.

As it happens, this recipe is surprisingly good when $\epsilon < 0.01$ and, for larger values, further terms in the series initiated here do not improve the accuracy. Like most asymptotic series, there is only a very small range of values of ϵ over which a two-term approximation isn't accurate enough but a three-term approximation is!

14.5.7 Ekman Layers and Stewartson Layers

There is a fascinating problem which arises in the study of rotating fluids, an informative version of which is

$$\left.\begin{aligned} &\epsilon \Delta\Delta\psi + v_z = 0 \\ &\epsilon \Delta v - \psi_z = 0 \end{aligned}\right\} \quad \text{in } -L < z < L,\ -\infty < x < \infty \tag{14.5-99}$$

where

$$\Delta \equiv \frac{\partial^2}{\partial x^2} + \frac{\partial^2}{\partial z^2} \quad \text{and } 0 < \epsilon \ll 1$$

The boundary conditions are

$$\begin{aligned} &\psi(x, \pm L, \epsilon) = \psi_z(x, \pm L, \epsilon) = 0 \\ &v(x, -L, \epsilon) = -v(x, L, \epsilon) = f(x) \end{aligned} \tag{14.5-100}$$

where $f(x)$ is any monotone increasing function for which $f(-\infty) = 0, f(\infty) = 1$, and for which the transition from zero to unity has a characteristic scale, λ. For example

$$f(x) = \frac{1}{2}\left[1 + \operatorname{erf}\left(\frac{x}{\lambda}\right)\right]$$

is a good special case to keep in mind.

In initiating the analysis, it is useful to think of λ as being of order unity even though we'll relax that constraint before the analysis is completed.

It is easily discovered that a conventional perturbation process fails and that a boundary layer structure must exist just above $z = -L$ and below $z = L$. Accordingly, we write (using the symmetry of the problem to minimize the number of unknown functions

$$\psi = \epsilon^{\beta}\Psi(x, z, \epsilon) + \epsilon^{\alpha}\chi\left(x, \frac{z + L}{\epsilon^{\nu}}, \epsilon\right) + \epsilon^{\alpha}\chi\left(x, \frac{L - z}{\epsilon^{\nu}}, \epsilon\right)$$

$$v = V(x, z, \epsilon) + w\left(x, \frac{z + L}{\epsilon^{\nu}}, \epsilon\right) - w\left(x, \frac{L - z}{\epsilon^{\nu}}, \epsilon\right)$$

Henceforth, we refer to either of $(z + L)/\epsilon^{\nu}$ or $(L - z)/\epsilon^{\nu}$ as η. The functions Ψ and χ are multiplied by powers of ϵ because we don't know how big they will be but, in view of the given boundary conditions, we can safely anticipate that V and w are of order unity.

With these substitutions, eq. (14.5-99) becomes

$$V_z + \epsilon^{1+\beta}\Delta\Delta\Psi + \epsilon^{1-4\nu+\alpha}\chi_{\eta\eta\eta\eta} + \epsilon^{-}w_{\eta} + O(\epsilon^{1-2\nu}) = 0 \qquad (14.5\text{-}101)$$

$$\epsilon^{\beta}\Psi_z - \epsilon\Delta V + \epsilon^{\alpha-\nu}\chi_{\eta} - \epsilon^{1-2\nu}w_{\eta\eta} + O(\epsilon) = 0 \qquad (14.5\text{-}102)$$

Since we expect (actually, we demand) that χ and w will die off exponentially rapidly with increasing η, eqs. (14.5-101) and (14.5-102) can be satisfied only if

$$V_z + \epsilon^{1+\beta}\Delta\Delta\Psi = 0 \qquad (14.5\text{-}103)$$

$$\epsilon^{\beta}\Psi_z - \epsilon\Delta V = 0 \qquad (14.5\text{-}104)$$

$$\epsilon^{1-4\nu+\alpha}\chi_{\eta\eta\eta\eta} + \epsilon^{-\nu}w_{\eta} \simeq 0 \qquad (14.5\text{-}105)$$

$$\epsilon^{\alpha-\nu}\chi_{\eta} - \epsilon^{1-2\nu}w_{\eta\eta} \simeq 0 \qquad (14.5\text{-}106)$$

We choose α and ν so that eqs. (14.5-105) and (14.5-106) reduce to

$$\chi_{\eta\eta\eta\eta} + w_{\eta} = 0 \qquad (14.5\text{-}107)$$

$$w_{\eta\eta} - \chi_\eta = 0 \tag{14.5-108}$$

i.e., $\alpha = \nu = \frac{1}{2}$.

Since w will be order unity, so will χ and, therefore, ψ must be of order $\epsilon^{1/2}$; thus, we must also choose $\beta = \frac{1}{2}$ and eqs. (14.5-103) and (14.5-104) become

$$V_z = 0 \tag{14.5-109}$$

$$\Psi_z = 0 \tag{14.5-110}$$

The appropriately decaying solutions of eqs. (14.5-107) and (14.5-108) are

$$w = A(x)\, e^{-\eta\sqrt{i}} + B(x)\, e^{-\eta\sqrt{-i}} \tag{14.5-111}$$

$$\chi = -A(x)\sqrt{i}\, e^{-\eta\sqrt{i}} - \sqrt{-i}\, B(x)\, e^{-\eta\sqrt{-i}} \tag{14.5-112}$$

whereas, (since V is odd in z)

$$V \equiv V(x) \equiv 0 \tag{14.5-113}$$

$$\Psi \equiv \Psi(x) \tag{14.5-114}$$

The boundary conditions imply that

$$A(x) + B(x) = f(x)$$

$$\Psi(x) - A(x)\sqrt{i} - B(x)\sqrt{-i} = 0$$

$$-iA(x) + iB(x) = 0$$

and it follows that

$$A(x) = B(x) = \frac{f(x)}{2} \tag{14.5-115}$$

$$\Psi(x) = \frac{f(x)}{\sqrt{2}} \tag{14.5-116}$$

The flow associated with this problem has velocity components which, in the x, y, and z directions are $-\psi_z$, v, ψ_x. Thus, the imposed boundary motion produces a flow in which fluid is transported in the x-direction in the boundary layer along the lower boundary toward the transition region of f; it flows vertically at speed $\epsilon^{1/2} f'(x)/\sqrt{2}$ toward the top boundary and then flows away from the updraft region in the boundary layer along the top plate.

The important thing to notice is that the structure of the vertical current is an exceedingly simple consequence of the boundary motion and has no structure imposed by the dynamics. However, if $f(x)$ were narrow enough, the dynamics would alter the structure and we turn now to the question: How narrow must $f(x)$ be in order that the foregoing analysis becomes inadequate?

The clue to the answer lies in eqs. (14.5-103) and (14.5-104). If the horizontal structure had a width, ϵ^μ, the appropriate horizontal variable would be $s = x/\epsilon^\mu$. Furthermore, since V in the foregoing was zero, it may be small in this extended analysis, so we write $V = \epsilon^p W(z, s)$.

If $\mu > \frac{1}{2}$, eqs. (14.5-105) and (14.5-106) would still contain the dominant contributions in the boundary layer regions but eqs. (14.5-103) and (14.5-104) would become $\epsilon^p W_z + \epsilon^{3/2-4\mu}\Psi_{ssss}$ and $\Psi_z - \epsilon^{1/2-2\mu+p} W_{ss}$. Thus, $\mu = \frac{1}{3}$ and $p = \frac{1}{6}$.

Accordingly, we now extend the family of functions $f(x)$ to include those for which $f'(0) = O(\epsilon^{-1/3})$ and the equations defining the problem are

$$\Psi_{ssss} + W_z = 0 \tag{14.5-117}$$

$$W_{ss} - \Psi_z = 0 \tag{14.5-118}$$

Here χ and w are still governed by eqs. (14.5-107) and (14.5-108) and the boundary conditions are given by eq. (14.5-100).

Equations (14.5-117) and (14.5-118) are complicated enough that it pays to introduce their Fourier Transforms (e.g., $\overline{\Psi}(z, \xi) = \int_{-\infty}^{\infty} \Psi(z, s)\, e^{-i\xi s}\, ds$) and these equations become

$$\overline{W}_z + \xi^4\, \overline{\Psi} = 0$$

$$\overline{\Psi}_z + \xi^2\, \overline{W} = 0$$

so that

$$\overline{W} = C(\xi) \sinh \xi^3 z$$

$$\overline{\Psi} = -\xi^{-1} C(\xi) \cosh \xi^3 z$$

Since w and χ are still described by eqs. (14.5-11) and (14.5-12), the boundary conditions (at $z = -1$) on the Fourier transforms of w, χ, W and Ψ become

$$-\epsilon^{1/6} C(\xi) \sinh \xi^3 + \overline{A}(\xi) + \overline{B}(\xi) = \overline{f}(\xi)$$

$$-\xi^{-1} C(\xi) \cosh \xi^3 - \overline{A}\sqrt{i} - \overline{B}\sqrt{-i} = 0$$

$$i\overline{A} - i\overline{B} + \epsilon^{1/2} \xi^2 C(\xi) \sinh \xi^3 = 0$$

Ignoring the contribution of order $\epsilon^{1/2}$ but not that of $\epsilon^{1/6}$ (the reader can ascertain that this is consistent with the earlier approximations) we obtain

$$C = \frac{-\xi \bar{f}(\xi)}{[\sqrt{2}\cosh \xi^3 + \epsilon^{1/6}\xi \sinh \xi^3]}$$

That is

$$\bar{\Psi} = \frac{\bar{F}(\xi)\cosh \xi^3 z}{\sqrt{2}\cosh \xi^3 \left[1 + \frac{\epsilon^{1/6}\xi}{\sqrt{2}}\tanh \xi^3\right]} \tag{14.5-119}$$

where $F(s) \equiv f(x)$.

This time the vertical velocity ψ_x has more structure than in the preliminary analysis but the integrated vertical flow is the same $\left(\int_{-\infty}^{\infty} \psi_x \, dx = \epsilon^{1/2} \lim_{\xi \to 0} (i\xi\bar{\Psi}) = \epsilon^{1/2} [f(\infty) - f(-\infty)/\sqrt{2}]\right)$

The structure of the flow can best be inferred by noting that $\Psi_s(s, x)$ is the convolution of three functions, $F_s(s)$, the inverse of $\cosh \xi^3 z/\cosh \xi^3$, and the inverse of $[1 + \epsilon^{1/6} 2^{-1/2} \xi \tanh \xi^3]^{-1}$. When $z = \pm 1$, the second of these is a delta function; elsewhere it has a width of order unity in the s-coordinate, unit area, and zero second moment, i.e., with $\bar{G}(\xi, z) = \cosh \xi^3 z/\cosh \xi^3$, $\int_{-\infty}^{\infty} s^2 G(s)\, ds = 0$.

The third function also has unit area, a width of order unity, zero second moment and a logarithmic singularity at the origin. Thus, if $F(s)$ is wide compared to unity, the convolution integral is for all practical purposes the convolution of F with two delta functions; that is $\Psi = F/\sqrt{2}$.

Alternatively, when F is narrow compared to each of the other functions, $F'(s)$ plays the role of a delta function and, *if the analysis is not invalid for such cases*, the structure is invariant for different functions $F(s)$.

Furthermore, that structure which is characterized by the zero second moment of Ψ_s, is such that there must be a backflow outside of the upflow region.

Whenever, $F'(s) \gg 1$, especially when $F'(s_0) > \epsilon^{1/6}$, there will necessarily be an interval in s near s_0 where the x-derivatives of w and x cannot be ignored in comparison with their z-derivatives and it would seem that a more refined procedure would be needed to treat such cases. If, in fact, one really wants to see the detailed structure near $(-L, s_0)$, such a refinement is necessary but if one cares only about those characterizing features of the flow which are described by knowing $\Psi(z, \infty) - \Psi(z, -\infty)$ for $L - |z| \gg \epsilon^{1/2}$ and by knowing $\int_{-\infty}^{\infty} s^2 \Psi_s \, ds$ and other such macroscopic features, no such refinement is necessary.

To see this one can note that the Fourier transform of the exact solution for ψ can be written

$$\bar{\psi}_s(z,\xi) = \frac{\begin{vmatrix} \cosh s_1 & \cosh s_2 & \cosh s_3 \\ s_1 \sinh s_1 & s_2 \sinh s_2 & s_3 \sinh s_3 \\ \cosh s_1 x & \cosh s_2 x & \cosh s_3 x \end{vmatrix}}{\begin{vmatrix} \cosh s_1 & \cosh s_2 & \cosh s_3 \\ s_1 \sinh s_1 & s_2 \sinh s_2 & s_3 \sinh s_3 \\ \dfrac{(s_1^2 - \xi^2)^2}{s_1} \sinh s_1 & \dfrac{(s_2^2 - \xi^2)^2}{s_2} \sinh s_2 & \dfrac{(s_3^2 - \xi^2)^2}{s_3} \sinh s_3 \end{vmatrix}}$$

where s_1, s_2, and s_3 are the three of the roots (i.e., the values of r) of $\epsilon^2(\xi^2 - r^2)^3 = r^2$ which lie close to, respectively, $\epsilon\,\xi^3$, $\sqrt{i/\epsilon}$, and $\sqrt{-i/\epsilon}$.

Once again, one can find (exactly) the net vertical flow and the second moment of ψ_s and once again, the total vertical flow is controlled entirely by $f(\infty) - f(-\infty)$ and by ϵ but this time the second moment is ϵ. But the width of Ψ is still $\epsilon^{1/3}$ (in the x-coordinate system) and if there were no downflow region, the second moment would have to be of order $\epsilon^{2/3}$. Hence, the structure which is revealed in simple form by the crude boundary layer analysis is correct and the interpretability of the description is much better than that which a more meticulous procedure would provide. We conclude that the analysis which leads to eq. (14.5-91) is, for almost all purposes, the optimal one and that the superficially inadequate features of the analysis and of the result really are of no importance.

In particular, of course, the presence of the $\epsilon^{1/6}$ factor in the denominator of eq. (14.5-91) is especially disconcerting to some. There is a procedure whereby this particular discomfort can be eliminated and by whose use the formalities are so carried out that one can continue to construct the later terms of the series. For this particular refinement the reader is referred to Ref. 14-6.

14.5.8 Convection vs Diffusion

Another instructive boundary layer problem arises when one studies the boundary value problem

$$\epsilon(U_{xx} + U_{yy}) - U_x = 0 \quad \text{outside of } \Gamma \qquad (14.5\text{-}120)$$

with $U = 1$ on Γ and $U \to 0$ (see Fig. 14.5-3) as $x^2 + y^2 \to \infty$.

This problem would arise if a fluid at temperature T_0 were to flow uniformly and unimpeded to the right, flowing through the object, Γ, whose temperature is $T_0 + T_1$. Near $x, y = (-1, 0)$ heat would diffuse to the left but with ϵ small, the convective motion would keep the layer of heated air thin. Accordingly, for the region just to the left of the object, we could write $U(x, y) = w(\eta, \theta)$, where, because the layer is thin, we define $\eta = (r - 1)/\epsilon^\gamma$ and $r^2 = x^2 + y^2$ and $\theta = \arctan y/x$.

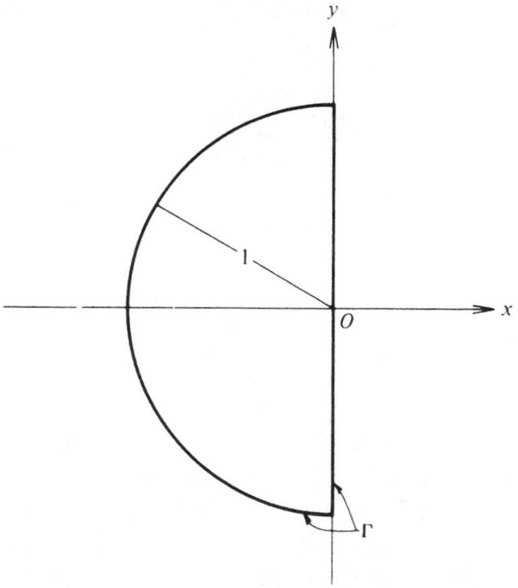

Fig. 14.5-3 Geometry of problem of section 14.5.8.

Equation (14.5-120) becomes

$$\epsilon^{1-2\gamma} w_{\eta\eta} + \frac{\epsilon^{1-\gamma}}{1+\epsilon^\gamma \eta} w_\eta + \epsilon(1+\epsilon^\gamma \eta)^{-2} w_{\theta\theta} - \frac{\cos\theta}{\epsilon^\gamma} w_\eta + \frac{\sin\theta}{1+\epsilon^\gamma \eta} w_\theta = 0 \tag{14.5-121}$$

With $\gamma = 1$ (the only choice for which the diffusive term and the convecting term can make comparable contributions) the dominant terms are

$$w_{\eta\eta} - \cos\theta\, w_\eta = 0 \tag{14.5-122}$$

and

$$w = e^{\eta \cos\theta} \tag{14.5-123}$$

is the solution which obeys the boundary condition on the semicircle and tends to zero (as it should) as $\eta \to \infty$, in $|\theta - \pi| < \pi/2$. However, for $|\theta - \pi|$ close enough to $\pi/2$, $\cos\theta$ is very small and the terms we ignored in eq. (14.5-121) are just as large [for the w of eq. (14.5-123)] as those we depended on in eq. (14.5-122). Thus in, a small range of θ near $\theta = \pi/2$, eq. (14.5-122) cannot be a suitable approximation and we must try a different representation of U. A broader set of possibilities arises when we write $U(x, y) = \psi(z, \xi)$, where $z = (r-1)/\epsilon^\alpha$ and $\xi = (\pi/2 - \theta)/\epsilon^k$ whereupon

$$\epsilon^{1-2\alpha}\psi_{zz} + \frac{\epsilon^{1-\alpha}}{1+\epsilon^{\alpha}z}\psi_z + \epsilon^{1-2k}(1+\epsilon^{\alpha}z)^{-2}\psi_{\xi\xi} - \frac{\sin(\epsilon^k\xi)}{\epsilon^{\alpha}}\psi_z - \frac{\cos\epsilon^k\xi}{\epsilon^k}\psi_\xi = 0 \tag{14.5-124}$$

It seems intuitively clear that, near the top of the circle, the tangential component of the convective process must be important and that, if ψ is to blend with w, the radial diffusion and the radial convection must be represented in the equation which is taken to govern ψ just as they were in the equation for w. Thus, in order that each of these three terms have the same coefficient (unity), we choose $1 - 2\alpha = k - \alpha = -k$, i.e., $\alpha = 2k = \frac{2}{3}$, and the dominant terms of eq. (14.5-124) become

$$\psi_{zz} - \xi\,\psi_z - \psi_\xi = 0 \tag{14.5-125}$$

The boundary conditions require that ψ blend smoothly with w when $-\xi \gg 1$, that $\psi(0, \xi) = 1$, and that $\psi \to 0$ as $z \to \infty$.

A reasonably exact treatment of this equation is rather elaborate and cumbersome but a very good description is forthcoming when one uses an approximation scheme which is generally useful in problems involving a balance of convection and diffusion. For this problem one replaces

$$\psi_{zz} - \xi\psi_z - \psi_\xi = 0$$

by

$$\psi_{zz} - f(\xi)\,\psi_\xi = 0 \tag{14.5-126}$$

using the argument that there must be a fictitious horizontal convective speed $f(\xi)$ which is equivalent (in the balance with the diffusion process) to the convective velocity whose components are $(1, \xi)$. The function $f(\xi)$ will be determined after the solution has been obtained by demanding that

$$\int_0^\infty f(\xi)\,\psi_\xi\,dz = \int_0^\infty (\xi\psi_z + \psi_\xi)\,dz \tag{14.5-127}$$

that is, by demanding that on the average, the replaced term has the same size as its replacement. Calling

$$s = \int_{-\infty}^{\xi} \frac{d\xi}{f} \tag{14.5-128}$$

we have

$$\psi_{zz} - \psi_s = 0 \tag{14.5-129}$$

so that in accord with the boundary conditions,

$$\psi = \text{erfc}\left(\frac{z}{2\sqrt{s}}\right) \tag{14.5-130}$$

Using eq. (14.5-127) we obtain

$$\frac{1}{f} \equiv \frac{ds}{d\xi} = 1 + \sqrt{\pi s}\ \xi \tag{14.5-131}$$

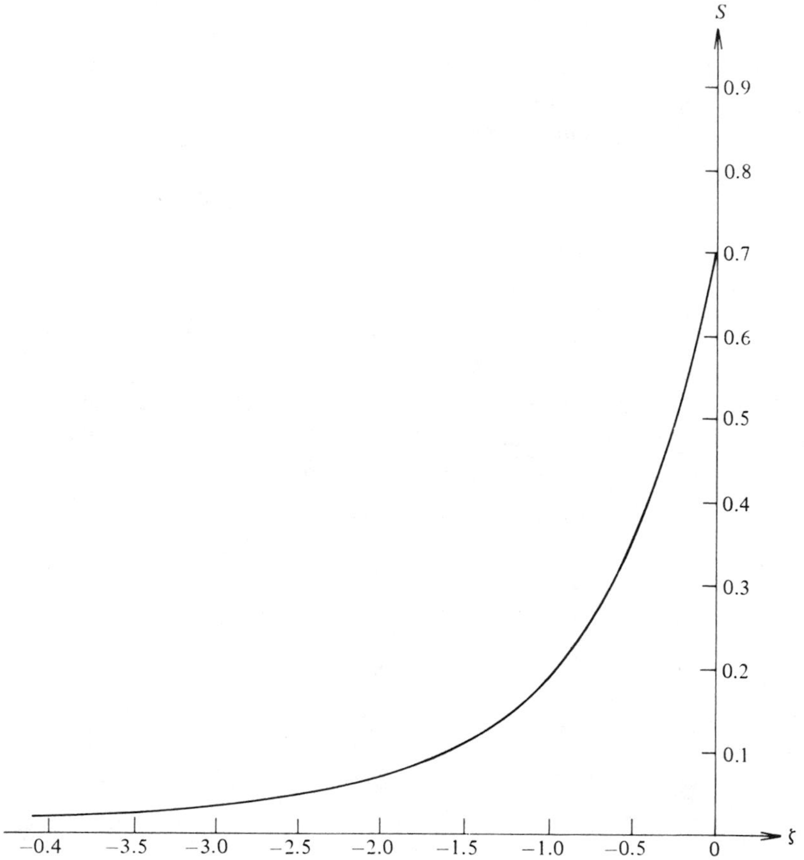

Fig. 14.5-4 Plot of S vs ξ with $S(0) = 0.71$ and $S(\xi) \sim 1/\pi\xi^2 + 4/\pi^2\xi^5 + 40/\pi^3\xi^8 + \cdots$.

However, it is clear that for large negative ξ, ψ should closely resemble $e^{\xi z}$. In particular, comparing the z-derivative at $z = 0$ of $e^{\xi z}$ and $\operatorname{erfc} z/2\sqrt{s}$, we get (at $z = 0$)

$$\frac{d}{dz}(e^{\xi z}) = \xi \quad \text{and} \quad \frac{d}{dz}\operatorname{erfc}\left(\frac{z}{2\sqrt{s}}\right) = -\frac{1}{\sqrt{\pi s}}$$

Thus, we want that solution of eq. (14.5-131) for which $s \to 1/\pi\xi^2$ as $\xi \to -\infty$.

It is also comforting to note that as $\xi \to 0$ the proper convective term is U_ξ so that $ds/d\xi \to 1$ as $\xi \to 0$ in accord with eq. (14.5-131).

The solution of eq. (14.5-131) is shown in Fig. (14.5-4) along with its asymptotic description. It, together with eqs. (14.5-130) and (14.5-123) describe $U(x, y)$ in the region $1 < r < \infty$, $\pi/2 < \theta < \pi$ and, by symmetry, in $\pi < \theta < 3\pi/2$ as well.

In $x > 0$ the solution is best obtained by transform methods using the foregoing results to provide boundary conditions on $x = 0$. Since perturbation procedures are not relevant there, we drop the problem at this point.

14.6 REMARKS

There are many research papers, review articles, and books which display various aspects of perturbation methods, and there is nothing in this section, except perhaps the details of some specific illustrative problems, which could not be inferred from the existing literature. However, the author hopes that he has displayed convincingly one facet of this topic.

The perturbation method appropriate to a given problem is not ordinarily a stereotyped member of a collection of such methods; in fact, the situation is quite the opposite; a surprising number of the problems which arise do not succumb to a routine analysis but require that the investigator be perceptive in anticipating the structure of the solution and that he be innovative in choosing the details of his procedure.

In support of this statement (and as a general summary) the following items may be informative.

The problems in sections 14.2.1, 14.3.1, and 14.4.4 (with the cubic term in u omitted and $\beta = 0$) each admit an ordinary expansion in powers of ϵ. The problems in 14.2.1 and 14.4.4 have solutions which are entire functions of ϵ but the standard series would fail to reveal the interesting features of 14.4.4 whereas it is uniformly valid and useful in 14.2.1. In section 14.3.1, the series fails to converge for $t > 4/3\epsilon$ and is thereby useless for a different reason.

In section 14.3.1 the multiscaling analysis proves to be useful because the phenomenon really has two scales—that of the oscillation and that of its decay. In section 14.4.3 multiscaling is needed again, but really for a different reason. The length scale, ϵ^{-1}, is not particularly important in the solution (although it is certainly present); the real reason that one needs the second variable stems from the transcendental dependence of the solution on ϵL.

Boundary layer problems admit a great variety of procedures. Those of sections 14.5.1, 14.5.5, and 14.5.7 admit additive descriptions whereas 14.5.8 involves additive descriptions with regard to the variable r (at a rather trivial level) for each of two ranges of θ but the description over all θ is more readily constructed by the blending type of analysis. In the rest of section 14.5, the blending analysis is more appropriate but requires, in several instances, a recognition of the fact that the fragmentation of the problem requires more than two regions with differing scales. In particular, the problems of sections 14.5.6 and 14.5.7 are of such a nature that it is valuable to include contributions which depend on ϵ within a term from which the ϵ dependence has presumably been 'scaled-out.'

Finally and repetitiously, then, the successful use of perturbation methods on other than routine problems requires a willingness to invent variations on various well-defined techniques in such a way that the construction adopted is well matched to the intrinsic structure of the function sought.

14.7 REFERENCES

14-1 Burkill, J. C., *The Theory of Ordinary Differential Equations*, Oliver and Boyd, Ltd., London, 1956.

14-2 Kevorkian, J. and Cole, J. D., *Perturbation Methods in Applied Mathematics*, Springer-Verlag, New York, 1981.

14-3 Cohen, H. G., "On the Subharmonic Synchronization of the Van der Pol Oscillator," *Acts du Colloque International des Vibrations Non Lineaires*, Du Ministre de L'Air, 1953.

14-4 Perng, D., "Wave Reflection by a Periodic Medium," *SIAM Journal on Applied Mathematics*, **22**, 280, March 1972.

14-5 Baum, H. R., "On the Weakly Damped Harmonic Oscillator," *Quarterly of Applied Mathematics*, **29** (4), Jan. 1972.

14-6 Carrier, G. F., "Another Singular Perturbation Device," Congres International des Mathematiciens, Nice, Sept. 1970.

14-7 Bender, C. M. and Orszag, S. A., *Advanced Mathematical Methods for Scientists and Engineers*, Chapters 7, 9, 10, 11, McGraw-Hill, New York, 1978.

15

Wave Propagation

Wilbert Lick*

15.0 INTRODUCTION

Wave propagation is an extremely broad and diverse field of study and includes such topics as vibrating strings and membranes, water waves, acoustics, waves in solids, electromagnetic waves, and probability waves in quantum mechanics. No effort will be made here to survey this field of wave propagation, but, consistent with the title of this handbook, selected results and methods of solution will be presented. More detailed and extensive treatments of various topics in wave propagation can be found in the references listed in the bibliography.

Instead of treating each physical system separately, as is done in most texts, the attempt will be made here to classify and treat systems on the basis of mathematical and physical similarities and characteristics. In the next section, elementary concepts of waves are introduced followed by a functional classification of both linear and nonlinear waves into categories of simple (nondispersive, nondiffusive), dispersive, and diffusive waves. In section 15.2, the basic equations of wave propagation for various physical systems are presented, and the systems are discussed in terms of the above categories.

Simple, dispersive, and diffusive waves are discussed in more detail in sections 15.3, 15.4, and 15.5, respectively. In these sections, selected topics are presented (1) to demonstrate basic mathematical methods, and (2) to illustrate the important and general physical similarities and differences among the various types of wave propagation. For convenience and clarity, the topics discussed are limited to a few physical systems. In section 15.3, the principal physical systems discussed are vibrating strings and waves in compressible fluids. In section 15.4, the main example chosen to illustrate a dispersive system is water waves while, in section 15.5, a diffusive system is illustrated mainly by topics concerned with waves in a compressible fluid including dissipation.

*Prof. Wilbert Lick, Dep't. of Mechanical Engineering, University of California, Santa Barbara, Calif. 93106.

15.1 GENERAL DEFINITIONS AND CLASSIFICATION OF WAVES

15.1.1 Elementary Concepts of Waves

A wave will be defined as a disturbance propagating through a medium. Since the concept of a disturbance implies energy, the basic characteristic of wave propagation is the transfer of energy through a system. In the following, simple examples of wave propagation will be discussed which will serve to define some of the more elementary concepts encountered in wave propagation although it must be realized that the definitions presented are sometimes provisional and will be modified when more complicated wave propagation problems are discussed.

Mathematically, the simplest example of wave propagation is that of a disturbance moving in one direction only, say the positive x-direction of a rectangular coordinate system, at constant speed c and with unchanging form. A physical example closely approximating this ideal situation is that of a disturbance traveling along a stretched, elastic string.

If, in the present example, a measure of the disturbance, say ψ, is represented at the time $t = 0$ by some arbitrary function $f(x)$, then the above description of the motion of the disturbance implies that the disturbance can be represented at a later time by $\psi(x, t) = f(\xi)$, where $\xi = x - ct$. That is, to an observer moving with the speed c in the positive x-direction, conditions remain unchanged. The argument $\xi = x - ct$ is called the *phase* of the wave function ψ. The speed c is called the *wave velocity* or, more precisely and more clearly, the *phase velocity* for reasons to be explained later (see section 15.3.1). Similarly a wave propagating in the negative x-direction at the constant speed c, with unchanging form, and with initial condition $g(x)$ can be represented at a later time by $\psi(x, t) = g(\eta)$, where $\eta = x + ct$.

For the wave traveling in the positive x-direction, the wave function satisfies the differential equation

$$\frac{\partial \psi}{\partial t} + c\frac{\partial \psi}{\partial x} = 0 \tag{15.1-1}$$

as can be verified by substitution. For a wave traveling in the negative x-direction, the wave function satisfies the equation

$$\frac{\partial \psi}{\partial t} - c\frac{\partial \psi}{\partial x} = 0 \tag{15.1-2}$$

A first-order equation cannot be derived which satisfies both of these wave functions. However, both functions do satisfy the same second-order equation

$$\frac{\partial^2 \psi}{\partial t^2} - c^2\frac{\partial^2 \psi}{\partial x^2} = 0 \tag{15.1-3}$$

In ξ, η coordinates, this equation reduces to

$$\frac{\partial^2 \psi}{\partial \xi \partial \eta} = 0 \tag{15.1-4}$$

of which the most general solution is $\psi = f(\xi) + g(\eta) = f(x - ct) + g(x + ct)$, where f and g are arbitrary functions. The solution consists of a superposition of two waves traveling in the positive and negative x-directions, respectively, each of which travels at constant speed c and whose waveform remains unchanged with time.

The generalization of eq. (15.1-3) to two and three dimensions is

$$\frac{\partial^2 \psi}{\partial t^2} - c^2 \nabla^2 \psi = 0 \tag{15.1-5}$$

where ∇^2 is the Laplacian operator. In rectangular coordinates, this equation takes the form

$$\frac{\partial^2 \psi}{\partial t^2} - c^2 \left(\frac{\partial^2 \psi}{\partial x^2} + \frac{\partial^2 \psi}{\partial y^2} + \frac{\partial^2 \psi}{\partial z^2} \right) = 0 \tag{15.1-6}$$

Solutions of this equation can be represented by

$$\psi(x, y, z, t) = f(\alpha x + \beta y + \gamma z \pm ct) \tag{15.1-7}$$

where f is an arbitrary function and α, β, and γ are arbitrary constants except for the restriction that $\alpha^2 + \beta^2 + \gamma^2 = 1$. These solutions are called *plane waves* and represent a wave in a three-dimensional medium propagating in one direction only. The disturbance at a particular time will be the same at all points of a plane perpendicular to the direction of propagation. This plane is called the *wave front*. In general, a wave front is defined as a surface at all points of which at any instant of time the phase of the disturbance has the same value. As in one-dimensional waves, the form of the wave represented by eq. (15.1-7) remains unchanged as the wave propagates.

All possible solutions of eqs. (15.1-5) and (15.1-6) can be expressed as linear combinations of solutions of the form given by eq. (15.1-7). However, other solutions are possible and convenient in certain cases. In general, these latter solutions do not have the property that the wave form remains unchanged with time.

In the above discussion, the phase velocity c has been assumed to be constant. This velocity is a function of the properties of the medium and therefore generally varies in a nonhomogeneous medium. Because of this variation in phase velocity, when waves impinge on a boundary separating two media with different properties and hence different wave velocities, complex wave motions result. Of the incoming disturbance, part is reflected. Part is transmitted into the second medium but is generally refracted or changed in direction. Also occurring in this situation is the phenomenon of diffraction, or the ability of waves to bend around the corners of obstacles. When the properties of the medium are continuously varying, the above phenomena are still present but become more complicated.

So far, only scalar waves have been discussed, i.e., ψ has magnitude only. More generally, vector waves are possible in which the wave function is a vector $\boldsymbol{\psi}$. The wave equation for $\boldsymbol{\psi}$ is then equivalent to three scalar equations, which of course are not necessarily independent. If the wave function is a vector and its direction is always parallel to the direction of propagation of the wave, the wave is called *longitudinal*. If the wave function is a vector and its direction is always at right angles to the direction of propagation of the waves, the wave is called *transverse*.

Irrotational and solenoidal vector wave functions can also be defined. A wave function vector is *irrotational* if its curl is zero, and it is *solenoidal* if its divergence is zero.

An important characteristic of the wave functions which satisfy eq. (15.1-5) is that they are linear. As a consequence, if $\psi_1, \psi_2, \ldots, \psi_N$ are each solutions, then so is any linear combination

$$\psi = \sum_{n=1}^{N} a_n \psi_n \tag{15.1-8}$$

where the a_n's are arbitrary constants. A great many time-dependent physical phenomena can be described to a good approximation by linear wave equations, although these equations generally are not as simple as those shown above.

More generally, the wave function and its derivatives appear as products in the governing equations. The equations are then nonlinear and solutions can not be superimposed. The mathematical methods of solution are then considerably more complex.

15.1.2 Classification of Wave Equations

As indicated previously, equations describing the propagation of disturbances or waves can be considerably more complicated than eq. (15.1-5). The physical behavior of waves can also be quite different from the behavior as described by eq. (15.1-5). Despite these differences, there are certain broad similarities among solutions of certain classes of equations describing waves.

So as to be able to discuss wave propagation problems readily, wave equations will be categorized as simple (nondispersive, nondiffusive), dispersive, or diffusive, classifications which will be seen not to be precise but to be useful. The discussion will follow that of Ref. 15-1. In order to classify wave equations, let us consider linear equations first and, for the present, consider the one-dimensional, time-dependent propagation of waves in a medium which is infinite in extent and has constant properties. It can be shown (see Chapter 11 and section 15.3.1) that the general solution of an initial value problem can be considered as the superposition of elementary solutions of the form $A \exp[i(kx - \omega t)]$, where the amplitude $A(k)$ and $\omega(k)$ are functions of the wave number k. In general, ω is a complex function.

The characteristics of the wave motion depend on the functional relation $\omega(k)$. In particular, the variation of the phase velocity $c \equiv \text{Re}\, \omega(k)/k$ with k indicates that harmonic waves of different wave number travel at different speeds and leads to dispersion of the different wave components. If $\omega(k)$ has an imaginary part, damping of a monochromatic wave and diffusion of a general wave system will be present.

These properties of ω lead to a classification of linear wave propagation problems as follows: (a) If ω is real and proportional to k, the phase speed is independent of wave number, dispersion and diffusion are absent, and the system is classified as simple, or nondispersive, nondiffusive; (b) If ω is real with $\omega''(k) \neq 0$, dispersion is present but diffusion is absent and the system is classified as dispersive; and (c) If ω is complex, both diffusion and dispersion are generally present. However, in the wave propagation problems considered here, diffusion is dominant, at least for large enough values of the independent variables, and therefore this type of system will be classified as diffusive.

Simple examples of the cases defined above are

Simple (nondispersive, nondiffusive) waves

$$\frac{\partial \psi}{\partial t} + \frac{\partial \psi}{\partial x} = 0 \qquad (15.1\text{-}9)$$

Dispersive waves

$$\frac{\partial \psi}{\partial t} + \frac{\partial \psi}{\partial x} = -\beta \frac{\partial^3 \psi}{\partial x^3} \qquad (15.1\text{-}10)$$

Diffusive waves

$$\frac{\partial \psi}{\partial t} + \frac{\partial \psi}{\partial x} = \delta \frac{\partial^2 \psi}{\partial x^2} \qquad (15.1\text{-}11)$$

It should be noted that the equations describing the propagation of waves in a physical system are not necessarily hyperbolic.

Equation (15.1-9) is similar to eq. (15.1-1) and describes a wave propagating in the positive x-direction at unit speed. The phase velocity c equals 1. Dispersion and diffusion are absent. The wave form remains unchanged with time.

Equation (15.1-10) describes a wave propagating in the positive x-direction but modified by dispersion (for β a small positive constant). The phase velocity, c, equals $1 - \beta k^2$. No damping or diffusion are present since ω is real. For an initial disturbance $S(-x)$, where $S(-x) = 0$ for $x > 0$ and $S(-x) = 1$ for $x < 0$, the solution to this equation can be written as an integral of the Airy function and indicates a wave traveling in the positive x-direction at unit speed. Ahead of the line $x = t$, ψ is monotonically decreasing to zero as x increases. The function ψ reaches a maximum slightly behind the line $x = t$ and then, as $x \to -\infty$, oscillates rapidly as a nearly uniform wave train with slowly decreasing amplitude and slowly changing frequency. This solution is somewhat atypical of solutions to linear dispersive equations. More generally for linear dispersive wave equations, an asymptotic expansion for large x and t can be obtained by the method of stationary phase and is of the form

$$\psi \sim \left[\frac{2\pi}{x\,|\omega''(k_0)|}\right]^{1/2} f(k_0) \cos\left(k_0 x - \omega t - \frac{\pi}{4} \operatorname{sgn} \omega''\right) \tag{15.1-12}$$

where f is determined from the initial conditions and $k_0(x, t)$ is the solution of $x = \omega'(k_0)t$. The above solution represents a nearly uniform wave train with slowly varying amplitude and frequency, which is also a characteristic feature of the solution of eq. (15.1-10).

Equation (15.1-11) describes a wave propagating in the positive x-direction with the wave form modified by diffusion. It is assumed that δ is a small positive constant. The phase velocity c equals 1, but ω has an imaginary part indicating damping of a monochromatic wave and diffusion of a nonmonochromatic wave. For an initial disturbance $S(-x)$, the solution to this equation is

$$\psi = \frac{1}{2} \operatorname{erfc}\left(\frac{x - t}{2\sqrt{\delta t}}\right) \tag{15.1-13}$$

indicating that the center of the wave travels in the positive x-direction at unit speed and the wave diffuses with a characteristic width proportional to $(\delta t)^{1/2}$. The same results hold true asymptotically for large x and t for an arbitrary initial disturbance.

The method of classification illustrated here can easily include two- and three-dimensional wave propagation problems and vector waves as long as in all cases the equations are linear. For an inhomogeneous medium, the system will be classified as if the properties of the medium were constant.

For nonlinear equations, no general classification is commonly accepted. However, from a functional point of view, nonlinear systems for a wide range of physical wave propagation problems can be categorized on the basis of the type of equation that they reduce to upon linearization. The equations below are simple extensions of eqs. (15.1-9) to (15.1-11) to include nonlinear effects and can be classified readily by this rule. It has been assumed that, upon linearization, ψ can be written as one plus a small perturbation from one.

1. The usual wave equation common in gas dynamics

$$\frac{\partial \psi}{\partial t} + \psi \frac{\partial \psi}{\partial x} = 0 \tag{15.1-14}$$

which describes waves propagating in the positive x-direction. The linearized version, eq. (15.1-9), is nondispersive, nondiffusive and therefore the above equation is also classified as nondispersive, nondiffusive (or simple).

2. The Korteweg-deVries equation

$$\frac{\partial \psi}{\partial t} + \psi \frac{\partial \psi}{\partial x} = -\beta \frac{\partial^3 \psi}{\partial x^3} \tag{15.1-15}$$

a nonlinear dispersive wave equation.

3. Burgers' equation

$$\frac{\partial \psi}{\partial t} + \psi \frac{\partial \psi}{\partial x} = \delta \frac{\partial^2 \psi}{\partial x^2} \tag{15.1-16}$$

a nonlinear diffusive wave equation.

These equations arise naturally in many wave propagation problems. Their solutions will be discussed in sections 15.3.7, 15.4.6, and 15.5.5. It will be seen that the character of the solutions and the methods of solution are distinctly different for each of these equations. They also may be significantly different from those of the related linear equation although retaining certain similarities.

15.2 PHYSICAL SYSTEMS AND THEIR CLASSIFICATION

15.2.1 Vibrating Strings

Probably the simplest example of wave propagation, both physically and mathematically, is that of waves on a stretched, elastic string. This case has been alluded to in section 15.1.1 where it served to introduce certain elementary definitions.

To be specific, consider a finite string of length L, of uniform density ρ, and, when in equilibrium, under tension T and lying along the x-axis. The string is disturbed from equilibrium by being pulled aside in the x, y-plane. The x and y components of the displacement of the element originally at $(x, 0)$ are denoted by ξ and η respectively.

The equations of motion in the horizontal and transverse directions result in two coupled nonlinear equations for ξ and η (see, for example, Ref. 15-2). However, if it is assumed that $\partial\xi/\partial x$ and $\partial\eta/\partial x$ are both small, then the equations can be linearized, are also uncoupled, and become

$$\frac{\partial^2 \eta}{\partial t^2} - \frac{T}{\rho}\frac{\partial^2 \eta}{\partial x^2} = 0 \tag{15.2-1}$$

$$\frac{\partial^2 \xi}{\partial t^2} - \frac{Y}{\rho}\frac{\partial^2 \xi}{\partial x^2} = 0 \tag{15.2-2}$$

where Y is Young's modulus. It can be seen that both displacements are propagated as simple waves without dispersion or diffusion. The phase velocity of the transverse component η is given by $\sqrt{T/\rho}$ while the phase velocity of the longitudinal component is $\sqrt{Y/\rho}$.

When the density of the string varies as a function of position, the basic equations describing the motion are of the same form as eqs. (15.2-1) and (15.2-2) except that $\rho = \rho(x)$. The solutions of this type of equation can be significantly different from those of eqs. (15.2-1) and (15.2-2) (see section 15.3.5).

If a frictional force proportional to the velocity of the string is considered, the equation of motion for the transverse displacement reduces to

$$\frac{\partial^2 \eta}{\partial t^2} - c^2 \frac{\partial^2 \eta}{\partial x^2} + R \frac{\partial \eta}{\partial t} = 0 \tag{15.2-3}$$

where $c^2 = T/\rho$ and R is the effective resistance per unit length per unit mass of string. Solutions to this equation are no longer of the simple wave type but are diffusive in character.

15.2.2 Vibrating Membrane

A slightly more complex problem is that of waves on a membrane, here approximated by a two-dimensional, perfectly flexible surface stretched by a uniform force per unit length T and with constant density or mass per unit area σ. It is assumed that the equilibrium position of the surface is a plane, say the x, y-plane in rectangular coordinates, and that a displacement $\xi(x, y)$ of an element of the surface occurs only in the z-direction.

In a similar manner as for a vibrating string, it is assumed that $\partial\xi/\partial x$ and $\partial\xi/\partial y$ are sufficiently small so that the equation of motion in the vertical direction can be linearized. In this approximation, the resulting equation for the displacement ξ is

$$\frac{\partial^2 \xi}{\partial t^2} - \frac{T}{\sigma} \nabla^2 \xi = 0 \tag{15.2.4}$$

where ∇^2 is the two-dimensional Laplacian. The above equation is of the simple wave type with the phase speed given by $\sqrt{T/\sigma}$.

If damping is considered, then the above equation is modified by an additional term proportional to $\partial\xi/\partial t$ and the motion is then diffusive in character.

15.2.3 Waves in Compressible Fluids

The equations of motion for a compressible fluid are extremely complex if a realistic description of the fluid is desired, e.g., if viscosity, heat conduction, mass diffusion, chemical reaction, etc. are considered. Only two idealized cases will be considered here. These are (1) an isentropic gas, and (2) a chemically reacting gas.

For the first case, it is assumed that the gas is compressible, the entropy of each fluid element remains constant, and all dissipative processes are neglected. With these assumptions, the equations describing the conservation of mass, momentum, and energy are

$$\frac{\partial \rho}{\partial t} + \boldsymbol{\nabla} \cdot \rho \mathbf{v} = 0 \tag{15.2-5}$$

$$\frac{D\mathbf{v}}{Dt} + \frac{1}{\rho} \boldsymbol{\nabla} p = 0 \tag{15.2-6}$$

$$\frac{Dh}{Dt} - \frac{1}{\rho} \frac{Dp}{Dt} = 0 \tag{15.2-7}$$

where

$$\frac{D}{Dt} = \frac{\partial}{\partial t} + \mathbf{v} \cdot \nabla \tag{15.2-8}$$

ρ is the density, $\mathbf{v}$ is the three dimensional velocity of the fluid, p is the pressure, and h is the enthalpy per unit mass. Equation (15.2-7) can be written alternatively as

$$\frac{DS}{Dt} = 0 \tag{15.2-9}$$

where S is the entropy per unit mass. In addition an equation of state is needed and will be assumed to be

$$p = \rho R T \tag{15.2-10}$$

where R is the gas constant and T is the temperature.

These equations are nonlinear and difficult to solve. Some methods of solution and characteristic results for the nonlinear problem are presented in sections 15.3.6 and 15.3.7. For the present, in order to classify the system, consider the one-dimensional, time-dependent wave propagation problem. If it is assumed that disturbances are small and nonlinear terms can be neglected, a single linear equation governing the motion can be derived and is

$$\frac{\partial^2 u}{\partial t^2} - a^2 \frac{\partial^2 u}{\partial x^2} = 0 \tag{15.2-11}$$

where u is the velocity of the fluid in the x-direction, and $a^2 = (\partial p/\partial \rho)_s$ and is the isentropic speed of sound evaluated for conditions in the undisturbed fluid.

The system is therefore of the simple wave type. For waves propagating in an inhomogeneous but stationary medium, the basic linear equation is of the same form as eq. (15.2-11) except that $a = a(x)$.

For a chemically reacting gas, the equations of motion take the same form as eqs. (15.2-5) to (15.2-7). However, all thermodynamic variables are now functions of the composition and a continuity equation for each component of the gas must be used to complete the system of equations.

For simplicity, only one nonequilibrium chemical reaction will be considered. In this case, the degree of nonequilibrium is determined by some parameter q. If dissociation of a diatomic gas is being considered, q is the mass fraction of one component, say Y_1. Since $Y_1 + Y_2 = 1$, Y_2 is determined if Y_1 is known.

An equation of state is assumed to be known in the form

$$h = h(p, \rho, q) \tag{15.2-12}$$

In equilibrium, $q = \tilde{q}(p, \rho)$ and $h = \tilde{h}[p, \rho, \tilde{q}(p \rho)] = \tilde{h}(p, \rho)$ where the symbol $\sim$ denotes the equilibrium value of the quantity at the local values of pressure and density.

In the absence of diffusion, the continuity or rate equation will be assumed to have the form

$$\frac{Dq}{Dt} = -\frac{q - \tilde{q}}{\tau} \tag{15.2-13}$$

where τ denotes the relaxation time for the reaction being considered and is assumed constant.

The rate of change of entropy can be written as

$$T\frac{DS}{Dt} = -\mu_P \frac{Dq}{Dt} \tag{15.2-14}$$

where μ_P has the character of a chemical potential. For dissociation of a diatomic gas $A_2 \to 2A$, where q is the mass fraction of the monatomic gas, $\mu_P = \mu_A - \mu_{A_2}$.

If nonequilibrium of an internal degree of freedom is being considered, say the vibrational degree of freedom of a diatomic gas, the above equations apply if q is taken to be T_i, the temperature of the internal degrees of freedom, and μ_P is taken to be $c_i(T_i - T)/T_i$, where c_i is the specific heat due to the internal degrees of freedom.

The above equations are nonlinear of course and even more complex than the ones for an isentropic fluid. Various approximate methods of solution will be discussed in section 15.5. For the present, let us note that for the one-dimensional, time-dependent problem when the disturbances are small, the resulting linear equation is

$$\frac{\partial}{\partial t}\left(\frac{\partial^2\phi}{\partial t^2} - a_f^2\frac{\partial^2\phi}{\partial x^2}\right) + \lambda\left(\frac{\partial^2\phi}{\partial t^2} - a_S^2\frac{\partial^2\phi}{\partial x^2}\right) = 0 \qquad (15.2\text{-}15)$$

where ϕ is a velocity potential defined by $u = \partial\phi/\partial x$, $p = -\rho_0\,\partial\phi/\partial t$, and where $a_f^2 = (\partial p/\partial\rho)_{S,q}$ is the frozen isentropic speed of sound calculated at constant composition and constant entropy, while $a_S^2 = (\partial p/\partial\rho)_{S,q=\tilde{q}}$ is the equilibrium isentropic speed of sound calculated at equilibrium composition and constant entropy. The parameter λ is inversely proportional to the relaxation time τ. The above equation is of the diffusive type and its solution will be discussed in sections 15.5.1 and 15.5.2.

15.2.4 Waves in Elastic Solids

Consider a homogeneous, isotropic, elastic medium in which the disturbances are small. Hooke's law, wherein each stress is proportional to the corresponding strain, will be assumed. Define a displacement vector $\mathbf{r} \equiv \mathbf{i}\xi + \mathbf{j}\eta + \mathbf{k}\zeta$ where ξ, η, and ζ are the components of the displacement of an element in a rectangular coordinate system.

With these approximations, the basic vector equation of motion for the displacement is

$$\rho\frac{\partial^2\mathbf{r}}{\partial t^2} = \left(B + \frac{4\mu}{3}\right)\nabla\nabla\cdot\mathbf{r} - \mu\nabla\times\nabla\times\mathbf{r} \qquad (15.2\text{-}16)$$

where ρ is the density, B the bulk modulus, and μ the shear modulus.

It is convenient to distinguish two limiting cases: (1) $\mathbf{r}$ is an irrotational vector for which $\nabla\times\mathbf{r} = 0$, and (2) $\mathbf{r}$ is a solenoidal vector for which $\nabla\cdot\mathbf{r} = 0$.

Consider first the case $\nabla\times\mathbf{r} = 0$. Then the above equation becomes

$$\frac{\partial^2\mathbf{r}}{\partial t^2} - \frac{1}{\rho}\left(B + \frac{4\mu}{3}\right)\nabla^2\mathbf{r} = 0 \qquad (15.2\text{-}17)$$

which is a simple wave equation for each component of $\mathbf{r}$. The wave propagation speed for each component of $\mathbf{r}$ is then $c = \sqrt{(B + 4\mu/3)/\rho}$. Because of the condition $\nabla\times\mathbf{r} = 0$, general solutions of this set of equations can be found in the form of plane waves with the propagated displacement being directed along the direction of propagation. The wave is therefore longitudinal. No shears are present.

Consider next the case $\nabla\cdot\mathbf{r} = 0$. Then eq. (15.2-16) becomes

$$\frac{\partial^2\mathbf{r}}{\partial t^2} - \frac{\mu}{\rho}\nabla^2\mathbf{r} = 0 \qquad (15.2\text{-}18)$$

which is also a simple wave equation for each component of $\mathbf{r}$. The wave propagation speed is now $c = \sqrt{\mu/\rho}$. Plane wave solutions are possible but in this case, because of the condition $\nabla \cdot \mathbf{r} = 0$, the propagated displacement is at right angles to the direction of propagation, i.e., a plane solenoidal elastic wave is transverse.

15.2.5 Electromagnetic Waves

The Maxwell equations relating the electric field intensity **E** and the magnetic induction **B** in a medium free of electric charges and currents are

$$\nabla \cdot \mathbf{E} = 0 \tag{15.2-19}$$

$$\nabla \cdot \mathbf{B} = 0 \tag{15.2-20}$$

$$\nabla \times \mathbf{E} = -\frac{1}{c}\frac{\partial B}{\partial t} \tag{15.2-21}$$

$$\nabla \times \mathbf{B} = \frac{\mu\epsilon}{c}\frac{\partial \mathbf{E}}{\partial t} \tag{15.2-22}$$

where c is the velocity of light in a vacuum, μ is the permeability, and ϵ is the dielectric constant. The Gaussian system of units has been used here where **E** is measured in electrostatic units and **B** in electromagnetic units.

From these equations, one can derive

$$\frac{\partial^2 \mathbf{E}}{\partial t^2} - \frac{c^2}{\mu\epsilon}\nabla^2 \mathbf{E} = 0 \tag{15.2-23}$$

$$\frac{\partial^2 \mathbf{B}}{\partial t^2} - \frac{c^2}{\mu\epsilon}\nabla^2 \mathbf{B} = 0 \tag{15.2-24}$$

which shows that both **E** and **B** propagate as simple waves with a wave speed given by $c/\sqrt{\mu\epsilon}$. Since **E** and **B** have zero divergence, the waves are solenoidal and the disturbances are at right angles to the direction of propagation of the wave. It can be shown that although **E** and **B** separately satisfy the simple wave equation, they are not propagated independently but are coupled and are always perpendicular to each other.

When Maxwell's equations are written for a continuous but conducting medium, **E** and **B** no longer have solutions of the simple wave type. Waves periodic in time are now attenuated with distance due to the conductivity of the medium. Dispersion is also present.

At very high frequencies, a continuum model is no longer valid and an atomic model of the medium must be considered. An analysis of the interaction between an electromagnetic wave and an atom shows strong absorption and dispersion when the frequency of radiation is approximately the same as one of the resonance frequencies of the atom. Anomalous dispersion, in which the group velocity (see section 15.4.1) is greater than the velocity of light in vacuum, results.

15.2.6 Water Waves

Surface waves on water are one of the most common and at the same time most intriguing examples of wave propagation. However, the mathematics of surface waves is quite complex and, because of this, certain physical and mathematical approximations have had to be made in order to obtain solutions to any but the most trivial problems. Most of the classic work on this problem has assumed that the water is incompressible, inviscid, and irrotational; these approximations will be made here.

Under these conditions, the continuity equation can be written as

$$\nabla^2 \phi = 0 \tag{15.2-25}$$

where $\phi(x, y, z, t)$ is the velocity potential and satisfies

$$\mathbf{v} = \boldsymbol{\nabla}\phi \tag{15.2-26}$$

where $\mathbf{v}$ is the velocity of a fluid element. In addition, Bernoulli's equation is valid throughout the fluid and can be written in terms of the velocity potential as

$$\frac{\partial \phi}{\partial t} + \frac{1}{2}\left[\left(\frac{\partial \phi}{\partial x}\right)^2 + \left(\frac{\partial \phi}{\partial y}\right)^2 + \left(\frac{\partial \phi}{\partial z}\right)^2\right] + \frac{p}{\rho} + gy = 0 \tag{15.2-27}$$

where g is the acceleration due to gravity and acts in the negative y-direction.

The boundary conditions are: (1) a kinematic condition on the free surface $y = \eta(x, z, t)$

$$\frac{\partial \phi}{\partial y} - \frac{\partial \phi}{\partial x}\frac{\partial \eta}{\partial x} - \frac{\partial \phi}{\partial z}\frac{\partial \eta}{\partial z} - \frac{\partial \eta}{\partial t} = 0 \tag{15.2-28}$$

which states that a fluid particle which is initially on the surface remains on the surface; (2) no normal velocity on fixed surfaces, so that

$$\frac{\partial \phi}{\partial n} = 0 \tag{15.2-29}$$

on these surfaces; and (3) the pressure is prescribed at a free surface, so that Bernoulli's equation can be written as

$$\frac{\partial \phi}{\partial t} + \frac{1}{2}\left[\left(\frac{\partial \phi}{\partial x}\right)^2 + \left(\frac{\partial \phi}{\partial y}\right)^2 + \left(\frac{\partial \phi}{\partial z}\right)^2\right] + \frac{p}{\rho} + g\eta = 0 \tag{15.2-30}$$

on $y = \eta(x, z, t)$.

The above equations are of course nonlinear. Their solution is complex due to the nonlinear terms appearing in the boundary conditions, and the fact that, in the case of free surface problems, the location of the free surface is not known *a priori* and has to be found as part of the solution. Because of this, progress in understanding wave propagation in water with a free surface has only been made by analyzing various limiting theories.

One limiting theory results from the assumption that the wave amplitudes are small. The basic equations are then linear. The theory has been extensively used for wave propagation studies. In this limit, Laplace's equation for ϕ, eq. (15.2-25), is still valid throughout the fluid. The appropriate boundary conditions are (1) $\partial\phi/\partial n = 0$ on a fixed surface, and (2) at a free surface, the linearized kinematic condition and Bernoulli's equation can be combined to form a single equation for ϕ, which is

$$\frac{\partial^2 \phi}{\partial t^2} + g \frac{\partial \phi}{\partial y} = 0 \tag{15.2-31}$$

and is to be applied at the location of the undisturbed free surface taken as $y = 0$. These equations are linear but dispersive in character. Solutions will be discussed in section 15.4.

A second limiting theory results when it is assumed that the depth of the water is small, say, by comparison with the wave length of the disturbance. Small amplitude disturbances are not assumed and the resulting lowest order approximation equations are nonlinear. The approximation is known as shallow water theory or the theory of long waves.

To lowest order, it can be shown that the pressure is hydrostatic. In two dimensions (x, y variations only), the horizontal momentum and continuity equations reduce to

$$\frac{\partial u}{\partial t} + u \frac{\partial u}{\partial x} = -g \frac{\partial \eta}{\partial x} \tag{15.2-32}$$

$$\frac{\partial}{\partial x} [u(\eta + h)] = -\frac{\partial \eta}{\partial t} \tag{15.2-33}$$

By a suitable transformation of dependent variables (Ref. 15-3), it can be shown that these equations are analogous to the equation describing one-dimensional, time-dependent wave propagation in a compressible fluid. In this limit, the waves described by these equations are not dispersive, but are of the simple wave type although nonlinear.

A third limiting case of considerable interest can be derived as a far-field expansion of the solution of an initial-value problem (Ref. 15-4). This expansion is valid for small disturbances and sufficiently large x and t, and takes into account first-order nonlinear and dispersive effects simultaneously. The governing equation is

$$\frac{\partial u}{\partial \tilde{t}} + 3u \frac{\partial u}{\partial \tilde{x}} = -\frac{1}{3} \frac{\partial^3 u}{\partial \tilde{x}^3} \tag{15.2-34}$$

where $\tilde{x} = x - t$, $\tilde{t} = \epsilon t$, and ϵ is small and is a measure of the amplitude of the waves. This is essentially the Korteweg-deVries equation, eq. (15.1-15). Its solution will be discussed in section 15.4.6.

15.2.7 Probability Waves in Quantum Mechanics

Consider a nonrelativistic system with one degree of freedom consisting of a particle of mass m restricted to motion along a fixed straight line, which we take as the x-axis. Assume that the system can be described as having a potential energy $V(x)$ throughout the region $-\infty < x < +\infty$.

For this system, the Schrodinger wave equation is

$$\frac{-h^2}{8\pi^2 m}\frac{\partial^2 \psi}{\partial x^2} + V\psi = \frac{-h}{2\pi i}\frac{\partial \psi}{\partial t} \tag{15.2-35}$$

where h is Planck's constant. The function $\psi(x, t)$ is called the Schrodinger wave function including the time, or the probability amplitude. This quantity is necessarily complex due to the appearance of $i \equiv \sqrt{-1}$ in the above equation. Denote the complex conjugate of ψ by ψ^*. The product $\psi^*\psi$ is then the probability distribution function, and $\psi^*\psi\, dx$ is the probability that the particle is in the region between x and $x + dx$ at time t.

Alternatively, the above equation can be written as a set of two equations, second-order in time and fourth-order in space, for two real quantities which are the real and imaginary parts of ψ.

Wavelike solutions of the above equation can be found. However, they are not of the simple wave type but are dispersive in character and lead, for example, to the spreading of an electron wave packet with time. If initially the wave packet has a width Δx_0, then after a time t the width of the packet will be on the order of $\Delta x = ht/2\pi m \Delta x_0$ where m is the mass of the electron. The uncertainty principle, of which the above is one example, is a direct consequence of the dispersive character of probability waves.

15.3 SIMPLE WAVES: NONDISPERSIVE, NONDIFFUSIVE

15.3.1 Basic Solutions for Linear Waves

In this section, only the basic solutions for the linear, simple wave equation, eq. (15.1-5), with constant coefficients will be discussed. In general, linear wave forms are neither periodic nor harmonic. Nevertheless, the study of periodic and, in particular, harmonic waves is important since (1) the behavior and analysis of harmonic waves is relatively simple; (2) many wave sources used in practice produce waves that are almost harmonic; and (3) when the principle of superposition holds, an arbitrary wave function can be represented as a sum of harmonic waves by the use of Fourier analysis.

Let us consider first harmonic waves for one-dimensional, time-dependent wave propagation in a homogeneous medium. The general solutions of the governing equation, eq. (15.1-3), can be written as $f_1(x - ct)$ and $f_2(x + ct)$ where f_1 and f_2 are arbitrary functions. Particular solutions of this type are

$$\begin{aligned} \psi &= A \cos [k(x \pm ct) + \epsilon] \\ &= A \cos [kx \pm \omega t + \epsilon] \end{aligned} \tag{15.3-1}$$

where A is the amplitude, ϵ is the phase constant, k is a constant called the *wave number*, and ω is the angular frequency and is equal to kc. The above solutions represent traveling periodic waves varying harmonically in space and time. The time period in which ψ varies through a complete cycle is $2\pi/\omega = T$. The reciprocal of the *period* T, or $\omega/2\pi = \nu$, is called the *frequency*. The spatial period of ψ is given by $2\pi/k = \lambda$ and is called the *wavelength*.

The above waves can also be represented as combinations of sine and cosine waves, i.e.,

$$\psi = B \cos k(x \pm ct) + C \sin k(x \pm ct) \tag{15.3-2}$$

where $B = A \cos \epsilon$ and $C = -A \sin \epsilon$. In fact, it can be easily shown that any combination of trigonometric waves, all having the same frequency but with arbitrary amplitudes and phase constants, can be reduced to a single trigonometric wave of the same frequency.

Waves of the above type can also be represented in complex form as

$$\psi = Be^{i(kx \pm \omega t)} \tag{15.3-3}$$

where either the real or imaginary part can represent a traveling harmonic wave. If $B = Ae^{i\epsilon}$, then the real part of the above expression corresponds to eq. (15.3-1).

As noted above, linear combinations of the above solutions are also solutions to the simple wave equation. In particular, if one adds two waves traveling in opposite directions with the same amplitude A and with phase constants which are of the same magnitude but opposite in sign, one obtains

$$\psi = 2A \cos kx \cos (\omega t + \epsilon) \tag{15.3-4}$$

This no longer represents a traveling wave but a standing harmonic wave. The disturbance does not propagate. At points for which $x = (2n - 1)\pi/2k$, where n is an integer, the disturbance vanishes for all t, while at points for which $x = n\pi/k$, the disturbance has maximum amplitude. These points are called nodes and anti-nodes respectively.

An alternate representation of a standing wave is given by

$$\psi = 2A \sin kx \sin (\omega t + \epsilon) \tag{15.3-5}$$

Linear combinations of standing waves can always be made to represent traveling waves.

The above solutions represent waves propagating in an infinite medium. In a finite medium, the effect of boundary conditions is to restrict the frequencies allowable for a solution to certain discrete numbers. As an example, consider a one-dimensional, finite medium $0 \leqslant x \leqslant l$ at the ends of which the conditions are that the disturbance is zero for all time. It follows from eqs. (15.3-4) and (15.3-5) that possible solutions of this problem are

$$\psi = \sum_n A_n \sin k_n x \sin (\omega_n t + \epsilon_n) \tag{15.3-6}$$

or

$$\psi = \sum_n (A_n \sin \omega_n t + B_n \cos \omega_n t) \sin k_n x \tag{15.3-7}$$

where k_n is no longer arbitrary but must satisfy the condition $k_n = n\pi/l$, where n is an integer.

The solution associated with a particular value of k_n is called a characteristic or normal mode. The characteristic or normal mode frequencies are then given by $\omega_n = n\pi c/l$ and the characteristic wavelengths by $\lambda_n = 2l/n$.

It can be seen that, for bounded media, the standing wave solution is the natural way to represent the solution because of the ease in handling the boundary conditions in contrast to a superposition of traveling waves.

The complete statement of the above problem usually requires that some initial conditions be satisfied. In general, a finite sum of normal mode solutions will not satisfy these initial conditions. However, it can be shown, by means of the Fourier theorem (see Chapter 11) that an infinite series of the above form will. For example, if at $t = 0$ conditions are specified such that $\psi(x, 0) = f_1(x)$ and $\partial\psi/\partial t(x, 0) = f_2(x)$, then, from eq. (15.3-7), it follows that

$$f_1(x) = \sum_{n=1}^{\infty} B_n \sin \frac{n\pi x}{l} \tag{15.3-8}$$

$$f_2(x) = \sum_{n=1}^{\infty} A_n \omega_n \sin \frac{n\pi x}{l} \tag{15.3-9}$$

Each of these series is a Fourier series and represents the corresponding function if the function and its derivative with respect to x are piecewise continuous in the interval $0 \leqslant x \leqslant l$. Then the series for f_1, for instance, converges to f_1, at all points of continuity and to $\frac{1}{2}[f_1(x+) + f_1(x-)]$ at points of discontinuity. Here $f_1(x+)$ means the value of the function at the discontinuity as the point is approached from the right, and $f_1(x-)$ means the value when approached from the left.

The A_n's and B_n's can be calculated by multiplying eqs. (15.3-8) and (15.3-9) by $\sin m\pi x/l$, where m is an integer, and integrating from 0 to l. The result is

$$A_n = \frac{2}{n\pi c} \int_0^l f_2(x) \sin \frac{n\pi x}{l}\, dx \tag{15.3-10}$$

$$B_n = \frac{2}{l} \int_0^l f_1(x) \sin \frac{n\pi x}{l}\, dx \tag{15.3-11}$$

The solution to the problem is then given by eq. (15.3-7) with the A_n's and B_n's given by the above.

The Fourier theorem allows an arbitrary function of a single variable in a bounded medium to be represented as an infinite series of harmonic waves. For an infinite medium, a function such as a wave pulse or bounded wave train can also be represented in terms of harmonic waves. This is made possible by the Fourier integral theorem (see Chapter 11 for a statement of the theorem and its use). The above type of analysis can be extended to media of higher dimensions. This is discussed in Chapter 11 and in the references listed in the bibliography.

15.3.2 Energy Transport

As noted in section 15.1.1, the concept of a disturbance or wave implies the transfer of energy. For a linear, simple wave in a homogeneous medium, the calculation of this energy transport is relatively straightforward. The procedure is simply to calculate the total energy density in the part of the medium being traversed by the wave, average over time, and then multiply by the velocity of propagation, which in this case is just the wave or phase speed.

As an example, let us consider the vibrating string. At any instant of time, the total energy of the string consists of kinetic and potential energy. Only transverse displacements will be considered in the following. The potential energy can be calculated as the work required to stretch the string by displacing it from its equilibrium position to the form $\eta(x, t)$. This work is given by the tension T times the change in length of the string, which is $\sqrt{1 + (\partial\eta/\partial x)^2}\, dx - dx$. It follows that the potential energy per unit length is

$$V = \frac{1}{2} T \left(\frac{\partial \eta}{\partial x}\right)^2 \tag{15.3-12}$$

if higher order terms in $\partial\eta/\partial x$ are neglected. The kinetic energy per unit length is given by

$$K = \frac{1}{2} \rho \left(\frac{\partial \eta}{\partial t}\right)^2 \tag{15.3-13}$$

where ρ is the density of the string.

For a harmonic wave with displacement given by

$$\eta = A \cos(\omega t - kx) \tag{15.3-14}$$

the average energy transported per unit time or power transmission is then

$$P = \tfrac{1}{2} \rho \omega^2 A^2 c \tag{15.3-15}$$

where $c^2 = T/\rho$.

In other more complicated cases such as wave propagation in inhomogeneous, dispersive, or diffusive media, or nonlinear wave propagation, the above procedure of calculating energy transport is no longer valid. The concept of the velocity of energy transport is very difficult in these cases and it is better to use the concept of energy flux, a quantity which is much easier to define and calculate and which accomplishes the same objectives.

15.3.3 Reflection and Refraction

To illustrate the effects of reflection and refraction, consider first the simple case of an infinite string, $-\infty < x < +\infty$, which has a discontinuous change in density ρ at $x = 0$. Since the phase speed is given by $c = \sqrt{T/\rho}$ where T is the tension in the string, this implies a discontinuous change in phase speed such that, say, $c = c_1$ for $-\infty < x < 0$ and $c = c_2$ for $0 < x < +\infty$.

For $x < 0$, assume a periodic wave propagating in the positive x-direction of the form

$$\psi = Ae^{i(\omega t - k_1 x)} \tag{15.3-16}$$

where ω is the angular frequency and $k_1 = \omega/c_1$. This wave will be incident on the junction at $x = 0$ where a partial reflection and transmission will occur. Write the reflected wave as

$$\psi = Be^{i(\omega t + k_1 x)} \tag{15.3-17}$$

and the transmitted wave as

$$\psi = Ce^{i(\omega t - k_2 x)} \tag{15.3-18}$$

where $k_2 = \omega/c_2$.

Continuity of ψ and $\partial\psi/\partial x$ are required in order that the displacements and restoring force are continuous across the junction. These conditions lead to the relations

$$\frac{B}{A} = \frac{c_2 - c_1}{c_1 + c_2} \tag{15.3-19}$$

$$\frac{C}{A} = \frac{2c_2}{c_1 + c_2} \tag{15.3-20}$$

A reflection coefficient R can be defined as the ratio of the energy reflected to the incident energy. Since energy is proportional to the amplitude squared (see section 15.3.2), it follows that

$$R = \left|\frac{B}{A}\right|^2 = \left(\frac{c_2 - c_1}{c_1 + c_2}\right)^2 \tag{15.3-21}$$

Similarly a transmission coefficient $\mathcal{T}$ can be defined as the ratio of the energy transmitted to the incident energy. Since energy must be conserved, and therefore the relation $R + \mathcal{T} = 1$ must be satisfied, it follows that

$$\mathcal{T} = \frac{4c_1c_2}{(c_1 + c_2)^2} \tag{15.3-22}$$

This last relation can also be derived by direct considerations, i.e., by calculating the rate of work done by the incident and transmitted waves at the junction. One can show that the incident energy is then $T\omega^2A^2/2c_1$ while the transmitted energy is $T\omega^2C^2/2c_2$. The above result for $\mathcal{T}$ then follows after the use of eq. (15.3-20).

A more complex example than the previous one is that of the reflection of a plane sound wave at a plane boundary separating two media (see Fig. 15.3-1). The density of the medium from which the wave is incident, the upper medium, will be denoted by ρ_1, the wave speed by c_1, and the wave function by ψ_1. Quantities in the lower medium will be denoted by the subscript 2. The angle of incidence is θ_1 and the angle of refraction is θ_2. It will be assumed that the normal to the wave front lies in the x, z-plane.

Introduce a potential function ψ such that $\mathbf{v} = \nabla\psi$ and $p = -\rho\partial\psi/\partial t$. From this definition and the basic equations for small disturbances in a compressible, isen-

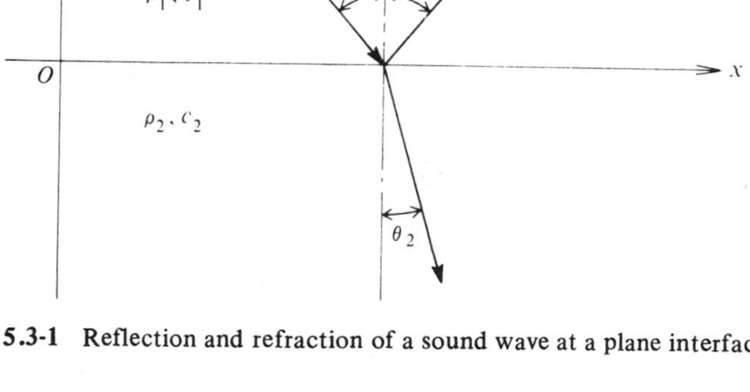

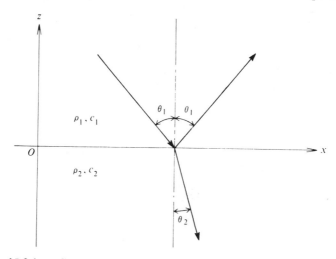

Fig. 15.3-1 Reflection and refraction of a sound wave at a plane interface.

tropic gas given in section 15.2.3, it follows that ψ satisfies the simple, linear wave equation.

For a wave periodic in time, the incident wave can then be written as

$$\psi = A \exp\left[-ik_1(x \sin\theta_1 - z\cos\theta_1)\right] \tag{15.3-23}$$

where A is the amplitude, $k_1 = \omega/c_1$ and is the wave number, and the factor $e^{i\omega t}$ has been omitted here and in the following. The reflected wave can be written as

$$\psi = B \exp\left[-ik_1(x \sin\theta_1 + z\cos\theta_1)\right] \tag{15.3-24}$$

and the refracted, or transmitted, wave can be written as

$$\psi = C \exp\left[-ik_2(x \sin\theta_2 - z\cos\theta_2)\right] \tag{15.3-25}$$

where B and C are amplitudes and $k_2 = \omega/c_2$ and is the wave number of the transmitted wave.

Across the interface separating the two media, the pressure and normal component of the particle velocity must be continuous. From these requirements and the definition of ψ, it follows that

$$\rho_1\psi_1 = \rho_2\psi_2 \tag{15.3-26}$$

$$\frac{\partial\psi_1}{\partial z} = \frac{\partial\psi_2}{\partial z} \tag{15.3-27}$$

at $z = 0$.

By the use of the above conditions and eqs. (15.3-23) to (15.3-25), it can be shown that

$$\sin\theta_1 = n \sin\theta_2 \tag{15.3-28}$$

where $n = k_2/k_1 = c_1/c_2$. This is the usual form of Snell's law.

The amplitudes of the reflected and transmitted waves are found to be

$$\frac{B}{A} = \frac{\rho_2 c_2 \cos\theta_1 - \rho_1 c_1 \cos\theta_2}{\rho_2 c_2 \cos\theta_1 + \rho_1 c_1 \cos\theta_2} \tag{15.3-29}$$

$$\frac{C}{A} = \frac{\rho_1}{\rho_2}\left(1 + \frac{B}{A}\right) \tag{15.3-30}$$

From eq. (15.3-29), it follows that the amplitude of the reflected wave becomes zero at the angle θ^*, where θ^* satisfies the relation

$$\sin \theta^* = \sqrt{\frac{m^2 - n^2}{m^2 - 1}} \tag{15.3-31}$$

where $m = \rho_2/\rho_1$. The angle θ^* is not necessarily a real angle, i.e., for arbitrary values of m and n, the amplitude of the reflected wave will not necessarily go to zero as θ is varied.

When $n < 1$ and therefore $c_2 > c_1$, and the angle of incidence satisfies the condition $\sin \theta > n$, total internal reflection will occur. The modulus of the reflected wave will be one. The amplitude of the refracted wave then decreases exponentially with distance from the interface in the negative z-direction.

Of course, the law of conservation of energy is valid and in the present case states that the energy transported to the interface by the incident wave is equal to the energy transported away from the interface by the reflected and refracted waves. Only the normal component of the energy flux needs to be considered in the calculation.

15.3.4 Diffraction

Diffraction effects arise whenever a wave is partially obstructed due to a change in the properties of the medium, either a discontinuous change, e.g., an obstacle, or a continuous change. When the wavelength of the disturbance is small by comparison with the dimensions of the obstacle or effective dimension of the inhomogeneous region, the use of ray theory (the geometrical optics approximation) is sufficient to describe the wave motion. In this limit, diffraction effects are absent. However, when the wavelength of the disturbance is comparable to or greater than the effective dimension of the obstacle, diffraction effects are important and in order to describe these effects a better approximation to the motion of the wave must be obtained.

In this section, only diffraction effects for linear, simple wave propagation in a homogeneous medium will be considered. Of course, one method of solution of this problem is to solve the wave equation with the appropriate boundary conditions directly either by analytic or numerical methods. This is an extremely difficult boundary-value problem and relatively few problems have been solved in this manner.

A powerful and approximate, but relatively accurate, method of solving diffraction problems is an integral procedure due to Kirchhoff. By starting with the wave equation and using Green's theorem, one can write the wave function ψ at any point P in a region V enclosed by a surface S approximately as

$$\psi(t) = \frac{1}{4\pi} \int_S \left\{ \frac{1}{r} \left[\frac{\partial \psi}{\partial n} \right]_- + \frac{ik}{r} \frac{\partial r}{\partial n} [\psi]_- \right\} ds \tag{15.3-32}$$

where r is the distance from the point P. In this relation, the approximation that the distance r is much greater than the wavelength of the disturbance has been made. The notation $[f]_-$ means that the quantity f is to be evaluated at the retarded time $t' = t - r/c$ and not at the time t.

The above relation, called the Kirchhoff theorem or Kirchhoff integral, states that the value of ψ at any point P may be expressed as a sum of contributions from each element on a surface enclosing P, each such contribution being determined by the value of the wave on the element in question at a time t' which is r/c earlier than the time t. This delay time is the time it would take for a wave to travel the distance r at the speed c. The above is a precise form of Huyghens' principle commonly found in elementary texts.

To illustrate the method of solution by means of the above integral and also to show some typical results, let us consider the case of a plane wave incident on an infinite plane screen containing a finite aperture. To find $\psi(t)$ at a point behind the screen, apply the above equation to a closed surface made up of the entire screen (including the aperture) and an infinite hemisphere behind the screen centered at the aperture.

Contributions to the integral over the hemisphere vanish. If it is assumed that there is no wave directly behind the screen, then $\psi = \partial\psi/\partial n = 0$ there, and the surface integral over the screen contributes nothing to the integral. The only contribution to the integral is that due to the wave incident on the aperture, which wave is assumed to be the same as if no screen were present.

Even for this simple case, for an arbitrary aperture, the evaluation of the Kirchhoff integral is difficult and the solution is quite complicated. To illustrate the general characteristics of the solution, consider a relatively simple case for which the solution has been found in analytic form, i.e., a plane wave incident on a rectangular aperture with sides of length a and b parallel to the x and y axes respectively.

When the wave number k becomes large (and therefore the wavelength becomes small), a first approximation to the integral can be found rather simply. The result states that the wave amplitude is identically zero in the shadow of the screen and the wave elsewhere is just a plane wave. This is just the well known geometrical optics limit or ray approximation.

In the opposite limit as $k \to 0$ or $\lambda \to \infty$, an asymptotic analysis gives

$$\psi = \frac{iAkab}{2\pi d} \frac{\sin u}{u} \frac{\sin v}{v} e^{i(\omega t - kd)} \tag{15.3-33}$$

where A is the amplitude of the incident wave, $u = kax/2d$, $v = kby/2d$, and d is the distance of the point P measured in a normal direction from the screen. The resulting diffraction pattern consists of an array of rectangles bounded by lines of zero intensity given by $x = n\lambda d/a$ and $y = m\lambda d/b$, where n and m are positive and negative integers but not zero. The intensity is maximum at the center. Within each rectangle there is a local maximum. The above limit is called Fraunhofer diffraction.

All other diffraction phenomena which are intermediate between the geometrical optics and Fraunhofer approximations, are classified as being of the Fresnel type.

For the above problem, a complete solution can be expressed in terms of Fresnel integrals.

In addition to the Kirchhoff integral procedure, another approximate but general method of solution of diffraction problems is by the use of the geometrical theory of diffraction (Refs. 15-37, 15-38, 15-39). This theory is an extension of geometrical optics (see section 15.35), accounts for diffraction in an approximate manner, and is valid asymptotically for small values of the wavelength.

The basic idea of the theory is to introduce diffracted rays which are similar to the usual rays of geometrical optics. It is assumed that the diffracted disturbance travels along these rays. The diffracted rays are produced by incident rays which hit edges, corners, or vertices of boundary surfaces, or which graze such surfaces. The initial amplitude and phase of the field on a diffracted ray is determined from the incident field with the aid of an appropriate diffraction coefficient. These diffraction coefficients are different for each type of diffraction, e.g., edge, vertex, etc. and must be determined from exact solutions or experiment. However, it is assumed that only the immediate neighborhood of the point of diffraction can affect the values of the diffraction coefficients and therefore these coefficients can be determined from relatively simple problems in which only the local geometrical and physical properties enter. Away from the diffracting surface, diffracted rays behave like ordinary rays.

The above ideas are an extension of Young's idea (in 1802) that diffraction is an edge effect, and can be tested readily for at least one simple problem where an exact solution is known, the diffraction of a plane wave by a semi-infinite screen with a straight edge. It can be shown that the solution to this problem (Ref. 15-37) consists of incident and reflected waves plus a third wave. When the direction of propagation of the incident wave is normal to the edge of the screen, this third wave is cylindrical with the edge of the screen as its axis and appears to come from the edge. This diffracted wave, obtained from the exact solution in the limit of small wavelength, is identical to the wave predicted by the geometrical theory of diffraction. Other more complicated problems also show good agreement between the exact and approximate theories.

The theory has been applied to diffraction problems in optics, water waves, elastic waves, and quantum-mechanical waves. As long as the wavelengths are small compared to the effective dimensions of the apparatus, the results are extremely accurate.

15.3.5 Inhomogeneous Media

The propagation of disturbances is considerably modified by inhomogeneities in the properties of the medium through which the disturbance is propagating. Discontinuous changes in the properties of the medium have been shown in the previous two sections to lead to reflection, refraction, and diffraction. These effects are also present when the properties of the medium change continuously.

To understand some of these phenomena as they occur in a medium with continuously varying properties and to see how some approximate solutions can be generated, let us consider first the one-dimensional time-dependent propagation of waves in an inhomogeneous medium. Assume that the governing equation is

$$\frac{\partial^2 \psi}{\partial t^2} - c^2(x)\frac{\partial^2 \psi}{\partial x^2} = 0 \tag{15.3-34}$$

where the wave speed c is now a function of position x. It is assumed that the inhomogeneity is effectively restricted to some region of extent L located near $x = 0$.

For periodic waves of frequency ω, the substitution $\psi(x, t) = e^{i\omega t}\,\phi(x)$ reduces the above equation to

$$\frac{d^2\phi}{dx^2} + \frac{\omega^2}{c^2}\phi = 0 \tag{15.3-35}$$

or, in dimensionless form

$$\frac{d^2\phi}{d\xi^2} + (kL)^2 g(\xi)\,\phi = 0 \tag{15.3-36}$$

where $\xi = x/L$, $k = \omega/c_\infty$, $g = c_\infty^2/c^2$, and c_∞ is the wave speed as $|x| \to \infty$. It is convenient to let $g = 1 + \epsilon f$ where f is $O(1)$ and tends to zero as $|x| \to \infty$ and ϵ is a constant and is a measure of the magnitude of the inhomogeneity.

An incident wave is specified propagating in the positive x-direction to the left of the inhomogeneity, i.e., at $x = -\infty$. The effect of the inhomogeneity in producing reflected and transmitted waves and distorting the incident wave in the region of the inhomogeneity is to be investigated.

Two dimensionless parameters appear naturally in the analysis, ϵ and kL, where kL represents the ratio of the characteristic length of the inhomogeneity to the wavelength of the incident wave. A common method of solution is a regular perturbation expansion of ϕ in terms of ϵ. It can be shown that this solution is only valid when $\epsilon kL \ll 1$.

In order to obtain solutions for large values of kL, the WKB method of solution is quite often used. The basic idea of this method is to assume a solution of eq. (15.3-36) in the form

$$\phi = \alpha(\xi)\exp\left[-ikL\beta(\xi)\right] \tag{15.3-37}$$

where α and β are real functions of ξ and are, hopefully, slowly varying. By substituting this expression into eq. (15.3-36) and equating real and imaginary parts, one obtains two simultaneous equations for α and β.

If it is assumed that the properties of the medium vary slowly in the sense that the properties do not change significantly in a wavelength, approximate solutions to these equations can be found readily and, to first order, are

$$\beta = \pm \int_0^\xi g^{1/2}(\xi)\,d\xi \tag{15.3-38}$$

$$\alpha = \Lambda/g^{1/4} \tag{15.3-39}$$

where Λ is some constant.

An approximate solution to eq. (15.3-36) is then

$$\phi = \frac{A_0}{g^{1/4}} \exp\left(-ikL \int_0^\xi g^{1/2}\,d\xi\right) \tag{15.3-40}$$

where A_0 is the magnitude of the incident wave. It can be seen that the incident wave is modified as it propagates through the inhomogeneity. However, the wave is completely transmitted through the medium and there are no reflected waves. Indeed, to any approximation the WKB procedure does not predict reflected waves, a not very satisfying result physically although valid asymptotically as $kL \to \infty$.

Although the WKB solution (and other methods such as two-timing and coordinate stretching) are limited because of this, these methods do suggest other dependent and independent variables which can be used to advantage (Ref. 15-5). For example, by transforming the independent variable $\eta = kx$, which appears naturally in the regular perturbation solution, to

$$\eta^* = \int_0^\eta g^{1/2}(\eta')\,d\eta' \tag{15.3-41}$$

which appears naturally in the WKB solution, a transformed equation is obtained. This equation can then be solved by a regular perturbation method. The resulting solution does predict reflected waves and can be shown to be valid when $\epsilon \ll 1$ for arbitrary values of kL.

Similarly, by transforming the dependent variable ϕ to $\phi^* = g^{1/4}\phi$ and solving the resulting equation by a regular perturbation method, one obtains a solution valid when $\epsilon/kL \ll 1$. Further solutions can be generated by extensions of this procedure to produce solutions valid when $\epsilon/(kL)^n \ll 1$, where $n = 2, 3, \ldots$.

A comparison of the results by these methods with an exact solution for the case when

$$g(x) = 1 + \frac{4\epsilon e^{mx}}{(1 + e^{mx})^2} \tag{15.3-42}$$

where m is a parameter, is shown in Fig. 15.3-2. Here only the first term of each

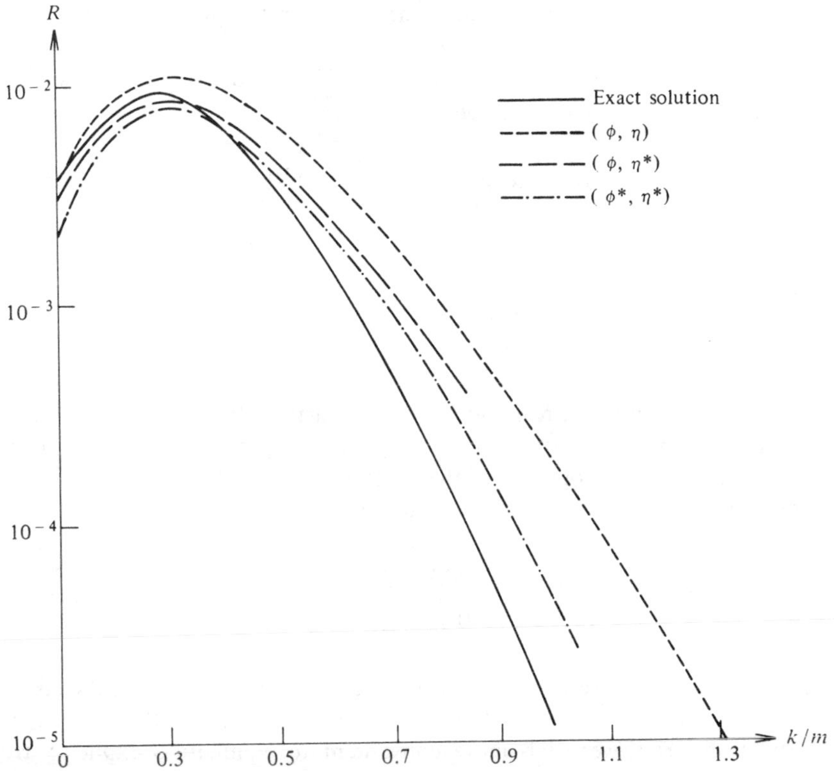

Fig. 15.3-2 Reflection coefficient for an inhomogeneous medium. The reflection coefficient R is defined as the square of the absolute value of the ratio of the amplitude of the reflected wave to the amplitude of the incident wave.

series has been used in calculating the results shown. It can be seen that, for fixed ϵ, the ϕ, η-solution is best when kL is small, the ϕ^*, η^*-solution is best when kL is large, and the ϕ, η^*-solution is a fairly good representation of the solution for all values of kL. This is in agreement with the convergence criteria of the solutions indicated above.

The above procedures are only useful when $g(x)$ is not close to zero anywhere in the region under consideration. When turning points, i.e., points at which $g(x)$ is zero, are present, different approximate solutions are necessary near the turning point. These solutions are then matched to solutions such as those above which are valid away from the turning point (Refs. 15-5, 15-6) in order to obtain a uniformly valid solution.

In two and three dimensions, the solution and description of wave propagation in an inhomogeneous medium is correspondingly more complex than in the one-dimensional case. In practice, the principal method of solution is by means of ray theory, which is analogous to (1) the WKB method described above for one-dimensional, time-dependent wave propagation in an inhomogeneous medium; and to

(2) the geometrical optics approximation described in the previous section in the discussion of diffraction by a screen.

The basic procedure of solution is similar to the WKB procedure described above, i.e., substitution of an assumed solution of the form

$$\psi = \alpha(x, y, z) e^{i[\omega t - k\beta(x, y, z)]} \tag{15.3-43}$$

into the basic equation

$$\frac{\partial^2 \psi}{\partial t^2} - c^2 \nabla^2 \psi = 0 \tag{15.3-44}$$

where $c = c(x, y, z)$, equating real and imaginary parts of the resulting equation to obtain two simultaneous equations for α and β, and approximate solution of these equations valid when the properties of the medium vary slowly.

The approximate equation for β is found to be

$$\left(\frac{\partial\beta}{\partial x}\right)^2 + \left(\frac{\partial\beta}{\partial y}\right)^2 + \left(\frac{\partial\beta}{\partial z}\right)^2 = n^2 \tag{15.3-45}$$

where $n = c_\infty / c$. The above is known as the eikonal equation and β as the eikonal. This equation must then be solved for β where $c(x, y, z)$, and therefore n, is a known function. However, it is more convenient to eliminate β explicitly from the above equation. It can be shown that this equation can be transformed to

$$\frac{\partial}{\partial s} n\hat{\mathbf{s}} = \nabla n \tag{15.3-46}$$

where $\hat{\mathbf{s}}$ is a unit vector normal to the wave front given by β = constant, and s is distance as measured along a curve, or ray, that is everywhere parallel to the local direction of $\hat{\mathbf{s}}$. From this and the initial direction of a ray, the entire family of rays can be constructed. Once the ray direction, and hence β, is known, α can be found and the solution is complete.

In general, the above equation must be solved numerically. However, when n is a function of one variable only, say y, the above equation reduces to

$$n \sin\theta = \text{constant} \tag{15.3-47}$$

where θ is the angle of incidence of the ray to the y axis. This is Snell's law, discussed previously in the case of reflection and refraction at a plane surface and now shown to be a valid approximation for a continuously varying (in one direction only) medium.

15.3.6 Nonlinear Wave Propagation and the Method of Characteristics

For linear and nonlinear equations of the hyperbolic type, the idea of characteristics plays an important part in both the physical interpretation of solutions and in obtaining numerical solutions. Characteristics, or characteristic coordinates, are defined as lines across which derivatives of the dependent variables may be discontinuous. The relations among the dependent variables along these curves are called the characteristic equations. Equivalently, it can be stated that the characteristic equations are relations which involve derivatives of the dependent variables in one direction only, i.e., the characteristic direction.

The linear simple wave equation, eq. (15.1-3), is of the hyperbolic type as are the equations of motion for (a) an isentropic, compressible gas; and (b) long water waves, among others. For the linear simple wave equation in a homogeneous medium, the general solution to the problem is given by $\psi = f_1(\xi) + f_2(\eta)$, where f_1 and f_2 are arbitrary functions. The characteristic coordinates are $\xi = x - ct$ and $\eta = x + ct$ and the characteristic equations are f_1 = constant on ξ = constant and f_2 = constant on η = constant.

For gas dynamics, the situation is somewhat more complicated. For compressible flows in which the entropy is constant along streamlines (but not necessarily constant throughout the flow), it can be shown (Ref. 15-7) that the characteristic directions are given by

$$\frac{dx}{dt} = u \pm a \tag{15.3-48}$$

$$\frac{dx}{dt} = u \tag{15.3-49}$$

corresponding to the positive and negative Mach lines and the streamlines, respectively. In these three directions, the corresponding characteristic equations are

$$dp \pm \rho a\, du = 0 \tag{15.3-50}$$

$$dS = 0 \tag{15.3-51}$$

For a flow which is isentropic throughout, i.e., homentropic flow, it is convenient to introduce the Riemann invariants r and s defined by

$$r = \frac{u}{2} + \frac{a}{\gamma - 1} \tag{15.3-52}$$

$$s = -\frac{u}{2} + \frac{a}{\gamma - 1} \tag{15.3-53}$$

where γ is the ratio of specific heats. The characteristic equations can then be written as r = constant on $dx/dt = u + a$ and s = constant on $dx/dt = u - a$.

To understand the usefulness and importance of characteristics, let us consider one-dimensional time-dependent disturbances in a compressible gas which is isentropic throughout. Assume that some initial data for the problem is given along a curve C (see Fig. 15.3-3). Consider some point $P(x, t)$ and the two characteristic curves $dx/dt = u \pm a$ drawn through P and which intersect the curve C at the points A and B. The domain of dependence of the point P is defined as the section A to B of the curve C, i.e., the solution at P is determined uniquely by the initial data on C between A and B and is not influenced by changes in the initial data outside of this section of C. Similarly the range of influence of a point P is the region in the positive t-direction from P bounded by the two characteristics $dx/dt = u \pm a$. Conditions at P can influence the entire solution within this region.

The theory of characteristics is particularly simple for an important class of problems called simple waves (not to be confused with nondispersive, nondiffusive waves). In the present context, simple waves are defined, for homogeneous isentropic flow for example, as waves in which either of the Riemann invariants, r or s, is constant throughout the region. This situation arises, for instance, when the disturbance is adjacent to a region of constant state. It can be shown that a basic property of simple waves is that the characteristics of one kind are straight lines in the x, t-plane.

A simple wave is called an expansion or rarefaction wave if the pressure and density of the gas decrease on crossing the wave while the wave is called a compression wave if the pressure and density increase. Although in a linear wave, the wave form remains unchanged, in a nonlinear wave, it is distorted. In particular, for a simple expansion wave, the velocity profile flattens out, while the velocity profile steepens in a compression wave.

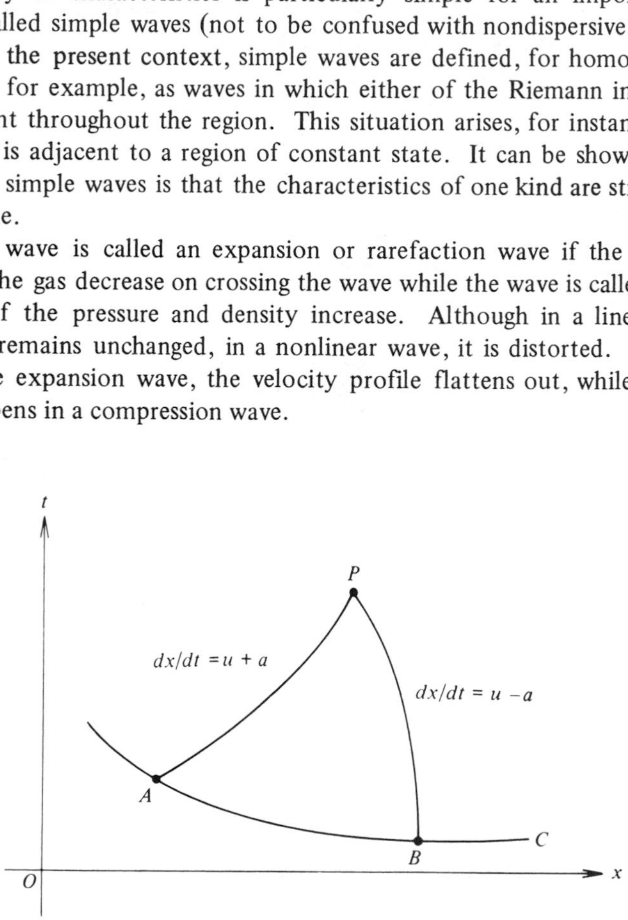

Fig. 15.3-3 Domain of dependence of a point P.

By this process, multivalued solutions will eventually be obtained for a compression wave, a physically unacceptable situation. In this multivalued region, the basic physical hypothesis of isentropic flow is no longer valid and irreversible processes must be considered. However, it is not necessary to consider the details of the flow in this region (see section 15.5.3 for a discussion of this problem). Rather it is sufficient to only satisfy the conservation equations for mass, momentum, and energy across this region, or shock wave. These conservation equations plus the condition that entropy must increase in an irreversible process lead to the Rankine-Hugoniot equations relating conditions across the shock. By the introduction of these shock waves, a complete single-valued, but not necessarily continuous, solution to the problem can be found.

In the next section, approximate analytic methods of solution for weakly nonlinear waves will be discussed. For strongly nonlinear waves, few analytic solutions are available and numerical methods must be employed to obtain solutions. One numerical procedure used is based on the characteristic coordinates and the characteristic relations valid along these lines, i.e., eqs. (15.3-50) and (15.3-51). By the use of finite differences, the characteristic coordinates are calculated, and therefore constructed in the x, t-plane, at the same time that the characteristic equations are integrated.

Although this procedure seems quite natural and is quite valuable for studying the general properties of the solution, an alternate procedure of using finite differences and a fixed rectangular net in the x, t-plane is more convenient. To insure convergence of this latter scheme, certain restrictions on the size of the grid intervals in the x, t-plane must be obeyed. In order to understand this, consider the grid shown in Fig. 15.3-4, where all dependent variables are known at points 1, 2, and 3, the characteristic lines through points 1 and 3 are given by $C+$ and $C-$, and P is the point at which the dependent variables are to be calculated. This procedure is only valid if P is within the region bounded by the x-axis and the lines $C+$ and $C-$, since this is the region of influence of the portion of the x-axis between 1 and 3. For points P outside of this region, it follows from the theory of characteristics discussed above that information from additional points on the x-axis would be

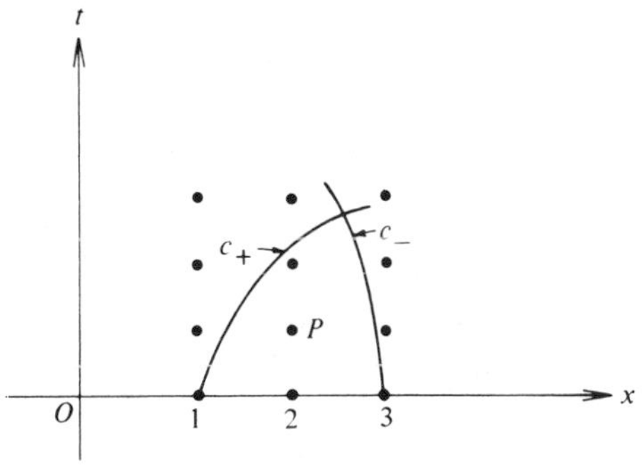

Fig. 15.3-4 Rectangular net for numerical calculations by finite differences.

needed to fully determine the solution at P. It can be seen that, for given Δx, these considerations restrict the range of Δt.

15.3.7 Weakly Nonlinear Waves: Analytic Methods of Solution

For weakly nonlinear waves in a compressible fluid, a regular perturbation analysis leads to nonuniformities in the solution due to cumulative nonlinear convective effects which are not accounted for by the regular perturbation procedure. Recently, approximate analytic methods of solution have been developed which do lead to uniformly valid solutions. These methods will be briefly discussed here (also, see Ref. 15-1 and references listed there).

The two main methods of use in this type of problem are (a) coordinate stretching procedures, and (b) the method of matched asymptotic expansions. Let us first consider coordinate stretching procedures. As noted in the previous section, the equations of motion describing the propagation of waves in an isentropic, compressible fluid are hyperbolic. The characteristics are the two sets of Mach lines and, for flows with entropy varying across particle path lines, the particle path lines themselves. The usual linear solution is equivalent to integrating the linearized characteristic equations along characteristic directions given by the undisturbed flow. For nonsteady, one-dimensional, homentropic flow, only two sets of characteristics, the Mach lines, are present, and the characteristic directions of the undisturbed flow are therefore $x \pm a_{so}t = \text{constant}$, where a_{so} is the isentropic speed of sound in the undisturbed flow. For a disturbed flow, the approximation that each set of characteristics consists of parallel straight lines is generally a poor approximation to the exact solution, except perhaps in a small, localized region, i.e., the nonlinear convective terms may be small everywhere compared with the linear terms, but their cumulative effects may be important. Higher approximations to the ordinary perturbation method do not give uniformly valid corrections to the linear solution.

To correct this defect of the ordinary perturbation method, researchers (Refs. 15-8 to 15-12) have developed several related methods that may be classified as coordinate stretching or coordinate perturbation methods. In these methods, the position variables are stretched, i.e., perturbed in an asymptotic series, at the same time as are the dependent variables such as velocity, pressure, etc.

The coordinate perturbation method as presented in Refs. 15-1 and 15-12 will be followed here. For simplicity, consider the one-dimensional, time-dependent flow of a fluid semi-infinite in extent and bounded on the left by a movable piston located at $x = 0$. The motion of the piston sends out waves in the positive x-direction.

Write the equations of motion in terms of two characteristic coordinates, say α and β, where α is constant on positive Mach lines and β is constant on negative Mach lines. A parameter f is defined such that f is constant on particle path lines. The dependent variables u, a, f, etc. and the position variables x and t are then expressed as perturbation series in powers of ϵ, $u(\alpha, \beta) = u_0(\alpha, \beta) + \epsilon u_1(\alpha, \beta) + \cdots$, etc., where ϵ is a small parameter of the order of the piston velocity divided by the isentropic speed of sound. By equating coefficients of like powers of ϵ to zero, one obtains an ordered set of linear partial differential equations. These equations can be solved successively to obtain $u_n(\alpha, \beta), a_n(\alpha, \beta), x_n(\alpha, \beta), t_n(\alpha, \beta)$, etc. to any order.

The zeroth approximations for u and a are given. The zeroth approximations for x, t, and f as functions of α and β can then be found from the zeroth-order equations. By inverting these relations, one obtains the zeroth approximation for the characteristics, which turn out to be the characteristics for the undisturbed flow, i.e., the Mach lines $x \pm a_{so}t$ = constant and the particle path lines x = constant. In successive steps, the nth approximation to $u(\alpha, \beta)$ and $a(\alpha, \beta)$ can be found, followed by the nth approximation to $x(\alpha, \beta)$, $t(\alpha, \beta)$, and $f(\alpha, \beta)$. At each step, an implicit relationship $u(x, t), a(x, t)$ can then be determined.

To complete the solution, shock waves must be included where the above procedure gives multivalued solutions. For a weak discontinuous shock, the Rankine-Hugoniot relations show that the shock speed is the average of the speeds of small disturbances (given by the slope of the characteristics) before and after the shock. This is sufficient to determine the shock location.

In the above analysis, at each step of the iteration procedure the characteristic lines are corrected and are generally curved and nonparallel in contrast to the usual linear theory. Reflected disturbances are predicted in second and higher approximations.

Also, a simple method of improving linear solutions of problems in which waves may propagate in both positive and negative x-directions is indicated. The procedure amounts to using the usual linear theory but simultaneously improving the location of all sets of characteristics, an extension of a hypothesis used by Whitham (Ref. 15-9) in an analysis of a similar problem.

A distinctly different approach to the problem of nonlinear wave propagation is the method of matched asymptotic expansions. Consider the same problem as above but for a homentropic fluid. First- and second-order theories based on the regular perturbation analysis have been calculated (see, for instance, Ref. 15-4). The second-order results show that the solution is nonuniformly valid when the wave front has traveled a distance $x/a_{so}T = 1/\epsilon$, where T is a characteristic time associated with the piston motion and ϵ is again of the order of the piston velocity divided by the isentropic speed of sound. That is, the solution becomes invalid at a time t_1 at which the deviation of the wave front from its location as predicted by the linear theory is comparable with the characteristic length $L = a_{so}T$. This time t_1 is of the order of $x_p/a_{so}\epsilon^2$, where x_p is a characteristic length associated with the piston motion so that $x_p = O(\epsilon a_{so}T)$.

By the method of matched asymptotic expansions (Refs. 15-4, 15-13), one attempts to find a solution valid in the region $a_{so}t/L > O(1/\epsilon)$, or far field, where the ordinary linear expansion, or near-field solution, is no longer valid. In order to describe the far field, a new time scale $t^* = \epsilon a_{so}t/L$ is suggested by the above estimates. A new length coordinate $x^* = x/L - a_{so}t/L$ moving at the isentropic speed of sound is also introduced.

By substituting these new coordinates into the basic equations of motion, expanding all dependent variables ϕ as $\phi = \phi_0 + \epsilon\phi_1 \ldots$, and equating terms $O(\epsilon)$ and $O(\epsilon^2)$ separately to zero, one eventually obtains the equation

$$\frac{\partial u}{\partial t} + \left(\frac{\gamma + 1}{2} u + a_{so}\right) \frac{\partial u}{\partial x} = 0 \qquad (15.3\text{-}54)$$

which, by a simple transformation of coordinates, can be reduced to eq. (15.1-14). The solution of this equation gives a first approximation to the far field and can then be matched with the linear, or near-field, solution to give a uniformly valid solution for all x, t. For simple problems, the above equation includes the linear equation and therefore no matching is necessary. In any case, the shock waves have to be fitted in as in the previous method.

For the problem of the propagation of waves predominantly in one direction in an isentropic fluid, both the method of matched asymptotic expansions and the coordinate-stretching method are equally simple and equally effective. In the lowest approximation, the solutions by the two methods are essentially the same. The method of matched asymptotic expansions is more widely applicable, although the coordinate-stretching method is more suitable for problems of waves propagating in more than one direction. In this latter method, it is also easier to find higher approximations.

15.4 DISPERSIVE WAVES

15.4.1 Dispersion and Group Velocity

In the present section, some of the features associated with dispersive waves, i.e., waves in which the phase velocity is a function of frequency, will be discussed. The main emphasis will be on water waves. Much of the analysis presented in section 15.4.1 to 15.4.4 will follow that of Stoker (Ref. 15-3). Also see Ref. 15-40 for an additional treatment of water waves. Interesting photographs and a discussion of water waves are given in Ref. 15-14.

To begin with, let us consider the special case of linear, two-dimensional, time-dependent surface waves propagating in water of depth h. Let x be the horizontal coordinate while y is the vertical coordinate measured from the undisturbed free surface, which is assumed to be plane. The basic equations (see section 15.2.6) which must be satisfied are

$$\nabla^2\phi = 0 \qquad (15.4\text{-}1)$$

throughout the flow field,

$$\frac{\partial \phi}{\partial y} = 0 \qquad (15.4\text{-}2)$$

at the bottom surface $y = -h$, and

$$\frac{\partial^2 \phi}{\partial t^2} + g \frac{\partial \phi}{\partial y} = 0 \tag{15.4-3}$$

at the free surface $y = 0$.

A solution of these equations which describes simple harmonic progressive waves is

$$\phi = A \cosh k(y + h) \cos (kx \pm \omega t + \epsilon) \tag{15.4-4}$$

where A is the amplitude, ϵ is the phase constant, and ω and k are related by

$$\omega^2 = gk \tanh kh \tag{15.4-5}$$

The surface displacement η is given by

$$\begin{aligned} \eta &= -\frac{1}{g} \frac{\partial \phi}{\partial t} \\ &= \pm \frac{A\omega}{g} \cosh kh \sin (kx \pm \omega t + \epsilon) \end{aligned} \tag{15.4-6}$$

The phase speed c is given by $c = \omega/k$, or

$$\frac{c^2}{gh} = \frac{\lambda}{2\pi h} \tanh \frac{2\pi h}{\lambda} \tag{15.4-7}$$

where λ is the wavelength. In general, it can be seen that the phase speed is a function of wavelength, and hence frequency, and therefore the waves are dispersive. In particular, in deep water as $h/\lambda \to \infty$, $c^2 \doteq g\lambda/2\pi$, while in shallow water where $h/\lambda \to 0$, $c^2 \doteq gh$, so that the wave speed is independent of frequency.

Within the present linear approximation, an arbitrary disturbance in the water can be considered as a superposition of waves of various amplitudes and wave lengths and hence different phase speeds. As time progresses, the form of any initial disturbance will change due to the dispersive quality of the waves. However, groups of waves with about the same phase speed will tend to travel together, although not necessarily at the phase speed of the waves.

To understand this more clearly, let us consider first the superposition of only two progressive waves given by

$$\phi_1 = A \sin [kx - \omega t] \tag{15.4-8}$$

$$\phi_2 = A \sin [(k + \delta k)x - (\omega + \delta\omega)t] \tag{15.4-9}$$

where δk and $\delta\omega$ are small quantities. Superposition of these two waves gives

$$\phi = B \sin (k'x - \omega' t) \tag{15.4-10}$$

where $k' = k + \delta k/2$, $\omega' = \omega + \delta\omega/2$, and B is a slowly varying function of x and t given by $2A \cos [\frac{1}{2} (x\delta k - t\delta\omega)]$.

The function ϕ can therefore be described as an almost harmonic wave of frequency k and phase speed ω/k whose amplitude varies harmonically with x and t (see Fig. 15.4-1). This modulation changes slowly with x and t and propagates in

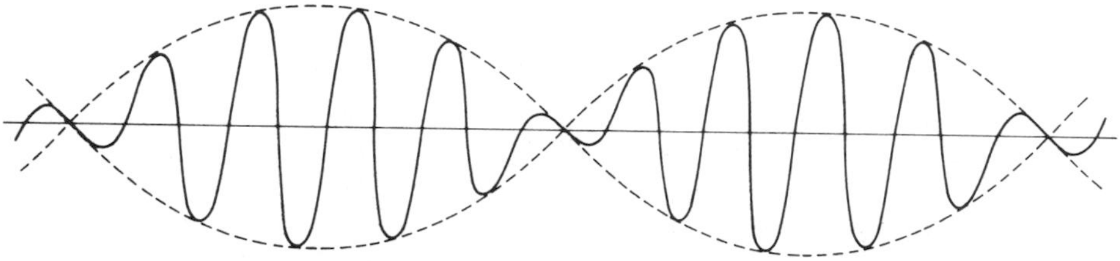

Fig. 15.4-1 The superposition of two sinusoidal waves with the same amplitudes but different frequencies.

the direction of the wave at the speed $\delta\omega/\delta k$, called the group velocity c_g. In the limit as δk and $\delta\omega$ become small

$$c_g = \frac{d\omega}{dk} = \frac{d(kc)}{dk} = c - \lambda \frac{dc}{d\lambda} \tag{15.4-11}$$

More generally, a superposition, by means of an integral, of infinitely many waves with amplitudes and wave lengths which vary over a small range can be considered. By means of the method of stationary phase, one can then show that the above relation for group velocity is still valid for describing the propagation speed of the disturbance.

For the present case where the phase speed is given by eq. (15.4-7), it follows that the group velocity is given by

$$_g = \frac{c}{2} \left(1 + \frac{4\pi h/\lambda}{\sinh 4\pi h/\lambda}\right) \tag{15.4-12}$$

Note that as $h/\lambda \to 0$, $c_g = c$, while for $h/\lambda \to \infty$, $c_g = c/2$.

15.4.2 Energy Transport

The calculation of the energy transfer or energy flux in a dispersive medium is more complex, because of the dispersive quality of the waves, than in a nondispersive, nondiffusive medium. As an example of a dispersive medium, consider again surface gravity waves in water within the linear approximation (e.g., see Refs. 15-3 and 15-40).

Let V be the volume occupied by the water and S be the surface enclosing the volume. The surface $S(t)$ may move independently of the fluid. The energy E of the water consists of kinetic energy and potential energy due to gravity and is given by

$$E = \rho \int_V \left\{ \frac{1}{2} \left[\left(\frac{\partial \phi}{\partial x} \right)^2 + \left(\frac{\partial \phi}{\partial y} \right)^2 + \left(\frac{\partial \phi}{\partial z} \right)^2 \right] + gy \right\} dV \qquad (15.4\text{-}13)$$

which can also be written, by means of Bernoulli's law, as

$$E = -\int_V \left(p + \rho \frac{\partial \phi}{\partial t} \right) dV \qquad (15.4\text{-}14)$$

The rate of change of energy dE/dt within V can be found by differentiation of the above equations and is

$$\frac{dE}{dt} = \int_S \left[\rho \frac{\partial \phi}{\partial t} \left(\frac{\partial \phi}{\partial n} - v_n \right) - p v_n \right] dS \qquad (15.4\text{-}15)$$

where v_n is the normal velocity of S taken positive in the direction outward from V and $\partial\phi/\partial n$ is the velocity component of the fluid in the same direction.

Some special cases are of interest. First, if a portion of the boundary S, say S_1, is moving with the fluid and therefore always consists of the same fluid particles, then $\partial\phi/\partial n$ and v_n are identical, and

$$\frac{dE}{dt} = -\int_{S_1} p v_n \, dS = -\int_{S_1} p \frac{\partial \phi}{\partial n} \, dS \qquad (15.4\text{-}16)$$

In addition, if the surface is a free surface on which the pressure vanishes, then it can be seen from the above that there is no contribution to the energy flux from this surface. Second, if a portion of the surface, say S_2, is fixed in space, then $v_n = 0$, and

$$\frac{dE}{dt} = \rho \int_{S_2} \frac{\partial \phi}{\partial t} \frac{\partial \phi}{\partial n} \, dS \qquad (15.4\text{-}17)$$

For the special case of a harmonic progressive wave in water of constant depth [in which case the potential is given by eq. (15.4-4)], the energy flux, or energy per unit breadth of wave averaged over the time, is given by

$$F = \frac{A^2 \rho \omega k h}{4} \left(1 + \frac{\sinh 2kh}{2kh}\right) \tag{15.4-18}$$

which, from eq. (15.4-12), can be written as

$$F = \frac{A^2 \rho \omega^2}{2g} \cosh^2 kh \cdot c_g \tag{15.4-19}$$

It can be shown from eq. (15.4-1) (by integrating over a wavelength and dividing by the wavelength) that the quantity multiplying c_g in the above formula is the average energy per unit length in the x-direction (neglecting the potential energy of the water when at rest). It can be seen from the above that energy is propagated in the direction of the wave with the group velocity c_g, which is always equal to or less than the phase velocity c. This is true for linear dispersive waves in general (and trivially true for linear simple waves where the group and phase velocities are the same), but is not true when diffusion is present.

15.4.3 Instantaneous Point Source

An interesting example which illustrates some of the effects of dispersion is the wave motion due to a point disturbance at the surface of the water. Only the linear problem will be discussed here. To be specific, let us consider the two dimensional, time-dependent motion in water of infinite depth due to an initial elevation concentrated at the origin, i.e., the motion due to a finite surface elevation $\eta(x, 0) = a$ extending from $x = -b/2$ to $x = +b/2$ in the limit as $b \to 0$ with ab constant. A unit value of ab will be assumed.

The basic equations for this problem are eqs. (15.4-1) through (15.4-3) with the initial condition as stated above. The solution to this problem can be obtained by means of Fourier transforms (Ref. 15-3) and is

$$\eta(x, t) = \frac{1}{\pi} \lim_{y \to 0} \int_0^\infty e^{\lambda y} \cos \lambda x \cos \left(\sqrt{g\lambda}\, t\right) d\lambda \tag{15.4-20}$$

Asymptotic evaluations of this integral valid for both small and large values of gt^2/x can be obtained. If one expands all terms in the integrand in power series and integrates, one obtains the formula

$$\eta = \frac{1}{\pi x} \left[\frac{gt^2}{2x} - \frac{1}{1 \cdot 3 \cdot 5} \left(\frac{gt^2}{2x}\right)^3 + \cdots\right] \tag{15.4-21}$$

This series converges for all values of $gt^2/2x$ but is only useful for small values of this quantity. Of more interest is the solution for large values of $gt^2/2x$. For this case, one can obtain an asymptotic, but divergent, series for η by the method of stationary phase. To a first approximation, the result is

$$\eta = \frac{\sqrt{g}}{2\sqrt{\pi}} \frac{t}{x^{3/2}} \cos\left(\frac{gt^2}{4x} - \frac{\pi}{4}\right) \tag{15.4-22}$$

It can be seen that in this approximation the motion is approximately simple harmonic with slowly varying wavelength and with the amplitude of the motion decreasing as $x^{-3/2}$ and increasing linearly with time.

The singularities in the above approximate solutions as $x \to 0$ and $t \to \infty$ are due to the concentrated singularity in η as $t \to 0$. Other initial conditions are possible which do not show either of these singularities.

Schematic diagrams of the motion are shown in Figs. 15.4-2 and 15.4-3. The motion shown in these figures can be interpreted in terms of the discussion of group and phase velocities given in section 15.4.1 as follows. The initial disturbance can be considered as a superposition of waves of all frequencies and hence wave-

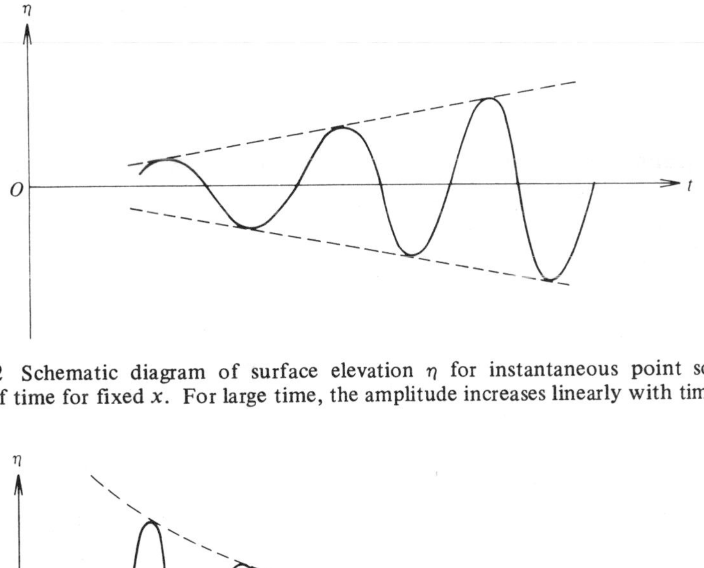

Fig. 15.4-2 Schematic diagram of surface elevation η for instantaneous point source as a function of time for fixed x. For large time, the amplitude increases linearly with time.

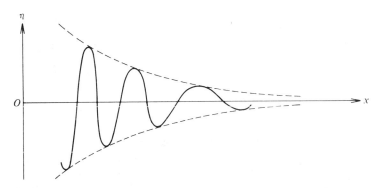

Fig. 15.4-3 Schematic diagram of surface elevation η for instantaneous point source as a function of distance x for fixed time. For large time, the amplitude decreases as $x^{-3/2}$.

lengths. The phase speed of each of these waves is given by eq. (15.4-7) which for deep water gives $c = \sqrt{g\lambda/2\pi}$ while the group velocity is $c_g = c/2$. It can be seen that the group and phase velocities increase with wavelength and therefore the waves with the longest wavelengths will tend to be at the front of the disturbance and waves with the shortest wavelengths at the back. Although each individual wave or phase travels at the phase speed, the speed of a group of waves, or the speed at which the location of waves of a particular wavelength travels, is given by the group speed.

15.4.4 Kelvin Ship Waves

An interesting application of the theory of water waves is to the calculation of the wave pattern due to a moving ship. This wave pattern is very distinctive and, for a ship in deep water moving in a straight line at constant speed, is restricted to a region behind the ship in the shape of a V. For deep water and to a first approximation, this shape is the same regardless of the size or speed of the ship.

Although more complex theories have been developed which approximate the ship more realistically, the analysis presented here will follow the work of Kelvin and its discussion by Stoker (Ref. 15-3). In this approximation, the ship is considered to be a continuous point impulse moving over the surface of water which is infinite in depth.

An instantaneous point impulse (in cylindrical geometry) is defined as the limit of an instantaneous impulse of uniform strength b applied to an area of radius $r \leqslant a$ in the limit as $a \to 0$ while $b \to \infty$ in such a way that the total impulse $\pi a^2 b$ is constant. In terms of the potential ϕ, the instantaneous impulse is given by $I = -\rho\phi(x, o, z; 0)$.

The surface elevation for the initial condition of an instantaneous point impulse can be found by the use of Fourier transforms and is

$$\eta(r, t) = \frac{-1}{2\pi\rho\sqrt{g}} \lim_{y \to 0} \int_0^\infty e^{\lambda y} J_0(\lambda r) \sin(\sqrt{g\lambda}\, t)\, \lambda^{3/2}\, d\lambda \qquad (15.4\text{-}23)$$

where J_0 is the Bessel function of zeroth order and the rest of the notation is the same as in previous sections. Asymptotic evaluation of this integral by means of the method of stationary phase gives

$$\eta = \frac{-gt^3}{2^{7/2}\pi\rho r^4} \sin \frac{gt^2}{4r} \qquad (15.4\text{-}24)$$

This expression can be shown to be valid for large values of the parameter $gt^2/4r$.

In order to obtain the wave pattern due to a moving ship, the procedure is to integrate the above equation along the course of the ship assuming that the strength of the impulse is constant. Let the course of the ship be specified by $x_1 = x_1(t)$, $z_1 = z_1(t)$, where $0 \leqslant t \leqslant T$. The parameter t is the time required for the ship to

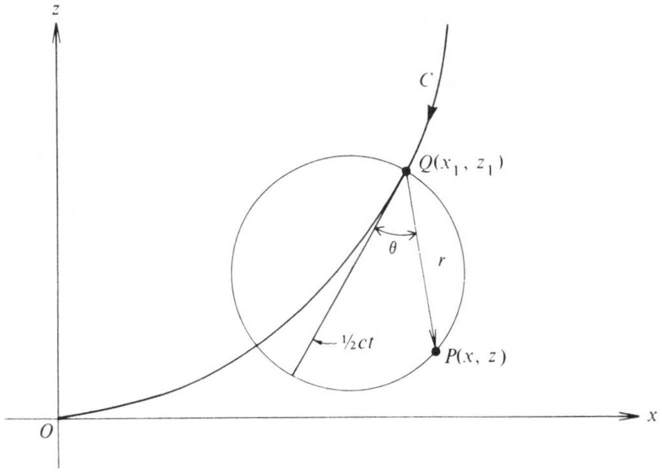

Fig. 15.4-4 Point P influenced by a given point Q. The path of the ship is given by the curve C. (After Stoker, Ref. (15.3)).

travel from some point $Q(x_1, z_1)$ to its present position, assumed to be at the origin (see Fig. 15.4-4). The surface displacement at some point $P(x, z)$ is to be calculated.

Under these conditions, integration of eq. (15.4-24) gives

$$\eta = A \int_0^T \frac{t^3}{r^4} \sin \frac{gt^2}{4r} \, dt \tag{15.4-25}$$

where A is a constant and $r = r(t)$ is the distance from P to Q.

This integral can be evaluated by the method of stationary phase. The resulting expression for the condition of stationary phase can be written as

$$r = \tfrac{1}{2} ct \cos \theta \tag{15.4-26}$$

where c is the speed of the ship and θ is the angle between $\mathbf{r}$ and the direction of the ship at any instant of time (see Fig. 15.4-4). This expression can be interpreted as follows. For a fixed point P, the above equation gives the point Q which is effective, within the stationary phase approximation, in causing the disturbance at P. The contributions due to all other instantaneous point impulses are presumably cancelled out because of mutual interference effects. For a fixed point Q, the points P which are influenced in this manner lie on a circle (influence circle) with a diameter which is tangent to the path of the ship at Q with Q being on this circle and at one end of the diameter.

The disturbance due to the ship is therefore restricted to the region bounded by the envelope of these influence circles. This region can be constructed rather easily for the particular case of a ship moving at constant speed in a straight line. By choosing several points Q along the path of the ship, drawing the influence circles and their envelope, one obtains Fig. 15.4-5. As can be seen, the envelope is a pair of straight lines with semi-angle which satisfies $\sin \phi = \frac{1}{3}$, so that $\phi \doteq 19°28'$. This angle and hence region of disturbance is independent of ship speed, as long as the speed is constant.

Further calculations show that the wave system is made up of two sets of waves, the diverging and transverse waves, the former at acute angles to the course of the ship and the latter at approximately right angles (see Fig. 15.4-6). Near the envelope of the disturbance, the usual lowest order approximation to the method of station-

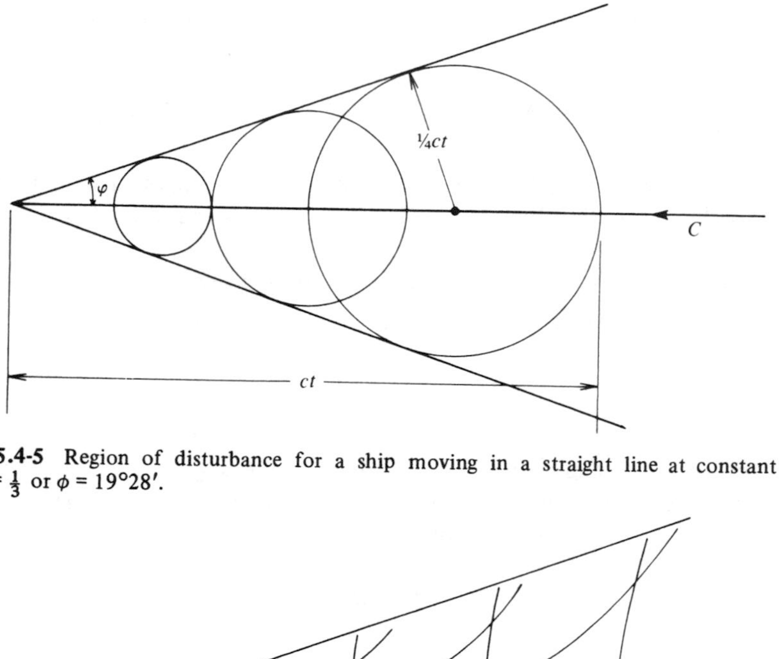

Fig. 15.4-5 Region of disturbance for a ship moving in a straight line at constant speed; $\sin \phi = \frac{1}{3}$ or $\phi = 19°28'$.

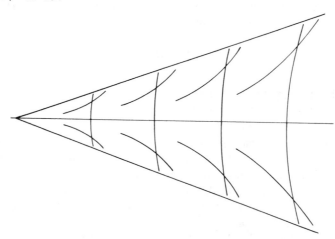

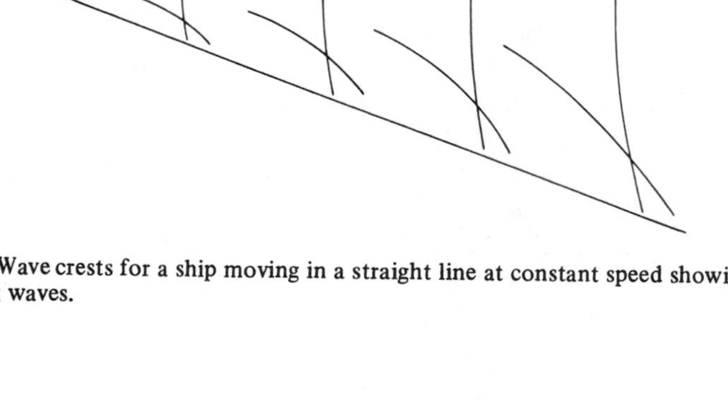

Fig. 15.4-6 Wave crests for a ship moving in a straight line at constant speed showing transverse and diverging waves.

ary phase is no longer valid and higher approximations must be considered in order to get reasonable results.

For a ship moving in shallow water, the wave motion is quite different from that described above. In particular, the general characteristics of the wave motion do depend on the speed of the ship. For shallow water of depth h and for a ship moving in a straight line at constant speed V, which is greater than $\sqrt{gh}$, it can be shown that the disturbances are again confined to a wedge-shaped region behind the ship but, in this case, the semi-angle is given by $\sin \phi = \sqrt{gh}/V$.

15.4.5 Wave Refraction: Ray Theory

Effects of dispersion are also quite significant in the refraction of water waves caused by obstacles or by a variable water depth. Of course, refraction is present in the absence of dispersion, e.g., for a monochromatic wave, but it will be seen that the amount of refraction depends on the wavelength of the waves and hence the amount of refraction is dispersive in character.

Let us consider the wave diffraction problem for a simple harmonic wave propagating from very deep water into shallower water. As long as the variations in bottom depth are sufficiently slow, ray theory is a good approximation and that theory will be discussed here. This theory assumes irrotational wave motion and conservation of energy between rays, an approximation which is only valid when the rays change direction and amplitude slowly in the distance of a wavelength. In addition, it is assumed that the phase and group speeds are given by eqs. (15.4-7) and (15.4-12) which were derived for constant depth.

With these assumptions, refraction diagrams can be readily constructed for any specified bottom topography. Consider Fig. 15.4-7 where a simple case of straight parallel depth contours is shown with the wave crests in deep water being straight and parallel and at an angle to the shore. Since the phase speed decreases as the depth decreases, waves at point A, for instance, will tend to travel more slowly than waves at point B. As a result, waves tend to turn parallel to shore as shown. The refraction diagram can be constructed by calculating the distance traveled during one wave period by the waves originally at various points on a crest, constructing orthogonals to the waves at each such point, and advancing each point

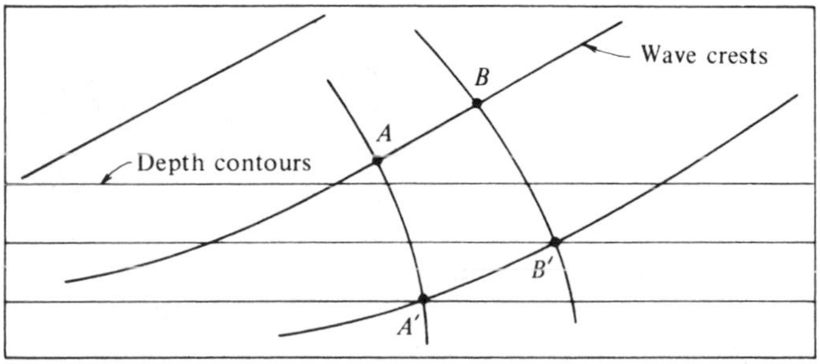

Fig. 15.4-7 Refraction diagram for beach with straight and parallel depth contours.

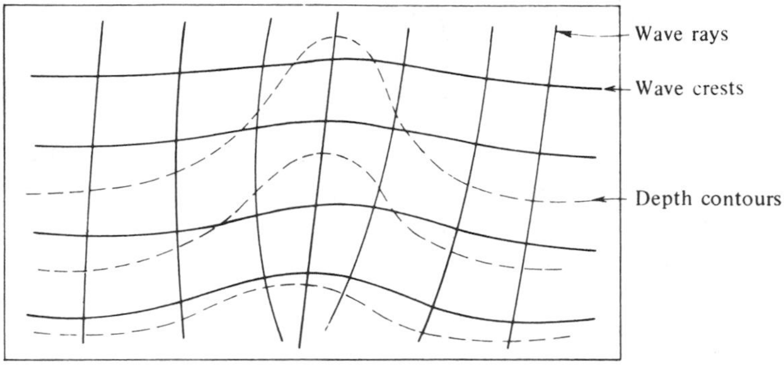

Fig. 15.4-8 Refraction diagram for submerged ridge.

the appropriate distance, i.e., point A moves to A', etc. The complete refraction pattern can be constructed in this manner with as much accuracy as needed. More refined methods of construction are available (Refs. 15-15 and 15-16) but the above procedure shows the basic ideas.

Figure 15.4-8 shows a slightly more complex situation, refraction by a submerged ridge (Ref. 15-17). Directly over the ridge, the wave speed is lower and the waves tend to lag behind while on either side of the ridge, the wave speed is greater and the waves tend to speed ahead. This causes a convergence of waves over the ridge and a divergence of waves to either side.

Wave heights can also be predicted adequately on the basis of the above theory. It was shown in sections 15.4.1 and 15.4.2 that for constant depth the energy flux is proportional to the square of the wave amplitude and also that the energy traveled at the group speed c_g. This is also assumed in the present case of slowly varying topography. The above assumptions plus the assumption of conservation of energy between rays leads to the relation

$$A^2 c_g S = \text{constant} \tag{15.4-27}$$

where A is the amplitude of the wave, c_g is the group velocity, and S is the distance between adjacent rays. Once the refraction diagram is constructed, S can be calculated; c_g is a function of the depth, which is known, and hence A can be calculated.

An interesting application of these ideas was to the calculation and later measurement of swell along a beach on Aruba, Netherlands Antilles (Ref. 15-16). Here it was observed that intermittently during the winter, large waves appeared on the eastern or lee side of Aruba on a particular beach although the prevailing winds were from the west. It was hypothesized that these waves were generated by a storm in the North Atlantic some time previous to the appearance of the swell at Aruba. The waves caused by this storm then propagated southward through the North Atlantic (at the group speed), were refracted as they passed between islands in the Caribbean Sea, were also refracted at the north-west tip of Aruba, and arrived on

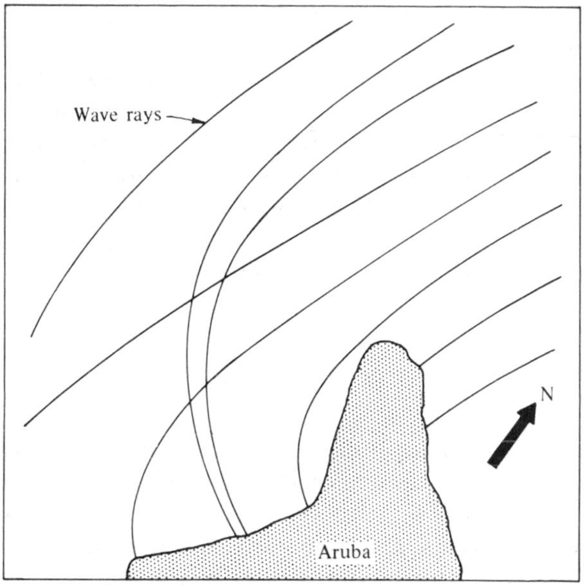

Fig. 15.4-9 Schematic refraction diagram for the shore off Aruba for waves with periods of 15 secs. coming from the North-East.

the leeward coast of Aruba (see Fig. 15.4-9). As can be seen, the rays converge along a particular section of the coast (where the highest waves were observed) and diverge elsewhere. It has been shown that only waves with 14 to 15 second periods are capable of doing this. Appearance of swell with this period has been correlated (with the appropriate time lag) with storms in the North Atlantic.

15.4.6 Nonlinear Wave Propagation

In the previous sections, only the linear theory of water waves has been disucssed. This was sufficient to illustrate many of the dispersive properties of water waves and dispersive wave effects in general. Dispersive effects are also present in non-linear water waves. However the theory of nonlinear water waves is extremely complex and no general method of solution of this problem has been devised.

Two limiting cases of water waves which take into account nonlinear effects were indicated in section 15.2.6, i.e., (1) shallow water theory, in which dispersive effects are not present, and (2) the wave motion governed by the Korteweg-deVries equation, repeated here for convenience

$$\frac{\partial u}{\partial \widetilde{t}} + 3u \frac{\partial u}{\partial \widetilde{x}} = -\frac{1}{3} \frac{\partial^3 u}{\partial \widetilde{x}^3} \tag{15.4-28}$$

This equation takes into account first order nonlinear convective effects and dispersive effects simultaneously. Other limiting cases in the theory of water waves are possible. For a discussion of these, see Refs. 15-3, 15-15, and 15-18.

Many of the recent significant advances in the understanding of dispersive effects in nonlinear waves have come from the study and solution of the above equation. Because of this and because of the importance of the Korteweg-deVries equation in physical situations, e.g., water waves, ion-acoustic waves, collisionless magnetohydrodynamic waves in a warm plasma, acoustic waves in an anharmonic crystal, and vortex breakdown in a rotating fluid, some of the work on this equation will be summarized here. For a more complete summary, see Ref. 15-1 and the references mentioned there.

The most common form of the Korteweg-deVries equation is

$$\frac{\partial u}{\partial t} + u\frac{\partial u}{\partial x} = -\beta\frac{\partial^3 u}{\partial x^3} \tag{15.4-29}$$

where β is a positive constant. It should be noted that the quantities u, x, and t can be rescaled to produce any desired coefficients before the terms in this equation.

Stationary solutions of this equation were first obtained by Korteweg and deVries in 1895. These solutions are of the form $u(x, t) = u(\xi)$, where $\xi = x - ct$ and c is some constant velocity. This substitution transforms the above equation into a third-order nonlinear ordinary differential equation for which closed-form analytic solutions are possible. Depending on the boundary conditions prescribed as $|x| \to \infty$, the solutions are either solitary or periodic waves propagating with constant velocity c relative to the medium.

For a solitary wave, or 'soliton', for which $u \to u_\infty$ as $|x| \to \infty$, u is of the form

$$u(\xi) = u_\infty + (u_0 - u_\infty)\,\text{sech}^2\left[\left(\frac{u_0 - u_\infty}{12\beta}\right)^{1/2}\xi\right] \tag{15.4-30}$$

where u_∞ and u_0 are arbitrary constants and u_0 is the amplitude of the wave relative to u_∞. The width of the wave is of the order of $[12\beta/(u_0 - u_\infty)]^{1/2}$ and is inversely proportional to the square root of the amplitude. The velocity of the wave is related to the amplitude by $c = u_\infty + (u_0 - u_\infty)/3$.

For periodic, or cnoidal waves, u can be written in terms of the Jacobian elliptic function. In two opposing limits, these cnoidal waves reduce to either (a) a sinusoidal wave of finite amplitude, or (b) a sequence of solitons each separated by a very large distance.

Numerical solutions (Refs. 15-19 to 15-22) have given us considerable insight into the behavior of nonstationary wave motion as governed by the Korteweg-deVries equation. Researchers have studied (a) the interactions of two or more solitons initially widely separated and of different amplitude and therefore different speeds of propagation, and (b) the formation and interaction of solitons as determined by the initial condition $u = \cos \pi x$ (with $\beta^{1/2} = 0.0222$).

The investigation of case (a) first showed that solitons are remarkably stable, i.e., as two solitons meet, they interact nonlinearly but emerge as solitons with amplitude and velocity unchanged. The motion in case (b) is considerably more complex. In this case, initially the third term of eq. (15.4-29) is negligible and the motion is strictly wave-like. A steepening of the wave form in front begins. Before a discontinuity occurs, the third, or dispersive, term of eq. (15.4-29) becomes significant, causing an oscillatory structure to form superimposed on the periodic waves of wavelength 2. These oscillations grow in amplitude and form solitons with an almost linear variation of amplitude with distance (per wavelength interval). These solitons then interact and periodically almost reconstruct the initial state.

Berezin & Karpman (Ref. 15-19) studied a different initial condition

$$u(x,0) = u_0 \exp\left(-\frac{x^2}{L^2}\right) \tag{15.4-31}$$

with quite interesting results. This initial condition attenuates as $|x| \to \infty$ in contrast to the periodic initial condition assumed in case (b) above. The character of the solutions was shown to depend on the parameter $\sigma = L(u_0/\beta)^{1/2}$. For σ greater than approximately 4, the initial perturbation decays into (a) two or more solitons propagating in the positive x-direction and (b) a rapidly oscillating wave train propagating in the negative x-direction but with negligible (and time-decreasing) amplitude. With increasing σ, the initial solution breaks up into a larger number of solitons. For σ slightly less than 4, the initial solution breaks up into one soliton moving in the positive x-direction and a rapidly oscillating wave packet moving in the negative x-direction but now with nonnegligible amplitude, energy, and momentum (see Fig. 15.4-10). For $\sigma << 4$, no solitons are present asymptotically for large time, and the solution consists only of a rapidly oscillating wave packet propagating

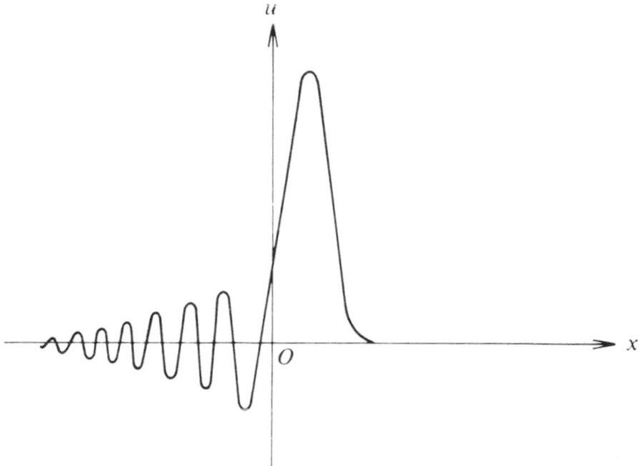

Fig. 15.4-10 Schematic diagram showing amplitude of dispersive wave for large time for initial condition $u = u_0 \exp(-x^2/L^2)$ and σ slightly less than 4. Solution consists of one soliton moving in positive x-direction and a rapidly oscillating wave packet moving in the negative x-direction.

in the negative x-direction. The values of σ at which the solution changes qualitatively were determined only approximately by numerical trial and error.

A method of solution applicable to general nonlinear dispersive wave systems has been developed by Whitham (Refs. 15-23 and 15-24). The procedure is valid when the system can be described in terms of nonuniform oscillating wave trains where the amplitude and wave number vary slowly in space and time, i.e., where the length and time scales of the variation in amplitude and wave number are large compared with the wavelength and period. Originally Whitham used an averaging technique analogous to the Krylov-Bogoliubov averaging procedure for ordinary differential equations in order to obtain approximate equations for the variations of amplitude and wave number. Subsequently, the desired equations were derived by application of an averaging procedure to the Lagrangian for the system and use of a variational principle based on this averaged Lagrangian.

This procedure can be applied to both the soliton and rapidly oscillating parts of the solution to the problem investigated by Berezin and Karpman (Ref. 15-19), for which a special case is shown in Fig. 15.4-10, to obtain asymptotic results for large t. In particular, it can be shown that, for large time and at one instant of time, the amplitude of the solitons increases linearly with distance, and the slope of the line of amplitude versus distance decreases inversely with time. In addition, a general asymptotic form of the oscillatory part of the solution can be determined.

A formal perturbation expansion that can be applied to general nonlinear dispersive wave equations has been developed by Luke (Ref. 15-25). Whitham's results are then obtained directly as the first approximation to the expansion.

15.5 DIFFUSIVE WAVES

15.5.1 Linear Analysis: A Simple Example

The main emphasis in the present section will be on diffusive waves in gas dynamics. In this case, diffusion is caused by some dissipative process, e.g., viscosity, thermal conductivity, chemical reaction, radiation, etc., or combination of processes. One simple wave propagation problem will be considered, the one-dimensional, time-dependent propagation of disturbances through a gas, semi-infinite in extent. It is assumed that the gas is bounded by a piston at the left and initially is in static equilibrium. For time $t > 0$, conditions on certain variables such as velocity and temperature of the piston will be changed to prescribed constant values different from their initial values. In general, this will cause a wave to be propagated to the right in the positive x-direction. This is called the signalling problem.

When no dissipative process is considered, the basic equation is just the simple linear wave equation, as mentioned in section 15.2.3. However, when dissipative processes are considered, the basic equations are considerably modified as is the corresponding wave motion. For each dissipative process or combination of processes, in the limit of small amplitude motions, a single equation governing the motion can always be found. This equation will have the general form

$$\sum_m \lambda_m \left(\frac{\partial}{\partial t} + c_{m1} \frac{\partial}{\partial x}\right)\left(\frac{\partial}{\partial t} + c_{m2} \frac{\partial}{\partial x}\right) \cdots \left(\frac{\partial}{\partial t} + c_{mn} \frac{\partial}{\partial x}\right) \phi = 0 \qquad (15.5\text{-}1)$$

where the λ_m's are known parameters, the c_{mn}'s are different wave speeds, and ϕ may be the velocity or a related quantity. The above equation is a simple generalization of an equation studied extensively by Whitham (Ref. 15-26) which was

$$\left(\frac{\partial}{\partial t} + c_1 \frac{\partial}{\partial x}\right)\left(\frac{\partial}{\partial t} + c_2 \frac{\partial}{\partial x}\right) \phi + \lambda \left(\frac{\partial}{\partial t} + a \frac{\partial}{\partial x}\right) \phi = 0 \qquad (15.5\text{-}2)$$

In order to illustrate the mathematical procedure of obtaining certain limiting solutions and to show the basic characteristics of diffusive wave motion, the above equation will be discussed first. In section 15.5.2, the linear wave motions due to specific dissipative processes will be considered briefly. A more complete discussion of diffusive waves is given in Ref. 15-27.

In eq. (15.5-2), it will be assumed that $\lambda > 0$, that $c_1 > a > 0$, and that $c_2 < 0$. These conditions insure stability of the solution and correspond closely to the physical problems encountered later. For a discussion of other cases, Ref. 15-26 should be consulted.

Appropriate boundary conditions for the signalling problem for the present case are

$$t = 0, \quad \phi = \frac{\partial \phi}{\partial t} = 0 \qquad (15.5\text{-}3)$$

$$x = 0, \quad \phi = \phi_0 \qquad (15.5\text{-}4)$$

$$x \to \infty, \quad \phi = 0 \qquad (15.5\text{-}5)$$

The solution to this problem can be found readily by Laplace transforms. By use of the inversion integral, one can write the solution as

$$\phi = \frac{\phi_0}{2\pi i} \int_\Gamma \frac{\exp(pt + \gamma x)}{p} \, dp \qquad (15.5\text{-}6)$$

where Γ is the path such that Re p is constant and is to the right of all singularities, $\gamma = -\frac{1}{2}\left[\delta_1 + \sqrt{\delta_1^2 - 4\delta_2}\right]$; $\delta_1 = [p(c_1 + c_2) + \lambda a]/c_1 c_2$; and $\delta_2 = p(p + \lambda)/c_1 c_2$. This equation can be numerically integrated but with great difficulty. Analytic solutions are desirable and can be obtained for certain limiting and important cases.

An approximate evaluation of the above integral for small time can be accomplished by substituting expansions for large p for the function γ. Large p corresponds to high frequencies and therefore this approximation is valid when the high frequency waves dominate, i.e., when t is small or near discontinuities in the wave form.

By this procedure, one finds that

$$\phi = \phi_0 H(t - x/c_1) \exp\left[\frac{-\lambda}{c_1}\left(\frac{c_1 - a}{c_1 - c_2}\right) x\right] \tag{15.5-7}$$

where $H(x)$ is the step function defined such that $H(x) = 1$ for $x > 0$ and $H(x) = 0$ for $x < 0$. Therefore, for small time, the wave propagates at the speed c_1 and decays exponentially with a characteristic length defined by

$$\frac{\lambda}{c_1}\left(\frac{c_1 - a}{c_1 - c_2}\right) x = 1; \quad \text{or } x = \frac{c_1}{\lambda}\left(\frac{c_1 - c_2}{c_1 - a}\right) \tag{15.5-8}$$

As λ increases, the wave decays more rapidly.

For large time, the form of the integral in eq. (15.5-6) suggests evaluation by the method of steepest descent. In this approximation, the dominant contributions to the integral come from the neighborhood of the saddle point and perhaps from any singularities enclosed by the contour path deformed to pass through the saddle point. We anticipate that the saddle point will be located near the origin. This is consistent with the idea that for long time the high frequency waves will have attenuated and the lower frequency waves are dominant. We then can approximate γ by an expansion for small p.

Evaluation of the resulting integral gives

$$\phi = \frac{\phi_0}{2} \operatorname{erfc}\left(\frac{x - at}{\sqrt{\beta t}}\right) \tag{15.5-9}$$

where $\beta = 4(c_1 - a)(a - c_2)/\lambda$. It can be seen from the above equation that for large time the main part of the wave propagates at the speed a and diffuses with a characteristic diffusion width defined by

$$\frac{x - at}{\sqrt{\beta t}} = 1; \quad \text{or } x - at = \left[\frac{4(c_1 - a)(a - c_2)t}{\lambda}\right]^{1/2} \tag{15.5-10}$$

As λ increases, the diffusion width decreases, i.e., the wave front becomes steeper. A schematic diagram of the motion is shown in Fig. 15.5-1.

The general result is then the following: (1) For small time, the highest order term of eq. (15.5-2) dominates. The lower order term produces an exponential damping of the wave described by the higher order term. (2) For long time, the lower order term dominates. The higher order term causes the wave described by the lower order term to diffuse. (3) As λ increases, the lower order term becomes dominant at an earlier time. In addition, it can be shown that the characteristics are

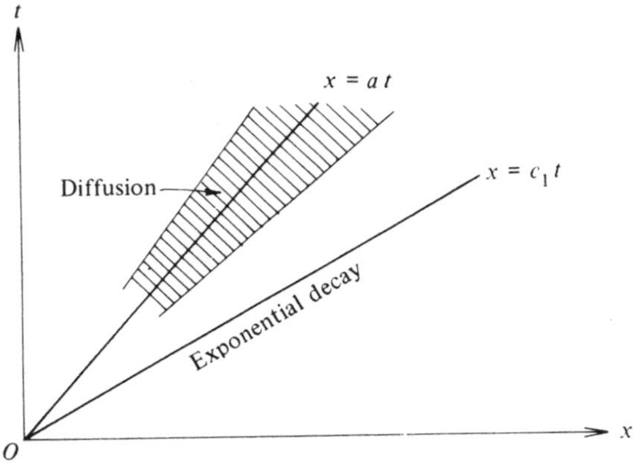

Fig. 15.5-1 Schematic diagram of wave motion. The wave propagating at the speed c_1 decays exponentially. As this wave decays, a wave propagating at the speed a forms and diffuses with a diffusion width proportional to $\sqrt{t}$.

given by the first term in eq. (15.5-2) and are the lines $x - c_1 t = \text{constant}$ and $x - c_2 t = \text{constant}$. The subcharacteristics are given by the second term in eq. (15.5-2) and are $x - at = \text{constant}$. As λ increases, ϕ and its derivatives change rapidly near $x = at$ but are still continuous.

15.5.2 General Characteristics of Linear Motion

A particular example of a dissipative process that results in an equation and solution very similar to that discussed in the previous section is the signalling problem for a chemically reacting mixture of gases (Refs. 15-28 and 15-29). The basic equations and assumptions for this problem were given in section 15.2.3. The linear equation for the velocity potential ϕ can be shown to be

$$\frac{\partial}{\partial t}\left(\frac{\partial^2 \phi}{\partial t^2} - a_f^2 \frac{\partial^2 \phi}{\partial x^2}\right) + \lambda \left(\frac{\partial^2 \phi}{\partial t^2} - a_s^2 \frac{\partial^2 \phi}{\partial x^2}\right) = 0 \tag{15.5-11}$$

where all quantities were defined in section 15.2.3. The proper boundary conditions for the signalling problem are

$$t = 0, \quad \phi = \frac{\partial \phi}{\partial t} = \frac{\partial^2 \phi}{\partial t^2} = 0 \tag{15.5-12}$$

$$x = 0, \quad \frac{\partial \phi}{\partial x} = u_p \tag{15.5-13}$$

$$x \to \infty, \quad \phi \to 0 \tag{15.5-14}$$

This problem can be solved readily by Laplace transforms and limiting solutions can be obtained in the manner described in section 15.5.1. For small time, a first approximation to the solution can be shown to be

$$u = u_p H(t - x/a_f) \exp\left[-\left(\frac{a_f^2 - a_s^2}{a_f^2}\right)\frac{\lambda x}{2a_f}\right] \tag{15.5-15}$$

The above equation indicates a wave which travels at the frozen speed of sound and decays exponentially with a decay length of

$$x = \frac{2a_f}{\lambda}\left(\frac{a_f^2}{a_f^2 - a_s^2}\right) \tag{15.5-16}$$

A decay time can be defined as the time it takes the wave to travel a distance corresponding to a decay length. This decay time is

$$t = \frac{x}{a_f} = \frac{2}{\lambda}\left(\frac{a_f^2}{a_f^2 - a_s^2}\right) = O(\tau) \tag{15.5-17}$$

where τ is the relaxation time for the reaction being considered. It can be shown that the entropy and composition are constant through the wave front given by $x \doteq a_f t$.

For large time, approximate evaluation of the inversion integral leads to

$$\frac{u}{u_p} = \frac{1}{2}\operatorname{erfc}\left[\frac{x - a_s t}{(2\delta t)^{1/2}}\right] \tag{15.5-18}$$

where $\delta = (a_f^2 - a_s^2)/\lambda$. The wave propagates at the isentropic speed of sound and diffuses with a diffusivity δ. In addition, it can be shown that the entropy is constant and the composition is given by its local equilibrium value. The general features of the motion are the same as shown in Fig. (15.5-1) with the substitution of a_f for c_1 and a_s for a.

Other dissipative processes, although leading to different basic equations, can be analyzed in the same way. For instance, if the effects of thermal radiation on the propagation of small disturbances in a gas are considered (Refs. 15-30 and 15-31), it can be shown that an approximate equation describing the flow can be written as

$$\frac{\partial^3}{\partial t \partial x^2}\left(\frac{\partial^2 \phi}{\partial t^2} - a_s^2 \frac{\partial^2 \phi}{\partial x^2}\right) + a_1^2 \frac{\partial^2}{\partial x^2}\left(\frac{\partial^2 \phi}{\partial t^2} - a_T^2 \frac{\partial^2 \phi}{\partial x^2}\right) - b_1^2 \frac{\partial}{\partial t}\left(\frac{\partial^2 \phi}{\partial t^2} - a_s^2 \frac{\partial^2 \phi}{\partial x^2}\right) = 0 \tag{15.5-19}$$

where ϕ is a velocity potential, $a_s^2 = (\partial p/\partial \rho)_s$, $a_T^2 = (\partial p/\partial \rho)_T$, and a_1 and b_1 are constant parameters which depend on the radiative and thermodynamic properties of the gas.

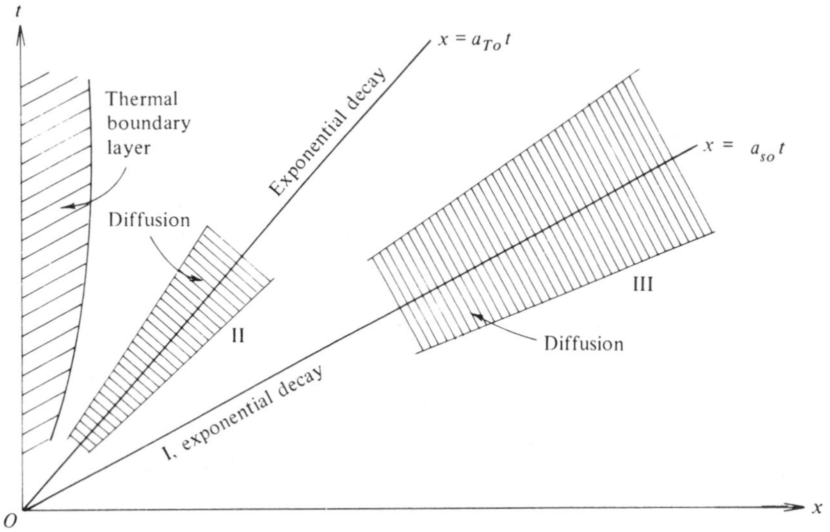

Fig. 15.5-2 Schematic diagram of wave motion showing the effects of thermal radiation.

The resulting motion is shown schematically in Fig. 15.5-2. Wave I propagates at the isentropic speed of sound and decays exponentially with the velocity and temperature discontinuous across the wave. Wave II propagates at the isothermal speed of sound, diffuses with a diffusion width proportional to $\sqrt{t}$ and eventually decays exponentially. Wave III propagates at the isentropic speed of sound, diffuses, but does not decay. The entropy is essentially constant through waves I and III. Disturbances are present in front of the waves due to radiation. A thermal boundary layer is present at $x = 0$.

In general, the basic characteristics for any type of dissipative wave motion can be anticipated by analogy with the results of section 15.5.1. When the linearized equation contains more than two terms, each pair of adjacent terms can be thought of as independent as a first guess at the solution.

Except for long time, the wave motion is generally different for each particular dissipative process. However, for long time, the wave motions are similar, are diffusive in character and depend on a certain diffusivity δ. These diffusivities can be added when different dissipative processes are considered, but only when the diffusing waves, considered separately, travel at the same speed.

15.5.3 Steady-State Shock Waves

Although the linear theory of wave propagation is quite useful, the convective effects are not included properly. In the signalling problem, for instance, no matter how small the perturbations at the piston, nonlinear convective effects will ultimately become important. For long time, the linear theory predicts a wave diffusing gradually with time, although the correct limiting solution for long time would give a compression wave which would be steady in a coordinate system moving with the wave and in which the diffusive effects would be exactly balanced by nonlinear convective effects.

This steady-state structure of the shock wave will be discussed in the present section while the time-dependent formation of this shock wave from some initial condition will be discussed in the following two sections. Many dissipative processes are potentially important in the study of shock waves in a real gas. However, because the presence of chemical reaction leads to several interesting features, that case will be discussed briefly here. See Ref. 15-27 for a more comprehensive presentation of dissipative effects on the structure of shock waves.

In the present discussion, only the case of vibrational nonequilibrium will be considered. The basic equations have been presented in section 15.2.3. It is assumed that the internal energy e is the sum of the energy of the active (translational plus rotational) degrees of freedom e_t and the energy of an internal (vibrational) degree of e_i, and that $e_t = c_{vt}T$ and $e_i = c_i T_i$, where the specific heats c_{vt} and c_i are constant. The equation of state is $p = \rho RT$ and the rate equation has the form

$$\frac{DT_i}{Dt} = -\frac{(T_i - T)}{\tau} \tag{15.5-20}$$

It is assumed that a state exists which is steady in a coordinate system moving at the shock speed U and that all variables are functions of $\xi = x - Ut$ only. The equations of conservation of mass, momentum, and energy, eqs. (15.2-5) to (15.2-7), are then integrable and become

$$\rho v = m \tag{15.5-21}$$

$$p + \rho v^2 = P \tag{15.5-22}$$

$$e + \frac{p}{\rho} + \frac{1}{2} v^2 = H \tag{15.5-23}$$

where m, P, and H are constants and are determined from the uniform conditions at $\xi = \pm\infty$. The velocity $v = U - u$ and is the fluid velocity in the coordinate system moving at the shock speed U. Hereafter, conditions at $\xi = \pm\infty$ will be denoted by the subscripts 0 and 1 respectively.

From these equations, one can derive the relations

$$c_i dT_i = \left(\frac{\gamma_t + 1}{\gamma_t - 1}\right)(v - v_c)\, dv = \frac{1}{(\gamma_t - 1)} \frac{(v^2 - a_f^2)}{v} dv \tag{15.5-24}$$

$$\frac{dv}{d\xi} = -\frac{1}{2\tau}\left(\frac{\gamma_t - 1}{\gamma_t + 1}\right)\left(\frac{\gamma + 1}{\gamma - 1}\right)\frac{(v_0 - v)(v - v_1)}{v(v - v_c)} \tag{15.5-25}$$

where $\gamma_t = c_{pt}/c_{vt}$; $a_f^2 = (\partial p/\partial \rho)_{s,T_i} = \gamma_t rT$; and $v_c = \gamma_t P/[(\gamma_t + 1)m]$. The velocities v_0 and v_1 are just the velocities at $\xi = \pm\infty$ and are determined from eqs. (15.5-21) to (15.5-23) with $T_i = T$. These velocities depend only on the conditions at $\xi = \pm\infty$, are independent of the dissipative processes occurring in between, and can be shown to be

$$v_{0,1} = \left(\frac{\gamma}{\gamma + 1}\right)\frac{P}{m} \pm \left\{\left[\left(\frac{\gamma}{\gamma + 1}\right)\frac{P}{m}\right]^2 - 2\left(\frac{\gamma - 1}{\gamma + 1}\right)H\right\}^{1/2} \qquad (15.5\text{-}26)$$

It can be seen that $v_0 \geqslant v_1$. By writing m, P, and H in terms of conditions at $\xi = \pm\infty$, it is easily shown that the flow is supersonic in front of the shock, $v_0 \geqslant a_{s0}$, and subsonic behind the shock, $v_1 \leqslant a_{s1}$. From the definition of v, v_0 is identical to the shock speed U. This choice of v_0 and v_1 insures that the entropy will increase across the shock. The usual jump conditions across a shock wave follow from the above equations.

Equation (15.5-25) can be integrated exactly to obtain

$$\frac{1}{2\tau}\left(\frac{\gamma_t - 1}{\gamma_t + 1}\right)\left(\frac{\gamma + 1}{\gamma - 1}\right)\xi = v - \frac{v_0(v_c - v_0)}{v_0 - v_1}\ln(v_0 - v) + \frac{v_1(v_c - v_1)}{v_0 - v_1}\ln(v - v_1)$$
$$(15.5\text{-}27)$$

As long as $v_0 < v_c$, the solution for v is a single-valued function of ξ. However, if $v_0 > v_c$ [and therefore $v_0 > a_{f0}$, since it can be shown from the previous equations and definitions that $(\gamma_t + 1)v(v - v_c) = v^2 - a_f^2$], the solution becomes double-valued (see Fig. 15.5-3). A suitable solution can be obtained by introducing a solution discontinuous in v but continuous in T_i. The vibrational temperature T_i must be continuous because otherwise $dT_i/d\xi$ would be infinite, which is not allowed by

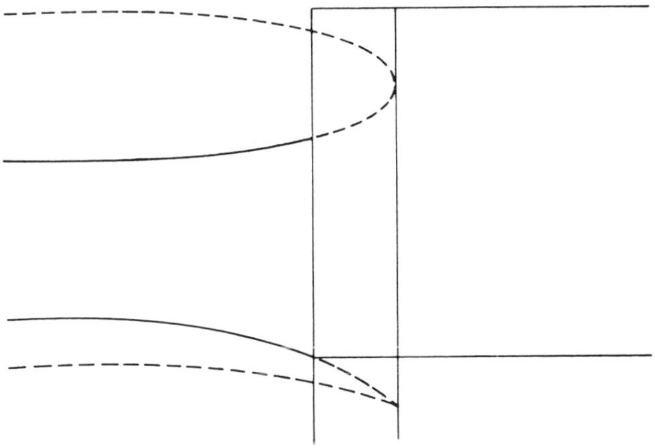

Fig. 15.5-3 Discontinuous velocity and vibrational temperature profiles for a shock wave in a gas with vibrational nonequilibrium. A discontinuous profile occurs when $U > a_{f0}$.

the rate equation. The jump conditions across this inner discontinuity may be found from eqs. (15.5-21) to (15.5-23) with T_i constant. It can be shown that the velocity ahead of the discontinuity is supersonic relative to the frozen speed of sound, while the velocity behind is subsonic relative to this same speed.

For an arbitrary dissipative process, an approximate solution valid for weak shocks can be found in the following manner. It can be shown (Ref. 15-27, also see section 15.5.5) that an approximate equation which governs the one-dimensional, time-dependent propagation of a weak, diffuse wave traveling in one direction only is Burgers' equation, which can be written in the present context as

$$\frac{\partial u}{\partial t} + \left[\left(\frac{\gamma + 1}{2}\right) u + a_{so}\right] \frac{\partial u}{\partial x} = \frac{\delta}{2} \frac{\partial^2 u}{\partial x^2} \tag{15.5-28}$$

where δ is a diffusivity which depends on the particular dissipative process or combination of processes being considered. For a chemical reaction, $\delta = (a_{fo}^2 - a_{so}^2)/\lambda$. Once δ is known, the steady-state structure of a shock wave within the above approximation can be found easily from the steady-state solution of the above equation. The solution is

$$\begin{aligned} \frac{u}{u_1} &= \left\{1 + \exp\left[\left(\frac{\gamma + 1}{2}\right) \frac{u_1}{\delta} \xi\right]\right\}^{-1} \\ &= \frac{1}{2}\left\{1 - \tanh\left[\left(\frac{\gamma + 1}{4}\right) \frac{u_1}{\delta} \xi\right]\right\} \end{aligned} \tag{15.5-29}$$

For weak enough shocks, the solutions when arbitrary dissipative processes are considered are generally continuous. However, for strong shocks, the solutions will generally permit discontinuities as long as viscosity is not considered. Discontinuities are possible when the characteristics travel at a speed intermediate between the speeds of the fluid at the front and rear of the shock. The speed of the characteristics can be predicted from the linear theory except that, in the nonlinear theory, the characteristics propagate relative to the fluid, which is moving, and the speed must be calculated using the local values of the thermodynamic variables.

15.5.4 Method of Characteristics

The method of characteristics for a compressible fluid without dispersion or diffusion present was treated in section 15.3.6. When diffusion is present, characteristics may or may not be present, i.e., the equations of motion may or may not be hyperbolic in character. For example, when viscosity is included in the analysis, the front of the disturbance is propagated at infinite speed, at least within the continuum gas hypothesis, and therefore the equations of motion must be parabolic in character. However, when chemical reactions with finite reaction times are included in the analysis and other dissipative processes are neglected, disturbances propagate at a finite rate of speed and therefore the equations of motion are hyperbolic in character.

This latter problem has many interesting features and so will be discussed briefly here. Again the one-dimensional, time-dependent motion will be considered. The characteristics for this case were investigated by Chu (Ref. 15-28) and Broer (Ref. 15-29) while, for the two-dimensional, steady case, the analogous problem was discussed in Ref. 15-32.

The basic equations for the present problem were given in section 15.2.3. With relatively little algebra and a little thermodynamics, these equations can be rewritten as

$$D_{\pm}\, p \pm \rho a_f\, D_{\pm}\, u = -\frac{a_f^2}{\tau}\left[\left(\frac{\partial h}{\partial q}\right)_{p,\rho} \Big/ \left(\frac{\partial h}{\partial \rho}\right)_{q,p}\right](q - \tilde{q}) \tag{15.5-30}$$

$$T\frac{DS}{Dt} = \frac{Q}{\tau}(q - \tilde{q}) \tag{15.5-31}$$

$$\frac{Dq}{Dt} = -\frac{(q - \tilde{q})}{\tau} \tag{15.5-32}$$

where $Q = (\partial h/\partial q)_{p,s}$ and

$$D_{\pm} = \frac{\partial}{\partial t} + (u \pm a_f)\frac{\partial}{\partial x} \tag{15.5-33}$$

$$\frac{D}{Dt} = \frac{\partial}{\partial t} + u\frac{\partial}{\partial x} \tag{15.5-34}$$

and the other terms are as defined in section 15.2.3.

It can be seen that these equations are in characteristic form and therefore define the characteristic directions. In eq. (15.5-30), the speeds defined are equal to $u \pm a_f$, i.e., disturbances propagate relative to the fluid in both directions at the frozen isentropic speed of sound. In eqs. (15.5-31) and (15.5-32), the speed defined is u or just the speed of the fluid itself.

By comparison of these equations and the characteristic equations for a compressible fluid without dissipation given in section 15.3.6, it can be seen that the characteristic relations and particularly the characteristic directions are different for the two cases. This fact has caused some difficulty in the physical and mathematical understanding of the problem. The difficulty is in the interpretation of the characteristic speeds when the motion is near equilibrium. In equilibrium, $q - \tilde{q}$ is exactly zero, and the discussion of section 15.3.6 applies. Rewriting eqs. (15.3-50) and

(15.3-51) for the equilibrium case in the present notation, one finds that

$$D'_{\pm} p \pm \rho a_s D'_{\pm} u = 0 \tag{15.5-35}$$

where $a_s^2 = (\partial p/\partial \rho)_{s,\, q=\tilde{q}}$ and

$$D'_{\pm} = \frac{\partial}{\partial t} + (u \pm a_s)\frac{\partial}{\partial x} \tag{15.5-36}$$

These are characteristic equations but with $u \pm a_s$ as the characteristic speeds. It can be seen that they are different from eqs. (15.5-30) which are valid for a chemically reacting gas with finite reaction times. It can be shown that these characteristic speeds do not change continuously from $u \pm a_f$ to $u \pm a_s$ as the flow goes toward equilibrium. Rather, as long as there is a finite reaction rate, no matter how small, $u \pm a_f$ is the characteristic speed.

The explanation to this seemingly discontinuous behavior is the following. Although the mathematical characteristics change discontinuously between the equilibrium and nonequilibrium situations, the physical characteristics or features of the flow do not. In near equilibrium when $u \pm a_f$ are the proper mathematical characteristics, although the front of the disturbance travels at the speed $u \pm a_f$, as equilibrium is approached more and more of the disturbance travels at the speed $u \pm a_s$. The motion therefore approaches the motion predicted by equilibrium theory in a smooth manner. This feature of the flow was discussed in section 15.5.2 for the linear case and will be discussed further in the next section for the nonlinear case.

15.5.5 Weakly Nonlinear Waves

For weakly nonlinear waves, approximate analytic methods of solution have recently been developed. These solutions can be used to easily characterize the flow and will be discussed briefly here. The diffusive effects due to viscosity and heat conduction will be discussed first followed by a discussion of effects due to chemical relaxation. The mathematical methods of solution described are similar to those presented in section 15.3.7. See Ref. 15-1 and the references listed there for a more comprehensive treatment of this problem.

Consider the one-dimensional, time-dependent signalling problem with the effects of viscosity and thermal conductivity included. In this problem, immediately after the piston is set in motion, the velocity gradients are very steep and it is expected that the diffusive terms are more important than the nonlinear convective terms. This implies that the linear theory is valid for some early time. As the wave propagates through the gas, viscosity and thermal conductivity cause the velocity gradients to ease. Eventually the nonlinear effects must become important since, for long time, it is expected that the exact solution would be a steady-state compression wave with the nonlinear convective effects balanced by the diffusive effects.

Let us introduce a dimensionless length $x^* = a_{s0}x/\nu$, a dimensionless time $t^* = a_{s0}^2 t/\nu$, and a dimensionless parameter ϵ that is of the order of a characteristic piston velocity divided by the isentropic speed of sound. The symbol ν denotes the kinematic viscosity. For long time, the linear theory predicts a wave progressing in the positive x-direction with a diffusion width of the same order as $\sqrt{t^*}$. The thickness of a weak, steady-state compression wave is of the order of $1/\epsilon$. The linear solution is therefore necessarily invalid when the diffusion thickness and the thickness of the compression wave are equal, i.e., when t^* is approximately $1/\epsilon^2$.

The method of matched asymptotic expansions can be used in a manner similar to that of section 15.3.7 to find a solution valid in a far-field region, $t^* > 1/\epsilon^2$, where the ordinary linear expansion is no longer valid, and then to match this far-field solution smoothly to the linear solution. The result is a uniformly valid expansion for the entire flow field.

In order to describe the far field properly, a new time scale $t^+ = \epsilon^2 t^*$ and a new length scale (moving with the compression wave) $x^+ = \epsilon(x^* - t^*)$ are suggested. By introducing these coordinates into the equations of motion, expanding all dependent variables in an asymptotic series in powers of ϵ, and equating coefficients of like powers of ϵ to zero, one can derive the equation

$$\frac{\partial u}{\partial t} + \left(\frac{\gamma + 1}{2} u + a_{so}\right) \frac{\partial u}{\partial x} = \frac{\delta}{2} \frac{\partial^2 u}{\partial x^2} \tag{15.5-37}$$

which is the first approximation for the flow in the far field and of course is just Burgers' equation. Here, $\delta = \nu + \alpha(\gamma - 1)\gamma$, and α is the thermal diffusivity.

The utility of Burgers' equation, which of course is nonlinear, is due principally to the fact that it can be solved exactly (Refs. 15-33 and 15-34) by use of a transformation that reduces it to the linear, one-dimensional, time-dependent heat equation. Introduce the quantities $w = \frac{1}{2}(\gamma + 1)u$ and $X = x - a_{so}t$. Equation (15.5-37) then reduces to

$$\frac{\partial w}{\partial t} + w \frac{\partial w}{\partial X} = \frac{\delta}{2} \frac{\partial^2 w}{\partial X^2} \tag{15.5-38}$$

Define a function ψ such that w and ψ are related by

$$\psi = \exp\left(\frac{1}{\delta} \int_x^\infty w \, dX\right) \tag{15.5-39}$$

$$w = -\frac{\delta}{\psi} \frac{\partial \psi}{\partial X} \tag{15.5-40}$$

With this substitution, one can reduce eq. (15.5-38) to

$$\frac{\partial \psi}{\partial t} = \frac{\delta}{2} \frac{\partial^2 \psi}{\partial X^2} \tag{15.5-41}$$

which is just the classical heat-conduction equation, for which solutions can readily be found.

Boundary or initial conditions, or both, are needed for both the near-field and far-field equations. For the signalling problem with constant piston velocity u_p for $t > 0$, boundary and initial conditions are prescribed for the linear near-field problem. An initial condition for the far-field equation is determined from matching with the linear solution and can be shown to be $u = u_p S(-x)$ for $t = 0$.

The solution to the signalling problem shows that the shock forms in a time of order $1/\epsilon^2$. The solution also gives the correct result, to the lowest order in shock strength, for the steady-state shock (see section 15.5.3).

For the case when other dissipative effects are present, alone or in combination, the method of matched asymptotic expansions can be used in essentially the same form as above, i.e., the far field is described by Burgers' equation and the near field by the linear theory. Of course, the linear theory is quite different in each case. The solution for the far field as given by Burgers' equation is limited to weak waves since it predicts a continuous shock structure in all cases. As noted in the previous section, this is not true for strong enough waves.

Implicit in the above arguments and analyses has been the assumption that the nonuniformity in the linear solution has been caused by diffusion of the wave front until the thickness of the wave front as predicted by linear theory becomes greater than the thickness of the shock wave as predicted by a steady-state analysis. For the problem of a chemically reacting fluid, this occurs in a dimensionless time $t/\tau = O(\epsilon^{-2})$, where τ is the relaxation time for the particular reaction being considered. The dimensional time is then $\tau/\epsilon^2 \equiv t_2$.

For the nonlinear, nondiffusive, nondispersive wave-propagation problem described in section 15.3.7, the nonuniformity of the linear solution was caused by a different mechanism, i.e., the nonuniformity was a result of nonlinear convection causing the wave front to deviate from its position as predicted by linear theory until this difference in position was comparable to or greater than the characteristic length $L = a_{s0}T$, where T is a characteristic time associated with the piston motion. This occurs in a dimensionless time $a_{s0}t/L = O(\epsilon^{-1})$ and a dimensional time $L/a_{s0}\epsilon = x_p/a_{s0}\epsilon^2 = t_1$, where x_p is a characteristic length associated with the piston motion and is of the order of the characteristic piston velocity multiplied by T.

For the problem of wave propagation in a chemically reacting fluid, the ratio of these times t_1/t_2 is $O(x_p/a_{s0}\tau)$. As this ratio becomes very large, the above analyses are valid. However, if τ is very large, t_1 could be less than t_2, indicating that the nonuniformity as described in section 15.3.7 would be of importance at an earlier time than the nonuniformity corrected by the use of Burgers' equation and described at the beginning of this section. An intermediate region must then be present where neither the far-field nor near-field solutions of the above method of matched asymptotic expansion are valid. Another way of looking at it is that in this intermediate region, nonlinear convective effects are important but the approximation of describing chemical reaction by a diffusion term (as in Burgers' equation) is not correct.

A uniformly valid solution for this type of problem has recently been obtained by the use of (a) an N-timing procedure, or the method of multiple scales (Ref. 15-35); and (b) the method of matched asymptotic expansions (Ref. 15-36). Similar results were obtained. By both methods it was found that (a) the near-field region is described by the linear theory; (b) the far-field is described by Burgers' equation; and (c) an intermediate expansion was needed in an intermediate region. In this intermediate region, the characteristics are no longer straight as predicted by linear theory but are now curved lines. The analyses show that there is a smooth transition from the linear solution, valid for short time, to a steady-state continuous shock wave, which is the correct solution (for weak enough shocks) as $t \to \infty$.

15.6 REFERENCES AND BIBLIOGRAPHY

15.6.1 References

15-1 Lick, W., "Nonlinear Wave Propagation in Fluids," *Annual Review of Fluid Mechanics*, Vol. 2, Annual Reviews, Inc., Palo Alto, California, 1970.

15-2 Lindsay, R. B., *Mechanical Radiation*, McGraw-Hill, New York, 1960.

15-3 Stoker, J. J., *Water Waves*, Wiley-Interscience, New York, 1957.

15-4 Kevorkian, J. and Cole, J. D., *Perturbation Methods in Applied Mathematics*, Springer-Verlag, N.Y., 1981.

15-5 Haq, A., and Lick, W., "The Propagation of Waves in Inhomogeneous Media," FTAS/TR-71-59, Case Western Reserve University, Cleveland, Ohio, 1971.

15-6 Carrier, G. F.; Krook, M.; and Pearson, C. E., *Functions of a Complex Variable*, McGraw-Hill, New York, 1966.

15-7 Courant, R., and Friedrichs, K. O., *Supersonic Flow and Shock Waves*, Wiley-Interscience, New York, 1948.

15-8 Lighthill, M. J., "A Technique for Rendering Approximate Solutions to Physical Problems Uniformly Valid," *Philadelphian Magazine* 7 (40), 1949.

15-9 Whitham, G. B., "The Flow Pattern of a Supersonic Projectile," *Comm. on Pure and Appl. Math.* 5, 1952.

15-10 Lin, C. C., "On a Perturbation Theory Based on the Method of Characteristics," *J. Math. Phys.* **33**, 1954.

15-11 Fox, P. A., "On the Use of Coordinate Perturbations in the Solution of Physical Problems," D. Sci. Thesis, Mass. Inst. Technol., Cambridge, Mass., 1953.

15-12 Lick, W., "Solution of Non-Isentropic Flow Problems by a Coordinate Perturbation Method," FTAS/TR-67-22, Case Western Reserve University, Cleveland, Ohio, 1967.

15-13 Van Dyke, M., *Perturbation Methods in Fluid Mechanics*, Academic Press, New York, 1964.

15-14 Tricker, R. A. R., *Bores, Breakers, Waves, and Wakes*, American Elsevier Publishing Co., New York, 1965.

15-15 Ippen, A. T., *Estuary and Coastline Hydrodynamics*, McGraw-Hill, New York, 1966.

15-16 Wilson, W. S.; Wilson, D. S.; and Michael, J. A., "Field Measurements of

Swell Near Aruba, N. A., submitted for publication to *Journal of the Waterways and Harbors*, 1971.

15-17 Munk, W. H., and Traylor, M. A., "Refraction of Ocean Waves," *The Journal of Geology*, **LV**, 1947.

15-18 Laitone, E. V., "Higher Approximations to Nonlinear Water Waves and the Limiting Heights of Cnoidal, Solitary and Stokes' Waves," *Beach Erosion Board TM. No. 133*, 1963.

15-19 Berezin, Y. A., and Karpman, V. L., "Nonlinear Evolution of Disturbances in Plasmas and Other Dispersive Media," *Soviet Phys. JETP*, **24**, 1967.

15-20 Zabusky, N. J., and Kruskal, M. D., "Interaction of 'Solitons' in a Collisionless Plasma and the Recurrence of Initial States," *Phys. Rev. Letters*, **15**, 1965.

15-21 Zabusky, N. J., "A Synergetic Approach to Problems of Nonlinear Dispersive Wave Propagation and Interaction," *Proc. Symp. Nonlinear Partial Differential Equations*, Academic Press, New York, 1967.

15-22 Zabusky, N. J., "Solitons and Bound States of the Time-Independent Schrodinger Equation," *Phys. Rev.*, **168**, 1968.

15-23 Whitham, G. B., "Nonlinear Dispersive Waves," *Proc. Roy. Soc. A.*, **283**, 1965.

15-24 Whitham, G. B., "A General Approach to Linear and Nonlinear Dispersive Waves Using a Lagrangian," *J. Fluid Mech.*, 22, 1965.

15-25 Luke, J. C., "A Perturbation Method for Nonlinear Dispersive Wave Problems," *Proc. Roy. Soc. A*, 1966.

15-26 Whitham, G. B., "Some Comments on Wave Propagation and Shock Wave Structure with Application to Magnetohydrodynamics," *Comm. Pure and App. Math.*, **XII**, 1959.

15-27 Lick, W., "Wave Propagation in Real Gases," in *Advances in Applied Mechanics*, Vol. 10, Academic Press, New York, 1967.

15-28 Chu, B. T., "Wave Propagation and the Method of Characteristics in Reacting Gas Mixtures with applications to Hypersonic Flow," Brown University, WADC TN 57-213, 1957.

15-29 Broer, L. J. F., "Characteristics of the Equations of Motion of a Reacting Gas." *J. Fluid Mech.*, 1958.

15-30 Baldwin, B. S., Jr., "The Propagation of Plane Acoustic Waves in a Radiating Gas," *NASA TR R-138*, 1962.

15-31 Lick, W., "The Propagation of Small Disturbances in a Radiating Gas," *J. Fluid Mech.* **18**, 1964.

15-32 Lick, W., "Inviscid Flow of a Reacting Mixture of Gases Around a Blunt Body," *J. Fluid Mech.* **7**, 1959.

15-33 Hopf, E., "The Partial Differential Equation $u_t + uu_x = uu_{xx}$." *Comm. Pure and App. Math.*, **3**, 1950.

15-34 Cole, J. D., "On a Quasi-Linear Parabolic Equation Occurring in Aerodynamics," *Quart. App. Math.* **9**, 1951.

15-35 Lick, W., "Two-Variable Expansions and Singular Perturbation Problems," *SIAM J. Appl. Math.* **17**, 1969.

15-36 Blythe, P. A., "Nonlinear Wave Propagation in a Relaxing Gas," *J. Fluid Mech.* **37**, 1969.

15-37 Sommerfeld, A., *Optics*, Academic Press, New York, 1954.

15-38 Keller, J. B., "The Geometrical Theory of Diffraction," *Proceedings of the Symposium on Microwave Optics*, Eaton Electronics Research Laboratory, McGill University, Montreal, Canada, 1953. Also, *Calculus of Variations and its Applications, Proceedings of Symposia in Applied Math*, edited by L. M. Graves, McGraw-Hill, New York, 1958.

15-39 Keller, J. B., "Geometrical Theory of Diffraction," *Journal of the Optical Society of America*, **52** (2), 1962.

15-40 Wehausen, J. V., and Laitone, E. V., "Surface Waves," *Handbuch der Physik*, **9**, Springer-Verlag, Berlin, 1960.

15.6.2 Bibliography

Born, M., and Wolf, E., *Principles of Optics*, The Macmillan Co., New York, 1964.

Brekhovskikh, L. M., *Waves in Layered Media*, Academic Press, New York, 1960.

Courant, R., and Friedrichs, K. O., *Supersonic Flow and Shock Waves*, Wiley-Interscience, New York, 1948.

Elmore, W. C., and Heald, M. A., *Physics of Waves*, McGraw-Hill, New York, 1969.

Lighthill, M. J., *Waves in Fluids*, Cambridge University Press, Cambridge, 1978.

Lighthill, M. J., "Viscosity Effects in Sound Waves of Finite Amplitude," *Surveys in Mechanics*, edited by Batchelor, G. K. and Davies, R. M., Cambridge University Press, Cambridge, 1956.

Lindsay, R. B., *Mechanical Radiation*, McGraw-Hill, New York, 1960.

Morse, P. M., and Ingard, K. U., *Theoretical Acoustics*, McGraw-Hill, New York, 1968.

Pauling, L., and Wilson, E. B., *Introduction to Quantum Mechanics*, McGraw-Hill, New York, 1935.

Pearson, J. M., *A Theory of Waves*, Allyn and Bacon, Boston, 1966.

Rayleigh, Lord, *Theory of Sound*, Dover Publications, New York, 1945.

Schiff, I., *Quantum Mechanics*, McGraw-Hill, 1955.

Stoker, J. J., *Water Waves*, Wiley-Interscience, New York, 1957.

Whitham, G. B., *Linear and Nonlinear Waves*, Wiley-Interscience, New York, 1974.

16

Matrices and Linear Algebra

Dr. Tse-Sun Chow*

16.1 PRELIMINARY CONSIDERATIONS

16.1.1 Notation

A *matrix* $\mathbf{A} = [a_{ij}]_{mn}$ is a rectangular array consisting of m rows and n columns of elements:

$$\mathbf{A} = \begin{bmatrix} a_{11} & a_{12} & \cdots & a_{1n} \\ a_{21} & a_{22} & \cdots & a_{2n} \\ \cdots & \cdots & \cdots & \cdots \\ a_{m1} & a_{m2} & \cdots & a_{mn} \end{bmatrix}$$

Here a_{ij} is the element in the ith row and the jth column. Unless otherwise stated, we will take the elements as being complex numbers, although they could be elements of any field or even of a more general structure. As a special case a_{ij} can be real for all i and j; **A** is then called a *real matrix*.

The *dimension* of **A** is $m \times n$ (or simply n, if $m = n$). When $m = n$ **A** is a *square matrix*, and we write $\mathbf{A} = [a_{ij}]_n$. In this case we also say the *order* of **A** is n, and its *diagonal elements* are a_{ii} $(i = 1, 2, \ldots, n)$. When $n = 1$, **A** is a *column matrix* or *column vector*, (or more specifically, m-dimensional column vector). When $m = 1$, **A** is a *row matrix*, or *row vector*. A vector will be denoted by a bold-faced small letter. To make a distinction between a column vector and a row vector, we write:

*Dr. Tse-Sun Chow, Boeing Computer Services, Inc., Seattle, Wash.

$$\mathbf{x} = [x_i]_m = \begin{bmatrix} x_1 \\ x_2 \\ \vdots \\ x_m \end{bmatrix}, \quad \mathbf{y} = \{y_i\}_n = [y_1, y_2, \ldots, y_n]$$

When $m = n = 1$, **A** is reduced to one element, or a *scalar*.

If $a_{ij} = 0$ for all i and j, then **A** is the *null matrix* **O**; similarly a *null vector* is a vector of which all the elements are zero, and is denoted by **o**.

16.1.2 Rules of Operations

1. Two matrices are *equal* if they have the same dimensions and if corresponding elements are equal. Thus if $\mathbf{A} = [a_{ij}]_{mn}$, $\mathbf{B} = [b_{ij}]_{mn}$, then $\mathbf{A} = \mathbf{B}$ if and only if $a_{ij} = b_{ij}$ for all i and j.

2. Let α be a complex number, and let $\mathbf{A} = [a_{ij}]_{mn}$. Then $\alpha\mathbf{A}$ is the matrix obtained by multiplying each element of **A** by α. Thus $\mathbf{C} = \alpha\mathbf{A}$ implies $\mathbf{C} = [c_{ij}]_{mn}$ with $c_{ij} = \alpha a_{ij}$ for all i and j.

3. Two matrices of the same dimensions may be *added*: if $\mathbf{A} = [a_{ij}]_{mn}$ and $\mathbf{B} = [b_{ij}]_{mn}$, then $\mathbf{C} = \mathbf{A} + \mathbf{B}$ means $\mathbf{C} = [c_{ij}]_{mn}$, with $c_{ij} = a_{ij} + b_{ij}$ for all i and j. Note that $\mathbf{A} + \mathbf{B} = \mathbf{B} + \mathbf{A}$. From (2) and (3), we have $\mathbf{A} - \mathbf{B} = [a_{ij} - b_{ij}]_{mn}$.

4. Let $\mathbf{A} = [a_{ij}]_{mn}$, $\mathbf{B} = [b_{ij}]_{np}$, i.e., **A** has the same number of columns as **B** has rows. Then the *product* **AB** is defined, and $\mathbf{C} = \mathbf{AB}$ means $\mathbf{C} = [c_{ij}]_{mp}$ with $c_{ij} = \sum_{k=1}^{n} a_{ik}b_{kj}$.

Example:

$$\mathbf{A} = \begin{bmatrix} 1 & 0 & -1 \\ 2 & i & 3 \end{bmatrix}, \quad \mathbf{B} = \begin{bmatrix} 1 & 0 & 4 \\ 2 & i & 0 \\ 3 & 1 & -i \end{bmatrix}, \quad \mathbf{AB} = \begin{bmatrix} -2 & -1 & 4+i \\ 11+2i & 2 & 8-3i \end{bmatrix}$$

If $\mathbf{A} = [a_{ij}]_{mn}$, $\mathbf{B} = [b_{ij}]_{np}$, $\mathbf{D} = [d_{ij}]_{np}$, then $\mathbf{A}(\mathbf{B} + \mathbf{D}) = \mathbf{AB} + \mathbf{AD}$. Also, if $\mathbf{E} = [e_{ij}]_{pq}$, then $\mathbf{A}(\mathbf{BE}) = (\mathbf{AB})\mathbf{E}$. Thus matrix multiplication is *distributive* and *associative*.

It is however not, in general *commutative*. For if $\mathbf{A} = [a_{ij}]_{mn}$ and $\mathbf{B} = [b_{ij}]_{nq}$, then **AB** is defined, but **BA** is not defined unless $q = m$. Moreover, if $q = m$, **AB** and **BA** are of different orders unless $m = n$. If **A** and **B** are square matrices of the same order, then they are said to *commute* if $\mathbf{AB} = \mathbf{BA}$.

Example:

$$\begin{bmatrix} 1 & 2 \\ 3 & 4 \end{bmatrix} \text{ and } \begin{bmatrix} 3 & -2 \\ -3 & 0 \end{bmatrix} \text{ commute}$$

The above rules of operations are motivated by the properties of linear algebraic equations. Thus a convenient way to display the equations

$$\begin{aligned} y_1 &= a_{11}x_1 + a_{12}x_2 + \cdots + a_{1n}x_n \\ y_2 &= a_{21}x_1 + a_{22}x_2 + \cdots + a_{2n}x_n \\ &\cdots\cdots\cdots\cdots\cdots\cdots\cdots\cdots \\ y_m &= a_{m1}x_1 + a_{m2}x_2 + \cdots + a_{mn}x_n \end{aligned}$$

is $\mathbf{y} = \mathbf{A}\mathbf{x}$, where $\mathbf{A} = [a_{ij}]_{mn}$ and $\mathbf{y}$, $\mathbf{x}$ are the column vectors: $\mathbf{y} = [y_i]_m$, $\mathbf{x} = [x_i]_n$. Moreover, if $\mathbf{z} = \mathbf{B}\mathbf{y}$, where $\mathbf{z} = [z_i]_s$, $\mathbf{B} = [b_{ij}]_{sm}$, then $\mathbf{z} = \mathbf{B}\mathbf{A}\mathbf{x}$.

16.1.3 Definitions

In the following, $\bar{a}_{ij}$ denotes the complex conjugate of a_{ij}, and $|a_{ij}|$ the positive square root of $a_{ij}\bar{a}_{ij}$, i.e., the absolute value of a_{ij}.

1. Let $\mathbf{A} = [a_{ij}]_{mn}$. The *conjugate* of $\mathbf{A}$ is $\bar{\mathbf{A}} = [\bar{a}_{ij}]_{mn}$. The *transpose* of $\mathbf{A}$ is $\mathbf{A}^T = [a'_{ij}]_{nm}$ where $a'_{ij} = a_{ji}$ for all i, j. The *conjugate transpose* of $\mathbf{A}$, or the *adjoint* of $\mathbf{A}$ is $\mathbf{A}^* = \bar{\mathbf{A}}^T$. The matrix obtained by replacing each element of $\mathbf{A}$ by its absolute value is written as $|\mathbf{A}| = [|a_{ij}|]_{mn}$.

2. Let $\mathbf{A} = [a_{ij}]_n$, a square matrix of order n. Then

A is symmetric if $\mathbf{A} = \mathbf{A}^T$.

A is skew-symmetric if $\mathbf{A} = -\mathbf{A}^T$.

A is hermitian if $\mathbf{A} = \mathbf{A}^*$.

A is skew-hermitian if $\mathbf{A} = -\mathbf{A}^*$.

A is diagonal if $a_{ij} = 0$ for $i \neq j$, in this case we write: $\mathbf{A} = \text{diag}\,\{a_{11}, a_{22}, \ldots, a_{nn}\}$.

A is banded of width $2s + 1$ if $a_{ij} = 0$ for $|i - j| > s$. When $s = 1$, A is tri-diagonal.

A is upper triangular if $a_{ij} = 0$ for $i > j$, and lower triangular if $a_{ij} = 0$ for $j > i$.

A is strictly upper triangular if $a_{ij} = 0$ for $i \geqq j$ and strictly lower triangular if $a_{ij} = 0$ for $j \geqq i$.

A is of upper Hessenberg form if $a_{ij} = 0$ for $i - j > 1$ with $a_{ij} = 1$ for $i - j = 1$, and of lower Hessenberg form if $a_{ij} = 0$ for $j - i > 1$ with $a_{ij} = 1$ for $j - i = 1$.

A is the identity matrix if it is a diagonal matrix such that each diagonal element is unity. We denote an identity matrix by $\mathbf{I}_n$, or simply $\mathbf{I}$ if its order is clear in the context and its jth column by $\mathbf{e}_j$.

A is normal if $\mathbf{A}\mathbf{A}^* = \mathbf{A}^*\mathbf{A}$. It is unitary if $\mathbf{A}\mathbf{A}^* = \mathbf{A}^*\mathbf{A} = \mathbf{I}$. If all the elements of a unitary matrix are real, then it is orthogonal.

A is indempotent if $\mathbf{A}^2 = \mathbf{A} \neq \mathbf{I}$. If r is the smallest integer such that $\mathbf{A}^r = \mathbf{O}$, where $\mathbf{A}^r = \mathbf{A} \cdot \mathbf{A} \cdots \mathbf{A}$ (r times), then $\mathbf{A}$ is nilpotent with index r.

The trace of $\mathbf{A}$, written tr $\mathbf{A}$, is the sum of all the diagonal elements.

3. The following are direct consequences of the above definitions.

The product of two upper (lower) triangular matrices is again upper (lower) triangular. The product of two unitary matrices is again unitary.

For any matrix $\mathbf{A}$, $\mathbf{A}^*\mathbf{A}$ and $\mathbf{A}\mathbf{A}^*$ are hermitian, and $\operatorname{tr}(\mathbf{A}^*\mathbf{A}) = \operatorname{tr}(\mathbf{A}\mathbf{A}^*) = \sum_i \sum_j |a_{ij}|^2$.

Let $\mathbf{A} = [a_{ij}]_{np}$ and $\mathbf{B} = [b_{ij}]_{pn}$, then $\operatorname{tr}(\mathbf{AB}) = \operatorname{tr}(\mathbf{BA})$.

When the product $\mathbf{AB}$ is defined, then $(\mathbf{AB})^* = \mathbf{B}^*\mathbf{A}^*$.

If $\mathbf{A}$ is hermitian, then $i\mathbf{A}$ is skew-hermitian. If $\mathbf{A}$ is skew-hermitian, then $i\mathbf{A}$ is hermitian.

If $\mathbf{A}$ is hermitian it can be expressed as $\mathbf{R}_1 + i\mathbf{R}_2$, where $\mathbf{R}_1$ is real symmetric and $\mathbf{R}_2$ real skew-symmetric. If $\mathbf{A}$ is skew-hermitian, it can be expressed as $\mathbf{R}_2 + i\mathbf{R}_1$.

Hermitian, skew-hermitian and unitary matrices are normal.

Any square matrices $\mathbf{A}$ can be expressed as the sum of a symmetric and a skew-symmetric matrix: $\mathbf{A} = \frac{1}{2}(\mathbf{A} + \mathbf{A}^T) + \frac{1}{2}(\mathbf{A} - \mathbf{A}^T)$; also as the sum of a hermitian and a skew-hermitian matrix: $\mathbf{A} = \frac{1}{2}(\mathbf{A} + \mathbf{A}^*) + \frac{1}{2}(\mathbf{A} - \mathbf{A}^*)$.

Examples:

(1) The matrix $\mathbf{G} = \sqrt{2/(n+1)}\,[g_{ij}]_n$, where $g_{ij} = (-1)^{i+j-1} \sin ij\pi/(n+1)$ is orthogonal. But $\mathbf{G} = \mathbf{G}^T$, hence $\mathbf{G}^2 = \mathbf{I}$.

(2) Let $1, \omega, \ldots, \omega^{n-1}$ be the roots of $x^n - 1 = 0$. The matrix $W = (1/\sqrt{n})\,[w_{ij}]_n$ where $w_{ij} = \omega^{(i-1)(j-1)}$ is unitary.

(3) Let $\mathbf{A} = [a_{ij}]_n$ be strictly upper triangular. Then $\mathbf{A}$ is nilpotent since $\mathbf{A}^s = \mathbf{O}$, for $s \leqslant n$.

(4) Let $\mathbf{A} = (1/n)\,[a_{ij}]_n$, where $a_{ij} = 1$ for all i, j, then $\mathbf{A}$ is indempotent.

16.1.4 Submatrices

A *submatrix* of $\mathbf{A} = [a_{ij}]_{mn}$ is the matrix of the remaining elements after certain rows and columns of $\mathbf{A}$ have been crossed out. Let $\sigma_1^r(i_k)$ denote the set of r integers $i_1, i_2, \ldots, i_r$ $(1 \leqslant i_1 < \cdots < i_r \leqslant m)$, and similarly let $\sigma_1^s(j_k)$ denote the set of s integers $j_1, j_2, \ldots, j_s$ $(1 \leqslant j_1 < \cdots < j_s \leqslant n)$. The submatrix of $\mathbf{A}$ obtained by deleting from $\mathbf{A}$ all the rows except the $i_k{}^{\text{th}}$ rows $(k = 1, 2, \ldots, r)$, and all the columns except the $j_k{}^{\text{th}}$ columns $(k = 1, 2, \ldots, s)$ will be denoted by $\mathbf{A}[\sigma_1^r(i_k)|\sigma_1^s(j_k)]$. The submatrix *complementary to* $\mathbf{A}[\sigma_1^r(i_k)|\sigma_1^s(j_k)]$ is the submatrix of $\mathbf{A}$ after deleting the $i_k{}^{\text{th}}$ rows $(k = 1, 2, \ldots, r)$ and the $j_k{}^{\text{th}}$ columns $(k = 1, 2, \ldots, s)$, and will be denoted by $\mathbf{A}[\sigma_1^r(i_k)|\sigma_1^s(j_k)]$. When $\mathbf{A}$ is square and $r = s$ and $i_k = j_k$ $(k = 1, 2, \ldots, r)$, then $\mathbf{A}(\sigma_1^r(i_k)|\sigma_1^s(j_k))$ is a *principal* submatrix; in this case it is written $\mathbf{A}(\sigma_1^r(i_k))$. A *leading* principal submatrix is when $i_k = k$, i.e., $\mathbf{A}(1, 2, \ldots, k)$.

Example: The Hankel matrix is defined as

$$\mathbf{H}_{k+1}^{(n-1)} = [h_{ij}]_{k+1}, \qquad \begin{cases} h_{ij} = a_{i+j+n-3}, & i \leqslant j \\ h_{ji} = h_{ij}, & i > j \end{cases}$$

Then

$$\mathbf{H}_k^{(n)} = \mathbf{H}_{k+1}^{(n-1)}\overline{(1|k+1)} = \mathbf{H}_{k+1}^{(n-1)}\overline{(k+1|1)}$$

$$\mathbf{H}_k^{(n-1)} = \mathbf{H}_{k+1}^{(n-1)}\overline{(k+1)} = \mathbf{H}_{k+1}^{(n-1)}(1, 2, \ldots, k)$$

$$\mathbf{H}_{k-1}^{(n+1)} = \mathbf{H}_{k+1}^{(n-1)}\overline{(1, k+1)} = \mathbf{H}_{k+1}^{(n-1)}(2, 3, \ldots, k)$$

16.2 DETERMINANTS

16.2.1 Notation and Definitions

The set of numbers: $\sigma_1^n(\pi_k) = (\pi_1, \pi_2, \ldots, \pi_n)$ is an even (odd) permutation of the set $\sigma_1^n(k) = (1, 2, \ldots, n)$ if the latter can be brought to the former by an even (odd) number of interchanges of adjacent elements. We associate the index $\rho[\sigma_1^n(\pi_k)]$ with $\sigma_1^n(\pi_k)$ such that $\rho[\sigma_1^n(\pi_k)] = 1$ or -1 according as $\sigma_1^n(\pi_k)$ is an even or odd permutation of $\sigma_1^n(k)$.

The *determinant* of a square matrix $\mathbf{A} = [a_{ij}]_n$ is a number associated with $\mathbf{A}$, defined as:

$$\det \mathbf{A} = \sum \rho[\sigma_1^n(\pi_k)] \prod_{i=1}^{n} a_{i\pi_i}$$

where the summation is extended to all sets of $\sigma_1^n(\pi_k)$ which are permutations of $\sigma_1^n(k)$.

By deleting the ith row and the jth column of $\mathbf{A}$, we obtain a submatrix of order $n - 1$, and its determinant is called the *minor* of a_{ij}. More generally, we call the determinant of any square submatrix of $\mathbf{A}$ by deleting some of its rows and columns a minor of $\mathbf{A}$.

16.2.2 Basic Properties

The following basic properties of determinants follow directly from the definitions:

1. If $\mathbf{A} = \mathbf{B}^T$, then $\det \mathbf{A} = \det \mathbf{B}$. Also, $\det \mathbf{A}^* = \overline{\det \mathbf{A}}$.
2. If $\mathbf{B}$ is obtained from $\mathbf{A}$ by the interchange of two rows (columns), then $\det \mathbf{A} = -\det \mathbf{B}$.
3. If any two rows (columns) of $\mathbf{A}$ are identical, then $\det \mathbf{A} = 0$.
4. If $\mathbf{B}$ is obtained from $\mathbf{A}$ by multiplying any row (column) by k, other rows (columns) remaining unchanged, then $\det \mathbf{B} = k \det \mathbf{A}$.

5. If **A**, **B**, **C** differ only in the ith row (column), such that each element of the ith row of **C** is the sum of the corresponding elements of **A** and **B**, then det **C** = det **A** + det **B**.
6. det (k**A**) = k^n det **A**, n being the dimension of **A**.
7. det (**AB**) = det **A** det **B**.

16.2.3 Laplace Expansion Theorem

The Laplace expansion theorem states: *for a given set* $\sigma_1^r(i_k) = (i_1, i_2, \ldots, i_k)$, $(1 \leqslant i_1 < i_2 < \cdots < i_k \leqslant n)$

$$\det \mathbf{A} = (-1)^{I_r} \sum (-1)^{J_r} \det \mathbf{A}[\sigma_1^r(i_k)|\sigma_1^r(j_k)] \det \mathbf{A}\overline{[\sigma_1^r(i_k)|\sigma_1^r(j_k)]}$$

where the summation is extended to all possible sets of $\sigma_1^r(j_k)$ *taken from* $\sigma_1^n(k)$, $(1 \leqslant j_1 < j_2 < \cdots < j_k \leqslant n)$ *and*

$$I_r = \sum_{k=1}^{r} i_k, \quad J_r = \sum_{k=1}^{r} j_k$$

16.2.4 Expansion by Rows or Columns

A special case of the Laplace expansion theorem is to expand a determinant by a certain row or column. With the notation of 16.1.4, expansion by the ith row or the jth column gives

$$\sum_{j=1}^{n} (-1)^{i+j} a_{ij} \det \mathbf{A}_{kj} = \delta_{ik} \det \mathbf{A}$$

$$\sum_{i=1}^{n} (-1)^{i+j} a_{ij} \det \mathbf{A}_{ik} = \delta_{jk} \det \mathbf{A}$$

where $\delta_{ik} = 0$ or 1 according as $i \neq k$ or $i = k$. The term $(-1)^{i+j} \det \mathbf{A}_{ij}$ is called the *cofactor* of a_{ij}.

16.2.5 Cauchy-Binet Formula

Let $\mathbf{C} = [c_{ij}]_n$ *and* $c_{ij} = \sum_{s=1}^{m} a_{is} b_{sj}$, $(m \geqslant n)$, *then*

$$\det \mathbf{C} = \sum \det \mathbf{A}[\sigma_1^n(k)|\sigma_1^n(i_k)] \cdot \det \mathbf{B}[\sigma_1^n(i_k)|\sigma_1^n(k)]$$

where the summation is extended to all $\sigma_1^n(i_k)$ *such that* $1 \leqslant i_1 < i_2 < \cdots < i_n \leqslant m$.

When $m = n$, the above reduces to the familiar determinant multiplication formula. When $m < n$, $\det \mathbf{C} = 0$.

16.2.6 Sylvester's Theorem

Let $\mathbf{B}^{(s)} = [b_{ij}^{(s)}]_{n-s}$, $(i, j = s+1, s+2, \dots, n)$ *and* $b_{ij}^{(s)} = \det \mathbf{A}[\sigma_1^s(k), i | \sigma_1^s(k), j]$, *then* $\det \mathbf{B}^{(s)} = \{\det \mathbf{A}[\sigma_1^s(k)]\}^{n-s-1} \cdot \det \mathbf{A}$.

Examples:

(1) With the notation of 16.1.4, consider the determinant

$$\Delta = \left|\begin{array}{c|c} \mathbf{H}_{k+1}^{(n-1)}(1, 2, \dots, k | 1, 2, \dots, k+1) & \mathbf{O} \\ \hline a_{n+k-1} \quad a_{n+k} \cdots a_{n+2k-1} & a_{n+k} \quad a_{n+k+1} \cdots a_{n+2k-2} \\ \mathbf{H}_{k+1}^{(n-1)}(2, 3, \dots, k+1 | 1, 2, \dots, k+1) & \mathbf{H}_{k-1}^{(n+1)} \end{array}\right|$$

Apply the Laplace expansion theorem according to the vertical partitioning and we obtain $\Delta = \det \mathbf{H}_{k+1}^{(n-1)} \det \mathbf{H}_{k-1}^{(n+1)}$. However, expansion by the horizontal partitioning leads to $\Delta = -[\det \mathbf{H}_k^{(n)}]^2 + \det \mathbf{H}_k^{(n-1)} \cdot \det \mathbf{H}_k^{(n+1)}$. Thus we have the identity

$$\det \mathbf{H}_{k+1}^{(n-1)} \cdot \det \mathbf{H}_{k-1}^{(n+1)} - \det \mathbf{H}_k^{(n-1)} \det \mathbf{H}_k^{(n+1)} + [\det \mathbf{H}_k^{(n)}]^2 = 0$$

(2) Let $\mathbf{A} = [a_{ij}]_n$, and consider the following matrix of dimension $2n$

$$\mathbf{B} = \begin{bmatrix} \mathbf{A} & \mathbf{I} \\ \mathbf{I} & \mathbf{O} \end{bmatrix}$$

It follows from the Sylvester theorem with $s = n$ and $\mathbf{B}^{(n)} = [b_{ij}^{(n)}]_n$, where

$$\begin{aligned} b_{ij}^{(n)} &= \det \mathbf{B}[\sigma_1^n(k), i | \sigma_1^n(k), j] \\ &= (-1)^{i+j} \det \mathbf{A}(1, \dots, i-1, i+1, \dots, n | 1, \dots, j-1, j+1, \dots, n) = (-1)^{i+j} A_{ij} \end{aligned}$$

then

$$(\det \mathbf{A})^{n-1} = \det \mathbf{B}^{(n)}$$

(3) The *Vandermonde* matrix is defined as: $\mathbf{V}(x_1, x_2, \dots, x_n) = [v_{ij}]_n$, $v_{ij} = x_j^{i-1}$, $(i, j = 1, 2, \dots, n)$.

$$\det \mathbf{V}(x_1, x_2, \dots, x_n) = \prod_{n \geqslant i > j \geqslant 1} (x_i - x_j)$$

(4) Define $\mathbf{D}(x, y) = [d_{ij}]_n, d_{ij} = 1/(x_i + y_j)$, then

$$\det \mathbf{D}(x, y) = \frac{\prod_{1 \leqslant i < j \leqslant n} (x_i - x_j)(y_i - y_j)}{\prod_{1 \leqslant i < j \leqslant n} (x_i + y_j)}$$

If $x_i = i$, $y_i = i - 1$, $\mathbf{D}(x, y)$ is the *Hilbert* matrix.

(5) The determinant of a skew-symmetric matrix of odd order is equal to zero, and that of even order is a perfect square of a polynomial function of its elements. The determinant of a skew-hermitian matrix of odd order is pure imaginary.

(6) The determinant of an orthogonal matrix is either +1 or − 1, and the modulus of the determinant of a unitary matrix is 1.

16.3 VECTOR SPACES AND LINEAR TRANSFORMATION

16.3.1 Introduction

A *vector space* is a set of vectors which are closed under the operation of scalar multiplication and vector addition. For a given set of vectors (column vectors or row vectors, 16.1.1), $\mathbf{z}_k = [z_{k1}, z_{k2}, \ldots, z_{kn}]$, $(k = 1, 2, \ldots, m)$, and a set of scalars $\alpha_1, \alpha_2, \ldots, \alpha_m$, the vector $\sum_{k=1}^{m} \alpha_k \mathbf{z}_k$ is a *linear combination* of $\mathbf{z}_1, \mathbf{z}_2, \ldots,$ $\mathbf{z}_m$. The totality of all such combinations obtained by varying $\alpha_1, \alpha_2, \ldots, \alpha_m$ forms a vector space spanned by $\mathbf{z}_1, \mathbf{z}_2, \ldots, \mathbf{z}_m$. In turn, the vectors $\mathbf{z}_1, \mathbf{z}_2, \ldots, \mathbf{z}_m$ are called the *generators* of the vector space.

The *inner product* of two n-dimensional vectors $\mathbf{a}, \mathbf{b}$ is defined as: $(\mathbf{a}, \mathbf{b}) = \sum_{k=1}^{n} a_k \bar{b}_k$. The inner product satisfies the following properties:

(i) $(\mathbf{a}, \mathbf{b}) = \overline{(\mathbf{b}, \mathbf{a})}$
(ii) $(\mathbf{a}, \mathbf{a}) = \|\mathbf{a}\|^2 > 0$ unless $\mathbf{a} = \mathbf{o}$, in which case $(\mathbf{o}, \mathbf{o}) = 0$
(iii) $(\mathbf{a} + \mathbf{b}, \mathbf{c} + \mathbf{d}) = (\mathbf{a}, \mathbf{c}) + (\mathbf{a}, \mathbf{d}) + (\mathbf{b}, \mathbf{c}) + (\mathbf{b}, \mathbf{d})$
(iv) $(k\mathbf{a}, \mathbf{b}) = k(\mathbf{a}, \mathbf{b})$

When $(\mathbf{a}, \mathbf{b}) = 0$, $\mathbf{a}, \mathbf{b}$ are said to be *orthogonal*.

A vector space with an inner product defined is called a *unitary space*. We have occasions to deal with a real vector space (i.e., when all the elements of vectors are real numbers); in this case (i) becomes $(\mathbf{a}, \mathbf{b}) = (\mathbf{b}, \mathbf{a})$. A real vector space with inner product so defined is called the *euclidean space*.*

*It will be understood from now on that $\|\mathbf{a}\|$ implies $\|\mathbf{a}\|_E$, see 16.8.1.

16.3.2 Linear Dependence

A set of n-dimensional vectors $\mathbf{z}_k$, $(k = 1, 2, \ldots, m)$ are *linearly dependent* when there exist scalars α_k, $(k = 1, 2, \ldots, m)$, not all zero, such that $\sum_{k=1}^{m} \alpha_k \mathbf{z}_k = 0$. The vector space generated by the linear combinations of $\mathbf{z}_k$, $(k = 1, 2, \ldots, m)$ is said to be of dimension s if it has s independent vectors and any set of $s + 1$ vectors in the vector space are linearly dependent. When $\mathbf{z}_k$ are linearly independent and $m = n$, the set $\mathbf{z}_k$, $(k = 1, 2, \ldots, m)$ forms a basis for the vector space of dimension n: $\mathbf{C}_n$, and any vector in $\mathbf{C}_n$ can be expressed uniquely as a linear combination of $\mathbf{z}_k$, $(k = 1, 2, \ldots, n)$.

A test for the linear dependence is the vanishing of the *Gramian:*

$$\Gamma(\mathbf{z}_1, \mathbf{z}_2, \ldots, \mathbf{z}_m) = \det\ [(\mathbf{z}_i, \mathbf{z}_j)]_m$$

The set of vectors $\mathbf{z}_k$, $(k = 1, 2, \ldots, m)$ are linearly dependent if and only if $\Gamma(\mathbf{z}_1, \mathbf{z}_2, \ldots, \mathbf{z}_m) = 0$. By applying the Cauchy-Binet formula (16.2.5) to $\Gamma(\mathbf{z}_1, \mathbf{z}_2, \ldots, \mathbf{z}_m)$ we see that it is the sum of squares of the moduli of determinants of dimension m. Hence $\Gamma(\mathbf{z}_1, \mathbf{z}_2, \ldots, \mathbf{z}_m) \geqslant 0$. For $m > n$, $\Gamma(\mathbf{z}_1, \mathbf{z}_2, \ldots, \mathbf{z}_m) = 0$ and the set $\mathbf{z}_k$ is always dependent.

Let $m = 2$, then $\Gamma(\mathbf{z}_1, \mathbf{z}_2) \geqslant 0$ gives $|(\mathbf{z}_1, \mathbf{z}_2)| \leqslant \{\|\mathbf{z}_1\|\ \|\mathbf{z}_2\|\}^{1/2}$ with equality holding if and only if $\mathbf{z}_1, \mathbf{z}_2$ are linearly dependent. This is the *Schwartz inequality*.

A general inequality for the Gramian is: $\Gamma(\mathbf{z}_1, \mathbf{z}_2, \ldots, \mathbf{z}_m) \leqslant \Gamma(\mathbf{z}_1, \ldots, \mathbf{z}_p) \cdot \Gamma(\mathbf{z}_{p+1}, \ldots, \mathbf{z}_m)$ where the equality is attained when $\mathbf{z}_1, \ldots, \mathbf{z}_p$ are simultaneously orthogonal to $\mathbf{z}_{p+1}, \ldots, \mathbf{z}_m$. By repeated application of the above, we obtain

$$\Gamma(\mathbf{z}_1, \mathbf{z}_2, \ldots, \mathbf{z}_m) \leqslant (\mathbf{z}_1, \mathbf{z}_1) \ldots (\mathbf{z}_m, \mathbf{z}_m)$$

and in case $m = n$, we have

$$\Gamma(\mathbf{z}_1, \mathbf{z}_2, \ldots, \mathbf{z}_n) = \det \mathbf{Z}^*\mathbf{Z} = |\det \mathbf{Z}|^2 \leqslant \prod_j \sum_i |z_{ij}|^2$$

which is known as *Hadamard's inequality*, $(\mathbf{Z} = [z_{ij}]_n)$.

16.3.3 Subspaces

A subset $\mathbf{V}$ of a vector space $\mathbf{C}_n$ is called a *subspace* of $\mathbf{C}_n$ if $\mathbf{V}$ is itself also a vector space. If $\mathbf{W}$ is also a subspace of $\mathbf{C}_n$, then the common part of $\mathbf{V}$ and $\mathbf{W}$, denoted by $\mathbf{V} \cap \mathbf{W}$, is also a subspace of $\mathbf{C}_n$. However, the set of all vectors of $\mathbf{V}$ and $\mathbf{W}$ is in general not a subspace.

Let $\mathbf{V} + \mathbf{W}$ represent the set of all vectors $\mathbf{v} + \mathbf{w}$ such that $\mathbf{v}$ belongs to $\mathbf{V}$ and $\mathbf{w}$ belongs to $\mathbf{W}$, then $\mathbf{V} + \mathbf{W}$ is also a subspace. Furthermore, $\dim \mathbf{V} + \dim \mathbf{W} = \dim (\mathbf{V} + \mathbf{W}) + \dim (\mathbf{V} \cap \mathbf{W})$, where $\dim \mathbf{V}$ is the dimension of the subspace $\mathbf{V}$.

Two subspaces **V** and **W** are complementary if $\mathbf{V} \cap \mathbf{W} = \mathbf{o}$, i.e., the only common vector of **V** and **W** is **o**, and $\mathbf{V} + \mathbf{W} = \mathbf{C}_n$. In particular, if $(\mathbf{v}, \mathbf{w}) = 0$ for every **v** in **V** and every **w** in **W**, then **V** is the *orthogonal complement* of **W** and vice versa.

16.3.4 Linear Transformation

A transformation $\mathcal{T}$ is said to be *linear* if it preserves addition and scalar multiplication of vectors

$$\mathcal{T}(\alpha_1 \mathbf{z}_1 + \alpha_2 \mathbf{z}_2) = \alpha_1 \mathcal{T}(\mathbf{z}_1) + \alpha_2 \mathcal{T}(\mathbf{z}_2)$$

where α_1, α_2 are scalars. Let the vector space $\mathbf{C}_n$ have the basis: $\mathbf{u}_1, \mathbf{u}_2, \ldots, \mathbf{u}_n$, and let $\mathcal{T}$ be the transformation that maps $\mathbf{C}_n$ into itself. If $\mathcal{T}(\mathbf{u}_i) = \sum_{s=1}^{n} t_{is}\mathbf{u}_s$ $(i = 1, 2, \ldots, n)$ we say the matrix of the transformation $\mathcal{T}$ with respect to $\mathbf{u}_1, \mathbf{u}_2, \ldots, \mathbf{u}_n$ is $\mathbf{T} = [t_{ij}]_n$. The identity transformation $\mathcal{I}$ is one which leaves every vector unchanged. The zero transformation $\mathcal{O}$ changes every vector to the zero (null) vector.

Example: Let any vector **z** in the *x*-*y* plane be rotated a given angle θ. Referred to the basis: $\mathbf{e}_1 = [1, 0]^T, \mathbf{e}_2 = [0, 1]^T$, we have $\mathcal{T}(\mathbf{e}_1) = \cos\theta\, \mathbf{e}_1 + \sin\theta\, \mathbf{e}_2, \mathcal{T}(\mathbf{e}_2) = -\sin\theta\, \mathbf{e}_1 + \cos\theta\, \mathbf{e}_2$. The matrix of the transformation is therefore

$$\begin{bmatrix} \cos\theta & \sin\theta \\ -\sin\theta & \cos\theta \end{bmatrix}$$

Let **z** be any vector in $\mathbf{C}_n$ which has the basis $\mathbf{u}_1, \mathbf{u}_2, \ldots, \mathbf{u}_n$ then $\mathbf{z} = \sum_{s=1}^{n} z_s\mathbf{u}_s$, where z_s are the components of **z**. Let $\mathcal{T}(\mathbf{z}) = \boldsymbol{\zeta} = \sum_{s=1}^{n} \zeta_s\mathbf{u}_s$, and similarly ζ_s be the components of $\boldsymbol{\zeta}$. If the matrix of the transformation is **T**, then $\mathcal{T}(\mathbf{u}_i) = \sum_{s=1}^{n} t_{is}\mathbf{u}_s$. Thus

$$\boldsymbol{\zeta} = \sum_s \zeta_s \mathbf{u}_s = \mathcal{T}\left(\sum_s z_s \mathbf{u}_s\right) = \sum_s z_s \mathcal{T}(\mathbf{u}_s) = \sum_{s,p} z_s t_{sp}\, \mathbf{u}_p$$

and $\zeta_s = \sum_p z_p t_{ps}$. If we write $\mathbf{z} = [z_1, z_2, \ldots, z_n]^T, \boldsymbol{\zeta} = [\zeta_1, \zeta_2, \ldots, \zeta_n]^T$, the effect of the transformation $\mathcal{T}$ is to multiply **z** by $\mathbf{T}^T$: $\boldsymbol{\zeta} = \mathbf{T}^T\mathbf{z}$.

For any transformation $\mathcal{T}$, the *adjoint* of $\mathcal{T}$, written $\mathcal{T}^*$, is defined by $(\mathcal{T}\mathbf{u}, \mathbf{v}) = (\mathbf{u}, \mathcal{T}^*\mathbf{v})$ for all vectors **u** and **v**. This agrees with the definition of adjoint for matrices (16.1.3), since the matrix of $\mathcal{T}^*$ is the conjugate transpose of the matrix of $\mathcal{T}$. Similarly a transformation $\mathcal{T}$ is *hermitian* if $\mathcal{T}^* = \mathcal{T}$; it is *normal* if $\mathcal{T}\mathcal{T}^* = \mathcal{T}^*\mathcal{T}$. A *unitary* transformation is one which preserves the inner product, $(\mathcal{T}\mathbf{u}, \mathcal{T}\mathbf{u}) = (\mathbf{u}, \mathbf{u})$.

16.3.5 Orthonormalization

Given a set of n-dimensional independent vectors: $\mathbf{z}_k$, $(k = 1, 2, \ldots, m)$ it is required to find the set ζ_k, $(k = 1, 2, \ldots, m)$ such that $(\zeta_i, \zeta_j) = \delta_{ij}$, $(i, j = 1, 2, \ldots, m)$. The classical Gram-Schmidt method yields

$$\begin{cases} \zeta_1 = \dfrac{\mathbf{z}_1}{\|\mathbf{z}_1\|} \\ \zeta_k' = \mathbf{z}_k - \displaystyle\sum_{s=1}^{k-1} (\mathbf{z}_k, \zeta_s)\zeta_s; \quad \zeta_k = \dfrac{\zeta_k'}{\|\zeta_k'\|} \qquad k = 2, 3, \ldots, m \end{cases}$$

The modified Gram-Schmidt method: let $\mathbf{z}_k^{(0)} = \mathbf{z}_k$, $(k = 1, 2, \ldots, m)$

$$\mathbf{z}_k^{(s)} = \mathbf{z}_k^{(s-1)} - (\mathbf{z}_k^{(s-1)}, \zeta_k)\zeta_s; \quad \zeta_s = \frac{\mathbf{z}_s^{(s-1)}}{\|\mathbf{z}_s^{(s-1)}\|} \qquad \begin{matrix} s = 1, 2, \ldots, m; \\ k = s+1, s+2, \ldots, m \end{matrix}$$

For $m > 2$, the methods of calculating ζ_s begin to differ. Although the two methods are theoretically equivalent, the modified method gives more accurate numerical answers; this is especially true if the given set $\mathbf{z}_k$ are approximately parallel so that large cancellations occur in the arithmetic process.

Let ζ_k, $(k = 1, 2, \ldots, m)$ form an orthonormal basis in $\mathbf{C}_n$. The matrix $\mathbf{Z}$ of which the columns are ζ_k is unitary, and $\mathbf{Z}^*\mathbf{Z} = \mathbf{I}$. Since $\mathbf{Z}^T$ is also unitary, the columns of $\mathbf{Z}^T$ likewise form an orthonormal basis in $\mathbf{C}_n$.

The orthonormalization process of Gram-Schmidt effectively expresses a given matrix $\mathbf{Z} = [\mathbf{z}_1, \mathbf{z}_2, \ldots, \mathbf{z}_n]$ as the product of $\mathbf{UT}$ where $\mathbf{U}$ is unitary and $\mathbf{T}$ is upper triangular with real nonnegative elements on the diagonal. Moreover, if $\mathbf{Z}$ is nonsingular, such decomposition is unique.

16.3.6 Biorthogonalization

Two different bases of vectors $\mathbf{z}_k$, ζ_k, $(k = 1, 2, \ldots, n)$ in $\mathbf{C}_n$ are said to be *biorthogonal* if $(\mathbf{z}_i, \zeta_j) = \delta_{ij}$. Given the basis $\mathbf{z}_k$, $(k = 1, 2, \ldots, n)$, the basis that is biorthogonal to it can be calculated by the following algorithms:

$$\begin{cases} \zeta_k^{(k)} = \dfrac{\zeta_k^{(k-1)}}{(\zeta_k^{(k-1)}, \mathbf{z}_k)} \\ \zeta_i^{(k)} = \zeta_i^{(k-1)} - (\zeta_i^{(k-1)}, \mathbf{z}_k)\zeta_k^{(k)} \qquad i, k = 1, 2, \ldots, n; i \neq k \end{cases}$$

The choice of $\zeta_k^{(0)} = \mathbf{z}_k$, $(k = 1, 2, \ldots, n)$ ensures all the denominators $(\zeta_k^{(k-1)}, \mathbf{z}_k)$

do not vanish in the process and the final calculated set of vectors $\zeta_k^{(n)}$, ($k = 1, 2, \ldots, n$) are the basis sought which is biorthogonal to the basis $\mathbf{z}_k$.

16.4 MATRICES

16.4.1 Rank

The *rank* of a matrix is the dimension of the largest square submatrix of which the determinant is not zero. The rank is also equal to the number of the independent columns or the number of independent rows. A square matrix is said to be nonsingular if its rank is equal to its dimension (order).

Let $\mathbf{A}, \mathbf{B}$ be of the same dimension, then Rank $(\mathbf{A} + \mathbf{B}) \leq$ Rank $\mathbf{A}$ + Rank $\mathbf{B}$ and Rank $(\mathbf{A} + \mathbf{B}) \geq |$Rank $\mathbf{A}$ - Rank $\mathbf{B}|$.

Frobenius inequality: if the produce $\mathbf{ABC}$ is defined, then Rank $(\mathbf{AB})$ + Rank $(\mathbf{BC}) \leq$ Rank $\mathbf{B}$ + Rank $(\mathbf{ABC})$. From this the *Sylvester's law* follows as a special case: *if the product of* $\mathbf{A}, \mathbf{B}$ *is defined, then Rank* $(\mathbf{AB}) \leq$ *min (Rank* $\mathbf{A}$, *Rank* $\mathbf{B}$), *and Rank* $(\mathbf{AB}) \geq$ *Rank* $\mathbf{A}$ + *Rank* $\mathbf{B} - n$, *n being the number of columns of* $\mathbf{A}$.

Multiplication of a matrix by a nonsingular matrix does not alter its rank. Let $\mathbf{A} = [a_{ij}]_{mn}$ be of rank r. Then $\mathbf{A}$ can be expressed as the product $\mathbf{BC}$, where $\mathbf{B} = [b_{ij}]_{mr}$, $\mathbf{C} = [c_{ij}]_{rn}$, and both $\mathbf{B}, \mathbf{C}$ are of rank r.

Let $\mathbf{A} = [a_{ij}]_{mn}$, then Rank $\mathbf{A}$ = Rank $(\mathbf{A}^*\mathbf{A})$ = Rank $(\mathbf{AA}^*)$.

16.4.2 Estimates of Rank

Three methods of rank estimation are as follows:

(i) Let $\mathbf{A} = [a_{ij}]_{mn}$ and form the sums

$$b_i(s) = \max_{j_1, j_2, \ldots, j_s \neq i} \{|a_{ij_1}| + |a_{ij_2}| + \cdots + |a_{ij_s}|\}$$

$$c_j(s) = \max_{i_1, i_2, \ldots, i_s \neq j} \{|a_{i_1 j}| + |a_{i_2 j}| + \cdots + |a_{i_s j}|\}$$

If either $|a_{ii}| > b_i(s)$ for s distinct rows of $\mathbf{A}$, or $|a_{jj}| > c_j(s)$ for s distinct columns of $\mathbf{A}$, then the rank of $\mathbf{A}$ is at least s.

(ii) Let $f_i = |a_{ii}| \Big/ \sum_{j=1}^{n} |a_{ij}|$, $g_j = |a_{jj}| \Big/ \sum_{i=1}^{n} |a_{ij}|$, $(i, j = 1, 2, \ldots, n)$ which are understood to be zero if both the numerator and the denominator are zero, then

$$\text{Rank } \mathbf{A} \geqslant \max \left\{ \sum_{i=1}^{n} f_i, \sum_{i=1}^{n} g_i \right\}$$

(iii) Let $F_i = |a_{ii}|^2 \Big/ \sum_{j=1}^{n} |a_{ij}|^2$, $G_j = |a_{jj}|^2 \Big/ \sum_{i=1}^{n} |a_{ij}|^2$, $(i, j = 1, 2, \ldots, n)$ which are understood to be zero if both the numerator and the denominator are zero, then

$$\text{Rank } \mathbf{A} \geqslant \max \left\{ \sum_{i=1}^{n} F_i, \sum_{j=1}^{n} G_j \right\}$$

16.4.3 Inverse

A nonsingular square matrix $\mathbf{A} = [a_{ij}]_n$ has the unique *inverse:* $\mathbf{A}^{-1} = [\alpha_{ij}]_n$, $\mathbf{A}^{-1}\mathbf{A} = \mathbf{A}\mathbf{A}^{-1} = \mathbf{I}$, where $\alpha_{ij} = (-1)^{i+j} \det \mathbf{A}_{ji} / \det \mathbf{A}$, $\mathbf{A}_{ji} = \mathbf{A}(1, \ldots, j-1, j+1, \ldots, n \mid 1, \ldots, i-1, i+1, \ldots, n)$. If $\mathbf{B}$ is also nonsingular of dimension n, then $(\mathbf{AB})^{-1} = \mathbf{B}^{-1}\mathbf{A}^{-1}$.

The *adjugate* of $\mathbf{A}$, (sometimes also called the *adjoint*, but this is entirely different from the definition given in 16.1.3), is the matrix: $\text{adj } \mathbf{A} = [(-1)^{i+j} \det \mathbf{A}_{ji}]_n$. Therefore $\mathbf{A} \cdot \text{adj } \mathbf{A} = (\det \mathbf{A}) \cdot \mathbf{I}$. The rank of adj $\mathbf{A}$ can be n, 1, 0 according as $\mathbf{A}$ is nonsingular, of rank $n - 1$, or less than $n - 1$.

Suppose the inverse of $\mathbf{B}$ is known and $\mathbf{A} = \mathbf{B} + \mathbf{uv}$, where $\mathbf{u}, \mathbf{v}$ are column and row vectors respectively. Then $\mathbf{A}^{-1} = \mathbf{B}^{-1} - \mathbf{B}^{-1}\mathbf{uvB}^{-1}/\gamma$ where $\gamma = 1 + \mathbf{vB}^{-1}\mathbf{u}$. In particular, if the matrix $\mathbf{uv}$ is zero everywhere except the kth row, then $\mathbf{A}^{-1} = \mathbf{B}^{-1} - \beta_k \mathbf{vB}^{-1}/(1 + (\mathbf{v}^T, \beta_k))$, where β_k is the kth column of $\mathbf{B}^{-1}$, ($\mathbf{v}$ is the kth row of $\mathbf{uv}$).

Example: Let $\mathbf{A} = [a_{ij}]_n$ be strictly upper triangular, then $\mathbf{A}^n = \mathbf{0}$, (16.1.3). We have then $(\mathbf{I} - \mathbf{A}) \sum_{i=1}^{n-1} \mathbf{A}^i = \mathbf{A} = \mathbf{I} - (\mathbf{I} - \mathbf{A})$, so that $\sum_{i=1}^{n-1} \mathbf{A}^i = (\mathbf{I} - \mathbf{A})^{-1} - \mathbf{I}$.

16.4.4 Block Operations

A matrix $\mathbf{A} = [a_{ij}]_{mn}$ can be partitioned into blocks such that each block is a submatrix of $\mathbf{A}$. For example, if $m = m_1 + m_2 + \cdots + m_r$, $n = n_1 + n_2 + \cdots + n_s$, $\mathbf{A}$ can be partitioned into $r \times s$ blocks of submatrices each having dimensions $m_i \times n_j$ ($i = 1, 2, \ldots, r$; $j = 1, 2, \ldots, s$). We can then write $\mathbf{A} = [\alpha_{ij}]_{rs}$ where α_{ij} represents a submatrix of $\mathbf{A}$. If $\mathbf{B} = [b_{ij}]_{nl}$ and if the rows of $\mathbf{B}$ are similarly partitioned as the columns of $\mathbf{A}$, then $\mathbf{AB}$ can be block multiplied. Thus if $\mathbf{B} = [\beta_{ij}]_{st}$ where $l = l_1 + l_2 + \cdots + l_t$ and β_{ij} is a submatrix of $\mathbf{B}$ of dimension $n_i \times l_j$ ($i = 1, 2, \ldots, s$; $j = 1, 2, \ldots, t$) then $\mathbf{AB} = \left[\sum_{k=1}^{s} \alpha_{ik}\beta_{kj} \right]_{rt}$. Of special interest are those partitioned matrices in which all the blocks on the diagonal are square.

Let a matrix be partitioned into 2×2 blocks: $\mathbf{A} = [a_{ij}]_n = [\alpha_{ij}]_2$. If α_{11} is nonsingular, then $\det \mathbf{A} = \det \alpha_{11} \cdot \det(\alpha_{22} - \alpha_{21}\alpha_{11}^{-1}\alpha_{12})$; if α_{22} is nonsingular, then $\det \mathbf{A} = \det \alpha_{22} \cdot \det(\alpha_{11} - \alpha_{12}\alpha_{22}^{-1}\alpha_{21})$. Now suppose $\mathbf{A}, \alpha_{11}, \alpha_{22}$ are all nonsingular, then

$$\mathbf{A}^{-1} = \begin{bmatrix} (\alpha_{11} - \alpha_{12}\alpha_{22}^{-1}\alpha_{21})^{-1} & -\alpha_{11}^{-1}\alpha_{12}(\alpha_{22} - \alpha_{21}\alpha_{11}^{-1}\alpha_{12})^{-1} \\ -\alpha_{22}^{-1}\alpha_{21}(\alpha_{11} - \alpha_{12}\alpha_{22}^{-1}\alpha_{21})^{-1} & (\alpha_{22} - \alpha_{21}\alpha_{11}^{-1}\alpha_{12})^{-1} \end{bmatrix}$$

16.4.5 Elementary Matrices

The *elementary* operations that can be performed on a matrix **A** are:

(i) Interchange the ith and j^{th} rows (columns) of **A**. This can be accomplished by premultiplying (postmultiplying) **A** by **P** which is obtained by interchanging the ith and j^{th} column of the identity matrix **I**. **P** is a special case of a *permutation matrix* of which the columns are the reordered columns of **I**. The inverse of a permutation matrix is equal to its transpose.

(ii) Multiply the ith row (column) of **A** by a nonzero scalar k.

(iii) Add to the ith row (column) of **A** by k times the jth row (column) of **A**.

Operations (ii) and (iii) are accomplished by premultiplying (postmultiplying) **A** by $\mathbf{Q}_{ij} = \mathbf{I} + \mathbf{E}_{ij}$, where $\mathbf{E}_{ij}$ is zero everywhere except its (i, j) entry which is k. Note that $\mathbf{Q}_{ij}^{-1} = \mathbf{I} - \mathbf{E}_{ij}$ $(i \neq j)$.

Both **P** and **Q** are *elementary matrices*, and their inverses are again elementary matrices. Every square nonsingular matrix can be expressed as the product of elementary matrices. Two matrices are *equivalent* if one can be obtained from the other by a sequence of elementary operations.

16.4.6 Elementary Unitary Matrices

Let $\mathbf{U}_{pq} = [u_{ij}]_n$ where $u_{ij} = \delta_{ij}$ except the four entries: $u_{pp}, u_{qq}, u_{pq}, u_{qp}$. These are defined as follows: $u_{pp} = e^{i\alpha} \cos\theta$, $u_{qq} = e^{i\delta} \cos\theta$, $u_{pq} = e^{i\beta} \sin\theta$, $u_{qp} = -e^{i\gamma} \sin\theta$ and $\alpha - \beta - \gamma + \delta = 0$ or any integral multiple of 2π. Then $\mathbf{U}_{pq}$ is a *unitary elementary matrix*. As a special case $u_{pp} = c$, $u_{qq} = \bar{c}$, $u_{pq} = -\bar{s}$, $u_{qp} = s$, where $|c|^2 + |s|^2 = 1$, which is sometimes referred to as a plane rotation.

The *elementary Householder matrix* is defined as: $\mathbf{H}^{(s)} = \mathbf{I} - 2\mathbf{w}^{(s)}\mathbf{w}^{(s)*}$, where $\mathbf{w}^{(s)T} = [0, \ldots, 0, w_{s+1}, \ldots, w_n]$ and $\mathbf{w}^{(s)*}\mathbf{w} = 1$. This matrix is also hermitian.

16.4.7 Linear Transformation and Change of Basis

Let $\mathbf{u}_1, \mathbf{u}_2, \ldots, \mathbf{u}_n$ be a basis for $\mathbf{C}_n$ and $\mathbf{v}_1, \mathbf{v}_2, \ldots, \mathbf{v}_n$ be another basis; these are related by: $\mathbf{v}_i = \sum_{s=1}^{n} \sigma_{is} \mathbf{u}_s$ $(i = 1, 2, \ldots, n)$. In matrix form the relation is $\mathbf{V} = \mathbf{SU}$ where $\mathbf{V} = [\mathbf{v}_1, \mathbf{v}_2, \ldots, \mathbf{v}_n]$, $\mathbf{U} = [\mathbf{u}_1, \mathbf{u}_2, \ldots, \mathbf{u}_n]$ and $\mathbf{S} = [\sigma_{ij}]_n$ which is nonsingular. Let $\mathbf{T} = [t_{ij}]_n$ be the matrix of the transformation $\mathscr{T}$ with respect to the basis $\mathbf{u}_1, \mathbf{u}_2, \ldots, \mathbf{u}_n$ and $\mathbf{R} = [r_{ij}]_n$ the matrix of $\mathscr{T}$ with respect to $\mathbf{v}_1, \mathbf{v}_2, \ldots, \mathbf{v}_n$. From $\mathscr{T}(\mathbf{v}_i) = \sum_s \sigma_{is} \mathscr{T}(\mathbf{u}_s)$ it follows $\sum_l r_{il} \mathbf{v}_l = \sum_s \sum_m \sigma_{is} t_{sm} \mathbf{u}_m$;

$$\sum_l \sum_m r_{il} \sigma_{lm} \mathbf{u}_m = \sum_s \sum_m \sigma_{is} t_{sm} \mathbf{u}_m$$

Thus $\mathbf{RSU} = \mathbf{STU}$ and the relation between **T** and **R** is given by: $\mathbf{T} = \mathbf{S}^{-1}\mathbf{RS}$.

16.4.8 Rank and Nullity

Suppose the matrix **T** of the transformation $\mathcal{T}$ with respect to the basis $\mathbf{u}_1, \mathbf{u}_2, \ldots, \mathbf{u}_n$ has rank r, then we say the *rank of transformation* $\mathcal{T}$ *is* r. From 16.4.7 we see that the rank of transformation remains the same when one basis changes to another. The transformed vectors $\boldsymbol{\zeta} = \mathcal{T}(\mathbf{z})$, $\mathbf{z}$ belonging to $\mathbf{C}_n$, also form a vector space which is called the *range* of the transformation. The dimension of the range is equal to the rank of the transformation, and is also r.

The set of vectors $\mathbf{z}$ of $\mathbf{C}_n$ such that $\mathcal{T}(\mathbf{z}) = \mathbf{o}$ obviously form a vector space, called the *null space* (or *kernel*) of the transformation $\mathcal{T}$. Its dimension s is called the *nullity* of $\mathcal{T}$. If the rank of $\mathcal{T}$ is r, then $s + r = n$, the dimension of the space $\mathbf{C}_n$.

Example: Consider the vector spaces $\mathbf{C}_2$ and $\mathbf{C}_3$. The matrix of transformation

$$\begin{bmatrix} a & a' \\ b & b' \\ c & c' \end{bmatrix}$$

which maps any vector in $\mathbf{C}_3$ into a vector in $\mathbf{C}_2$ has rank 2. Then the kernel of transformation is the one-dimensional vector space: $k(bc' - b'c)$, $k(ca' - c'a)$, $k(ab' - a'b)$.

16.5 LINEAR SYSTEM OF EQUATIONS

16.5.1 A General Theorem

Let $\mathbf{A} = [a_{ij}]_{mn}$, $\mathbf{x} \in \mathbf{C}_n$, $\mathbf{b} \in \mathbf{C}_m$, *the equation* $\mathbf{Ax} = \mathbf{b}$ *has a solution if and only if the augmented matrix* $[\mathbf{A}, \mathbf{b}]$ *has the same rank as* $\mathbf{A}$. If $\mathbf{A}$ and $[\mathbf{A}, \mathbf{b}]$ are of rank r, then there are $n - r + 1$ independent solutions for $\mathbf{x}$. If $\mathbf{b} = \mathbf{0}$, then the homogeneous equation $\mathbf{Ax} = \mathbf{0}$ has $n - r$ independent solutions for $\mathbf{x}$, disregarding the trivial solution $\mathbf{x} = \mathbf{0}$.

16.5.2 Gauss Elimination and LR Decomposition

Let $\mathbf{A} = [a_{ij}]_n$ be square and assume $a_{ii} \neq 0$. Then premultiplication of $\mathbf{A}$ by a suitable elementary matrix $\mathbf{Q}_1$ (16.4.5): $\mathbf{Q}_1 = \mathbf{I} + \mathbf{E}_{21}$ with $k = a_{21}/a_{11}$ reduces the element at the position (2, 1) to zero. In like manner, continued premultiplication by suitably chosen $\mathbf{Q}_2, \mathbf{Q}_3, \ldots, \mathbf{Q}_{n-1}$ will reduce the elements at the positions (3, 1), (4, 1), . . . , (n, 1) successively to zero. The element at the (2, 2) position of the reduced matrix is now $(a_{11}a_{22} - a_{12}a_{21})/a_{11} = \det \mathbf{A}(1, 2)/\det \mathbf{A}(1)$, and if it is not zero, all the elements at the positions (3, 2), (4, 2), . . . , (n, 2) can be similarly reduced to zero by premultiplication using suitable **Q**'s. The process can be continued as long as all the elements at the diagonal positions (3, 3), (4, 4), . . . , (n - 1, n - 1) in the subsequent steps are nonzero, and **A** is ultimately reduced to the upper triangular form. Let $\Delta_k = \det \mathbf{A}(1, 2, \ldots, k)$,

($k = 1, 2, \ldots, n$), then

$$\left(\prod_i \mathbf{Q}_i\right)\mathbf{A} = \begin{bmatrix} \Delta_1 & * & * & \cdots & * \\ & \dfrac{\Delta_2}{\Delta_1} & * & \cdots & * \\ & & \ddots & & \vdots \\ & & & \ddots & * \\ 0 & & & & \dfrac{\Delta_n}{\Delta_{n-1}} \end{bmatrix} \tag{16.5-1}$$

Since $\prod_i \mathbf{Q}_i$ is lower triangular, $\mathbf{L} = \left(\prod_i \mathbf{Q}_i\right)^{-1}$ is also lower triangular. Thus every square matrix can be decomposed into the product **LR**, where **L** is lower triangular with all diagonal elements equal to unity, and **R** is upper triangular, provided the principal minors det $\mathbf{A}(1, 2, \ldots, k)$, ($k = 1, 2, \ldots, n - 1$) do not vanish. Moreover, this decomposition is unique.

Let eq. (16.5-1) be written as

$$\mathbf{A} = \mathbf{L}\ \mathrm{diag}\left(\Delta_1, \frac{\Delta_2}{\Delta_1}, \ldots, \frac{\Delta_n}{\Delta_{n-1}}\right)\mathbf{R}'$$

where $\mathbf{R}'$ is upper triangular with all diagonal elements equal to unity. If **A** is symmetric, then $\mathbf{A}^T = \mathbf{A} = \mathbf{R}'^T\ \mathrm{diag}\left(\Delta_1, \frac{\Delta_2}{\Delta_1}, \ldots, \frac{\Delta_n}{\Delta_{n-1}}\right)\mathbf{L}^T$ and comparison with eq. (16.5-1) shows that $\mathbf{L} = \mathbf{R}'^T$. Furthermore, if $\Delta_k > 0$, ($k = 1, 2, \ldots, n$), then by writing $\mathbf{J} = \mathbf{L}\ \mathrm{diag}\ (\sqrt{\Delta_1}, \sqrt{\Delta_2/\Delta_1}, \ldots, \sqrt{\Delta_n/\Delta_{n-1}})$ it follows $\mathbf{A} = \mathbf{J}\mathbf{J}^T$. This the *Cholesky decomposition*.

The solution of the equation $\mathbf{Ax} = \mathbf{b}$ is facilitated by decomposing **A** into **LR** form. With $\mathbf{LRx} = \mathbf{b}$ and $\mathbf{y} = \mathbf{Rx}$ one first solves $\mathbf{Ly} = \mathbf{b}$ for **y** by forward substitution, and then solves $\mathbf{Rx} = \mathbf{y}$ by back substitution.

The diagonal element at the (r, r) position in the rth step is commonly referred to as the *pivot* in the elimination process. Evidently if the pivot is zero, the method breaks down. Partial pivoting is a strategy whereby this difficulty is overcome by rearranging the remaining $n - r$ rows such that the new pivot now has the largest absolute value of the $n - r$ elements in this column. (If all the $n - r$ elements are zero, then the first r columns of **A** are linearly dependent). Complete pivoting implies a rearrangement of the remaining $n - r$ rows and $n - r$ columns such that the new pivot has the largest absolute value of all the $(n - r) \times (n - r)$ elements.

16.5.3 Householder's Triangularization

Assume after $r - 1$ steps **A** is reduced to $\mathbf{A}_r$, ($\mathbf{A}_1 = \mathbf{A}$), which is already triangularized in its first r columns. Now, (16.4.6)

$$\mathbf{H}^{(r)}\mathbf{A}_r = [\mathbf{I} - 2\mathbf{w}^{(r)}\mathbf{w}^{(r)*}] \begin{bmatrix} \mathbf{A}_r^{(1)} & \mathbf{A}_r^{(2)} \\ 0 & \mathbf{A}_r^{(4)} \end{bmatrix} = \begin{bmatrix} \mathbf{A}_r^{(1)} & \mathbf{A}_r^{(2)} \\ 0 & \hat{\mathbf{A}}_r^{(4)} \end{bmatrix}$$

where $\mathbf{A}_r^{(1)}$ is in upper triangular form. Let the first column of $\mathbf{A}_r^{(4)}$ be $\mathbf{a}^T = (a_{r+1,r+1}, a_{r+2,r+1}, \ldots, a_{n,r+1})$; if $\mathbf{w}$ is specified as

$$\begin{cases} \mathbf{w}^T = \dfrac{1}{2K}\left(0, \ldots, 0, a_{r+1,r+1}\left(1 + \dfrac{S}{|a_{r+1,r+1}|}\right), a_{r+2,r+1}, \ldots, a_{n,r+1}\right) \\ |K|^2 = \dfrac{1}{2}(S^2 + S\,|a_{r+1,r+1}|), \\ S^2 = \mathbf{a}^*\mathbf{a}, \end{cases}$$

then the first column of $\hat{\mathbf{A}}_r^{(4)}$ will assume the form: $(k, 0, \ldots, 0)^T$ where $k = -Sa_{r+1,r+1}/|a_{r+1,r+1}|$. The process then continues until the entire matrix is triangularized. Note that k is complex if $a_{r+1,r+1}$ is complex.

16.5.4 Gauss-Jordan Elimination and the Product Form of Inverse

The Gauss-Jordan method reduces a given matrix $\mathbf{A} = [a_{ij}]_n$ to a permutation matrix. Consider the sth column of $\mathbf{A}$: $\mathbf{a}_s$, and if $a_{\sigma s} \neq 0$, it is possible to reduce this column to the σth column of the identity matrix $\mathbf{e}_\sigma$ by the following operations. First premultiply by $\mathbf{Q}_\sigma = \mathbf{I} + \mathbf{E}_{\sigma\sigma}$ with $k = 1/a_{\sigma s} - 1$, as in section 16.4.5, then the elements at the remaining positions $(\sigma', s), (\sigma' = 1, 2, \ldots, n; \sigma' \neq \sigma)$ can be reduced to zero by premultiplication of $\mathbf{Q}_{\sigma' s} = \mathbf{I} + \mathbf{E}_{\sigma' s}$ with $k = -a_{\sigma' s}$. The whole operation can be done by a single premultiplication of the matrix $\mathbf{Q}_1 = \left(\prod_{\sigma'} \mathbf{Q}_{\sigma' s}\right) \mathbf{Q}_\sigma$, which differs from the identity only in the σth column

$$\mathbf{Q}_1 = \begin{bmatrix} 1 & \cdots & -a_{1s}/a_{\sigma s} & & \\ & 1 & \cdots -a_{2s}/a_{\sigma s} & & \\ & & \ddots \quad \vdots & & \\ & & 1/a_{\sigma s} & & \\ & & \vdots & \ddots & \\ & & -a_{ns}/a_{\sigma s} & \cdots & 1 \end{bmatrix}$$

Next a similar operation $\mathbf{Q}_2$ on the rth column of $\mathbf{A}$ is performed to transform it to $\mathbf{e}_\rho\ (a_{\rho r} \neq 0)$. This transformation leaves $\mathbf{e}_\sigma$, the previous transformed vector unaltered, and after n such transformations, $\mathbf{A}$ is reduced to a permutation matrix $\mathbf{P}$. Thus $\mathbf{Q}_n\mathbf{Q}_{n-1} \ldots \mathbf{Q}_1\mathbf{A} = \mathbf{P}$, and $\mathbf{A}^{-1} = \mathbf{P}^T\mathbf{Q}_n\mathbf{Q}_{n-1} \ldots \mathbf{Q}_1$, which is generally referred to as the *product form of the inverse* of $\mathbf{A}$.

16.5.5 Sufficient Conditions for Nonsingularity; Irreducibility

(i) Let $\mathbf{A} = [a_{ij}]_n$, and if $|a_{ii}| - \sum_{\substack{j=1 \\ \neq i}}^{n} |a_{ij}| > 0, (i = 1, 2, \ldots, n)$, then $\mathbf{A}$ is nonsingular. Such matrices are *strictly diagonally dominant.*

(ii) If $|a_{ii}||a_{jj}| - \sum_{\substack{l=1 \\ \neq i}}^{n} |a_{il}| \sum_{\substack{k=1 \\ \neq j}}^{n} |a_{jk}| > 0, (i, j = 1, 2, \ldots, n)$, then $\mathbf{A}$ is nonsingular.

Obviously matrices exist that satisfy the conditions of (ii) but not of (i), but there are more inequalities to be tested in (ii).

A matrix $\mathbf{A}$ is *reducible* (or *decomposable*) if there exists a permutation matrix $\mathbf{P}$ such that

$$\mathbf{P}^T\mathbf{A}\mathbf{P} = \begin{bmatrix} \mathbf{A}_{11} & \mathbf{A}_{12} \\ 0 & \mathbf{A}_{22} \end{bmatrix}$$

where $\mathbf{A}_{11}, \mathbf{A}_{22}$ are square. Otherwise $\mathbf{A}$ is *irreducible.*

When $\mathbf{A}$ is irreducible, then the conditions in (i) can be relaxed to: if $|a_{ii}| - \sum_{\substack{j=1 \\ \neq i}}^{n} |a_{ij}| \geqslant 0\ (i = 1, 2, \ldots, n)$, with at least one strict inequality, then $\mathbf{A}$ is nonsingular. Such matrices are *irreducibly diagonally dominant.* Similarly (ii) can be relaxed to: if $\mathbf{A}$ is irreducible and $|a_{ii}||a_{jj}| - \sum_{\substack{l=1 \\ \neq i}}^{n} |a_{il}| \sum_{\substack{k=1 \\ \neq j}}^{n} |a_{jk}| \geqslant 0, (i, j = 1, 2, \ldots, n)$ with at least one strict inequality, then $\mathbf{A}$ is nonsingular.

Similar conditions can be formulated based on the columns of $\mathbf{A}$ by considering $\mathbf{A}^T$ instead of $\mathbf{A}$.

16.5.6 Condition Numbers

For a square matrix $\mathbf{A}$, the *condition number* $C_\phi(\mathbf{A})$ is defined by $\phi(\mathbf{A})\,\phi(\mathbf{A}^{-1})$, where ϕ refers to a certain norm (16.8.1). This number gives a bound on the sensitivity of changes in the solution of the equation: $\mathbf{Ax} = \mathbf{b}$ with respect to changes of $\mathbf{b}$. If $\mathbf{A}$ is assumed to hold constant, then the changes in $\mathbf{x}$: $\delta\mathbf{x}$ are related to changes in $\mathbf{b}$: $\delta\mathbf{b}$ by

$$\frac{\phi(\delta\mathbf{x})}{\phi(\mathbf{x})} \leqslant C_\phi(\mathbf{A})\,\frac{\phi(\delta\mathbf{b})}{\phi(\mathbf{b})}$$

Similarly, the condition number $C_\phi(\mathbf{A})$ also gives a measure of changes in $\mathbf{x}$ to changes in $\mathbf{A}$. If $\mathbf{b}$ is assumed to be constant and $\delta\mathbf{A}$ denotes changes in $\mathbf{A}$, then

$$\frac{\phi(\delta \mathbf{x})}{\phi(\mathbf{x}+\delta \mathbf{x})} \leqslant C_\phi(\mathbf{A}) \frac{\phi(\delta \mathbf{A})}{\phi(\mathbf{A})}$$

When ϕ is the euclidean norm E, then $C_E(\mathbf{A})$ is just the ratio of the largest to the least singular values of $\mathbf{A}$, (16.6.1). If $\mathbf{A}$ is hermitian or real symmetric, then $C_E(\mathbf{A})$ is the ratio of its largest to the least eigenvalues. If, in addition, $\mathbf{A}$ is positive definite, then $C_E(\mathbf{A}+k\mathbf{I}) < C_E(\mathbf{A})$, for $k > 0$.

16.5.7 Matrix Scaling

Let $\mathbf{D}_1$, $\mathbf{D}_2$ be nonsingular diagonal matrices. Then $\mathbf{D}_1\mathbf{A}\mathbf{D}_2$ is *diagonally equivalent* to $\mathbf{A}$ and such transformation is called *matrix scaling*. From a practical point of view it is desirable to scale a matrix to improve its condition, if such diagonal matrices can be found.

Let $\mathbf{A}$ be irreducible and the norm under consideration be the e-norm, (16.8.1). Let ω be the maximum real eigenvalue of the positive matrix $|\mathbf{A}||\mathbf{A}^{-1}|$, and $\mathbf{z} = (z_1, z_2, \ldots, z_n)^T$ the corresponding eigenvector: $|\mathbf{A}||\mathbf{A}^{-1}|\, \mathbf{z} = \omega\mathbf{z}$, (16.10.1). Similarly let $|\mathbf{A}^{-1}||\mathbf{A}|\, \mathbf{w} = \omega\mathbf{w}$. Then $\min_{\mathbf{D}_1,\mathbf{D}_2} C_e(\mathbf{D}_1\mathbf{A}\mathbf{D}_2)$ is achieved when $\mathbf{D}_1 = \mathrm{diag}\,(1/z_1, 1/z_2, \ldots, 1/z_n)$, $\mathbf{D}_2 = \mathrm{diag}\,(w_1, w_2, \ldots, w_n)$.

16.5.8 Iterative Methods

The principal feature of any *iterative method* is the construction of a sequence of approximate solutions which ultimately converge to the solution. Let the successive iterates $\mathbf{x}^{(0)}, \mathbf{x}^{(1)}, \ldots$ be constructed according to $\mathbf{x}^{(k)} = \mathbf{G}\mathbf{x}^{(k-1)} + \mathbf{f}$, where $\mathbf{x}^{(0)}$ is any initial vector. For the process to converge it is necessary and sufficient that the spectral radius of $\mathbf{G}$ be less than one, (16.6.1).

For the solution of the linear system of equations $\mathbf{A}\mathbf{x} = \mathbf{b}$, let $\mathbf{A} = \mathbf{D} - \mathbf{L} - \mathbf{U}$, where $\mathbf{D}$ is the diagonal of $\mathbf{A}$, $\mathbf{L}$ and $\mathbf{U}$ are respectively strictly lower and upper triangular matrices. Then starting with an arbitrary $\mathbf{x}^{(0)}$ the various iterative methods for solving $\mathbf{A}\mathbf{x} = \mathbf{b}$ are

(i) Jacobi method:

$$\mathbf{x}^{(k)} = \mathbf{D}^{-1}(\mathbf{L}+\mathbf{U})\,\mathbf{x}^{(k-1)} + \mathbf{D}^{-1}\mathbf{b}; \qquad (k = 1, 2, \ldots)$$

(ii) Gauss-Seidal method:

$$\mathbf{x}^{(k)} = (\mathbf{D}-\mathbf{L})^{-1}\mathbf{U}\,\mathbf{x}^{(k-1)} + (\mathbf{D}-\mathbf{L})^{-1}\mathbf{b}; \qquad (k = 1, 2, \ldots)$$

(iii) Successive over-relaxation (SOR) method:

$$\mathbf{x}^{(k)} = (\mathbf{D}-\omega\mathbf{L})^{-1}\{(1-\omega)\,\mathbf{D} + \omega\mathbf{U}\}\mathbf{x}^{(k-1)} + \omega(\mathbf{D}-\mathbf{L})^{-1}\mathbf{b}; \qquad (k = 1, 2, \ldots)$$

where ω is the relaxation parameter.

The Jacobi or Gauss-Seidal methods are convergent for any initial $\mathbf{x}^{(0)}$ if $\mathbf{A}$ is strictly or irreducibly diagonally dominant. The SOR method converges for any initial $\mathbf{x}^{(0)}$ if $\mathbf{A}$ is hermitian and positive definite.

16.5.9 SOR Method for Block Tri-diagonal Matrices

Let $\mathbf{A}$ be block tri-diagonal (with square diagonal blocks), and write $\mathbf{A} = \tilde{\mathbf{D}} - \tilde{\mathbf{L}} - \tilde{\mathbf{U}}$ where $\tilde{\mathbf{D}}$ is block diagonal, and $\tilde{\mathbf{L}}$ and $\tilde{\mathbf{U}}$ are respectively strictly lower and upper block triangular matrices. The block SOR method for solving $\mathbf{Ax} = \mathbf{b}$ is then given by

$$\mathbf{x}^{(k)} = (\tilde{\mathbf{D}} - \omega\tilde{\mathbf{L}})^{-1}\{(1-\omega)\tilde{\mathbf{D}} + \omega\tilde{\mathbf{U}}\}\mathbf{x}^{(k-1)} + \omega(\tilde{\mathbf{D}} - \tilde{\mathbf{L}})^{-1}\mathbf{b}; \qquad (k = 1, 2, \ldots)$$

and the block Jacobi and Gauss methods can be similarly defined. The proper choice of the relaxation parameter ω can greatly accelerate the convergence of the SOR method.

Let the eigenvalues of $\mathbf{D}^{-1}(\tilde{\mathbf{L}} + \tilde{\mathbf{U}})$ be denoted by θ and the spectral radius $\hat{\theta}$. If θ's are real (for example, if $\mathbf{A}$ is positive definite hermitian, see section 16.9.2) and if $\hat{\theta} < 1$, then the block SOR method converges for $0 < \omega < 2$. Under these conditions convergence is fastest (relaxation is optimal) if $\omega = 2/(1 + \sqrt{1 - \hat{\theta}^2})$ (optimal relaxation factor), for which the spectral radius of $(\tilde{\mathbf{D}} - \omega\tilde{\mathbf{L}})^{-1}\{(1-\omega)\tilde{\mathbf{D}} + \omega\tilde{\mathbf{U}}\}$ is $(1 - \sqrt{1 - \hat{\theta}^2})/(1 + \sqrt{1 - \hat{\theta}^2})$.

16.6 EIGENVALUES AND THE JORDAN NORMAL FORM

16.6.1 Definitions and Basic Results

The following definitions and results will be of use in the discussion of section 16.6:

1. Let $\mathbf{A}$ be square of dimension n, and $\mathbf{Az} = \lambda\mathbf{z}$ (or $\mathbf{uA} = \lambda\mathbf{u}$), then λ is an *eigenvalue* of $\mathbf{A}$ and $\mathbf{z}$ (or $\mathbf{u}$) the corresponding *eigenvector* (*eigenrow*). The eigenvalues of $\mathbf{A}$ are the n roots of the *characteristic equation*: $\det(\lambda\mathbf{I} - \mathbf{A}) = \lambda^n + p_1\lambda^{n-1} + \cdots + p_n = 0$. The polynomial: $F(\lambda) = \det(\lambda\mathbf{I} - \mathbf{A})$ is called the *characteristic polynomial* of $\mathbf{A}$ and $\lambda\mathbf{I} - \mathbf{A}$ the *characteristic matrix*. The largest modulus of the eigenvalues of $\mathbf{A}$ is called the *spectral radius*.

2. The product of all the eigenvalues of $\mathbf{A}$ is equal to $\det \mathbf{A}$. $\mathbf{A}$ is singular if and only if at least one of its eigenvalues is zero.

3. If $\mathbf{A}$ is nonsingular and λ one of its eigenvalues, then $1/\lambda$ is an eigenvalue of $\mathbf{A}^{-1}$.

4. If $f(\mathbf{A})$ is a polynomial of $\mathbf{A}$, then the eigenvalues of $f(\mathbf{A})$ are $f(\lambda_i)$.

5. The sum of all the eigenvalues of $\mathbf{A}$ is equal to $\operatorname{tr}\mathbf{A}$.

6. Every matrix satisfies its own characteristic equation (*Cayley Hamilton Theorem*).

7. If the n eigenvalues of $\mathbf{A}$ are distinct, then $\mathbf{A}$ has n independent eigenvectors. Furthermore, if $\mathbf{Az}_i = \lambda_i\mathbf{z}_i$, $\mathbf{w}_j^*\mathbf{A} = \lambda_j\mathbf{w}_j^*$ then $(\mathbf{z}_i, \mathbf{w}_j) = 0$ for $i \neq j$.

8. If **A** has multiple eigenvalues, **A** may have less than n eigenvectors. However, if **A** is normal, hermitian, or real symmetric, then **A** must have n independent eigenvectors. Besides, the eigenvalues of hermitian or real symmetric matrices are real.

9. Let $\mathbf{A} = [a_{ij}]_{mn}$, $\mathbf{B} = [b_{ij}]_{nm}$, then the nonzero eigenvalues of **AB** are the same as those of **BA**. If $m > n$, then **AB** has $m - n$ more zero eigenvalues than **BA**.

10. Let λ_i be the eigenvalues of **A**, then $\overline{\lambda_i}$ are the eigenvalues of $\mathbf{A}^*$. If the eigenvalues of $\mathbf{A}^*\mathbf{A}$ are σ_i^2, then in general $\sigma_i^2 \neq \lambda_i\overline{\lambda_i}$, unless **A** is normal. The *singular values* of **A** are defined as σ_i $(\sigma_i \geqslant 0)$. Furthermore, the following inequality holds for any **A**: $\min \sigma_i \leqslant |\lambda_i| \leqslant \max \sigma_i$.

11. The *field of values* of **A** is defined as the collection of all values $(\mathbf{Ax}, \mathbf{x})$ where $\mathbf{x}^*\mathbf{x} = 1$. The field of values of **A** includes the eigenvalues of **A**, and if **U** is unitary, the field of values of **A** is also the field of values of $\mathbf{U}^*\mathbf{AU}$. If **A** is normal, then its field of values is the same as the convex hull of its eigenvalues.

12. If $\mathbf{A} = \mathbf{S}^{-1}\mathbf{BS}$, then **A** is *similar* to **B**, and the eigenvalues of **A** are the same as those of **B**. **A** is said to be obtained from **B** by a *similarity transformation*. If **S** is unitary, then **A** is *unitarily similar* to **B**.

13. The eigenvalues of a skew-hermitian matrix are either purely imaginary or zero. The eigenvalues of a unitary matrix have modulus one. A nilpotent matrix has all eigenvalues equal to zero. An indempotent matrix has eigenvalues equal to either one or zero.

14. If $\sigma_i(\mathbf{A})$ are the singular values of **A**, and $\lambda_i(\mathbf{A})$ the eigenvalues, then

$$\min \sigma_i(\mathbf{A}) \cdot \min \sigma_i(\mathbf{B}) \leqslant |\lambda(\mathbf{AB})| \leqslant \max \sigma_i(\mathbf{A}) \cdot \max \sigma_i(\mathbf{B}) \cdot$$

$$\min \sigma_i(\mathbf{A}) \cdot \min \sigma_i(\mathbf{B}) \leqslant \sigma(\mathbf{AB}) \leqslant \max \sigma_i(\mathbf{A}) \cdot \max \sigma_i(\mathbf{B}) \cdot$$

$$\min \sigma_i(\mathbf{A}) - \max \sigma_i(\mathbf{B}) \leqslant |\lambda(\mathbf{A} + \mathbf{B})| \leqslant \max \sigma_i(\mathbf{A}) + \max \sigma_i(\mathbf{B}) \cdot$$

$$\min \sigma_i(\mathbf{A}) - \max \sigma_i(\mathbf{B}) \leqslant \sigma(\mathbf{A} + \mathbf{B}) \leqslant \max \sigma_i(\mathbf{A}) + \max \sigma_i(\mathbf{B}) \cdot$$

16.6.2 Polynomial Matrices

Let $\mathbf{A} = [a_{ij}(\lambda)]_{mn}$ where $a_{ij}(\lambda)$ is a polynomial of λ. If the highest degree of $a_{ij}(\lambda)$ is N, then **A** is a *polynomial matrix* of degree N. For polynomial matrices the elementary operations (16.4.5) are now

(i) Interchange any two rows (columns).
(ii) Multiply any row (column) by a nonzero scalar constant.
(iii) Add to any row (column) by another row (column) after being multiplied by an arbitrary polynomial $k(\lambda)$.

Again, two matrices are *equivalent* if one can be obtained from the other by a sequence of such elementary operations. Suppose the rank of **A** is r. Then by means of a chain of elementary operations, it can be shown that **A** is equivalent to the diagonal form:

$$\Lambda(\lambda) = \begin{bmatrix} i_1(\lambda) & & & & & & \\ & i_2(\lambda) & & & & & \\ & & \ddots & \cdots & \cdots & \cdots & \\ & & & i_r(\lambda) & & & \\ & & & & 0 & & \\ & & \cdots & \cdots & \cdots & \ddots & \\ & & & & & & 0 \end{bmatrix} \qquad (16.6\text{-}1)$$

where $i_k(\lambda)$ divides $i_{k+1}(\lambda)$, $(k = 1, 2, \ldots, r - 1)$ and the leading coefficient of each $i_k(\lambda)$ is unity. The polynomials $i_k(\lambda)$, $(k = 1, 2, \ldots, r)$ are uniquely determined and are the *invariant polynomials* of **A**.

Let $\mathbf{D}_p(\lambda)$ be the greatest common divisor of all the minors of order p of $\mathbf{A} = [a_{ij}(\lambda)]_{mn}$ (of rank r), $(p = 1, 2, \ldots, r)$, such that the leading coefficient of each $D_p(\lambda)$ is unity. Then in the sequence $D_r(\lambda), D_{r-1}(\lambda), \ldots, D_1(\lambda)$ each $D_k(\lambda)$ divides $D_{k+1}(\lambda)$, $(k = 1, 2, \ldots, r - 1)$. If $\psi_k(\lambda) = D_k(\lambda)/D_{k-1}(\lambda)$ $(k = 1, 2, \ldots, r)$, $(D_0(\lambda) = 1)$, then both $\psi_k(\lambda), D_k(\lambda)$ remain invariant by elementary operations on **A**. But **A** is equivalent to $\Lambda(\lambda)$, therefore $\psi_k(\lambda) = i_k(\lambda)$, $(k = 1, 2, \ldots, r)$.

16.6.3 Minimal Polynomial

For every vector **z** in $\mathbf{C}_n$ there is a unique *annihilating polynomial* of the lowest degree $P(\lambda)$ (the coefficient of the term of the highest degree of $P(\lambda)$ being one), called the *minimal polynomial*, such that $P(\mathbf{A})\,\mathbf{z} = \mathbf{o}$. Let $e_1, e_2, \ldots, e_n$ be the basis vectors of $\mathbf{C}_n$, and their respective minimal polynomials be $P_1(\lambda), P_2(\lambda), \ldots, P_n(\lambda)$. Then the least common multiple of $P_1(\lambda), P_2(\lambda), \ldots, P_n(\lambda)$, say $\psi(\lambda)$, annihilates the entire space $\mathbf{C}_n$, and $\psi(\mathbf{A}) = \mathbf{0}$. $\psi(\lambda)$ is also called the minimal polynomial of **A**.

The characteristic polynomial, $F(\lambda) = \det(\lambda\mathbf{I} - \mathbf{A})$, also annihilates **A**, $F(\mathbf{A}) = \mathbf{0}$, but it may not be the minimal polynomial $\psi(\lambda)$. It is always divisible by $\psi(\lambda)$. If $F(\lambda)$ is identical with $\psi(\lambda)$, **A** is *nonderogatory*; otherwise *derogatory*.

Example: For the matrix

$$\mathbf{A} = \begin{bmatrix} 1 & 1 & & & \\ & 1 & & & \\ & & 1 & & \\ & & & 2 & \\ & & & & 2 \end{bmatrix}; \qquad \lambda\mathbf{I} - \mathbf{A} = \begin{bmatrix} \lambda - 1 & -1 & & & \\ & \lambda - 1 & & & \\ & & \lambda - 1 & & \\ & & & \lambda - 2 & \\ & & & & \lambda - 2 \end{bmatrix}$$

$F(\lambda) = \det(\lambda\mathbf{I} - \mathbf{A}) = (\lambda - 1)^3(\lambda - 2)^2$, the minimal polynomial for e_1, e_2 is

$(\lambda - 1)^2$, for $\mathbf{e}_3$ is $\lambda - 1$, and for $\mathbf{e}_3, \mathbf{e}_4$ is $\lambda - 2$. Consequently the minimal polynomial for $\mathbf{C}_5$ is $(\lambda - 1)^2(\lambda - 2)$. It also follows from $\lambda\mathbf{I} - \mathbf{A}$ that $D_5(\lambda) = (\lambda - 1)^3(\lambda - 2)^2, D_4(\lambda) = (\lambda - 1)(\lambda - 2), D_3(\lambda) = \lambda - 1, D_2(\lambda) = 1, D_1(\lambda) = 1$, and $\psi(\lambda) = (\lambda - 1)^2(\lambda - 2)$, the minimal polynomial of $\mathbf{A}$.

16.6.4 The Jordan Normal Form

Let

$$\mathbf{L} = \left[\begin{array}{cccccc} 0 & 1 & - & - & - & 0 \\ 0 & 0 & 1 & 0 & 0 & 0 \\ \hline 0 & 0 & 0 & 0 & 0 & 1 \\ -p_n & -p_{n-1} & - & - & - & -p_1 \end{array}\right] \tag{16.6-2}$$

Then the characteristic polynomial $D_n(\lambda) = \det(\lambda\mathbf{I} - \mathbf{L}) = \lambda^n + p_1\lambda^{n-1} + \cdots + p_n$ and $D_{n-1}(\lambda) = \cdots = D_1(\lambda) = 1$. For this matrix the minimal polynomial therefore coincides with its characteristic polynomial. Now if $D_n(\lambda) = (\lambda - \lambda_1)^{s_1}(\lambda - \lambda_2)^{s_2} \ldots (\lambda - \lambda_r)^{s_r}$ where $\lambda_1, \lambda_2, \ldots, \lambda_r$ are distinct, and $\mathbf{z}_{ij} = (1/(j-1)!)(d/d\lambda_i)^{j-1}[1, \lambda_i, \ldots, \lambda_i^{n-1}]^T$, $(i = 1, 2, \ldots, r; j = 1, 2, \ldots, s_i)$, and $T = [\mathbf{z}_{11}, \ldots, \mathbf{z}_{1s_1}, \mathbf{z}_{21}, \ldots, \mathbf{z}_{2s_2}, \ldots, \mathbf{z}_{rs_r}]$, then $\mathbf{L}$ referred to the basis $\mathbf{z}_{11}, \ldots, \mathbf{z}_{1s_1}, \mathbf{z}_{21}, \ldots, \mathbf{z}_{2s_2}, \ldots, \mathbf{z}_{rs_r}$ will have the *Jordan normal form*

$$\mathbf{T}^{-1}\mathbf{L}\mathbf{T} = \begin{bmatrix} \mathbf{J}(\lambda_1) & & \\ & \mathbf{J}(\lambda_2) & \\ \cdots & \cdots & \cdots \\ & & \mathbf{J}(\lambda_r) \end{bmatrix}; \quad \mathbf{J}(\lambda_i) = \begin{bmatrix} \lambda_i & 1 & & & & \\ & \lambda_i & 1 & & & \\ & & \cdot & \cdot & & \\ & & & \cdot & \cdot & \\ & & & & \lambda_i & 1 \\ & & & & & \lambda_i \end{bmatrix} \tag{16.6-3}$$

each *Jordan box* $\mathbf{J}(\lambda_i)$ being of dimensions $s_i \times s_i$, $(i = 1, 2, \ldots, r)$; besides, the following relations holds:

$$(\mathbf{L} - \lambda_i\mathbf{I})\,\mathbf{z}_{i1} = \mathbf{o}; \qquad (i = 1, 2, \ldots, r)$$

and

$$(\mathbf{L} - \lambda_i\mathbf{I})\,\mathbf{z}_{ij} = \mathbf{z}_{i,j-1}; \qquad (i = 1, 2, \ldots, r; j = 2, 3, \ldots, s_i)$$

The vectors $\mathbf{z}_{i1}$ $(i = 1, 2, \ldots, r)$ are the eigenvectors of $\mathbf{L}$.

If $P(\lambda) = \lambda^m + p_1\lambda^{m-1} + \cdots + p_m$ is the minimal polynomial of $\mathbf{z}$ in $\mathbf{C}_n$ then $\mathbf{z}$, $\mathbf{Az}, \ldots, \mathbf{A}^{m-1}\mathbf{z}$ are necessarily independent and form an m-dimensional *cyclic subspace* which is invariant with respect to $\mathbf{A}$.

If $\psi(\lambda)$ is the minimal polynomial of the entire space $\mathbf{C}_n$, then there exists a vector $\mathbf{g}_1$ whose minimal polynomial is also $\psi(\lambda)$. If $m = n$, then $\mathbf{g}_1, \mathbf{Ag}_1, \ldots, \mathbf{A}^m \mathbf{g}_1$ span $\mathbf{C}_n$. If $m < n$, then there exists another vector $\mathbf{g}_2$ with minimal polynomial $\psi_2(\lambda)$ such that $\psi_1(\lambda)$ is divisible by $\psi_2(\lambda)$ and no vector in the cyclic space generated by $\mathbf{g}_2$ can lie in the cyclic space generated by $\mathbf{g}_1$. In general $\mathbf{C}_n$ can always be decomposed into cyclic spaces $\mathbf{S}_i$, with minimal polynomial: $\psi_i(\lambda) = \lambda^{m_i} + p_{i1}\lambda^{m_i-1} + \cdots + p_{im_i}$ $(i = 1, 2, \ldots, t)$, $\mathbf{C}_n = \sum_{i=1}^{t} \mathbf{S}_i$, such that $\psi_i(\lambda)$ is the minimal polynomial of $\mathbf{C}_n$ and each $\psi_i(\lambda)$ is divisible by $\psi_{i+1}(\lambda)$, $(i = 1, 2, \ldots, t-1)$.

The matrix $\mathbf{A}$ referred to the basis $\mathbf{g}_1, \mathbf{Ag}_1, \ldots, \mathbf{A}^{m_i-1}\mathbf{g}_1, \mathbf{g}_2, \ldots, \mathbf{A}^{m_2-1}\mathbf{g}_2, \ldots, \mathbf{A}^{m_t-1}\mathbf{g}_t$ is equal to

$$
\begin{aligned}
&[\mathbf{g}_1, \ldots, \mathbf{A}^{m_1-1}\mathbf{g}_1, \ldots, \mathbf{A}^{m_t-1}\mathbf{g}_t]^{-1}\mathbf{A}[\mathbf{g}_1, \ldots, \mathbf{A}^{m_1-1}\mathbf{g}_1, \ldots, \mathbf{A}^{m_t-1}\mathbf{g}_t] \\
&= [\mathbf{g}_1, \ldots, \mathbf{A}^{m_1-1}\mathbf{g}_1, \ldots, \mathbf{A}^{m_t-1}\mathbf{g}_t]^{-1}[\mathbf{Ag}_1, \ldots, \mathbf{A}^{m_1}\mathbf{g}_1, \ldots, \mathbf{A}^{m_t}\mathbf{g}_t] \\
&= \begin{bmatrix} \mathbf{L}_1^T & & \\ & \mathbf{L}_2^T & \\ \cdots & \cdots & \cdots \\ & & \mathbf{L}_t^T \end{bmatrix}
\end{aligned}
\tag{16.6-4}
$$

where $\mathbf{L}_i$ is of the form given by eq. (16.6-2). If $\psi_i(\lambda) = (\lambda - \lambda_1)^{s_{i1}}(\lambda - \lambda_2)^{s_{i2}} \ldots (\lambda - \lambda_r)^{s_{ir}}$ $(i = 1, 2, \ldots, t)$, where $s_{1j} \geqslant s_{2j} \geqslant \cdots \geqslant s_{tj}$, $(j = 1, 2, \ldots, r)$ then $\mathbf{A}$ is similar to

$$
\mathbf{J} = \begin{bmatrix}
\mathbf{J}_1(\lambda_1) & & & & & & \\
& \ddots & & & & & \\
& & \mathbf{J}_1(\lambda_r) & & & & \\
& & & \mathbf{J}_2(\lambda_1) & & & \\
& & & & \ddots & & \\
& & & & & \mathbf{J}_2(\lambda_r) & \\
& & & & & & \ddots \\
& & & & & & & \mathbf{J}_t(\lambda_r)
\end{bmatrix}
\tag{16.6-5}
$$

where each $\mathbf{J}_i(\lambda_j)$ is of dimensions $s_{ij} \times s_{ij}$. The irreducible factors $(\lambda - \lambda_k)^{s_{ij}}$ of $\psi_i(\lambda)$ are called the *elementary divisors* $(s_{ij} > 0)$.

16.6.5 Deductions from the Jordan Form

For each Jordan box of **A** in the Jordan form there is one and only one eigenvector of **A** corresponding to the eigenvalue of the box. **A** has as many independent eigenvectors as the number of Jordan boxes. **A** is *defective* if it has fewer than n independent eigenvectors. Suppose $\mathbf{z}_{j1}$ is the eigenvector corresponding to λ_j of the box $\mathbf{J}_i(\lambda_j)$ (of dimensions $s_{ij} \times s_{ij}$). Then $(\mathbf{A} - \lambda_j\mathbf{I})\,\mathbf{z}_{j1} = \mathbf{o}, (\mathbf{A} - \lambda_j\mathbf{I})\,\mathbf{z}_{jk} = \mathbf{z}_{j,k-1}\,(k = 2, 3, \ldots, s_{ij}$. The vectors $\mathbf{z}_{j1}, \mathbf{z}_{j2}, \ldots, \mathbf{z}_{js_{ij}}$ are the columns of $\mathbf{S}$: $\mathbf{S}^{-1}\mathbf{AS} = \mathbf{J}$.

If each of the Jordan boxes of **A** consists of only one element, then **A** is similar to a diagonal matrix, and is said to be of *simple structure.*

The rank of $\lambda_i\mathbf{I} - \mathbf{A}$ is equal to $n - c_i$, where c_i is the number of Jordan boxes with the eigenvalues λ_i, or the number of independent eigenvectors of **A** corresponding to λ_i. A matrix is derogatory if for some $i, c_i > 1$; non-derogatory if $c_i = 1$ for all i, (compare 16.6.3). The rank of **A** is at least equal to the number of its nonzero eigenvalues.

Every matrix is similar to a matrix with arbitrarily small off-diagonal elements. Let $\mathbf{D} = \text{diag}\,(1, \delta, \ldots, \delta^{n-1})$, then $(\mathbf{SD})^{-1}\mathbf{A}(\mathbf{SD}) = \mathbf{D}^{-1}\mathbf{JD}$ which has arbitrarily small off-diagonal elements if δ is so.

16.6.6 Schur's Theorem

Schur's theorem states that *every matrix is unitarily similar to a triangular matrix*. It is noted that the triangular matrix is not unique.

From this theorem it follows that if **A** is normal, then $\mathbf{TT}^* = \mathbf{T}^*\mathbf{T}$ where **T** is the triangular matrix in question. Comparison of the corresponding elements on both sides of this equation shows that **T** is actually diagonal. A normal matrix is therefore of simple structure, and its eigenvectors are mutually orthogonal. In fact, a matrix is unitarily similar to a diagonal matrix if and only if it is normal. The circulant matrix $\mathbf{C} = [c_{ij}]_n$ where $c_{ij} = c_{i+1,j+1}$ (indices mod n) is normal. With **W** defined in 16.1.3, it can be shown $\mathbf{W}^*\mathbf{CW} = \text{diag}\,(\lambda_1, \lambda_2, \ldots, \lambda_n)$, where $\lambda_i = \sum_{i=1}^{n} \omega^{(i-1)(j-1)}\,c_{ij}$.

Another consequence of the theorem is the following: Let $\mathbf{A} = [a_{ij}]_n$ have the eigenvalues $\lambda_1, \lambda_2, \ldots, \lambda_n$, then $\sum_{i=1}^{n} |\lambda_i|^2 \leqslant \sum_{i,j=1}^{n} |a_{ij}|$ with equality if and only if **A** is normal. This follows from the fact that: $\text{tr}\,(\mathbf{T}^*\mathbf{T}) = \text{tr}\,(\mathbf{U}^*\mathbf{A}^*\mathbf{UU}^*\mathbf{AU}) = \text{tr}\,(\mathbf{A}^*\mathbf{A})$, and leads to the estimate: $|\lambda_i| \leqslant n \max_{i,j} |a_{ij}|$.

Further deductions can be made by letting $\mathbf{G} = (\mathbf{A} + \mathbf{A}^*)/2 = [g_{ij}]_n$, $\mathbf{H} = (\mathbf{A} - \mathbf{A}^*)/2i = [h_{ij}]_n$, then $(\mathbf{T} + \mathbf{T}^*)/2 = \mathbf{U}^*\mathbf{GU}$, $(\mathbf{T} - \mathbf{T}^*)/2i = \mathbf{U}^*\mathbf{HU}$. Thus $\sum_{i=1}^{n} |\text{Re}(\lambda_i)|^2 \leqslant \sum_{i,j} |g_{ij}|^2$, $\sum_{i=1}^{n} |\text{Im}\,(\lambda_i)|^2 \leqslant \sum_{i,j} |h_{ij}|^2$, and this leads to the estimate: $|\text{Re}(\lambda_i)| \leqslant n \max_{i,j} |g_{ij}|$, $|\text{Im}(\lambda_i)| \leqslant n \max_{i,j} |h_{ij}|$. If **A** is real then the diagonal elements of **H** are zeros and λ_i occur in conjugate pairs, so a better estimate can be obtained for $|\text{Im}(\lambda_i)|$: $|\text{Im}(\lambda_i)|^2 \leqslant [n(n-1)/2] \max_{i,j} |h_{ij}|^2$.

16.6.7 Lanczos Decomposition Theorem

Let $\mathbf{A} = [a_{ij}]_{mn}$ *and be of rank r. Then there exist*: $\mathbf{X} = [x_{ij}]_{mr} = [\mathbf{x}_1, \mathbf{x}_2, \ldots, \mathbf{x}_r]$. $\mathbf{Y} = [y_{ij}]_{nr} = [\mathbf{y}_1, \mathbf{y}_2, \ldots, \mathbf{y}_r]$ *where* $(\mathbf{x}_i, \mathbf{x}_j) = \delta_{ij}$, $(\mathbf{y}_i, \mathbf{y}_j) = \delta_{ij}$ $(i, j = 1, 2, \ldots, r)$ *and* $\mathbf{\Lambda} = \text{diag}\ \{\sigma_1, \sigma_2, \ldots, \sigma_r\}$, *where σ's are the singular values of* **A** *such that* $\mathbf{A} = \mathbf{X\Lambda Y}^*$.

This decomposition theorem follows from the fact that $\mathbf{A}^*\mathbf{A}$ is hermitian and of the same rank as **A**, so that there exists a unitary matrix **U** such that $\mathbf{U}^*\mathbf{A}^*\mathbf{AU} = (\mathbf{AU})^*(\mathbf{AU}) = \text{diag}\ \{\sigma_1^2, \ldots, \sigma_r^2, 0, \ldots, 0\}$. By normalizing the columns of **AU** it is seen that $\mathbf{x}_i$ are the orthonormalized eigenvectors of $\mathbf{AA}^*$ corresponding to the eigenvalues σ_i^2, $(i = 1, 2, \ldots, r)$, and $\mathbf{y}_i$ are the orthonormalized eigenvectors of $\mathbf{A}^*\mathbf{A}$ corresponding to the eigenvalues σ_i^2, $(i = 1, 2, \ldots, r)$.

16.6.8 Functions of a Matrix

Let $f(z)$ be a polynomial of z, and **A** have the eigenvalues λ_i of multiplicities s_i $(i = 1, 2, \ldots, k)$. Let **J** be one Jordan block of **A**, then

$$f(\mathbf{J}) = \begin{bmatrix} f(\lambda_i) & \dfrac{f'(\lambda_i)}{1!} & \dfrac{f''(\lambda_i)}{2!} & \cdots \\ & f(\lambda_i) & \dfrac{f'(\lambda_i)}{1!} & \cdots \\ 0 & & f(\lambda_i) & \cdots \\ & & & \ddots \end{bmatrix} \tag{16.6-6}$$

The function $f(\mathbf{A})$ is thus similar to a block diagonal matrix with blocks $f(\mathbf{J})$ and $f(\mathbf{A})$ has the eigenvalues $f(\lambda_i)$ with multiplicities s_i $(i = 1, 2, \ldots, k)$.

Let $g(z) = a_0 + a_1 z + a_2 z^2 + \cdots$ have the radius of convergence r and the spectral radius of **A** be $\rho(\mathbf{A})$. The matrix power series $g(\mathbf{A}) = a_0\mathbf{I} + a_1\mathbf{A} + a_2\mathbf{A}^2 + \cdots$ is convergent if and only if (a) $\rho(\mathbf{A}) \leqslant r$ and (b) for every eigenvalue λ_i of **A** of multiplicity s_i for which $|\lambda_i| = r$, the power series $f^{(s_i-1)}(\lambda_i)$ is also convergent. If the z-series is convergent in the whole complex plane, then the matrix power series is convergent for arbitrary **A**. We can therefore write

$$e^{\mathbf{A}} = \mathbf{I} + \mathbf{A} + \frac{1}{2!}\mathbf{A}^2 + \cdots$$

$$\ln(\mathbf{I} + \mathbf{A}) = \mathbf{A} - \tfrac{1}{2}\mathbf{A} + \tfrac{1}{3}\mathbf{A}^2 - \cdots \qquad [\rho(\mathbf{A}) < 1]$$

16.6.9 Roots of a Matrix

Let $\mathbf{X}$ be an mth root of $\mathbf{A}$, where $\mathbf{A}$ is of the nth order and is nonsingular, then $\mathbf{X}$ satisfies the equation: $\mathbf{X}^m = \mathbf{A}$, and commutes with $\mathbf{A}$.

(i) Let $\mathbf{J} = \mathbf{S}^{-1}\mathbf{A}\mathbf{S}$, where $\mathbf{J}$ is of the Jordan form: $\mathbf{J} = \{\mathbf{J}_1(\lambda_1), \mathbf{J}_2(\lambda_2), \ldots\}$ and $\mathbf{J}_1(\lambda_1)$ is a Jordan box of eigenvalue λ_1, etc. Then

$$\mathbf{X} = \mathbf{S}\mathbf{K}\,\{\mathbf{J}_1(\lambda_1)^{1/m}, \mathbf{J}_2(\lambda_2)^{1/m}, \ldots\}\,\mathbf{K}^{-1}\mathbf{S}^{-1}$$

where $\mathbf{K}$ is a matrix that commutes with $\mathbf{J}$ and $\mathbf{J}_1(\lambda_1)^{1/m} \cdots$ can be found by expanding $\mathbf{J}_1(\lambda_1)^{1/m} = (\lambda_1 \mathbf{I} + \mathbf{C}_1)^{1/m}$, since this series terminates, $\mathbf{C}_1$ being strictly upper triangular.

(ii) If $\mathbf{A}$ is nonderogatory, then $\lambda_1, \lambda_2, \ldots$ are all distinct, and $\mathbf{K}$ commutes with $\{\mathbf{J}_1(\lambda_1)^{1/m}, \mathbf{J}_2(\lambda_2)^{1/m}, \ldots\}$. In this case

$$\mathbf{X} = \mathbf{S}\,\{\mathbf{J}_1(\lambda_1)^{1/m}, \mathbf{J}_2(\lambda_2)^{1/m}, \ldots\}\,\mathbf{S}^{-1}$$

and every root of $\mathbf{A}$ can be expressed as a polynomial of $\mathbf{A}$.

(iii) The mth root of a singular matrix does not always exist.

Example:

(1) Consider the Jordan box

$$\mathbf{J}(\lambda) = \begin{bmatrix} \lambda & 1 & & \\ & \lambda & 1 & \\ & & \lambda & 1 \\ & & & \lambda \end{bmatrix} = \lambda\mathbf{I} + \mathbf{C}, \qquad (\lambda \neq 0)$$

where $\mathbf{C}$ is strictly upper triangular and has only the elements unity on the first super diagonal. Hence

$$\mathbf{J}(\lambda)^{1/2} = (\lambda\mathbf{I} + \mathbf{C})^{1/2} = \lambda^{1/2}\left(\mathbf{I} + \frac{1}{2\lambda}\mathbf{C} - \frac{1}{8\lambda^2}\mathbf{C}^2 + \frac{1}{16\lambda^3}\mathbf{C}^3\right)$$

$$= \lambda^{1/2}\begin{bmatrix} 1 & \dfrac{1}{2\lambda} & -\dfrac{1}{8\lambda^2} & \dfrac{1}{16\lambda^3} \\ & 1 & \dfrac{1}{2\lambda} & -\dfrac{1}{8\lambda^2} \\ & & 1 & \dfrac{1}{2\lambda} \\ & & & 1 \end{bmatrix}$$

(2) Let

$$\mathbf{A} = \begin{bmatrix} \lambda & 1 & & \\ & \lambda & 1 & \\ & & \lambda & \\ & & & \mu \end{bmatrix} \qquad (\lambda \neq 0, \mu \neq 0)$$

then the square root of **A** is, depending on whether $\lambda \neq \mu$ or $\lambda = \mu$:

$$\begin{bmatrix} \alpha \begin{bmatrix} 1 & \frac{1}{2\lambda} & -\frac{1}{8\lambda^2} \\ & 1 & \frac{1}{2\lambda} \\ & & 1 \end{bmatrix} & \\ & \beta \end{bmatrix}, \qquad (\alpha^2 = \lambda, \beta^2 = \mu \text{ and } \lambda \neq \mu)$$

$$\begin{bmatrix} a & b & c & e \\ & a & b & \\ & & a & \\ & & d & f \end{bmatrix} \begin{bmatrix} \alpha \begin{bmatrix} 1 & \frac{1}{2\lambda} & -\frac{1}{8\lambda^2} \\ & 1 & \frac{1}{2\lambda} \\ & & 1 \end{bmatrix} & \\ & \alpha \end{bmatrix} \begin{bmatrix} a & b & c & e \\ & a & b & \\ & & a & \\ & & d & f \end{bmatrix}^{-1},$$

$$(\alpha^2 = \lambda = \mu \text{ and } a, b, c, d, e, f \text{ arbitrary})$$

(3) Let

$$\mathbf{A} = \begin{bmatrix} 0 & 1 & 0 \\ 0 & 0 & 1 \\ 0 & 0 & 0 \end{bmatrix}.$$

Suppose it has a square root **X**. We reduce **X** to the Jordan form: $\mathbf{X} = \mathbf{SJS}^{-1}$. Since all the eigenvalues of **X** are zero, **J** can assume one of the following forms:

$$\begin{bmatrix} 0 & & \\ & 0 & \\ & & 0 \end{bmatrix}, \begin{bmatrix} 0 & 1 & \\ & 0 & \\ & & 0 \end{bmatrix}, \begin{bmatrix} 0 & 1 & \\ & 0 & 1 \\ & & 0 \end{bmatrix}$$

The first form shows $\mathbf{X} = \mathbf{O}$, and the second form leads to $\mathbf{X}^2 = \mathbf{O}$. In the third one

$$\mathbf{X}^2 = \mathbf{S}\begin{bmatrix} 0 & 0 & 1 \\ 0 & 0 & 0 \\ 0 & 0 & 0 \end{bmatrix}\mathbf{S}^{-1} = \mathbf{S}\begin{bmatrix} 1 & 0 & 0 \\ 0 & 0 & 1 \\ 0 & 1 & 0 \end{bmatrix}\begin{bmatrix} 0 & 1 & 0 \\ 0 & 0 & 0 \\ 0 & 0 & 0 \end{bmatrix}\begin{bmatrix} 1 & 0 & 0 \\ 0 & 0 & 1 \\ 0 & 1 & 0 \end{bmatrix}\mathbf{S}^{-1}$$

Since the Jordan form of $\mathbf{X}^2$ is different from $\mathbf{A}$, we conclude $\mathbf{A}$ has no square root.

16.7 ESTIMATES AND DETERMINATION OF EIGENVALUES

16.7.1 Gershgorin-type Estimates of Eigenvalues

If G_i is the ith *Gershgorin disk*: $|z - a_{ii}| \leqslant \sum_{\substack{j=1 \\ \neq i}}^{n} |a_{ij}|$ in the complex plane of z for the matrix $\mathbf{A} = [a_{ij}]_n$, then the eigenvalues of $\mathbf{A}$ lie in the union of G_i, $(i = 1, 2, \ldots, n)$.

If there is a particular disk isolated from the remaining disks, then this isolated disk contains exactly one eigenvalue of $\mathbf{A}$. More generally, if the union of k disks is isolated from the union of the remaining $n - k$ disks, then the union of k disks contains exactly k eigenvalues, counting multiple eigenvalues.

Let O_{ij} be the *Cassini ovals* defined by $|z - a_{ii}||z - a_{jj}| \leqslant \sum_{\substack{l=1 \\ \neq i}}^{n} |a_{il}| \sum_{\substack{k=1 \\ \neq j}}^{n} |a_{jk}|$ then the eigenvalues of $\mathbf{A}$ lie in the union of the $n(n-1)/2$ ovals, $(i, j = 1, 2, \ldots, n)$. It is noticed that the union of the ovals are contained in the union of the Gershgorin disks, or at most they touch at the boundaries. If $\sum_{\substack{l=1 \\ \neq i}}^{n} |a_{il}| \sum_{\substack{k=1 \\ \neq j}}^{n} |a_{jk}| < |a_{ii} - a_{jj}|/2$, then the oval separates into two ovals centered at a_{ii} and a_{jj}.

16.7.2 Tridiagonalization

Since it takes comparatively less effort to evaluate the eigenvalues of a tridiagonal matrix, methods have been devised to transform a given matrix (square) to the tridiagonal form while at the same time all the eigenvalues are preserved. The methods of Givens and Householder employ a finite sequence of unitarily similar transformations, which when applied to hermitian matrices, will reduce them to the tridiagonal form. When applied to arbitrary matrices, they will be reduced to the Hessenberg form. The algorithms of Lanczos (finite) reduce an arbitrary matrix to the tridiagonal form.

16.7.3 Given's Method

In this method a sequence of plane rotational similarity transformations are applied to $\mathbf{A} = [a_{ij}]_n$ such that a zero once created by one similarity transformation will not be destroyed by a subsequent transformation. The zeros are created in the following order: $(3, 1), (4, 1), \ldots, (n, 1); (4, 2), (5, 2), \ldots, (n, 2); (5, 3), \ldots,$

$(n, 3), \ldots, (n, n-2)$ by using rotational similarity transformations: $\mathbf{U}_{31}, \mathbf{U}_{41}, \ldots,$ $\mathbf{U}_{41}, \mathbf{U}_{42}, \ldots, \mathbf{U}_{n,n-2}$ (16.4.6). Suppose $\mathbf{A}$ is already reduced to $\mathbf{A}_r$ of the form

$$\mathbf{A}_r = \begin{bmatrix} \mathbf{A}_r^{(1)} & \mathbf{A}_r^{(2)} \\ \mathbf{A}_r^{(3)} & \mathbf{A}_r^{(4)} \end{bmatrix} \tag{16.7-1}$$

where $\mathbf{A}_r^{(1)}$ is of $r \times r$ and $\mathbf{A}_r^{(3)}$ is zero everywhere except the $(r+1, r)^{\text{th}}$ entry. Let the first column of $\mathbf{A}_r^{(4)}$ be $(a_{r+1,r+1}, a_{r+2,r+1}, \ldots, a_{nr})^T$, then putting: $\mathbf{U}_{q,r+1} = [u_{ij}]_n$, $u_{ij} = \delta_{ij}$ except $u_{r+2,r+2} = c$, $u_{qq} = \bar{c}$, $u_{r+2,q} = -\bar{s}$, $u_{q,r+2} = s$ where $c = a_{r+2,r+1}/\theta$, $s = a_{q,r+1}/\theta$ if $\theta = \sqrt{|a_{r+2,r+1}|^2 + |a_{q,r+1}|^2} \neq 0$ and $c = 1, s = 0$ if $\theta = 0$, $(q = r+3, r+4, \ldots, n)$, in the product $\mathbf{U}_{n,r+1}^* \cdots \mathbf{U}_{r+3,r+1}^* \mathbf{A}_r \mathbf{U}_{r+3,r+1} \cdots \mathbf{U}_{n,r+1}$ zeros are created in the positions: $(r+3, r+1), \ldots, (n, r+1)$. If $\mathbf{A}$ is hermitian, the final result is a real tridiagonal matrix.

16.7.4 Householder's Method

Here the similarity transformations are applied in the sequence: $\mathbf{H}^{(1)}, \mathbf{H}^{(2)}, \ldots,$ $\mathbf{H}^{(n-2)}$ (16.4.6), and zeros are created in the positions: (3, 1), (4, 1), ..., (n, 1); (4, 2), (5, 2), ..., (n, 2), ..., (n − 2, n). After $r-1$ steps $\mathbf{A}$ is reduced to $\mathbf{A}_r$ $(\mathbf{A}_1 = \mathbf{A})$, so that

$$\mathbf{A}_r = \begin{bmatrix} \mathbf{A}_r^{(1)} & \mathbf{A}_r^{(2)} \\ \mathbf{A}_r^{(3)} & \mathbf{A}_r^{(4)} \end{bmatrix} \tag{16.7-2}$$

where $\mathbf{A}_r^{(1)}$ is of dimensions $r \times r$ and is in Hessenberg form, $\mathbf{A}_r^{(3)}$ consists of all zeros except the rth column. Let the rth column of $\mathbf{A}_r^{(3)}$ be $\mathbf{a}^T = (a_{r+1,r}, a_{r+2,r}, \ldots, a_{nr})$, then in the next step $\mathbf{A}_{r+1} = \mathbf{H}^{(r)*}\mathbf{A}_r\mathbf{H}^{(r)}$, $\mathbf{H}^{(r)} = \mathbf{I} - 2\,\mathbf{w}^{(r)}\,\mathbf{w}^{(r)*}$, where

$$\begin{cases} \mathbf{w}^T = \dfrac{1}{2K}\left(0, \ldots, 0, a_{r+1,r}\left(1 + \dfrac{S}{|a_{r+1,r}|}\right), a_{r+2,r}, \ldots, a_{nr}\right) \\ |K|^2 = \frac{1}{2}\,(S^2 + S|a_{r+1,r}|) \\ S^2 = \mathbf{a}^*\mathbf{a} \end{cases} \tag{16.7-3}$$

this column is reduced to $(-S\,a_{r+1,r}/|a_{r+1,r}|, 0, \ldots, 0)^T$. The process can then be continued. If $\mathbf{A}$ is hermitian, the result is a tridiagonal hermitian matrix.

16.7.5 Lanczos's Algorithms

Let $\mathbf{A} = [a_{ij}]_n$; by starting with two arbitrary vectors ξ_1, η_1, two systems of vectors: ξ_i, η_i $(i = 2, 3, \ldots)$ are constructed according to the following recursive relations:

$$\begin{cases} \xi_2 = \mathbf{A}\,\xi_1 - \gamma_1\xi_1, \\ \eta_2 = \mathbf{A}^*\eta_1 - \bar{\gamma}_1\eta_1, \end{cases} \quad \begin{cases} \xi_{l+1} = \mathbf{A}\xi_l - \gamma_l\xi_l - \beta_{l-1,l}\xi_{l-1} \\ \eta_{l+1} = \mathbf{A}^*\eta_l - \bar{\gamma}_l\eta_l - \bar{\beta}_{l-1,l}\eta_{l-1} \end{cases} \quad (l = 2, 3, \ldots)$$

(16.7-3)

where

$$\gamma_l = \frac{(\mathbf{A}\xi_l, \eta_l)}{(\xi_l, \eta_l)}; \quad \beta_{l-1,l} = \frac{(\mathbf{A}\xi_l, \eta_{l-1})}{(\xi_{l-1}, \eta_{l-1})} \tag{16.7-4}$$

It can be shown that $\xi_i, \eta_i (i = 1, 2, \ldots)$ are biorthogonal. The above formulas are equivalent to the application of the biorthogonalization algorithm of (16.3.6) to the two systems: $\xi_1, \mathbf{A}\xi_1, \mathbf{A}^2\xi_1, \ldots; \eta_1, \mathbf{A}^*\eta_1, \mathbf{A}^{*2}\eta_1, \ldots$. The algorithms have to terminate if one of the denominators in eq. (16.7-4) becomes zero; however, it can be shown that if the degree of the minimal polynomial of $\mathbf{A}$ is m, then the algorithms terminate in at most m steps because $\xi_{m+1} = \eta_{m+1} = \mathbf{0}$, and it is always possible to choose ξ_1, η_1 such that $(\xi_l, \eta_l) \neq 0$, $(l = 2, 3, \ldots, m - 1)$. If $m = n$, and the algorithms terminate normally, then $\mathbf{A} = \mathbf{XSX}^{-1}$, where $\mathbf{X} = [\xi_1, \xi_2, \ldots, \xi_n]$, $\mathbf{S}$ is tridiagonal, $\mathbf{S} = [s_{ij}]_n$, $s_{ii} = \beta_{ii}$, $(i = 1, 2, \ldots, n)$, $s_{i,i+1} = \beta_{i,i+1}$, $s_{i+1,i} = 1$, $(i = 1, 2, \ldots, m - 1)$, $s_{ij} = 0$, $(|i - j| > 1)$.

If the algorithms break down before the nth step, it is possible, in certain cases, to continue it without going back to change ξ_1, η_1; if $m < n$, it is also possible to continue the process.

16.7.6 QR Algorithm

It is generally recognized that the QR algorithm is probably one of the most effective methods ever developed for the evaluation of matrix eigenvalues. It is closely related to the LR algorithm discovered earlier.

A given matrix $\mathbf{A}(=\mathbf{A}_1)$ is first decomposed as the product $\mathbf{Q}_1\mathbf{R}_1$ where $\mathbf{Q}_1$ is unitary and $\mathbf{R}_1$ upper triangular. The order of multiplication is then reversed to form $\mathbf{R}_1\mathbf{Q}_1 = \mathbf{A}_2$. The process is repeated on $\mathbf{A}_2$: $\mathbf{A}_2 = \mathbf{Q}_2\mathbf{R}_2$, $\mathbf{R}_2\mathbf{Q}_2 = \mathbf{A}_3 = \mathbf{Q}_3\mathbf{R}_3$, etc. A sequence of unitarily similar matrices: $\mathbf{A}_1, \mathbf{A}_2 = \mathbf{Q}_1^*\mathbf{A}_1\mathbf{Q}_1, \mathbf{A}_3 = \mathbf{Q}_2^*\mathbf{A}_2\mathbf{Q}_2, \ldots$ are then constructed. It has been shown that the following basic results justify the algorithm:

(i) The QR transformation preserves the band width of hermitain matrices. It also preserves the upper Hessenberg form; thus an arbitrary matrix may be first reduced to the upper Hessenberg form to which the QR algorithm is then applied with considerable savings in computation.

(ii) If $\mathbf{A}$ is nonsingular with eigenvalues all of distinct modulus, then as $k \to \infty$, $\mathbf{A}_k$ converges to an upper triangular matrix, the diagonal elements of which are the eigenvalues of $\mathbf{A}$. If $\mathbf{A}$ has a number of eigenvalues of equal modulus, then certain principal submatrix of $\mathbf{A}_k$ need not converge but their eigenvalues will converge to these eigenvalues of equal modulus in the limit.

16.8 NORMS

16.8.1 Vector and Matrix Norms

The *norm* of a vector: $||\mathbf{x}||$ is a scalar satisfying the following properties:

(i) $||\mathbf{x}|| > 0$ unless $\mathbf{x} = \mathbf{0}$,

(ii) $\|k\mathbf{x}\| = |k|\,\|\mathbf{x}\|$, k being a scalar,
(iii) $\|\mathbf{x}+\mathbf{y}\| \leqslant \|\mathbf{x}\| + \|\mathbf{y}\|$.

Examples: (vector norms)
(1) The Hölder norm: $\|\mathbf{x}\|_{(p)} = \left\{\sum_j |x_j|^p\right\}^{1/p}$, $(p \geqslant 1)$.
(2) The e-norm*: $\|\mathbf{x}\|_e = \max_j |\mathbf{x}_j|$.
(3) The e'-norm: $\|\mathbf{x}\|_{e'} = \sum_j |\mathbf{x}_j|$, being a special case of (1), $p = 1$.
(4) The euclidean norm: $\|\mathbf{x}\|_E$, being a special case of (1), $p = 2$.

The norm of a square matrix satisfies the following properties:

(i) $\|\mathbf{A}\| > 0$ unless $\mathbf{A} = \mathbf{0}$,
(ii) $\|k\mathbf{A}\| = |k|\,\|\mathbf{A}\|$ for any scalar k,
(iii) $\|\mathbf{A}+\mathbf{B}\| \leqslant \|\mathbf{A}\| + \|\mathbf{B}\|$, and
(iv) $\|\mathbf{AB}\| \leqslant \|\mathbf{A}\|\,\|\mathbf{B}\|$.

Examples: (matrix norms, for $\mathbf{A} = [a_{ij}]_n$)
(1) $\|\mathbf{A}\|_M = n \max_{i,j} |a_{ij}|$.
(2) The Hölder norm: $\|\mathbf{A}\|_{(p)} = \left\{\sum_{i,j} |a_{ij}|^p\right\}^{1/p}$, $(1 \leqslant p \leqslant 2)$.
(3) The euclidean norm: $\|\mathbf{A}\|_E = \{\text{tr}\,(\mathbf{A}^*\mathbf{A})\}^{1/2}$, being a special case of (2), $p = 2$. If $\mathbf{U}$ is unitary, then $\|\mathbf{UA}\|_E = \|\mathbf{A}\|_E = \|\mathbf{A}^*\|_E = \|\mathbf{AU}\|_E$.
(4) The expansion norm: $\|\mathbf{A}\|_{\varphi,\psi} = \max_{\|\mathbf{x}\|_\psi = 1} \|\mathbf{Ax}\|_\varphi$.

16.8.2 Consistent and Subordinate Norms

If for a particular norm $\|\mathbf{Ax}\| \leqslant \|\mathbf{A}\|\,\|\mathbf{x}\|$ for all $\mathbf{A}$ and $\mathbf{x}$, the matrix norm is *consistent* with the vector norm. Given a vector norm φ, the matrix norm constructed according to $\|\mathbf{A}\|_\varphi = \max_{\|\mathbf{x}\|_\varphi = 1} \|\mathbf{Ax}\|_\varphi$ (16.8.1, expansion norm $\varphi = \psi$) is said to be *subordinate* to the vector norm. It is always consistent with the vector norm, and is no greater than any other matrix norms consistent with the vector norm.

Examples:
(1) The matrix norm subordinate to the vector euclidean norm is its largest singular value, sometimes called the *spectrum norm*, $\|\mathbf{A}\|_2 = \max_i \sqrt{\lambda_i(\mathbf{A}^*\mathbf{A})} = \max_i \sigma_i(\mathbf{A})$. If $\mathbf{U}$ is unitary, then $\|\mathbf{U}^*\mathbf{AU}\|_2 = \|\mathbf{A}\|_2$.
(2) The matrix norm subordinate to the vector e-norm is $\|\mathbf{A}\|_e = \max_i \sum_j |a_{ij}|$.
(3) The matrix norm subordinate to the vector e'-norm is $\|\mathbf{A}\|_{e'} = \max_j \sum_i |a_{ij}|$.

*The e-norm may also be interpreted as a special case of (1), for $p = \infty$, also called the *maximum norm*.

16.8.3 Inequalities

Following is a list of norm inequalities:

1. Vector norms–Let $\mathbf{x} \in \mathbf{C}_n$, then

$$\begin{cases} \|\mathbf{x}\|_e \leqslant \|\mathbf{x}\|_E \leqslant \|\mathbf{x}\|_{e'} \\ \|\mathbf{x}\|_e \leqslant \|\mathbf{x}\|_E \leqslant \sqrt{n}\,\|\mathbf{x}\|_e \\ \dfrac{1}{\sqrt{n}}\|\mathbf{x}\|_{e'} \leqslant \|\mathbf{x}\|_E \leqslant \|\mathbf{x}\|_{e'} \end{cases}$$

2. Matrix norms–Let $\mathbf{A} = [a_{ij}]_n$, then

$$\begin{cases} \dfrac{1}{n}\|\mathbf{A}\|_M \leqslant \|\mathbf{A}\|_{(p)} \leqslant n^{(2/p)-1}\,\|\mathbf{A}\|_M & (1 \leqslant p \leqslant 2) \\ \|\mathbf{A}\|_{(p)} \leqslant \|\mathbf{A}\|_{(1)} \leqslant n^{2-2/p}\,\|\mathbf{A}\|_{(p)} & (1 \leqslant p \leqslant 2) \end{cases}$$

$$\begin{cases} \dfrac{1}{\sqrt{n}}\|\mathbf{A}\|_2 \leqslant \|\mathbf{A}\|_e; \quad \|\mathbf{A}\|_{e'} \leqslant \sqrt{n}\,\|\mathbf{A}\|_2 \\ \dfrac{1}{n}\|\mathbf{A}\|_M \leqslant \|\mathbf{A}\|_e; \quad \|\mathbf{A}\|_{e'} \leqslant \|\mathbf{A}\|_M \\ \dfrac{1}{\sqrt{n}}\|\mathbf{A}\|_E \leqslant \|\mathbf{A}\|_e; \quad \|\mathbf{A}\|_{e'} \leqslant \sqrt{n}\,\|\mathbf{A}\|_E \end{cases}$$

$$\begin{cases} \dfrac{1}{n}\|\mathbf{A}\|_M \leqslant \|\mathbf{A}\|_2 \leqslant \|\mathbf{A}\|_M \\ \dfrac{1}{\sqrt{n}}\|\mathbf{A}\|_E \leqslant \|\mathbf{A}\|_2 \leqslant \|\mathbf{A}\|_E \\ \|\mathbf{A}\|_2 \leqslant \{\|\mathbf{A}\|_e\|\mathbf{A}\|_{e'}\}^{1/2} \end{cases}$$

$$\begin{cases} \dfrac{1}{n}\|\mathbf{A}\|_M \leqslant \|\mathbf{A}\|_E \leqslant \|\mathbf{A}\|_M \\ \|\mathbf{A}\|_2 \leqslant \|\mathbf{A}\|_E \leqslant \sqrt{n}\,\|\mathbf{A}\|_2 \end{cases}$$

The spectral radius of a matrix cannot exceed any of its norms.

16.8.4 Boundedness of Matrix Powers

The $\lim_{m \to \infty} \mathbf{A}^m = 0$ if and only if $\rho(\mathbf{A}) < 1$ ($\rho(\mathbf{A})$, the spectral radius of $\mathbf{A}$). $\mathbf{A}^m$ remains bounded if and only if $\rho(\mathbf{A}) \leqslant 1$, and corresponding to all eigenvalues of $\mathbf{A}$ of modulus one, there are as many independent eigenvectors. A sufficient condition for $\lim_{m \to \infty} \mathbf{A}^m = 0$ is that $\|\mathbf{A}\|_\varphi < 1$, where φ can be any matrix norm.

Let $\mathbf{A} = [a_{ij}]_n$, and s the maximum multiplicity of its eigenvalues in the minimal polynomial of $\mathbf{A}$. Then there exists a constant $K > 0$ depending only on $\mathbf{A}$ such that $\left\{\sum_{i=1}^{n} |\lambda_i|^{2m}\right\}^{1/2} \leqslant \|\mathbf{A}^m\|_E \leqslant K m^{s-1} \left\{\sum_{i=1}^{n} |\lambda_i|^{2m}\right\}^{1/2}$. If all $\lambda_i = 0$, then $\|\mathbf{A}^m\|_E = 0$ for $m \geqslant s$. Furthermore, if $\mathbf{A}$ is nonsingular, then $n\|\mathbf{A}\|_E^{-1} \leqslant \|\mathbf{A}^{-1}\|_E \leqslant |\det \mathbf{A}|^{-1} (1/n)^{(n-2)/2} \|\mathbf{A}\|_E^{n-1}$, where the equality is attained when $\mathbf{A}$ is a constant multiple of a unitary matrix.

If the field of values of $\mathbf{A}$ lie in the unit disk, $|(\mathbf{Ax}, \mathbf{x})| \leqslant 1$, $\|\mathbf{x}\|_E = 1$, then there exists a constant K such that $\|\mathbf{A}^m\|_E \leqslant K$, $(m = 1, 2, \ldots)$.

16.9 HERMITIAN FORMS AND MATRICES

16.9.1 Quadratic and Hermitian Forms

The expression $\mathbf{z}^*\mathbf{Hz}$ where $\mathbf{H}$ is hermitian (square and of order n) is a *hermitian form*, which is real for all $\mathbf{z}$. If $\mathbf{z} = \mathbf{Qu}$, where $\mathbf{Q}$ is nonsingular, then $\mathbf{z}^*\mathbf{Hz} = \mathbf{u}^*\mathbf{Q}^*\mathbf{HQu} = \mathbf{u}^*\mathbf{Ku}$, and $\mathbf{H}$ and $\mathbf{K} = \mathbf{Q}^*\mathbf{HQ}$ are said to be *conjunctive*. The real counterpart of hermitian forms is the *quadratic form*: $\mathbf{x}^T\mathbf{Ax}$ where $\mathbf{A}$ can be taken as real symmetric. Again $\mathbf{A}$ and $\mathbf{B}$ are said to be *congruent* if there is a nonsingular $\mathbf{Q}$ (real) such that $\mathbf{B} = \mathbf{Q}^T\mathbf{AQ}$. Quadratic forms are special cases of hermitian forms.

Since $\mathbf{H}$ is unitarily similar to a diagonal matrix, we have: $\mathbf{z}^*\mathbf{Hz} = \mathbf{u}^*\mathbf{U}^*\mathbf{HUu} = \mathbf{u}^*\mathbf{Du} = \sum_i \lambda_i |u_i|^2 = \sum_i (\pm)(\sqrt{|\lambda_i|}\, u_i)^2$, and the hermitian form is reduced to a sum of squares. The eigenvalues of $\mathbf{H}$ are given by λ_i, $(i = 1, 2, \ldots, n)$. Let the number of positive and negative eigenvalues be respectively denoted by p and q, then the rank of $\mathbf{H}$ or of the hermitian form is $p + q$, the *index* of $\mathbf{H}$ or of the hermitian form is p and $p - q$ is its *signature*. Conjunctive hermitian matrices have the same rank and signature.

The hermitian form $\mathbf{z}^*\mathbf{Hz}$ is *positive (negative) definite* if it is positive (negative) unless $\mathbf{z} = \mathbf{0}$. It is *positive (negative) semi-definite* if it is nonnegative (nonpositive) for all $\mathbf{z} \neq \mathbf{0}$. If neither definite or semi-definite, it is *indefinite*. The matrices $\mathbf{A}^*\mathbf{A}$, $\mathbf{AA}^*$ are positive semi-definite for all $\mathbf{A}$.

A positive (negative) definite hermitian form has all positive (negative) eigenvalues. A positive (negative) semi-definite hermitian form has all non-negative (non-positive) eigenvalues, and its rank less than n. An indefinite form has both positive and negative eigenvalues. Let the leading principal minors of $\mathbf{H}$ be $D_i = \det \mathbf{H}(1, 2, \ldots, i)$, then $\mathbf{z}^*\mathbf{Hz}$ is positive definite if and only if $D_i > 0$, $(i = 1, 2, \ldots, n)$; $\mathbf{z}^*\mathbf{Hz}$ is negative definite if and only if $(-1)^i D_i > 0$, $(i = 1, 2, \ldots, n)$; $\mathbf{z}^*\mathbf{Hz}$ is positive semi-definite if and only if $\det \mathbf{H} = 0$ and *all* principal minors of $\mathbf{H}$ are nonnegative: $\det \mathbf{H}(i_1, i_2, \ldots, i_s) \geqslant 0$, $(i_1, i_2, \ldots, i_s = 1, 2, \ldots, n; s = 1, 2, \ldots, n)$; and $\mathbf{z}^*\mathbf{Hz}$ is negative semi-definite if and only if $\det \mathbf{H} = 0$ and

$$(-1)^s \det \mathbf{H}(i_1, i_2, \ldots, i_s) \geqslant 0, (i_1, i_2, \ldots, i_s = 1, 2, \ldots, n; s = 1, 2, \ldots, n)$$

Example: Let

$$\mathbf{H} = \begin{bmatrix} 1 & 0 & 0 & 1 \\ 0 & 1 & 1 & 0 \\ 0 & 1 & 1 & 0 \\ 1 & 0 & 0 & 0 \end{bmatrix}$$

then det $\mathbf{H}(1) = 1$, det $\mathbf{H}(1, 2) = 1$, det $\mathbf{H}(1, 2, 3) = 0$, det $\mathbf{H}(1, 2, 3, 4) = 0$. But $\mathbf{H}$ has eigenvalues: $0, 2, (1 \pm \sqrt{5})/2$, so that $\mathbf{H}$ is not positive semi-definite. In fact det $\mathbf{H}(1, 4) = -1 < 0$. The condition that every leading principal minor of $\mathbf{H}$ is nonnegative is not enough to guarantee that $\mathbf{H}$ be positive semi-definite.

Two hermitian forms $\mathbf{z}^*\mathbf{Hz}$ and $\mathbf{z}^*\mathbf{Kz}$ can be simultaneously reduced to the sum of squares respectively as $\sum_i \lambda_i |u_i|^2$ and $\sum_i |u_i|^2$ where λ_i are the roots of det $(\mathbf{H} - \lambda\mathbf{K}) = 0$.

Let $\mathbf{H} = [h_{ij}]_n$, $\mathbf{G} = [g_{ij}]_n$ be positive definite hermitian. Then $\mathbf{J} = [h_{ij} + g_{ij}]_n$, $\mathbf{K} = [h_{ij}g_{ij}]_n$ are also positive definite hermitian.

16.9.2 Eigenvalues of Hermitian Matrices

Let $\mathbf{H}$, $\mathbf{G}$ be two hermitian matrices, and one of them nonsingular, then the eigenvalues of $\mathbf{HG}$ are all real. Similarly, let $\mathbf{H}$ be hermitian and $\mathbf{S}$ skew-hermitian, and one of them nonsingular, then the eigenvalues of $\mathbf{HS}$ are all real.

Root separation theorem: let $\mathbf{H} = [h_{ij}]_n$ *be hermitian with eigenvalues* $\lambda_1 \leqslant \lambda_2 \leqslant \cdots \leqslant \lambda_n$, *and the eigenvalues of any of its principal submatrix of order* $n - 1$ *be* $\mu_1 \leqslant \mu_2 \leqslant \cdots \leqslant \mu_{n-1}$, *then* $\lambda_1 \leqslant \mu_1 \leqslant \lambda_2 \leqslant \mu_2 \leqslant \cdots \leqslant \mu_{n-1} \leqslant \lambda_n$. This leads to the following estimate of the eigenvalues of $\mathbf{H}$: let $\sigma_k^{(\nu)}$ be the sum of the moduli of all off-diagonal elements of the kth row of any principal submatrix of order ν, and $h_k^{(\nu)}$ its kth diagonal element, then $\min_k \{h_k^{(n-\nu+1)} - \sigma_k^{(n-\nu+1)}\} \leqslant \lambda_\nu \leqslant \max_k \{h_k^{(\nu)} + \sigma_k^{(\nu)}\}$. A special case is $\lambda_1 \leqslant \min_i \{h_{ii}\}$ and $\lambda_n \geqslant \max_i \{h_{ii}\}$.

Let the eigenvalues of a positive definite hermitian matrix $\mathbf{H}$ be ordered $\lambda_1 \leqslant \lambda_2 \leqslant \cdots \leqslant \lambda_n$ and $\zeta_1, \zeta_2, \cdots, \zeta_k$ be k orthonormal vectors. Let $\Gamma(\xi_1, \xi_2, \ldots, \xi_k | \eta_1, \eta_2, \ldots, \eta_k) = \det [\gamma_{ij}]_k$, $\gamma_{ij} = (\xi_i, \eta_i)$. Then, as ζ's vary:

$$\max \Gamma(\zeta_1, \zeta_2, \ldots, \zeta_k | \mathbf{H}\zeta_1, \mathbf{H}\zeta_2, \ldots, \mathbf{H}\zeta_k) = \lambda_{n-k+1}\lambda_{n-k+2} \ldots \lambda_n$$

$$\min \Gamma(\zeta_1, \zeta_2, \ldots, \zeta_k | \mathbf{H}\zeta_1, \mathbf{H}\zeta_2, \ldots, \mathbf{H}\zeta_k) = \lambda_1\lambda_2 \ldots \lambda_k \qquad (k \leqslant n)$$

Also,

$$\max \prod_{i=1}^{k} (\zeta_i, \mathbf{H}\zeta_i) = \{(\lambda_{n-k+1} + \lambda_{n-k+2} + \cdots + \lambda_n)/k\}^k$$

$$\min \prod_{i=1}^{k} (\zeta_i, \mathbf{H}\zeta_i) = \lambda_1\lambda_2 \cdots \lambda_k \qquad (k \leqslant n)$$

The following is true for any hermitian $\mathbf{H}$:

$$\max \sum_{i=1}^{k} (\zeta_i, \mathbf{H}\zeta_i) = \lambda_{n-k+1} + \lambda_{n-k+2} + \cdots + \lambda_n$$

$$\min \sum_{i=1}^{k} (\zeta_i, \mathbf{H}\zeta_i) = \lambda_1 + \lambda_2 + \cdots + \lambda_k$$

Let $\mathbf{A}$ be arbitrary, with eigenvalues $|\lambda_1| \leqslant |\lambda_2| \leqslant \cdots \leqslant |\lambda_n|$, and singular values $\sigma_1 \leqslant \sigma_2 \leqslant \cdots \leqslant \sigma_n$. Then

$$|\lambda_n \lambda_{n-1} \cdots \lambda_{n-k+1}| \leqslant \sigma_n \sigma_{n-1} \cdots \sigma_{n-k+1} \qquad (k = 1, 2, \ldots, n)$$

$$|\lambda_n| + |\lambda_{n-1}| + \cdots + |\lambda_{n-k+1}| \leqslant \sigma_n + \sigma_{n-1} + \cdots + \sigma_{n-k+1} \qquad (k = 1, 2, \ldots, n)$$

Let $\mathbf{H}_1$, $\mathbf{H}_2$ be hermitian with eigenvalues respectively $\lambda_1 \leqslant \lambda_2 \leqslant \cdots \leqslant \lambda_n$ and $\mu_1 \leqslant \mu_2 \leqslant \cdots \leqslant \mu_n$; and let $\mathbf{H}_3 = \mathbf{H}_1 + \mathbf{H}_2$ with eigenvalues $\nu_1 \leqslant \nu_2 \leqslant \cdots \leqslant \nu_n$. Then (a) $\sum_i |\lambda_i - \mu_i|^2 \leqslant \|\mathbf{H}_1 - \mathbf{H}_2\|_E^2$; and (b) $\mu_1 \leqslant \nu_i - \lambda_i \leqslant \mu_n$ $(i = 1, 2, \ldots, n)$.

If $\mathbf{H}$ is positive definite $\lambda_1 \leqslant \lambda_2 \leqslant \cdots \leqslant \lambda_n$, and x, y orthonormal, then $|\mathbf{x}^*\mathbf{H}\mathbf{y}|^2/(\mathbf{x}^*\mathbf{H}\mathbf{x})(\mathbf{y}^*\mathbf{H}\mathbf{y}) \leqslant (\lambda_1 - \lambda_n)^2/(\lambda_1 + \lambda_n)^2$, and $1 \leqslant (\mathbf{A}\mathbf{x}, \mathbf{x})(\mathbf{A}^{-1}\mathbf{x}, \mathbf{x}) \leqslant \frac{1}{4}\{(\lambda_1/\lambda_n)^{1/2} + (\lambda_n/\lambda_1)^{1/2}\}^2$.

16.9.3 Polar Decomposition

If $\mathbf{U}$ is unitary, then there exists a hermitian matrix $\mathbf{F}$ such that $\mathbf{U} = e^{iF}$. In fact, since $\mathbf{W}^*\mathbf{U}\mathbf{W} = \text{diag}\,\{e^{i\theta_1}, e^{i\theta_2}, \ldots, e^{i\theta_n}\}$, where θ's are real and $\mathbf{W}$ is unitary, we can write $\mathbf{F} = \mathbf{W}\,\text{diag}\,\{\theta_1, \theta_2, \ldots, \theta_n\}\,\mathbf{W}^*$.

For every square matrix $\mathbf{A}$ there exist two positive semi-definite hermitian matrices $\mathbf{H}$ and $\mathbf{K}$ and two unitary matrices $\mathbf{U}$ and $\mathbf{V}$ such that $\mathbf{A} = \mathbf{HU} = \mathbf{VK}$, where $\mathbf{H} = (\mathbf{AA}^*)^{1/2}$, $\mathbf{K} = (\mathbf{A}^*\mathbf{A})^{1/2}$, and if $\mathbf{A}$ is nonsingular, $\mathbf{U}$, $\mathbf{V}$ are uniquely determined: $\mathbf{U} = \mathbf{H}^{-1}\mathbf{A}$, $\mathbf{V} = \mathbf{K}^{-1}\mathbf{A}^*$. Moreover, $\mathbf{A}$ is normal if and only if $\mathbf{HU} = \mathbf{UH}$ and $\mathbf{KV} = \mathbf{VK}$. We can then write $\mathbf{A} = \mathbf{H}e^{iF} = e^{iG}\mathbf{K}$, where $\mathbf{F}$, $\mathbf{G}$ are hermitian.

16.10 MATRICES WITH REAL ELEMENTS

16.10.1 Nonnegative Matrices

A matrix is *positive* $(\mathbf{A} > \mathbf{0})$ or *nonnegative* $(\mathbf{A} \geqslant \mathbf{0})$ if its elements $a_{ij} > 0$ or $a_{ij} \geqslant 0$ for all i, j.

Perron's theorem: If $\mathbf{A} = [a_{ij}]_n > \mathbf{0}$, *it has a (real) positive eigenvalue* ω *which is larger than the modulus of any other eigenvalue. This eigenvalue* ω *is simple, and the corresponding eigenvector* $\mathbf{z} > \mathbf{0}$.

Frobenius theorem: If $\mathbf{A} = [a_{ij}]_n \geqslant \mathbf{0}$ *and is irreducible, then* $\mathbf{A}$ *has a positive eigenvalue* ω *which is simple, and if* λ_i *is any of the remaining eigenvalues, then*

$\omega \geqslant |\lambda_i|$. This eigenvalue ω also increases if any a_{ij} increases. The eigenvector **z** corresponding to ω is also positive. In case there are h eigenvalues having the same modulus ω: $\omega_1 \lambda_1, \ldots, \lambda_{h-1}$, then they are the roots of the equation: $\lambda^h - \omega^h = 0$. If $h > 1$ then there is a permutation matrix **P** such that

$$\mathbf{P}^T\mathbf{A}\mathbf{P} = \begin{bmatrix} 0 & A_{12} & 0 & \cdots & 0 \\ 0 & 0 & A_{23} & \cdots & 0 \\ & & & \cdot & \\ \cdots & \cdots & \cdots & \cdots & \cdots \\ & & & \cdot & \\ & & & & A_{h-1,h} \\ A_{h1} & 0 & 0 & \cdots & 0 \end{bmatrix}$$

where the zero diagonal blocks are all square. If $h = 1$, **A** is *primitive*; if $h > 1$ **A** is *cyclic of index h*.

It is noticed that the nonzero eigenvalues of **A** must be contained in the roots of $\det(\lambda^h \mathbf{I} - \mathbf{A}_{12}\mathbf{A}_{23} \cdots \mathbf{A}_{h1}) = 0$; thus if λ is an eigenvalue of **A**, $\lambda = e^{i(2\pi/h)\theta}$ $(\theta = 1, 2, \ldots, h-1)$ are also eigenvalues of **A**.

Let $\mathbf{A} = [a_{ij}]_n \geqslant 0$ and irreducible. Let $R_i = \sum_{j=1}^{n} a_{ij}$, $R = \max_i R_i$, $r = \min_i R_i$, then $r \leqslant \omega \leqslant R$. For a better estimate, let $m_i = \min_{j \neq i} a_{ji}$, then

$$\omega \leqslant \min_i \{R - m_i + a_{ii} + \sqrt{(R - a_{ii} - m_i)^2 + 4m_i(R - a_{ii})}\}/2$$

$$\omega \geqslant \max_i \{r - m_i + a_{ii} + \sqrt{(r - a_{ii} - m_i)^2 + 4m_i(R - a_{ii})}\}/2$$

Let $\mathbf{A} \geqslant \mathbf{0}$ and irreducible, then **A** cannot have two linearly independent positive eigenvectors.

A general nonnegative matrix $\mathbf{A} \geqslant \mathbf{0}$ always has a nonnegative eigenvalue $\omega \geqslant |\lambda_i|$, where λ_i are the remaining eigenvalues of **A**. The eigenvector corresponding to ω is also nonnegative. If ω is simple, and if the corresponding eigenvectors for both **A** and $\mathbf{A}^T$ are positive, then **A** is irreducible.

If $\mathbf{A} \geqslant \mathbf{0}$, then $\mu\mathbf{I} - \mathbf{A}$ is nonsingular and $(\mu\mathbf{I} - \mathbf{A})^{-1} > \mathbf{0}$, for $\mu > \omega$. Moreover, all the principal minors of $\mu\mathbf{I} - \mathbf{A}$ are positive.

A (square) matrix is *monotone* if $\mathbf{Az} \geqslant \mathbf{0}$ implies $\mathbf{z} \geqslant \mathbf{0}$. If **A** is monotone, then $\mathbf{A}^{-1}$ exists and $\mathbf{A}^{-1} \geqslant \mathbf{0}$.

Let $\mathbf{Z} = \text{diag}\,\{z_1, z_2, \ldots, z_n\}$ where $(z_1, z_2, \ldots, z_n)^T$ is the positive eigenvector corresponding to ω of an irreducible $\mathbf{A} \geqslant \mathbf{0}$. Then $\mathbf{Z}^{-1}\mathbf{AZ} = \omega\mathbf{S}$, where $\mathbf{S} \geqslant \mathbf{0}$ and each row sum of **S** is one. **S** is called a *stochastic* matrix. A *doubly stochastic* matrix is a nonnegative matrix of which all the row sums equal to one as well as the column sums. If $\mathbf{A} > \mathbf{0}$, then there exist two diagonal matrices $\mathbf{D}_1$ and $\mathbf{D}_2$ with positive diagonal elements such that $\mathbf{D}_1\mathbf{A}\mathbf{D}_2$ is doubly stochastic. This doubly stochastic matrix is uniquely determined, and $\mathbf{D}_1$ and $\mathbf{D}_2$ are unique up to a scalar factor.

16.10.2 M-matrices

Let $\mathbf{A} = [a_{ij}]_n$ and $a_{ij} \leqslant 0\ (i \neq j)$, then $\mathbf{A}$ is called an *M-matrix* if $\mathbf{A}$ is nonsingular and $\mathbf{A}^{-1} \geqslant \mathbf{0}$. *M*-matrices have the following properties:

(i) $a_{ii} > 0\ (i = 1, 2, \ldots, n)$ and if $\mathbf{B} = \mathbf{I} - \mathbf{DA}$, $\mathbf{D} = \operatorname{diag}\{1/a_{11}, 1/a_{22}, \ldots, 1/a_{nn}\}$, $\mathbf{B} \geqslant 0$ and the spectral radius of $\mathbf{B}$ is less than one.
(ii) The real part of all the eigenvalues of an *M*-matrix is positive.
(iii) All the principal minors of *M*-matrices are positive.
(iv) If an *M*-matrix is also symmetric, then it is also positive definite. It is called a *Stieljes matrix*.

16.10.3 Oscillatory Matrix

A matrix $\mathbf{A} = [a_{ij}]_n$ is *totally nonnegative* (*totally positive*) if all its minors of any order are nonnegative (positive), [i.e., if $\mathbf{C}_r(\mathbf{A}) \geqslant \mathbf{0}\ (\mathbf{C}_r(\mathbf{A}) > \mathbf{0})$, $r = 1, 2, \ldots, n$, (16.13.1)]. If a matrix is totally nonnegative, and if there is an integer $p > 0$ such that $\mathbf{A}^p$ is totally positive, then $\mathbf{A}$ is *oscillatory*.

A totally nonnegative matrix is oscillatory if and only if $\mathbf{A}$ is nonsingular, and all elements of $\mathbf{A}$ in the principal diagonal, the first super- and first subdiagonal are positive ($a_{ij} > 0$ for $|i - j| \leqslant 1, i, j = 1, 2, \ldots, n$).

The fundamental theorem on oscillatory matrices is as follows. An oscillatory matrix of order n always has n distinct positive eigenvalues $\lambda_1 > \lambda_2 > \cdots > \lambda_n > 0$. The eigenvector $\mathbf{z}_1$ associated with λ_1 has nonzero coordinates of like sign; the eigenvector $\mathbf{z}_2$ associated with λ_2 has exactly one variation of sign in its coordinates and in general the eigenvector $\mathbf{z}_i$ has exactly $i - 1$ variations of sign in its coordinates.

16.10.4 Criteria for Stability of Routh-Hurwitz and Schur-Cohn

1. *Routh-Hurwitz criteria.* Let $f(x) = a_0x^n + a_1x^{n-1} + \cdots + a_{n-1}x + a_n$, $(a_0 > 0)$ be a polynomial with real coefficients, and form the matrix $\mathbf{A} = [a_{ij}]_n$, $a_{ij} = a_{2i-j}$, $(i, j = 1, 2, \ldots, n, a_k = 0 \text{ if } k < 0)$

$$\mathbf{A} = \begin{bmatrix} a_1 & a_0 & 0 & 0 & 0 & 0 & 0 & 0 & 0 \\ a_3 & a_2 & a_1 & a_0 & 0 & 0 & & \cdots & \\ a_5 & a_4 & a_3 & a_2 & a_1 & a_0 & & \cdots & \\ a_7 & a_6 & a_5 & a_4 & a_3 & a_2 & & \cdots & \\ \cdots & \cdots & \cdots & \cdots & \cdots & \cdots & \cdots & \cdots & \cdots \end{bmatrix}$$

then the roots of $f(x)$ will lie in the left half of the complex plane if and only if $\det \mathbf{A}(1) > 0$, $\det \mathbf{A}(1, 2) > 0, \ldots, \det \mathbf{A}(1, 2, \ldots, n) > 0$. It has been shown that the above criteria are equivalent to the requirement that the matrix $\mathbf{C} = [c_{ij}]_n$ be positive definite, where

$$c_{ij} = \begin{cases} 0, & i+j \text{ odd} \\ \sum_{k=1}^{i} (-)^{i+k} a_{k-1} a_{i+j-k}, & j \geqslant i \\ c_{ji}, & j < i \end{cases}$$

The relations between the principal minors of **A** and **C** are given by the following formula: det $\mathbf{C}(1, 2, \ldots, k) = a_0$ det $\mathbf{A}(1, 2, \ldots, k)$ det $\mathbf{A}(1, 2, \ldots, k-1)$ $(k = 1, 2, \ldots, n)$, if det $\mathbf{A}(0) = 1$, Ref. 16-21.

2. *Schur-Cohn criteria.* The roots of $f(x) = a_0 x^n + a_1 x^{n-1} + \cdots + a_{n-1} x + a_n$ will lie within the unit circle if and only if the following $2n$ conditions are satisfied: $f(1) > 0$, $(-)^n f(-1) > 0$; det $(\mathbf{X}_i + \mathbf{Y}_i) > 0$, det $(\mathbf{X}_i - \mathbf{Y}_i) > 0$ $(i = 1, 2, \ldots, n-1)$, where

$$\mathbf{X}_i = \begin{bmatrix} a_0 & a_1 & \cdots & a_{i-1} \\ & a_0 & \cdots & a_{i-2} \\ & & & \vdots \\ 0 & & a_0 & a_1 \\ & & & a_0 \end{bmatrix}; \quad \mathbf{Y}_i = \begin{bmatrix} & & & a_n \\ & 0 & a_n & a_{n-1} \\ & & & \vdots \\ a_n & a_{n-1} & \cdots & a_{n-i+1} \end{bmatrix}$$

Simplification in the evaluation of these determinants has been given in Ref. 16-23.

A symmetric matrix formulation was given in Ref. 16-22: the roots of $f(x)$ will lie within the unit circle if and only if the matrix $\mathbf{B} = [b_{ij}]_n$ is positive definite, where $b_{ij} = \sum_{k=0}^{\min(i,j)} (a_{i-k} a_{jk} - a_{n+k-i} a_{n+k-j})$ $(i, j = 1, 2, \ldots, n)$. The connection between **B** and **C** was formally established in Ref. 16-21, by using the transformation $z = (1 + w)/(1 - w)$ which maps the left half of w-plane into the interior of the unit circle of the z-plane.

16.11 GENERALIZED INVERSE

16.11.1 Definition

Let $\mathbf{A} = [a_{ij}]_{mn}$ and the rank of **A** be r; then it can be shown the system of matrix equations: $\mathbf{AXA} = \mathbf{A}$, $\mathbf{XAX} = \mathbf{X}$, $\mathbf{AX} = (\mathbf{AX})^*$ and $\mathbf{XA} = (\mathbf{XA})^*$ have the unique solution $\mathbf{X} = \mathbf{A}^+$, which is the *generalized inverse* of **A**. When **A** is square and of full rank, then $\mathbf{A}^{-1}$ satisfies the above system of equations, so that the generalized inverse is indeed an extension of the usual definition of matrix inverse to singular and rectangular matrices.

16.11.2 Properties of A^+

The properties of the generalized inverse are as follows:

(i) $\mathbf{A}^{++} = \mathbf{A}$.
(ii) $\mathbf{A}^{*+} = \mathbf{A}^{+*}$.
(iii) $(\mathbf{A}^*\mathbf{A})^+ = \mathbf{A}^+\mathbf{A}^{+*}$, $(\mathbf{A}\mathbf{A}^*)^+ = \mathbf{A}^{+*}\mathbf{A}^+$.
(iv) Rank $(\mathbf{A}^+)$ = Rank $(\mathbf{A}\mathbf{A}^+)$ = Rank $(\mathbf{A})$.
(v) $\mathbf{A}^+\mathbf{A}\mathbf{A}^* = \mathbf{A}^*$ and $\mathbf{A}^*\mathbf{A}\mathbf{A}^+ = \mathbf{A}^*$.
(vi) If $\mathbf{U}, \mathbf{V}$ are unitary, then $(\mathbf{UAV})^+ = \mathbf{V}^*\mathbf{A}^+\mathbf{U}^*$.
(vii) Let $\mathbf{A}$, $\mathbf{B}$ be any two matrices such that $\mathbf{AB}$ is defined. Then if $\mathbf{B}_1 = \mathbf{A}^+\mathbf{AB}$, $\mathbf{A}_1 = \mathbf{AB}_1\mathbf{B}_1^+$, the relation holds: $(\mathbf{AB})^+ = \mathbf{B}_1^+\mathbf{A}_1^+$.
(viii) $\mathbf{AA}^+$, $\mathbf{A}^+\mathbf{A}$ are indempotent.

16.11.3 Methods for Computing Generalized Inverse

(i) Let $\mathbf{B} = [b_{ij}]_{nr}$, and of rank $r \leqslant n$. Then $\mathbf{B}^*\mathbf{B}$ is nonsingular, and $\mathbf{B}^+ = (\mathbf{B}^*\mathbf{B})^{-1}\mathbf{B}^*$. Similarly if $\mathbf{C} = [c_{ij}]_{rn}$, and of rank $r \leqslant n$, then $\mathbf{C}^+ = \mathbf{C}^*(\mathbf{CC}^*)^{-1}$.

(ii) Let $\mathbf{A} = [a_{ij}]_{mn}$ and of rank r. Then $\mathbf{A} = \mathbf{PQ}$, $\mathbf{P}$ of dimensions $m \times r$ and $\mathbf{Q}$ of dimensions $r \times n$ and both $\mathbf{P}$, $\mathbf{Q}$ are of rank r, (16.4.1). From (i), $\mathbf{P}^+ = (\mathbf{P}^*\mathbf{P})^{-1}\mathbf{P}^*$; $\mathbf{Q}^+ = \mathbf{Q}^*(\mathbf{QQ}^*)^{-1}$, $\mathbf{P}^+\mathbf{P} = \mathbf{I}$, $\mathbf{QQ}^+ = \mathbf{I}$, and $\mathbf{A}^+ = \mathbf{Q}^+\mathbf{P}^+ = \mathbf{Q}^*(\mathbf{QQ}^*)^{-1}(\mathbf{P}^*\mathbf{P})^{-1}\mathbf{P}^*$, [16.11.2 (vii)].

(iii) Let $\mathbf{A} = [a_{ij}]_{mn}$, of rank r, $\mathbf{B}$ be any $m \times r$ matrix of independent columns of $\mathbf{A}$, and $\mathbf{C}$ any $r \times n$ matrix of independent rows of $\mathbf{A}$. Then $\mathbf{A}^+ = \mathbf{C}^+(\mathbf{B}^+\mathbf{AC}^+)^{-1}\mathbf{B}^+$. It is noticed that $\mathbf{B}^+\mathbf{AC}^+$ is of dimensions $r \times r$ and of rank r. $\mathbf{B}^+$ and $\mathbf{C}^+$ can be calculated as in (i).

16.11.4 Least Square Solution of Ax = b

If the equation $\mathbf{Ax} = \mathbf{b}$ has a solution, then $\mathbf{x} = \mathbf{A}^+\mathbf{b} + (\mathbf{I} - \mathbf{A}^+\mathbf{A})\mathbf{c}$, where $\mathbf{c}$ is arbitrary.

Let $\mathbf{A} = [a_{ij}]_{mn}$ and $\mathbf{b} \in \mathbf{C}_m$. If $\mathbf{x} = \mathbf{A}^+\mathbf{b}$ and $\mathbf{z} \in \mathbf{C}_n$, $\mathbf{z}$ arbitrary, then $\|\mathbf{Ax} - \mathbf{b}\|_E \leqslant \|\mathbf{Az} - \mathbf{b}\|_E$; furthermore $\|\mathbf{x}\|_E < \|\mathbf{z}\|_E$, for any $\mathbf{z} \neq \mathbf{x}$ such that $\|\mathbf{Ax} - \mathbf{b}\|_E = \|\mathbf{Az} - \mathbf{b}\|_E$. This shows that: (a) $\mathbf{A}^+\mathbf{b}$ is the solution of $\mathbf{Ax} = \mathbf{b}$ in case $m = n$ and $\mathbf{A}$ is nonsingular; (b) $\mathbf{A}^+\mathbf{b}$ is the solution of least euclidean norm, if $\mathbf{Ax} = \mathbf{b}$ has a solution; and (c) $\mathbf{A}^+\mathbf{b}$ is the least squares approximation, whenever $\mathbf{Ax} = \mathbf{b}$ fails to have a solution.

The equation $\mathbf{AXB} = \mathbf{C}$ has a solution if and only if $\mathbf{AA}^+\mathbf{CB}^+\mathbf{B} = \mathbf{C}$, in which case the general solution is $\mathbf{X} = \mathbf{A}^+\mathbf{CB}^+ + \mathbf{Y} - \mathbf{A}^+\mathbf{AYBB}^+$, where $\mathbf{Y}$ is arbitrary. If $\mathbf{AA}^+\mathbf{CB}^+\mathbf{B} = \mathbf{C}$, then the best approximation of $\mathbf{AXB} = \mathbf{C}$ is $\mathbf{X} = \mathbf{A}^+\mathbf{CB}^+$ in the sense that for any other $\mathbf{Z}$: $\|\mathbf{AZB} - \mathbf{C}\|_E \leqslant \|\mathbf{AXB} - \mathbf{C}\|_E$, and if $\|\mathbf{AZB} - \mathbf{C}\|_E = \|\mathbf{AXB} - \mathbf{C}\|_E$ and $\mathbf{Z} \neq \mathbf{X}$, then $\|\mathbf{X}\|_E < \|\mathbf{Z}\|_E$.

16.12 COMMUTING MATRICES

16.12.1 Solutions of AX = XA

The solutions of this equation give all matrices commuting with $\mathbf{A} = [a_{ij}]_n$. Let $\mathbf{A}$ be reduced to the (upper) Jordan form, $\mathbf{J} = \mathbf{S}^{-1}\mathbf{AS}$, with Jordan boxes $\mathbf{J}_1, \mathbf{J}_2, \ldots, \mathbf{J}_s$

on the diagonal; then writing $\tilde{\mathbf{X}} = \mathbf{S}^{-1}\mathbf{X}\mathbf{S}$, we have $\mathbf{J}\tilde{\mathbf{X}} = \tilde{\mathbf{X}}\mathbf{J}$. We partition the matrix $\tilde{\mathbf{X}}$ by rows and columns as the dimensions of the Jordan boxes $n_1, n_2, \ldots, n_s$: $\tilde{\mathbf{X}} = [\tilde{\mathbf{x}}_{ij}]_s$. Then $\mathbf{J}_1\tilde{\mathbf{x}}_{11} = \tilde{\mathbf{x}}_{11}\mathbf{J}_1$, $\mathbf{J}_1\tilde{\mathbf{x}}_{12} = \tilde{\mathbf{x}}_{12}\mathbf{J}_2, \ldots, \mathbf{J}_2\tilde{\mathbf{x}}_{21} = \tilde{\mathbf{x}}_{21}\mathbf{J}_1$, etc. The forms of $\tilde{\mathbf{x}}_{ij}$ will be determined by the following rules:

(i) In the case $\mathbf{J}_i\tilde{\mathbf{x}}_{ii} = \tilde{\mathbf{x}}_{ii}\mathbf{J}_i$, we have

$$\tilde{\mathbf{x}}_{ii} = \begin{bmatrix} x_1 & x_2 & \cdots & x_{n_i} \\ & & & \cdot \\ & & & \cdot \\ & & & \cdot \\ & & & x_2 \\ & 0 & & x_1 \end{bmatrix} \tag{16.12-1}$$

(ii) In the case $\mathbf{J}_i\tilde{\mathbf{x}}_{ik} = \tilde{\mathbf{x}}_{ik}\mathbf{J}_k$, we have

$$\tilde{\mathbf{x}}_{ik} = \begin{bmatrix} x_1 & x_2 & \cdots & x_{n_k} \\ & & & \cdot \\ & & & \cdot \\ & & & \cdot \\ & & & x_2 \\ 0 & & & x_1 \end{bmatrix}; \quad \tilde{\mathbf{x}}_{ik} = \begin{bmatrix} x_1 & x_2 & \cdots & x_{n_i} \\ & & & \cdot \\ & & & \cdot \\ & & & \cdot \\ & & & x_2 \\ 0 & & & x_1 \end{bmatrix} \tag{16.12-2}$$

$$\text{if } n_i \geqslant n_k \qquad\qquad \text{if } n_i \leqslant n_k$$

If the invariant polynomials of $\mathbf{A}$ are $i_1(\lambda), i_2(\lambda), \ldots, i_r(\lambda)$, and the degree of these polynomials are respectively $d_1 \leqslant d_2 \leqslant \cdots \leqslant d_r$, then the number of linearly independent solutions of the equation $\mathbf{AX} = \mathbf{XA}$ is equal to $d_r + 3d_{r-1} + \cdots + (2r-1)d_1$. Moreover, if $\mathbf{A}$ is nonderogatory, then every solution $\mathbf{X}$ can be expressed as a polynomial of $\mathbf{A}$ and conversely, (Ref. 16-5, Vol. I, p. 222).

16.12.2 General Theorems on Commuting Matrices

If $\mathbf{AB} = \mathbf{BA}$, then $\mathbf{A}$, $\mathbf{B}$ have at least one common eigenvector. Let $\mathbf{Az} = \lambda\mathbf{z}$ so that $\mathbf{z}$ is an eigenvector of $\mathbf{A}$. If the minimal polynomial of $\mathbf{B}$ is $f(\mathbf{B}) = (\mathbf{B} - \mu\mathbf{I})\varphi(\mathbf{B})$, then $\varphi(\mathbf{B})\mathbf{z}$ is an eigenvector of $\mathbf{B}$; and $(\mathbf{B} - \mu\mathbf{I})\varphi(\mathbf{B})\mathbf{z} = \mathbf{0}$. But $\mathbf{A}\varphi(\mathbf{B})\mathbf{z} = \varphi(\mathbf{B})\mathbf{Az} = \lambda\varphi(\mathbf{B})\mathbf{z}$, showing that $\varphi(\mathbf{B})\mathbf{z}$ is also an eigenvector of $\mathbf{A}$.

An induction proof of the following theorem can be constructed, using the common eigenvector of $\mathbf{A}$, $\mathbf{B}$: *if* $\mathbf{AB} = \mathbf{BA}$, *then* $\mathbf{A}$, $\mathbf{B}$ *can be simultaneously reduced to the triangular form by a similarity transformation* $\mathbf{S}$; *moreover*, $\mathbf{S}$ *can be chosen as a unitary matrix.*

If $\mathbf{AB} = \mathbf{BA}$ and $\mathbf{A}$, $\mathbf{B}$ are both normal, then $\mathbf{A}$, $\mathbf{B}$ can be simultaneously reduced to the diagonal form by the same similarity transformation which is unitary.

16.12.3 Partitioned Matrices with Commuting Submatrices

Let $\mathbf{A} = [\mathbf{A}_{ij}]_m$ where $\mathbf{A}_{ij} = [a_{rs}^{(ij)}]_n$ so that $\mathbf{A}$ is of dimensions $mn \times mn$. Such partitioned matrices occur quite often, for example, in the difference approximations of partial differential equations; and in many cases the submatrices $\mathbf{A}_{ij}$ are commuting among themselves: $\mathbf{A}_{ij}\mathbf{A}_{kl} = \mathbf{A}_{kl}\mathbf{A}_{ij}$ for all i, j, k, l. Such matrices have the following properties:

(i) The mn eigenvalues of $\mathbf{A}$ are given by the eigenvalues of the matrices $[\lambda_s(\mathbf{A}_{ij})]_m$, $(s = 1, 2, \ldots, n)$, where $\lambda_s(\mathbf{A}_{ij})$ is the sth eigenvalue of $\mathbf{A}_{ij}$, when the eigenvalues of $\mathbf{A}_{ij}$ are arranged in a certain order.

(ii) $\det \mathbf{A} = \det \left\{ \sum \rho[\sigma_1^n(\pi_k)] \prod_{i=1}^{n} \mathbf{A}_{i\pi_i} \right\}$, where $\sum \rho[\sigma_1^n(\pi_k)] \prod_{i=1}^{n} \mathbf{A}_{i\pi_i}$ is the expansion of $\det \mathbf{A}$ if each submatrix $\mathbf{A}_{ij}$ is considered as a scalar.

The proof of both (i) and (ii) depends on the property that a set of commuting matrices can be simultaneously brought to the triangular form by a similarity transformation (16.12.2), and a subsequent permutation of rows and columns of the transformed matrix.

Example:

(1) Let $\mathbf{U} = [\mathbf{A}_{kl}]_n$, $\mathbf{V} = [\mathbf{B}_{kl}]_n$ in which

$$\mathbf{A}_{kl} = \begin{bmatrix} \alpha_{kl} & \beta_{kl} \\ -\beta_{kl} & \alpha_{kl} \end{bmatrix}, \quad \mathbf{B}_{kl} = \begin{bmatrix} \alpha_{kl} & i\beta_{kl} \\ i\beta_{kl} & \alpha_{kl} \end{bmatrix}, \tag{16.12-3}$$

and $\mathbf{W} = [\alpha_{kl} + i\beta_{kl}]_n$, $i = \sqrt{-1}$, α_{kl}, β_{kl} real. Then the submatrices $\mathbf{A}_{kl}$ commute among themselves; so also do $\mathbf{B}_{kl}$. The eigenvalues of $\mathbf{A}_{kl}$, $\mathbf{B}_{kl}$ are both $\alpha_{kl} \pm i\beta_{kl}$. Hence the eigenvalues of $\mathbf{U}$ and $\mathbf{V}$ are the collection of the eigenvalues of the matrices $\mathbf{W}$ and $\overline{\mathbf{W}}$. Therefore, $\det \mathbf{U} = \det \mathbf{V} = |\det \mathbf{W}|^2$.

(2) Let $\mathbf{A} = [a_{ij}]_n$, $\mathbf{B} = [b_{ij}]_m$, $\mathbf{C} = [\mathbf{C}_{ij}]_n$, where $\mathbf{C}_{ij} = a_{ij}\mathbf{I} + \delta_{ij}\mathbf{B}$. Then $\mathbf{C}_{ij}$ commute among themselves and if the eigenvalues of $\mathbf{A}$, $\mathbf{B}$ are λ_k, μ_l respectively, the eigenvalues of $\mathbf{C}$ are those of $[a_{ij} + \mu_l\delta_{ij}]_n$, i.e., $\lambda_k + \mu_l$ $(k = 1, 2, \ldots, n;\ l = 1, 2, \ldots, m)$.

16.12.4 Kronecker Product

Let $\mathbf{A} = [a_{ij}]_{mn}$, $\mathbf{B} = [b_{ij}]_{m'n'}$, then the *Kronecker product* (or *direct product*) of $\mathbf{A}$, $\mathbf{B}$ is given by $\mathbf{A} \otimes \mathbf{B} = [\mathbf{F}_{ij}]_{mn}$, where $\mathbf{F}_{ij} = a_{ij}\mathbf{B}$. When the matrix $\mathbf{B}$ is written out in full in $\mathbf{A} \otimes \mathbf{B}$, the dimensions of the latter is $mm' \times nn'$. We note the following properties:

(i) $\mathbf{A} \otimes \mathbf{B} \otimes \mathbf{C} = (\mathbf{A} \otimes \mathbf{B}) \otimes \mathbf{C} = \mathbf{A} \otimes (\mathbf{B} \otimes \mathbf{C})$.

(ii) $\mathbf{A} \otimes (\mathbf{B} + \mathbf{C}) = \mathbf{A} \otimes \mathbf{B} + \mathbf{A} \otimes \mathbf{C}$.

(iii) If $\mathbf{C} = [c_{ij}]_{nk}$, $\mathbf{D} = [d_{ij}]_{n'k'}$, then $(\mathbf{A} \otimes \mathbf{B})(\mathbf{C} \otimes \mathbf{D}) = \mathbf{AC} \otimes \mathbf{BD}$.

(iv) $(\mathbf{A} \otimes \mathbf{B})^T = \mathbf{A}^T \otimes \mathbf{B}^T$, $(\mathbf{A} \otimes \mathbf{B})^* = \mathbf{A}^* \otimes \mathbf{B}^*$.

(v) If $\mathbf{A}$ and $\mathbf{B}$ are nonsingular, so is $\mathbf{A} \otimes \mathbf{B}$ and $(\mathbf{A} \otimes \mathbf{B})^{-1} = \mathbf{A}^{-1} \otimes \mathbf{B}^{-1}$. .

(vi) $\operatorname{tr}(\mathbf{A} \otimes \mathbf{B}) = (\operatorname{tr} \mathbf{A})(\operatorname{tr} \mathbf{B})$.

(vii) If $\mathbf{A} = [a_{ij}]_{mn}$, $\mathbf{B} = [b_{ij}]_{kl}$, then $\operatorname{Rank}(\mathbf{A} \otimes \mathbf{B}) = \operatorname{Rank} \mathbf{A} \cdot \operatorname{Rank} \mathbf{B}$.

(viii) If $\mathbf{A}$ and $\mathbf{B}$ are both normal, hermitian, positive definite, or unitary, then $\mathbf{A} \otimes \mathbf{B}$ is also normal, hermitian, positive definite, or unitary.

(ix) If $\mathbf{A} = [a_{ij}]_n$, $\mathbf{B} = [b_{ij}]_m$, then in the product $\mathbf{A} \otimes \mathbf{B} = [\mathbf{F}_{ij}]_n = [a_{ij}\mathbf{B}]_n$, $\mathbf{F}_{ij}$ commute among themselves. The eigenvalues of $\mathbf{A} \otimes \mathbf{B}$ are therefore the eigenvalues of the matrices $[a_{ij}\mu_l]_n$ where μ_l $(l = 1, 2, \ldots, m)$ are the eigenvalues of $\mathbf{B}$. Thus if the eigenvalues of $\mathbf{A}$ are λ_k $(k = 1, 2, \ldots, n)$, then the eigenvalues of $\mathbf{A} \otimes \mathbf{B}$ are $\lambda_k \mu_l$ $(k = 1, 2, \ldots, n; l = 1, 2, \ldots, m)$.

(x) $\det(\mathbf{A} \otimes \mathbf{B}) = \det \{\rho[\sigma_1^n(\pi_k)] (a_{1\pi_1}\mathbf{B})(a_{2\pi_2}\mathbf{B}) \cdots (a_{n\pi_n}\mathbf{B})\} = \det \{\mathbf{B}^n \det \mathbf{A}\} = (\det \mathbf{A})^m (\det \mathbf{B})^n$.

16.13 COMPOUND MATRICES

16.13.1 Definition and Properties

Let $\mathbf{A} = [a_{ij}]_{mn}$ and $1 \leqslant r \leqslant \min(m, n)$. From the set of numbers $(1, 2, \ldots, m)$, take r numbers to form $M = \binom{m}{r}$ subsets $(p_1, p_2, \ldots, p_r)$, $1 \leqslant p_1 < p_2 < \cdots < p_r \leqslant m$ and order these sets lexicographically as $\sigma_1, \sigma_2, \ldots, \sigma_M$. Similarly choose r numbers from $(1, 2, \ldots, n)$ and form $N = \binom{n}{r}$ subsets $(q_1, q_2, \ldots, q_r)$, $1 \leqslant q_1 < q_2 < \cdots < q_r \leqslant n$ and order the subsets lexicographically as $\tau_1, \tau_2, \ldots, \tau_N$. The rth *compound* of $\mathbf{A}$: $\mathbf{C}_r(\mathbf{A}) = [\alpha_{ij}]_{MN}$ is then defined as the matrix such that $\alpha_{ij} = \det \mathbf{A}(\sigma_i | \tau_j)$. For example, if $m = 6$, $n = 5$, $r = 4$, then $\sigma_1 = (1, 2, 3, 4)$, $\sigma_2 = (1, 2, 3, 5)$, $\sigma_3 = (1, 2, 3, 6)$, $\sigma_4 = (1, 2, 4, 5), \ldots$, etc.; and $\alpha_{11} = \det \mathbf{A}(1, 2, 3, 4 | 1, 2, 3, 4)$, $\alpha_{34} = \det \mathbf{A}(1, 2, 3, 6 | 1, 3, 4, 5)$, $\alpha_{45} = \det \mathbf{A}(1, 2, 4, 5 | 2, 3, 4, 5)$, etc.

The following properties follow from the definition:

(i) The compound of the transpose of a matrix is equal to the transpose of the compound of the matrix: $\mathbf{C}_r(\mathbf{A}^T) = (\mathbf{C}_r(\mathbf{A}))^T$.

(ii) If $\mathbf{AB} = \mathbf{G}$, $\mathbf{A}$ of dimensions $m \times n$, $\mathbf{B}$ of dimensions $n \times l$, then $\mathbf{C}_r(\mathbf{A})\,\mathbf{C}_r(\mathbf{B}) = \mathbf{C}_r(\mathbf{G})$, $[r \leqslant \min(m, n, l)]$.

(iii) If $\mathbf{A}$ is square and nonsingular, then $\mathbf{C}_r(\mathbf{A}^{-1}) = [\mathbf{C}_r(\mathbf{A})]^{-1}$.

(iv) If $\mathbf{A}$ is unitary, normal, hermitian, or positive definite, so is $\mathbf{C}_r(\mathbf{A})$.

(v) If $\mathbf{A} = [a_{ij}]_{mn}$ and of rank k, then the rank of $\mathbf{C}_r(\mathbf{A})$ is $\binom{k}{r}$ $[\binom{k}{r} = 0$ if $k < r]$.

(vi) If $\mathbf{A}$ is square and of order n, then $\det[\mathbf{C}_r(\mathbf{A})] = (\det \mathbf{A})^{\binom{n-1}{r-1}}$.

16.13.2 Eigenvalues and Eigenvectors

If $\mathbf{A}$ is square with eigenvalues λ_i, then the eigenvalues of $\mathbf{C}_r(\mathbf{A})$ are $\Lambda_i [i = 1, 2, \ldots, \binom{n}{r}]$, where $\Lambda_1 = \lambda_1\lambda_2 \cdots \lambda_r$, $\lambda_1\lambda_2 \cdots \lambda_{r-1}\lambda_{r+1}$, $\Lambda_3 = \lambda_1\lambda_2 \cdots \lambda_{r-1}\lambda_{r+2}$, etc. Moreover, if $\mathbf{A}$ is of simple structure, then $\mathbf{C}_r(\mathbf{A})$ is also of simple structure and the eigenvectors of $\mathbf{C}_r(\mathbf{A})$ are given by the columns of $\mathbf{C}_r(\mathbf{Z})$, where $\mathbf{Z} = [\mathbf{z}_1, \mathbf{z}_2, \ldots, \mathbf{z}_n]$, $\mathbf{z}_i$ being the ith eigenvector of $\mathbf{A}$ for λ_i.

16.13.3 Generalized Sylvester's Theorem

Taking the tth compound of $\mathbf{B}^{(s)}$ in section 16.2.6, using 16.13.1 (vi), and applying the Sylvester's theorem to each element of $\mathbf{C}_t[\mathbf{B}^{(s)}]$, we obtain the following generalized result:

$$\det \mathbf{G}^{(s)(t)} = \{\det \mathbf{A}(\sigma_1^s(k))\}^{\binom{n-s-1}{t}} \{\det \mathbf{A}\}^{\binom{n-s-1}{t-1}}$$

where $\mathbf{G}^{(s)(t)}$ is of order $\binom{n-s}{t}$ and a typical element of $\mathbf{G}^{(s)(t)}$ is $\det \mathbf{A}[\sigma_1^s(k), i_1, \ldots, i_t | \sigma_1^s(k), j_1, \ldots, j_t)$, $(i_1, i_2, \ldots, i_t)$, $(j_1, j_2, \ldots, j_k)$ being any t elements of $(s+1, s+2, \ldots, n)$ arranged in lexicographical order.

16.14 HANDLING LARGE SPARSE MATRICES

16.14.1 Introduction

Sparse matrices are those in which the nonzero elements amount to only a very small percentage of the total number of elements. Techniques have been developed to take advantage of the large number of zeros to avoid unnecessary computation.

Sparse matrices occur frequently in practice, indeed such problems as structural analysis, network flow analysis, difference approximations to differential equations, etc., all lead to sparse matrices. We shall present some brief outline of sparse matrix techniques for the two main problem areas in matrix analysis: eigenvalue problems and matrix inverses.

The principal techniques employed in handling large sparse matrices are permutations of rows and columns to transform them to some standard form so that the problems are reduced to the determination of eigenvalues and inverses of several matrices of smaller dimensions. Whereas different row and column permutations are permissible for the inverse problems, the same row and column permutations are required in order to preserve the eigenvalues.

16.14.2 The Eigenvalue Problem

If it is possible to transform a given matrix to a block triangular form by (the same) row and column permutations, then the work of computing the eigenvalues of the original matrix is reduced to that of finding eigenvalues of the diagonal blocks.

Let $\mathbf{A} = [a_{ij}]_n$ be the given matrix, then construct the boolean counterpart $\mathbf{B} = [b_{ij}]_n$ where $b_{ij} = 1$ if $a_{ij} \neq 0$, and $b_{ij} = 0$ if $a_{ij} = 0$. Using boolean rules of multiplication and addition (i.e., $0 \cdot 0 = 0$, $1 \cdot 0 = 0$, $1 \cdot 1 = 1$, $1 + 0 = 1$, $1 + 1 = 1$), it can be shown that $\mathbf{I} + \mathbf{B} \leqslant (\mathbf{I} + \mathbf{B})^2 \leqslant \cdots \leqslant (\mathbf{I} + \mathbf{B})^t = (\mathbf{I} + \mathbf{B})^{t+1} = \cdots (t \leqslant n - 1)$, where $\mathbf{I}$ is the identity and t is the smallest index such that further multiplication will not alter the matrix elements. The matrix $\mathbf{R} = [r_{ij}]_n = (\mathbf{I} + \mathbf{B})^t$ is called the *reachability matrix*. If $r_{ij} \neq 0$ for some i, j $(i \neq j)$, then it implies that there exist at least one sequence of indices $i, i_1, i_2, \ldots, i_s, j$ such that $a_{ii_1}, a_{i_1 i_2}, \ldots, a_{i_s j}$ are all nonzero. If there exist at least another sequence $j, j_1, j_2, \ldots, j_t, i$ such that a_{jj_1}, $a_{j_1 j_2}, \ldots, a_{j_t i}$ are all nonzero, the points i, j are *strongly connected.* The extraction of the maximal subset of indices which are strongly connected from $(1, 2, \ldots, n)$ forms the basis of the algorithms of determining the row and column permutations.

From $\mathbf{R}$, the reachability matrix just calculated we form the *elementwise product*, $\mathbf{R} \times \mathbf{R}^T = [r_{ij}]_n \times [r_{ji}]_n = [r_{ij} r_{ji}]_n$. Let the nonzero elements of the first row of $\mathbf{R} \times \mathbf{R}^T$ be identified at the positions: $(1, \alpha_1)$, $(1, \alpha_2), \ldots, (1, \alpha_p)$. Then $S_\alpha = (\alpha_1, \alpha_2, \ldots, \alpha_p)$ forms one subset of indices which are strongly connected. Delete the $\alpha_1, \alpha_2, \ldots, \alpha_p^{\text{th}}$ rows and columns from $\mathbf{R} \times \mathbf{R}^T$ and repeat the procedure for the next row, etc., until the set $(1, 2, \ldots, n)$ is decomposed into $S_\alpha, S_\beta, \ldots$. Reorder $S_\alpha, S_\beta, \ldots$ so that they are consistent with the reachability matrix $\mathbf{R}$ in the sense if $r_{ij} = 1$ and $i \in S_\alpha, j \in S_\beta$ then S_α precedes S_β. Call the new ordered subsets: $\overline{S}_\alpha, \overline{S}_\beta, \ldots$. Then by permuting the rows and columns according to $\overline{S}_\alpha, \overline{S}_\beta, \ldots$, the given matrix will be reduced to block triangular form, Ref. 16-30.

Example:

(1)

$$\mathbf{A} = \begin{bmatrix} 1 & 0 & 0 & 0 & 5 & 0 & 0 \\ 0 & 4 & 0 & 2 & 0 & 0 & 0 \\ 0 & 0 & 1 & 0 & 0 & 0 & 2 \\ 0 & 0 & 0 & 4 & 0 & 0 & 3 \\ 2 & 0 & 3 & 0 & 1 & 3 & 0 \\ 0 & 0 & 3 & 2 & 0 & 2 & 0 \\ 0 & 0 & 1 & 0 & 0 & 0 & 4 \end{bmatrix}; \quad \mathbf{C} = \mathbf{I} + \mathbf{B} = \begin{bmatrix} 1 & 0 & 0 & 0 & 1 & 0 & 0 \\ 0 & 1 & 0 & 1 & 0 & 0 & 0 \\ 0 & 0 & 1 & 0 & 0 & 0 & 1 \\ 0 & 0 & 0 & 1 & 0 & 0 & 1 \\ 1 & 0 & 1 & 0 & 1 & 1 & 0 \\ 0 & 0 & 1 & 1 & 0 & 1 & 0 \\ 0 & 0 & 1 & 0 & 0 & 0 & 1 \end{bmatrix}$$

$$\mathbf{R} = \mathbf{C}^2 = \mathbf{C}^4 = \begin{bmatrix} 1 & 0 & 1 & 1 & 1 & 1 & 1 \\ 0 & 1 & 1 & 1 & 0 & 0 & 1 \\ 0 & 0 & 1 & 0 & 0 & 0 & 1 \\ 0 & 0 & 1 & 1 & 0 & 0 & 1 \\ 1 & 0 & 1 & 1 & 1 & 1 & 1 \\ 0 & 0 & 1 & 1 & 0 & 1 & 1 \\ 0 & 0 & 1 & 0 & 0 & 0 & 1 \end{bmatrix}; \quad \mathbf{R} \times \mathbf{R}^T = \begin{bmatrix} 1 & 0 & 0 & 0 & 1 & 0 & 0 \\ 0 & 1 & 0 & 0 & 0 & 0 & 0 \\ 0 & 0 & 1 & 0 & 0 & 0 & 1 \\ 0 & 0 & 0 & 1 & 0 & 0 & 0 \\ 1 & 0 & 0 & 0 & 1 & 0 & 0 \\ 0 & 0 & 0 & 0 & 0 & 1 & 0 \\ 0 & 0 & 1 & 0 & 0 & 0 & 1 \end{bmatrix}$$

$$S_\alpha = (1, 5); \quad S_\beta = (2); \quad S_\gamma = (3, 7); \quad S_\delta = (4); \quad S_\epsilon = (6)$$

Reorder S to obtain: $\overline{S}_\alpha = (1, 5), \overline{S}_\beta = (2), \overline{S}_\gamma = (6), \overline{S}_\delta = (4), \overline{S}_\epsilon = (3, 7)$. After permutation of rows and columns in the order: $(1, 5, 2, 6, 4, 3, 7)$, we obtain

$$\begin{bmatrix} 1 & 5 & 0 & 0 & 0 & 0 & 0 \\ 2 & 1 & 0 & 0 & 0 & 0 & 0 \\ 0 & 0 & 4 & 0 & 2 & 0 & 0 \\ 0 & 0 & 0 & 2 & 2 & 3 & 0 \\ 0 & 0 & 0 & 0 & 4 & 0 & 3 \\ 0 & 0 & 0 & 0 & 0 & 0 & 2 \\ 0 & 0 & 0 & 0 & 0 & 1 & 0 \end{bmatrix}$$

Although the determination of the reachability matrix **R** can be speeded up by calculating the powers $(\mathbf{I}+\mathbf{B})^2$, $(\mathbf{I}+\mathbf{B})^4, \ldots$ until $(\mathbf{I}+\mathbf{B})^{2^t} = (\mathbf{I}+\mathbf{B})^{2^{t+1}}$, the process is nevertheless slow if n is large. Since $2^{t-1} \leqslant n-1$, the number of multiplications required to calculate R is approximately $n^3 \log_2 n$. Define:

(a) *Forward matrix multiplication*—The product **A** by **A** by forward multiplication is carried out in such a way that the product element immediately replaces the corresponding element of **A** as soon as it is calculated, multiplication being performed in the order: $(1,1), (1,2), \ldots, (1,n), (2,1), \ldots, (2,n), \ldots, (n,1), \ldots, (n,n)$, (written: $\mathbf{A} \overset{Ⓕ}{\times} \mathbf{A}$).

(b) *Backward matrix multiplication*—The product element immediately replaces the corresponding element of **A** as soon as it is calculated, but in the reverse order as (a): $(n,n), (n,n-1), \ldots, (n,1), (n-1,n), \ldots, (n-1,1), \ldots, (1,n), \ldots, (1,1)$, (written: $\mathbf{A} \overset{Ⓑ}{\times} \mathbf{A}$).

It has been shown that, using boolean rules of multiplication and addition, the reachability matrix is the result of one forward multiplication of **I** + **B** by **I** + **B** (let the product be **D**), and then followed by one backward multiplication of **D** by **D**, Ref. 16-31. The number of multiplications required is thus $2n^3$ and represents considerable savings in computation when n is large.

Example:

(2) Let **A** be given as in (1), then

$$\mathbf{D} = (\mathbf{I}+\mathbf{B}) \overset{Ⓕ}{\times} (\mathbf{I}+\mathbf{B}) = \begin{bmatrix} 1 & 0 & F & 0 & 1 & F & 0 \\ 0 & 1 & 0 & 1 & 0 & 0 & F \\ 0 & 0 & 1 & 0 & 0 & 0 & 1 \\ 0 & 0 & F & 1 & 0 & 0 & 1 \\ 1 & 0 & 1 & F & 1 & F & F \\ 0 & 0 & 1 & 1 & 0 & 1 & F \\ 0 & 0 & 1 & 0 & 0 & 0 & 1 \end{bmatrix}; \quad \mathbf{D} \overset{Ⓑ}{\times} \mathbf{D} = \begin{bmatrix} 1 & 0 & F & B & 1 & F & B \\ 0 & 1 & B & 1 & 0 & 0 & F \\ 0 & 0 & 1 & 0 & 0 & 0 & 1 \\ 0 & 0 & F & 1 & 0 & 0 & 1 \\ 1 & 0 & 1 & F & 1 & F & F \\ 0 & 0 & 1 & 1 & 0 & 1 & F \\ 0 & 0 & 1 & 0 & 0 & 0 & 1 \end{bmatrix}$$

The letters F and B indicate new nonzero elements formed in the forward and backward multiplication process respectively.

A very efficient procedure was given in Ref. 17-32 consisting essentially of finding loops in tracing the indices of the nonzero elements of the given matrix and eliminating rows and columns in succession. To illustrate, using example (1), the procedures follow:

(a) First remove all the nonzero diagonal elements of the given matrix (replace them by zeros). If there is a row of all zero elements, it is removed as well as the corresponding column. Repeat it until every row has some nonzero elements.

(b) Starting with row 1 we trace the indices of the nonzero elements until a loop is found, thus (1, 5), (5, 1), i.e., 1, 5, 1. We "collapse" row 5 into row 1 to form a new row 1 which consists of all the nonzero elements of row 5 and row 1 minus its diagonal element. The new element in row 1 is indicated by +. At the same time write 1 on the left side of row 5 to indicate row 5 has been collapsed into row 1. Similarly column 5 is collapsed into column 1.

(c) Remove row 5 and column 5.

(d) Start with row 1 again look for another loop; this time it is (1, 3), (3, 7), (7, 3), i.e., 1, 3, 7, 3 showing that 3, 7, 3 form a loop. We collapse row 7 into row 3 (no diagonal element), and then remove row 7.

(e) Row 3 and row 4 are now all zeros, so they are crossed out as well as column 3 and column 4. Since row 7 has been collapsed into row 3, the first set that is freed is (3, 7). Next is (4).

(f) In the remaining matrix row 2 and row 6 are now all zeros. The next sets freed are then (2), (6), and finally (1, 5), since row 5 has been collapsed into row 1.

(g) The rows and columns are eliminated in the following order: (3, 7), (4), (2), (6), (1, 5). Permutation of rows and columns in this order leads to a lower block triangular form. To obtain an upper block triangular form, we can either start with $\mathbf{A}^T$ or reverse the role of rows and columns in the above procedure.

$$
\begin{array}{cc|ccccccc|}
 & & 1 & 2 & 3 & 4 & 5 & 6 & 7 \\
\hline
 & 1 & 0 & 0 & 0 & 0 & 1 & 0 & 0 \\
 & 2 & 0 & 0 & 0 & 1 & 0 & 0 & 0 \\
 & 3 & 0 & 0 & 0 & 0 & 0 & 0 & 1 \\
 & 4 & 0 & 0 & 0 & 0 & 0 & 0 & 1 \\
 & 5 & 1 & 0 & 1 & 0 & 0 & 1 & 0 \\
 & 6 & 0 & 0 & 1 & 1 & 0 & 0 & 0 \\
 & 7 & 0 & 0 & 1 & 0 & 0 & 0 & 0 \\
\hline
\end{array}
,\qquad
\begin{array}{cc|ccccccc|}
 & & 1 & 2 & 3 & 4 & 5 & 6 & 7 \\
\hline
 & 1 & 0 & 0 & + & 0 & 1 & + & 0 \\
 & 2 & 0 & 0 & 0 & 1 & 0 & 0 & 0 \\
 & 3 & 0 & 0 & 0 & 0 & 0 & 0 & 1 \\
 & 4 & 0 & 0 & 0 & 0 & 0 & 0 & 1 \\
1 & 5 & 1 & 0 & 1 & 0 & 0 & 1 & 0 \\
 & 6 & 0 & 0 & 1 & 1 & 0 & 0 & 0 \\
 & 7 & 0 & 0 & 1 & 0 & 0 & 0 & 0 \\
\hline
\end{array}
$$

(a) (b)

$$
\begin{array}{cc|cccccc|}
 & & 1 & 2 & 3 & 4 & 6 & 7 \\
\hline
 & 1 & 0 & 0 & + & 0 & + & 0 \\
 & 2 & 0 & 0 & 0 & 1 & 0 & 0 \\
 & 3 & 0 & 0 & 0 & 0 & 0 & 1 \\
 & 4 & 0 & 0 & 0 & 0 & 0 & 1 \\
 & 6 & 0 & 0 & 1 & 1 & 0 & 0 \\
3 & 7 & 0 & 0 & 1 & 0 & 0 & 0 \\
\hline
\end{array}
,\qquad
\begin{array}{c|ccccc|}
 & 1 & 2 & 3 & 4 & 6 \\
\hline
1 & 0 & 0 & + & 0 & + \\
2 & 0 & 0 & 0 & 1 & 0 \\
3 & 0 & 0 & 0 & 0 & 0 \\
4 & 0 & 0 & 0 & 0 & 0 \\
6 & 0 & 0 & 1 & 1 & 0 \\
\hline
\end{array}
,\qquad
\begin{array}{c|ccc|}
 & 1 & 2 & 6 \\
\hline
1 & 0 & 0 & + \\
2 & 0 & 0 & 0 \\
6 & 0 & 0 & 0 \\
\hline
\end{array}
$$

(c) (d) (e)

16.14.3 Inverses of Sparse Matrices

In this case the row and column permutations need not be the same. Thus it is required to find two permutation matrices **P** and **Q** such that **PAQ** is in block triangular form. The equation $\mathbf{Ax} = \mathbf{b}$ is then equivalent to $\mathbf{PAQ} \cdot \mathbf{Q}^T \mathbf{x} = \mathbf{Pb}$, and since $\mathbf{PAQ}$ is already in block triangular form, the solution $\mathbf{Q}^T \mathbf{x}$ can be found by *LR* decomposition of the diagonal blocks. It is noticed that certain matrices can be reduced to block triangular form by different row and column permutations but not by the same row and column permutations.

To determine $\mathbf{P}$ and $\mathbf{Q}$ we first find a nonzero term in the expansion of det $\mathbf{A}$. As is suggested in Ref. 16-34, this can be done by an algorithm of Hall (16-35) or that of Ford and Fulkerson (16-36). The rows (or columns) of $\mathbf{A}$ are permuted so that the entries of this nonzero term all lie on the principal diagonal. Suppose $\mathbf{P}_1$ is the permutation matrix to accomplish this, so that $\mathbf{P}_1 \mathbf{A}$ has nonzero diagonal elements. Any of the methods described in 16.14.2 can then be applied to $\mathbf{P}_1 \mathbf{A}$ to reduce it to block triangular form: $\mathbf{Q}^T \mathbf{P}_1 \mathbf{AQ}$; the product $\mathbf{Q}^T \mathbf{P}_1$ thus gives the permutation matrix $\mathbf{P}$.

16.15 REFERENCES

16-1 Aitken, A. C., *Determinants and Matrices*, Oliver and Boyd, Edinburgh, 1954.

16-2 Barnett, S., and Storey, C., *Matrix Methods in Stability Theory*, Barnes-Noble, New York, 1971.

16-3 Bellman, R., *Introduction to Matrix Analysis*, McGraw-Hill, New York, 1960.

16-4 Fadeev, D. K., and Fadeeva, V. N., *Computational Methods of Linear Algebra*, W. H. Freeman, San Francisco, 1963.

16-5 Gantmacher, F. R., *The Theory of Matrices*, vols. I and II, Chelsea, New York, 1959.

16-6 Householder, A. S., *The Theory of Matrices in Numerical Analysis*, Blaisdell Publishing Co., New York, 1965.

16-7 MacDuffee, C. C., *The Theory of Matrices*, Chelsea, New York, 1946.

16-8 Marcus, M., and Minc, H., *Survey of Matrix Theory and Matrix Inequalities*, Prindle, Weber, and Schmidt, Ltd., Boston, 1969.

16-9 Todd, J., *Survey of Numerical Analysis*, McGraw-Hill, New York, 1962.

16-10 Varga, R. S., *Matrix Iterative Analysis*, Prentice Hall, Englewood Cliffs, New Jersey, 1962.

16-11 Wedderburn, J. H. M., *Lectures on Matrices*, American Math. Society, New York, 1954.

16-12 Wilkinson, J. H., *The Algebraic Eigenvalue Problem*, Clarendon Press, Oxford, 1965.

16-13 Hestenes, M. R., "Inversion of Matrices by Biorthogonalization and Related Results," *J. SIAM*, **6**, 51–90, 1958. (Section 16.3.4.)

16-14 Fan, K., and Hoffman, A. J., *Lower Bounds for the Rank and Location of the Eigenvalues of a Matrix*, National Bureau of Standards, Applied Mathematics Series 39, 117–130, 1954. (Section 16.4.2.)

16-15 Bauer, F. L., "Optimally Scaled Matrices," *Numer. Math.*, **5**, 73–87, 1963. (Section 16.5.7.)

16-16 Schwerdtfeger, H., "Direct Proof of Lanczos Decomposition Theory," *Amer. Math. Monthly*, **67**, 855–860, 1960. (Section 16.6.7.)

16-17 Lax, P. D., and Wendroff, B., "Difference Schemes for Hyperbolic Equations with High Order of Accuracy," *Comm. Pure and Appl. Math.*, **17**, 381–398, 1964. (Section 16.8.4.)

16-18 Sinkhorn, R., "A Relationship Between Arbitrary Positive Matrices and Doubly Stochastic Matrices," *Ann. Math. Statistics*, **35**, 876–879, 1964. (Section 16.10.1.)

16-19 Brauer, A., "The Theorems of Ledermann and Ostrowski on Positive Matrices," *Duke Mathematics Journal*, **24**, 265–274, 1957. (Section 16.10.1.)

16-20 Cohn, A., "Ueber die Anzahl der Wurzeln einer Algebraischen Gleichung in einem Kreise," *Mathematical Algorithms*, **14**, 110–148, 1922. (Section 16.10.4.)

16-21 Ralston, A., "A Symmetric Matrix Formulation of the Hurwitz-Routh Stability Criterion," *IRE Trans. Auto, Control*, **7**, 50–51, 1962. (Section 16.10.4.)

16-22 Wilf, H. S., "A Stability Criterion for Numerical Integration," *J. Assoc. Comp. Mach.*, **6**, 363–365, 1959. (Section 16.10.4.)

16-23 Jury, E. I., and Bharucha, B. H., "Notes on the Stability Criterion for Linear Discrete Systems," *IRE Trans. Auto. Control*, **AC-6**, 88–90, 1961. (Section 16.10.4.)

16-24 Penrose, R., "A Generalized Inverse for Matrices," *Proc. Camb. Philo. Soc.*, **51**, 406–413, 1961. (Section 16.11.1.)

16-25 Greville, T. N. E., "Note on the Generalized Inverse of a Matrix Product," *SIAM Review*, **8**, 518–521, 1966. (Section 16.11.1.)

16-26 Drazin, M. A.; Dungey, J. W.; and Gruenberg, K. W., "Some Theorems on Commuting Matrices," *J. London Math. Soc.*, **26**, 221–228, 1951. (Section 16.12.2.)

16-27 Afriat, S. N., "Composite Matrices," *Quart. J. Math.*, Second Series, **5**, 81–98, 1954. (Section 16.12.3.)

16-28 Brenner, J. L., "Expanded Matrices from Matrices with Complex Elements," *SIAM Review*, **3**, 165–166, 1961. (Section 16.12.3.)

16-29 Gott., E., "A Theorem on Determinants," *SIAM Review*, **2**, 288–291, 1960. (Section 16.12.3.)

16-30 Harary, F., "A Graph Theoretical Approach to Matrix Inversion by Partitioning," *Numer. Math.*, **7**, 255–259, 1959. (Section 16.14.2.)

16-31 Hu, T. C., "Revised Matrix Algorithms for Shortest Paths," *SIAM J. Appl. Math.*, **15**, 207–218, 1967. (Section 16.14.2.)

16-32 Steward, D. V., "Partitioning and Tearing Systems of Equations," *J. SIAM Numer. Anal.*, Series B, **2**, 345–365, 1965. (Section 16.14.2.)

16-33 Dulmage, A. L., and Mendelsohn, N. S., "Two Algorithms for Bipartite Graphs," *J. SIAM*, **11**, 183–194, 1963. (Section 16.14.3.)

16-34 —, "On the Inverse of Sparse Matrices," *Math. Comp.*, **16**, 494–496, 1962. (Section 16.14.3.)

16-35 Hall, M., "An Algorithm for Distinct Representatives," *Amer. Math. Monthly*, 716–717, 1956. (Section 16.14.3.)

16-36 Ford, L. R., and Fulkerson, D. R., "A Simple Algorithm for Finding Maximal Network Flows and an Application to Hitchcock Problems," *Canad. J. Math.*, **9**, 210–218, 1957. (Section 16.14.3.)

16-37 Tewarson, R. P., "Row Column Permutation of Sparse Matrices," *The Computer Journal*, **10**, 300–305, 1967. (Section 16.14.3.)

16-38 —, "Computations with Sparse Matrices," *SIAM Review*, **12**, 527–543, 1970. (Section 16.14.3.)

16-39 Dongarra, J. J., Moler, C. B., Bunch, J. R., and Stewart, G. W., *LINPACK User's Guide*, Soc. Indust. Appl. Math., Phila., 1979.

16-40 Smith, B. T., Boyle, J. M. Dongarra, J. J., Garbow, B. S., Ikebe, Y., Klema, V. C., and Moler, C. B., *Matrix Eigensystem Routines, EISPACK Guide*, Springer-Verlag, Berlin, 1976.

17

Functional Approximation

Robin Esch*

17.0 INTRODUCTION

17.0.1 The Fundamental Problem

A fairly general statement of the fundamental mathematical problem of approximation theory can be given as follows:

1. Let $\Phi = \{f(x)\}$ be a set of functions from which the function to be approximated is drawn (frequently one is concerned with establishing results that are valid for *any* $f \in \Phi$). For example one may be interested in the set $C[0, 1]$ of all functions continuous on $0 \leqslant x \leqslant 1$.
2. Let $G = \{g(x)\}$ be the set of all approximating functions which one decides to admit (for example frequently $G = P_n$ = the set of all polynomials of degree n or less is taken); naturally G is ordinarily a much smaller set than Φ.
3. A measure of 'goodness of approximation' is needed. Let $\|\cdot\|$ be a measure of the 'size' of a function, so that the smallness of $\|f(x) - g(x)\|$ can serve as a measure of how well $g(x)$ approximates $f(x)$. For example one may take the *Chebyshev norm* on $[0, 1]$:

$$\|\phi(x)\| = \max_{0 \leqslant x \leqslant 1} |\phi(x)|$$

   Clearly then if $\|f(x) - g(x)\|$ is small $g(x)$ is 'close to' $f(x)$ in a meaningful sense.

Then the fundamental problem may be stated: for any $f(x) \in \Phi$, does a function $g^*(x) \in G$ exist which minimizes $\|f(x) - g(x)\|$? If so what are its properties, and how may it be computed?

*Prof. Robin Esch, Dep't. of Mathematics, Boston University, Boston, Mass.

A solution $g^*(x)$ has the obvious property that, for all $g(x) \in G$

$$\| f(x) - g^*(x) \| \leqslant \| f(x) - g(x) \|$$

We will frequently denote this minimum error measure by

$$\lambda^* = \| f(x) - g^*(x) \|$$

A variety of such problems result, as one can take various choices for Φ, G and $\| \cdot \|$. In any such problem mathematicians are interested in existence (whether $g^*(x)$ exists), uniqueness (whether only one $g^*(x)$ exists), characterization properties, and other interesting properties of $g^*(x)$, and of course in how to compute $g^*(x)$. Many other matters are of interest, for example convergence (in polynomial approximation, $G = P_n$, this concerns how the error and its derivatives depend on n), the relationship between approximations derived with different error criteria, etc.

17.0.2 Applied Approximation Problems

In applied work one is less likely to be concerned with establishing results for a very general class of $f(x)$. More often one is dealing with a single known function, or a family of functions of similar known character. Thus the applied mathematician usually knows much more about the character of $f(x)$ than the pure mathematician would be willing to assume. The choice of error measure may be clearly dictated, or it may be somewhat arbitrary, i.e., nearly equally satisfactory results may be obtained with two different error measures. Frequently the greatest practical difficulty is in defining and handling the set G of admissible approximations. This is obviously crucial, for good results are not going to be obtained if G simply does not contain functions which are 'close' to $f(x)$. Frequently in practice side conditions or constraints are present which reduce the size of G in more or less complicated ways. For example, in polynomial approximation it may be necessary that certain coefficients be nonnegative, or it may be important that no two roots lie too close to each other. This sort of implicit curtailment of the set G of admissible approximations can greatly increase the difficulty of an approximation problem.

Let us cite some examples of approximation problems arising in practice.

1. A digital computer subroutine to compute $\sin x$ to machine accuracy is required. It is easy to give $\sin x$ for any x in terms of its values in $0 \leqslant x \leqslant \pi/2$, so the problem becomes: find an approximation to $\sin x$ which can be easily evaluated by an automatic computer, and which deviates from $\sin x$ by no more than some tolerance (say 10^{-7}, or in double precision work, 10^{-16}) in $0 \leqslant x \leqslant \pi/2$. Here no element of generality with regard to the set Φ of $f(x)$ is present; we are concerned with just one specific $f(x)$. A Chebyshev norm is clearly indicated (if we construct an approximation $g(x)$ using some other error criterion, say least mean square error, we would have to test *a posteriori* to see that the accuracy requirement is fulfilled). The choice of the set G of

admissible approximations is difficult, as the set of functions which are easy (inexpensive) for a digital computer to evaluate is not well-defined. In current practice rational functions are usually chosen, as a) they are easy for a digital computer to evaluate; b) they yield high-accuracy approximations to well-behaved functions such as sin x; and c) satisfactory methods for computing best rational function approximations are well-known. It is suspected that better results (less running time and storage) could be obtained by enlarging the class of admissible approximations, to include also for example compositions of rational functions, but the best approximation problem then is beyond the reach of current established practice.

2. Subroutines for the evaluation of 100 functions of a single variable x are to be prepared for use by an airborne digital computer. These functions, which might represent for example vehicle dynamic characteristics, are defined by test measurements, plus certain knowledge about their characteristics (it might be known for example that the functions have value unity at $x = 0$ and die out exponentially as $x \to \infty$).

   Here as in the previous example extreme generality in the set Φ of $f(x)$ is not present; even though many $f(x)$ are to be treated, it is probably known from physical considerations that they all have certain characteristics and a certain minimum level of good behavior. The set of admissible approximations must again be easily computable. Since the approximations are to be determined from data given at only a finite set of discrete abscissas, the error criterion must involve only those points; some sort of norm over a finite set of discrete points is indicated. The possibility of experimental error may make a least-mean-square criterion preferable to the Chebyshev strategy of minimizing the maximum error magnitude. The question of how best to make use of the known properties of the functions may be difficult if limiting the set of admissible approximations to functions with those properties is not convenient.

3. An algorithm for generating a given function in an analog computer is to be devised. The set of functions which are easy for analog computers to compute are the solutions of constant coefficient differential equations, i.e., sums and products of exponentials, sinusoidal functions, and polynomials. One is thus led to consider exponential approximations, i.e., approximating functions of the form e.g.,

$$\sum_{k=1}^{n} a_k e^{b_k x}$$

   Because the undetermined coefficients $b_1, b_2, \ldots$ appear nonlinearly in this expression, the resulting exponential approximation problem is difficult, though not beyond the reach of current techniques.

4. An electric filter is to be designed whose power spectrum approximates to within a given tolerance a specified curve, and where cost is to be minimized.

This filter might be intended for example to match certain standard control equipment to the characteristics of a specific aircraft. Transfer functions of filters are rational functions, so this is a rational approximation problem (however if time delays are present exponential factors make their appearance). A Chebyshev type error criterion is probably indicated, although special provision must be made if the prescribed ideal curve has vertical portions (as in band-pass filters). The minimum cost criterion is a tough one to handle; a high order rational function whose denominator has several well-separated negative real roots may be cheaper to fabricate than a lower order rational function whose denominator roots are all complex conjugate pairs (as in the Chebyshev filters).

5. A curve is to be passed through specified points in the plane in a maximally smooth manner. In the linearized case this reduces to a spline interpolation problem, which fits into the domain of ideas of approximation theory. The nonlinear case, in which maximizing smoothness may mean minimizing the integral of the square of the curvature, is a difficult nonlinear problem. If the independent variable is time, the dependent variable in such a problem might represent a coordinate of a vehicle; Newton's laws imply that no derivative of lower than second order of such a coordinate can be discontinuous.

Approximation theory will frequently be of great practical value in such applied problems, but frequently ad hoc methods and ingenuity will be required in addition.

17.1 NORMS AND RELATED MEASURES OF ERROR

17.1.1 Approximation on [0, 1]

As remarked earlier a key part of an approximation problem is the choice of the measure of goodness of approximation. We discuss first the case of a finite one-dimensional interval, which (by means of a linear transformation) may without loss of generality be taken as the interval $0 \leqslant x \leqslant 1$, denoted $[0, 1]$.

Let p be a number in the interval $1 \leqslant p < \infty$. Then the L_p norm of a function $\phi(x)$ is given by

$$L_p(\phi) = \| \phi \|_p = \left[\int_0^1 | \phi(x) |^p \, dx \right]^{1/p} \tag{17.1-1}$$

For continuous $\phi(x)$, the limit of $L_p(\phi)$ as $p \to \infty$ is[17-45]:

$$L_\infty(\phi) = \| \phi \|_\infty = \max_{0 \leqslant x \leqslant 1} | \phi(x) | \tag{17.1-2}$$

The L_∞ norm is also called the *Chebyshev norm*, the *uniform norm*, or the *maximum norm* [The term 'uniform' stems from the fact that, if for a sequence of functions $g_1(x), g_2(x), \ldots, \| f(x) - g_n(x) \|_\infty \to 0$, then the sequence converges uniformly to

$f(x)]$. The L_∞, or Chebyshev, approximation is often referred to as *minimax* approximation, since it minimizes the maximum error magnitude.

If we minimize

$$\int_0^1 |e(x)|^p \, dx \tag{17.1-3}$$

it is equivalent to minimizing $L_p(e)$; the technicality that, because of omission of the $1/p$-power, eq. (17.1-3) is not a norm, is unimportant in this respect. Technically, a norm $\|\cdot\|$ is a functional which, for any two functions $f(x)$ and $g(x)$, and any scalar α satisfies the conditions

$$\begin{aligned} &\|f(x)\| \geqslant 0 \\ &\|f(x)\| = 0 \quad \text{if and only if} \quad f(x) = 0 \\ &\|\alpha f(x)\| = |\alpha| \cdot \|f(x)\| \\ &\|f(x) + g(x)\| \leqslant \|f(x)\| + \|g(x)\| \end{aligned} \tag{17.1-4}$$

The last two hold for (17.1-1) but not for (17.1-3). The last condition is the famous triangle inequality, in the L_p case known as the *Minkowski inequality*[17-19]; it is easy to prove for the special cases $p = 1$ and ∞, and, by use of the Schwarz inequality, not difficult for $p = 2$.

These cases, $p = 1, 2$, and ∞, are by far the most important cases in practical work. They each have a simple interpretation. Minimizing the L_1 norm of $e(x) = [f(x) - g(x)]$ is equivalent to minimizing the area between the two curves. Minimizing $L_\infty(e)$ means minimizing the maximum error magnitude; this is Chebyshev, or minimax, approximation. Minimizing $L_2(e)$, or equivalently minimizing

$$[L_2(e)]^2 = \int_0^1 e(x)^2 \, dx \tag{17.1-5}$$

is the familiar least-mean-square error strategy, analytically attractive because absolute value signs can be omitted [thus it is easy to differentiate (17.1-5) with respect to parameters appearing in $e(x)$], and because the resulting equations are usually linear.

Regions in which errors are large contribute relatively more heavily to $L_2(e)$ than to $L_1(e)$. This effect increases in $L_p(e)$ as p increases, so that, for large p a good approximation to $L_p(e)$ would be obtained by including in the integral only the regions close to the points at which $|e(x)|$ attains its maximum. In the limit where p goes to infinity only the maximum error magnitude counts.

17.1.2 Weight Functions; Semi-Infinite and Infinite Regions

There may be reasons for weighting errors in some regions of the interval more heavily than errors in other regions. Or weight functions may arise because they occur in orthogonality properties of approximating functions being used. If $w(x)$ is any integrable function obeying

$$w(x) > 0 \qquad 0 \leqslant x \leqslant 1$$

then the generalization of (17.1-1)

$$L_p(\phi; w) = \left[\int_0^1 w(x) \, |\phi(x)|^p \, dx \right]^{1/p} \tag{17.1-6}$$

is still a norm. The outer radical may be omitted in minimization.

If the interval is semi-infinite ($0 \leqslant x < \infty$) or infinite ($-\infty < x < \infty$), weight functions are needed in L_2 approximation; otherwise the integrals involved will exist for too small a class of functions. The most usual norms are

$$\left[\int_0^\infty e^{-x} \, [\phi(x)]^2 \, dx \right]^{1/2} \tag{17.1-7}$$

and

$$\left[\int_{-\infty}^\infty e^{-x^2} \, [\phi(x)]^2 \, dx \right]^{1/2} \tag{17.1-8}$$

in these cases. In the finite interval [0, 1] the most usual weight function is

$$w(x) = x^\alpha (1 - x)^\beta \tag{17.1-9}$$

where α and β are constants $\geqslant -\frac{1}{2}$. The case $w(x) = [x(1 - x)]^{-1/2}$, associated with the name Chebyshev, is appropriate when the ends of the interval are to be weighted more heavily than the middle; it will be discussed in section 17.4.3.

17.1.3 The Case of Several Dimensions; Arbitrary Domains

The approximation problem increases greatly in difficulty when we go from one to several dimensions. However, there is no difficulty in generalizing the error criteria mentioned above. Suppose a function $f(\mathbf{x})$, defined on an n-dimensional space with Cartesian coordinates $\mathbf{x} = (x_1, x_2, \ldots, x_n)$ is to be approximated by a member $g(\mathbf{x})$ of a set of admissible approximations, over some domain R. The L_p norm of the error $e(\mathbf{x}) = [f(\mathbf{x}) - g(\mathbf{x})]$ is a straightforward generalization of (17.1-1)

$$L_p(e) = \left[\int_R |e(\mathbf{x})|^p \, dx \right]^{1/p} \tag{17.1-10}$$

The L_∞ norm

$$L_\infty(e) = \max_{\mathbf{x}\in R} |e(\mathbf{x})| \tag{17.1-11}$$

is the limit as $p \to \infty$ of (17.1-10), with two possible exceptions which are purely technical; if $e(\mathbf{x})$ has some point discontinuities of measure zero, or R contains some isolated regions of measure zero.

17.1.4 Discrete Point Sets

We may decide, as a matter of computational convenience, to look at an error at only a finite set of discrete points x_i, $i = 1, \ldots, N$; or we may be forced to do this, as the function $f(x)$ to be approximated may be given at only such a finite set of points. A discrete analogy to the continuous L_p norm (17.1-1) is

$$L_p(\phi) = \left[\sum_{i=1}^{N} |\phi(x_i)|^p\right]^{1/p} \tag{17.1-12}$$

The cases $p = 1, 2, \infty$ are again the ones commonly employed in practical work. The $p = \infty$ case (or more properly the limit as $p \to \infty$ case) is

$$L_\infty(\phi) = \max_{i=1,N} |\phi(x_i)| \tag{17.1-13}$$

Inclusion of positive weights w_i is frequently desirable; for example in the L_2 case the weighted norm

$$L_2(w; \phi) = \left[\sum_{i=1}^{N} w_i [\phi(x_i)]^2\right]^{1/2} \tag{17.1-14}$$

is frequently employed.

In cases where the values of $f(x_i)$ are given by measurements which contain normally distributed errors, a statistical argument may be given that the L_2 norm is the most appropriate.[17-37]

17.1.5 The Sobolev Norms; Smoothness

In cases where the smallness of derivatives, as well as functional values, is important, the following generalization of the L_p and L_∞ norms are useful (for simplicity we take the interval [0, 1]):

$$\|\phi(x)\|_{W_q^s} \equiv \left[\sum_{j=0}^{s} \int_0^1 \left|\frac{d^j\phi}{dx^j}\right|^q dx\right]^{1/q} \tag{17.1-15}$$

$$\| \phi(x) \|_{W_\infty^s} \equiv \sum_{j=0}^{s} \max_{0 \leqslant x \leqslant 1} \left| \frac{d^j \phi}{dx^j} \right| \tag{17.1-16}$$

These *Sobolev norms* are useful in obtaining error bound and convergence information for smooth approximations, such as spline functions and related piecewise functions, where continuity of derivatives or convergence of derivative value, as well as functional value, is important. The Besov norms are a related development.[17-46, 17-47]

17.2 RELATIONSHIP BETWEEN APPROXIMATION ON A CONTINUUM AND ON A DISCRETE POINT SET

Suppose it is required to compute the best L_∞ [0, 1] approximation $g^* \in G$ to an arbitrary continuous $f(x)$. As usual let λ^* denote the L_∞ error norm of g^*

$$\lambda^* = \| f - g^* \| = \max_{0 \leqslant x \leqslant 1} | f(x) - g^*(x) | \tag{17.2-1}$$

One technique for handling such problems is to first discretize them by agreeing to look at the error at only a finite set of m points X, typically of the form

$$X\colon\ 0 = x_1 < x_2 < x_3 < \cdots < x_m = 1 \tag{17.2-2}$$

Methods such as linear programming then become applicable (and often turn out to be efficient even for quite large m, e.g., $m = 1000$).

Denote the L_∞ error norm on the discrete point set X by

$$\lambda_X(g) = \| f(x) - g(x) \|_X = \max_{i=1,\ldots,m} | f(x_i) - g(x_i) | \tag{17.2-3}$$

(this is a norm on X but of course is not a norm on [0, 1]), and denote the best approximation in this norm by g_X^*

$$\lambda_X^* = \| f - g_X^* \|_X = \inf_{g \in G} \lambda_X(g) \tag{17.2-4}$$

Suppose now that we have solved the discrete problem, i.e., computed g_X^* and its error norm λ_X^* on X, and we want to know how good an approximation this yields to the solution g^* of the continuous problem.

First we note that bounds on the continuous least error norm λ^* are immediately available; λ_X^* itself is a lower bound, since $\lambda_X^* = \| f - g_X^* \|_X \leqslant \| f - g^* \|_X$ (since not even g^* surpasses g_X^* on X!) $\leqslant \| f - g^* \| = \lambda^*$.

Furthermore, the error norm over [0, 1] of g_X^* furnishes an upper bound to λ^* (this of course is true for any admissible approximation); thus we have

$$\lambda_X^* \leqslant \lambda^* \leqslant \| f(x) - g_X^*(x) \| = M \tag{17.2-5}$$

In the majority of practical computations, if these bounds are close [usually $M \leqslant (1.001)\,\lambda_X^*$ is more than close enough] g_X^* may be taken as an adequately good approximation to g^*, and the job is done. The lower bound in (17.2-5) is available at no cost from the discrete approximation problem computation; the upper bound M requires further computation, as the continuous interval must be searched. However, this also constitutes an independent final check, usually considered mandatory in careful work.

We present further discussion to study the effects of the choice of the discrete point set X. Denote the error curve of the computed approximation by

$$e_X(x) = f(x) - g_X^*(x) \tag{17.2-6}$$

and let $\tilde{x}$ be a point where $|e_X(x)|$ attains its maximum value M on $[0, 1]$

$$|e_X(\tilde{x})| = |f(\tilde{x}) - g_X^*(\tilde{x})| = \|f(x) - g_X^*(x)\| = M$$

Also let $\tilde{y}$ be a nearest point of the discrete point set X. Then the following maneuver yields a lower bound on λ_X^*:

$$\begin{aligned} M = \|f - g_X^*\| &= |f(\tilde{x}) - g_X^*(\tilde{x})| \\ &= |f(\tilde{x}) - f(\tilde{y}) + f(\tilde{y}) - g_X^*(\tilde{y}) + g_X^*(\tilde{y}) - g_X^*(\tilde{x})| \\ &\leqslant |[f(\tilde{x}) - g_X^*(\tilde{x})] - [f(\tilde{y}) - g_X^*(\tilde{y})]| + |f(\tilde{y}) - g_X^*(\tilde{y})| \\ &\leqslant |e_X(\tilde{x}) - e_X(\tilde{y})| + \lambda_X^* \end{aligned}$$

Combining this with (17.2-5), we have then

$$M - \Delta \leqslant \lambda_X^* \leqslant \lambda^* \leqslant M \tag{17.2-7}$$

where

$$\Delta = |e_X(\tilde{x}) - e_X(\tilde{y})| \tag{17.2-8}$$

Thus to achieve good results we will wish to choose the point set X so as to force Δ to be small.

This affords another *a posteriori* check: locate a point $\tilde{x}$ at which the computed error (17.2-6) attains its maximum magnitude on $[0, 1]$; locate a nearest point $\tilde{y}$ of the discrete point set X; and then compute Δ by taking the difference of the computed error curve at $\tilde{x}$ and at $\tilde{y}$.

Ordinarily the approximating functions $g(x)$ are continuous and smooth, so that the continuity properties of $e_X(x)$ hinge on the continuity properties of f. If f is continuous on $[0, 1]$ Δ will be small if $|\tilde{x} - \tilde{y}|$ is small. Recalling the definition of $\tilde{x}$ and $\tilde{y}$, we see that good results will be obtained, i.e., Δ will be small in (17.2-7), if the discrete point set X is dense in the neighborhood of $\tilde{x}$.

As a practical matter, it is necessary to choose X suitably dense in the neighborhood of each anticipated extremum of the error curve. It is interesting to note however that X need not be dense everywhere. In fact let us consider a sequence of calculations, on discrete point sets $X_1, X_2, X_3, \ldots$. For continuous $f(x)$, $\lambda^*_{X_i}$ will converge to λ^* provided $\tilde{x}_i - \tilde{y}_i \to 0$. This can be achieved even if the number of points in X_i is bounded for all i.

In discussing convergence, and showing that not only does $\lambda^*_{X_i} \to \lambda^*$ but also $g^*_{X_i} \to g^*$, it is desirable to replace Δ by something less dependent on the specific point set X_i; Rivlin and Cheney do so (see Ref. 17-36, and 17-37, section 1.3).

A further insight afforded by this discussion concerns the case where $f(x)$ is smooth, i.e., $f' \in C[0, 1]$. Then [assuming $g(x)$ smooth as is ordinarily the case], the error curve will also be smooth, and therefore horizontal at interior extrema (see Fig. 17.5-1 for an example of such an error curve). If $\tilde{x}$ is an interior point ($\Delta = 0$ if $\tilde{x}$ is an end point as we would then have $\tilde{y} = \tilde{x}$), then Δ–given by eq. (17.2-8)–will be proportional to $(\tilde{x} - \tilde{y})^2$. This indicates that density of the point set X is not so crucial for smooth extrema as for cusps of the error curve, or regions where oscillation is rapid, as is intuitively apparent. The discretization approach works best for functions which are smooth everywhere, which fortunately is the usual case.

So far we have been assuming the L_∞ error norm. It is also true in the L_p norms that the best approximation on a finite point set X approaches the best approximation on the continuum [0, 1], provided that weights are included corresponding to the intervals $\Delta x_i = x_{i+1} - x_i$ between the discrete points of X. This follows in an obvious manner from the definition of the Riemann integral. For a discussion of the L_2 and L_1 cases see Rice,[17-31] section 2.5, and Rivlin,[17-35] p. 80.

17.3 EXISTENCE OF BEST APPROXIMATIONS

The proving of existence of best approximation–that is, given some set Φ of functions to be approximated and some set G of approximating functions, the proving that for *any* $f \in \Phi$ a $g^* \in G$ exists which minimizes the error norm $\| f - g \|$ under consideration–is frequently a fascinating mathematical problem. However in practical work the existence question in its general form can usually be bypassed, since one is usually dealing with rather specific functions with known properties. We now outline some of the considerations involved in proving existence.

For an arbitrary $f \in \Phi$, to each $g \in G$ corresponds a nonnegative number

$$\lambda = \lambda(g) = \| f - g \|$$

This infinite set of nonnegative real numbers has a greatest lower bound λ^*; and each neighborhood of λ^* includes a λ of some $g \in G$. Thus there must be an infinite sequence $\{g_1, g_2, \ldots\}$ with corresponding error norms $\{\lambda_1, \lambda_2, \ldots\}$ which approach λ^*. This much is true in full generality, i.e., approximations g with error norms arbitrarily close to λ^* always exist.

The usual technique in proving existence is to show that the sequence $\{g_1, g_2, \ldots\}$ can be taken in a compact subset of G (often this involves just putting a suitably large upper bound on the magnitude of g). A closed set of points includes all its limit points, and the idea of a compact set is a related idea. A function set S is compact if it has the following property: every infinite sequence in S possesses a subsequence which approaches a limit in S. Thus if $\{g_1, g_2, \ldots\}$ comes from a compact subset of G a subsequence approaches a limit g^* in that subset, and g^* must have error norm λ^*, completing the demonstration of existence.

Rivlin gives an efficient proof which covers all cases where G is a linear space and all error norms (see Ref. 17-35, p. 1). Thus for linear approximating functions, i.e., approximating functions of the form

$$g(x) = \sum_{k=1}^{n} a_k \Phi_k(x) \tag{17.3-1}$$

where $\Phi_1(x), \Phi_2(x), \ldots, \Phi_n(x)$ are given functions of x, and in particular for polynomials (but without side conditions) existence holds for any error norm. Rational functions are not of the linear form (17.3-1) and a separate proof is needed (see, Ref. 17-35, p. 120).

Existence may be disturbed in an obvious manner by appending strong inequalities as side conditions. Thus if G is the set of constants < 0 there exists no best approximation to $f(x) = 1$ (in just about any error norm). In exponential approximation lack of closure of the set G can be less obvious: consider approximating functions of the form

$$g(x) = a_1 e^{a_2 x} + a_3 \tag{17.3-2}$$

where a_1, a_2, a_3 are any finite real numbers. Then, since

$$g = a_1 [1 + a_2 x + O(a_2^2 x^2)] + a_3$$

if we set $a_1 + a_3 = 0$, $a_1 a_2 = 1$ and let $a_2 \to 0$ we have $g \to x$. But in this process $a_1 \to \infty$, and so no best approximation to $f(x) = x$ exists. The trouble is that the function x is a limit of the set G but not in G. To obtain existence in general the set (17.3-2) must be closed by appending all straight lines (Rice, Refs. 17-31, 17-32, 17-34).

17.4 L_2 OR LEAST-MEAN-SQUARE APPROXIMATION

Approximation in the L_2 norm is computationally in many ways the easiest sort of approximation (in fact it was the only widely used sort before the advent of automatic computers). When random errors are present in the data to be approximated a statistical argument may be given that L_2 approximation is most appropriate (see for example the Gauss-Markov theorem in Ref. 17-37, p. 14). The L_2 approximation of exact data also has many properties which are frequently desirable, and the introduction of weight functions adds a dimension of flexibility that is often of use. Much of the theory is of course very classical, being found in almost identical form for example in the theory of Fourier series. We start with the simplest case.

17.4.1 L_2 Polynomial Approximation on a Finite Interval; Legendre Polynomials

Consider least-mean-square approximation to an arbitrary $f(x)$ on a finite interval, which we take to be $[-1, 1]$. This is in conformity with most references in this context; but, moreover, there is an important *symmetry* consideration that makes this a wise choice: since evenness and oddness about the midpoint are important features of many functions that arise, putting the origin at the center aids recognition of symmetry properties and yields algebraic simplifications. It is no loss of generality to take this interval, as the linear transformation

$$x = \frac{2s - a - b}{b - a} \tag{17.4-1}$$

maps the arbitrary finite interval $a \leqslant s \leqslant b$ into $-1 \leqslant x \leqslant 1$.

The set Φ from which the arbitrary function $f(x)$ to be approximated is drawn might be the set of all functions continuous on $[0, 1]$ (the continuity requirement can be relaxed greatly if desired). The set of approximating functions will be P_n, the set of all polynomials of degree n or less, and the approximation criterion is the L_2 error norm or equivalently its square the mean square error

$$Q = \int_{-1}^{1} [f(x) - p_n(x)]^2 \, dx \tag{17.4-2}$$

The fact that Q is analytic in the unknown coefficients a_i (since it does not contain absolute value signs), and that the problem may therefore be attacked in straightforward fashion by dealing with the Euler minimization equations, which moreover are linear in the unknowns, is of course of decisive importance.

First we illustrate a procedure which, though perfectly satisfactory in easy problems, is for several reasons disadvantageous in problems of some difficulty or when n is large. We represent $p_n(x)$ in Maclaurin form

$$p_n(x) = a_0 + a_1 x + a_2 x^2 + \cdots + a_n x^n \tag{17.4-3}$$

so that what we are doing is tantamount to approximation by a linear combination of functions drawn from the set

$$\{1, x, x^2, \ldots, x^n\} \tag{17.4-4}$$

The Euler equations for minimizing (17.4-2)

$$\frac{\partial Q}{\partial a_i} = 0; \quad i = 0, 1, \ldots, n \tag{17.4-5}$$

yield the $(n + 1)$ by $(n + 1)$ linear system of *normal equations*

$$\mathbf{Aa} = \mathbf{b} \tag{17.4-6}$$

where

$$\mathbf{a} = \begin{bmatrix} a_0 \\ a_1 \\ \cdot \\ \cdot \\ \cdot \\ a_n \end{bmatrix}; \quad (\mathbf{b})_i = \int_{-1}^{1} x^i f(x)\, dx; \quad (\mathbf{A})_{ij} = \int_{-1}^{1} x^{i+j}\, dx \qquad i, j = 0, 1, \ldots, n \tag{17.4-7}$$

Since Q is quadratic in any of the a_i, and increases as a_i grows large either positively or negatively, this clearly yields a unique minimum of Q and the approximation problem is in principle solved.

As an aside we note that, in more technical language, the fact that Q has a unique minimum in the $(n + 1)$-dimensional space $\mathbf{a}$ is associated with the fact that it is a positive-definite quadratic form

$$Q = \mathbf{a}^T \mathbf{Aa} - 2\mathbf{a}^T \mathbf{b} + \int_{-1}^{1} [f(x)]^2\, dx \tag{17.4-8}$$

where $\mathbf{A}$ is a positive-definite real-symmetric matrix. Whenever a linear system is obtained by writing the Euler minimization equations for a nonnegative quadratic expression such as (17.4-2), the positive definiteness properties of the matrix are easily investigated by such a line of argument.

There are several disadvantages in this approach:

1. It is necessary to compute the elements of the $(n + 1)$ by $(n + 1)$ matrix $\mathbf{A}$ and solve the linear system (17.4-6). In this simple case, because of symmetry, half of the elements of $\mathbf{A}$ are zero, and the others may be expressed by a simple formula; nevertheless $\mathbf{A}$ is still a dense matrix. In more general L_2 approximation with linear approximating functions, although $\mathbf{A}$ retains its real symmetric property, computation of its elements is laborious.
2. Moreover, when n is fairly large, the matrix $\mathbf{A}$ becomes poorly conditioned, necessitating extra care in error control. In fact, had we taken the interval $[0, 1]$ instead of $[-1, 1]$, $\mathbf{A}$ would have turned out in this polynomial case to be the Hilbert matrix, $\mathbf{A}_{ij} = 1/(i + j - 1)$, a famous example of an ill-conditioned matrix.

3. The resulting coefficients a_i depend on the degree of approximation n; if the result is not sufficiently accurate, and consequently a decision is made to increase n, all coefficients a_i must be recomputed from scratch (this is known as *nonfinality of coefficients*).
4. Important theoretical results (Bessel's inequality, etc.) do not emerge easily in this formulation, nor do certain important physical insights (in the alternative approach the component functions entering the linear approximating function may turn out to be eigenfunctions or normal modes of a physical system under study, and as such provide the key to understanding its behavior).

In the preferred procedure we represent our approximating $p_n(x)$ in the alternative form

$$p_n(x) = b_0 L_0 + b_1 L_1(x) + b_2 L_2(x) + \cdots + b_n L_n(x) \tag{17.4-9}$$

where the *Legendre polynomials* $L_k(x)$ have the properties

$L_k(x)$ is a polynomial of degree k

$$\int_{-1}^{1} L_j(x) L_k(x)\, dx = 0, \qquad j \neq k \tag{17.4-10}$$

$$L_k(1) = 1 \tag{17.4-11}$$

We express eq. (17.4-10) by saying "the set $\{L_k(x)\}$ is orthogonal with respect to integration from -1 to 1, with weight unity."

These may be taken as the defining properties of the Legendre polynomials. Actually it is somewhat more convenient to take in place of eq. (17.4-10) the requirement

$$\int_{-1}^{1} L_k(x) p(x)\, dx = 0 \tag{17.4-12}$$

where $p(x)$ is any polynomial of degree less than k; the orthogonality condition (17.4-10) is an immediate consequence. The first few Legendre polynomials are $L_0 = 1, L_1 = x, L_2 = \frac{3}{2}x^2 - \frac{1}{2}, L_3 = \frac{5}{2}x^3 - \frac{3}{2}x$, and the general formula

$$L_k(x) = \frac{1}{2^k k!} \frac{d^k}{dx^k} [x^2 - 1]^k \tag{17.4-13}$$

is readily derived by setting $L_k(x) = c\, d^k/dx^k\, G_{2k}(x)$ in eq. (17.4-12).

Many more facts about these important functions may be found, e.g., in Ref. 17-8, pp. 82–87. However here we need only the orthogonality property, the fact

$$\int_{-1}^{1} [L_k(x)]^2\, dx = \frac{2}{2k+1} \qquad (17.4\text{-}14)$$

and the knowledge that any polynomial of degree n can be uniquely expressed in the form of eq. (17.4-9), i.e., represented as a linear combination of Legendre polynomials.

The set $\{L_0, L_1, \ldots, L_n\}$ may be regarded as resulting from applying the Gram-Schmidt orthogonalization process to the set $\{1, x, x^2, \ldots, x^n\}$ (cf. section 17.4-2).

Substituting (17.4-9) into the mean-square error (17.4-2), the Euler minimization equations

$$\frac{\partial Q}{\partial b_i} = 0; \qquad i = 0, 1, \ldots, n$$

yield the $(n + 1)$ by $(n + 1)$ system of normal equations

$$\sum_{k=0}^{n} b_k \int_{-1}^{1} L_k(x) L_j(x)\, dx = \int_{-1}^{1} f(x) L_j(x)\, dx; \qquad j = 0, 1, \ldots, n \qquad (17.4\text{-}15)$$

which because of the orthogonality property is diagonal, and yields a separate formula for each of the b_j

$$b_j = \frac{\displaystyle\int_{-1}^{1} f(x) L_j(x)\, dx}{\displaystyle\int_{-1}^{1} [L_j(x)]^2\, dx} = \frac{2k+1}{2} \int_{-1}^{1} f(x) L_j(x)\, dx \qquad (17.4\text{-}16)$$

This is called the Fourier formula for the coefficients. The fact that it does not depend on n is known as the *finality of coefficients* property: If a decision to increase n is made no recomputation need be made of coefficients previously determined. The final answer one obtains by this process is of course the same polynomial that the approach of eq. (17.4-6) yields, except for round-off errors and errors of approximate numerical integration. It may be convenient to transform the answer back into Maclaurin form (17.4-3) for ease of later evaluation.

It is rewarding to express the mean-square error (17.4-2) in terms of the Fourier coefficients (17.4-16). Let those be denoted by b_j as in eq. (17.4-16), and consider the mean-square error of an arbitrary p_n approximation

$$p_n(x) = \sum_{k=0}^{n} c_k L_k(x)$$

This is

$$Q = \int_{-1}^{1} \left[f(x) - \sum_{k=0}^{n} c_k L_k(x) \right]^2 dx$$

Using eqs. (17.4-10), (17.4-14), (17.4-16) and completing the square, this can be written

$$Q = \int_{-1}^{1} [f(x)]^2\, dx - \sum_{k=0}^{n} \frac{2}{2k+1} b_k^2 + \sum_{k=0}^{n} \frac{2}{2k+1} [c_k - b_k]^2 \qquad (17.4\text{-}17)$$

This confirms very explicitly the earlier claim that a minimum is uniquely obtained by the choice $c_k = b_k$, $k = 0, 1, \ldots, n$. With this choice we have the optimal least-square error

$$Q^* = \int_{-1}^{1} [f(x)]^2\, dx - \sum_{k=0}^{n} \frac{2}{2k+1} b_k^2 \qquad (17.4\text{-}18)$$

This shows that the approximation cannot diverge as n grows large, as Q^* can never increase. Since Q^* is nonnegative we have *Bessel's inequality*

$$\int_{-1}^{1} [f(x)]^2\, dx \geqslant \sum_{k=0}^{\infty} \frac{2}{2k+1} b_k^2$$

Parseval's Identity or the *completeness relation*

$$\int_{-1}^{1} [f(x)]^2\, dx = \sum_{k=0}^{\infty} \frac{2}{2k+1} b_k^2$$

expressing the fact that $Q^* \to 0$ as $n \to \infty$, is a deeper result (see, e.g., Ref. 17-8). This is true for a very general class of $f(x)$; for example if $f(x)$ is bounded on $[-1, 1]$, has at most a finite number of discontinuities and a finite number of maxima and minima, then $Q^* \to 0$ as $n \to \infty$. This L_2 convergence is called *convergence in the mean*, as the mean-square error goes to zero. At any point

$$p(x) \to \lim_{\epsilon \to 0} \frac{f(x+\epsilon) + f(x-\epsilon)}{2}$$

so pointwise convergence occurs at points where $f(x)$ is continuous, and uniform convergence occurs within any closed interval on which $f(x)$ is continuous. [This is a slightly weaker but practically much more useful result than the Weierstrass theorem on approximation of continuous functions by polynomials, section (18.4.1.1)]. The rate of convergence depends on the continuity properties of the derivatives of f. These facts apply in general to L_2 approximation by any complete set of functions, such as Fourier trigonometric series or expansions in other orthogonal polynomials.

From the normal equation (17.4-16) it follows that the error of the optimal $L_2 P_n$ approximation p^* is orthogonal to any $p_n(x)$:

$$\int_{-1}^{1} p_n(x)\,[f(x) - p^*(x)]\,dx = 0 \tag{17.4-19}$$

This may be considered the *error characterization property* of best L_2 polynomial approximations.

17.4.2 General Linear L_2 Approximation; Gram-Schmidt Orthogonalization

Let us consider now the L_2 approximation problem on $[-1, 1]$ where the approximating function is a linear combination of a linearly independent set of n given functions $\{\phi_1(x), \phi_2(x), \ldots, \phi_n(x)\}$

$$G(x; \mathbf{a}) = \sum_{k=1}^{n} a_k \phi_k(x) \tag{17.4-20}$$

Minimization of the mean-square error

$$Q = \int_{-1}^{1} [f(x) - G(x; \mathbf{a})]^2\,dx \tag{17.4-21}$$

leads to the n by n system of Euler equations

$$\frac{\partial Q}{\partial a_j} = 0; \quad j = 1, 2, \ldots, n$$

or

$$\mathbf{A}\mathbf{a} = \mathbf{b} \tag{17.4-22}$$

where

$$\mathbf{a} = \begin{pmatrix} a_1 \\ a_2 \\ \cdot \\ \cdot \\ \cdot \\ a_n \end{pmatrix}; \quad (\mathbf{b})_k = \int_{-1}^{1} f(x)\,\phi_k(x)\,dx; \quad (\mathbf{A})_{jk} = \int_{-1}^{1} \phi_j(x)\,\phi_k(x)\,dx$$

As in the previous section Q is identifiable as the positive-definite quadratic form

$$Q = \mathbf{a}^T \mathbf{A} \mathbf{a} - 2\mathbf{a}^T \mathbf{b} + \int_{-1}^{1} [f(x)]^2 \, dx \tag{17.4-23}$$

and **A** is a positive-definite real-symmetric matrix.

If the set of functions $\{\phi_k(x)\}$ were orthogonal, the off-diagonal elements of **A** would be zero and eq. (17.4-22) would reduce to the Fourier formula for the coefficients

$$a_j = \frac{\displaystyle\int_{-1}^{1} f(x)\,\phi_j(x)\,dx}{\displaystyle\int_{-1}^{1} [\phi_j(x)]^2\,dx} \tag{17.4-24}$$

To secure this advantage it may be worth carrying out the Gram-Schmidt orthogonalization process on the set $\{\phi_k(x)\}$. We describe this process first in its simplest form.

Choose $v_1(x)$ as $\phi_1(x)$ normalized, so that the integral of its square is unity

$$v_1(x) = \frac{\phi_1(x)}{\left[\displaystyle\int_{-1}^{1} [\phi_1(x)]^2\,dx\right]^{1/2}}$$

Next take

$$u_2(x) = \phi_2(x) - c_1 v_1(x) \tag{17.4-25}$$

and choose the constant c_1 such that u_2 is orthogonal to v_1

$$c_1 = \int_{-1}^{1} v_1(x)\,\phi_2(x)\,dx \tag{17.4-26}$$

and finally normalize $u_2(x)$ to obtain $v_2(x)$

$$v_2(x) = \frac{u_2(x)}{\left[\displaystyle\int_{-1}^{1} [u_2(x)]^2\,dx\right]^{1/2}}$$

Continuing,

$$u_3(x) = \phi_3(x) - d_1 v_1(x) - d_2 v_2(x)$$

$$d_1 = \int_{-1}^{1} v_1(x)\phi_3(x)\,dx; \quad d_2 = \int_{-1}^{1} v_2(x)\phi_3(x)\,dx$$

etc. In this fashion an orthogonal set $\{v_1(x), v_2(x), \ldots\}$ is built up.

This procedure may work well if the functions of the set $\{\phi_1(x), \phi_2(x), \ldots\}$ are strongly linearly independent and arranged in some logical order [in the polynomial case, eq. (17.4-4), this procedure can't be improved upon]. However sometimes care needs to be taken in selecting the order in which the function $\phi_k(x)$ are brought into the orthogonal set; this is particularly true when the $\phi_k(x)$ are in no logical order, and when some sort of errors may be present, or when linear dependencies may be present.

An alternative procedure requires some sort of criterion for determining the 'size' of functions; computation of the L_2 norm or its square would do, but often some less laborious criterion is quite satisfactory. The function $v_1(x)$ is chosen as the 'biggest' ϕ_k, and then normalized (if desired). Before proceeding all other $\phi_k(x)$ are made orthogonal to $v_1(x)$

$$\tilde{\phi}_k(x) = \phi_k(x) - c_k v_1(x); \qquad k = 2, 3, \ldots$$

$$c_k = \int_{-1}^{1} \phi_k(x) v_1(x)\,dx \tag{17.4-27}$$

Then $v_2(x)$ is chosen as the biggest of the remaining $\phi_k(x)$, etc.

The evil which this procedure is designed to avoid is bringing into the orthogonal set $\{v_1(x), v_2(x), \ldots, v_j(x)\}$ a $\phi_{j+1}(x)$ which is nearly linearly dependent on the functions already in the set. That would involve a

$$u_{j+1} = \phi_{j+1} - c_1 v_1 - c_2 v_2 - \cdots - c_j v_j$$

which was small, involving the difference between nearly equal quantities; the relative error in the normalized v_{j+1} could then be quite large, and subsequent calculations would be affected. The above procedure insures that such effects, if they must occur, are postponed to as late a stage in the process as possible and prevented from affecting the accuracy of the early part of the process. Perhaps this trick is even more important in orthogonalizing sets of vectors, as in dealing with L_2 approximation over discrete point sets.

17.4.3 Weighted L_2 Polynomial Approximation

The mean square error criterion may be generalized by the inclusion of a weight function $w(x)$, i.e., we may proceed to minimize

$$Q = \int_{-1}^{1} w(x)[f(x) - G(x)]^2 \, dx \tag{17.4-28}$$

where $w(x)$ is some weight function obeying $w(x) > 0$ in $[-1, 1]$. We may do this because we are more concerned with certain parts of the interval than with others, and therefore wish to weigh errors more heavily in the critical regions. However, perhaps more frequently than not, the weighted error criterion is thrust upon us because we are dealing with approximating functions which obey a weighted orthogonality condition

$$G(x) = \sum_{k=1}^{n} a_k \phi_k(x) \tag{17.4-29}$$

$$\int_{-1}^{1} w(x) \phi_k(x) \phi_j(x) \, dx = 0; \quad j \neq k \tag{17.4-30}$$

We can express this by saying that ϕ_k and ϕ_j are "w-orthogonal." In this case if we use the error criterion (17.4-28) the Euler equations reduce to the Fourier formula for the coefficients

$$a_j = \frac{\displaystyle\int_{-1}^{1} w(x) f(x) \phi_j(x) \, dx}{\displaystyle\int_{-1}^{1} w(x) [\phi_j(x)]^2 \, dx} \tag{17.4-31}$$

and all of the results of section (17.4.2) readily generalize, the factor $w(x)$ making its appearance inside every integral.

For example, the Chebyshev polynomials

$$T_k(x) = \cos(k \cos^{-1} x) \tag{17.4-32}$$

are orthogonal under integration -1 to 1 with weight $w(x) = 1/\sqrt{1 - x^2}$. This weight function weighs the end of the interval heavily compared to the middle. The truncated Chebyshev series expansion of $f(x)$

$$\sum_{k=0}^{n} c_k T_k(x) \tag{17.4-33}$$

$$c_k = \int_{-1}^{1} \frac{f(x) T_k(x)}{\sqrt{1 - x^2}} \, dx \Bigg/ \int_{-1}^{1} \frac{[T_k(x)]^2}{\sqrt{1 - x^2}} \, dx$$

is the best P_n approximation to $f(x)$ in the weighted L_2 sense with the weight function $1/\sqrt{1-x^2}$; it has no necessary connection to the best P_n approximation in the L_∞ or Chebyshev sense; however, since L_2 approximation errors tend to be greater near the ends of the interval, (17.4-33) is usually better in the L_∞ sense than the best unweighted L_2 approximation [the truncated Legendre series expansion (17.4-9)].

Both the Legendre polynomials and the Chebyshev polynomials are special cases of Jacobi polynomials (see Ref. 17-8, p. 90).

For the semi-infinite interval $0 \leqslant x < \infty$ a weight factor is needed to make the integrals involved converge for a wide class of functions. The usual weight function is $w(x) = e^{-x}$, as in the norm (17.1-7). The corresponding orthogonal polynomials are the Laguerre polynomials[17-8]

$$p_n(x) = e^x \frac{d^n}{dx^n} (x^n e^{-x}) \tag{17.4-34}$$

Similarly, for the doubly infinite interval $-\infty < x < \infty$ the usual weight function is $w(x) = e^{-x^2}$, as in the norm (17.1-8) and the corresponding orthogonal polynomials are the Hermite polynomials

$$p_n(x) = (-1)^n e^{x^2} \frac{d^n}{dx^n} e^{-x^2} \tag{17.4-35}$$

For any positive weight function (say for simplicity over a finite interval) an orthogonal set of polynomials exists, and can be determined by applying Gram-Schmidt orthogonalization [with weight $w(x)$] to the set $\{1, x, x^2, \ldots\}$. However there is an easier approach: Define the $w(x)$-weighted inner product by

$$\langle f(x), g(x) \rangle = \int_{-1}^{1} w(x) f(x) g(x)\, dx$$

Then take

$$p_0(x) = 1; \quad p_1(x) = x - \frac{\langle x, 1 \rangle}{\langle 1, 1 \rangle}$$

these two being $w(x)$-orthogonal. Now assume that the first k members of a set of w-orthogonal polynomials have been computed, and write

$$p_{k+1}(x) = (x - \alpha_k) p_k(x) - \beta_k p_{k-1}(x) \tag{17.4-36}$$

It is easy to see that, under the inductive hypothesis, $p_{k+1}(x)$ is w-orthogonal to

$p_j(x)$ for $j < k - 1$. It will also be orthogonal to $p_{k-1}(x)$ and $p_k(x)$ if we choose

$$\alpha_k = \frac{\langle xp_k, p_k \rangle}{\langle p_k, p_k \rangle}; \quad \beta_k = \frac{\langle xp_k, p_{k-1} \rangle}{\langle p_{k-1}, p_{k-1} \rangle} \tag{17.4-37}$$

which enables the set to be built up by recursion.

17.4.4 L_2 Approximation on Discrete Point Sets

Let

$$X: \ -1 \leqslant x_1 < x_2 < \cdots < x_m \leqslant 1$$

be a set of points in [0, 1], and let $w_i > 0$, $i = 1, \ldots, m$ be prescribed weights. Suppose we are given m values $(f_1, f_2, \ldots, f_m)$, which might simply be measured or otherwise given data, or they might be values of some function $f(x)$ sampled at the points $x_1, x_2, \ldots, x_m$. Suppose it is desired to approximate these values by a linear function of the form

$$G(x; \mathbf{a}) = \sum_{k=1}^{n} a_k \phi_k(x) \tag{17.4-38}$$

(since it will only be necessary to know values of the $\phi_k(x)$ at the points x_i these also could be just vectors of m values rather than functions of a continuous variable). Then a best L_2 weighted approximation minimize the weighted mean-square error

$$Q_X = \sum_{j=1}^{m} w_j \left[f_j - \sum_{k=1}^{n} a_k \phi_k(x_j) \right]^2 \tag{17.4-39}$$

in analogy with eq. (17.4-1); $Q_X^{1/2}$ is a norm of the error. All formulas of the previous sections apply, with summations replacing integrations. In particular, if the set of functions is w-orthogonal

$$\sum_{k=1}^{m} w_j \phi_p(x_j) \phi_q(x_j) = 0 \qquad p \neq q \tag{17.4-40}$$

then the normal equations for minimization of Q_X decouple into Fourier formulas for the coefficients

$$a_k^* = \frac{\sum_{j=1}^{m} w_j f_j \phi_k(x_j)}{\sum_{j=1}^{m} w_j [\phi_k(x_j)]^2} \tag{17.4-41}$$

(It may be convenient to normalize the functions $\phi_k(x)$ so that these denominators become unity.)

In section (18.4.4.1) of this book, Forsythe's method for constructing polynomials which are w-orthogonal over an arbitrary set X with arbitrary weights w_i is presented; this is the discrete analog of the method of eq. (17.4-36) for the construction of sets of polynomials that are w-orthogonal over a continuum.

Since the orthogonal polynomials for discrete point sets X depend on the particular spacing of the points x_i, one does not have orthogonal polynomial sets of such general utility as in the context of continuous intervals. However two cases may be mentioned: The Gram polynomials and the Chebyshev polynomials are each orthogonal on a discrete point set X with unity weight; in the former case the points in X are equally spaced, and in the latter they are the zeros of a higher order Chebyshev polynomial. (The summation orthogonality property of Chebyshev polynomials is related to a similar property of sets of sinusoidal functions.)

The classical reference on orthogonal polynomials, for both the continuous and discrete cases, is Szego.[17-44]

17.5 THEORY OF CHEBYSHEV APPROXIMATION

Many methods are based on the characterization properties of good Chebyshev approximations and best Chebyshev approximations, and the recognition and evaluation of such good and best approximations often relies on that theory. For maximum clarity we outline the theory first for the simplest case, unweighted polynomial L_∞ approximation.

17.5.1 Chebyshev Approximation on a Finite Interval by Polynomials

We choose the interval [0, 1] as is customary (although if the functions under treatment have symmetry or antisymmetry properties the interval [-1, 1] may be preferable). We let the function $f(x)$ to be approximated be drawn from the set of all functions continuous on [0, 1], choose the set P_n of polynomials of degree n or less as the set of best approximations, and take the Chebyshev or L_∞ error norm

$$\lambda = \lambda(p_n) = \|f(x) - p_n(x)\|_\infty = \max_{0<x<1} |f(x) - p_n(x)| \tag{17.5-1}$$

We seek to investigate $p_n^*(x)$ such that, for all $p_n(x)$

$$\lambda^* = \|f(x) - p_n^*(x)\| \leqslant \|f(x) - p_n(x)\| \tag{17.5-2}$$

$p_n^*(x)$ is known to exist, as discussed in section 17.3.

The fundamental property of polynomials which is decisive in the theory follows from the fundamental theory of roots (zeros) of polynomials: $p_n(x)$ has exactly n roots, counting multiplicity and admitting complex roots, so that in particular it cannot have more than n. This last statement is equivalent to the possibility and uniqueness of polynomial interpolation; for, given any $n+1$ distinct abscissas $x_0, x_1, \ldots, x_n$, the system of $n+1$ equations

$$p_n(x_j) = \sum_{k=0}^{n} a_k x_j^k = 0; \qquad j = 0, \ldots, n \tag{17.5-3}$$

can have only the zero solution $a_0 = a_1 = \cdots = a_n = 0$. Thus the matrix of this system (called a *Vandermonde matrix*) is nonsingular, and the nonhomogeneous system

$$p_n(x_j) = \sum_{k=0}^{n} a_k x_j^k = f_j; \qquad j = 0, \ldots, n \tag{17.5-4}$$

where $f_0, f_1, \ldots, f_n$ are any given ordinates, has a unique solution. In generalizations of the linear theory this condition is called the *Haar condition*, and sets of functions which obey it are called *Chebyshev sets*.

Now consider any $p_n(x)$, and construct a *set of critical points* $[t_0, t_1, t_2, \ldots]$ of $p_n(x)$ as follows: Denote its error norm λ as in eq. (17.5-1), and let t_0 be the smallest value of $x \geqslant 0$ for which the error attains the magnitude λ, say with positive sign

$$e(t_0) = f(t_0) - p_n(t_0) = \lambda$$

[If $e(t_0)$ is negative simply reverse all signs in the following argument.] Then let t_1 be the first value of $x > t_0$ at which $e(t_1) = -\lambda$ (if there is any), and let z_1 be the largest zero of $e(x)$ less than t_1. Then let t_2 be the first value of x greater than t_1 at which $e(t_2) = +\lambda$ and z_2 the largest zero of $e(x)$ less than t_2, etc. Continuing in this fashion we construct the critical point set $\{t_0, t_1, t_2, \ldots\}$ and a nested set of zeros $\{z_1, z_2, \ldots\}$. Figure 17.5-1 illustrates this process for a specific example, showing some unusual situations which could occur.

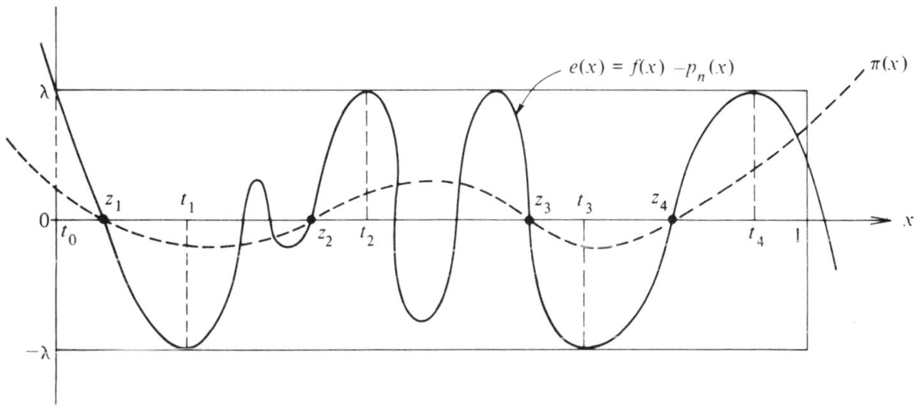

Fig. 17.5-1 Sample error curve.

Now let us examine the assumption that a critical set of points $\{t_0, t_1, \ldots, t_q\}$ with $q \leqslant n$ suffices to cover the range $0 \leqslant x \leqslant 1$. In this case

$$\pi(x) = \prod_{k=1}^{q} (x - z_k) \tag{17.5-5}$$

(shown as a dashed line on Fig. 17.5-1) is a polynomial of degree not exceeding n, so that

$$\tilde{p}_n = p_n + \epsilon\pi \tag{17.5-6}$$

is an admissible approximation. Now the error of this approximation is

$$f - \tilde{p}_n = [f - p_n] - \epsilon\pi$$

and by examining Fig. 17.5-1 we see that, if ϵ is properly chosen, this error will have norm less than λ. For by the subtraction of $\epsilon\pi$, each of the error extrema of p_n will be pulled back a little toward the horizontal axis.

Thus may be proven the first part of Chebyshev's *characterization theorem*: a best L_∞ polynomial approximation to a continuous $f(x)$ possesses a set of $(n + 2)$ critical points, i.e., equioscillates (at least) $n + 1$ times.

It is efficient to show next that $p^*(x)$ is unique. Assume there are two best P_n approximations $S_n(x)$ and $T_n(x)$ and consider the admissible approximation

$$Q_n(x) = \frac{S_n(x) + T_n(x)}{2}$$

At all points in $0 \leqslant x \leqslant 1$ we have

$$|f(x) - Q_n(x)| = \left|\frac{f - S_n}{2} + \frac{f - T_n}{2}\right| \leqslant \left|\frac{f - S_n}{2}\right| + \left|\frac{f - T_n}{2}\right|$$

$$\leqslant \frac{1}{2}\|f - S_n\|_\infty + \frac{1}{2}\|f - T_n\|_\infty = \lambda^*$$

so that $Q_n(x)$ is also a best approximation. (As an aside we remark that this type of argument is valid for any error norm and shows that the set of best approximations is convex.) By the previous result then the error of Q_n equioscillates on some set of $n + 2$ critical points t_i in $[0, 1]$

$$f(t_i) - Q_n(t_i) = \frac{f(t_i) - S_n(t_i)}{2} + \frac{f(t_i) - T_n(t_i)}{2} = (-1)^i\lambda \qquad i = 0, 1, \ldots, n + 1$$

where $|\lambda| = \lambda^*$. But since $\lambda^*/2$ is the greatest magnitude that either $\frac{1}{2}(f - S_n)$ or $\frac{1}{2}(f - T_n)$ can attain in [0, 1], this can only be true if

$$\frac{f(t_i) - S_n(t_i)}{2} = \frac{f(t_i) - T_n(t_i)}{2} = \frac{(-1)^i\lambda}{2}; \qquad i = 0, 1, \ldots, n+1$$

from which we conclude that the polynomials $S_n(x)$ and $T_n(x)$ are equal at $n + 2$ distinct points. Therefore they are identically equal, proving uniqueness.

The final step in the fundamental theory is to prove that a p_n with the equioscillation property must be p_n^*. Let $p_n(x)$ be a polynomial obeying

$$f(t_i) - p_n(t_i) = (-1)^i\lambda; \qquad i = 0, 1, \ldots, n+1$$

where $0 \leqslant t_0 < t_1 < \cdots < t_{n+1} \leqslant 1$, and assume that

$$|\lambda| = \|f - p_n\|_\infty > \lambda^* \tag{17.5-7}$$

Then

$$R_n(x) = [f(x) - p_n(x)] - \{f(x) - p_n^*(x)\}$$

is a polynomial of degree n which alternates in sign on the $n + 2$ critical points, since the curly bracket has magnitude strictly less than $|\lambda|$. That however is impossible since it would imply that $R_n(x)$ would have $n + 1$ distinct roots, nested among the critical points $t_0, \ldots, t_{n+1}$. Therefore eq. (17.5-7) cannot be true and $\|f - P_n\|_\infty = \lambda^*$. Therefore p_n is a best approximation, and, because the best approximation is unique, p_n is *the* best approximation.

This completes the proof of the *Fundamental Characterization and Uniqueness Theorem: p_n^*, the best Chebyshev p_n approximation to a continuous $f(x)$ on [0, 1], is unique; $p_n(x)$ is p_n^* if and only if its error equioscillates on a set of $n + 2$ critical points $0 \leqslant t_0 < t_1 < \cdots < t_{n+1} + 1$:*

$$f(t_i) - p_n(t_i) = (-1)^i\lambda \tag{17.5-8}$$

where

$$|\lambda| = \|f(x) - p_n(x)\|_\infty \tag{17.5-9}$$

In a few very easy problems (e.g., best P_1 approximation to $f(x) = \sqrt{x}$ in [0, 1]) p^* may be calculated directly from the characterization property. Most problems require iterative methods, which however still may be based on the characterization property (section 17.6).

Any p_n of course yields an upper bound on λ^*. A lower bound is given by an important *Theorem of de la Vallée Poussin*: *If $p_n(x)$ takes on alternating error* *values*

$$f(t_k) - p_n(t_k) = (-1)^k \lambda_k; \qquad k = 0, 1, \ldots, n+1 \tag{17.5-10}$$

on a set of n + 2 points

$$0 \leqslant t_0 < t_1 < \cdots < t_{n+1} \leqslant 1 \tag{17.5-11}$$

where the λ_k are all of the same sign, then

$$\lambda^* \geqslant \min_k |\lambda_k| \tag{17.5-12}$$

This may be proven by assuming the contrary, whereupon the polynomial

$$Q_n(x) = [f(x) - p_n^*(x)] - [f(x) - p_n(x)] = p_n(x) - p_n^*(x)$$

has alternating signs on the $n + 2$ point (17.5-11), and therefore has $n + 1$ distinct zeros, which is impossible.

Many iterative computational methods obtain on each iteration a polynomial with the property (17.5-10). Such methods therefore have the attractive feature that at each step the norm of the current approximating polynomial (a 'good approximation,' if you will) can be compared to the current lower bound to λ^*, giving a good basis for the decision as to when to terminate the iterations.

17.5.2 Extension to Chebyshev Sets

The general linear approximation function is of the form

$$G(x; \mathrm{a}) = \sum_{k=1}^{n} a_k \phi_k(x) \tag{17.5-13}$$

where the $\phi_k(x)$ are arbitrary functions of x (usually assumed linearly independent to avoid the nuisance of having to remove linear dependencies in proving existence, etc.) Equation (17.5-13) is *linear* in the unknown parameters $a_1, a_2, \ldots, a_n$; the given functions $\phi_1(x), \ldots, \phi_n(x)$ may of course be arbitrary nonlinear functions of x. Polynomials are a special case of eq. (17.5-13) with $\phi_k(x) = x^{k-1}$ (note the unfortunate redefinition of n).

The fundamental uniqueness and characterization theorem does not apply to (17.5-13), as can be seen from the example $n = 1, \phi_1(x) = x, f(x) = 1$. The proof does not go through because the fundamental property of polynomials, discussed at eq. (17.5-3), is not necessarily possessed by the general linear approximating function (17.5-13). A restriction on the set of function $\{\phi_k(x)\}$ is needed, and the appropriate restriction is to Chebyshev sets. $\{\phi_1(x), \ldots, \phi_n(x)\}$ is a *Chebyshev set* on the interval $[a, b]$ if it obeys a condition often called the *Haar condition*, which may be stated in three equivalent forms [note the analogy to eqs. (17.5-3), (17.5-4)]:

1. No linear combination of form (17.5-13) has more than $n - 1$ zeros in $[a, b]$ unless it vanishes identically
2. If $x_1, \ldots, x_n$ are n arbitrary but distinct abscissas in $[a, b]$ and $f_1, \ldots, f_n$ are arbitrary numbers, then the interpolation problem

$$\sum_{k=1}^{n} a_k \phi_k(x_j) = f_j; \quad j = 1, \ldots, n \tag{17.5-14}$$

   has a unique solution $a_1, \ldots, a_n$.
3. The determinant of the system (17.5-14) is nonzero for any n distinct values $x_1, \ldots, x_n$ in $[a, b]$.

With the restriction to Chebyshev sets the characterization theory goes through just as with polynomials, and we have the *Fundamental Theorem*: $G(x; \mathbf{a}^*)$, *the best Chebyshev approximation to an arbitrary continuous* $f(x)$ *on* $[0, 1]$, *is unique*; $G(x; \mathbf{a})$ *is* $G(x; \mathbf{a}^*)$ *if and only if its error equioscillates on a set of* $n + 1$ *critical point* $0 \leqslant t_0 < t_1 < \cdots < t_n \leqslant 1$:

$$\begin{aligned} f(t_i) - G(t_i; \mathbf{a}) &= (-1)^i \lambda \\ |\lambda| &= \|f(x) - G(x; \mathbf{a})\| \end{aligned} \tag{17.5-15}$$

(See, for example, Ref. 17-31, section 3.2, or Ref. 17-18, section 7.5.) The de la Vallée Poussin result likewise generalizes: *If an approximation obeys*

$$f(t_k) - A(t_k; \mathbf{a}) = (-1)^k \lambda_k \tag{17.5-16}$$

where the λ_k *are all of one sign, then*

$$\min_k |\lambda_k| \leqslant \lambda^* \tag{17.5-17}$$

Unfortunately most sets of functions $\{\phi_k(x)\}$ are not Chebyshev sets, so it must be admitted that this generalization is of more theoretical than practical value. The set $\{1, x, x^2, \ldots, x^{n-1}\}$ is a Chebyshev set on any interval $[a, b]$ (the polynomial case), but if one power of x is omitted the property fails; $\{x, x^2\}$ is not a Chebyshev set on $[0, 1]$; $\{1, x^2\}$ is on $[0, 1]$ but is not on $[-1, 1]$. The other well-known type of Chebyshev set includes trigonometric polynomials, with the range of x appropriately restricted (Ref. 17-31, p. 55). The example of approximation of x^2 by the set $\{x, e^x\}$, which is not a Chebyshev set, is given by Handscomb et al. (Ref. 17-18, p. 69); the best Chebyshev approximation on $[0, 2]$ has just two error extrema.

Failure of the Chebyshev set property disables the methods of section 17.6, but not the linear programming approach of 17.7.

17.5.3 Extension to Rational Functions

Though rational functions

$$R(x) = R_{p,q}(x) = \frac{P_p(x)}{Q_q(x)} = \frac{\sum_{k=0}^{p} a_k x^k}{\sum_{k=0}^{q} b_k x^k} \tag{17.5-18}$$

are not linear in the coefficients b_k of the denominator polynomial, the characterization theory generalizes to rational approximating functions, though with some complications. There are $p + q + 2$ parameters in (17.5-18), but since numerator and denominator may be multiplied by the same constant, one constraint may be arbitrarily imposed, so there are in fact

$$n = p + q + 1 \tag{17.5-19}$$

degrees of freedom. (We usually prefer to take $b_0 = 1$; since $x = 0$ is included in the region of interest this is always legitimate. Sometimes, however, the constraint $\sum_{k=0}^{q} b_k^2 = 1$ is convenient.)

For generality we include a positive weight function $w(x)$ in the error norm

$$\lambda = \lambda(R) = \| w(x)[f(x) - R(x)] \|_\infty$$
$$= \max_{0 \leqslant x \leqslant 1} |w(x)[f(x) - R(x)]| \tag{17.5-20}$$

(This could have been done in the previous two sections; indeed polynomial approximation is a special case of rational function approximation.)

The best rational approximation $R^*(x)$ to an arbitrary $f(x)$ that minimizes (17.5-20) exists and is unique (see Refs. 17-1, 17-31, 17-35); the complications that arise in the characterization property are associated with the fact that in irreducible form (i.e., with all common factors of numerator and denominator cancelled off), the numerator and denominator of R^* may be of degree less than p and q respectively. Let R^* in irreducible form be

$$R^*(x) = \frac{\sum_{k=0}^{p-\alpha} a_k^* x^k}{\sum_{k=0}^{q-\beta} b_k^* x^k} \tag{17.5-21}$$

Then the *defect* of R^* is defined as

$$d = \min(\alpha, \beta) \qquad (17.5\text{-}22)$$

and $R^*(x)$ equioscillates on a set of $n + 1 - d = p + q + 2 - d$ critical points, i.e., we have the theorem: *A rational function $R(x)$ with defect d is the best approximation R^* if and only if there exist (at least) $n + 1 - d$ abscissas*

$$0 \leqslant t_1 < t_2 < \cdots < t_{n+1-d} \leqslant 1 \qquad (17.5\text{-}23)$$

such that

$$w(t_i)\,[f(t_i) - R(t_i)] = (-1)^i \lambda \qquad i = 1, \ldots, n + 1 - d \qquad (17.5\text{-}24)$$

$$|\lambda| = \|w(x)\,[f(x) - R(x)]\|_\infty$$

If its defect d is >0 the solution R^* is called *degenerate*. In practical work true degeneracy is unusual and the assumption of nondegeneracy often not restrictive.[17-27]

17.5.4 Rational Function Interpolation

It is efficient at this point to discuss interpolation by rational functions, required in some computational approaches. As we saw at eq. (17.5-19) the right number of conditions to impose in a rational function interpolation problem is $n = p + q + 1$, so the fundamental problem may be stated

$$R(x_i) = \frac{P_p(x_i)}{Q_q(x_i)} = f_i; \qquad i = 1, 2, \ldots, n \qquad (17.5\text{-}25)$$

where $x_1, \ldots, x_n$ are n arbitrary but distinct given abscissas, $f_1, \ldots, f_n$ are n arbitrary numbers, with $n = p + q + 1$.

A practical procedure for solving such a problem is as follows: first solve the homogeneous linear system obtained by multiplying out the denominators of eqs. (17.5-25)

$$P_p(x_i) - f_i Q_q(x_i) = 0; \qquad i = 1, 2, \ldots, n \qquad (17.5\text{-}26)$$

This always has nontrivial solutions for the unknown $\{a_j, b_j\}$ since there are one more unknowns than equations. Furthermore in the solution Q cannot be identically zero, as then eq. (17.5-26) would imply that P_p has too many roots.

The next step is to check whether Q is zero at any of the interpolation points x_i. If not, the solution to (17.5-25) has been found. The solution is then unique, to within multiplication of numerator and denominator by the same factor.

If $Q(x_i) = 0$ for some x_i, eq. (17.5-26) implies $P(x_i) = 0$ also. Thus P and Q have the common factor $(x - x_i)$ which may be divided off. The result almost certainly will not obey eq. (17.5-25) at $x = x_i$, and in this case (17.5-25) has no solution.

That this situation is unusual may be seen as follows: Suppose $Q(x_i) = 0$ at $m > 0$ interpolation points, say $x_1, x_2, \ldots, x_m$. The solution of eqs. (17.5-26) then gives a rational function R which satisfies the remaining $n - m$ of conditions (17.5-25). But the m factors $(x - x_1)(x - x_2) \ldots (x - x_m)$ are common to P and Q and may be cancelled in $R = P/Q$, so R involves in fact only $n - 2m$ degrees of freedom (it equals a polynomial of degree $p - m$ divided by a polynomial of degree $q - m$). This R solves the interpolation problem

$$R(x_i) = f_i; \qquad i = m + 1, \ldots, n$$

which contains m more conditions than R has degrees of freedom.

For further discussion, including proofs of uniqueness, see Ref. 17-35, p. 132.

17.5.5 Extension to Other Nonlinear Approximating Functions

Rice has extended the characterization theory approach to nonlinear problems through the idea of varisolvence—a local generalization of the unique interpolation property of Chebyshev sets.[17-29, 17-33] The most important practical application of this theory is in exponential approximation, i.e., in the use of approximating functions of the form

$$G(x) = \sum_{k=1}^{m} a_k e^{b_k x} \tag{17.5-27}$$

(The theory applies also to rational functions but they perhaps are more easily handled by other methods.) The characterization properties of best approximations of the form (17.5-27) are complicated by special cases, corresponding to the fact that the closure of the set (17.5-27) contains polynomials multiplied by exponentials, (cf 17-32). The real set of interest is the set of all solutions to all constant coefficient, ordinary differential equations of the form

$$\prod_{k=1}^{m} (D - b_k)\, G(x) = 0 \tag{17.5-28}$$

this being the closure of the set (17.5-27).

17.6 CHEBYSHEV APPROXIMATION METHODS BASED ON CHARACTERIZATION PROPERTIES

The methods discussed in this section, for Chebyshev approximation on a continuous finite interval which we take to be [0, 1], include the methods most used to date. Let the approximating function be denoted

$$G(x; \mathbf{a}); \quad \mathbf{a} = \begin{pmatrix} a_1 \\ a_2 \\ \cdot \\ \cdot \\ \cdot \\ a_n \end{pmatrix} \tag{17.6-1}$$

and denote the (possibly weighted) error by

$$e(x) = w(x)\,[f(x) - G(x; \mathbf{a})] \tag{17.6-2}$$

where $w(x)$ is a given positive weight function which may be taken $\equiv 1$ if the uniformly weighted case is desired.

The methods of this section rely on a full characterization property, i.e., the best approximation G^* is assumed to have a weighted error which equioscillates on a set of $n + 1$ abscissas (critical points)

$$0 \leqslant t_0 < t_1 < \cdots < t_n \leqslant 1 \tag{17.6-3}$$

$$e(t_k) = (-1)^k \lambda \tag{17.6-4}$$

$$|\lambda| = \lambda^* = \max_{0 \leqslant x \leqslant 1} |w(x)\,[f(x) - G^*(x; \mathbf{a})]| \tag{17.6-5}$$

The methods thus are applicable for the following types of approximation:

1. Polynomial approximation–

$$G(x; \mathbf{a}) = \sum_{k=1}^{n} a_k\, x^{k-1} = P_{n-1}(x) \tag{17.6-6}$$

(In this case n, the number of parameters in the approximation, is one less than the n in section 17.5.1, which there denoted the degree of the polynomial).

2. Linear approximation with a Chebyshev set–

$$G(x; \mathbf{a}) = \sum_{k=1}^{n} a_k\, \phi_k(x) \tag{17.6-7}$$

where $\{\phi_1(x), \phi_2(x), \ldots, \phi_n(x)\}$ is a Chebyshev set [see eq. (17.5-14)].

3. Rational approximation–

$$G(x; \mathbf{a}) = \frac{\sum_{j=0}^{p} a_j\, x^j}{\sum_{k=0}^{q} b_k\, x^k} \tag{17.6-8}$$

where $p + q + 1 = n$. If $b_0 = 1$ is taken one may take $\mathbf{a}^T = (a_0, a_1, \ldots, a_p, b_1, \ldots, b_q)$.

4. To a limited extent to other situations when characterization results have been established, such as simple exponential approximation–

$$G(x; \mathbf{a}) = a_1 a_2^x + a_3; \quad \text{or} \quad a_1 e^{a_3 x} + a_2 e^{a_4 x}; \text{ etc.} \tag{17.6-9}$$

In each case a weight function $w(x)$ can be included in the error as in eq. (17.6-2). In the last two cases the characterization property is complicated by special cases as noted earlier.

The methods of this section start with an externally supplied starting approximation $G(x; \mathbf{a}^{(0)})$, which (although of course it will not equioscillate) must have a full complement of n oscillations. The typical iteration examines the error curve of the kth iterate $G(x; \mathbf{a}^{(k)})$ and uses its properties in some manner to construct a 'better' next iterate $G(x; \mathbf{a}^{(k+1)})$. There must be a set of $n + 1$ abscissas

$$0 \leqslant t_0^{(k)} < t_1^{(k)} < \cdots < t_n^{(k)} \leqslant 1 \tag{17.6-10}$$

on which the kth error curve oscillates

$$e^{(k)}[t_j^{(k)}] = (-1)^j \lambda_j^{(k)} \tag{17.6-11}$$

where the $\lambda_j^{(k)}$ are all of one sign. This requirement must be met by the externally supplied starting approximation $G(x; \mathbf{a}^{(0)})$. A de la Valée Poussin theorem gives a lower bound of λ^* at each step

$$\lambda^* \geqslant \max_j |\lambda_j^{(k)}| \tag{17.6-12}$$

These methods are sometimes called *ascent methods* since the lower bound (17.6-12) is raised on each iteration. One motivation behind the methods is to *level* the error curve, i.e., to make modifications in **a** so that the relative extrema of $e(x)$, which at the start have various different magnitudes, become more equal in magnitude on each iteration.

17.6.1 Generalized Remes Methods on a Continuum

Methods of Remes type are the most used of this family of methods and also perhaps among the most complicated. They are based on iteratively adjusting a set of abscissas as in eq. (17.6-10), which hopefully converge to a critical point set on which G^* equioscillates. Methods of this type are sometimes called *exchange methods*, particularly when one is working on a finite point set, as will be explained later.

At the start of the $(k + 1)$ st iteration one has available a set of $n + 1$ abscissas $t_0^{(k)}, t_1^{(k)}, \ldots, t_n^{(k)}$ as in eq. (17.6-10). The first step is to solve the following system for $\mathbf{a}^{(k+1)}$ and $\lambda^{(k+1)}$

$$e^{(k+1)}(t_j^{(k)}) = w(t_j^{(k)}) [f(t_j^{(k)}) - G(t_j^{(k)}; \mathbf{a}^{k+1})] = (-1)^j \lambda^{(k+1)} \qquad j = 0, 1, \ldots, n \tag{17.6-13}$$

This is an $(n + 1)$ by $(n + 1)$ linear system if $G(x; \mathbf{a})$ is linear in the components of $\mathbf{a}$ [cases 1 and 2 above]; otherwise, as in rational approximation, it is nonlinear. Note that it is more complicated than an interpolation problem, as the values which $G(x; \mathbf{a}^{(k+1)})$ must take on at the $n + 1$ given abscissas depend on the unknown $\lambda^{(k+1)}$. In the rational function case of eq. (17.6-8), there are typically $q + 1$ solutions, one of which is the desired one; the others yield rational functions with poles in [0, 1], totally unsuitable as approximations to $f(x)$.

After this step, by the appropriate de la Vallée Poussin theorem, $|\lambda^{(k+1)}|$ is a lower bound on λ^*, and of course $\|G(x; \mathbf{a}^{(k+1)})\|_\infty$ is an upper bound. The iterations may be terminated if the upper and lower bounds are sufficiently close.

The second step is to examine the error curve $e^{(k+1)}(x)$ of the new iterate $G(x; \mathbf{a}^{(k+1)})$. This is known to take on values as given by eq. (17.6-13), but these of course are not its relative extrema, except in the final stages when the iterations have converged. A typical behavior of the error curve is shown in Fig. (17.6-1). The error is searched, and its new relative extrema located (shown as vertical dashed lines on Fig. (17.6-1). This is of course a nontrivial computational task.

In a conservative version of the process which is usually called the *one-for-one* exchange process, a single exchange is made to obtain the new set of abscissas $t_0^{(k+1)}, \ldots, t_n^{(k+1)}$. The maximum error magnitude is located, at t_m in Fig. 17.6-1, and t_m is exchanged with the neighboring $t_j^{(k)}$ for which the error has the same sign. In the case shown in Fig. 17.6-1 we would have

$$\begin{aligned} t_j^{(k+1)} &= t_j^{(k)}; \quad j \neq 2 \\ t_2^{(k+1)} &= t_m \end{aligned} \tag{17.6-14}$$

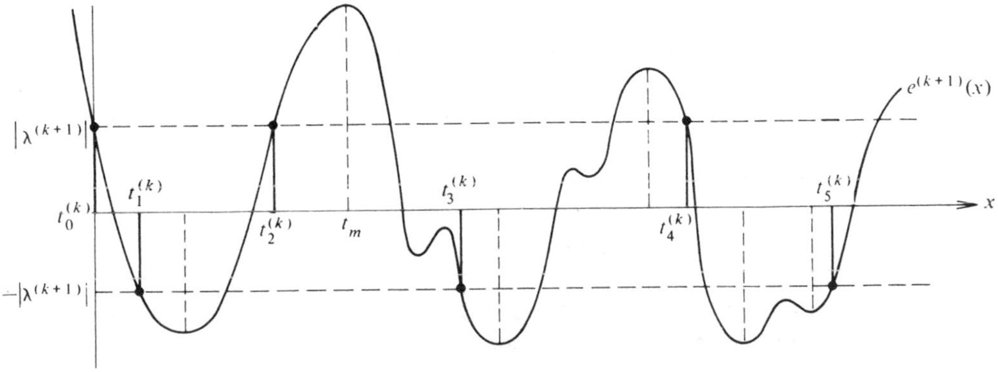

Fig. 17.6-1 Critical point exchange process.

(An obvious other type of choice, involving renumbering, is made if $t_m < t_0$ or $t_m > t_n$.)

In a more powerful version, which may be called the *multiple exchange process*, every member of the set of points $\{t_0^{(k)}, \ldots, t_n^{(k)}\}$ is moved to a nearby extremum to obtain the set $\{t_0^{(k+1)}, \ldots, t_n^{(k+1)}\}$. Sometimes it is preferred to move each $t_j^{(k)}$ to the nearest extremum of the right sign (note $t_5^{(k)}$ in Fig. 17.6-1) because theoretical results have been proven under this assumption; however if this choice omits the global maximum, an exception should be made.

Apparently then the starting requirement is not an initial approximation $G(x; \mathbf{a}^{(0)})$ but merely a set of points $\{t_0^{(0)}, \ldots, t_n^{(0)}\}$. However for a number of reasons an actual initial approximation $G(x; \mathbf{a}^{(0)})$ is desirable, especially in the rational function case. Theoretical assurances are then available, and as might be suspected computational difficulties are encountered if the initial guess is not sufficiently good (cf. Ref. 17-27).

17.6.2 Remes (or Exchange) Methods on a Discrete Point Set

The situation is simplified if one is concerned with only a discrete set of abscissas

$$X:\ x_1, x_2, \ldots, x_m \tag{17.6-15}$$

even if m is quite large (say, $m > 1000$). The error norm (17.6-5) may be replaced by the maximum (weighted) error magnitude over the discrete set

$$\lambda_X^* = \max_{x_i \in X} |w(x_i)\,[f(x_i) - G(x_i; \mathbf{a})]| \tag{17.6-16}$$

This situation may come about because (1) we decide to discretize the problem, for convenience of calculation (see section 17.2); (2) we have data at only a finite set of points; or (3) we are dealing from the start with a discrete situation, such as m inconsistent equations in $n < m$ unknowns.

The set of abscissas (17.6-10) must then be chosen from the m points (17.6-15); in this context such a set is called a *reference*, and there are only the finite number $\binom{n}{m}$ of possible choices.

A *levelled reference function* corresponding to the reference

$$x_0 \leqslant t_0 < t_1 < \cdots < t_n \leqslant x_m \tag{17.6-17}$$

is, in the present context, a solution to

$$e(t_j) = w(t_j)\,[f(t_j) - G(t_j; \mathbf{a})] = (-1)^j \lambda \tag{17.6-18}$$

i.e., a solution to eq. (17.6-13). (A more complicated sign alternation rule is given in the literature for the Chebyshev set case,[17-6, 17-43] but rarely occurs in practice.)

Of all approximations, the maximum error magnitude over the reference is least for the levelled reference function (Stiefel, Ref. 17-43; this and subsequent statements are made for the linear Chebyshev set approximations only; the possible presence of poles complicates the rational function case.)

In a typical *one-for-one exchange process* the maximum error magnitude over the entire point set X of a levelled reference function is located: say it occurs at x_m. If x_m is in the reference the solution has been found. Otherwise x_m is exchanged with a member of the reference so as to preserve the error alternation property, and a new levelled reference function calculated. The reference deviation λ is strictly monotone increasing in this process, so the same reference can never recur; and since there are a finite number of references the process must terminate after a finite number of steps with $\lambda = \lambda_X^*$.

Theoretically quite a large number of steps could be required, but in practice typically only a few more than n exchanges are required, even when m is 100 or more times larger than n. A similar phenomenon is encountered in the simplex method in linear programming, and indeed this process has striking similarities to the simplex method, which is also an exchange process in which a scalar merit function changes in monotone fashion. However the Chebyshev set property required here is very much stronger than the requirements for the simplex method (section 17.7). To trace the precise relationship between the exchange process and the simplex method see Powell (Chapter 8 in Handscomb, Ref. 17-18); also see Stiefel (Ref. 17-43) and Cheney (Ref. 17-6).

What is the relationship between the best approximation $G(x; \mathbf{a}_X^*)$ on the discrete set X and the best approximation $G(x; \mathbf{a}^*)$ on the continuous interval $[0, 1]$? By the de la Vallée Poussin theorem λ_X^* furnishes a lower bound on λ^*, and if the trouble is taken to search for the maximum magnitude on $[0, 1]$ of the error curve of $G(x; \mathbf{a}_X^*)$, an upper bound is obtained. The crucial matter is that the point set X must have points close to the critical points of $G(x; \mathbf{a}^*)$; theoretically it does not matter whether X is dense elsewhere (see section 17.2). However, since these points are unknown a priori, and since the labor of the process is so weakly dependent on the total number of points in X, it is usually most expedient to choose X fairly dense over the whole interval. If $f(x) \in C^1$ the error curve will be horizontal at its interior extrema, and slight misses will have only a small effect. Of course, in very careful work one might even follow the calculation with a second computation on a discrete point set X' chosen to have points densely bunched about each error extremum, the locations of which are approximately known as a result of the first calculation.

The single-exchange process for approximation on the continuum $[0, 1]$ discussed in the previous section may be considered an exchange process on a discrete point set X where the members of X are chosen during the computations. All the theorems which guarantee success of the process remain valid. This may partially explain why the single exchange process is sometimes used when the more powerful multiple exchange process is available.

17.6.3 Comments on Generalized Remes Methods

Although the characterization theory is classical, Chebyshev approximation, in contradistinction to L_2 approximation, is computationally laborious, and development of computational methods had to await the advent of modern automatic computation. The first successes, computational and also theoretical (convergence proofs, etc.), were with Remes type methods, which are obviously intimately related to the characterization properties. As a short bibliography we cite the following references: 17-7, 17-14, 17-24, 17-26, 17-27, and 17-28. Perhaps the single most valuable reference is Ralston, 17-27, who describes in detail a complete state-of-the-art computer program for rational approximation by the Remes method and discusses many important attendant theoretical matters, some proven by Ralston himself.

One advantage of the Remes approach is certainly that a great deal is known about it theoretically. Another is the swift ultimate convergence (of the multiple exchange process), typical of generalized Newton's methods. This is associated with the fact that, at the solution $G(x; \mathbf{a}^*)$ with critical point set $\{t_0^*, t_1^*, \ldots, t_n^*\}$, perturbation of the form $t_k = t_k^* + \delta t_k$ in interior critical points in eq. (17.6-13) produce only second-order perturbations in $\mathbf{a}$ and λ (note that the error curve is horizontal at the interior critical points).

Among disadvantages we must list: it is computationally quite laborious. The solution of eqs. (17.6-13), and the search for error extrema, and the obtaining of a sufficiently accurate starting approximation are all difficult computational tasks. Explicit provision may be needed for cases where the error curve has extra oscillations ('nonstandard cases'), and degeneracy or near-degeneracy in rational function approximation causes trouble (although this is rare in practice; cf. Ref. 17-27). Finally, this approach does not generalize to approximation by non-Chebyshev sets, such as spline-function approximation.

17.6.4 Direct Parameter Adjustment

Simpler than the Remes methods is an iterative method which directly adjusts the parameters of the approximation [the components of $\mathbf{a}$ in $G(x; \mathbf{a})$] on the basis of an examination of the error curve. Let the kth (weighted) error be given by

$$E(x; \mathbf{a}) = w(x)\,[f(x) - G(x; \mathbf{a})]$$

(we suppress the superscript k for simplicity of notation); this might have an appearance similar to the solid curve in Fig. 17.6-2. It is required that there be $n + 1$ relative extrema of the error curve, as with the Remes methods; for simplicity we assume the *normal* case, in which there are no more than this number of oscillations.

Now consider the effect of perturbing each component of $\mathbf{a}$, i.e., consider the error curve $E(x; \mathbf{a} + \delta\mathbf{a})$, which might appear as the dotted curve in Fig. 17.6-2.

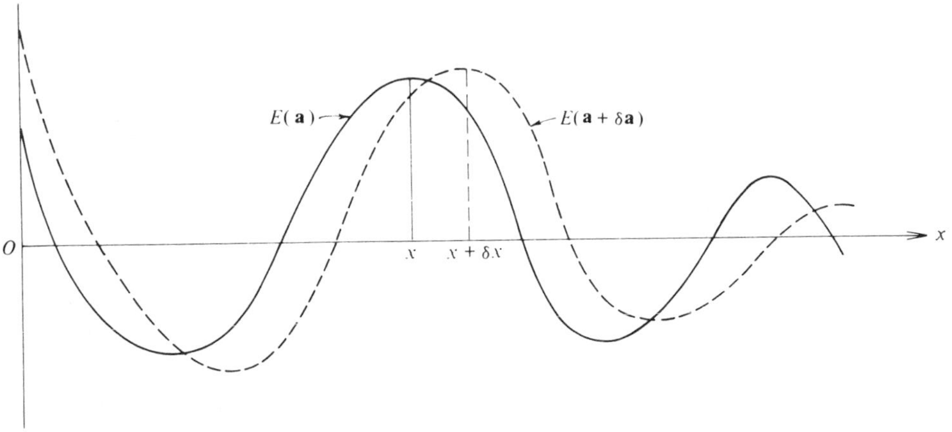

Fig. 17.6-2 Perturbation of error curve.

The effect of small variations in a and x is given by

$$\delta E = E(x + \delta x; \mathbf{a} + \boldsymbol{\delta}\mathbf{a}) - E(x; \mathbf{a})$$

$$= \sum_{k=1}^{n} \left[\frac{\partial E}{\partial a_k}\right]_{x;\mathbf{a}} \delta a_k + \left[\frac{\partial E}{\partial x}\right]_{x;\mathbf{a}} \delta x + O(|\boldsymbol{\delta}\mathbf{a}|^2, \delta x^2) \qquad (17.6\text{-}19)$$

Now suppose we consider a variation such that x is at a relative extremum of $E(x; \mathbf{a})$, and $x + \delta x$ is at the corresponding relative extremum of $E(x + \delta x; \mathbf{a} + \boldsymbol{\delta}\mathbf{a})$, as shown in Fig. 17.6-2. Then $\delta x = 0$ if the extremum in question lies at an end point (as at $x = 0$ in Fig. 17.6-2), and $[\partial E/\partial x]_{x;\mathbf{a}}$ is zero for an interior extremum (such as the case pictured in Fig. 17.6-2; we assume $f \in C^1$), so that the second term in eq. (17.6-19) is zero. Thus we may write for the jth extremum

$$\delta E_j = \sum_{k=1}^{n} \left.\frac{\partial E}{\partial a_k}\right|_{t_j;\mathbf{a}} \delta a_k + O(|\boldsymbol{\delta}\mathbf{a}|^2, \delta x^2) \qquad (17.6\text{-}20)$$

where t_j is the abscissa of the jth extremum of $E(x; \mathbf{a}), j = 0, 1, \ldots, n$.

The parameter adjustment algorithm attempts to level the error curve on each step by neglecting the second order terms in eq. (17.6-20), and defining $\boldsymbol{\delta}\mathbf{a}$ by the equations

$$E_j + \delta E_j = E_j + \sum_{k=1}^{n} \left[\frac{\partial E}{\partial a_k}\right]_{t_j;\mathbf{a}} \delta a_k = (-1)^j \lambda; \quad j = 0, 1, \ldots, n \qquad (17.6\text{-}21)$$

These are $n + 1$ linear equations in the $n + 1$ unknowns $(\delta a_1, \delta a_2, \ldots, \delta a_n, \lambda)$; the coefficients $[\partial E/\partial a_k]_{t_j;\mathbf{a}}$ do not involve the function $f(x)$ to be approximated, but merely first derivatives of $G(x; \mathbf{a})$, and subroutines for their evaluation may be written which are applicable for approximating arbitrary $f(x)$.

Though on each iteration the error curve must be searched for relative extrema, just as with the Remes methods, and the requirements for initial extrema are just as stringent as with Remes methods, the job of solving the system (17.6-13), which is nonlinear in the rational function case, is obviated. We note from the derivation (based on neglecting quantities that in the final stages of convergence are of the second order of smallness) that this method is of generalized Newton method type, and ultimate convergence is correspondingly quadratic.

This method is especially attractive when algebraic side conditions are placed on the unknown components of $\mathbf{a}$ (such as for example $a_1 = 1$, $a_5 = a_3$, etc.) as long as these do not destroy the characterization property (the optimal approximation must equioscillate as many times as there are degrees of freedom remaining in $\mathbf{a}$). Such side conditions simply reduce the size of the linear system (17.6-21) to be solved on each step.

17.6.5 A Collocation or Zero-Adjustment Method

Many methods have been tried which iteratively adjust the zero error point; perhaps the most successful is Maehly's "second direct method,"[17-24] which we now outline.

Since the best approximation has an error which equioscillates on a critical set of $n + 1$ points, there must be n distinct zeros of its error curve nested among the $n + 1$ oscillating extrema. At the start of the typical k^{th} iteration n points z_1, $z_2, \ldots, z_n$, supposed to approximate the zero error points of $G(x; \mathbf{a}^*)$, are available. The first step is to interpolate $G(x; \mathbf{a}^{(k)})$ to $f(x)$at these collocation points, i.e., to solve for $\mathbf{a}^{(k)}$ the system

$$G(z_j; \mathbf{a}^k) = f(z_j); \quad j = 1, \ldots, n \tag{17.6-22}$$

These equations are linear even in the rational function case, though in exceptional cases rational function interpolation fails (see section 17.5.4). Thus solution of (17.6-22) is fundamentally an easier task than solving the system (17.6-13) of the Remes methods. Furthermore the points $z_1, \ldots, z_n$ are perforce zero error points, and unless by bad luck some zeros are not simple zeros the error curve will have the required number of oscillations.

Next some algorithm is needed for improving the values $z_1, \ldots, z_n$ based on an examination of the error curve. Maehly's approach was to write

$$E(x) = G(x) \prod_{k=1}^{n} (x - z_k) \tag{17.6-23}$$

thus factoring off explicitly the n known zeros, and to consider the result of a perturbation δz_k in each of the z_k's and δx in x

$$\delta E = E(x + \delta x; \mathbf{z} + \delta \mathbf{z}) - E(x; \mathbf{z})$$

$$= \sum_{k=1}^{n} \frac{\partial E}{\partial z_k} \delta z_k + \frac{\partial E}{\partial x} \delta x + O(|\delta \mathbf{z}|^2, \delta x^2) \tag{17.6-24}$$

At an error extremum the second term in (17.6-24) is zero for the same reason as in the previous section (assuming $f \in C'$ so that interior error extrema are horizontal points); substituting in (17.6-23) then, at an extremum t_j E undergoes the change

$$\delta E = \sum_{k=1}^{n} \left[\frac{\partial G}{\partial z_k} \cdot \frac{E}{G} - \frac{E}{x - z_k} \right]_{x=t_j} + O(|\delta \mathbf{z}|^2, \delta x^2) \tag{17.6-25}$$

For purposes of deriving an iterative algorithm, the second order terms are neglected, and the terms involving derivatives of G are also; the latter is partially justified by the hope that most of the effect of perturbing the z_k will enter eq. (17.6-23) explicitly through the second factor, and the implicit dependence through the $G(x)$ factor will be relatively less important.

The result is the following estimate for the perturbation in the error E_j originally located at $x = t_j$:

$$\frac{\delta E_j}{E_j} = \delta(\log |E_j|) \doteq - \sum_{k=1}^{n} \frac{\delta z_k}{t_j - z_k} \tag{17.6-26}$$

We will be tending to level the error curve if we set, for the jth error extremum

$$\log |E_j| + \delta \log |E_j| = \log |\lambda| \qquad j = 0, 1, \ldots, n$$

which leads to the linear system

$$\sum_{k=0}^{n} \frac{\delta z_k}{t_j - z_k} + \log |\lambda| = \log |E_j| \qquad j = 0, 1, \ldots, n \tag{17.6-27}$$

for the $n + 1$ unknowns $(\delta z_1, \ldots, \delta z_n, \log |\lambda|)$.

In the early stages of iteration a provision is necessary to prevent the perturbed collocation points $z_k + \delta z_k$ from getting out of order or passing outside of the interval $[0, 1]$; an empirical rule that limits the motion of each collocation point to 99% of its original distance from a neighbor or endpoint has been found satisfactory.

17.7 USE OF LINEAR PROGRAMMING IN CHEBYSHEV APPROXIMATION

Linear programming has the great advantage that it does not depend on characterization properties; consequently it is more generally applicable than the methods of the previous section. Even when characterization properties are available, the linear programming approach maintains an advantage, in that it is immune to troubles caused by extra oscillations in the error curve, or degeneracy or near-degeneracy of characterization such as can occur in rational function or exponential approximation. Furthermore, since linear programming is a standard problem of widespread importance, efficient computer subroutines (using the simplex method) are widely available.

It is necessary to discretize the problem a priori, i.e., agree to look at the error at only a discrete set of m points

$$X:\ 0 = x_1 < x_2 < \cdots < x_m = 1 \tag{17.7-1}$$

when m might be, for example, 100 or 1000. The discussion of section 17.2 gives theoretical justification for this discretization. In this feature of the linear programming approach a positive and a negative consideration approximately cancel out. For on the one hand it is always necessary to verify *a posteriori* that the point set X was chosen densely enough in the critical regions to make the effects of the discretization negligible. This disadvantage is traded for the advantage that the computational job of searching over a continuum for the extrema of error curves is otherwise avoided. This problematical feature of Remes and other methods based on characterization is about as problem-dependent as choice of the discrete point set X is.

17.7.1 Linear Approximation

We consider first Chebyshev approximation of an arbitrary $f(x)$ on $[0, 1]$ by linear approximating functions as in eq. (17.6-7)

$$G(x; \mathrm{a}) = \sum_{k=1}^{n} a_k \phi_k(x) \tag{17.6-7}$$

We take the discrete L_∞ error norm over X

$$\lambda_X = \lambda_X(G) = \max_{i=1,\ldots,m} |f(x_i) - G(x_i; \mathrm{a})| \tag{17.7-2}$$

and we wish to compute $\mathbf{a}^*$ which makes $\lambda_X = \lambda_X(\mathrm{a})$ take on its minimum value λ_X^*.

This may be phrased as the linear programming problem: Minimize λ, subject to the $2m$ inequality constraints–

$$\begin{aligned} \lambda + \left[\sum_{k=1}^{n} a_k \phi_k(x_i) - f(x_i)\right] &\geqslant 0 \\ \lambda - \left[\sum_{k=1}^{n} a_k \phi_k(x_i) - f(x_i)\right] &\geqslant 0 \qquad i = 1, \ldots, m \end{aligned} \tag{17.7-3}$$

(where the $n + 1$ variables $\lambda, a_1, a_2, \ldots, a_n$ are not sign restricted). Two linear inequalities are thus required to express each of the absolute value inequalities $\lambda \geqslant |i^{\text{th}} \text{ error}|$. This is one standard form of linear programming problems (see section 20.5.1 of this book, and also references 17-3, 17-17). We may cast it in matrix language by defining

$$\mathbf{u} = [\lambda, a_1, a_2, \ldots, a_n]$$

$$\mathbf{b} = \begin{bmatrix} 1 \\ 0 \\ 0 \\ \cdot \\ \cdot \\ \cdot \\ \cdot \\ 0 \end{bmatrix}; \quad \mathbf{c} = \begin{bmatrix} f(x_1) \\ -f(x_1) \\ f(x_2) \\ -f(x_2) \\ \cdot \\ \cdot \\ \cdot \\ f(x_m) \\ -f(x_m) \end{bmatrix};$$

$$\mathbf{C} = \begin{bmatrix} 1 & 1 & 1 & 1 & \cdots & 1 & 1 \\ \phi_1(x_1) & -\phi_1(x_1) & \phi_1(x_2) & -\phi_1(x_2) & \cdots & \phi_1(x_m) & -\phi_1(x_m) \\ \phi_2(x_1) & -\phi_2(x_1) & \phi_2(x_2) & -\phi_2(x_2) & \cdots & \phi_2(x_m) & -\phi_2(x_m) \\ \cdot & \cdot & \cdot & \cdot & & \cdot & \cdot \\ \cdot & \cdot & \cdot & \cdot & & \cdot & \cdot \\ \cdot & \cdot & \cdot & \cdot & & \cdot & \cdot \\ \cdot & \cdot & \cdot & \cdot & & \cdot & \cdot \\ \phi_n(x_1) & -\phi_n(x_1) & \phi_n(x_2) & -\phi_n(x_2) & \cdots & \phi_n(x_m) & -\phi_n(x_m) \end{bmatrix} \quad (17.7\text{-}4)$$

whereupon the linear programming problem may be stated:

$$\left.\begin{array}{l} \text{minimize } \mathbf{ub} \\ \mathbf{uC} \geqslant \mathbf{c}^T \\ (\mathbf{u} \text{ not sign restricted}) \end{array}\right\} \quad (17.7\text{-}3')$$

Since m is usually much greater than n, and since the constraints in the dual problem are equalities rather than inequalities, it is much more efficient to solve the dual of this problem. It is most conventional to call (17.7-3′) the dual of the following *primal* linear programming problem:

$$\left.\begin{aligned}&\text{maximize } \mathbf{c}^T\mathbf{w}\\&\mathbf{C}\mathbf{w} = \mathbf{b}\\&\mathbf{w} \geqslant \mathbf{0}\end{aligned}\right\} \qquad (17.7\text{-}5)$$

here

$$\mathbf{w}^T = (u_1, v_1, u_2, v_2, \ldots, u_m, v_m) \qquad (17.7\text{-}6)$$

is a new vector of $2m$ unknowns whose meaning is not immediately apparent. It follows from the theory of linear programming that

$$\max_{\mathbf{w}} \mathbf{c}^T\mathbf{w} = \min_{\mathbf{u}} \mathbf{ub}$$

and, by standard computational techniques, when the primal problem (17.7-5) is solved the solution $\mathbf{u}$ to the dual problem (17.7-3′) can be obtained with negligible additional labor.

Feasible solutions to (17.7-5) (vectors $\mathbf{w}$ obeying the second and third of conditions (17.7-5) obviously exist (e.g., take $u_1 = v_1 = \frac{1}{2}$ and all other components of $\mathbf{w}$ zero), so the fundamental theorem of linear programming states that a optimal solution $\mathbf{w}^*$ exists which is *basic*, which means that no more than $n + 1$ of the m components of $\mathbf{w}^*$ are nonzero. The significance of that turns out to be the following: the error achieves its maximum magnitude at each point corresponding to a nonzero component of $\mathbf{w}^*$, with one sign or the other depending on whether it is a u_i or a v_i which is nonzero. Thus a partial characterization property, holding for all types of linear approximation, comes out of linear programming theory. However sign alternation of the error cannot be inferred in general.

We may emphasize how much weaker the conditions on the set $\{\phi_1(x), \ldots, \phi_n(x)\}$ are for success of the linear programming approach than for methods based on characterization as follows: Linear programming requires only that the matrix C be of rank $n + 1$, i.e., that there exist some set of $n + 1$ columns of C that are linearly independent. Linear independence of the set $\{\phi_i(x)\}$ on X is sufficient for that. The Haar condition, eq. (17.5-14), is a very much stronger condition on the set $\{\phi_i(x)\}$: If we strike out every other column of the matrix C, the Haar condition requires that *every* set of n columns of the result be linearly independent.

The simplex method, the most efficient method for solving linear programming problems, requires a starting basic feasible solution, i.e., a solution of the second and third of eqs. (17.7-5) with no more than $n + 1$ nonzero components. There is a standard linear programming technique for generating such a starting basic feasible solution, which involves introducing and then eliminating $n + 1$ additional 'fictitious' variables. However in practice this conservative approach is usually unnecessary, if use is made of the interpretation of the components of the unknown vector $\mathbf{w}$ mentioned above. A plausible error curve with n oscillations is assumed, and a basic solution with $n + 1$ corresponding nonzero elements of $\mathbf{w}$ taken to get the simplex method started. Of course when several similar problems are being solved, or linear programming is being used in an iterative method for handling nonlinear problems, a more 'educated' choice of the initial basic solution may be available.

Starting from a random initial basic solution, the simplex method typically requires slightly over n steps or exchanges to arrive at the optimal solution. The number of exchanges depends about linearly on the number of rows of the matrix C, and only very weakly on the number of columns; increasing m from 100 to 1000 typically adds only about two exchanges to the process.

Higher than single precision arithmetic is recommended for this type of calculation on 36-bit word machines (and for other best Chebyshev approximation calculations as well), as otherwise round-off errors are likely to interfere with the various exchange and convergence criteria involved.

It should finally be mentioned that the linear programming approach easily accommodates such complications as linear equality or inequality constraints on the components of the unknown vector **a**.

17.7.2 Rational Approximation by Linear Programming

The $L_\infty[0, 1]$ rational function approximation problem can be phrased as the following nonlinear programming problem:

Minimize λ, subject to the constraints

$$\begin{aligned} &\lambda + \left[\frac{\sum_{j=0}^{p} a_j x_i^j}{1 + \sum_{j=1}^{q} b_j x_i^j} - f(x_i)\right] \geqslant 0 \\ &\lambda - \left[\frac{\sum_{j=0}^{p} a_j x_i^j}{1 + \sum_{j=1}^{q} b_j x_i^j} - f(x_i)\right] \geqslant 0 \qquad i = 1, \ldots, m \end{aligned} \tag{17.7-7}$$

This can be handled by the local linearization and iteration technique discussed in the next section; however an iteration technique due to Grabner[17-15] works so well that it deserves special mention. We first write the constraints (17.7-7) in the form

$$\lambda \pm \frac{1}{1 + \sum_{j=1}^{q} b_j x_i^j}\left[\sum_{j=0}^{p} a_j x_i^j - f(x_i)\left(1 + \sum_{j=1}^{q} b_j x_i^j\right)\right] \geqslant 0$$

Now an iterative method is set up in which the factor to the left of the square bracket (including the denominator polynomial) is treated as a weighting factor, using old values of the unknowns $b_1, b_2, \ldots, b_q$. Putting in iteration numbers, the $(k+1)^{\text{st}}$ iterates $[\lambda^{(k+1)}, a_0^{(k+1)}, \ldots, a_p^{(k+1)}, b_1^{(k+1)}, \ldots, b_q^{(k+1)}]$ are obtained from the kth by solving the linear program:

Minimize $\lambda^{(k+1)}$, subject to

$$\left(1 + \sum_{j=1}^{q} b_j^{(k)} x_i^j\right) \lambda^{(k+1)} + \sum_{j=0}^{p} a_j^{(k+1)} x_i^j - f(x_i) \left(1 + \sum_{j=1}^{q} b_j^{(k+1)} x_i^j\right) \geqslant 0$$

$$\left(1 + \sum_{j=1}^{q} b_j^{(k)} x_i^j\right) \lambda^{(k+1)} - \sum_{j=0}^{p} a_j^{(k+1)} x_i^j + f(x_i) \left(1 + \sum_{j=1}^{q} b_j^{(k+1)} x_i^j\right) \geqslant 0$$

$$i = 1, \ldots, m \qquad (17.7\text{-}8)$$

To get started it usually suffices to take $b_j^{(0)} = 0, j = 1, \ldots, q$. The constraint matrix C changes in only the first row from iteration to iteration [the coefficients of $\lambda^{(k+1)}$ in eq. (17.7-8)—see eqs. (17.7-3′)] so its updating is not unduly laborious. If the final basis of each iteration is used as the starting basis of the simplex calculation in the next, the numbers of exchanges required in the simplex method calculations are greatly reduced. Again the appending of side conditions on the coefficients a_j, b_j is a simple matter.

17.7.3 Nonlinear Approximation by Local Linearization and Use of Linear Programming

The generalized Newton's method technique of local linearization and iteration is one of the most useful tools of the numerical analyst for handling nonlinear problems. We present such a method for nonlinear L_∞ approximation. Rather than deal in generalities the technique will be illustrated for the exponential approximating function

$$G(x) = a_1 e^{b_1 x} + a_2 e^{b_2 x} \qquad (17.7\text{-}9)$$

The ideas are so simple that their application to other problems will be obvious.

As before, we discretize the problem, i.e., agree to look at the error on only a finite (but if necessary large) set of points X: $0 = x_1 < x_2 < \cdots < x_m = 1$. At the end of the kth iteration approximate values $a_1^{(k)}, a_2^{(k)}, b_1^{(k)}, b_2^{(k)}$ have been computed.

For purposes of computing the next iterate we linearize about these values. Note that only the two parameters b_1, b_2 appear nonlinearly in (17.7-9), so it is only necessary to linearize with respect to these two of the four parameters. We have

$$\begin{aligned} G(x) &= a_1 e^{[b_1^{(k)} + \delta b_1]x} + a_2 e^{[b_2^{(k)} + \delta b_2]x} \\ &\doteq a_1 e^{b_1^{(k)} x}[1 + \delta b_1 x] + a_2 e^{b_2^{(k)} x}[1 + \delta b_2 x] \qquad (17.7\text{-}10) \\ &= a_1 e^{b_1^{(k)} x} + d_1 x e^{b_1^{(k)} x} + a_2 e^{b_2^{(k)} x} + d_2 x e^{b_2^{(k)} x} \qquad (17.7\text{-}11) \end{aligned}$$

where

$$\left.\begin{aligned} d_1 &= a_1 \delta b_1 \\ d_2 &= a_2 \delta b_2 \end{aligned}\right\} \tag{17.7-12}$$

The function (17.7-11) is linear in the parameters a_1, a_2, d_1, d_2 $b_1^{(k)}$ and $b_2^{(k)}$ being given numbers at this point in the calculations, and the linear programming technique of section 17.7.1 above may be applied to find a best L_∞ approximation of the form (17.7-11) to an arbitrary $f(x)$.

The resulting iterative process will ordinarily diverge, the trouble being that the solution for the best approximation of form (17.7-11) will usually yield values of δb_1 or δb_2 large enough so that the approximation (17.7-10) is poor. The process may be converted to a convergent one by another of the standard tricks of the numerical analyst, to limit the size of $\delta b_1, \delta b_2$, so that the values of b_1, b_2 remain within a range within which the linearization has some accuracy.

We would therefore like to impose the constraints

$$\left.\begin{aligned} |\delta b_1| &= \left|\frac{d_1^{(k+1)}}{a_1^{(k+1)}}\right| \leqslant \epsilon \\ |\delta b_2| &= \left|\frac{d_2^{(k+1)}}{a_2^{(k+1)}}\right| \leqslant \epsilon \end{aligned}\right\} \tag{17.7-13}$$

where ϵ is an empirical parameter which we intend to vary experimentally to obtain convergence. However the constraints (17.7-13) are somewhat inconvenient so we take instead

$$\left.\begin{aligned} |d_1^{(k+1)}| &\leqslant \epsilon\, |a_1^{(k)}| \\ |d_2^{(k+1)}| &\leqslant \epsilon\, |a_2^{(k)}| \end{aligned}\right\} \tag{17.7-14}$$

where the right-hand sides are known numerical values. When written in linear form this yields four linear inequalities which are added to the linear programming problem to be solved to obtain the $(k + 1)^{\text{st}}$ iterates.

The linear program solver encounters trouble if either b_1 or b_2 gets very close to zero, or if b_1 and b_2 approach each other, as then the rows of the C matrix (17.7-4) approach linear dependence. In particular trouble is likely to occur if one or both of the initial guesses $b_1^{(0)}$, $b_2^{(0)}$ has the wrong sign. However often, and especially in cases where exponential approximation of the form (17.7-9) is appropriate, the process converges to the optimal best approximation.

In addition to exponential approximations,[17-12] this technique has been successfully applied to other nonlinear problems, including the determination of optimal coordinate system rotation,[17-25] and best L_∞ approximation by compositions of polynomials.[17-41]

17.8 L_1 APPROXIMATION

A classical theory exists for approximation in the L_1 norm by linear approximating functions, i.e., approximating functions of the form

$$G(x; \mathrm{a}) = \sum_{j=1}^{n} a_j \phi_j(x) \tag{17.8-1}$$

where the $\phi_j(x)$ are arbitrary given functions. We take the interval $[-1, 1]$ for simplicity (although any bounded subset of the real numbers may be taken as the domain of approximation), and as usual we assume that the function $f(x)$ to be approximated is continuous (although integrability is really all that is needed). Unity weight function is taken for simplicity (a positive weight function can be included in the L_1 norm if desired).

It is desired then to find $\mathbf{a}^*$ which minimizes

$$\lambda(\mathrm{a}) = \|f(x) - G(x; \mathrm{a})\|_1 = \int_{-1}^{1} |f(x) - G(x; \mathrm{a})|\, dx \tag{17.8-2}$$

Analogously to the characterization properties (17.4-19) for L_2 approximation and (17.5-8) for L_∞ approximation, we have the L_1 *Characterization Property*: *If for any* $G(x; \mathrm{a})$ *it is true that*

$$\int_{-1}^{1} G(x; \mathrm{a}) \operatorname{sgn}\, [f(x) - G(x; \mathrm{a}^*)]\, dx = 0 \tag{17.8-3}$$

Then $G(x; \mathrm{a}^*)$ *is a best* L_1 *approximation to* $f(x)$ *on* $[-1, 1]$. Here sgn is the sign function, defined by

$$\operatorname{sgn}\, [y(x)] = \begin{cases} -1 & y(x) < 0 \\ 0 & y(x) = 0 \\ 1 & y(x) > 0 \end{cases}$$

To show this, consider the L_1 error norm of any admissible approximation

$$\lambda(\mathrm{a}) = \|f(x) - G(x; \mathrm{a})\|_1 = \int_{-1}^{1} |f(x) - G(x; \mathrm{a})|\, dx$$

$$= \int_{-1}^{1} [f(x) - G(x; \mathrm{a})] \operatorname{sgn}\, [f(x) - G(x; \mathrm{a})]\, dx$$

$$\geqslant \int_{-1}^{1} [f(x) - G(x;\mathbf{a})] \operatorname{sgn} [f(x) - G(x;\mathbf{a}^*)]\, dx$$

$$= \int_{-1}^{1} [f(x) - G(x;\mathbf{a}^*)] \operatorname{sgn} [f(x) - G(x;\mathbf{a}^*)]\, dx$$

[where we have used (17.8-3)]

$$= \int_{-1}^{1} |f(x) - G(x;\mathbf{a}^*)|\, dx = \lambda(\mathbf{a}^*)$$

Thus no **a** yields error norm smaller than that of $\mathbf{a}^*$.

The characterization condition (17.8-3) is not in its most general form (a necessary and sufficient condition). The latter is complicated by the possibility that $f(x) = G(x;\mathbf{a}^*)$ on a set of measure greater than zero. Fortunately this rarely occurs; if it does not, (17.8-3) is necessary as well as sufficient. See Rice[17-31] for the discussion of the general case and Rivlin[17-35] for the general polynomial approximation discussion.

If the set of functions $\phi_j(x)$ is a Chebyshev set [see eq. (17.5-14)], $\mathbf{a}^*$ is unique.[17-31] Furthermore, the characterization condition (17.8-3) frequently allows computation of the best approximation by a simple interpolation process. We illustrate this for the case of polynomial approximation

$$G(x;\mathbf{a}) = \sum_{j=1}^{n} a_j x^{j-1} \tag{17.8-4}$$

The basic idea is to find a sign function $s(x)$ [a function for which $\operatorname{sgn} s(x) = s(x)$] such that for all **a**

$$\int_{-1}^{1} \left[\sum_{j=1}^{n} a_j x^{j-1}\right] s(x)\, dx = 0 \tag{17.8-5}$$

and then to attempt to choose **a** such that

$$\operatorname{sgn}\left[f(x) - \sum_{j=1}^{n} a_j x^{j}\right] = s(x) \tag{17.8-6}$$

If successful this will yield a best approximation $G(x;\mathbf{a}^*)$ by (17.8-3).

Setting $x = \cos\theta$ changes (17.8-5) into the form

$$\int_{0}^{\pi} \sum_{j=1}^{n} a_j (\cos\theta)^{j-1} \sin\theta\, s(\cos\theta)\, d\theta = 0 \tag{17.8-7}$$

Furthermore it is easily established by induction that $\cos^p \theta \sin \theta$ is expressible in the form

$$\cos^p \theta \sin \theta = \sum_{k=1}^{p} c_k \sin k\theta$$

Thus (17.8-7) may be expressed

$$\int_0^{\pi} \left[\sum_{j=1}^{n} b_j \sin j\theta \right] s(\cos \theta)\, d\theta = 0 \tag{17.8.8}$$

Now one of the many orthogonality properties of the sinusoidal functions is used (Ref. 17-31, p. 110)

$$\int_0^{\pi} \sin m\theta \operatorname{sgn} [\sin (n+1)\theta]\, d\theta = 0; \qquad m = 1, 2, \ldots, n \tag{17.8-9}$$

In the interior of $[0, \pi]$ the sign changes of this sign function are at

$$\theta = \frac{k\pi}{n+1}; \qquad k = 1, 2, \ldots, n$$

Therefore if $s(x)$ has sign changes at the points

$$x_k = \cos \frac{k\pi}{n+1}; \qquad k = 1, 2, \ldots, n \tag{17.8-10}$$

(and only these points), (17.8-8) will hold.

To attempt to achieve (17.8-6), then, we calculate **a** by solving the interpolation problem

$$\sum_{j=1}^{n} a_j x_k^j = f(x_k); \qquad k = 1, \ldots, n \tag{17.8-11}$$

where x_k is given by (17.8-10). We then inspect the error curve; if it has no zeroes at points other than the x_k (and simple zeroes at the x_k), then (17.8-3) will hold and we have found the best approximation.

The fact that the interpolation points (17.8-10) are independent of $f(x)$ at first seems remarkable; it is essentially associated with the fact that small perturbations yield changes in the L_1 error norm, which we recall is the area between $f(x)$ and $G(x; \mathbf{a})$, which depend to first order only on the location of the zeroes of the error curve and not on its shape. This may be visualized from elementary considerations in the $n = 0$ and $n = 1$ cases.

Thus we are assured that, in the Chebyshev function case, the best L_1 approximation *oscillates* at least n times, just as in the L_∞ case it *equioscillates* at least n times. The non-normal case of more than n oscillations causes trouble, from a computational viewpoint, in both cases.

If the above interpolation procedure is inapplicable, because of failure of the Chebyshev set property, or fails, because of the occurrence of a non-normal case with more than n oscillations, it is recommended that the problem be solved in discrete form, by minimizing

$$\sum_{i=1}^{m} w_i \left| f(x_i) - G(x_i; \mathrm{a}) \right| \tag{17.8-12}$$

If the discrete point set $x_1, \ldots, x_m$ is not equally spaced a corresponding variation in the positive weights w_i is taken. Such minimization problems are easily formulated as linear programming problems which are readily solved by standard means (Ref. 17-35, p. 78).

17.9 PIECEWISE APPROXIMATION WITHOUT CONTINUITY AT THE JOINTS

When a function to be approximated has different character in different regions, or when high accuracy is otherwise hard to obtain, it is frequently profitable to break up the domain of the independent variable into subregions and employ a separate approximation in each subregion. (This is even more widely necessary when working in more than one dimension, i.e., in surface approximation, etc.) When the resulting composite approximation is not required to be continuous across the boundaries of the subregions, the one-dimensional theory is in good shape, due to Lawson.[17-21, 17-22] We present it for the following fundamental problem:

Let $f(x)$ be a given aribtrary continuous function to be approximated over some given finite interval $\alpha \leqslant x \leqslant \beta$, and suppose we decide to subdivide the interval $[\alpha, \beta]$ into m subintervals by a partition π:

$$\pi: \alpha = u_0 \leqslant u_1 \leqslant u_2 \leqslant \cdots \leqslant u_m = \beta \tag{17.9-1}$$

The abscissas u_j are called *joints* or *knots*, and optimal locations are to be found for the interior joints $u_1, u_2, \ldots, u_{m-1}$. Ordinarily no two adjacent joints would coincide; however the existence discussion is facilitated if weak inequalities are taken in (17.9.1).]

In the typical i^{th} subinterval $f(x)$ is to be approximated by a member of the set of admissible approximations G_i, in the norm $\|\cdot\|_i$; G_i, and even the error norm, may vary with i (though this would be unusual in practice). It is assumed that a subroutine is available to solve the best approximation problem in each interval, i.e., to compute g_i^* and $\lambda^*(u_{i-1}, u_i)$ where

$$\lambda^*(u_{i-1}, u_i) = \|f(x) - g_i^*(x)\|_i = \min_{g_i \in G_i} \|f(x) - g_i(x)\|_i \tag{17.9-2}$$

when u_{i-1} and u_i are given.

The problem is to compute the optimal joint locations $u_1^*, \ldots, u_{m-1}^*$, and the least global error norm Λ^*, where

$$\Lambda^* = \min_{u_1, \ldots, u_{m-1}} [\max_i \lambda^*(u_{i-1}, u_i)] \tag{17.9.3}$$

When the L_∞ or maximum norm

$$\|f(x) - g_i(x)\|_i = \max_{u_{i-1} \leqslant x \leqslant u_i} |f(x) - g_i(x)|$$

is used in (17.9-2), this is a min-max-min-max problem.

It will be assumed that $\lambda^*(u_{i-1}, u_i)$ is a continuous function of both arguments, as is ordinarily the case; for example this holds for rational or general linear approximation in the L_∞ or L_2 norms.[17-11, 17-21] Furthermore, $\lambda^*(u_{i-1}, u_i)$ is (weakly) monotone increasing in its second argument and monotone decreasing in its first; i.e., if $\Delta > 0$ then

$$\left.\begin{aligned} \lambda^*(u_{i-1}, u_i + \Delta) &\geqslant \lambda^*(u_{i-1}, u_i) \\ \lambda^*(u_{i-1} - \Delta, u_i) &\geqslant \lambda^*(u_{i-1}, u_i) \end{aligned}\right\} \tag{17.9-4}$$

This follows because the best approximation on $[u_{i-1}, u_i + \Delta]$ is a candidate for approximation on the smaller interval $[u_{i-1}, u_i]$.

The case $m = 2$ of one interior joint now becomes simple (see Fig. 17.9-1); if we plot $\lambda^*(\alpha, u_1)$ and $\lambda^*(u_1, \beta)$ versus u_1 on the same graph, the former is (weakly) monotone increasing, the latter (weakly) monotone decreasing, and

$$\Lambda^* = \min_{u_1} \{\max [\lambda^*(\alpha, u_1), \lambda^*(u_1, \beta)]\}$$

is obtained by choosing $u_1 = u_1^*$ at a point where the two curves cross. Ordinarily there is only one such point and we have a unique solution (Fig. 17.9-1a). The error norms are then equal or *leveled* in the optimal solution, and we may imagine searching for the solution by moving u_1 in such a manner as to *level* the error norms in the two intervals.

There is however the possibility of nonuniqueness if one or both of the two curves is horizontal in the critical neighborhood, as shown in Fig. (17.9-1b). In this case solutions do not have to be leveled, although a leveled solution always exists.

Turning to the general m-interval case, we note that, by virtue of the continuity of $\lambda^*(u_{i-1}, u_i)$,

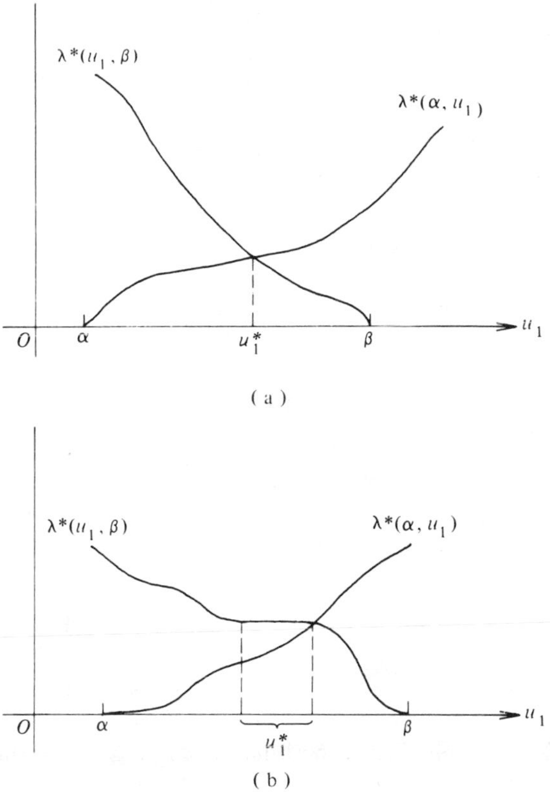

Fig. 17.9-1 Optimal location of one interior joint. a) Unique solution; and b) nonunique solution.

$$\Lambda(u_1, u_2, \ldots, u_{m-1}) = \max_{i=1,\ldots,m} \lambda^*(u_{i-1}, u_i)$$

is a continuous function on the compact subset of $(m-1)$-dimensional space defined by (17.9-1); this ensures existence of Λ^* and $u_1^*, \ldots, u_{m-1}^*$.

Now consider any two m-partitions

$$\begin{aligned} \pi_1:\ \alpha = u_0 \leqslant u_1 \leqslant \cdots \leqslant u_m = \beta \\ \pi_2:\ \alpha = v_0 \leqslant v_1 \leqslant \cdots \leqslant v_m = \beta \end{aligned} \tag{17.9-5}$$

We will show that

$$\min_i \lambda^*(u_{i-1}, u_i) \leqslant \max_i \lambda^*(v_{i-1}, v_i) \tag{17.9-6}$$

If π_1 and π_2 are identical this is trivially true; if not, let j be the smallest index for which $u_j \neq v_j$. If $u_j < v_j$ we have by (17.9-4)

$$\lambda^*(u_{j-1}, u_j) \leqslant \lambda^*(v_{j-1}, v_j) = \lambda^*(u_{j-1}, v_j)$$

From which (17.9-6) follows. If $u_j > v_j$, let k be the first index greater than j for which $u_k \leqslant v_k$. Then $[u_{k-1}, u_k] \subset [v_{k-1}, v_k]$ so, again by (17.9-4), we have

$$\lambda^*(u_{k-1}, u_k) \leqslant \lambda^*(v_{k-1}, v_k)$$

which again implies (17.9-6).

Now suppose π_2 in (17.9-5) is the optimal partition. Then (17.9-6) becomes

$$\min_i \lambda^*(u_{i-1}, u_i) \leqslant \max_i \lambda^*(u_{i-1}^*, u_i^*) = \Lambda^*$$

Hence any partition π_1 gives us bounds on Λ^*

$$\min_i \lambda^*(u_{i-1}, u_i) \leqslant \Lambda^* \leqslant \max_i \lambda^*(u_{i-1}, u_i) \tag{17.9-7}$$

A *leveled* partition [one for which the $\lambda^*(u_{i-1}, u_i)$ are all equal] is thus a solution, since the upper and lower bounds in (17.9-7) coincide.

The existence of leveled solutions, under the assumption of continuity of $\lambda^*(u_{i-1}, u_i)$, may be shown by proving inductively that for an m-partition of (α, β), Λ^* is a (weakly) monotonic increasing function of β. We have already shown this for $m = 1$ and the proof for $m = 2$ follows readily from inspection of Fig. (17.9-1). Assuming the desired result to hold for some arbitrary m, we regard the $(m + 1)$-partition problem as a two-interval problem with a single interior joint separating m intervals on the left from one interval on the right. The leveled solution is at an intersection as in Fig. (17.9-1), and Λ^* must increase (or in exceptional cases remain fixed but not decrease) as the right boundary is increased.

It remains to discuss computation of the solution. Processes patterned after the above inductive line of reasoning turn out to be unfeasible (except possibly for very small m), because of the large amount of computation involved. A better procedure is to guess joint locations and iteratively adjust them so as to tend to level the error norms by means of some algorithm which uses the values $\lambda^*(u_{i-1}, u_i)$. Lawson gives such a method,[17-22] based on the assumption that $\lambda^*(u_{i-1}, u_i)$ can be usefully approximated by an expression of the form

$$\lambda^*(u_{i-1}, u_i) \doteq k_i [u_i - u_{i-1}]^{\alpha_i} \tag{17.9-8}$$

where k_i is an undetermined parameter, and α_i is an integer equal to the number of degrees of freedom in G_i, the set of admissible approximations for the ith interval. On each iteration the constants k_i are evaluated using 'old' values of the joints, and then 'new' values of the joints $u_i^{(\text{new})}$ and a new estimate $\Lambda^{(\text{new})}$ for Λ^* are computed from the equations

$$k_i[u_i^{(\text{new})} - u_{i-1}^{(\text{new})}]^{\alpha_i} = \Lambda^{(\text{new})}; \qquad i = 1, \ldots, m \tag{17.9-9}$$

Solving for $u_i^{(\text{new})} - u_{i-1}^{(\text{new})}$ and adding yields

$$\beta - \alpha = \sum_{i=1}^{m} k_i^{-1/\alpha_i} [\Lambda^{(\text{new})}]^{1/\alpha_i} \tag{17.9-10}$$

which is easily solved for $\Lambda^{(\text{new})}$ if the α_i are all the same; otherwise Newton's method or the false position method may be employed. (See Ref. 17-11 or 17-22 for further discussion.)

17.10 APPROXIMATION BY SPLINES AND RELATED SMOOTH PIECEWISE FUNCTIONS

17.10.1 Introduction

As in the previous section, we consider the problem of approximating a given $f(x)$ over a finite interval $[\alpha, \beta]$, and we divide the interval into m subregions by the partition

$$\pi: \alpha = u_0 \leqslant u_1 \leqslant \cdots \leqslant u_m = \beta \tag{17.10-1}$$

Now however the approximating functions are required to have certain continuity properties at the joints u_j.

A spline function $s(x)$ of degree n with the joints (17.10-1) may be defined by the following two properties:

(i) In each interval $u_{i-1} \leqslant x \leqslant u_i$, $i = 1, \ldots, m$, $s(x)$ is a polynomial of degree not exceeding n;
(ii) at each interior joint $s(x)$ and its first $n - 1$ derivatives are continuous, i.e., (17.10-2)

$$s(x) \in C^{n-1} [\alpha, \beta] \tag{17.10-3}$$

Various modifications of this definition occur in the literature, e.g., the definition is often extended to cover the entire real axis; in the odd n case $s(x)$ is often then defined to be a polynomial of degree $(n - 1)/2$ in $(-\infty, \alpha]$ and $[\beta, \infty)$.

The odd n splines have an important variational property stated below at eq. (17.10-11), which may be interpreted as a maximal smoothness condition. The $n = 3$ case (the *cubic spline*) is especially important; it occurs in linearized thin beam theory, and in fact the term "spline" derives from a drafting tool used by naval architects.

Other types of piecewise functions result when the continuity requirement (17.10-3) is modified; for example, the *piecewise cubic Hermite function* obeys (17.10-2) with $n = 3$, but with a C^1 continuity requirement.

The allowing of discontinuities in high order derivatives results in a character which will be termed *local* [see below at eq. (17.10-13)]; roughly this means that constraints placed on $s(x)$ in one region have only small effects in regions far removed. As a result splines are much better suited to approximating nonanalytic (and many analytic) functions than are analytic approximating functions. This is not surprising when one recalls that the shape of an analytic function in any small region determines its shape everywhere. The escape from the tyranny of analyticity is tied up with many favorable properties of spline functions, including good convergence properties, accuracy of derivative approximation, and stability in the presence of round-off errors. Some authors like to view the use of splines as a middle ground between classical methods, which employ an analytic approximation over the whole domain of a problem, and numerical finite difference methods, which in essence break the domain into very many small intervals.

The extension of piecewise approximation techniques to two and three dimensions, and their employment in partial differential equation problems, has given rise to a family of methods known as *finite element methods*, currently of great importance in engineering analysis.

17.10.2 Approximation of $f(x)$ by a Spline

We consider now the problem of approximating an arbitrary $f(x)$ by an $s(x)$ obeying (17.10-2) in the L_∞ or L_2 norm.

We assume first that the joints (17.10-1) are given *a priori*. Then $s(x)$ is a linear approximating function. This is most easily seen from the representation

$$s(x) = \sum_{j=0}^{n} a_j x^j + \sum_{k=1}^{m-1} \frac{c_k}{n!}(x - u_k)_+^n \tag{17.10-4}$$

where the function $(x)_+^n$ is defined by

$$(x)_+^n = \begin{cases} x^n & x > 0 \\ 0 & x \leqslant 0 \end{cases} \tag{17.10-5}$$

(This itself is a spline function of degree n, with a single joint at $x = 0$.) In eq. (17.10-4) the first sum gives the polynomial of degree n to which $s(x)$ is equal in the first interval $[\alpha, u_1]$, and c_k equals the discontinuity in the nth derivative at the kth joint.

The spline $s(x)$ is then of the linear form (17.5-13)

$$s(x) = \sum_{j=0}^{n+m-1} a_j \phi_j(x)$$

$$\phi_j(x) = \begin{cases} x^j & j = 0, \dots, n \\ (x - u_{j-n})_+^n & j = n+1, \dots, n+m-1 \end{cases}$$

However this set of functions $\{\phi_j(x)\}$ is not a Chebyshev set, as unique interpolation to arbitrary data is not always possible. [Consider for example the case $n = 1$, $m = 2$; then the interpolation problem $s(x_i) = f_i$, $i = 1, 2, 3$, has a unique solution only if $x_1 < u_1 < x_3$.] Rice has given a characterization theory,[17-30] which however is complicated.

Thus in the problem of L_∞ approximation of an arbitrary $f(x)$, the methods of section 17.6, which are based on characterization and therefore assume the Chebyshev set property, cannot be applied with full assurance that they will succeed. The linear programming method of section 17.6 is perfectly applicable, however, and is the method of choice for this problem.[17-10, 17-40]

Approximation in the L_2 norm is straightforward.

Similar remarks apply to related approximating functions with lower continuity. If (17.10-3) is replaced by $s(x) \in C^r$ then we have the linear form

$$s(x) = \sum_{j=0}^{n} a_j x^j + \sum_{j=0}^{n-r-1} \sum_{k=1}^{m-1} \frac{C_{j,k}}{(n-j)!} (x - u_k)_+^{n-j} \tag{17.10-6}$$

in which $c_{j,k}$ equals the discontinuity in the $(n - j)^{\text{th}}$ derivative at the kth joint.

So far the joints have been considered fixed *a priori*. If they are variable, and are included among the parameters to be optimized, then the problem becomes nonlinear and much harder.[17-10, 17-40]

In the variable joint case the set of $s(x)$ typically has limit functions of lower continuity, which must be appended to obtain general existence of best approximations. For example, the set of quadratic splines in $[-1, 1]$ with two interior joints has the limit function $f(x) = |x|$ which is only C^0; the two joints have coalesced at $x = 0$ in the limiting process.

For an authoritative account of recent developments in spline approximation, including documented computer programs, see Ref. 17-48.

17.10.3 Interpolation by Splines

If we regard the joints as fixed, then we see from (17.10-4) that $s(x)$ contains $n + m$ independent parameters or degrees of freedom, so the general interpolation problem should contain $n + m$ conditions

$$s(x_j) = f_j; \qquad j = 1, \ldots, n + m \tag{17.10-7}$$

Here the x_j are given distinct abscissas, and the f_j are arbitrary given ordinates.

Schoenberg and Whitney 17-39 have shown that a solution exists for arbitrary f_j if and only if the interpolation points and joints are related by the conditions

$$x_i < u_i < x_{i+n+1}; \qquad i = 1, \ldots, m - 1 \tag{17.10-8}$$

In particular then, as can be deduced from elementary considerations, there must be at least one and at most $n + 1$ interpolation points to the left of u_1, there must be no more than $n + 1$ interpolation points in any interval $[u_{i-1}, u_i]$, etc.

Since the problem is linear, (17.10-8) also guarantees uniqueness.

In most spline interpolation problems that arise in practice, however, the interpolation points coincide with the joints [and there is no problem about satisfying (17.10-8)]. Perhaps the most frequently occurring cases are the following two:

1. cubic spline with given first derivative at the end points–

$$\begin{cases} s(u_k) = f_k & k = 0, \ldots, m \\ s'(u_0) = f_0' \\ s'(u_m) = f_m' \end{cases} \tag{17.10-9}$$

2. cubic spline with zero second derivative at the end points–

$$\begin{cases} s(u_k) = f_k & k = 0, \ldots, m \\ s''(u_0) = 0 \\ s''(u_m) = 0 \end{cases} \tag{17.10-10}$$

The method of solving these equations to determine $s(x)$ is given in section 18.4.3.2 of this book; see also Refs. 17-2, 17-5, and 17-16.

These cubic splines have the following variational property: of all functions $y(x) \in C^2 [\alpha, \beta]$ obeying conditions (17.10-9), $s(x)$, the spline function obeying (17.10-9), is the *smoothest* in the sense that it minimizes

$$\int_\alpha^\beta [y''(x)]^2 \, dx \tag{17.10-11}$$

Similarly of all $y(x) \in C^2 [\alpha, \beta]$ obeying $y(u_k) = f_k$, $k = 0, \ldots, m$, the spline function defined by (17.10-10) is the smoothest in that it minimizes (17.10-11).

In thin beam theory, with $y(x)$ denoting the deflection of the beam and with $|y'(x)|$ small (so that the curvature may be set equal to y''), eq. (17.10-11) is proportional to the strain energy, i.e., the work necessary to deform the beam into the shape $y(x)$. A variational form of the equilibrium conditions is that the strain energy be minimized. Thus, to linear approximation, if a beam is constrained by conditions (17.10-9) or (17.10-10) it will assume the shape of the corresponding cubic spline. (The end conditions in (17.10-10) correspond to zero applied torque at the ends.) Thus we see the connection between cubic splines and the draftsman's spline, and we see the appropriateness of cubic spline interpolation to such mechanical design problems as the laying out of ship hull contours.

Starting with $n = 3$, all odd-order spline functions interpolated to given data in this manner obey a similar variational condition,[17-9] with the square of the derivative of order $(n + 1)/2$ appearing in the integrand.

A *cardinal spline*, for example corresponding to the problem (17.10-10), is a spline for which all data on the right-hand side are zero except for one of the f_k

$$s_j(u_k) = \delta_{k,j} = \begin{cases} 0 & k \neq j \\ 1 & k = j \end{cases} \tag{17.10-12}$$

Then (17.10-10) has the solution in Lagrangian form

$$s(x) = \sum_{j=0}^{m} f_j s_j(x) \tag{17.10-13}$$

which facilitates study of the effects of perturbing one of the values f_j. The amplitude of the cardinal spline $s_j(x)$ is found to die out exponentially as one moves away from the jth interpolation point,[17-5] in contrast to the situation with polynomial interpolation. This is the basis for the characterization of spline interpolation as *local*. Cubic spline interpolation is only approximately local, as exponentially small but nonzero effects are felt far away from a perturbation. Cubic Hermite interpolation and linear ($n = 1$) spline interpolation are examples of strictly local interpolating functions. As seen from the Lagrange interpolation formula (section 18.4.2.1), polynomial interpolation has the opposite property: perturbations at an interpolation point x_k result in large changes in the interpolating polynomial far away from x_k.

Of great interest are the convergence properties of an interpolated spline function as $\Delta_k = u_k - u_{k-1}$ is made small. As is well-known, polynomial interpolation can fail to converge even for functions of the class C^∞ (the famous Runge example of $f(x) = 1/(1 + x^2)$ in $[-5, 5]$ is discussed in Steffenson.[17-42]) In contrast it has been proven that, for $f(x) \in C^4$, if $[\max_k |\Delta_k|]\ [\min_k |\Delta_k|]^{-1}$ remains bounded, then as $||\Delta|| = \max_k |\Delta_k| \to 0$ cubic spline interpolation has the convergence property

$$|f^{(j)}(x) - s^{(j)}(x)| \leqslant K||\Delta||^{4-j} \tag{17.10-14}$$

for $j = 0, 1, 2, 3$ (Birkhoff and deBoor;[17-4] sharper results are also proven. See also Ref. 17-2). Thus we have uniform convergence of derivatives up to order three. We see that, again in contrast to polynomial interpolation, spline interpolation (and approximation) is suitable when we wish to compute derivatives.

17.11 REFERENCES

17-1 Achiesser, N. I., *Theory of Approximation*, Ungar, New York, 1956.

17-2 Ahlberg, J. H., Nilson, E. N., and Walsh, J. L., *The Theory of Splines and Their Applications*. Academic Press, New York, 1967.

17-3 Arden, D. N., "The Solution of Linear Programming Problems," 263–279, *Mathematical Methods for Digital Computers*, vol. 1, eds., A. Ralston and H. S. Wilf, Wiley, New York, 1965.

17-4 Birkhoff, G., and deBoor, C., "Error Bounds for Cubic Spline Interpolation," *Journal of Math. and Mech.*, **13**, 827–835, 1964.

17-5 Birkhoff, G., and deBoor, C., "Piecewise Polynomial Interpolation and Approximation," 164–190, *Approximation of Functions*, ed. Garabedian, Elsevier, Amsterdam, 1965.

17-6 Cheney, E. W., *Introduction to Approximation Theory*, McGraw-Hill, New York, 1966.

17-7 Cheney, E. W., and Southard, T. H., "A Survey of Methods for Rational Approximation with a Particular Reference to a New Method Based on a Formula of Darboux," *SIAM Review*, **5**, 219–231, 1963.

17-8 Courant, R., and Hilbert, D., *Methods of Mathematical Physics*, vol. 1, Wiley-Interscience, New York, 1953.

17-9 deBoor, C., "Best Approximation Properties of Spline Functions of Odd Degree," *Journal of Math. and Mech.*, **12**, 747–749, 1963.

17-10 Esch, R. E., and Eastman, W. L., "Computational Methods for Best Spline Function Approximation," *Journal of Approximation Theory*, **2**, 85–96, 1969.

17-11 Esch, R. E., and Eastman, W. L., "Computational Methods for Best Approximation," *Wright-Patterson Air Force Base Report* SEG-TR-67-30, Dayton, Ohio, 1967.

17-12 Esch, R. E., and Eastman, W. L., "Computational Methods for Best Approximation and Associated Numerical Analysis," *Wright-Patterson Air Force Base Report* ASD-TR-68-37, Dayton, Ohio, 1968.

17-13 Forsythe, G. E., "Generation and Use of Orthogonal Polynomials for Data Fitting with a Digital Computer," *J. SIAM*, **5**, 74–88, 1957.

17-14 Fraser, W., and Hart, J. F., "On the Computation of Rational Approximations to Continuous Functions," *C.A.C.M.*, **5**, 401–403, 1962.

17-15 Grabner, C., "Application of an Iterative Linear Programming Algorithm for Obtaining Generalized Rational Function Approximations," International Symposium on Mathematical Programming, London School of Economics, London, 1964.

17-16 Greville, T. N. E., ed., *Theory and Applications of Spline Functions*, Academic Press, New York, 1969.

17-17 Hadley, G., *Linear Programming*, Addison-Wesley, Reading, Mass., 1962.

17-18 Handscombe, D. C., ed., *Methods of Numerical Approximation*, Pergamon, Oxford, 1966.

17-19 Hardy, G. H., Littlewood, J. E., and Polya, G., *Inequalities*, Cambridge University Press, Cambridge, 1964.

17-20 Jackson, D., *The Theory of Approximation*, American Math Society, New York, 1930.

17-21 Lawson, C. L., "Characteristic Properties of the Segmented Rational Minimax Approximation Problem," *Numerische Mathematik*, **6**, 293–301, 1964.

17-22 Lawson, C. L., "Segmented Rational Minimax Approximation, Characteristic Properties and Computational Methods," *Jet Propulsion Laboratory Technical Report 32-579*, Pasadena, California, 1963.

17-23 Lorentz, G. G., *Approximation of Functions*, Holt, Rinehart and Winston, New York, 1966.

17-24 Maehly, H. J., "Methods for Fitting Rational Approximations, Parts II and III," *J.A.C.M.*, **10**, 257–277, 1963.

17-25 Michaud, R. O., "Best Rotated Approximation," *Boston University Math Research Report 71-16*, Boston, 1971.

17-26 Murnaghan, F. D., and Wrench, J. W., "The Approximation of Differentiable Functions by Polynomials," *David Taylor Model Basin Report 1175*, 1958.

17-27 Ralston, A., "Rational Chebyshev Approximation," *Mathematical Methods for Digital Computers*, vol. II, eds. A. Ralson and H. S. Wilf, Wiley, New York, 1967.

17-28 Ralston, A., "Rational Chebyshev Approximation by Remes Algorithms," *Numerische Math* **7**, 322–330, 1965.

17-29 Rice, J. R., *The Approximation of Functions, Vol. II–Advanced Topics*, Addison-Wesley, Reading, Mass., 1969.

17-30 Rice, J. R., "Characterization of Chebyshev Approximations by Splines," *SIAM Journal of Numerical Analysis* **4**, 557–565, 1967.

17-31 Rice, J. R., *The Approximation of Furctions, Vol. I–Linear Theory*, Addison-Wesley, Reading, Mass., 1964.

17-32 Rice, J. R., "Chebyshev Approximation by Exponentials," *J. SIAM*, **10**, 149–161, 1962.

17-33 Rice, J. R., "The Characterization of Best Nonlinear Tchebycheff Approximations," *Trans. Am. Math Soc.*, **96**, 322–340, 1960.

17-34 Rice, J. R., "Chebyshev Approximation by $ab^x + c$," *J. SIAM*, **8**, 691–702, 1960.

17-35 Rivlin, T. J., *An Introduction to the Approximation of Functions*, Blaisdell, Waltham, Mass., 1969.

17-36 Rivlin, T. J., and Cheney, E. W., "A Comparison of Uniform Approximations on an Interval and a Finite Subset Thereof," *SIAM Journal of Numerical Analysis* **3**, 311–320, 1966.

17-37 Scheffé, H., *The Analysis of Variance*, Wiley, New York, 1959.

17-38 Schoenberg, I. J. ed., *Approximations with Special Emphasis on Spline Functions*, Academic Press, New York, 1969.

17-39 Schoenberg, I. J., and Whitney, A., "On Pólya Frequency Functions III: The Positivity of Translation Determinants with an Application to the Interpolation Problem by Spline Curves," *Tr. Am. Math Soc.* **74**, 246–259, 1953.

17-40 Schumaker, L. L., "On Computing Best Spline Approximations," *Mathematics Research Center Report No. 833*, Madison, Wisconsin, 1968.

17-41 Spuria, A. J., "Best Approximation by Polynomial Composition," *Boston University Math Research Report*, May, 1970.

17-42 Steffenson, J. F., *Interpolation*, Chelsea, New York, 1950.

17-43 Stiefel, E. L., "Numerical Methods of Chebyshev Approximation," 217–232, *On Numerical Approximation*, ed. U. Langer, Wisconsin Press, 1959.

17-44 Szego, G., *Orthogonal Polynomials*, American Math Society, Providence, R.I., 1939.

17-45 Taylor, A. E., *Introduction to Functional Analysis*, Wiley, New York, 1958.

17-46 Varga, R. S., "Error Bounds for Spline Interpolation," 367–388, *Approximations with Special Emphasis on Spline Functions*, ed. I. J. Schoenberg, Academic Press, New York, 1969.

17-47 Varga, R. S., *Functional Analysis and Approximation Theory in Numerical Analysis*, SIAM, Philadelphia, 1971. (No. 3 of the Regional Conference Series in Applied Mathematics.)

17-48 deBoor, Carl, *A Practical Guide to Splines*, Appl. Math. Series no. 27, Springer-Verlag, New York, 1978.

18

Numerical Analysis

A. C. R. Newbery*

18.0 INTRODUCTION

The basic responsibilities of the numerical analyst are:

1. To design computer programs for the solution of numerical problems.
2. When reliable computer programs are available, to select a program which is optimally matched to a given problem.
3. To design and implement tests which will verify that a given program is behaving according to specifications, and which will give a clear indication of any weaknesses in the program.
4. To provide error estimates associated with the computerized solution of any numerical problem.

Of course there are other peripheral activities, but the above four points represent the main core. The last of the four is the most basic because it is essential to the other three. The design, selection, and validation of computer programs all presuppose an understanding of the way in which errors arise and are propagated.

In this chapter it will be the general policy to let the references look after the detailed specification of algorithms, proofs of convergence etc. We shall concentrate mainly on drawing attention to the existence, merits and demerits of the various algorithms which may be available for a given purpose. In addition to the specific references given in section 10.1, a special mention should be made of the N.B.S. Handbook which section 10.2 assigns for further reading. It provides the details and numerical constants for many of the basic formulas, particularly in quadrature and interpolation.

18.1 GENERAL INFORMATION ON ERROR ANALYSIS

18.1.1 Sources of Error

These may be:

a. Modelling error, i.e., a mismatch between the physical problem and its mathe-

*Prof. A. C. R. Newbery, Dep't. of Computer Science, University of Kentucky, Lexington, Ky.

matical formulation. For instance the parameters involved in the mathematical definition of a satellite orbit, e.g., the effect of the gravitational anomalies of the earth, are not known to perfect accuracy. Consequently some imperfections of the mathematical model are inevitable.

b. Reformulation error. A given problem may be unsolvable on a digital computer, but it may be possible to formulate a related problem which is solvable, and whose solution approximates that of the given problem. This situation arises from the fact that there is no number continuum on a digital computer—the numbers form a finite discrete set. Consequently any problem, such as a differential equation, which presupposes the existence of a continuum has to be reformulated. In this case the reformulated problem will be a difference equation.

c. Computational errors. This represents the cumulative effect of using finite-precision arithmetic on a lengthy problem. The term is here taken as *excluding* all logical mistakes and machine malfunctions. Only errors of type (b) and (c) will be considered here.

18.1.2 Floating-Point Arithmetic

This is the mode in which almost all scientific computing is performed. It means that every number is represented to a fixed number of significant digits. The digits are generally binary digits (= bits) but in the case of the IBM 360/370 series they are hexadecimal digits.

18.1.2.1 Round-Off Error (also Known as Number-Truncation Error) A real number correctly rounded to β significant bits may differ from the same number rounded to $\beta + 1$ bits by as much as $\pm\frac{1}{2}$ in the last (i.e., the βth) position; hence there may be that much difference between the number stored in the machine and the real number which it is supposed to represent. The difference between a real number r and its machine representation r' obeys the inequality $|r - r'| \leqslant \epsilon |r'|$, where $\epsilon = 2^{-\beta}$, and the number r' correctly represents r to β significant bits.

18.1.2.2 Loss of Significant Bits The most common cause of loss of significance is the subtraction of nearly equal numbers. If two numbers x, y are represented to β significant bits and they agree to the first k of these bits, then the number of significant digits in the difference $x - y$ is $\beta - k$. This means that although x and y were correctly rounded machine representations, the error in the machine representation of $x - y$ could be as great as $\pm 2^{-(\beta - k)}|x - y|$. Frequently a loss of significance due to this cause can be avoided by careful programming. For example the expression $9 - \sqrt{80}$ can be reformulated as $1/(9 + \sqrt{80})$ so as to avoid the subtraction of nearly equal numbers. Similarly one should beware of writing expressions like $1 - \cos x$ if there is a possibility that $\cos x \simeq 1$. It can be replaced by $2 \sin^2 (x/2)$. See Ref. 18-28.

18.1.2.3 Error Propagation for Basix Floating-Point Arithmetic In the following the symbol $M(\cdot)$ denotes "the machine representation of." Thus $M(x)$ is the machine representation of a real number x. It follows from section 18.1.2.1 that $|M(x) - x| \leqslant \epsilon |M(x)| = \epsilon |x| + O(\epsilon^2)$. For positive numbers x, y combined by floating point operations the following inequalities hold:

$$
\begin{array}{lll}
M(x+y) & \text{is in the range} & (x+y)(1 \pm 2\epsilon) \\
M(x-y) & \text{is in the range} & (x-y) \pm \epsilon(|x| + |y| + |x-y|) \\
M(xy) & \text{is in the range} & xy(1 \pm 3\epsilon) \\
M(x/y) & \text{is in the range} & (x/y)(1 \pm 3\epsilon)
\end{array}
$$

The last two statements are first-order approximate, but in scientific computation one may reasonably assume that $\epsilon < 10^{-6}$, so that one can justify the neglect of ϵ^2 in comparison with 1.

18.1.2.4 Nonaxioms of Floating-Point Arithmetic Certain axioms and theorems of number fields fail to hold when numbers are combined by the rules of floating-point arithmetic. In particular the following are *FALSE*:

(i) Distributive law: $a(b + c) = ab + ac$
(ii) Associative law: $(a + b) + c = a + (b + c)$
(iii) Cancellation law: If $a + \delta = a$ then $\delta = 0$. (The converse of this, howeve .s true.)

18.1.3 Interval Arithmetic

It was observed in section 18.1.2.3 that a machine number $M(x)$ actually represents any real number x in the interval $M(x)(1 \pm \epsilon)$. Thus, what is thought of as being a number is in reality not a number but an *interval*. It is possible to perform arithmetic on intervals; for example the "sum" of two intervals is an interval defined by the uprounded floating-point sum of the two upper bounds and the downrounded sum of the lower bounds. The subject is discussed in Ref. 18-47. So far the subject has not proved to be of much value in scientific computation. The reasons for this include

1. There is no generally satisfactory way of dealing with correlated errors. Thus if three intervals are related by $I_3 = I_1$ "$-$" I_2, where "$-$" denotes subtraction in the interval arithmetic sense, then it will not follow from $I_2 = I_1$ that $I_3 = 0$. Hence when one has finally computed bounds to a solution, one has no assurance that these bounds are attainable or even remotely approachable.
2. The techniques are only applicable to problems of type (c) as in section 18.1.1. It will be observed that error estimation in this area, particularly in linear algebra, can often be performed by methods which are more economical and more realistic.

18.1.4 Condition of a Problem

Solving a computational problem may be viewed as mapping an ordered set or vector of input (data) numbers onto a vector of outputs (the solution). For instance if one is solving the $n \times n$ linear system $Ax = b$, the input vector consists of the $n^2 + n$ elements of A and b, and the output consists of the n elements of x. In general there is an input vector u and an output vector v which may be of different dimension. Let it be assumed that $u \to v$ and $u + \delta u \to v + \delta v$. If it is true that for some small vector δu, $\|\delta v\|/\|\delta u\|$ is very large, the problem is said to be ill-conditioned. It means that a small perturbation in the data is consistent with a large variation in the solution.

Since the data are generally not known to perfect precision, this is a sign that the computed solution has to be viewed with distrust. There is no unique way of assigning condition numbers to problems, since different choices of norm will be appropriate to different situations. However, the condition number C will represent, or approximate, or be an upper bound for the quantity $\|\delta v\|/\|\delta u\|$, where δu is an infinitesimal vector whose direction is chosen to maximize the quotient of norms. Specific examples will follow in section 18.2.7.

18.1.5 Backward Error Analysis

Given a problem for which D and S represent the (ordered) data set and solution set of numbers, it may be difficult to put a bound on the 'forward' error norm $\|S_c - S\|$, which represents the difference between the computed solution S_c and the true solution S. It is often easier, and equally useful, to prove that S_c is the exact solution corresponding to a perturbed data set D_p where a bound on $\|D_p - D\|$ can be stated. To take a trivial example, let x be an exact zero of the polynomial $P(x)$; an approximation x_1 is found such that $P(x_1) = h$. It is obvious that x_1 is an exact zero of the perturbed polynomial $P(x) - h$. Lanczos' τ-method Ref. 18-43 for solving differential equations is an example of this approach, whereby instead of finding an approximate solution to a given problem, he finds an exact solution to a problem which differs from the given one by a function with a computable (small) bound.

18.1.6 Richardson Process

This is a technique for error estimation, useful principally in the area of quadrature and differential equations. It presupposes the existence of an approximate method whose error is of the form Mh^k, where M is (effectively) constant, h is a known parameter, generally a step-size, and k is also known. The error may then be estimated by means of a consistency test. For example the error associated with the trapezoidal quadrature formula Ref. 18-61 is $-(h^3/12)f''(z)$ where the function $f(x)$ was to be integrated over an interval of length h, and z is a point within this interval. If it may be assumed that $f''(z)$ is approximately constant, then the error is of the form Mh^3 and we can estimate the unknown pseudo-constant M. Let Q denote the result of applying the formula once over the whole range; let Q_l, Q_r

denote the results of applying it over the left-half and right-half ranges, and let T, T_l, T_r be the (unknown) true areas. Then $T = Q - Mh^3$, $T_l = Q_l - M(h/2)^3$ and $T - T_l - T_r = 0 = Q - Q_l - Q_r - Mh^3(1 - \frac{1}{8} - \frac{1}{8})$. Hence we estimate that the error $Mh^3 = (4/3)(Q - Q_l - Q_r)$. The estimate is exact if M is truly constant. In general when errors are of the form Mh^k, the error estimate is $(Q - Q_l - Q_r)\, 2^{k-1}/(2^{k-1} - 1)$. If it is known that the quantity M, while not constant, has the property that its maximum and minimum over the range of integration do not differ by more than a known quantity δ, then it can be shown that the Richardson error estimate is correct within

$$\pm\delta h^k/(2^{k-1} - 1)$$

Richardson-type error estimates are useful in many situations; they often have to be designed with a specific application in mind. It should be noted that parameter *halving* is not an essential feature. One can also estimate errors by observing the discrepancy between two solutions that were computed with step-sizes that differed in a ratio other than 1 : 2.

18.1.7 Mean Value Theorem: $f(x + \delta) = f(x) + \delta f'(z), z \in (x, x + \delta)$

This theorem is useful for the estimation of function-evaluation errors that are attributable to argument errors. For instance if the relative error in a machine number x is ϵ, i.e., $\delta = \epsilon x$, then the relative error in the function-evaluation can be taken to be $\delta f'(x)/f(x)$, since there is ordinarily no important error introduced by evaluating the derivative at x rather than at the neighboring point z. The ratio of relative error on output to that on input then simplifies to $xf'(x)/f(x)$. For instance, when $f(x) = \sin x$, this ratio has a limiting value of 1 at $x = 0$, but the ratio becomes unbounded near $x = \pi$. Argument error is only *one* of the sources of function-evaluation error. The other two are arithmetical error section 18.1.2.3 and approximation error. The latter ordinarily denotes the result of truncating an infinite series or continued fraction for the purpose of approximating a function.

18.2 LINEAR EQUATION SYSTEMS

This section will not attempt to present enough detail to enable a reader to construct his own programs. It is mainly concerned with the considerations to be borne in mind when selecting a method for a given problem, and for this purpose there is some consideration of data errors and computational errors. The program libraries of section 18.9.5 are all strong in linear algebra. The typical user should seldom have to do anything more than data preparation.

18.2.1 Matrix Inversion

The first thing that needs to be said about matrix inversion is that there are very few situations in applied mathematics where an explicit matrix inverse is ever strictly required. One of these is discussed in section 18.2.8.2; also there are some implementations of the Simplex algorithm and of nonlinear equation-solving algorithms where matrix inversion is called for.

In general if one has a set of linear equation systems $AX = B$, it is convenient to write $X = A^{-1}B$, but this notational convenience does *not* imply that the explicit inverse A^{-1} should ever be computed. As observed in Ref. 18-20, it is at least equally economical, and generally more so, to compute X by triangular decomposition (section 18.2.2.1). If B is the identity matrix then the two methods are indistinguishable.

If an inverse is needed, then Gaussian elimination with scaling and full or partial pivoting is recommended Ref. 18-70.

18.2.2 Matrix Decomposition

Generally decomposition is the first step toward solving an equation system, but it occurs also in connection with eigenvalue problems. There are basically two kinds of decomposition—triangular and orthotriangular—but there are many alternative implementations of each kind.

18.2.2.1 Triangular Decomposition This means expressing a square matrix A as a product LU, where L is lower triangular with unit diagonal elements and U is upper triangular. (For simplicity we have assumed that any necessary row-interchanges in A were done ahead of time.) The equation-solving procedure then consists of a forward substitution and a back substitution as described in Ref. 18-20. The old Gaussian elimination schemes belong to this class. One performs row operations on A to convert it into an upper triangular matrix U', but the row operations are equivalent to premultiplying A by a lower triangular matrix L'. Hence one obtains $L'A = U'$, or $A = L'^{-1}U'$ so that this is a disguised form of LU decomposition. The LU decomposition as described in Ref. 18-20 is the most efficient known method for solving general linear equation systems of the form $AX = B$ where A is $n \times n$, X and B are $n \times m$ for arbitrary positive integers n, m.

18.2.2.2 Choleski Decomposition This algorithm Ref. 18-70 is closely related to that of the previous section but it is restricted to symmetric, positive definite matrices. Any such matrix H can be put in the form $H = LL^T$, where L does not necessarily have unit diagonal elements. It can be shown Ref. 18-62 that no pivoting is necessary in this case. Also if H is banded with half bandwidth k (i.e., there are never more than k nonzero elements of H either to the left or right of any diagonal element), then L will have a diagonal and k subdiagonals. Elsewhere it will be zero. If $k \ll n$ then the storage requirements for L are clearly far smaller than what would be needed to store the full inverse matrix H^{-1}.

18.2.2.3 Orthotriangularization This is a process whereby a given matrix A is represented in the form ϕU, where ϕ is an orthogonal matrix. Since ϕ has the property that $\phi^T = \phi^{-1}$, we see that the problem $AX = B$ reduces to $X = A^{-1}B =$

$U^{-1}\phi^T B$. We can define $Y = \phi^T B$ by matrix multiplication, then we can obtain $X = U^{-1} Y$ by back-substitution through the upper triangular matrix U.

In chapter 16 it was observed that the classical Gram-Schmidt algorithm for orthotriangularization is unstable, and it should be rejected. The 'modified Gram-Schmidt' process is useable. It generates an upper triangular matrix U' such that $AU' = \phi$, so that mathematically $U' = U^{-1}$ of the previous paragraph, but the computation is stable. Equation-solving algorithms based on stable orthotriangularization are given by Björck, Ref. 18-3 and Householder, Ref. 18-70).

Orthotriangularization is about twice as expensive as LU decomposition in terms of operation counts, but this is largely offset by the greater logical complexity of the latter method. Orthotriangularization has additional 'fringe benefits' relative to LU, as will be noted in sections 18.2.2.4 and 18.2.7.

18.2.2.4 Overdetermined Linear Systems If A is an $m \times n$ matrix of rank n with $m > n$, and it is required to 'solve' $Ax \simeq b$ in the sense that we are to minimize $h^T h$ where $h = Ax - b$, then we may proceed as follows:

1. Determine an upper triangular $n \times n$ matrix U' and an $m \times n$ matrix ϕ whose columns are orthonormal, such that $AU' = \phi$
2. The required solution is given as before by $x = U'\phi^T b$.

18.2.3 Perturbation Methods

The basis for these is Woodbury's formula Ref. 18-34. The formula states that

$$(A - USU')^{-1} = A^{-1} - A^{-1} UTU'A^{-1}$$

where

$$T^{-1} + S^{-1} = U'A^{-1}U$$

It is assumed that A^{-1} is already known either explicitly or in the form of a decomposition, and that USU' denotes some sparse perturbation of A. Three examples follow.

1. The perturbation involves one column only—the jth. (This situation arises with the simplex algorithm for linear programming.) In this case U is an $n \times 1$ matrix representing the perturbation; S can be taken as the scalar 1, and U' is a $1 \times n$ matrix with 1 in the jth position and zeros elsewhere. In Ref. 18-60 Powell discusses a variant of this procedure with improved stability properties.
2. Let there be two perturbed columns, the jth perturbed by u and the kth by v. We can take U to be an $n \times 2$ matrix whose columns are u, v, S is the 2×2 identity matrix and U' has e_j and e_k for its first and second rows.
3. Let USU' be a symmetric perturbation consisting of a vector u in the jth row *and* j^{th} column. (The perturbation to a_{jj} is interpreted half as a row and half as a column perturbation.) Let U denote an $n \times 2$ matrix whose first column is u

and whose second is e_j; let $U' = U^T$ and let S be the reversal matrix of order two, i.e., $s_{12} = s_{21} = 1, s_{11} = s_{22} = 0$. It can be verified that USU' then has the required form, so that the Woodbury formula becomes applicable.*

In order for the Woodbury formula to be competitive it is essential that the perturbations should be confined to a small number of rows and/or columns; otherwise the cost of computing T becomes too great. It is not essential that the explicit inverse A^{-1} be known; an LU or ϕU decomposition of A will serve the purpose.

18.2.4 Iterative Refinement

Let the system $Ax = b$ be solved by the LU method of section 18.2.2.1 and let the computed solution be the vector x_1. On checking back we might find $Ax_1 - b = h_1$, where h_1 is not exactly a zero vector because the LU decomposition was not executed to perfect precision. Assuming that the vector h_1 is significant, (a clarification of this is given in section 18.2.8.3), then we may assume that the computed solution x_1 was wrong and that the true solution is a vector $x_1 - e_1$. If $A(x_1 - e_1) = b$, then $Ae_1 = h_1$ and $e_1 = A^{-1}h_1$. We can 'solve' this last equation for e_1 using the LU decomposition we already have. Since this decomposition is not perfect we shall not get a perfect value of e_1, nor will our corrected 'solution' of $x_1 - e_1$ be perfect either, but it is reasonable to expect it will be better than the uncorrected first solution. In Ref. 18-20 there is a careful analysis of the method describing:

1. the situations in which this 'reasonable' expectation is justified.
2. the warning signals to look for if the successive refinements are not converging.
3. the signals which indicate that the iteration has reached 'noise level,' i.e., additional iterations without augmented wordlength cannot be expected to produce additional accuracy.
4. a means of estimating the condition (section 18.2.7) of the matrix by studying the behavior of successive iterates.

Iterative refinements based on an orthotriangular decomposition can be executed in an analogous fashion both for the exactly determined and the overdetermined case. Estimation of condition number is implemented in the linear equation solvers contained in the libraries of section 18.9.5.

18.2.5 Iterative Methods

The most powerful iterative methods for solving linear equation systems are given in Chapter 16 and Ref. 18-69. Many of these are special-purpose methods designed for use specifically with the kinds of algebraic system which arise from the discretization of partial differential equations.

In this section we shall only record some of the facts relating to gradient methods, while noting that such methods, if useful at all, should be considered primarily as a

*I am indebted to Mr. H. Rosenfeld of the Boeing Company for this account of symmetric perturbation.

means of refining a nearly correct solution, rather than for computing a solution *ab initio*.

Given the problem $Ax = b$, let y be an approximate solution; let e be the error in y and let h be the associated residual, so that

$$Ay - b = h, \quad x = y - e, \quad \text{and} \quad Ae = h$$

1. The gradient of h^2 is parallel to $A^T h$. This means that the scalar h^2 varies as the vector y varies; among all possible directions in which y could vary, the direction $A^T h$ is the one which will make h^2 vary most steeply.

2. The gradient of $||h||_1$, i.e., of the scalar

$$\sum_1^n |h_i|$$

is parallel to $A^T s$ where $s_i = \text{sign}\,(h_i)$.

3. Given an arbitrary vector u, define $w = A^T u$ and let y be replaced by $y - \alpha w$. The effect on h and e will depend on the choice of scalar α. We can minimize either e^2 or h^2 with respect to α by setting

$$\alpha_e = h \cdot u/w^2 \quad \text{or} \quad \alpha_h = w^2/(Aw)^2$$

The vector u will ordinarily be chosen to equal either h or s as defined above.

18.2.6 Finite Iterative Methods

These methods are iterative in form but they have the property that 'mathematically' (i.e., in the absence of any rounding errors) they have to terminate in a finite number of steps. There are several different versions applicable to the various types of problem, including least-squares solution of over-determined systems etc. This material is summarized in two articles, one by Fischbach and one by Hestenes in Ref. 18-10.

18.2.7 Scaling and Condition Numbers

It is shown in Ref. 18-70 that if we are given the system $Ax = b$, but the b vector is subjected to a perturbation δb then the corresponding perturbation δx in the solution vector obeys the inequality $||\delta x||/||x|| \leqslant C(A)\,||\delta b||/||b||$, where $C(A) = ||A||\ ||A^{-1}||$ and is called the 'condition number' of the matrix A. Although the numerical value of $C(A)$ varies with the choice of norm, the foregoing definition and inequality hold for any vector norm and subordinate matrix norm. For practical purposes there are only three norm-pairs that are of interest, viz.,

$$||x||_1 = \sum_1^n |x_i|, \qquad ||A||_1 = \max_j \sum_{i=1}^n |a_{ij}|$$

$$|||x||_2 = \sqrt{\sum x_i^2}, \quad ||A||_2 = (\text{max eigenvalue of } A^TA)^{1/2}$$

$$||x||_\infty = \max_i |x_i|, \quad ||A||_\infty = \max_i \sum_{j=1}^{n} |a_{ij}|$$

It is clear that, regardless how one chooses to define the norm, a large value of $C(A)$ implies that the system is very sensitive to small perturbations.

The value of $C(A)$ may be materially affected by the manner in which the problem is scaled. For example if we take the third norm above, and row-scale the matrix in such a way that every absolute row sum shall equal the largest absolute row sum of A, this is equivalent to premultiplying A by a diagonal matrix D all of whose elements are $\geqslant 1$ and generally some are > 1. The condition number $C(DA)$ is $||DA||\ ||A^{-1}D^{-1}||$. By construction $||DA|| = ||A||$, but since the elements of D^{-1} are all $\leqslant 1$ it is clear that $||A^{-1}D^{-1}|| \leqslant ||A^{-1}||$. Hence, $C(DA) \leqslant C(A)$.

If orthotriangularization is used, so that $AU' = \phi$, then $A^{-1} = (\phi U'^{-1})^{-1} = U'\phi^T$ and $||A^{-1}|| \leqslant ||U'||\ ||\phi^T||$. This will yield an upper bound for $C(A)$.

Attempting to find an optimal scaling is generally not worth the labor, and one should be content to find a simple scaling that is not unduly bad. Wilkinson's suggestion Ref. 18-70 for avoiding any gross misscaling of A is to equilibrate, by which he means scale each column of A by an integer power of 2 in such a way that its element of largest magnitude should be in the interval $[\frac{1}{2}, 1)$.

If a matrix is row- and/or column-scaled for the purpose of inversion, this means that A was replaced by D_1AD_2. What emerges from the inversion process will therefore be $D_2^{-1}A^{-1}D^{-1}$. This output must be premultiplied by D_2 and postmultiplied by D_1 to yield the required value of A^{-1}.

18.2.8 Error Analysis

This subject is treated in great detail in Ref. 18-70. In the following synopsis it is assumed that the equation system is $Ax = b$; the quantities $\delta A, \delta x, \delta b$ denote small perturbations in A, x, b respectively; they could be interpreted as data errors or number truncation errors.

18.2.8.1 Variation of x with A

$$||\delta x|| / ||x|| \leqslant \frac{C(A)\ ||\delta A|| / ||A||}{1 - C(A)\ ||\delta A|| / ||A||}$$

provided

$$||A^{-1}||\ ||\delta A|| < 1, \quad \text{i.e.,} \quad \frac{C(A)\ ||\delta A||}{||A||} < 1$$

18.2.8.2 Variation of x with b

$$||\delta x|| = ||A^{-1}\delta b|| \leqslant ||A^{-1}|| \; ||\delta b|| = C(A) \, ||\delta b|| / ||A||$$

Since $\delta x = A^{-1}\delta b$ we can put termwise bounds on the elements of δx if we have A^{-1} explicitly and know termwise bounds on δb.

This leads to

$$|\delta x_i| \leqslant \sum_{j=1}^{n} |a'_{ij}\delta b_j|$$

where $A^{-1} = [a'_{ij}]$.

With the help of the explicit inverse one can also answer questions of the form: Given the requirement that the x_i have to be accurate to $p\%$, what is the largest percentage inaccuracy we can tolerate in the b_i while guaranteeing to meet the requirement?

*18.2.8.3 **Computational Errors*** Suppose the computed solution to the system is y, and we observe a residual vector h defined by $Ay - b = h$. Ordinarily we expect all components h_i to be small, but since the b_i are not necessarily small, the h_i may be the result of subtracting a pair of larger, nearly-equal numbers, and may therefore not be significant. The following observations should be made:

1. Let ϵb_i denote "the magnitude of one rounding error in b_i," i.e., $\epsilon b_i = 2^{-\beta}|b_i|$ for a binary machine with β-bit mantissa; then if $|h_i| < \epsilon b_i$ it is insignificant, even if it is known that b_i is *not* subject to rounding error. If, for example $h_i = 0$, that means that the dot product of the ith row of A with y exactly equalled b_i. The dot product is then subject to rounding error at least as big as ϵb_i, even if b_i itself is exact.

2. The rounding errors in the dot product *can* be far greater than ϵb_i. The dot product is the algebraic sum of n numbers of possibly varying sign. Some of these numbers may be vastly greater than the algebraic sum; consequently the sum, if accumulated in single precision, may be grossly inaccurate.

3. In view of the above remarks it is recommended that the dot products should be accumulated in double precision. One makes the reasonable (but not strictly rigorous) assumption that the double precision sum will round correctly to single precision. One considers 'noise level' to be $2\epsilon b_i$. If all the h_i are below noise level, then there is no point in continuing with any iterative method.

*18.2.8.4 **Relations Between the Residual and Data Errors*** If we find that $Ay - b = h$ and h is significant, then y is the exact solution to the problem $Az = b + h$, where we assume a data error of size h_i in the given value b_i.

We can also attribute the observed h_i to hypothetical errors δ_{ij} in the elements a_{ij} obeying the equation

$$\sum_{j=1}^{n} \delta_{ij} y_j = h_i$$

This is one equation in n unknowns $\delta_{ij}, j = 1, 2, \ldots, n$. There may be a solution for which all the δ_{ij} are small enough in magnitude that they are all within the radius of uncertainty of the associated element a_{ij}. In this case the problem has been correctly solved to within the accuracy warranted by the quality of the data. This form of *a posteriori* error analysis is discussed in Ref. 18-53. It has the merit that it can be implemented without any knowledge of the condition number.

18.2.9 Selection of Method

Given the problem $AX = B$, the choice of method will depend firstly on whether B is a single or multiple right-hand side; secondly it depends on the nature of A. The following are suggested guidelines.

1. If A is square, dense, and unstructured the choice is between LU and ortho-triangularization. The choice is discussed by Wilkinson Ref. 18-70, but he expresses no strong preference either way. In either case accuracy can generally be improved by iterative refinement.

2. If A is symmetric and positive-definite then Choleski decomposition is clearly recommended. For the symmetric indefinite case the recommended method is given in Ref. 18-6. Iterative refinement can be used.

3. If A is sparse and B is a single vector, the methods of Chapter 16 may be applicable, particularly if A is diagonal-dominant. One should also consider the methods of section 18.2.6. These finite iterative methods can solve a general $n \times n$ system for the cost of $2n$ matrix-vector multiplications. This is about six times the cost of LU decomposition on a dense matrix, so that one reaches break-even point economically when the matrix is one-sixth dense. However, densities down to 1% are commonly encountered. Conjugate gradient methods, unlike LU decomposition, do not involve any operations on the matrix A. Whereas the LU decomposition will require n^2 locations to store the decomposition of A, the conjugate gradient method will only need to store the few nonzero elements of A plus a small number of vector arrays. On occasions this storage economy will make the difference between an in-core and an out-of-core problem. Some sparse matrix programs are available in Ref. 18-33.

4. For least-squares problems the method of section 18.2.2.4 is generally recommended. However, if A is sparse and B is a single vector, there is a finite-iterative method which may be competitive.

5. For sparse matrices which differ only in a few rows or columns from a matrix whose inverse or decomposition is known or easily computable the perturbation methods of section 18.2.3 are appropriate.

18.3 EIGENVALUE AND LAMBDA-MATRIX PROBLEMS

18.3.1 Relevant Theorems

The reader is referred to Ref. 18-34 for a comprehensive treatment of this topic.

18.3.1.1 Gersgorin's Theorem The largest eigenvalue-modulus of A does not exceed $\|A\|_1$ nor $\|A\|_\infty$, where these norms are defined in section 18.2.7. An extended version states that every eigenvalue of A lies in the union of n disks with centers a_{ii} and radius

$$\sum_{j \neq i}^{n} |a_{ij}|; \quad \text{for } i = 1, 2, \ldots, n$$

For additional extensions see Ref. 18-34. The theorems apply also to complex matrices.

18.3.1.2 If A has eigenvalues λ_i and eigenvectors v_i, then

(i) A^{-1} has eigenvalues λ_i^{-1} and eigenvectors v_i

(ii) For any algebraic polynomial P, $P(A)$ has eigenvalues $P(\lambda_i)$ and eigenvectors v_i. A similar statement holds for the rational function $R(A) \equiv P(A)Q^{-1}(A)$. The 'constant term' of a polynomial has to be interpreted as a scalar multiple of the identity matrix.

(iii) If $P(x)$ is a polynomial whose n roots are λ_i, then $P(A) \equiv 0$, and P is called the "Characteristic polynomial" of A.

18.3.1.3 Biorthogonality Theorem If A has distinct eigenvalues λ_i, then the eigenvectors v_i of A and u_i of A^T form a 'biorthogonal system'. This means that $(u_i, v_j) = 0$ unless $i = j$, i.e., unless both vectors correspond to the same eigenvalue. The (generally complex) scalar product of complex vectors x, y is defined by

$$(x, y) = \sum_{r=1}^{n} \bar{x}_r y_r$$

where the bar denotes complex conjugate.

If A is real-symmetric then all its eigenvalues and vectors are real, and if tne eigenvalues are distinct, all eigenvectors will be orthogonal.

18.3.1.4 Similarity Transform If S is an arbitrary nonsingular matrix then the eigenvalues of SAS^{-1} are the same as those of A. The eigenvectors are those of A premultiplied by S.

18.3.1.5 The Sum and Product of Eigenvalues The sum of the eigenvalues of A equals the trace of A, i.e., the sum of diagonal elements. The product of the eigenvalues equals $(-1)^n$ Det (A).

18.3.2 Hermitian Matrices

A Hermitian matrix is one which equals its own conjugate transpose. Its eigenvalues are real and its eigenvectors orthogonal. The recommended method for solving the eigenproblem is presented by Ortega, Ref. 18-62, together with the associated error analysis. One would not be advised to depart from this procedure unless one were seeking only a very small subset of the eigenvectors. In the latter case inverse iteration might be preferable.

18.3.3 The Rayleigh Quotient

If x is an estimated eigenvector of a real matrix A, then one can find a 'best' associated eigenvalue by minimizing the squared Euclidean norm $\|Ax - \lambda x\|_2^2$ with respect to λ. The value of λ which minimizes this quadratic function is known as the 'Rayleigh quotient' $R = x^T Ax/x^T x$. It can be shown that when x is real and A is real symmetric, then R has to be in the closed interval defined by the greatest and least eigenvalue of A.

18.3.3.1 Inverse Iteration If x is an approximate eigenvector of A and R is the Rayleigh quotient, we can attempt to define a vector y by $(A - RI)y = x$. The near-singularity of $A - RI$ may cause y to be very large, but it will in general be a better approximant to the eigenvector than x. After scaling y down, a new Rayleigh quotient can be computed and the process repeated. The process is recommended by Wilkinson Ref. 18-70 as a means of determining an eigenvector when the eigenvalue is known. He studies the convergence of the process and generalizes the Rayleigh quotient to complex matrices.

18.3.4 Non-Hermitian Matrices

Solving an eigenproblem of this type involves at least two phases. First the matrix is similarity-transformed into Hessenberg form; secondly the eigenvalues of the Hessenberg matrix are determined; lastly the eigenvectors, if required, will also be computed. Wilkinson Ref. 18-70 cites three acceptable algorithms for phase 1 and two, namely the *QR* method and Laguerre's method, for phase 2. Inverse iteration is recommended for the third phase. There are several good computer programs in the public domain for this purpose. Since Laguerre's method requires the first and second derivative of Det $(A - \lambda I)$ with respect to λ the formulas of section 18.3.5.3 are appropriate.

18.3.5 Lambda-Matrix Problems

If B is a square matrix, each of whose elements is a function of a single parameter λ. It may be required to determine values of λ such that Det $(B) = 0$. It also may be required to find a nontrivial vector v associated with one of these λ-values such that $Bv = 0$. In the following discussion elements are assumed to be at least continuous functions of λ. On occasions the context will indicate an assumption of differentiability. Regardless how the roots λ_i may be computed, the associated vectors v_i can be computed by inverse iteration.

18.3.5.1 ***Muller's Method*** This is undoubtedly the simplest method to use. Wilkinson, Ref. 18-70, states that it has worked well for him when the elements b_{ij} are low-degree polynomials in λ.

18.3.5.2 ***Other Methods*** Methods like Newton's and Laguerre's require one or two derivatives of Det (B) with respect to λ. Their merits are studied empirically by Lancaster in Ref. 18-42.

18.3.5.3 ***Derivative of a Lambda-Matrix*** The formulas for this are derived in Ref. 18-42. Let Δ denote Det (B), then

$$\frac{\Delta'(\lambda)}{\Delta(\lambda)} = \text{Tr } [B^{-1}(\lambda)B'(\lambda)] \tag{18.3-1}$$

and

$$\frac{d}{d\lambda}\left(\frac{\Delta'(\lambda)}{\Delta(\lambda)}\right) = \text{Tr } [B^{-1}B'' - (B^{-1}B')^2] \tag{18.3-2}$$

In the above formulas the derivative matrices B', B'' are defined by termwise differentiation and Tr stands for 'trace', i.e., "the sum of the diagonal elements of."

In the simple case corresponding to the standard eigenvalue problem, we have

$$\frac{d}{d\lambda}[\text{Det }(A - \lambda I)] = -\text{Det }(A - \lambda I) \cdot [\text{Tr }(A - \lambda I)^{-1}] \tag{18.3-3}$$

18.3.5.4 $Av = \lambda Bv$ Given an equation of this form, where A, B are symmetric and B is positive-definite, the recommended method is to factor B into LL^T by Choleski decomposition section 18.2.2.2. Then the problem Det $(A - \lambda B) = 0$ can be replaced by Det $L^{-1}(A - \lambda B){L^{-1}}^T = 0$, i.e., Det $(L^{-1}A{L^{-1}}^T - \lambda I) = 0$. This gives us a standard eigenproblem with the symmetric matrix $L^{-1}A{L^{-1}}^T$. It can be solved by the method of section 18.3.2. If u_j are the eigenvectors of $L^{-1}A{L^{-1}}^T$, then they are related to those of the original system by $v_j = {L^{-1}}^T u_j$.

If the problem is as above except that A is positive-definite, while B is not, the problem should be reformulated as $Bv = \mu Av$. The problem can be solved as above by factoring A; the resulting eigenvalues μ_i are reciprocals of the corresponding eigenvalues λ_i of the original problem.

18.3.6 Error Analysis

The following results are mostly abstracted from Ref. 18-70.

18.3.6.1 Symmetric Matrix If $Au - \lambda u = h$, where A is symmetric and $u^T u = 1$, then there is at least one true eigenvalue λ_i of A such that

$$|\lambda_i - \lambda| \leqslant \sqrt{h^T h}$$

18.3.6.2 Symmetric Perturbation If A and B are symmetric with eigenvalues λ_i, μ_i, then the matrix $C = A + B$ has eigenvalues y_i where

$$|y_i - \lambda_i| \leqslant \max_j |\mu_j|$$

This result holds for all symmetric perturbations, but is most useful when B is small of the same order as the rounding errors in A. All eigenvalues of B can be bounded using Gersgorin's theorem, section 18.3.1.1 and hence we can bound the errors in an eigenvalue attributable to data errors in A.

18.3.6.3 Unsymmetric Matrix If Det $(A - \lambda I) = \Delta$, then there is at least one eigenvalue λ_i of A such that

$$|\lambda_i - \lambda| \leqslant |\Delta|^{1/n}$$

and also

$$|\lambda_i - \lambda| \leqslant n / |\mathrm{Tr}\,(A - \lambda)^{-1}|$$

These results follow by combining sections 18.7.6, 18.3.5.3.

18.4 APPROXIMATION AND INTERPOLATION

18.4.1 Relevant Theorems

18.4.1.1 Weierstrass' Theorem Let $f(x)$ be continuous on a finite interval $[a, b]$ and let $\epsilon > 0$ be given. Corresponding to arbitrarily small ϵ there exists an n such that one can construct an nth degree polynomial $p_n(x)$ with the property that $|f(x) - p_n(x)| \leqslant \epsilon$ for all $x \in [a, b]$.

18.4.1.2 Extensions and Remarks on Weierstrass' Theorem It should not be assumed that a polynomial $p_n(x)$ which uniformly approximates $f(x)$ in the above sense will necessarily have the property that $p_n'(x)$ approximates $f'(x)$, (assuming $f'(x)$ to be defined). However polynomials *can* be constructed to provide simultaneous uniform approximation to $f(x)$ and its derivatives of arbitrarily high order, provided only that $f(x)$ is sufficiently differentiable Ref. 18-11. For uniform approximation of $f(x)$ by trigonometric polynomials see Ref. 18-67.

18.4.1.3 ***The Runge Phenomenon*** If $\{q_n(x)\}$ is a sequence of polynomials interpolating $f(x)$ at an increasingly dense set of points in $[a, b]$, it can *not* be assumed that the sequence will uniformly approximate $f(x)$. An example, due to Runge, shows that a sequence so constructed can diverge Ref. 18-67. This theorem constitutes a warning that results obtained from high-order polynomial interpolation should be viewed with some suspicion.

18.4.1.4 ***Chebyshev's Theorem*** In the notation of section 18.4.1.1, if we fix n and then seek the smallest positive ϵ for which the inequality holds, then the following statements can be made: (1) The polynomial $p_n(x)$ of best approximation is unique; and (2) the difference $f(x) - p_n(x)$ will attain its maximum magnitude of ϵ at least $n + 2$ times in $[a, b]$ with alternating signs Ref. 18-67.

18.4.2 Polynomial Interpolation

18.4.2.1 ***Lagrange's Formula*** Given $n + 1$ points (x_i, f_i), $i = 0, \ldots, n$ with the x_i distinct, there exists a unique polynomial of degree $\leqslant n$ passing through those points. The usual textbook representation of this polynomial $y(x)$ is as follows:

Let

$$\pi(x) \equiv \prod_0^n (x - x_i); \quad \pi_j(x) \equiv \pi(x)/(x - x_j)$$

then

$$y(x) = \sum_{i=0}^n f_i \pi_i(x)/\pi_i(x_i)$$

and the remainder is given by

$$f(x) - y(x) = \pi(x) f^{(n+1)}(z)/(n+1)!$$

where z is an abscissa somewhere in the continuous interval defined by the greatest and least of x and the x_i. Differentiability of $f(x)$ is required only for the validity of the remainder term, and not for the constructibility of $y(x)$.

18.4.2.2 Newton's Formula An alternative method for constructing and representing the same interpolant $y(x)$ is

$$y(x) = \sum_{i=0}^{n} a_i Q_i(x)$$

where

$$Q_0 = 1; \quad Q_i = \prod_{0}^{i-1} (x - x_j)$$

and

$$a_0 = f_0; \quad a_j = \frac{\left[f_j - \sum_{0}^{j-1} a_i Q_i(x_j) \right]}{Q_j(x_j)}; \qquad j \geqslant 1$$

This formulation is more economical to implement than that of section 18.4.2.1; moreover it has the merit that one is not constrained to preassign the number of points that will be incorporated into the interpolation. Since the a_i and Q_i are defined recursively, it is always possible to incorporate additional data points when they become available.

18.4.2.3 Osculatory Interpolation Sometimes an interpolant is required to match slopes and higher derivatives in addition to ordinates. It is then called 'osculatory.' Supposing we have $n + 1$ such mixed conditions, then we can guarantee to find an nth degree polynomial matching the conditions, provided that (a) whenever the kth derivative is specified at x_i, all lower-order derivatives are also specified there, and (b) at a given x_i only *one* derivative of a given order may be specified. (Clearly we cannot allow two distinct ordinates or slopes to be specified at a single abscissa.) The case where an ordinate f_k and a slope f_k' are given at each of $n + 1$ points is known as *Hermite interpolation* (Ref. 18-32). Let $l_i(x) = \pi_i(x)/\pi_i(x_i)$ where $\pi_i(x)$ is defined in section 18.4.2.1. Define

$$h_i(x) = [1 - 2l_i'(x)\,(x - x_i)]\; l_i^2(x)$$

and

$$\bar{h}_i(x) = (x - x_i)\, l_i^2(x)$$

Then the required osculatory interpolant is

$$y(x) = \sum_0^n h_i(x) f_i + \sum_0^n \bar{h}_i(x) f_i'$$

An algorithm for osculatory interpolation in terms of trigonometric polynomials is given in Ref. 18-52.

18.4.2.4 Finite Difference Methods There is a great profusion of such methods using forward, central or backward differences on equispaced data and using divided differences otherwise. The methods which are (often incorrectly) attributed to Gregory, Laplace, Bessel etc. are all mathematically equivalent to constructing the polynomial interpolant and evaluating it at a given point; however some economy is achieved by bypassing the explicit construction of the polynomial. The methods differ from each other partly by the way the interpolation points are sequenced after they have been chosen. Details may be found in Ref. 18-32 and most other numerical analysis texts.

18.4.3 Spline Interpolation

The definition of a spline function Ref. 18-25 is: Given a strictly increasing sequence of real numbers $x_1, x_2, \ldots, x_n$, a *spline function*, $S(x)$, of degree m with *knots* $x_1, x_2, \ldots, x_n$ is a function defined on the entire real line having the following two properties:

1. In each interval (x_i, x_{i+1}) for $i = 0, 1, \ldots, n$ (where $x_0 = -\infty$ and $x_{n+1} = \infty$), $S(x)$ is given by some polynomial of degree m or less.
2. $S(x)$ and its derivatives of orders $\leqslant m - 1$ are continuous everywhere.

A 'natural' spline of odd degree $2k - 1$ is a spline with the property that in the two external intervals, $(-\infty, x_1)$ and (x_n, ∞), it has degree $\leqslant k - 1$ rather than the maximum admissible degree of $2k - 1$. It is not required to be represented by the *same* polynomial in the two external intervals. A *knot* is a point of transition from one polynomial to another. The knots are commonly taken to coincide with data points, although this is not formally required.

18.4.3.1 Smoothness Property of Natural Splines Given n points (x_i, f_i) and a positive integer $k < n$, consider the problem of finding an interpolating spline function $g(x)$ of degree $2k - 1$ with the property that among all such interpolants the roughness, R, is given as

$$R = \int_{x_1}^{x_n} [g^{(k)}(x)]^2 \, dx$$

and shall be minimal. The solution to this problem is known (Ref. 18-25) to be the unique interpolating natural spline of degree $2k - 1$, with knots at x_i.

In the important case of $k = 2$, the roughness as defined above is approximately proportional to the integral of squared curvature (assuming that $|g'(x)|$ does not fluctuate greatly). This smoothness property accounts for the prevalence of cubic splines in engineering applications.

18.4.3.2 Construction of Cubic Splines Assuming that the knots are at the nodes, a cubic spline is defined by n points (x_i, y_i) together with two boundary conditions. There are three standard types of boundary conditions: (i) y'_1, y'_n given, (ii) y''_1, y''_n given and (iii) $y'_1 = y'_n, y''_1 = y''_n$. Case (iii) specifies a "periodic spline." In all cases we find that each continuity equation relates three points (usually neighboring points). This gives rise to a linear equation system which is tridiagonal in cases (i), (ii) and almost tridiagonal in case (iii). All the systems are strongly diagonal-dominant and hence very well conditioned. A compendium of FORTRAN programs and formulas can be found in Ref. 18.65. The coverage extends to nonpolynomial splines and to surface fitting. Further information can be found in Ref. 18.13.

18.4.3.3 Spline Interpolation on a Closed Contour Given the problem of constructing a smooth closed curve through a given sequence of points (x_i, y_i), there are two standard approaches: (1) Define a discrete variable s_i denoting cumulative chord length, i.e., $s_1 = 0$, $s_{i+1} = s_i + \sqrt{(x_{i+1} - x_i)^2 + (y_{i+1} - y_i)^2}$. Fit separate periodic splines to the point sets (s_i, x_i) and (s_i, y_i), with s being regarded as the independent variable. We then abandon the restriction that s_i is discrete, and we can define x and y for all s in the range. (2) A second approach is to pick an origin inside the closed contour, translate to polar coordinates (θ, r) and fit a periodic spline to the data points (θ_i, r_i). The second approach is more restrictive. Choosing an origin may require human judgment; moreover, if the contour is nonconvex, it may be impossible to find an origin such that r is uniquely defined for a given θ. The first approach, known as the "parametric spline," is more general, and will handle self-intersecting curves. It has the demerit that one can not directly answer the question "what is y when $x = X$?" One has to solve a cubic equation to determine s when $x = X$, then calculate y.

18.4.4 Least-Squares Approximation by Polynomials

Given m points (x_i, f_i) with x_i distinct, and m positive weights w_i, it is required to construct a polynomial $q_n(x)$ of degree $n \leqslant m - 1$ such that the scalar quantity

$$E = \sum_{i=1}^{m} w_i [q_n(x_i) - f_i]^2$$

shall be minimized with respect to the class of nth degree polynomials.

18.4.4.1 ***Forsythe's Method*** The basic method recommended for the above problem, given by Forsythe in Ref. 8-19 and based on orthogonal polynomials, is as follows:

1. Define the inner product of the orthogonality relation to be the weighted sum over the m abscissas x_i, i.e., $[p(x), f]$ is defined to mean

$$\sum_{i=1}^{m} w_i p(x_i) f_i$$

for any function $p(x)$.

2. Define a sequence of polynomials $\{p_k(x)\}$ by

$$p_0(x) = 1, \quad p_1(x) = x - \alpha_1$$

for $k = 1, 2, \ldots,$

$$p_{k+1}(x) = (x - \alpha_{k+1}) p_k(x) - \beta_k p_{k-1}(x)$$

where

$$\alpha_{k+1} = \frac{(p_k, x p_k)}{(p_k, p_k)}, \quad \beta_k = \frac{(p_k, x p_{k-1})}{(p_{k-1}, p_{k-1})}$$

3. The polynomial of degree n which best approximates f is then given by

$$q_n(x) = \sum_{k=0}^{n} c_k p_k(x)$$

where

$$c_k = \frac{(f, p_k)}{(p_k, p_k)}$$

18.4.4.2 Notes on Forsythe's Method For many applications it will not be necessary to get any of the polynomials p_k or q_n in explicit algebraic form. If we store $p_k(x_i)$ for each k and i, this information will get us through all stages of the algorithm, including the calculation of the c_k. Then if we wish to evaluate $q_n(x)$ for some argument x possibly different from all x_i, we can recursively evaluate $p_0(x)$, $p_1(x), \ldots$ from their definitions using the known values of α_k, β_k. This storage convention for the polynomials $p_k(x)$ is wasteful because, for example we are using m storage locations to represent the monic polynomial $p_1(x)$, when one location would have been sufficient. The practice is, however, commonly accepted because it is convenient to program and economical to run.

It can be shown that $(p_k, xp_{k-1}) = (p_k, x^k) = (p_k, p_k)$. Since the self-inner products of all the polynomials p_k will be needed in the second and third stages of the above algorithm, it is generally preferable to redefine β_k as $(p_k, p_k)/(p_{k-1}, p_{k-1})$ rather than the way it was given in the previous section.

18.4.4.3 Absolute Constraints If it is required to approximate a function in such a way that some constraints must be met *exactly* while others are to be met only in a least-squares sense, then a method due to Hayes' Ref. 18-29 is recommended.

Let $f(x)$ be approximated by a function of the form $y(x) + \pi(x)g(x)$, where

1. $y(x)$ is an interpolant meeting all the absolute constraints exactly, e.g., it could be the Lagrangian or osculatory interpolant;

2. $\pi(x)$ is a function satisfying homogeneous constraints corresponding to those of $y(x)$, i.e., if the absolute constraints imposed values for y_3, y_7, y_7' then we must construct $\pi(x)$ so that $\pi(x_3) = \pi(x_7) = \pi'(x_7) = 0$. $\pi(x)$ must not vanish at any other abscissa x_i.

3. $g(x)$ is a polynomial to be determined in a least-squares sense.

The form of the approximant is such that it is guaranteed to meet all the absolute constraints exactly and it only remains to determine the polynomial $g(x)$ in such a way as to minimize

$$E = \sum{}' w_i \left[y(x_i) + \pi(x_i)g(x_i) - f_i\right]^2$$

where $\sum{}'$ denotes summation over the abscissas for which there is *no* absolute constraint.

Equivalently we have to minimize

$$E = \sum{}' w_i \pi^2(x_i) \left[g(x_i) - \left(\frac{f_i - y(x_i)}{\pi(x_i)}\right)\right]$$

This is now precisely the form of problem that can be treated by Forsythe's method.

(See section 18.4.4.1.) All that is needed is some pre-editing of the data; the weights are changed from w_i to $w_i \pi^2 (x_i)$, and the approximand and range of summation are also modified as indicated.

It is not essential that $y(x)$ or $\pi(x)$ should be defined as algebraic polynomials, although these are the simplest functions to construct according to the requirements. On occasions, if there is some need to control the asymptotic behavior of the approximant, one could for example construct $y(x)$ and $\pi(x)$ to be of the form e^{-x^2} times a polynomial.

18.4.5 Trigonometric Interpolation and Curve-Fitting

All the techniques described above with reference to algebraic polynomials have their counterparts in terms of trigonometric polynomials. In particular interpolation on ordinates, osculatory interpolation, constrained and unconstrained least-squares approximation can all be performed in terms of trigonometric or hyperbolic polynomials. The algorithms are given in Ref. 18-52. In some cases, as we shall note in section 18.4.9.1, we can solve a trigonmetric interpolation or (possibly constrained) curve-fitting problem by using the software for the corresponding algebraic problem. These cases include (a) the pure cosine series, (b) the pure sine series, and (c) the mixed sine-cosine series defined on equispaced nodes.

18.4.5.1 Fast Evaluation of Trigonometric Polynomials The cosine series $C(\theta) = \sum_{r=0}^{n} c_r \cos r\theta$ can be economically evaluated while calculating only one cosine, namely $\cos \theta$. By using the identity

$$\cos n\theta = 2 \cos \theta \cos(n-1)\theta - \cos(n-2)\theta$$

we can express the term $c_n \cos n\theta$ as a weighted sum of $\cos(n-1)\theta$ and $\cos(n-2)\theta$. Hence $C(\theta)$ can been reduced to an $(n-1)^{\text{th}}$ degree polynomial with the coefficients c_{n-1}, c_{n-2} appropriately modified. Successive further reductions leave us ultimately with a linear expression that we can evaluate directly. The recursion is unstable when $|\cos \theta| \simeq 1$, but in Ref. 18-24 a stabilized recursion is given for sine and cosine series.

In the case of the discrete Fourier transform we need to evaluate a Fourier series at n equispaced argument values. Let the (complex) coefficients be a_r. We have to evaluate $\sum_{1}^{n} a_r e^{i(r-1)\theta}$ for $\theta = 0, -2\pi/n, -4\pi/n, \ldots, -2(n-1)\pi/n$. This is equivalent to premultiplying the complex vector a by a matrix M whose $k-j^{\text{th}}$ element is $m_{kj} = \omega^{(k-1)(j-1)}$, where $\omega = e^{-2\pi i/n}$.

If M were a general matrix, the matrix-vector multiplication would ordinarily require n^2 complex multiplications. However, since every element of M is an nth root of unity, the matrix can contain only n distinct entries instead of the usual n^2. If we used a general-purpose matrix-vector multiplier, we would therefore expect to be doing a lot of redundant multiplications. The "Fast Fourier Transform" is a technique for minimizing the redundant labor in the situation. The economies are greatest when n is an exact power of 2. The work is then proportional to $n \log_2 n$. A program and discussion are given in Ref. 18-14.

18.4.5.2 Trigonometric Interpolation by Prony's Method If the abscissas of $2m$ points (x_i, f_i) are equispaced, then Prony's method yields an interpolant to $f(x)$ in the form $y(x) = \sum_{r=1}^{m} C_r e^{\alpha_r x}$. The quantities α_r are determined to be zeros of a real polynomial; whenever we have a complex conjugate pair of zeros, the two corresponding terms in the summation can be rewritten as a pair of real trigonometric terms. The result will generally be a combination of real exponential and trigonometric terms. If it happens to be purely trigonometric, it will not generally be a Fourier series since the frequencies do not have to be in arithmetic progression. Although the method is formally an interpolation algorithm, its principal value is in applications to periodicity-searching. Details are given in Ref. 18-32.

18.4.6 Rational Interpolation and Approximation

Although many cases are known in which a rational function will yield a far more economical approximant than a polynomial, there is no known algorithm which will guarantee to construct the rational function of given total degree (numerator plus denominator) which 'best' approximates an arbitrary continuous or discrete function in some sense (least-squares, minimax or L_∞). The output of any program which purports to construct a 'best' rational approximation has to be carefully checked. In particular its behavior at or near its poles should be investigated. By 'near' we mean to include real abscissas which are close to complex poles. Reference 18-7 surveys the field.

Rational functions can be converted into continued-fraction form and vice-versa. The latter form is about twice as economical to evaluate but is commonly more sensitive to rounding errors.

18.4.6.1 Padé Approximant This can be used in connection with an approximand whose Taylor expansion is known. If we let $T_k(x)$ denote the kth degree polynomial obtained by truncating the Taylor series, then we may construct a rational function $P_r(x)/P_s(x)$ such that the coefficients of $P_s(x)T_k(x)$ agree with those of P_r as far as possible. There are effectively $r + s + 1$ degrees of freedom in the coefficients of P_s, P_r since the constant term of P_s is normalized to one. Consequently we can hope to match up to $r + s + 1$ terms of the Taylor expansion, but there is no guarantee that this can be done. The linear equations defining the coefficients are simple in structure but cannot be guaranteed nonsingular. An algorithm is given in Ref. 18-29.

18.4.6.2 Thiele's Method This algorithm generates a continued-fraction interpolant to a function defined on a point set. Confluent forms and other variants exist and are described in Ref. 18-32.

18.4.7 Surface Fitting

Given a set of points in space (x_i, y_i, z_i) one may wish to determine a function f such that $z_i \simeq f(x_i, y_i)$. Unless the points (x_i, y_i) have some kind of structure to help us, (e.g., they might form a rectangular grid), the problem may be computationally quite difficult.

18.4.7.1 Unstructured Case One can set up $f(x, y)$ as a polynomial in two variables, i.e., a linear combination of terms of the form $x^p y^q$ for various (small) integer pairs p, q. In order to determine the associated coefficients a_{pq}, one can solve the over-determined linear equation system in a (weighted) least-squares sense. Just as normal equations in the one-variable case tend to be highly ill-conditioned, so we shall have to expect the same trouble in this case. By choosing an origin in the $x - y$ plane somewhere close to the centroid of the points (x_i, y_i) we shall do something to alleviate the conditioning problem. For instance it is to be expected that $x^0 y^0$ and $x^1 y^0$, i.e., one and x, will be among the functions in the linear combination approximating $f(x, y)$. Hence one column of the matrix will be all ones and another will consist of the numbers x_i. If all the x_i are in the range $[100, 101]$ then those two columns will be nearly parallel, and the system is ill-defined. By moving the origin to the centroid we shall make the two columns orthogonal, so at least we can shut out trouble from that particular source. The resulting over-determined linear system can be solved by the methods discussed in sections 18.2.2.3, 18.2.2.4.

18.4.7.2 Structured Cases The most highly structured case is where the points (x_j, x_j) form a rectangular grid. Hayes, Ref. 18-29, indicates a way of handling this problem with or without constraints by means of orthogonal polynomials. Birkhoff and de Boor, Ref. 18-22, construct a two-dimensional spline interpolant. Hayes also considers a less structured case where the points $(x_i y_i)$ lie on a small number of parallel lines.

18.4.8 Smoothing

For some purposes it may be essential for an approximant to have certain overall smoothness properties, and one may be willing to trade in some measure of accuracy to attain this. The quality of approximant is not necessarily well described by the measure of its error. For example an approximant based on the classical Fourier method will commonly have very small errors of rapidly alternating sign. This is fine for some purposes, but it had better not be used to produce tapes for numerically controlled tools.

There is necessarily some vagueness and subjectivity in the concept of smoothness. Commonly accepted definitions of roughness include the integral of squared curvature or squared second or higher derivative. Also, the integral may be replaced by a sum or weighted sum and derivatives by divided differences. The concept of roughness is therefore context-dependent. Having decided on a way to quantify it, one usually tries to minimize the roughness while adjusting the data within the limits of the probable error. Powell, Ref. 18-29, and Reinsch, Ref. 18-63, proposed algorithms for accomplishing this by use of spline approximation. The σ-smoothing described by Lanczos in Ref. 18-43 has the effect of damping the high-frequency oscillations of a Fourier series.

18.4.9 Selection of Method

In matching a method to a problem there are many questions which need to be asked. First concerning the data: How accurate are they? Are they periodic? Could there be any (derivative) discontinuities in the function which generated them? Any poles, and if so, where? What are their asymptotic behaviors? Second, concerning the purpose of the proposed interpolation or approximation: Does one simply want to evaluate ordinates, and if so, it is a one-shot affair or an often-repeated process? Does one intend to evaluate derivatives or integrals instead of, or in addition to, ordinates? How much accuracy is required, how much can be guaranteed and to what extent will the computed approximation be sensitive to small errors or variations in the data?

18.4.9.1 Polynomial Interpolation Algorithms If the approximand has poles or horizontal asymptotes, the use of any unmodified polynomial approximant is generally unwise. The existence of the Runge phenomenon section 18.4.1.3 was demonstrated by producing a sequence of polynomial interpolants to $1/(1 + x^2)$. Since this function is asymptotically zero, one might expect to encounter difficulty when approximating it with polynomials which are asymptotically infinite. The phenomenon was first demonstrated around 1900. There have been many demonstrations since, but these were mostly unintentional.

It should be remembered that a polynomial interpolation or approximation algorithm can be used for much broader purposes then plain polynomial approximation. For example if $y(x)$ is thought to have a logarithmic singularity at an abscissa X, one can approximate the data $(x_i, y_i/\ln|X - x_i|)$ and derive a polynomial $P(x)$ such that $y(x) \simeq P(x) \ln|X - x|$. One can also scale the independent variable instead of, or in addition to, the dependent variable. For example if it is thought that $y(x)$ would be well represented by a cosine polynomial in x, this is equivalent to the assumption that $y \simeq P(z)$, where P is an algebraic polynomial and $z = \cos x$. Given the points x_i, y_i one constructs the values $z_i = \cos x_i$; then in order to 'read' the interpolant at a point X, one defines $Z = \cos X$ and one obtains the ordinate Y by using the fact that y is well represented as a polynomial in z.

Another point that should be remembered in connection with polynomial interpolation is that there is *no* tendency for perturbation effects to be localized. As can be seen from section 18.4.2.1, if the ordinate f_k is perturbed by δ_k, then the effect of this at a point x is $\delta_k \pi_k(x)/\pi_k(x_k)$. There is no tendency for the multiplier of δ_k to become smaller as x becomes more remote from x_k; on the contrary it tends to infinity as x becomes large. This behavior is often entirely incompatible with that of the physical system that one is supposed to be modelling.

Supposing that one nevertheless decides that polynomial interpolation is indicated, one is then faced with a large set of alternative implementations, no one of which is best for all situations. The reader is referred to Krogh, Ref. 18-40, for some help in making the decision.

18.4.9.2 Use of Rational Functions On occasions rational functions or continued fractions may represent an extremely powerful way of approximating a mathematical constant or function. Their value, however, in a general engineering environment seems to be limited. This is partly because the algorithms are unreliable, and partly because the user is seldom willing to automate the process of locating the poles of a function. In the absence of precise information it is often preferred to guess where they are and what their nature is, and then to proceed, perhaps iteratively, by the method of the previous section.

18.4.9.3 Use of Splines Spline interpolation is preferred whenever smoothness is important, and this includes specifically cases where one wishes to evaluate (higher) derivatives. Another important, and usually advantageous, feature is that the effect of an ordinate perturbation is strongly localized. It is shown by Curtis, Ref. 18-29, that with equispaced knots the perturbation effect decays by a factor of -0.268 per interval.

18.4.9.4 Choice of Norm Having made the decision whether the approximant is to be polynomial, rational, spline or trigonometric, the next decision is the choice of norm. We can interpolate, i.e., make the approximant pass exactly through the data points; we can fit in a least squares sense or in a minimax sense.

Interpolation—other than spline interpolation—is generally not recommended for estimation of derivatives. The interpolant must meet, and will ordinarily cross, the interpoland at each data point. The two curves may have substantially different derivatives at these points, but the situation is better midway between data points.

If the choice is between minimax and least squares, the advantage of convenience and reliability is strongly with the latter. Powell, in Ref. 18-58, shows that if the data are continuous and the degree of polynomial approximant is moderate, then there can never be a great difference in quality between the three types of polynomial approximants: (1) True minimax; (2) approximate minimax based on interpolation at Chebyshev nodes; and (3) least-squares. While these theorems do not say anything directly about functions defined on a point set, the message seems to be that the ordinary practitioner should seldom prefer minimax to least-squares. To pursue the minimax means a lot of extra work for very little potential gain.

18.4.9.5 Acceptance Testing At some point a decision must be made on whether a given approximation is "adequate," which often means "as good as is warranted by the quality of the data." The size of the least-squares error is one important guide, but it can tell us nothing about small systematic errors. If these are a matter of concern, then one should also perform a runs test as follows: At each abscissa x_i in monotonic order, determine whether $f(x_i)$ exceeds the approximant $\tilde{f}(x_i)$. If so, record a + sign, otherwise record a −. Having obtained a sequence of m signs, count the number of switches from + to − or vice versa. Ideally, if there are no systematic errors, the probability of a switch should be .5. Hence if the number of switches differs substantially (binomial distribution) from $(m - 1)/2$ we should distrust it, just as we would distrust a supposedly honest coin that landed heads substantially more or less than half the time.

18.4.10 Cluster Analysis

This topic is commonly regarded as belonging to statistics; however, it has a strong enough curve-fitting content to make it worth mentioning. A cluster is a grouping of objects that are "close" according to some context-dependent measure. For instance, if the objects are points in the unit square, there may be concentrations of them, just as a village is concentration of houses. A cluster-analysis problem might be to identify such "villages" or to determine whether two contiguous villages should properly be regarded as a single larger village. If not, what is the demarcation line between them? In taxonomy one identifies a species of mammal by collating its measurements with certain flexibly defined standards. In order to be identified as a rat the mammal has to fall inside a certain range with respect to several dimensions. These are admissible ranges of length, weight, color, tail-length as a fraction of total length etc. At the same time there is much redundancy in our criteria. If the rat has been painted blue and its tail is missing, we may still correctly identify it. Commonly the data will be subjected to a factor analysis to reduce (or at least identify) the redundancies before a cluster analysis is applied. We are still left with a set of points in a space that may be of quite high dimension. The task is to define a relatively small number of clusters and to assign every point to one such cluster in such a way that the variance among points in each cluster shall be small. The concept of "variance" implies a measure of "distance" from a mean. The "distances" used in this area are often non-Euclidean and may be asymmetric. In principle this is a combinatorial task of great complexity, and there have to be algorithms which yield approximate solutions within acceptable time limits. Refs. 18-66 and 18-1 deal with the algorithmic and conceptual aspects, respectively.

18.5 QUADRATURE AND INTEGRAL EQUATIONS

The simplest form of quadrature problem is that of constructing an equality

$$\int_a^b f(x)\,dx = \sum_{r=1}^{n} w_r f(x_r) + T \qquad (18.5\text{-}1)$$

where the truncation error T is as small as possible, i.e., the nodes x_r and the weights w_r are to be chosen in such a way as to make the approximation as close as possible. In the simplest case, a, b will be finite and $f(x)$ will be smooth. A large (some would say disproportionately large) amount of development work has been done on this problem, and there are plenty of programs available based on Romberg or Gauss-Legendre quadrature which will perform well on this type of problem. Once we depart from the simple problem, the list of possible complications is long, and since these can be present in numerous mixes and combinations there appears to be no hope that anyone will ever write a general purpose quadrature program which truly lives up to its name. The following is a list of the more important complications:

1. Either a and/or b is infinite.
2. The function $f(x)$ has integrable singularities in the range, but we may not know what kind they are or where they are located.
3. a and b are points on an open or closed contour.
4. We may not be free to assign the nodes x_r; we have to do the best we can with the given ordinates.
5. This could be a volume or hypervolume integral problem.
6. The integrand may be highly oscillatory.

Unless some analytic information about the integrand is available there can be no possibility of obtaining any quantitative error bound. The approximation is based on a finite sample of points drawn from an infinite population. If we have literally no knowledge of what can happen between the sample points, then there can be no bound to the possible error in the approximation, and it makes no sense to claim that one method is better than another. In order to establish an error bound, one needs to have some information about the integrand that is valid throughout the continuum over which it is being integrated. For example, if we have a bound on the magnitude of the first or a higher derivative in the range, we can then establish error bounds for the quadrature.

18.5.1 Simple Quadrature Problems

Problems of the simplest type described above can be effectively handled either by formulas of Gauss-Legendre type or of Romberg type. The relative merits of these two approaches are set out very fairly by Wilf in Ref. 18-62.

18.5.2 Polynomial-Based Quadratures

Most of the established methods rely at least partly on the assumption that the integrand can be locally approximated by a polynomial or a sequence of polynomials. They differ from each other principally in their mode of sampling.

18.5.2.1 The Newton-Cotes Formulas The n-point member of this family is derived by sampling the integrand at n equispaced points including both endpoints of the range of integration. By constructing and integrating the n-point Lagrangian interpolant one arrives at a formula in the form of eq. (18.5-1). The weights and the truncation error are recorded in numerous textbooks. The best known formulas in the family are the Trapezoidal rule ($n = 2$) and Simpson's rule ($n = 3$).

18.5.2.2 Composite Newton-Cotes Formulas If the function is known for example, at five equispaced points including the endpoints, one would not necessarily apply the five-point formula. Instead one could apply the trapezoidal rule on each of the four intervals separately and add the results; or one could apply Simpson's rule twice, once on the first pair of intervals and once on the second. The composite form of Simpson's rule is sometimes known as the 'parabolic rule,' and it can be used for any odd number of ordinates.

18.5.2.3 Quadrature in Terms of Endpoint Derivatives If some of the higher derivatives of the integrand are known at the two endpoints, this information may be used to refine the composite trapezoidal rule approximation. The process is known as 'end correction.' See Ref. 18-12. An extreme version of this, based on the plain trapezoidal rule and using no interior ordinates but only derivatives at the endpoints, is known as 'quadrature in terms of end-data.' Discussion and formulas are presented in Ref. 18-43.

18.5.2.4 Gauss-Legendre Quadrature This method is also based on integrating an n-point polynomial interpolant; however the x_r are no longer equispaced. They are chosen so as to make the formula exact for integrating polynomials of as high a degree as possible. Since there are $2n$ degrees of freedom x_r, w_r in the specification of the formula, one expects to be able to make it satisfy $2n$ conditions, e.g., to be exact for all polynomials of degree 0 through $2n - 1$. This expectation is realized, and the truncation error is of the form $H_n D^{2n} f(z)$, where H_n is a tabulated numerical constant, and D is the differential operator, so that D^{2n} annihilates all polynomials of degree $\leqslant 2n - 1$ as required.

18.5.2.5 Other Variants of Gaussian Quadrature If one of the x_r is to be an endpoint, a or b, the number of freedoms is reduced by one, and the form of the truncation error is adjusted accordingly. This is known as 'Radau quadrature.' If both endpoints are nodes, but the other nodes x_r are assigned optimally (in the sense of yielding exactness for as high a degree of polynomial as possible) then we have 'Lobatto quadrature.' Gauss-Laguerre formulas are designed for integrands of the form $e^{-x} P_{2n-1}(x)$ over the range $[0, \infty]$, and Gauss-Hermite formulas are designed for integrands of the form $e^{-x^2} P_{2n-1}(x)$ over the range $[-\infty, \infty]$, where $P_{2n-1}(x)$ denotes an arbitrary polynomial of degree $\leqslant 2n - 1$. The nodes, weights and truncation error coefficients for all these formulas are widely tabulated.

In general, any integrand of the form $P_{2n-1}(x)W(x)$ can be exactly integrated over a finite range by a formula of the type of eq. (18.5-1) where the w_r and x_r depend on the (positive) weight function $W(x)$ and on the range of integration. For further detail see Ref. 18-37. Positivity of $W(x)$ is a sufficient but not necessary condition for the existence of the formula. The best-known member of this class is the Gauss-Chebyshev formula

$$\int_{-1}^{1} \frac{f(x)}{(1-x^2)^{1/2}}\,dx = \frac{\pi}{n}\sum_{r=1}^{n} f\left(\cos\frac{2r-1}{2n}\pi\right) + T$$

and T has the form $H_n D^{2n} f(z)$. The distinctive feature of the formula is that all the weights w_r are equal.

18.5.3 Transformation of Problems

Since most standard formulas presuppose that the range of integration has been normalized to $[-1, 1]$, we commonly have to change the variable in eq. (18.5-1) thus

$$\int_a^b f(x)\,dx = \frac{(b-a)}{2}\int_{-1}^{1} f\left(\frac{a+b}{2} + t\,\frac{b-a}{2}\right)dt$$

On occasions it may be desirable to transform an infinite or semi-finite range into a finite range. The principal motivation for this is that one cannot directly apply the Richardson error estimation technique (section 18.1.6) to an infinite range. For instance one can use the transformation

$$\int_0^\infty f(x)\,dx = 2\int_{-1}^{1} f\left(\frac{1+t}{1-t}\right)\cdot\frac{dt}{(1-t)^2}$$

This transformation gives us a singularity at $t = 1$, and we shall have to apply our knowledge of the limiting behavior of $f\left(\frac{1+t}{1-t}\right)\Big/(1-t)^2$ as $t \to 1$. Presumably some knowledge of this is available, otherwise we had no right to assume that the problem as originally posed had a solution.

18.5.4 Error Estimation

Two sources of error need to be considered. Firstly the ordinates $f(x_r)$ could each be wrong by as much as $\pm\delta_r$. A reference to eq. (18.5-1) then indicates that the error due to this source alone could be as great as

$$\sum_1^n |w_r|\,\delta_r$$

It is also evident from eq. (18.5-1) that $\sum w_r \simeq b - a$, because if this were not so the formula would not perform well on the very simple problem when $f(x) = 1$. If all the w_r are positive, we can effectively say that the error due to inaccurate function-evaluation does not exceed $|b - a| \max \delta_r$. On the other hand if the w_r are of varying sign, then the constraint $\sum w_r \simeq b - a$ does not impose any bound on individual magnitudes of w_r, and without such a bound we cannot bound the effect of errors in function evaluation. It is therefore wise to use formulas with positive weights only. This includes all Gaussian formulas, but not all Newton-Cotes formulas involving more than seven points.

Estimates of truncation error can most conveniently be made by means of a Richardson process (section 18.1.6). Some such estimation technique is built into every nontrivial quadrature program.

Determining rigorous bounds for the truncation error is an extensive topic. There is no shortage of scholarly articles about it, but there is, unfortunately, a severe shortage of useable techniques. The formal truncation error term for n-point Gaussian quadrature is only useable in the unlikely event that one can obtain a bound for the magnitude of the $2n^{\text{th}}$ derivative of the integrand. If there is any analytic information available, it is most likely to take the form of a bound on the magnitude of the slope. If this bound is M then the *a priori* error bound for the composite trapezoidal rule is

$$\frac{M}{4} \sum_{r=1}^{n-1} (x_{r+1} - x_r)^2$$

Secrest showed in Ref. 18-64 that a generally better *a posteriori* bound could be given in the same situation, namely

$$\sum_{r=1}^{n-1} \left[\frac{M}{4} (x_{r+1} - x_r)^2 - \frac{(f_{r+1} - f_r)^2}{4M} \right] \tag{18.5-2}$$

In the given situation this bound is attainable (but only by a function with a discontinuous second derivative); furthermore there is no other formula that can guarantee to produce a smaller error. The same reference also gives an *a posteriori* bound in terms of a bound on the magnitude of the second derivative.

18.5.5 Adaptive Quadrature

This expression denotes an automation of some of the decision-making that is involved in the quadrature process, the decisions being made "at run time," i.e., using information that becomes available as the computation progresses. Any quadrature over a large domain will usually have to be split into a set of subproblems defined on subdomains. The subdomains will be of varying size, being relatively large in regions where the integrand is thought to be smooth. The judgment will normally be based on a Richardson-type process (section 18.1.6). If the estimated error in a subdomain exceeds the tolerance, then the subdomain size must be reduced accordingly. Programming a computer to adjust subdomain sizes appropriately is one relatively simple part of the adaptive quadrature concept. One may go further and attempt to identify singularities and choose appropriate formulas. Further discussion may be found in Ref. 18-45.

18.5.6 Selecting a Method

If the data points are randomly spaced, one can use the composite trapezoidal rule with Richardson's error estimation. One can use low-order polynomial or trigonometric or other interpolation in the composite mode and integrate the interpolant. High order algebraic interpolation should be avoided because of its sensitivity to data errors as noted in section 18.4.9.1. For equispaced data the options are the same but the implementations are easier—particularly the parabolic rule and also trigonometric quadrature if the range is a full period. For cases where the nodes are not preassigned, all the advice one can give boils down to one sentence: "Make the best possible use of any knowledge you can get about the analytic behavior of the function." In particular, if the function has any unpolynomial-like properties, one should beware of using polynomial-based quadratures, such as Gauss-Legendre. Such properties include low-order derivative discontinuities, horizontal asymptotes and high-frequency oscillation. If the location of a derivative discontinuity is known, then one should use a composite formula in such a way that the approximant will also have a derivative discontinuity at that point. If the derivative discontinuity takes the form of a pole, then it is best for the user to state the nature of the pole and assign a quadrature formula appropriately; e.g., the Gauss-Chebyshev formula of section 18.5.2.5 will handle one kind of integrable singularity. A computer can be programmed to locate a pole quite accurately, but it can only make a very tentative and fallible identification of the nature of the pole. If all that the reader could see was a sequence of points drawn with uncertain accuracy on a graph, could he make a confident distinction between a radical and a logarithmic singularity? Could he even distinguish between an integrable and a nonintegrable singularity? There is no reason to expect a computer to be any smarter than the reader in this respect. Automated detection and processing of singularities provides some insurance against gross blunders, but it is not a substitute for careful analysis.

In the case of an integrand with horizontal asymptotes whose effect becomes evident within the range of integration, the high-order Gaussian formulas will not necessarily perform badly, but they are likely to be inefficient, because in effect they use a high-order polynomial to approximate a curve that is locally almost linear. A composite Simpson's rule program is likely to be preferable.

For rapidly oscillating integrands, Filon's method, Ref. 18-12, may be appropriate; however if the period varies in a known nonconstant manner it will often be preferable to integrate each loop separately; an appropriate method for each loop being Lobatto quadrature, since the ordinates at the end of each loop are zero. The accuracy of such a computation is very hard to ensure, because the final result is the algebraic sum (perhaps near-zero) of many loop areas which have alternating signs.

In summary it is clear that one needs a large variety of programs in order to handle the various different problems. All programs should be adaptive to the extent that they monitor the error bounds or error estimates, and they adjust the density of sampling accordingly. Additional adaptivity beyond this is of questionable value, becuase it tends to increase the bulk and the running costs of the program; it pays off only when there is a high premium on the provision of idiot-proof programs.

18.5.7 Multidimensional Quadrature

It has to be admitted that this is a very underdeveloped area, and only in simple cases can we hope to proceed without a lot of custom-made programming. Suppose we are integrating over a rectangle of sides a, b and we are to compute

$$\int_0^a \int_0^b f(x,y)\, dy\, dx$$

For any arbitrary x-value we can perform the inner integration by one of the methods previously considered. If the result of doing this at a point x_i is g_i, then the outer integration is equivalent to calculating

$$\int_0^a g(x)\, dx$$

where $g(x)$ is a function which we can evaluate at any point x_i. This can also be done by one of the previously considered methods. We can extend this approach to the case where b is not constant, but is a known function of x. Extensions into higher-dimensional cases are accessible, provided the domains of integration are hypercubes, hyperspheres, or geometric figures which transform into these by a change of coordinate system. As a last resort one could try a Monte Carlo approach. The subject is discussed in Refs. 18-12 and 18-62. The former reference also discusses contour integration.

18.5.8 Integral Equations

An integral equation is one in which the unknown function occurs under an integral sign. The function and/or its derivatives may also occur elsewhere in the equation. The function may be multivariate, and it may occur nonlinearly. The domain of integration will not necessarily be constant or finite. In short, this area includes some very intractable problems. We shall here illustrate the collocation approach to a simple problem. Let

$$f(x) = g(x) + \int_a^b f(t)\, k(x, t)\, dt \qquad (18.5\text{-}3)$$

The unknown function is f. All other symbols represent known functions or variables. We assume $f(x) \simeq \sum_1^n A_r \phi_r(x)$, where the ϕ_r are basis functions, e.g., $\phi_r(x) = x^{r-1}$. Let x_1 be a selected value of x, and substitute into eq. (18.5-3), obtaining

$$\sum_1^n A_r \phi_r(x_1) = g(x_1) + \sum_1^n A_r \int_a^b \phi_r(t)\, k(x_1, t)\, dt$$

Calculating the integral by quadrature, we now have a linear equation relating the A_r. Repeat the process for $x_2, x_3, \ldots$ until there are at least n equations, so that the A_r can be calculated by solving a (possibly overdetermined) linear equation system. A comprehensive, numerically oriented approach to general problems in this area will be found in Ref. 18-2. A report on available software is given by Delves in Ref. 18-38.

18.6 ORDINARY DIFFERENTIAL EQUATIONS

A comprehensive numerical treatment is given in Ref. 18-41.

18.6.1 Classification of Problems

Given an independent scalar variable x and a vector variable y, the most general form of differential equation is $G(x, y, y') = 0$, where G denotes an arbitrary (nonlinear) vectorial combination of x and the various components of the vectors y and y'. If G is an implicit function it may be possible to rewrite the problem in explicit form. Some transformations of this sort are discussed in Ref. 18-8. The numerical algorithms to be discussed below are designed for explicit problems, i.e., problems of the form $y' = f(x, y)$. Some algorithms are further restricted to cases where f is linear, so that the problem is of the form $y' = Ay + b$ where the elements of the vector b and the square matrix A are functions of x only.

A general problem as described above will not necessarily have a unique solution (or indeed *any* solution). In most practical situations, it will be required to impose side conditions of the form $F(x, y, y') = 0$ in order that the solution be uniquely specified. The vectorial function F specifies a relation between a *finite number* of abscissas, ordinates, and derivatives. This is in contrast with the function G above which relates these quantities over a *continuum* of values x. If the side conditions involve only one abscissa x then the problem is called an initial-value problem. Otherwise it is a boundary-value problem.

A differential equation involving higher derivatives can be reformulated as a system of first-order equations. For instance, the problem $y'' = f(x, y, y')$ can be written as $y' = z$, $z' = f(x, y, z)$.

Differential eigenproblems are defined similarly to algebraic ones. A commonly occurring form is $y'' = \lambda By$, where the elements of B are functions of x only. One is seeking those values of λ for which the equation has a nontrivial solution.

A differential equation of the form $y' = Ay$ is called *stiff* if it has the property that over a large range of relevant x-values, the matrix A has eigenvalues all in the left half-plane and whose real parts differ widely. Special-purpose methods are required in order to handle such problems effectively.

A description of some of the more important algorithms will now be given.

18.6.2 Runge-Kutta Methods

Given the problem $y' = f(x, y)$ where y_0 is given at x_0, one can define a recursion in the form

$$y_{n+1} = y_n + \sum_1^m w_j k_j$$

which can be used to generate $y_1, y_2, \ldots, y_r, \ldots$; the latter represent approximate ordinates on the solution curve at $x_1, x_2, \ldots, x_r, \ldots$, where $x_r = x_0 + rh$ and h is a locally constant step-size. In the above recursion the limit m (commonly $m = 4$) depends on which of the numerous Runge-Kutta variants one is using. The quantities k_j represent $hf(x, y)$ evaluated at various points (x, y) specified by the formula, and the weights w_j are also specified. The parameter values specified by the formula are generally calculated with a view to making the computed solution match the Taylor series solution to as many terms as are permitted by the degrees of freedom (number of parameters) in the algorithm.

In the classical Runge-Kutta formulas each successive k_j was explicitly defined in terms of lower-subscripted k-values. These are known as explicit Runge-Kutta formulas. More recently implicit formulas have been studied, where each k-value depends on each other one. Implementation of such formulas is clearly more difficult in view of the generally nonlinear interdependence of the k's; however the disadvantage may be offset by the acquisition of free parameters.

18.6.3 Multistep Methods

These methods have the form

$$\sum_0^k a_j y_j = h \sum_0^k b_j f(x_j, y_j) \tag{18.6-1}$$

where the quantities a_j, b_j and k are characteristic of the particular multistep method which one chose, except that $a_k = 1$. If $b_k = 0$ the formula is a *predictor*, otherwise it is called a *corrector*. In either case the formula is used to define y_k in terms of other y and y' values. [It is customary to denote $f(x_j, y_j)$ by y'_j.] In the case of a corrector formula the quantity y_k appears on both sides of the equation, since it is involved implicitly in y'_k. We therefore have a (generally nonlinear) equation system defining y_k. It can be shown that for sufficiently small h the system can be solved by iteration; however for stiff systems this sufficiently small h may be too small to be workable; however, stiff problems require special purpose methods in any case.

When y_k has been determined from eq. (18.6-1), one can use the same formula, or perhaps a different one to determine y_{k+1} in terms of y'_{k+1} and earlier values of y, y'. The usual strategy is to use a predictor in the form of eq. (18.6-1) to get a trial value of y_k; this determines a trial value of $y'_k = f(x_k, y_k)$. Then using a corrector version of eq. (18.6-1) put the trial value of y'_k on the right and obtain a 'corrected' value of y_k. This would be a PEC method (Predict-evaluate-correct). The literature contains many studies of iterated corrector formulas such as PECEC etc. Formulas of this type are sometimes called linear multistep formulas because only linear combinations of y, y' occur. In most texts a multistep formula is assumed to be linear unless otherwise stated. The coefficients a_j, b_j are ordinarily chosen so as to make the formula give exact results whenever the true solution curve corresponds to a polynomial up to a specified degree L. Such a formula is said to be "of order L" and its truncation error is of the form $Th^{L+1}D^{L+1}Y(z)$, where T is a numerical constant specified by the formula, Y is the true solution, z is an abscissa in the range $[x_0, x_k]$, and D is the differential operator, so that, as required, D^{L+1} annihilates all polynomials of degree $\leqslant L$.

Since a corrector formula has $2k + 1$ parameters a_j, b_j one expects to be able to make it satisfy $2k + 1$ conditions, e.g., to be exact for all polynomials of degree zero through $2k$. This can indeed be done, but it has been proved that for $k > 2$ such formulas are necessarily unstable. For a compact statement of known results on stability (most of which are due to Dahlquist) the reader is referred to Ref. 18-44. A more detailed treatment is contained in Ref. 18-30. An unstable formula is one with the property that the effect of small errors, e.g., rounding errors, will grow exponentially as the computation proceeds.

18.6.4 Extrapolatory Methods

These methods are based on Richardson-type extrapolations (section 18.1.6). The version in most common use is by Bulrisch and Stoer and is based on rational approximation (Ref. 18-5). It is considered to be among the best methods for nonstiff problems.

18.6.5 Other Methods

The previous three sections cover the most commonly used algorithms, but there are at least three more worth mentioning.

18.6.5.1 Off-Step Methods These resemble the multistep methods of section 18.6.3 except that the x_j are not necessarily equispaced. The freedom to assign abscissas at will provides more free parameters, which can be used to obtain greater precision without loss of stability; however, one pays for this in terms of an increased number of function evaluations. More detail is given in Ref. 18-44.

18.6.5.2 Block Multistep Methods These methods have 'been around' a long time. Booth (1957) in Ref. 18-4 ascribes them to Picard. In 1967 Newbery gave a convergence analysis and observed that the high-accuracy formulas in this family were more economical than the corresponding formulas in the Runge-Kutta family, Ref. 18-51.

One needs k independent formulas each in the format of eq. (18.6-1). The jth of these formulas expresses y_j in terms of

$$y_0 \quad \text{and} \quad \{y_i'\}, \qquad i = 0, 1, \ldots, k$$

One predicts a whole vector of y-values $y_1, y_2, \ldots, y_k$ using any convenient predictor, e.g.

$$y_j \simeq y_{j-1} + hy_{j-1}', \qquad j = 1, 2, \ldots, k$$

Then the whole vector is corrected by cyclic application of the k independent formulas mentioned above.

The foregoing description applies strictly to a single first-order equation. If we have a system with several dependent variables, then each y is itself a vector, and for "vector of y-values" one should read "matrix of y-values."

18.6.5.3 Cyclic Multistep Methods Donelson and Hansen, Ref. 18-16, have studied the possibility of using cyclically a battery of multistep formulas of the type of eq. (18.6-1). Thus the formula which defines y_k in terms of y_k' and earlier values of y, y' will not be reused to define y_{k+1} in terms of y_{k+1}' etc. They use a sequence of different formulas all in the same format. By this device one can obtain higher precision without loss of stability than is possible by repeated reuse of any one formula of the type of eq. (18.6-1).

18.6.6 Error Estimation

If one is using a predictor-corrector method, it is appropriate to choose a pair of the same order; then by observing the discrepancy between the predicted and corrected values of an ordinate one can estimate the error in either one. In this case the error estimation is economical, because the data needed are by-products of the regular computation.

Error estimation for Runge-Kutta methods is less economical. Again one uses some variant of Richardson's method, but this time the data are not provided free.

In the extrapolation techniques, error estimation and correction are usually built in as an integral part of the algorithm.

18.6.7 Step-Size Change

A well-constructed program will monitor the error estimates, and if they go above a prescribed tolerance the step-size will be reduced. Conversely if the errors are substantially below tolerance level, the step-size will be increased so as to save time and reduce accumulated round-off error. Of course the tolerance should never be set below the level that is warranted by the word length; that would be like asking for seven significant figures from a six-figure machine. A good program will check to ensure that the user-specified tolerance is reasonable and, if not, replace it by the tightest tolerance that the word length will permit. Changing the step-size with a multistep formula is relatively expensive, because one needs k points recomputed with the new spacing. Consequently a multistep program will ordinarily be written in such a way that a step-size increase will not be undertaken unless the errors are very substantially below tolerance level. It is more economical to run at a somewhat suboptimal step-size rather than to incur the expense of changing it. With Runge-Kutta methods the step-size change is trivial, and this enables them to run at a more nearly optimal step-size than is possible for multistep methods.

18.6.8 Choice of Method for Initial-Value Problems

For the nonstiff problem there are good implementations of Runge-Kutta, multistep and extrapolation methods available in the program libraries. A block multistep method received a good report from Skappel, Ref. 18-15, but there is a lack of direct evidence comparing it with its competitors. A cyclic multistep method with enlarged stability region is reported in Ref. 18-46. Some systematic tests on the more readily available programs are given in Ref. 18-35. For the stiff problem there are again three principal contenders, and the choice should be made according to the geometry of the eigenvalues. The details can be found in Ref. 18-17.

18.6.9 Boundary-Value Problems

18.6.9.1 Algebraization If the range of integration is subdivided into n small (not necessarily equal) intervals, then every derivative occurring in the equation or the boundary conditions can be approximated by a finite divided difference. This is equivalent to replacing y_i' or y_i'' by the first or second derivative of a parabola defined by y_i and its two closest neighbors. This in turn is equivalent to replacing the original differential system by an approximating algebraic system. A linear differential system converts to a linear algebraic system and nonlinear converts to nonlinear. In both cases the algebraic system will be sparse; in the linear case a compensation can be made for the errors introduced by conversion to finite differences. See Fox, Ref. 18-21.

18.6.9.2 ***Shooting Methods*** Suppose the system has, say, 10 degrees of freedom, i.e., it can be expressed as a system of 10 first-order differential equations. The boundary conditions are specified at a number of points no greater than 10 and no fewer than 2 (otherwise it would be an initial-value problem), and let $B_j(y) = 0$ denote the jth boundary condition; it may involve all 10 components of the vector y at any one specific abscissa x_j. Let the range of integration be $[A, Z]$. Since we do not know the side conditions at A, we guess all 10 initial values at A and let the 10 guesses be represented by a vector g. Having made the guess at 10 scalars, we solve the *initial value* problem from A to Z by a method suitably selected, and we expect to find that the computed solution $\tilde{y}$ does *not* exactly satisfy the 10 boundary conditions; however we can define a 10-dimensional residual vector h whose components are $B_1(\tilde{y}), B_2(\tilde{y}), \ldots, B_{10}(\tilde{y})$. Since we represented our ten guesses as a vector g, we have a mechanism for mapping g onto h. The problem is to determine that value of the input vector g which will cause the output vector h to vanish. If the differential equation system and its boundary conditions are linear, then the mapping of g to h is (in principle) linear. Although the linearity may become a little 'roughed up' in practice, at least we should expect no difficulty, because the problem will not be highly nonlinear. On the other hand, if the given differential system is nonlinear, then the mapping will be likewise, and there is no way of telling how difficult the problem can be, if it is indeed solvable at all. Methods for solving nonlinear equation systems are discussed in section 18.8.

18.6.9.3 ***Selecting a Method for Boundary-Value Problems*** As far as convenience is concerned, the advantage is very clearly with the shooting method, rather than algebraization. For nonlinear problems one needs a nonlinear algebraic equation solver driven by an initial-value solver. The latter is driven by a derivative evaluator. The last component is the only one that should need programming, since the first two should be in the standard library. The dimension of the nonlinear algebraic system is the same as the number of boundary conditions. Unfortunately the shooting method is often highly unstable; that is to say that a small change in the assumed initial conditions may cause a large change in the computed function at some of the boundary points. One is, so to speak, shooting with a highly temperamental rifle. The alternative is to algebraize. One again has a nonlinear algebraic system, but its dimension is now a multiple of the number of steps into which the interval was divided. One may not know this figure in advance, but it can go into the thousands on just a second-order, two-point boundary-value problem. The equation system is sparse (often banded), but the storage requirements and programming effort are still likely to be far greater than with the shooting method. Often the best advice will be "Shoot first and ask questions afterwards." The algebraization of a nonlinear boundary-value problem is discussed in Ref. 18-56.

18.7 NONLINEAR FUNCTIONS OF ONE VARIABLE

A general reference for this area is Ref. 18-30.

18.7.1 Localization Theorems for Polynomial Zeros

18.7.1.1 Descartes' Rule If

$$P(x) \equiv \sum_0^n p_r x^r$$

has all real coefficients p_r, then the number N_p of positive zeros does not exceed the number of sign changes in the sequence $p_n, p_{n-1}, \ldots, p_0$. If N_p is less than the number of sign changes, then it is less by an even number. A similar relation holds between the number of negative zeros and the coefficients of $P(-x)$.

18.7.1.2 If z_i are the zeros of $P(x)$, then

$$\sum z_i = -\frac{p_{n-1}}{p_n}$$

$$\sum_{i \neq j} z_i z_j = \frac{p_{n-2}}{p_n}, \ldots, \prod_1^n z_i = (-1)^n \frac{p_0}{p_n}$$

18.7.1.3 If z_i are zeros of $P(x)$, then $1/z_i$ are zeros of

$$P^*(x) \equiv \sum_0^n p_r x^{n-r}$$

18.7.1.4 Sturm Sequence A Sturm sequence relative to an nth degree real polynomial $P(x)$ is a sequence of functions $f_1(x), \ldots, f_m(x)$ which indicate the number of real zeros of $P(x)$ in an interval $[a, b]$ in the following way: *Let $V(a)$ denote the number of sign changes in the sequence $f_1(a), \ldots, f_m(a)$ and similarly define $V(b)$. Then $P(x)$ has exactly $|V(a) - V(b)|$ real zeros in $[a, b]$.* Sturm's theorem establishes that *the f_i defined as follows constitute a Sturm sequence for $P(x)$: $f_1(x) = P(x)$, $f_2(x) = P'(x)$, and thereafter $f_{i+1}(x)$ is the negative of the remainder polynomial arising from the synthetic division of $f_{i-1}(x)$ by $f_i(x)$.*

18.7.1.5 Gersgorin's Theorem for Polynomials *If $P(x)$ is a monic polynomial (i.e., $p_n = 1$) then a bound on the magnitude of all its zeros is*

$$\max\left[1, \sum_0^{n-1} |p_r|\right]$$

and another bound is max $[|p_0|, 1 + |\underset{0<r<n}{p_r}|]$. The theorem holds for complex zeros and complex coefficients.

18.7.1.6 Rouché's Theorem If $f(z)$ and $g(z)$ are regular within and on a closed contour C and $|g(z)| < |f(z)|$ on C then $f(z)$ and $g(z) + f(z)$ have the same number of zeros inside C.

18.7.1.7 If X is a zero of $P(x)$, and $g(x)$ is a polynomial, all of whose coefficients are small relative to the mean of $|p_r|$, then the difference δX between X and a zero of the perturbed polynomial $P(x) + g(x)$ is given to a first approximation by $|\delta X| \simeq |g(X)/P'(X)|$. It follows that clustered zeros are generally more sensitive to coefficient perturbation than isolated zeros. It also follows that zeros of small magnitude are more sensitive to variations in the low-order coefficients and are generally insensitive to variations in the high-order coefficients.

18.7.2 Order of Convergence

An iterative method for approximating a zero X of a function $f(x)$ is a procedure for advancing from the ith iterate x_i to x_{i+1}. Generally the iteration function is of the form $x_{i+1} = g(x_i)$, though the function g may also involve several arguments $x_i, x_{i-1}, \ldots, x_{i-k}$. If the errors $e_i = X - x_i$ are related by $e_{i+1} = O(e_i^p)$ then the convergence is said to be of pth order. If $X = g(X)$ then the order of convergence is at least linear (i.e., first order). If in addition $g^{(j)}(X) = 0$ for $j = 1, 2, \ldots, k$ but $g^{(k+1)}(X) \neq 0$ the order is said to be $(k + 1)^{\text{st}}$. These conditions do not guarantee convergence but they indicate the rapidity of convergence when it occurs.

18.7.3 Algorithms for Determining Zeros of Functions of One Variable

A comprehensive treatment of this subject is given by Traub, Ref. 18-68. Some of the more important methods are:

18.7.3.1 Newton's Method for Approximating a Zero of $f(x)$ If x_i is the ith iterate, then $x_{i+1} = x_i - f(x_i)/f'(x_i)$. The method has second-order convergence to an isolated zero; otherwise it is first order. When $f(x)$ is an nth degree polynomial it can be proved that $f(x)$ has at least one zero in or on a complex-plane disc with center at x_i and radius $n\,|f(x_i)/f'(x_i)|$. Newton's method remains valid in complex arithmetic, but problems can arise. Suppose we attempt to find a complex zero of a real polynomial; if the initial approximation is real, then every subsequent approximation has to be real, and we can never succeed. Muller's method (section 18.7.3.3) does not have this difficulty about getting away from the real axis.

18.7.3.2 Bairstow's Method (Ref. 18-32) This method is applicable only to polynomials and is primarily useful for finding real quadratic factors of a real polynomial. It has second-order convergence to isolated zeros.

18.7.3.3 Muller's Method (Ref. 18-68) This method is based on three-point parabolic interpolation and is of order 1.84 for nonmultiple zeros. Since each iteration only requires one new function evaluation and no derivative, it is generally among the most economical and popular methods for zero-finding, both for polynomials and for arbitrary continuous functions of one variable. The comments in Ref. 18-68 should be carefully noted before programming, since there are formulations which are mathematically correct but computationally unsatisfactory.

18.7.4 Implementation of Polynomial Root-Finding Algorithms

If one uses the common synthetic division algorithm of Ref. 18-61 to evaluate a polynomial $P(x)$ at an argument X of large magnitude, then one may anticipate serious error build-up, since it is possible for the error at each stage of the recursion to be $|X|$ times its value at the previous stage. If $P(x)$ has a zero X of large magnitude, one is advised *not* to attempt to approximate it directly. Instead one can apply section 18.7.1.3 and seek the small zero $1/X$ of the polynomial $P^*(x)$.

18.7.4.1 Polynomial Deflation When a zero of a polynomial has been found, one ordinarily factors it out and then proceeds to find a zero of the reduced (deflated) polynomial. Since the reduction process involves some rounding errors, the computed coefficients of the deflated polynomial may be slightly different from their proper mathematical values. Since it was observed in section 18.7.1.7 that a small perturbation of the coefficients may induce an indefinitely large perturbation of the roots, it is clear that the deflation process can create problems. In Ref. 18-71 there is an extensive discussion of this, including some alternatives to deflation. In summary, the problem is not generally severe provided that the reduced polynomial is always defined either by factoring the smallest root from $P(x)$ or by factoring the reciprocal of the largest root from $P^*(x)$.

18.7.4.2 Specification of Tolerances Most algorithms for locating zeros of polynomials involve generating a sequence of iterations $x_1, x_2, \ldots,$ and terminating the iterations when either $|x_{i+1} - x_i| < \tau_1(|x_{i+1}| + |x_i|)$ or $|P(x_i)| < \tau_2$, where τ_1, τ_2 are tolerances. The tightest reasonable value for τ_1 is $2^{-\beta}$ and for τ_2 it is $|p_0|2^{-\beta}$, where β significant binary digits are carried in floating-point arithmetic. This does not imply that it is impossible to satisfy a tighter tolerance; it means that there is no reason for believing that a tighter tolerance will produce greater accuracy. In practice it is common to 'back off' from these tightest reasonable tolerances by about a factor of 4.

18.7.4.3 Initialization The first estimate x_0 for a zero of a polynomial is often essentially a blind guess. One approach that is commonly satisfactory is to note from section 18.7.1.2 that the product of the zeros z_i is given by $\prod|z_i| = |p_0/p_n|$; consequently there exists at least one zero z_1 such that $|z_1| \leqslant |p_0/p_n|^{1/n}$. Moreover, by 'trading' the $P(x)$ problem into the $P^*(x)$ problem if necessary by section 18.7.1.3, one can arrange that $|z_1| \leqslant |p_0/p_n|^{1/n} \leqslant 1$. We are then certain that the origin is at most one unit away from the smallest zero in the complex plane.

18.7.4.4 Correction Limitation Some algorithms are liable to give 'wild' corrections if the initial approximation is not good. This is true of Newton's method, section 18.7.3.1, if $f'(x_0) \simeq 0$. A program should provide for overruling the mathematical algorithm in this situation.

Many algorithms which normally have superlinear convergence may be slowed down to linear convergence in the presence of a multiple root. It is sometimes helpful to apply the algorithm not to $P(x)$ but to $P(x)/P'(x)$. The latter function has no repeated zeros; the fact that it has singularities may be an acceptable price to pay for this.

18.7.5 Acceleration

Another way of dealing with the problem raised in section 18.7.4.4 is by applying the Aitken δ^2 process of Ref. 18-68.

If a sequence $\{x_i\}$ is converging linearly, then an estimate $\tilde{x}$ of the value to which it is converging is given by

$$\tilde{x} = x_{i+2} - \frac{(x_{i+2} - x_{i+1})^2}{x_{i+2} - 2x_{i+1} + x_i}$$

It is essential for the validity of the extrapolation that the sequence should be *converging linearly*. Since one may not be sure that this condition is satisfied, it is advisable to test $|P(\tilde{x})|$ against $|P(x_{i+2})|$ before accepting the extrapolation.

18.7.6 Error Analysis for Polynomial Zeros

The accuracy to which a polynomial zero can be computed is limited by the accuracy of the polynomial evaluation procedure. In the following circumstances a bound for evaluation error can be stated: (1) There are no data errors, i.e., the polynomial coefficients as stored are assumed 'correct'; (2) The usual synthetic division procedure is used for evaluation of $P(x)$ at $x = \alpha$; and (3) The rounding errors for each arithmetical operation obey a relation of the form $|N - N^*| \leqslant \epsilon N^*$, where N, N^* are the true and computed results of the operation and ϵ is a known positive constant. The evaluation error is then bounded by

$$\tilde{P}(|\alpha|)\ [\epsilon + n\sigma]/(1 - n\sigma)$$

where $\sigma = \epsilon(2 - \epsilon)$ and $\tilde{P}$ denotes a polynomial whose coefficients are those of P set positive (or replaced by their moduli if complex).

If condition (1) above is not satisfied, it is simple to compute a bound for the error attributable to this cause, provided one knows bounds for the errors in the stored coefficients.

Given that the evaluation error of a function $f(x)$ at an argument x is $\pm b$, then if X is a computed zero of $f(x)$ the radius of uncertainty in the computed zero will be approximately $|b/f'(X)|$. Further, if $f(x)$ is an nth degree polynomial and it is known that $|f(X)| \leqslant b$, then $f(x)$ has at least one zero in a circle with center at X and radius $n\,|b/f'(X)|$. These results are corollaries of section 18.7.3.1. It can also be shown that there is at least one zero within a distance of $|b/p_n|^{1/n}$ from X.

18.7.7 Deflation of Nonpolynomial Zeros

Since it is generally not possible by analytic means to factor out known zeros from an arbitrary function $f(x)$, the usual practice is to apply the zero-finding algorithm to $f(x)/\prod(x - X_i)$ where the product is taken over all previously determined zeros X_i.

18.8 NONLINEAR EQUATION SYSTEMS AND OPTIMIZATION

Methods for solving nonlinear equation systems fall mainly into two categories:

1. Generalizations of Newton's method, section 18.7.3.1 for solving a nonlinear equation in a single variable. The secant method is here regarded as a variant of Newton's method, where derivatives are replaced by difference quotients.
2. Generalizations of the gradient methods for solving (perhaps in a least-squares sense) a system of linear equations.

The problem of minimizing a scalar function of several variables is more general and basically more difficult. Here we are generally trying to find the zero of a gradient vector rather than a residual vector. The gradient vector will not be explicitly given; it will have to be inferred from the successive iterates. If we find a point where the gradient is zero, it will not necessarily be an extremum. It could be a saddlepoint. If the minimization is subject to constraints, then the solution may be at a point where the gradient is *not* zero.

18.8.1 Generalization of Newton's Method

Let the system be written in the form $G(x) = 0$, where x is the n-dimensional vector to be determined, and G is a set of m nonlinear functions $g_i(x)$; let $y_1, y_2, \ldots$ be successive vector approximations to x. Then if $m = n$ we have, according to Newton's method

$$y_{k+1} = y_k - J^{-1}G(y_k)$$

where J is the Jacobian

$$[J_{ij}] = [\partial g_i/\partial y_j] \tag{18.8-1}$$

Newton's method is seldom used in its pure form. Computing the Jacobian may be unacceptably expensive or even impossible. When the Jacobian is available, it may be ill conditioned. That means that the computed corrections may be far larger than what one can justify under the heading of local extrapolation. Nevertheless the formula is important, since it underlies several modified-Newton methods that are in regular use.

18.8.2 Generalization of the Secant Method

In the single-variable case the secant method uses the slope of a secant defined by two points for the purpose of estimating the derivative in the neighborhood of those points. It is an estimate based on the assumption that the mapping is (approximately) linear. If the mapping really is linear, then we obtain immediate convergence. Correspondingly with an $n \times n$ problem, we can construct $n + 1$ residual vectors h_i, and hence determine where the zero of the function would be if the mapping were linear.

With the notation of the previous section we define y_j to be the jth (vector) approximation to the true solution x of the system $G(x) = 0$ and we define h_j to be the (vector) residual associated with y_j, i.e., $G(y_j) = h_j$.

We then define a vector

$$\tilde{y} = y_{n+1} + \sum_{1}^{n} p_r(y_r - y_{n+1}) \tag{18.8-2}$$

where the p_r are undetermined scalars. We then argue that *if the mapping were linear* it would be the case that

$$G(\tilde{y}) \text{ would equal } h_{n+1} + \sum_{1}^{n} p_r(h_r - h_{n+1}) \tag{18.8-3}$$

We then note that in general the right side of the quasi-equality (18.8-3) can be reduced to zero by suitable choice of the p_r. When this suitable choice is reinserted into eq. (18.8-2) the vector $\tilde{y}$, so defined, can be taken as the next approximation; the next set of $n + 1$ approximations will consist of $\tilde{y}$ and n of the previously used $n + 1$ approximations. After renumbering, the process is repeated.

At each stage we must solve a linear equation system to calculate $\bar{p}$. Each matrix is closely related to its predecessor, and the most efficient algorithms take advantage of this. Instead of solving each system *ab initio*, one can solve more cheaply by updating either an inverse or a decomposition of the required matrix. In the next section we discuss some difficulties that may arise.

18.8.2.1 Pitfalls of the Generalized Secant Method Apart from the obvious difficulties that may arise from making a linear approximation to a perhaps highly nonlinear system, it should be noted from the previous section that the method will break down if either of the vector sets $\{h_r - h_{n+1}\}$ or $\{y_r - y_{n+1}\}$, $r = 1, 2, \ldots, n$, becomes dependent. Dependence of the former set means that the p_r are not defined; dependence of the latter set implies that all successive approximations are going to be in the subspace spanned by the vectors $\{y_r - y_{n+1}\}$. If the solution happens not to be in this subspace, then it can never be found. If the linear system which supposedly defines the p_r is singular the gradient of the residual of the system may still be calculated (see section 18.2.5). The Jacobian of the system is approximated by HY^{-1}, where the columns of H are $h_r - h_{n+1}$ and those of Y are $y_r - y_{n+1}$. Some methods define an iteration of the form

$$y_{n+2} = y_{n+1} - (\alpha I + \tilde{J}^T \tilde{J})^{-1} \tilde{J}^T h_{n+1} \tag{18.8-4}$$

where $\tilde{J}$ is the approximate Jacobian as defined above, and α is a scalar parameter to be suitably chosen. If $\alpha = 0$ the method reverts to Newton's method with approximate Jacobian, i.e.

$$y_{n+2} = y_{n+1} - \tilde{J}^{-1} h_{n+1}$$

When α is taken to be very large, the direction of the correction gets to be more like the direction $-\tilde{J}^T h_{n+1}$, which would be specified by the gradient method. For positive α it can be guaranteed that the matrix inverse in eq. (18.8-4) is well defined. It remains important to ensure that the revised Y matrix does not have (nearly) dependent columns.

It may be seen from eq. (18.8-4) that if $\tilde{J}^T h_{n+1} = 0$ the iteration becomes static. There could be two causes of this; either $h_{n+1} = 0$, so that the solution is found, or $\tilde{J}^T$ is singular; in the latter case the gradient of $h_{n+1}^T h_{n+1}$ is approximately zero. This can be inferred from the fact that the true gradient is parallel to $J^T h_{n+1}$; hence assuming we are satisfied that $\tilde{J}$ is an adequate approximation to J we would conclude that we had reached the bottom of a valley. Possibly the problem has no solution, but if it has, the solution cannot be reached by any downhill method starting at the point y_{n+1}. Most computer programs will cut off iterations in this situation. They will return with either a zero or a local minimum, which is characterized by a zero gradient.

18.8.3 Generalizations of Gradient Methods

We have already seen how the gradient concept may be called on to help out when a Newton-type method breaks down. Its merit lies in the fact that the gradient is always well defined even when the Jacobian is singular or rectangular. It will often happen that the number of equations, m, differs from the number of variables, n. Of particular importance is the case $m = 1$, $n > 1$, i.e., the minimization-of-functionals problem. In general we can let the index r in eqs. (18.8-2) and (18.8-3) go from 1 to k, where k is not necessarily related to m, n. It should normally be at least as great as max (m, n). This will lead to

$$\tilde{y} = y_{n+1} - \lambda (YY^T)^{-1} YH^T h_{n+1}$$

where Y is a matrix whose columns are $y_r - y_{n+1}$; H has columns $h_r - h_{n+1}$, $r = n - k + 1, n - k + 2, \ldots, n$; and λ is a parameter to be determined. If λ is determined by linear search for a minimum of $\|G(\tilde{y})\|_2$ then we have a nonlinear analog of the Cauchy gradient method for linear equations. Since this method is of ill repute in linear algebra, it is unlikely to be of direct value; however, there are several nonlinear versions of the conjugate gradient method which will solve the functional minimization problem in a finite number of steps when the functional is a positive definite quadratic form. The methods are effective also on other types of functional, but the guarantee of termination in a finite number of steps is restricted to quadratic forms. See Ref. 18-55.

18.8.4 One-Dimensional Search

Equation-solvers and functional-minimizers will both generally need to execute a one-dimensional search. A direction will have first been computed, possibly by Newton's method or a gradient method, but we must still decide how far to go in this direction. We shall search along the line for a minimum of the functional. (In the case of equation-solvers the functional is here defined to be the residual norm.) Parabolic approximation is the most commonly used technique. One calculates the function at three points on the line; one finds the point where the interpolating parabola has a minimum; one replaces one of the three original points by the newly found point and proceeds iteratively.

18.8.5 Unconstrained Minimization

The problems of minimizing and maximizing a functional are essentially equivalent, because the minimum of $G(y)$ occurs at the same y-value as the maximum of $-G(y)$. A program designed primarily for rectangular equation-solving will also handle these problems provided that it adopts the convention that it shall exit when it encounters a point where the residual norm has zero gradient. Conversely, a program designed for functional minimization can be used for equation-solving, because one can define the functional to be the residual norm.

18.8.6 Deflation, Constraints and Penalty Functions

Suppose a vector solution s_1 has been found, so that $G(s_1) = 0$. If we now want a second solution we generate a sequence of vectors $\{y_i\}$ according to the method of our choice, but we have to take steps to avoid finding the solution s_1 over again. In order to steer the iterative sequence away from s_1, one can replace the given nonlinear system $G(x)$ by $G(x)/\|x - s_1\|$ which has the same zeros other than s_1. Then, having found s_2, one again redefines the system as $G(x)/(\|x - s_1\| \, \|x - s_2\|)$ etc. If the original system had an analytic Jacobian, the modified systems will probably become too complicated for the analytic Jacobian to be usable. The same numerical deflation technique can be used if one unintentionally found a nonzero minimum when one was looking for a zero. The intention in all cases is the same: to prevent the iteration sequence from heading in that direction again.

Another situation in which we need to head off the iteration sequence from some undesired direction is when we have a minimization problem subject to inequality constraints. Here the problem is to minimize $G(x)$ subject to constraints $g(x) \leqslant 0$, where g is a (possibly) nonlinear vectorial mapping of the vector x; the requirement means that *every* element of the vector $g(x)$ must be $\leqslant 0$. One way to handle this is by constructing a penalty function, possibly in the form $p^T g$, i.e., a weighted sum of the elements of the vector $g(y)$; each weight listed in the vector p will be zero whenever the g-element which it multiplies is $\leqslant 0$, but a large value is assigned to the weight when the corresponding g element is positive; hence $p^T g$ is nonnegative, and it takes on a large value whenever one or more of the constraints are violated. We then say the problem is to minimize $G(x) + p^T g(x)$, and the presence of the penalty function tends to prevent the iteration from heading in directions such that the constraints are violated. In certain simple situations the constraints may be satisfied by a change of variable. For example, if it is required that $|x_i| \leqslant 1$, we could redefine x_i as $\cos \theta_i$.

18.8.6.1 Lagrangian Multipliers When we have equality constraints, our first thought is to reduce the dimension of the problem if possible. If this is not practicable, we may proceed as follows by the method of Lagrangian multipliers. Let $G(x)$ denote a scalar function of the n-dimensional vector x, and let $g(x)$ be an m-dimensional (generally nonlinear) mapping of x. It is required to minimize $G(x)$ subject to the constraint that $g(x) = 0$. Let

$$\tilde{G}(x) = G(x) + \sum_{1}^{m} s_r g_r(x)$$

The variables x_i, s_r now form an augmented set of $n + m$ variables. To determine these we have to solve $n + m$ equations, namely $\partial \tilde{G}/\partial x_i = 0$, $i = 1, 2, \ldots, n$ and $g_r = 0, r = 1, 2, \ldots, m$.

18.8.7 Approximation of the Jacobian

If one is attempting to construct a finite-difference approximation to a Jacobian matrix, it is important to realize that high accuracy is virtually unattainable on this problem. In effect one is repeatedly using two scalar ordinates y_1, y_0 with the horizontal spacing h to approximate the slope at some intermediate abscissa. Errors arise from two sources: firstly there are inevitable inaccuracies in the computed ordinates y_1, y_0, and secondly even if we could obtain the secant slope exactly this could differ from the exact tangent slope by as much as $hF''/2$, where F'' denotes $\max |f''(x)|$ in the interval of length h. In many implementations the bound is hF'' because for economy one often uses the less accurate forward difference rather than a central difference. Independently it can be verified that if y_0, y_1 are subject to errors $\pm\delta_0$, $\pm\delta_1$ respectively, then the error in the computed secant slope could be as great as $(|\delta_1| + |\delta_0|)/h$. Assuming that these errors are additive, we get a combined error bound E in the form

$$E = \frac{A}{h} + Bh$$

where A, B are positive constants. If we make the unlikely assumption that we have enough information to compute an optimum h in the sense of minimizing E, this leads to $h = \sqrt{A/B}$ and

$$E_{\min} = 2\sqrt{AB} \quad \text{or} \quad \sqrt{2F''(|\delta_0| + |\delta_1|)} \qquad (18.8\text{-}5)$$

under the assumption of central differences. The presence of the square root in eq. (18.8-5) is obviously damaging. Besides, this bound presupposes the use of an optimum step-size, and one will seldom be lucky enough to guess anywhere near the optimum.

18.8.8 General Observations

Methods requiring an analytic Jacobian or Hessian are seldom of use in an engineering environment. The information one needs for computing them may be lacking, or the cost may be unacceptable. When numerical deflation is needed, even an initially simple Jacobian will become intractable. The HSL library, Ref. 18-33, has a particularly strong reputation in the area of optimization and nonlinear equation systems.

18.9 MISCELLANEOUS TOPICS

18.9.1 Elementary Functions

A user ordinarily relies on the manufacturer to provide satisfactory subroutines for $\sin x$, $\ln x$ etc. It is important to note that even when the manufacturer does the best possible job (and some do not) there have to be weaknesses in these routines when applied to large arguments. For instance in approximating $\sin x$ for a large x the first step is to reduce x by $N\pi/4$ where N is an integer such that $y = (x - N\pi/4) \in [0, \pi/4)$. The problem is thereby replaced by that of finding sine (or cos) of y and possibly reversing the sign. The quantity y is called the 'reduced range' argument, but it cannot be computed accurately because $\pi/4$ cannot be stored accurately. If $\tilde{y}$ is the machine value of y, we shall have $|\tilde{y} - y| = N\epsilon$, where ϵ is the representational error in $\pi/4$. Hence instead of sine (or cos) of y we shall get an approximation to sine (or cos) of $y \pm N\epsilon$. The relative errors in this become unbounded for sines near $y \simeq 0$. Similar problems exist and are unavoidable for ln, exp, etc. These range-reduction errors are not connected with the errors in the reduced-range approximation. The latter errors, due to truncation of an infinite series or continued fraction, are superimposed on the former.

18.9.2 Linear Recursion Formulas

If a sequence of scalars or vectors $\{y_r\}$ obeys a recursion of the form

$$y_{r+1} = a_{r+1}\, y_r + b_{r+1}\, y_{r-1} \cdots + p_{r+1}\, y_{r-k} + c_{r+1}$$

then it is clear that any errors in $y_r, y_{r-1}, \ldots, y_{r-k}$ will be propagated into y_{r+1} and hence into $y_{r+2}, \ldots,$ and it is important to know the extent to which the effect of an error can grow as it is propagated throughout a computation. The solution of ordinary differential equations by multistep methods provides an example. The evaluation of Bessel functions by three-term recursion is discussed in Ref. 18-28. This is a classical example of a case where the use of a mathematically correct recursion, without regard to its computational aspects, can quickly lead to totally incorrect results. The problem is discussed in a more general setting by Oliver (Ref. 18-54) and Gautschi (Ref. 18-23). The general conclusion is that if it is possible to trade an initial value problem into a boundary-value problem, the stability is likely to be improved, even though there may be some cost in terms of convenience and computational complexity. In discussing boundary-value problems we already noted in section 18.6.9.3 that the stability was likely to be impaired if we tried (by shooting) to reformulate the system as an initial-value problem. Intuitively it seems plausible that if you can anchor an approximation at two or more distinct points, this will be better than just one point. Experience generally bears this out.

18.9.3 Scalar Products

The scalar product of two vectors forms such a large ingredient of many programs that it will commonly be found desirable to have a special fast-running assembly-coded subroutine for it. Apart from the speed factor, it may also be important, as noted in section 18.2.8.3, to run the scalar products in higher precision than the rest of the computation.

In Ref. 18-72 the following device for improving the speed of a scalar product is suggested:

Let x, y be the n-dimensional vectors. Let

$$\zeta = \sum_{j=1}^{[n/2]} x_{2j-1} x_{2j} \quad \text{and} \quad \eta = \sum_{j=1}^{[n/2]} y_{2j-1} y_{2j}$$

where $[n/2]$ denotes the integer part of $n/2$, then let

$$D = \sum_{j=1}^{[n/2]} (x_{2j-1} + y_{2j})(x_{2j} + y_{2j-1}) - \zeta - \eta.$$

The scalar product $x \cdot y$ then equals D if n is even and $D + x_n y_n$ if n is odd. It is only in the case of multiple scalar products, e.g., matrix-matrix multiplication that the device is profitable. In this case the computation of the various ζ, η values may be regarded as a small overhead cost, while the cost of D is only about half that of a regular scalar product.

18.9.4 Monte Carlo Methods

Any system which is of professional interest to applied mathematicians has to be describable, at a given instant, by a set of numbers. An orbiting satellite, for example, is characterized by position velocity and acceleration vectors. It will occasionally happen that the relevant numbers or parameters of equations which relate them are not exactly ascertainable. All we can do is obtain a theoretical or empirical frequency distribution for them. These frequency distributions do not lend themselves to regular calculation like ordinary numbers–imagine the task of inverting a matrix whose elements are not numbers but frequency distributions. A problem of this kind has to be solved by simulation.

We set up n^2 random number generators, one for each element of the matrix; we generate and invert a large number of matrices and from this large finite sample of inverses we draw some inferences concerning the frequency distributions of the inverse elements. Strictly within the realm of numerical analysis one does not frequently encounter the need to solve a problem all the way by Monte Carlo methods, but they may be useful for the empirical estimation of the sensitivity of a given operator. For instance if G is a nonlinear vector mapping of a vector x it may be of interest to have some rough idea of the max of

$$\| G(x + \delta x) - G(x) \| / \| \delta x \|$$

for various small vectors δx. A simple way is to take a look at a couple of dozen. For further information see Ref. 18-26.

18.9.5 Computer Program Libraries

There are now several good mathematical software program libraries availble to the user. In many cases the programs were written by the world's leading experts. The typical user is strongly advised against thinking that be can do better by writing his own programs. Without wishing to imply any order of merit, we can endorse all the following.

a. IMSL, Ref. 18-36.
b. HSL, Ref. 18-33. This library is particularly strong in the area of optimization and nonlinear equation systems.
c. EISPACK, Ref. 18-49. This is a program package restricted to eigenproblems only.
d. LINPACK, Ref. 18-50. This package is restricted to linear equations.
e. NAG, Ref. 18-48. This is a complete library offered in Algol as well as Fortran versions.

18.9.6 Acceptance Tests for Computer Programs

On occasions one may need to verify that a given computer program is meeting its specifications. The nature and extent of testing depends on the type of program and on the environment in which it is to be used. To make even a simple program like a quadratic equation solver totally idiot-proof is not an entirely simple task, and it has to be paid for in terms of leadtime, bulk and running time. One should take a careful look at one's environment before deciding how much idiot-proofing is appropriate. The following is a checklist of points to consider when one is validating a computer program. It is not suggested that every item is applicable in every situation, and in some cases a point may be legitimately ignored at the user's discretion.

1. Do you know what kind of performance you can reasonably expect on this test? If you establish that the solution is correct to five significant figures, do you have some way of judging whether this result is good, bad, or indifferent? If you are not able to make some such judgment, what benefit was derived from performing the test?

2. Have you checked out all the special situations which are known by theory to be potentially troublesome? e.g. equal, clustered or equimodular roots in the case of polynomial factoring.

3. Make sure that data errors do not obscure a test that is designed to investigate computational accuracy. Using a Hilbert matrix to test linear equation solvers and using the explicit inverse to obtain the 'true' result is an example of this error. Since the Hilbert matrix cannot be exactly machine-stored, we cannot learn much from observing the difference between the computed solution and the true solution. The difference is partly due to data errors and partly due to computational errors and there is no simple way of separating the two effects.

4. Did you run a complete spectrum of problems ranging from trivial to impossible? Remember that a polynomial rootfinder which gets into some kind of logical hang-up when the polynomial is of degree one can be a nuisance when it is embedded in a large program. Nobody thinks the failure could possibly have occurred just there, so much time is spent looking for and possibly 'correcting' nonexistent errors elsewhere in the program. At the other end of the spectrum, a linear equation solver should be able to give warning of ill-conditioning, and it should cut out if the ill-conditioning is so severe as to place all the allegedly significant figures in doubt.

5. Check the error returns for singular matrices, polynomials with high-order coefficient zero or with negative degree, etc.

6. Did you perform enough tests of each type, so that your results have some statistical significance? When you claim that the singular-matrix error return has been 'checked out,' do you mean that it worked once on a 2×2 case, or did it work fifty times on matrices of dimension 1×1 through 50×50?

18.10 REFERENCES AND BIBLIOGRAPHY

18.10.1 References

18-1 Anderberg, M. R., *Cluster Analysis for Applications*, Academic Press, New York, 1973.

18-2 Baker, C. T. H., *The Numerical Treatment of Integral Equations*, Clarendon, Oxford, 1977.

18-3 Björck, A., "Solving Linear Least Squares Problems by Gram-Schmidt Orthogonalization," *Nordisk Tidskr. Informationsbehandlung*, vol. 7, 1-21, 1967.

18-4 Booth, A. D., *Numerical Methods*, Butterworth, London, 1957.

18-5 Bulirsch, R., and Stoer, J., Numerical Treatment of Ordinary Differential Equations by Extrapolation Methods," *Numer. Math.*, **8**, 1-13, 1966.

18-6 Bunch, J. R., and Parlett, B. N., "Direct Methods for Solving Symmetric Indefinite Systems of Linear Equations," *SIAM J. Numer. Anal.*, **8**, 639-655, 1971.

18-7 Cody, W. J., "A Survey of Practical Rational and Polynomial Approximation of Functions," *SIAM Rev.*, **12**, 400-423, 1970.

18-8 Collatz, L., *The Numerical Treatment of Differential Equations*, Springer, Berlin, 1960.

18-9 Crane, P. C., and Fox, P. A., *A Comparative Study of Computer Programs for Integrating Differential Equations*, Bell Labs, Murray Hill, N.J., 1969.

18-10 Curtis, J. H. (ed.), *Proc. Symp. Appl. Math.*, Amer. Math. Soc., vol. 6, McGraw-Hill, New York, 1956.

18-11 Davis, P. J., *Interpolation and Approximation*, Blaisdell, New York, 1963.

18-12 Davis, P. J., and Rabinowitz, P., *Numerical Integration*, Blaisdell, Waltham, Mass., 1967.

18-13 de Boor, C., *A Practical Guide to Splines*, Springer, New York, 1978.

18-14 de Boor, C., *Elementary Numerical Analysis*, McGraw-Hill, New York, 1980

18-15 Dold, A., and Eckmann, E. (ed.), *Lecture Notes in Mathematics, Conference on the Numerical Solution of Differential Equations*, Dundee, Scotland, June 1969, Springer, New York, 1969.

18-16 Donelson, J., and Hansen, E., "Cyclic composite multistep predictor-corrector methods," *SIAM J. Numer. Anal.*, **8**, 137-157, 1971.

18-17 Enright, W. H., Hull, T. E., and Lindberg, B., "Comparing Numerical Methods for Stiff Systems of O.D.E.s.," *BIT*, **15**, 10-48, 1975.

18-18 Fletcher, R., *Methods for the Solution of Optimization Problems,* Technical Report HL. 70/5927, Atomic Energy Research Establishment, Harwell, England, 1970.

18-19 Forsythe, G. E., "Generation and use of orthogonal polynomials for data fitting with a digital computer," *SIAM J.*, **5**, 74-88, 1957.

18-20 Forsythe, G. E., and Moler, C. B., *Computer Solution of Linear Algebraic Systems*, Prentice Hall, Englewood Cliffs, N.J., 1967.

18-21 Fox, L. (ed.), *Numerical Solution of Ordinary and Partial Differential Equations*, Pergamon, London, 1962.

18-22 Garabedian, H. L. (ed.), *Approximation of Functions*, Proc. Symp. on Approximation of Functions, Elsevier, N.Y., 1964.

18-23 Gautschi, W., "Computational Aspects of Three-Term Recurrence Relations," *SIAM Rev.*, **9**, 24-82, 1967.

18-24 Gentleman, W. M., "An Error Analysis of Goertzel's (Watt's) Method for Computing Fourier Coefficients," *Comput. J.*, **12**, 160-165, 1969.

18-25 Greville, T. N. E. (ed.), *Theory and Applications of Spline Functions*, Academic Press, New York, 1969.

18-26 Hammersley, J. M., and Handscomb, D. C., *Monte Carlo Methods*, Methuen, London, 1964.
18-27 Hamming, R. W., *Introduction to Applied Numerical Analysis*, McGraw-Hill, New York, 1971.
18-28 Hart, J. F. (et al.), *Computer Approximations*, Wiley, New York, 1968.
18-29 Hayes, J. G. (ed.), *Numerical Approximation to Functions and Data*, Athlone Press, London, 1970.
18-30 Henrici, P., *Applied and Computational Complex Analysis*, Wiley, New York, 1974.
18-31 Henrici, P., *Discrete Variable Methods in Ordinary Differential Equations*, Wiley, New York, 1962.
18-32 Hildebrand, F. B., *Introduction to Numerical Analysis*, McGraw-Hill, New York, 1956.
18-33 Hopper, M. J., Harwell Subroutine Library Document R.7477, Atomic Research Establishment, 1973.
18-34 Householder, A. S., *The Theory of Matrices in Numerical Analysis*, Blaisdell, New York, 1964.
18-35 Hull, T. E., et al., "Comparing Numerical Methods for Ordinary Differential Equations," *SIAM J. Numer. Anal.*, **9**, 603–637, 1972.
18-36 IMSL, International Mathematical and Statistical Libraries, Inc., Houston, 1979.
18-37 Isaacson, E., and Keller, H. B., *Analysis of Numerical Methods*, Wiley, New York, 1966.
18-38 Jacobs, D. (ed.), *Numerical Software Needs and Availability*, Academic Press, New York, 1978.
18-39 Kowalik, J., and Osborne, M. R., *Methods for Unconstrained Optimization Problems*, Elsevier, New York, 1968.
18-40 Krogh, F. T., "Efficient Algorithms for Polynomial Interpolation and Numerical Differentiation," *Math. Comput.*, **24**, 185–190, 1970.
18-41 Lambert, J. D., *Computational Methods in Ordinary Differential Equations* Wiley, New York, 1973.
18-42 Lancaster, P., *Lambda-Matrices and Vibrating Systems*, Pergamon, Oxford, 1966.
18-43 Lanczos, C., *Applied Analysis*, Prentice-Hall, Englewood Cliffs, N.J., 1956.
18-44 Lapidus, L., and Seinfeld, J. H., *Numerical Solution of Ordinary Differential Equations*, Academic Press, N.Y., 1971.
18-45 Lyness, J. N., and Kaganove, J. J., "A Technique for Comparing Automatic Quadrature Routines," *Comput. J.*, **20**, 170–177, 1977.
18-46 Mihelcic, M., "A( )-Stable Cyclic Composite Multistep Methods of Order 5," *Computing*, **20**, 267–272, 1978.
18-47 Moore, R. E., *Interval Analysis*, Prentice-Hall, Englewood-Cliffs, N.J., 1966.
18-48 NAG (Numerical Algorithms Group), *Mathematical Subroutine Library*, Oxford University Computing Centre, 1973+.
18-49 National Energy Software Center, *EISPACK*, Argonne National Lab.,
18-50 National Energy Software Center, *LINPACK*, Argonne National Lab., 1979+.
18-51 Newbery, A. C. R., "Convergence of Successive Substitution Procedures," *Math. Comput.*, **21**, 489–490, 1967.

18-52 Newbery, A. C. R., "Trigonometric Interpolation and Curve-Fitting," *Math. Comput.*, **24**, 869–876, 1970.

18-53 Oettli, W., and Prager, W., "Compatibility of Approximate Solutions of Linear Equations with Given Error Bounds for Coefficients and Right-hand Sides," *Numer. Math.*, **6**, 405–409, 1964.

18-54 Oliver, J., "The Numerical Solution of Linear Recurrence Relations," *Numer. Math.*, **11**, 349–360, 1968.

18-55 Ortega, J. M., and Rheinboldt, W. C., *Iterative Solution of Nonlinear Equations in Several Variables*, Academic Press, New York, 1970.

18-56 Pearson, Carl E., "On Non-linear Ordinary Differential Equations of Boundary Layer Type," *Studies in Appl. Math.*, **47**, 351–358, 1968.

18-57 Powell, M. J. D., *A Fortran Subroutine for Solving Systems of Nonlinear Algebraic Equations*, H. M. Stationery Office, London, 1968.

18-58 Powell, M. J. D., "On the Maximum Errors of Polynomial Approximations Defined by Interpolation and by Least Squares Criteria," *Comput. J.*, **9**, 404–407, 1967.

18-59 Powell, M. J. D., "A Survey of Numerical Methods for Unconstrained Optimization," *SIAM Rev.*, **12**, 79–97, 1970.

18-60 Powell, M. J. D., "A Theorem on Rank-One Modifications to a Matrix and Its Inverse," *Comput. J.*, **12**, 288–290, 1969.

18-61 Ralston, A., *A First Course in Numerical Analysis*, McGraw-Hill, New York, 1965.

18-62 Ralston, A., and Wilf, H. S., *Mathematical Methods for Digital Computers, Vol. II*, Wiley, New York, 1967.

18-63 Reinsch, C. H., "Smoothing by Spline Functions," *Numer. Math.*, **10**, 177–183, 1967.

18-64 Secrest, D., "Numerical Integration of Arbitrarily Spaced Data and Estimation of Errors," *SIAM J. Numer. Anal.*, **2**, 52–68, 1965.

18-65 Späth, H., *Spline Algorithms for Curves and Surfaces*, translated by W. D. Hoskins and H. W. Sager, Utilias Mathematica, Winnipeg, 1974.

18-66 Späth, H., *Cluster Analysis Algorithms*, translated by Ursula Bull, Ellis Horwood, New York, 1980.

18-67 Todd, J. (ed.), *A Survey of Numerical Analysis*, McGraw-Hill, New York, 1962.

18-68 Traub, J. F., *Iterative Methods for the Solution of Equations*, Prentice-Hall, Englewood Cliffs, N.J., 1964.

18-69 Varga, R. S., *Matrix Iterative Analysis*, Prentice-Hall, Englewood Cliffs, N.J., 1962.

18-70 Wilkinson, J. H., *The Algebraic Eigenvalue Problem*, Clarendon, Oxford, 1965.

18-71 Wilkinson, J. H., *Rounding Errors in Algebraic Processes*, H. M. Stationery Office, London, 1963.

18-72 Winograd, S., "A New Algorithm for Inner Product," *IEEE Trans. Computers*, 693–694, 1968.

18.10.2 Bibliography

Abramowitz, M., and Stegun, I. A. (ed.) *Handbook of Mathematical Functions*, National Bureau of Standards, Washington, D.C., 1964.

19

Mathematical Models and Their Formulation

Frederic Y. M. Wan*

19.1 MATHEMATICAL MODELING

19.1.1 The Continuum Mechanics Model of Matter

With a few exceptions, the chapters of this handbook are concerned with mathematical methods useful in the quantitative analysis of problems in science and engineering.† An important and challenging aspect of any quantitative study of a real-life phenomenon is the formulation of mathematical problems which are relevant to a better understanding of the phenomenon and to which these mathematical methods can be applied. Real-life phenomena are usually too complex to be analyzed quantitatively without idealization and simplification. It is just not feasible or practical to follow the individual motions of trillions of molecules in a cubic centimeter of air or the evolution of billions of stars in a typical galaxy. For many practical purposes, however, information about a body of matter (or a galaxy) can be obtained by treating the collection of molecules (or stars) in that body as a "continuous medium" having properties, such as density, velocity, etc., that vary smoothly throughout the body. In this *continuum model*, the equilibrium or motion of the body under external forces and torques, for example, may be taken as a consequence of Euler's laws of mechanics for *continuous* media. The mathematical methods described in this handbook may now be used to deduce from Euler's law an initial/boundary-value problem for differential equations that governs the mechanical behavior of the continuum. Section 4.7 of this handbook gives a sample derivation of some relevant differential equations of this mathematical model, widely known as *continuum mechanics*, for the study of the mechanics of deformable bodies of matter.

*Prof. Frederic Y. M. Wan, Dep't. of Mathematics and Institute of Applied Mathematics and Statistics, University of British Columbia, Vancouver, B.C. V6T 1Y4, Canada.

†Here, science and engineering are to be taken in the most general context. For example, we would include in science, the quantitative aspects of archaeology, commerce, etc.

For more than two centuries, countless macroscopic phenomena involving mechanical behavior of matter (solid, fluid, or gas) have been explained, and the effects of any change in their setting have been accurately assessed by the continuum mechanics model, with the help of mathematical methods described in this handbook and elsewhere. This article is not concerned with the contents of the continuum mechanics model; nor is a knowledge of the model a prerequisite for the rest of the chapter. (The construction of the mathematical models of this chapter will start from first principles.) We merely wish to identify the principles operating in the development of this extremely successful mathematical model to be used in the construction of new mathematical models of current interest.

Beyond the process of idealization and application of the laws of mechanics already discussed above, another important aspect in the model development is the process of simplification. In actual applications of the continuum mechanics model, it is usually very difficult or costly to extract useful information about the phenomena from this sophisticated mathematical model without some simplifications. Beyond those resulting naturally from a certain uniformity or symmetry inherent in a particular problem, simplifications of the model through judicious approximations are often made for special classes of problems, based on the analyst's understanding of the phenomenon, retaining only those features in the model expected to be most relevant. Beam, plate, and shell theories in solid mechanics and water waves (see section 15.2.6) and lubrication theory in fluid mechanics are examples of such simplifications for different classes of problems to which the general continuum mechanics model applies. For specific problems where one of these simplified theories is appropriate, further idealization and approximation may still be possible (and sometimes necessary!) to make progress toward its solution. We saw in section 15.2.6 the examples of small amplitude approximation, shallow water theory, and far-field expansion (leading to the Korteweg-deVries equation) for water waves. At other times, simplification, idealization, and approximation with very little mathematical or scientific justification may have to be made to bring the continuum mechanics model to bear on a problem. A real-life example of this situation will be described in section 19.6. The discussion there illustrates why applied mathematicians must have a good command of the scientific and engineering aspects of the phenomenon under investigation if they are to be successful in mathematical modeling. It is simply not enough to be adroit in mathematical methods alone.

19.1.2 Construction of Mathematical Models

The various steps in the development of the continuum mechanics model of matter, briefly described in section 19.1.1, are basic to constructing mathematical models for any real-life situations. The principal steps consist of

1. the idealization of the actual phenomenon of interest (such as space-time continuum mentioned in section 19.1.1)
2. the introduction of mathematical quantities (such as the density and velocity fields for continuous media) for an adequate description of the behavior of the idealized phenomenon

3. the application of the appropriate fundamental scientific principles (such as Euler's laws of mechanics for continuous media) to relate the various descriptive quantities

Typically, one arrives at a mathematical problem at the end of these steps; its solution is expected to provide useful information about the phenomenon under investigation. In the continuum mechanics model, the mathematical problem takes the form of an initial/boundary-value problem for differential equations. For other phenomena, the resulting mathematical models may not be continuous or deterministic, and they may not involve differential equations.

Whatever form the mathematical model may take, it is imperative that the relevant mathematical problems be *properly posed*. That is,

4. the mathematical problems associated with the model should be checked for consistency; whenever appropriate, they should also be checked for determinacy and stability with respect to small perturbations

A more detailed discussion of properly posed problems for models involving partial differential equations can be found in section 9.2.2 of this handbook.

If necessary and possible, the analyst should undertake some

5. further simplifications of the mathematical model by judicious approximations

before attempting to solve a particular problem or a class of problems. The implications of the solution obtained, tested against the available data for the actual phenomenon, may in turn suggest some

6. improvements of the mathematical model including the possibility of further simplifications.

In other words, there should be feedback and interplay between the model and the real-life situation being modeled.

Beyond the six basic steps outlined above, there is no set recipe for mathematical model construction in general. In the article "The Process of Applied Mathematics," Maynard Thompson (Ref. 19-1) supports that view: "There is today no treatise which provides a definitive discussion of the theory and practice of model building. It is not even clear that such an undertaking is a reasonable one. . . . The repeated successes of ceratin individuals in creating and developing mathematical models for scientific phenomena indicate quite clearly that model building can be learned. It is not clear that it can be taught in the same way as most academic subjects." Therefore, it is not surprising that all available applied mathematics texts and monographs use the case study approach in their discussion of general mathematical modeling, usually preceded by a few pages of preliminary remarks on general principles, such as the six basic steps outlined above (see Refs. 19-1–19-7 and other references cited therein). Given the nature of the subject, our discussion on mathematical models and their construction cannot deviate substantially from this conventional course. However, our viewpoints on modeling, also to be demonstrated by specific examples, are intended to complement those already stated elsewhere. Also, our examples are chosen to avoid duplication whenever possible.

To the extent that expertise on mathematical modeling is acquired principally through studying specific examples of past successes (and personal involvement in current modeling activities), the simple and elegant formulation of the continuum mechanics model (as well as the Lagrange-Hamilton-Jacobi theory for particle dynamics discussed briefly in sections 8.1.3, 13.1.1, and 13.1.2) requires some comments. No one interested in mathematical modeling should be lulled by this model into expecting comparable simplicity and elegance in all mathematical models and similar ease in their formulation. No less a scientific giant than Newton himself grappled unsuccessfully with the development of a continuum model for the motion of a "resisting medium." Nearly a century had elapsed before the breakthrough finally came, with the publication of Euler's laws of mechanics. A compact and polished formulation, such as the one partially described in sections 4.7.1–4.7.4, is the culmination of another two centuries of embellishment. Therefore, it is reasonable to expect the first efforts to formulate an appropriate mathematical model for a new area of inquiry to resemble work on the macroscopic behavior of matter in the pre-Euler (or even pre-Newton) era. The absence of a reasonably complete set of simple unifying fundamental laws for many new areas, such as the biological and social sciences, is often responsible for this unfortunate resemblance. In extreme cases, nothing is known about the phenomenon being modeled other than the possible dependence of some of its descriptive quantities on the others.

19.1.3 Survey of Contents

The one advantage the present generation of analysts has over its predecessors lies in the large collection of mathematical methods and models that accumulated over the years. At the very least, a knowledge of the mathematical techniques discussed in this handbook and elsewhere and of the successful models described in the literature should suggest the kind of mathematical models for a new phenomenon with which progress can be made. Among the mathematical methods available, *dimensional analysis* is unusual in that it can be helpful even if little is known about the phenomenon to be analyzed. It is also unusual in its wide range of applicability, from proving Pythagoras' theorem, Ref. 19-8,* to the derivation of similarity laws in continuum mechanics, Ref. 19-9. Section 19.2 discusses some elementary aspects of dimensional analysis and some of its applications to mathematical model formulation, including cases when little is known about the fundamental laws governing the actual phenomenon to be analyzed.

In most modeling situations, however, some fundamental laws of nature governing the phenomenon under investigation are known, and they are to form the basis of a mathematical model. As an extreme case, the mechanical, electromagnetic, and optical phenomena of matter at the macroscopic level are accurately determined by about a dozen fundamental laws. More commonly however, only some of the numerous important mechanisms governing the phenomenon of interest are identified and understood. Two consequences of this unfavorable condition, which prevails in biological and social sciences, are immediate. The analyst cannot expect

*The author is grateful to Professor George Bluman of UBC for a preprint of Ref. 19-8 from which the writing of section 19.2 has benefitted considerably.

any mathematical model to be an adequate (idealized) representation of the phenomenon modeled; it would not and could not have included all important features of the phenomenon. Such an incomplete model, in turn, can normally provide the analyst only with useful qualitative information about the phenomenon. Both are in direct contrast to our experience with the continuum mechanics model, which has given many engineers in the last two centuries design data for actual constructions.

Incomplete models are still useful, as the information obtained from them often offers some insight into some aspects of the actual phenomena. To the extent that no model in such disciplines as the social and biological sciences can be comprehensive, a successful study of complex social and biological processes usually begins with the simplest but still nontrivial mathematical model. More elaborate models, which include additional effects and features, are formulated and analyzed after a thorough understanding of the simpler models. The choice of model sequence is usually suggested by the relative importance of the various underlying mechanisms; the available mathematical methods for analyzing different types of models may also have some influence on the choice. The history of science weighs heavily in favor of the "from the simple to the elaborate" approach for scientific inquiries in previously untouched or little-explored areas. Unifying fundamental principles and comprehensive mathematical models are built on knowledge gained from investigations of more limited scope. The current polished form of continuum mechanics notwithstanding, the buckling of an elastic column was understood long before the formulation of the Cauchy-Navier three-dimensional theory of elasticity. Even Newton and Euler benefited from the partial research results obtained by themselves and others. The "simple-to-elaborate" approach is so important in mathematical modeling (and the trend toward generality in published research is so prevailing) that some work on urban land economics will be described in section 19.3 to illustrate the approach.

Comprehensive models tend to give rise to new, more specialized, and more difficult mathematical problems for which available mathematical methods may not be useful or applicable. In contrast, the mathematical problems associated with the simple and less complete models are often found to be simlar in structure to previously solved problems. The simple models of urban land economics in section 19.3 for example are similar in mathematical structure to problems in continuum mechanics and may therefore be solved by similar mathematical methods. The process of *scaling*, so useful in continuum mechanics and other mathematical models, can be introduced here in a completely different setting. Appropriately scaled, the solution of the mathematical problems in section 19.3 may be obtained by the perturbation methods of Chapter 14.

For a given phenomenon, often more than one possible mathematical model with comparable levels of descriptive and predictive power, may be formulated. The preference for a particular model is sometimes determined by the analyst's scientific and mathematical expertise, as well as the established practice of closely related areas. Also, a certain approach may seem more natural than others in a particular setting. It is important however for an analyst to be flexible and receptive to other possible approaches to the same problem. Occasionally, a less direct approach may lead to a more tractable mathematical problem than a more natural or established model. This is demonstrated in section 19.4 by some recent developments on a fundamental problem in forest management. The results presented there also illustrate how insight into an important class of problems may be obtained from qualitative results based on less than comprehensive models of the real-life situation.

Serious students and practitioners of quantitative methods of analysis today are well aware of the fact that linear models are much easier to analyze than nonlinear models. As a first approximation, it is often worthwhile to make model simplifications that remove the nonlinearities in the model. The tentative results from the first approximation linear model may then be used to estimate the effects of the neglected nonlinearities in the original model; those having a significant effect are then restored to the model for a more accurate solution. Unfortunately, the a posteriori consistency check is not always reliable. The deformation of a steadily rotating shallow thin elastic toroidal shell, discussed in section 19.5, demonstrates this point. The results based on a linear (bending) theory show no significant effects from the neglected nonlinear terms; nevertheless, these results are completely unacceptable no matter how small the inertia loading (or rotating speed) may be. Though linearity is an extremely desirable feature in a mathematical model and should be exploited whenever possible, its implications not only should be examined for mathematical consistency, but also must be tested against scientific reality.

Model simplifications with very little mathematical or scientific justifications are often necessary for many models to be useful. A real-life example of this situation will be reported in section 19.6. The example illustrates why there is more to mathematical modeling than a dexterity in mathematics and science.

Finally, we return, in section 19.7, to the theme "mathematical methods and concepts know no temporal barriers or scientific boundaries." We will have touched on this theme in section 19.3, but it deserves more emphasis and a more striking illustration. A simple model for the study of the effect of the 200-mile fishing limit turns out to have some well-known mathematical structures. The possible existence of an equilibrium fish population is identical to the problem of (Euler) column buckling. The problem of temporal stability of the equilibrium population has the same form as the initial/boundary-value problem for one-dimensional heat conduction. Thus, the practitioners of mathematical modeling, like other professionals, can also benefit from a knowledge of past accomplishments.

19.2 GROPING IN THE DARK

19.2.1 A Simple Example

Suppose a bomb is dropped vertically from a height h above ground level, and we are interested in its time in flight u before hitting the ground, assuming zero initial velocity. Even if we know little or nothing about the laws of physics that determine the behavior of a falling body, we might intuitively expect u to depend on the initial height h, the weight mg, and the shape s of the bomb. By treating the mass of the bomb m and the gravitational acceleration g as independent parameters, we can express this dependence by the mathematical expression*

$$u = f(h, m, g, s) \tag{19.2-1}$$

If this is the extent of our knowledge of the situation, then f is an unknown function to be determined by suitable experiments.

In the extreme case where the initial height, the weight, and the shape (as well as the material) of the bomb have already been decided, we need only to perform one experiment for one specific value of each of the parameters h, m, and s, as g is a known number once we specify the units for length and time. However, the cost of performing an experiment for this set of specific values may be too expensive or too dangerous, as in the case of megaton bombs. It would be desirable (if not imperative) that we find a more attractive alternative.

More often, we need the flight time u for a range of values of each of the parameters, to arrive at a final choice of one or more sets of parameter values for actual applications. In that case, we might perform experiments using, say 10 appropriate values for each of h, m, and s, leading to a total of 10^3 experiments and an expenditure of $10^3 \times D$ dollars, if one experiment costs D dollars. We might then use some kind of interpolation to estimate the function f for other values of the three parameters (far from a simple task in itself). Again, it would be desirable to find a more attractive alternative.

To find an alternative to performing costly and dangerous experiments, we obseve that all quantities in the relation (19.2-1) have as their dimension some combination of mass $\equiv M$, length $\equiv L$, and time $\equiv T$. With the dimension of Z denoted by $[Z]$, we have

$$\begin{gathered} [u] = T, \quad [m] = M, \quad [h] = L \\ [g] = LT^{-2}, \quad [s] = 1 \end{gathered} \tag{19.2-2}$$

We call T, M, and L the *fundamental dimensional units* (fdu) of our problem and describe s as *dimensionless*. Note that, although the shape parameter s has no dimension, $[s]$ may be taken as the combination $T^{\alpha_1}M^{\alpha_2}L^{\alpha_3}$ with $\alpha_1 = \alpha_2 = \alpha_3 = 0$. From (19.2-2), we have $[h/g] = T^2$ so that $u/\sqrt{h/g}$ is dimensionless. The numerical value of a dimensionless quantity is unaffected by the choice of unit system, whether it be MKS, cgs, or the British foot-pound-second system, i.e., it is *invariant under arbitrary scalings of fdu's*. We next divide both sides of (19.2-1) by $\sqrt{h/g}$ to get

$$\frac{u}{\sqrt{h/g}} = \frac{f(h, m, g, s)}{\sqrt{h/g}} \tag{19.2-3}$$

If we expect the relation (19.2-1) to remain unchanged for any choice of unit system, then the right side of (19.2-3) must be invariant under an arbitrary scaling

*More technical items, such as air drag, are not included in this first model for simplicity; they may be included as an improved model if the present model turns out to be inadequate.

of any of the *fdu*, just as the left side must be. Therefore, the right side must be a function of only dimensionless combinations of the quantities appearing there. But the only possible dimensionless combination of h, g, m, and s is s itself; we must have then

$$\frac{u}{\sqrt{h/g}} = F(s) \qquad \text{or} \qquad u = \sqrt{\frac{h}{g}}\, F(s) \tag{19.2-4}$$

where F is another unknown function.

The reduction of (19.2-1) to (19.2-4) is significant both qualitatively and quantitatively. We now know that the mass of the bomb is irrelevant to the time of impact and that the dependence of impact time on initial height (and gravitational acceleration) is completely determined, both without dropping a single bomb or any other object from the tower of Pisa! To obtain the impact time for a particular bomb, we have only to perform a single inexpensive and nondestructive experiment, by dropping any object of the same shape from a reasonable height that permits accurate measurements.

In the context of the mathematical modeling process discussed in section 19.1.2, the above development has effectively idealized the bomb as a point mass with the effect of its volume summarized by the shape parameter s (step 1). The wealth of material on the mechanics of falling bodies suggests that this is a reasonable first attempt to idealize for the purpose of a quantitative study. A set of measurable quantities $\{u, h, m, g, s\}$ has also been introduced for a description of the idealized model (step 2). We have tacitly assumed that the set is adequate, if not complete, for the study of our point mass model. In contrast to the validity of the point mass idealization, this assumption must be tested. Also, a dimensional constant, such as the one in Newton's inverse square law, might have been left out in the relation (19.2-1) (step 3). The reduced relation (19.2-4) may be used to check these assumptions with minimal cost. If different objects with the same shape yield different values for $F(s)$, then we must reexamine the model for possible improvements. We may expand the number of descriptive quantities upon which the impact time depends; we may include the mass density and material composition of the bomb, the effect of air drag, the variation of atmospheric properties with height, etc. (step 6).

Whether we use the reduced relation (19.2-4) for validating the mathematical model of the falling bomb or deducing information about it for an optimal effect in real-life application, the reduction from (19.2-1) to (19.2-4) is obviously invaluable, and the method of reduction deserves a closer examination for possible application in other situations. In the next section, we distill the basic ideas of dimensional analysis used in this simple example and formulate a more general procedure for a wider range of applications in mathematical modeling.

19.2.2 Dimensional Analysis

In any quantitative study of a real-life phenomenon, we have, after step 2 of the mathematical modeling process (see section 19.1.2), identified a set of measurable scalar quantities (field variable and parameters), say $\{u, W_1, W_2, \ldots, W_n\}$, that describes an idealized and simplified version of the phenomenon. The set is assumed to be complete within the framework of the model contemplated, and the dimension of all quantities in the set is known and denoted by $\{[u], [W_1], \ldots, [W_n]\}$. For simplicity, suppose the model aims to determine the single scalar quantity u, and suppose, on the basis of our meager knowledge about the phenomenon, we know only that u is a function of all the Ws

$$u = f(W_1, W_2, \ldots, W_n) \tag{19.2-5}$$

so that the mathematical problem at the end of step 3 of our modeling process is to find out as much about f as possible.

Our experience with the simple example of section 19.2.1 suggests that we first identify the fdu's, say $L_1, L_2, \ldots, L_m$, of the problem. The number and types of fdu's vary with problems (but typically $m < n$). For example, we have $\{L_1 = M =$ mass, $L_2 = L =$ length, $L_3 = T =$ time$\}$ for mechanical problems and $\{L_1 = D =$ dollars, $L_2 = P =$ labor population, $L_3 = T =$ time, $L_4 = G =$ goods$\}$ in some economic problems. We now postulate* that all descriptive quantities in the mathematical models have dimensions that are products of powers of the fdu's, i.e., for any quantity Z we have

$$[Z] = L_1^{a_1} L_2^{a_2} \ldots L_m^{a_m} \tag{19.2-6}$$

for m real numbers $a_1, \ldots, a_m$. The m-tuple $\mathbf{a} = (a_1, a_2, \ldots, a_m)$ is called the *dimension vector* of Z. Z is said to be *dimensionless* if and only if $[Z] = 1$, i.e., its dimension vector is the null vector. *The numerical value of a dimensionless quantity is unaffected by a change of units of measurement.*

In dimensionless analysis, expression (19.2-5) is assumed to remain unchanged also for any system of measurement units. A change from one system of units to another involves a scaling (or rescaling) of each fdu, which in turn induces a scaling of the descriptive quantities of the model. For example, in changing from cgs to MKS units, $L_1 = M =$ mass is scaled by 10^{-3}, $L_2 = L =$ length is scaled by 10^{-2}, and $L_3 = T =$ time is unchanged. In turn, this induces a scaling of a quantity such as energy E with $[E] = ML^2T^{-2}$ by $10^{-3}(10^{-2})^2 = 10^{-7}$. Expression (19.2-5) is therefore assumed to be invariant under an arbitrary scaling of any fdu.

With no loss in generality, let $W_1, W_2, \ldots, W_m$ have independent dimensions, and call them the *primary quantities*. The dimensions of the *secondary quantities* $W_{m+1}, \ldots, W_n$ can then be expressed in terms of the dimensions of the primary quantities in the form

$$[W_{m+j}] = [W_1]^{a_{m+j,1}} \ldots [W_m]^{a_{m+j,m}} \qquad (j = 1, 2, \ldots, n - m) \tag{19.2-7}$$

*This postulate is sometimes deduced as a theorem from what may be considered more intuitive hypotheses (e.g., Refs. 19-5 and 19-9). But it is equally easy to verify the present postulate for a specific problem, since the dimensions of all quantities involved are known.

Given that the relation (19.2-5) is to remain *unchanged* for any system of measurement units, the quantity u must also be a secondary quantity, so that

$$[u] = [W_1]^{a_1} \dots [W_m]^{a_m} \tag{19.2-8}$$

Otherwise, the dimension of u would be independent of those of $W_1, \dots, W_m$, and u would depend not only on $W_1, \dots, W_n$, i.e., the set of descriptive quantities for the model is incomplete.

We now form the dimensionless combinations

$$\begin{aligned} \pi_j &= \frac{W_{m+j}}{W_1^{a_{m+j,1}} \dots W_m^{a_{m+j,m}}} \qquad (j = 1, 2, \dots, n - m) \\ \pi &= \frac{u}{W_1^{a_1} \dots W_m^{a_m}} \end{aligned} \tag{19.2-9}$$

and write (19.2-5) as

$$\begin{aligned} \pi &= \frac{f(W_1, \dots, W_n)}{W_1^{a_1} \dots W_m^{a_m}} \\ &= \frac{f(W_1, \dots, W_m, W_1^{a_{m+1,1}} \dots W_m^{a_{m+1,m}} \pi_1, \dots, W_1^{a_{n,1}} \dots W_m^{a_{n,m}} \pi_{n-m})}{(W_1^{a_1} \dots W_m^{a_m})} \\ &\equiv F(W_1, \dots, W_m, \pi_1, \dots, \pi_{n-m}) \end{aligned} \tag{19.2-10}$$

where F is some new unknown function. Because $W_1, W_2, \dots, W_m$ are of independent dimensions and $\pi, \pi_1, \dots, \pi_{m-n}$ are dimensionless, and because (19.2-5) (and therefore (19.2-10)) is *unchanged* for any system of units of measurements, we may change from one system of measurement units to another, which leaves the numerical values of all Ws and πs unchanged except for an arbitrary change in W_1. It follows that $\partial F/\partial W_1 \equiv 0$ or π does not depend on W_1. Similar arguments give $\partial F/\partial W_k \equiv 0$, $k = 1, 2, \dots, m$, so that we have

$$\pi = \Phi(\pi_1, \pi_2, \dots, \pi_{n-m}) \tag{19.2-11}$$

or in terms of the original quantities

$$\begin{aligned} u &= f(W_1, \dots, W_n) \\ &= W_1^{a_1} \dots W_m^{a_m} \Phi\left[\frac{W_{m+1}}{W_1^{a_{m+1,1}} \dots W_m^{a_{m+1,m}}}, \dots, \frac{W_n}{W_1^{a_{n,1}} \dots W_m^{a_{n,m}}}\right] \end{aligned} \tag{19.2-12}$$

This is the basic result in the theory of dimensional analysis and is known as the Buckingham Π-theorem (see Ref. 19-8 for references to E. Buckingham's original publications). The theorem asserts that a relation such as (19.2-5) in a mathematical model, which is unchanged for any system of measurement units, can be rewritten as a relation among dimensionless combinations of the original quantities appearing in (19.2-5), and that the number of dimensionless combinations involved is equal to the difference between the number of original quantities and the number of fdu's for the model.

The above basic result in dimensional analysis is intuitively obvious and had been used by Fourier, Maxwell, Reynolds, and Rayleigh, just to name a few, long before the explicit formulation and formal proof of the Π-theorem. This basic result is also useful when more is known about the model. We show in a later section how it may be used to reduce the solution of a partial differential equation to that of an ordinary differential equation. More general versions of the theorem (e.g., Ref. 19-5), as well as group theoretical treatments of the subject (e.g., Refs. 19-10 and 19-11), have expanded the usefulness of dimensional analysis to a larger class of problems, notably in differential equations. Interested readers are referred to Refs. 19-5 and 19-10 through 19-16 (and references cited there) for these developments.

19.2.3 The Atomic Explosion of 1945

To emphasize the importance of dimensional analysis in applications beyond the textbook variety, we sketch in this section a dimensional analysis argument for Sir G. I. Taylor's astounding deduction of the approximate energy released by the first atomic explosion in New Mexico. Taylor carried out his analysis in 1941. The motion picture records of the explosion by J. E. Mack, declassified in 1947, provided the experimental data needed for completing the deduction. However, the amount of energy released by the blast apparently was still classified in 1947!

An atomic explosion is idealized and simplified as the release of a large amount of energy E from a "point." The explosion sends off an expanding spherical fireball whose (outer) surface corresponds to a powerful shock wave. Let R be the spherical shock wave radius, which increases with time. It is not unreasonable to expect this increase to vary with the amount of energy E released by the explosion, the initial (or ambient) air density ρ_0, and the initial (or ambient) air pressure p_0. We assume that these quantities are adequate for a proper description of the model and write

$$R = f(t, E, \rho_0, p_0) \tag{19.2-13}$$

The fdu's in this case are L_1 = mass, L_2 = length, and L_3 = time, and the dimensions of the five quantities involved are

$$\begin{gathered}[R] = L_2, \quad [t] = L_3, \quad [E] = L_1 L_2^2 L_3^{-2} \\ [\rho_0] = L_1 L_2^{-3}, \quad [p_0] = L_1 L_2^{-1} L_3^{-2}\end{gathered} \tag{19.2-14}$$

By the second part of the Π-theorem, there is only one possible dimensionless combination among the four quantities t, E, ρ_0, and p_0 on the right side of (18.2-13), which may be taken as

$$\pi_1 = p_0\left[\frac{t^6}{E^2\rho_0^3}\right]^{1/5} \tag{19.2-15}$$

and R may be made dimensionless by

$$\pi = R\left[\frac{\rho_0}{Et^2}\right]^{1/5} \tag{19.2-16}$$

By the first part of the Π-theorem, we may write (19.2-13) as

$$\pi = F(\pi_1) \quad \text{or} \quad R = \left[\frac{Et^2}{\rho_0}\right]^{1/5} F(\pi_1) \tag{19.2-17}$$

where F is an unknown function of π_1.

In cgs units, we have $\rho_0 = 1.25 \times 10^{-3}$ (g/cm^3) and $p_0 = 10^6$ (g/cm-sec^2). The released energy E is expected to be a very large number; an explosion of 50 pounds of TNT corresponds to an E of the order of 10^{15} ergs (=g-cm^2/sec^2). An atomic explosion is much more powerful than that, e.g., $E = 9 \times 10^{20}$ ergs for a 20-kiloton explosion. From these figures, we see that π_1 is a small number, if t is not more than a second. For $\pi_1 < 0.01$ (or $t < 0.07$ sec for the 20-kiloton explosion), $F(\pi_1)$ may be approximated by $F(0)$, so that (19.2-17) is approximately

$$R \simeq \left[\frac{E}{\rho_0}\right]^{1/5} F(0)t^{2/5} \tag{19.2-18}$$

which is the formula derived by Taylor (Ref. 19-17). Experiments using light explosives can be conducted to determine $F(0)$, which turns out to be approximately 1.

The approximate relation (19.2-18) may be written as

$$\frac{5}{2}\log_{10}(R) = \log_{10}(t) + \frac{1}{2}\log_{10}\left[\frac{E}{\rho_0}\right] \tag{19.2-19}$$

If $(5/2)\log_{10}(R)$ is plotted against $\log_{10}(t)$ for a fixed value of E/ρ_0, (19.2-19) gives a straight line with unit slope. An actual record of the radius of the spherical fire ball for different times after the explosion may be similarly plotted to verify the unit slope linear relation between $(5/2)\log_{10}(R)$ and $\log_{10}(t)$. If the linear relation is confirmed, we can then get from the same plot of the data an estimate of the corresponding value of E.

Mack's motion picture and other declassified photographic data provided Taylor with the information needed to validate the approximate relation (19.2-19) and then to estimate the energy released for that first atomic explosion. His estimate turns out to be remarkably accurate. It is also rather amazing that a straight line log-log plot of the data was predicted theoretically by Taylor more than four years before the actual explosion! As we can see from Refs. 19-9 and 19-18, dimensional analysis has become increasingly more useful in the study of explosions since that time.

19.2.4 Fundamental Solutions of the One-Dimensional Heat Equation

So far, we have applied dimensional analysis to situations where the analyst does not know the scientific laws and principles that govern the phenomenon being investigated. In this section, we discuss how, when more information about the phenomenon is available, dimensional analysis may help to attain a complete understanding of the idealized phenomenon without any experiment on the phenomenon.

For heat conduction along a straight thin bar, the energy in any segment of the bar must be conserved, so that the rate of change of thermal energy in a bar segment must equal the net rate at which heat energy enters the segment. From this conservation law, a second-order partial differential equation has been deduced in section 9.1.1 of this handbook for the temperature distribution $u(x, t)$ in the idealized bar at time t and location x. We consider here an infinitely long bar of uniform properties oriented in such a way that its central axis coincides with the x-axis. For $t < 0$, the entire bar is at some uniform ambient temperature, which we will take to be zero (degrees). At the instant $t = 0$, a heat pulse of strength Q is applied instantaneously at the location $x = 0$ of the bar. Of interest is the subsequent diffusion of that heat pulse for later times $t > 0$. The partial differential equation (PDE) of one-dimensional heat conduction applies to the above situation in the form

$$\rho c u_t - k u_{xx} = Q\delta(t)\,\delta(x) \qquad (-\infty < x < \infty, t > 0-) \qquad (19.2\text{-}20)$$

where ρ is the density, c is the specific heat, k is the thermal conductivity, Q is the amount of heat added per cross-section area, $\delta(\cdot)$ is the Dirac delta function, and $(\ )_y \equiv \partial(\ )/\partial y$. Equation (19.2-20) is supplemented by the auxiliary condition

$$u(x, t) \equiv 0 \qquad (t \leqslant 0-) \qquad (19.2\text{-}21)$$

We also expect to have

$$\lim_{|x| \to \infty} u(x, t) = 0 \qquad (0 < t < \infty) \qquad (19.2\text{-}22)$$

as it takes finite time for the heat pulse to diffuse through a finite distance. The mathematical problem defined by (19.2-20), (19.2-21), and (19.2-22) completely determines the temperature distribution $u(x, t)$ throughout the bar for all $t > 0$. An explicit solution of this problem may be obtained with the help of dimensional analysis.

There are five fdu's in the above heat conduction problem, namely

$$M = \text{mass}, \quad L = \text{length}, \quad T = \text{time}$$
$$C = \text{heat (in calories)}, \quad D = \text{temperature (in degrees)}$$

Correspondingly, the dimensions of the variables and parameters in the problem are

$$\begin{aligned}[]
[\rho] = ML^{-3}, \quad [c] = CD^{-1}M^{-1}, \quad [k] = CD^{-1}T^{-1}L^{-1} \\
[Q] = CL^{-2}, \quad [u] = D, \quad [x] = L, \quad [t] = T
\end{aligned} \tag{19.2-23}$$

Evidently, the solution u may depend on all six remaining quantities:

$$u = f(x, t, \rho, c, k, Q) \tag{19.2-24}$$

By the second part of the Π-theorem, we can have only one dimensionless combination of the six quantities on the right, which we take in the form

$$\pi_1 = \frac{x}{2\sqrt{\kappa t}} \tag{19.2-25}$$

where $\kappa \equiv k/\rho c$ is the thermal diffusivity with $[\kappa] = L^2 T^{-1}$. Since

$$\pi = \frac{2\sqrt{\kappa t}}{Q/\rho c} u \tag{19.2-26}$$

is also dimensionless, we have from the second part of the Π-theorem that

$$\pi = F(\pi_1) \qquad \text{or} \qquad u = \frac{Q}{\rho c} \frac{F(x/2\sqrt{\kappa t})}{2\sqrt{\kappa t}} \tag{19.2-27}$$

Thus, the application of dimensional analysis reduces the solution for u to a form involving only an unknown function of one variable, which is a specific combination of the two independent variables x and t (and the parameter κ). The actual form of F can now be determined by a direct substitution of (19.2-27) into the PDE (19.2-20).

With (19.2-27) and the fact that $\delta(t) = 0$ for $t > 0$, the PDE (19.2-20) for $t > 0$ becomes an ordinary differential equation (ODE) for $F(\pi_1)$:

$$F'' + 2\pi_1 F' + 2F = 0 \qquad (-\infty < \pi_1 < \infty) \tag{19.2-28}$$

where $(\ )' \equiv d(\ )/d\pi_1$. Rewritten as $F'' + 2(\pi_1 F)' = 0$, the ODE (19.2-28) may be integrated once immediately to give

$$F' + 2\pi_1 F = c_0 \tag{19.2-29}$$

where c_0 is an arbitrary constant of integration. The first-order linear ODE has an integrating factor $e^{\pi_1^2}$ and can be formally integrated once more to get $F(\pi_1)$. Unfortunately, the integral of $e^{\pi_1^2}$ is unbounded at both end limits, $\pi_1 \to \pm\infty$, so that a more detailed analysis (as in Ref. 19-10) is necessary to obtain the correct solution. Here we simply appeal to the physical fact that, with only a single point source heat pulse, the temperature field $u(x, t)$ and therefore $F(\pi_1)$ should be nonnegative everywhere and at all times (with $F \to 0$ as $|x| \to \infty$ for all $t > 0$). A sketch of the direction field for $F(\pi_1)$ on the basis of (19.2-29) shows that this is not possible unless $c_0 = 0$.

With $c_0 = 0$, the first-order ODE (19.2-29) yields

$$F(\pi_1) = c_1 e^{-\pi_1^2} \tag{19.2-30}$$

The new constant of integration c_1 is determined by the strength of the heat pulse at $t = 0$, a piece of information in the PDE (19.2-20) that we have not used. We exhaust the content of (19.2-20) by integrating it over the entire bar and up to time t (which is equivalent to the global conservation of energy) to get

$$\begin{aligned}\int_{-\infty}^{t}\int_{-\infty}^{\infty} [\rho c u_t - k u_{xx}]\, dx\, dt &= Q \int_{-\infty}^{t}\int_{-\infty}^{\infty} \delta(t)\,\delta(x)\, dx\, dt \\ &= QH(t)\end{aligned} \tag{19.2-31}$$

where $H(t)$ is the unit step function (equal to 1 for $t > 0$ and 0 for $t < 0$). With $u = 0$ for $t < 0$ (and $u_x \to 0$ as $|x| \to \infty$), it follows from (19.2-31) and (19.2-30) that

$$\rho c \int_{-\infty}^{\infty} u(x, t)\, dx = \frac{Qc_1}{2\sqrt{\kappa t}} \int_{-\infty}^{\infty} e^{-x^2/4\kappa t}\, dx = Q \tag{19.2-32}$$

or $c_1 = 1/\sqrt{\pi}$ (here π is the number 3.14159 . . .), so that

$$u = \frac{Q}{\rho c\sqrt{\pi}} \frac{e^{-x^2/4\pi t}}{2\sqrt{\kappa t}} H(t) \tag{19.2-33}$$

with $\kappa \equiv k/\rho c$. This solution of the initial-value problem (19.2-20)-(19.2-22) is known as the *fundamental solution* of the one-dimensional heat (or diffusion) equation.

The fundamental solution (19.2-33) is important beyond the specific problem of an infinitely long bar subject to a heat pulse at a point. It plays a significant role in the solution of many diffusion problems in one dimension. For example, the solution of

$$\rho c v_t - k v_{xx} = g(x, t) \qquad (-\infty < x < \infty, t > 0)$$

$$v(x, 0) = 0 \qquad (-\infty < x < \infty)$$

$$\lim_{|x| \to \infty} v(x, t) = 0 \qquad (t > 0)$$

where $g(x, t)$ is a prescribed (integrable) distributed heat source density, is

$$v(x, t) = \int_0^t \int_{-\infty}^{\infty} U(x - \xi, t - \tau)\, g(\xi, \tau)\, d\xi\, d\tau \qquad (19.2\text{-}34)$$

where $U(x - \xi, t - \tau)$ is the fundamental solution with the unit strength heat pulse to be located at the position ξ along the bar and to occur at the instant τ (instead of the origin of both space and time), i.e.,

$$U(x - \xi, t - \tau) = \frac{1}{\rho c \sqrt{\pi}} \frac{e^{-(x-\xi)^2/4\kappa(t-\tau)}}{2\sqrt{\kappa(t - \tau)}} H(t - \tau) \qquad (19.2\text{-}35)$$

(We will not digress to derive the solution (19.2-34) and will note only that it can be verified by direct substitution.) Similar fundamental solutions in two and three dimensions may also be obtained with the help of dimensional analysis.

19.2.5 Similarity

It should be evident by now that dimensionless quantities have a significant role in mathematical model formulation and the extraction of useful information from the model. In a dimensionless formulation of the model, it is possible for two or more different realizations of the modeled phenomenon to have the same set of values for all the dimensionless quantities; different numerical values of the same physical quantities associated with the phenomenon may give the same set of numerical values for particular dimensionless combinations of the physical quantities. For example, $\{x = 4$ meters and $t = 1$ sec$\}$ and $\{x = 8$ meters and $t = 4$ sec$\}$ give the same value for π_1 in (19.2-25) as long as κ is kept fixed. Two realizations are said to be *similar* if all dimensionless quantities in the model have the same values for both, e.g., two falling bodies modeled by (19.2-1) with the same shape. This concept of similarity is most important in experiments on test models, smaller versions of the actual item (quite unrelated to mathematical models of a phenomenon). Experimental results on test models can be related to the actual item only if they are similar.

The concept of similarity is also important in establishing the existence and obtaining the so-called *similarity solution(s)* of initial/boundary-value problems (IBVP) associated with a particular mathematical model, as in the case of one-dimensional heat conduction discussed earlier. If we expect the spatial distribution of a property of the modeled phenomenon at different times to be similar, we may then seek a similarity solution of the IBVP by way of a dimensionless combination of time and spatial coordinates, such as π_1 in (19.2-25). Such a dimensionless combination is called a *similarity variable*. It may also happen that a similarity solution is established with the help of dimensional analysis and thereby reveals a useful similarity structure inherent in the modeled phenomenon.

In connection with mathematical models that can be formulated as an IBVP in partial differential equations, the concept of similarity has in recent years extended the determination of similarity solutions beyond what can be accomplished within the framework of dimensional analysis. Working directly with a dimensionless form of the IBVP, invariance consideration and group theoretical methods have been instrumental in finding similarity solutions not tractable by dimensional analysis alone (Ref. 19-10). In another direction, similarity solutions are found to be important for the description of the "intermediate asymptotic" behavior of a phenomenon. Loosely speaking, this is the behavior of the phenomenon in the range where the solution of the IBVP no longer depends on the details of the initial or boundary conditions, but is still far from being in a limiting state. For a finite bar of length L heated locally over a small interval ϵ, an intermediate asymptotic solution is appropriate at a time when the heat is diffused up to a distance l away from the source with $\epsilon \ll l \ll L$ and the temperature of the bar is sought at point $\epsilon \ll x < l$.

In mathematical modeling situations where no IBVP can be properly formulated, various kinds of similarity methods have also found important applications. Aside from the examples discussed in sections 19.2.1 and 19.2.3, the theory of turbulence, the theory of phase transitions, and quantum field theory are among those that have benefited from these methods (Ref. 19-9).

19.3 FROM THE SIMPLE TO THE ELABORATE

19.3.1 Urban Land Economics

Cities all over the world grow larger and more important in the 20th century. As they spread across the countryside, urban blight, traffic congestion, housing shortages, diminished open space, environmental pollution, slums, poverty, and other urban problems become more acute. Governments and social sciences have become increasingly concerned with them. To deal effectively with these problems, the forces behind them must be uncovered, delineated, and controlled. Urban economics is concerned with the economic forces behind many of our urban problems; a typical introductory text on the subject (e.g., Ref. 19-19) applies the tools of economics, such as theories of production, consumer behavior, equilibrium, etc., to analyze all the urban problems listed above and more. The internal structure of cities and specific urban problems are so complex that only a modest beginning has been made toward a complete understanding of them. The task is complicated by the fact that, to varying degrees, individual urban problems are interrelated. To make progress, broad areas of intimately related urban issues are often isolated for independent economic analyses, with the interaction between areas treated subsequently. The situation is analogous to the design study of an airplane: the wings, the fuselage, the jet engines, etc., are first isolated for analysis, and the information obtained is used as an input to a global study of the entire structure.

Urban land economics is a broad area within urban economics that is concerned with the relation of land values to land uses within a city. The subject is of particular interest to public policy makers. Many urban public expenditures, such as the construction of roads and government buildings and the creation of public parks and other recreational areas, involve the acquisition of land. With the market price not always a true indicator of the land's actual worth (called the *shadow price* or *shadow rent*) of land in a city, it is prudent that governments at different levels have an understanding of both prices and their differences to arrive at proper (or economically efficient) decisions on expenditures involving land. A difference between the market price and the shadow price of land is often a consequence of the fact that private individuals do not pay for the social costs they create. For example, except for a few places in the United States where tolls are collected from commuters, individuals do not pay for the traffic congestion they cause and inflict upon others. A government may also wish to intervene in private decisions that are based on the market price if it differs significantly from the shadow price, e.g., by changing the price structure of public transportation or imposing a commuter tax. Again, an understanding of urban land values is necessary for a proper decision.

In this section we describe a group of results in urban land economic pertaining to residential land values. About 80% of all privately developed land in major cities in the United States is directed toward residential use (Ref. 19-20). The economic theory of residential land therefore deserves as much attention and effort from economists as commercial or agricultural lands. In fact, the values of land of all three types are interrelated, and the study of their proper allocation for different purposes must ultimately be unified. However, even a comprehensive study of residential land values and uses alone is not possible, because the real-life situation is far too complex to be effectively idealized and not all operating forces with a significant influence on the eventual outcomes are economic in nature. Under the circumstances, the study of residential land values has proceeded with the formulation and analysis of some very simple models that capture a number of the dominant economic features involved, with the number of features increasing as the models become more and more elaborate. As such, the economic analyses of residential land values provide a good illustration of the "from the simple to the elaborate" approach in mathematical modeling, an approach that seems to require renewed emphasis in view of the recent trend in teaching continuum mechanics by starting with very elaborate models.

In selecting the economic theory of residential land values to illustrate a good working principle in mathematical modeling, we also have in mind the fact that the analysis of the models within that theory to be presented below will enable us to bring out some additional features of dimensional analysis not already discussed in the last section. These include scaling and dimensionless formulation, which in turn show how the perturbation methods of Chapter 14, developed originally for solid and fluid mechanics, are also useful in the social sciences. Such mathematical problems as free-boundary problems in differential equations, which arise naturally in the study of the mechanics of continuous media, are also shown to appear in social science by our models in residential land economics.

19.3.2 Household Preference and Utility Function

Serious attempts to understand the economics of residential land began in 1926 with a brief qualitative analysis by R. M. Haig (Ref. 19-21). However, the skeleton of the standard model for the study of the equilibrium distribution of residential land values (or rents) and population densities emerged only in W. Alonso's 1960 doctoral dissertation entitled "A Model of the Urban Land Market: Location and Densities of Dwellings and Business." An expanded version of the material in the dissertation can be found in Ref. 19-21, in which a review of earlier investigations can be found.

The standard model for residential land allocations considers a featureless, monocentric, circular city consisting of a circular central business district (CBD) surrounded by a circular annular region of roads and houses. Because we pay no attention to natural geographical or topographical features, one unit of land differs from another only in its distance from the CBD. We denote the radius of the CBD by R_c (miles) and the radius of the outer boundary of the residential district by R_e. A population of N_0 *identical* families lives in the city; each household uses a certain amount of housing space s (square miles) at a distance X from the CBD and consumes a certain amount c (bags) of a simple composite consumption good per annum. Each household sends one commuter traveling to work in the CBD to earn an identical annual wage Y (dollars). The standard model is concerned only with traveling to and from the CBD by the workers, which results in a transportation cost of t dollars per annum (which depends on the distance traveled) for each household. Traveling within the CBD is assumed to be free, so that $t(X = R_c) = 0$. The land values are determined (primarily) by differential accessibility to the CBD in a manner to be described below. The radial road system is densely and evenly distributed in the circumferential direction, so that we effectively have complete polar symmetry.*

Let $r(X)$ (dollars/mile^2) be the land value or rent per unit area of housing space and p (dollars/bag) be the price of a bag of consumption good. Each household uses its annual wage to pay for housing, consumption goods, and travel cost (to and from the CBD), so that

$$pc + rs + t = Y \tag{19.3-1}$$

*The polar symmetry stipulation may be removed by considering only a pie-shaped sector of the city with a (small) sectoral angle θ.

Each of the N_0 families chooses a location X, an amount of land s, and an amount of consumption goods c, which give the household the most satisfaction subject to the budget constraint (19.3-1). As we shall see, the land rent per unit area, which varies with location, is determined by the household decisions on location, housing space, and the amount of consumption goods.

The satisfaction of a household is given by a utility index U. In the absence of natural geographical features and socially determined neighborhood characteristics, it is difficult to know whether it is better to live near the CBD or far away from it, given that the costs of transportation (including commuting time) have already been included in the budget equation (19.3-1). For simplicity and a lack of information, we will assume that location has no direct effect on the utility of a household (though it definitely has an indirect effect through (19.3-1)).

The modern theory of utility is an indispensable tool for the social sciences. Though not without serious limitations, the theory developed by von Neumann and Morgenstern and a modified version suggested by R. D. Luce (in which preference is assumed to be probabilistic) are on solid mathematical foundation. Readers are referred to Ref. 19-22 for a detailed discussion of the von Neumann-Morgenstern theory, including a discussion of experimental determinations of utility functions, as well as a sketch of the Luce theory. We limit ourselves here to listing a few basic properties of utility functions relevant to our analyses of the residential land values and uses.

The utility function $U(c, s)$ (which we will take to be twice differentiable) assigns a real number to each point in the (c, s)-plane (*the commodity plane*) to reflect the household preferences. A point (c, s) in the commodity plane corresponds to a specific combination of c amount of consumption goods and s amount of land for housing and is called a *commodity bundle*. The numerical values of $U(c, s)$ give the ranking by the household preferences, with $U(c, s) \geqslant U(c', s')$ whenever the bundle (c, s) is at least as satisfactory or desirable as (c', s'). To the extent that the utility function reflects only preference rankings, it is *ordinal;* hence a utility function is determined only up to a one-one monotone transformation.

A curve in the commodity plane along which U is a constant is called an *indifference curve*. The household is indifferent to all commodity bundles along an indifference curve in the sense that it considers each as desirable as all the others. A household always prefers more of each commodity if the other is not reduced, i.e., $\partial u/\partial c > 0$ and $\partial u/\partial s > 0$. Along an indifference curve, we have

$$dU = U_c\, dc + U_s\, ds = 0 \tag{19.3-2}$$

with $U_z \equiv \partial U/\partial z$, so that

$$\frac{dc}{ds} = -\frac{U_s}{U_c} < 0 \tag{19.3-3}$$

Therefore, indifference curves are convex to the origin of the commodity plane (Fig. 19.3-1). Note that this property is invariant with respect to a one-one monotone transformation. We also expect $U_{cc} < 0$ and $U_{ss} < 0$, as the additional satisfaction to be derived from an extra amount of a commodity diminishes with the amount of the commodity already in possession. The addition assumption $U_{cc}U_{ss} - U_{cs}^2 > 0$ is usually made, so that U is concave with a negative definite Hessian matrix in the entire (first quadrant of the) commodity plane.

For a fixed X, if $r(X)$, $t(X)$, and p are known, then the graph of (19.3-1) in the commodity plane would be a straight line with a negative slope. In that case, the bundle $(\bar{c}, \bar{s})$ that maximizes $U(c, s)$ subject to the budget constraint (19.3-1) would be the point on a particular indifference curve where the budget line (19.3-1) is tangent to that curve. (As shown in Fig. 19.3-1, an indifference curve with lower utility intersects the budget line at two distinct points, and one with a higher utility value does not intersect the budget line at all.) For the residential land problem, the situation is more complicated, for at least $r(X)$ is not known and is to be determined as a part of the solution process.

19.3.3 Distance Cost of Transportation

Until the early seventies, the cost of transportation was taken to depend only on the distance from the CBD. From available data, we can calculate the distance cost per mile per annum τ_d, from which we get the total distance cost of transportation

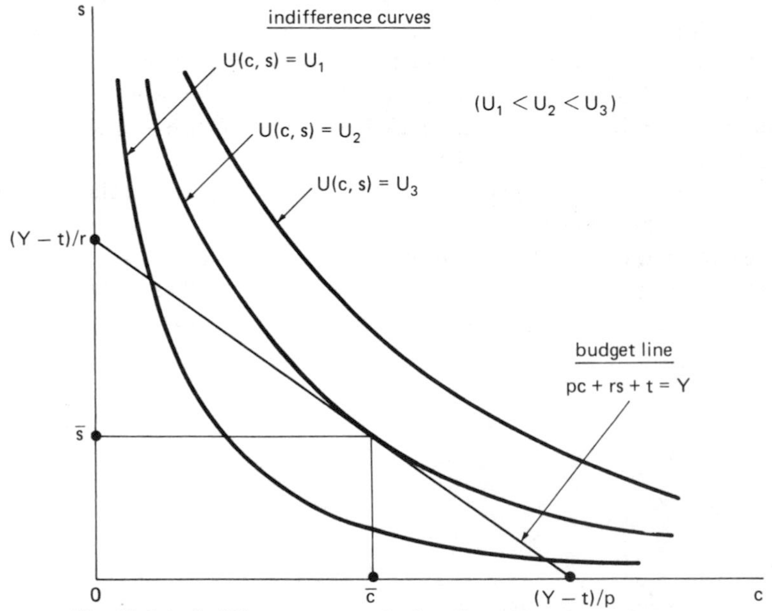

Fig. 19.3-1 Indifference curves, budget line and consumer's optimum.

per annum

$$t(X) = t_d(X) \equiv \int_{R_c}^{X} \tau_d \, dX \tag{19.3-4}$$

because transportation is free within the CBD. Note that we have $t_d = \tau_d(X - R_c)$ if the travel cost per mile per annum τ_d is a constant.

The household optimum is now obtained from the first-order necessary conditions (see section 20.2.2)

$$\left.\frac{\partial L}{\partial c}\right|_{(\bar{c},\bar{s})} = [U_c + \lambda p]_{(\bar{c},\bar{s})} = 0, \quad \left.\frac{\partial L}{\partial s}\right|_{(\bar{c},\bar{s})} = [U_s + \lambda r]_{(\bar{c},\bar{s})} = 0 \tag{19.3-5}$$

where the Lagrangian function L is

$$L(c, s, \lambda) \equiv U(c, s) + \lambda [pc + rs + t - Y] \tag{19.3-6}$$

with λ being a Lagrange multiplier and $t = t_d$. Together with (19.3-1), the conditions (19.3-5) determine $\bar{c}$, $\bar{s}$, and $\bar{\lambda}$ in terms of $Y - t(X)$, p, and the yet unknown land rent distribution $r(X)$. We may then use the results to eliminate $\bar{c}$ and $\bar{s}$ from the utility function to get

$$U(\bar{c}, \bar{s}) = \bar{U}(r(x), Y - t(X), p) \tag{19.3-7}$$

To illustrate, we consider a Cobb-Douglas-type utility function often used in urban land economics for households in residential districts:

$$U(c, s) = U(\xi), \qquad \xi = s^{\sigma} c^{1-\sigma} \tag{19.3-8}$$

where $0 < \sigma < 1$, and $U(\cdot)$ is monotone increasing and strictly concave. The first-order necessary conditions for a maximum U yield the stationary point

$$\bar{c} = \frac{1 - \sigma}{p}(Y - t), \qquad \bar{s} = \frac{\sigma}{r}(Y - t) \tag{19.3-9}$$

and the corrsponding stationary value of U is

$$U(\bar{c}, \bar{s}) = U\left(\frac{(1 - \sigma)^{1-\sigma}\sigma^{\sigma}}{p^{1-\sigma} r^{\sigma}}(Y - t)\right) \equiv \bar{U}(r(X), Y - t(X), p) \tag{19.3-10}$$

The monotone increasing and strictly concave properties of $U(\cdot)$ ensure that the stationary value is in fact a maximum.

Because all N_0 families are identical in our model city, they must have the same taste and therefore share the same utility function. For them to settle into an equilibrium dwelling pattern, all households must have attained the same utility (value); otherwise the less satisfied households would move to a more satisfactory location with a higher utility, bidding up the land rent there in the process, etc., to reach equilibrium. Therefore, we have in equilibrium (and with $t(R_c) = 0$)

$$\overline{U}(r(X), Y - t(X), p) = \overline{U}(r(R_c), Y, p) \tag{19.3-11}$$

say, which determines $r(x)$ in terms of $Y - t(X)$ up to an unknown constant $r_c \equiv r(R_c)$, the unit area land rent at the edge of the CBD. For the Cobb-Douglas-type utility function (19.3-8), we have

$$r^\sigma = \frac{r_c^\sigma}{Y} [Y - t(X)] \tag{19.3-12}$$

To determine the remaining unknown constant r_c, we let $N(X)$ be the number of households living outside the circle of radius X. Evidently, we must have

$$N(R_c) = N_0, \qquad N(R_e) = 0 \tag{19.3-13a, b}$$

because the entire population lives outside the CBD, and no family lives beyond the city limits. In an annular ring in the residential district extending from X to $X + dX$, a fraction $b(X)$ of the land is used for housing, and the remaining land area ($= [1 - b(X)]\, 2\pi X\, dX$) is for roads. We have for that ring*

$$2\pi b(X)\, X\, dX = -\overline{s}\, dN = -\frac{\sigma}{r(X)} [Y - t(X)]\, dN \tag{19.3-14}$$

since the amount of land area occupied by the households in the ring (of area $2\pi X\, dX$) must equal the total amount of land area in the ring allocated for housing. From (19.3-14) and (19.3-13a) we get

$$\begin{aligned} N(X) &= N_0 - 2\pi \int_{R_c}^{X} \frac{b(Z)}{\overline{s}} Z\, dZ \\ &= N_0 - \frac{2\pi}{\sigma} \int_{R_c}^{X} \frac{r(Z)\, b(Z)}{Y - t(Z)} Z\, dZ \end{aligned} \tag{19.3-15}$$

The condition (19.3-13b) gives

$$N_0 = \frac{2\pi}{\sigma} \int_{R_c}^{R_e} \frac{r(Z)\, b(Z)}{Y - t(Z)} Z\, dZ \tag{19.3-16}$$

it determines r_c, which appears as a parameter in $r(X)$. For the utility function

*To be concrete, we will henceforth work with the general class of Cobb-Douglas utility functions (19.3-8).

(19.3-8), the relation (19.3-12) gives

$$r_c = \frac{\sigma Y N_0 / 2\pi}{\displaystyle\int_{R_c}^{R_e} [1 - t(Z)/Y]^{\alpha} b(Z) Z \, dZ}, \qquad \alpha + 1 \equiv \frac{1}{\sigma} \qquad (19.3\text{-}17)$$

At this point, we have completed the formulation of the standard model for residential land uses. The model is sufficiently simple that we have effectively obtained the solution of the relevant mathematical problem in the process of formulating the model. Once τ_d and $b(X)$ are prescribed, it is only a matter of evaluating two integrals to get all the information about the phenomenon contained in this model.

Before leaving it, we note that the rent function $r(X)$ determined from this model (as well as from all subsequent improved models) is the (equilibrium) *market rent*. This is what an individual family would pay for a unit area of housing space. With the equilibrium distribution of market rent, lot size, and composite good consumption, it is now possible to compute the short-term shadow rent distribution by considering what happens when another fraction of land is added for housing at a particular location. The long-term shadow rent distribution can also be obtained with a more elaborate calculation. We will not pursue these economic issues in this article and will refer readers to Refs. 19-24 and 19-25 for a report on these calculations. Instead, we note a rather peculiar feature of this simple model that can be seen from (19.3-17). Suppose the fraction of land allocated for housing is the same at every ring so that $b(X) \equiv b_c$. The model gives a well-defined rent profile for $0 < b_c < 1$ and continues to do so as b_c tends to unity. But with $b(X) \equiv 1$, the entire residential district is allocated for housing, and there is no road to get to and from the CBD! With b_c sufficiently close to (but still less than) unity, gigantic traffic congestions should have sent the rent for lands near the CBD skyrocketing. Evidently, the assumption that only distance (and fixed) costs of transportation are important in calculating the travel cost is inappropriate for heavy-traffic situations. The effect of traffic congestion must be included in the model if it is to give useful insight to an admittedly complicated economic problem.

19.3.4 Congestion Cost of Transportation

The effect of traffic congestion was included in the model for the first time by R. M. Solow (Ref. 19-26), who took the household transportation cost per annum $t(X)$ to consist of two parts: the known distance cost $t_d(X)$ of (19.3-4) and a congestion cost, which depends on the traffic density. Recall that $N(X)$ is the number of households outside the ring of radius X; it also gives the number of commuters crossing that ring each day to get to the CBD. The total road width at the same ring is $2\pi[1 - b(X)]X$; therefore the quantity $N(X)/2\pi[1 - b(X)]X$ is a measure of the traffic density at the ring of radius X. The congestion cost per person-mile (per annum) τ_c of the round-trip travel of the CBD at the ring of radius X must depend on this traffic density measure, i.e., $\tau_c = \tau_c(N(X)/2\pi X[1 - b(X)])$.

To be concrete, we consider in this article only the case where

$$\tau_c = a\left[\frac{N(X)}{2\pi X\{1 - b(X)\}}\right] \tag{19.3-18}$$

where $a \geqslant 0$ is a constant of proportionality. We have for this case

$$\begin{aligned} t(X) &= \int_{R_c}^{X} [\tau_d + \tau_c]\, dX \\ &= t_d(X) + \frac{a}{2\pi}\int_{R_c}^{X} \frac{N(Z)}{Z[1 - b(Z)]}\, dZ \end{aligned} \tag{19.3-19}$$

or in differentiated form

$$\frac{dt}{dX} = \tau_d + \frac{a}{2\pi}\,\frac{N}{X(1 - b)} \qquad (R_c < X < R_e) \tag{19.3-20}$$

again with free transportation inside the CBD so that

$$t(R_c) = 0 \tag{19.3-21}$$

The travel cost now depends on the population distribution described by $N(X)$. We know already from (19.3-15) that $N(X)$ itself depends on $t(X)$ through the location decisions of individual households. This mutual dependence is transparent if we write (19.3-15) in differentiated form:

$$\begin{aligned} \frac{dN}{dX} &= -\frac{2\pi X}{\sigma}\,\frac{r(X)\, b(X)}{Y - t(X)} \qquad (R_c < X < R_e) \\ &= -\frac{2\pi r_c}{\sigma Y} X b(X)\left[1 - \frac{t(X)}{Y}\right]^{\alpha} \end{aligned} \tag{19.3-22}$$

with

$$\alpha + 1 = \frac{1}{\sigma} \tag{19.3-23}$$

For a prescribed $b(X)$, (19.3-20) and (19.3-22) form a pair of coupled first-order ODE for N and t with an unknown parameter r_c. This second-order system and the three auxiliary conditions (19.3-13) and (19.3-21) define a two-point boundary value problem (BVP) in ODE (see Chapter 6).* It determines $N(X)$, $t(X)$, and r_c, completely. Once we have these, we can calculate the remaining economic quantities of interest, e.g., r, $\bar{c}$, $\bar{s}$, and λ, in this model from (19.3-12), (19.3-9), and (19.3-5). In contrast to the simpler model of section (19.3.3), we have completed the model formulation in this case and still have the task of solving a mathematical problem whose solution is to give all the information about the model phenomenon contained in the improved model and, one hopes, some insight to the phenomenon itself.

19.3.5 Dimensionless Formulation and Perturbation Methods

Before we attempt to obtain the solution of the boundary-value problem (BVP) defined by (19.3-20), (19.3-22), (19.3-13), and (19.3-21), we observe, on the basis of the expression for $r\bar{s}$ in (19.3-9), that the constant σ, introduced by the utility function (19.3-8), is the fraction of the household income net transportation cost spent on housing. This fraction is usually less than $\frac{1}{2}$, so that $\alpha + 1 = 1/\sigma > 2$ or $\alpha > 1$. Thus, the BVP for N and t is *nonlinear*, unless the households spend exactly one-half of their after-transportation-cost income on housing. In the United States and Canada, the figure is about $\frac{1}{4}$, so that $\alpha \cong 3$; linearization is therefore not in the cards. Numerical solution of nonlinear BVPs is difficult, and the success of iterative methods for these problems often depends on a good initial guess. It behooves us to first seek a good approximate solution of the problem.

From the form of the BVP, it appears that N and t should depend on X, as well as the parameters R_c, R_e, a, τ_d (taken to be a constant), α (or σ), Y, N_0, and whatever parameters were introduced by the prescribed land allocation function $b(X)$, while the constant r_c depends only on the parameters. The discussion of section 19.2 suggests that we may apply dimensional analysis to simplify the dependence. Instead of working with $t(X, R_c, R_e, a, \tau_d, \alpha, Y, N_0, \ldots)$ etc., we apply the basic idea of dimensional analysis directly to the BVP. Evidently, N and X may be made dimensionless by setting

$$x = \frac{X}{R_c}, \qquad u(x) = \frac{N(X)}{N_0} \tag{19.3-24}$$

Though $t(X)/Y$ is dimensionless, it is customary to work with a dimensionless form of the after-transportation-cost household income

$$w(x) = \frac{1}{Y}\,[Y - t(X)] = 1 - \frac{t(X)}{Y} \tag{19.3-25}$$

*By introducing a new differential equation $dr_c/dX = 0$, the problem becomes a standard BVP for a third-order system of ODE.

With (19.3-24) and (19.3-25), we may write the BVP for t and N as

$$\begin{cases} \dfrac{dw}{dx} = -\epsilon^2\left[\eta + (1-\eta)\dfrac{u}{x(1-b)}\right] & (19.3\text{-}26) \\ & (1 < x < R \equiv R_e/R_c) \\ \dfrac{du}{dx} = -(\alpha+1)\,\nu_c x b w^{\alpha} & (19.3\text{-}27) \end{cases}$$

with

$$w(1) = 1, \quad u(1) = 1, \quad u(R) = 0 \tag{19.3-28}$$

where

$$\epsilon^2\eta \equiv \epsilon_d^2 = \frac{\tau_d R_c}{Y}, \quad \epsilon^2(1-\eta) \equiv \epsilon_c^2 = \frac{aN_0}{2\pi Y} \tag{19.3-29}$$

It is not difficult to check that $\epsilon^2 = \epsilon_d^2 + \epsilon_c^2$, η, ν_c, and R are all dimensionless parameters and that the numerical values of w, u, η, and b never exceed unity and are nonnegative. It follows from (19.3-26)-(19.3-28) that the dimensionless solution $u(x)$, $w(x)$, and ν_c (the dimensionless rent per unit area at the edge of the CBD) depend only on the dimensionless parameters ϵ^2, η, R, α, and whatever dimensionless parameters introduced by b. (For conciseness, we will not use a different symbol for $b(x)$.) Given that time does not appear explicitly in our problem, the conclusion is consistent with the π-theorem.

Now $\epsilon^2 = \epsilon_d^2 + \epsilon_c^2$ is of the order of the fraction of household wage spent on transportation (including both distance and congestion costs) and must be (considerably) smaller than unity. Treating ϵ^2 as a small parameter, we may use the perturbation method of section 14.2 to obtain a regular perturbation solution. With all three unknowns expanded as regular perturbation series in powers of ϵ^2,

$$\{u(x;\epsilon), w(x;\epsilon), \nu_c(\epsilon)\} = \sum_{n=0}^{\infty} \{u_n(x), w_n(x), \nu_n\}\epsilon^{2n} \tag{19.3-30}$$

the nonlinear BVP is reduced to a sequence of *linear* BVPs for the coefficients of the series that admit explicit solutions in terms of elementary functions or quadratures (see Ref. 19-27). We emphasize that it is not meaningful to say something is (numerically) small or large, unless it is a dimensionless quantity; otherwise, a change of measurement units could alter its order of magnitude. As such, a dimensionless formulation of the mathematical problem is essential if a perturbation or asymptotic analysis is contemplated.

We pointed out earlier in this section that housing costs consume about a quarter of the annual household income net travel cost in the United States and Canada, and therefore σ should be about $\frac{1}{4}$. As our rather stark model does not allow for a separate accounting of land and building costs, the choice of $\sigma = \frac{1}{4}$ reflects an implicit assumption that more housing space always begets a bigger and more expensive house. However, a more general model in Ref. 19-27, which allows for separate expenditures on land and building, leads again to the same BVP with $r\bar{s}$ being exclusively the expenditure on land acquisition. In that case, σ is considerably smaller, say $\frac{1}{10}$ or less. With $\alpha = 0(10)$, it is found in Ref. 19-27 that successive terms of the regular perturbation series (19.3-30) contain increasingly higher powers of α. Unless $\alpha\epsilon^2 \ll 1$, the perturbation solution is not at all useful. In any event, the accuracy of the same truncated series for a fixed ϵ^2 decreases as α increases, and a different approximate method of solution is required if α is sufficiently large.

To seek an appropriate simple approximate solution for $\alpha \gg 1$, we observe that $w(x) < 1$ for $x > 1$, as there is always some travel cost for households away from the edge of the CBD. Therefore, we have $w^\alpha \ll 1$ for $x > 1$ if $\alpha \gg 1$, so that the nonlinear term in (19.3-27) is important only when $x \cong 1$ and $u(x)$ has a large gradient of order α there, dropping off abruptly as we move away from the CBD. These observations suggest that there is a layer phenomenon adjacent to $x = 1$, and the method of matched asymptotic expansions (see section 14.5) may be used to obtain an approximate solution for $\alpha \gg 1$.

With $w^\alpha \ll 1$ away from the CBD for $\alpha \gg 1$ and therefore negligible there, a leading term *outer solution* is straightforward. It should be noted that there is no reduction of the order of the system associated with the outer solution, which can be made to satisfy the only boundary condition at $x = R$, namely, $u(R) = 0$. The remaining constant of integration will be needed for later matching with the *inner solution*. For the inner solution near $x = 1$, the structure of (19.3-27) suggests that the stretched variable $Z = \alpha(x - 1)$ is appropriate. Also, we have $w(1) = 1$ and $w(x) \lesssim 1$ for $0 < Z = 0(1)$ however large α may be. This suggests that we write the inner solution for w as $1 - \hat{w}(Z)/\alpha$ with $w^\alpha = (1 - \hat{w}/\alpha)^\alpha$ tending to $e^{-\hat{w}(Z)}$ as $\alpha \to \infty$. We get in this way a leading term inner solution for w and u (in terms of elementary functions) that satisfies the two boundary conditions $u(x = 1) = w(x = 1) = 1$. The unknown parameter ν_c and the unknown constant of integration in the leading term outer solution are then determined by matching of the inner and outer solutions in some intermediate region where both solutions are valid.

Without getting into details, which can be found in Ref. 19-27, we have tried to indicate how the regular perturbation method and the method of matched asymptotic expansions also can be useful in social science and how a formulation of the relevant mathematical problems in terms of dimensionless variables and parameters help us to see the applicability of these methods. When a truncated perturbation or asymptotic solution is not sufficiently accurate, it may still be useful as an initial guess for an iterative numerical solution scheme, such as those described in section 18.6 for BVP (see Ref. 19-27).

A word of caution is appropriate at this point. The selection of the scaled (dimensionless) variables (19.3-24) and (19.3-25) is rather straightforward for this problem. Such is not always the case, as shown by a simple textbook example in Ref. 19-5 and by many research problems in journal articles including one to be discussed in section 19.5. Often, experience in modeling and, more importantly, an insight into the modeled phenomenon are the only valuable guides to the correct choice of scaling.

19.3.6 An Expanding Metropolis and Public Ownership

In this and the next two sections, Solow's equilibrium model for residential land allocations will be modified and made more elaborate to include other effects of interest. In this subsection, we show how two realistic economic features may be incorporated into the basic equilibrium model. In section 19.3.7, we will discuss how a city planner may choose the land allocation function $b(x)$ to optimize some benefit measure. In section 19.3.8, we will show how we may remove the unrealistic assumption of a homogeneous population.

19.3.6.1 A Free-Boundary Problem Some cities have natural boundaries, such as mountains or seas; others have well-defined political boundaries fixed by legislative or legal actions. However, a metropolis that has a fixed population but is free to expand geographically reaches its equilibrium size when the land value at the outer edge of the annular residential district is equal to the *agricultural rent* r_A (dollars/mile^2), i.e., the worth of the land as farmland, so that

$$r(X = R_e) = r_A \qquad \text{or} \qquad \nu_c w^{\alpha+1}(R;\epsilon) = \nu_A \tag{19.3-31}$$

where $\nu_A \equiv 2\pi R_c^2 r_A / YN_0$. For such a city, the dimensionless outer boundary is unknown and must be determined as a part of the solution by (19.3-31), giving us a free-boundary problem in ODE.

For moderate values of α, so that a regular perturbation solution is appropriate, the expansions (19.3-30) must now be supplemented by an expansion for R, which depends on ϵ, as all other unknowns of the problem:

$$R(\epsilon) = \sum_{n=0}^{\infty} R_n \epsilon^{2n} \tag{19.3-32}$$

With the relation

$$w(R(\epsilon);\epsilon) = w\Big|_{\epsilon=0} + \epsilon \frac{dw}{d\epsilon}\Big|_{\epsilon=0} + \frac{1}{2}\epsilon^2 \frac{d^2w}{d\epsilon^2}\Big|_{\epsilon=0} + \cdots \tag{19.3-33}$$

where

$$
\begin{aligned}
w\Big|_{\epsilon=0} &= w(R(0);0) = w_0(R_0)\\
\frac{dw}{d\epsilon}\Big|_{\epsilon=0} &= \left[\frac{\partial w}{\partial R}\frac{dR}{d\epsilon} + \frac{\partial w}{\partial \epsilon}\right]_{\epsilon=0} = w_1(R_0) + R_1 w_0'(R_0)\\
\frac{d^2w}{d\epsilon^2}\Big|_{\epsilon=0} &= \left[\frac{\partial^2 w}{\partial R^2}\left(\frac{dR}{d\epsilon}\right)^2 + 2\frac{\partial^2 w}{\partial\epsilon\partial R}\frac{dR}{d\epsilon} + \frac{\partial w}{\partial R}\frac{d^2R}{d\epsilon^2} + \frac{\partial^2 w}{\partial \epsilon^2}\right]_{\epsilon=0}\\
&= 2\left[\frac{1}{2}R_1^2 w_0''(R_0) + R_1 w_1'(R_0) + R_2 w_0'(R_0) + w_2(R_0)\right]
\end{aligned}
$$

(19.3-34)

etc., the condition (19.3-31) gives a sequence of conditions in which $R_0, R_1, R_2, \ldots$, etc., appear explicitly as multiplicative factors. This sequence supplements those previously obtained for the fixed-boundary problem (but now with $u(R(\epsilon);\epsilon)$ also expanded in the form (19.3-33) and (19.3-34)) and is exactly what is needed for determining $R_0, R_1, R_2, \ldots$, etc. Actual solutions, including numerical results obtained in this way, are reported in Ref. 19-27.*

19.3.6.2 Public Ownership Up to now, our model does not account for the rent payments annually collected from the households; effectively, we have an absentee landlord who takes the total payment out of the city and spends it somewhere else. In reality, some portion of that total payment (a portion collected as property tax for example) is returned to the households in the form of services, subsidies, and grants. We consider here the situation where the rent payments are collected by the city and a fraction $\beta(\leqslant 1)$ of the total payment is redistributed in equal shares as cash dividends to all families in the city; the $\beta = 1$ case corresponds to *public ownership*. We see from (19.3-14) that the actual amount D of a divident share is given by

$$D = -\frac{\beta}{N_0}\int_1^R r(x)\,s(x)\,dN \qquad (19.3\text{-}35)$$

(as the population increment is negative for a postive dx). This dividend must be added to the household income, so that, in equilibrium, Y in (19.3-1) should be replaced by $y \equiv Y + D \equiv Y(1 + \Delta)$ with $\Delta = D/Y$. Correspondingly, we have for the

***There is a misprint in $dw/d\epsilon$ at $\epsilon = 0$ in Ref. 19-27, where R_1 in the second term on the right is incorrectly printed as R_0. But the numerical results of Tables 2 and 3 are correct.**

Cobb-Douglas-type utility functions the relations

$$\bar{c} = \frac{1-\sigma}{p}(y-t), \qquad \bar{s} = \frac{\sigma}{r}(y-t)$$
$$r^{\sigma} = r_c^{\sigma}\left(1 - \frac{t}{y}\right) \equiv r_c^{\sigma}\hat{w}(x), \qquad \hat{w} \equiv 1 - \frac{t}{y} \tag{19.3-36}$$

If τ_d is a constant, the dimensionless BVP (19.3-26), (19.3-27), and (19.3-28) for the absentee landlord model is now replaced by

$$\left\{ \begin{aligned} \frac{d\hat{w}}{dx} &= -\frac{\epsilon^2}{1+\Delta}\left[\eta + (1-\eta)\frac{u}{x(1-b)}\right] && (19.3\text{-}37) \\ & && (1 < x < R) \\ \frac{du}{dx} &= -\frac{(\alpha+1)}{1+\Delta}\nu_c x b(x)\hat{w}^{\alpha} && (19.3\text{-}38) \end{aligned} \right.$$

with

$$\hat{w}(1) = 1, \quad u(1) = 1, \quad u(R) = 0 \tag{19.3-39}$$

In this new BVP, the parameters ν_c, ϵ, and η are as previously defined in (19.3-29), while the new dimensionless dividend parameter Δ is D/Y. The dimensionless outer city boundary R is either given or determined by the agricultural rent through (19.3-31).

As D (and hence Δ) is defined in terms of r, $\bar{s}$, and N by (19.3-35), it must be determined as a part of the solution. With (19.3-35) written as

$$\Delta = \beta\nu_c \int_1^R \hat{w}^{\alpha+1}(x)\, b(x) x \, dx \tag{19.3-40}$$

we may, in principle, solve (for the prescribed R case) the two first-order ODE (19.3-37) and (19.3-38) along with the first two initial conditions of (19.3-39) as an initial-value problem (IVP) with ν_c and Δ as parameters. The last condition of (19.3-39) and the integrated condition (19.3-40) are then used to determine ν_c and Δ. This procedure has been implemented numerically in Ref. 19-24 and elsewhere.

19.3.7 Cost-Benefit and Second-Best Allocation

Solow's pioneering work and subsequent developments (see Ref. 19-28 and other research articles cited therein) have consistently shown that residential lands near the CBD are undervalued by the market land price, because private individuals (or households) do not pay for the social costs generated by traffic congestions. An optimal, i.e., economically efficient, allocation of residential land, through a particular choice of the allocation function, $b_{OP}(x)$, can be achieved by imposing corrective commuter tolls (not the same as tolls for paying off construction costs). But such tolls are extremely unpopular and have been levied in only a few places in the United States and Canada. It is considered politically suicidal for any local government to institute commuter tolls, not to mention the possible congestions they may create. Nevertheless, analyses such as Ref. 19-28, delineating the differences between the optimal land allocation $b_{OP}(x)$ and the allocation $b_{CB}(x)$ by a cost-benefit criterion based on market land price (henceforth called the *cost-benefit allocation*) for general utility functions, are of some interest.

Given that congestion tolls are not levied, it is more important to know whether or not the cost-benefit allocation leads to the highest possible common household utility. Therefore, there is considerably more interest in the so-called *second-best* allocation $b_{SB}(x)$, the allocation of residential land that gives the highest utility U under the constraint of no congestion tolls. The research progress in this direction is limited because the mathematical problem associated with the determination of the second-best allocation is more complex. Until recently, only some partial results and numerical solutions have been obtained for this problem. In this section, we will formulate the mathematical model for equilibrium land allocations by the cost-benefit criterion and by the second-best criterion.

19.3.7.1 The Cost-Benefit Allocation Suppose land allocation for roads and housing is made by a cost-benefit consideration based on the market land price. An additional (marginal) unit area of land used for roads would reduce the total transportation cost; it also reduces the total revenue from land rents. Land will continue to be taken away from housing for roads when the benefit from reduced transportation cost exceeds the cost of lost land revenue. The allocation process reaches an equilibrium when the marginal benefit from more land for roads equals the marginal cost of land lost to residential uses priced at the market rent. For the congestion cost function (19.3-18), this leads to the condition

$$a\left[\frac{N}{2\pi X(1-b)}\right]=r \tag{19.3-41}$$

The cost-equals-benefit condition (19.3-41) determines the equilibrium cost-benefit allocation distribution b_{CB}. For Solow's absentee ownership model, outlined in section 19.3.4 and 19.3.5, we have

$$b_{CB}=1-\sqrt{\frac{a}{r}\,\frac{N}{2\pi X}}=1-\sqrt{\frac{\epsilon^2(1-\eta)}{v_c w^{\alpha+1}}\,\frac{u}{x}} \tag{19.3-42}$$

With (19.3-42), we essentially have the cost-benefit residential land allocation model independently formulated by A. J. Robson (Ref. 19-29) and Y. Kanemoto (Ref.

19-28). In principle, the process of determining the equilibrium residential land values and uses from this model consists of using (19.3-42) to eliminate $b(x)$ from (19.3-26) and (19.3-27) and then solving the resulting BVP for v_c, $u(x)$, and $w(x)$ (or $\hat{w}(x)$ and Δ), as well as R if the city boundary is not prescribed. In actual fact, the one condition (19.3-42) eliminates both u and b from (19.3-26), leaving a first-order separable ODE for $w(x)$. An exact solution of the BVP in terms of quadratures is therefore immediate (Refs. 19-28 and 19-29). Accurate approximate perturbation solutions have also been obtained in Refs. 19-30 and 19-31. For later references, we give here a two-term perturbation solution for the cost-benefit allocation function in the fixed boundary and absenteee landlord case:

$$b_{\text{CB}}(x;\epsilon) = 1 - \epsilon\left[\frac{R^2 - x^2}{x}\sqrt{\frac{(\alpha+1)(1-\eta)}{2(R^2-1)}}\right] + 0(\epsilon^2) \qquad (19.3\text{-}43)$$

Note that all perturbation series are now in powers of ϵ instead of ϵ^2, as in section 19.3.5 where $b(x)$ is prescribed.

It is also worth mentioning that the relevant BVP for the optimal allocation (when congestion tolls are levied) has the same mathematical structure as the one for the cost-benefit allocation and therefore also admits an exact solution.

19.3.7.2 The Second-Best Solution Though the mathematical problem associated with the second-best allocation model is much more difficult, the formulation of the model is just as straightforward as the cost-benefit allocation model. Recall from (19.3-11) that the common household utility $\overline{U}$ for the absentee landlord case is a function of the unit area land rent at the edge of CBD, r_c. But the development of subsections 19.3.4 and 19.3.5 shows that the value of r_c depends on $b(x)$. We may therefore use $b(x)$ as a policy control variable to maximize $\overline{U}$ (see also Ref. 19-28). The choice of b, which gives a maximum $\overline{U}$, is called the second-best allocation and is denoted by b_{SB}. The dependence of r_c on b is through the BVP (19.3-26)–(19.3-28) for a city with fixed boundaries; the condition (19.3-26)-(19.3-28) assume the role of equality constraints in the optimization problem. In addition, there are also the inequality constraints

$$0 \leqslant b, w, u \leqslant 1 \qquad (19.3\text{-}44)$$

which follow from the definition of the three quantities.

A systematic procedure to deduce the allocation function for the above second-best allocation model follows from the Pontryagin maximum principle applied to the optimal control problem in this model (Ref. 19-30). This principle (see sections 20.3.7 and 20.3.3) requires that the second-best allocation $b_{\text{SB}}(x)$ (for the class of Cobb-Douglas-type utility functions (19.3-8)) be determined in terms of v_c, $w(x)$, $u(x)$, and two auxiliary quantities (adjoint variables) $\phi(x)$ and $\psi(x)$ by

$$v_c x w^\alpha \psi - \epsilon^2(1-\eta)\frac{\phi u}{x(1-b)^2} = 0 \qquad (19.3\text{-}45)$$

The four functions $\phi(x)$, $\psi(x)$, $w(x)$, and $u(x)$ must be the unique solution of the BVP defined by the fourth-order system of ODE, consisting of

$$\phi' = -\alpha \nu_c x b \psi w^{\alpha-1} \tag{19.3-46}$$

$$\psi' = -\epsilon^2(1-\eta)\frac{\phi}{x(1-b)} \tag{19.3-47}$$

along with (19.3-26) and (19.3-27) and the four boundary conditions (19.3-28) and

$$\psi(R) = 0 \tag{19.3-48}$$

The remaining unknown parameter ν_c is then determined by the integral condition

$$\frac{\mu(\nu_c)}{\nu_c} = (\alpha+1)\int_1^R b\psi w^\alpha x\,dx \tag{19.3-49}$$

where

$$\mu(\nu_c) \equiv \frac{\sigma^\sigma(1-\sigma)^{1-\sigma}Y^2}{2\pi r_c^{\sigma+1}R_c^2}\,\overline{U}^{\cdot}(\sigma^\sigma(1-\sigma)^{1-\sigma}Y/r_c^\sigma) \tag{19.3-50}$$

In contrast to the cost-benefit allocation case, an exact solution is not available in terms of elementary functions or quadratures. However, accurate truncated perturbation solutions have been obtained in Refs. 19-30 and 19-31 and confirmed by accurate numerical solutions (with prescribed error tolerances) also reported there. It is clear from the definition of the second-best allocation that we must have $\overline{U}(b = b_{CB}) \leqslant \overline{U}(b = b_{SB})$. A significant result from the perturbation solution is the profile of $b_{SB}(x)$ itself. A two-term perturbation solution (for absentee ownership and prescribed boundaries) is

$$b_{SB}(x;\epsilon) = 1 - \epsilon\left[\frac{R^2 - x^2}{x}\sqrt{\frac{\alpha(1-\eta)}{2(R^2-1)}}\right] + 0(\epsilon^2) \tag{19.3-51}$$

Together with (19.3-43), we have

$$\frac{1 - b_{SB}(x)}{1 - b_{CB}(x)} = \sqrt{\frac{\alpha}{\alpha+1}}\,[1 + 0(\epsilon)] \tag{19.3-52}$$

The ratio (19.3-52) indicates that, except for higher order terms in ϵ (which is small compared to unity), the cost-benefit criterion based on market land price allocates more land for roads than the second-best allocation, resulting in no higher (and usually lower) common household utility!

19.3.8 An Inhomogeneous Population

Alonso's landmark development recognized for the first time the significance of both location and size of housing space in economic models of residential land uses. The introduction of congestion cost (by Solow) into Alonso's model leads to the standard basic theory of the subject. More general descriptive and normative models formulated more recently have introduced more realistic features (including an open city with migration, etc.) into the basic model and allowed for interaction with other aspects of urban economics. Most of these generalizations and extensions can be found in Kanemoto's monograph (Ref. 19-28). The few extensions of the basic model already described are more than enough to illustrate the theme of this section. However, with the highly unrealistic restriction of a homogeneous population, models for residential land economics cannot and will not be taken seriously. The results from any of these models will not be meaningful unless this restriction is removed. As the last item of this section, we will remove this restriction.

Within the framework of the Alonso-Solow model, households can be different from each other through their taste and preference or through their annual wage. For simplicity, we treat these two items separately, each for a two-class population. First, we consider a population of households with the same utility function, but with some households getting a high wage Y_h and others getting the lower wage Y_l. It will be evident that the case of identical income but two different utility functions can be similarly treated. In fact, the analysis can be extended to an inhomogeneous population of multiple income classes and with multiple household utility preferences.

19.3.8.1 Two Income Classes Consider a population of two income classes with identical tastes, N_l households each with an annual wage Y_l and N_h households each with an annual wage $Y_h > Y_l$. We are interested here in the equilibrium population density pattern and the rent profile for a given $b(x)$. Within each class, the equilibrium analysis is as before. In addition, we will show that the entire low-income class will live in a ring adjacent to the CBD and the high-income class will occupy the outer ring with more housing space per household. The remaining new task will be to find the boundary separating the two residential zones. This is done concurrently with the determination of other economic variables in the model.

The two income classes do not mix; otherwise, we have in equilibrium

$$U\left(\frac{(1-\sigma)^{1-\sigma}\sigma^{\sigma}(Y_j - t)}{p^{1-\sigma}r^{\sigma}}\right) = U\left(\frac{(1-\sigma)^{1-\sigma}\sigma^{\sigma}Y_j}{p^{1-\sigma}r_c^{\sigma}}\right) \tag{19.3-53}$$

from (19.3-10) (for a Cobb-Douglas utility function) for both $j = l$ and $j = h$. They give different profiles for the same equilibrium unit area rent distribution $r(x)$! Similar arguments also rule out a multizone pattern. Therefore, the two income classes live in two different zones with a single circular boundary between the zones at $X = R_i$.

Within each zone, previous analyses for a single homogeneous population apply, giving a smoothly decreasing rent profile (convex to the origin) in each zone and a previously not used identity

$$r'\bar{s}^{(j)} + t' = 0 \qquad (j = l, h) \tag{19.3-54}$$

The rent profile must also be continuous across the boundary at R_i between the two zones. Otherwise, those living on the high-rent side of a jump discontinuity would move across to achieve a higher utility, for the transportation cost is continuous across the boundary. However, the rent profile cannot be smooth across the boundary; in fact it must be steeper on the side closer to the CBD. If this were not so, each class could do no worse (and much better if it is steeper the other way) by moving into the other's zone.

Finally, the steeper side of the rent profile at R_i must belong to the low-income class. To see this, we recall that $r\bar{s}^{(j)} = \sigma(Y_j - t)$, so that, at R_i, the rich spend more on housing. Now, if r is continuous across R_i, it follows that $\bar{s}^{(h)} > \bar{s}^{(l)}$ and then, from (19.3-54), that the rent gradient on the side of the poor must be steeper. Hence, the poor stay in an inner zone close to the CBD, while *the rich stay out in the suburbs*.

For the poor in the inner zone, $R_c \leqslant X \leqslant R_i$, we have for the absentee landlord case,

$$\left.\begin{aligned} r = r_c\left(1 - \frac{t}{Y_l}\right)^{\alpha+1} &\equiv r_c w_l^{\alpha+1} \\ r\bar{s}^{(l)} = \sigma Y_l w_l, \qquad \bar{c}^{(l)} &= (1 - \sigma) Y_l w_l \end{aligned}\right\} \tag{19.3-55}$$

$$\frac{dw_l}{dX} = -\frac{\tau_d}{Y_l} - \frac{aN}{2\pi Y_l X(1 - b)}, \qquad \frac{dN}{dX} = -\frac{2\pi r_c}{\sigma Y_l} X b w_l^{\alpha} \tag{19.3-56}$$

with

$$\left.\begin{aligned} w_l(R_c) = 1, \qquad w_l(R_i) = 1 - \frac{t_i}{Y_l}, \qquad t_i \equiv t(R_i) \\ N(R_c) = N_l + N_h \equiv N_0, \qquad N(R_i) = N_h \end{aligned}\right\} \tag{19.3-57}$$

For the rich in the suburbs, $R_i \leqslant X \leqslant R_e$, we have

$$\left.\begin{aligned} r = r_i\left(\frac{Y_h - t}{Y_h - t_i}\right)^{\alpha+1} &\equiv r_i w_h^{\alpha+1}, \\ r_i \equiv r(R_i) = r_c\left(1 - \frac{t_i}{Y_l}\right)^{\alpha+1} & \\ r\bar{s}^{(h)} = \sigma(Y_h - t_i) w_h, \qquad \bar{c}^{(h)} &= (1 - \sigma)(Y_h - t_i) w_h \end{aligned}\right\} \tag{19.3-58}$$

$$\frac{dw_h}{dX} = -\frac{\tau_d}{Y_h - t_i} - \frac{aN}{2\pi(Y_h - t_i)X(1 - b)}, \quad \frac{dN}{dX} = -\frac{2\pi r_i}{\sigma(Y_h - t_i)} Xbw_h^{\alpha} \tag{19.3-59}$$

with

$$w_h(R_i) = 1, \quad N(R_i) = N_h, \quad N(R_e) = 0 \tag{19.3-60}$$

The analysis leading to these results is the same as before, supplemented by some continuity considerations. Conceptually, the BVP (19.3-59) and (19.3-60), with r_i expressed in terms of r_c and $t_i \equiv t(R_i)$ by (19.3-58), determines w_h, N, and t_i in $R_i \leqslant X \leqslant R_e$ with r_c and R_i as parameters. The two ODEs (19.3-56) and the four boundary conditions in (19.3-57) determine w_l and N in $R_c \leqslant X \leqslant R_i$ and the two parameters r_c and R_i. Perturbation, asymptotic, and numerical methods may be used to obtain the solution of this and other related problems, much as before.

19.3.8.2 Two Utility Classes The analysis of a population of two classes of households with different Cobb-Douglas utility functions is nearly the same as that for two income classes. Consider N_m households with $\sigma = \sigma_m$ and N_l households with $\sigma = \sigma_l < \sigma_m$. In equilibrium, the entire class of households that would spend a bigger fraction σ_m of their after-travel-cost income on housing will live in an outer ring of the residential district, while those who prefer to spend less will live in an inner ring adjacent to the CBD. They do not mix; otherwise the two different common household utility values would give different rent profiles, which is unacceptable. By the same argument, the rent profile is continuous across the interzone boundary at R_i, with a steeper gradient on the side of the boundary near the CBD. The steeper side of the rent profile at R_i is occupied by households with $\sigma = \sigma_l$, now by way of $r\bar{s}^{(j)} = \sigma_j(Y - t)$, $j = l, m$. Hence, those who would spend more after-travel income on housing stay in the suburbs; the others stay in an inner zone adjacent to the CBD.

For those in the inner zone, $R_c \leqslant X \leqslant R_i$, we have for the absentee landlord case,

$$\left.\begin{aligned} r = r_c\left(1 - \frac{t}{Y}\right)^{\alpha_l+1} &\equiv r_c w_l^{\alpha_l+1} \\ r\bar{s}^{(l)} = \sigma_l Y w_l, \quad \bar{c}^{(l)} &= (1 - \sigma_l) Y w_l \end{aligned}\right\} \tag{19.3-61}$$

$$\frac{dw_l}{dX} = -\frac{\tau_d}{Y} - \frac{aN}{2\pi YX(1 - b)}, \quad \frac{dN}{dX} = -\frac{2\pi r_c}{\sigma_l Y} Xbw_l^{\alpha_l} \tag{19.3-62}$$

with

$$\left.\begin{aligned} w_l(R_c) = 1, \quad w_l(R_i) = 1 - \frac{t_i}{Y}, \quad t_i \equiv t(R_i) \\ N(R_c) = N_l + N_m \equiv N_0, \quad N(R_i) = N_m \end{aligned}\right\} \tag{19.3-63}$$

For those who live in the outer ring, $R_i \leqslant X \leqslant R_e$, we have

$$\left.\begin{aligned} r &= r_i\left(\frac{Y-t}{Y-t_i}\right)^{\alpha_m+1} \equiv r_i w_m^{\alpha_m+1}, \\ r_i &\equiv r(R_i) = r_c\left(1-\frac{t_i}{Y}\right)^{\alpha_l+1} \\ r\overline{s}^{(m)} &= \sigma_m(Y-t_i)w_m, \quad \overline{c}^{(m)} = (1-\sigma_m)(Y-t_i)w_m \end{aligned}\right\} \tag{19.3-64}$$

$$\frac{dw_m}{dX} = -\frac{\tau_d}{Y-t_i} - \frac{aN}{2\pi(Y-t_i)X(1-b)}, \quad \frac{dN}{dX} = -\frac{2\pi r_i}{\sigma_m(Y-t_i)} Xbw_m^{\alpha_m} \tag{19.3-65}$$

with

$$w_m(R_i) = 1, \quad N(R_i) = N_m, \quad N(R_e) = 0 \tag{19.3-66}$$

Perturbation, asymptotic, and numerical solutions of this problem are again feasible.

With the two special cases treated above, it should be clear how we could proceed with more general situations. At the same time, the results for the two special cases suggest that the interaction between differential utility and differential income among households would be rather complex in general. Hence, the insight into their effect on population density pattern and rent distribution gained from the study of the two limiting situations above is invaluable, confirming once more the merits of going "from the simple to the elaborate."

19.4 TRY A DIFFERENT FORMULATION

19.4.1 When to Cut a Tree?

Forestry is a major industry in the Pacific Northwest, and proper forest management is of major concern to government and to the forest industry, albeit for different reasons. With their profit motive, private logging companies are keenly interested in the economics, more specifically the financial return, of their operations. In this section, we will describe some results on a long-standing problem in forestry economics to illustrate an important working principle in mathematical modeling. When faced with a difficult mathematical problem from a mathematical model, it is prudent to be flexible. A different formulation may lead to a new and solvable problem.

To an economist, a forest of trees is a stock of capital that increases in value with tree age. For example, a typical stand of 110-year-old British Columbia Douglas fir was worth \$1,000 (after harvesting cost) for timber production in 1967, while a 30-year-old stand had no net commercial value (Ref. 19-32). With the average (or marginal) yield rate and relative growth rate of the trees diminishing with tree age A and with a positive real discount rate (net inflation) for future income, the "when to cut a tree" question has long been a fundamental problem in forestry economics. Given the net commercial value V of a tree (before the onset of biological decay) as a monotone increasing concave function of tree age, and with a constant real interest rate δ for discounting future revenue, the optimal cutting time of the tree T_{IF} is easily determined by the condition that $e^{-\delta T} V(A)$ attains its maximum value at $T = T_{IF}$. With $T = A + T_0$ where $T_0 \leqslant 0$ is the germination time of the tree, the condition yields the so-called Fisher's rule (Refs. 19-33–19-35)

$$\frac{V'(A_{IF})}{V(A_{IF})} = \delta, \qquad T_{IF} = A_{IF} + T_0 \tag{19.4-1}$$

The optimal cutting time is when the relative yield rate equals the discount rate.

Fisher's rule applies only to a single cutting (of a "once-and-for-all forest"); it must be modified to account for the opportunity cost of not logging sooner for an "ongoing forest," where replanting is assumed to be done immediately after cutting. If all biological and economical factors remain unchanged over time, the optimal cutting age A_{MF}, for trees in an ongoing forest is the so-called *Faustmann rotation*, determined by the condition (Refs. 19-34–19-37):

$$\frac{V'(A_{MF})}{V(A_{MF})} = \frac{\delta}{1 - e^{-\delta A_{MF}}} \tag{19.4-2}$$

The Faustmann formula (19.4-2) is obtained by choosing a sequence of cutting times $\{T_1, T_2, \ldots\}$ to maximize the present value of all future net revenues (Ref. 19-33)

$$P \equiv e^{-\delta T_1} V(A_1) + e^{-\delta T_2} V(A_2) + \cdots = \sum_{k=1}^{\infty} e^{-\delta T_k} V(A_k) \tag{19.4-3}$$

where $A_k = T_k - T_{k-1}$, with T_0 being the (uniform) germination time of the initial forest. An alternate maximum rent formulation, which leads to the same result, is given by P. Samuelson in Ref. 19-35, where the effect of labor as an input to production, as well as nonsteady state considerations, can also be found. The effect of forest-thinning is discussed by C. W. Clark in Ref. 19-33.

In this article, we are mainly concerned with recent attempts to improve upon Fisher's result for a once-and-for-all forest and Faustmann's result for an ongoing forest in two directions. When applied to an entire forest, it is implicit in these two results that harvesting of the whole forest or any part of it can be instantaneous. In other words, there is sufficient equipment and a sufficiently large labor force to log the entire forest in a relatively short time interval compared with the time scale for a significant loss in interest on the delayed profit. For a large forest, this assumption is not realistic, and an upper limit to the available harvesting effort and therefore the harvesting rate should be imposed on the optimal solution.

In arriving at Fisher's and Faustmann's results, harvesting costs have either been ignored or taken in a form so that $V(A)$ could be regarded as the net commercial value or profit. However, average harvesting cost (per tree say) does vary substantially with the rate of harvest. At low harvesting rates, fixed cost facilities for cutting and transportation are not used to their full capacity, giving a high unit harvesting cost. At high harvesting rates, available facilities and manpower are used beyond their normal capacity, resulting in more frequent equipment breakdown and overtime pay. Therefore, as a function of the harvesting rate, the unit harvesting cost is U-shaped and cannot be absorbed in $V(A)$.

The effect of an upper bound on harvesting rate and a U-shaped dependence of unit harvesting cost on harvesting rate have been investigated in several articles (Refs. 19-38–19-40). With regard to the effect of a rate-dependent unit harvesting cost, only some partial results have been obtained, principally because the mathematical problems associated with these improved models are too difficult to yield a complete solution. With regard to the effect of an upper bound on harvesting rate, the mathematical problems associated with the improved models are also difficult, but it is possible to deduce their solution by arguments based on insight. In the next section, we indicate the nature of the difficulties with a brief summary of the models developed by Heap and Neher (Ref. 19-38), which are natural extensions of the conventional formulation. In subsequent developments, we show how a different approach to the same problem, which allows for an additional degree of realism (namely, trees are cut in a certain order imposed by laws and government regulations or by economic considerations), leads to parallel models that are completely tractable mathematically. As such, the results serve as an illustration for another good working principle in mathematical modeling: *when the going gets tough with one model, try a different formulation.*

19.4.2 Conventional Formulation with Two New Effects

19.4.2.1 The Once-and-for-All Forest When there is an upper bound to the harvesting rate $h(t)$, say $h(t) \leqslant m$, the harvesting of an entire forest (or any part of it) has to be spread over a finite time interval, say $t_s \leqslant t \leqslant t_e$. In that case, the present value of the discounted future profit is given by

$$P \equiv \int_{t_s}^{t_e} [p(t) - c(h)]\, h e^{-\delta t}\, dt \qquad (19.4\text{-}4)$$

where p is the price function for unit harvest, and c is the rate-dependent harvesting cost for unit harvest. (In the conventional formulation, c does not depend explicitly on t for simplicity.) If the unit cost function c is independent of h, then $p(t) - c$ is identical to the net commercial value of a unit harvest $V(A)$. Again, δ is the real interest rate (apart from inflation, and assumed to be constant over time). The logging company is to choose a starting time t_s and a harvesting rate $h(t)$ (which determines the completion time t_e) to maximize P subject to

(i) $$0 \leqslant h(t) \leqslant m \tag{19.4-5}$$

(ii) $$\int_{t_s}^{t_e} h(t)\,dt = F_0 \tag{19.4-6}$$

The second condition is simply a statement of the fact that the logging of the entire forest is completed by t_e.

To put the problem in the conventional form of an optimal control problem, we define an accumulated harvest function $F(t)$ by

$$\frac{dF}{dt} = h(t), \qquad (t_s \leqslant t) \tag{19.4-7}$$

Evidently, we have

$$F(t_s) = 0, \qquad F(t_e) = F_0 \tag{19.4-8a, b}$$

The methods of dynamic optimization described in subsections 20.3.3 and 20.3.4 (suitably modified to cover the case of a free starting time) may now be applied to obtain the conditions that determine the optimal harvest policy $\{\bar{t}_s, \bar{h}(t)\}$ (see Ref. 19-38). We merely wish to point out that the solution of the relevant free-end-point BVP defined by these conditions is not at all straightforward, though it is made less complicated in Ref. 19-38 by a separate treatment of the two effects and by the assumption of a uniform distribution of tree ages.

19.4.2.2 The Ongoing Forest For a repeatedly harvested forest, it is natural to attempt an extension of the formulation that yields the Faustmann rotation. This is the approach used in Ref. 19-38 in which the present value of the total future net profit P is maximized with

$$P = \sum_{n=1}^{\infty} \int_{t_{sn}}^{t_{en}} [p(A_n(t)) - c(h_n)]\, h_n e^{-\delta t}\, dt \tag{19.4-9}$$

In this expression, (t_{sn}, t_{en}) is the time interval of the nth harvest with harvesting rate $h_n(t)$, and $A_n(t)$ is the age of the trees being cut at time t during the nth harvest. Evidently, (19.4-9) is a straightforward extension of (19.4-3) and (19.4-4). The logging company is to choose a sequence of starting times $\{t_{sn}\}$ and a sequence of harvesting rates $\{h_n(t)\}$ to maximize P subject to the constraints

(i) $$0 \leqslant h_n(t) \leqslant m \qquad (n = 1, 2, \ldots) \tag{19.4-10}$$

(ii) $$\int_{t_{sn}}^{t_{en}} h_n(t)\,dt = F_0 \qquad (n = 1, 2, \ldots) \tag{19.4-11}$$

In addition, it is important to note that the tree age function A_n is not an independent quantity. The portion of the forest cut during the nth harvest from t_{sn} to t must be the same as that portion cut during the $(n-1)$th harvest from $t_{s(n-1)}$ to $t - A_n(t)$. We have, therefore the additional set of constraints

$$\text{(iii)} \qquad \int_{t_{s(n-1)}}^{t-A_n(t)} h_{n-1}(Z)\,dZ = \int_{t_{sn}}^{t} h_n(Z)\,dZ, \qquad (n = 2, 3, \ldots) \qquad (19.4\text{-}12)$$

For the application of the methods of dynamic optimization, we again transform (19.4-11) and (19.4-12) into a more suitable form. As in a single-harvest case, we introduce a sequence of accumulated harvest functions $\{F_n(t)\}$ by

$$\frac{dF_n}{dt} = h_n(t), \qquad F_n(t_{sn}) = 0, \qquad F_n(t_{en}) = F_0 \qquad (19.4\text{-}13)$$

for $n = 1, 2, 3, \ldots$. The conditions (19.4-13) for $\{F_n\}$ replace (19.4-11). To get a local form of (19.4-12), we differentiate both sides of that expression with respect to t to get (after some rearrangement)

$$\frac{dA_n}{dt} = \frac{h_{n-1}(t - A_n) - h_n(t)}{h_{n-1}(t - A_n)}, \qquad (n = 2, 3, \ldots) \qquad (19.4\text{-}14a)$$

with

$$A_n(t_{sn}) = t_{sn} - t_{s(n-1)}, \qquad A_n(t_{en}) = t_{en} - t_{e(n-1)} \qquad (n = 2, 3, \ldots)$$
$$(19.4\text{-}14b)$$

Unfortunately, the conventional maximum principle in optimal control theory does not apply to problems with state equations in the form of functional-differential equations, where state variables appear as delays in the argument of h_k.

By an ad hoc argument suggested by experience and insight, the determination of the optimal harvest policy for the case of rate-independent unit harvest costs (so that $c_n(h)$, $n = 1, 2, 3, \ldots$, are constant) has been reduced (Ref. 10-38) to an equivalent once-and-for-all forest problem, so that the results already obtained for that case apply immediately. For more general situations, only some partial results are obtained in Ref. 19-38, as well as in more recent efforts such as Refs. 19-39 and 19-40.

19.4.3 The Once-and-for-All Forest with Ordered Site Access

Inherent in the conventional model of forest logging is an assumption that loggers may cut trees from any part of the forest. This is not a serious restriction if the initial tree age distribution of the forest is uniform. In forests with a nonuniform initial age distribution, there may be requirements or regulations that dictate the order in which trees are to be cut. For example, some jurisdictions require that trees in a forest be cut in the order of their age, the oldest one to be cut first. In the absence of such laws, it may be physically necessary and economically prudent to cut trees in the order of their distance from one or more logging camp sites. In short, tree logging is necessarily ordered by practical and regulatory considerations; random access to tree sites does not occur in reality.

In this article, we consider only the situation where a single logging crew is to cut a (uniformly) narrow strip of tree stumps along a prescribed path winding through the entire forest. In that case, the position of any tree site can be described by the arc length s along the path to the site. For a continuous model of the forest, the discrete tree stands of the forest are smeared out over their respective assigned areas. Except for cases of sharp discontinuities in the initial age distribution, the actual distribution of tree age is replaced by a continuous approximation (see Ref. 19-41). The commercial value of the stumpage at different sites may be different because of a nonuniform age distribution, different growth conditions, market price differences at different cutting times, etc.

Let $T(s) \geqslant 0$ be the time at which the tree site at location s along the path is harvested in the future, $T = 0$ being now. The initial age distribution of the trees in the forest is denoted by $-T_0(s)$ with $T_0(s) \leqslant 0$ being the germination time distribution of the current trees. At cutting time, the tree stand at s will be $A(s) \equiv T(s) - T_0(s)$ years old. By construction, $T' \equiv dT/ds$ is nonnegative along the path with $T' = 0$ only if instantaneous harvesting is possible (with an unbounded harvest rate), as T' is a measure of the time consumed in logging a particular tree site and $1/T'$ is therefore a measure of the harvesting rate $h(s)$.

Let $p(s, T, A, T')\, ds$ and $c(s, T, A, T')\, ds$ be the commercial price and the harvesting (cutting, shipping, etc.) cost of the timber from the incremental strip $(s, s + ds)$ of the logging path. That the price per unit site harvested at location s and the harvesting cost per unit site at that location may depend on tree age A and absolute time has been discussed in the literature of forestry economics; the dependence on absolute time reflects in part the fluctuation of the lumber and labor markets. The possible dependence of p and c on location is not unexpected; trees may grow faster at one site than another, and they may be more difficult to log at some locations because of the geography and topography. We have already discussed how harvesting cost may change with the harvesting rate. If the logging company has any degree of monopolistic power, lumber price may also be affected by the rate of harvest.

The present value of the discounted future net revenue for the tree stumpage along an incremental path $(s, s + ds)$ is $e^{-\delta T(s)} [p - c]\, ds$. For a path so normalized that it is of unit length, the present value of the discounted future net revenue for the entire forest is given by

$$P \equiv \int_0^1 (p - c)\, e^{-\delta T}\, ds \qquad (19.4\text{-}15)$$

The management problem for the logging company is to choose a harvest schedule $T(s)$ for the forest so that this present value is a maximum. The maximization is subject to the constraint on the harvest rate, which takes the form of $T' \geqslant \tau (\geqslant 0)$ in our model. At this point, we have effectively completed our model formulation. The solution of the optimization problem will offer the insight this model can provide for the real phenomenon.

Note that in our formulation, the harvest rate is $F_0 ds/dT$, where F_0 is the amount of forest to cut. A change of the independent variable from s to T transforms (19.4-15) back to the expression for P (19.4-4) used in the conventional model. (We have taken $F_0 = 1$ and will continue to do so.)

19.4.4 The Optimal Harvest Policy—The Once-and-for-All Forest

In order to bring the available tools in modern control theory to bear on the solution of the optimization problem described in the previous section, we introduce a new control variable u by the defining equation (of state)

$$\frac{dT}{ds} \equiv u \tag{19.4-16}$$

and write the present value of future net revenue P as

$$P = \int_0^1 e^{-\delta T} V(s, T, A, u)\, ds, \qquad V \equiv p - c \tag{19.4-17}$$

where $A = T - T_0$. The maximum principle (see section 20.3-7) now requires that $u(s)$ be chosen to maximize the Hamiltonian

$$\mathcal{H} \equiv V(s, T, A, u)\, e^{-\delta T} + \lambda u \tag{19.4-18}$$

subject to the equation of state (19.4-16), the equation for the adjoint variable $\lambda(s)$

$$\frac{d\lambda}{ds} = -\frac{\partial \mathcal{H}}{\partial T} = -[V_T + V_A - \delta V]\, e^{-\delta T}, \qquad (\;)_y \equiv \frac{\partial(\;)}{\partial y} \tag{19.4-19}$$

the transversality conditions,

$$\lambda(0) = \lambda(1) = 0 \tag{19.4-20}$$

and the inequality constraints

$$u \geqslant \tau\ (\geqslant 0), \qquad T(0) \geqslant 0 \tag{19.4-21}$$

By allowing τ to be positive, we include the possibility of an imposed maximum feasible harvesting rate. If there is no such imposed upper limit, then $\tau = 0$ and

(19.4-21) simply reflects the fact that, in our model, the tree sites are ordered for the purpose of harvesting, as is usually the case in reality.

(A) When the inequality constraints (19.4-21) are not binding, we have an interior solution for the optimal control problem given by

$$\frac{\partial \mathcal{H}}{\partial u} = e^{-\delta T} \frac{\partial V}{\partial u} + \lambda = 0 \tag{19.4-22}$$

Equations (19.4-22) and (19.4-16) may be used to eliminate λ and u from (19.4-19) and (19.4-20) to get a second-order differential equation for T in the interval (0, 1) and one boundary condition for T at each end of the interval. This two-point boundary value may then be solved to get the optimal harvest time for different tree sites, denoted by $T(s)$.*

(B) When the first inequality constraint is binding, we have then a corner solution with

$$u(s) \equiv \frac{dT}{ds} = \tau \tag{19.4-23}$$

which can be integrated to give

$$T(s) = \tau s + t_0 \tag{19.4-24}$$

where the constant of integration t_0 will be determined presently. Upon substituting (19.4-23) and (19.4-24) into (19.4-19), we get

$$\lambda(s) = -\int_0^s [(V_T + V_A - \delta V) e^{-\delta T}]_{T=\tau s + t_0} \, ds \tag{19.4-25}$$

where the condition $\lambda(0) = 0$ has been used to eliminate a new constant of integration. The remaining transversality condition $\lambda(1) = 0$ becomes

$$\int_0^1 [(V_T + V_A - \delta V) e^{-\delta T}]_{T=\tau s + t_0} \, ds = 0 \tag{19.4-26}$$

and serves as a condition for the determination of the constant t_0.

By way of (A) or (B), we have in effect found a (and usually the only) candidate $T(s)$ for the optimal policy. That it is optimal may be verified by checking the appropriate concavity conditions or using arguments similar to those in Ref. 19-38. The systematic procedure for finding $T(s)$ in the present formulation may be com-

*Strictly speaking, $T(s)$ so determined satisfies only the necessary condition for a maximum P. For brevity, we call it the optimal policy, with the understanding that optimality is still to be demonstrated. In most cases (e.g., when V is strictly concave in u and T), this is straightforward.

pared with the corresponding derivation in Ref. 19-38. The conceptual and computational simplicity of our solution procedure can be attributed to the fact that the end points in the optimal control problem are now fixed. Although solutions for problems with various types of nonuniform and possibly discontinuous initial age distributions and unit harvesting costs have been obtained by the procedure outlined above and analyzed for their implications (Ref. 19-41), we confine ourselves in this article to a brief discussion of two specific cases with a uniform initial age distribution treated in Ref. 19-38.

19.4.5 Fisher's Rule and a Maximum Harvesting Rate

19.4.5.1 Fisher's Age Suppose the price (per unit tree site) is a function of tree age only, $p = p(A)$, and the cost (per unit site) is a constant, $c = c_0$. Because $\partial V/\partial u = \partial(p - c)/\partial u \equiv 0$, we have from (19.4-22)

$$\lambda \equiv 0 \tag{19.4-27}$$

for an interior solution. The solution (19.4-27) satisfies the two transversality conditions (19.4-20), and reduces (19.4-19) (with $V_T \equiv 0$) to

$$\frac{\dot{V}(A)}{V(A)} \equiv \frac{p(A)}{p(A) - c_0} = \delta \tag{19.4-28}$$

where a dot indicates differentiation with respect to the argument of the function. With $A = T - T_0$, the condition (19.4-28) is Fisher's rule (19.4-1) in a slightly different notation.

For a monotone increasing, concave function $p(A)$ (with $p - c_0 > 0$ for sufficiently large A), we let $\bar{A}(\delta)$ denote the root of (19.4-28). When there is no limit to the harvesting rate so that $\tau = 0$ and the initial age distribution of the forest is *uniform* ($T_0(s) = T_0$ = constant), then the "optimal" policy is $T(s) = T_0 + \bar{A}(\delta)$. It requires the tree site at s to be logged when the tree stump there reaches the Fisher age $\bar{A}(\delta)$. If $T_0 + \bar{A}(\delta) < 0$, logging should have been done before now; therefore we log immediately.

When the initial age distribution of the forest $-T_0(s)$ is nonuniform, the situation is more complicated. We must distinguish and treat separately the three cases (i) $T_0'(s) \geqslant 0$, $0 \leqslant s \leqslant 1$, (ii) $T_0'(s) \leqslant (0)$, $0 \leqslant s \leqslant 1$, and (iii) T_0' changes sign over the interval $0 \leqslant s \leqslant 1$. The analyses for these three cases and others are reported in Ref. 19-41.

19.4.5.2 A Bounded Feasible Harvest Rate In reality, a logging company is usually faced with a maxium feasible harvesting rate corresponding to a positive lower bound τ on $T'(s)$. Whether the optimal harvesting policy is given by the interior solution (19.4-22) now depends on whether the inequality constraint $T' > \tau > 0$ is binding. Again, we limit ourselves in this section to the case $p = p(A)$ and $c = c_0$, so that we have $\partial\mathcal{H}/\partial u \equiv 0$ and $\lambda \equiv 0$ in the case of an interior solution. The corresponding harvesting schedule, $T(s) = \bar{A}(\delta) + T_0(s)$, is again a consequence of Fisher's rule (19.4-28). We consider here only uniform initial age distributions, so that $T_0'(s) \equiv 0$. For this case, we have $T'(s) = T_0'(s) \equiv 0$ for the interior solution, so that the inequality constraint is binding, and the "optimal" harvest schedule is $T'(s) = \tau$ or

$$T = \tau s + t_0 \tag{19.4-29}$$

for $0 \leqslant s \leqslant 1$. For simplicity, we discuss only the case $t_0 \geqslant 0$. The adjoint variable is then given by (19.4-25), and the transversality conditions are satisfied by a value of t_0 determined by the integral condition (19.4-26). For the class of p and c considered here, (19.4-26) becomes

$$[p(\tau + t_0 - T_0) - c_0]\, e^{-\delta(\tau + t_0 - T_0)} = [p(t_0 - T_0) - c_0]\, e^{-\delta(t_0 - T_0)} \tag{19.4-30}$$

Therefore, the *optimal harvest schedule is to harvest at maximum feasible rate $1/\tau$ (or minimum time per tree site τ) throughout the entire forest (see (19.4-29)) starting at t_0 when the present value of the net revenue of the first tree cut equals that of the last tree cut.* It is not difficult to show $t_0 < \bar{A}(\delta) + T_0$. This "optimal" harvesting policy is identical to that obtained in Ref. 19-38, as it should be. Just like all other cases previously considered and to be investigated, the solution procedure for the "optimal" policy here is a systematic consequence of our formulation and the conventional maximum principle (for fixed initial and terminal s) without the need for special treatment.

When the initial age distribution of the forest is not uniform, the situation is more complicated, but similar to the unlimited harvest rate case. A separate treatment of the three cases, (i) $T_0'(s) \geqslant \tau$, (ii) $T_0'(s) \leqslant \tau$, and (iii) $T_0'(s) - \tau$ changes sign, can be found in Ref. 19-41.

19.4.6 Unit Harvest Cost Varying with Harvest Rate—Single Harvest

Instead of c being a constant, we consider in this section average (per unit site) harvest cost functions that depend only on T' with the conventional U-shaped graph, i.e., $c(T') > 0$ is convex in T' with a minimum at $T' = \tau_{min} > 0$ (so that $\min c(T') = c(\tau_{min})$). This class of average cost functions includes both the effect of a fixed cost component important at a low harvest rate and an overload cost component important at a high harvest rate.

With $p = p(A) \equiv p(T - T_0)$ as before, we have from the Hamiltonian $\partial\mathcal{H}/\partial u = -e^{-\delta T}\dot{c}(u) + \lambda$, where $u = T'$ and a dot on top of a function indicates differentiation with respect to its argument. An interior solution of the optimal control problem requires (see (19.4-22))

$$\lambda(s) = e^{-\delta T}\dot{c}(u) = e^{-\delta T(s)}\dot{c}(T'(s)) \tag{19.4-31}$$

with the transversality conditions $\lambda(0) = \lambda(1) = 0$ satisfied by taking

$$T'(0) = T'(1) = \tau_{\min} \tag{19.4-32}$$

(No other choice is possible, as c has a unique stationary point and $e^{-\delta T}$ never vanishes.) As such, we have reproduced and extended the principal result of Ref. 19-38 for the same class of problems, namely, *harvesting should start and end with a harvest rate giving a minimum per unit site harvest cost*. If there is an upper bound to the feasible harvest rate, the interior solution (19.4-31) with $T'(0) = T'(1) = \tau_{\min}$ may not be appropriate, and the inequality constraint on the control $u \equiv T'$ may be binding. We shall return later to a discussion of the optimal solution for the case of limited harvest capacity.

From (19.4-31) and the qualitative behavior of the class of $c(T')$ of interest here, we see that T' remains positive along the entire logging path for the interior solution, independent of the initial age distribution. Therefore, the optimal harvest schedule is determined by inserting (19.4-31) into the differential equation (19.4-19) for the adjoint variable λ and solving the resulting second-order ODE for T with (19.4-32) as boundary conditions. As an alternative solution process, we may solve (19.4-31) for T' to get the unique solution

$$T' = f(\lambda e^{\delta T}) \tag{19.4-33}$$

(because $\dot{c}$ is a monotone increasing function of its argument), and then write (19.4-19) as

$$\lambda' = -\{\dot{p}(T - T_0) - \delta p(T - T_0) + \delta c(f(\lambda e^{\delta T}))\}\, e^{-\delta T} \tag{19.4-34}$$

The second-order system of two first-order equations, (19.4-33) and (19.4-34), and the two transversality conditions, $\lambda(0) = \lambda(1) = 0$, define a two-point boundary-value problem for $\lambda(s)$ and $T(s)$.

We note that (19.4-34) and $\lambda(0) = 0$ yield

$$\lambda(s) = -\int_0^s \{\dot{p}(T - T_0) - \delta\,[p(T - T_0) - c(T')]\}\, e^{-\delta T}\, ds \tag{19.4-35}$$

The condition $\lambda(1) = 0$ then gives

$$\frac{\displaystyle\int_0^1 \dot{p}(T - T_0)\, e^{-\delta T}\, ds}{\displaystyle\int_0^1 [p(T - T_0) - c(T')]\, e^{-\delta T}\, ds} = \delta \tag{19.4-36}$$

Whether we have a uniform initial age distribution or a maximum feasible harvest rate, *the optimal policy, according to (19.4-36), logs the forest on a schedule that makes the relative discounted marginal yield of the forest equal the discount rate, with the first and last tree site logged at the rate* $1/\tau_{min}$ *(for a minimum unit harvest cost).* For simplicity, we have tacitly assumed $T(s) \geqslant 0$; otherwise, the entire path segment $0 \leqslant s \leqslant \bar{s}$, where $T(s) < 0$ should be clear-cut immediately.

For a uniform initial age distribution so that $T_0(s)$ is a constant, the system (19.4-33) and (19.4-34) is autonomous and admits a first integral. For important classes of $c(T')$, an exact solution of the BVP for λ and T is given in Ref. 19-41. When there is an upper limit to the harvesting capacity, (a lower bound τ on T'), the optimal harvest schedule depends on the sign of $T' - \tau$. The optimal harvest schedule continues to be the interior solution defined by the two-point boundary-value problem (19.4-33), (19.4-34), and (19.4-20) if $T' \geqslant \tau$ for the entire forest. The situation is more complicated if $T' - \tau$ is negative for some s. For example, if $\tau_{min} < \tau$, then the inequality constraint $T' \geqslant \tau$ is binding for an initial segment of the logging path $0 \leqslant s \leqslant \bar{s}$, so that we have

$$T(s) = \tau s + t_0 \tag{19.4-37}$$

and

$$\begin{aligned}\lambda(s) &= -\int_0^s \{\dot{p}(\tau s + t_0 - T_0) - \delta p(\tau s + t_0 - T_0) + \delta c(\tau)\} e^{-\delta(\tau s + t_0)}\, ds \\ &= -\frac{1}{\tau}\left[\{p(\tau s + t_0 - T_0) - c(\tau)\} e^{-\delta(\tau s + t_0)}\right]_0^s\end{aligned} \tag{19.4-38}$$

there. However, for $\bar{s} \leqslant s \leqslant 1$, the condition (19.4-31) is admissible, and the optimal harvest policy satisfies (19.4-33) and (19.4-34) with $\lambda(1) = 0$. The two unknown parameters t_0 and $\bar{s}$ are determined by the continuity of λ and T at the junction $s = \bar{s}$. A similar procedure for determining the optimal harvest schedule applies when $T' - \tau$ becomes negative in one or more segments of $[0, 1]$, which may or may not include an end point.

19.4.7 The Ongoing Forest with Ordered Site Access

Unless there is an abundance of forests, a harvested forest should be replanted for future lumber supply. Clearly, the longer the logging of the existing forest is delayed, the longer it takes to acquire revenue from future harvests. The significance of the opportunity cost associated with not logging sooner (than the Fisher age, say) was recognized by Faustmann, who examined the optimal harvest policy for a forest to be harvested and replanted repeatedly.

Analyses of harvest policy for ongoing forests are meaningful only for long-term planning over a span of centuries, given that replanted trees are of no net commercial value during the first few decades after germination. The fluctuation of price and cost with chronological time should be significant over such a planning period and should be included in a realistic mathematical model. On the other hand, the incorporation of such fluctuations in a model of ongoing forests is certain to make the mathematical problems much less tractable, as we shall soon see. Therefore, it is that much more important for us to seek as simple a mathematical formulation of the model problem as we possibly can. A formulation for an ongoing forest similar to that of subsection 19.4.3 is even more attractive than the conventional formulation of subsection 19.4.2 from this viewpoint.

Within the framework of our formulation, we let $T_k(s)$ be the time (measured from now) at which the tree site at location s along the logging path is harvested during the kth harvest, $k = 1, 2, \ldots$. The initial age distribution of the trees in the forest is again denoted by $-T_0(s)$ with $T_0(s) \leqslant 0$ being the germination time distribution of the existing trees. The tree at location s will be $A_k(s) \equiv T_k(s) - T_{k-1}(s)$ years old when it is logged during the kth harvest. By construction, $T_k' \equiv dT_k/ds$ is nonnegative along the path with $T_k' = 0$ only if instantaneous harvesting is possible (with unlimited harvest capacity), as T_k' is a measure of the time consumed in logging a particular tree site during the kth harvest and $1/T_k'$ is therefore a measure of the harvest rate h_k at the location s for the kth harvest.

Similar to the case of "once-and-for-all forests," we let $p_k\, ds$ and $c_k\, ds$ be the commercial price and harvesting cost of the timber from the kth harvest over the incremental path strip $(s, s + ds)$. For reasons already explained, both p_k and c_k may vary with location s, logging time T_k, and tree age A_k, as well as current and previous harvest rates, $1/T_j', j = 1, 2, \ldots, k$. The present value of the discounted future net revenue for tree stumpage along an incremental path $(s, s + ds)$ from the kth harvest is $e^{-\delta_k T_k(s)}[p_k - c_k]\, ds$ where δ_k is the constant discount rate at the time of the kth harvest. The present value of the discounted future net revenue from the entire forest at the end of the Nth harvest is

$$P_N \equiv \sum_{k=1}^{N} \left[\int_0^1 (p_k - c_k)\, e^{-\delta_k T_k}\, ds \right] \qquad (19.4\text{-}39)$$

where N is ∞ if the forest is to be harvested repeatedly for the whole future.* The management problem for the logging company is to choose a sequence of harvest schedules $\{T_1, T_2, \ldots\}$ for the forest so that P_N is a maximum. With $dT_k/ds = 1/h_k$, $p_k \equiv p$ and $C_k \equiv c_k h_k$, the expression (19.4-39) reduces to the conventional expression for Refs. 19-34–19-39 when unit site price and harvest cost are identical for all harvests.

To apply the conventional maximum principle to the optimal control problem in our model for an ongoing forest, we introduce a new set of controls by the defining

***For the ongoing forest problem to be meaningful, P_N must remain bounded as $N \to \infty$.**

equations (of state)

$$T'_k \equiv u_k, \qquad (k = 1, 2, \ldots) \tag{19.4-40}$$

and write P_N as

$$P_N = \sum_{k=1}^{N} \int_0^1 e^{-\delta_k T_k} V_k(s, T_k, A_k, u_1, u_2, \ldots, u_k)\, ds \tag{19.41}$$

where $V_k \equiv p_k - c_k$ is the net revenue per unit path length. The maximum principle now requires that u_k be chosen to maximize the Hamiltonian

$$\mathcal{H} \equiv \sum_{k=1}^{N} [e^{-\delta_k T_k} V_k + \lambda_k u_k] \tag{19.4-42}$$

subject to the equations of state (19.4-40), the equations for the adjoint variables (discounted shadow prices for the different harvests), $\lambda_1, \lambda_2, \ldots,$

$$\lambda'_k = -\frac{\partial \mathcal{H}}{\partial T_k} = -\left[\frac{\partial V_k}{\partial T_k} + \frac{\partial V_k}{\partial A_k} - \delta_k V_k\right] e^{-\delta_k T_k} + \frac{\partial V_{k+1}}{\partial A_{k+1}} e^{-\delta_{k+1} T_{k+1}}$$
$$(k = 1, 2, \ldots, N) \tag{19.4-43}$$

(with $V_{N+1} \equiv 0$), the transversality conditions

$$\lambda_k(0) = \lambda_k(1) = 0 \qquad (k = 1, 2, \ldots) \tag{19.4-44}$$

and the inequality constraints

$$u_k \geqslant \tau_k (\geqslant 0), \quad T_1(0) \geqslant 0, \quad T_k(0) \geqslant T_{k-1}(1) \qquad (k = 1, 2, \ldots, N)$$
$$\tag{19.4-45}$$

By allowing τ_k to be positive, we include the possibility of a maximum feasible harvest rate for each harvest reflecting a limited harvest capacity. If harvest capacity is unlimited, then we have $\tau_k = 0$ and (19.4-45) simply reflects the fact that the tree sites are ordered for the purpose of logging. (We also assume $T_k(s) \geqslant T_{k-1}(s) \geqslant \cdots \geqslant T_1(s) \geqslant 0$ for simplicity.)

When the inequality constraints (19.4-45) are not binding, we have an interior solution for the optimal control problem given by

$$\frac{\partial \mathcal{H}}{\partial u_k} = \lambda_k + \sum_{j=1}^{N} e^{-\delta_j T_j} \frac{\partial V_j}{\partial u_k} = 0, \qquad (k = 1, 2, \ldots, N) \tag{19.4-46}$$

The conditions (19.4-46) and (19.4-40) may be used to eliminate λ_k and u_k from (19.4-43) and (19.4-44) to get a system of N second-order ODEs for T_k, $k = 1, 2, \ldots, N$, and one set of N boundary conditions at each end of the logging path. This two-point boundary value problem may then be solved to get the optimal harvest schedule for each harvest.*

At the other extreme, when the inequality constraints on u_k in (19.4-45) are binding, we have a corner solution with

$$u_k \equiv T_k' = \tau_k \qquad \text{or} \qquad T_k(s) = \tau_k s + t_k \tag{19.4-47}$$

where the constants of integration t_k, $k = 1, 2, \ldots$, are to be determined by (19.4-43) and (19.4-44).

Intermediate situations with some of the inequality constraints on u_k in (19.4-45) being binding are also possible and must be dealt with separately in a manner to be indicated in the subsequent sections. Regardless of whether one or more of (19.4-45) are binding, we get from (19.4-43) and the transversality conditions $\lambda_k(0) = 0$ (see (19.4-44))

$$\lambda_k(s) = -\int_0^s \left[\left(\frac{\partial V_k}{\partial A_k} + \frac{\partial V_k}{\partial T_k} - \delta_k V_k\right) e^{-\delta_k T_k} - \frac{\partial V_{k+1}}{\partial A_{k+1}} e^{-\delta_{k+1} T_{k+1}}\right] ds$$

$$(k = 1, 2, \ldots) \tag{19.4-48}$$

The remaining transversality conditions, $\lambda_k(1) = 0$ of (19.4-44) gives

$$\int_0^1 \left[\left(\frac{\partial V_k}{\partial A_k} + \frac{\partial V_k}{\partial T_k}\right) e^{-\delta_k T_k} - \frac{\partial V_{k+1}}{\partial A_{k+1}} e^{-\delta_{k+1} T_{k+1}}\right] ds = \delta_k \int_0^1 V_k e^{-\delta_k T_k}\, ds \tag{19.4-49}$$

Thus, under the optimal policy, the *discounted net gain* through time marginal yield of the entire forest (for not harvesting) *equals the opportunity cost* consisting now of the sum of the time marginal yield of the replanted forest and the interest earned on discounted net revenue of the harvested forest.

Observe that in our formulation of the problem, tree ages and harvest schedules are related in a natural way by $A_k(s) = T_k(s) - T_{k-1}(s)$. These simple relations replace the integral conditions (19.4-12) in the formulation of Ref. 19-38, which leads to functional-differential equations of state (19.4-14) for the optimal control problem. Because of our choice of space instead of time as the independent variable, no functional-differential equation appears in our formulation, and the maximum principle is directly applicable to the new optimal control problem.

*We assume the various concavity conditions are satisfied, so that the necessary conditions for optimality are also sufficient.

19.4.8 A Finite Harvest Sequence and the Faustmann Rotation

Suppose the net revenue per unit tree site for the kth harvest, $V_k \equiv p_k - c_k$, is a monotone increasing concave function of tree age only, and the constant discount rate is the same for all harvests, $\delta_k = \delta$, $k = 1, 2, \ldots$. Then the system (19.4-46) reduces to

$$\lambda_k \equiv 0 \qquad (k = 1, 2, \ldots, N) \tag{19.4-50}$$

and the transversality conditions (19.4-44) are trivially satisfied. The differential system (19.4-43) for λ_k now becomes an algebraic system of N simultaneous nonlinear equations for $A_k(s)$ (and therefore the optimal harvest schedules $T_k(s)$), $k = 1, 2, 3, \ldots, N$:

$$\dot{V}_k(A_k) - \delta V_k(A_k) = \dot{V}_{k+1}(A_{k+1})\, e^{-\delta A_{k+1}} \qquad (k = 1, 2, 3, \ldots, N) \tag{19.4-51}$$

with $V_{N+1}(A) \equiv 0$.

The system (19.4-51) may be solved by noting that the Nth equation

$$\dot{V}_N(A_N) - \delta V_N(A_N) = 0 \tag{19.4-52}$$

involves only one unknown A_N and its unique solution is the well-known Fisher age $\alpha_N(\delta)$ (denoted by $\bar{A}(\delta)$ earlier):

$$A_N(s) \equiv T_N(s) - T_{N-1}(s) = \alpha_N(\delta), \qquad \frac{\dot{V}_N(\alpha_N(\delta))}{V_N(\alpha_N(\delta))} = \delta \tag{19.4-53}$$

Having determined $A_N(s)$, the $(N-1)$th equation

$$\dot{V}_{N-1}(A_{N-1}) - \delta V_{N-1}(A_{N-1}) = \dot{V}_N(\alpha_N)\, e^{-\delta \alpha_N} \tag{19.4-54}$$

involves only one unknown and may be solved to get the unique solution $\alpha^*_{N-1}(\delta)$ for $A_{N-1}(s)$. The process is repeated to get $A_{N-2} = \alpha^*_{N-2}, A_{N-3} = \alpha^*_{N-3}, \ldots$, with the first equation giving $A_1(s) = \alpha^*_1(\delta)$. These results for the tree age distribution during the different harvests are then used to determine the optimal harvest schedules $\{T_k(s)\}$:

$$T_1(s) = T_0(s) + \alpha^*_1(\delta) \quad T_2(s) = T_1(s) + \alpha^*_2(\delta) = T_0(s) + \alpha^*_1(\delta) + \alpha^*_2(\delta)$$

$$T_k(s) = T_{k-1}(s) + \alpha^*_k(\delta) = T_0(s) + \sum_{j=1}^{k} \alpha^*_j(\delta), \qquad (k = 3, \ldots, N-1) \tag{19.4-55}$$

$$T_N(s) = T_{N-1}(s) + \alpha_N(\delta) = T_0(s) + \sum_{j=1}^{N-1} \alpha^*_j(\delta) + \alpha_N(\delta)$$

For the case of a uniform initial age distribution and unlimited harvesting capacity, the optimal harvest policy is to clear-cut the entire forest instantaneously when the trees reach the age $\alpha_k^*(\delta)$ during the kth harvest ($k = 1, 2, \ldots, N-1$) and when they reach the Fisher age $\alpha_N(\delta)$ for the last harvest (see Fig. 19.4-1). (For simplicity, we will discuss only the $T_0(s) + \alpha_1^*(\delta) \geqslant 0$ case.) This policy is the only one that satisfies the necessary conditions for optimality. As $N \to \infty$, it can be shown rigorously that α_1^* (and therefore any α_k^*) tends to the Faustmann age A_{MF} defined by (19.4-2). (See Ref. 19-42.) It can also be verified formally that A_{MF} is a solution of the infinite system of equations (19.4-51) (with $N = \infty$).

When there is a maximum feasible harvest rate so that $\tau_k > 0$ for all harvests, the inequality constraint $T_1' \geqslant \tau_1$ is binding for a uniform initial age distribution ($T_0'(s) \equiv 0$). For this case, we have $T_1(s) = \tau_1 s + t_1$, where t_1 is a constant of integration to be determined by

$$\frac{d\lambda_1}{ds} = -[\dot{V}_1(T_1 - T_0) - \delta V(T_1 - T_0)]\, e^{-\delta T_1} + \dot{V}_2(T_2 - T_1)\, e^{-\delta T_2} \tag{19.4-56}$$

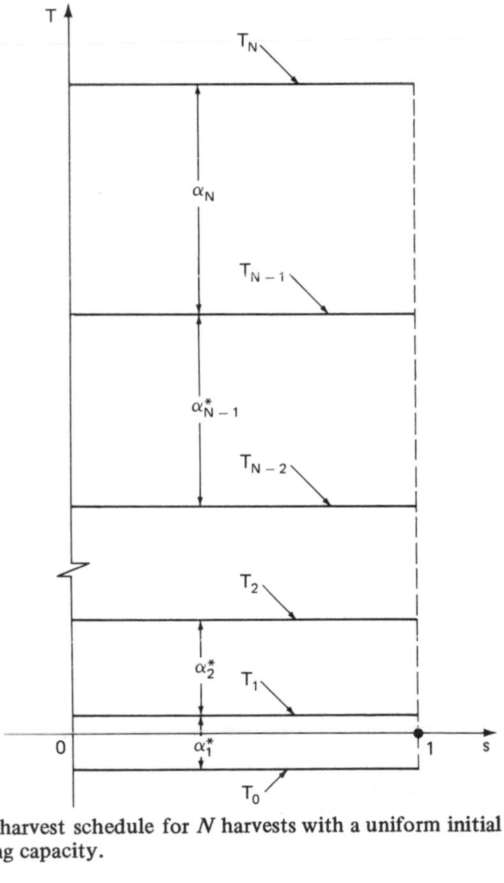

Fig. 19.4-1 Optimal harvest schedule for N harvests with a uniform initial age distribution and an unlimited harvesting capacity.

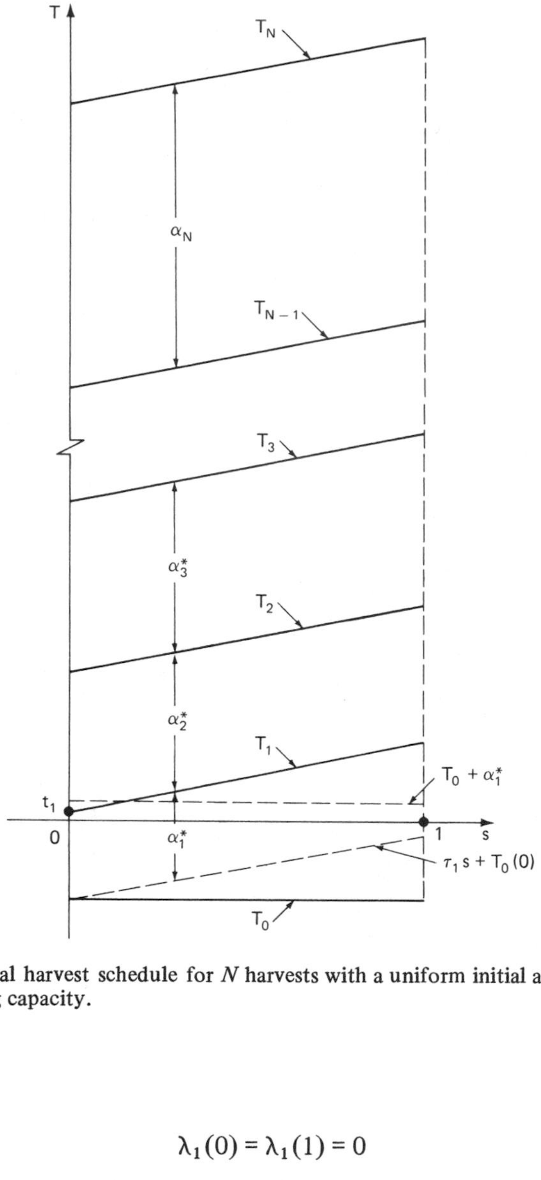

Fig. 19.4-2 **Optimal harvest schedule for *N* harvests with a uniform initial age distribution and a limited harvesting capacity.**

and

$$\lambda_1(0) = \lambda_1(1) = 0 \tag{19.4-57}$$

which follows from (19.4-43) and (19.4-44) with $k = 1$. The ODE for λ_1 (19.4-56) involves $T_2(s)$, the yet unknown harvest schedule for the second harvest. It is evident from the form of (19.4-43) (when $T_k' \equiv \tau_k$) and (19.4-51) (when $\lambda_k \equiv 0$) that $T_k(s)$, $k = 2, 3, \ldots, N$, and t_1 are in general determined simultaneously. To illustrate, consider the specific case with $\tau_k \geqslant \tau_{k+1}$, $k = 1, 2, \ldots, N$, so that the maximum feasible harvesting rate increases with time.* In that case, the inequality constraints are not binding for $k \geqslant 2$, so that we have $\lambda_k \equiv 0$, $k = 2, 3, \ldots, N$. It follows from the analysis leading up to (19.4-55)

$$\begin{aligned} A_N &\equiv T_N(s) - T_{N-1}(s) = \alpha_N(\delta) \\ A_k &\equiv T_k(s) - T_{k-1}(s) = \alpha_k^*(\delta), \qquad (k = N-1, \ldots, 2) \end{aligned} \tag{19.58}$$

In particular, we have $T_2(s) = T_1(s) + \alpha_2^*(\delta) = \tau_1 s + t_1 + \alpha_2^*(\delta)$, so that (19.4-56) and (19.4-57) may be written as a single condition for t_1:

$$\frac{\int_0^1 \{\dot{V}_1(T_1 - T_0) - \dot{V}_2(\alpha_2^*)\, e^{-\delta\alpha_2^*}\}\, e^{-\delta T_1}\, ds}{\int_0^1 V_1(T_1 - T_0)\, e^{-\delta T_1}\, ds} = \delta \tag{19.4-59}$$

According to (19.4-59), the optimal first harvest policy $T_1(s)$ in this case is to log at the maximum feasible rate starting at the instant t_1 when the discounted marginal yield (of the first harvest) net the same yield of the second harvest, as a fraction of the present value of the first harvest, equals the discount rate. The optimal harvest schedule for the remaining harvests is given by (see (19.4-55))

$$T_k(s) = T_{k-1}(s) + \alpha_k^*(\delta) = T_1(s) + \sum_{j=2}^{k} \alpha_j^*(\delta) \tag{19.4-60a}$$

for $k = 2, 3, \ldots, N-1$, and

$$T_N(s) = T_{N-1}(s) + \alpha_N(\delta) = T_1(s) + \sum_{j=1}^{N-1} \alpha_j^*(\delta) + \alpha_N(\delta) \tag{19.4-60b}$$

With the forest logged at the same constant rate τ_1 for all harvests, a limited harvest capacity gives a sustained yield of lumber over a finite interval of time. With α_k^*, $k = 1, 2, 3, \ldots$, tending to the Faustmann age as $N \to \infty$, the above results are identical to those obtained in Ref. 19-38 (see Fig. 19.4-2).

Although more complicated combinations of maximum feasible harvest rates for different harvests, as well as different types of nonuniform initial age distributions, can be handled similarly (Ref. 19-42), the advantages of the present formulation of the ongoing forest model over the conventional formulation should be quite evident by this time. It avoids the necessity of dealing with a free-endpoint optimal con-

*The (smallest) maximum feasible harvest rate τ_1 is assumed to be sufficiently large, so that it takes less than the Faustmann rotation A_{MF} to log the entire forest.

trol problem involving functional differential equations of state and allows more general situations (with time fluctuation on price, cost, feasible harvest rate, discount rate, etc.) to be treated by the same systematic procedure.

19.4.9 Unit Harvest Cost Dependent on Harvest Rate

Up to now, the merits of the new formulation may be considered as merely simplifications, embellishments, and improvements of earlier results. It is for the case of a harvest-rate-dependent unit harvest cost that the new formulation makes its unique contribution. Suppose unit price functions again depend only on tree age, but now the unit harvest cost functions depend only on the *current* harvesting rate. As pointed earlier, the graphs of these unit cost functions are positive and U-shaped, each with a single minimum at $T_k' = \tau_{\min}^{(k)} > 0$, so that $\dot{c}_k(\tau_{\min}^{(k)}) = 0$ and $\min\,[c_k(T_k')] = c_k(\tau_{\min}^{(k)})$. For this class of problems, we have from (19.4-42)

$$\lambda_k = e^{-\delta_k T_k} \dot{c}_k(u_k), \qquad u_k = T_k' \tag{19.4-61}$$

$k = 1, 2, 3, \ldots$, for an interior solution; the conditions (19.4-61) reduce to (19.4-51) if c_k is independent of harvesting rates.

We see from (19.4-61) that the transversality conditions $\lambda_k(0) = \lambda_k(1) = 0$ (see (19.4-44)) are satisfied by the interior solution with

$$T_k'(0) = T_k'(1) = \tau_{\min}^{(k)} > 0, \qquad (k = 1, 2, \ldots) \tag{19.4-62}$$

No other possibility exists, as each $c_k(u_k)$ has a unique stationary point and $e^{-\delta_k T_k}$ is always positive. With (19.4-62), we have reproduced and extended property (A) obtained in Ref. 19-38 by a partial maximization procedure: *For an optimal harvest policy (with no restriction on harvest rates), the* kth *harvest should start and end with the harvest rate* $1/\tau_{\min}^{(k)}$ *that gives a minimum unit site harvest cost for that harvest.* Aside from allowing for ordered site access and an upper limit on harvesting rate, we know that this property also holds for a finite harvest sequence and more general price and cost functions. The property has as its economic content a zero discounted shadow price for the first and last tree cut.

From (19.4-61) and the fact that $\dot{c}_k \to -\infty$ as T_k' tends to zero from above, we see that the harvest rate must remain finite and positive along the entire logging path for the interior solution, independent of the initial age distribution of the forest. Provided that $T_k'(s)$ is not less than its lower bound $\tau_k \geqslant 0$ (with $1/\tau_k$ being the upper bound on harvest rate), the optimal harvest policy must satisfy the system (19.4-61) and the equations for the adjoint variables (discounted shadow prices) (19.4-43), which simplify to read

$$\lambda_k' = -[\dot{p}_k(A_k) - \delta_k\{p_k(A_k) - c_k(T_k')\}]\, e^{-\delta_k T_k} + \dot{p}_{k+1}(A_{k+1})\, e^{-\delta_{k+1} T_{k+1}}, \quad (k = 1, 2, \ldots) \tag{19.4-63}$$

where $A_k = T_k - T_{k-1}$, and the transversality conditions are in the form of (19.4-62).

To determine the above interior solution, our experience with the rate-independent unit cost functions suggests that we begin with the problem of planning for a finite number of harvests, so that we have (19.4-61)-(19.4-63) for $k = 1, 2, \ldots, N$ with $p_{N+1} \equiv 0$. For a prescribed $T_0(s)$, these $2N$ simultaneous first order ODEs and $2N$ boundary conditions define a two-point boundary-value problem that can be solved for the $2N$ unknowns $\{\lambda_1, \ldots, \lambda_N\}$ and $\{T_1, \ldots, T_N\}$ by available methods. If $\ddot{p}_1, \ldots, \ddot{p}_N$ are bounded, we expect (and it can be shown) that, for a fixed k, the sequence $\{T_k(s; N)\}$ tends to a limit as $N \to \infty$. It follows that we may calculate the schedule for as many harvests of an ongoing forest as we wish and as accurately as we wish by solving the above BVP for a sufficiently large N. Though to solve the BVP (19.4-61), (19.4-63), and (19.4-62) (or (19.4-44)) for a large N is at best a very costly computational problem, the present formulation at least yields a systematical procedure for calculating the harvest sequence for an ongoing forest that satisfies the necessary conditions for optimality. In practice, it is difficult (if not impossible) to anticipate developments in the distant future; it would be rather unrealistic and meaningless to seek an optimal policy for more than a century. For an optimal policy for a less-than-ten harvest sequence, the solution process for the two-point BVP is definitely manageable; this is true even for more general classes of $\{p_k\}$ and $\{c_k\}$, such as those appearing in (19.4-41).

If $\tau_{\min}^{(k)} < \tau_k$ ($1/\tau_k$ being the maximum feasible harvest rate for the kth harvest) or if it should turn out that the solution of the two-point BVP is such that $T_k' < \tau_k$ for some range of s values and for one or more harvests, some or all inequality constraints would be binding, and the optimal harvest schedules would have to be obtained by a procedure similar to those described earlier for similar situations.

19.4.10 Identical Price and Cost Functions for All Harvests

As the last item on the subject of optimal harvest schedules, we wish to relate our results for an ongoing forest to the partial results for a Pareto optimum obtained in Ref. 19-38.* For this purpose, we specialize the two-point BVP defined by (19.4-61)-(19.4-63) to the case where we have $\delta_k = \delta$, $p_k = p(A_k)$, and $c_k = c(T_k')$, $k = 1, 2, 3, \ldots$. Upon using (19.4-61) to eliminate λ_k from (19.4-63) for this case, we get

$$\ddot{c}(T_k')\,T_k'' = -\dot{p}(A_k) + \delta\,[\,p(A_k) + T_k'\dot{c}(T_k') - c(T_k')] + \dot{p}(A_{k+1})\,e^{-\delta A_{k+1}} \qquad (k = 1, 2, 3, \ldots) \tag{19.4-64}$$

With h_k and $C(h_k)$ of Ref. 19-38 identified as $1/T_k'$ and $c(1/T_k')/T_k'$ in our formulation and with

$$\frac{dh_k}{dt} = \frac{d}{ds}\left(\frac{1}{T_k'}\right)\frac{ds}{dT_k} = -\frac{T_k''}{(T_k')^3} \tag{19.4-65}$$

it is straightforward to verify that (19.4-64) is identical to the corresponding condition (21) for a Pareto optimum obtained there. Therefore, all properties correctly

*The optimal policy for maximizing the discounted net revenue for a single harvest with all other harvests kept fixed.

deduced in Ref. 19-38 from the Pareto optimal solution also hold for the actual policy for the same restricted class of $\{p_k\}$, $\{c_k\}$, and $\{\delta_k = \delta\}$ considered here.

19.5 LINEARIZE WITH CARE

19.5.1 Steadily Rotating Elastic Shells of Revolution

Any analyst familiar with the methods of applied mathematics summarized in this handbook and elsewhere cannot help but be aware of the fact that *linear* problems are generally more tractable mathematically than nonlinear ones. At least as a first approximation, it is usually worthwhile to start with a linear model or to simplify an inherently nonlinear model by removing nonlinearities through appropriate restrictive assumptions, e.g., small-amplitude response. The tentative results from a linear or linearized model may then be used to estimate the neglected nonlinear effects. In practice, the nonlinear effects are to be included or restored only if estimates of these effects *based on the solution of the linear or linearized models* turn out to be comparable in magnitude to the contribution from effects retained in the linear theory. Unfortunately, such a back-check for consistency is not always reliable; experience and insight often offer better guides to the adequacy of a linear theory. In this section, we will illustrate the need for scientific insight in the use of linear models by the example of a steadily rotating thin elastic shell of revolution.

Loosely speaking, a shell structure is a layer of material bounded by two curved surfaces separated by a distance h, the shell thickness, which is small compared with the overall dimension of the structure. Shell structures occur everywhere in our daily lives, from beer cans and food containers to aircraft fuselages and biological cell membranes. Whatever form the applications of shell structures may take, analyses of the structural integrity of shells are difficult to perform and will continue to challenge the current and future generations of mechanicians and applied mathematicians. In view of the difficulties, shell structures are designed for applications with as simple a shape as possible, consistent with their functions. Commonly encountered shell shapes include cylindrical, conical, spherical, and other shells of revolution.

Among the many modern applications of shell structures are outer space stations contemplated for deep space explorations. For stability, they are to rotate steadily about some axis, usually an axis of symmetry of the shell structure. Some space stations contemplated are effectively thin toroidal shells of revolution, a hollow donut similar to the inner tube of bicycle or car tires. The structural properties of such a station have been the subject of numerous design studies, starting in the early fifties, e.g., Refs. 19-43, 19-44, and 19-45. In the context of mathematical modeling, the structural analysis of toroidal shells provides an example that demonstrates how deceptively adequate a linear model of a phenomenon may look. In actual fact, the linear model for a rotating toroidal shell could never give the correct qualitative behavior of the shell no matter how small the external load (the inertia force due to steady rotation) and the shell response may be! For this demonstration, it suffices to consider the simpler problem of the stresses and deformation of a steadily rotating *shallow* portion of a toroidal shell of revolution with a circular cross section. The shell portion is in the shape of a washer.

The structural analysis of a rotating toroidal cap and of other shells is instructive for another reason. Shells are three-dimensional solid bodies and, in principle, should be analyzed by the use of three-dimensional models of continuous media (see section 4.7). Unfortunately, for most shell problems, it is not possible to obtain by analytical methods the structural behavior of the shell from such a model; it is also too costly and impractical to get it by numerical methods. Simplifications of the continuum mechanics models for shells, leading to what is known as *thin shell theory* today, had already been made by G. Kirchhoff and A. E. H. Love back in the 19th century, employing ideas introduced even earlier by L. Euler and the Bernoulli brothers for simpler structures (Ref. 19-46). With the developments of the last two centuries in mathematics and mechanics, we now know that the classical theory of thin elastic shells may be deduced as the leading term outer-solution of a matched asymptotic expansion solution of the original three-dimensional elasticity problem. To deduce a thin shell model for the analysis of rotating toroidal shells from the continuum mechanics model would be well beyond the scope of this article. For our purpose, it is sufficient to outline the standard shallow shell model of Ref. 19-47 for axisymmetric deformations of thin elastic shells of revolution, using as a point of departure certain geometrical properties (known as the Euler-Bernoulli hypothesis) of the outer-solution. Such a model will be useful in many applications beyond the analysis of a rotating toroidal cap, including a real-life application to be discussed in section 19.6.

19.5.2 Axisymmetric Deformation of Shallow Shells of Revolution

In cylindrical coordinates (r, θ, z), a surface of revolution may be characterized by the fact that the axial coordinate z of a point on the surface is a function of the radial coordinate r, the radial distance from the axis of revolution. We consider in this article, shell structures with a middle surface defined by $z = Z(r)$ and with a thickness h, which may vary only in the meridional direction, i.e., h may be a function of r. The shells are subject to only axisymmetric external force and moment intensities (such as uniform pressure and gravity loading) that induce only axisymmetric radial and axial deformation, so that measures of displacement, strain, and stress (see section 4.7) are independent of the angular coordinate θ.

Let $u(r)$ and $w(r)$ be the radial and axial displacement components, respectively, of points on the middle surface of a shell of revolution. Normal strain components are defined in terms of u and w as relative changes of length. With Fig. 19.5-1 giving a sketch of an elemental cross section of the shell along any meridian before and after deformation, the circumferential strain component, e_θ, and the meridional strain component, e_r, of a surface at a distance ζ from the middle surface are taken in the form (Ref. 19-47)

$$e_\theta = \epsilon_\theta + \zeta\kappa_\theta, \qquad e_r = \epsilon_r + \zeta\kappa_r \tag{19.5-1,2}$$

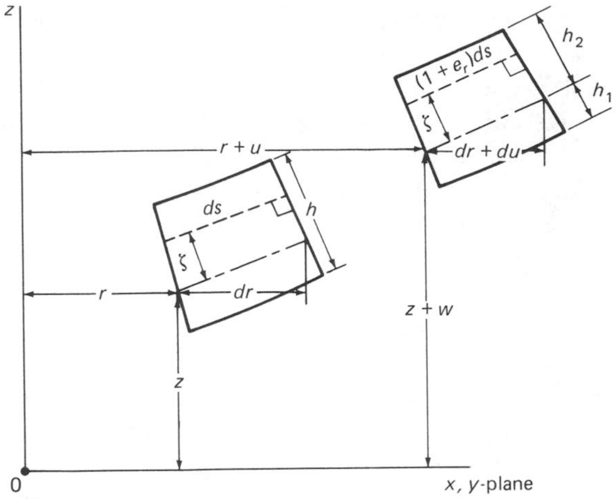

Fig. 19.5-1 Axisymmetric displacement and strain components of thin shells of revolution.

where ϵ_θ and ϵ_r are the midsurface strain components given by

$$\epsilon_\theta = \frac{u}{r}, \qquad \epsilon_r = u' + z'w' + \frac{1}{2}(w')^2 \tag{19.5-3,4}$$

and κ_θ and κ_r are the midsurface curvature changes given by

$$\kappa_\theta = -\frac{w'}{r}, \qquad \kappa_r = -w'' \tag{19.5-5,6}$$

where primes indicate differentiation with respect to r. To obtain the simplified approximate strain-displacement relations (19.5-1) and (19.5-2), we have limited our consideration to shells of revolution that are *thin* and *shallow*, so that we can make the following two approximations:

1. *The Thin Shell Approximation* (Euler-Bernoulli hypothesis): The normals to the undeformed middle surface are deformed, without extension, into normals to the deformed middle surface (in particular $|\zeta z'| \ll r$).
2. *The Shallow Shell Approximation:* The difference between meridional slope and the (small) sloping angle may be disregarded ($z' = \tan \xi \cong \xi$).

Also, we are interested here only in infinitesimal strain problems, so that $|\epsilon_r| \ll 1$, $|\epsilon_\theta| \ll 1$, $|h\kappa_\theta| \ll 1$, and $|h\kappa_r| \ll 1$.

The straining of the deformable shell medium induces internal reactions within the medium (in the form of stress components) to resist the distortion from its natural state. For the class of linearly isotropic shell problems of interest here, two induced stress components σ_r and σ_θ are given in terms of e_r and e_θ by two generalized Hooke's laws. With the ζ-dependence of e_r and e_θ (and therefore σ_r and

σ_θ) known explicitly through the thinness approximation, it is desirable to eliminate the explicit appearance of ζ in the model by working with weighted averages of σ_r and σ_θ across the shell thickness. We introduce stress resultants, N_r and N_θ, and stress couples, M_r and M_θ, by the integrated relations

$$(N_r, N_\theta) = \int_{-h/2}^{h/2} (\sigma_r, \sigma_\theta)\, d\zeta, \qquad (M_r, M_\theta) = \int_{-h/2}^{h/2} (\sigma_r, \sigma_\theta)\zeta\, d\zeta \tag{19.5-7,8}$$

For the stress resultants and couples, four stress-strain relations of the form (see Ref. 19-47)

$$\epsilon_r = A(N_r - \nu N_\theta), \qquad \epsilon_\theta = A(N_\theta - \nu N_r) \tag{19.5-9a,b}$$

$$M_r = D(\kappa_r + \nu\kappa_\theta), \qquad M_\theta = D(\kappa_\theta + \nu\kappa_r) \tag{19.5-10a,b}$$

may be obtained from the generalized Hooke's law relating the e's and σ's. For a homogeneous material, we have in terms of Young's modulus E, Poisson's ratio ν, and h

$$A = \frac{1}{Eh}, \qquad D = \frac{Eh^3}{12(1-\nu^2)} \tag{19.5-11}$$

With the resultants and couples, we have effectively idealized the three-dimensional shell body as a two-dimensional surface (which is usually taken to be the middle surface of the shell) endowed with mechanical properties that are the two-dimensional analogues of the three-dimensional properties. The resultants and couples themselves are scalar fields defined on the surface, but they do not vary in the circumferential direction for axisymmetric problems.

The stress resultants and couples of the shell must be in equilibrium with external surface force and moment intensities (Fig. 19.5-2) for any portion of the midsurface. Force and moment equilibrium equations for the class of problems of interest here may be taken in the form

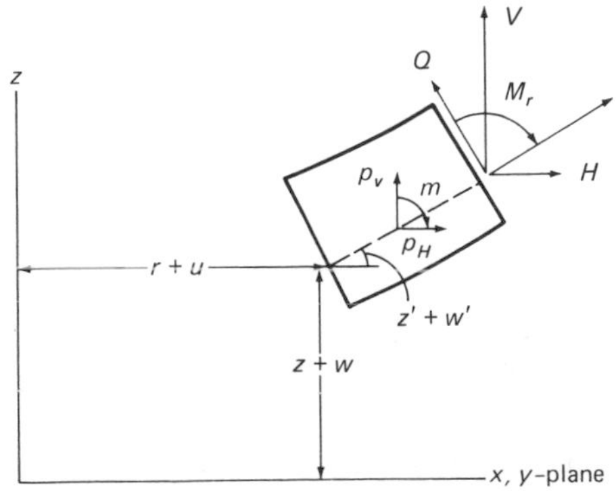

Fig. 19.5-2 Stress resultants, moment resultants and surface load intensities for axisymmetric bending and stretching of thin shells of revolution.

$$(rH)' - N_\theta + rp_H = 0, \qquad (rV)' + rp_V = 0 \tag{19.5-12,13}$$

$$(rM_r)' - M_\theta + rQ + rm = 0 \tag{19.5-14}$$

where p_H, p_V and m are surface load intensities. Consistent with the shallow shell approximation, the radial and axial stress resultants, H and V, are related to the meridional stress resultant N_r and a transverse shear resultant Q by the relations

$$Q = V - (z' + w')H, \qquad N_r = H - (z' + w')V \cong H \tag{19.5-15a,b}$$

The systems (19.5-3)-(19.5-6), (19.5-9), (19.5-10), and (19.5-13)-(19.5-15) are 13 equations for the 13 unknowns u, w, ϵ_r, ϵ_θ, κ_r, κ_θ, N_r, N_θ, Q, H, V, M_r, and M_θ. As a set of ODE, it is sixth-order. With six suitably prescribed auxiliary conditions, the system determines all field quantities in the model.

19.5.3 A Boundary-Value Problem for rH and w'

With load intensities p_H, p_V, and m being known functions of r, the ODE (19.5-13) can be integrated immediately to give

$$rV(r) = \frac{F_0}{2\pi} + \int_{r_i}^{r} p_V(x)x\,dx \tag{19.5-16}$$

where F_0 is a constant of integration, and r_i is the radial coordinate of some reference edge of the surface of revolution. The ODE (19.5-12) can be used to express N_θ in terms of $\psi \equiv rH$

$$N_\theta = \psi' + rp_H, \qquad H = \frac{\psi}{r} \tag{19.5-17}$$

while (19.5-15) gives Q and N_r in terms of rV and ψ

$$Q = V - (\xi + \phi)\frac{\psi}{r}, \qquad N_r = H = \frac{\psi}{r} \tag{19.5-18,19}$$

where we have set

$$\phi = w', \qquad \xi = z' \tag{19.5-20a,b}$$

Upon the introduction of (19.5-5) and (19.5-6) into (19.5-10), we get

$$M_r = -D\left(\phi' + \frac{\nu}{r}\phi\right), \qquad M_\theta = -D\left(\nu\phi' + \frac{1}{r}\phi\right) \tag{19.5-21a,b}$$

The expressions (19.5-21) and (19.5-19) reduce the moment equilibrium equation (19.5-14) to a second-order ODE for ϕ and ψ. For a shell of constant thickness and uniform material properties, this equation takes the form

$$D\left[\phi'' + \frac{1}{r}\phi' - \frac{1}{r^2}\phi\right] - \frac{1}{r}(\xi + \phi)\psi = -V + m \tag{19.5-22}$$

To get a second equation for ϕ and ψ, we write (19.5-3) as $u = r\epsilon_\theta$ and use it to eliminate u from (19.5-4) to get a compatibility equation (see also subsection 4.7.4), $\epsilon_r = (r\epsilon_\theta)' + (z' + \frac{1}{2}\phi)\phi$. Now ϵ_r and ϵ_θ may be expressed in terms of ψ upon substituting (19.5-17) and (19.5-19) into (19.5-9). The resulting expressions may in turn be used to write the compatibility equation as an equation for ψ and ϕ. For a shell of uniform thickness and material properties, this equation takes the form

$$A\left[\psi'' + \frac{1}{r}\psi' - \frac{1}{r^2}\psi\right] + \frac{1}{r}\left(\xi + \frac{1}{2}\phi\right)\phi = -A\,(rp_H)' - (1 + \nu)Ap_H \tag{19.5-23}$$

At this point, we have exhausted the content of the original sixth-order simultaneous system of 13 equations. The reduction process results in three uncoupled subsystems:

1. the simple equation (19.5-16), which determines $V(r)$ up to a constant F_0
2. the fourth-order system of two second-order ODEs (19.5-22) and (19.5-23) for ϕ and ψ
3. the single equation (19.5-20a), which determines w up to a constant w_0

$$w(r) = w_0 + \int_{r_i}^{r} \phi(x)\,dx \tag{19.5-24}$$

The three subsystems should be solved in the order they are listed. These subsystems also indicate the appropriate boundary conditions for the problem. Evidently, one condition must involve the resultant axial force for the determination of F_0. A second condition must fix the vertical position of some edge of the shell to determine w_0. The remaining four conditions may be prescribed in terms of the radial stress resultant H and bending moment M_r or in terms of the radial distplacement component u and meridional change of slope $\phi = w'$ or some combinations of these quantities.

As a specific example of a BVP in the class of problems for which the model developed in subsection (19.5.2) is appropriate, consider a frustum of shallow shell of revolution with an inner edge at $r = r_i > 0$ and an outer edge at $r = r_0 > r_i$. The shell is steadily rotating about its axis of revolution with a constant angular rotating speed ω, resulting in an outward radial inertia force intensity as its only external load. In that case, we have $p_H = \rho h\omega^2 r$, $p_V \equiv 0$, and $m \equiv 0$, where ρ is the volume mass density of the shell material, and $rV = 0$. (The fact that there is no resultant axial force acting on any part of the shell requires that F_0 be set equal to zero.) With V completely determined, we may now simplify (19.5-22) and (19.23) to

$$\left\{\begin{aligned} &D\left[\phi'' + \frac{1}{r}\phi' - \frac{1}{r^2}\phi\right] - \frac{1}{r}(\xi + \phi)\psi = 0 && (19.5\text{-}25)\\ &A\left[\psi'' + \frac{1}{r}\psi' - \frac{1}{r^2}\psi\right] + \frac{1}{r}\left(\xi + \frac{1}{2}\phi\right)\phi = -(3+\nu)A\rho h\omega^2 r && (19.5\text{-}26)\end{aligned}\right.$$

The shell is free of any edge load at both its edges, so that

$$r = r_i, r_0: \qquad N_r = \frac{\psi}{r} = 0, \qquad M_r = -D\left(\phi' + \frac{\nu}{r}\phi\right) = 0 \tag{19.5-27}$$

Having $\phi(r)$, we then determine $w(r)$ by (19.5-24) up to a constant, as the stress and strain of the shell are unaffected by a vertical rigid translation. We may fix the shell in space by setting $w(r_i) = 0$, say. All stress, strain, and curvature change measures are determined by ψ and ϕ through (19.5-17)-(19.5-19), (19.5-21), (19.5-9), and (19.5-10). Finally u is given by (19.5-3).

The BVP is nonlinear because of the $\phi\psi$ term in (19.5-25) and the ϕ^2 term in (19.5-26). There is no other nonlinearity in the problem. For a given shell of revolution, $z' \equiv \xi$ is known, e.g., $z' = \xi_0$ (a constant) for a conical shell and $z' = (r - a)/R$ for a toroidal cap where a is the center line radius and R is the radius of the (circular) cross section.

19.5.4 Linear Model of Steadily Rotating Shallow Shells

Unless there is some kind of instability lurking around, it is a cardinal rule in particle, rigid body, and (classical) continuum mechanics that, for a sufficiently small external load, the response is small in amplitude, and nonlinear effects in the relevant mathematical model associated with products or positive powers of the response may be neglected with no serious loss of qualitative or quantitative accuracy. Several centuries of scientific and engineering successes resulting from its applications have made it a platitude to enunciate the rule as a good working principle in mathematical modeling. In fact, there is an alarmingly excessive (if not total) reliance on past successes and a posteriori consistency arguments for linearization in mathematical modeling today. The rotating shell problem indicates that there can be exceptions to this general rule.

For a linear model of axisymmetric bending and stretching of shallow shells of revolution, we have only to omit the two quadratic terms in the fourth-order system (19.5-22) and (19.5-23) for ϕ and ψ. For a steadily rotating shallow shell of revolution, these equations become

$$\left\{\begin{aligned} &D\left[\phi_L'' + \frac{1}{r}\phi_L' - \frac{1}{r^2}\phi_L\right] - \frac{1}{r}\xi\psi_L = 0\\ &A\left[\psi_L'' + \frac{1}{r}\psi_L' - \frac{1}{r^2}\psi_L\right] + \frac{1}{r}\xi\phi_L = -(3+\nu)A\rho h\omega^2 r\end{aligned}\right. \tag{19.5-28}$$

where we denote by a subscript L the solution of the linear (bending) model. The free-edge (boundary) conditions in (19.5-27) remain unchanged, as they do not contain any nonlinear term. The linear BVP defined by (19.5-28) and (19.5-27) may be solved by a number of available methods, once the shape function $\xi(r)$ is prescribed. We note in particular that the BVP uncouples into two simpler problems if $\xi(r) \equiv 0$, i.e., if the shell is in fact a flat plate. The unique solution for the plate bending problem, defined by the first ODE in (19.5-28) with $\xi \equiv 0$ and the second boundary condition in (19.5-27) for both edges, is the trivial solution $\phi_L \equiv 0$. The unique solution for the plate stretching (or generalized plane stress) problem, defined by the remaining half of (19.5-27) and (19.5-28) is just the well-known solution for a rotating circular disc, Ref. 19-46.

If ξ does not vanish identically, but $|\xi|$ is sufficiently small, the shell behavior is not expected to be qualitatively different from the rotating disc solution. Intuitively, we expect the structure to be shell-like only if its highest point rises significantly above the lowest point or, more correctly (as large or small is meaningful only for dimensionless combinations of parameters), if the thickness-to-rise ratio is small compared with unity. This turns out to be the case, as we shall see from a dimensionless form of the BVP. For this dimensionless form, we introduce the dimensionless quantities

$$x = \frac{r}{r_0}, \qquad \xi(r) = \xi_0 s(x)$$
$$\phi_L(r) = \bar{\phi}_L f_L(x), \qquad \psi_L(r) = \bar{\psi}_L g_L(x) \tag{19.5-29}$$

where ξ_0 is a known constant, and $\bar{\phi}_L$ and $\bar{\psi}_L$ are as yet undetermined amplitude factors, all chosen so that s, f_L, and g_L are $0(1)$ quantities. For a shallow conical shell, we have $s(x) \equiv 1$. For a toroidal cap, we may take $\xi_0 = a/R$, so that $s(x) = \delta x - 1$, where $\delta = r_0/a$, with $1 < \delta < 2$, and $0 < r_i/a < 1$. Consistent with the shallowness approximation, we have $\xi_0 r_0$ as the differential rise of the two edges, so that $h/(\xi_0 r_0)$ is of the order of the thickness-to-rise ratio if r_i is not nearly r_0.

In terms of the dimensionless quantities in (19.5-29) and with $(\ )^{\cdot} \equiv d(\ )/dx$, the two ODEs in (19.5-28) become

$$\frac{D\bar{\phi}_L}{r_0 \xi_0 \bar{\psi}_L}\left[f_L^{\cdot\cdot} + \frac{1}{x} f_L^{\cdot} - \frac{1}{x^2} f_L\right] - \frac{s}{x} g_L = 0 \tag{19.5-30}$$

$$\frac{A\bar{\psi}_L}{r_0 \xi_0 \bar{\phi}_L}\left[g_L^{\cdot\cdot} + \frac{1}{x} g_L^{\cdot} - \frac{1}{x^2} g_L\right] + \frac{s}{x} f_L = -\frac{A\rho h \omega^2 r_0^2}{\xi_0 \bar{\phi}_L}(3+\nu)x \tag{19.5-31}$$

For the shell not to be platelike, the only term with $\xi(r)$ as a multiplicative factor in both (19.5-30) and (19.5-31) must not be small compared with other terms of the same equation. It follows that we must take

$$\bar{\phi}_L = \frac{A\rho h \omega^2 r_0^2}{\xi_0} \tag{19.5-32}$$

while the remaining two dimensionless combinations must be $0(1)$ at most. If $\bar{\psi}_L$ is chosen so that one of them is unity, then the other would be

$$\epsilon^4 = \frac{DA}{r_0^2 \xi_0^2} = \frac{h^2}{12(1 - \nu^2) r_0^2 \xi_0^2} \ll 1 \tag{19.5-33}$$

Our experience with singular perturbation problems (see chapter 14) rules out either choice of $\bar{\psi}_L$ and suggests instead

$$\bar{\psi}_L = \frac{r_0 \xi_0 \bar{\phi}_L}{A} \epsilon^2 = \rho h \omega^2 r_0^3 \epsilon^2 \tag{19.5-34}$$

so that (19.5-30) and (19.5-31) take on a more symmetric form,

$$\epsilon^2 \left[\ddot{f}_L + \frac{1}{x} \dot{f}_L - \frac{1}{x^2} f_L \right] - \frac{s}{x} g_L = 0 \tag{19.5-35}$$

$$\epsilon^2 \left[\ddot{g}_L + \frac{1}{x} \dot{g}_L - \frac{1}{x^2} g_L \right] + \frac{s}{x} f_L = -(3 + \nu) x \tag{19.5-36}$$

while the boundary conditions (19.5-27) become

$$x = x_i, 1: \qquad \epsilon^2 g_L = 0, \qquad \epsilon^4 \left[\dot{f}_L + \frac{\nu}{x} f_L \right] = 0 \tag{19.5-37}$$

With $0 < x_i \leqslant x \leqslant 1$ and $\epsilon^2 \ll 1$, the *linear* BVP is in the form of a singular perturbation problem. If $s(x) \neq 0$, its solution may be taken as a linear combination of a smoothly varying (*outer*) solution and two rapidly varying layer solutions. The leading-term outer solution is the *linear membrane* solution

$$g_{LM} \equiv 0 \qquad f_{LM}(x) = -(3 + \nu) \frac{x^2}{s} \tag{19.5-38}$$

which is dominant (at least) away from the edges. The layer (edge bending) solutions contribute significantly only in a small interval ($0(\epsilon)$ compared with the shell span) adjacent to one edge (and one solution for each edge), decaying rapidly to zero a short distance away from the edge. With g_{LM} itself satisyfing the boundary conditions $g_L(x_i) = g_L(1) = 0$ and the remaining conditions to be satisfied involve $\dot{f}$, the layer solutions must be $0(\epsilon)$ compared to the linear membrane solution. Therefore, the exact solution of the linear model, when $s(x) \neq 0$ (which is the case for conical and spherical shells, for example), is the linear membrane solution (19.5-38) except for $0(\epsilon)$ terms.

For shells with $s(x_t) = 0$ for some x_t inside the interval $(x_i, 1)$, e.g., $s(1/\delta) = 0$ for toroidal caps with $\delta = a/r_0$, the leading-term outer solution is no longer the linear membrane solution, although it tends to the latter away from the "turning point" x_t of the ODEs. For a toroidal cap, this leading-term outer solution is a combination of Airy and Lommel functions (Refs. 19-48, 19-49, and 19-50). The qualitative behavior of shells with turning points in the ODEs (19.5-35) and (19.5-36) (corresponding to various kinds of flat points with a horizontal tangent along the shell meridians) has been thoroughly analyzed and fully documented ever since R. A. Clark's pioneering work on that subject.

With the above solution of the linear bending model, (19.5-35)-(19.5-37), we may now estimate the contribution of the neglected terms, $\phi\psi/r$ in (19.5-25) and $\phi^2/2r$ in (19.5-26). For shells with no horizontal meridional slope, the neglected nonlinear terms are $0(\overline{\phi}_L/\xi_0)$ compared to the most dominant term retained in the same ODE. The contributions of the nonlinear terms, as estimated by the solution of the linearized model, are insignificant whenever the rotating speed is sufficiently slow that $\overline{\phi}_L/\xi_0 \ll 1$, i.e., the magnitude of the change in meridional slope must be small compared with the magnitude of the undeformed slope. For shells with a horizontal meridional slope at one or more locations, the dominant term retained in the ODEs is no longer the one with $s(x)$ as a multiplicative factor, at least not near a turning point. An estimate based on the solution of the linear model suggests the more stringent requirement of $\overline{\phi}_L/\xi_0 \ll \epsilon^{4/3}$. In either case, the consistency criterion for the adequacy of a linear model can always be met by a sufficiently small rotating speed, as ϵ is a measure of the shell geometry and does not involve ω. In other words, there is always a range of positive ω, possibly depending on shell geometries, for which the solution of the linear problem provides a quantitatively and qualitatively accurate approximation of the solution of the nonlinear model.

Experience with the linear bending solution of toroidal shells indicates that the solution ϕ_L changes sign near the turning point $x_t = 1/\delta$. Actual computation shows, as we shall see later, that the corresponding deformed meridional slope $\xi + \phi_L$ is always negative for some interval to the far side of the turning point away from the axis of revolution. Given that the inertia force loading is radially outward and increasing with distance from the axis of revolution, this negative deformed meridional slope is *qualitatively incorrect* however small the amplitude of the negative deformed slope; the radially outward centrifugal force should always keep any flattened portion of the shell from turning downward (or upward)! To get some insight into the actual shell deformation, we describe in the next subsection the nonlinear membrane solution obtained in Ref. 19-51 for the toroidal cap.

19.5.5 Nonlinear Membrane Solution

We now return to the original nonlinear BVP defined by (19.5-25)-(19.5-27) and consider the situation when the shell is "very thin."* Given that D is proportional to h^3, we may, to a first approximation, neglect terms in (19.5-25) with D as a multiplicative factor, leaving us with

$$\frac{1}{r}(\xi + \phi_{NM})\psi_{NM} = 0 \qquad (19.5\text{-}39)$$

*This statement will be restated in a meaningful and correct form later, when we work with a dimensionless form of the nonlinear BVP.

where we have denoted by a subscript NM this so-called *nonlinear membrane* approximation. Equation (19.5-39) may be satisfied in two ways:

1. $\psi_{NM} \equiv 0$: The corresponding ϕ_{NM} is now determined by (19.5-26) to be

$$\phi_{NM} = -\xi + \sqrt{\xi^2 + 2\xi\phi_{LM}} \tag{19.5-40}$$

where $\phi_{LM} = \bar{\phi}_L f_{LM} = -(3+\nu)A\rho h\omega^2 r^2/\xi$. Evidently, this solution is not applicable around a turning point, because the quantity inside the square root is negative there.

2. $\xi + \phi_{NM} \equiv 0$: The corresponding solution ψ_{NM} is now determined by (19.5-26), which may now be written as

$$A\left[\psi''_{NM} + \frac{1}{r}\psi'_{NM} - \frac{1}{r^2}\psi_{NM}\right] = \frac{\xi^2}{2r} - (3+\nu)A\rho h\omega^2 r \tag{19.5-41}$$

and may be solved along with the boundary condition $\psi_{NM} = 0$ at $r = r_i$ and $r = r_0$. This is merely a rotating disc problem with a part of the original inertia force intensity "used up" to flatten the shell into a disc so that the deformed slope is horizontal (as required by $\xi + \phi_{LM} \equiv 0$). Intuitively, we do not expect this solution to be appropriate for very small ω, for the amplitude of the shell response ϕ should also be small in that case.

It appears that both types of nonlinear membrane solutions should be rejected for a toroidal cap with a sufficiently small ω. However, it was recognized (see Ref. 19-51) that each may apply to a different region of the solution domain and that a suitable combination of these solutions does provide a good first approximation for the exact solution of the original nonlinear model except for layer phenomena, with the accuracy of the approximate solution improved with decreasing thickness (more precisely, as $\epsilon \to 0$).

For a toroidal cap, we must have for this combination

$$\phi_{NM} = -\xi \qquad (r_{ti} < r < r_{to}) \tag{19.5-42}$$

along with a ψ_{NM} from (19.5-41), for some interval (r_{ti}, r_{to}) containing the turning point $r_t = r_0 x_t$ in the interior. For a fixed $\omega > 0$, the right side of (19.5-40) is complex only for $r_{ci} < r < r_{co}$ where r_{ci} and r_{co} are the two roots of $\xi^2(r) + \xi(r)\phi_{LM}(r) = \xi^2(r) - 2(3+\nu)A\rho h\omega^2 r^2 = 0$ with $r_{ci} < r_t < r_{co}$. The limitation on the range of (19.5-42) requires $r_{ti} \leqslant r_{ci}$ and $r_{to} \geqslant r_{co}$. Outside the interval (r_{ti}, r_{to}), it is possible to take

$$\psi_{NM} \equiv 0, \quad \phi_{NM} = -\xi + \sqrt{\xi^2 + 2\xi\phi_{LM}}$$
$$(r_i < r < r_{ti} \leqslant r_{ci}, \quad r_{co} \leqslant r_{to} < r < r_0) \tag{19.5-43}$$

The solution (19.5-41)–(19.5-43) satisfies the condition on ψ in (19.5-27) at both edges, r_i and r_0, but not the condition on ϕ at these edges. In addition, ϕ_{NM} and

ψ_{NM} are generally not continuous across the transition points r_{ti} and r_{to} where one solution takes over from the other. We may use the two constants of integration from the solution of (19.5-41) and the two yet unspecified parameters r_{ti} and r_{to} to meet some of the expected continuity requirements. Now, ψ_{NM} can be made continuous across r_{ti} and r_{to} by using $\psi_{NM}(r_{ti}) = \psi_{NM}(r_{to}) = 0$ as two boundary conditions for the second order ODE (19.5-41) and thereby fixing the two constants of integration. We can make ϕ_{NM} continuous across the transition points by taking $r_{ti} = r_{ci}$ and $r_{to} = r_{co}$. Alternatively, we can choose $r_{ti} < r_{ci}$ and $r_{to} > r_{co}$ to make ψ'_{NM}, and thereby the radial displacement u, continuous across the transition points. We choose the latter on the grounds that a discontinuity in the deformed slope can be removed by a layer solution associated with the shell's bending stiffness, which has been ignored completely in the above nonlinear membrane solution (as we have effectively set $D = 0$ to get (19.5-39)). The same bending stiffness also introduced another layer solution of much smaller amplitude at each edge to allow for the satisfaction of the moment-free condition on ϕ at the two edges.

The above nonlinear membrane solution gives a deformed configuration for the toroidal cap with a deformed meridional slope everywhere not as steep as the undeformed slope, i.e., $|\xi + \phi| \leqslant |\xi|$. This is consistent with the radially outward inertia loading, which works to flatten the shell with a strength in direct proportion to the distance away from the axis of revolution. Moreover, once a portion of the shell is flattened, it will stay flat as the radial load prevents it from sloping in either direction. A toroidal cap has a horizontal slope at its crown ($r = a$) and an increasingly steeper slope away from the crown in both directions. For a sufficiently high rotating speed, the most shallow region of the cap around $r = a$ begins to flatten first; this flattened region spreads in both directions as ω increases further. Therefore, except for the moment-free boundary conditions at the edges and the continuity of ϕ across the junction r_{ti} and r_{to} (to be satisfied with the addition of bending layers), the nonlinear membrane solution given by (19.5-42) and (19.5-43) is physically more acceptable than the linear solution of the last subsection, even though the latter meets the small-amplitude consistency requirements for sufficiently small ω. We shall see in the next subsection that this nonlinear membrane solution is in fact an accurate approximation of the exact solution for $\epsilon \ll 1$, except for the layer solutions mentioned.

19.5.6 Exact Solution for Toroidal Caps

For the actual solution of the nonlinar BVP for a rotating toroidal cap, we tentatively scale, as we did for the linear case, to get

$$\epsilon^2\left[f^{\cdot\cdot} + \frac{1}{x}f^{\cdot} - \frac{1}{x^2}f\right] - \frac{1}{x}(s + \kappa f)g = 0 \tag{19.5-44}$$

$$\epsilon^2\left[g^{\cdot\cdot} + \frac{1}{x}g^{\cdot} - \frac{1}{x^2}g\right] + \frac{1}{x}\left(s + \frac{1}{2}\kappa f\right)f = -(3+\nu)x \tag{19.5-45}$$

where

$$\kappa = \frac{\bar{\phi}_L}{\xi_0} = \frac{A\rho h\omega^2 r_0^2}{\xi_0^2} \tag{19.5-46}$$

While linearization suggests itself if $\kappa \ll 1$ and $f = 0(1)$ at most, the results for the linear (bending) theory and nonlinear membrane model indicate that it may not be appropriate to do so.

In view of the structure of (19.5-44) and (19.5-45) for $\epsilon^2 \ll 1$, an approximate solution of the nonlinear problem may be obtained by the method of matched asymptotic expansions. Unlike the linear problem, it is now possible to have more than one type of outer solution, depending on the magnitude of g. We limit ourselves here to the interesting case of $\kappa = 0(1)$ at most.* If g is $0(1)$ at most, then we have a perturbation series solution in powers of ϵ^2 with the leading term being (19.5-40). On the other hand, g may be large, of order ϵ^{-2} (ψ should have been scaled by $A\rho h\omega^2 r_0^3$ and not by $\bar{\psi}_L \equiv A\rho h\omega^2 r_0^3 \epsilon^2$), so that we either rescale g or seek a regular perturbation solution for $\epsilon^2 g$. Either way, we get as a leading term solution the other nonlinear membrane solution $\phi_{NM} = -\xi$. In fact, with ψ of order $A\rho h\omega^3 r_0^3$, the leading term perturbation solution allows for both types of nonlinear membrane solutions. The correct choice of the outer solution is to be complemented by appropriate inner solutions. Evidently, one inner solution is needed in a narrow region adjacent to each edge, as the outer solution cannot satisfy the traction free boundary conditions (19.5-27), which may be written in the dimensionless form

$$x = x_i, 1: \quad \epsilon^2 g = 0, \quad \epsilon^4 \left[f^{\cdot} + \frac{\nu}{f} f \right] = 0 \tag{19.5-47}$$

The construction of these edge-zone-type inner solutions is well understood (Ref. 19-47). For an outer solution of the form (19.5-42) and (19.5-43), an additional inner solution is needed in the interior to bridge the two different types of outer solution for two adjacent portions of the shell. Such an inner solution exists under favorable load and geometric conditions (Ref. 19-51).

We will not pursue a discussion on the construction of the layer solutions here. Instead, we give in Fig. 19.5-3 the graph of $(\phi + \xi)/\xi_0$ for a toroidal cap, $s(x) = \delta x - 1$, with a typical set of load and geometric parameter values ($\epsilon^2 = 10^{-4}, x_i = \frac{1}{3}$, $\delta = r_0/a = 1.5$, $\nu = 0.3$, and $\kappa = 0.014281$). This and similar solutions for other sets of parameter values are obtained numerically by a spline-collocation method for the nonlinear BVP accurate to at least five significant figures. For the solution given in Fig. 19.5-3, the parameter κ characterizing the nonlinear effects is $0(10^{-2})$; assuming that ϕ and ψ have been properly scaled by $\bar{\psi}_L$ and $\bar{\phi}_L$, it strongly suggests that the nonlinear effects are negligible. An estimate of the magnitude of the nonlinear terms using κf_L shows $|\kappa f_L| < 0.26$. Yet, the linear bending solution, also shown in Fig. 19.5-3, is quantitatively and qualitatively incorrect for a shell span occupying slightly more than the middle third portion of the toroidal cap

*For $\kappa \gg 1$, we have essentially a rotating disc problem with a small perturbation after rescaling.

surrounding the turning point of the ODE at $r = a$. In contrast, the nonlinear membrane solution given by (19.5-42) and (19.5-43) (also shown in Fig. 19.5-3) is indistinguishable from the exact solution except for layer phenomena of the type described earlier. According to both, the deformed shell has a flattened region extending over the middle third of the shell span having the turning point nearly at the center. The linear bending solution, on the other hand, describes a deformed configuration that has a wrinkle around the turning point and extending over the same middle third region. For $\kappa = 0.02892$ (roughly doubling the first value but still very small), the corresponding solutions for the linear and nonlinear model show an enlargement of the region of discrepancy from one-third to more than one-half of the shell span. Although the magnitude and region of this discrepancy diminish with decreasing load magnitude (for rotating speed), the wrinkle in the linear bending solution persists. Physically, the shell perfers to resist the straining from the radially outward inertia force intensity by its stretching stiffness through the nonlinear membrane shell action; the linear theory strips the shell of this option

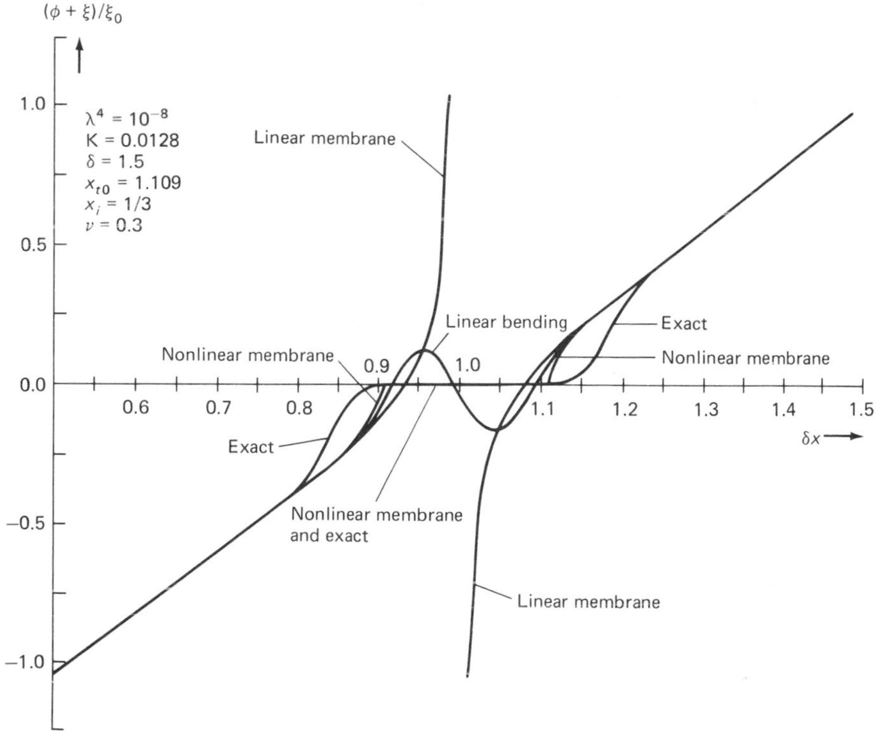

Fig. 19.5-3 The normalized deformed meridional slope of a steadily rotating shallow elastic toroidal cap by linear membrane, linear bending, nonlinear membrane and exact (nonlinear bending) models.

and forces it to respond to the straining by developing its bending actions and hence the wrinkling. To the extent that the linear and nonlinear theory model two intrinsically different shell actions, the linear theory cannot be a first approximation for the nonlinear theory for an inertia load of any magnitude.

19.6 STEPPING BEYOND REALITY

19.6.1 A Corrugated Lid for a Rigid Container

Many mathematical models are useful beyond the study of the phenomenon for which the models were originally developed. We will see in the next section that a model for a new phenomenon may have the same mathematical structure as an existing model for a completely different phenomenon and may therefore be analyzed by similar mathematical techniques. In this section, we show, by way of an example, that an existing mathematical model, when viewed from a proper perspective, may also be an appropriate model for a seemingly unrelated class of problems.

Most household kitchens have a few plastic food containers, essentially cannisters fitted with a detachable lid. Among those commercially available are some with a polarly corrugated, circular lid fitted to an effectively rigid cylindrical cannister Fig. 19.6-1. The lid consists of a corrugated annular "plate" attached to a small thin flat disc at its inner edge and to a *ring lip* at the outer edge. The corrugations are periodic in the circumferential direction with coplanar troughs. The roofs of the corrugations slope radially downward toward the center so that their height (measured from the plane of the troughs) increases with the radial distance from the center. The corrugated lid is fitted to the rigid cannister by pushing down on the center of the small inner circular disc, which results in a radial contraction of the lid, so that the thin wall of the cannister fits into the ring lip of the lid. The polar corrugations make the lid more flexible and facilitate the (air-tight) fitting with the cannister.

To design and produce such food containers, it is important to know the magnitude of the push-down force required to fit the ring lip to the cannister and the sealing pressure after the fitting, as functions of various design parameters, e.g., the number and shape of the corrugations, the material properties and the thickness of the corrugated "plate," the material properties, thickness and size of the inner disc, and structural properties of the lip, etc. A tractable mathematical model of the lid is therefore needed to obtain these two load-deformation relations as well as other design information, such as the life expectancy of the lip with regard to fatigue failure, etc.

To attempt an accurate solution of the above structural mechanics problem by a (three-dimensional) continuum mechanics model would be unrealistic. For a more tractable model, we observe that the corrugated annular portion of the lid is essentially a thin layer and the corrugation height is small compared with the overall dimension of the lid. These structural features suggest that a shell or plate model would be appropriate. With the radial slanting of the corrugations and the observed coupling between bending and stretching actions (as the push-down force does

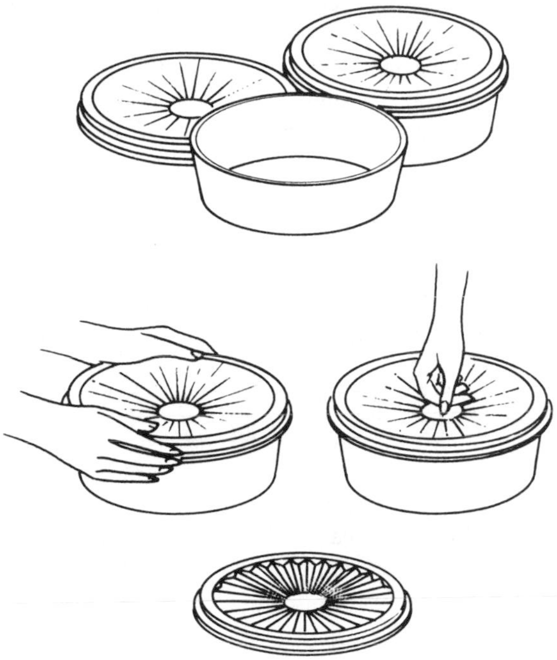

Fig. 19.6-1 Rigid cylindrical cannister with a corrugated circular lid.

induce a radial contraction), we should attempt a shell model. Though the loading of interest is axisymmetric, the structure is not naturally a shell of revolution; a more general shell model than the one in subsection 19.5.2 seems to be necessary for the analysis of the lid. However, with sufficiently numerous (closely spaced) corrugations in the actual design, it is intuitively reasonable to expect effective polar symmetry in the corrugated structure if we average all structural properties over a period of the periodic corrugations. Therefore, we treat the corrugated structure as a conical shell of revolution with a midsurface located halfway between the roof and the trough of the corrugations. The equivalent conical shell is shallow, as the slanting of the roofs is slight. Because of the polar corrugations, the shell is more flexible in the circumferential direction than the radial direction. The shell model of subsection 19.5.2 must be modified to reflect this simple directional perference in structural properties, known as *orthotropy*.

Orthotropic shell theories have been formulated for various types of orthotropic materials (e.g., Ref. 19-47); they are not applicable to our equivalent conical shell whose orthotropicity is geometric in origin (induced by the corrugations). An appropriate orthotropic shell theory for our equivalent shell will be developed in subsections 19.6.2 and 19.6.3. Two BVPs directly relevant to design studies will be formulated in subsection 19.6.4.

The design data generated by the seemingly crude model of the actual structure are in excellent agreement with available experimental data. This is rather unexpected, as the step from a thin corrugated structure to an equivalent shallow conical shell of revolution has no real mathematical or scientific justification. It is a part of the process of idealization and simplification so important to the construction of useful mathematical models. Unfortunately, there is no recipe for this process; experience and past successes are the only guides available.

19.6.2 Axisymmetric Deformation of Shallow Orthotropic Shells of Revolution

For a material that is polarly orthotropic in the tangent plane of the midsurface of a shell of revolution, the relevant stress-strain relations for the shell medium are

$$e_\theta = \frac{\sigma_\theta}{E_\theta} - \frac{\sigma_r}{E_{\theta r}}, \qquad e_r = \frac{\sigma_r}{E_r} - \frac{\sigma_\theta}{E_{r\theta}} \tag{19.6-1}$$

where the elastic moduli E_θ, E_r, and $E_{r\theta} = E_{\theta r}$ may be functions of the thickness coordinate ζ and the radial coordinate r, i.e., the material may be inhomogeneous. The relations (19.6-1) may be inverted to give

$$\sigma_\theta = E_\theta^* e_\theta + E_{\theta r}^* e_r, \qquad \sigma_r = E_r^* e_r + E_{r\theta}^* e_\theta \tag{19.6-2}$$

where

$$\begin{aligned} E_\theta^* &= \frac{E_\theta}{1 - \nu^2}, & E_r^* &= \frac{E_r}{1 - \nu^2} \\ E_{r\theta}^* &= E_{\theta r}^* = \frac{\nu^2 E_{r\theta}}{1 - \nu^2}, & \nu^2 &= \frac{E_\theta E_r}{E_{r\theta}^2} \end{aligned} \tag{19.6-3}$$

To obtain a set of stress-strain relations for stress resultants and stress couples, we again use (19.5-1) and (19.5-2) in (19.6-2) and integrate them across the shell thickness. In the most general situation treated in Ref. 19-47, each of the stress resultants and couples is related by an algebraic relation to all four quantities ϵ_r, ϵ_θ, κ_r, and κ_θ. We consider here only the class of orthotropic materials with all four moduli being even functions of the thickness coordinate, so that the four relations uncouple into two groups

$$N_\theta = C_\theta \epsilon_\theta + C_{\theta r} \epsilon_r, \qquad N_r = C_r \epsilon_r + C_{r\theta} \epsilon_\theta \tag{19.6-4}$$

and

$$M_\theta = D_\theta \kappa_\theta + D_{\theta r} \kappa_r, \qquad M_r = D_r \kappa_r + D_{r\theta} \kappa_\theta \tag{19.6-5}$$

where the stretching and bending stiffness factors, C and D,

$$\{C, D\} = \int_{-h/2}^{h/2} \{1, \zeta\} E^* \, d\zeta \tag{19.6-6}$$

may be functions of r. For a homogeneous medium, we have

$$C_r = \frac{E_r h}{1 - \nu^2}, \quad C_\theta = \frac{E_\theta h}{1 - \nu^2}, \quad C_{r\theta} = C_{\theta r} = \frac{\nu^2 E_{r\theta} h}{1 - \nu^2} \tag{19.6-7}$$

$$D_r = \frac{E_r h^3}{12(1 - \nu^2)}, \quad D_\theta = \frac{E_\theta h^3}{12(1 - \nu^2)}, \quad D_{r\theta} = D_{\theta r} = \frac{\nu^2 E_{r\theta} h^3}{12(1 - \nu^2)} \tag{19.6-8}$$

The relations (19.6-4) will be used in their inverted forms

$$\epsilon_\theta = A_\theta N_\theta + A_{\theta r} N_r, \qquad \epsilon_r = A_r N_r + A_{r\theta} N_\theta \tag{19.6-9}$$

where

$$\begin{aligned} A_\theta &= \frac{1}{C_\theta (1 - \nu_s^2)}, & A_r &= \frac{1}{C_r (1 - \nu_s^2)} \\ A_{r\theta} = A_{\theta r} &= -\frac{\nu_s^2}{C_{r\theta} (1 - \nu_s^2)}, & \nu_s^2 &= \frac{C_{r\theta}^2}{C_\theta C_r} \end{aligned} \tag{19.6-10}$$

For a homogeneous material, (19.6-10) becomes

$$A = \frac{1}{E_\theta h}, \quad A_r = \frac{1}{E_r h}, \quad A_{r\theta} = A_{\theta r} = -\frac{1}{E_{r\theta} h} \tag{19.6-11}$$

Whether the material is homogeneous or not, the important feature of the stress-strain relations (19.6-4) (or (19.6-9)) and (19.6-5) relevant to later developments is that the orthotropicity of the shell has its origin in the material orthotropicity of the shell medium, as reflected in (19.6-1).

With (19.6-5) and (19.6-9) taking the place of (19.5-9) and (19.5-10), the set of differential equations of equilibrium (19.5-12) to (19.5-14) and strain-displacement relations, (19.5-3) to (19.5-6), can again be reduced to three uncoupled subsystems. Two of these remain as before, with (19.5-16) determining the resultant axial force $2\pi r V$ and (19.5-24) determining the axial displacement component, w, respectively. The third subsystem, namely the two coupled second order ODEs for ϕ and ψ, now take the form

$$\phi'' + \frac{1}{r}\left(1 + r\frac{D_r'}{D_r}\right)\phi' - \frac{1}{r^2}\left(\frac{D_\theta}{D_r} - r\frac{D_{r\theta}'}{D_r}\right)\phi - \frac{1}{rD_r}(\xi + \phi)\psi$$
$$= \frac{1}{D_r}(-V + m) \tag{19.6-12}$$

$$\psi'' + \frac{1}{r}\left(1 + r\frac{A_\theta'}{A_\theta}\right)\psi' - \frac{1}{r^2}\left(\frac{A_r}{A_\theta} - r\frac{A_{r\theta}'}{A_\theta}\right)\psi + \frac{1}{rA_\theta}\left(\xi + \frac{1}{2}\phi\right)\phi$$
$$= -\frac{1}{A_\theta}(A_\theta r p_H)' - \left(1 + \frac{A_{r\theta}}{A_\theta}\right)p_H \tag{19.6-13}$$

where a prime again indicates differentiation with respect to r. The system (19.6-12) and (19.6-13), along with four suitably prescribed boundary conditions, determines ϕ and ψ. We may then compute all other stress and displacement quantities associated with this subsystem from (19.5-17) to (19.5-19),

$$M_r = -D_r\phi' - \frac{1}{r}D_{r\theta}\phi, \quad M_\theta = -D_{\theta r}\phi' - \frac{1}{r}D_\theta\phi \tag{19.6-14}$$

and

$$u = r\epsilon_\theta = rA_\theta\psi' + A_{\theta r}\psi + r^2A_\theta p_H \tag{19.6-15}$$

19.6.3 Orthotropicity of the Corrugated Lid

Before we can apply the shallow orthotropic shell model developed in the last subsection, we have to obtain the stretching and bending stiffness factors $\{C\}$ and $\{D\}$ of the equivalent orthotropic shell for our corrugated structure. It is important to realize that they cannot be obtained from the definition of these factors (19.6-6) for a genuine orthotropic material. The material for our lid is in fact homogeneous and isotropic! However, it is physically clear that the lid is more flexible in both stretching and bending in the circumferential direction. The source of this property of preferred stiffness direction is the lid's geometry, not its material, and we must somehow translate this geometrical effect into an equivalent material orthotropy.

The linearly increasing corrugation height with radial distance from the lid center suggests that we model the lid as a conical shell (or revolution) with a middle surface at midheight of the corrugations. As the maximum corrugation height h_0 is small compared with the radial dimension r_0, $h_0/r_0 \ll 1$, the equivalent shell is shallow with a midsurface meridional slope function $\xi(r) \equiv \xi_0 = h_0/2r_0$, so that $s(x) \equiv 1$. In view of the difference in both the bending and stretching stiffness in the radial and circumferential direction, we take the stress-strain relations for the equivalent conical shell to be in the form of the orthotropic relations (19.6-5) and (19.6-9). For a genuine homogeneous and isotropic shell, its bending stiffness factors are effectively the product of $E/(1 - \nu^2)$ and the appropriate moment of inertia I per unit arc length along a tangent curve, meridional or circumferential. The key to obtaining the bending stiffness factors of our equivalent orthotropic shell (of a homogeneous and isotropic material) is to use an effective moment of inertia per unit arc length of the cross section for one period of the periodic corrugations, Ref. 19-52. For example, we have for the radial bending stiffness factor

$$D_r = \frac{1}{l} \frac{EI_r}{1 - \nu^2} \qquad (19.6\text{-}16)$$

where I_r is the bending moment of inertia and depends on the actual cross-sectional shape of the corrugations. Because the corrugations taper toward the axis of revolution, I_r changes with r.

Similarly, the stretching stiffness factor $1/A_r$ for a homogeneous and isotropic shell is the product of E and the cross-sectional area per unit arc length in the circumferential direction. For the equivalent orthotropic shell, we use instead an effective area per unit arc length of the cross section for one period of the periodic corrugations. This effective area also changes with r for a tapered corrugation.

The other stiffness factors, D_θ, $D_{r\theta}$, A_θ, and $A_{r\theta}$, can also be obtained using the relevant effective stiffness of a single corrugation, i.e., one period of the periodic corrugated annular region. We do not need the actual expressions for the stiffness factors there. It should be pointed out, however, that the determination of A_θ is far more complicated and challenging than the above derivations for D_r and A_r.

19.6.4 Two Boundary-Value Problems

Two boundary-value problems are relevant to the design of the corrugated lid for the essentially rigid cylindrical cannister. To put the lid on the cannister, a vertical force of magnitude P is applied to the small inner disc of the lid, inducing a large enough radial contraction for the lip of the lid to be fitted to the cannister. Of interest here is the relation between the push-down force and the uniform radial contraction at the shell edge. This relation depends on the various geometric and material parameters of the lid and gives the magnitude of P needed to fit the lip to the cannister. After the lid is fitted on, the push-down force will be removed. Without the vertical force, the lid would expand to return to its undeformed configuration, but it is prevented from doing so by the rigid wall of the cannister. The radial stress distribution induced in the lid by the radial constraint exerts a pressure on the cannister wall and thereby seals it. Of interest here is the relation between the radial displacement of the lid at the edge caused by the constraining wall and the sealing pressure induced. How this relation changes with the geometrical and material parameters of the lid will provide us with the design information needed for a secure sealing.

In this section, we formulate two boundary-value problems for the equivalent shallow orthotropic conical shell relevant to the two load-deformation relations described above. In contrast to the problem of a steadily rotating toroidal cap, a linear model of the shell is adequate for these two problems.

19.6.4.1 Radial Contraction Induced by an Axial Force For this phase of the problem, the only external load is an axial force of magnitude P applied at the center of the inner circular disc of the lid. Therefore, we have

$$2\pi r V(r) = -P = \text{constant} \qquad (r > r_i) \tag{19.6-17}$$

With ξ equal to a constant ξ_0 for a shallow conical shell, the governing differential equations for a linear model may be obtained from (19.6-12) and (19.6-13) in the form

$$\begin{aligned} \phi'' + \frac{1}{r}\left(1 + r\frac{D_r'}{D_r}\right)\phi' - \frac{1}{r^2}\left(\frac{D_\theta}{D_r} - r\frac{D_{r\theta}'}{D_r}\right)\phi - \frac{\xi_0}{rD_r}\psi &= \frac{P}{2\pi r D_r} \\ \psi'' + \frac{1}{r}\left(1 + r\frac{A_\theta'}{A_\theta}\right)\psi' - \frac{1}{r^2}\left(\frac{A_r}{A_\theta} - r\frac{A_{\theta r}'}{A_\theta}\right)\psi + \frac{\xi_0}{rA_\theta}\phi &= 0 \end{aligned} \tag{19.6-18}$$

The outer edge of the annular corrugated structure is attached to a ring lip that is considerably stiffer than the corrugated structure. The effect of such an edge stiffener is treated in the theory of plates or shells as an elastic support (Ref. 19-52)

$$r = r_0: \qquad N_r = -k_u u + k_c \phi, \qquad M_r = -k_c u + k_\phi \phi \tag{19.6-19}$$

where the stiffness coefficients k_u, k_c, and k_ϕ are determined by the shape of the stiffener.

The inner edge of the annular corrugated structure is attached to a homogeneous, isotropic, circular, flat, elastic disc of radius r_i and uniform thickness h_d. When h and h_d are of the same order of magnitude, the flat disc also acts as a stiffener for the corrugated structure, with the stiffening effect characterized by an elastic support at the $r = r_i$

$$r = r_i: \qquad N_r + k_{ud} u = 0, \qquad M_r - k_{\phi d}\phi = 0 \tag{19.6-20}$$

with

$$k_{ud} = \frac{Eh_d}{r_i(1 - \nu^2)}, \qquad k_{\phi d} = \frac{Eh_d^3}{12(1 - \nu^2)r_i} \tag{19.6-21}$$

Note that, in contrast to the effect of a ring stiffener of arbitrary cross section, the bending and stretching actions are uncoupled in the elastic support associated with the inner disc.

The linear BVP defined by (19.6-18), (19.6-19), and (19.6-20) determines ϕ and ψ with P as a parameter. The value of the midsurface radial displacement component at $r = r_0$ calculated from ψ gives a load-deflection relation between the radial contraction of the corrugated lid at $r = r_0$ and the push-down force P.

*19.6.4.2 **Sealing Pressure Induced by Radial Contraction*** For this phase of the problem, there is no distributed or point load in the shell interior, so that $V = m = p_H = 0$. The governing differential equations for this phase are therefore (19.6-18) with $P = 0$. The boundary conditions at the inner edge $r = r_i$ remain as given by (18.6-20). At the outer edge $r = r_0$, the shell is subject to a prescribed inward radial (midsurface) displacement of magnitude δ, so that

$$r = r_0: \qquad u = -\delta, \qquad M_r = k_c\delta + k_\phi\phi \tag{19.6-22}$$

The BVP defined by (19.6-18) (with $P = 0$), (19.6-20), and (19.6-22) determines ϕ and ψ with δ as a parameter. The values of N_r and M_r at the edge $r = r_0$ may then be computed from (19.5-19) and (19.6-14), respectively. The sealing pressure is then given by the maximum value of the compressive radial stress at $r = r_0$. This radial stress is the sum of the membrane and bending stresses associated with N_r and M_r, respectively (see (19.5-7) and (19.5-8)). The effect of the sealing pressure is measured by the frictional force between the lip of the lid and the cannister induced by the radial stress.

The numerical solutions of the two linear BVPs described above are straightforward for a very modest computing cost. Such solutions have been obtained for the available lids of various sizes and corrugated configurations. Predictions from the two overall load deformation relations based on these numerical solutions are nearly indistinguishable from the experimental data for the same lids. This good agreement is rather amazing considering the tenuous scientific basis for the contrived modeling of the geometric orthotropy by an equivalent material orthotropy. Without this rather artificial mathematical model, the analysis of the actual lid design would have been very difficult and costly at best.

19.7 WHY REINVENT THE WHEEL?

19.7.1 The 200-Mile Fishing Limit

It has been known for some time that some of the more valued fish stocks off the east coast of Canada and the United States (especially cod, haddock, and redfish) are severely depleted by overfishing, and a "20-year program of intensive experimental management," including the possibility of "a complete closure of east coast fisheries," has been proposed to bring them back up toward the region's carrying capacity (Ref. 19-53). With the recent establishment of a 200-mile offshore zone for regulated fishing, it is now theoretically possible to impose such a fishing moratorium. But fish do swim in and out of the regulated fishing ground, and foreign fishing fleets may station themselves just outside the political boundary to catch whatever fish that cross the boundary. How effective will a moratorium be under those circumstances? Can the fish stock be rebuilt up to the region's carrying capacity? Or will a moratorium prove to be totally ineffective because of the fish movement?

As a first step toward answering these and other questions associated with the establishment of the 200-mile limit, D. Ludwig (Ref. 19-54) modeled the natural growth of the fish population within a region where fishing is not allowed by an initial-boundary-value problem (IBVP) for the nonnegative fish density u. Assuming a random motion for the fish population in a region adjacent to a long and straight coastline, Ludwig's model requires the fish density $u(x, t)$ to satisfy the spatially one-dimensional reaction-diffusion equation

$$u_t = \nu^2 u_{xx} + \Gamma f(u) \qquad (0 < x < l, t > 0) \tag{19.7-1}$$

where x measures the distance from shore, with $x = l$ being the outer boundary of the regulated fishing zone, and Γ is a real constant chosen so that the normalized *growth rate* function, $f(u)$, has a unit derivative (with respect to its argument) at the origin, i.e., $f'(0) = 1$. The rate of diffusion of fish in the x-direction is assumed to be proportional to the gradient of the fish density with a constant of proportionality ν^2. This constant has a role similar to the thermal diffusivity κ in the one-dimensional heat conduction problems of section 19.2 and is called the *diffusion coefficient.*

With $\nu^2 = 0$, (19.7-1) is the usual ODE characterizing the population growth in population dynamics. Typical growth rate functions include the familiar *logistic growth* model and the *depensation* model. For logistic growth, we have $\Gamma > 0$ and $f(u) = u(1 - u/u_c)$, where the positive constant u_c is the stable steady-state population density in conventional logistic growth and is often taken to be the uniform density of the maximum sustainable population, i.e., the *carrying capacity*, of a region. For the depensation model, we have $\Gamma < 0$ and $f(u) = u(1 - u/u_c)(1 - u/u_i)$, where u_c and $u_i < u_c$ are known positive constants. For simplicity, we will restrict our discussion in this article to the class of $f(u)$ analytic in a finite neighborhood of $u = 0$. This class includes most growth rates we encountered in the fishery literature, Ref. 19-34.

Just as the thermal diffusivity κ varies with the material of the bar, the diffusion coefficient ν^2 takes on different values for different fish habitats and must be measured for a particular habitat. In principle, an estimate of the value of ν^2 may be obtained from empirical data based on the physical interpretation of the diffusion term $\nu^2 u_{xx}$. As the rate of diffusion of fish is taken to be proportional to the gradient of the fish density, the rate of diffusion toward a location x from the population farther offshore is approximately proportional to $[u(x + h) - u(x)]/h$. Similarly, the rate of diffusion from the location x toward the shore is proportional to $[u(x) - u(x - h)]/h$. Thus the net rate of increase of the fish density at x is given by $[u(x + h) - 2u(x) + u(x - h)]/h^2$, which tends to u_{xx} in the limit as $h \to 0$. Therefore, measurements of the fish density at selective locations can be made to approximately determine ν^2 as long as complicating factors, such as schooling (Ref. 19-34), are absent.

In Ref. 19-54, Ludwig discussed the worst possible situation, where fish are harvested as soon as they go outside the political boundary $x = l$, so that

$$u(l, t) = 0 \tag{19.7-2}$$

Much of our discussion here is also concerned with this situation, though more realistic models will also be formulated. Since there is no flux of fish onshore, we have

$$u_x(0, t) = 0 \tag{19.7-3}$$

Given the distribution of fish density in the region $0 \leqslant x \leqslant l$ at the start of the fishing moratorium, say

$$u(x, 0) = u_0(x) \qquad (0 \leqslant x \leqslant l) \tag{19.7-4}$$

eqs. (19.7-1) to (19.7-4) determine the fish density $u(x, t)$ in the region of no fishing for some time thereafter. What is of interest to the fishery managers, however, is whether the fish density evolves in time toward some equilibrium (steady-state) density. In other words, are there time-independent solutions of the boundary value problem (19.7-1) to (19.7-3), and if so, are they stable?

With $f(0) = 0$, it is clear that $u(x) \equiv 0$ is an equilibrium state of the system. It is also not difficult to obtain the nontrivial equilibrium population densities whenever they exist (as we will do later). On the other hand, it is much more difficult to decide whether a particular equilibrium density is stable; that is, starting with a density close to the equilibrium density, does it remain close to the equilibrium density thereafter, or does it evolve away from it as time goes on? Global stability theorems for the time-independent solution of the PDE (19.7-1) have been obtained by Aronson and Weinberger (Ref. 19-55) for a slightly different set of boundary conditions with the help of a maximum principle. These theorems can be extended to cover our problem. If we begin less ambitiously by asking only about local stability,* then the relevant mathematical problem is the same as the one-dimensional heat conduction in a finite bar and can be solved by an elementary method. In fact, the problem of determining nontrivial equilibrium densities itself is mathematically the same as Euler's *Elastica* (Ref. 19-46) and can be handled in exactly the same way. The methods for these two classes of classical problems also apply, either directly or with some modifications, to more realistic models proposed herein (see also Ref. 19-56). It is not an exaggeration to say that many emerging problems in mathematical modeling may be new in appearance, but their mathematical structures may still be the same or similar to problems already treated successfully in the past. The general class of the 200-mile fishing limit problems is only one of the many examples. The practitioners of applied mathematics should be forever ready to take another page from history.

*Local stability analyses are usually easier, and suggest the appropriate global stability results that may not be apparent.

19.7.2 The Local Stability of the Trivial State

With $f(0) = 0$, the trivial state $u(x, t) \equiv 0$ is an equilibrium density, i.e., a time independent solution of the BVP (19.7-1) to (19.7-3). To analyze the local stability of this equilibrium state, consider an initial density distribution $u_0(x)$ near the trivial state, i.e., $u_0 \ll 1$. For such a $u_0(x)$, we expect $u(x, t) \simeq v(x, t)$ initially where $v(x, t)$ is the solution of the linearized IBVP

$$v_t = \nu^2 v_{xx} + \Gamma v \qquad (0 < x < l, t > 0) \tag{19.7-5}$$

$$v_x(0, t) = v(l, t) = 0 \qquad (t > 0) \tag{19.7-6}$$

$$v(x, 0) = u_0(x) \qquad (0 \leqslant x \leqslant l) \tag{19.7-7}$$

since $f(0) = 0$ and $f'(0) = 1$. The method of separation of variables gives

$$\begin{aligned} v(x, t) &= \sum_{n=0}^{\infty} V_n \cos\left(\frac{\lambda_n x}{l}\right) e^{(\alpha^2 - \lambda_n^2)\nu^2 t/l^2} \\ &= V_0 \cos\left(\frac{\pi x}{2l}\right) e^{(1-\pi^2/4\alpha^2)\Gamma t} + V_1 \cos\left(\frac{3\pi x}{2l}\right) e^{(1-9\pi^2/4\alpha^2)\Gamma t} + \cdots \end{aligned} \tag{19.7-8}$$

where $\alpha^2 \equiv \Gamma l^2/\nu^2$, $\lambda_k = \pi(2k+1)/2$, and the constants V_k, $k = 0, 1, 2, \ldots$, are determined by the given initial distribution $u_0(x)$.

Evidently, we have $v(x, t) \to 0$ as $t \to \infty$ if $\Gamma < 0$, as in the case of the depensation model. In this case, the trivial equilibrium state is (locally) asymptotically stable. On the other hand, if $\Gamma = \gamma^2 > 0$, as in the logistic growth case, the steady-state behavior of $v(x, t)$ is dominated by the leading term of the series (19.7-8). If $\alpha \equiv \gamma l/\nu$ is smaller than $\pi/2$, then again $v(x, t)$ tends to 0 as $t \to \infty$. The fish population will head toward extinction if it ever gets "too small." If $\alpha > \pi/2$, the trivial equilibrium state is unstable, and the fish stock, if left undisturbed within the region of regulated fishing, will grow with time beyond the range of applicability of a linearized analysis. The above conclusions merely tell us in more precise terms what we should have expected all along, namely, the fish population will grow in time if the reproduction rate of the fish population is large compared with the rate at which fish leave the regulated fishing zone. The value of our analysis is that it tells us, in the case of $\Gamma = \gamma^2 > 0$, how large γ^2 (or how small ν^2) has to be for the fish stock to increase in spite of attrition due to fish movement across the political boundary. For a given fish population so that both $\gamma > 0$ and ν are fixed constants, our results for the case $\Gamma > 0$ suggest how far we have to extend the offshore fishing moratorium, i.e., the location of the boundary $x = l$, in order for such a moratorium to be effective.

The above theoretical results and those to be obtained in the subsequent development show the importance of having a numerical value for the diffusion coefficient. Without at least an order of magnitude estimate of ν^2, the theoretical result on an effective moratorium is useless.

19.7.3 Nontrivial Equilibrium Densities

We consider henceforth only the case $\alpha^2 > (\pi/2)^2$, so that the trivial state is unstable. For this case, the initially depleted fish stock will grow with the help of a fish moratorium. But this growth cannot be indefinite, because there is a limit to the size of the fish stock the region can carry. It is not difficult to see that, unlike the case of no fish movement, nontrivial solutions of $f(u) = 0$ are not equilibrium fish densities for our problems because they do not satisfy the boundary condition $u(l, t) = 0$. Therefore, we expect that there must be one or more equilibrium densities that are not uniformly distributed over the interval $(0, l)$.

An *equilibrium fish density* $U(x)$ is a solution of the BVP

$$\nu^2 U_{xx} + \gamma^2 f(U) = 0, \qquad U_x(0) = U(l) = 0 \tag{19.7-9}$$

where we have set $\Gamma \equiv \gamma^2 > 0$. It is the same mathematical problem encountered in the *Elastica* with the column clamped at $x = l$ and simply supported at $x = 0$. A first integral of the above nonlinear ODE gives

$$U_x \equiv \frac{dU}{dx} = -\frac{\gamma}{\nu}\,[F(U_0) - F(U)]^{1/2} \tag{19.7-10}$$

where $dF/dU \equiv 2f(U)$, and where $U_0 \equiv U(0)$ is an unknown constant. The negative square root was chosen because we must have $U(l) = 0$. Incidentally, the fact that $U_x = 0$ when $U = U_0$ has been used to fix the constant of integration in (19.7-10).

The first-order ODE (19.7-10) is separable and can be integrated to give

$$x = \frac{\nu}{\gamma} \int_U^{U_0} \frac{dY}{\sqrt{F(U_0) - F(Y)}} \tag{19.7-11}$$

where we have made use of the fact $U(0) = U_0$ to fix the constant of integration. The relation (19.7-11) may be inverted to give a nontrivial equilibrium density distribution $U(x)$. To determine the yet unknown constant U_0, we use the fact that $U(l) = 0$ to get from (19.7-11)

$$\alpha \equiv \frac{\gamma l}{\nu} = \int_0^{U_0} \frac{dY}{\sqrt{F(U_0) - F(Y)}} \tag{19.7-12}$$

The relation (19.7-12) may be inverted to give U_0 as a function of $\alpha \equiv \gamma l/\nu$. For a discussion of local stability, we will need a few more details about U.

To carry out the analysis beyond this point, we will work out the specific case of *logistic growth* with $f(U) = U(1 - U/u_c)$. A similar analysis for the general case can be found in Ref. 19-56. For the logistic growth case, eqs. (19.7-11) and (19.7-12) take the form

$$x = \frac{\nu}{\gamma} \int_U^{U_0} \frac{dY}{\sqrt{(U_0^2 - Y^2) - (2/3u_c)(U_0^3 - Y^3)}}$$
$$= \frac{\nu}{\gamma} \int_w^1 \frac{dy}{\sqrt{(1 - y^2) - (2a/3)(1 - y^3)}} \tag{19.7-13}$$

and

$$\alpha \equiv \frac{\gamma l}{\nu} = \int_0^1 \frac{dy}{\sqrt{(1 - y^2) - (2a/3)(1 - y^3)}} \tag{19.7-14}$$

respectively, with $a \equiv U_0/u_c$, and $w \equiv U/U_0$. Note that α, a, and w are dimensionless quantities. It is not difficult to see that the integral in (19.7-14) is a monotone increasing function of the parameter a in the range $0 \leqslant a < 1$. The value of α is $\pi/2$ at $a = 0$ and tends to infinity as a tends to unity. Hence, there is exactly one nontrivial equilibrium fish density given by (19.7-13) and (19.7-14) for all $\alpha > \pi/2$ (and the only equilibrium density for $\alpha < \pi/2$ is the trivial state $U(x) \equiv 0$). The bifurcation from the trivial state occurs at $\alpha = \pi/2$, precisely the value above which the trivial state was found earlier to be unstable.

For α slightly larger than $\pi/2$, say $\alpha = \frac{1}{2}\pi(1 + \epsilon)$ with $0 < \epsilon \ll 1$, we can obtain an approximate expression for a and for the nontrivial equilibrium density in terms of elementary functions. Since $a(\alpha = \frac{1}{2}\pi) = 0$, we expand a in a Taylor series in powers of ϵ for the case $\alpha = \frac{1}{2}\pi(1 + \epsilon)$

$$a = a_1 \epsilon + a_2 \epsilon^2 + \cdots \tag{19.7-15}$$

and the integral (19.7-14) can then be written as

$$\frac{\pi}{2}(1 + \epsilon) = \int_0^1 \left[1 + \frac{a_1 \epsilon}{3} \frac{1 - v^3}{1 - v^2} + 0(\epsilon^2)\right] \frac{dv}{\sqrt{1 - v^2}}$$
$$= \frac{\pi}{2} + \frac{2}{3} a_1 \epsilon + 0(\epsilon^2) \tag{19.7-16}$$

which can be solved to give

$$a_1 = \tfrac{3}{4}\pi \tag{19.7-17}$$

While the remaining coefficients a_k, $k = 2, 3, \ldots$, in (19.7-15) can also be obtained from terms in (19.7-16) involving higher powers of ϵ, it suffices for our purpose to know

$$a = \tfrac{3}{4}\pi\epsilon + 0(\epsilon^2) \tag{19.7-18}$$

and, correspondingly, from (19.7-13)

$$U(x) = U_0 w = \frac{3}{4}\pi u_c \epsilon \cos\left(\frac{\pi x}{2l}\right)[1 + 0(\epsilon)] \tag{19.7-19}$$

The expression (19.7-19) gives an adequate first approximation for the nontrivial equilibrium density when $\alpha \gtrsim \frac{1}{2}\pi$. This expression will be useful in a local stability analysis of the nontrivial equilibrium state near bifurcation in the next section. Note that it is often easier to obtain higher order terms in the parametric series for $U(x;\epsilon)$ by directly seeking a perturbation solution of the BVP (19.7-9) in the form

$$U(x;\epsilon) = \tilde{U}_0(x) + \epsilon\tilde{U}_1(x) + \epsilon^2\tilde{U}_2(x) + \cdots \tag{19.7-20}$$

keeping in mind the series expansion (19.7-15) for a and $\alpha \equiv \gamma l/\nu = \frac{1}{2}\pi(1+\epsilon)$.

19.7.4 Local Stability near Bifurcation

To see whether a nontrivial equilibrium fish density is stable, we consider an initial fish density $u_0(x) \equiv U(x) + v_0(x)$ with $|v_0(x)| \ll U(x)$. We are interested in the time evolution of the fish density $u(x,t) \equiv U(x) + v(x,t)$ starting from the small perturbation from the equilibrium state. Since $|v_0(x)| \ll U(x)$, we expect to have $|v(x,t)| \ll U(x)$, at least for a while, so that we can linearize the PDE (19.7-1) in $v(x,t)$ to get

$$v_t = \nu^2 v_{xx} + \gamma^2 q(x) v, \qquad q(x) \equiv f'(U(x)) \tag{19.7-21}$$

while the associated boundary and initial conditions for v follow directly from (19.7-2) to (19.7-4) and the definition of $v(x,t)$

$$v_x(0,t) = v(l,t) = 0 \qquad (t > 0) \tag{19.7-22}$$

$$v(x,0) = v_0(x) \qquad (0 \leqslant x \leqslant l) \tag{19.7-23}$$

The solution of the linear initial-boundary-value problem (IBVP) for $v(x,t)$ may be obtained by the method of separation of variables

$$v(x,t) = \sum_{k=0}^{\infty} \overline{V}_k \phi_k(x)\, e^{(\alpha^2 - \overline{\lambda}_k^2)\nu^2 t/l^2} \tag{19.7-24}$$

where $\{\bar{\lambda}_k^2\}$ and $\{\phi_k(x)\}$ are the eigenvalues (ordered in increasing magnitude) and eigenfunctions, respectively, of the eigenvalue problem

$$l^2\phi^{\cdot\cdot} + [\alpha^2(q-1) + \bar{\lambda}^2]\phi = 0, \qquad \phi^{\cdot}(0) = \phi(l) = 0 \tag{19.7-25}$$

with $(\ )^{\cdot} = d(\ )/dx$. The constants $\bar{V}_k$, $k = 0, 1, 2, \ldots$, in (19.7-24) are determined by the initial condition (19.7-23). The local stability of the nontrivial equilibrium state $U(x)$ evidently depends on the sign of $(\alpha^2 - \bar{\lambda}_0^2)$, which can be found by an elementary method when the system is near bifurcation.

To illustrate, take again the logistic growth case where $f(u) = u(1 - u/u_c)$, so that $q - 1 = -2U/u_c$. Near bifurcation, i.e., when $\alpha = \frac{1}{2}\pi(1+\epsilon)$ for $0 < \epsilon \ll 1$, we have $U/u_c = \frac{3}{4}\pi\epsilon\cos(\pi x/2l)\,[1 + 0(\epsilon)]$ (see eq. (19.7.19)), so that 19.7-25 becomes

$$\begin{gathered} l^2\phi^{\cdot\cdot} + \left[\bar{\lambda}^2 - \frac{3}{2}\pi\epsilon\lambda_0^2\cos\left(\frac{\pi x}{2l}\right)\{1 + 0(\epsilon)\}\right]\phi = 0 \\ \phi^{\cdot}(0) = \phi(l) = 0 \end{gathered} \tag{19.7-26}$$

where $\lambda_0^2 = (\pi/2)^2$ is the lowest eigenvalue for the limiting case with $\epsilon = 0$. For local stability, it suffices to seek a perturbation solution of the first eigenvalue $\bar{\lambda}_0^2$ and the corresponding eigenfunction $\phi_0(x)$

$$\begin{gathered} \bar{\lambda}_0^2 = \Lambda_0^2[1 + \epsilon\mu_1 + \epsilon^2\mu_2 + \cdots] \\ \phi_0(x) = \Phi_0(x) + \epsilon\Phi_1(x) + \epsilon^2\Phi_2(x) + \cdots \end{gathered} \tag{19.7-27}$$

Upon substituting (19.7-27) into (19.7-26), we get

$$\begin{gathered} [l^2\Phi_0^{\cdot\cdot} + \Lambda_0^2\Phi_0] + \epsilon\left[l^2\Phi_1^{\cdot\cdot} + \Lambda_0^2\Phi_1 + \left\{\Lambda_0^2\mu_1 - \frac{3}{2}\pi\lambda_0^2\cos\left(\frac{\pi x}{2l}\right)\right\}\Phi_0\right] + 0(\epsilon^2) = 0 \\ \Phi_0^{\cdot}(0) + \epsilon\Phi_1^{\cdot}(0) + 0(\epsilon^2) = 0, \qquad \Phi_0(l) + \epsilon\Phi_1(l) + 0(\epsilon^2) = 0 \end{gathered} \tag{19.7-28}$$

From (19.7-28), we get

$$l^2\Phi_0^{\cdot\cdot} + \Lambda_0^2\Phi_0 = 0, \qquad \Phi_0^{\cdot}(0) = \Phi_0(l) = 0 \tag{19.7-29}$$

for the leading term solution Λ_0^2 and $\Phi_0(x)$, and

$$l^2\Phi_1^{\cdot\cdot} + \Lambda_0^2\Phi_1 = \Phi_0(x)\left[\lambda_0^2\frac{3}{2}\pi\cos\left(\frac{\pi x}{2l}\right) - \mu_1\Lambda_0^2\right], \qquad \Phi_1^{\cdot}(0) = \Phi_1(l) = 0 \tag{19.7-30}$$

for the first-order correction terms μ_1 and $\Phi_1(x)$, etc.

The solution of the eigenvalue problem (19.7-29) is straightforward with

$$\Lambda_0^2 = \left(\frac{\pi}{2}\right)^2 \equiv \lambda_0^2, \quad \text{and} \quad \Phi_0(x) = \cos\left(\frac{\pi x}{2l}\right) \tag{19.7-31}$$

as the lowest eigenvalue and the corresponding eigenfunction, respectively (keeping in mind that the series (19.7-27) are for the lowest eigenvalue of (19.7-25) and the associated eigenfunction).

With (19.7-31), the general solution of the ODE for Φ_1 that satisfies the condition $\dot{\Phi}_1(0) = 0$ is

$$\Phi_1(x) = c_0 \cos\left(\frac{\pi x}{2l}\right) + \frac{\pi}{4}\left[3 - \cos\left(\frac{\pi x}{2l}\right)\right] - \frac{\mu_1}{2}\left[\cos\left(\frac{\pi x}{2l}\right) + \frac{\pi x}{2l}\sin\left(\frac{\pi x}{2l}\right)\right] \tag{19.7-32}$$

The remaining condition in (19.7-30), $\Phi_1(l) = 0$, requires

$$\pi - \frac{\mu_1}{2}\frac{\pi}{2} = 0 \qquad \text{or} \qquad \mu_1 = 4 \tag{19.7-33}$$

(The undetermined constant c_0 is available for a suitable normalization of the approximate eigenfunction.) Therefore, we have

$$\bar{\lambda}_0^2 = \left(\frac{\pi}{2}\right)^2 [1 + 4\epsilon + 0(\epsilon^2)] \tag{19.7-34}$$

and

$$\alpha^2 - \bar{\lambda}_0^2 = -\frac{\pi^2}{2}\epsilon[1 + 0(\epsilon)] \tag{19.7-35}$$

so that, near bifurcation, the nontrivial equilibrium density $U(x)$ for the logistic growth case is locally (asymptotically) stable.

19.7.5 Finite Rate of Fishing Effort

Up to now, we have effectively assumed the fishing effort at the political boundary $x = l$ to be unlimited, so that $u(x, t) \equiv 0$ for $x \geqslant l$. We can now comment briefly on the more realistic case of a finite rate of fishing effort E beyond $x = l$.

With a fish harvest rate density $h_0(u, E)$ (often taken to be proportional to Eu) in the region $x > l$, the dynamics of the fish population outside the political boundary is evidently governed by

$$u_t = \nu^2 u_{xx} + \Gamma_0 f_0(u) - h_0(u, E) \qquad (l < x < L) \tag{19.7-36}$$

where we have allowed for a different natural growth rate function for the fish population beyond $x = l$, and where $x = L$ is a line beyond which the fish population would not survive because of unfavorable environment, e.g., the edge of the continental shelf, so that

$$u(L, t) = 0 \tag{19.7-37}$$

If no such ecological boundary exists, then L extends to infinity while the total fish population remains finite. At $x = l$, the condition $u(l, t) = 0$ is now replaced by continuity conditions on u and u_x. Given the initial fish population density, we have an initial boundary value problem for $u(x, t)$ in the larger domain $0 < x < L$.

The analysis of the above more realistic problem is evidently more complex. However, the methods for obtaining nontrivial equilibrium densities and for performing local stability analysis, originally developed for the study of column buckling and one-dimensional heat conduction of a finite rod, respectively, may still be used for this new problem, either as it stands or transformed into a two-component vector field problem on the interval (0, 1). Two special cases of the new problem deserve further discussion.

19.7.5.1 Concentrated Fishing Effort at $x = l$ For the special case of $\Gamma_0 f_0 = \Gamma f$ and concentrated fishing effort at $x = l$, resulting in a total harvest rate $H_0(u(l), E_0)$, we have a simpler problem consisting of the PDE (19.7-1) for $0 < x < L$, the boundary conditions (19.7-3) and (19.7-27), the continuity condition $u(l-, t) = u(l+, t)$, the jump condition

$$\nu^2 u_x(l-, t) + H_0(u(l), E_0) = \nu^2 u_x(l+, t) \tag{19.7-38}$$

and a prescribed initial fish density distribution (19.7-4). With some modifications, methods used for Ludwig's model earlier are again applicable to this simplified problem.

When the fish population is severely harvested at the boundary $x = l$, we often have $|u_x(l+, t)| \ll |u_x(l-, t)|$. In that case, we may neglect the $u_x(l+, t)$ term in (19.7-38) to obtain an initial boundary-value problem consisting of (19.7-1), (19.7-3), (19.7-4), and (19.7-38) with the right-hand side replaced by zero. For a constant fishing effort rate E_0 and with H_0 proportional to $u(l)$, so that $H_0 = c_l E_0 u$, the simplified boundary condition at $x = l$ becomes

$$\nu^2 u_x(l, t) + c_l E_0 u(l, t) = 0 \tag{19.7-39}$$

The IBVP, (19.7-1), (19.7-3), (19.7-4), and (19.7-39), is akin to the IBVP (19.7-1) to (19.7-4) already analyzed and is analogous to the elastica hinged at $x = 0$ and elastically supported at $x = l$. By the methods used there, it is not difficult to show that the trivial equilibrium state $u \equiv 0$ is asymptotically stable for $\Gamma < \nu^2 \alpha_0^2 / l^2$, where α_0 is the smallest positive root of

$$\cot(z) = z\mu \qquad \text{with} \qquad \mu \equiv \frac{\nu^2}{E_0 c_l} \tag{19.7-40}$$

and is unstable otherwise. For $0 < \mu < \infty$, we have $\alpha_0 < \pi/2$, and the difference decreases with increasing E_0. In other words, the threshold for extinction, as expected, is reduced by a decrease of fishing effort rate. Furthermore, an asymptotically stable nontrivial equilibrium state corresponding to (19.7-11) exists for $\Gamma \equiv \gamma^2 > \nu^2 \alpha_0^2 / l^2$, and a bifurcation from the trivial state occurs at $\gamma l / \nu = \alpha_0$. As $E_0 \to 0$ (and therefore $\mu \to \infty$), we have $\alpha_0 \to 0$, so that we recover the expected results for the natural growth of an unharvested fish population.

19.7.5.2 Fixed Mortality Rate beyond the Political Boundary Another model deals with the situation where the fish population in the region $l < x < L$ is so heavily harvested that the population density declines at a fixed rate $s^2 > 0$. In that case, we have

$$u_t = \nu^2 u_{xx} - s^2 u \qquad (l < x < L) \tag{19.7-41}$$

With (19.7-41), the nontrivial equilibrium fish population density U in the outer region is

$$U = \begin{cases} c \sinh h\left(\dfrac{s}{\nu}\,[L - x]\right) & (l \leqslant x \leqslant L) \\[2ex] c\, e^{-sx/\nu} & (l \leqslant x < \infty) \end{cases} \tag{19.7-42}$$

where the unknown constant c is determined as a part of the solution of the BVP in the inner region, consisting of the ODE in (19.7-9), the no-flux condition (19.7-3), and the continuity conditions on U and U_x at $x = l$.

As in subsections 19.7.3, a first integral of the ODE in (19.7-9) that satisfies the no-flux condition at shore (19.7-3) is

$$U_x = -\frac{\gamma}{\nu} \sqrt{F(U_0) - F(U)} \qquad (0 < x < l) \tag{19.7-43}$$

where $dF/dU = 2f(U)$, and U_0 is the as yet unknown equilibrium population density at $x = 0$. The first-order, separable ODE (19.7-43) for $U(x)$ can be integrated immediately to give

$$x = \frac{\nu}{\gamma} \int_{U}^{U_0} \frac{dY}{\sqrt{F(U_0) - F(Y)}} \qquad (0 < x < l) \tag{19.7-44}$$

The continuity of U at $x = l$ requires

$$\alpha \equiv \frac{\gamma l}{\nu} = \int_{U_l}^{U_0} \frac{dY}{\sqrt{F(U_0) - F(U)}} \tag{19.7-45}$$

where

$$U_l \equiv U(l) = c\, e^{-sl/\nu} \tag{19.7-46}$$

A second condition for the two unknown parameters U_0 and c comes from the continuity of U_x at $x = l$, which may be simplified to read

$$\frac{\gamma}{s} \sqrt{F(U_0) - F(U_l)} = ce^{-sl/\nu} \tag{19.7-47}$$

The two conditions (19.7-45) and (19.7-47), together with (19.7-46), determine U_0 and c and therewith the nontrivial equilibrium fish density for $0 < x < \infty$ by way of (19.7-44) and (19.7-42). As in Ludwig's model, a nontrivial equilibrium density does not exist for all possible geometric and growth parameters. For the logistic growth model, we found in subsection 19.7.3 that $\alpha \equiv \gamma l/\nu$ must be greater than $\pi/2$ for $U_0 > 0$. A similar condition for α can be determined when the fishing effort beyond the region of no fishing is finite. For this purpose, we eliminate the unknown c from (19.7-46) and (19.7-47) to get for the logistic growth model

$$F(U_0) - F(U_l) = (U_0^2 - U_l^2) - \frac{2}{3u_c}(U_0^3 - U_l^3) = \frac{s^2}{\gamma^2} U_l^2 \tag{19.7-48}$$

which determine U_l in terms of U_0 and the parameter s, γ, and u_c with

$$U_l(U_0 = 0) = 0.$$

The condition (19.7-45) may be written in the dimensionless form

$$\alpha \equiv \frac{\gamma l}{\nu} \int_{w_l}^{1} \frac{dy}{\sqrt{(1 - y^2) - \frac{2}{3} a(1 - y^3)}} \tag{19.7-49}$$

where $w_l = U_l/U_0$, and $a = U_0/u_c$. For $a = 0$, we have from (19.7-49)

$$\int_{w_0}^{1} \frac{dy}{\sqrt{1-y^2}} = \frac{\pi}{2} - \sin^{-1}(w_0) \equiv \frac{\pi}{2} - \phi(\gamma, s) \qquad (19.7\text{-}50)$$

where w_0 is given by (19.7-48) with $a = 0$

$$0 < w_0 = \frac{\gamma}{\sqrt{s^2 + \gamma^2}} < 1 \qquad (19.7\text{-}51)$$

so that $\sin\phi = \gamma/\sqrt{s^2 + \gamma^2}$. Note that $\phi \to 0$ as $s \to \infty$, and we recover the limiting result obtained in subsection 19.7.3.

For a small positive value of a, we have from (19.7-48)

$$w_l = w_0\{1 - \tfrac{1}{3}a(1 - w_0^3) + \cdots\} \qquad (19.7\text{-}52)$$

so that the integral in (19.7-49) is again a monotone increasing function of the parameter a in the range $0 \leqslant a < 1$. Hence, there is a unique nontrivial equilibrium density given by (19.7-44) and (19.7-42) for $\alpha > \frac{1}{2}\pi - \phi$, and the only equilibrium density for $\alpha < \frac{1}{2}\pi - \phi$ is the trivial state $U(x) \equiv 0$. For a particular fish habitat, a finite equilibrium fish population is possible for a smaller region of fishing moratorium than what is required by Ludwig's model, when the harvest rate outside the region is bounded. With $\phi \to 0$ monotonically as $s \to \infty$, this smaller region increases in size as the harvest rate beyond the political boundary increases. The local stability of nontrivial equilibrium densities in the neighborhood of the trivial state can again be analyzed by the method of subsections 19.7.4 and 19.7.5.

Instead of a total ban on fishing inside the region $0 < x < l$, the case of a regulated fishery is also of interest. Whether the regulation is by a catch quota or by a maximum allowed fishing effort (such as the number of vessel-days), the fish population in $0 < x < l$ is harvested at a density rate h, so that a term $-h$ is to be added to (19.7-1). Otherwise, the structure of all the models considered in this subsection and subsection 19.7.1 remains unchanged, and the models themselves can be similarly analyzed.

19.8 BETTER ROBUST THAN REALISTIC

A large part of the mathematical modeling process described in the preceding sections is effectively the well-known "scientific method" for the theoretical aspect of the physical sciences. Articles on mathematical modeling such as this merely try to distill the essence of this time-proven method for applications in areas of quantitative studies beyond the physical sciences, adding a few useful working principles from mathematics and engineering in the process. Mathematical models developed in this way are diametrically opposite to *simulation models* for the same phenomena. In the mechanics of deformable bodies of matter for example, a simulation model would attempt to be as "realistic" as possible by working at the molecular level. For many complex phenomena, the two types of models complement each other.

A mathematical model does not have to mirror closely the "real" phenomenon being modeled; but it should be robust. In many applications, continuum mechanics is sufficiently robust to be used for generating actual design data, though it is not particularly realistic from a molecular perspective. Note that "reality" itself may be rather elusive and often has meaning only in a relative sense. Molecules are "really" complicated structures of atoms. Atoms are themselves tiny solar systems of electrons in orbit about a nucleus. A tiny solar system is only a statistical average model for the behavior of elementary particles. Indications are that these elementary particles may be made up of still more fundamental units, and so on.

On the other hand, no mathematical model can be robust for every realization of a particular phenomenon. Newtonian mechanics has been most successful for describing and predicting the motion of most planets in the solar system. Yet, to account for the advance of the perihelion of the planet Mercury, we need the more general theory of relativistic mechanics.

Faced with the above two constraints, it would appear that our goal in modeling a phenomenon should be to formulate as simple a model as possible that is still robust for the class of realizations to be analyzed by the model. The "simple-to-elaborate" approach to mathematical modeling of section 19.3 is particularly significant in this regard. The experience gained from a sequence of increasingly more sophisticated models usually makes it possible to identify those features of a phenomenon that must be retained for the robustness of the model (and others that may be neglected because they have only secondary effects). To seek the most comprehensive model in a first attempt, without any regard for the actual scope of the immediate objectives or for the possibility of some preliminary insight into the phenomenon, usually leads to mathematically untractable models. Often, the application of such a comprehensive model to a particular situation would resemble the use of relativity theory to explain the curving of a baseball.

19.9 REFERENCES

19-1 Thompson, M. D., et al., *Case Studies in Applied Mathematics*, MAA Publication (Comm. on the Undergrad. Program in Math.), 1976.

19-2 von Kármán, T., and Biot, M. A., *Mathematical Methods in Engineering*, McGraw-Hill, New York, 1940.

19-3 Genin, J., and Maybee, J. S., *Introduction to Applied Mathematics, vol. 1*, Holt, Rinehart and Winston, New York, 1970.

19-4 Pollard, H., *Applied Mathematics: An Introduction*, Addison-Wesley, Reading, Mass., 1972.

19-5 Lin, C. C., and Segel, L. A., *Mathematics Applied to Deterministic Problems in the Natural Sciences*, Macmillan, New York, 1974.

19-6 Lancaster, P., *Mathematics: Models of the Real World*, Prentice-Hall, Englewood Cliffs, N.J., 1976.

19-7 Haberman, R., *Mathematical Models*, Prentice-Hall, Englewood Cliffs, N.J., 1977.

19-8 Bluman, G. W., "Dimensional Analysis, Symmetry and Modelling," Appl. Math. Notes, *6*, 122–135, 1981.

19-9 Barenblatt, G. I., *Similarity, Self-Similarity, and Intermediate Asymptotics*, Consultants Bureau (Div. Plenum), New York, 1979.

19-10 Bluman, G. W., and Cole, J. D., *Similarity Methods for Differential Equations,* Springer-Verlag, New York-Heidelberg-Berlin, 1974.

19-11 Birkhoff, G., *Hydrodynamics* (2nd ed.), Princeton Univ. Press, Princeton, N.J., 1960.

19-12 Bridgman, P. W., *Dimensional Analysis* (rev. ed.), Yale Univ. Press, New Haven, Conn., 1931 (paperback ed., 1963).

19-13 Sedov, L. J., *Similarity and Dimensional Methods in Mechanics* (4th ed.), Academic Press, New York, 1959.

19-14 de Jong, F. J., *Dimensional Analysis for Economists*, North-Holland Publishing, Amsterdam, 1967.

19-15 Becker, H. A., *Dimensionless Parameters. Theory and Method*, Halsted Press (div. Wiley), New York, 1976.

19-16 Kurth, R., *Dimensional Analysis and Group Theory in Astrophysics*, Pergamon Press, Oxford-New York, 1972.

19-17 Taylor, G. I., "The Formation of a Blast Wave by a Very Intense Explosion, II: The Atomic Explosion of 1945," *Proc. Roy. Soc.* A, **201**, 175, 1950.

19-18 Baker, W. E., *Explosions in Air*, Univ. of Texas Press, Austin, Tex., 1973.

19-19 Mills, E. S., *Urban Economics*, Scott, Foresman, Glenview, Ill., 1972.

19-20 Bartholomew, H., *Land Uses in American Cities*, Harvard Univ. Press, Cambridge, Mass., 1955.

19-21 Haig, R. M., "Toward an Understanding of the Metropolis," *Quart. J. Econ.*, **40**, 421–423, 1926.

19-22 Alonso, W., *Location and Land Use*, Harvard Univ. Press, Cambridge, Mass., 1964.

19-23 Luce, R. D., and Raiffa, H., *Games and Decisions*, Wiley, New York, 1957.

19-24 Solow, R. M., "Congestion Cost and the Use of Land for Streets," *Bell J. Econ. Manag. Sci.*, **4**, 602–618, 1973.

19-25 Arnott, R. J., and MacKinnon, J. G., "Market and Shadow Land Rents with Congestion," *Am. Econ. Rev.*, **68**, 588–600, 1978.

19-26 Solow, R. M., "Congestion, Density and the Use of Land in Transportation," *Swedish J. Econ.*, **74**, 161–173, 1972.

19-27 Wan, F. Y. M., "Perturbation and Asymptotic Solutions for Problems in the Theory of Urban Land Rent," *Studies Appl. Math.*, **56**, 219–239, 1977.

19-28 Kanemoto, Y., *Theories of Urban Externalities*, North-Holland, Amsterdam-New York, 1980. (Also, "Cost-Benefit Analysis and the Second Best Land Use for Transportation," *J. Urban Econ.*, **4**, 483–503, 1977.)

19-29 Robson, A. J., "Cost-Benefit Analysis and the Use of Urban Land for Transportation," *J. Urban Econ.*, **3**, 180–191, 1976.

19-30 Wan, F. Y. M., "Accurate Solutions for the Second Best Land Use Problem, I: Absentee Ownership," I.A.M.S. Tech. Report 79-30, Univ. of British Columbia, July 1979.

19-31 Wan, F. Y. M., "Accurate Solutions for the Second Best Land Use Problem, II: Public Ownership," I.A.M.S. Tech. Report 83-20, Univ. of British Columbia, 1983.
19-32 Pearse, P., "The Optimum Forest Rotation," *Forestry Chron.*, **2**, 178–195, 1967.
19-33 Fisher, I., *The Theory of Interest*, Macmillan, New York, 1930.
19-34 Clark, C. W., *Mathematical Bioeconomics*, Wiley, New York, 1976.
19-35 Samuelson, P. A., "Economics of Forestry in an Evolving Society," *Econ. Inqu.*, **XIV**, 466–492, 1976.
19-36 Faustmann, M., "Berechnung des Werthes, welchen Weldboden sowie nach nicht haubare Holzbestande fur die Weldwirtschaft besitzen," *Allgemeine Forst und Jagd Zeitung*, **25**, 441, 1849.
19-37 ——, "On the Determination of the Value Which Forest Land and Immature Stands Pose for Forestry," in *Martin Faustmann and the Evolution of Discounted Cash Flow*, M. Gane (ed.)., Oxford Institute Paper No. 42, 1968, Oxford.
19-38 Heaps, T., and Neher, P. A., "The Economics of Forestry When the Rate of Harvest Is Constrained," *J. Environ. Econ. Manage.*, **6**, 297–319, 1979.
19-39 Davidson, R., and Hellsten, M., "Optimal Forest Rotation With Costly Planting and Harvesting," presented at the Fifth Canadian Conference on Economic Theory in Vancouver, B.C., May 1980.
19-40 Heaps, T., "The Forestry Maximum Principle," presented at the Can. Appl. Math. Soc. Annual Meeting (Montreal), May 1981.
19-41 Wan, F. Y. M., and Anderson, K., "Ordered Site Access and Age Distribution in Harvesting Once-and-for-all Forests," *Stud. Appl. Math.*, **65**, 1983.
19-42 Anderson, K., and Wan, F. Y. M., "Finite and Infinite Sequences Harvests of Ongoing Forests," I.A.M.S. Tech. Rep. 81-14, Univ. British Columbia, 1981.
19-43 Flugge, W., and Riplog, P. M., "A Large Deformation Theory of Shell Membranes," Tech. Rep. No. 102, Div. of Eng. Mech., Stanford Univ., Sept. 1956.
19-44 Johnson, M. W., "On the Dynamics of Shallow Elastic Membranes," Theory of Thin Elastic Shells, (ed. W. T. Koiter), Proc. IUTAM Symp. Delft, 1959.
19-45 Simmonds, J. G., "The Finite Deflection of a Normally Loaded Spinning Elastic Membrane," *J. Aerospace Sci.*, **29**, 1180–1189, 1962.
19-46 Love, A. E. H., *A Treatise on the Mathematical Theory of Elasticity* (4th ed.), Dover Publications, New York, 1944.
19-47 Reissner, E., "Symmetric Bending of Shallow Shells of Revolution," *J. Math. Mech.*, **7**, 121–140, 1958.
19-48 Clark, R. A., "On the Theory of Thin Elastic Toroidal Shells," *J. Math. Phys.*, **29**, 146–178, 1950.
19-49 ——, "Asymptotic Solutions of Elastic Shell Problems," *Asymp. Soln. of ODE & Appl.* (ed. C. H. Wilcox), Wiley, New York, 1964, 185–209.
19-50 Seaman, W. J., and Wan, F. Y. M., "Lateral Bending and Twisting of Thin-Walled Curved Tubes," *Stud. Appl. Math.*, **53**, 73–89, 1974.
19-51 Reissner, E., and Wan, F. Y. M., "Rotating Shallow Elastic Shells of Revolution," *J. SIAM*, **13**, 333–352, 1965. (Also, *T. V. Kármán in Memoriam*, SIAM Publication, Philadelphia, 1965, 159–178.)

19-52 Timoshenko, S., and Woinowsky-Krieger, S., *Theory of Plates and Shells* (2nd ed.), McGraw-Hill, New York, 1959.
19-53 Larkin, P. A., "Scientific Technology Needs for Canadian Shelf-seas Fisheries," Interim Report, Fisheries Research Board of Canada, Ottawa, Feb. 1975.
19-54 Ludwig, D., "Some Mathematical Problems in the Management of Biological Resources," *Appl. Math. Notes*, **2**, 39–56, 1976.
19-55 Aronson, D. G., and Weinberger, H. F., "Nonlinear Diffusion in Population in Mathematics, vol. 446 (*Partial Differential Equations and Related Topics*), Springer, Berlin, 1975, 5–49.
19-56 Wan, F. Y. M., "Bifurcation Theory and the Two Hundred Mile Fishing Limit," *Appl. Math. Notes*, **4**, 74–87, 1977.

20

Optimization Techniques

Juris Vagners*

20.1 INTRODUCTION

20.1.1 Optimization in Perspective

In this chapter, we will present some fundamental elements of the diverse field covered by the term "optimization theory." Optimization encompasses such areas as the theory of ordinary maxima and minima (section 20.2.1), calculus of variations (20.3.2), linear and nonlinear programming (20.5.1, 2), dynamic programming (20.5.3), maximum principles (20.3.7), discrete and continuous games (20.6), and differential games of varying degrees of complexity (20.6). The literature on these topics, and variations thereof, is staggering and unfortunately scattered throughout the various publications by engineers, mathematicians and systems analysts. In a necessarily brief treatment such as given in this chapter, many aspects of optimization theory are neglected. However, we have endeavored to present sufficiently complete references and biographical sources to more complete works to guide the interested reader in the majority of cases.

In view of the complexity of the field, it appears desirable to provide an overview of how the primary topics of optimization theory are interrelated and perhaps indicate a possible hierarchy. Two views of such a hierarchy are advanced by Isaacs[20-1] and Ho[20-2] in recent survey papers on differential game theory. Isaacs views overall optimization theory in the context of generalized differential games and Ho terms the general theory "generalized control theory." The hierarchy implied is the same in both cases.

Central to all problems in optimization theory are the concepts of *payoff*, *controllers* or *players*, *system*, and *information sets*. In order to define what one means by a solution to an optimization problem, the concept of payoff must be defined and the controllers or players identified. If there is only one person on

*Prof. Juris Vagners, Dep't. of Aeronautics and Astronautics, University of Washington, Seattle, Wash.

whose decisions the outcome of some particular process depends and the outcome can be described by a single quantity, then the meaning of payoff (and hence solution to the optimization problem) and controller or player is clear. In the literature the payoff is known by many names: measure or index of performance, criterion function, cost function, performance criterion, return function, figure of merit are some of the more common terms. If, however, there is more than one criterion–a situation we often face in our daily lives–then we require rational and systematic methods of performing trade-offs and the concept of a solution becomes blurred. Similarly, if we introduce more decision-making persons (controllers) into a given process (i.e., n-person games), we find that there may be no single, generally accepted concept of a solution. The concept of controllers or players is not altered by these complications, we merely have to identify the decision making or purposeful agents in any process.

A system can be defined as a collection of entities governed by rules of operation such that the outputs exhibited by the collection are influenced by the inputs and the rules of operation. The state of a system at any arbitrary instant of time is usually defined by a finite set of numbers. Knowledge of the state and the inputs constitutes an information set, thus for multiple players there may be multiple information sets. These sets may be complete or incomplete depending on whether or not all players have access to the same information. In addition, these sets may be perfect (deterministic) or imperfect (noisy). Obviously, the nature and quality of the information available drastically influences the control actions during an optimization process.

Based on the characteristics and number of criteria, controllers and information sets, we can establish a hierarchy of optimization problems and indicate where in the structure the topics of this chapter lie. The simplest of course is the problem of parameter optimization which includes the classical theory of maxima and minima, linear and nonlinear programming. In parameter optimization there is one (deterministic or probabilistic) criterion, one controller, one complete information set and the system state is described by static equations and/or inequalities in the form of linear or nonlinear algebraic or difference equations.

On the next rung of complexity we might consider optimization problems for dynamic systems where the state is defined by ordinary or partial differential equations. These can be thought of as the limiting cases of multistage (static) parameter optimization problems where the time increment between steps tends to zero. In this class, developed extensively in recent years as optimal control, we encounter the classical calculus of variations problems and their extension through various maximum and optimality principles (i.e., Pontryagin's Minimum Principle, section 20.3.7, and the dynamic programming principle of optimality section 20.5.3). We are still concerned with one criterion, controller and information set but have the added dimension in that the problem is dynamic and might be deterministic or stochastic.

The next level would introduce two controllers (players) with a single conflicting criterion. Here we encounter elementary or finite matrix game theory where the controls and payoff matrix are discrete, and continuous game theory where the controls and payoff are continuous functions. Each player at this level has complete information regarding the payoff for each strategy but may or may not have knowledge of his opponent's strategy. Such games are known as zero-sum games since the sum of the payoffs to all players for each move is zero—what one player gains the other loses. If the order in which the players act does not matter, i.e., the minmax and maxmin (section 20.6.2) of the payoff are equal (this minmax is termed the *Value* of the game and is unique), the optimal strategies of the players are unaffected by knowledge or lack of knowledge of each other's strategy. A solution to this game involves the Value and at least one optimal strategy for each player.

An interesting extension here is Danskin's Max-Min Theory[20-3] wherein the maxmin does not equal minmax but one player selects his strategy after, and with full knowledge of, his opponent's selection. This problem differs from classical game theory when minmax does not equal maxmin since there it is assumed that the information on the current state available to each player is incomplete. In the classical case, game theory tells us to play probabilistically (mixed strategy) and the solution to the game consists of finding the minimax of the expected value of the payoff, the optimal mixed strategy and the best probability distribution of decisions for each player.

Next, we can consider extensions to dynamical systems where the state is governed by differential equations, we have one conflicting criterion and two players, i.e., zero-sum, two-player differential games. The information available to each player might be complete or incomplete. In cases with complete information, the finite game concept of a solution is directly applicable. For the incomplete information case, it is reasonable to expect mixed strategies to form the solution but not much is known of solution methods or whether a solution always exists in the finite game theoretic sense.

In the last group or uppermost rung of the hierarchy we identify a class of optimization problems where the concept of a solution is far from clear. To this class belong multiple criteria one player games with complete information,[20-4, 5] nonzero sum (either or both players may lose or gain) two-player games with complete or incomplete information[20-6, 7, 8] and finally, *n*-person, multiple criteria, multiple information set games with complete or incomplete information.[20-9] In the case of multiple players, various candidates for the concept of a solution have been offered, cf. the equilibrium point theory of Nash[20-10] and the concept of a pareto-optimal set.[20-2] For further discussion of *n*-person differential games, see the work by Case.[20-11]

In this chapter we will be discussing the elements of optimization for deterministic problems with primary focus on one player games or optimal control theory, hence we offer some references to bibliographical sources for the other problems outlined in the hierarchy. For the basic foundations of differential games, the classic book by Isaacs[20-12] is strongly recommended. A survey of differential games is the paper by Berkovitz;[20-13] bibliographical information may be found in Refs. 20-24, 20-15. Stochastic optimization problems in the context of control theory are briefly reviewed by Paiewonsky[20-16] as part of an overall review of optimal control theory and practice. This 1965 paper lists 362 entries while concentrating primarily on flight mechanics and control. The text by Pierre[20-17] offers extensive references. For current work in all areas, one might refer to the *Journal of Optimization Theory and Applications* (*JOTA*), the *SIAM Journal on Control*, the *International Journal of Control*, and the *IEEE Transactions on Automatic Control* as initial sources.

20.1.2 On Solution Techniques

A general classification of solution techniques in optimization is difficult since each particular problem requires its own peculiar blend of methods. Nevertheless, two general classes in optimal control theory can be identified: a) "indirect" methods where the optimal solution is determined primarily from necessary conditions based on differential or functional properties of the optimum; and b) "direct" methods requiring the direct use of the payoff and constraints in systematic, recursive search for the optimum. This division is somewhat artificial in that the two concepts can be jointly utilized in a comprehensive optimization scheme.

The direct methods are the simplest to identify. After formulation of the payoff and constraints, a scheme must be devised which will maximize (minimize) the payoff in the admissible space allowed by the constraints. In this group we find linear programming, nonlinear programming or search techniques, and dynamic programming. Linear programming involves optimization of a linear criterion function (which may be stochastic) subject to linear constraints; extensive references and discussion of the various methods may be found in Pierre[20-17] and Künzi, et al.[20-18]

As implied by the title, nonlinear programming, which we take to be synonymous with search procedures, attempts to find the extremum of a payoff that is a nonlinear (real valued) function of one or more parameters that may or may not be subject to constraints. Search techniques, whether sequential or nonsequential, discrete or continuous, may be divided into two classes: a) those that utilize derivatives of the criterion function, and b) those that do not. Possibly the best known (and oldest) of class (a) techniques is the method of gradients or method of steepest descent, dating back to Cauchy. Search techniques are too numerous to discuss here, extensive references to search techniques are given by Leon[20-19] and Pierre.[20-17] See also Gill et al.[20-57]

In solving problems of ordinary maxima and minima, whether constrained or not, application of the necessary conditions for an optimum reduces the problem to one of root-finding. Computation in complicated cases then proceeds via some appropriate search technique.

Dynamic programming[20-20] is the general term assigned to solution of multistage (discrete or continuous) decision processes leading to the extremum of a payoff. In the context of direct methods, dynamic programming implies imbedding the given problem in a family of problems with the same terminal or initial point (or hyper surface) determining the optimum return function (see section 20.4.1) by solving a partial differential equation and then finding the minimizing (maximizing) state and control solutions via direct solution of the state and control equations. The primary drawback to this approach is computer storage limitation, storage of the extremal fields for even moderately complicated problems can overrun current machine capacities.

The class of "indirect" methods is (primarily) based on the variational formulation of optimization processes. The variational formulation of dynamic optimization processes (usually) leads to nonlinear, two-point boundary-value problems which rarely allow analytical solution. The problem then is to find the state, control and adjoint vectors satisfying four conditions for an optimal solution: the state or system equations, adjoint equations, optimality or necessary conditions and the initial and final boundary conditions for the state and adjoint variables. Numerical solutions of such problems all involve either "flooding" or iterative techniques. Flooding techniques generate a dense field of extremals, or candidate optimum solutions, satisfying the prescribed boundary conditions at either the initial or terminal point using the unspecified conditions there as parameters. One attempts to generate the field in such a manner that the boundary conditions at the other end are included in the field, interpolation for an extremal satisfying the specific boundary conditions then yields the optimal solution to the problem. This approach falls in the realm of dynamic programming for boundary value problems.

Other direct methods for solution of two-point boundary value problems involve changing the variational problem into one of ordinary maxima and minima. One approximates the unknown function (payoff) by a truncated series and then maximizes over the coefficients of the series. Two well-known techniques are those of Galerkin and Rayleigh-Ritz.

Iterative solutions to two-point boundary value problems usually involve some scheme of successive linearization, i.e. a nominal solution is chosen that satisfies none, one, two or three of the conditions for an optimum solution and then this nominal is modified to satisfy the remaining condition(s). As noted by Bryson and Ho,[20-21] there are fifteen possible combinations; only the methods of neighboring extremals (section 20.4.2), gradient methods (section 20.5.4) and quasi-linearization (section 20.5.5) have been used extensively.

The neighboring extremal method nominal solutions satisfy all but the boundary conditions, iterative techniques are then used to improve the estimates of the unspecified initial (terminal) conditions so as to satisfy specified terminal (initial) conditions. The drawback to these techniques is the extreme sensitivity of the process to the changes in boundary conditions often necessitating educated guessing to establish good starting values for the estimated parameters and encourage solution convergence.

Gradient methods satisfy the state equations and the adjoint equations but not the boundary or optimality conditions. One may use either first order or second order gradient methods or a combination of both in a given problem to take advantage of the good features of each method.

Quasi-linearization methods satisfy the optimality conditions, may or may not satisfy the boundary conditions, but do not satisfy the state and adjoint equations.

In most cases these methods involve guessing nominal state and adjoint vectors satisfying as many boundary conditions as possible and then solving for the corresponding control vector from the optimality condition. Successive linearization of the state and adjoint equations about the nominal is then applied to improve the state and adjoint vectors (the control vector is determined from the optimality condition) until a solution is reached.

20.1.3 Notation

In the following sections we will use the vector and matrix notation now current in the optimization literature. In this section we give a brief outline of the conventions used in the main body of the chapter; for details of linear algebra and matrix operation see Chapter 16 of this Handbook. A criticism often leveled at the use of vector and matrix (or "state-space") notation is that the gains in brevity are outweighed by the loss of interpretation for the unaccustomed reader. Since the majority of current papers and texts in the field use vector notation, it is advisable for anyone interested in optimization to become familiar with it. One of the best aids to gaining familiarity with the notation is expansion of the vector formulae in components for two-dimensions. The reader is urged to do so whenever uncertainty over a specific relationship arises.

All lower case letters will denote a vector, i.e.,

$$x = \begin{bmatrix} x_1 \\ x_2 \\ \cdot \\ \cdot \\ \cdot \\ x_n \end{bmatrix}$$

where $x_1, x_2, \ldots, x_n$ are scalar components of x. The letters i, j, k will be used as indexing variables taking on values 0, 1, 2, . . . , for example, x_i will mean the ith component of the vector x. The basis (cf. Chapter 16) is always assumed to be Cartesian. A scalar variable in this convention is simply a "vector" of one component. The scalar variable t will be used for the independent variable time with subscripts indicating specific instants, for example t_0, the initial time.

When upper case letters are used to indicate matrices, the convention for the elements is as follows:

$$A = \begin{bmatrix} a_{11} & a_{12} & \cdots & a_{1n} \\ a_{21} & & & \\ \cdot & & & \\ \cdot & \cdots & \cdots & \cdots \\ \cdot & & & \\ a_{m1} & \cdots & \cdots & a_{mn} \end{bmatrix}$$

The elements $a_{i,j}$ of A are scalar or real variables.

A superscript T will be used to indicate the transpose of a vector or matrix, so that the scalar product of two vectors will be written as

$$x^T y = x_1 y_1 + x_2 y_2 + \cdots + x_n y_m$$

The vector product in this notation is

$$xy^T = \begin{bmatrix} x_1 y_1 & x_1 y_2 & \cdots & x_1 y_m \\ x_2 y_1 & & & \\ \vdots & & \cdots\cdots\cdots & \\ x_n y_1 & \cdots & \cdots & x_n y_m \end{bmatrix}$$

(Note that x and y need not have the same dimension.) The norm of a vector x is thus defined by $x^T x = \|x\|^2$.

In differentiating or integrating vectors or matrices we operate on each component

$$\dot{x} = \frac{dx}{dt} = \begin{bmatrix} \frac{dx_1}{dt} \\ \frac{dx_2}{dt} \\ \vdots \\ \frac{dx_n}{dt} \end{bmatrix} = \begin{bmatrix} \dot{x}_1 \\ \dot{x}_2 \\ \vdots \\ \dot{x}_n \end{bmatrix}; \quad \int x \, dt = \begin{bmatrix} \int x_1 \, dt \\ \int x_2 \, dt \\ \vdots \\ \int x_n \, dt \end{bmatrix}$$

and

$$\dot{A} = \frac{dA}{dt} = \begin{bmatrix} \frac{da_{11}}{dt} & \frac{da_{12}}{dt} & \cdots & \frac{da_{1n}}{dt} \\ \frac{da_{21}}{dt} & & & \\ \vdots & & \cdots\cdots\cdots & \\ \frac{da_{m1}}{dt} & & \cdots\cdots & \frac{da_{mn}}{dt} \end{bmatrix}$$

The gradient of a scalar function J is defined to be a row vector

$$\nabla J = J_x = \frac{\partial J}{\partial x} = \left[\frac{\partial J}{\partial x_1}\ \frac{\partial J}{\partial x_2} \cdots \frac{\partial J}{\partial x_n}\right] = [J_{x_1} J_{x_2} \cdots J_{x_n}]$$

We will also need the second derivative of a scalar function with respect to vector arguments which we define as

$$J_{xy} = \frac{\partial}{\partial y}(J_x^T) = \frac{\partial}{\partial y}(J_{x^T}) = \begin{bmatrix} \frac{\partial^2 J}{\partial x_1 \partial y_1} & \frac{\partial^2 J}{\partial x_1 \partial y_2} & \cdots & \frac{\partial^2 J}{\partial x_1 \partial y_m} \\ \frac{\partial^2 J}{\partial x_2 \partial y_1} & & & \\ \vdots & \cdots & \cdots & \cdots \\ \frac{\partial^2 J}{\partial x_n \partial y_1} & \cdots & \cdots & \frac{\partial^2 J}{\partial x_n \partial y_m} \end{bmatrix}$$

20.2 PARAMETER OPTIMIZATION

In this section we will consider the following general class of problems*:

Minimize (maximize) the scalar criterion $M(y)$, where y is an l-vector, subject to the constraint conditions $e(y) = 0$, e an s-vector, and $f(y) \geqslant 0$, f an m-vector. The fundamental problem thus is to find the vector y^0 in the admissible or feasible region defined by the constraint relations, where M attains a minimum (maximum) value.

It is instructive to consider the problem in two dimensions. For this case, $M(y_1, y_2)$ represents a surface over a region in the y_1, y_2-plane defined by the constraint curves. One possible case for inequality constraints is illustrated in Fig. 20.2-1. The feasible region is defined as the region bounded by the y_1, y_2-axes and the arc of a circle of radius one. The y_1, y_2-axes form part of the feasible region but the circular arc is not included. The behavior of the surface $M(y_1, y_2)$ is indicated by contour lines of constant values of M. The admissible points y may be further limited by imposing equality constraints $e(y) = 0$.

Considering Fig. 20.2-1, we note that at the points B, D, and F the gradient of the surface vanishes, $\partial M/\partial y = 0$; such points are called *stationary points*. A stationary point can be a maximum (F), a minimum (B), or neither (D), in which case it is called a *saddle point.*

Within the interior of the admissible region, the surface M has a minimum at the point B, a saddle point at D and another minimum at C which is the lowest point of a ravine defined by the curve EG (some partial derivatives are discontinuous). By direct comparison, we see that the function attains the lowest value, or *absolute minimum*, at the point B. The point C is termed a *relative minimum* by

*These are also referred to as *mathematical programming problems* in the current literature.

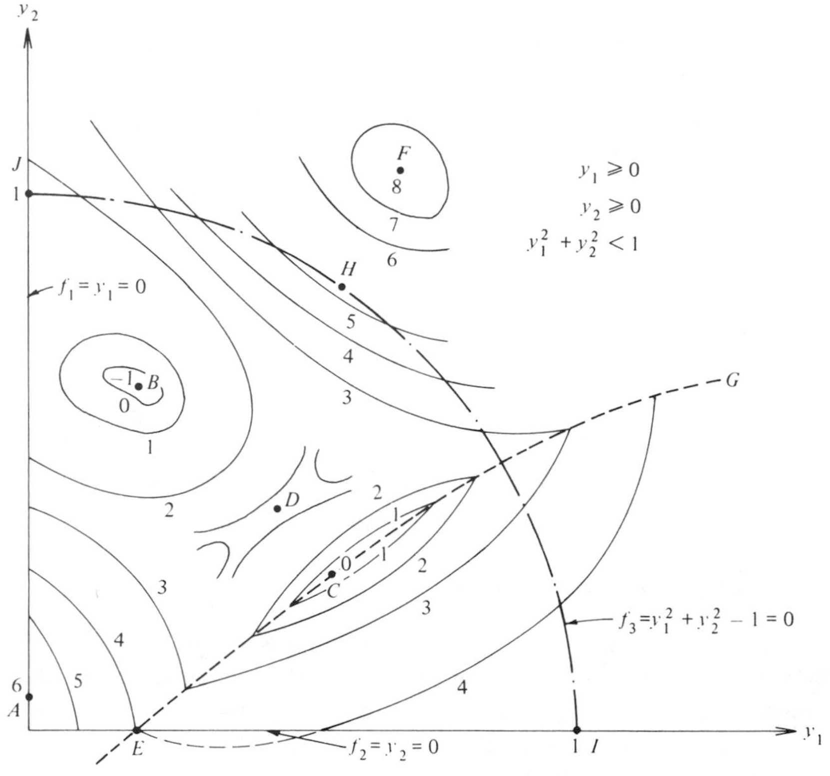

Fig. 20.2-1

virtue of the fact that it is lower than all points in a small, finite, neighborhood of it. The boundaries formed by the y_1, y_2 axes are part of the admissible region and we find the maximum of M at A. Points on the arc of radius one are not admissible, so that the highest point along this arc, H, can be approached as closely as desired but the value never achieved. In this case, by comparing the values of M just outside the boundary, we see that $M(A)$ is greater than the limiting value $M(H)$ and hence A is the *absolute maximum* (F lies outside the admissible region); $M(H)$ is thus termed a *relative maximum*. Based on these results, we see that in general all points on or near the boundaries should be examined; in so doing we find relative maxima at I and J.

From this example, it is clear that direct solution of the general problem in l-dimensional space can be very difficult indeed. We therefore consider some special cases and the theory developed for their solution.

20.2.1 Maxima and Minima of Real Functions

For continuous functions $M(y)$ over a closed and bounded set, the existence of a maximum and minimum is guaranteed by the *theorem of Weierstrass: such functions possess a largest and a smallest value either in the interior or on the boundary of the set.* This knowledge does not, however, give us a method for searching for the extrema. We must usually locate all the relative or local extrema and then compare the values of M there.

In order to ascertain whether a relative extremum is achieved at a point y^0 we must analyze the effects of small displacements from y^0. To do so, we expand $M(y)$ in a Taylor series about y^0

$$M(y) = M(y^0) + M_y\big|_{y^0}\,\Delta y + \tfrac{1}{2}\Delta y^T M_{yy}\big|_{y^0}\,\Delta y + R \qquad (20.2\text{-}1)$$

where R includes all terms whose order in Δy is higher than second. If y^0 is a stationary point, then the first order term must vanish for arbitrary Δy

$$M_y\big|_{y^0} = 0 \qquad (20.2\text{-}2)$$

For, if this were not so, we could increase or decrease the value of M by moving along a line of some nonzero $\partial M/\partial y_i$.

The restrictions on eq. (20.2-2) are that all Δy be possible (so that the point y^0 cannot lie on the boundary of the admissible set) and that all components of M_y be continuous. Under these conditions, the solution of the set of algebraic equations (20.2-2) gives the location of all interior maxima, minima, and saddle points.

To determine the nature of an extremum point, we examine the matrix of second derivatives M_{yy} evaluated at y^0. For, if y^0 is to be a maximum for example, then the second order terms in Δy cannot yield a higher value for M than $M(y^0)$. Thus *for a maximum the matrix M_{yy} must be negative definite, for a minimum, positive definite* (cf., Chapter 16). If $M_{yy} = 0$, further investigation is required—one may have a saddle point, maximum or minimum. The sufficient conditions for an isolated maximum (minimum) over an open set thus are

$$\nabla M = M_y = 0$$

and

$$\begin{aligned} \nabla^2 M &= M_{yy} > 0 \quad \text{local minimum} \\ \nabla^2 M &= M_{yy} < 0 \quad \text{local maximum} \end{aligned} \qquad (20.2\text{-}3)$$

The situation when the vector y is constrained, either by equality or inequality conditions, is examined in the next section.

20.2.2 Constraints and Accessory Conditions

The constrained problem may be posed as follows: Find the vector y^0 that minimizes* $M(y)$ subject to the constraint conditions

*In the sequel we shall examine only the minimization problem with the understanding that to maximize M, minimize $-M$.

$$e(y) = 0 \qquad e \ \ s\text{-vector} \tag{20.2-4a}$$

$$f(y) \geqslant 0 \qquad f \ \ m\text{-vector} \tag{20.2-4b}$$

where f, e, and y have different dimensions in general. Problems for which M, e, or f exhibit some nonlinearity are usually referred to as *nonlinear programming problems* in the current literature. When M, e, and f are all linear functions of y, we have a *linear programming* problem. For such problems, the minimum, if it exists, *must* occur on the boundary of the admissible set, for the curvature of M is zero everywhere. Such problems are discussed in more detail in section 20.5.1. The nonlinear programming problem may be further classified as convex, concave, separable, quadratic and/or factorable depending on possible special characteristics of the functions M, e, and f. For this discussion, any form of nonlinearity of M, e, f is allowable subject only to continuity and differentiability requirements; for a discussion of the special classifications noted see the text and cited references of Fiacco and McCormick.[20-22] The results stated here will thus apply to *relative* or *local* solutions of the constrained problem, although for special properties of the problem functions, conditions for *global* or absolute solutions are attainable. For example, in the convex programming problem where $M(y)$ is convex, the $f_i(y)$ concave and $e_j(y)$ linear, a local solution is also the global solution.

In order to present a unified treatment of the general constrained problem, and the various subcases such as the problem with equality constraints only or inequality constraints only, we introduce the function $\mathcal{H}$ defined as

$$\mathcal{H}(y, \mu, \nu) = M(y) - \mu^T f(y) + \nu^T e(y) \tag{20.2-5}$$

In mathematical programming literature, this is known as the *Lagrangian* function and the vectors μ, ν are called *generalized Lagrange multipliers* or *dual variables*. In the control literature μ, ν are known as *adjoint vectors*.

For problem functions M, f, and e which are all once-continuous differentiable, the *first-order necessary conditions* for a local minimum may be stated as follows:

If a) the point y^0 satisfies the constraints (20.2-4a) and (20.2-4b); b) the problem functions M, f, e are all differentiable at y^0; and c) no vector λ exists such that $\lambda^T \nabla M(y^0) < 0$, $\lambda^T \nabla e_i(y^0) = 0$, $i = 1, 2, \ldots, s$ and $\nabla^T f_j(y^0) \geqslant 0$ for all j for which $f_j(y^0) = 0$, then there exist vectors μ^0, ν^0 such that (y^0, μ^0, ν^0) satisfies

$$f(y^0) \geqslant 0 \tag{20.2-6a}$$

$$e(y^0) = 0 \tag{20.2-6b}$$

$$\mu^{0T} f(y^0) = 0 \tag{20.2-6c}$$

$$\mu_j^0 \begin{cases} \geqslant 0 & \text{if } f_j(y^0) = 0 \\ = 0 & \text{if } f_j(y^0) > 0 \end{cases} \tag{20.2-6d}$$

$$\mathcal{H}_y(y^0, \mu^0, \nu^0) = 0 \tag{20.2-6e}$$

and (20.2-6a)-(20.2-6e) are the necessary conditions that y^0 be a local minimum.

Condition (c), or the existence of finite Lagrange multipliers μ^0, ν^0 is usually assured by imposing qualifications on the constraints, such as the Kuhn-Tucker first-order constraint qualification.[20-23] The constraint qualification is designed to rule out situations for which at the minimum point, ∇M is not equal to any finite linear combination of ∇e_i and ∇f_j, where e_i and f_j are all the effective constraints at y^0. However, the constraint qualification is not *necessary* that finite Lagrange multipliers exist, whereas condition (c) is. A simple example for which the constraints do not satisfy the constraint qualification yet condition (c) holds and a local minimum exists at the point $y^0 = (1, 0)$ is

Minimize $M = y^2$ subject to $y_1 \geqslant 0, y_2 \geqslant 0$ and $(1 - y_1)^3 - y_2 \geqslant 0$

For proof of the first-order necessary conditions, see Ref. 20-22.

There may also occur situations—termed *abnormal*—when the function $\mathcal{H}$ takes the form

$$\mathcal{H}(y, \mu, \nu) = \nu_0 M(y) - \mu^T f(y) + \nu^T e(y) \tag{20.2-7}$$

and, for a solution to exist, ν_0 must be chosen as zero. A simple geometrical interpretation of the abnormal case is possible in the case $e(y)$ is a two-vector, there are no binding inequality constraints and y is three-dimensional. In this case, the surfaces $e_1 = 0$ and $e_2 = 0$ have a point P in common where the normals coincide. For higher dimensions of y, e, and binding constraints f_j, we note that abnormality occurs whenever all second order determinants of the matrix ∇g vanish, $g = (e, f_j)^T$, all j such that $f_j = 0$.

An elementary example of the abnormal problem follows.

Example: Find the maximum of

$$M(y_1, y_2) = y_1 - 2y_2^2 \tag{20.2-8a}$$

subject to

$$e_1(y_1, y_2, y_3) = y_1^3 + y_2^2 + y_3^2 = 0 \tag{20.2-8b}$$

Forming $\nu_0 M_y + \nu^T e_y$ we have

$$\nu_0 + 3\nu_1 y_1^2 = 0 \quad \text{(a)}$$

$$-4y_2\nu_0 + 2\nu_1 y_2 = 0 \quad \text{(b)}$$

$$2\nu_1 y_3 = 0 \quad \text{(c)} \qquad (20.2\text{-}9a)$$

and the constraint

$$y_1^3 + y_2^2 + y_3^2 = 0 \quad \text{(d)} \qquad (20.2\text{-}9b)$$

to solve for y_1^0, y_2^0, y_3^0, ν_0^0 and ν_1^0. From (c) we have either ν_1^0 or y_3^0 or both equal to zero. If $\nu_1^0 = 0$, then $\nu_0^0 = 0$ so we must choose $y_3^0 = 0$. From (b) $y_2^0 = 0$ *or* $\nu_1^0 = 2\nu_0^0$, in which case (a) yields $\nu_0^0 + 6\nu_0^0 y_1^{02}$ or $\nu_0^0 = 0$ and $\nu_1^0 = 0$, thus $y_2^0 = 0$. Then from the constraint (d) $y_1^0 = 0$, which implies $\nu_0^0 = 0$ but $\nu_1^0 \neq 0$. Attempting to solve the problem by omitting ν_0 we find no solution among real numbers.

Solutions to constrained problems are still attainable even if condition (c) of the first-order necessary conditions is not satisfied. In such problems, we must then admit infinite Lagrange multipliers.

For the unconstrained problem we were able to state the necessary and sufficient conditions for an isolated local extremum of a twice-differentiable function as $M_y = 0$ and $M_{yy} \gtrless 0$.

Similarly, we can utilize the curvature of the problem functions to establish sufficiency for isolated extrema of constrained problems. However, we note that the term "isolated" is required, since for general problems we cannot obtain necessary *and* sufficient conditions for a point to be an extremum utilizing information only at the point.

The sufficient conditions that a point y^0 be an isolated local minimum are that:

For twice-differentiable functions M, e, f, the gradients of the binding constraints ∇e and ∇f_j, all j such that $f_j = 0$, be linearly independent, there exist vectors μ^0, ν^0 such that (y^0, μ^0, ν^0) satisfies (20.2-6a)–(20.2-6e) and for every nonzero vector z satisfying $z^T \nabla e = 0$, $z^T \nabla f_k = 0$ for all k for which $\mu_k^0 > 0$ and $z^T \nabla f_i \geqslant 0$ for all remaining f_i of the binding constraints f_j,

$$z^T \mathcal{H}_{yy}(y^0, \mu^0, \nu^0) z > 0 \qquad (20.2\text{-}10)$$

or the matrix $\mathcal{H}_{yy}(y^0, \mu^0, \nu^0)$ be positive definite.

Consider now the special case of the general constrained problem where the dimension l of y is greater than the dimension of the equality constraint vector e and no inequality constraints exist. In this case we can decompose y into two vectors x and u of dimension s and $n = l - s$ respectively with $l \geqslant s$. The choice of x and u is arbitrary but must be such that $e(x, u) = 0$ determines x given u, i.e., e_x^{-1} is nonsingular. This is the typical formulation of control problems—x is called the state vector and u the control or decision vector. With this decomposition the necessary conditions for a stationary value of $M(x, u) = M(y)$ become

$$\mathcal{H}_x = 0; \quad \mathcal{H}_u = 0; \quad e(x, u) = 0 \qquad (20.2\text{-}11)$$

The sufficient conditions require that the differential change of $\mathcal{H}$ to second order away from the nominal solution (x^0, u^0) as defined by (20.2-11) be positive for a local minimum. Expanding $\mathcal{H}$ to second order in dx, du yields

$$d\mathcal{H} = (\mathcal{H}_x, \mathcal{H}_u)\begin{pmatrix} dx \\ du \end{pmatrix} + \tfrac{1}{2}(dx^T, du^T)\begin{pmatrix} \mathcal{H}_{xx} & \mathcal{H}_{xu} \\ \mathcal{H}_{ux} & \mathcal{H}_{uu} \end{pmatrix}\begin{pmatrix} dx \\ du \end{pmatrix} + R \quad (20.2\text{-}12)$$

where R contains third and higher order terms in dx and du. By virtue of eq. (20.2-11), the first term of (20.2-12) vanishes. Then, since we also require $de = 0$ at the stationary point, we can express dx as a function of du

$$de = e_x\, dx + e_u\, du + 2^{\text{nd}} \text{ and higher order terms} = 0$$

$$dx = -e_x^{-1}\, e_u\, du + \text{higher order terms} \quad (20.2\text{-}13)$$

we find

$$d\mathcal{H} = \tfrac{1}{2} du^T[-e_u^T(e_x^T)^{-1}, I]\begin{bmatrix} \mathcal{H}_{xx} & \mathcal{H}_{xu} \\ \mathcal{H}_{ux} & \mathcal{H}_{uu} \end{bmatrix}\begin{bmatrix} -e_x^{-1}(e_u) \\ I \end{bmatrix} du + R \quad (20.2\text{-}14)$$

where I is the identity matrix. Since $d\mathcal{H} = dM + \nu^T de$

$$dM = \tfrac{1}{2} du^T \left(\frac{\partial^2 M}{\partial u^2}\right)_{e=0} du + R \quad (20.2\text{-}15a)$$

with

$$\left(\frac{\partial^2 M}{\partial u^2}\right)_{e=0} = \mathcal{H}_{uu} - \mathcal{H}_{ux}\, e_x^{-1}\, e_u - e_u^T(e_x^T)^{-1}\,\mathcal{H}_{xu} + e_u^T(e_x^T)^{-1}\,\mathcal{H}_{xx}\, e_x^{-1}\, e_u \quad (20.2\text{-}15b)$$

Thus, the matrix $(\partial^2 M/\partial u^2)_{e=0}$ must be positive definite for a local minimum, a condition which can be directly checked for a given problem.

20.2.3 Neighboring Optimal Solutions

Often it is of interest to determine how much does the optimum change when one changes the criterion function $M(y)$ and the constraint functions $e(y)$ and $f(y)$. Here we shall consider the special case where $y^T = (x, u)$; for a discussion of the more general mathematical programming problem, see Ref. 20-22.

Assume that the constraints are changed slightly so that $e(x, u) = de$, a constant infinitesimal vector. Then, examine the necessary conditions for stationarity with $x = x^0 + dx$, $u = u^0 + du$ and $e(x, u) = de$

$$d\mathcal{H}_x^T = \mathcal{H}_{xx}\, dx + \mathcal{H}_{xu}\, du + e_x^T\, d\nu = 0$$

$$d\mathcal{H}_u^T = \mathcal{H}_{ux}\, dx + \mathcal{H}_{uu}\, du + e_u^T\, d\nu = 0 \quad (20.2\text{-}16)$$

$$de = e_x\, dx + e_u\, du$$

where the partials are evaluated at x^0, u^0. Solving these equations for du in terms of the known de we find:

$$du = -\left[\mathcal{H}_{uu} + e_u^T(e_x^T)^{-1}\,\mathcal{H}_{xx}\,e_x^{-1}\,e_u - e_u^T(e_x^T)^{-1}\,\mathcal{H}_{xu} - \mathcal{H}_{ux}\,e_x^{-1}\,e_u\right]^{-1} \cdot \left[\mathcal{H}_{ux} - e_u^T(e_x^T)^{-1}\,\mathcal{H}_{xx}\right] e_x^{-1}\,de \quad (20.2\text{-}17)$$

or

$$du = -\left(\frac{\partial^2 M}{\partial u^2}\right)_{e=0}^{-1} \cdot \left[\mathcal{H}_{ux} - e_u^T(e_x^T)^{-1}\,\mathcal{H}_{xx}\right] e_x^{-1}\,de \quad (20.2\text{-}18)$$

Therefore, if the point (x^0, u^0) is a local extremum, the existence of neighboring optimal solutions is guaranteed. The corresponding changes in the state and adjoint vectors are

$$dx = e_x^{-1}\,de - e_x^{-1}\,e_u\,du \quad (20.2\text{-}19)$$

$$d\nu = -(e_x^T)^{-1}\left[\mathcal{H}_{xx}\,dx + \mathcal{H}_{xu}\,du\right] \quad (20.2\text{-}20)$$

20.2.4 Significance of the Lagrange Multipliers

The physical significance of the adjoint vector—or vector of Lagrange multipliers—can now be seen. Existence of neighboring optimal solutions under small constant changes of the constraints is guaranteed if the nominal stationary point is a local extremum. The corresponding changes of control du and the state and adjoint vectors dx and $d\nu$ are given by eqs. (20.2-18)-(20.2-20). If we express the differential changes in M in terms of de we have

$$\begin{aligned} dM &= \frac{\partial M}{\partial e}\,de + \frac{1}{2}\,de^T\,\frac{\partial^2 M}{\partial e^2}\,de + \cdots \\ &= -\nu^T\,de + \frac{1}{2}\,de^T\,\frac{\partial^2 M}{\partial e^2}\,de + \cdots \end{aligned} \quad (20.2\text{-}21)$$

Since the partials are evaluated at the extremum, we have

$$\nu^T = -\frac{\partial M}{\partial e}(x^0, u^0) \quad (20.2\text{-}22)$$

i.e., the adjoint vector is the sensitivity of the extremum value of M to changes in the constraints.

Examples:

(1) A typical problem from the ordinary calculus is to find the scalar control u that minimizes the criterion

$$M = \frac{1}{2}(x^2 + u^2) \tag{20.2-23}$$

subject to the constraint

$$e(x, u) = k - ux = 0 \tag{20.2-24}$$

where x is a scalar parameter and k a positive constant. Forming the function $\mathcal{H}$, we have

$$\mathcal{H} = M + \nu^T e = \frac{1}{2}(x^2 + u^2) + \nu(k - ux) \tag{20.2-25}$$

and the necessary conditions are

$$\begin{aligned} \mathcal{H}_x &= x - \nu u = 0 \\ \mathcal{H}_u &= u - \nu x = 0 \\ k - ux &= 0 \end{aligned} \tag{20.2-26}$$

Hence $x = \nu u$, $u = \nu x \Rightarrow \nu = \pm 1$ and $u^0 = \pm\sqrt{k}$. The sufficiency condition (20.2-15) is clearly satisfied:

$$\frac{\partial^2 M}{\partial u^2} = [-e_u^T(e_x^T)^{-1}, I] \begin{bmatrix} \mathcal{H}_{xx} & \mathcal{H}_{xu} \\ \mathcal{H}_{ux} & \mathcal{H}_{uu} \end{bmatrix} \begin{bmatrix} -e_x^{-1}(e_u) \\ I \end{bmatrix}$$

$$\frac{\partial^2 M}{\partial u^2} = \left[-\frac{x}{u}, 1\right] \begin{bmatrix} 1 & -\nu \\ -\nu & 1 \end{bmatrix} \begin{bmatrix} -\dfrac{x}{u} \\ 1 \end{bmatrix}$$

$$\left.\frac{\partial^2 M}{\partial u^2}\right|_{x^0, u^0} = 4 > 0 \tag{20.2-27}$$

(2) Another typical parameter optimization problem is the simplified sailboat problem proposed by Bryson.[20-21] Assume that a cat-rigged sailboat (one sail) moves at a constant velocity and we wish to determine the heading and sail setting angles for maximum velocity and maximum upwind velocity for a given wind direction. The force equilibrium is shown in Fig. 20.2-2.

The sail setting is determined by the angle θ. The sail angle of attack is α and the heading angle is ψ. We assume that the magnitude of the sail force S and the magnitude of the drag force D are given in terms of the relative velocity V_r and boat velocity V as

$$\begin{aligned} S &= c_1 V_r^2 \sin\alpha \\ D &= c_2 V^2 \end{aligned} \tag{20.2-28}$$

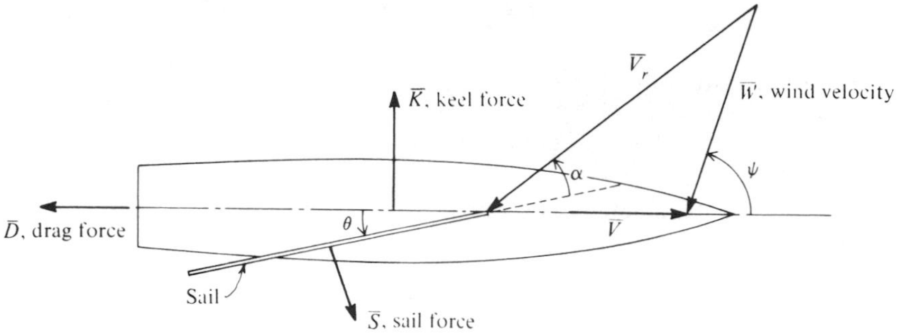

Fig. 20.2-2

For determining maximum velocity for a given heading, the state variables are α, V_r, V, the control is θ for fixed ψ. The constraint equations from equilibrium are

$$\begin{aligned} e_1(V, V_r, \alpha, \theta) &= V^2 - \mu^2 V_r^2 \sin\alpha \sin\theta = 0 \\ e_2(V_r, \theta, \alpha) &= W \sin\psi - V_r \sin(\alpha + \theta) = 0 \\ e_3(V_1 V_r) &= V_r^2 - (V^2 + W^2 + 2VW\cos\psi) = 0 \end{aligned} \qquad (20.2\text{-}29)$$

where $\mu^2 = c_1/c_2$ and W is the magnitude of wind velocity. The function $\mathcal{H}$ for the problem is $\mathcal{H} = M + \nu^T e$ or

$$\mathcal{H} = V + \nu_1(V^2 - \mu^2 V_r^2 \sin\alpha \sin\theta) + \nu_2[W\sin\psi - V_r \sin(\theta + \alpha)] + \nu_3[V_r^2 - (V^2 + W^2 + 2VW\cos\psi)] \qquad (20.2\text{-}30)$$

so the necessary conditions yield

$$\begin{aligned} \mathcal{H}_\alpha &= -\nu_1 \mu^2 V_r^2 \cos\alpha \sin\theta - \nu_2 V_r \cos(\theta + \alpha) = 0 \\ \mathcal{H}_V &= 1 + 2\nu_1 V - 2\nu_3(V + W\cos\psi) = 0 \\ \mathcal{H}_{V_r} &= -2\nu_1 \mu^2 V_r \sin\alpha \sin\theta - \nu_2 \sin(\theta + \alpha) + 2\nu_3 V_r = 0 \\ \mathcal{H}_\theta &= -\nu_1 \mu^2 V_r^2 \sin\alpha \cos\theta - \nu_2 V_r \cos(\theta + \alpha) = 0 \end{aligned} \qquad (20.2\text{-}31)$$

From $\mathcal{H}_\alpha$ and $\mathcal{H}_\theta$ we find $\cos\alpha \sin\theta - \sin\alpha \cos\theta = \sin(\theta - \alpha) = 0$ or $\theta = \alpha$, i.e., the maximum velocity is attained if the sail setting is half the relative wind inclination to the centerline of the boat (usually obtained from telltales or masthead wind indicators).

The maximum velocity is attained running before the wind, $\psi = 180°$. From the first of (20.2-29) we find, with $\theta = \alpha$

$$V^2 = \mu^2 V_r^2 \sin^2 \theta \tag{20.2-32}$$

hence V_{max} is attained for $\theta = 90°$.

From the third of (20.2-29) we have

$$V_r^2 = (W - V)^2; \quad W > V$$

so that

$$V = \mu(W - V)$$

or

$$V_{max} = W \frac{\mu}{1 + \mu} \tag{20.2-33}$$

To maximize the upwind velocity, the payoff is $V \cos \psi$ and we have the additional state variable ψ. The function $\mathcal{H}$ becomes $\mathcal{H} = V \cos \psi + \nu^T e$ with the change in $\mathcal{H}_V$ as given by

$$\mathcal{H}_V = \cos \psi + 2\nu_1 V - 2\nu_3(V + W \cos \psi) = 0 \tag{20.2-34}$$

and the additional necessary condition

$$\mathcal{H}_\psi = -V \sin \psi + \nu_2 W \cos \psi + 2\nu_3 VW \sin \psi = 0 \tag{20.2-35}$$

The conclusion that $\theta = \alpha$ remains unaltered. To determine the value of θ and the heading angle ψ, the constraint eqs. (20.2-29) give

$$\sin^2 \psi(1 - \mu^2 \sin^2 \theta) = \sin^2 2\theta + \mu \sin \theta \sin 2\theta \sin 2\psi \tag{20.2-36}$$

From (20.2-31) the ν's are

$$\nu_2 = -\frac{\nu_1 V^2 \cot \theta}{V_r \cos 2\theta} \tag{20.2-37}$$

$$\begin{aligned} \nu_3 &= \frac{\cos \psi + 2V\nu_1}{2(V + W \cos \psi)} \\ \nu_1 &= -\frac{V_r^2 \cos \psi}{2(V + W \cos \psi)} \left[\frac{1}{\dfrac{V_r^2 V}{V + W \cos \psi} - V^2 \left(1 - \dfrac{\tan 2\theta}{2 \tan \theta}\right)} \right] \end{aligned} \tag{20.2-38}$$

Using (20.2-35) we can determine

$$\tan^2 \psi = \frac{\tan 2\theta}{2 \tan \theta} \left[1 - \mu^2 \sin^2 \theta \left(1 - \frac{\tan 2\theta}{2 \tan \theta}\right)\right]^{-1} \tag{20.2-39}$$

For small θ, $\psi \cong 45°$ and $\theta^2 \cong [(\mu + 2)^2 + 4]^{-1}$. The maximum upwind velocity $(V \cos \psi)_{max}$ is $W/4\mu$.

20.3 DYNAMIC OPTIMIZATION, NECESSARY CONDITIONS

By "dynamic optimization" we shall mean any optimization problem where the state of the system is continuous and determined either by ordinary or partial differential equations. The bulk of the results will be developed for ordinary differential equations, in section 20.3.8, we shall show how distributed parameter systems, governed by partial differential equations, may be included in the theory. Optimization problems for continuous systems are properly *calculus of variations* problems; developments in optimal control theory such as Pontryagin's Minimum Principle have extended some of the classical calculus of variations results, primarily in regard to control constraints.

When considering continuous systems we may be faced with two different types of problems: (a) optimize the response history of the controlled variables to disturbances of the system to be controlled, or (b) optimize a single payoff function or performance criterion. In the first type of problem we wish to minimize the error with which the system achieves a known, desired time history of the controlled variables. If this desired value is fixed, we have the classical regulator problem; if changing due to dynamics of its own, we have the servomechanism problem. The typical problems of a process control engineer fall into this class. We will not discuss the necessary theory for problems of type (a) here but refer the interested reader to the text by Oldenburger.[20-24]

When optimizing the value of a single payoff function, we do not consider the entire response curve but only the value of some quantity associated with the response curve. For such problems we may speak of either *analogic* or *proper* applications. For example, in the variational formulation of classical mechanics, we are concerned with analogic applications; we search for a configuration of a system characterized by M whereby M is minimized, rather than the value of M_{min} itself. We may simply illustrate this case by considering the variational problem we instinctively solve each day: What is the shortest distance between two points? If we *have* to go from point A to point B anyway, then we are primarily interested in the qualitative result that the shortest path under no constraints is a straight line. The actual value of the minimum distance (and hence perhaps the time) is of secondary value, although readily available once we know the motion must take place in a straight line. As an example of a proper application we might consider the problem of computing the minimum fuel required for a spacecraft to execute a desired orbit change maneuver—we need to know the least amount of fuel we must take along. The two classes of applications do not differ in analytical development, rather, in the importance attached to the result.

20.3.1 The General Problem

In the succeeding sections we will consider the following general problem:

Given the initial value of an n-dimensional state vector

$$x(t_0) = x_0 \tag{20.3-1}$$

with x subject to the differential constraints

$$\dot{x} = f(x, u, t) \tag{20.3-2}$$

where $u(t)$ is an m-dimensional control vector, $u \in U(t)$, a control domain independent of x, find $u(t)$ such that the scalar criterion

$$M = \varphi[x(t_f), t_f] + \int_{t_0}^{t_f} h(x, u, t)\, dt \tag{20.3-3}$$

achieves a minimum (maximum) over the time interval $[t_0, t_f]$ with q terminal constraints

$$\psi[x(t_f), t_f] = 0 \tag{20.3-4}$$

Both the state and control vector may be subject to inequality and/or equality constraints and the final time t_f might be either fixed or free.

The state equations (20.3-2) are allowed to be discontinuous at a finite number of points t_i, $t_0 < t_i < t_f$, and the state might be required to satisfy interior boundary conditions at times t_j.

The usual calculus of variations problems where the initial state and time are allowed to be free and may appear in φ can be easily included in the results to follow by minor extensions of the arguments.

20.3.2 The Problems of Mayer, Bolza and Laqrange

In the classical calculus of variations three types of problems are identified—those of Mayer, Bolza and Lagrange. The distinction rests with the form of the payoff (or functional) employed

$$M = \varphi(x_f, t_f, x_0, t_0) + \int_{t_0}^{t_f} h(x, \dot{x}, t)\, dt \qquad \text{(Bolza)} \tag{20.3-5a}$$

$$M = \varphi(x_f, t_f, x_0, t_0) \qquad \text{(Mayer)} \quad (20.3\text{-}5b)$$

$$M = \int_{t_0}^{t_f} h(x, \dot{x}, t)\, dt \qquad \text{(Lagrange)} \quad (20.3\text{-}5c)$$

[In these expressions and in the sequel x_f, x_0 denote $x(t_f)$ and $x(t_0)$.] We could, instead of the Bolza form, use either the Mayer or Lagrange form. To change the Bolza problem into the Mayer form, for example, we introduce an $n + 1^{\text{st}}$ variable such that

$$\dot{x}_{n+1} = h(x, \dot{x}, t)$$

so

$$x_{n+1}(t_f, t_0) = \int_{t_0}^{t_f} h(x, \dot{x}, t)\, dt$$

and

$$M = \varphi(x_f, t_f, x_0, t_0) + x_{n+1}(t_f, t_0)$$

$$M = \varphi(x_f, t_f, x_0, t_0); \qquad x \text{ an } n + 1 \text{ vector}$$

Equivalently, we could deal with only the Lagrange form. In the calculus of variations it is assumed that $h(x, \dot{x}, t)$ are real valued functions of class C^2.* Thus, the problem of Mayer can be written as

$$M = \varphi(x_f, t_f, x_0, t_0) = \int_{t_0}^{t_f} \frac{d\varphi}{dt}\, dt$$

$$M = \int_{t_0}^{t_f} \left[\frac{\partial \phi}{\partial t}(x, t) + \left(\frac{\partial \phi}{\partial x} \right)^T \dot{x} \right] dt$$

which is the Lagrange form. A similar device may be employed for the problem of Bolza.

20.3.3 The Necessary Conditions: No Terminal Constraints, Fixed Terminal Time

To introduce the basic concepts, let us consider the following problem from the classical calculus of variations. Given a scalar function of two scalar variables y

*Functions twice continuously differentiable are denoted by the symbol C^2.

and t

$$h = h(y, \dot{y}, t) \tag{20.3-6}$$

and the definite integral

$$M = \int_{t_0}^{t_f} h(y, \dot{y}, t)\, dt \tag{20.3-7}$$

with boundary conditions

$$y(t_0) = \alpha; \quad y(t_f) = \beta \tag{20.3-8}$$

find $y = y(t)$, twice continuously differentiable (C^2), such that M attains a stationary value in the interval $t_0 \leqslant t \leqslant t_f$. A typical problem of this type can be illustrated as shown in Fig. 20.3-1.

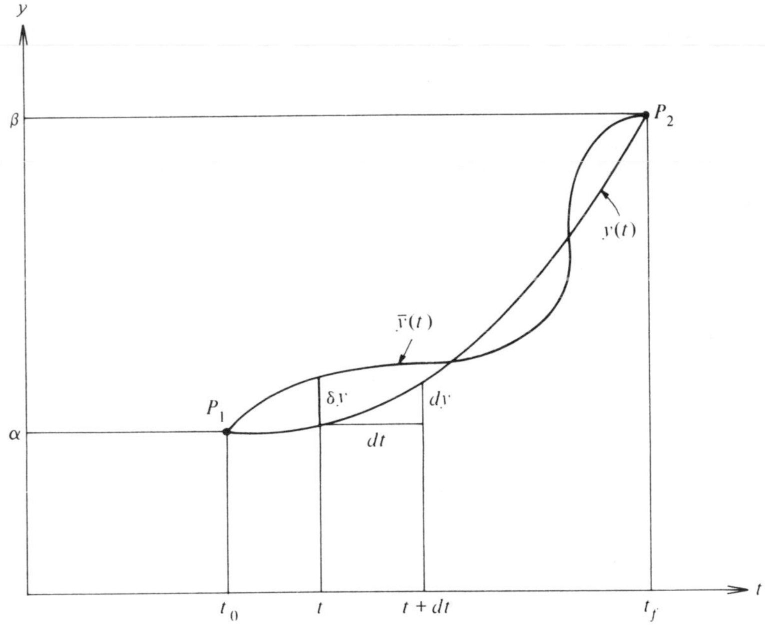

Fig. 20.3-1

We *assume* that the function $y(t)$ yields an extremum (or at least stationary) value of M. In order to prove that we do have a stationary value, we must evaluate M for a modified function $\bar{y}(t)$ and show that the rate of change of M due to the change in the function y becomes zero. Initially we consider trial functions $\bar{y}(t)$

such that $|\bar{y} - y| \leqslant \epsilon_1$ and $|\dot{\bar{y}} - \dot{y}| \leqslant \epsilon_2$ with ϵ_1 and ϵ_2 small. This restriction will result in a *weak extremum* of M and $\bar{y} - y = \delta y$ is a *weak variation.*

Let us write the trial function $\bar{y}(t)$ as

$$\bar{y}(t) = y(t) + \epsilon\eta(t) \tag{20.3-9}$$

where $\eta(t)$ is a new (arbitrary continuous) function and ϵ is a small real number. Comparing $\bar{y}$ and y at the same value of t we find

$$\delta y = \bar{y} - y = \epsilon\eta \tag{20.3-10}$$

The difference between the d-process of ordinary calculus and the δ-process of the calculus of variations is easily visualized from Fig. 20.3-1. The δ-process is characterized by two features:

(a) It is an *infinitesimal* change as $\epsilon \to 0$.
(b) It is a *virtual* change—made in any arbitrary manner.

Next we ask for the change of the integral M caused by the variation δy. Note that the integrand h is a given function of $y, \dot{y}$ and t, hence the *functional dependence* is not altered by the variation process. The increment in M is

$$\Delta M = M(y + \delta y, \dot{y} + \delta\dot{y}, t) - M(y, \dot{y}, t) \tag{20.3-11}$$

If the first term of eq. (20.3-11) possesses finite derivatives of all orders with respect to ϵ in an open neighborhood of $\epsilon = 0$, a Maclaurin series expansion with respect to ϵ of the first term will give

$$\Delta M = \frac{\partial M}{\partial \epsilon}(y + \delta y, \dot{y} + \delta\dot{y}, t)\bigg|_{\epsilon=0} \epsilon + \frac{\partial^2 M}{\partial \epsilon^2}(y + \delta y, \dot{y} + \delta\dot{y}, t)\bigg|_{\epsilon=0} \frac{\epsilon^2}{2} + \cdots \tag{20.3-12}$$

By convention, the first term of (20.3-12) is called the *first variation* of M denoted by δM, and the next term as the second variation denoted by $\delta^2 M$. Expanding the first term yields

$$\begin{aligned} \frac{\delta M}{\epsilon} &= \left[\frac{\partial}{\partial \epsilon}\int_{t_0}^{t_f} h(y + \delta y, \dot{y} + \delta\dot{y}, t)\,dt\right]\bigg|_{\epsilon=0} \\ \frac{\delta M}{\epsilon} &= \int_{t_0}^{t_f} (h_y\eta + h_{\dot{y}}\dot{\eta})\,dt \end{aligned} \tag{20.3-13}$$

The second term of the integrand of eq. (20.3-13) may be integrated by parts to give

$$\frac{\delta M}{\epsilon} = h_{\dot{y}}\eta \Big|_{t_0}^{t_f} + \int_{t_0}^{t_f} \left[h_y - \frac{d}{dt}(h_{\dot{y}})\right]\eta \, dt \qquad (20.3\text{-}14)$$

Since we are integrating between fixed limits, all paths—real as well as varied—start and end at the same points, we have $\eta(t_0) = \eta(t_f) = 0$. Furthermore, if M is to have a stationary value for $y(t)$ as assumed, then $\delta M/\epsilon$ must be zero for arbitrary $\eta(t)$. The *necessary condition* for a stationary value of M thus becomes

$$h_y - \frac{d}{dt}(h_{\dot{y}}) = 0 \qquad (20.3\text{-}15)$$

This equation is known as the *Euler-Lagrange* equation.

Let us now examine the optimization problem posed in section 20.3.1 but still under no terminal constraints and for fixed final time. In order to account for the differential constraints, eq. (20.3-2), we introduce the adjoint vector λ and form the augmented criterion

$$J = \varphi(x_f, t_f) + \int_{t_0}^{t_f} [h(x, u, t) + \lambda^T(t)\ \{f(x, u, t) - \dot{x}\}]\ dt \qquad (20.3\text{-}16)$$

For convenience, we introduce the *Hamiltonian* defined as

$$\mathcal{H}(x, u, \lambda, t) = h(x, u, t) + \lambda^T(t) f(x, u, t) \qquad (20.3\text{-}17)$$

and integrate the last term of eq. (20.3-16) by parts to obtain

$$J = \varphi(x_f, t_f) - \lambda^T x \big|_{t_0}^{t_f} + \int_{t_0}^{t_f} \{\mathcal{H}(x, u, \lambda, t) + \dot{\lambda}^T(t)\, x\}\ dt \qquad (20.3\text{-}18)$$

Now consider the *variation* of J due to (weak) variations in the control vector $u(t)$ for fixed t_0, t_f

$$\delta J(= \delta M) = [\varphi_x - \lambda^T]\ \delta x \big|_{t_f} + [\lambda^T \delta x]\big|_{t_0} + \int_{t_0}^{t_f} \{[\mathcal{H}_x + \dot{\lambda}^T]\ \delta x + \mathcal{H}_u \delta u\}\ dt \qquad (20.3\text{-}19)$$

We could determine the δx corresponding to a given δu but it is simpler to *choose*

$$\dot{\lambda}^T = -\mathcal{H}_x \tag{20.3-20}$$

and (the boundary condition)

$$\lambda^T(t_f) = \varphi_x|_{t_f} \tag{20.3-21}$$

so that

$$\delta J = [\lambda^T \delta x]|_{t_0} + \int_{t_0}^{t_f} [\mathcal{H}_u \delta u] \, dt \tag{20.3-22}$$

In particular, note that

$$\lambda^T(t_0) = \frac{\partial J}{\partial x_0} \tag{20.3-23}$$

for fixed $u(t)$. Since t_0 is arbitrary, the adjoint functions $\lambda(t)$ are also called *influence functions* on J of variations in $x(t)$. The functions $\mathcal{H}_u$ are called *impulse response functions.*

As before δJ must be zero for arbitrary δu (satisfying the weak variation restrictions which we will relax later), hence

$$\mathcal{H}_u = 0 \quad \text{for all } t_0 \leqslant t \leqslant t_f \tag{20.3-24}$$

Solution of eq. (20.3-24) gives the optimal control u^0; if we have $u^0 = u^0(t)$, it is known as the *control function* or open loop control, if $u^0 = u^0(x, t)$, it is called the *control law* or *closed loop control.*

To solve for $u(t)$ then, such that J achieves a stationary value, we must solve the *two-point boundary value problem* defined by the necessary conditions

$$\dot{x} = f(x, u, t) = \mathcal{H}_\lambda^T \tag{20.3-25a}$$

$$\dot{\lambda} = -\mathcal{H}_x^T \tag{20.3-25b}$$

subject to the mixed boundary conditions

$$x(t_0) = x_0 \tag{20.3-25c}$$

$$\lambda(t_f) = \varphi_x^T|_{t_f} \tag{20.3-25d}$$

with control determined from

$$\mathcal{H}_u = 0 \tag{20.3-25e}$$

Equations (20.3-25b) and (20.3-25e) are the *Euler-Lagrange* equations for our control problem.

Note the similarity of equations (20.3-25a, b) to the Hamiltonian equations of motion in mechanics, where x would be the generalized coordinates and λ the generalized momenta. As in mechanics, we can determine a *first integral* of our boundary value problem if $\mathcal{H}$ is not an explicit function of time. For then

$$d\mathcal{H} = \mathcal{H}_x\, dx + \mathcal{H}_u\, du + \mathcal{H}_t\, dt + d\lambda^T f$$

or

$$\begin{aligned} \dot{\mathcal{H}} &= \mathcal{H}_x \dot{x} + \mathcal{H}_u \dot{u} + \mathcal{H}_t + \dot{\lambda}^T f \\ \dot{\mathcal{H}} &= \mathcal{H}_t + \mathcal{H}_u \dot{u} + (\mathcal{H}_x + \dot{\lambda}^T) f \end{aligned} \tag{20.3-26}$$

The last two terms of eq. (20.3-26) vanish; $\mathcal{H}_u$ if $u(t)$ is optimal and the coefficient of f by virtue of eq. (20.3-25b), so that if t does not appear explicitly in $\mathcal{H}$, $\mathcal{H}_t = 0$ and

$$\mathcal{H} = \text{constant} \tag{20.3-27}$$

This constancy of the Hamiltonian along the optimum path can often be exploited in solution of problems for which the integral exists.

The conditions of eq. (20.3-25) do not guarantee that the solution is in fact a local minimum, additional necessary conditions must be satisfied. We shall postpone examination of the sufficient conditions until section 20.4, and turn our attention to the extension necessary for the first necessary conditions of the general problem posed in section 20.3.1.

First, we consider an example of the class of problems discussed in this section, the Zermelo problem.

Example: A water-borne craft (ship, swimmer) must travel through a region of strong currents, which are known functions of the position. The problem is to determine a steering program for minimum time travel between two points. We introduce the fixed coordinates $x = (x_1, x_2)^T$ and express the speed of the craft as $V(x)$ and the currents as $w = [u(x), v(x)]^T$. The equations of motion are

$$\dot{x}_1 = V(x) \cos \theta + u(x) \tag{20.3-28a}$$

$$\dot{x}_2 = V(x) \sin \theta + v(x) \tag{20.3-28b}$$

where θ is the steering angle relative to the fixed x_1 direction. The payoff is the elapsed time or

$$M = \int_{t_0}^{t_f} dt = (t_f - t_0) \tag{20.3-29}$$

We may choose the time origin as $t_0 = 0$ for convenience. The Hamiltonian is

$$\mathcal{H} = 1 + \lambda_1 [V(x) \cos \theta + u(x)] + \lambda_2 [V(x) \sin \theta + v(x)] \tag{20.3-30}$$

Since time dost not appear explicitly in $\mathcal{H}$, we have $\mathcal{H}$ = constant or $\mathcal{H} = 0$ since the constant must be zero for a minimum time problem.

The control equation is

$$\mathcal{H}_\theta = V[-\lambda_1 \sin \theta + \lambda_2 \cos \theta] = 0; \quad \Rightarrow \tan \theta = \frac{\lambda_2}{\lambda_1} \tag{20.3-31}$$

From $\mathcal{H}$ we can determine the Euler-Lagrange equations

$$\dot{\lambda}_1 = -\mathcal{H}_{x_1} = -[\lambda_1 (V_{x_1} \cos \theta + u_{x_1})] + \lambda_2 (V_{x_1} \sin \theta + v_{x_1})] \tag{20.3-32a}$$

$$\dot{\lambda}_2 = -\mathcal{H}_{x_2} = -[\lambda_1 (V_{x_2} \cos \theta + u_{x_2})] + \lambda_2 (V_{x_2} \sin \theta + v_{x_2})] \tag{20.3-32b}$$

and from $\mathcal{H}$ and the control equation

$$\lambda_1 = -\frac{\cos \theta}{V + u \cos \theta + v \sin \theta} \tag{20.3-33a}$$

$$\lambda_2 = -\frac{\sin \theta}{V + u \cos \theta + v \sin \theta} \tag{20.3-33b}$$

The governing equation for the steering angle θ then becomes

$$\dot{\theta} = \sin^2 \theta \, v_{x_1} + \sin \theta \cos \theta \, (u_{x_1} - v_{x_2}) - \cos^2 \theta \, u_{x_2} + V_{x_1} \sin \theta - V_{x_2} \cos \theta \tag{20.3-34}$$

which is determined by substitution of (20.3-33) into (20.3-32). We must solve $\dot{\theta}$, $\dot{x}_1$, $\dot{x}_2$ simultaneously to get the minimum time paths; to have the trajectory pass through two given points A and B the correct value of θ at the starting point must be selected.

If the speed of the craft is constant, V_B, and the current varies "from shore" as some function $f(x_2)$ so that $v = 0$, $u = f(x_2)$ then

$$\dot{\lambda}_1 = -\mathcal{H}_{x_1} = 0 \qquad \Rightarrow \lambda_1 = \text{constant} \tag{20.3-35a}$$

$$\dot{\lambda}_2 = -\mathcal{H}_{x_2} \qquad \Rightarrow \dot{\lambda}_2 = -\lambda_1 \frac{df(x_2)}{dx_2} \tag{20.3-35b}$$

The second of equations (20.3-35) can be written as

$$\frac{d\lambda_2}{dx_2} = \frac{\dot{\lambda}_2}{\dot{x}_2} = -\frac{\lambda_1 f'(x_2)\,[\lambda_1^2 + \lambda_2^2]^{1/2}}{\lambda_2 V_B}$$

which integrates to

$$[\lambda_1^2 + \lambda_2^2]^{1/2}\Big|_{\lambda_{2_0}}^{\lambda_2} = -\frac{\lambda_1}{V_B}\,[f(x_2) - f(x_{2_0})]$$

but $\cos\theta = \lambda_1/[\lambda_1^2 + \lambda_2^2]^{1/2}$ so

$$\sec\theta_0 - \sec\theta = \frac{1}{V_B}\,[f(x_2) - f(x_{2_0})] \tag{20.3-36}$$

and since $x_2(t_f)$ is free, we have $\lambda_2(t_f) = 0$ hence $\theta(t_f) = 0$.

20.3.4 The Necessary Conditions: Terminal State Constraints, Free Terminal Time

To account for terminal state constraints $\psi(x_f, t_f) = 0$ we introduce the Lagrange multipliers ν_0, ν^T with ν a $(q \times 1)$ vector, not all zero, and form the augmented criterion as

$$J = \nu_0 M + \nu^T \psi + \int_{t_0}^{t_f} \lambda^T(t)\,\{f(x,u,t) - \dot{x}\}\,dt$$

$$J = [\nu_0 \varphi(x_f, t_f) + \nu^T \psi(x_f, t_f)] + \int_{t_0}^{t_f} \{\nu_0 h(x,u,t) + \lambda^T(t)\,[f(x,u,t) - \dot{x}]\}\,dt \tag{20.3-37}$$

where the multiplier ν_0 has been introduced to accommodate abnormal situations. (Abnormality will be discussed in detail later, after the necessary conditions for a stationary value of J have been determined.) The terminal time t_f is unspecified and may be thought of as an additional control variable which can be adjusted to find a minimum (maximum) of J while satisfying the terminal state constraints.

To derive the necessary conditions we consider the change in J due to control variations δu (and hence δx) and differential changes in the terminal time t_f. The

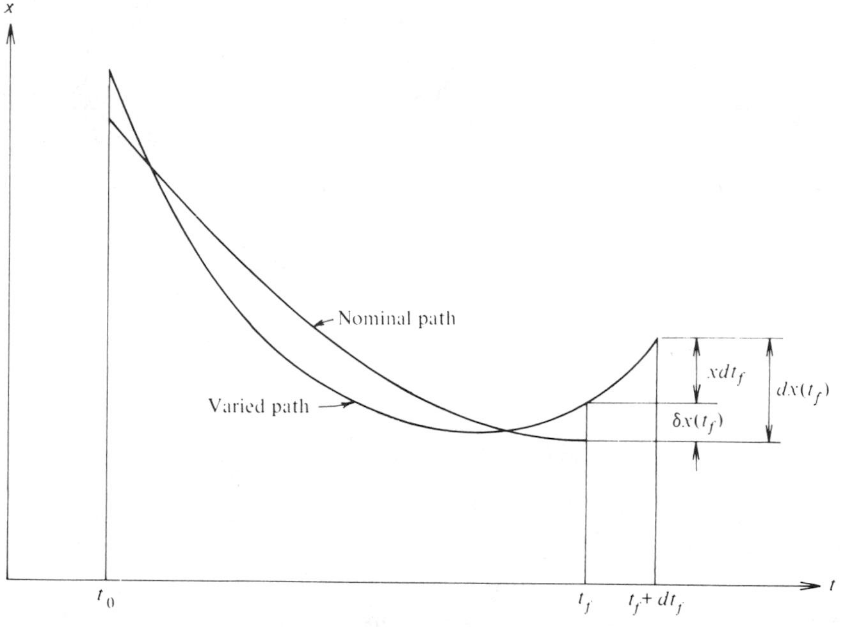

Fig. 20.3-2

situation at the terminus of two comparison paths may be visualized geometrically for one state variable x as shown in Fig. 20.3-2.

Thus

$$dx(t_f) = \delta x(t_f) + \dot{x}(t_f)\, dt_f \tag{20.3-38}$$

Forming the (differential) change in J due to dt_f and δu (in the integrand t *does not* vary)

$$dJ = \left[\nu_0 \frac{\partial \varphi}{\partial t} + \nu^T \frac{\partial \psi}{\partial t}\right]\Bigg|_{t_f} dt_f + \left[\nu_0 \frac{\partial \varphi}{\partial x} + \nu^T \frac{\partial \psi}{\partial x}\right]\Bigg|_{t_f} dx_f$$

$$+ \int_{t_0}^{t_f} \left\{ \nu_0 \frac{\partial h}{\partial x} \delta x + \nu_0 \frac{\partial h}{\partial u} \delta u + \lambda^T \left[\frac{\partial f}{\partial x} \delta x + \frac{\partial f}{\partial u} \delta u - \delta \dot{x}\right]\right\} dt$$

$$+ \nu_0 h|_{t_f} dt_f \tag{20.3-39}$$

where the last term arises from the variable limit of integration t_f. Integrating the $\delta \dot{x}$ term of the integrand by parts yields

$$dJ = \left[\nu_0 \frac{\partial \varphi}{\partial t} + \nu_0 h + \nu^T \frac{\partial \psi}{\partial t}\right]\Bigg|_{t_f} dt_f + \left[\nu_0 \frac{\partial \varphi}{\partial x} + \nu^T \frac{\partial \psi}{\partial x}\right]\Bigg|_{t_f} dx_f$$

$$+ \lambda^T \delta x|_{t_0} - \lambda^T \delta x|_{t_f} + \int_{t_0}^{t_f} \left\{\left[\nu_0 \frac{\partial h}{\partial x} + \lambda^T \frac{\partial f}{\partial x} + \dot{\lambda}^T\right] \delta x\right.$$

$$\left. + \left[\nu_0 \frac{\partial h}{\partial u} + \lambda^T \frac{\partial f}{\partial u}\right] \delta u \right\} dt \tag{20.3-40}$$

Introducing eq. (20.3-38) and the definition of the Hamiltonian, $\mathcal{H} = \nu_0 h + \lambda^T f$

$$dJ = \left[\nu_0 \frac{\partial \varphi}{\partial t} + \nu_0 h + \nu^T \frac{\partial \psi}{\partial t} + \lambda^T \dot{x}\right]\Bigg|_{t_f} dt_f + \left[\nu_0 \frac{\partial \varphi}{\partial x} + \nu^T \frac{\partial \psi}{\partial x} - \lambda^T\right]\Bigg|_{t_f} dx_f$$

$$+ \lambda^T \delta x|_{t_0} + \int_{t_0}^{t_f} \{[\mathcal{H}_x + \dot{\lambda}^T]\, \delta x + \mathcal{H}_u \delta u\}\, dt \tag{20.3-41}$$

Now, *choose* $\lambda(t)$ such that

$$\dot{\lambda}^T = -\mathcal{H}_x^T \tag{20.3-42}$$

$$\lambda^T(t_f) = \left[\nu_0 \frac{\partial \varphi}{\partial x} + \nu^T \frac{\partial \psi}{\partial x}\right]\Bigg|_{t_f} \tag{20.3-43}$$

and, since $x(t_0) = x_0$ is given, $\delta x(t_0) = 0$, and dJ becomes

$$dJ = \left[\nu_0 \frac{\partial \varphi}{\partial t} + \nu_0 h + \nu^T \frac{\partial \psi}{\partial t} + \lambda^T \dot{x}\right]\Bigg|_{t_f} dt_f + \int_{t_0}^{t_f} \mathcal{H}_u \delta u\, dt \tag{20.3-44}$$

Thus, for $dJ = 0$, we have the following *set of first necessary conditions:*

$$\dot{x} = f(x, u, t) = \mathcal{H}_\lambda^T \qquad n \text{ differential equations} \tag{20.3-45a}$$

$$\dot{\lambda} = -\mathcal{H}_x^T \qquad n \text{ differential equations} \tag{20.3-45b}$$

$$\mathcal{H}_u = 0 \qquad m \text{ algebraic equations} \tag{20.3-45c}$$

with

$$x(t_0) = x_0, \quad \lambda^T(t_f) = \left[\nu_0 \frac{\partial \varphi}{\partial x} + \nu^T \frac{\partial \psi}{\partial x}\right]\Bigg|_{t_f} \tag{20.3-45d}$$

$$\psi(x_f, t_f) = 0 \qquad q \text{ algebraic equations} \tag{20.3-45e}$$

and the additional (*transversality*) condition

$$\left[\nu_0 \frac{\partial \varphi}{\partial t} + \nu^T \frac{\partial \psi}{\partial t} + \nu_0 h + \lambda^T \dot{x}\right]\Bigg|_{t_f} = 0$$

or

$$\begin{gathered}\left[\nu_0 \frac{\partial \varphi}{\partial t} + \nu^T \frac{\partial \psi}{\partial t} + \nu_0 h + \left(\nu_0 \frac{\partial \varphi}{\partial x} + \nu^T \frac{\partial \psi}{\partial x}\right) f\right]\Bigg|_{t_f} = 0 \\ \left[\nu_0 \frac{d\varphi}{dt} + \nu_0 h + \nu^T \frac{d\psi}{dt}\right]\Bigg|_{t_f} = 0\end{gathered} \tag{20.3-45f}$$

Should some $x_i(t_0)$ be *not* prescribed, we must choose $\lambda_i(t_0) = 0$; since there exists an optimum $x_i^0(t_0)$, small arbitrary changes of $x_i(t_0)$ around this value must give $dJ = 0$.

Example: To illustrate the effects of end constraints, consider a variation of the Zermelo problem. For constant boat speed V_B and a current that increases linearly from shore, i.e., $u(x) = kx_2$, we ask for maximum distance traveled downstream in a fixed time interval. In addition, we require that the boat return to shore, $x_2(t_f) = 0$, at a specified terminal time t_f. We introduce the terminal constraint

$$\psi(x_f, t_f) = x_2(t_f) = 0 \tag{20.3-46}$$

and the state equations

$$\begin{aligned}\dot{x}_1 &= V_B \cos\theta + kx_2 \\ \dot{x}_2 &= V_B \sin\theta\end{aligned} \tag{20.3-47}$$

with initial conditions

$$x_1(t_0) = x_2(t_0) = 0 \tag{20.3-48}$$

The Hamiltonian is

$$\mathcal{H} = \lambda^T f = \lambda_1(V_B \cos\theta + kx_2) + \lambda_2 V_B \sin\theta \tag{20.3-49}$$

since the payoff is $M = \varphi(x_f, t_f) = x_1(t_f)$. The adjoint equations become

$$\dot{\lambda}_1 = 0; \quad \dot{\lambda}_2 = -k\lambda_1 \tag{20.3-50}$$

with boundary conditions

$$\begin{aligned} \lambda^T(t_f) &= \varphi_{x_f} + \nu^T \psi_{x_f} \\ \lambda_1(t_f) &= 1; \quad \lambda_2(t_f) = \nu_2 \end{aligned} \tag{20.3-51}$$

Hence λ_1 = constant = 1 and $\dot{\lambda}_2 = -k$, so that $\lambda_2(t) = k(t_f - t) + \nu_2$. Now, the control $\theta(t)$ must maximize the Hamiltonian which may be written as

$$\mathcal{H} = V_B(\lambda_1 \cos\theta + \lambda_2 \sin\theta) + \lambda_1 k x_2$$

so that we have $\max_\theta \mathcal{H}$ if

$$\sin\theta = \frac{\lambda_2}{[\lambda_1^2 + \lambda_2^2]^{1/2}}; \quad \cos\theta = \frac{\lambda_1}{[\lambda_1^2 + \lambda_2^2]^{1/2}} \tag{20.3-52}$$

or, since $\lambda_1 = 1$, $\lambda_2 = \tan\theta$. The control then is

$$\tan\theta_0 - \tan\theta = k(t - t_0) \tag{20.3-53}$$

The angle θ_0 is determined by $\theta_0 = -\theta_f$ or $2\tan\theta_0 = k(t_f - t_0)$.

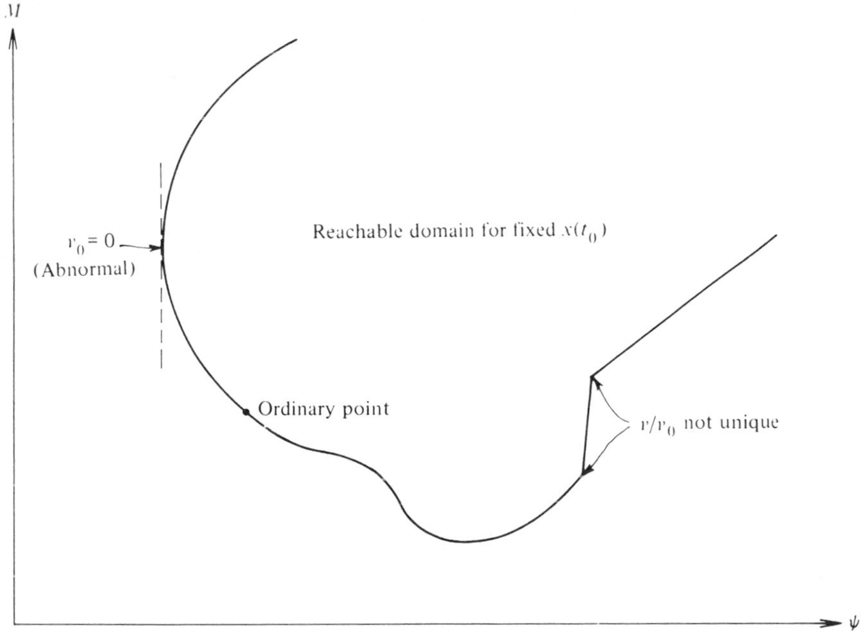

Fig. 20.3-3

Let us now return to the question of abnormality and the interpretation of the multipliers ν_0 and ν^T. The necessary conditions expressed by equations (20.3-45a) to (20.3-45f) for minimizing the criterion M subject to q terminal constraints $\psi(x_f, t_f) = \psi_D$ (with ψ_D a given constraint level) state that there exist $(q + 1)$ multipliers ν_0, ν^T not necessarily unique, such that $\nu_0 \delta M + \nu^T \delta \psi \geqslant 0$ for all possible variations in control and/or final time t_f. The reachable domain formed by M and ψ for fixed initial state $x(t_0)$ is locally convex. Thus, if ν^T is unique, the ratio ν^T/ν_0 is the sensitivity of $\min M$ to changes in the constraint level ψ_D, $-\partial M_{\min}/\partial \psi_D$. If ν^T is not unique, then ν^{T-}/ν_0 and ν^{T+}/ν_0 are the left and right hand limits (unique) and are the sensitivities on the left and right of the ambiguity point. When $\partial M_{\min}/\partial \psi_D \to \infty$, we have the abnormal case and we must choose $\nu_0 = 0$; for normal cases, we may set $\nu_0 = 1$. This geometrical interpretation is depicted in Fig. 20.3-3 for one constraint function ψ. Further discussion of abnormality in the classical calculus of variations may be found in the text by Bliss.[20-25]

20.3.5 The Necessary Conditions with Path Constraints

Up to now we have discussed optimization problems where only end-point constraints may appear; in this section we extend the theory to include equality and/or inequality *path constraints.* A path constraint may apply over the entire interval of interest and may involve state and/or control variables. These constraints may be dealt with by appropriately introducing additional Lagrange multipliers and adjoining the constraints to the variational Hamiltonian.

First, we consider the *isoperimetric constraints* of the calculus of variations. In this case, we must find the stationary value of the performance criterion subject to an additional integral constraint

$$G(t_f) = \int_{t_0}^{t_f} g(x, u, t)\, dt \tag{20.3-54}$$

As was done in section 20.3.2 to change the Bolza problem into the Mayer form, we introduce an additional state variable x_{n+1} such that

$$\dot{x}_{n+1} = g(x, u, t) \tag{20.3-55}$$

with boundary conditions $x_{n+1}(t_0) = 0$ and $x_{n+1}(t_f) = G(t_f)$. Then, with a (constant) Lagrange multiplier μ, we form the augmented Hamiltonian

$$\mathcal{H} = h(x, u, t) + \lambda^T(t) f(x, u, t) + \mu g(x, u, t) \tag{20.3-56}$$

So the only modifications to the previous results are in the Euler-Lagrange equations which become

$$\dot{\lambda}^T = -\frac{\partial \mathcal{H}}{\partial x} = -h_x - \lambda^T f_x - \mu g_x \tag{20.3-57}$$

$$\frac{\partial \mathcal{H}}{\partial u} = h_u + \lambda^T f_u + \mu g_u = 0 \qquad (20.3\text{-}58)$$

Suppose now that we have an *equality constraint** involving the state and control variables of the form

$$C(x, u, t) = 0 \qquad (20.3\text{-}59)$$

where $\partial C/\partial u \neq 0$ for any u. We introduce the Lagrange multiplier $\mu(t)$ and form

$$\mathcal{H} = h(x, u, t) + \lambda^T(t) f(x, u, t) + \mu(t)\, C(x, u, t) \qquad (20.3\text{-}60)$$

so that the Euler-Lagrange equations become

$$\frac{\partial \mathcal{H}}{\partial u} = h_u + \lambda^T f_u + \mu C_u = 0 \qquad (20.3\text{-}61)$$

$$\dot{\lambda}^T = -h_x - \lambda^T f_x - \mu C_x \qquad (20.3\text{-}62)$$

If the constraint C is $C(u, t)$ only, eqs. (20.3-61) and (20.3-62) remain valid with the absence of the C_x term in (20.3-62). Suppose that C is a function of state variables and time only

$$C(x, t) = 0 \qquad (20.3\text{-}63)$$

In this case, control of $C(x, t)$ can only be achieved through its time derivatives: one takes repeated time derivatives of $C(x, t)$ until an expression depending explicitly on u is obtained when $f(x, u, t)$ is substituted for $\dot{x}$. If this occurs on the qth derivative then

$$\frac{d^q C(x, t)}{dt^q} = S(x, u, t) = 0 \qquad (20.3\text{-}64)$$

and $S(x, u, t)$ plays the same role as (20.3-59). The constraint $C(x, t)$ is termed a *q^{th}-order state variable equality constraint.* In addition, the conditions

$$C(x, t) = 0, \frac{dC(x, t)}{dt} = 0, \ldots, \frac{d^{q-1}}{dt^{q-1}} C(x, t) = 0 \qquad (20.3\text{-}65)$$

must be added as a set of boundary conditions either at t_0 or t_f.

*There may be r such constraint relations; however, r must be less than m, the dimension of u, in order for an optimization problem to exist.

The situation is modified somewhat if we have an inequality constraint of the form

$$C(x, u, t) \leqslant 0 \tag{20.3-66}$$

We again adjoin C to the Hamiltonian to form

$$\mathcal{H} = h(x, u, t) + \lambda^T(t) f(x, u, t) + \mu(t)\, C(x, u, t) \tag{20.3-67}$$

with the additional qualification that

$$\mu(t) \begin{cases} \geqslant 0 & \text{if } C = 0 \\ = 0 & \text{if } C < 0 \end{cases} \tag{20.3-68}$$

The corresponding Euler-Lagrange equations then become

$$\left.\begin{aligned} \frac{\partial \mathcal{H}}{\partial u} &= h_u + \lambda^T f_u + \mu C_u = 0 \\ \dot{\lambda}^T = -\frac{\partial \mathcal{H}}{\partial x} &= -h_x - \lambda^T f_x - \mu C_x \end{aligned}\right\} \quad C = 0 \tag{20.3-69a}$$

$$\left.\begin{aligned} \frac{\partial \mathcal{H}}{\partial u} &= h_u + \lambda^T f_u = 0 \\ \dot{\lambda}^T = -\frac{\partial \mathcal{H}}{\partial x} &= -h_x - \lambda^T f_x \end{aligned}\right\} \quad C < 0 \tag{20.3-69b}$$

For the case $C < 0$, the constraint is not binding and the optimum control is determined from $\partial \mathcal{H}/\partial u = 0$ only. When $C = 0$, the control $u(t)$ and the adjoint vector $\mu(t)$ are determined jointly from $C = 0$ and $\partial \mathcal{H}/\partial u = 0$. As before, we require that $\partial C(x, u, t)/\partial u \neq 0$ for any u, but the case $C(u, t) \leqslant 0$ is included.

When the control is absent, we have the *state variable inequality constraint*

$$C(x, t) \leqslant 0 \tag{20.3-70}$$

As in the equality constraint case, we take successive time derivatives until on the qth derivative, the control appears explicitly upon the substitution of $f(x, u, t)$ for $\dot{x}$. The governing equations are formally the same as when $C = C(x, u, t)$ with

$$\frac{d^q}{dt^q} C(x, t) = S(x, u, t) \tag{20.3-71}$$

replacing $C(x, u, t)$ and the Hamiltonian

$$\mathcal{H} = h(x, u, t) + \lambda^T(t) f(x, u, t) + \mu(t) S(x, u, t) \tag{20.3-72}$$

In addition, we require $S(x, u, t) = 0$ and $\mu(t) \geqslant 0$ on the constraint boundary $C(x, t) = 0$ and $\mu = 0$ when the constraint is not effective, $C(x, t) < 0$. Furthermore, since control of the constraint $C(x, t)$ can only be obtained by changing the qth time derivative, to keep the system on the constraint one must also satisfy the tangency requirements at the junctions of constrained and unconstrained arcs

$$C(x, t) = 0, \quad \frac{dC(x, t)}{dt} = 0, \ldots, \frac{d^{q-1}}{dt^{q-1}} C(x, t) = 0 \tag{20.3-73}$$

The eqs. (20.3-73) form a set of interior boundary conditions which may be taken into account by regarding the total time interval $[t_0, t_f]$ as composed of subintervals $[t_0, t_i]$, $[t_i, t_j]$, $[t_j, t_f]$ where t_i is the entry and t_j the exit time from the constraint. The problem thus is broken down into unconstrained and constrained arcs, with (20.3-73) representing terminal conditions for the interval $[t_0, t_i]$. Thus, we may adjoin (20.3-73) to the Hamiltonian with a (constant) multiplier vector ζ* [determined so that (20.3-73) are satisfied] and obtain from the first variation

$$\lambda^T(t_i^-) = \lambda^T(t_i^+) + \zeta^T N_x |_{t_i} \tag{20.3-74}$$

$$\mathcal{H}(t_i^-) = \mathcal{H}(t_i^+) - \zeta^T N_t |_{t_i} \tag{20.3-75}$$

where

$$N(x, t) \triangleq \begin{bmatrix} C(x, t) \\ \dfrac{dC}{dt}(x, t) \\ \cdots\cdots\cdots \\ \dfrac{d^{q-1}}{dt^{q-1}} C(x, t) \end{bmatrix}$$

*Recent results reported by Jacobson, D. H.; Lele, M. M.; and Speyer, J. L. in "New Necessary Conditions of Optimality for Control Problems with State-Variable Inequality Constraints," *Journal of Mathematical Analysis and Applications,* **35**, (2), August 1971, indicate that the necessary conditions as given here underspecify the behavior of the adjoint variables at the junctions of constrained and internal arcs. In the reference, it is shown that the multipliers ζ and $\mu(t)$ are *related* along the constrained arc and furthermore, that for *odd* order constraints with $q \geqslant 3$, the optimal arc can only *touch* the boundary but not remain on the boundary for any finite time.

The adjoint vector λ and the Hamiltonian thus are discontinuous at the entry point to the constraint and continuous at the exit point.

Example: As an example of a problem with state constraints, consider the following: $x_1(t)$ to minimize $\int_{t_0}^{t_f} \ddot{x}_1^2\, dt$ subject to $x_1(t_f) = 0$ and the state constraint

$$C(x, t) = x_1 - W(t_f - t) \leqslant 0 \tag{20.3-76}$$

$$W > 0, \quad \forall\, t \leqslant t_f$$

with $x(t_0) = 0, \dot{x}(t_0) = v_0 > 1$.

We introduce an additional state variable x_3 defined by

$$x_3 = \int_{t_0}^{t_f} \ddot{x}_1^2\, dt = \int_{t_0}^{t_f} u^2\, dt \tag{20.3-77}$$

where $u = \ddot{x}_1$. Then, we maximize $\varphi(x_f, t_f) = -x_3(t_f)$ subject to the terminal constraints on $x_1(t_f)$, t_f. The state equations are

$$\dot{x}_1 = x_2; \quad \dot{x}_2 = u; \quad \dot{x}_3 = u^2 \tag{20.3-78}$$

with the constraints

$$\begin{aligned} C(x, t) &= x_1 - W(t_f - t) \leqslant 0 \\ \dot{C}(x, t) &= x_2 + W = 0 \qquad \text{on } C \\ \ddot{C}(x, t) &= u = 0 \qquad \text{on } C \end{aligned} \tag{20.3-79}$$

The Hamiltonian is

$$\mathcal{H} = \lambda_1 x_2 + \lambda_2 u + \lambda_3 u^2 \tag{20.3-80}$$

From the terminal conditions $\nu_0 \varphi_{x_f} + \nu^T \psi_{x_f} = \lambda^T(t_f)$ we find

$$\lambda_1(t_f) = \nu_1; \quad \lambda_2(t_f) = 0; \quad \lambda_3(t_f) = -1 \tag{20.3-81}$$

but since x_3 does not appear in $\mathcal{H}$, $\lambda_3 = -1$ for all time.

Thus, if $C < 0$ then $\max_u \mathcal{H}$ implies $\lambda_2 = 2u$. If $C = 0$, and stays thus, then $u = 0$, $x_2 = -W$ and $\mathcal{H} = -\lambda_1 W$. Hence

$$\mathcal{H}^+ = -\lambda_1^+ W \tag{20.3-82}$$

$$\mathcal{H}^- = -\lambda_1^- W + 2(u^-)^2 - (u^-)^2 = -\lambda_1^- W + (u^-)^2 \tag{20.3-83}$$

From the general theory

$$\mathcal{H}^- = \mathcal{H}^+ - \zeta_1 C_t - \zeta_2 \dot{C}_t = \mathcal{H}^+ - \zeta_1 W \tag{20.3-84}$$

$$\lambda_1^- = \lambda_1^+ + \zeta_1 C_{x_1} + \zeta_2 \dot{C}_{x_1} = \lambda_1^+ + \zeta_1 \tag{20.3-85}$$

So

$$-\lambda_1^- W + (u^-)^2 = -W(\lambda_1^- - \zeta_1) - \zeta_1 W \tag{20.3-86}$$

hence $(u^-)^2 = 0$ or $\ddot{x}_1^- = 0$.

On the unconstrained arc for $t < t'$, where t' is the time of entry onto the constraint, $\dot{\lambda}_1 = 0$ or λ_1 = constant, and $\dot{\lambda}_2 = -\lambda_1$. But $\lambda_2 = 2u$, so that $\dot{\lambda}_2 = 2\dot{u} = 2\dddot{x}_1$. This yields $\dddot{x}_1 = -\lambda_1/2$ = constant and the solution is, in general, a cubic of the form

$$x_1(t) = a_3(t - t_0)^3 + a_2(t - t_0)^2 + a_1(t - t_0) + a_0 \tag{20.3-87}$$

From the initial conditions we have $a_0 = 0$ and $a_1 = v_0$. Evaluating x_1, $\dot{x}_1$ and $\ddot{x}_1$ at t':

$$\begin{aligned} x_1(t') &= a_3(t' - t_0)^3 + a_2(t' - t_0)^2 + v_0(t' - t_0) = W(t_f - t') \\ \dot{x}_1(t') &= [x_2(t')] = 3a_3(t' - t_0)^2 + 2a_2(t' - t_0) + v_0 = -W \\ \ddot{x}_1(t') &= [u(t')] = 6a_3(t' - t_0) + 2a_2 = 0 \end{aligned} \tag{20.3-88}$$

These equations yield

$$t' - t_0 = \frac{3W}{W + v_0}(t_f - t_0) \tag{20.3-89}$$

Since $(t' - t_0)/(t_f - t_0) \leqslant 1$, the path reaches the constraint before t_f if $W < v_0/2$. For $W > \frac{1}{2} v_0$, $x_1(t_f) = a_3(t_f - t_0)^2 + a_2(t_f - t_0) + v_0 = 0$ at the end point, and since $\lambda_2(t_f) = 0$ [$x_2(t_f)$ free] we have $u(t_f) = 0$ or $u(t_f) = 6a_3(t_f - t_0) + 2a_2 = 0$. Solving these relationships for a_3 and a_2 we have

$$a_3 = \frac{v_0}{2(t_f - t_0)^2}; \quad a_2 = -\frac{3v_0}{2(t_f - t_0)}$$

or

$$x = \frac{v_0(t - t_0)^3}{2(t_f - t_0)^2} - \frac{3v_0(t - t_0)^2}{2(t_f - t_0)} + v(t - t_0) \tag{20.3-90}$$

20.3.6 Corner Conditions

The derivation of the necessary conditions given in section 20.3.3 assumes that the extremal vector $y(t)$ was twice continuously differentiable. We can admit extremal vectors with discontinuous derivatives at a finite number of points by considering the solution to be composed of sub-arcs between discontinuities along each of which the Euler-Lagrange equations hold. At each discontinuity t_i integrate the Euler-Lagrange equations across the discontinuity, i.e., from $t_i - \epsilon$ to $t_i + \epsilon$

$$\int_{t_i-\epsilon}^{t_i+\epsilon} \left[\frac{d}{dt}\lambda^T + \mathcal{H}_x\right] dt = 0$$

$$\lambda^T(t_i + \epsilon) - \lambda^T(t_i - \epsilon) + \int_{t_i-\epsilon}^{t_i+\epsilon} \mathcal{H}_x \, dt = 0 \tag{20.3-91}$$

$$\int_{t_i-\epsilon}^{t_i+\epsilon} \left[\frac{d\mathcal{H}}{dt} - \frac{\partial \mathcal{H}}{\partial t}\right] dt = 0$$

$$\mathcal{H}(t_i + \epsilon) - \mathcal{H}(t_i - \epsilon) - \int_{t_i-\epsilon}^{t_i+\epsilon} \mathcal{H}_t \, dt = 0 \tag{20.3-92}$$

If $\mathcal{H}$ and the partial derivatives are *bounded*, then as $\epsilon \to 0$ the integrals vanish and we have

$$\lambda^T(t_i^-) = \lambda^T(t_i^+) \tag{20.3-93}$$

$$\mathcal{H}(t_i^-) = \mathcal{H}(t_i^+) \tag{20.3-94}$$

On either side of t_i the control is optimal, $\mathcal{H}_u = 0$, so

$$\mathcal{H}_u(t_i^-) = \mathcal{H}_u(t_i^+) \tag{20.3-95}$$

Equations (20.3-93) to (20.3-95) are a generalized version of the *Weierstrass-Erdmann* corner conditions of the classical calculus of variations. In the calculus of variations, the criterion is of the form

$$M = \int_{t_0}^{t_f} F(y, \dot{y}, t) \, dt \tag{20.3-96}$$

and the Euler-Lagrange equations are

$$\frac{d}{dt} F_{\dot{y}} - F_y = 0 \tag{20.3-97}$$

which can also be written in the form

$$\frac{d}{dt}\left[F_{\dot{y}}\dot{y} - F\right] = -F_t \tag{20.3-98}$$

Integration over the discontinuity and the limit $\epsilon \to 0$ yields

$$F_{\dot{y}}(t_i^-) = F_{\dot{y}}(t_i^+) \tag{20.3-99}$$

$$(F_{\dot{y}}\dot{y} - F)\Big|_{t_i^-} = (F_{\dot{y}}\dot{y} - F)\Big|_{t_i^+} \tag{20.3-100}$$

The conditions given by eqs. (20.3-93) to (20.3-95) hold for problems without path constraints or problems with control variable inequality constraints $[C(u, t) \leqslant 0]$ only. For problems with state variable inequality constraints $[C(x, t) \leqslant 0]$, at the entry point to the constraint the corner conditions are

$$\lambda^T(t_i^-) = \lambda^T(t_i^+) + \zeta^T N_x \tag{20.3-101}$$

$$\mathcal{H}(t_i^-) = \mathcal{H}(t_i^+) - \zeta^T N_t \tag{20.3-102}$$

$$\mathcal{H}_u(t_i^-) = \mathcal{H}_u(t_i^+) \tag{20.3-103}$$

where $N(x, t) = 0$ are the q tangency constraints $C(x, t) = 0$, $d/dt\ C(x, t) = 0, \ldots$ $d^{q-1}/dt^{q-1}\ C(x, t) = 0$ and ζ is a q vector of (constant) Lagrange multipliers.* At the exit point conditions, eqs. (20.3-93) to (20.3-95) hold.

The relationships given by eqs. (20.3-101) to (20.3-103) also apply to problems with q interior boundary conditions $N[x(t_i), t_i] = 0$ where $t_0 < t_i < t_f$ with no modifications. If the system equations are discontinuous at some interior time $t_0 < t_1 < t_f$, i.e.,

$$\dot{x} = f(x, u, t) \qquad t < t_1 \tag{20.3-104}$$

$$\dot{x} = \tilde{f}(x, u, t) \qquad t_1 < t \tag{20.3-105}$$

with t_1 determined from an interior (scalar) boundary condition of the form $\psi[x(t_1), t_1] = 0$, then eqs. (20.3-101) through (20.3-103) apply with

$$\mathcal{H}^{(1)}(t_1^-) = h + \lambda^T f = \mathcal{H}^{(2)}(t_1^+) - k\psi_t = h + \lambda^T \tilde{f} - k\psi_t \tag{20.3-106}$$

For a discussion of the problem when there are discontinuities in the state variables x as well as in the system equations, see the derivation by Bryson and Ho,[20-21] p. 106.

*See last footnote.

20.3.7 Strong Control Variations, Pontryagin's Minimum Principle

The necessary conditions considered in the previous sections were determined subject to the weak variation restriction on the control, i.e., $\delta u \leqslant \epsilon_1$ and $\delta \dot{u} \leqslant \epsilon_2$ with ϵ_1 and ϵ_2 small. Let us now consider what modifications, if any, are required when we admit *strong variations*, i.e., only $\delta u \leqslant \epsilon$. In the classical calculus of variations these considerations lead to the Weierstrass E condition. We shall show that when the region of control U is an open set of the vector space of u_i, $i = 1, 2, \ldots, m$, the Weierstrass E condition is equivalent to the weak form of Pontryagin's Minimum Principle.* If U is closed and bounded, then the Weierstrass condition no longer holds. The necessary extension to include closed and bounded regions U is contained in the strong form of Pontryagin's Minimum Principle.

We will first derive the Weierstrass E condition. For simplicity of argument we consider the (normal) fixed end point problem with

$$J = \int_{t_0}^{t_f} g(t, y, \dot{y}, \lambda)\, dt \tag{20.3-107}$$

where $g(t, y, \dot{y}, \lambda) = h(t, y, \dot{y}) + \lambda^T \Phi(t, y, \dot{y})$ with $\Phi(y, y, \dot{y}) = 0$ the differential constraints, and assume that we have found the minimizing vector $y(t)$. Then, introduce a comparison function which equals the extremal solution everywhere except in some interval $t_1 \leqslant t \leqslant t_2$ with $t_0 \leqslant t_1 \leqslant t \leqslant t_2 \leqslant t_f$. In an interval of ϵ width we allow some other function $\tilde{y}(t)$ such that $\dot{y}(t_1) \neq \dot{\tilde{y}}(t_1)$. The situation can be visualized as in Fig. 20.3-4.

In the interval $t_1 \leqslant t \leqslant t_1 + \epsilon$ we allow a strong variation, but only a weak variation in $t_1 + \epsilon \leqslant t \leqslant t_2$ returning to the extremal solution. The composite comparison solution is

$$\begin{aligned} y(t) &\text{ for } \quad t_0 \leqslant t \leqslant t_1, \quad t_2 \leqslant t \leqslant t_f \\ \tilde{y}(t) &\text{ for } \quad t_1 \leqslant t \leqslant t_1 + \epsilon \\ y(t, \epsilon) &\text{ for } \quad t_1 + \epsilon \leqslant t \leqslant t_2 \end{aligned} \tag{20.3-108}$$

Continuity of y and $\tilde{y}$ (but not their rates) requires that

$$\begin{aligned} y(t_1) &= \tilde{y}(t_1) \\ \tilde{y}(t_1 + \epsilon) &= y(t_1 + \epsilon, \epsilon) \\ y(t_2, \epsilon) &= y(t_2) \end{aligned} \tag{20.3-109}$$

*In Pontryagin's original work,[20-26] the negative of the Hamiltonian was used, hence it was called the Maximum Principle.

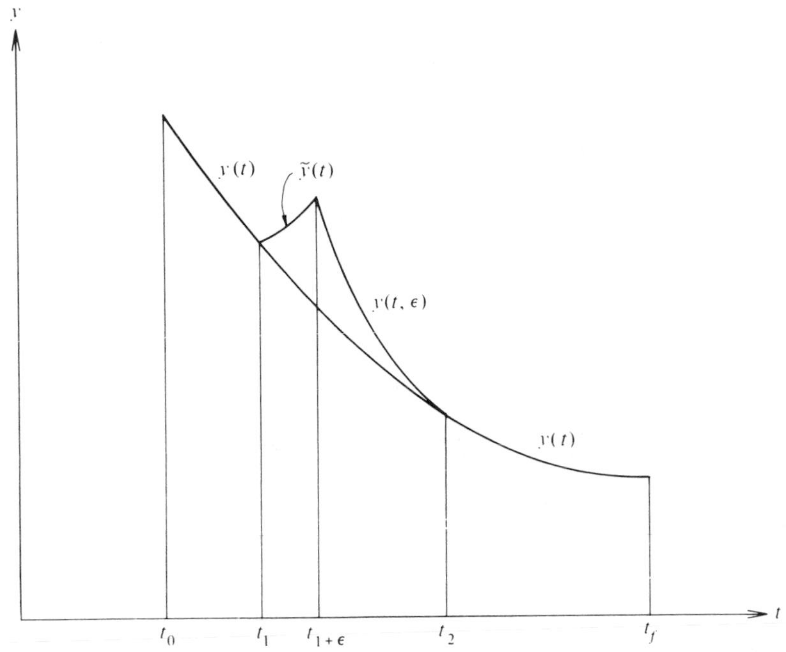

Fig. 20.3-4

The integral, eq. (20.3-107) then breaks down as

$$J = \int_{t_0}^{t_1} g(t, y, \dot{y}, \lambda)\, dt + \int_{t_1}^{t_1+\epsilon} g(t, \tilde{y}, \dot{\tilde{y}}, \lambda)\, dt$$

$$+ \int_{t_1+\epsilon}^{t_2} g[t, y(t, \epsilon), \dot{y}(t, \epsilon), \lambda]\, dt + \int_{t_2}^{t_f} g(t, y, \dot{y}, \lambda)\, dt \quad (20.3\text{-}110)$$

Since $y(t)$ minimizes J, we must have $J(\epsilon) - J(0) > 0$. The condition that $y(t)$ still be the extremal under strong variations thus is $dJ(0) = (dJ/d\epsilon)\big|_{\epsilon=0}\, \epsilon \geqslant 0$. Note that

$$d\tilde{y}(t_1 + \epsilon) = dy(t_1 + \epsilon, \epsilon) = \frac{\partial y}{\partial t}\frac{dt}{d\epsilon}\, d\epsilon + \frac{\partial y}{\partial \epsilon}\, d\epsilon \quad (20.3\text{-}111)$$

so that at $\epsilon = 0$

$$\frac{d\tilde{y}(t_1)}{dt} = \frac{\partial y(t_1, 0)}{\partial t} + \frac{\partial y(t_1, 0)}{\partial \epsilon}$$

or

$$\frac{\partial y(t_1, 0)}{\partial \epsilon} = \dot{\tilde{y}}(t_1) - y(t_1) \tag{20.3-112}$$

Since $y(t_2)$ is extremal, $\partial y/\partial \epsilon\ (t_2, 0) = 0$. Thus

$$\frac{dJ(\epsilon)}{d\epsilon} = g[t_1 + \epsilon, \tilde{y}(t_1 + \epsilon), \dot{\tilde{y}}(t_1 + \epsilon), \lambda] - g[t_1 + \epsilon, y(t_1 + \epsilon), \dot{y}(t_1 + \epsilon), \lambda]$$

$$+ \int_{t_1+\epsilon}^{t_2} \frac{\partial}{\partial \epsilon} g(t, y(t, \epsilon), \dot{y}(t, \epsilon), \lambda)\, dt$$

$$= [g(t, \tilde{y}, \dot{\tilde{y}}, \lambda) - g(t, y, \dot{y}, \lambda)]\Big|_{t_1+\epsilon} + \int_{t_1+\epsilon}^{t_2} \left[g_y \frac{\partial y}{\partial \epsilon} + g_{\dot{y}} \frac{\partial \dot{y}}{\partial \epsilon} \right] dt \tag{20.3-113}$$

Integrating the last term by parts yields

$$\frac{dJ(\epsilon)}{d\epsilon} = [g(t, \tilde{y}, \dot{\tilde{y}}, \lambda) - g(t, y, \dot{y}, \lambda)]\Big|_{t_1+\epsilon} + \frac{\partial g}{\partial \dot{y}(t, \epsilon)} \frac{\partial y(t, \epsilon)}{\partial \epsilon}\Bigg|_{t_1+\epsilon}^{t_2}$$

$$- \int_{t_1+\epsilon}^{t_2} \left[\frac{d}{dt} g_{\dot{y}} - g_y \right] \frac{\partial y(t, \epsilon)}{\partial \epsilon}\, dt \tag{20.3-114}$$

Evaluating eq. (20.3-114) at $\epsilon = 0$, the integrand vanishes since then $y(t, \epsilon)$, $\dot{y}(t, \epsilon) \to y(t)$, $\dot{y}(t)$, the extremal solution for which the Euler equations are satisfied. Therefore, with eq. (20.3-112) and the fact that $\partial y(t_2, 0)/\partial \epsilon = 0$, we find

$$\frac{dJ(0)}{d\epsilon} = [g(t, \tilde{y}, \dot{\tilde{y}}, \lambda) - g(t, y, \dot{y}, \lambda)]\Big|_{t_1} - g_{\dot{y}}\Big|_{t_1} [\dot{\tilde{y}}(t_1) - \dot{y}(t_1)] \tag{20.3-115}$$

For extremals to be minimizing $dJ(0) \geqslant 0$ or $[dJ(0)/d\epsilon]\ \epsilon \geqslant 0$ and since $\epsilon > 0$, $dJ(0)/d\epsilon \geqslant 0$. This quantity is usually denoted by E and is known as the Weierstrass excess or E function in the classical calculus of variations. Since t_1 was arbitrary, we require $E(t) \geqslant 0$ for $y(t)$ to be minimizing under strong variations. Extension of eq. (20.3-115) to abnormal arcs and to variable end points may be made by similar arguments.

In order to cast the result, eq. (20.3-115) into the control formulation, let us introduce the notation

$$\begin{aligned} y_i &= x_i \qquad i = 1, 2, \ldots, n \\ \dot{y}_{n+j} &= u_j \qquad j = 1, 2 \ldots m \end{aligned} \tag{20.3-116}$$

$$\Phi = f(x, u, t) - \dot{x}$$

Then

$$g(t, y, \dot{y}, \lambda) = h + \lambda^T(f - \dot{x}) = \mathcal{H}(x, u, t, \lambda) - \lambda^T \dot{x} \qquad (20.3\text{-}117)$$

If $x(t)$, $u(t)$, and $\lambda(t)$ are some functions and $\tilde{u}(t)$ is the strong variation at time t_i (arbitrary) with $\dot{\tilde{x}} = f(x, \tilde{u}, t)$, then the E function becomes

$$E = [\mathcal{H}(x, \tilde{u}, t, \lambda) - \mathcal{H}(x, u, t, \lambda) - \lambda^T(\dot{\tilde{x}} - \dot{x})]\Big|_{t_i} + (\tilde{u} - u)\frac{\partial}{\partial u}[\mathcal{H}(x, u, t, \lambda) - \lambda^T \dot{x}]\Big|_{t_i} \qquad (20.3\text{-}118)$$

$$E = [\mathcal{H}(x, \tilde{u}, t, \lambda) - \mathcal{H}(x, u, t, \lambda)] + (\tilde{u} - u)\frac{\partial \mathcal{H}}{\partial u}(x, u, \lambda, t)\Big|_{t_i} \qquad (20.3\text{-}119)$$

Since $\partial\mathcal{H}/\partial u$ vanishes at each interior point of U (i.e., all control variations must be admissible) if $u(t)$ is minimizing, then the E condition requires that

$$E = \mathcal{H}(x, \tilde{u}, t, \lambda) - \mathcal{H}(x, u, t, \lambda) \geqslant 0 \qquad (20.3\text{-}120)$$

or

$$\mathcal{H}(x, \tilde{u}, t, \lambda) \geqslant \mathcal{H}(x, u, t, \lambda) \qquad (20.3\text{-}121)$$

Equation (20.3-121) is the principal result of *Pontryagin's Minimum Principle*.

We now state *Pontryagin's Minimum Principle* which is applicable for *any topological* control space U; specifically, it includes the case when the control $u(t)$ may assume a number of discrete levels $u_1, u_2, \ldots, u_k$.

Let $u(t)$ be an admissible control (defined over the interval $t_0 \leqslant t \leqslant t_f$, measurable and bounded) and let $x(t)$ be the corresponding trajectory satisfying

$$\dot{x} = f(x, u, t); \quad x(t_0) = x_0 \qquad (20.3\text{-}122)$$

subject to

$$\psi(x_f, t_f) = 0 \qquad (20.3\text{-}123)$$

In order that $u(t)$ and $x(t)$ be optimal for $t_0 \leqslant t \leqslant t_f$, i.e., minimize (maximize) the criterion

$$M = \varphi(x_f, t_f) + \int_{t_0}^{t_f} h(x, u, t)\, dt \qquad (20.3\text{-}124)$$

it is necessary that there exist a nonzero vector function ν_0, ν^T, with $\nu_0 \geqslant 0$ such that $u(t)$ minimizes (maximizes) the Hamiltonian

$$\mathcal{H} = \nu_0 h + \lambda^T f \tag{20.3-125}$$

i.e.

$$\mathcal{H}(x, \tilde{u}, t, \lambda) \underset{(\leqslant)}{\geqslant} \mathcal{H}(x, u, t, \lambda) \tag{20.3-126}$$

for any admissible control variation $\tilde{u}(t)$ and where x, λ obey the Hamiltonian system

$$\dot{x} = \mathcal{H}_\lambda^T \tag{20.3-127}$$

$$\dot{\lambda} = -\mathcal{H}_x^T \tag{20.3-128}$$

subject to

$$\lambda^T(t_f) = [\nu_0 \varphi_x + \nu^T \psi_x]\Big|_{t_f} \tag{20.3-129}$$

For proof of the Minimum Principle under the general assumptions on the control space U, see the text by Pontryagin, et al.[20-26] We note that if $\mathcal{H}$ is considered as a function $\mathcal{H}(x, u, \lambda, t)$, then it follows from the canonical equations (20.3-127), (20.3-128) that $\dot{\mathcal{H}} = \mathcal{H}_t + \mathcal{H}_u \dot{u}$, and $\mathcal{H}_u \dot{u}$ vanishes whenever U is independent of t. If u lies in the interior of U (classical case) then $\mathcal{H}_u = 0$, if u lies on the boundary of U then $\mathcal{H}_u \dot{u} = 0$. The first integral of the two-point boundary value problem defined by the Minimum Principle is $\mathcal{H}$ = constant, whenever h, f, and U are independent of t.

20.3.8 Optimization of Distributed Parameter Systems

The concepts of the previous sections can be extended to optimization problems where the system is described by partial differential equations, integral equations or other functional relationships. In this section, we will present the minimum principle for systems described by a certain class of partial differential equations; for extensions to integral and integro-differential state equations, see the text by Butkovskiy.[20-27]

We assume that the controlled process is defined over a domain $D(0 \leqslant \xi \leqslant a, 0 \leqslant \eta \leqslant b)$ by

$$\frac{\partial^2 Q}{\partial \xi \partial \eta} = f(\xi, \eta, Q, Q_\xi, Q_\eta, u) \tag{20.3-130}$$

where Q and f are n-vectors and u is the m-dimensional (bounded) control $u(\xi, \eta)$, which may be piecewise continuous in some convex closed region R of the m-dimensional control space U. The boundary conditions for $Q = Q(\xi, \eta)$ are assumed of the form

$$Q(\xi, 0) = X(\xi); \quad Q(0, \eta) = Y(\eta); \quad Q(0, 0) = 0 \tag{20.3-131}$$

The criterion function M may be given in different ways analogous to the Bolza, Mayer, or Lagrange problems in the classical calculus of variations. It is convenient here to choose M as

$$M = c^T Q(a, b) \tag{20.3-132}$$

where c is a constant vector. Other forms of M may be reduced to eq. (20.3-132) by various devices. For example, if

$$M = \iint_D f_0(\xi, \eta, Q, Q_\xi, Q_\eta, u)\, d\xi\, d\eta \tag{20.3-133}$$

introduction of a new function $Q_0(\xi, \eta)$ defined by

$$\frac{\partial^2 Q_0}{\partial \xi \partial \eta} = f_0(\xi, \eta, Q, Q_\xi, Q_\eta, u) \tag{20.3-134}$$

subject to $Q_0(0, \eta) = Q_0(\xi, 0) = 0$ reduces eq. (20.3-133) to eq. (20.3-132).

We introduce the Hamiltonian defined by

$$\mathcal{H}(\xi, \eta, Q, \Lambda, Q_\xi, Q_\eta, u) = \Lambda^T(\xi, \eta) f(\xi, \eta, Q, Q_\xi, Q_\eta, u) \tag{20.3-135}$$

where $\Lambda(\xi, \eta)$ is the adjoint vector determined from

$$\frac{\partial^2 \Lambda}{\partial \xi \partial \eta} = -\mathcal{H}_Q + \frac{\partial}{\partial \xi}\frac{\partial \mathcal{H}}{\partial Q_\xi} + \frac{\partial}{\partial \eta}\frac{\partial \mathcal{H}}{\partial Q_\eta} \tag{20.3-136}$$

with boundary conditions

$$\begin{aligned} \Lambda_\xi &= \frac{\partial \mathcal{H}}{\partial Q_\eta} \quad \text{at } \eta = b \\ \Lambda_\eta &= \frac{\partial \mathcal{H}}{\partial Q_\xi} \quad \text{at } \xi = a \\ \Lambda(a, b) &= c \end{aligned} \tag{20.3-137}$$

The necessary conditions are given by the Minimum Principle:

If $u^0(\xi, \eta)$ is the optimum control and $Q^0(\xi, \eta)$, $\Lambda^0(\xi, \eta)$ are the corresponding state and adjoint vectors satisfying eqs. (20.3-130, 131) and eqs. (20.3-136, 137) respectively, then

$$\mathcal{H}(\xi, \eta, Q, \Lambda, Q_\xi, Q_\eta, u) \geqslant \mathcal{H}(\xi, \eta, Q^0, \Lambda^0, Q_\xi^0, Q_\eta^0, u^0) \qquad (20.3\text{-}138)$$

for any admissible control $u(\xi, \eta)$.

For discussion of various special forms of eq. (20.3-130) and inclusion of state constraints see Butkovskiy.[20-27] Some problems from mathematical physics and minimization of multiple integrals depending on a function of several variables and its first-order partial derivatives are discussed by Gelfand and Fomin.[20-28]

20.4 EXTREMAL FIELDS, SUFFICIENCY CONDITIONS

The conditions we have determined in the previous sections are *necessary* for optimality, we now consider the development of *sufficient* conditions that $u(t)$ be optimal. If only one solution satisfies the first necessary conditions, eq. (20.3-45a-f), either it is the optimum solution or an optimum does not exist in the class of functions considered. If many solutions satisfy the necessary conditions, further criteria must be developed to determine the optimum. Central to these criteria are the concepts of *extremal fields*.

To introduce the necessary ideas, consider the family of solutions to the simple harmonic oscillator equation

$$\ddot{x} + x = 0 \qquad x(0) = a, \quad \dot{x}(0) = 0$$

The solution $x = a \cos t$ is graphically portrayed in Fig. 20.4-1 for various values of a.

At $t = \pi/2$, $3\pi/2$ the slope function is not unique. For the domain indicated by $D_1 (0 \leqslant t < t_1 < \pi/2)$ each point is pierced by one and only one solution curve; the solutions constitute a proper field or simply a *field*. For the domain $D_2 (\pi/2 \leqslant t \leqslant \pi)$ the solutions form a central field, i.e., at $t = \pi/2$ and only $\pi/2$ of D_2 the slope function is not unique (the point $t = \pi/2$ is known as a *pencil point*). Over the domain $D_3 (\pi/2 \leqslant t \leqslant 2\pi)$ the family of solutions does not form a field since the slope function is not unique at $t = \pi/2$ *and* $t = 3\pi/2$.

These concepts are readily generalized to higher dimensional spaces. A family of solutions forms a field over a simply connected multidimensional domain D if at each point of D there exists only one solution curve.

20.4.1 Hamilton-Jacobi Theory

Solution of the optimization problem as formulated in section 20.3.1, will yield one optimal control $u(t)$ which will take the system from the (specified) initial point $x(t_0) = x_0$ to the terminal hypersurface $\psi(x_f, t_f) = 0$. To determine the opti-

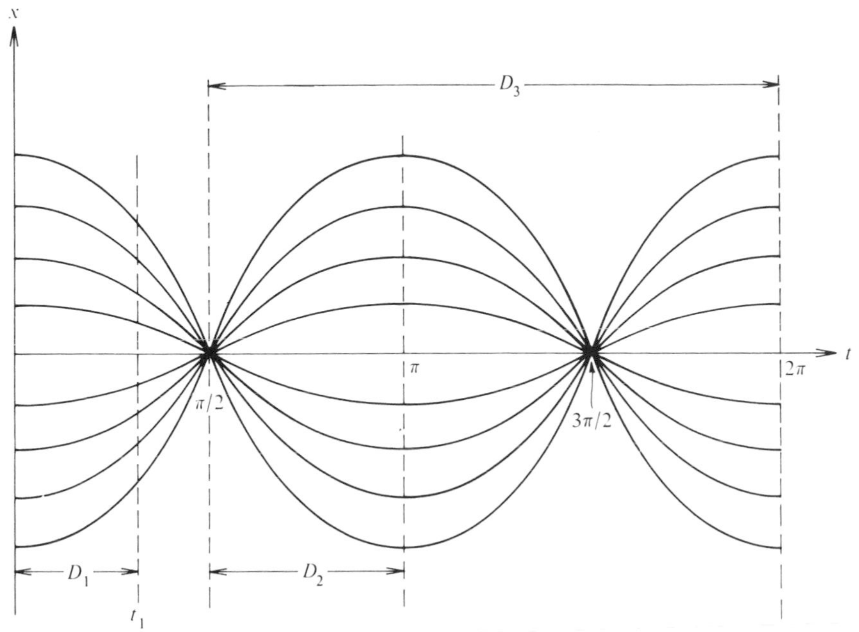

Fig. 20.4-1

mal control from any other initial point (x, t), we have to solve a new optimization problem with the new initial conditions. The resulting family of solutions from all possible initial points (x, t) forms a field of extremals if it satisfies the criteria for a proper field. One may generate the field by straightforward solution of many optimization problems, but this is tedious. Often one is interested in the optimum solution when the initial and/or terminal conditions are changed slightly; this problem of neighboring extremal paths is considered in section 20.4.2. An alternate approach to defining the entire extremal field is a generalization of *Hamilton-Jacobi* theory.

We assume that for each point (x, t) of the domain, there exists an associated unique optimal control law*

$$u^0 = u^0(x, t) \tag{20.4-1}$$

which is a closed-loop optimal control law. Then for each specified initial point (x_0, t_0), terminal hypersurface and associated optimal control law $u^0(x, t)$, there exists a unique value of the criterion or payoff function. We denote this value by $M^0(x, t)$, which is often termed the optimal return function

$$M^0(x, t) = \min_{u(t)} \left\{ \varphi(x_f, t_f) + \int_t^{t_f} h[x(\tau), u(\tau), \tau] \, d\tau \right\} \tag{20.4-2}$$

*This assumption implies that no conjugate points, cf. section 20.4.2, exist, i.e., a proper field of extremals exists.

subject to the boundary condition

$$M^0(x,t) = \varphi(x,t) \quad \text{on} \quad \psi(x,t) = 0 \tag{20.4-3}$$

The governing equation for M^0 is usually derived as follows. Divide the time interval $[t, t_f]$ into $[t, t + \Delta t]$ and $[t + \Delta t, t_f]$ where Δt is small. Then

$$M^0(x,t) = \min_{u(t)} \left[\varphi(x_f, t_f) + \int_t^{t+\Delta t} h\, d\tau + \int_{t+\Delta t}^{t_f} h\, d\tau\right] \tag{20.4-4}$$

Since Δt is small the first integral may be approximated as $h[x(t), u(t), t]\ \Delta t + O(\Delta t^2)$ so that

$$M^0(x,t) = \min_{u(t)} \left\{\varphi(x_f, t_f) + h[x(t), u(t), t]\ \Delta t + O(\Delta t^2) + \int_{t+\Delta t}^{t_f} h\, d\tau\right\} \tag{20.4-5}$$

By definition

$$\left[\varphi(x_f, t_f) + \int_{t+\Delta t}^{t_f} h\, d\tau\right] \tag{20.4-6}$$

is the optimal return function $M^0(x + \Delta x, t + \Delta t)$ if $u = u^0$, the optimal control, is used from $t + \Delta t$ to t_f. The initial state $x + \Delta x$ results from $u(t)$ over time Δt as determined by the system equations $\dot{x} = f(x, u, t)$

$$\Delta x = f[t, x(t), u(t)]\ \Delta t + O(\Delta t^2) \tag{20.4-7}$$

Substituting these relationships in eq. (20.4-5) we get

$$M^0(x,t) = \min_{u(t)} \{h[x(t), u(t), t]\ \Delta t + O(\Delta t^2) + M^0(x + \Delta x, t + \Delta t)\} \tag{20.4-8}$$

where the minimization with respect to $u(t)$ is now over the interval t to $t + \Delta t$. Expanding $M^0(x + \Delta x, t + \Delta t)$ about $M^0(x,t)$ (assuming that M^0 is a continuous, twice differentiable function over the entire x, t space of interest):

$$M^0(x,t) = \min_{u(t)} \left\{h[x(t), u(t), t]\ \Delta t + O(\Delta t^2) + M^0(x,t) + \frac{\partial M^0}{\partial x} f(x,u,t)\, \Delta t + \frac{\partial M^0}{\partial t} \Delta t + O(\Delta t^2)\right\} \tag{20.4-9}$$

Since $M^0(x, t)$ and $\partial M^0/\partial t \ \Delta t$ are not functions of $u(t)$ they may be taken outside the minimization operation. Dividing by Δt and taking the limit $\Delta t \to 0$

$$0 = \frac{\partial M^0}{\partial t}(x, t) + \min_{u(t)} \left[\frac{\partial M^0}{\partial x} f(x, u, t) + h(x, u, t) \right] \tag{20.4-10}$$

This is known as the *Hamilton-Jacobi-(Bellman) equation* for M^0 whose solution, with appropriate boundary conditions, leads to the optimal control law $u^0(x, t)$. The generalization of the classical Hamilton-Jacobi theory to multistage systems and combinatorial problems is due to Bellman[20-20] who calls the overall theory dynamic programming (cf. section 20.5.3).

To cast eq. (20.4-10) into a more familiar form, recall that we had found for fixed ψ

$$\lambda^T(t_0) = \frac{\partial J}{\partial x_0} = \frac{\partial M}{\partial x_0} \tag{20.4-11}$$

From the definition of $M^0(x, t)$ it is clear that along an optimal trajectory

$$\lambda^T = \frac{\partial M^0}{\partial x} \tag{20.4-12}$$

Hence eq. (20.4-10) may be written in the form

$$\begin{gathered} \frac{\partial M^0(x, t)}{\partial t} + \min_{u(t)} \left[\mathcal{H} \left(x, \frac{\partial M^0}{\partial x}, u, t \right) \right] = 0 \\ \frac{\partial M^0(x, t)}{\partial t} + \mathcal{H}^0 \left(x, \frac{\partial M^0}{\partial x}, t \right) = 0 \end{gathered} \tag{20.4-13}$$

where

$$\mathcal{H}^0 \left(x, \frac{\partial M^0}{\partial x}, t \right) = \min_{u(t)} \mathcal{H} \left(x, \frac{\partial M^0}{\partial x}, u, t \right) \tag{20.4-14}$$

Solution of eq. (20.4-13) is most effectively accomplished via the method of characteristics (cf. Chapter 8) which is equivalent to determining the field of extremals. In practice, analytical solution of eq. (20.4-14) in all but trivial cases is usually impossible. Thus one resorts to numerical techniques, in general utilizing recursion techniques of dynamic programming, cf. section 20.5.4.

20.4.2 Neighboring Extremals, Weak Sufficient Conditions

Assume that a nominal extremal path has been found with the associated control $u(t)$. Consider then small perturbations about the nominal path resulting from small perturbations in the initial state $\delta x(t_0)$ and terminal hypersurface $\delta\psi$. For simplicity, we consider the fixed end-time problem. The existence of a *neighboring extremal path*—a path that satisfies the first necessary conditions, eq. (20.3-45a–f), for the new initial state and terminal constraints—will be guaranteed if the nominal path may be embedded in a field of extremals. Furthermore, if we can show that the second order terms in the expansion of the criterion about the nominal are greater than zero for any $\delta u(t) \neq 0$ resulting from $\delta x(t_0)$, $\delta\psi$, then the local minimality of the nominal path is guaranteed.

The perturbations $\delta x(t_0)$ and $\delta\psi$ will give rise to perturbations $\delta x(t)$, $\delta u(t)$, $\delta\lambda(t)$, and dv which are determined by linearizing eqs. (20.3-45a–f) about the nominal

$$\delta\dot{x} = f_x \delta x + f_u \delta u \tag{20.4-15a}$$

$$\delta\dot{\lambda} = -\mathcal{H}_{xx}\delta x - \mathcal{H}_{xu}\delta u - \mathcal{H}_{x\lambda}\delta\lambda \tag{20.4-15b}$$

$$0 = \mathcal{H}_{uu}\delta u + \mathcal{H}_{ux}\delta x + \mathcal{H}_{u\lambda}\delta\lambda \tag{20.4-15c}$$

$$\delta\lambda(t_f) = [(\nu_0\varphi_{xx} + \nu^T\psi_{xx})\,\delta x + \psi_x^T dv]\Big|_{t_f} \tag{20.4-15d}$$

$$\delta\psi = [\psi_x\delta x]\Big|_{t_f}; \quad \text{fixed } t_f \tag{20.4-15e}$$

$$\delta x(t_0) \quad \text{given} \tag{20.4-15f}$$

Since the coefficients of these equations are evaluated on the nominal path, the system, eqs. (20.4-15a, f), forms a *linear* two-point boundary value problem for $\delta x(t)$, $\delta\lambda(t)$, $\delta u(t)$ and dv. The necessary control change $\delta u(t)$ is given in terms of δx and $\delta\lambda$ by (since $\dot{x} = f = \lambda^T$)

$$\delta u(t) = -\mathcal{H}_{uu}^{-1}[\mathcal{H}_{ux}\delta x + f_u^T\delta\lambda] \tag{20.4-16}$$

An additional necessary condition for the existence of a neighboring extremal path, emerges from eq. (20.4-16): $\mathcal{H}_{uu}$ *must be nonsingular over* $[t_0, t_f]$. This is the ***strengthened Legendre-Clebsch condition*** in the calculus of variations. Using eq (20.4-16) we may write the equations for $\delta\dot{x}$ and $\delta\dot{\lambda}$ as

$$\begin{bmatrix} \delta\dot{x} \\ \delta\dot{\lambda} \end{bmatrix} = \begin{bmatrix} C_1(t) & -C_2(t) \\ -C_3(t) & -C_1^T(t) \end{bmatrix} \begin{bmatrix} \delta x \\ \delta\lambda \end{bmatrix} \tag{20.4-17}$$

where C_i are the matrices defined by

$$\begin{aligned} C_1(t) &= f_x - f_u\mathcal{H}_{uu}^{-1}\mathcal{H}_{ux} \\ C_2(t) &= f_u\mathcal{H}_{uu}^{-1}f_u^T \\ C_3(t) &= \mathcal{H}_{xx} - \mathcal{H}_{xu}\mathcal{H}_{uu}^{-1}\mathcal{H}_{ux} \end{aligned} \tag{20.4-18}$$

Equations (20.4-17) could be integrated subject to the boundary conditions, eq. (20.4-15d-f), adjusting $d\nu$ so that the desired $\delta\psi$ would be met. The usual approach is to seek solutions for $\delta\lambda(t)$ and $\delta\psi$ of the form

$$\delta\lambda(t) = A_1(t)\,\delta x(t) + A_2(t)\,d\nu \tag{20.4-19}$$

$$\delta\psi = A_2^T(t)\,\delta x(t) + A_3(t)\,d\nu \tag{20.4-20}$$

where the matrices $A_i(t)$ must satisfy the boundary conditions implied by eq. (20.4-15d-f).

$$A_1(t_f) = [\nu_0 \varphi_{xx} + \nu^T \psi_{xx}]|_{t_f} \tag{20.4-21a}$$

$$A_2(t_f) = [\psi_x^T]|_{t_f} \tag{20.4-21b}$$

$$A_3(t_f) = 0 \tag{20.4-21c}$$

Thus, it is simpler to determine the governing differential equations for $A_i(t)$ and *integrate backwards* with the *initial conditions* $A_i(t_f)$ as given by eq. (20.4-21). The differential equations are [from eq. (20.4-17)]

$$[-C_3 - C_1^T A_1 - A_1 C_1 + A_1 C_2 A_1 - \dot{A}_1]\,\delta x - [(C_1^T - A_1 C_2) A_2 + \dot{A}_2]\,d\nu = 0 \tag{20.4-22}$$

$$[\dot{A}_2^T + A_2^T (C_1 - C_2 A_1)]\,\delta x + [-A_2^T C_2 A_2 + \dot{A}_3]\,d\nu = 0 \tag{20.4-23}$$

These expressions must hold for arbitrary δx, $d\nu$ hence the coefficients vanish, so

$$\dot{A}_1 = -C_3 - C_1^T A_1 - A_1 C_1 + A_1 C_2 A_1 \tag{20.4-24}$$

$$\dot{A}_2 = -(C_1^T - A_1 C_2) A_2 \tag{20.4-25}$$

$$\dot{A}_3 = A_2^T C_2 A_2 \tag{20.4-26}$$

The $\dot{A}_i$ equations with boundary conditions, eq. (20.4-21) may be integrated backwards to t_0. Known the values $A_3(t_0)$, $A_2(t_0)$, we can then solve for the appropriate $d\nu$ to produce the desired $\delta\psi$ from eq. (20.4-20):

$$d\nu = A_3^{-1}(t_0)[\delta\psi - A_2^T(t_0)\,\delta x(t_0)] \tag{20.4-27}$$

Thus we have another necessary condition for the neighboring extremal path: $A_3(t_0)$ *must be nonsingular* for $d\nu$ to exist for all possible $\delta\psi$. Since t_0 is arbitrary and functions $A_3(t)$, $A_2(t)$ are determined from the backward integration at any time $t_0 \leqslant t \leqslant t_f$, we can find the control $u(t)$ necessary for a neighboring extremal path from any point t, $\delta x(t)$ to $\delta\psi$ near the nominal from eqs. (20.4-16), (20.4-19), and (20.4-27)

$$\delta u(t) = -\mathcal{H}_{uu}^{-1}\{[\mathcal{H}_{ux} + f_u^T(A_1 - A_2A_3^{-1}A_2^T)]\,\delta x + f_u^T A_2 A_3^{-1}\,\delta\psi\} \tag{20.4-28}$$

The remaining necessary condition then is that $(A_1 - A_2A_3^{-1}A_2^T)$ remain finite for $t_0 \leqslant t < t_f$, i.e., there is no conjugate point (neighboring extremals cross) in the interval (the Jacobi condition of the calculus of variations).

In summary, the sufficient conditions that a solution to eqs. (20.3-5a–f) be a weak local minimum (for small δx, δu) are

$$\mathcal{H}_{uu} > 0 \qquad t_0 \leqslant t \leqslant t_f \tag{20.4-29}$$

$$A_3 < 0 \qquad t_0 \leqslant t < t_f \tag{20.4-30}$$

$$A_1 - A_2A_3^{-1}A_2^T \qquad \text{finite for } t_0 \leqslant t < t_f \tag{20.4-31}$$

For a compact proof of this assertion see Bryson and Ho,[20-21] p. 182–183.

Example: A simple example in which a focal or conjugate point occurs is the problem of finding the shortest distance between a point and a great circle on a sphere. The coordinates for the problem are θ, the latitude, and φ, the longitude with φ_i indicating a particular meridian or desired terminal great circle. If R is the radius of the sphere, the elemental distance ds on the surface is given by

$$ds = [R^2(d\theta)^2 + R^2\cos^2\theta\,(d\varphi)^2]^{1/2} \tag{20.4.32}$$

Thus, we must find $u(\varphi)$ to minimize

$$M = \int_0^{\varphi_i} [u^2 + \cos^2\theta]^{1/2}\, d\varphi \tag{20.4-33}$$

with $u = d\theta/d\varphi$. Let us assume the initial condition $\theta(0) = 0$. The first necessary conditions yield $u = 0$, $\theta = 0$ as the candidate optimal solution.

If we consider neighboring paths and expand M about the nominal $u = 0$, $\theta = 0$ path

$$\delta M = M - \varphi_i = \frac{1}{2}\int_0^{\varphi_i} (u^2 - \theta^2)\, d\theta + \cdots \tag{20.4-34}$$

The Hamiltonian for this accessory minimum problem is

$$\mathcal{H} = \tfrac{1}{2}(u^2 - \theta^2) + \lambda u \tag{20.4-35}$$

with the Euler-Lagrange equations

$$\begin{gathered}\frac{d\lambda}{d\varphi} = -\frac{\partial \mathcal{H}}{\partial \theta} = \theta; \quad \lambda(\varphi_i) = 0; \quad \theta(\varphi_i) \text{ free} \\ 0 = \frac{\partial \mathcal{H}}{\partial u} = u + \lambda\end{gathered} \tag{20.4-36}$$

The governing equation for θ then becomes

$$\frac{d^2\theta}{d\varphi^2} + \theta = 0; \quad \theta(0) = 0, \quad \left(\frac{d\theta}{d\varphi}\right)_{\varphi_i} = 0 \tag{20.4-37}$$

with general solution $\theta(\varphi) = A \sin \varphi + B \cos \varphi$. From initial condition $\theta(0) = 0$, $B = 0$, and from $(d\theta/d\varphi)_{\varphi_i} = 0$, $A = 0$. However, we note that if $\varphi_i = \pi/2$ then the terminal condition is satisfied by any A. The initial point 0 is said to be *conjugate* to $\varphi_i = \pi/2$. Similar problems arise if one asks for minimum time paths on the surface of a sphere at constant speed.

20.4.3 Strong Sufficient Conditions

The results of the previous section assume that we admit only weak variations. Sufficient conditions for a strong minimum require that the conditions of section 20.4.2 be met and the strengthened Weierstrass condition hold

$$\mathcal{H}\left(x, \frac{\partial M^0}{\partial x}, \tilde{u}, t\right) > \mathcal{H}\left(x, \frac{\partial M^0}{\partial x}, u, t\right) \quad \text{for all } t \tag{20.4-32}$$

This is nothing more than the principal result of the Pontryagin Minimum Principle (cf. section 20.3.7), i.e., the control u^0 minimizes the Hamiltonian along the optimal path.

20.4.4 Singular Arcs

Under certain conditions, the second derivative $\mathcal{H}_{uu}$ may vanish identically, for example, when one or more components of the control vector enter the Hamiltonian linearly. When this happens, neither the Minimum Principle nor the classical calculus of variations provide adequate tests for optimality. Such problems are called singular and the extremals along which $\mathcal{H}_{uu} = 0$ are called *singular arcs*.

Let us assume that one component u_1 of u appears linearly in the Hamiltonian. The Minimum Principle requires that either u_1 is on the boundary of U or

$$\mathcal{H}_{u_1}(x, \lambda, -, u_2, u_3, \ldots, u_m, t) = 0 \tag{20.4-39}$$

Since u_1 is absent from eq. (20.4-39), we cannot find the singular control directly in terms of x and λ. All other u_i may be determined in the usual manner. It may be shown[20-21,29] that on the singular arc the control u_1 may be determined by evaluating the even number of time derivatives of $\mathcal{H}_{u_1}$

$$\frac{d^{2q}\mathcal{H}_{u_1}}{dt^{2q}} = V(x, \lambda, t) + W(x, \lambda, t)u_1 \tag{20.4-40}$$

where $2q$ is the number of differentiations required before u_1 appears explicitly. If the control is to remain on the arc, all lower time derivatives of $\mathcal{H}_{u_1}$ must be zero and u_1 chosen so that eq. (20.4-40) is zero. Thus

$$u_1 = -\frac{V(x, \lambda, t)}{W(x, \lambda, t)} \tag{20.4-41}$$

and

$$\frac{d^{j-1}}{dt^{j-1}}\mathcal{H}_{u_1}(x, \lambda, t, u_2, \ldots, u_m) = 0, \quad j = 1, \ldots, 2q \tag{20.4-42}$$

All singular arcs must lie on a $2(n\text{-}q)$-dimensional manifold—the *singular surface*—in the $2n$-dimensional space of x, λ.

Note that the Legendre-Clebsch convexity condition is trivially satisfied along a singular arc. A generalized Legendre-Clebsch condition[20-29] requires that in addition

$$(-1)^i \frac{\partial}{\partial u_1}\left[\frac{d^{2i}}{dt^{2i}}\mathcal{H}_{u_1}\right] \geqslant 0 \tag{20.4-43}$$

with the strict inequality holding for $i = q$. The integer q is referred to as the order of singularity.

The problem of joining singular and nonsingular arcs is in general quite complex. If the control appears linearly in the Hamiltonian and is bounded above and below, it may be shown by expanding $\mathcal{H}_{u_1}$ in a Taylor series near the junction time t_1 that the necessary condition for transition is

$$\frac{\partial}{\partial u_1}\left[\frac{d^{2i}}{dt^{2i}}\mathcal{H}_u\right]_{t_1} \leqslant 0 \quad i = 1, 2, \ldots, q \tag{20.4-44}$$

Furthermore, it can be shown[20-30] that there is also a "chattering" solution to the junction problem, i.e., the control switches at infinite frequency from upper to lower control bound.

The sufficiency conditions for nonsingular arcs were determined from second variations about an extremal arc (the solution of the accessory problem in the calculus of variations). For singular arcs, the accessory problem itself is singular and results analogous to the sufficiency conditions are very recent. The principal references are Jacobson and Speyer[20-31,32] for nonnegativity of the singular second variation and a Jacobi-type (conjugate point) condition.

Example: An example of a singular arc occurs in a simplified version of the minimum time airplane climb problem. If h is the altitude, v the speed, T the thrust, D the drag, g gravity, and γ the flight path angle, the state equations become (neglecting induced drag which depends on lift or $\dot{\gamma}$):

$$\begin{aligned} \dot{h} &= v \sin \gamma \\ \dot{v} &= T - D(h, v) - g \sin \gamma \end{aligned} \tag{20.4-45}$$

If we consider $\sin \gamma = u$, the control, with $|u| \leqslant 1$, we must minimize the Hamiltonian

$$\mathcal{H} = 1 + \lambda_h uv + \lambda_v [T - D(h, v) - gu] \tag{20.4-46}$$

In this minimum time problem, $\mathcal{H} = 0$, and we require

$$\min_u \{1 + u(\lambda_h v - \lambda_v g) + \lambda_v [T - D(h, v)]\} = 0 \tag{20.4-47}$$

The singular arc is defined by

$$\lambda_h v - \lambda_v g = 0 \tag{20.4-48}$$

since then $\mathcal{H}$ is independent of u. The adjoint equations are

$$\begin{aligned} \dot{\lambda}_v &= -u\lambda_h + \lambda_v \frac{\partial D}{\partial v} \\ \dot{\lambda}_h &= \lambda_v \frac{\partial D}{\partial h} \end{aligned} \tag{20.4-49}$$

Hence

$$\frac{d}{dt} [\lambda_h v - \lambda_v g] = \frac{T - D}{v} (\lambda_h v - \lambda_v g) + \lambda_v \left[v \frac{\partial D}{\partial h} + g \left(\frac{T - D}{v} - \frac{\partial D}{\partial v} \right) \right] \tag{20.4-50}$$

so that on the singular arc

$$1 - \lambda_v(D - T) = 0$$

$$v\frac{\partial D}{\partial h} - g\left(\frac{D - T}{v} + \frac{\partial D}{\partial v}\right) = 0 \qquad (20.4\text{-}51)$$

If the drag is of the form $D = Av^2 e^{-kh}$ where A is a scalar constant, and T is constant, the singular arc is

$$e^{kh} = \frac{Av^2}{gT}(kv^2 + 3g) \qquad (20.4\text{-}52)$$

Typical trajectories in h, v-space might appear as shown in Fig. 20.4-2.

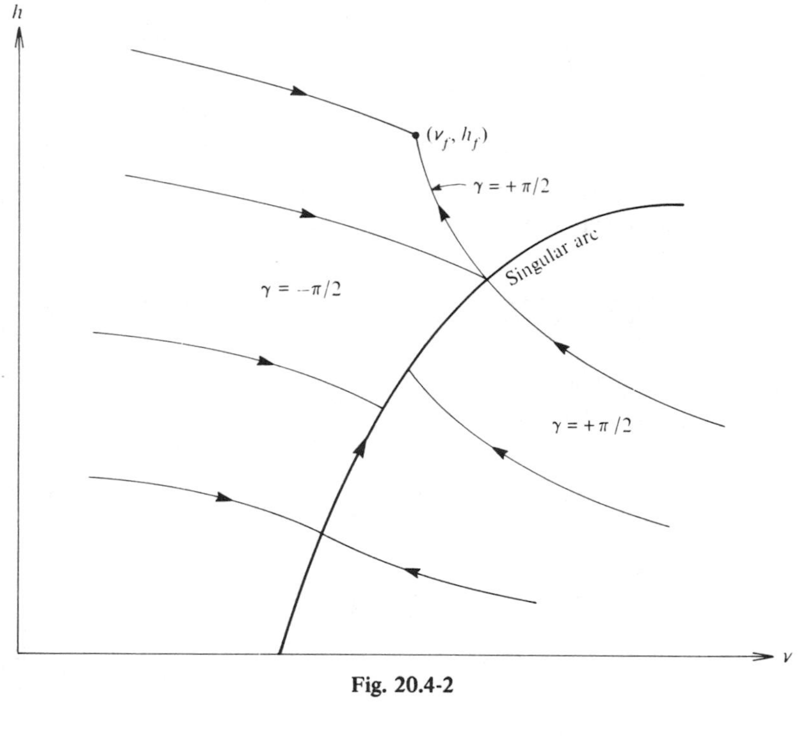

Fig. 20.4-2

20.5 COMPUTATIONAL TECHNIQUES

Whereas the basic theory of optimization can be rather compactly presented, a concise description of available computational techniques is not possible due to the extensive number of methods available. However, we can identify three major classes of computational algorithms: linear programming, nonlinear programming or search techniques, and numerical solution of two-point boundary value problems. Extensive literature is available on all three classes and most major computer systems have general codes available for solution of standard types of problems. The proliferation of numerical solution techniques is primarily due to the fact that each optimization problem has its own peculiarities and hence commands "customized" solutions.

For linear programming problems, the definitive text is by Dantzig;[20-33] other useful texts are by Hadley,[20-34] Künzi,[20-18] who lists some explicit codes, and Pierre[20-17] who gives an extensive list of references to the literature. In section 20.5.1, we present the salient features of linear programming problems and introduce the pertinent terminology.

Nonlinear programming problems, from a computational point of view, are usually divided into unconstrained and constrained problems. General texts on the subject are Hadley,[20-35] Pierre[20-17] and Zoutendijk.[20-36] Comparisons of some current methods have been performed by Leon[20-37] and Box.[20-38] In section 20.5.2, we briefly outline some of the more useful techniques for unconstrained and constrained nonlinear programming problems. A good source of current numerical techniques and comparisons for this class of optimization techniques is the *Computer Journal.*

Finally, we discuss some of the aspects of the numerical solution of dynamic optimization problems. The principal emphasis is on the solution of the two-point boundary value problem which results from the necessary conditions for an optimum. An excellent discussion of various algorithms based on the variational theory is given in the text by Bryson and Ho.[20-21]

Each numerical solution of an optimization problem will have its own unique numerical analysis subproblems, for example, iteration schemes, numerical integration of differential equations and computer storage manipulation problems. Material relevant to such numerical analysis problems may be found in Chapters 18 and 19 of this *Handbook.*

20.5.1 Linear Programming

If the performance index and the constraints are linear functions of the state variables of the form

$$J = a^T x \tag{20.5-1}$$

$$Ax \underset{(=)}{\geqslant} b \quad b \geqslant 0 \tag{20.5-2}$$

$$x \geqslant 0 \tag{20.5-3}$$

where A is an $m \times n$ matrix of rank m, x is an n-vector, b an m-vector, $m < n$, the problem is called a *linear programming problem* in standard form. The coefficients

a are known as *value coefficients.* Almost all linear programming problems may be cast into the standard form via simple transformations.

Since the curvature of J is zero everywhere, the minimum (maximum), if it exists, must occur on the boundary defined by the constraints. These constraints, eq. (20.5-2) and eq. (20.5-3), define a convex set R_c so that the minimum (maximum) is attained at one (at least) of the vertices of R_c. Thus for numerical solution, the search for extrema is narrowed to the vertices of R_c which are determined by the *basic solutions* of eq. (20.5-3) resulting from setting $n - m$ of the x_i's equal to zero. In the remaining m x_i's are forced to be nonzero, the basic solution in nondegenerate; furthermore, if a basic solution contains no negative x_i's, it is termed a basic feasible solution.

The principal solution techniques for linear programming are variations of the simplex algorithm.[20-33] The key idea of the simplex algorithm is to establish an initial basic feasible solution and then systematically proceed to another neighboring basic feasible solution which is guaranteed to have a lower (higher) value of J. The means of establishing the initial basic feasible solution may be quite involved since one must solve the m-equations (20.5-3) and the n-m equations $x_i = 0$ simultaneously. For some approaches to the problem the interested reader is referred to Künzi, et al.[20-18] and Pierre.[20-17] To proceed to a neighboring vertex which will have a smaller (larger) value of J, one usually uses some version of steepest descent (ascent) or method of feasible directions.[20-36]

Corresponding to any linear programming problem, called the *primal problem* (either max or min), is a *dual problem* (min or max). If the primal problem is defined by

Minimize

$$J = a^T x$$

Subject to

$$Ax \underset{(=)}{\geqslant} b$$

$$x \geqslant 0$$

Then the dual problem is defined by

Maximize

$$I = \alpha^T b$$

Subject to

$$A^T \alpha \underset{(=)}{\leqslant} a$$

$$\alpha \geqslant 0$$

We note that if both the primal and dual problems have feasible solutions, then both have optimal solutions and

$$\min J = \max I$$

The principal utility of the dual and primal formulation, in view of the noted equality of the optimal solutions, is that one or the other may be more easily solved. Simplification in solution usually arises from the dimension of the unknown matrix, for example, if the primal problem has two unknowns x and ten inequality constraints characterized by b, then the dual problem has ten unknowns and two inequality constraints. In the simplex algorithm, the *basis matrix*, defined as the $m \times m$ matrix formed from some m columns of the constraint matrix A, must be inverted at each step. In the illustrative primal problem, we have a 10×10 basis matrix while the dual only has a 2×2 basis matrix with apparent computational advantages.

20.5.2 Nonlinear Programming

If the performance index and/or constraints in a parameter optimization problem are nonlinear, real-valued functions, we have a *nonlinear programming problem*. The performance index may be given analytically or determined experimentally, may be deterministic or stochastic and subject to constraints or unconstrained. The literature on numerical solution of such problems is extensive, Pierre[20-17] lists 111 references in his text covering papers from many different fields. The two principal categories of numerical nonlinear programming methods are (1) methods of feasible directions, and (2) penalty function methods. Constrained optimization problems may sometimes be transformed into unconstrained problems, cf. Box,[20-38,22] hence reducing their difficulty. We will first discuss the unconstrained problem.

The most straightforward approach to solving unconstrained parameter optimization problems would be to solve the system of equations resulting from the necessary conditions (20.2-2). For techniques of solving systems of equations, see Chapter 18 of this *Handbook*. We note however, that such a direct approach may be exceedingly difficult due to the nature of the criterion function, furthermore, the necessary equations only yield stationary points. Thus other techniques are required which will search for the optimum in a systematic manner.

Probably the best known search techniques are those based on the gradient of the performance index. The gradient of a function, f_x, points in the direction of maximum rate of increase of $f(x)$. Thus at any given point x^i, one can compute $f_x(-f_x)$ and step in the direction of steepest ascent (descent) to a new point x^{i+1} with f_x scaled by some step size $\alpha^i > 0$ so that $x^{i+1} = x^i \underset{(-)}{+} \alpha^i f_x$. The process will converge to a local maximum (minimum) of f for sufficiently well-behaved functions f if α^i are chosen so that $f(x^{i+1}) \underset{(<)}{>} f(x^i)$. The step length satisfying this condition may be found by minimizing the function $g(\alpha) = f(x^i - \alpha^T f_x |_i)$; for a discussion of this technique, called the method of optimum gradients see Curry.[20-39] The procedure is repeated until some stop criteria are met, for example, $|f_x| < \epsilon$ or $|f(x^{i+1}) - f(x^i)| < \eta$ where ϵ and η are pre-set limits. Since f_x may not necessarily point toward the extremum point when the equivalue contours of the criterion function are unsymmetrical, and $f_x \to 0$ as the extremum is approached, more efficient techniques are often desirable.

To overcome these difficulties, use may be made of second order terms in the expansion of the criterion function. Expansion of f about the extremum x^0 yields

$$f(x) = f(x^0) + \tfrac{1}{2}(x - x^0)^T f_{xx}|_{x^0} (x - x^0) + \cdots \tag{20.5-4}$$

Near the optimum, the function thus behaves like a pure quadratic. Methods that utilize the second-order term are known as second-order gradient methods and are designed to minimize (maximize) a quadratic function of n variables in n steps. The most powerful of these in the literature appears[20-38] to be the method of Fletcher and Powell[20-40] known as the *Davidon-Fletcher-Powell* (DFP) method.

The central idea is the use of conjugate directions. The general quadratic function can be expressed as

$$f(x) = a + b^T x + \tfrac{1}{2} x^T Q x \tag{20.5-5}$$

where Q is a symmetric, positive-definite matrix. The minimizing x^0 is determined as

$$x^0 = -Q^{-1} b \tag{20.5-6}$$

Given a point $\tilde{x}$ and a direction vector e, a constant vector β can be found such that

$$x^0 = \tilde{x} + \beta^T e \tag{20.5-7}$$

If e^1 and e^2 are two direction vectors in n-space, they are called *Q-conjugate direction* vectors if

$$e^{1^T} Q e^2 = 0 \tag{20.5-8}$$

Whenever e is Q-conjugate, the constants β are determined from

$$\beta = -(b + Q\tilde{x})^T \frac{e}{e^T Q e} \tag{20.5-9}$$

The iterative minimization procedure then starts at $\tilde{x}$ say, and minimizes f down the e^i-directions, $i = 1, 2, \ldots, n$, where e^i are Q-conjugate. Successive points x^i

are determined from

$$x^{i+1} = x^i + \beta^{iT} e^i \tag{20.5-10}$$

with β^i defined by eq. (20.5-9) at the ith point ($i^{\text{th}} e$).

The DFP method chooses H^1 as any positive definite matrix and at each successive step i, $e^i = -H^i f_x |_{x^i}$. Then α^i is chosen by minimizing $f(x^i + \alpha^{iT} e^i)$

$$x^{i+1} = x^i + \alpha^{iT} e^i \tag{20.5-11}$$

$$H^{i+1} = H^i + A^i + B^i \tag{20.5-12}$$

where A^i, B^i are defined by

$$C^i = \alpha^{iT} e^i; \quad \Delta^i = f_x(x^{i+1}) - f_x(x^i)$$

$$A^i = \frac{C^i C^{iT}}{C^{iT} \Delta^i} \tag{20.5-13}$$

$$B^i = -\frac{H^i \Delta^i \Delta^{iT} H^i}{\Delta^{iT} H^i \Delta^i} \tag{20.5-14}$$

The matrix H thus is updated at each step i and supplies the current direction vector e^i by multiplying the current gradient vector.

For problems where the function f cannot be determined analytically but is defined by experimental data, the techniques of minimizing a function without evaluation of the derivatives as proposed by Powell[20-41] and Zangwill[20-42] are useful.

For constrained problems, some version of feasible directions[20-36] or penalty function method is usually employed. Methods of feasible directions or gradient projection methods[20-43,44] use the same basic philosophy as the unconstrained optimization techniques but must be so constructed as to cope with inequality restrictions. In essence, one chooses an initial point that does not violate any constraint and then finds a direction (the feasible direction) such that a (small) move in that direction improves the criterion but violates no constraints. This process is repeated until no feasible direction at the current point can be found that improves the criterion. One such method is due to Zoutendijk,[20-36] briefly as follows.

Consider minimizing $f(x)$ subject to $g(x) \geqslant 0$, an m-vector. Assume that a starting point x^1 has been chosen which satisfies the constraints. Let $g_j(x^1) = 0$ where g_j are the binding constraints at x^1. To find a feasible direction e, a small move along e from x^1 makes no binding constraint negative, i.e.,

$$\left. \frac{d}{d\alpha} g_j(x^1 + \alpha^T e) \right|_{\alpha=0} = \nabla g_j^T(x^1) e \geqslant 0 \tag{20.5-15}$$

where ∇ is the gradient. In addition, the feasible vector has the property that

$$\frac{d}{d\alpha} f(x^1 + \alpha^T e)\bigg|_{\alpha=0} = \nabla f^T(x^1)e < 0 \tag{20.5-16}$$

In searching for the best direction vector e along which to move, one should minimize eq. (20.5-16). However, in presence of nonlinear constraints, one might encounter zigzagging along the (curved) boundary. Such problems can be avoided; for the necessary modifications and computation details, see Zoutendijk.[20-36]

The problem of minimizing $f(x)$ subject to $g(x) \geqslant 0$ can be approached via penalty function methods. One can define the penalty function $W(x)$

$$W(x) = \begin{cases} 0 & x \geqslant 0 \\ \infty & x < 0 \end{cases} \tag{20.5-17}$$

and consider the minimization of the modified function

$$F(x) = f(x) + W[g(x)] \tag{20.5-18}$$

subject to no constraints. The techniques of unconstrained optimization may then be employed. If difficulties are encountered because of discontinuities of $F(x)$, one can choose penalty functions $W(x)$ with continuity properties necessary to alleviate the problems. However, peculiarities in the topography of the modified function may still create convergence difficulties.

To overcome these problems, the *Fiacco-McCormick Method*[20-22] is widely used. In this approach, define the function

$$P(x, \xi) = f(x) + \xi \sum_{i=1}^{m} \frac{1}{g_i(x)} \tag{20.5-19}$$

with $\xi > 0$. Choose $\xi^1 > 0$ and x^1 strictly inside the constraint set and minimize $P(x, \xi^1)$ subject to no constraints starting from x^1. The minimum depends on ξ^1 and hence we must repeat the process for a sequence ξ^i. Since

$$\xi \sum_{i=1}^{m} \frac{1}{g_i(x)}$$

is the penalty function, decreasing ξ will concentrate the minimization more on $f(x)$. The sequence of minimizing points $x^0(\xi^j)$ can approach the constraint boundaries if such approach improves f. As $\xi \to 0$, the minimizing point $x^0(\xi)$ must approach the minimum of the original constrained problem. For additional discussion of this method as well as proof of the validity of the process, see Ref. 20-22.

Let us now consider what is involved in working with the necessary conditions of sections 20.2, 20.3 directly. In this case, one guesses the initial values of the state, control and adjoint vectors and computes $\mathcal{H}_x$, $\mathcal{H}_u$ and the constraint f. Then, linearize the necessary conditions eqs. (20.2-14) about the initial guess x^i, u^i and λ^i

$$\mathcal{H}_x^i + \mathcal{H}_{xx}^i\, dx + \mathcal{H}_{xu}^i\, du + (f_x^T)^i\, d\lambda = 0 \tag{20.5-20}$$

$$\mathcal{H}_u^i + \mathcal{H}_{ux}^i\, dx + \mathcal{H}_{uu}^i\, du + (f_u^T)^i\, d\lambda = 0 \tag{20.5-21}$$

$$f^i + f_x^i\, dx + f_u^i\, du = 0 \tag{20.5-22}$$

Solution of eqs. (20.5-20) to (20.5-22) yields the increments dx, du and $d\lambda$, so the new values are

$$x^{i+1} = x^i + dx; \quad u^{i+1} = u^i + du; \quad \lambda^{i+1} = \lambda^i + d\lambda \tag{20.5-23}$$

Scaling factors may be introduced to regulate the magnitudes of dx, du and $d\lambda$.

All these techniques will converge (if at all) to local extrema. Since there may be more than one extremum in the domain of interest, means must be devised to search the entire domain for relative extrema which are then compared to determine the global or absolute extrema. The nature of each stationary point determined by the gradient procedures must be checked via the sufficiency condition eq. (20.2-19).

20.5.3 Dynamic Programming

A computational technique enjoying wide popularity for both nonlinear programming problems and dynamic optimization problems is the *dynamic programming* method. The method was developed initially to cope with sequential decision stochastic programming problems which occur in inventory theory. Subsequently, the basic ideas have been extended to cope with a wide class of problems.

These problems may be characterized as multistage decision processes which are Markovian: i.e., after i decisions the effect of the remaining decision sequence upon the total return function (cf. section 20.4.1) depends only on the state of the system at stage i and the subsequent decisions. Implementation of dynamic programming is based on the *principle of optimality* due to Bellman:[20-20]

> If u^i is an optimal policy sequence resulting in the state sequence x^i for a given dynamic programming problem with initial state x^1, then u^{i+1} is an optimal policy sequence for the same criterion and final state but initial state x^2.

Thus the dynamic programming approach imbeds the given optimization problem, whether continuous or discrete, in a class of similar problems with different initial conditions. This process requires the solution of the Hamilton-Jacobi-(Bellman) partial differential eq. (20.4-10) or difference equation to generate the field of extremals emanating backwards from the terminal point. For the partial differential equation this may be accomplished by the *method of characteristics*, for the theoretical basis of this method see Courant and Hilbert,[20-45] and for numerical techniques, Chapter 19 of this *Handbook.*

Except for special cases, solution of optimization problems for continuous systems where the dimension of the state variables is greater than say, three, is impractical using the straightforward dynamic programming approach due to computer storage limitations. In order to circumvent the storage problems, modifications to the conventional method of storing all computed values of the optimum return function are sought. One may use finite series approximations to the return function, for example, polynomial approximation.[20-40] Other devices may be introduced, such as the use of Lagrangian multipliers, iterative techniques in conjunction with standard dynamic programming over a restricted domain of policy or control space or state space or both, cf. Ref. 20-20, and the so-called state-increment dynamic programming[20-47] method which utilizes slow and rapid-access storage of a particular computer system in a manner to decrease rapid-access storage requirements.

We note that in contrast to the storage problems associated with continuous and discrete many variable problems, dynamic programming in direct application is very useful for combinatorial or allocation problems where there are only few possible choices of control at each step of the decision sequence.

20.5.4 Gradient Procedures

As noted in section 20.1.2, the variational formulation of optimization problems leads to the solution of two-point boundary value problems or a succession of two-point boundary value problems in case of interior point constraints. Numerical solution of such problems may be accomplished via the dynamic programming approach of generating central fields of extremals (fields emanating from either initial or terminal points) when computer storage problems are not encountered. The alternative to such flooding procedures involves some type of iteration. A general class of iteration procedures for which the state and adjoint equations are satisfied by the nominal solution (via numerical integration), but the optimality and boundary conditions are not, is formed by *gradient methods.* Gradient methods involve iterative algorithms for improving the control history $u(t)$ in order to approach the optimality conditions and prescribed boundary conditions.

As an example of gradient methods, consider minimizing

$$M = \varphi(x_f, t_f) + \int_{t_0}^{t_f} h(x, u, t)\, dt \tag{20.5-24}$$

subject to q terminal constraints

$$\psi(x_f, t_f) = \psi_D \tag{20.5-25}$$

where ψ_D is a given constraint level. Since q x_k's at t_f are defined by eq. (20.5-25), we may consider $\varphi(x_f, t_f)$ to be a function of the remaining $n - q$ components of

x. For fixed initial conditions, along *any* trajectory the variation of M due to variations in control $\delta u(t)$ and terminal time δt_f is given by

$$\delta M = (h + \dot{\varphi})\Big|_{t_f} \delta t_f + \int_{t_0}^{t_f} (h_u + \lambda_{(M)}^T f_u)\, \delta u \, dt \tag{20.5-26}$$

where the differential constraints $\dot{x} = f(x, u, t)$ have been adjoined to M with the adjoint vector $\lambda_{(M)}^T$ chosen to satisfy

$$\dot{\lambda}_{(M)}^T + \lambda_{(M)}^T f_x = -h_x \tag{20.5-27}$$

$$\lambda_{(M)}^T (t_f) = \varphi_{x_f} \tag{20.5-28}$$

with the first q components of $\lambda_{(M)} (t_f)$ zero.

Since each $\psi_j(x_f, t_f) = \psi_{jD}$ may be identified with $\varphi(x_f, t_f)$ for a new performance index $M_j^* = \psi_j(x_f, t_f)$ with $h(x, u, t) = 0$, we can form the first variation of ψ_j along any trajectory as [after adjoining $\dot{x} = f(x, u, t)$]

$$\delta \psi_j = \dot{\psi}_j \Big|_{t_f} \delta t_f + \int_{t_0}^{t_f} \lambda_{(\psi)j}^T f_u \, \delta u \, dt \tag{20.5-29}$$

which can be written compactly as a q-vector

$$\delta \psi = \dot{\psi} \,|_{t_f} \delta t_f + \int_{t_0}^{t_f} \lambda_{(\psi)}^T f_u \, \delta u \, dt \tag{20.5-30}$$

where $\lambda_{(\psi)}^T$ is understood to be $q \times n$ and satisfies

$$\dot{\lambda}_{(\psi)}^T + \lambda_{(\psi)}^T f_x = 0 \tag{20.5-31}$$

$$\lambda_{(\psi)}^T (t_f) = \psi_x |_{t_f} \tag{20.5-32}$$

We now guess a nominal control history $u(t)$, compute $x(t)$ from the state equations (numerical integration) and given initial conditions $x(t_0) = x_0$ to find $\psi\,[x(t_f), t_f]$. This nominal trajectory will usually fail to satisfy the local optimality conditions $\mathcal{H}_u = 0$ in addition to not meeting the prescribed terminal constraint level $\psi(x_f, t_f) = \psi_D$. We find the adjoint quantities $\lambda_{(M)}(t)$ and $\lambda_{(\psi)}(t)$ by backwards integration (numerical) of the adjoint equations (20.5-27) and (20.5-31), using $x(t_f)$ from state equation integration to determine the boundary conditions.

Since eqs. (20.5-26) and (20.5-30) are *linearized*, i.e., valid only for small $\delta u(t)$, there is no minimum for δM subject to constraints on $\delta \psi$. To resolve this difficulty, introduce a *quadratic penalty function* in $\delta u(t)$

$$P = \tfrac{1}{2} \int_{t_0}^{t_f} \delta u^T(t)\, W(t)\, \delta u(t)\, dt \tag{20.5-33}$$

where $W(t)$ is a positive definite, possibly time-varying $m \times m$ weighting matrix. We may then find an improved trajectory by determining $\delta u(t)$ and δt_f to minimize (maximize)*

$$\delta \overline{M} = P \underset{(-)}{+} \delta M \tag{20.5-34}$$

subject to $\delta\psi$ = desired $\delta\psi_f$.

We may adjoin $\delta\psi$ to $\delta\overline{M}$ with suitable multipliers μ and minimize

$$\delta \overline{J} = P + \delta M - \mu^T \delta\psi \tag{20.5.35}$$

The first variation of (20.5-35) must vanish, which yields

$$\delta u^T(t) = [h_u(t) + [\lambda^T_{(M)}(t) - \mu^T \lambda^T_{(\psi)}(t)]\, f_u(t)]\, W^{-1}_{(t)} \tag{20.5-36}$$

where μ and δt_f are determined by

$$(h + \dot{\varphi})\big|_{t_f} - \mu^T \dot{\psi}(t_f) = 0 \tag{20.5-37}$$

and

$$\delta\psi_f = \dot{\psi}\big|_{t_f} \delta t_f + \int_{t_0}^{t_f} \lambda^T_{(\psi)} f_u W^{-1} (h_u^T + f_u^T \lambda_{(M)} - f_u^T \lambda_{(\psi)} \mu)\, dt \tag{20.5-38}$$

or

$$\begin{bmatrix} I_{\psi\psi} & -\dot{\psi}(t_f) \\ \dot{\psi}(t_f) & 0 \end{bmatrix} \begin{bmatrix} \mu \\ \delta t_f \end{bmatrix} = \begin{bmatrix} I_{\psi M} - \delta\psi_f \\ h(t_f) + \dot{\varphi}(t_f) \end{bmatrix} \tag{20.5-39}$$

where

$$I_{\psi\psi} = \int_{t_0}^{t_f} \lambda^T_{(\psi)} f_u W^{-1} f_u^T \lambda_{(\psi)}\, dt = I^T_{\psi\psi} \tag{20.5-40}$$

*This approach is due to John V. Breakwell; for other (related) gradient techniques see Bryson and Ho, Ref. 20-21, Chapter 7.

$$I_{\psi M} = \int_{t_0}^{t_f} \lambda_{(\psi)}^T f_u W^{-1} (h_u^T + f_u^T \lambda_{(M)})\, dt \qquad (20.5\text{-}41)$$

The above integrals are computed backwards along the nominal trajectory along with the adjoints $\lambda_{(\psi)}(t)$ and $\lambda_{(M)}(t)$ which are stored, and then the improved control $u(t) + \delta u(t)$ is used to generate a new trajectory with presumably smaller terminal errors. To control the magnitude of the control change $\delta u(t)$, the weighting function $W(t)$ is adjusted. The whole process is then iterated until improvement in M and the terminal errors $\delta\psi$ are negligible. The convergence of this process depends on the first guess nominal path and the choice of the weighting function $W(t)$.

In general however, first-order gradient algorithms may show poor convergence near the optimal solution. To improve the convergence, second-order gradient algorithms are often used. These methods are based on expanding the criterion function to second order along the nominal trajectory and hence require that the nominal solution be "convex" over the entire time period: $\mathcal{H}_{uu} > 0$ for minimization, $\mathcal{H}_{uu} < 0$ for maximization. If it is difficult to pick a nominal convex solution, a first-order method such as we have described may be used to start the whole process. By choice of the matrix $W(t)$ we can insure that the nominal path generated is convex. For an excellent, detailed discussion of a second-order transition-matrix algorithm[20-48] and a backward-sweep algorithm see Bryson and Ho, Ref. 20-21, pp. 229–232. For an account of a successive sweep method, its relation to dynamic programming, and references to related iterative algorithms for solving two-point boundary value problems, see the paper by McReynolds.[20-29]

20.5.5 Neighboring Extremal Methods and Quasilinearization

Another widely used class of iterative procedures is characterized by successive solution of *linear* two-point boundary value problems derived by linearizing the system and adjoint equations about a nominal path. *Neighboring extremal methods*[20-50,51] involve guessing the unspecified boundary conditions to generate a nominal path, which satisfies the state and adjoint equations as well as the optimality conditions, and then iterating to improve the unspecified boundary conditions so as to satisfy the given boundary conditions. The primary difficulty with these methods is obtaining a reasonably accurate set of initial conditions to start the process. This starting sensitivity is due to the nature of the Euler-Lagrange equations; small changes in the starting values of λ result in large errors in the terminal values. The difficulty is usually circumvented by using a first-order gradient procedure to generate "reasonable" initial values of λ and then applying a neighboring extremal method.

Quasilinearization methods have been proposed by many authors under different names such as differential variations,[20-52] quasilinearization[20-53] and generalized Newton-Raphson.[20-54] The essential features of these methods involve choosing nominal state and adjoint functions $x(t)$ and $\lambda(t)$ to satisfy as many boundary conditions as possible and then determining the corresponding control $u(t)$ from the optimality conditions. The system and adjoint equations are linearized about this nominal and the iteration proceeds via solution of successive linear two-point boundary value problems until the exact system and adjoint equations are satisfied to the desired accuracy.

20.6 ELEMENTS OF GAME THEORY

Modern mathematical treatment of problems in interest conflict, or game theory as such problems are now termed, has its foundations in the classical text by von Neumann and Morgenstern.[20-55] The theory and problems that evolved subsequently deal with discrete and continuous games characterized by the (finite) game matrix concept. An excellent review of the scope of classical game theory, its assumptions and limitations is contained in the book by Luce and Haiffa.[20-10]

In the early 1950's, the scope of game theory was extended to dynamic game processes, where the players are confronted with lengthy continuous (or discrete) sequences of decisions. The theory was given a unifying exposition under the name of differential games by Isaacs in his text published in 1965.[20-12] Subsequent to the publication of Isaacs' text, the developments of optimal control theory and differential game theory have become unified, hence a brief exposition of the elements of game theory is fitting in this chapter.

The applications of game theory, now generalized to include dynamic processes, are many. The scope of problems ranges from parlor games, such as bridge, poker, chess, to warfare to economic policy to social behavior. In many of these areas, elegant theorems are possible but often techniques of solution are non-existent. Such is the case in any problem where the criteria are difficult to establish other than by subjective assumptions. Discussion of the logical difficulties encountered in formulating game theory problems is available in Ref. 20-10. An excellent presentation of mathematical methods and theory of games is available in the text by Karlin.[20-56]

In this section, we will introduce the key concepts of game theory and the elementary formulation of some classes of games. For the interested reader, we recommend the cited references.

20.6.1 Characterization of Games

Common to all game theory are the concepts of *strategy*, the *payoff* of a game, and *players*. The *players* in a game are the decision-making entities, whether individuals, governments or "fate". Each of the players may have a prescribed course of action, or sequence of decisions, in a given situation or may have to choose among many alternatives at each step based either on complete knowledge or incomplete knowledge of the current and/or past events. A prescription of decision for each conceivable situation in a given game is known as a *pure strategy*. A *mixed* (or *randomized*) *strategy* involves a probability mixture of pure strategies, i.e., if u_1

and u_2 are two pure strategies, a mixed strategy chooses u_1 with probability p_1 and u_2 with probability p_2.

A game is defined by the rules of the game and characterization of the players. The rules of the game consist of a finite *game tree* or *connected graph* depicting the alternatives available at each move, a labeling of each move which identifies the player taking that move, a probability distribution for each chance move, a specification of what each player may know at the time of each move and an assigned outcome or payoff for each possible sequence of moves.

The players must be identified and their *utility function* specified; the utility function represents a player's preference numerically. A *rational* player will always choose the alternative he prefers, i.e., the one with the largest utility. For an axiomatic development of the modern concepts of utility theory and how they enter game theory, see Chapter 2 of Luce and Haiffa.[20-10] For the elementary concepts it suffices to assume that each player can identify his preference among a particular set of alternatives and that he will consistently choose his preference. This means that for each player we can identify a *payoff function*, as we have used it in control theory, defined over the end points of the game tree, i.e., the terminal points of all decision sequences.

20.6.2 Two-Player Games

Two-player games form the most extensively studied class of games. If we assume that player 1 and player 2 each has available a set of pure strategies u and v respectively represented by

$$u = (u_1, u_2, \ldots, u_n) \tag{20.6-1}$$

$$v = (v_1, v_2, \ldots, v_m) \tag{20.6-2}$$

and denote the payoff for the strategy choice u_i for player one and v_j for player two by J_{ij}, the game may be represented by the *game matrix* (*payoff matrix*)

Player 2, PURE STRATEGY v

Player 1, PURE STRATEGY u

$$\begin{array}{c|ccccc}
 & v_1 & v_2 & \cdots v_j & \cdots & v_m \\
\hline
u_1 & J_{11} & J_{12} & \cdots J_{1j} & \cdots & J_{1m} \\
u_2 & J_{21} & J_{22} & \cdots J_{2j} & \cdots & J_{2m} \\
\cdots & \cdots & \cdots & \cdots & \cdots & \cdots \\
u_i & J_{i1} & J_{i2} & \cdots J_{ij} & \cdots & J_{im} \\
\cdots & \cdots & \cdots & \cdots & \cdots & \cdots \\
u_n & J_{n1} & J_{n2} & \cdots J_{nj} & \cdots & J_{nm}
\end{array} \tag{20.6-3}$$

A choice by player one of strategy u_i is equivalent to choice of row i and a choice of strategy v_j by player 2 equivalent to the choice of column j.

Classical game theory in normal form assumes each player has a preference ordering over the outcomes, and each player knows the matrix (20.6-3), his opponent's preference pattern for the outcome of the game, and if the outcome involves chance, the probabilities of each possibility. An elementary game of this type known to all is tic-tac-toe, another, considerably more complicated in dimension but not in type, is chess.

A further classification of games is possible according to the player's preference as to the outcome(s). If the desires of player one are directly opposite those of player two, we have a *strictly competitive* game. Such win-lose games are termed *zero-sum two-player games* and a wealth of results in game theory are available. For example, we suppose that each J_{ij} of eq. (20.6-3) represents a certain number of dollars that player 2 must pay player 1. Assuming that both are "rational"—player one wishes to win as much as possible and player two to lose as little as possible—the game strategy is clear. If player 1 moves first, he should pick the row with the *largest minimum*, since he knows that player 2 will subsequently choose the column with the minimum. If player 2 moves first, he should choose the column with the smallest maximum, since player one will subsequently pick the row with the maximum. If the game matrix has an entry which is both the minimum of its row and maximum of its column, the entry is known as an *equilibrium pair* or *minimax solution* of the game. For an equilibrium pair

$$\max_{1} \min_{2} J_{ij} = \min_{2} \max_{1} J_{ij} \tag{20.6-4}$$

where the order of play is 1, 2 on the left and 2, 1 on the right. Such equilibrium pairs need not be unique, however, all equilibrium pairs give rise to the same payoff.

It is easily shown that not all zero-sum or strictly competitive games have an equilibrium pair, for example, the game matrix

$$\begin{bmatrix} 3 & 1 \\ 2 & 4 \end{bmatrix} \quad \text{as opposed to} \quad \begin{bmatrix} 1 & 2 \\ 3 & 4 \end{bmatrix}$$

Thus, the question remains: Can one tell in advance from the game tree structure whether or not the game has a minimax solution? It can be shown (von Neumann and Morgenstern)[20-55] that for any game for which the player has complete and unambiguous knowledge of the previous moves optimal (minimax and maximin) strategies exist.

If an equilibrium pair does not exist, then the order of play makes a difference. The minimax principle of von Neumann and Morgenstern asserts that by randomizing, i.e., using mixed strategies, the minimax and maximin are equal on an expected value basis, i.e.,

$$E(\max_{1} \min_{2} J_{ij}) = E(\min_{2} \max_{1} J_{ij}) \tag{20.6-5}$$

A mixed strategy for player one is a probability distribution over n points; if player one has n pure strategies $(u_1, u_2 \ldots u_n)$ then a mixed strategy is $(p_1u_1, p_2u_2 \ldots p_mu_n)$ where $p_i \geqslant 0$ and $\sum_{i=1}^{n} p_i = 1$. For player two, we have m pure strategies $(v_1, v_2 \ldots v_m)$ and the mixed strategy $(q_1v_1, q_2v_2, \ldots, q_mv_m)$ with $q_i \geqslant 0$, $\sum_{i=1}^{m} q_i = 1$. The mixed strategies isolated by the minimax principle form an equilibrium pair. Existence of pairs of equilibrium strategies is thus guaranteed for all two-player zero-sum games and the payoff $E(\min\max J_{ij}) = V$ corresponding to the equilibrium pair is known as the *Value* of the game. (Pure strategies are included in this definition of Value.)

A number of techniques for solution of two-player games are discussed in the Appendices of Luce and Haiffa.[20-55] The most broadly applicable approach reduces the game to a linear programming problem. To do so, the payoff matrix is scaled so that the value of the game $V > 0$. The optimal strategies p_i, q_j and the value V are given by

$$p_i = Vx_i; \quad i = 1, 2, \ldots, n \tag{20.6-6}$$

$$q_j = V\alpha_j; \quad j = 1, 2, \ldots, m \tag{20.6-7}$$

$$V = \frac{1}{J_{\min}} = \frac{1}{I_{\max}} \tag{20.6-8}$$

where $V, J_{\min}, I_{\max}, x_i$ and α_j are determined from the solution of the dual linear programming problems

$$J = x_1 + x_2 + \cdots + x_n = \min \tag{20.6-9}$$

$$J_{1j}x_1 + J_{2j}x_2 + \cdots + J_{nj}x_n \geqslant 1; \quad j = 1, 2, \ldots, m \tag{20.6-10}$$

$$x_i \geqslant 0; \quad i = 1, 2, \ldots, m \tag{20.6-11}$$

$$I = \alpha_1 + \alpha_2 + \cdots + \alpha_m = \max \tag{20.6-12}$$

$$J_{i1}\alpha_1 + J_{i2}\alpha_2 + \cdots + J_{im}\alpha_m \leqslant 1; \quad i = 1, 2, \ldots, n \tag{20.6-13}$$

$$-\alpha_j \leqslant 0; \quad j = 1, 2, \ldots, m \tag{20.6-14}$$

Thus far, the discussion has been focused on games where the choice of strategies u, v was discrete. If u and v are continuous, then we have a continuous payoff function $J(u, v)$ rather than the payoff matrix. A solution of the game entails finding functions u^0 v^0 such that

$$J(u^0, v) \leqslant J(u^0, v^0) \leqslant J(u, v^0) \tag{20.6-15}$$

This is equivalent to a parameter optimization problem with two controls and the sufficient conditions for u^0, v^0 are

$$J_u = J_v = 0 \tag{20.6-16}$$

$$J_{uu} > 0; \quad J_{vv} < 0 \tag{20.6-17}$$

Any u^0, v^0 satisfying eqs. (20.6-16) and (20.6-17) is a *game-theoretic saddle point.* In cases where $\min_u \max_v J \neq \max_v \min_u J$, a mixed strategy is again used to equalize the difference.

It is clear that, although important, two-player zero-sum games only cover a small fraction of possible interest conflict situations. Other situations such as two-player *nonzero sum*—where the preferences of the players are not necessarily opposed—with or without cooperation, and multiplayer (n-player) games are covered in the (extensive) literature.[20-10,55,56]

20.6.3 Differential Games

The extension of classical game theory to include the dynamical case results in the theory of differential games.[20-13] An alternate viewpoint is to take optimal control theory and add an opposing controller—thus we get "extended" two-sided calculus of variations problems. We may generally define a differential game by the state equations

$$\dot{x} = f(x, u, v, t); \quad x(t_0) = x_0 \tag{20.6-18}$$

with terminal constraints

$$\psi(x_f, t_f) = 0 \tag{20.6-19}$$

and a performance criterion

$$M = \varphi(x_f, t_f) + \int_{t_0}^{t_f} h(x, u, v, t)\, dt \tag{20.6-20}$$

The problem is to find functions $u^0(x, t)$, $v^0(x, t)$ such that

$$M(u^0, v) \leqslant M(u^0, v^0) \leqslant M(u, v^0) \tag{20.6-21}$$

for all other closed loop strategies $u(x, t)$, $v(x, t)$. Formally, no new ideas are required for the first-order necessary conditions for a stationary value of M. We define a Hamiltonian by adjoining the differential constraints and find the necessary conditions as

$$\mathcal{H} = h + \lambda^T f \tag{20.6-22a}$$

$$\dot{x} = f(x, u, v, t) \tag{20.6-22b}$$

$$\dot{\lambda} = -\mathcal{H}_x^T \tag{20.6-22c}$$

$$x(t_0) = x_0 \tag{20.6-22d}$$

$$\lambda^T(t_f) = \varphi(x_f, t_f) + \mu^T \psi(x_f, t_f) \tag{20.6-22e}$$

$$\mathcal{H}_u = \mathcal{H}_v = 0 \tag{20.6-22f}$$

or

$$\mathcal{H}^0 = \max_v \min_u \mathcal{H} \tag{20.6-22g}$$

The solution of the two-point boundary value problem defined by eq. (20.6-22) must satisfy the game-theoretic saddle point condition, eq. (20.6-21). If $M_{uv} = 0$, the problem is called *separable* and a saddle point always exists. For nonseparable problems, the strategies u^0, v^0 defined by eq. (20.6-22) may still be useful if a specification is made that one player will play first.

20.7 REFERENCES

20-1 Isaacs, R., "Differential Games: Their Scope, Nature, and Future," *Journal of Optimization Theory and Applications*, (*JOTA*) **3** (5), 283–295, 1969.

20-2 Ho, Y. C., "Differential Games, Dynamic Optimization, and Generalized Control Theory," Survey Paper, *JOTA*, **6** (3), 179–209, 1970.

20-3 Danskin, J., *The Theory of Max-Min*, Springer-Verlag, New York, 1967.

20-4 Zadek, L. A., "Optimality and Non-Scalar Valued Performance Criteria," *IEEE Transactions on Automatic Control*, **AC-8**, No. (1), 1963.

20-5 Da Cunha, N. O., and Polak, E., "Constrained Minimization Under Vector-Valued Criteria in Finite-Dimensional Spaces, *Memorandum No. ERL-M188*, University of California at Berkeley, 1966.

20-6 Case, J. H., "Towards a Theory of Many Players Differential Games," *SIAM Journal on Control*, **7** (2), 1969.

20-7 Starr, A. W., and Ho, Y. C., "Nonzero-Sum Differential Games," *JOTA*, **3** (3), 1969.

20-8 Starr, A. W., and Ho, Y. C., "Further Properties of Nonzero-Sum Differential Games," *JOTA*, **3** (4), 1969.

20-9 Harsanyi, J., "Games with Incomplete Information Played by Bayesian Players, Parts I, II, III," Management Science, **14**, 3, 5, & 7, 1967 & 1968.

20-10 Luce, R. D., and Haiffa, H., *Games and Decisions*, John Wiley & Sons, New York, 1957.

20-11 Case, H. H., "Equilibrium Points in N-Person Differential Games," University of Michigan, Department of Industrial Engineering, *TR No. 1967-1*, 1967.

20-12 Isaacs, R., *Differential Games*, John Wiley and Sons, New York, 1965.

20-13 Berkovitz, L. D., "A Survey of Differential Games" in *Mathematical Theory of Control*, ed. Balakrishnan, A. V. and Neustadt, L. W., Academic Press, New York, 1967.

20-14 "Collected Bibliography on Differential Games," *Proceedings of the First International Conference on the Theory and Applications of Differential Games*, ed. Ho, Y. C. and Leitmann, G., Amherst, Massachusetts, 1969.

20-15 *Proceedings of the First International Conference on the Theory and Applications of Differential Games*, ed. Ho, Y. C. and Leitmann, G., Amherst, Massachusetts, 1969.

20-16 Paiewonsky, B., "Optimal Control: A Review of Theory and Practice," *AIAA Journal*, **3**, 1985–2006, November 1965.

20-17 Pierre, D. A., *Optimization Theory with Applications*, John Wiley & Sons, New York, 1969.

20-18 Künzi, H. P., Tzschah, H. G., and Zehnder, C. A., *Numerical Methods of Mathematical Optimization*, Academic Press, New York, 1968.

20-19 Leon, A., "A Classified Bibliography on Optimization," in *Recent Advances in Optimization Techniques*, ed. Lavi, A., and Vogl, T. P., John Wiley & Sons, New York, 1966.

20-20 Bellman, R., *Dynamic Programming*, Princeton University Press, Princeton, New Jersey, 1957.

20-21 Bryson, A., and Ho, Y. C., *Applied Optimal Control*, Hemisphere Publishing Co., Wash., D.C., 1975.

20-22 Fiacco, A. V., and McCormick, G. P., *Nonlinear Programming: Sequential Unconstrained Minimization Techniques*, John Wiley and Sons, New York, 1968.

20-23 Kuhn, H. W. and Tucker, A. W., "Nonlinear Programming," in *Proceedings of the Second Berkeley Symposium on Math., Stat., and Probab.*, ed. Neyman, J., Berkeley, California, 481–492, 1950.

20-24 Oldenburger, R., *Optimal Control*, Holt, Rinehart and Winston, New York, 1966.

20-25 Bliss, G. A., *Lectures on the Calculus of Variations*, University of Chicago Press, 1945.

20-26 Pontryagin, L. S., Boltyanskii, V. G., Gamkrelidze, R. V., and Mischchenko, E. F., *The Mathematical Theory of Optimal Processes*, Wiley-Interscience, New York, 1962.

20-27 Butkovskiy, A. G., *Distributed Control Systems*, American Elsevier Publishing Co., New York, 1969.

20-28 Gelfand, I. M., and Fomin, S. V., *Calculus of Variations*, Prentice-Hall, Inc., Englewood Cliffs, New Jersey, 1963.

20-29 Robbins, H. M., "A Generalized Legendre-Clebsch Condition for the Singular Cases of Optimal Control," *IBM Journal of Research and Development*, **11**, 361, 1967.

20-30 Kelley, H. J., Kopp, R. E., and Moyer, A. G., "Singular Extremals," Chapter 3, in *Optimization–Theory and Applications*, ed. Leitman, G., Academic Press, 1966.

20-31 Jacobson, D. H., and Speyer, J. L., "Necessary and Sufficient Conditions for Optimality for Singular Control Problems: A Transformation Approach," *TR No. 69-24*, Harvard University, 1969.

20-32 Jacobson, D. H., and Speyer, J. L., "Necessary and Sufficient Conditions for Optimality of Singular Control Problems: A Limit Approach," *TR No. 604*, Harvard University, 1970.

20-33 Dantzig, G. B., *Linear Programming and Extensions*, Prentice-Hall, Englewood Cliffs, New Jersey, 1961.

20-34 Hadley, G., *Linear Programming*, Addison-Wesley, Reading, Massachusetts, 1962.

20-35 Hadley, G., *Nonlinear and Dynamic Programming*, Addison-Welsey, Reading, Massachusetts, 1964.

20-36 Zoutendijk, G., *Method of Feasible Directions*, Elsevier Publishing Co., London, 1961.

20-37 Leon, A., "A Comparison Among Eight Known Optimization Procedures," *Recent Advances in Optimization Techniques*, Ed. Vogl, T. P. and Lavi, A., John Wiley & Sons, New York, 1966.

20-38 Box, J. M., "A Comparison of Several Current Optimization Methods, and the Use of Transformations in Constrained Problems," *The Computer Journal*, **9**, 1966.

20-39 Curry, H., "Methods of Steepest Descent for Nonlinear Minimization Problems," *Quarterly of Applied Mathematics*, **2**, 1954.

20-40 Fletcher, R., and Powell, M. J. D., "A Rapidly Convergent Descent Method for Minimization," *The Computer Journal*, **6**, July 1963.

20-41 Powell, M. J. D., "An Efficient Method for Finding the Minimum of a Function of Several Variables without Calculating Derivatives," *The Computer Journal*, **7**, July 1964.

20-42 Zangwill, W. I., "Minimizing a Function without Calculating Derivatives," *The Computer Journal*, **10**, November 1967.

20-43 Rosen, J. B., "The Gradient Projection Method for Nonlinear Programming: Part I–Linear Constraints," *SIAM Journal of Applied Mathematics*, **8**, 1960.

20-44 Rosen, J. B., "The Gradient Projection Method for Nonlinear Programming: Part II–Nonlinear Constraints," *SIAM Journal*, **9**, 1961.

20-45 Courant, R., and Hilbert, D., *Methods of Mathematical Physics*, Vol. II, Chapter 2, Wiley-Interscience, New York, 1953.

20-46 Bellman, R., and Dreyfus, S. E., "Functional Approximation and Dynamic Programming," *Mathematical Tables and Other Aids to Computation*, Vol. 13, October 1959.

20-47 Larson, R. E., "An Approach to Reducing the High-Speed Memory Requirement of Dynamic Programming," *Journal of Mathematical Analysis and Applications*, **11**, July 1965.

20-48 Kelley, H. J., Kopp, R., and Moyer, G., "A Trajectory Optimization Technique Based on the Second Variation," in *Progress in Astronautics*, Vol. 14, Academic Press, New York, 1964.

20-49 McReynolds, S. R., "The Successive Sweep Method and Dynamic Programming," *Journal of Mathematical Analysis and Applications*, **19**, 1967.

20-50 Breakwell, J. V., Speyer, J. L., and Bryson, A. E., "Optimization and Control of Nonlinear Systems Using the Second Variation," *SIAM Journal of Control*, Series A, **1**, 1963.

20-51 Kelley, H. J., Kopp, R., and Moyer, G., "A Trajectory Optimization Technique Based Upon the Theory of the Second Variation," *AIAA Astrodynamics Conference*, Yale University, New Haven, Connecticut, August 19–21, 1963.

20-52 Hestenes, M. R., "Numerical Methods of Obtaining Solutions of Fixed End Point Problems in the Calculus of Variations," The RAND Corporation, *Memorandum No. RM-102*, 1949.

20-53 Bellman, R., and Kalaba, R. E., *Quasilinearization and Boundary Value Problems*, American Elsevier Publishing Co., New York, 1965.

20-54 Kemuth, P., and McGill, R., "Two-Point Boundary Value Problem Techniques," *Advances in Control Systems*, Vol. 3, Chapter 2, Academic Press, New York, 1966.

20-55 von Neumann, J. and Morgenstern, O., *Theory of Games and Economic Behavior*, Princeton University Press, Princeton, New Jersey, 1944.

20-56 Karlin, S., *Mathematical Methods and Theory in Games, Programming and Economics*, Vol. I, II, Addison-Wesley, Reading, Massachusetts, 1959.

20-57 Gill, P., Murray, W., and Wright, M., *Practical Optimization*, Academic Press, N.Y., 1981.

21

Probability and Statistics

L. Fisher*

21.0 INTRODUCTION

Probability and statistical theory arose historically when it was noticed that certain regularities were observed in games of chance and in human mortality (Refs. 21-1, 21-2, and 21-3). The philosophical foundations of probability are a matter of debate to the present day. The "frequency" school holds that the probability of an event is the fraction of times the event would occur in indefinitely many independent trials repeating exactly the same situation. The "Bayesian" school holds that a probability is a subjective measurement of the strength of one's belief that an event will occur. In addition, every position between the two views given above seems to be occupied.

21.1 PROBABILITY SPACES

During the early development of probability theory it was not appreciated that the subject could be formalized mathematically. Work was often a mixture of mathematics dealing with situations with a finite number of equally likely outcomes and intuitive discussion. A. N. Kolmogorov, in 1933 (Ref. 21-65), placed probability within the mainstream of axiomatic mathematics by formalizing the field as a branch of measure theory. Statistics was developing along different pathways by and large. It was Harald Cramér (Ref. 21-66) who in 1945 placed statistical theory upon the axiomatic structure of probability theory.

The formal mathematical structure is difficult to comprehend unless one understands the situation being modeled. One has in mind a "chance situation." For example, hands of poker may be dealt or an operation with a certain chance of operative mortality may be performed. The probability of an event may be thought of as the proportion of the time that the event would occur if the experiment or

*Prof. Lloyd Fisher, Dep't. of Biostatistics, University of Washington, Seattle, Wash.

situation could be repeated an indefinite number of times. To formalize this one needs to consider all possible outcomes. Events are then defined as subsets of the possible outcomes; the probability is then a number defined for each event.

21.1.1 Definitions

The mathematical formulation of proability theory begins with a set S, consisting of all possible outcomes of the situation at hand, and certain subsets of S, to be called events, for which a probability will be defined.

If one has a sequence of events, it would be useful to be able to speak about the new event that at least one of the events in the sequence occurs. Further, if one talks about an event occurring, it should be possible to talk about the event *not* occurring. By requiring the set of possible events to be a sigma-field, these aims are met.

Definition 21.1.1: A collection $\mathfrak{F}$ of subsets of a set S is a sigma-field on S if

i) ϕ (the empty set) is in $\mathfrak{F}$.
ii) If $A_1, A_2, \ldots$ is a sequence of sets in $\mathfrak{F}$ then $U_{i=1}^{\infty} A_i$ is in $\mathfrak{F}$.
iii) If A is in $\mathfrak{F}$ then A^c is in $\mathfrak{F}$, where A^c, A complement, consists of all elements of S that are not in A.

The next definition adds the probabilities to the individual events. Since the probability is a proportion, it must always be a number between 0 and 1. Property i) below assures that all possible outcomes are being considered. Finally, if disjoint events are considered, the proportion of the time that one of them occurs is the sum of the proportions of the time that the individual disjoint events occur.

Definition 21.1.2: A probability space $(S, \mathfrak{F}, P)$ is a triple where $\mathfrak{F}$ is a sigma-field on S and $P(A)$ is defined for A in $\mathfrak{F}$. P satisfies the following properties:

i) $P(S) = 1$
ii) $0 \leqslant P(A) \leqslant 1$ for each A in $\mathfrak{F}$.
iii) If $A_1, A_2, A_3, \ldots$ is a sequence of mutually exclusive elements of $\mathfrak{F}$ then

$$P(U_{i=1}^{\infty} A_i) = \sum_{i=1}^{\infty} P(A_i)$$

or

$$P(A_1 \cup A_2 \cup \cdots) = P(A_1) + P(A_2) + \cdots$$

P is called a probability measure.

Example: Let $S = \{s_1, \ldots, s_n\}$ be a finite set of n objects and $\mathfrak{F}$ consist of all subsets of S. A probability measure is characterized by $p_i = P(\{s_i\})$, $i = 1, \ldots, n$. The p_i must satisfy $p_i \geqslant 0$ and $p_1 + \cdots + p_n = 1$.

For example, if S consists of all possible poker hands then each point in S has probability $1/\binom{52}{5}$ under ideal shuffling. Here $\binom{n}{k} = n!/(n-k)!k!$ where $0 \leqslant k \leqslant n$ and n and k are integers with $0! = 1, j! = 1 \cdot 2 \cdot 3 \cdots j$; $\binom{n}{k}$ is a binomial coefficient, the number of subsets of size k that can be formed from n objects. Reference 21-4 gives an excellent (though demanding) introduction to finite probability spaces.

21.1.2 Conditional Probability and Independence

Conditional probabilities are defined to take into account new information. If we know that an event A occurred, then the probability that an event B occurs should be changed. The fraction of the time that B occurs when A occurs should be the conditional probability. Translated into probabilities, this gives the next definition.

Definition 21.1.3: Let A and B be events with $P(A) > 0$ then $P(B|A) = P(A \cap B)/P(A)$ is the conditional probability of B given A.

The conditional probability of B given A allows us to update the probability that B will occur when the information that A has occurred is received. For example, the probability of all heads in five flips of a fair coin is $\frac{1}{32}$, but the probability of all heads given that the first four flips give heads is $\frac{1}{2}$.

Definition 21.1.4: Events A and B are independent or statistically independent if $P(A \cap B) = P(A)P(B)$. More generally, events of a finite or countable family $\{A_1, A_2, \ldots\}$ are independent if for any finite subcollection $A_{i_1}, \ldots, A_{i_n}$

$$P(A_{i_1} \cap \cdots \cap A_{i_n}) = P(A_{i_1})P(A_{i_2}) \cdots P(A_{i_n})$$

Note that if $P(A) > 0$ and $P(B) > 0$ and if A and B are independent events then $P(A) = P(A|B)$ and $P(B) = P(B|A)$. Also, if $P(A) = P(A|B)$ then $P(A)\,P(B) = P(A \cap B)$ and A and B are independent events. These facts motivate the definition of independence. If A is independent of B, then knowledge of B should not affect the probability of A occurring, i.e., $P(A) = P(A|B)$. In experimentation a common assumption is that the outcome of one trial of an experiment does not affect the outcome of another trial. Mathematically, one models this by letting the outcomes be independent events (or more exactly the event that the first outcome takes on a specific value is independent of the event that the second outcome takes on a specific value).

21.1.3 Probability Measures on R^n

Most statistical studies and probabilistic models involve the recording of a set of numbers. The following definitions prepare the way for this most important situation.

Definition 21.1.5: R^n is n-dimensional Euclidean space, the space of n-tuples of real numbers. For $\mathbf{x}$ and $\mathbf{y}$ in R^n we write $\mathbf{x} \leqslant \mathbf{y}$ if $\mathbf{x} = (x_1, \ldots, x_n)$, $\mathbf{y} =$

$(y_1, \ldots, y_n)$ and $x_i \leqslant y_i$, $i = 1, 2, \ldots, n$. For each $\mathbf{x}$ in R^n let $I_{\mathbf{x}}$ be the set of $\mathbf{y}$ such that $\mathbf{y} \leqslant \mathbf{x}$. The Borel field of subsets of R^n is the smallest sigma-field of subsets of R^n containing all sets of the form $I_{\mathbf{x}}$. A probability measure on R^n is a probability measure on the Borel sigma-field.

Definition 21.1.6: The distribution function F of a probability measure on R^n is defined by $F(\mathbf{x}) = P(I_{\mathbf{x}})$

The distribution function uniquely determines P (and conversely).

Two general types of probability measures on R^n cover most of the situations encountered in practice.

Definition 21.1.7: A probability measure, P, on R^n is discrete if there is a finite or countable set of points $\{\mathbf{x}_1, \mathbf{x}_2, \ldots\}$ such that $\sum_i P(\{\mathbf{x}_i\}) = 1$. The function $p(\mathbf{x}) = P(\{\mathbf{x}\})$ is the probability mass function.

Definition 21.1.8: A probability measure, P, on R^n is continuous if there is a non-negative function f such that for each $\mathbf{x}$, $P(I_{\mathbf{x}}) = \int_{I_{\mathbf{x}}} f(\mathbf{z})\, d\mathbf{z}$. The function f is called a probability density function.

For a discrete probability measure the probability of a set A is $\sum_{\mathbf{x} \text{ in } A} p(\mathbf{x})$. For a continuous probability measure $P(A) = \int_A f(\mathbf{x})\, d\mathbf{x}$. The integral is an n-dimensional integral.

21.2 RANDOM VECTORS AND RANDOM VARIABLES

21.2.1 Definitions and the Induced Probability Measure on R^n

In probabilistic and statistical reasoning one is often concerned with a number or set of numbers associated with the outcome of an experiment. Even data that is not numerical may often be coded numerically for convenience.

Definition 21.2.1: An m-dimensional random vector is a function $\mathbf{X}(s)$ defined on a probability space $(S, \mathcal{F}, P)$ such that for each $\mathbf{x}$ in R^m, $\{s\colon \mathbf{X}(s) \text{ in } I_{\mathbf{x}}\}$ is in $\mathcal{F}$.

The intuition behind the definition is as follows: with each outcome of the experiment, say s, we associate m numbers $\mathbf{X}(s)$. Since we want to be able to talk about such things as the probability that $\mathbf{X}(s) \leqslant \mathbf{x}$ we must require that the set of s with $\mathbf{X}(s) \leqslant \mathbf{x}$ be an event so that the probability is defined.

It often is the case that we are only interested in the numbers $\mathbf{X}(s)$ and not in the original space S. In this case we can reduce the problem to considering only R^m and its Borel subsets.

Definition 21.2.2: Let $\mathbf{X}$ be an m-dimensional random vector on $(S, \mathcal{F}, P')$. The induced probability measure P on the Borel sigma-field of R^m is defined by

$$P(A) = P'(\{s\colon\ \mathbf{X}(s) \in A\})$$

If P is a discrete measure we say that $\mathbf{X}$ is a discrete random vector. If P is a continuous measure then $\mathbf{X}$ is a continuous random vector. If $m = 1$ then $\mathbf{X} = X$ is called a random variable.

We may think of a random vector as a collection of m random variables, i.e., one random variable for each coordinate.

Definition 21.2.3: The distribution of a random vector refers to its induced probability measure expressed in any way feasible. For a discrete random vector it is enough to specify $p(\mathbf{x})$, the probability mass function. For a continuous random vector it is enough to give the probability density function to determine the measure.

The idea of independent random vectors or random variables is an essential idea in statistical applications. It is a mathematical formulation of the intuitive idea that several determinations of a number are unrelated or have no effect on each other.

Definition 21.2.4: Random vectors $\mathbf{X}_1, \ldots, \mathbf{X}_k$ all defined on the same probability space are independent if for any Borel sets $A_1, A_2, \ldots, A_k$ the events $\{\mathbf{X}_1 \in A_1\}, \ldots, \{\mathbf{X}_k \in A_k\}$ are independent events. If the $\mathbf{X}_1, \ldots, \mathbf{X}_k$ all have the same distribution they are called identically distributed.

If $\mathbf{X}_1, \ldots, \mathbf{X}_k$ are defined on the same probability space we may think of the $\mathbf{X}_i$'s as comprising one vector $\mathbf{X} = (\mathbf{X}_1', \ldots, \mathbf{X}_k')'$. From each $\mathbf{X}_i$ we have a distribution function $F_i(\mathbf{x}_i)$. For $\mathbf{X}$ we also have a distribution function $F(\mathbf{x}_1, \ldots, \mathbf{x}_k)$. Then the $\mathbf{X}_1, \ldots, \mathbf{X}_k$ are independent if and only if

$$F(\mathbf{x}_1, \ldots, \mathbf{x}_k) = F_1(\mathbf{x}_1) \cdot F_2(\mathbf{x}_2) \ldots F_k(\mathbf{x}_k)$$

If $\mathbf{X}$ is a discrete random vector with probability mass function $p(\mathbf{x}_1, \ldots, \mathbf{x}_k)$ then each $\mathbf{X}_i$ is discrete with probability mass function $p_i(\mathbf{x}_i) = \sum_{\mathbf{x}_1} \cdots \sum_{\mathbf{x}_{i-1}} \sum_{\mathbf{x}_{i+1}} \cdots \sum_{\mathbf{x}_k} p(\mathbf{x}_1, \ldots, \mathbf{x}_k)$. The $\mathbf{X}_i$ are independent if and only if

$$p(\mathbf{x}_1, \ldots, \mathbf{x}_k) = p_1(\mathbf{x}_1) \cdots p_k(\mathbf{x}_k)$$

If $\mathbf{X}$ is a continuous random vector with probability density function $f(\mathbf{x}_1, \ldots, \mathbf{x}_k)$ then $\mathbf{X}_i$ is a continuous random vector with probability density function

$$f_i(\mathbf{x}_i) = \int \cdots \int f(\mathbf{x}_1, \ldots, \mathbf{x}_k)\, d\mathbf{x}_1 \cdots d\mathbf{x}_{i-1}\, d\mathbf{x}_{i+1} \cdots d\mathbf{x}_k$$

The $\mathbf{X}_i$ are independent if and only if $f(\mathbf{x}_1, \ldots, \mathbf{x}_k) = f_1(\mathbf{x}_1) \cdots f_k(\mathbf{x}_k)$ (except possibly for values of $\mathbf{x}_1, \ldots, \mathbf{x}_k$ which lie in a set whose volume is zero).

The importance of these comments lies not in the fact that one usually checks that random vectors are independent but that one uses the above to construct the probability distribution of independent random vectors.

Example: Find the probability density function for three independent random variables x_1, x_2, x_3 which each have probability density function $(2\pi)^{-1/2} \times \exp(-x^2/2)$.

Solution:

$$f(x_1, x_2, x_3) = \prod_{i=1}^{3} (2\pi)^{-1/2} \exp(-x_i^2/2)$$

$$= (2\pi)^{-3/2} \exp\left(-\sum_{i=1}^{3} x_i^2/2\right)$$

21.2.2 Expectations, Moments and Generating Functions

Definition 21.2.5: Let $\mathbf{X}$ be a random vector with distribution function F. If for some function h, $\int h(\mathbf{y})\, dF(\mathbf{y})$ exists (as a Lebesgue-Stieltjes integral) then the quantity is called the expected value of $h(\mathbf{X})$ and denoted by $E[h(\mathbf{X})]$.

For a discrete random vector with probability mass function $p(\mathbf{x})$,

$$E[h(\mathbf{X})] = \sum_{\mathbf{x}} h(\mathbf{x})\, p(\mathbf{x})$$

For a continuous random vector with probability density function $f(\mathbf{x})$

$$E[h(\mathbf{X})] = \int h(\mathbf{x}) f(\mathbf{x})\, d\mathbf{x}$$

In speaking of a set of random variables $X_1, X_2, \ldots, X_n$ defined on the same probability space we speak of their joint distribution when we speak of the probability distribution of $\mathbf{X} = (X_1, \ldots, X_n)'$.

Examples:

(1) Let X and Y be random variables with probability density function

$$f(x,y) = \begin{cases} 1 & 0 \leqslant x \leqslant 1 \text{ and } 0 \leqslant y \leqslant 1 \\ 0 & \text{otherwise} \end{cases}$$

Find $E(XY)$.

Solution:

$$E(XY) = \int_{-\infty}^{+\infty} \int_{-\infty}^{+\infty} xyf(x,y)\,dx\,dy = \int_0^1 \int_0^1 xy \cdot 1 \cdot dx\,dy = \frac{1}{4}$$

(2) Let X be a discrete random variable with $p(x) = e^{-\lambda} \lambda^x / x!$, for $x = 0, 1, 2, \ldots, p(x) = 0$ for all other x. Find $E(e^{tX})$.

Solution:

$$E(e^{tX}) = \sum_{x=0}^{\infty} \frac{e^{tx} e^{-\lambda} \lambda^x}{x!} = e^{-\lambda} \sum_{x=0}^{\infty} \frac{(\lambda e^t)^x}{x!} = e^{-\lambda} e^{\lambda e^t} = e^{\lambda(e^t - 1)}$$

Certain expected values have proved to be useful indices for giving a rough idea of the form of a probability distribution. The following definition lists some of these expectations.

Definition 21.2.6: In this definition it is assumed that all expectations given exist.

(a) The expected value or mean value of random variable X is $E(X)$ and is often denoted by μ.

(b) The variance of a random variable X is $E\{[X - E(X)]^2\}$ and is often denoted by σ^2 or Var (X). The standard deviation of a random variable X is σ, the square root of the variance of X.

(c) The covariance of two jointly distributed random variables X and Y is $E\{[X - E(X)]\ [Y - E(Y)]\}$ denoted by Cov (X, Y). The correlation or correlation coefficient of X and Y is

$$\frac{E\{[X - E(X)]\ [Y - E(Y)]\}}{(E\{[X - E(X)]^2\}\ E\{[Y - E(Y)]^2\})^{1/2}}$$

assuming that X and Y have positive variances. That is, the correlation of X and Y is the covariance of X and Y divided by the product of their standard deviations.

(d) The rth moment of a random variable X is $E(X^r)$, denoted by μ_r'.

(e) The rth central moment of a random variable X is $E[(X - \mu)^r]$, denoted by μ_r.

(f) If $\mathbf{X} = (X_1, \ldots, X_n)'$ is a random vector, its mean vector is $\boldsymbol{\mu} = [E(X_1), \ldots, E(X_n)]'$ and its covariance matrix Σ is an $n \times n$ matrix with i, j^{th} entry

$$\Sigma_{ij} = \text{Cov}\,(X_i, X_j)$$

(g) The moment generating function of a random vector $\mathbf{X} = (X_1, \ldots, X_n)'$ is

$$M(\mathbf{t}) = E(e^{\mathbf{t}' \cdot \mathbf{X}}) = E(e^{\sum_{j=1}^{n} t_j X_j})$$

if it is defined for all t with $|t_1| + \cdots + |t_n| < \epsilon$ for some $\epsilon > 0$.

(h) The characteristic function of $\mathbf{X} = (X_1, \ldots, X_n)'$ is (with $i^2 = -1$)

$$\phi(\mathbf{t}) = E(e^{i\mathbf{t}' \cdot \mathbf{X}}) = E\left[\cos\left(\sum_{j=1}^{n} t_j X_j\right)\right] + iE\left[\sin\left(\sum_{j=1}^{n} t_j X_j\right)\right]$$

THEOREM 21.2.1: *Assume all expectations involved exist.*

(*a*) *The characteristic function of a random vector* $\mathbf{X}$ *always exists and uniquely determines the distribution of* $\mathbf{X}$.

(*b*) *If the moment generating function of* $\mathbf{X}$ *exists it uniquely determines the distribution of* $\mathbf{X}$.

(c) *Jointly distributed random vectors* $\mathbf{X}$ *and* $\mathbf{Y}$ *are independent if and only if their joint characteristic function is the product of the two individual characteristic functions:*

$$\phi_{(\mathbf{X}', \mathbf{Y}')'}(\mathbf{t}_1, \mathbf{t}_2) = \phi_{\mathbf{X}}(\mathbf{t}_1)\, \phi_{\mathbf{Y}}(\mathbf{t}_2)$$

(d) $E[ah(\mathbf{X}) + bg(\mathbf{Y})] = aE[h(\mathbf{X})] + bE[g(\mathbf{Y})]$ *for any functions h and g.*

(e) *If* $\mathbf{X}$ *and* $\mathbf{Y}$ *are independent random vectors*

$$E[h(\mathbf{X})\, g(\mathbf{Y})] = E[h(\mathbf{X})]\, E[g(\mathbf{Y})]$$

(f) *If k is a nonnegative integer and $E(X^k)$ exists and $\phi(t)$ is the characteristic function of X then*

$$i^k \mu'_k = i^k E(X^k) = \left.\frac{d^k \phi(t)}{dt^k}\right|_{t=0}$$

(g) *If the moment generating function of X exists then all positive moments of X exist and for nonnegative integers k*

$$\mu'_k = E(X^k) = \left.\frac{d^k M(t)}{dt^k}\right|_{t=0}$$

(h) *Let $\mathbf{X} = (X_1, \ldots, X_n)'$ have characteristic function $\phi(t_1, \ldots, t_n)$ and let $k_1, \ldots, k_n$ be nonnegative integers such that $E(X_1^{k_1} \cdots X_n^{k_n})$ exists then*

$$i^{k_1+\cdots+k_n} E(X_1^{k_1} \cdots X_n^{k_n}) = \frac{\partial^{k_1+\cdots+k_n}}{\partial t_1^{k_1} \cdots \partial t_n^{k_n}} \phi(t_1, \ldots, t_n)\Big|_{t_1=\cdots=t_n=0}$$

(i) *If the moment generating function of $\mathbf{X} = (X_1, \ldots, X_n)'$ exists then for any nonnegative integers $k_1, \ldots, k_n$*

$$E(X_1^{k_1} \cdots X_n^{k_n}) = \frac{\partial^{k_1+\cdots+k_n}}{\partial t_1^{k_1} \cdots \partial t_n^{k_n}} M(t_1, \ldots, t_n)\Big|_{t_1=\cdots=t_n=0}$$

(j)

$$\operatorname{Var}(X) = \mu'_2 - (\mu'_1)^2$$

(k) *If X and Y are independent random variables,*

$$\operatorname{Var}(X + Y) = \operatorname{Var}(X) + \operatorname{Var}(Y)$$

(l) $\operatorname{Var}(aX + b) = a^2 \operatorname{Var}(X)$ *for constants a and b.*

(m) *If X has characteristic function $\phi(t)$, then $aX + b$ has characteristic function $e^{ibt}\,\phi(at)$.*

(n) *Let $\mathbf{X}^{n \times 1}$ and $\mathbf{Y}^{n \times 1}$ be independent random vectors, then the characteristic function of the sum is the product of the characteristic functions: $\phi_{\mathbf{X}+\mathbf{Y}}(\mathbf{t}) = \phi_{\mathbf{X}}(\mathbf{t})\,\phi_{\mathbf{Y}}(\mathbf{t})$ and vice versa.*

21.2.3 Specific Discrete and Continuous Distributions

In this section we describe specific probability distributions. The following tables present the distributions after which some of the uses of these distributions are remarked upon.

Table 21.2-1 Discrete Probability Distributions.

A. One-Dimensional Distributions

Name	Parameters	Nonzero Values of the Probability Mass Function, $p(k)$	Mean	Variance	Moment Generating Function
Binomial	n a positive integer, $0 \leqslant p \leqslant 1$	$\binom{n}{k} p^k (1-p)^{n-k}$ $k = 0, 1, 2, \ldots, n$	np	$np(1-p)$	$(pe^t + 1 - p)^n$
Poisson	$\lambda > 0$	$e^{-\lambda} \lambda^k / k!$ $k = 0, 1, 2, 3, \ldots$	λ	λ	$e^{\lambda(e^t - 1)}$
Geometric	$0 < p \leqslant 1$	$p(1-p)^{k-1}$ $k = 1, 2, 3, \ldots$	$\frac{1}{p}$	$\frac{(1-p)}{p^2}$	$\frac{pe^t}{1 - e^t(1-p)}$
Hyper-geometric	n, N, M positive integers, $n \leqslant N + M$	$\binom{N}{k}\binom{M}{n-k} \Big/ \binom{N+M}{n}$ $k = 0, 1, 2, \ldots, \min(N, n)$	$\frac{nN}{N+M}$	$\frac{NMn(N+M-n)}{(N+M)^2 (N+M-1)}$	$\frac{\binom{M}{n}}{\binom{N+M}{n}} F(-n, -N; M-n+1; e^t)$ [See 1 below]
Negative Binomial	r positive $0 < p \leqslant 1$	$\binom{k+r-1}{r-1} p^r (1-p)^k$ $k = 0, 1, 2, \ldots$	$\frac{r(1-p)}{p}$	$\frac{r(1-p)}{p^2}$	$p^r[1 - (1-p)e^t]^{-r}$

Logarithmic	$0 < q < 1$	$\dfrac{-q^k}{[k \log (1-q)]}$ $k = 1, 2, 3, \ldots$	$\dfrac{-q}{(1-q) \log (1-q)}$	$\dfrac{-q[\log (1-q) + q]}{[\log (1-q)]^2 (1-q)^2}$	$\dfrac{\log (1-qe^t)}{\log (1-q)}$
Uniform on $1, \ldots n$	positive integer n	$\dfrac{1}{n}$, $k = 1, 2, \ldots, n$	$\dfrac{(n+1)}{2}$	$\dfrac{(n^2-1)}{12}$	$\dfrac{e^t(1-e^{nt})}{n(1-e^t)}$

B. n-Dimensional Distributions

			Mean	Variances and Covariances	
Multinomial	$p_i \geqslant 0$, $i = 1, \ldots, n$ $\sum_{i=1}^{n} p_i = 1$ m a positive integer	$p(k_1, \ldots, k_n) = \binom{m}{k_1, \ldots, k_n} p_1^{k_1} \cdots p_n^{k_n}$, k_i, $i = 1, \ldots, n$ nonnegative integers, $\sum_{i=1}^{n} k_i = m$ [see 2 below]	$E(k_i) = mp_i$	$\mathrm{Cov}(k_i, k_j) = -mp_ip_j$, $i \neq j$ $\mathrm{Var}(k_i) = mp_i(1-p_i)$	$(p_1 e^{t_1} + \cdots p_n e^{t_n})^m$

1. F is the hypergeometric function, see Ref. 21-5.

2. $\binom{m}{k_1, \ldots, k_n} = \dfrac{m!}{k_1!\, k_2! \cdots k_n!}$ is the multinomial coefficient.

Table 21.2-2 Continuous Probability Distributions.

A. One-Dimensional Distributions

Name	Parameters	Probability Density Functions	Mean	Variance	Moment Generating Function
Normal or Gaussian	$-\infty < \mu < \infty$ $\sigma > 0$	$\frac{1}{\sqrt{2\pi}\,\sigma} e^{-(x-\mu)^2/2\sigma^2}$	μ	σ^2	$e^{t\mu+(\sigma^2 t^2/2)}$
Exponential	$\lambda > 0$	$\frac{1}{\lambda} e^{-x/\lambda}, \quad x > 0$ $0, \quad x \leqslant 0$	λ	λ^2	$(1 - \lambda t)^{-1}$
Gamma	$\alpha > 0$ $\beta > 0$	$\frac{1}{\Gamma(\alpha)\beta^\alpha} x^{\alpha-1} e^{-x/\beta}, \quad x > 0$ $0, \quad x \leqslant 0$	$\beta\alpha$	$\beta^2\alpha$	$(1 - \beta t)^{-\alpha}$
Chi-square or χ^2	n a positive integer	$\frac{x^{(n-2)/2} e^{-x/2}}{2^{n/2}\,\Gamma(n/2)}, \quad x > 0$ $0, \quad x \leqslant 0$	n	$2n$	$(1 - 2t)^{-n/2}$
F or Snedecor's F	m and n positive integers, called degrees of freedom.	$\frac{\Gamma\left(\frac{m+n}{2}\right)\left(\frac{m}{n}\right)^{m/2} x^{(m-2)/2}}{\Gamma\left(\frac{m}{2}\right)\Gamma\left(\frac{n}{2}\right)\left(1+\frac{m}{n}x\right)^{(m+n)/2}}, \quad x > 0$ $0, \quad x \leqslant 0$	$\frac{n}{n-2}, \quad n > 2$	$\frac{2n^2(m+n-2)}{m(n-2)^2(n-4)}, \quad n > 4$	

t or Students' t	n a positive integer, called degrees of freedom.	$\dfrac{\Gamma\left(\frac{n+1}{2}\right)}{\sqrt{n\pi}\,\Gamma\left(\frac{n}{2}\right)\left(1+\frac{x^2}{n}\right)^{(n+1)/2}}$	$0, n > 1$	$\dfrac{n}{n-2}, \quad n > 2$	Characteristic Function: See Ref. 21-64, pp. 74 and 375.
Beta	$\alpha > 0$ and $\beta > 0$	$\dfrac{x^{\alpha-1}(1-x)^{\beta-1}}{B(\alpha,\beta)}, \quad 0 \leqslant x \leqslant 1$ $0, \quad$ other x (see note 1 below)	$\dfrac{\alpha}{\alpha+\beta}$	$\dfrac{\alpha\beta}{(\alpha+\beta)^2(\alpha+\beta+1)}$	$M(\alpha, \alpha+\beta, t)$ (see note 2 below)
Cauchy	$\beta > 0$	$\dfrac{1}{\pi\beta[1+(x/\beta)^2]}$	does not exist	does not exist	Characteristic Function $e^{-\beta\|t\|}$
Uniform	$a < b$	$\dfrac{1}{b-a}, \quad a < x < b$ $0, \quad$ elsewhere	$\dfrac{a+b}{2}$	$\dfrac{(b-a)^2}{12}$	$\dfrac{e^{bt}-e^{at}}{t(b-a)}$
Lognormal	$\alpha > 0$ $-\infty < \beta < +\infty$	$\dfrac{\alpha}{\sqrt{2\pi}\,x} e^{-(1/2)(\alpha \log x+\beta)^2}, \quad 0 < x < \infty$ $0, \quad x \leqslant 0$	$e^{(1/2\alpha^2)-(\beta/\alpha)}$	$e^{[(1/\alpha^2)-(2\beta/\alpha)]}(e^{(1/\alpha^2)}-1)$	

Table 21.2-2 (Continued)

B. n-Dimensional Distributions

			Mean Vector	Covariance Matrix	
Normal or Gaussian, $N(\boldsymbol{\mu}, \Sigma)$	$\boldsymbol{\mu}$ an $n \times 1$ vector and Σ an $n \times n$ positive definite matrix	$\frac{1}{(2\pi)^{n/2}\lvert\Sigma\rvert^{-(1/2)}} e^{-(1/2)(\mathbf{x}-\boldsymbol{\mu})'\Sigma^{-1}(\mathbf{x}-\boldsymbol{\mu})}$ where $\lvert\Sigma\rvert$ is the determinant of Σ. $(\mathbf{x}-\boldsymbol{\mu})'$ is the transpose of $\mathbf{x}-\boldsymbol{\mu}$	$\boldsymbol{\mu}$	Σ	$e^{t'\boldsymbol{\mu}+(1/2)t'\Sigma t}$
Wishart	Let $n = k^2$ and $\boldsymbol{\Sigma}$ be an $k \times k$ positive definite matrix, m a positive integer $\geqslant k$.	Denote the $\mathbf{x}$ vector by $x_{ij},\ i,j = 1, 2, \ldots, k;\ x_{ij} = x_{ji}$. $\frac{\lvert\boldsymbol{\Sigma}^{-1}\rvert^{m/2}\lvert\mathbf{x}\rvert^{(1/2)(m-k-1)} e^{-(1/2)\sum_{i,j=1}^{k}(\boldsymbol{\Sigma}^{-1})_{i,j}x_{i,j}}}{2^{km/2}\pi^{k(k-1)/4}\Gamma\left(\frac{m}{2}\right)\Gamma\left(\frac{m-1}{2}\right)\cdots\Gamma\left(\frac{m-k+1}{2}\right)}$ for x_{ij} such that the $k \times k$ matrix with entries x_{ij} is positive definite. This is the probability density function over that $\frac{k}{2}(k+1)$ dimensional region; 0, elsewhere. The absolute value $\lvert A\rvert$ of a matrix is its determinant.	$m\boldsymbol{\Sigma}$		Let $c_{ij} = (\boldsymbol{\Sigma}^{-1})ij/2$, $\lvert d_{ij}\rvert$ denote the determinant of a $k \times k$ matrix with entries d_{ij} $\frac{\lvert c_{ij}\rvert^{m/2}}{\lvert c_{ij} - \epsilon_{ij}t_{ij}\rvert^{m/2}}$ where $\epsilon_{ij} = \begin{cases} 1 & \text{if } i = j \\ \frac{1}{2} & \text{if } i \neq j \end{cases}$

1. For $B(\alpha, \beta)$ see Chapter 2.
2. M is Kummer's function see Ref. 21-5.

When a moment generating function exists the characteristic function may be found by substituting it for $\mathbf{t}$.

We now discuss briefly the importance of some of these distributions.

Binomial: Let an experiment be independently performed n times where p is the probability of a specified event A happening. If X denotes the number of times A occurs then X has a binomial distribution with parameters n and p. Among the uses of the binomial distribution are quality control, acceptance sampling, public opinion polls, drug trials, and testing for E.S.P.

Poisson: If we consider the number of atomic particles decaying in a given time (from a large number of particles) this number will have approximately a Poisson distribution. The binomial distribution is approximated nicely by a Poisson distribution if n is large and p is small (see section 21.2.4).

Geometric: Independently repeat a task where p is the probability of success until a success occurs. $p(k)$ is the probability the first success occurs on the kth try.

Multinomial: Suppose we independently try an experiment m times. Also suppose that one of n possible outcomes results each time (and only one). If p_i is the probability of the ith outcome then $p(k_1, \ldots, k_n)$ is the probability that outcome one occurs k_1 times, outcome two occurs k_2 times, ..., where $k_1 + \cdots + k_n = m$. If $n = 2$ we have the binomial distribution.

Normal or Gaussian: If a random vector is the sum of many independent small contributions then the random vector will be approximately normal (see section 21.2.4). For this reason many phenomena occurring in nature have approximately a normal distribution. Also for this reason errors are often modeled as normally distributed. Further, for many statistical procedures a large sample leads to procedures that may be based on the normal distribution (see section 21.4.5 on Robustness).

Exponential: Suppose that we are waiting for an event that does not have a memory in the sense that if it has not occurred by a given time then the distribution of the time to occurrence is the same as the original distribution of the time to occurrence. Then this time to occurrence has an exponential distribution. This distribution is often used as the model of the time to failure of a component in an electronic device. Thus it is a cornerstone of reliability theory. The "memoryless property" is described by saying that "an old object is the same as a new one."

The *Chi-square*, F, and t distributions are all related to the normal distribution and will be considered later. The relations are as follows: a) Let $X_1, \ldots, X_n$ be independent $N(0, 1)$ random variables. [$N(0, 1)$ means that they are normal with mean 0 and variance one.] Then $\sum_{i=1}^{n} X_i^2$ has a χ^2 distribution with n degrees of freedom; b) Let X and Y be independent χ^2 random variables with m and n degrees of freedom, respectively. Then $(X/m)/(Y/n)$ is an F random variable with m and n degrees of freedom. [Note that the order in which we state m and n is important!]; c) Let X be a $N(0, 1)$ random variable and Y be a χ^2 random variable with n degrees of freedom then $X/\sqrt{Y/n}$ is a t random variable with n degrees of freedom.

The following results are useful in the distribution theory of statistics and may be verified using the results of section 21.2.1 or Theorem 21.2.1. We use the following abbreviations: r.v. = random variable; i.i.d. = independent identically distributed; ind. = independent, r.vec. = random vector.

(a) Let $\mathbf{X}$ and $\mathbf{Y}$ be ind. $N(\mu_1, \Sigma_1)$ and $N(\mu_2, \Sigma_2)$ r.vec.'s respectively; then $\mathbf{X} + \mathbf{Y}$ is $N(\mu_1 + \mu_2, \Sigma_1 + \Sigma_2)$.

(b) Let $\mathbf{X}$ be a $N(\mu, \Sigma)$ r.vec. in n dimensions then $(\mathbf{X} - \mu)' \Sigma^{-1} (\mathbf{X} - \mu)$ is a χ^2 random variable with n-degrees of freedom if Σ^{-1} exists.

(c) If X is a $N(\mu, \sigma^2)$ r.v. then $(X - \mu)/\sigma$ is a $N(0, 1)$ r.v.

(d) If X and Y are ind. χ^2 r.v.'s with m and n degrees of freedom then $X + Y$ is a χ^2 r.v. with $m + n$ degrees of freedom. Also $X/(X + Y)$ has a beta distribution with parameters $\alpha = m$ and $\beta = n$.

(e) If X has a t distribution with k degrees of freedom then $(1 + t^2/k)^{-1}$ has a beta distribution with $\alpha = k/2$ and $\beta = \frac{1}{2}$.

21.2.4 Limit Theorems

THEOREM 21.2.2 (*Strong Law of Large Numbers*): *Let $X_1, X_2, X_3, \ldots$ be a sequence of i.i.d. r.v.'s with (finite) mean μ; then*

$$P\left(\lim_{n \to \infty} \sum_{i=1}^{n} X_i/n = \mu\right) = 1$$

(*Weak Law of Large Numbers*): *For every $\epsilon > 0$,*

$$\lim_{n \to \infty} P\left(\left|\left(\sum_{i=1}^{n} X_i/n\right) - \mu\right| > \epsilon\right) = 0$$

THEOREM 21.2.3: (*Poisson approximation to the binomial distribution.*) *Let X_λ be a Poisson r.v. with parameter λ and for each n let X_n be a binomial r.v. with parameters n and p_n. If $\lim_{n \to \infty} np_n = \lambda$ then*

$$\lim_{n \to \infty} P(X_n = k) = P(X_\lambda = k), \qquad k = 0, 1, 2, \ldots$$

Definition 21.2.7: Let $F, F_1, F_2, F_3, \ldots$ be distribution functions all on R^n. We say that F_k, $k = 1, 2, \ldots$ converges in distribution to F if for each $\mathbf{x}$ at which $F(\mathbf{x})$ is continuous $\lim_{k \to \infty} F_k(\mathbf{x}) = F(\mathbf{x})$. If $\mathbf{X}, \mathbf{X}_1, \mathbf{X}_2, \ldots$ are n-dimensional random vectors we say that the $\mathbf{X}_k$ converge in distribution to $\mathbf{X}$ if the corresponding distribution functions converge.

THEOREM 21.2.4 (*Central Limit Theorem*): *Let* $\mathbf{X}_1, \mathbf{X}_2, \ldots$ *be i.i.d. random vectors with mean vector* μ *and covariance matrix* Σ. *For each k let* $\overline{\mathbf{X}}_k = \sum_{i=1}^{k} \mathbf{X}_i/k$. *Then* $\sqrt{k}(\overline{\mathbf{X}}_k - \mu)$ *converges in distribution to a* $N(\mathbf{0}, \Sigma)$ *distribution.*

In one dimension the Central Limit Theorem can be written in the following form. Let $X_1, X_2, \ldots$ be i.i.d. r.v.'s with mean μ and variance $\sigma^2 > 0$. Let $-\infty \leqslant a \leqslant b \leqslant +\infty$ then

$$\lim_{k \to \infty} P\left(a \leqslant \sqrt{k}\left(\sum_{i=1}^{k} X_i/k - \mu\right)\bigg/\sigma \leqslant b\right) = \int_a^b \frac{1}{\sqrt{2\pi}} e^{-x^2/2}\, dx$$

21.3 DESCRIPTIVE STATISTICS

The statistical emphasis in this chapter is on mathematical statistics. Statistics is used to help in understanding the world. Even when using mathematical statistics one is almost never absolutely certain that all assumptions hold (and thus the trend to nonparametric statistics to reduce the number of assumptions needed.) When dealing with human populations (e.g., sociology, political science, epidemiology) "clean" experiments almost never are possible. Statistics then, in practice, is an aid to common sense (and a restraint that keeps personal prejudices from leading to misinterpretation of data). An understanding of statistics cannot substitute for, or overcome, a lack of knowledge in the area being studied. To gain an understanding of an area descriptive statistics can be very useful. Indeed a persuasive argument can be advanced that descriptive statistics is more important than mathematical statistics. In this chapter we treat descriptive statistics only briefly because of the smaller mathematical content. Reference 21-6 is excellent concerning the pitfalls of descriptive statistics. See also Ref. 21-67.

Consider a set of numbers $x_1, \ldots, x_n$. Think of choosing a number at random (i.e., each number has probability $1/n$ of being chosen). The *mean* of these numbers is $\bar{x} = \sum_{i=1}^{n} x_i/n$. The *variance* is $\sum_{i=1}^{n} (x_i - \bar{x})^2/n$. A *median* of these numbers is any value m such that at least $\frac{1}{2}$ of the x_i's are less than or equal to m and at least $\frac{1}{2}$ of the x_i's are greater than or equal to m. A *histogram* of the data is given by dividing the line into intervals and counting the number of the x_i in each interval. Over each interval a line is drawn whose height is proportional to the number of points with value in the interval times the length of the interval. Figure 21.3-1 illustrates the histogram.

21.4 STATISTICAL INFERENCE

21.4.1 Basic Idea of Statistical Inference

Suppose that we have two teams we consider to be of equal ability who play a series of twenty games. We might model the situation by assuming the number

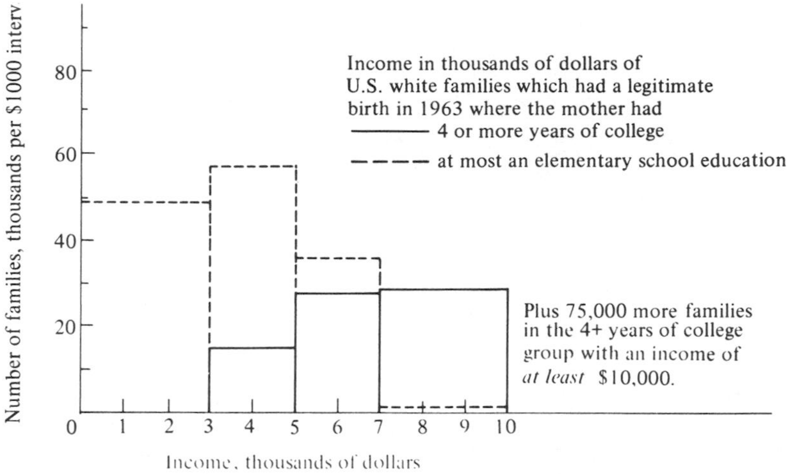

Fig. 21.3-1 Histogram for income level of U.S. white families which had a legitimate birth in 1963 where the mother had:

——————— 4 or more years of college

-------- at most an elementary school education

(Source of numbers: National Center for Health Statistics, Series 22, Number 6, "Educational Attainment of Mother and Family Income: White Legitimate Births.")

of wins by one team, say team A, to be a binomial distribution with parameters $n = 20$ and $p = \frac{1}{2}$. If in fact team A wins 19 of 20 games we might be led to revise our opinion that the two teams are of the same strength. Under the model there is a positive (though very small) probability that team A will win 19 of 20 games. Thus we observe an outcome that can happen if the teams are of equal strength but the small probability of the event being this one-sided leads us to suspect that a larger value of p would have been appropriate.

Roughly speaking this example illustrates the difference between probability and statistics. In probability the emphasis is on working with fixed probability distributions. In statistics the assumption is that we are observing one of a number of possible probability distributions and are trying to infer information as to the correct state of affairs. In the example above we might infer a larger value of p is needed. Thus, *statistical inference* is the process of inferring facts about the underlying distribution from observing random vectors with the distribution.

21.4.2 Parametric Models

Before proceeding with a statistical inference problem the data is often assumed to come from one a given set of distributions. If the class of distributions is naturally described by a finite set of parameters (usually a subset of Euclidean space) then the problem has a parametric model. In the sections to follow the idea will become clearer through examples.

21.4.3 Estimation and Confidence Sets

We will consider the situation where we observe $\mathbf{X}$, a random vector, whose distribution comes from one of a delimited number of possible distributions. In statistical problems even when the distribution of the observed vector $\mathbf{X}$ is completely specified by some parameters it may often be the case that some or all of the parameters are unknown. As we cannot calculate a random vector which explicitly uses an unknown quantity we must restrict ourselves to using only certain random vectors. A *statistic* is a random vector which depends only on observed (or to be observed) random vectors and known quantities. It is often of interest to estimate unknown parameters.

Definition 21.4.1: A statistic $\mathbf{X}$ is an *unbiased* estimate of a parameter vector $\boldsymbol{\theta}$ if $E_{\theta}(\mathbf{X}) = \boldsymbol{\theta}$. (The expected value of a vector is the vector of the expected values.) If X is a r.v. then the *bias* in estimating θ is $E_{\theta}(X - \theta)$. E_{θ} denotes the expected value when X has a probability distribution with θ the true parameter value.

The equation $E(\mathbf{X}) = \boldsymbol{\theta}$ must hold for all possible distributions of $\mathbf{X}$. The bias may change with the distribution of X so that it is a function of the possible distributions for X and not one fixed number.

Example: Let $X_1, \ldots, X_n$ be observed i.i.d. r.v.'s with finite mean μ and variance σ^2. Then $\overline{X} = \sum_{i=1}^{n} X_i/n$, called the sample mean, is an unbiased estimate of μ and $s^2 = \sum_{i=1}^{n} (X_i - \overline{X})^2/(n - 1)$, called the sample variance, is an unbiased estimate of σ^2.

It is also desirable to get more and more accurate estimates as the amount of observed data increases.

Definition 21.4.2: Let $\mathbf{X}_1, \mathbf{X}_2, \mathbf{X}_3, \ldots$ be a sequence of i.i.d. observed k-dimensional random vectors. Suppose that for each n, $\mathbf{S}_n$ depending on $\mathbf{X}_1, \ldots, \mathbf{X}_n$ is an estimate of a vector $\boldsymbol{\theta}$. $\mathbf{S}_n$ is a consistent sequence of estimates for $\boldsymbol{\theta}$ if for each $\epsilon > 0$

$$\lim_{n \to \infty} P(\| \mathbf{S}_n - \boldsymbol{\theta} \| < \epsilon) = 1$$

where $\| \mathbf{S}_n - \boldsymbol{\theta} \|$ is the Euclidean distance between the two vectors.

In the last example $\overline{X}$ and s^2 are consistent estimates of μ and σ^2. In practice, one is not interested in the fact that when the number of observations is extremely large the estimates are quite good with a high probability (as in the definition of consistency). It is desirable to have some idea of the reliability of an estimate based upon the data at hand. It is usually impossible to estimate a parameter with a number in such a way that the probability of the two being equal is very large. What can be done is to choose a set such that with a high probability the parameter is in the set. This idea is formalized in confidence sets.

Definition 21.4.3: Let $P(- \mid \theta)$ denote a probability when the distribution of the observed random vector has θ as the value of a parameter vector. Let θ be an m-dimensional vector and let $\mathbf{S}(\mathbf{X})$ be an m-dimensional set that may found from the observed random vector $\mathbf{X}$ and known quantities. $\mathbf{S}(\mathbf{X})$ is a β *confidence set* if

$$P(\theta \text{ is in } \mathbf{S}(\mathbf{X}) \mid \theta) \geqslant \beta$$

for all possible distributions of $\mathbf{X}$ under consideration.

If the parameter vector is one-dimensional and $S(X)$ is always an interval then $S(X)$ is a *confidence interval.*

A fairly subtle logical point occurs here. The equation in the last situation refers to the situation before $\mathbf{X}$ is observed. In practice $\mathbf{X}$ will have a fixed probability distribution and θ will thus have a specific value, say θ. When we observe $\mathbf{X}$ then "θ is in $\mathbf{S}(\mathbf{X})$" is either a true or false statement which has probability one or zero. *Before* observing $\mathbf{X}$ the probability that θ will be in $\mathbf{S}(\mathbf{X})$ may be β, $0 < \beta < 1$. The interpretation of the *degree of confidence* β is that we've constructed a set $\mathbf{S}(\mathbf{X})$ in such a way that in many trials of the same situation θ would be in $\mathbf{S}(\mathbf{X})$ a fraction β of the time. In any one experiment we do not know which actually happened.

Example: Let $X_1, \ldots, X_n$ be i.i.d. r.v.'s with unknown mean and variance, μ and σ^2. Assuming that n is large we can use the Central Limit Theorem to construct a confidence interval of (approximately) degree of confidence 0.9 for μ.

Let z be such that

$$\int_{-z}^{z} \frac{1}{\sqrt{2\pi}} e^{-x^2/2}\, dx = 0.9$$

Then

$$0.9 \cong P\left(-z \leqslant \frac{\sqrt{n}(\overline{X} - \mu)}{s} \leqslant z\right)$$

$$= P\left(\overline{X} - \frac{zs}{\sqrt{n}} \leqslant \mu \leqslant \overline{X} + \frac{zs}{\sqrt{n}}\right)$$

Thus

$$S(\mathbf{X}) = \left[\overline{X} - \frac{zs}{\sqrt{n}}, \overline{X} + \frac{zs}{\sqrt{n}}\right]$$

is an appropriate confidence interval where the brackets denote a closed interval. (Here we've gone slightly beyond the Central Limit Theorem in using s instead of σ, but this can be justified.)

There is a standard method of choosing estimates of unknown parameters which has quite wide applicability and has very nice properties for large sample sizes (see Ref. 21-7). This is the method of *maximum likelihood estimates.* Suppose that we observe a random vector **X** that is assumed to come from one of a family of distributions indexed by a parameter vector θ. We will assume that **X** is either a discrete or continuous random vector. After observing $\mathbf{X} = \mathbf{x}$ we define the likelihood function $l(\theta)$ to be: a) $l(\theta) = f(\mathbf{x} | \theta)$ when **X** is a continuous random vector and $f(\mathbf{x} | \theta)$ is the density function when θ is the true parameter value; and b) $l(\theta) = p(\mathbf{x} | \theta)$ when **X** is discrete and $p(\mathbf{x} | \theta)$ is the probability mass function when θ is the parameter value. If we can choose $\hat{\theta}$ so that $l(\hat{\theta}) \geqslant l(\theta)$ for all allowable parameter values θ then $\hat{\theta}$ is the maximum likelihood estimate of θ. Note that $\hat{\theta}$ is a function of the value of **X**. The rationale of the maximum likelihood estimate results from the idea that $l(\theta)$ is the probability of observing $\mathbf{X} = \mathbf{x}$ when θ is the parameter vector (in the discrete case). Thus we pick $\hat{\theta}$ to maximize the probability of observing $\mathbf{X} = \mathbf{x}$.

Examples: Let X be a binomial r.v. with parameters n and p; n known. We may assume $X = k$ an integer from 0 through n (if not we have the wrong model). Then $l(p) = \binom{n}{k} p^k (1 - p)^{n-k}$. Taking the derivative and setting it equal to zero we find that $\hat{p} = X/n$. Let $X_1, \ldots, X_n$ be independent normal random variables with mean μ and variance σ^2. If $X_i = x_i$ the likelihood function is

$$l(\mu, \sigma^2) = (2\pi)^{-n/2} e^{-\sum_{i=1}^{n} (X_i - \mu)^2/\sigma^2}$$

Setting the partial derivatives of log $l(\mu, \sigma^2)$ equal to zero one finds $\hat{\mu} = \overline{X}$ and $\hat{\sigma}^2 = \sum_{i=1}^{h} (X_i - \overline{X})^2/n$.

If a parametric model can be constructed maximum likelihood estimates often may be found numerically even when analytical forms are not feasible.

21.4.4 Hypothesis Testing

Suppose that someone enters a poker game (feeling that the game is fair). If this person is consistently beaten by flushes and full houses he may feel that his "luck" is bad. If this continues, however, he will probably come to the conclusion that he is being cheated. This conclusion will be reached even though under a reasonable mathematical model (e.g., all hands dealt independently, in a given deal all possibilities equally likely) the events which occurred had a positive probability. The reasoning is that even though there is a positive probability of such events occurring the probability is so small that it may be discounted. This is an informal example of what is called a hypothesis test. We now proceed to formalize the situation.

In hypothesis testing a data vector **X** will be observed. Based upon the values taken by **X** the experimenter would like to decide whether or not to reject a hypothesis to be taken as the status quo or working model unless the evidence is sufficiently convincing to reject the hypothesis. In order to set this into a mathematical framework the hypothesis must be stated mathematically, that is it must specify something about the distribution of **X**. In a hypothesis testing situation many mistakes occur because the translation from a hypothesis about the "real world" to a statement about the distribution of a random vector is faultily carried out. We won't discuss the point further as our emphasis is on the mathematical side of statistics but in practice it must be born constantly in mind.

Definition 21.4.4: A *hypothesis testing situation* is one in which a random vector **X** is to be observed and the distribution of **X** must come from some specified set of possibilities. A hypothesis is the statement that **X** has a distribution from some subset of the possible distributions. Often the hypothesis will be identified with the subset of distributions specified. If there is a working hypothesis that will require "solid" evidence before it is rejected this hypothesis will be called the *null hypothesis* and denoted by H_0. In a situation where we are only interested in two hypotheses, a null hypothesis and another hypothesis and the two hypotheses are disjoint we call the other hypothesis the *alternative hypothesis* and denote it by H_A. If a hypothesis completely specifies the distribution of **X** (that is, consists of only one distribution) it is called a *simple hypothesis.* Otherwise it is a composite hypothesis.

In this chapter we restrict consideration to the case where there are two hypotheses, H_0 and H_A, which give all possibilities and do not overlap. A hypotheses test is specified by deciding which outcomes for **X** will lead us to reject the null hypotheses. As an aid in giving definitions we introduce the following notation. The possible probability distributions for **X** will be indexed by θ in Θ. Then $\Theta = \Theta_0 \cup \Theta_A$ where Θ_0 corresponds to the indices of distributions possible under the null hypotheses and Θ_A for the alternative hypotheses.

Definition 21.4.5: Consider testing H_0 versus H_A. A *hypothesis test* consists of the set of possible outcomes for **X** for which we reject H_0. This set, say C, is called the *critical region* of the test. If the distribution of **X** comes from H_0, but **X** takes a value in C so that the null hypotheses is rejected then a *type I error* is said to be made. If H_A holds and H_0 is not rejected a *type II error* has been made. A test is said to have *significance level* β if

$$P_\theta(\mathbf{X} \text{ is in } C) \leqslant \beta$$

for each $\theta \in \Theta_0$. P_θ denotes the probability measure when θ is the "true" index value for the probability distribution for **X**. The *power function* of a hypothesis test is a function of each possible distribution for **X** and is defined to be the probability of rejecting H_0, that is, the probability that **X** is in C.

Roughly speaking the significance level of a test is the fraction of the time that the null hypothesis will be rejected when it is true. The value of β quoted for a test is usually chosen as small as possible, but in complex situations the smallest possible value for β is often difficult to find. In constructing a hypothesis test it is usual to decide on β first and then to find an appropriate critical region so that the significance level is as desired. In speaking about the power of a test one means, roughly speaking, how well the power function behaves (i.e., we desire the power function close to one when H_A holds and close to zero when H_0 holds).

Examples:

(1) Let $X_1, \ldots, X_{20}$ be i.i.d. normal random variables with unknown mean μ and unknown variance σ^2. Test $H_0: \mu = 0$ versus $H_A: \mu > 0$ at a .05 significance level. Since we are worried about the mean it makes sense to base the decision on $\overline{X}$. Since we are concerned about a mean which is too large (H_A) we want to reject H_0 when $\overline{X}$ is large. If $X_1, \ldots, X_n$ are i.i.d. $N(\mu, \sigma^2)$ r.v.'s then $t = \sqrt{n}(\overline{X} - \mu)/s$ has a t-distribution so that in our problem if H_0 is true $\sqrt{20}\ \overline{X}/s$ has a t-distribution with 19 degrees of freedom. From tables (for example, Refs. 21-8, 21-9, or 21-69) we find that if Y is a t r.v. with 19 degrees of freedom

$$P(Y \geqslant 1.7291) = .05$$

Thus an appropriate critical region is the set of points such that $\sqrt{20}\ \overline{X}/s \geqslant 1.7291$.

(2) Let X be a binomial r.v. with $n = 24$ and p unknown. Find an appropriate critical region at significance level .10 for testing $H_0: p = \frac{1}{2}$ versus $H_A: p \neq \frac{1}{2}$. Since we are worried about all values of p not equal to $\frac{1}{2}$ we will reject H_0 when the number of successes is too large or too small. It seems reasonable to choose a critical region as symmetrical as possible working in from both ends of the distribution. By using tables we find that the critical region

$$\{0, 1, 2, 3, 4, 5, 6, 7, 8, 17, 18, 19, 20, 21, 22, 23, 24\}$$

has probability .0640 under H_0, but that if we try to add 16 to the critical region the significance level exceeds .10.

(3) Let X be a binomial r.v. with $n = 20$. Test $H_0: p \leqslant \frac{1}{2}$ versus $H_A: p > \frac{1}{2}$ at significance level .20. Plot the power function (as a function of p). We want to reject when X is too large. From tables

$$\{13, 14, 15, 16, 17, 18, 19, 20\}$$

has probability .1316 when $p = \frac{1}{2}$ (and less than this probability when $p < \frac{1}{2}$) and if we try to add 12 to the critical region the probability jumps to .2517 when $p = \frac{1}{2}$, above the required significance level. The power function is

$$f(p) = P(\text{Reject } H_0 \mid p) = P(X \geqslant 13 \mid p)$$
$$= \sum_{k=13}^{20} \binom{20}{k} p^k (1 - p)^{20-k}$$

See Fig. 21.4-1.

From these examples we see that when a hypothesis test is based on a random variable that the critical region often has the form of an interval or union of intervals, in this case a special terminology has been introduced.

Definition 21.4.6: Suppose that a *hypothesis test* is based on a statistic Y. If the critical region is of the form $(c, +\infty)$, $[c, +\infty)$, $(-\infty, c)$ or $(-\infty, c]$ for some number c then the test is a *one-tailed test* and the critical region is a *one-tailed critical region*. If the critical region is of the form $(-\infty, c) \cup (d, +\infty)$ or some modification of this having one or both endpoints included the test is said to be a *two-tailed test* and the critical region a *two-tailed critical region*. When one or two-tailed critical regions are used the end points of the intervals are called *critical values*.

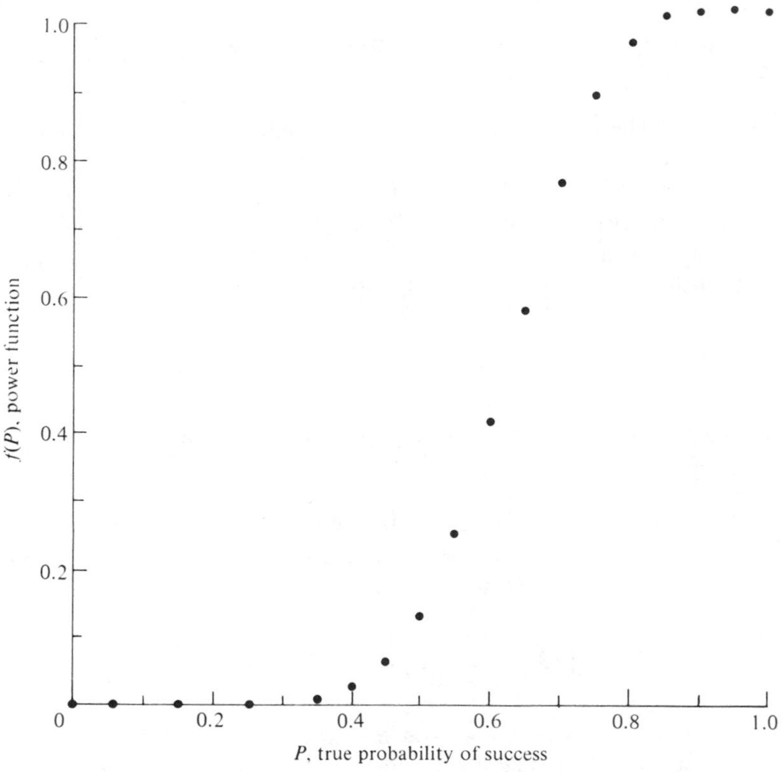

Fig. 21.4-1 The power function, $f(p)$, as a function of p, the true probability of success, for ex. 3, Definition 21.4.5.

In a specific situation the question of finding an appropriate test arises. This question has been studied extensively in the statistical literature. References 21-10, 21-11, 21-14, and 21-18 give a good introduction (at an advanced mathematical level) to such topics. This selection ends with two results in this direction.

THEOREM 21.4.1: *(Neyman-Pearson lemma). Consider testing H_0 versus H_A where both hypotheses are simple hypotheses. Suppose that* $\mathbf{X}$ *is a continuous or discrete random vector. Let $g_0(g_A)$ be the probability density function or probability mass function when $H_0(H_A)$ is true. Suppose that C is a critical region such that there is a constant ξ such that if* $\mathbf{x}$ *is in C, $g_0(\mathbf{x}) \leqslant \xi \cdot g_A(\mathbf{x})$; if* $\mathbf{x}$ *is not in C, $g_0(\mathbf{x}) \geqslant \xi \cdot g_A(\mathbf{x})$. Then among all critical regions with significance level $P(X \text{ in } C|H_0)$, C minimizes $P(X \text{ is not in } C|H_A)$.*

This theorem illustrates the general principle that there is a trade-off between the probabilities of type I and type II errors. If they cannot both be lowered to the levels desired then more data must be collected, the experiment at hand is not appropriate to the aims desired. There is a cost to gathering data so that an experimental design is usually a compromise between conflicting aims: low probabilities of type I and II errors and low cost in performing the experiment or collecting data.

Example: Let $X_1, \ldots, X_n$ be i.i.d. exponential random variables. Find the Neyman-Pearson critical regions for testing $H_0: \lambda = 1$ versus $H_A: \lambda = 3$. We may assume that all the $X_i = x_i$ are positive (or we have the wrong model). Then

$$g_0(x_1, \ldots, x_n) = e^{-\sum_{i=1}^{n} x_i} \text{ and } g_A(x_1, \ldots, x_n) = \left(\frac{1}{3}\right)^n e^{-\sum_{i=1}^{n} x_i/3}$$

By taking logs the critical regions satisfy

$$\bar{x} = \sum_{i=1}^{n} x_i/n \geqslant 3 \log(\xi/3^n)/2n$$

Thus, we reject H_0 when the sample mean is large which is intuitively reasonable as the test is that the true mean is 1 against an alternative of 3.

The Neyman-Pearson approach has an analog in the case where the hypotheses may be composite. This is the *likelihood ratio test* approach. Suppose that $\mathbf{X}$ is discrete or continuous with likelihood function $l(\theta)$. The null hypotheses is θ in Θ_0. The likelihood ratio statistic is

$$\lambda = \frac{\max\limits_{\theta \text{ in } \Theta_0} l(\theta)}{\max\limits_{\theta \text{ in } \Theta_0 \cup \Theta_A} l(\theta)}$$

(Strictly speaking the maximum should be replaced by the supremum if the maximum is not attained.) The critical region is specified by $\lambda \leqslant \xi$ for a fixed constant ξ.

Example: Let $X_1, \ldots, X_n$ be i.i.d. $N(\mu, \sigma^2)$ r.v.'s. Find the likelihood ratio statistic for testing H_0: $\mu = 1$ versus H_A: $\mu \neq 1$ where σ^2 is also unknown. The parameter vector is $\theta = (\mu, \sigma^2)'$. Under H_0 the likelihood function is maximized by $\hat{\sigma}^2 = \sum_{i=1}^{n} (X_i - 1)^2/n$ (μ is restricted to be 1). Under H_A the likelihood function is maximized by $\tilde{\mu} = \bar{x}$ and $\tilde{\sigma}^2 = \sum_{i=1}^{n} (x_i - \bar{x})^2/n$. From this the likelihood ratio statistic is found to be

$$\lambda = \{1 + [\sqrt{n}\,(\bar{x} - 1)/s]^2/(n - 1)\}^{-n/2}$$

λ increases as $\sqrt{n}\,(\bar{x} - 1)/s$ decreases in absolute value. Thus, the likelihood ratio test is based on $\sqrt{n}\,(\bar{x} - 1)/s$ which has a t distribution with $n - 1$ degrees of freedom under the null hypothesis.

Often in statistical literature only the null hypothesis is specified. In this case it is understood that the alternative is "anything other than the null hypothesis."

It will be noticed that in many of the examples i.i.d. random vectors have been used. The model is prevalent because it may be used to represent (independent) repetitions of the same experiment.

Definition 21.4.7: A *random sample of size n* (or a sample of size n) is a set of n i.i.d. random vectors.

The term random sample is not always used this way in the literature and care must be taken to look at the context in which the term is used in order to be certain of its meaning in any particular usage.

In this section the formalism of hypothesis testing has been considered. Consideration is now turned to some of the more practical aspects of using estimation or hypothesis testing procedures for statistical inference. Often a sample is taken from a finite population—i.e., the random vector is associated with one of a finite number of objects. In particular, studies dealing with human populations usually arise in this way. It is an obvious but crucial point that when sampling from a finite population the statistical inference is valid only for the population being sampled. Any extension involves some assumption such as: the physical properties of matter are the same anywhere. The assumptions we use in statistical techniques are crucial. For example, repeated measurements with a biased measuring instrument will not compensate in any way for the fixed bias. In this case, the error of measurement is not independent between experimental trials.

21.4.5 Relationship between Hypothesis Testing and Confidence Sets

Suppose that a parameter vector θ is being considered. Suppose $\mathbf{S}(\mathbf{X})$ gives a β-confidence set; then if a test of significance level 1-β is desired for the hypothesis

$\theta = \theta_0$ this may be accomplished by using $\{\mathbf{x}: \theta_0$ is not in $\mathbf{S}(\mathbf{x})\}$ as the critical region. Conversely suppose that for each θ we have a test of significance level α (i.e., where the null hypothesis is that θ is the correct value of the parameter vector). Given that $\mathbf{X} = \mathbf{x}$ form a confidence set consisting of the θ such that $\mathbf{x}$ is not in the critical region of the test then we have a $1 - \alpha$ confidence set. From this it is seen that hypothesis testing and confidence set theory are essentially the same subject.

21.4.6 Efficiency, Power, and Robustness

In estimation problems one would like an estimate that has high probability of being close to the true value of the quantity being estimated. If we have two unbiased estimates of a real valued quantity then their relative efficiency is defined as the ratio of the two variances. The efficiency of an unbiased estimate in certain situations is the ratio of a theoretical lower bound for the variance (the Cramer-Rao lower bound, Refs. 21-7, 21-12, 21-68, and 21-69) to the variance. In hypothesis testing the power function describes the characteristics of the test. If one test has a power function that is better than the power function for a second test the first test is said to be more powerful than the second. (In this context better means that the first power function is less than or equal to the second when the null hypothesis is true and greater than or equal to the second when the alternative hypothesis is true, with at least one strict inequality.) Much of the development of statistics has been concerned with finding efficient and powerful procedures. In these situations there are assumptions (e.g., i.i.d. random normal vectors.) In practice, of course, all the assumptions are very rarely satified. A trend in modern statistics has been towards finding robust procedures. Robustness is a semimathematical concept. A procedure is robust if it still works "fairly well" when the assumptions are "not quite" satisfied. Or one procedure is more robust than another if its performance holds over a wider class of models. Or a type of robustness results if a test behaves well as the sample size increases. Section 21.7 returns to the topic of robustness.

21.5 THE GENERAL LINEAR MODEL

21.5.1 The Model

The section concerns some applications of one form of the general univariate linear model. Independent normal random variables are observed having means which are a linear combination of p unknown parameters. The random variables are assumed to have the same variance. Thus $Y_i, i = 1, 2, \ldots, n$ are observed

$$Y_i = x_{i1}\beta_1 + x_{i2}\beta_2 + \cdots + x_{ip}\beta_p + e_i \qquad i = 1, 2, \ldots, n \qquad (21.5\text{-}1)$$

where the x_{ij}'s are known constants, the $\beta_1, \ldots, \beta_p$ are unknown parameters and the e_i are i.i.d.$N(0, \sigma^2)$ random variables where σ^2 is unknown.

It is more natural to write the equations (21.5-1) in matrix form. Let

$$\mathbf{Y} = \begin{pmatrix} Y_1 \\ \vdots \\ Y_n \end{pmatrix}; \quad \beta = \begin{pmatrix} \beta_1 \\ \vdots \\ \beta_p \end{pmatrix}; \quad \mathbf{e} = \begin{pmatrix} e_1 \\ \vdots \\ e_n \end{pmatrix}, \quad X = \begin{pmatrix} x_{11} & x_{12} & \cdots & x_{1p} \\ x_{21} & & & \\ \cdots & \cdots & \cdots & \cdots \\ x_{n1} & \cdots & \cdots & x_{np} \end{pmatrix} \tag{21.5-2}$$

Then $\mathbf{Y} = X\beta + \mathbf{e}$ where $\mathbf{X}$ is a known $n \times p$ matrix, β is an unknown $p \times 1$ vector and e is $N(\mathbf{0}, \sigma^2 I)$ where σ^2 is unknown. Equivalently, $\mathbf{Y}$ is $N(X\beta, \sigma^2 I)$.

Maximum likelihood estimation of the unknown β and σ^2 gives the following. Any $\hat{\beta}$ which satisfies the following is a maximum likelihood estimate for β.

$$\sum_{i=1}^{n} \left(Y_i - \sum_{j=1}^{n} x_{ij}\hat{\beta}_j \right)^2 = \min_{\beta} \sum_{i=1}^{n} \left(Y_i - \sum_{j=1}^{n} x_{ij}\beta_j \right)^2 \tag{21.5-3}$$

Thus, the maximum likelihood estimates minimize the sum of the squares of the deviation (or *residual*) between Y_i and its estimated expected value $\sum_{j=1}^{n} x_{ij}\hat{\beta}_j$. The maximum likelihood estimates give an example of a least squares fit. That is, forgetting normality, a reasonable way to fit a vector $\mathbf{Y}$ with a vector of the form $X\beta$ is to minimize the sum of squares of the true minus fitted values. Let $||\cdot||^2$ represent the length of a vector in Euclidean space. If $Z = \begin{pmatrix} z_1 \\ \vdots \\ z_n \end{pmatrix}$ then $||Z||^2 = \sum_{i=1}^{n} z_i^2$. The least squares or maximum likelihood estimates of the β's satisfy

$$||\mathbf{Y} - X\hat{\beta}||^2 = \min_{\beta} ||\mathbf{Y} - X\beta||^2 \tag{21.5-4}$$

Another way of thinking of $\hat{\beta}$ is that it is chosen to make $X\beta$ as close as possible to $\mathbf{Y}$ (in n-dimensional Euclidean space).

β is a least squares estimate if and only if it satisfies the normal equations (where X' is the transpose of X)

$$X'X\beta = X'\mathbf{Y}$$

The maximum likelihood estimates are unique for all possible $\mathbf{Y}$ if and only if X has rank p. In this case

$$\hat{\beta} = (X'X)^{-1}X'\mathbf{Y} \tag{21.5-5}$$

In this case $\hat{\beta}$ is a $N[\beta, \sigma^2(X'X)^{-1}]$ random vector when β is the true value of the parameter vector.

Here, σ^2 may be estimated by using the fact that $\|\mathbf{Y} - X\hat{\beta}\|^2$ has a distribution equal to σ^2 times a Chi-square random variable with $n - r$ degrees of freedom where r is the rank of the matrix X. Let $s^2 = \|\mathbf{Y} - X\hat{\beta}\|^2/(n - r)$. Then s^2 is an unbiased estimate of σ^2 and a confidence set for β (when X has rank p) with degree of confidence α is given by

$$\{\beta: (\hat{\beta} - \beta)'X'X(\hat{\beta} - \beta) \leqslant ps^2 F_{1-\alpha;p,n-p}\}$$

where $F_{1-\alpha;p,n-p}$ is such that if Z is an F random variable with p and $n - p$ degrees of freedom, $P(Z \geqslant F_{1-\alpha;p,n-p}) = 1 - \alpha$.

In many situations the parameter vector β is subject to some linear restrictions, say, $H\beta = \mathbf{0}$ where H is a $k \times p$ matrix of rank k and $\mathbf{0}$ the $k \times 1$ zero vector. The estimation and confidence procedures may be extended to this case (see Refs. 21-10, 21-13, 21-14, 21-15, 21-70, and 21-74).

Hypothesis testing problems usually come in the following form: β is restricted as above, $H\beta = \mathbf{0}$. The null hypothesis is that some further linear restrictions hold, say, $M\beta = \mathbf{0}$. Another way of stating this is that β belongs to some subspace, say V, of p-dimensional Euclidean space. The null hypothesis is that β belongs to $V_0 \subseteq V$ where V_0 is a subspace of V. Let the dimension of V be s and the dimension of V_0 be r ($p \geqslant s > r \geqslant 0$). There are now maximum likelihood estimates under two situations. Define $\hat{\beta}$ and $\tilde{\beta}$ by

$$\begin{aligned} \|\mathbf{Y} - X\hat{\beta}\|^2 &= \min_{\beta \text{ in } V} \|\mathbf{Y} - X\beta\|^2 \\ \|\mathbf{Y} - X\tilde{\beta}\|^2 &= \min_{\beta \text{ in } V_0} \|\mathbf{Y} - X\beta\|^2 \end{aligned} \qquad (21.5\text{-}6)$$

Define F as follows:

$$F = \frac{(\|\mathbf{Y} - X\tilde{\beta}\|^2 - \|\mathbf{Y} - X\hat{\beta}\|^2) \cdot (n - s)}{\|\mathbf{Y} - X\hat{\beta}\|^2 (s - r)} \qquad (21.5\text{-}7)$$

Under the null hypothesis F has an F-distribution with $s - r$ and $n - s$ degrees of freedom and the null hypothesis is tested by rejecting when F is too large. This test is equivalent to the likelihood ratio test. In the remainder of the section some examples of the general linear model are considered.

21.5.2 Multiple Regression Analysis

In the simplest situation a quantity x_i is set or measured and the random variable Y_i is a linear function of functions of x_i plus a random error. The important point here is that for the linear model to hold things must be linear in the unknown parameters but not in terms of the known quantity x_i. Thus, all of the following may use the general linear model. Here the e_i are i.i.d. $N(0, \sigma^2)$ random variables, $i = 1, 2, \ldots, n$.

$$Y_i = \beta_1 + \beta_2 x_i + e_i$$

$$Y_i = \beta_1 + \beta_2 x_i + \beta_3 x_i^2 + \cdots + \beta_{k+1} x_i^k + e_i; \quad k \text{ a positive integer}$$

$$Y_i = \beta_1 + \beta_2 \cos x_i + \beta_3 \sin x_i + \beta_4 \cos 2x_i + \beta_5 \sin 2x_i + \beta_6 x_i + \beta_7 x_i^2 + e_i$$

$$Y_i = \beta_1 + \beta_2 e^{x_i} + e_i$$

(21.5-8)

The point of these equations is that the model is linear in the unknown β_i's, but not the x_i's. An excellent reference for regression problems is Ref. 21-16. In particular this reference gives useful suggestions for determining the form of the regression equation. The variances of the error terms have been assumed the same for each observation. This assumption and that of normality may be evaluated by looking at the residuals (the Y_i minus the least-squares fit). References 21-16 and 21-18 give further details. If a polynomial in the X_i is desired and an upper bound on the degree of the polynomial is available there are sequential procedures for determining the degree of the polynomial (see, for example, Ref. 21-16).

In multiple regression the assumption is that Y is a linear combination of the values of p quantities $x_1, \ldots, x_p$ which may result from a polynomial in one quantity x (i.e., $x_1 = 1, x_2 = x, \ldots, x_p = x^{p-1}$ so $\sum_{i=1}^{m} \beta_i x^{i-1}$). The $x_1, \ldots, x_p$ may be p distinct quantities (e.g., time, temperature, pressure). The $x_1, \ldots, x_p$ may be used to represent a more complicated function of smaller number of more "fundamental" quantities (e.g., $\beta_1 x_1 + \beta_2 x_2 + \cdots + \beta_6 x_6 = \beta_1 + \beta_2 x + \beta_3 x^2 + \beta_4 z + \beta_5 z^2 + \beta_6 xz$). If at the ith observation the values of $x_1, \ldots, x_p$ are denoted $x_{i1}, \ldots, x_{ip}$ then the model is

$$Y_i = \beta_1 x_{i1} + \cdots + \beta_p x_{ip} + e_i$$

the general linear model.

Some off-shoots of the multiple regression technique are discussed in section 21.6.1.

21.5.3 Analysis of Variance

One of the most useful statistical techniques was originally developed by R. A. Fisher and called the analysis of variance (ANOVA). The term arises because the variability in data in certain situations may be divided up into a sum of parts the various parts each accounting for some of the variance. Thus, there is an analysis of the variance. The mathematics is presented in Refs. 21-72 and 21-15. Precalculus introductions are Refs. 21-20 and 21-52. In addition most experimental design texts are by and large expositions of the ANOVA. In an article of this type an extended discussion is not feasible, but this section closes with some examples.

Examples:

(1) *One-way classification*—Consider a quantity that may take any one of S states. At the jth state n_j observations are taken. The model then is

$$Y_{ji} = \beta_j + e_{ji}; \qquad i = 1, \ldots, n_j, \; j = 1, \ldots, S$$

where the e_{ji} are i.i.d. $N(0, \sigma^2)$ r.v.'s. For notational convenience let a dot in place of a subscript denote averaging over that subscript. For example,

$$\beta. = \sum_{j=1}^{S} \beta_j/S, \quad Y_j. = \sum_{i=1}^{n_j} Y_{ji}/n_j, \quad \overline{Y} = \sum_{j=1}^{S} \sum_{i=1}^{n_j} Y_{ji} \bigg/ \sum_{j=1}^{S} n_j$$

Without loss of generality the model may be rewritten as

$$Y_{ji} = \mu + \alpha_j + e_{ji}$$

where $\alpha. = 0$. ($\mu = \beta.$, $\alpha_j = \beta_j - \beta.$)
Consider testing the hypothesis that

$$\alpha_1 = \alpha_2 = \cdots = \alpha_S = 0$$

This is the hypothesis that all S levels of the quantity being set have the same effect on the observed variable Y. Another way of stating the hypothesis is that random samples from S normal populations having the same variance are taken and that the hypothesis is that all S populations have the same mean.

The F statistic for this hypothesis turns out to be

$$F = \frac{\displaystyle\sum_{j=1}^{S} n_j (Y_j. - \overline{Y})^2 \cdot \left(\sum_{j=1}^{S} n_j - S \right)}{\displaystyle\sum_{j=1}^{S} \sum_{i=1}^{n_j} (Y_{ji} - Y_j.)^2 (S - 1)}$$

(2) *Two-way classification* ($m > 2$ *observations per cell*)—Consider an outcome that is effected by the levels of two quantities the first having R possible levels and the second having C possible levels. Assume that there are m observations at each possible setting with i.i.d. $N(0, \sigma^2)$ errors.

$$Y_{ijk} = \beta_{ij} + e_{ijk}; \qquad i = 1, \ldots, R; \; j = 1, \ldots, C; \; k = 1, \ldots, m$$

This may be rewritten as follows:

$$Y_{ijk} = \mu + \alpha_i + \delta_j + \gamma_{ij} + e_{ijk}$$

where $\alpha_{.} = \delta_{.} = \gamma_{i.} = \gamma_{.j} = 0$. ($\mu = \beta_{..}$, $\alpha_i = \beta_{i.} - \beta_{..}$, $\delta_j = \beta_{.j} - \beta_{..}$, $\gamma_{ij} = \beta_{ij} - \beta_{i.} - \beta_{.j} + \beta_{..}$). The γ_{ij}'s are called the interaction terms; α_i is the average effect when quantity one is at level i.

Consider testing the null hypothesis that there is no interaction (that is, $\gamma_{ij} = 0$ for all i and j). This hypothesis is the same as saying that the effects of the two factors are additive. The F statistic is found to be

$$F = \frac{m(m-1)RC \sum_{i=1}^{R} \sum_{j=1}^{C} (Y_{ij.} - Y_{i..} - Y_{.j.} + Y_{...})^2}{(R-1)(C-1) \sum_{i=1}^{R} \sum_{j=1}^{C} \sum_{k=1}^{m} (Y_{ijk} - Y_{ij.})^2}$$

which under the null hypothesis is an F random variable with $(R-1)(C-1)$ and $(m-1)RC$ degrees of freedom.

For more complex ANOVA techniques the reader is referred to the references mentioned above.

21.6 SOME OTHER TECHNIQUES OF MULTIVARIATE ANALYSIS

21.6.1 Correlation, Partial and Multiple Correlation Coefficients

In this section the concept of a conditional distribution is needed. Roughly speaking, if $X_1, \ldots, X_n$ are random variables then the conditional distribution of $X_1, \ldots, X_k$ given that $X_{k+1} = x_{k+1}, \ldots, X_n = x_n$ is the distribution of $X_1, \ldots, X_k$ "updated" by taking into account the values of the other X_i's. That is, for any Borel subset A of k-dimensional Euclidean space the conditional probability distribution assigns a measure to the set A equal to

$$P[(X_1, \ldots, X_k) \text{ in } A \mid X_{k+1} = x_{k+1}, \ldots, X_n = x_n]$$

(The "roughly speaking" phrase results from the fact that the conditioning event may have probability zero so that a more sophisticated concept of conditional probability is needed.)

If $X_1, \ldots, X_n$ are continuous then the conditional distribution of $X_1, \ldots, X_k$ is continuous with density function

$$f(x_1, \ldots, x_k) = \frac{f(x_1, \ldots, x_k, \ldots, x_n)}{\int_{-\infty}^{+\infty} \cdots \int_{-\infty}^{+\infty} f(x_1, \ldots, x_n)\, dx_1 \cdots dx_k} \tag{21.6-1}$$

whenever $X_{k+1} = x_{k+1}, \ldots, X_n = x_n$ is such that

$$\int_{-\infty}^{+\infty} \cdots \int_{-\infty}^{+\infty} f(x_1, \ldots, x_n)\, dx_1 \cdots dx_k > 0$$

In particular (see e.g., Ref. 21-21) let $(X_1, \ldots, X_n)$ be $N(\mu, \Sigma)$ where Σ is positive definite. The distribution of

$$\mathbf{X} = \begin{pmatrix} x_1 \\ \cdot \\ \cdot \\ \cdot \\ x_k \end{pmatrix} \text{ given that } \begin{pmatrix} X_{k+1} \\ \cdot \\ \cdot \\ \cdot \\ X_n \end{pmatrix} = \begin{pmatrix} x_{k+1} \\ \cdot \\ \cdot \\ \cdot \\ x_n \end{pmatrix} = \mathrm{x} \text{ and } \Sigma = \begin{pmatrix} \Sigma_{11} & \Sigma_{12} \\ \Sigma_{21} & \Sigma_{22} \end{pmatrix}$$

where Σ_{11} is the $k \times k$ covariance matrix of the unconditional distribution of $\mathbf{X}$ is $N[\mu_1 + \Sigma_{12}\Sigma_{22}^{-1}(\mathrm{x} - \mu_2), \Sigma_{11} - \Sigma_{12}\Sigma_{22}^{-1}\Sigma_{21}]$ where

$$\mu_1 = \begin{pmatrix} \mu_1 \\ \cdot \\ \cdot \\ \cdot \\ \mu_k \end{pmatrix} \text{ and } \mu_2 = \begin{pmatrix} \mu_{k+1} \\ \cdot \\ \cdot \\ \cdot \\ \mu_n \end{pmatrix}$$

In this section the term best linear predictor will be used. By the best linear predictor of Y by $X_1, \ldots, X_m$ we mean the linear combination $\sum_{i=1}^{m} a_i X_i$ which minimizes $E[(Y - \sum_{i=1}^{m} a_i X_i)^2]$. It is assumed that all random variables used in this section have a finite variance and are adjusted to have mean zero. (An equivalent procedure is to allow the linear predictors to have a constant term so that the mean of Y is subtracted out.)

Correlation Coefficient—The correlation coefficient of X and Y has been previously defined as

$$\rho_{X,Y} = \frac{\mathrm{Cov}\,(X, Y)}{\sqrt{\mathrm{Var}\,(X)\,\mathrm{Var}\,(Y)}} \tag{21.6-2}$$

Let $Z = a + bX$ be the best linear predictor for Y in terms of X. Then

$$\mathrm{Var}\,(Y) = \mathrm{Var}\,(Y - Z) + \mathrm{Var}\,(Z)$$

Thus, the variance of Y is equal to the variance explained by X plus the residual variation left in Y. Then

$$\rho_{X,Y}^2 = \mathrm{Var}\,(Z)/\mathrm{Var}\,(Y) \tag{21.6-3}$$

The correlation coefficient squared is the fraction of the variance of Y that can be explained by X (or visa versa).

Partial Correlation Coefficient—Often in studying multivariate statistics (that is statistics involving more than one random variable) it is desired to study the relationship between some random variables after somehow trying to adjust out or eliminate the effect of other variables that interfere. The partial correlation coefficient is the correlation between two random variables after adjusting out the effect of other variables. There are two useful interpretations of the partial correlation coefficient.

Let $X_1, X_2, \ldots, X_n$ be random variables and let Y be the best linear predictor for X_1 in terms of $X_3, \ldots, X_n$ and Z be the best linear predictor for X_2 in terms of $X_3, \ldots, X_n$. The partial correlation coefficient of X_1 and X_2 after adjusting for $X_3, \ldots, X_n$ is defined to be the correlation coefficient of $X_1 - Y$ and $X_2 - Z$. It is denoted by $\rho_{1,2.3,4,\ldots,n}$.

The sense in which the variables $X_3, \ldots, X_n$ have been adjusted for is to take out their linear predictive ability. If they are related to X_1 and X_2 in an extremely nonlinear fashion this technique would appear less useful than otherwise.

Suppose now that the random vector $(X_1, \ldots, X_n)'$ has a multivariate normal distribution. From the facts on conditional distributions given above it is seen that conditionally upon fixed values of $X_3, \ldots, X_n$ the conditional covariance and variances of X_1 and X_2 do not depend upon which values $X_3, \ldots, X_n$ take. Thus the conditional correlation coefficient of X_1 and X_2 does not depend upon which values the conditioning variables take. This value is also the partial correlation coefficient. For normal random variables the partial correlation coefficient is the correlation when the adjusting variables take on fixed values. For other multivariate distributions the conditional values of the correlation coefficient may change with the values taken on by the conditioning $X_3, \ldots, X_n$.

Multiple Correlation Coefficient—The multiple correlation coefficient between a random variable X_1 and a set of random variables $X_2, \ldots, X_n$ is defined as follows. Let Z be the best linear predictor for X_1 in terms of $X_2, \ldots, X_n$. The square of the multiple correlation coefficient is $\text{Cov}(X_1, Z)/\text{Var}(X_1)$. The multiple correlation coefficient, denoted by $\rho_{1(2,\ldots,n)}$, is the positive square root of $\text{Cov}(X_1, Z)/\text{Var}(X_1)$. The square of the multiple correlation coefficient is often denoted by R^2 (capital R squared); R^2 is the fraction of the variability of X_1 that may be explained (in a linear fashion) by $X_2, \ldots, X_n$.

Partial Multiple Correlation Coefficient—Consider random variables $X_1, X_2, \ldots, X_k, X_{k+1}, \ldots, X_n$. It might be desired to find the strength of the relationship between X_1 and $X_2, \ldots, X_k$ after adjusting for the effect of $X_{k+1}, \ldots, X_n$. Let $Z_1, \ldots, Z_k$ be $X_1, \ldots, X_k$ after subtracting out the best linear predictors in terms of $X_{k+1}, \ldots, X_n$. The multiple correlation coefficient of Z_1 with $Z_2, \ldots, Z_n$ is the partial multiple correlation coefficient, denoted by $\rho_{1(2,\ldots,k)\cdot k+1,\ldots,n}$. In other words, $\rho^2_{1(2,\ldots,k)\cdot k+1,\ldots,n}$ is the fraction of the residual variance left in X_1 after using $X_{k+1},\ldots, X_n$ (to predict linearly) that may be explained by $X_2,\ldots,X_k$.

Example: To illustrate these ideas a topic from multiple regression is used. Consider the relationship of X_1 to $X_2, \ldots, X_n$. To see which variable has the most effect on X_1 select the X_j which maximizes ρ_{1j}^2, that is, select the variable which explains as much of the variance of X_1 as possible. Using a foreward selection procedure after selecting the first X_j, say X_{j_0} choose the next most important random variable, say X_k to maximize $\rho_{1(j_0,k)}^2$. The foreward selection procedure then keeps adding random variables in succession. It is not necessarily true that the first two random variables selected maximize $\rho_{1(j,k)}^2$ where $j \neq k$, $j, k = 2, \ldots, n$. The computation time becomes prohibitive however, if all possible combinations are considered at each step.

In practice the partial and multiple correlations are not known but must be estimated. The distribution theory of such estimates will not be discussed here. In this foreward selection procedure "partial F tests" may be applied to see if the new variables added account for more variation than would be expected by chance. (see Ref. 21-16).

21.6.2 Principal Component Analysis

Let $X_1, \ldots, X_n$ have covariance matrix Σ. Let

$$Y = \sum_{i=1}^{n} a_i X_i, \qquad \mathbf{a} = \begin{pmatrix} a_1 \\ \cdot \\ \cdot \\ \cdot \\ a_n \end{pmatrix}$$

Then $\text{Var}(Y) = \mathbf{a}'\Sigma\mathbf{a}$. If $\mathbf{a}'\mathbf{a} = 1$ the vector $\mathbf{a}$ may be thought of as giving a direction in n-dimensional Euclidean space. Choose $\mathbf{a}^1$ subject to $\mathbf{a}'\mathbf{a} = 1$ to maximize $\mathbf{a}'\Sigma\mathbf{a}$. The random variable $Y_1 = \sum_{i=1}^{n} a_i^1 X_i$ is the *first principal component*; Y_1 is the linear combination of the X_i which has the largest possible variance (subject to $\mathbf{a}'\mathbf{a} = 1$). The $\text{Var}(Y_1)$ is the largest eigenvalue of Σ. If we observe $X_i = x_i$, $i = 1, 2, \ldots, n$ then $y = \sum_{i=1}^{n} a_i^1 x_i$ is the *score* of the first principal component. Similarly, the second principal component is $\sum_{i=1}^{n} a_i^2 X_i$ where $\mathbf{a}^2$ maximizes $\mathbf{a}'\Sigma\mathbf{a}$ subject to $\mathbf{a}^{2\prime}\mathbf{a}^2 = 1$ and $\mathbf{a}^{2\prime}\mathbf{a}^1 = 0$.

In general, let $\lambda_1 \geqslant \lambda_2 \geqslant \cdots \geqslant \lambda_n$ be the eigenvalues of Σ with associated eigenvectors $\mathbf{a}^1, \ldots, \mathbf{a}^n$ where $\mathbf{a}^{j\prime}\mathbf{a}^j = 1$ and the $\mathbf{a}^j$'s are orthogonal. The jth principal component is $\sum_{i=1}^{n} a_i^j X_i$. The principal components are unique if the eigenvalues are distinct; λ_j is the variance of the jth principal component.

The rationale behind the method is that the directions of greatest variability represent the most important part of the data. This assumption needs careful scrutiny and is probably not valid in many cases.

In dealing with real data the principal components are estimated from the sample covariance matrix. The scores on the first two principal components are often plotted in a *scatter diagram.* The scatter diagram is the best two dimensional representation of the data in the sense of maximizing the sum of the squares of the lengths of the projections of the data onto the two dimensional subspace. (This holds after adjusting the data to have mean zero.) See Refs. 21-22, 21-23, and 21-73 for principal component analysis.

21.6.3 Factor Analysis

Factor analysis is a model which is used to reduce the dimensionality of data to a smaller number of underlying factors. Suppose that one is observing a random sample of n-dimensional vectors $\mathbf{X}$. It may be that the n-components are functions of only $k < n$ underlying variables plus a random error term. In factor analysis the relationship is supposed linear and the random variables normally distributed. The model is

$$\mathbf{X} = \mu + A\mathbf{Y} + \mathbf{e} \tag{21.6-4}$$

where μ is a $n \times 1$ mean vector, A is an $n \times k$ matrix of constants (factor loadings), $\mathbf{Y}$ is $N(\mathbf{0}, I)$ where I is the $k \times k$ identity matrix and $\mathbf{e}$ is independent of $\mathbf{Y}$ and $N(\mathbf{0}, D)$ where D is an $n \times n$ diagonal matrix.

These assumptions do not uniquely define the model in terms of the observed distribution of $\mathbf{X}$. Let T be a $k \times k$ matrix of rank n then

$$\mathbf{X} = \mu + (AT^{-1})(T\mathbf{Y}) + \mathbf{e}$$

If T is an orthogonal matrix $T\mathbf{Y}$ will be $N(\mathbf{0}, I)$ and the model

$$\mathbf{X} = \mu + A^*\mathbf{Y}^* + \mathbf{e}$$

where $A^* = AT^{-1}$, $\mathbf{Y}^* = T\mathbf{Y}$ satisfies the model given above. There are various procedures for introducing assumptions that make the model unique, for estimating the quantities involved and for testing the adequacy of the model for a given k. Reference 21-74 is perhaps the best reference on factor analysis. See also Refs. 21-22, 21-23, and 21-24.

21.6.4 Cluster Analysis

Cluster analysis is not one specific technique but refers to the problem of trying to locate or identify groups or clusters of "like" objects from data. Often the number of clusters is not known (or indeed if in any reasonable sense there are clusters.) There are many methods advanced for clustering. References 21-25, 21-26, and 21-27 deal further with the subject.

21.6.5 Discriminant Analysis and Logistic Regression

In discriminant analysis the problem of classifying new data points in one of several populations is considered. In this case the distributions of the populations are either assumed known or some "training" data is supplied to help in estimating the characteristics of the population.

The most commonly used method of discriminant analysis is that of *linear discriminant analysis.* Consider classifying a new n-dimensional data point into one of two populations which each have a normal distribution, say $N(\mu_1, \Sigma)$ and $N(\mu_2, \Sigma)$ respectively. By using the Neyman-Pearson lemma it can be shown that in order to minimize the probability of an error the classification (upon observing $\mathbf{x}$) should be based upon

$$L(x) = (\mu_1' - \mu_2') \Sigma^{-1} \mathbf{x} \tag{21.6-5}$$

which is called the linear discriminant function. The decision rule is to classify into population one (mean μ_1) if $L(x) \geqslant c$ and into population two if $L(x) < c$. The constant c is chosen to reflect the frequencies with which the two populations occur and the costs of making the various types of errors. In practice μ_1, μ_2, and Σ are replaced by estimates. References 21-28, 21-14, and 21-22 deal with this. Another way of thinking of the linear discriminant function $L(X) = \sum_{i=1}^{n} b_i X_i$ is that it is the linear combination of the co-ordinates of $\mathbf{X}$ which has the means (under the two distributions) as many standard deviations apart as possible.

A less restrictive model, conditional upon the $\mathbf{x}$ values, assumes that the probability of falling into one group is given by the logistic function

$$P(\text{Group } 1 \mid \mathbf{x})) = e^{c + a'x}/(1 + e^{c + a'x}) \tag{21.6-6}$$

The coefficients c and $\mathbf{a}$ are estimated by the method of maximum likelihood. References 21-75 and 21-76 discuss these issues further. Bayes' theorem applied to two multivariate normal populations gives a logistic function as in (21.6-6), where the coefficients $\mathbf{a}$ are the linear discriminant coefficients of (21.6-5).

The topics of cluster analysis and discriminant analysis fall into the area called pattern recognition. The term pattern recognition is more prevalent in the engineering literature while the terms cluster analysis and discriminant analysis are more usual in the statistical literature.

21.7 PARAMETRIC, NONPARAMETRIC, AND DISTRIBUTION-FREE STATISTICAL TESTS

In this section further hypothesis testing situations are considered.

21.7.1 Nonparametric and Distribution-Free Statistics

In previous parts of this chapter parametric situations have been considered. That is, the distribution of the observed statistics has been characterized by some parameters. In order to increase the robustness of tests it is desirable to have hypothesis tests that work when faced with a wide variety of distributions. The intent is to extend the statistical inference to a class of distributions wider than some parametric family. These extensions constitute *nonparametric* statistics. Some of the precedures are illustrated in the remainder of section 21.7. It is easy to set up tests in many situations that seem reasonable for a wide variety of distributions but are such that the distribution varies for each possibility in such a way that finding critical regions for a given significance level or confidence intervals at a given degree of confidence is impossible. It is desirable to be able to find a test statistic whose distribution (under some null hypothesis) is the same for a wide variety of distributions. The distribution of the test statistic is then free of the underlying distribution (if it is in some class). Such a statistic is called *distribution-free*.

21.7.2 Two-sample Problem

The above ideas will be illustrated on the two-sample problem. Suppose that we observe a sample of size $n, X_1, \ldots, X_n$, of random variables from a population with distribution function F. Let $Y_1, \ldots, Y_m$ be a sample of size m, independent of the first sample, from a population with distribution function G. It is desired to compare the distributions of two populations. Situations that are appropriate to two sample techniques include comparing: two drugs, average lifetime of two alternative electronic components, incomes of two different socio-economic groups, response times with or without a drug and maximum amount of stress that can be taken by some structure.

Consider first a parametric approach. A difference in the mean values might be tested. If the X_i's are $N(\mu_1, \sigma^2)$ and the Y_j's are $N(\mu_2, \sigma^2)$ and the variance is unknown then

$$T = \frac{\overline{X} - \overline{Y}}{\sqrt{\dfrac{(n-1)\,s_x^2 + (m-1)\,s_y^2}{n+m-2}\left(\dfrac{1}{n} + \dfrac{1}{m}\right)}} \tag{21.7-1}$$

has a t-distribution with $n + m - 2$ degrees of freedom under the null hypothesis that $\mu_1 = \mu_2$. [Also, $\overline{X} = \sum_{i=1}^{n} X_i/n$, $\overline{Y} = \sum_{j=1}^{m} Y_j/m$, $s_x^2 = \sum_{i=1}^{n} (X_i - \overline{X})^2/(n-1)$, $s_y^2 = \sum_{j=1}^{m} (Y_j - \overline{Y})^2/(m-1)$.] The hypothesis test then is based on a t-distribution.

The assumption of normality and equal variances may be relaxed asymptotically for if the variances are finite then if $\mu_1 = \mu_2$

$$Z = \frac{\overline{X} - \overline{Y}}{\sqrt{\dfrac{s_x^2}{n} + \dfrac{s_y^2}{m}}} \tag{21.7-2}$$

is asymptotically $N(0, 1)$.

Now a nonparametric approach is considered. Suppose that F and G are continuous distribution functions. Consider the null hypothesis that $F = G$, that is the two distributions are the same. With a continuous distribution $X_1, \ldots, X_n$, $Y_1, \ldots, Y_m$ will be $n + m$ distinct values (with a probability equal to one.) Let the $n + m$ values be rearranged in increasing order

$$Z_{(1)} < Z_{(2)} < \cdots < Z_{(n+m)} \tag{21.7-3}$$

For $X_1, \ldots, X_n$ let $R_i = k$ if $X_i = Z_{(k)}$. That is, $R_i = k$ if the $X_i{}^{\text{th}}$ observation is the kth largest of the $n + m$ numbers. The R_i's are called the rank order statistics or the rank statistics. Under the null hypothesis of $F = G$ the random variables $X_1, \ldots, X_n$, $Y_1, \ldots, Y_m$ constitute a sample with distribution function F. By symmetry it is clear that $\{R_1, \ldots, R_n\}$ is equally likely to be any subset of $\{1, 2, \ldots, n + m\}$. For a given subset $(R_1, \ldots, R_n)$ is equally likely to be any permutation of the numbers. Thus, under the null hypothesis ($F = G$ a continuous distribution function), any statistic calculated from $(R_1, \ldots, R_n)$ has a distribution which is "free" of the distribution F.

A distribution of the form $\sum_{i=1}^{n} c_i f(R_i)$ is called a linear rank statistic (where f is an arbitrary function with domain of definition $\{1, 2, \ldots, n + m\}$). Suppose that the alternative that is under consideration is that one sample "tends" to have larger values than the other (e.g., $F \geqslant G$ or $G \geqslant F$ with some values with strict inequality). Then a reasonable statistic is the Wilcoxon two-sample statistic

$$W = \sum_{i=1}^{n} R_i \tag{21.7-4}$$

Tables for the critical values of W under the null hypothesis may be found in Refs. 21-32 and 21-77. The statistic is asymptotically normal when min (m, n) is large.

One virtue of rank order tests is that they may be designed to be sensitive to a parametric family if it is so desired. Suppose that a test is desired which is sensitive to data which comes from a normal distribution. Instead of working with the ranks it would be desirable to use the order statistics corresponding to $X_1, \ldots, X_n$. In order to make the test distribution free it is desirable to use a function of the rank statistics. From the rank statistics the X_i's may be estimated. Suppose the X_i's come from a $N(0, 1)$ sample. If $R_i = k$ then X_i is the kth smallest among $n + m$ $N(0, 1)$ random variables. Thus one would "expect" about $k/(n + m + 1)$ of the probability of a $N(0, 1)$ distribution to be less than or equal to the value of X_i corresponding to $R_i = k$. Estimate X_i by $f(R_i)$ where

$$f(x) = \Phi^{-1}\left(\frac{x}{n + m + 1}\right) \quad \text{and} \quad \Phi(z) = \frac{1}{\sqrt{2\pi}} \int_{-\infty}^{z} e^{-y^2/2}\, dy \tag{21.7-5}$$

$\Phi(x)$ is the $N(0, 1)$ distribution function so that $f(x)$ is that value such that if X is a $N(0, 1)$ random variable then

$$P[X \leqslant f(x)] = \frac{x}{n+m+1}$$

The linear rank order statistic

$$v = \sum_{i=1}^{n} f(R_i) \qquad (21.7\text{-}6)$$

is the van der Waerden statistic. In general it is clear that from the rank statistics there is no hope of estimating the mean and variance of a normal population from which the X_i's might have come. Thus, the van der Waerden statistic is appropriate in general if the data comes from normal distribution. For large n and m, the van der Waerden test has essentially the same power as the t-test of eq. (21.7-1). However, for *any other* shape of the distribution F if G is F shifted by some amount then the van der Waerden test is better than the t-test for large n and m. Thus this distribution free statistic behaves well in the parametric situation and also offers additional protection if the parametric assumptions are not satisfied.

Suppose now that the alternative hypothesis of interest is $F \neq G$. The interest is then whether the populations are the same against the alternative that they differ in some (unspecified) way. One method of approaching this is to estimate the two distribution functions F and G. Fix x, $F(x) = P(X_i \leqslant x)$. An obvious estimate of $F(x)$ is $\hat{F}(x) =$ (number of $X_i \leqslant x, i = 1, \ldots, n)/n$. Estimate F at all x in this manner. Denote the estimate for G by $\hat{G}$. Under the null hypothesis $F = G$ these two quantities are estimating the same quantity. A reasonable test statistic is

$$K = \underset{-\infty < x < \infty}{\text{maximum}} \, |\hat{F}(x) - \hat{G}(x)| \qquad (21.7\text{-}7)$$

where K is the (two-sided) Kolmogorov-Smirnov statistic. It is easy to show that K is a function of the rank statistics and so is distribution-free. (It is not a linear rank statistic.) Since K protects against all possible ways of $F \neq G$ it has fairly low power against many specific types of alternative hypotheses. An introduction to rank tests is given in Refs. 21-29 and 21-78. Reference 21-30 gives an exposition at a very mathematical level. Reference 21-31 discusses a wide range of distribution-free statistical methods. Recent tables for using the Wilcoxon and Kolmogorov-Smirnov statistics are given in Ref. 21-32. It should be emphasized that there is a wide range of nonparametric procedures that will not be touched upon in this chapter.

21.7.3 χ^2 (Chi-square) Tests

The importance of the χ^2-distribution arises from its relationship with the normal distribution. In this section some limit theorems involving the χ^2-distribution are stated and then examples of its usage are given. (The limit theorems hold because there are underlying random vectors that have a limiting normal distribution.)

THEOREM 21.7.1: *Let $(n_1, \ldots, n_k)$ be a multinomial random vector with parameters n and $p_1, \ldots, p_k$ all $p_i > 0$. Then as $n \to \infty$ (the p_i's fixed)*

$$x^2 = \sum_{i=1}^{k} \frac{(n_i - np_i)^2}{np_i} \tag{21.7-8}$$

converges in distribution to a χ^2 random variable with $k - 1$ degrees of freedom.

Note that np_i is the expected number of times that the ith outcome occurs. For this reason the variable x^2 is often given as

$$x^2 = \sum \frac{(\text{observed} - \text{expected})^2}{\text{expected}}$$

where "observed" denote the number of times the outcome occurs and "expected" is the expected number of occurrences.

Example: The χ^2 test is often used for testing *goodness-of-fit* problems. A goodness-of-fit problem is one where a model is hypothesized and the question of whether or not the model "fits" the data is considered. For example, suppose that a spinning wheel in a carnival is being observed and that it can stop on any of fifty numbers. It is desired to test that $p_i = \frac{1}{50}$ where p_i is the probability of stopping on the ith number. Let n_i be the number of times the ith outcome occurred in n spins of the wheel then

$$x^2 = \sum_{i=1}^{50} \frac{(n_i - n/50)^2}{n/50}$$

is asymptotically χ^2 with 49 degrees of freedom. If in fact the true probabilities are $\tilde{p}_i$ where some of the $\tilde{p}_i$ differ from the p_i then on the average the value of x^2 will be too large and thus the critical region for x^2 should be a one-tailed critical region of the form $(c, +\infty)$. Since the χ^2 distribution only holds in the limit the question arises as to how large n should be for the approximation to be reasonably valid. A common rule of thumb is that the expected value in each cell (i.e., of each outcome) should be at least 5. (In some cases fewer observations may be needed. See Ref. 21-33.) For very small significance levels the number of observations should be higher.

In many cases of testing goodness-of-fit the distribution to be fitted depends upon unknown parameters. In this case it is necessary to estimate the parameters and this changes the limiting distribution of the quantity analogous to x^2.

THEOREM 21.7.2: *Let $(n_1, \ldots, n_k)$ be a multinomial random vector whose probabilities $p_1(\theta), \ldots, p_k(\theta)$ depend upon an $s \times 1$ parameter vector θ, $s < k - 1$. Let $\hat{\theta}$ be an estimate of θ and*

$$X^2 = \sum_{i=1}^{n} \frac{[n_i - np_i(\hat{\theta})]^2}{np_i(\hat{\theta})} \tag{21.7-9}$$

Suppose that the $p_i(\theta)$'s in some neighborhood of the true parameter value θ_0 satisfy:

a. $p_i(\theta) \geqslant c > 0$ *(c a fixed constant).*
b. *The p_i's have continuous first and second partial derivatives with respect to the parameters (θ_j's).*
c. *The $k \times s$ matrix with entries $\partial p_i/\partial \theta_j$ has rank s.*
1. *Let the $\hat{\theta}$ be chosen by the method of maximum likelihood used on $(n_1, \ldots, n_k)$ or by the method of minimum χ^2 estimation [choose $\hat{\theta}$ to minimize (21.7-8)] then assuming that $\hat{\theta}$ converges to θ_0 in probability X^2 has a limiting χ^2 distribution with $k - s - 1$ degrees of freedom.*
2. *Consider the goodness-of-fit problem where the outcome space is partitioned into k cells and the number of outcomes in cell i is n_i. Let the parameters θ be estimated by maximum likelihood estimates using the entire data set. The limiting distribution of X^2 has a distribution function between the distribution functions of χ^2 random variables with $k - 1$ and $k - s - 1$ degrees of freedom (under conditions as set forth in Reference 21-34).*

Statement (1) of the theorem may be paraphrased by stating that one degree of freedom is lost for each parameter estimated.

Example: (Independence in two-way contingency tables.) Consider a sample taken from a population for which two traits are measured, the first trait having *r* possibilities and the second trait have *c* possibilities. If the sample size is *n* the data may be represented in a two-way contingency table.

	Trait 2			
Trait 1	n_{11}	$n_{12} \cdots$	n_{1c}	$n_{1\cdot}$
	n_{21}			
	$\cdots$	$\cdots$	$\cdots$	
	$n_{r1} \cdots$	$\cdots$	n_{cc}	$n_{r\cdot}$
	$n_{\cdot 1} \cdots$	$\cdots$	$n_{\cdot c}$	$n_{\cdot\cdot} = n$

where n_{ij} is the number of cases where the first trait had the ith possible outcome and the second trait had the jth possible outcome; $n_{i\cdot} = \sum_{j=1}^{c} n_{ij}$, etc. Let p_{ij} denote the probability that an outcome corresponds to the i, j^{th} cell. If the two traits are independent then there exist numbers $p_i \geqslant 0$, $i = 1, \ldots, r$, $\sum_{i=1}^{r} p_i = 1$ and $p_j' \geqslant 0$, $j = 1, \ldots, c$, $\sum_{j=1}^{c} p_j' = 1$ such that $p_{ij} = p_i p_j'$. Under independence the maximum likelihood estimates for the p_i and p_j' terms are $\hat{p}_i = n_{i\cdot}/n$ and $\hat{p}_j' = n_{\cdot j}/n$. Since the p_j's and p_i''s sum to one, the conditions of (1) of Theorem 21.7.7 are satisfied if $s = r - 1 + c - 1$ and the parameter vector θ is $(p_1, \ldots, p_{r-1}, p_1', \ldots, p_{c-1}')$. The number of possible outcomes is $r \cdot c = k$. If $p_{ij} = p_i p_j' > 0$ holds then

$$X^2 = \sum_{i=1}^{r} \sum_{j=1}^{c} \frac{[n_{ij} - (n_{i.} n_{.j}/n)]^2}{(n_{i.} n_{.j}/n)}$$

has a limiting χ^2 distribution with $k - s - 1 = rc - (c - 1 + r - 1) - 1 = (r - 1)(c - 1)$ degrees of freedom.

Example: Consider testing the goodness-of-fit for a sample $X_1, \ldots, X_n$ to a normal distribution with unknown parameters. Suppose that $(-\infty, +\infty)$ is partitioned into k intervals and that the probability of a point falling into the interval I is estimated by

$$p_I = \int_I \frac{1}{\sqrt{2\pi} s} e^{-(y-\bar{X})^2/2s^2}\, dy$$

where $\bar{X} = \sum_{i=1}^{n} X_i/n$ and $s^2 = \sum_{i=1}^{n} (X_i - \bar{X})^2/n$ are the maximum likelihood estimates for μ and σ^2. Let n_I = number of X_i in the interval I. Then $X^2 = \sum_I (n_I - np_I)^2/np_I$ has a limiting distribution between a χ^2 distribution with $k - 1$ and $k - 3$ degrees of freedom. Reference 21-34 indicates that assuming X^2 has a χ^2 distribution with $k - 3$ degrees of freedom can lead to a serious underestimate of the probability of a type I error.

Reference 21-35 gives much more information on the χ^2 distribution.

21.7.4 Conditional Tests

Some tests are best thought of in a conditional manner. After observing the data part of the data is assumed as known and the conditional distribution of the remaining portion of the data is used for significance testing. From a purely logical point of view such tests are not different—i.e., certain outcomes lead to rejection of a null hypothesis. Psychologically, however, there is a difference. Two examples of conditional tests are given.

Examples:

(1) (McNemar's test.) Suppose that it is desired to study the relationship between smoking and cancer. One might ask cancer patients if they smoke and then compare the rate with the population as a whole. This might be criticized on the grounds that the patients are not a sample from the population and may differ in many traits that predispose a person toward cancer. One may try to get around this by matching each cancer patient with someone from the population as a whole who has the same values on variables of interest e.g., sex, age, socio-

economic variables. Then ask the matched person if he (or she) smokes. The null hypothesis is that for each pair the probability of being a smoker is the same (but the value may vary from pair to pair.) The outcomes of the pairs may then be recorded in a 2 X 2 table.

		Match	
		Smoker	Non-Smoker
Cancer Patient	Smoker	n_1	n_2
	Non-Smoker	n_3	n_4

where n_1 is the number of pairs for which both members of the pair smoked, etc. The pairs for which both members smoked or did not smoke give no information as to whether smoking and cancer are more likely to occur together. If a pair disagrees, however, under the null hypothesis either of the two ways of disagreeing is equally likely. Thus, knowing $n_2 + n_3, n_2$ is a binomial random variable with $n = n_1 + n_2$ and $p = \frac{1}{2}$. Tests based on the conditional distribution of n_2 (conditional on $n_1 + n_2$) are called McNemar's test.

(2) (Fisher's Exact Test for 2 X 2 tables.) Consider two populations being tested for a trait with two possible outcomes, Yes and No. Let n_1 be the sample size from population 1 and n_2 the sample size from population 2. Record the data in a 2 X 2 table.

		Yes	No	
Population	1	Y_1	N_1	n_1
	2	Y_2	N_2	n_2
		Y	N	

That is, $Y_1 + N_1 = n_1$, $Y_2 + N_2 = n_2$, $Y_1 + Y_2 = Y$, and $N_1 + N_2 = N$. Suppose that the null hypothesis that populations 1 and 2 have the same probability of a yes or no reply holds. Suppose further that the values of n_1, n_2, Y and N are given but not the values of Y_1, Y_2, N_1 and N_2. It is easy to see that knowing Y, N, n_1 and n_2 only one of Y_1, Y_2, N_1, and N_2 is needed to specify the rest so that we need consider only Y_1. Conditionally upon knowing Y, N, n_1 and n_2 what is the distribution of Y_1? Of the Y yes replies in the $n_1 + n_2$ samples any arrangement is equally likely. From this it follows that Y_1 has a hypergeometric distribution

$$P(Y_1 = k) = \frac{\binom{Y}{k}\binom{N}{n_1 - k}}{\binom{Y+N}{n_1}} \qquad k = \max(0, Y - n_2), \ldots, \min(n_1, Y) \quad (21.7\text{-}10)$$

From this distribution appropriate one- and two-tailed tests may be devised. Appropriate tables are Refs. 21-36 and 21-37.

21.7.5 Robust Estimates, Trimming, and Winsorization

Consider estimating the mean of a distribution from a sample. If the sample is normally distributed the sample mean is the best method of estimation. However, even if the distribution is "nearly" normal, but not quite, there are often better methods of estimation (see Ref. 21-38 which gives an excellent introduction to these topics). Two such techniques used are trimming and Winsorization. Trimming refers to dropping the tail observations from data. Winsorization brings the tail observations in toward the center of the data.

Definition 21.7.1: Let $X_1, \ldots, X_n$ be numbers from a random sample. Let $p = 100\,k/n$ where k is a nonnegative integer, $p < 50$. Let $Z_{(1)} \leqslant Z_{(2)} \leqslant \cdots \leqslant Z_{(n)}$ be the order statistics. The *p% trimmed mean* is

$$\overline{X}_T = \sum_{j=k+1}^{n-k} Z_{(j)}/(n-2k) \tag{21.7-12}$$

The *p% Winsorized mean* is

$$\overline{X}_w = \left[(k+1)(Z_{(k+1)} + Z_{(n-k)}) + \sum_{j=k+2}^{n-k-1} Z_{(j)}\right]\Big/ n \tag{21.7-13}$$

The trimmed mean drops the bottom and top $p\%$ of the observations from the computation and uses the remaining observations to compute the mean. The Winsorized mean takes the largest and smallest $p\%$ of the sample values and moves them into the nearest value left. That is, $Z_{(1)}, \ldots, Z_{(k)}$ are "moved" into $Z_{(k+1)}$ and $Z_{(n-k+1)}, \ldots, Z_{(n-k)}$ are "moved" to $Z_{(n-k)}$.

Reference 21-38 shows that a trimmed mean is superior to the usual sample mean under certain departures from normality if p is small (say 1 to 6%) and suffers very little loss of efficiency when the distribution actually is normal. The subject of robust estimation is under rapid development at the present time and robust and nonparametric statistics may change standard statistical procedure somewhat in the future. Often a median is as useful (or a more useful) measure of location than the mean. In this case nonparametric confidence intervals and tests exist.

Considerable work has been done on what happens when the mathematical assumptions underlying a statistical procedure are violated. A statistical procedure is robust if it holds its performance well even when the underlying assumptions are violated. Robust estimates in complex situations, such as multiple regression models, have been developed. Reference 21-79 gives a comprehensive account of the theory. Computer software for robust estimation is just beginning to make its impact felt at this writing.

21.7.6 *P*-values

In hypothesis testing the appropriate significance level to use is often difficult to determine and is frequently set arbitrarily. The values 10%, 5%, 1% and 0.1% have gained respect through historical accident, but in most cases there are few objective reasons for choosing a given significance level. Often a test is set at a 5% level and the data is such that the null hypothesis would be rejected even at the 0.001% level. Should this data merely be reported as rejecting at the 5% level? It is clear that it is not "fair" to select the significance level after observing the data. An alternative approach is to give the smallest possible significance level for which the null hypothesis would be rejected. This value is called the *P*-value.

Definition 21.7.2: Consider a hypothesis testing situation based on a statistic $\mathbf{X}$. Suppose that as the significance level, α, increases the critical region increases (i.e., at a lower value of the significance level the critical region is contained in the critical region corresponding to a higher value of the significance level). If $\alpha = 0$ suppose that the critical region is empty and if $\alpha = 1$ suppose that the critical region is the whole outcome space. Let C_α denote the critical region at significance level α. If $\mathbf{X} = \mathbf{x}$ the *P-value* is

$$\text{infimum } \{\alpha \colon x \in C_\alpha\}$$

If the *P*-value is β then the data is such that the null hypothesis would be rejected at significance level α if $\alpha > \beta$ and not rejected if $\alpha < \beta$. Thus the *P*-value allows the reader of a result to determine what would happen with his own favorite significance level. It is also a reasonable indicator of the amount of evidence that the data gives against a hypothesis.

Examples:

(1) Suppose that under a null hypothesis a random variable X has a $N(0, 1)$ distribution. The critical regions are chosen as symmetric two-tailed critical regions. If $X = -2.18$ what is the *P*-value? As $P(X \leqslant -2.18) = 0.03788$ the *P*-value is $2 \times 0.03788 = 0.07576$. If the critical region were one-sided of the form (c, ∞) the *P*-value would be $P(X > -2.18) = 0.98537$.

(2) Suppose that the Kolmogorov-Smirnov statistic K, eq. (21.7-7), is computed for two samples each of size 20 and is found to be $\frac{8}{20}$. Then $P(K \leqslant \frac{8}{20}) = 0.9665$ (from Ref. 21-9). The *P*-value is thus 0.0335.

21.8 BAYESIAN STATISTICS AND DECISION THEORY

21.8.1 Bayes' Theorem

Suppose that probabilities for some events are known. If later some new information comes to hand how does this change the probabilities of the events occurring? Bayes' theorem deals with this problem. Several versions of Bayes' theorem are given in this section.

THEOREM 21.8.1: *Let $A_1, A_2, \ldots, A_n$ be disjoint events whose union is the entire probability space. Let B be an event with $P(B) > 0$ then*

$$P(A_i \mid B) = \frac{P(B \mid A_i)P(A_i)}{\sum_{j=1}^{n} P(B \mid A_j)P(A_j)} \tag{21.8-2}$$

where $P(B \mid A_j)P(A_j) = 0$ if $P(A_j) = 0$.

Bayes' theorem allows the $P(A_i \mid B)$ to be expressed in terms of the original probability of the A_j's and the probability of B given the A_j's.

Example: Suppose a person reaches in a pocket which has a fair coin and a two-headed coin, the probability being $\frac{1}{2}$ of grabbing each. If eight flips of the coin selected are all heads what is the probability that the fair coin is being used? B is the event of eight heads. A_1 is the event the fair coin is chosen, $P(A_1) = \frac{1}{2}$. A_2 is the event the two-headed coin is chosen, $P(A_2) = \frac{1}{2}$. $P(B \mid A_1) = \frac{1}{2^8}$, $P(B \mid A_2) = 1$. By Bayes' theorem

$$P(A_1 \mid B) = \frac{\frac{1}{2^8} \cdot \frac{1}{2}}{\frac{1}{2^8} \cdot \frac{1}{2} + 1 \cdot \frac{1}{2}} = (1 + 2^8)^{-1}$$

THEOREM 21.8.2: *A probability distribution is given over a set of θ's (θ is chosen at random). A random vector $\mathbf{X}$ is to be observed where the distribution of $\mathbf{X}$ is specified by θ. The following give the distribution of θ conditionally upon observing that $\mathbf{X} = \mathbf{x}$.*

A. *Let θ take on a countable (or finite) number of possibilities $\{\theta_1, \theta_2, \ldots\}$. Let $p_i = P(\{\theta_i\})$. Let $\mathbf{X}$ be a discrete random vector with probability mass function $p(\mathbf{x} \mid \theta)$ when θ is the true value.*

$$P(\theta_j \mid \mathbf{X} = \mathbf{x}) = \frac{p(\mathbf{x} \mid \theta_j)p_j}{\sum_i p(\mathbf{x} \mid \theta_i)p_i} \tag{21.8-4}$$

B. *Let θ be discrete as above and $\mathbf{X}$ a continuous random vector with conditional probability density function $f(\mathbf{x} \mid \theta)$. Then*

$$P(\theta_j \mid \mathbf{X} = \mathbf{x}) = \frac{f(\mathbf{x} \mid \theta_j)p_j}{\sum_i f(\mathbf{x} \mid \theta_i)p_i} \tag{21.8-5}$$

C. *Let $\theta = \boldsymbol{\theta}$ be a continuous random parameter in k-dimensional Euclidean space with probability density function $g(\boldsymbol{\theta})$. Let $\mathbf{X}$ be a discrete random vector with conditional probability mass function $p(\mathbf{x} \mid \boldsymbol{\theta})$. Then the conditional distribution of θ after observing $\mathbf{X}$ is continuous with probability density function*

$$f(\theta \mid \mathbf{X}=\mathbf{x}) = \frac{p(\mathbf{x} \mid \theta) g(\theta)}{\int_{R^k} p(\mathbf{x} \mid \tilde{\theta}) g(\tilde{\theta})\, d\tilde{\theta}} \tag{21.8-6}$$

D. *Let $\theta = \theta$ be a continuous random parameter in k-dimensional Euclidean space with probability density function $g(\theta)$. Let* **X** *be a continuous random vector with conditional probability density function $f(\mathbf{x} \mid \theta)$. The conditional distribution of θ after observing* **X** *is continuous with probability density function*

$$f(\theta \mid \mathbf{x}) = \frac{f(\mathbf{x} \mid \theta) g(\theta)}{\int_{R^k} f(\mathbf{x} \mid \tilde{\theta}) g(\tilde{\theta})\, d\tilde{\theta}} \tag{21.8-7}$$

Example: Assume a uniform distribution on p, $0 \leqslant p \leqslant 1$. Let X (given p) be a binomial random variable with parameters n and p. Find the distribution of p given that $X = k$ ($k = 0, 1, 2, \ldots, n$); (C) of the above theorem applies.

$$f(p \mid X = k) = \frac{\binom{n}{k} p^k (1-p)^{n-k}}{\int_0^1 \binom{n}{k} \tilde{p}^k (1-\tilde{p})^{n-k}\, d\tilde{p}}$$

The conditional distribution of p is known as a beta distribution.

In the above the initial probability distribution of θ is called the *a priori* or *prior* distribution. (Prior referring to the fact the distribution is prior to observing the value of **X**.) The distribution of θ after the **X** statistic has been observed is called the *a posteriori* or *posterior* distribution.

21.8.2 Bayesian Statistics

The starting point for Bayesian statistics is that a prior distribution over the possibilities at hand may be found. This requirement has been the subject of continuing debate. The foundations of subjective probability theory may be found in Refs. 21-39 and 21-40. Some articles giving reasons that Bayesian statistics may be preferable to the hypothesis testing approach are Refs. 21-41 and 21-42. An intermediate approach arguing that the likelihood function should be used for statistical inference is Ref. 21-43.

The prior distribution is both the strength and weakness of Bayesian statistics. By specifying a prior distribution information other than the experiment or data at hand may be brought to bear on reaching a reasonable conclusion. On the other hand it may be extremely difficult in a particular case to take subjective feelings and refine them to the point where a prior distribution is specified. The form of prior distributions are often chosen to make the calculation of posterior distributions feasible without numerical integration or summation. Such *conjugate* priors (see Ref. 21-44) hopefully have enough variety so that the "true" prior distribution can be approximated fairly well. The sensitivity of the posterior distribution to the form of the prior distribution has much to do with how precisely the prior distribution needs to be specified.

The Bayesian parallels to the procedures discussed previously are now considered. For estimation of parameters both point estimates and confidence sets have been discussed. A Bayesian analysis yields a (posterior) probability distribution for the unknown parameters. A set such that the probability of the parameters being in the set is α may be found. The interpretation of a $100\alpha\%$ confidence set is different. A confidence set is formed in such a way that if the same procedure were independently repeated many times a fraction α of the time the parameter will be in the set. The Bayesian set of probability α does not depend upon many independent repetitions for its interpretation. With the uncertainty of the present situation the posterior distribution measures the belief that the parameter values fall into various sets.

The analog of a point estimate is to choose some quantity from the posterior distribution to summarize the distribution over the parameter values. Possibilities include the mean vector of the distribution and mode of the distribution.

Example: Suppose that a prior distribution for a parameter μ is $N(0, 1)$. Observe $X_1, \ldots, X_n$ i.i.d. $N(\mu, 1)$ random variables. Find: (a) the posterior distribution of μ; (b) the shortest interval whose probability of containing μ is α; and (c) the mean and mode of the posterior distribution. From eq. (21.8-7) the density of the posterior distribution is proportional to

$$e^{\frac{-(n+1)}{2}\left(\mu-\frac{n\overline{X}}{n+1}\right)^2}$$

so that the posterior distribution is $N[n\overline{X}/(n+1), 1/(n+1)]$. From this it follows that the answer to (b) is

$$\left(\frac{n\overline{X}}{n+1} - \frac{Z_{\alpha/2}}{\sqrt{n+1}}, \frac{n\overline{X}}{n+1} + \frac{Z_{\alpha/2}}{\sqrt{n+1}}\right)$$

where

$$(2\pi)^{-(1/2)} \int_{Z_{\alpha/2}}^{\infty} e^{-x^2/2}\, dx = \alpha/2$$

and for (c) both parts are answered by $n\overline{X}/(n+1)$.

In a hypothesis testing situation a Bayesian approach will yield probabilities of the null hypothesis or the alternative being true. The analysis will be left at this stage unless there is a need to act upon the possibilities. In this case the consequences of the actions will be weighed along with the probabilities in making a decision.

21.8.3 Decision Theory

Decision theory rests upon the idea that statistical observations are taken in order to take some sort of action. The effect of the action is going to depend upon the true state of affairs and, of course, the purpose of taking statistical observations is to infer something about the true state of affairs. These ideas are formalized below.

Definition 21.8.1: A decision theory problem consists of the following elements:

1. A set $\mathcal{S}$ of the possible states of nature.
2. A set $\mathcal{A}$ of possible actions available.
3. A loss function $L(s, a)$ defined for s in $\mathcal{S}$ and a in $\mathcal{A}$ giving the loss if action a is taken when s is the true state of nature. $L(\cdot, \cdot)$ is a real-valued function.
4. A statistic $\mathbf{X}$ to be observed whose distribution varies with the state of nature s; E_s will denote the expected value when s is the true state of nature.

Definition 21.8.2: A *decision rule* is a function from the range of possible values of $\mathbf{X}$ into the action set $\mathcal{A}$.

In other words after observing $\mathbf{X}$ the action $a = d(\mathbf{X})$ is going to be taken. The fundamental problem of decision theory is how to choose the decision rule in a reasonable manner.

Definition 21.8.3: Associated with each state of nature s and decision rule d the average loss is called the *risk function*.

$$R(s, d) = E_s[L(s, d(X))] \tag{21.8-10}$$

(It is assumed that $R(s, d) > -\infty$ for all pairs s and d.)

One desire in decision theory is to make the risk small. The problem encountered is that a decision rule that works well for one state of nature may be terrible for another state of nature. Suppose that one takes the pessimistic view that the state of nature will be the worst possible case for the decision rule being used. In this case there are gains to allowing the decision rule to randomized, that is, after observing $\mathbf{X}$ a random (according to a specified distribution dependent on $\mathbf{X}$) action is taken.

Definition 21.8.4: A probability distribution on the set of decision rules, d, such that

$$E_{\tilde{d}}[R(s, \tilde{d})] < \infty \quad \text{for } s \in \mathcal{S}$$

is called a randomized decision rule. Let D denote the space of randomized decision rules. Extend the definition of risk to include randomized decision rules by defining the risk of a randomized decision rule $\tilde{d}$ to be $R(s, \tilde{d}) = E_{\tilde{d}}[R(s, \tilde{d})]$.

By placing probability one on one nonrandomized decision rule we see that the randomized decision rules include the nonrandomized decision rules that have finite risk function. (It is assumed that this can be done. We have not discussed the requirements to insure the $L[a, \tilde{d}(X)]$ is a measurable function so that the integral may be defined.)

How might a decision rule be selected? Clearly a decision rule d_1 in D is preferable to d_2 in D if $R(s, d_1) \leqslant R(s, d_2)$ for all $s \in \mathcal{S}$ with strict inequality for at least one s. In this case we'll say that d_1 is better than d_2.

Definition 21.8.5: A decision rule d in D is *admissible* if there is no d' in D which is better than d. Otherwise a decision rule is *inadmissible*.

One case where there is some hope of finding one best rule is when there is a prior distribution on the states of nature $\mathcal{S}$.

Definition 21.8.6: Let P be a probability measure on $\mathcal{S}$. The *Bayes' risk* of a decision rule $d \in D$ is $R(P, d) = E\,[R(s, d)]$. A decision rule d^* is a Bayes decision rule with respect to prior distribution P if

$$R(P, d^*) = \inf_{d \in D} R(P, d) \tag{21.8-14}$$

Another case where one decision rule may be selected from among the decision rules is when one is worried about the worst possible state of nature for the decision rule used. Suppose that after a decision rule d is chosen the worst possible state of nature holds, then the risk is $\sup_{s \in \mathcal{S}} R(s, d)$. It is then desired to minimize this quantity.

Definition 21.8.7: A decision rule d^* in D is a *minimax* decision rule if

$$\sup_{s \in \mathcal{S}} R(s, d^*) = \inf_{d \in D} \sup_{s \in \mathcal{S}} R(s, d) \tag{21.8-15}$$

Example: Consider two states of nature $\{s_1, s_2\}$, two possible actions $\{a_1, a_2\}$. Suppose that a random variable X is to be observed which takes on only the values 0 and 1. The distribution of X and the loss function are as follows:

$P(X = k \mid s_i)$	$i = 1$	$i = 2$
$k = 0$	$\frac{1}{2}$	$\frac{1}{4}$
$k = 1$	$\frac{1}{2}$	$\frac{3}{4}$

$L(s_i, a_j)$	$j = 1$	$j = 2$
$i = 1$	0	2
$i = 2$	1	0

There are four nonrandomized decision rules.

$$d_1(0) = a_1, d_1(1) = a_1; \quad d_2(0) = a_1, d_2(1) = a_2$$

$$d_3(0) = a_2, d_3(1) = a_1; \quad d_4(0) = a_2, d_4(1) = a_2$$

It can be checked that any decision rule is characterized by two numbers p, r; $0 \leqslant p, r \leqslant 1$ where p is the probability of taking action a_1 if $X = 0$ and r is the probability of taking action a_1 when $X = 1$. It is then found that if d corresponds to p and r the risks are

$$R(s_1, d) = 2 - r - p$$

$$R(s_2, d) = \frac{p}{4} + \frac{3r}{4}$$

Find all Bayes rules and a minimax rule.

If the prior distribution has probability θ that the true state of nature is s_1 then the Bayes risk of d is

$$\theta(2 - r - p) + (1 - \theta)\left(\frac{p}{4} + \frac{3r}{4}\right) = 2\theta + p\left(\frac{1}{4} - \frac{5\theta}{4}\right) + r\left(\frac{3}{4} - \frac{7\theta}{4}\right)$$

Since p and r may be varied independently the Bayes decision rules are

$$\begin{array}{lll} p = 0, r = 0 & \text{if} & 0 \leqslant \theta < \frac{1}{5} \\ 0 \leqslant p \leqslant 1, r = 0 & & \theta = \frac{1}{5} \\ p = 1, r = 0 & & \frac{1}{5} < \theta < \frac{3}{7} \\ p = 1, 0 \leqslant r \leqslant 1 & & \theta = \frac{3}{7} \\ p = 1, r = 1 & & \frac{3}{7} < \theta \leqslant 1 \end{array}$$

The minimax decision rule d^* is $p = 1$, $r = \frac{3}{7}$. To verify this it must be verified that

$$\max_{i=1,2} R(s_i, d^*) = \min_{\substack{0 \leqslant p \leqslant 1 \\ 0 \leqslant r \leqslant 1}} \max_{i=1,2} R[s_i, d(p, r)]$$

or

$$\max\left(2 - 1 - \frac{3}{7}, \frac{1}{4} + \frac{3 \cdot 3}{4 \cdot 7}\right) = \frac{4}{7} = \min_{r,p} \max\left(2 - r - p, \frac{p}{4} + \frac{3r}{4}\right)$$

Suppose $\max(2 - r - p, p/4 + 3r/4) \leqslant \frac{4}{7}$ then $2 - r - p \leqslant \frac{4}{7}$ or $r + p \geqslant \frac{10}{7}$. To minimize $p/4 + 3r/4$ for a fixed value of $r + p$ it is desirable to take p as large as possible, $p = 1$. Then $r \geqslant \frac{3}{7}$ and $p/4 + 3r/4 \geqslant \frac{1}{4} + [3(\frac{3}{7})]/4 = \frac{4}{7}$. So that $p = 1$, $r = \frac{3}{7}$ is the (unique) minimax decision rule.

It is desirable when possible to restrict attention to admissible decision rules (provided that they can be located.)

Definition 21.8.18: In a decision theory problem a set $\tilde{D} \subseteq D$ of decision rules is a *complete class* if for any d in D there is a $\tilde{d}$ in $\tilde{D}$ at least as good as d. [$\tilde{d}$ is at least as good as d if for every $s \in \mathcal{S}$, $R(s, \tilde{d}) \leqslant R(s, d)$.] $\tilde{D}$ is a *minimal complete class* if $\tilde{D}$ is complete but the removal of any element from $\tilde{D}$ gives a set which is not a complete class.

Parts of the following theorem hold under more general conditions, see Ref. 21-45. A fairly elementary introduction to decision theory is given in Ref. 21-47.

THEOREM 21.8.3: *Let $\mathcal{S}$ and $\mathcal{A}$ be finite sets. Then:*

a. *Let P be a prior distribution which puts positive probability on each $s \in \mathcal{S}$. Let d be a Bayes decision rule with respect to P then d is an admissible decision rule.*
b. *The set of Bayes decision rules is a complete class.*
c. *The set of admissible Bayes decision rules is a minimal complete class.*
d. *Any admissible decision rule is a Bayes decision rule.*

Example: What is the minimal complete class for the last example? The Bayes rules were calculated previously. It remains to eliminate the Bayes rules that are not admissible. From the last theorem the only rules whose admissibility needs to be considered correspond to $\theta = 0$ and $\theta = 1$. But these rules are known to be admissible as they also correspond to other values of θ. Thus all of the Bayes rules given in (21.8-15) constitute the minimal complete class.

In the context of the last theorem it is seen that someone behaving in an admissible manner is behaving in Bayesian fashion (knowingly or not).

Much of statistics may be subsumed under a decision theory approach. For example, when discussing estimation of a real parameter μ if the allowable actions are to pick a real number and the loss is $L(\mu, a) = (a - \mu)^2$ then the expected loss or risk is $E\{[a(x) - \mu]^2\}$ which is the mean square error. The references cited previously (21-45, 21-46) are appropriate references.

21.9 CONCLUDING REMARKS

Because of spatial limitations many topics have been treated quite briefly or not at all. Among the topics omitted entirely were experimental design (see Refs. 21-48, 21-49, and 21-50), reliability theory (Refs. 21-51, 21-52, and 21-81), survival analysis (Ref. 21-80), survey sampling (Refs. 21-53 and 21-54), time series analysis (Refs. 21-55, 21-56, 21-57, and 21-58), and stochastic processes including Markov chains

(Refs. 21-59, 21-60, 21-61, 21-62, and 21-63). By and large the theory and rationale behind the choice of a statistical test has been slighted (i.e., such topics as efficiency, asymptotic efficiency, invariance and unbiasedness.) The following references are appropriate: 21-7, 21-10, 21-11, 21-12, 21-13, 21-14, and 21-69.

The purpose of statistics is to aid in gaining knowledge of the "world" in order to make decisions and advance science. Statistical methods serve as a guide to intuition and guard against improper conclusions. In many (indeed most) situations there are of necessity points arising that do not fall within the logical framework of mathematical statistics. Statistics should not be viewed as a straight-jacket that implies only "perfect" experiments be considered. The mathematical theory is to be used in conjunction with good judgment to the best of one's ability. If questions as to appropriateness of data (or a mathematical model for the data) arise then intellectual honesty requires that the questions be considered and reported in conjunction with the mathematical treatment of the data. It is often claimed that statistics "proves" the obvious. This is somewhat true, however, one of the main values of statistics is when it does *not* prove what is "obviously" true but warns that indeed the data is not sufficient for the "obvious" conclusion. In a situation where the mathematical statistical theory and common sense give conflicting answers then extreme caution and review are needed before deciding which side is correct.

21.10 REFERENCES

21-1 Todhunter, I., *A History of the Mathematical Theory of Probability* (1st ed. 1865), Chelsea Publishing Co., New York, 1965.

21-2 Newman, J. R., *The World of Mathematics*, vol. 2, Simon and Schuster, New York, 1956.

21-3 Newman, J. R., *The World of Mathematics*, vol. 3, Simon and Schuster, New York, 1956.

21-4 Feller, W., *An Introduction to Probability Theory and its Applications*, vol. 1, 3rd ed., John Wiley and Sons, New York, 1968.

21-5 Abraham, M., and Stegun, I. A. (ed.), *Handbook of Mathematical Functions*, National Bureau of Standards, Wash., D.C., 1964.

21-6 Huff, D., *How to Lie with Statistics*, W. W. Norton, New York, 1954.

21-7 Wilks, S. S., *Mathematical Statistics*, John Wiley and Sons, New York, 1962.

21-8 Beyer, W. H. (ed.), *CRC Handbook of Tables for Probability and Statistics*, The Chemical Rubber Company, Cleveland, 1966.

21-9 Owen, D. B., *Handbook of Statistical Tables*, Addison-Wesley, Reading, Mass., 1962.

21-10 Lehmann, E. L., *Testing Statistical Hypotheses*, John Wiley and Sons, New York, 1959.

21-11 Kendall, M. G., and Stuart, A., *The Advanced Theory of Statistics*, vol. II, 2nd ed., Hafner, Darien, Conn., 1967.

21-12 Hogg, R. V., and Craig, A. T., *Introduction to Mathematical Statistics*, 3rd ed., Macmillan, New York, 1970.

21-13 Ferguson, T. S., *Mathematical Statistics, A Decision Theory Approach*, Academic Press, New York, 1967.

21-14 Rao, C. R., *Linear Statistical Inference and Its Applications*, John Wiley and Sons, New York, 1965.

21-15 Scheffé, H., *The Analysis of Variance*, John Wiley and Sons, New York, 1959.

21-16 Draper, N. R., and Smith, H., *Applied Regression Analysis*, John Wiley and Sons, New York, 1966.

21-17 Acton, F. S., *Analysis of Straight-Line Data*, Dover, New York, 1966.

21-18 Dempster, A. P., *Elements of Continuous Multivariate Analysis*, Addison-Wesley, Reading, Mass., 1969.

21-19 Anderson, T. W., *An Introduction to Multivariate Statistical Analysis*, John Wiley and Sons, New York, 1958.

21-20 Guenther, W. C., *Analysis of Variance*, Prentice-Hall, Englewood Cliffs, New Jersey, 1964.

21-21 Graybill, F. A., *An Introduction to Linear Statistical Models*, McGraw-Hill, New York, 1961.

21-22 Seal, H. L., *Multivariate Statistical Analysis for Biologists*, Methuen, London, 1968.

21-23 Morrison, D. F., *Multivariate Statistical Methods*, McGraw-Hill, New York, 1967.

21-24 Lawley, P., and Maxwell, A. E., *Factor Analysis as a Statistical Method*, Butterworth, London, 1963.

21-25 Jardine, N., and Sibson, R., *Mathematical Taxonomy*, John Wiley and Sons, New York, 1971.

21-26 Sokal, R. R., and Sneath, P. H. A., *Principles of Numerical Taxonomy*, Freeman, San Francisco, 1963.

21-27 Cormack, R. M., "A Review of Classification," (with a discussion), *Journal of the Royal Statistical Society*, Series B, **134**, 321–367, 1971.

21-28 Kendall, M. G., and Stuart, A., *The Advanced Theory of Statistics*, vol. III, Hafner, Darien, Conn., 1966.

21-29 Hájek, J., *A Course in Nonparametric Statistics*, Holden-Day, San Francisco, 1969.

21-30 Hájek, J., and Sidek, Z., *Theory of Rank Tests*, Academic Press, New York, 1967.

21-31 Bradley, J. V., *Distribution-Free Statistical Tests*, Prentice-Hall, Englewood Cliffs, New Jersey, 1968.

21-32 *Selected Tables in Mathematical Statistics, vol. I*, Markham, Chicago, 1970.

21-33 Craddock, J. M., and Flood, C. R., "The Distribution of the χ^2 Statistic in Small Contingency Tables," *Applied Statistics*, **19**, 173–181, 1970.

21-34 Chernoff, H., and Lehman, E. I., "The Use of Maximum Likelihood Estimates in χ^2 Tests for Goodness of Fit," *Annals of Mathematical Statistics*, **25**, 579–586, 1954.

21-35 Lancaster, H. O., *The Chi-Squared Distribution*, John Wiley and Sons, New York, 1969.

21-36 Finney, D. J.; Latscha, R.; Bennet, B.; and Hsu, P., *Tables for Testing Significance in a 2 $\times$ 2 Contingency Table*, Cambridge University Press, Cambridge, 1963.

21-37 Lieberman, G. J., and Owen, D. B., *Tables of the Hypergeometric Probability Distribution*, Stanford University Press, Stanford, 1961.

21-38 Tukey, J. W., "A Survey of Sampling from Contaminated Distributions," in Olkin, I., et al. (eds.), *Contributions to Probability and Statistics*, Stanford University Press, Stanford, 1961.

21-39 Savage, L. J., *The Foundations of Statistics*, John Wiley and Sons, New York, 1954.

21-40 Kyburg, H. E., Jr., and Smokler, H. E. (eds.), *Studies in Subjective Probability*, John Wiley and Sons, New York, 1964.

21-41 Cornfield, J., "Sequential Trails, Sequential Analysis, and the Likelihood Principle," *The American Statistician*, **20**, 18–23, 1966.

21-42 Cornfield, J., "The Bayesian Outlook and its Application" (with discussion), *Biometrics*, **25**, 617–657, 1969.

21-43 Birnbaum, A., "On the Foundations of Statistical Inference," *Journal of the American Statistical Association*, **57**, 269–326, 1962.

21-44 Raiffia, H., and Schlaifer, R., *Applied Statistical Decision Theory*, M.I.T. Press, Cambridge, Mass., 1968.

21-45 Blackwell, D., and Girshick, M. A., *Theory of Games and Statistical Decision*, John Wiley and Sons, New York, 1954.

21-46 Wald, A., *Statistical Decision Functions*, John Wiley and Sons, New York, 1950.

21-47 Chernoff, H., and Moses, L. E., *Elementary Decision Theory*, John Wiley and Sons, New York, 1959.

21-48 Cochran, W. G., and Cox, G. M., *Experimental Designs*, John Wiley and Sons, New York, 1957.

21-49 Cox, D. R., *Planning of Experiments*, John Wiley and Sons, New York, 1958.

21-50 Fisher, R. A., *The Design of Experiments*, 8th ed., Oliver and Boyd, Edinburgh, 1966.

21-51 Barlow, R. E., and Proschan, F., *Mathematical Theory of Reliability*, John Wiley and Sons, New York, 1965.

21-52 Gnedenko, B. V.; Belayev, Y. K.; and Solovyev, A. D., *Mathematical Methods of Reliability Theory* (translated by Scripta Technica), Academic Press, New York, 1969.

21-53 Deming, E. D., *Some Theory of Sampling*, John Wiley and Sons, New York, 1950, Dover, New York, 1966.

21-54 Kish, L., *Survey Sampling*, John Wiley and Sons, New York, 1965.

21-55 Anderson, T. W., *The Statistical Analysis of Time Series*, John Wiley and Sons, New York, 1971.

21-56 Hannan, E. J., *Multiple Time Series*, John Wiley and Sons, New York, 1970.

21-57 Box, G. E. P., and Kenkins, G. M., *Time Series Analysis: Forecasting and Control*, Holden-Day, San Francisco, 1970.

21-58 Jenkins, G. M., and Watts, D. G., *Spectral Analysis and Its Applications*, Holden-Day, San Francisco, 1968.

21-59 Doob, J. L., *Stochastic Processes*, John Wiley and Sons, New York, 1953.

21-60 Karlin, S., *A First Course in Stochastic Processes*, Academic Press, New York, 1966.

21-61 Parzen, E., *Stochastic Processes*, Holden-Day, San Francisco, 1962.

21-62 Kemeny, J. G., and Snell, J. L., *Finite Markov Chains*, Van Nostrand Reinhold, New York, 1960.

21-63 Kemeny, J. G., Snell, J. L., and Knapp, A. W., *Denumerable Markov Chains*, Van Nostrand Reinhold, New York, 1966.

21-64 Kendall, M. G., and Stuart, A., *The Advanced Theory of Statistics*, vol. I, 3rd ed., Hafner, Darien, Conn., 1969.

21-65 Kolmogorov, A. N., *Foundations of the Theory of Probability*, translated by Nathan Morrison, Chelsea Publishing Co., New York, 1950.

21-66 Cramér, Harald, *Mathematical Models of Statistics*, Princeton Univ. Press, Princeton, 1946.

21-67 Fisher, L., and van Belle, G., *Biostatistics: A Methodology for the Biological Sciences*, Duxbury, in press.

21-68 Cox, D. R., and Hinkley, D. V., *Theoretical Statistics*, Chapman and Hall, London, 1974.

21-69 Bickel, P. J., and Doksum, K. A., *Mathematical Statistics: Basic Ideas and Selected Topics*, Holden-Day, San Francisco, 1977.

21-70 Searle, S. R., *Linear Models*, John Wiley and Sons, New York, 1971.

21-71 Seber, G. A. F., *Linear Regression Analysis*, John Wiley and Sons, New York, 1977.

21-72 Fisher, L., and McDonald, J., *Fixed Effects Analysis of Variance*, Academic Press, New York, 1978.

21-73 Timm, N. H., *Multivariate Analysis with Applications in Education and Psychology*, Brooks/Cole, 1975.

21-74 Gorsuch, R. L., *Factor Analysis*, W. B. Saunders, Co., Philadelphia, 1974.

21-75 Cox, D. R., *Analysis of Binary Data*, Metheun & Co., Ltd., London, 1970.

21-76 Breslow, N. E., and Day, N. E., *Statistical Methods in Cancer Research. Volume 1–The Analysis of Case-Control Studies*, International Agency for Research on Cancer, Lyon, 1980.

21-77 Odeh, R. E., Owen, D. B., Birnbaum, Z. W., and Fisher, L., *Pocket Book of Statistical Tables*, Marcel Dekker, Inc., New York, 1977.

21-78 Lehmann, E. L., *Nonparametrics: Statistical Methods Based on Ranks*, Holden-Day, San Francisco, 1975.

21-79 Huber, P., *Robust Statistics*, John Wiley and Sons, New York, 1981.

21-80 Kalbfleisch, J. D., and Prentice, R. L., *The Statistical Analysis of Failure Time Data*, John Wiley and Sons, New York, 1980.

21-81 Mann, N. R., Schafer, R. E., and Singpurwalla, N. D., *Methods for Statistical Analysis of Reliability and Life Data*, John Wiley and Sons, New York, 1974.

21-82 Winer, B. J., *Statistical Principles in Experimental Design*, 2nd ed., McGraw-Hill, New York, 1971.

Index